Chemical Energy from Natural and Synthetic Gas

Chemical Energy from Natural and Synthetic Gas

Yatish T. Shah

CRC Press
Taylor & Francis Group
Boca Raton London New York

CRC Press is an imprint of the
Taylor & Francis Group, an **informa** business

CRC Press
Taylor & Francis Group
6000 Broken Sound Parkway NW, Suite 300
Boca Raton, FL 33487-2742

First issued in paperback 2019

© 2017 by Taylor & Francis Group, LLC
CRC Press is an imprint of Taylor & Francis Group, an Informa business

No claim to original U.S. Government works

ISBN-13: 978-1-4987-3802-6 (hbk)
ISBN-13: 978-0-367-87424-7 (pbk)

Visit the Taylor & Francis Web site at
http://www.taylorandfrancis.com

and the CRC Press Web site at
http://www.crcpress.com

Dedication

This book is dedicated to my family: Mary, James, Rebecca, Jonathan, Heather, Keith, Laura, and my eight grandchildren.

Contents

Preface

As the world is engaged in realigning the energy and fuel portfolio, which is currently fossil fuel dominated, to one that is more balanced between fossil, nuclear, and renewable energies, one source of fuel that is considered as playing the role of not only a "bridge or transition fuel" but also an "ultimate or end fuel" is gas. While for several decades coal and oil have been the dominant forces for heating, electrical power generation, and fuel for transportation industry, recent concerns about global warming and carbon emissions from these sources have led to more push toward the use of renewable energy sources such as biomass, waste, solar, wind, geothermal, and water, as well as more recognition of carbon-free nuclear energy. Unfortunately, the commercial development of energy and fuel from renewables will take time to economically compete with existing fossil energy resources and their infrastructure.

The development of renewable and nuclear energy at the large scale faces many challenges. Biomass has low mass and energy density compared to coal and oil. They are difficult to transport long distances and they are not as easy to store as coal and oil. The same is true for waste. In general, municipal solid waste is also highly heterogeneous and it can only be used from local sources. For these reasons, the development of stand-alone, large-scale, and sustainable power or fuel plants from these sources is highly problematic. Small-scale plants cannot compete with the economy of scale of fossil fuel plants. Furthermore, the infrastructure for their storage and transport is not well developed. Their penetration in the energy and fuel industry may require a different strategy than one used for fossil fuels. This different strategy is addressed in my previous book *Energy and Fuel Systems Integration*, CRC Press, New York (2015). Similarly, both solar and wind energy are time and location dependent and cannot provide large-scale sustainable power without backup power supply by fossil or nuclear energy or backup energy storage. Their large-scale implementation will also require hybrid energy system strategy. The possible hybrid structures for this purpose are also discussed in my previous book. Solar and wind energy are not highly energy efficient and their infrastructures also need to be further developed. The large-scale commercial experience for power generation from these sources of energy is in its early stages of development.

While enhanced geothermal system has an enormous potential for its role in energy and fuel industry, the needed infrastructure and commercial experience for it to become a reality is also still at the development stage. Finally, as shown in my book *Water for Energy and Fuel Production*, CRC Press, New York (2014), water also has an enormous potential for providing unlimited supply of energy and fuel; however, both infrastructure and commercial experience to tap this still needs to be developed. Nuclear energy, in some parts of the world, still suffers from social and political acceptance and, as shown in my previous book *Energy and Fuel Systems Integration*, CRC Press, New York (2015), it will require a different strategy (more toward helping renewable energy penetration and its use for nonelectrical applications) to gain more public acceptance.

While the world must pursue, more aggressively, obtaining energy and fuels from renewable sources, the resolutions of the issues outlined in the previous paragraph will take time. While renewable sources of energy and fuel have distinct advantages over fossil fuels, their commercial production must economically compete with fossil fuels. In the meantime, one fuel that has the potential to be a true "transition or bridge fuel" or even the "ultimate or end fuel" is gas. This book illustrates this point by examining all the roles of natural and synthetic gas in the energy and fuel industry.

Gaseous compounds containing carbon and/or hydrogen are what make gas a source of energy and fuel. These compounds, such as methane, ethane, propane, butane (and other volatile hydrocarbons), syngas (mixture of hydrogen and carbon monoxide), and hydrogen have high heating values and can be used to generate energy and other types of fuels. These fuels can be basically obtained from two sources: natural and man-made. Gas from a natural source (obtained from underground) recovered by conventional or unconventional method is called natural gas. This gas predominantly contains methane as a source of energy and fuel. Man-made gas is called "synthetic gas" and it can be obtained by three different methods, as detailed later. The composition of synthetic gas varies depending on the process used to produce it. While natural gas only comes from nonrenewable sources, synthetic gas can come from both nonrenewable and renewable sources.

Just like coal and oil, natural gas is a fossil fuel obtained from underground, from the bottom of the ocean or from an arctic environment. In the past, natural gas was recovered from relatively shallow and easy-to-access natural gas reservoirs by conventional drilling techniques. Just like conventional oil well, this source of natural gas is rapidly depleting. Natural gas often contains oil, and this type of gas is called associated gas. Natural gas can also be obtained from stand-alone pure gas reservoirs, called nonassociated gas.

Recent developments in and success of the process of hydraulic fracturing and horizontal (or directional) drilling have allowed successful access to gas trapped in deeper (even more than 15,000 ft) and tighter and more compressed matrices underground. These techniques have allowed us to access unconventional gas such as deep gas, tight gas, gas from geo-pressurized zones, shale gas, and gas from coal bed methane. In the United States, the biggest revolution occurred in the production of shale gas. Along with these unconventional gas sources, gas hydrates (mixture of water and methane) are obtained from the ocean floor and in arctic conditions. All these combined unconventional gas resources have vast potential, and our improved ability to tap them has led us into the "gas age." With recent successes, the United States is now the world's leading producer of natural gas. This expanding supply has allowed us to replace coal by natural gas in power plants and diesel oil in large vehicles by Liquid Natural Gas (LNG), Liquid Petroleum Gas or Liquid Propane Gas (LPG), and Compressed Natural Gas (CNG). This "shale gas" revolution has allowed us to replace more harmful coal and oil by gas, which has made it a "transition or bridge fuel." Thus, we are slowly transitioning from a coal and oil–based economy to a natural gas–based economy.

Natural gas or methane is cleaner than coal and oil. It contains significantly larger hydrogen to carbon ratio (4) compared to the ones for coal (less than 1) and refined oil (around 1.2–1.6). It does not contain other harmful chemicals that cannot be handled upfront and are prevented from emission into the environment. Unlike coal, natural gas can be used for both large- and small-scale (like micro turbines and engines) power applications in a convenient manner. While natural gas has low mass and energy density compared to oil, LNG, LPG, and CNG have been found to be good substitutes for diesel oil in large vehicles and their use is both economically and environmentally competitive and gaining ground. A gallon of CNG has about 25% of the energy content of a gallon of diesel fuel and LNG has 60% of the volumetric energy density of diesel fuel. LPG has a typical specific calorific value of 46.1 MJ/kg compared with 42.5 MJ/kg for fuel oil and 43.5 MJ/kg for premium-grade gasoline. However, its energy density per volume (26 MJ/L) is lower than either that of gasoline or fuel oil. Its density (about 0.5–0.58 kg/L) is lower than that of gasoline (about 0.71–0.77 kg/L). All old coal-based power plants in the United States are gradually being replaced by those operated by natural gas. Natural gas usage in power plants is expected to double by 2040 and surpass that of coal. Similarly, the use of natural gas is expected to surpass that of oil by 2025.

Synthetic gas can be produced by three distinct methods: (a) thermal gasification of all carbonaceous materials or refining of oil to produce hydrocarbons such as propane, butane, etc.; (2s) anaerobic digestion of cellulosic waste; and (3) hydrothermal processes involving either gasification in the presence of steam, sub- and supercritical water, and/or hydrogen for all carbonaceous materials or water dissociation to produce hydrogen. These three methods produce synthetic gas of different compositions.

Unlike coal- and petroleum-based oil, gas can be produced synthetically by gasification and reforming of both nonrenewable and renewable carbonaceous materials. The feedstock for the production of synthetic gas can be coal, oil, biomass, waste, or a mixture of these. The thermal gasification of coal is a commercially proven technology. The refining of oil to produce hydrocarbons such as propane, butane, etc., is also a commercial technology. The gasification of other feedstock (like biomass, waste, etc.) is also being aggressively developed at both small and large scales. Cogasification of coal/biomass/waste mixtures is gaining momentum due to its impact on carbon emission into the environment.

Unlike natural gas, synthetic gas contains many fuel components, such as all volatile hydrocarbons, hydrogen, carbon monoxide, and other gaseous impurities depending on the nature of feedstock and gasification conditions. Synthetic gas produced in this manner has been given many names, such as "producer gas," "town gas," "wood gas," "syngas," "water gas," etc., depending on its composition. Producer gas can also be described as high-, medium-, or low-Btu gas depending on its methane and nitrogen concentrations. Thus, synthetic gas can replace natural gas for heating and power production. The most useful form of synthetic gas for liquid fuel production is "syngas," which is a mixture of hydrogen and carbon monoxide that can be converted to a variety of liquid fuels and chemicals by the well-recognized Fischer–Tropsch synthesis, iso-synthesis, oxo-synthesis, and methanol and mixed alcohol production processes. Often, gas produced by the gasification technology is called by the generic name "synthesis gas," which is synonymous to "syngas." Unlike natural gas, the sources for synthesis gas are unlimited.

Both methane and hydrogen can also be produced by the biological process of "anaerobic digestion," which can be carried out in the absence of oxygen and with the help of suitable microorganisms. The gas produced by this method is called "biogas" or "bio-hydrogen." With the use of methanogenic bacteria, the "biogas" produced mainly contains methane and carbon dioxide. Landfill gas is a type of "biogas" largely containing methane and carbon dioxide. Anaerobic digestion processes can also produce bio-hydrogen with the help of appropriate microorganisms. In the presence of methanogenic bacteria, the produced "biogas" has a methane concentration of about 55% (the remainder being mostly carbon dioxide) as opposed to conventional natural gas that has a methane concentration of about 95%. The methane concentration in "biogas" is, however, very similar to that in "shale gas." Biogas can be refined to produce "bio-syngas," which is very similar to syngas produced from thermal gasification technologies.

The third type of synthetic gas is hydrogen or gas concentrated with hydrogen produced by two separate methods. Gas containing a high concentration of hydrogen can be produced by gasification in the presence of steam and sub- and supercritical water with or without hydrogenation of all carbonaceous materials. These processes are generally considered as "hydrothermal gasification." The second method involves the dissociation of water by electrolysis and photocatalytic, photobiological, thermal, thermochemical, and other novel methods. These methods generally produce pure hydrogen. Many consider hydrogen to be the "ultimate fuel" because it contains no carbon. The world will be much safer and cleaner if all the energy is provided from carbon-free sources. Thus, technological developments and commercialization of all types of "synthetic gas production," and in particular hydrogen production, will make large-scale synthetic gas production the end game (not just a transition game like natural gas) for energy and fuel industry.

Natural and synthetic gas can be converted to syngas with the desired composition of hydrogen and carbon monoxide by the process of "gas reforming" so that the mixture can be used to produce liquid fuels, fuel additives, and chemicals via Fischer–Tropsch synthesis, iso-synthesis, oxo-synthesis, and others. Gas reforming is one of the most important technologies for natural and synthetic gas and its further development can transform the role of gas from a "bridge fuel" to more of an "end fuel" through the production of syngas and hydrogen from methane and other carbonaceous materials. One type of gas reforming, namely, dry reforming, which involves the use of carbon dioxide to convert synthetic gas or carbonaceous materials to syngas, can also be an answer to reduce carbon dioxide emission. The book critically evaluates the effectiveness of various available technologies for gas reforming. Further developments in various types of reforming processes will further accentuate the role of "synthetic gas" as the "ultimate fuel."

Unlike coal and oil, natural gas, and various types of synthetic gas, can be easily cleaned up and upgraded to the desired level so that using them for power, heat, and liquid fuel applications does not result in harmful emissions into the environment. Both natural and synthetic gases also need to be cleaned to prepare them for storage and the transport. The book evaluates all the available technologies to clean and upgrade gas coming from different sources.

One of the reasons natural gas is becoming so important in the energy and fuel industry is that its storage and transportation infrastructure is well established and it is constantly expanding. The concept of gas grid, analogous to smart electrical grid, is being developed and this will make gas, heat, and electricity the dominant future forces of the energy and fuel industry. Natural gas storage technologies in all its forms (natural gas, LNG, LPG, CNG) are well established on a regional basis and progress is constant. While the natural gas infrastructure can be used for hydrogen in small quantities (5 to 15 vol%), the storage and transportation infrastructure for hydrogen is still at the research stage. Once that is developed, gas will truly become the "ultimate fuel." The book evaluates storage and transportation options for natural gas, syngas, and hydrogen.

The book also evaluates various end usages of natural gas, syngas, and hydrogen. Various gas-to-liquid fuel technologies and the role of hydrogen in refinery industries are assessed. Hydrogen is the most valuable commodity in the fuel industry. The book evaluates the role of methane, syngas, and hydrogen in large- and small-scale power production. Over next twenty years, gas will be the most dominant fuel in the power industry, surpassing coal and oil by a large margin. ExxonMobil predicts that by 2040, gas will generate 30% of the total electricity as opposed to its current value of 20%. The use of gas in small-scale power generation is rapidly rising. The use of gas for various heating purposes will also rapidly expand. Another area where the use of gas is rapidly expanding is the vehicle industry where LNG, LPG, and CNG are replacing gasoline and diesel fuels, and the use of fuel cell is on rise. The use of hydrogen for both stationary and mobile fuel cells is gaining momentum. The book evaluates all of these applications of gas in a critical manner.

In summary, this book differs from numerous other books on natural gas, synthesis gas, and hydrogen published previously in that it presents the unified and collective role of gas in the energy and fuel industry. It addresses both the "transition" as well as the "end game" role of gas. Most people believe hydrogen and electricity will be the pivotal sources of energy in the future. Syngas chemistry for the production of liquid fuels and chemicals has a vast future because syngas can be obtained from both natural gas and man-made synthetic gas. The development of smart gas grid will make the use of gas for heat, power, and liquid fuel unavoidable in the years to come.

This comprehensive book on natural and synthetic gas will be useful to all researchers involved in the development of new technologies for energy and fuel. It will be a good reference for graduate courses on energy and fuel and can serve as a graduate-level text on the subject of gas as a source of fuel and energy.

Author

Dr. Yatish T. Shah received his BSc in chemical engineering from the University of Michigan, Ann Arbor, Michigan and MS and ScD in chemical engineering from Massachusetts Institute of Technology, Cambridge, Massachusetts. He has more than 40 years of academic and industrial experience in energy-related areas. He was chairman of the Department of Chemical and Petroleum Engineering at the University of Pittsburgh, dean of the College of Engineering at the University of Tulsa and Drexel University, chief research officer at Clemson University, and provost at Missouri University of Science and Technology, University of Central Missouri, and Norfolk State University. He was also a visiting scholar at Cambridge University in the United Kingdom and a visiting professor at the University of California, Berkley, and Institut für Technische Chemie I der Universität Erlangen, Nürnberg, Germany. Dr. Shah has written six books related to energy: *Gas-Liquid-Solid Reactor Design* (published by McGraw-Hill, 1979), *Reaction Engineering in Direct Coal Liquefaction* (published by Addison-Wesley, 1981), *Cavitation Reaction Engineering* (published by Plenum Press, 1999), *Biofuels and Bioenergy—Processes and Technologies* (published by CRC Press, 2012), *Water for Energy and Fuel Production* (published by CRC Press, 2014) *and Energy and Fuel Systems Integration* (CRC Press, 2015). He has also published more than 250 refereed reviews, book chapters, and research technical publications in the areas of energy, environment, and reaction engineering. He is an active consultant to numerous industries and government organizations in the energy areas.

1 Introduction

1.1 INTRODUCTION

There are 10 sources of energy and fuel on earth. These are coal, oil, gas, biomass, waste, nuclear energy, solar energy, wind energy, geothermal energy, and water. They can also be divided into three categories: fossil fuels (coal, oil, and natural gas), nuclear energy, and renewables (biomass, waste, solar energy, wind energy, geothermal energy, and water). Currently, about 85% of our energy needs are supplied by fossil fuels. Among fossil fuels, coal and oil have been the dominant sources for our fuel supply for power and transportation, respectively. In recent years, due to environmental concerns, while the use of coal for power production is declining, the use of gas for power production is on the rise. Use of natural gas in transportation industry is also slowly increasing.

Three fossil fuels, namely, coal, oil, and natural gas, were formed many hundreds of millions of years ago before the Age of Dinosaurs—hence the name "fossil fuels." The age they were formed is called the Carboniferous (named after carbon) Period. It was part of the Paleozoic Era. The Carboniferous Period occurred from about 360 to 286 million years ago. At the time, the land was covered with swamps filled with huge trees, ferns, and other large leafy plants. The water and seas were filled with algae. Besides trees and vegetables, the remains of animals were also buried in the ground.

Natural gas is a fossil fuel formed when layers of buried plants, gases, and animals are exposed to intense heat and pressure over thousands of years. The energy that the plants originally obtained from the sun is stored in the form of chemical bonds in natural gas. Natural gas is a nonrenewable resource because it cannot be replenished on a human time frame. Natural gas is the gas component of coal, shale, oil, and water (in the form of clathrate hydrates under certain temperature and pressure conditions) formation. It is found with coal as *coalbed methane* (CBM), with oil as *associated gas*, with shale matrix as *shale gas* and as solid crystalline *inclusion* compounds, and with water as gas hydrates in arctic conditions and at the bottom of the sea. It can be used to generate heat, power, and liquid fuels and chemicals. It can be used for both static and mobile applications. Energy in 6000 cubic feet (ft^3) of natural gas is equivalent to 1 barrel of oil.

While the high-velocity gas generates wind energy, this book is focused on the conversion of chemical energy from gas to heat, electricity, and liquid fuels. This chemical energy is generally obtained from natural gas recovered from underground or synthetic gas produced from a variety of nonrenewable and renewable sources.

The useful chemical constituents of natural and synthetic gas are largely methane, syngas (mixture of hydrogen and carbon monoxide), and hydrogen. Other lower hydrocarbons, ethane, and particularly propane and butane, are also used as fuels. Olefins such as ethylene and propylene are often used as raw materials to produce chemicals, polymers, etc. In this book we will mainly focus on various aspects of chemical constituents such as methane, syngas, hydrogen, propane (liquefied petroleum gas [LPG]), and butane and their roles in the energy and fuel industry. Collectively, these are called gaseous fuels. The scope of this book is to describe in detail the production, cleaning, upgrade, storage, and transport of gaseous fuels that occurred naturally and the ones that are manmade. The book also illustrates versatile applications of natural and synthetic gas for a variety of end products.

While both coal and oil contain a large number of aliphatic, olefin, and aromatic hydrocarbons, which can provide energy, the processing of coal and oil for heat and power productions also results in the production of harmful by-products and emissions. On the other hand, gas containing volatile hydrocarbons (mostly methane), syngas, and hydrogen provide fuel for energy in a cleaner form.

Unlike coal and oil, natural or synthetic gas can be pretreated to produce high-quality pure methane, syngas, or hydrogen fuels, which can be used for numerous downstream applications. Both methane and hydrogen have high hydrogen to carbon ratios than coal or oil, making them cleaner and more efficient fuels. While the processing of coal and oil generates a significant amount of carbon dioxide, which is harmful to the environment, natural gas produces much lower carbon dioxide than coal during power production and much less harmful emissions than oil in heat and transportation applications. Harmful environmental effect by burning syngas is even lower and there is no carbon emission during the use of hydrogen.

The average emissions rates in the United States from natural gas–fired power plants are 1135 lb/MWh of carbon dioxide, 0.1 lb/MWh of sulfur dioxide, and 1.7 lb/MWh of nitrogen oxides [1]. Thus, compared to the average air emissions from coal-fired generation, natural gas produces half as much carbon dioxide, less than a third as much nitrogen oxides, and 1% as much sulfur oxides at the power plant [1]. All new power plants in the United States are planning to use natural gas instead of coal. An Massachusetts Institute of Technology (MIT) study [2] indicates that the use of natural gas will be nearly doubled by 2040. Other projections shown in Figure 1.1 [3] indicate that natural gas will compete well with oil in the future.

The increased share of natural gas in the global energy mix is not sufficient on its own to put the world on a carbon emission path consistent with an average global temperature rise of no more than 2°C. As shown in Table 1.1, natural gas is still not as good as renewable fuels and nuclear energy for its carbon emission. In the long term, more use of renewable energy and carbon-free nuclear energy is needed. The large-scale hydrogen production from natural or synthetic gas can, of course, make gas the ultimate source of fuel and energy along with renewable energy. At present time, nuclear power, renewable energy, and carbon capture and sequestration are relatively expensive next to gas. Until we make renewable energies and hydrogen commercially cheaper compared to fossil fuels, natural gas will remain the "transition fuel."

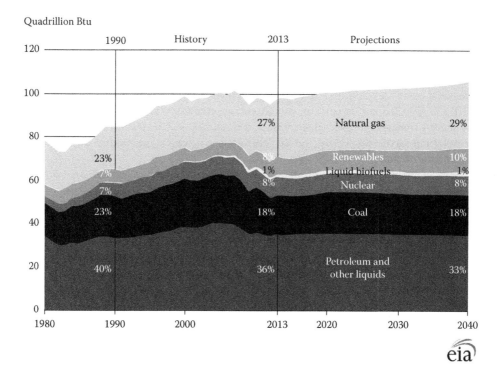

FIGURE 1.1 Energy consumption projection by fuel type. (From U.S. Energy Information Administration, Annual energy outlook 2015, EIA, Washington, DC, 2015, www.eia.gov/forecasts/aeo/, accessed on April, 2015.)

TABLE 1.1

Relative Ranges of Greenhouse Gas Emissions from Different Electricity Generation Technologies

Substance	Mean (% Relative to Lignite)	(Lower/Upper) Range (Tons of CO_2 eq/GWh)
Lignite	100	790/1372
Coal	84.2	756/1310
Oil	69.5	547/935
Natural gas	47.3	362/891
Solar PV	8.1	13/731
Biomass	4.3	10/101
Nuclear	2.75	2/130
Hydroelectric	2.5	2/237
Wind	2.5	6/124

Sources: Marano, J.J. and Ciferno, J.P., Life-cycle greenhouse-gas emissions inventory for Fischer–Tropsch fuels, Energy and Environmental Solutions, LLC, for the U.S. Department of Energy, National Energy Technology Laboratory, Pittsburgh, PA, 2001; IEA Statistics, *CO₂ Emissions from Fuel Combustion*, 2011 edn., OECD/IEA, Paris, France, 2011; Shah, Y., *Energy and Fuel Systems Integration*, CRC Press, Taylor & Francis Group, New York, 2015.

In general, natural gas is a mixture of several hydrocarbon gases, including methane (between 70% and 90%), ethane, propane, butane, and pentane, as well as carbon dioxide, nitrogen, and hydrogen sulfide. The composition of natural gas can vary widely, depending on the gas field. Natural gas is referred to as "wet" when hydrocarbons other than methane (particularly natural gas liquids [NGLs]) are present and "dry" when it is almost pure methane. Natural gas is also called "sour" when it contains significant amounts of hydrogen sulfide and "sweet" when it contains no sulfur.

Global gas demand was estimated at 3500 billion cubic meters (bcm) in 2013, which is up 1.2% from 2012 levels [4]. Gas demand has increased by around 800 bcm over the last decade, or 2.8% per year. Gas has a 21% share in the global primary energy mix, behind oil and coal. For comparison, 50 bcm of natural gas is roughly equivalent to 7% of U.S. consumption in 2012 or slightly more than Turkey's entire annual consumption in 2012. The United States, Russia, China, and Iran are the world's largest consumers of gas. The largest producers were Russia, the United States, Canada, Qatar, and Iran although this picture has changed in recent years due to the boom in shale gas production in the United States. It is important to note that Chinese gas consumption almost doubled over 2007–2012 and rose 9% in 2013 to reach nearly 120 bcm, while U.S. gas production increased by more than one-quarter over the same period to reach 688 bcm [4]. The United States consumed 19.7 million cubic feet (mcf) of natural gas in 1999, nearly all of which came from domestic production. Five states—Texas, Louisiana, Alaska, New Mexico, and Oklahoma—hold more than 85% of U.S. natural gas reserves. This picture is also significantly changed in recent years due to shale gas revolution.

In order to forecast the supply of natural gas, analysts mostly tend to refer to proven gas reserves, that is, volumes have been discovered and can be produced economically with existing technology at current gas prices. Worldwide proven conventional gas reserves are estimated at around 190 trillion cubic meters (tcm) or about 56 times the current annual global gas production. However, recoverable gas resources, that is, volumes that analysts are confident will be discovered or technology developed to produce them, are much larger, with recoverable conventional resources estimated at around

440 tcm [4–7]. Estimated recoverable unconventional resources (excluding methane hydrates) are around 240 tcm. Altogether, this would last around 220 years, based on the current rates of gas consumption [4–7].

Worldwide, many different units (for gas and associated energy supply and consumption) are used by different countries, which sometimes makes it difficult to reconcile data. At the Energy Information Administration (EIA) [8,9], natural gas statistics are given in bcm (for volume) and in terajoules (for energy). Worldwide, units such as kilowatt hours (kWh), kilocalories (kcal), million British thermal units (MBtu or MMbtu), therms (th), million tons of oil equivalent (mtoe), billion cubic feet (bcf), or billion cubic feet per day (bcf/d) are used. Data on liquefied natural gas (LNG) are given in tons or in bcm. It is worth noting that 1 m³ of LNG has much more energy than 1 m³ of gas at atmospheric pressure, due to their different physical states.

While the recent trend is to replace some of the fossil energy usage by renewable energy sources, within fossil energy the usage of coal and oil is also more substituted by gas both for heat and power productions. As mentioned earlier, this transformation is occurring due to more favorable impact by gas than coal or oil on environment. For example, the use of gas instead of coal can reduce CO_2 emission by more than 50%–60% and in some cases as high as 80%. The use of LNG, LPG, and compressed natural gas (CNG) in heavy vehicle industry reduces CO_2 emissions significantly compared to diesel fuel or gasoline. The automobile industry is making significant advances in the use of hydrogen in fuel cell–driven hybrid cars. It is clear that the use of gas has all the advantages that coal has in power industry with substantially less harmful effect on the environment. Gas can also be used for power generation at all scales. The major disadvantage of gas is its lower mass and energy density. While gas possesses higher heating value per unit mass than any other fossil fuels, its low mass and energy density require very large volume or high pressure, both of which can be problematic from the economic and operational points of view.

Gas is a very versatile fuel on both supply and demand sides. Natural gas (which predominantly contains methane) can be obtained from underground conventional reservoirs as well as numerous unconventional reservoirs. A significant amount of conventional gas is also trapped with oil (associated gas). Pure conventional gas is called "nonassociated" gas. The unconventional gas can be (1) tight gas, (2) deep gas, (3) gas from geopressurized zones, (4) shale gas, (5) CBM, or (6) gas hydrates. We have an enormous amount of reserve in these unconventional gas resources. Unlike coal, gas can also be generated by man-made processes, both from nonrenewable sources like coal and oil and renewable sources like biomass, waste, and water. In fact, water is our largest source of hydrogen gas. In fact, the man-made processes are flexible enough that they can generate gas from all types of carbonaceous materials, including all types of waste. The availability of huge feedstock and new technology developments make the availability of gas as the source of fuel and energy for a much longer time than underground coal or oil. On the demand side, gas can be used for all types of power, heat, and transportation applications, at large or small scale. Gas can be used for all industrial and manufacturing energy needs. Gas can also substitute oil and chemicals by numerous gas to liquid (GTL) conversion processes and gas (in the forms of LNG, LPG, and CNG) can substitute gasoline or diesel in the vehicle industry. Finally, gas (hydrogen) can be used in the rapidly expanding fuel cell industry. Thus, gas is truly a very versatile fuel.

Gas is a cleaner fuel because its impurities can be managed and removed. While we have many ranks of coal with different hydrogen to carbon ratios, mineral matters, and chemical reactivities and we have crude oils with different compositions and different degrees of reactivity and composition, the major fuel content of gas is methane (and sometimes other volatile hydrocarbons such as propane and butane), syngas (which is a mixture of hydrogen and carbon monoxide), and hydrogen. These three chemicals, namely, methane, hydrogen, and syngas, play the most important roles in the energy and fuel industry. These chemical substances with large hydrogen to carbon ratios give more precise control over their conversion to liquid fuel and heat and power generation. While the chemistry of coal and oil is very complex, the chemistry of gaseous fuel is more precise and controllable. The impurities in natural and synthetic gas are all detectable, removable, and manageable.

1.1.1 ADVANTAGES AND DISADVANTAGES OF GAS AS FUEL

Gaseous fuel possesses several inherent advantages and disadvantages compared to other fuels like coal, oil, biomass, waste, and water.

1. Collectively (natural and synthetic) gaseous fuel is in abundant supply. Since gas can be found from natural sources and can be generated from all carbonaceous fuels and water, unlike coal or oil, it can last almost indefinitely. Multiple sources of supply and multiple possibilities of its composition make it not only "bridge fuel" but also "ultimate fuel."

2. It is a cleaner fuel than coal, oil, and, to some extent, biomass and waste. The detailed chemical compositions of all gas components can be measured in quantitative terms. Its purification technologies are available and it can be used to any downstream applications without significant harm to the environment.

3. While the composition of refined oil, biooil, or synthetic gas produced from coal, oil, biomass, or waste depends on the nature of the feedstock, the production of liquid fuels and chemicals produced from syngas only depends on the syngas composition and a quantitative relationship between syngas composition and liquid fuel production can be found.

4. The burning of natural gas in power plants, in general, produces about 50%–60% less CO_2 emission compared to what is produced for coal combustion. A relative comparison of CO_2 emission from coal-, oil-, and gas-based power plants is described in Table 1.1. With proper optimization and the use of combined-cycle and cogeneration techniques, CO_2 emission in gas power plants can be as low as 80% of the emission in coal-based power plants. The coal-based power plants can reduce CO_2 emission by using carbon sequestration technology, but to this date it adds significant cost to the power production.

5. While coal, biomass, or waste combustion can produce other types of S-, Cl-, and N-based pollutants or alkalis and heavy metals in ash, the purified natural gas, syngas, or hydrogen do not produce any other pollutants. Gas cleaning technologies both at high and low temperatures are available.

6. Coal is very aromatic and possesses low values of hydrogen to carbon ratio (about 0.8), while refined oil has hydrogen to carbon ratio of about 1.2. The natural gas on the other hand has hydrogen to carbon ratio of 4 and pure hydrogen has no carbon. Larger concentration of hydrogen in the fuel always results in low-carbon emission irrespective to the thermochemical process used.

7. Coal is normally used for power production. Oil is normally used for heat and in vehicle industry. Oil can also be used for chemical and material industry. Gas, on the other hand, can be used for power and heat production and in the chemical and material industry. LNG, LPG and CNG can be used for mobile applications. Gas is thus the most versatile fuel. Gas (either methane or hydrogen) possesses the highest heating value per unit mass compared to coal or oil.

8. The biggest disadvantage of gaseous fuel compared to coal and oil is its low mass and energy density. For example, 4 kg of hydrogen occupies a volume of 49 m³. Similarly, 4 kg of methane will occupy a volume of about 6.1 m³. These numbers make their storage in small usages like vehicles problematic. The use of natural and synthetic gases requires good storage and transportation infrastructure.

9. CNG has about 25% of energy content of a gallon of diesel fuel and LNG has 60% of volumetric energy density of diesel fuel. Thus, these products from natural gas have bright future for use in larger vehicles. The conversions of LNG and CNG from natural gas, however, add cost to the fuels. Both LNG and CNG are cleaner than diesel oil for larger vehicles. NGL, a by-product of natural gas recovery, can also be used as fuel.

10. The infrastructure of storage and transportation of natural gas is well established. Gas grid and its development make the use of gas more convenient and efficient. Storage and

transportation of hydrogen, particularly for mobile applications, on the other hand, is still at the research and development stage.

11. The advances in steam and dry reforming technologies to convert methane to syngas and hydrogen make natural gas even more very versatile fuel. Since reforming requires a significant amount of heat input, new methods of providing heat such as solar, nuclear, and microwave heat can further accelerate the use of reforming technologies. The success of dry reforming can further decarbonize the use of gas as a source of fuel. The use of plasma in the reforming process offers some interesting possibilities.

12. Unlike coal power plants, the use of gas is equally efficient and convenient at both large and small scales. The use of gas in microturbines and engines is rapidly expanding.

13. Recent successes in hydraulic fracturing and horizontal drilling have rapidly expanded the supply of natural gas. This has lowered the gas price to the level that it has become very attractive all over the world.

1.2   TYPES OF NATURAL GAS AND THEIR MARKETS AND METHODS OF PRODUCTION

Natural gas is traditionally consumed in residential, industrial, transportation, and commercial sectors, mostly for heating, as raw materials for fertilizer production, and power generation. The power sector is by far the largest user of natural gas with around 40% of global gas demand as the fuel contributes to meeting incremental power demand and produces less CO_2 than coal. Industrials use roughly 24% of the total gas consumption and the residential/commercial sector 22%. Other uses include the energy and chemical industry (10%), for example, by GTL plants and losses. A more recent and rapidly developing sector is that of transport, which is expected to represent 9% of the incremental demand over 2012–2018, reaching almost 100 bcm by 2018 or 2.5% of the total gas use.

Natural gas largely contains methane (55%–95%) with some carbon dioxide; water; helium; other lower hydrocarbons such as ethane, propane, butane, and ethylene; and some hydrogen sulfide. The exact composition depends on its source and the nature and location of the reservoir. Natural gas is obtained from underground by both conventional and unconventional methods. Natural gas is often used, stored, and transported as CNG (3000–3600 psig) to reduce its volume and improve efficiency of its storage and transport [10,11].

LNG is natural gas that has been liquefied for transport (very often by ship and sometimes by truck). Depending on its exact composition, natural gas becomes liquid at approximately −162°C at atmospheric pressure. This liquefied state enables the natural gas to be shrunk to 1/600th of its original volume. To be transported, natural gas is first liquefied in an LNG liquefaction plant. This requires all heavier hydrocarbons to be removed from the natural gas, which leaves only pure methane, which is transported in specialized LNG carriers. The largest LNG carrier, called Q-Max, can transport 264,000 m^3 of LNG (or around 0.15 bcm). Finally, LNG is regasified. LNG trade represented around 320 bcm, or 9% of the global gas demand in 2012 [12–15].

The largest producer of LNG is Qatar, whose liquefaction capacity is roughly one-quarter of the global LNG liquefaction capacity as of late 2013. Qatar had a massive expansion of its capacity, which has increased by 63 bcm since early 2009 to reach 105 bcm. Indonesia, Malaysia, Australia, and Algeria are also significant LNG exporters. Russia and Yemen began exporting in 2009, Peru in 2010, and Angola in 2013. Papua New Guinea started exporting in 2014. Australia is set to become the second-largest LNG exporter behind Qatar by 2016—seven projects are committed or under construction, representing over 80 bcm of new capacity. The United States is also rapidly expanding its LNG export since the shale gas revolution [10–15].

LPG (or autogas) is a term used to describe a group of hydrocarbon-based gases derived from crude oil and/or natural gas. Natural gas purification produces about 55% of all LPG, while crude oil refining produces about 45%. LPG is mostly propane, butane, or a mix of the two. Varieties of LPG bought and sold include mixes that are primarily propane (C_3H_8), primarily butane (C_4H_{10}), and, most commonly, mixes including both propane and butane. It also includes ethane, ethylene, propylene, butylene, isobutene, and isobutylene; these are used primarily as chemical feedstock rather than fuel.

LPG becomes a liquid at normal pressure and a temperature of −42°C or at normal temperatures under a pressure of about 8 atm (standard units equivalent to ordinary atmospheric pressure at sea level and 0°C). Separating the economic impact of LPG is problematic because it is derived from both oil and natural gas [10,15].

There are different types of natural gas prices: wholesale prices (such as hub prices, border prices, and city gate) and end-user prices, which differ depending on the customer served (industrial or household). Prices typically include the cost of gas supplies; transmission, distribution, and storage costs; as well as the retailer's margin and taxes. End-user prices and wholesale prices vary widely across regions. Some wholesale gas prices are linked to oil prices, through an indexation present in long-term supply contracts in Continental Europe and OECD Pacific, but this represents only one-fifth of the global gas demand. Following the economic crisis in 2009, lower demand, and availability of cheaper sources of gas (LNG), many long-term contracts have been renegotiated in Europe so that they include now a partial spot indexation. However, long-term import prices in Asia remain closely linked to oil prices and stand therefore at much higher levels than in Europe or North America. This is becoming an issue for many Asian countries that are trying to move away from oil indexation.

Gas-to-gas competition (spot prices) can be found in North America, the United Kingdom, and parts of Continental Europe and represents one-third of the global gas demand. However, there is no such thing as a trading hub in Asia and it would take years to create one as highlighted in an International Energy Agency (IEA) special report [16]—*Developing a natural gas trading hub in Asia*. Prices in many other regions are often regulated: they can be set below costs, at cost of service, or be determined politically, reflecting perceived public needs. The IEA follows mostly the reported market prices in Europe, North America, and the OECD Pacific region. All these pricing regulations affect the market for natural gas. Pricing for synthetic gas depends on its supply and demand and in general not as regulated as natural gas, although in some parts pricing can be subsidized by local government such as the subsidy for hydrogen in California.

1.2.1 Conventional versus Unconventional Sources of Natural Gas

Traditionally, natural gas has been recovered using conventional drilling techniques [17–19]. These techniques have been useful to recover *associated* (with oil) and *nonassociated* (dry) gas from shallow gas reserves. In recent years, new technologies such as *fracking* and *horizontal and multidirectional drilling* have allowed recovery of natural gas from deeper and tighter unconventional gas reservoirs. As mentioned earlier, there are six types of unconventional gas: shale gas (found in shale deposits); CBM, also known as coal seam gas in Australia (extracted from coalbeds); tight gas (trapped underground in impermeable rock formations); deep gas (found in very deep [more than 3 miles] gas reservoirs); gas from geopressurized zones (sediments with trapped gas–water solutions at high pressure and temperature); and gas hydrates (generally found in arctic conditions or at the bottom of the sea).

Deep and ultradeep natural gas exists in deposits very far underground, beyond *conventional* drilling depths. This gas is typically 15,000 ft or deeper underground, quite a bit deeper than conventional gas deposits, which are traditionally only a few thousand feet deep at most. Deep drilling, exploration, and extraction techniques have substantially improved, making drilling for deep gas economical. However, deep gas is still more expensive to produce than conventional natural gas, and therefore economic conditions have to be such that it is profitable for the industry to extract from these sources.

Tight gas is stuck in a very tight formation underground [20–22], trapped in unusually impermeable, hard rock or in a sandstone or limestone formation that is unusually impermeable and nonporous (tight sand). Tight gas makes up a significant portion of the nation's natural gas resource

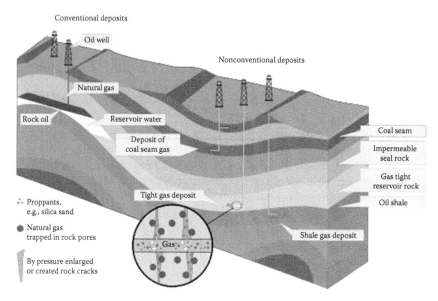

FIGURE 1.2 An illustration of tight gas compared to other types of deposits. (From Tight gas, Wikipedia, the free encyclopedia, 2015, https://en.wikipedia.org/wiki/Tight_gas; also from Shale gas, Wikipedia, the free encyclopedia, 2015, https://en.wikipedia.org/wiki/shale_gas.)

base, with the *Energy Information Administration* (EIA) [21–23] estimating that as of January 1, 2000, 253.83 trillion cubic feet (tcf) of technically recoverable deep natural gas exists in the United States. This represents over 21% of the total recoverable natural gas in the United States. An illustration of tight gas compared to other types of deposits is given in Figure 1.2.

Natural gas can also exist in shale deposits, which formed 350 million of years ago. Shale is a very-fine-grained sedimentary rock, which is easily breakable into thin, parallel layers. It is a very soft rock, but it does not disintegrate when it becomes wet. These shales can contain natural gas, usually when two thick, black shale deposits "sandwich" a thinner area of shale. Shale basins are scattered across the United States. As of November 2008, the FERC estimated [24] there was 742 tcf of technically recoverable shale gas resources in the United States. Shale represents a large and growing share of the U.S. recoverable resource base.

Coal, another fossil fuel, is formed underground under similar geologic conditions as natural gas and oil. Many coal seams also contain natural gas, either within the seam itself or the surrounding rock. This CBM is trapped underground and is generally not released into the atmosphere until coal mining activities unleash it. Today, however, CBM has become a popular unconventional form of natural gas. In June 2009, the Potential Gas Committee estimated that 163 tcf of technically recoverable CBM existed in the United States, which made up 7.8% of the total natural gas resource base [25–29].

Geopressurized zones are natural underground formations that are under unusually high pressure for their depth. These areas are formed by layers of clay that are deposited and compacted very quickly on top of more porous, absorbent material such as sand or silt. Water and natural gas that are present in this clay are squeezed out by the rapid compression of the clay, and these enter the more porous sand or silt deposits and are stored there under very high pressure (hence the term "geopressure"). Geopressurized zones are typically located at 10,000–25,000 ft below the surface of the earth and they contain methane and warm water solution under high pressure. Most of the geopressurized natural gas in the United States is located in the Gulf Coast region. The experts estimate that anywhere from 5,000 to 49,000 tcf of natural gas may exist in these areas [30,31].

Methane hydrates are made up of a lattice of frozen water, which forms a sort of "cage" around molecules of methane. These hydrates look like melting snow and were first discovered in permafrost regions of the Arctic. However, research on methane hydrates has revealed that they also exist at the bottom of the ocean. Experts estimated natural gas from 7,000 to over 73,000 tcf that may exist as gas hydrates. In fact, the USGS estimates that methane hydrates may contain more organic carbon than the world's coal, oil, and conventional natural gas combined. Hydrates are, however, spread over a large region and difficult to extract efficiently [30,31].

While different techniques are applied, depending on the type of gas being extracted, one method that has served well is hydraulic fracturing (also known as fracking) accompanied by horizontal or multidirectional drilling [32]. The process of horizontal or multidirectional drilling allows bypassing of impervious rocks, thus facilitating deep drilling. The horizontal drilling [17–19,32] (also see Chapter 2) also allows recovery of gas from shallow gas reservoirs without having multiple drilling wells. In the process of hydraulic fracturing, a well is drilled with porous ends, and large volumes of water (mixed with some sand and chemicals) are injected underground under high pressure to create cracks in the rock that remains open by sand. The chemicals facilitate loosening of trapped gas. The recovered trapped gas flow into the well bore so it can be produced. In the United States, tight gas has been produced for more than four decades and CBM for more than two decades [25–29]. Production of shale gas in the United States began much more recently and has rapidly increased from 2005 onward. In 2012, shale gas represented over a third of U.S. gas production, and its contribution to overall natural gas production in the United States is rapidly expanding.

Technological advances over the last two decades, particularly concerning hydraulic fracturing and soaring gas prices in the early 2000s, prompted drillers in the United States to pursue development of unconventional gas types more vigorously. The impressive growth in production in the United States over the last few years has actually led to a significant decline in gas prices, but output still remains robust and other countries have been prompted to explore whether they could enjoy similar results. Importing countries are keen to explore this, because if they are able to produce significant volumes of unconventional gas, they would have greater energy security and more energy independence—hence reducing a country's reliance on costly energy imports. Alternatively, some producers would be able to export more gas.

Canada already produces tight gas, CBM, and small amounts of shale gas. Although years behind the United States, some countries have made notable strides forward in exploring their own unconventional gas sources. Australia has shown good CBM potential, which is already being produced in small quantities. But Australia's future success most likely lies in projects that focus on producing LNG from CBM. China, India, and Indonesia have also produced small amounts of unconventional gas and are aggressively looking at ways to increase their respective volumes, notably among which China has very ambitious shale gas targets, that is, between 60 and 100 bcm by 2020.

Despite strong interest throughout Europe, there are areas of public concern—such as high population densities and issues surrounding the potentially damaging environmental impact by recovery of unconventional gas—which are hampering progress. A few other countries, such as Argentina, Algeria, and Mexico, seem also to have large shale gas potential. In the United States, of all unconventional sources of gas, "shale gas" recovery has made the most rapid progress. New shale gas fields could reconfigure the national map of gas producers and consumers. Gas production in the Marcellus shale and other burgeoning gas fields in the northeast, stretching from New England through the Great Lake states, are set to rise 78% by 2030. Under a carbon price regime, the researchers said gas production matches increasing gas consumption.

There are concerns about the environmental impact of unconventional gas production, notably in terms of land use, water use, and the potential for the contamination of drinking water and methane and other air emissions. There are also concerns about increasing minor earthquakes in the regions where a significant amount of hydraulic fracturing has taken place. In May 2012, a special *World Energy Outlook* report [33] looked at the environmental impact of unconventional gas production, including analysis of the "Golden Rules" that are needed to support a potential "Golden Age of Gas." So far, various organizations, including the Environmental Protection Agency (EPA) in the United States, have given green light to methods used for the recovery of unconventional gas. Cautions have been given to the practices that do not follow all safe protocols.

1.2.2 UNCONVENTIONAL OIL VERSUS GAS

As world conventional oil and gas supply diminishes, more efforts are being made to recover more oil and gas through unconventional means. Efforts are being made to recover the last parts of conventional oils, shale oils, bitumen, and heavy oils. The technologies that are used to recover these types of unconventional oils are more harmful to the environment, are more expensive, and involve more difficult upgrading strategies for recovered oils. The heavy oils from Venezuela and bitumen from Canada are more expensive to refine, require more hydrogen, and cause more harmful effects to the environment.

Unconventional gas, on the other hand, such as CBM, tight gas, deep gas and gas from geopressurized zones, shale gas, and gas from gas hydrates are very similar to natural gas (predominantly methane). While the hydraulic fracturing technology (i.e., fracking) [34] used to recover unconventional gas is controversial and may have some harmful environmental effects, these effects can be somewhat minimized by several precautionary actions. Unlike unconventional oils, the cleaning and upgrading of unconventional gas is not significantly different than that of conventional gas. They do not cause significantly more harmful effects to the environment. All unconventional gas is predominantly methane. Both conventional and unconventional gases are evaluated in more detail in Chapter 2.

1.3 SYNTHESIS GAS BY THERMAL GASIFICATION

Fundamentally, gaseous fuels are of two types: natural and synthetic or man-made. As shown in the last section, natural gaseous fuel, commonly known as natural gas, is obtained from underground by conventional or unconventional methods. Natural gas, irrespective of its source, predominantly contains methane and it is a nonrenewable source. Synthetic gas, on the other hand, is man-made and can be produced by three different methods: thermal gasification, anaerobic digestion, and hydrothermal gasification/water dissociation [35–38]. Unlike natural gas, the composition of synthetic gas can vary significantly depending on its method of production, nature of feedstock, and the prevailing operating conditions. Synthetic gas can have methane, propane, butane, syngas, or hydrogen as its predominant constituents. Synthetic gas can be produced from both nonrenewable and renewable feedstocks. It should be noted that oil refinery processes also produce methane, propane, and butane and they are a major source of liquid propane.

One of the major technologies for the production of synthesis gas is *thermal gasification* [35–38]. This technology produces synthetic gas of different compositions depending on the nature of the feedstock and operating conditions. It is generally carried out in the presence of air (or oxygen) and sometime carbon dioxide. When it is carried out in the presence of steam (or sub- and supercritical water) and/or hydrogen, it is known as "hydrothermal gasification."

The feedstocks for thermal gasification are normally coal, biomass, waste, oil shale, or their mixtures [35–38]. Coal gasification is a commercial technology used widely all over the world. In recent years, this technology has included a process for carbon capture to reduce its emission of carbon dioxide. The gasification of biomass is rapidly developing at a commercial scale and it can produce more hydrogen in product than coal gasification. Waste (municipal solid waste [MSW]) gasification is now considered to be a viable alternative to the landfill option and it is used both at small and large scales all over the world. Oil shale gasification is also practiced in some countries like Estonia, Turkey, and the former Soviet Union (see Chapter 3). In recent years, due to concerns on carbon dioxide emissions, efforts are also made to use mixed feedstock such as coal/biomass, coal/waste, and coal/oil shale [36].

The chemistry of thermal gasification varies with the nature of the feedstock. Subbituminous and bituminous coals are considered to be most suitable for coal gasification due to their favorable compositions of hydrogen, carbon, and oxygen. Biomass and MSW can be very heterogeneous and the gasification chemistry of these feedstocks will depend on the nature and composition of the particular biomass/waste used. Oil shale contains a significant amount of kerogen, which has an impact on the gasification product. In cogasification, the chemistry of gasification depends on the nature of the mixture. All of these factors are discussed in detail in Chapter 3.

Besides the feedstock type, two other parameters that affect product yield and composition during thermal gasification are feed preparation and the nature of gasifier and its operating conditions. Significant efforts have been made to prepare coal and biomass/waste for their effective gasification and to improve the quality of product gas. These feed preparation methods are discussed in detail in Chapter 3. Fundamentally, three types of gasifiers, namely, fixed (or moving) bed, fluidized bed, and entrained bed, and their numerous modifications are used in the thermal gasification depending on the nature of feedstock and operating conditions. The advantages and disadvantages of these different types of gasifiers (reactors) are also discussed in detail in Chapter 3.

Thermal gasification can produce either the so-called producer gas or syngas depending on the nature of the operating conditions. Producer gas contains significant amounts of methane and nitrogen (diluent) and these contents determine the Btu content of the gas. High methane concentration and low nitrogen concentration give high-Btu gas, while low methane concentration and high nitrogen concentration give low-Btu gas. At high temperature (>1200°C), syngas is produced, which largely contains hydrogen and carbon monoxide. Both producer gas and syngas can be generated from coal, biomass/waste, or their mixtures.

As shown in the following, producer gas and syngas have different downstream applications [35–37].

1.3.1 PRODUCER GAS

If the gasification product contains significant amounts of noncombustible gases such as nitrogen and carbon dioxide and a significant amount of methane, such a mixture is called "producer gas."

Producer gas is a generic term referring to the following:

- *Wood gas*: gas produced in a gasifier to power cars with ordinary internal combustion engines
- *Town gas*: manufactured gas, originally produced from coal, for sale to consumers and municipalities

Producer gas is a mixture of combustible (hydrogen, methane, and carbon monoxide) and noncombustible (nitrogen and carbon dioxide) gases. The heating value of producer gas varies from 4.5 to 6 MJ/m³ depending upon its constituents. Similar to syngas, producer gas is also produced by the gasification of carbonaceous materials such as coal or biomass. When atmospheric air is used as a gasification agent, the producer gas consists mostly of carbon monoxide, hydrogen, nitrogen, carbon dioxide, and methane. Depending on the methane and nitrogen content, the producer gas is called high-Btu, medium-Btu, or low-Btu gas. The typical composition of producer gas is illustrated in Table 1.2 ([35–37]; also see Chapter 3).

Applications of producer gas
Besides its use in power industry, producer gas is also used

1. As fuel for hot-air generators of the kind used to produce hot air in industries such as those involved with production of fertilizer and cement
2. As fuel for heating water in a number of industrial applications
3. As fuel for melting of glass in the production of artifacts
4. As fuel in food processing industry to provide heat for drying vegetables and seeds and for ovens in bakeries

TABLE 1.2

Typical Composition of Producer Gas

Carbon monoxide	18%–22%
Hydrogen	13%–19%
Methane	1%–5%
Heavier hydrocarbons	0.2%–0.4%
Carbon dioxide	9%–12%
Nitrogen	45%–55%
Water vapor	4%

Sources: Modified and adapted from Lee, S., Gasification of coal, Chapter 2, in: Lee, S. et al., eds., *Handbook of Alternative Fuel Technology*, Taylor & Francis Group, New York, 2007, pp. 26–78; Shah, Y., *Energy and Fuel Systems Integration*, CRC Press, Taylor & Francis Group, New York, 2015; Shah, Y.T., *Water for Energy and Fuel Production*, CRC Press, New York, 2014.

1.3.2 SYNGAS OR BIOSYNGAS

When thermal gasification of coal and biomass is carried out at very high temperatures and in the presence of pure oxygen, syngas (which mainly consists of hydrogen and carbon monoxide) is produced. When the feedstock used is biomass, the syngas is often called "biosyngas." Biosyngas contain very little tar and nitrogen. *Syngas* is used as a fuel source or as an intermediate for the production of liquid fuels, fuel additives, and chemicals.

Syngas, or *synthesis gas*, is a fuel gas mixture consisting primarily of hydrogen, carbon monoxide, and very often some carbon dioxide. The name comes from its use as intermediates in creating synthetic natural gas (SNG) and raw material for the production of ammonia or methanol. Syngas is also used as raw material for liquid fuels and fuel additives via Fischer–Tropsch (FT), methanol, or oxo- or isosyntheses. Syngas is combustible and often used as a fuel of internal combustion engines. It has less than half the energy density of natural gas. Syngas is used as an intermediate in the industrial synthesis of hydrogen. It can be also burnt and is used as a fuel source for numerous industrial applications and it is used as a fuel in power generation using integrated gasification combined-cycle (IGCC) power plant. Syngas is less harmful to the environment than natural gas. It can also be produced by gas reforming of methane and other hydrocarbons. This subject is treated in detail in Chapter 6.

There are numerous variations to the conventional thermal gasification examined in the literature. When coal is difficult to mine, the technology of underground gasification has been developed. In certain parts of the world this technology is now commercialized. The provision of heat during gasification can also be more efficiently implemented using solar energy or plasma. Efforts have been made to understand solar and plasma gasification. Efforts have also been made to carry out gasification in molten media (of salts or metals) environment. Energy efficiency of gasification has also been improved by novel process of indirect gasification where pyrolysis and gasification processes are separated from combustion process. Finally, efforts have also been made to improve gasification by use of cheap and disposable catalysts. All of these variations to conventional thermal gasification are discussed in detail in Chapter 3.

1.4 BIOGAS, BIOMETHANE, AND BIOHYDROGEN PRODUCTION BY ANAEROBIC DIGESTION

The second major method for synthetic gas is anaerobic digestion process to convert cellulosic biomass and waste into biogas or biohydrogen. This process occurs naturally in landfills to produce methane (also known as biogas). In recent years, the method is used to synthetically generate biogas (which can be upgraded to biomethane) and biohydrogen by manipulating and controlling operating conditions of anaerobic digestion process.

Methanogenic archaea are responsible for all biological sources of methane. When methane-rich gases are produced by the anaerobic decay of nonfossil organic matter (biomass or cellulosic waste), these are referred to as biogas. Natural sources of biogas include swamps, marshes, and landfills. Excluding water vapor, about half of landfill gas is methane and most of the rest is carbon dioxide, with small amounts of nitrogen, oxygen, and hydrogen, and variable trace amounts of hydrogen sulfide and siloxanes. If the gas is not removed, the pressure may get so high that it works its way to the surface, causing damage to the landfill structure, unpleasant odor, vegetation die-off, and explosion hazard. The gas can be vented to the atmosphere and flared or burned to produce electricity or heat [37–39].

The biogas can be produced by natural processes or by man-made synthetic processes. Synthetic biogas is produced by anaerobic digestion of organic materials from waste that otherwise goes to landfills or other primary, secondary, or tertiary treatment systems. This method is more efficient than just capturing the landfill gas it produces. The method is more controlled and flexible compared to the natural biogas generation process in landfill materials. This type of synthetic biogas can be produced and optimized from agricultural waste materials such as sewage sludge and manure by way of anaerobic digesters.

The method can also be used to produce hydrogen instead of methane by controlling the various steps of the anaerobic digestion process and utilizing process conditions and microbes that favor the production of hydrogen. Anaerobic lagoons produce biogas from manure, while biogas reactors can be used for manure or plant parts. In normal conditions, like landfill gas, synthetic biogas is mostly methane and carbon dioxide, with small amounts of nitrogen, oxygen, and hydrogen. However, with the exception of pesticides, there are usually lower levels of contaminants. As shown in Chapter 4, unlike methane produced in landfill, methane produced by synthetic processes can be optimized by manipulations of substrates, digester design, operating conditions, and nature of microbes. In recent years, the use of multiple substrates (codigestion) has optimized the methane production rate and efficiency of the space within the treatment plants. Codigestion has become a standard commercial practice in synthetic biogas production from numerous different types of substrates [36]. Chapter 4 also describes in detail the anaerobic digestion process to produce hydrogen with the use of different types of microbes.

Landfill gas or synthetic biogas cannot be distributed through utility natural gas pipelines unless it is cleaned up to less than 3% CO_2, and a few parts per million H_2S, because CO_2 and H_2S corrode the pipelines. The presence of CO_2 will lower the energy level of the gas below requirements for the pipeline. Siloxanes in the gas will also form deposits in gas burners and need to be removed prior to entry into any gas distribution or transmission system. Consequently, it may be more economical to burn the gas on-site or within a short distance of the landfill using a dedicated pipeline. Water vapor is often removed, even if the gas is burned on-site. If low temperatures condense water out of the gas, siloxanes can be lowered as well because they tend to condense out with the water vapor. Other nonmethane components may also be removed to meet emission standards to prevent fouling of the equipment or for environmental considerations. The upgrading of biogas produces biomethane (containing 95%–100% methane), which can be inserted into storage and transportation infrastructure of natural gas. Another approach is to co-fire landfill gas with natural gas that improves combustion and lowers emissions.

Synthetic biogas and landfill gas are already sources of energy for heat and power production. In some areas, their use, however, could be greatly expanded in the future. Gas generated in sewage treatment plants is commonly used to generate electricity. For example, the Hyperion sewage plant in Los Angeles burns 8 mcf (230,000 m^3) of biogas per day to generate power. New York City utilizes biogas to run equipment in the sewage plants, to generate electricity, and in boilers. Using sewage gas to make electricity is not limited to large cities. The city of Bakersfield, California, uses cogeneration at its sewer plants. California has 242 sewage wastewater treatment plants, 74 of which have installed anaerobic digesters. The total biopower generation from the 74 plants is about 66 MW [39].

While anaerobic digestion is a slow process, it is a low-temperature and most-energy-efficient process. The treatment of waste by anaerobic digestion is very popular in Europe and many parts of Asia. Germany, Denmark, and Holland have significant numbers of large-scale commercial biogas operations. These operations are being constantly optimized with the use of suitable mixtures of substrates, efficient digester designs, and new types of microbes. New microbes are also investigated to produce biohydrogen by anaerobic process. All of these topics are described in detail in Chapter 4.

1.5 HYDROGEN PRODUCTION BY HYDROTHERMAL GASIFICATION AND WATER DISSOCIATION

While synthetic gas production by thermal gasification produces synthesis gas and anaerobic digestion largely produces methane, the production of hydrogen or gas concentrated in hydrogen is very important for the energy and fuel industry. Hydrogen can be produced by numerous methods listed as follows [35–44]:

1. Gasification of carbonaceous materials by steam with or without hydrogen
2. Hydrothermal gasification of biomass/waste in subcritical water
3. Supercritical water gasification of all carbonaceous materials
4. Water dissociation by electrolysis
5. Photocatalytic and photobiological dissociation of water
6. Thermal and thermochemical dissociation of water
7. Novel methods for water dissociation
8. Production of hydrogen by anaerobic digestion
9. Production of hydrogen by steam reforming of carbonaceous materials including methane
10. Production of hydrogen by aqueous phase reforming of carbohydrates
11. Production of hydrogen by supercritical water reforming

As mentioned earlier, Chapter 4 discusses biohydrogen formation by anaerobic digestion (method 8). Chapter 5 examines hydrogen and hydrogen-rich gas productions by hydrothermal gasification and water dissociation (methods 1–7 listed above). Hydrogen production by various reforming processes (methods 9–11) is examined in Chapter 6.

The gasification in the presence of steam, water, supercritical water, and/or hydrogen produces gas rich in hydrogen from of all carbonaceous materials (coal, conventional and unconventional oils, biomass, waste, etc.). Chapter 5 discusses and describes all options for hydrothermal gasification. Hydrothermal gasification in supercritical water is a novel approach that evolved in the 1970s and has made significant progress in recent years. While this method can be applied to all carbonaceous feedstock, its application to biomass and waste has been found to be more promising for the production of hydrogen-rich gas. The process has been demonstrated at the pilot scale.

The term "hydrothermal gas" [37,38] needs to be separated from hydrogen produced by "hydrogasification," a process of thermal gasification of carbonaceous materials in the presence of hydrogen. While hydrogasification generally produces gas rich in hydrogen, it is an expensive process and often requires a catalyst to get desirable product distribution. The main actor in the production of "hydrothermal gas" is water in its different forms (steam, water, supercritical water, etc.) reacting with different types of carbonaceous materials. The commercial success of this process is very desirable because of unlimited resource of water and all carbonaceous materials, particularly biomass and wastes. Hydrothermal gasification can be carried out in the presence of hydrogen to further improve the production of hydrogen. This method is also discussed in Chapter 5.

Hydrogen can also be produced by water dissociation [36,37]. Currently, three fundamental processes for water dissociation are being pursued: electrolysis, photocatalytic and photobiological dissociation, and thermal and thermochemical water dissociation. Within each category, numerous variations have been examined. For example, the process of electrolysis can be carried out at high temperature and high pressure with the use of nuclear heat, plasma, etc. All of these variations are examined in Chapter 5. Thermochemical dissociation methods involve two, three, four, and even five step reactions. In fact, there are at least 400 different types of thermochemical reaction options examined in the literature. Chapter 5 evaluates some of more important two, three, and four step options. Numerous catalysts and microbes for thermocatalytic and thermobiological methods are also assessed in Chapter 5. There are also minor efforts in numerous novel technologies for water dissociation. While many options for water dissociation have great potential, most processes are at the laboratory or demonstration levels. The uses of solar and nuclear heat for water dissociation and hydrothermal gasification have also been examined, some of which are discussed in Chapter 5.

Unlike processes described in Chapters 3 and 4, hydrothermal gasification processes in sub- and supercritical waters as well as various methods for water dissociation are as yet not fully commercialized. Since hydrogen is the "ultimate fuel," future commercial success in the technologies discussed in Chapter 5 can have a very significant impact on future landscape of the energy and fuel industry.

1.6 GAS REFORMING

One reason why gas is so valuable compared to coal and oil is that not only clean gas can be used for smaller- and larger-scale power application, but also it can also be used to produce various types of liquid fuels such as middle distillate, gas oil, gasoline, diesel, and jet fuel through various syngas conversion processes such as the FT process and iso- and oxosyntheses. Methane can also be converted to methanol and higher alcohols. Methanol then can be converted to other fuels and fuel additives like dimethyl ether (DME), gasoline, and olefins. The chemistry for conversion of GTL fuels requires right composition of syngas (hydrogen to CO ratio) in the gas. This ratio is achieved by steam, dry (CO_2), partial oxidation (POX), and tri-reforming processes. Subcritical water reforming can also be used to convert cellulosic materials to hydrogen and syngas. Supercritical water reforming can be used to convert all carbonaceous materials in the hydrogen-rich gas. Gas reforming is a very powerful tool to prepare gas of the required composition for various downstream operations [35–38,40,42].

As mentioned earlier, there are numerous methods for gas reforming. Chapter 6 examines all different types of gas reforming processes. The most basic gas reforming is the water gas shift reaction (henceforth denoted as WGS reaction) that balances hydrogen to carbon monoxide ratio in the syngas via reversible reaction between carbon monoxide and water to produce hydrogen and carbon dioxide. A mixture of only carbon monoxide and hydrogen is called water gas. Water gas is generally used for the production of hydrogen. Since both constituents of water gas are combustible gases, they can be used as input to gas turbine for power production. WGSR is used to remove carbon monoxide from water gas to get pure hydrogen for the fuel cell applications. Carbon monoxide is produced from the reduction of carbon dioxide and it is generally 18%–22% on volume basis. Its octane number is 106 but its burning velocity is low. It is very toxic in nature. Hydrogen is also the product of reduction process in gasification. Its octane number is in the range of 60–66 but its burning velocity is very high. Hence, it increases the burning velocity of producer gases.

WGSR describes the reaction of carbon monoxide and water vapor to form carbon dioxide and hydrogen [42]:

$$CO + H_2O \rightleftharpoons CO_2 + H_2 \tag{1.1}$$

WGSR was discovered by the Italian physicist Felice Fontana in 1780. Before the early twentieth century, hydrogen was obtained by reacting steam under high pressure with iron to produce iron, iron oxide, and hydrogen. With the development of industrial processes that required hydrogen, such as the Haber–Bosch ammonia synthesis, the demand for a less expensive and more efficient method of hydrogen production was needed. As a resolution to this problem, WGSR was combined with the gasification of coal to produce pure hydrogen. As the idea of hydrogen economy gains popularity, the focus on hydrogen as a replacement fuel source for hydrocarbons is increasing.

WGSR is an important industrial reaction that is used in the manufacture of ammonia, hydrocarbons, methanol, and hydrogen. It is also often used in conjunction with steam reformation of methane and other hydrocarbons. In the FT process, WGSR is one of the most important reactions used to balance H_2/CO ratio. It provides a source of hydrogen at the expense of carbon monoxide, which is important for the production of high-purity hydrogen for use in ammonia synthesis.

The equilibrium of this reaction shows significant temperature dependence and the equilibrium constant decreases with an increase in temperature. Thus, optimum conversion of CO into hydrogen requires an optimum use of both faster kinetics at higher temperature and favorable thermodynamics at lower temperature. In industrial practice, this is achieved by carrying out WGSR in two stages: high-temperature shift (HTS) and low-temperature shift (LTS) with interstage cooling in between. HTS uses sulfur-resistant iron oxide–chromium oxide catalyst and converts carbon monoxide to 2%–4% level in the exit gas concentration. LTS uses copper catalyst. A sulfur removal guard chamber is used in between HTS and LTS reactors to protect the copper catalyst from sulfur poisoning. LTS brings CO concentration to less than 1% in the product gas. While both the HTS and LTS catalysts are commercially available, their formulations are continued to be further developed. All of these aspects of WGSR are discussed in detail in Chapter 6.

1.6.1 Steam, Dry, and Tri-Reforming

Reforming of natural gas or other carbonaceous materials to produce syngas or hydrogen is one of the very important reactions in the energy and fuel industry. Gas reforming is different from hydrocarbon reforming to produce higher octane number fuel in refining industry. While steam reforming of natural gas is still the most prominent method to produce hydrogen, in recent years numerous other methods of gas reforming have evolved [36–38,42,45]. This subject is also discussed in detail in Chapter 6.

Steam reforming of natural gas—sometimes referred to as steam methane reforming (SMR)—is the most common method of producing commercial bulk hydrogen. Hydrogen is used in the industrial synthesis of ammonia and other chemicals [5]. At high temperatures (700°C–1100°C) and in the presence of a metal-based catalyst (nickel), steam reacts with methane to yield carbon monoxide and hydrogen:

$$CH_4 + H_2O \rightleftharpoons CO + 3H_2 \tag{1.2}$$

Additional hydrogen can be recovered by a lower-temperature gas-shift reaction with the carbon monoxide produced (Equation 1.1).

The first reaction is strongly endothermic (consumes heat, $\Delta H_r = 206$ kJ/mol), while the second reaction is mildly exothermic (produces heat, $\Delta H_r = -41$ kJ/mol). SMR is approximately 65%–75% efficient. The United States produces nine million tons of hydrogen per year, mostly with steam reforming of natural gas. The worldwide ammonia production, using hydrogen derived from steam reforming, was 109 million metric tons in 2004 [42].

Carbon dioxide reforming (also known as "dry reforming") is a method of producing synthesis gas (mixtures of hydrogen and carbon monoxide) from the reaction of carbon dioxide with hydrocarbons such as methane. Synthesis gas is conventionally produced via the steam reforming reaction. In recent years, increased concerns on the contribution of greenhouse gases to global warming have increased interest in the replacement of steam as reactant with carbon dioxide [45].

The dry reforming reaction may be represented by

$$CO_2 + CH_4 \rightarrow 2H_2 + 2CO \tag{1.3}$$

Thus, two greenhouse gases are consumed and useful chemical building blocks, hydrogen and carbon monoxide, are produced. A challenge to the commercialization of this process is that the hydrogen that is produced tends to react with the carbon dioxide. For example, the following reverse WGSR typically proceeds with a lower activation energy than the dry reforming reaction itself:

$$CO_2 + H_2 \rightarrow H_2O + CO \tag{1.4}$$

Unlike steam and dry reforming, partial oxidation (*POX*) is an exothermic chemical reaction. It occurs when a substoichiometric fuel–air mixture is partially combusted in a reformer, creating a hydrogen-rich syngas, which can then be put to further use, for example, in a fuel cell. A distinction is made between *thermal POX* (TPOX) and *catalytic POX* (CPOX).

POX is a technically mature process in which natural gas or a heavy hydrocarbon fuel (heating oil) is mixed with a limited amount of oxygen in an exothermic process [46].

- General reaction equation (without catalyst, TPOX):

$$C_nH_m + \frac{2n+m}{4}O_2 \rightarrow nCO + \frac{m}{2}H_2O \tag{1.5}$$

- General reaction equation (with catalyst, CPOX):

$$C_nH_m + \frac{n}{2}O_2 \rightarrow nCO + \frac{m}{2}H_2 \tag{1.6}$$

POX can also be carried out with oil or coal and these reactions can be represented as [46]

- Possible reaction equation (heating oil):

$$C_{12}H_{24} + 6O_2 \rightarrow 12CO + 12H_2 \tag{1.7}$$

- Possible reaction equation (coal):

$$C_{24}H_{12} + 12O_2 \rightarrow 24CO + 6H_2 \tag{1.8}$$

The formulas given for coal and heating oil show only a typical representative of these highly complex mixtures. Water is added to the process for getting both the extreme temperatures and extra control on the formation of soot. *TPOX* reactions, which are dependent on the air–fuel ratio, proceed at temperatures of 1200°C and above. In *CPOX* the use of a catalyst reduces the required temperature to around 800°C–900°C [46].

The choice between TPOX and CPOX depends on the sulfur content of the fuel being used. CPOX can be employed if the sulfur content is below 50 ppm. Higher sulfur content can poison the catalyst, so the TPOX procedure is used for such fuels [46]. However, recent research shows that CPOX is possible with sulfur contents up to 400 ppm [46].

Tri-reforming process is a three-step reaction process. It avoids the separation step and has the promise of being cost efficient for producing industrially useful synthesis gas. Tri-reforming refers to simultaneous reforming of oxidative CO_2–steam from natural gas. It is a synergetic combination of endothermic CO_2 reforming (Equation 1.9), steam reforming (Equation 1.10), and exothermic oxidations of methane (Equations 1.11 and 1.12). Combining CO_2 reforming,

$$CH_4 + CO_2 \rightarrow 2CO + 2H_2 \quad [\Delta H^\circ = 247.3 \text{ kJ/mol}] \tag{1.9}$$

$$CH_4 + H_2O \rightarrow CO + 3H_2 \quad [\Delta H^\circ = 206.3 \text{ kJ/mol}] \tag{1.10}$$

$$CH_4 + \frac{1}{2}O_2 \rightarrow CO + 2H_2 \quad [\Delta H^\circ = -35.6 \text{ kJ/mol}] \tag{1.11}$$

$$CH_4 + 2O_2 \rightarrow CO_2 + 2H_2O \quad [\Delta H^\circ = -880 \text{ kJ/mol}] \tag{1.12}$$

steam reforming and POX can yield syngas with the desired H_2/CO ratios for methanol synthesis, FT synthesis, isosynthesis, oxosynthesis, etc.

Steam reforming is widely used in industry for making H_2 [36–38,42,45–48]. When CO-rich syngas for oxosynthesis and syngas with a H_2/CO ratio of 2 are needed for FT synthesis and methanol synthesis, steam reforming alone cannot give the desired H_2/CO ratio [36–38,42,45–48]. Steam reforming gives a H_2/CO ratio of 3, which is too high and thus needs to import CO_2 for making syngas with H_2/CO ratios of 2 or lower [36–38,42,45].

The CO_2 reforming (dry reforming) of methane has attracted considerable attention worldwide [36–38,42,45–48]. CO_2 reforming is 20% more endothermic than steam reforming; however, it is necessary to adjust the H_2/CO ratio for making MeOH or FT syngas. Two industrial processes that use this reaction are SPARG [45] and Calcor [45].

Carbon formation in the CO_2 reforming of methane is a major problem (Equations 1.14 and 1.15), particularly at elevated pressures [36–38,42,45].

$$CH_4 \rightarrow C + 2H_2 \quad [\Delta H^\circ = 74.9 \text{ kJ/mol}] \tag{1.13}$$

$$2CO \rightarrow C + CO_2 \quad [\Delta H^\circ = -172.2 \text{ kJ/mol}] \tag{1.14}$$

$$C + CO_2 \rightarrow 2CO \quad [\Delta H^\circ = 172.2 \text{ kJ/mol}] \tag{1.15}$$

$$C + H_2O \rightarrow CO + H_2 \quad [\Delta H^\circ = 131.4 \text{ kJ/mol}] \tag{1.16}$$

$$C + O_2 \rightarrow CO_2 \quad [\Delta H^\circ = -393.7 \text{ kJ/mol}] \tag{1.17}$$

When CO_2 reforming is coupled with steam reforming (Equations 1.9 and 1.10), this problem can be mitigated effectively (Equations 1.5 and 1.16). This carbon formation in CO_2 reforming can also be reduced by adding oxygen (Equation 1.17).

Direct POX of methane to produce syngas [36–38,42,45–48] and partial combustion of methane for energy-efficient autothermal syngas production [36–38,42,45–48] are being explored. These reactions are important, but the *CPOX is difficult to control* [36–38,42,45–48]. *The major operating problems in CPOX* are the overheating and hot spot formation caused by the exothermic nature of the oxidation reactions. Consequently, coupling the exothermic reaction with an endothermic reaction could solve this problem [36–38,42,45–48].

The combination of dry reforming with steam reforming can accomplish two important missions: to produce syngas with the desired H_2/CO ratios and to mitigate the carbon formation that is significant for dry reforming. Integrating steam reforming and POX with CO_2 reforming could dramatically reduce or eliminate carbon formation on reforming catalyst, thus increasing catalyst life and process efficiency. Therefore, the proposed tri-reforming can solve two important problems that are encountered in individual processing. Incorporating oxygen in the reaction also generates heat in situ that can be used to increase energy efficiency; oxygen also reduces or eliminates the carbon formation on the reforming catalyst. The tri-reforming can be achieved with natural gas and flue gases using the waste heat in the power plant and the heat generated in situ from oxidation with the oxygen that is present in flue gas. Chapter 6 discusses steam, dry, and tri-reforming processes in full detail.

1.6.2 Reforming in Sub- and Supercritical Water

The classic work done by Dumesic and coworkers ([37], also see Chapter 6) on aqueous phase reforming of carbohydrates (C:O 1:1) in subcritical water to produce hydrogen is now advanced to the level that hydrogen can be produced from many cellulosic materials by the bioforming process that is being developed by Virent Co., Madison, WI [37]. This process along with the basic principles of aqueous phase reforming is described in detail in Chapter 6.

Reforming of carbonaceous materials can also be carried out in supercritical water. Supercritical water offers many advantages to the catalytic reforming process. Literature studies [36–38,45] show that reforming under supercritical conditions may allow the same level of performance for syngas production, catalyst activity, and stability at a lower temperature than what is required for the current industrial reforming processes. Osada et al. [41] showed that at 400°C and 420°C temperatures and at 400 atm pressure, the yields of hydrogen and carbon monoxide increased with the water density during POX of nC_{16}. Supercritical conditions also improve catalyst performance by providing increased pore accessibility, better product selectivity, and more resistance to the coke deposition. The effects of supercritical water conditions on the reforming of diesel oil were examined by Pinkwart et al. [49] and the reforming of liquid hydrocarbon fuels by Shekhawat et al. [50]. Both studies showed similar results of increased hydrogen production and lowering of the required reaction temperature to obtain the same level of performance. These and many other literature studies for reforming in supercritical water are examined in Chapter 6.

1.6.3 Novel Methods for Reforming

In recent years, plasma has also been used to provide heat and carry out gas reforming in an efficient manner. Both nonthermal and thermal plasmas have been investigated. Different configurations of plasma and plasma reactors have also been investigated. Plasma can be used as a catalyst or can be accompanied by another solid catalyst. Hybrid plasma catalytic process has shown significant potential. Plasma reforming needs to be further examined both at laboratory scale and at large scale that involves sound procedure for plasma reactor scale-up. This subject is also discussed in Chapter 6.

Since both steam and dry reforming are endothermic processes and require high (about 800°C) temperature, significant efforts have been made to provide heat by novel techniques. Three methods investigated in the literature are microwave-assisted heating (using carbon or oxides of carbon), solar energy, and waste nuclear heat. These three methods have shown significant success at the laboratory scale and need to be developed at a larger scale. The use of cogeneration methods for solar nuclear and geothermal energy has shown great potential in the industry. These methods along with other novel reforming processes are discussed in detail in Chapter 6.

Finally, the effectiveness of gas reforming at both large and small scales requires novel reactor designs, particularly if microwave-, solar-, nuclear-, or plasma-based heating processes are involved. The efficient use of heat transfer is an important design problem for the reforming process. Chapter 6 also evaluates the effectiveness of various novel reforming reactors.

1.7 GAS PURIFICATION AND UPGRADING

Gas is a cleaner fuel than coal or oil because all of its chemical components can be well detected, so their removal technologies can be easily identified, monitored, and controlled. The need and extent of gas cleaning depends on its sources and its end usages. Generally, natural gas transported through a pipeline is at least 95%–99% pure methane. Significant impurities in this stream can cause problems in pipelines and downstream operations. The shale gas coming out of ground often contains only 55%–60% methane and it contains very large concentration of carbon dioxide and significant amounts of volatile higher hydrocarbons such as ethane, propane, butane, olefins, and acetylenes. These need to be separated by appropriate purification processes. Also, the composition of shale gas can vary from location to location. Tight gas, deep gas, and gas from geopressurized zones can have different gas compositions depending on the characteristics of the reservoir. All of these need to be prepared as pipeline quality gas before their transport to the desired locations.

While the impurities and their levels to some extent depend on the nature and location of natural gas, in general natural gas contains the following set of impurities [51]:

1. *Heavier gaseous hydrocarbons*: Ethane (C_2H_6), propane (C_3H_8), normal butane (n-C_4H_{10}), isobutane (i-C_4H_{10}), pentanes, and even higher-molecular-weight hydrocarbons. When processed and purified into finished by-products, all of these are collectively referred to as "NGLs."
2. *Acid gases*: Carbon dioxide (CO_2), hydrogen sulfide (H_2S), and mercaptans such as methanethiol (CH_3SH) and ethanethiol (C_2H_5SH).
3. *Other gases*: Nitrogen (N_2) and helium (He).
4. *Water*: Water vapor and liquid water. Also dissolved salts and dissolved gases (acids).
5. *Liquid hydrocarbons*: Perhaps some natural gas condensate (also referred to as "casing-head gasoline" or "natural gasoline") and/or crude oil.
6. *Mercury*: Very small amounts of mercury, primarily in elemental form, but chlorides and other species are possibly present.
7. *Naturally occurring radioactive material*: Natural gas may contain radon, and produced water may contain dissolved traces of radium, which can accumulate within piping and processing equipment. This can render piping and equipment radioactive over time.

The raw natural gas must be purified to meet the quality standards specified by the major pipeline transmission and distribution companies. Those quality standards vary from pipeline to pipeline and are usually a function of a pipeline system's design and the markets that it serves. In general, the standards require the following [51]:

1. Natural gas is within a specific range of heating value (caloric value). For example, in the United States, it should be about 1035 ± 5% Btu per cubic foot of gas at 1 atm and 60°F (41 MJ ± 5% per m³ of gas at 1 atm and 15.6°C).
2. Natural gas is delivered at or above a specified hydrocarbon dew-point temperature (below which some of the hydrocarbons in the gas might condense at pipeline pressure forming liquid slugs that could damage the pipeline). Dew-point adjustment serves the reduction of the concentration of water and heavy hydrocarbons in natural gas to such an extent that no condensation occurs during the ensuing transport in the pipelines. Natural gas should also be free of particulate solids and liquid water to prevent erosion, corrosion, or other damage to the pipeline.
3. Natural gas should be dehydrated of water vapor sufficiently to prevent the formation of methane hydrates within the gas processing plant or subsequently within the sales gas transmission pipeline. A typical water content specification in the United States is that gas must contain no more than 7 lb of water per million standard cubic feet (mscf) of gas [51].

4. Natural gas contains no more than trace amounts of components such as hydrogen sulfide, carbon dioxide, mercaptans, and nitrogen. The most common specification for hydrogen sulfide content is 0.25 grain H_2S per 100 ft^3 of gas, or approximately 4 ppm. Specifications for CO_2 typically limit the content to no more than 2% or 3% [51].

5. Mercury level in natural gas is at less than detectable limits (approximately 0.001 ppb by volume) primarily to avoid damaging equipment in the gas processing plant or the pipeline transmission system from mercury amalgamation and embrittlement of aluminum and other metals.

All the impurities in natural gas obtained at wellhead are removed in natural gas processing plants, which are either located right at the wellhead or skid mounted so they can be moved around in different locations. In some cases, raw natural gas is transported to a centralized natural gas processing plant for its transformation to pipeline quality gas. Ethane, propane, and butane are the primary heavy hydrocarbons extracted at a natural gas processing plant, but other petroleum gases, such as isobutane, pentanes, and normal gasoline, also may be processed. Numerous methods are available to treat these contaminants and these are described in detail in Chapter 7.

The composition of synthetic gas produced from nonrenewable or renewable sources by any one of the three methods described earlier depends on the chemical properties of original feedstock and the nature of the operating conditions. The synthetic gas generally contains methane, hydrogen, carbon monoxide, carbon dioxide, water, nitrogen, oxygen, as well as several other pollutants. The pollutants generally found in synthetic gas are particulates, tar (various polyaromatic hydrocarbons), sulfur, nitrogen, chlorine compounds, alkali and heavy metals, fly ash, etc., depending on the overall gasification conditions. These pollutants must be removed to the acceptable level before the gas can be used for power, heat, or liquid fuel productions. The maximum allowable concentrations of contaminants in syngas for downstream applications are outlined in Table 1.3 [36–38,52] and in Chapter 7.

The strategies for the contaminant removals from producer gas, syngas, or hydrothermal gas are threefold. Some of the contaminants such as sulfur, chlorine, alkaline matters, and heavy metals can be removed by feed pretreatment. The method of pretreatment depends on the nature of feedstock and their inherent composition. For example, sulfur from coal can be removed by physical

---

TABLE 1.3

Allowable Concentrations of Contaminants in Syngas for Catalytic Synthesis

Syngas Contaminants	Contaminant Specification
$H_2S + COS + CS_2$	<1 ppmv
$NH_3 + HCN$	<1 ppmv
$HCl + HBr + HF$	<10 ppbv
Alkali metals (Na + K)	<10 ppbv
Particles (soot, ash)	"Almost completely removed"
Organic components (viz., tar)	Removed to a level at which condensation occurs upon compression to FT synthesis pressure (25–60 bar)
Heteroorganic components (incl. S, N, O)	<1 ppmv
CO_2, N_2, CH_4, and larger hydrocarbons	"Soft maximum" of 15 vol% that has been identified (the lower, the better)

Sources: Ratafia-Brown, J. et al., Assessment of technologies for co-converting coal and biomass to a clean syngas—Task 2 report (RDS), DOE/NETL-403.01.08 Activity 2 Report, Department of Energy, Washington, D.C., May 10, 2007; Shah, Y.T., Biomass to liquid fuel via Fischer–Tropsch and related syntheses, Chapter 12, in: Lee, J.W., ed., *Advanced Biofuels and Bioproducts*, Springer Book Project, Springer Publ. Co., New York, September 2012, pp. 185–207; Shah, Y., *Energy and Fuel Systems Integration*, CRC Press, Taylor & Francis Group, New York, 2015.

and chemical cleaning of coal that results in less generation of sulfur compounds in the product gas. Same things can be done for biomass where feed pretreatment can reduce a number of impurities [36–38]. Various feed pretreatment methods are described in Chapters 3 and 7.

The second strategy is to manipulate operating conditions for gasification so as to minimize the formation of contaminant compounds in the product gas. This manipulation can involve the design of the gasifier, operating temperature and pressure, and the use of mixture (such as coal and biomass) where synergistic chemical reactions can affect the level of contaminants in the product. For example, tar-free, methane-free, and nitrogen-free syngas can be produced by an oxygen-blown, high-temperature entrained bed reactor. High temperature (temperature greater than approximately 1200°C) produces clean syngas with no methane. High temperature also causes slagging that carries ash in the molten stage. All tars can also be decomposed to syngas at very high temperatures. While gasifying coal–biomass mixtures, mineral matters in biomass can act as catalyst for the minimization of several contaminants. These and other issues related to this strategy are also described in Chapter 7.

Finally, there are numerous technologies available to treat syngas at low or high temperatures to remove different types of contaminants. These technologies are briefly described in Chapter 7.

The purification and upgrading of biogas and biohydrogen often follows the same processes as those for natural gas since biogas (or biomethane) is generated at low temperatures. The major impurity in biogas is carbon dioxide. The methods to clean and upgrade biogas are also described in Chapter 7.

1.8   GAS STORAGE AND TRANSPORTATION INFRASTRUCTURE (GAS GRID)

The constant, steady, and sustainable supply of natural gas requires its storage and transport infrastructure. There are numerous methods of gas storage (mostly underground) that are both natural and man-made. The storage is dispersed in various regions across the country. Underground gas storage (UGS) facilities largely contribute to the reliability of gas supplies to consumers. They level off daily gas consumption fluctuations and meet the peak demand during winter [36].

Basically, there are three types of UGS facilities: depleted reservoirs, aquifer, and salt caverns. Storage facilities in a depleted field or aquifer feature large capacities but low flexibility. Depleted reservoirs can be large and contain the major fraction of natural gas storage. Aquifers are underground porous, permeable rock formations that act as natural water reservoirs. However, in certain situations, these water containing formations may be reconditioned and used as natural gas storage facilities. Injection and extraction of gas are much faster in the storage facilities that were built in rock salt caverns (though they are inferior to the UGS facilities built in depleted fields in terms of capacity). Salt caverns are ideally impermeable reservoirs. It is not difficult to build an underground salt caverns, though it's a long process. Wells are drilled in a fitting bed of rock salt. Afterward, water is pumped into them and a cavity of the required size is washed out in the salt bed. A salt dome is not only impermeable to gas: salt is capable of "self-healing" fissures and fractures [11,15,53–57].

Gas can also be stored in a liquefied state. This is the most expensive of all storage options, but this solution is applicable when it is impossible to build other storage facilities near large consumers. During the construction of an underground storage facility, a portion of gas is trapped in the reservoir to build up the required pressure. This gas is called "a buffer gas." Its volume is about half of the total gas injected in the storage facility. Gas that will be later extracted from a UGS facility is called "active" or "working" gas. All of these issues are discussed in detail in Chapter 8.

Natural gas pipelines are used to move gas from the field to consumers. Gas produced from onshore and offshore facilities is transported via gathering systems and inter- and intrastate pipelines to residential, commercial, industrial, and utility companies.

A natural gas pipeline uses pressure from compressors to move the gas through the pipeline. There were 185,744 miles of interstate natural gas pipelines in the United States in 1997 and more than 300,000 miles in 2012. Most natural gas pipelines are operated, managed, and controlled by complex and fully automated computer systems [36].

An interesting aspect of natural gas pipelines is the introduction of odorants into the gas system. Natural gas is almost odorless as it comes from the well or processing facility. If the gas is destined for use as fuel in homes or industries, a chemical called "mercaptan" is added to give the gas a distinctive odor so that people can easily smell it when its concentration in air reaches 1%. Gas and air mixed in this concentration are not hazardous, but a mixture containing 5% gas is explosive. The odor is used to detect leaks in the pipelines.

Transporting natural gas thousands of miles through pipelines is the safest method of transportation. The transportation system for natural gas consists of a complex network of pipelines, designed to quickly and efficiently transport natural gas from its origin to areas of high natural gas demand. There are three major types of pipelines along the transportation route: the gathering system, the interstate pipeline system, and the distribution system. The gathering system consists of low-pressure, small-diameter pipelines that transport raw natural gas from the wellhead to the processing plant.

Pipelines can be classified as interstate or intrastate. Interstate pipelines are similar to the interstate highway system: they carry natural gas across state boundaries and, in some cases, across the country. Transmission pipes can measure anywhere from 6 to 48 in. in diameter. The actual pipeline consists of a strong carbon steel material, engineered to meet vigorous standards. The pipe is also covered with a specialized coating to ensure that it does not corrode once placed in the ground. The purpose of the coating is to protect the pipe from moisture, which causes corrosion and rusting.

Natural gas is highly pressurized as it travels through an interstate pipeline. To ensure that the natural gas flowing through any one pipeline remains pressurized, compression of this natural gas is required periodically along the pipe. This is accomplished by compressor stations, usually placed at 40–100 mile intervals along the pipeline. The natural gas enters the compressor station, where it is compressed by either a turbine, motor, or engine [11,15,53–57].

In addition to compressing natural gas to reduce its volume and push it through the pipe, metering stations are placed periodically along interstate natural gas pipelines. These stations allow pipeline companies to monitor the natural gas in their pipes. The flow is controlled by numerous valves on the pipes. The flow is controlled by the centralized gas control stations that collect, assimilate, and manage data received from monitoring and compressor stations along the pipe.

CNG is transported at high pressure, typically above 200 bars. Compressors and decompression equipment are less capital intensive and may be economical in smaller unit sizes than liquefaction/regasification plants. CNG trucks and carriers may transport natural gas directly to end users or to distribution points such as pipelines [11].

In order to transport natural gas in areas not served by pipelines, the gas is LNG to reduce its volume. When the gas is liquefied, it shrinks to 1/600 of its gaseous volume. Tankers equipped with pressurized, refrigerated, and insulated tanks are used to transport NGLs such as *LNG* and LPG. Natural gas is liquefied at the destination point and transported by special LNG cryogenic tankers to its destination. In order to liquefy the gas, its temperature is lowered to $-259°F$ ($-162°C$). Natural gas is kept in refrigerated and insulated tanks to maintain in its liquefied state during transport. At the delivery point, the LNG is regasified and charged into a gas pipeline system.

LNG carriers transport LNG across oceans, while tank trucks can carry LNG or CNG and LPG over shorter distances. Sea transport using CNG carrier ships that are now under development may be competitive with LNG transport in specific conditions. LNG is the preferred form for long-distance, high-volume transportation of natural gas, whereas pipeline is preferred for transport for distances up to 4000 km (2485 miles) over land and approximately half that distance offshore [57].

In the past, the natural gas that was recovered in the course of recovering petroleum could not be profitably sold and was simply burned at the oil field in a process known as flaring. Flaring is now illegal in many countries. Additionally, higher demand in the last 20–30 years has made production of gas associated with oil economically viable. As a further option, the gas is now sometimes reinjected into the formation for enhanced oil recovery by pressure maintenance as well as miscible or immiscible flooding. The chosen option among transportation, conservation, reinjection, or flaring of natural gas associated with oil is primarily dependent on proximity to markets (pipelines), economics of transport, and regulatory restrictions.

A "master gas system" was invented in Saudi Arabia in the late 1970s, ending any necessity for flaring. Satellite observation, however, shows that flaring and venting are still practiced in some gas-extracting countries. Floating LNG (FLNG) is an innovative technology designed to enable the development of offshore gas resources that would otherwise remain untapped because transferring this gas by pipeline to land-based storage or using land-based LNG operation is uneconomical and environmentally harmful. Many gas and oil companies are considering the economic and environmental benefits of FLNG ([57], also see Chapter 8).

Shell floats hull for the world's largest FLNG facility. The 488 m long hull of Shell's Prelude FLNG facility has been floated out of the dry dock at the Samsung Heavy Industries yard in Geoje, South Korea, where the facility is currently under construction. Once complete, Prelude FLNG will be the largest floating facility ever built. It will unlock new energy resources offshore and produce approximately 3.6 million tons of LNG per annum to meet the growing demand. FLNG will allow Shell to produce natural gas at sea, turn it into LNG, and then transfer it directly to the ships that will transport it to customers. It will enable the development of gas resources ranging from clusters of smaller more remote fields to potentially larger fields via multiple facilities where, for a range of reasons, an onshore development is not viable. This can mean faster, cheaper, more flexible development and deployment strategies for resources that were previously uneconomic or constrained by technical or other risks.

Besides being transported as gas, compressed gas, and liquefied gas, natural gas can also be transported as gas hydrates. This method has been found convenient and economical, particularly for small quantities of stranded gas. The storage and transport of LPG follow processes similar to the ones for LNG [15].

Currently, about 85% of natural gas is stored in underground reservoirs and transported by pipeline infrastructure. CNG is stored both underground or above the ground vessels and transported on smaller scales for mobile applications by various methods of road transportation. LNG and LPG are stored in special vessels and transported for mobile applications by road in specially designed tankers. About 11% of total natural gas is transported across sea in ships and about 4% of natural gas is transported by roads. These numbers will change as more and more natural gas is used for mobile applications [36]. These different types of storage and transportation issues for natural gas are discussed in detail in Chapter 8.

The requirements for storage and transport of synthetic gas are somewhat different. The methods for storage and transport of syngas are derived from those for natural gas, hydrogen, and carbon monoxide. Syngas is also often first converted to SNG and stored until it is needed. SNG can then be converted back to syngas. The storage and transportation methods used for natural gas can be easily adapted for SNG.

Biogas (or biomethane) is generally found in smaller quantities and often used locally unless it is accumulated in large quantities by combining several waste treatment facilities. This accumulated biogas can be purified as biomethane and can then be stirred and transported with natural gas.

While small amount of hydrogen (5–15 vol%) can be transported within natural gas pipeline, a larger quantity of hydrogen storage and transport will require its own independent infrastructure [36]. Hydrogen is even lighter than natural gas and creates embrittlement in materials. Significant research efforts are being carried out to explore various physical and chemical methods to store hydrogen. Efforts are also being made to transport hydrogen by pipelines or in liquid form. Since one of the major uses of hydrogen is in fuel cell for automobiles, significant research has been carried out to evaluate various storage methods that can work for mobile applications.

Along with storage and transportation technologies of natural gas, such as LNG, CNG, and LPG and gas hydrates [11,15,53–57], the storage and transportation methods for syngas, biogas (biomethane), and hydrogen are examined in detail in Chapter 8. Finally, this chapter also examines in detail the ever-growing importance of gas grid that along with heat grid and smart electrical grid manages and controls complex gas distribution system. Smart gas grid is a catalyst that will further promote the use of gas for its variety of applications [36].

1.9 GAS TO LIQUID FUELS AND FUEL ADDITIVES

Natural and synthetic gas consumption patterns, across nations, vary based on its access and local market needs. Countries with large reserves or synthetic gas generation capability tend to handle the raw material natural gas and synthetic gas more generously, while countries with scarce or lacking resources tend to be more economical. Despite the considerable findings, the predicted availability of the natural gas reserves or the synthetic gas generation capabilities have hardly changed except in few countries like the United States.

As mentioned earlier, natural and synthetic gas is a very versatile fuel both on the supply and demand sides. Usages of natural and synthetic gas are very wide. They can be

1. Fuel for industrial heating
2. Fuel for manufacturing and various commodity productions
3. Fuel for the operation of public and industrial power stations
4. Fuel for small-scale engines, turbines, and microturbines
5. Household fuel for cooking, heating, and providing hot water
6. Fuel for environment-friendly CNG or LNG vehicles
7. Raw materials for fuel cells
8. Raw material for alcohols and fuel additives
9. Fuel for large-scale fuel production using GTL process (e.g., to produce sulfur-free and aromatic-free diesel and gasoline)
10. Raw materials for various chemicals, polymers, and other consumer products as well as numerous refinery operations for upgrading crude oil, heavy oils, coal liquids, biooils, etc.

Natural gas plays a very significant role in liquid fuel and fuel additive industry. Both syngas and hydrogen can be used to produce liquid fuels or fuel additives in a number of different ways [58,59]:

1. From syngas to methanol
2. From methanol to DME, gasoline, and olefins
3. Mixed alcohol synthesis
4. FT synthesis
5. Isosynthesis
6. Oxosynthesis
7. Syngas fermentation
8. Role of hydrogen in the production of liquid fuels in refineries

The chemistry for the first six catalytic methods is well established and most of the conversion processes are practiced commercially. Different types of syngas conversion processes mentioned earlier require different types of catalyst, support, and promoters as well as different conditions of temperatures and pressures. Chapter 9 fully describes the basics and commercialization of all of these syngas conversion processes. It should be noted that syngas can also be converted to various chemicals; however, in this book, we have focused on syngas conversion for energy and fuel industry alone. Syngas can also be converted to alcohols via fermentation technique and this process is also described in detail in Chapter 9. Finally, the role of hydrogen in oil refinery to produce gasoline, diesel, jet fuel, naphtha, fuel oil, etc., from crude oils is also illustrated in detail in Chapter 9.

While syngas and hydrogen can produce a host of liquid fuels and fuel additives, it is once again worth mentioning that on the supply side there are numerous ways to produce syngas and hydrogen.

Besides direct production of syngas and hydrogen from various thermal and hydrothermal gasification methods, syngas of required composition can also be produced from any type of natural gas via gas reforming. For example, steam reforming of methane produces syngas with high hydrogen to carbon monoxide ratio of 3. Dry reforming produces syngas with hydrogen to carbon monoxide ratio of 1. POX produces the hydrogen to carbon monoxide ratio of 2. Thus, by a suitable combination of these three reforming techniques along with WGSR, one can produce syngas of the required composition from methane and other hydrocarbons. Different paths for syngas conversion such as FT synthesis, isosynthesis, oxosynthesis, methanol, and mixed alcohol production require syngas with different H_2/CO ratios that can be accommodated by these reforming processes. Thus, the supply side of syngas is also very versatile.

1.10   GAS FOR HEAT, ELECTRICITY, AND MOBILE APPLICATIONS

Besides liquid fuel and fuel additive productions, the major use of natural and synthetic gas is for electricity, heat, and mobile applications. Natural gas is extensively used for home, industrial, and process heating and it can effectively replace coal and oil for a variety of heating applications. Along with methane, both syngas and hydrogen can be used to generate electricity at large and small scales. Since the conversion of gas to electricity is a much cleaner process than that of coal to electricity, more and more utility scale heat and power plants now utilize gas instead of coal. Finally, hydrogen and natural gas are utilized in the vehicle industry in two ways: (1) LNG, CNG, and LPG serve as replacements for diesel and gasoline and (2) hydrogen is used in fuel cells for hybrid cars. Chapter 10 examines in detail these three application areas.

In power plants, the burning of natural gas produces nitrogen oxides and carbon dioxide, but it results to lower quantities when compared to burning of coal or oil. Emissions of sulfur dioxide and mercury compounds from burning natural gas are negligible. The burning of natural gas in combustion turbines requires very little water and it does not produce substantial amounts of solid wastes. Due to all these advantages, natural gas is a better choice than coal for power generation and its use is rapidly expanding. Methane is, however, a potent greenhouse gas and it should not be leaked into the atmosphere due to leakages during its underground recovery or due to its incomplete combustion during power production.

The EIA estimated that between 2009 and 2015, approximately 97 gigawatts (GW) of new electricity capacity was added in the United States. Of this, over 20%, or approximately 21.5 GW, was by natural gas. According to the EIA, natural gas–fired electricity generation is expected to account for 80% of all added electricity generation capacity by 2035. An MIT study [2] indicates that the use of natural gas for power plants will be nearly doubled by 2040.

Natural gas can be used to generate electricity in a variety of ways. The electricity can be generated in a large centralized plant or in a number of distributed plants. While a centralized plant offers economy of scale, distributed plants offer more flexibility options for the use of gas in various stranded locations. Distributed plants also provide backup to centralized plant in case of emergency needs or unusual peak power demand. Until recently, methods of generating power have been discussed in the context of large, centralized power plants. Both economy and regulations have favored that mode. However, with technological advancements, there is a trend toward what is known as "distributed generation." Distributed generation refers to the placement of individual, smaller-sized electric generation units at residential, commercial, and industrial sites of use. These small-scale power plants, which are primarily powered by natural gas, operate with small gas turbine or combustion engine units or natural gas fuel cells.

Distributed generation can take many forms, from small, low-output generators used, to back up the supply of electricity obtained from the centralized electric utilities, to larger, independent generators that supply enough electricity to power an entire factory. Distributed generation is attractive because it offers electricity that is more reliable, more efficient, and cheaper than purchasing power from a centralized utility. Distributed generation also allows for increased local price control over the electricity supply and cuts down on electricity losses during transmission. Due to its ample supply, well-designed infrastructure, and the environmental benefits, natural gas is one of the leading choices for on-site power generation. There are a number of ways in which natural gas may be used on-site to generate electricity. Fuel cells, gas-fired reciprocating engines, industrial natural gas–fired turbines, and microturbines are some of the more popular means of using natural gas for on-site electricity needs. Distributed power generation also allows better use of renewable resources.

Chapter 10 evaluates the pros and cons of centralized versus distributed power generation as well as on how both of these options can be managed by the use of gas, heat, and electrical grids and microgrids.

1.10.1 Gas Turbines and Microturbines

The most basic natural gas–fired electric generation consists of a steam generation unit, where fossil fuels are burned in a boiler to heat water and produce steam that then turns a turbine to generate electricity. Natural gas may be used for this process, although these basic steam units are more typical of large coal or nuclear generation facilities. Unfortunately, these basic steam generation units have fairly low energy efficiency. Typically, only 33%–35% of the thermal energy used to generate the steam is converted into electrical energy in these types of units.

Gas turbines [60,61] and combustion engines are also used to generate electricity. In these types of units, instead of heating steam to turn a turbine, hot gases from burning fossil fuels (particularly natural gas) are used to turn the turbine and generate electricity. Gas turbine and combustion engine plants are traditionally used primarily for peak load demands, as it is possible to quickly and easily turn them on. These plants have increased in popularity due to advances in gas turbine technology and the availability of natural gas.

The combustion (gas) turbines being installed in many of today's natural gas–fueled power plants are complex machines, but they basically involve three main sections:

1. *The compressor*, which draws air into the engine, pressurizes it, and feeds it to the combustion chamber at speeds of hundreds of miles per hour.
2. *The combustion system*, which is typically made up of a ring of fuel injectors that inject a steady stream of fuel into combustion chambers where it mixes with the air. The mixture is burned at temperatures of more than 2000°F. The combustion produces a high-temperature, high-pressure gas stream that enters and expands through the turbine section.
3. *The turbine* is an intricate array of alternate stationary and rotating aerofoil-section blades. As hot combustion gas expands through the turbine, it spins the rotating blades. The rotating blades perform a dual function: they drive the compressor to draw more pressurized air into the combustion section, and they spin a generator to produce electricity.

Land-based gas turbines are of two types: (1) heavy frame engines and (2) aeroderivative engines. Heavy frame engines are characterized by lower pressure ratios (typically below 20) and tend to be physically large. Pressure ratio is the ratio of the compressor discharge pressure and the inlet air pressure. Aero derivative engines are derived from jet engines, as the name implies, and operate at very high compression ratios (typically in excess of 30). Aeroderivative engines tend to be very compact and are useful where smaller power outputs are needed. As large frame turbines have higher power outputs, they can produce larger amounts of emissions and must be designed to achieve low emissions of pollutants, such as NO_x.

One key to a turbine's fuel-to-power efficiency is the temperature at which it operates. Higher temperatures generally mean higher efficiencies, which in turn can lead to more economical operation. Gas flowing through a typical power plant turbine can be as hot as 2300°F, but some of the critical metals in the turbine can withstand temperatures only as hot as 1500°F–1700°F. Therefore, air from the compressor might be used for cooling key turbine components, thus reducing ultimate thermal efficiency.

One of the major achievements of the Department of Energy's advanced turbine program was to break through previous limitations on turbine temperatures, using a combination of innovative cooling technologies and advanced materials. The advanced turbines that emerged from the department's research program were able to boost turbine inlet temperatures to as high as 2600°F—nearly 300° hotter than in previous turbines, and achieve efficiencies as high as 60%.

Gas is also very useful to generate electricity and heat in small engines and microturbines. This distributed heat and power generation for both static and mobile use is rapidly expanding. Gas-driven, micro- and intermediate-scale turbines are much cleaner than the ones using oil.

Microturbines are scale-down versions of industrial gas turbines. As their name suggests, these generating units are very small (Figure 1.3) and typically have a relatively small electric output. These types of distributed generation systems have the capacity to produce from 25 to 500 kW of electricity and are best suited for residential- or small-scale commercial units or for mobile applications.

Advantages to microturbines include a very compact size (about the same size as a refrigerator), a small number of moving parts, light weight, low cost, and increased efficiency. Using new waste heat recovery techniques in combined cycles or cogeneration, microturbines can achieve energy efficiencies of up to 80%.

The advances in both gas turbine technology and microturbine technology have immensely facilitated the use of natural gas in both large-scale and small-scale power generations. The large

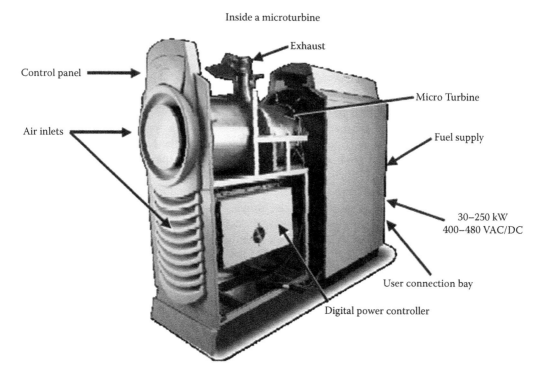

FIGURE 1.3 Microturbine with major assemblies. (From Staunton, R.H. and Ozpineci, B., Microturbine power conversion-technology review, ORNL/TM-2003/74, Oak Ridge National Laboratory, Oak Ridge, TN, U.S. Department of Energy under contract no. DE-AC05-00OR22725, April 8, 2003.)

gas turbines operated by gas have replaced the use of coal and steam turbine in large-scale power plants. Microturbines have facilitated the use of gas instead of oil in small-scale power generation. Chapter 10 discusses in detail our present state of the art of gas turbines and microturbines for electricity generation.

1.10.2 Combined Cycles and Cogeneration

Another way to boost thermal efficiency of power generation is to install a recuperator or heat recovery steam generator (HRSG) to recover energy from the turbine's exhaust. A recuperator captures waste heat in the gas turbine exhaust system to preheat the compressor discharge air before it enters the combustion chamber. A HRSG also generates steam by capturing heat from the turbine exhaust. The boilers that generate steam using waste turbine exhaust heat are also known as HRSGs. High-pressure steam from these boilers can be used to generate additional electric power with steam turbines, a configuration called a combined cycle [62]. Thus, thermodynamic efficiency of a power plant can be significantly enhanced by using two cycles in series. A simple cycle gas turbine can achieve energy conversion efficiencies ranging between 20% and 35%. With the higher temperatures achieved in the Department of Energy's turbine program, future hydrogen- and syngas-fired gas turbine combined-cycle plants are likely to achieve efficiencies of 60% or more. When waste heat is captured from these systems for heating or industrial purposes, the overall energy cycle efficiency could approach 80%.

In recent years, power is generated using coal gasification, nuclear energy, solar energy or geothermal energy combined with the concept of combined cycle. In these cases, processes are designed to integrate coal gasification unit or nuclear power or solar power unit with combined cycles to obtain better thermal efficiency. These processes are designated as Integrated gas combined cycle (IGCC), Integrated nuclear combined cycle (INCC), Integrated solar combined cycle (ISCC) or Integrated geothermal combined cycle (IGECC) [63] respectively. More and more existing gasification plants are converted to integrated combined-cycle gasification plants. These plants will become more environmentally acceptable by the introduction of a carbon capture unit in the process to reduce emission of carbon dioxide.

Industrial natural gas–fired turbines operate on the same concept as the larger centralized gas turbine generators discussed earlier. However, instead of being located in a centralized plant, these turbines are located in close proximity to where the electricity being generated and used. Industrial turbines—producing electricity through the use of high-temperature, high-pressure gas to turn a turbine that generates a current—are compact, lightweight, easily started, and simple to operate. This type of distributed generation is commonly used by medium- and large-sized establishments, such as universities, hospitals, commercial buildings, and industrial plants and can achieve efficiency up to 58%. This number will go up with the use of advanced materials that can withstand higher temperatures.

One of the advantages of distributed power generation is that waste heat coming out of turbine exhaust can also be used for local heat need such as district heating, industrial process heating, and desalination by operation commonly known as cogeneration or combined heat and power generation. When waste heat is used both for cooling and heating, the operation is called trigeneration or polygeneration [64]. Just like combined cycles, cogeneration significantly improves the overall thermal efficiency of power generation. Since heat cannot be transported at long distances, this concept is more useful for distributed power generation where both power and heat needs are local. It is difficult to use all waste heat for local needs for large centralized power plants. Both combined cycles and cogeneration methods to improve the thermal efficiency of gas-based power plants are discussed in detail in Chapter 10.

1.10.3 Gas Use for Mobile Applications

Natural gas is also being increasingly used for vehicle applications. There are two major uses of gas in vehicles. First, the use of CNG, LNG, and LPG as fuel in larger vehicles is rapidly increasing.

These fuels are replacing diesel because they have higher densities and are cleaner to burn. More efforts are being made to use CNG, LNG, and LPG in smaller cars. Chapter 10 examines this usage in detail.

Second, hydrogen-based fuel cell is becoming an increasingly important technology for the generation of electricity. They are much like rechargeable batteries, except that instead of using an electric recharger, they use fuel, such as natural gas or hydrogen, to generate electric power even when they are in use. Fuel cell converts chemical energy into electrical energy. Hydrogen is the most desired source of fuel. Fuel cells for distributed generation offer a multitude of benefits and are an exciting area of innovation and research for distributed generation applications. Chapter 10 also examines in detail the application of gas for mobile purposes.

From the details outlined in the remaining nine chapters, it will be very clear that natural and synthetic gas will play a more and more important role in the energy and fuel industry because its supply and demand are abundant and versatile with abundant availability of storage and transportation infrastructure. Collectively, not only both natural and synthetic gas will be important "transition fuel," but also with advancements in commercial hydrogen technology, it will become the "ultimate fuel."

REFERENCES

1. Natural gas and the environment (2015), Naturalgas.org, naturalgas.org/environment/naturalgas/, accessed September 2015.
2. Moniz, E.J., Jacoby, H.D., and Meggs, A.J.M., MIT study on the future of natural gas, an interdisciplinary MIT study. MIT Energy Initiative, Cambridge, MA, 104pp. (2013), web.mit.edu/.../Natural_Gas_Study.
3. Annual energy outlook 2015, EIA, U.S. Energy Information Administration, Washington, DC (2015), www.eia.gov/forecasts/aeo/, accessed April 2015.
4. Global natural gas markets overview: A report prepared by Leidos, Inc., under contract to EIA, U.S. Energy Information Administration, Washington, DC (August 2014), https://www.eia.gov/.../global_gas.pd, accessed August 2014.
5. List of countries by natural gas proven reserves, Wikipedia, the free encyclopedia (2015), https://en.wikipedia.org/.../List_of_countries_by_natural_gas, last modified October 04, 2016.
6. International energy statistics, Energy Information Administration, Department of Energy, Washington, DC (2015), www.eia.gov, accessed September 2015.
7. Dudley, R., BP statistical review of world energy, a BP report, 65th ed. (June 2016), British Petroleum, U.K., https://www.bp.com/.../bp-statistical-review-of-world-energy-2015-full-r.
8. Natural gas statistics IEA, Paris, France (2015), www.iea.org/statistics/topics/naturalgas/, accessed 2015.
9. Natural gas statistics at IEA, OECD/IEA, Paris, France (2014).
10. Liquefied Petroleum Gas (LPG), Liquefied Natural Gas (LNG) and Compressed Natural Gas (CNG), Envocare Ltd., Pune, India (March 21, 2007). Retrieved 2008-09-03.
11. Compressed natural gas, Wikipedia, the free encyclopedia (2015), https://en.wikipedia.org/wiki/Compressed_natural_gas, last modified October 09, 2016.
12. World LNG Report—2014 edition, International Gas Union, News, views and knowledge on gas worldwide', a report sponsored by Total, c/ Statoil, Fornebu, Norway (2014).
13. Global natural gas and LNG demand a report by EY, London, U.K. (2013).
14. Liquefied natural gas: Understanding the basic facts, a report prepared by the U.S. Department of Energy (DOE) in collaboration with the National Association of Regulatory Utility, USDOE and National Energy Technology Laboratory, Washington, DC, DOE/FE-0489 (August 2005).
15. Liquefied petroleum gas, Wikipedia, the free encyclopedia (2015), https://en.wikipedia.org/wiki/Liquefied_petroleum_gas, last modified September 19, 2016.
16. Natural gas prices and taxes, IEA report, OECD/IEA, Paris, France (2014), www.iea.org/Statistics.
17. Drilling, Wikipedia, the free encyclopedia (2015), https://en.wikipedia.org/wiki/Drilling, last modified October 13, 2016.
18. Natural gas extraction (2015), Naturalgas.org, naturalgas.org/naturalgas/extraction/.
19. Pad drilling and rig mobility lead to more efficient drilling, a report in Today in Energy by U.S. Energy Information Administration, Washington, DC (September 11, 2012).
20. Tight gas reservoirs, Petowiki, published by SPE, Washington, DC (2013), petrowiki.org/Tight_gas_reservoirs, accessed September 2013.

21. U.S. Energy Information Administration, Energy production, imports, and exports, Annual Energy Outlook 2015, EIA, Washington, DC (2015), https://www.eia.gov/.../section_energ, accessed September 2015.

22. Tight gas, Wikipedia, the free encyclopedia (2015), https://en.wikipedia.org/wiki/Tight_gas, accessed September 04, 2016.

23. World shale gas resources-analysis and projections, EIA, Washington, DC (2015), https://www.eia.gov/.../worldshalegas, accessed September 2015.

24. Natural gas market national overview, a report by Federal Energy Regulatory Commission, Washington, DC (October 2009), https://www.ferc.gov/overnight.

25. Al-Jubori, A., Boyer, C., Bustos, O., Pashin, J., and Wary, A., Coalbed methane: Clean energy for the world, *Oilfield Review*, 21 (2), 4–13 (Summer, 2009).

26. Franklen, M., Methane to markets - Global overview of policies affecting coal mine methane (CMM) recovery and utilization, U.S. EPA report, Washington, DC (August 30, 2010).

27. Coalbed methane proved reserves and production, U.S. DOE Energy Information Administration, Washington, DC, http://tonto.eia, doe.gov/dnav/ng/ng_enr_cbm_a_EPG0_r52_Bcf_a.htm, Accessed March 1, 2009.

28. For more on coalbed methane: Ayoub, J., Colson, L., Hinkel, J., Johnston, D., and Levine, J., Learning to produce coalbed methane, *Oilfield Review*, 3 (1), 27–40 (January 1991).

29. Anderson, J. et al., Producing natural gas from coal, *Oilfield Review*, 15 (3), 8–31 (Autumn 2003).

30. Heyward, T., BP Statistical Review of World Energy (June 2008), http://www.bp.com/liveassets/bp_internet/globalbp/globalbp_uk_english/reports_and_publications/statistical_energy_review_2008/STAGING/local_assets/downloads/pdf/statistical_review_of_world_energy_full_review_2008.pdf, Accessed February 13, 2009.

31. Unconventional natural gas resources (2015), NaturalGas.org, naturalgas.org/overview/unconventional-ng-resources, accessed September 2015.

32. Seale, R., Athans, J., and Themig, D., An effective horizontal well completion and stimulation system, *Journal of Canadian Petroleum Technology*, 46 (12), 12 (2007).

33. Birol, F., World energy outlook 2012, IEA report, Paris, France (November 2012).

34. Natural gas extraction-hydraulic fracturing, EPA report (February 2016), https://www.epa.gov/hydraulicfracturing, U.S. Environmental Protection Agency, Washington, DC (2016).

35. Lee, S., Gasification of coal, Chapter 2, in Lee, S., Speight, J., and Loyalka, S., (eds.), *Handbook of Alternative Fuel Technology*. Taylor & Francis Group, New York, pp. 26–78 (2007).

36. Shah, Y., *Energy and Fuel Systems Integration*. CRC Press, Taylor & Francis Group, New York (2015).

37. Shah, Y.T., *Water for Energy and Fuel Production*. CRC Press, New York (2014).

38. Lee, S. and Shah, Y.T., *Biofuels and Bioenergy: Technologies and Processes*. CRC Press, Taylor & Francis Group, New York (September 2012).

39. Biogas, Wikipedia, the free encyclopedia (2015), https://en.wikipedia.org/wiki/Biogas, last modified October 12, 2016.

40. Gasification, Wikipedia, the free encyclopedia (2015), https://en.wikipedia.org/wiki/Gasification.

41. Osada, M., Sato, T., Watanabe, M., Adschiri, T., and Arai, K., Low-temperature catalytic gasification of lignin and cellulose with a ruthenium catalyst in supercritical water. *Energy Fuels*, 18 (2), 327–333 (2004).

42. Steam reforming, Wikipedia, the free encyclopedia (2015), https://en.wikipedia.org/wiki/Steam_reforming.

43. Worley, M. and Yale, J., Biomass gasification technology assessment consolidated report, NREL technical monitor: Abhijit Dutta, subcontract report NREL/SR-5100-57085, Contract no. DE-AC36-08G028308, NREL, Golden, CO (November 2012).

44. Taylor, R., Howes, J., and Bauen, A., Review of technologies for gasification of biomass and wastes, Final report, NNFCC, Bioeconomy Consulting Co., York, OK (2009).

45. Shah, Y.T. and Gardner, T., Dry reforming of hydrocarbon feedstocks, *Catalysis Reviews: Science and Engineering*, 54, pp. 476–536; CRC Press, Taylor & Francis Group, New York (September 2014).

46. Partial oxidation, Wikipedia, the free encyclopedia (2015), https://en.wikipedia.org/wiki/Partial_oxidation.

47. Carbon dioxide reforming, Wikipedia, the free encyclopedia (2015), https://en.wikipedia.org/wiki/Carbon_dioxide_reforming.

48. Walker, D.M., Catalytic tri-reforming of biomass-derived syngas to produce desired H_2:CO ratios for fuel applications, Graduate Theses and Dissertations, Department of Chemical and Biomedical Engineering, University of South Florida, Tampa, FL (2012), http://scholarcommons.usf.edu/etd/4250.

49. Pinkwart, K., Bayha, T., Lutter, W., and Krausa, M., Gasification of diesel oil in supercritical water for fuel cells, *Journal of Power Sources*, 136, 211–214 (2004).

50. Shekhawat, D., Berry, D., Gardner, T., and Spivey, J., Catalytic reforming of liquid hydrocarbon fuels for fuel cell applications, *Catalysis*, 19, 184 (2006).

51. Natural gas processing, Wikipedia, the free encyclopedia (2015), https://en.wikipedia.org/wiki/Natural-gas_processing, last modified July 19, 2016.

52. Ratafia-Brown, J., Haslbeck, J., Skone, T., and Rutkowski, M., Assessment of technologies for co-converting coal and biomass to a clean syngas—Task 2 report (RDS), DOE/NETL-403.01.08 Activity 2 Report, Department of Energy, Washington, DC (May 10, 2007).

53. Natural gas storage, Wikipedia, the free encyclopedia (2015), https://en.wikipedia.org/wiki/Natural_gas_storage, last modified June 29, 2016.

54. Hydrogen storage, Wikipedia, the free encyclopedia (2015), https://en.wikipedia.org/wiki/Hydrogen_storage, last modified September 06, 2016.

55. Hydrogen pipeline transport, Wikipedia, the free encyclopedia (2015), https://en.wikipedia.org/wiki/Hydrogen_pipeline_transport, last modified September 07, 2016.

56. The transportation of natural gas (2015), naturalgas.org, naturalgas.org/naturalgas/transport, accessed September 2015.

57. Liquefied natural gas, Wikipedia, the free encyclopedia (2015), https://en.wikipedia.org/wiki/Liquefied_natural_gas, last modified October 15, 2016.

58. Shah, Y.T. and Perrotta, A., Catalysts for Fischer–Tropsch and isosynthesis, *I&EC Product Research and Development*, 15, 123 (1976).

59. Shah, Y.T., Biomass to liquid fuel via Fischer–Tropsch and related syntheses, Chapter 12, in Lee, J.W. (ed.), *Advanced Biofuels and Bioproducts*. Springer Book Project, Springer Publ. Co., New York, pp. 185–207 (September 2012).

60. Gas turbine, Wikipedia, the free encyclopedia (2015), https://en.wikipedia.org/wiki/Gas_turbine, last modified September 20, 2016.

61. Capstone turbine, Wikipedia, the free encyclopedia (2015), https://en.wikipedia.org/wiki/Capstone_Turbine, last modified July 21, 2016.

62. Combined cycle, Wikipedia, the free encyclopedia (2015), https://en.wikipedia.org/wiki/Combined_cycle.

63. Integrated gasification combined cycle, Wikipedia, the free encyclopedia (2015), https://en.wikipedia.org/wiki/Integrated_gasification_combined_cycle, last modified August 15, 2016.

64. Cogeneration, Wikipedia, the free encyclopedia (2015), https://en.wikipedia.org/wiki/Cogeneration, last modified October 15, 2016.

65. Staunton, R.H. and Ozpineci, B., Microturbine power conversion-technology review, ORNL/TM-2003/74, Oak Ridge National Laboratory, Oak Ridge, TN, U.S. Department of Energy under contract no. DE-AC05-00OR22725 (April 8, 2003).

66. Marano, J.J. and Ciferno, J.P., Life-cycle greenhouse-gas emissions inventory for Fischer–Tropsch fuels, Energy and Environmental Solutions, LLC, for the U.S. Department of Energy, National Energy Technology Laboratory, Pittsburgh, PA (2001).

67. IEA Statistics, *CO$_2$ Emissions from Fuel Combustion*, 2011 edn. OECD/IEA, Paris, France (2011).

68. Shale Gas, Wikipedia, the free Encyclopedia, 2015, https://en.wikipedia.org/wiki/shale_gas (last modified August 18, 2016).

2 Natural Gas

2.1 INTRODUCTION

Natural gas is a fossil fuel that formed when layers of buried plants, gases, and animals were exposed to intense heat and pressure over thousands of years. The energy that the plants originally obtained from the sun is stored in the form of chemical bonds in natural gas. Natural gas obtained from underground is a nonrenewable resource because it cannot be replenished in the human time frame [1–10]. Natural gas is colorless, shapeless, and odorless in its pure form, and when burned, it gives off a great deal of energy with fewer emissions than many other sources. Compared to other fossil fuels, natural gas burns cleaner and emits lower levels of potentially harmful by-products into the air. Natural gas is a combustible mixture of hydrocarbon gases. While natural gas is composed primarily of methane, it can also include ethane, propane, butane, and pentane. The composition of natural gas can vary widely; Table 2.1 [2,10] illustrates the typical makeup of natural gas before it is refined.

There are many different theories as to the origins of natural gas. The most widely accepted theory says that fossil fuels are formed when organic matter (such as the remains of a plant or animal) is compressed under the earth, at very high pressure for a very long time. This is referred to as thermogenic methane. Similar to the formation of oil, thermogenic methane is formed from organic particles that are covered in mud and other sediment. Over time, more and more sediment and mud and other debris are piled on top of the organic matter. These sediments and debris put a great deal of pressure on the organic matter and compressed it. This compression, combined with high temperatures found deep underneath the earth, broke down the carbon bonds in the organic matter. As one gets deeper and deeper under the earth's crust, the temperature gets higher and higher. At low temperatures (shallower deposits), more oil is produced relative to natural gas [1–11]. At higher temperatures, however, more natural gas is created, as opposed to oil. That is why natural gas is usually associated with oil in deposits that are 1–2 miles below the earth's crust. Deeper deposits, very far underground, usually contain primarily natural gas and, in many cases, pure methane [9].

Natural gas can also be formed through the transformation of organic matter by tiny microorganisms. This type of methane is referred to as biogenic methane. Methanogens, tiny methane-producing microorganisms, chemically break down organic matter to produce methane. These microorganisms are commonly found in areas near the surface of the earth that are void of oxygen. These microorganisms also live in the intestines of most animals, including humans. Formation of methane in this manner usually takes place close to the surface of the earth, and the methane produced is usually lost into the atmosphere. In certain circumstances, however, this methane can be trapped underground, recoverable as natural gas. An example of biogenic methane is landfill gas. Waste-containing landfills produce a relatively large amount of natural gas from the decomposition of the waste materials that they contain. New technologies allow this gas to be harvested and used to add to the supply of natural gas. The gas produced in this way largely contains methane (about 55%–60%) and carbon dioxide (about 30%–35%) [1–11]. This type of gas is called "biogas" and in recent years it has been produced synthetically to gain energy out of a variety of aqueous wastes. This type of synthetically produced gas is examined in detail in Chapter 4.

A third way in which methane (and natural gas) may be formed is through abiogenic processes. Extremely deep into the earth's crust, hydrogen-rich gases and carbon molecules exist. As these gases gradually rise toward the surface of the earth, they interact with minerals that also exist underground, in the absence of oxygen. This interaction results in a reaction that forms elements and compounds found in the atmosphere (including nitrogen, oxygen, carbon dioxide, argon, and water).

TABLE 2.1
Typical Composition of Natural Gas

Name	Formula	Mole %
Methane	CH_4	70–90
Ethane	C_2H_6	1.8–5.1
Propane	C_3H_8	0.1–1.5
iso-Butane	C_4H_{10}	0.01–0.3
n-Butane	C_4H_{10}	0.01–0.3
iso-Pentane	C_5H_{12}	Trace–0.14
n-Pentane	C_5H_{12}	Trace–0.04
Hexanes plus	C_6H_{14}	Trace–0.06
Carbon dioxide	CO_2	0–8
Oxygen	O_2	0.01–0.2
Nitrogen	N_2	1.3–5.6
Hydrogen sulfide	H_2S	0–5
Rare gases	A, He, Ne, Xe	Trace
Hydrogen	H_2	Trace–0.02
Specific gravity		0.57–0.62

Source: Modified and adapted from Composition of natural gas, Naturalgas.org, Retrieved July 14, 2012.

If these gases are under very high pressure as they move toward the surface of the earth, they are likely to form methane deposits, similar to thermogenic methane [1,3–11].

Natural gas was used by the Chinese around 500 BC [1,7]. They discovered a way to transport gas seeping from the ground in crude pipelines of bamboo to where it was used to boil sea water to extract the salt [6]. The world's first industrial extraction of natural gas started at Fredonia, New York, in 1825 [7]. By 2009, 66 trillion cubic meters (tcm) (or 8%) had been used out of the total 850 tcm of estimated remaining recoverable reserves of natural gas [8]. Based on an estimated 2015 world consumption rate of about 3.4 tcm of gas per year, the total estimated remaining economically recoverable reserves of natural gas would last 250 years at current consumption rates. An annual increase in usage of 2%–3% could result in currently recoverable reserves lasting significantly less, perhaps as few as 80–100 years [8]. These estimates are tentative because they do not consider any future development on the recoverable unconventional natural gas.

In the nineteenth century, natural gas was usually obtained as a by-product of producing oil, since the small, light gas carbon chains came out of solution as the extracted fluids underwent pressure reduction from the reservoir to the surface. Until early twentieth century, such unwanted gas was usually burned off at oil fields ("flaring") [12–14]. Today, unwanted gas (or stranded gas without a market) associated with oil extraction often is returned to the reservoir with "injection" wells while awaiting a possible future market or to repressurize the formation, which can enhance extraction rates from other wells. In regions with a high natural gas demand (such as the United States), pipelines are constructed when it is economically feasible to transport gas from a well site to an end consumer. A "master gas system" was invented in Saudi Arabia in the late 1970s, ending any necessity for flaring. Satellite observation, however, shows that flaring and venting are still practiced in some gas-extracting countries [12–14].

Because natural gas is not a pure product, as the reservoir pressure drops when nonassociated gas is extracted from a field under supercritical (pressure/temperature) conditions, the higher molecular weight components may partially condense upon isothermal depressurizing—an effect called retrograde condensation. The liquid thus formed may get trapped as the pores of the gas reservoir get depleted. One method to deal with this problem is to reinject dried gas free of condensate to maintain the underground pressure and to allow re-evaporation and extraction of condensates [10]. More frequently, the liquid condenses at the surface, and one of the tasks of the gas plant is to collect this condensate. The resulting liquid is called natural gas liquid (NGL) and it has a commercial value [10].

Natural gas is found in deep underground rock formations or associated with other hydrocarbon reservoirs such as coalbeds, shale deposits, and as methane clathrates (see Section 2.10). Petroleum is another resource and fossil fuel found in proximity to, and with, natural gas. Natural gas can be "associated" (found in oil fields), or "nonassociated" (isolated in natural gas fields). It sometimes contains a significant amount of ethane, propane, butane, and pentane—heavier hydrocarbons removed for commercial use prior to the methane being sold as a consumer fuel or chemical plant feedstock (see Chapters 9 and 10). Nonhydrocarbons such as carbon dioxide, nitrogen, helium (rarely), water, and hydrogen sulfide must also be removed before the natural gas can be stored and transported [15] and used for downstream applications. The technologies for these and other types of impurity removals are discussed in detail in Chapter 7.

Natural gas is often broken into two categories: conventional and nonconventional. This breakup is largely based on the convenience and economics of its recovery process. While both associated and nonassociated gas are considered as conventional natural gas, tight gas, deep gas, coalbed methane, shale gas, gas from geopressurized zones, and gas hydrates are considered as nonconventional gas (see Sections 2.5 through 2.10). The differentiation between conventional and unconventional natural gas is somewhat arbitrary because when technology and favorable economics for some of the unconventional gases are developed, they will become conventional gases. This chapter evaluates in detail both conventional and unconventional gases. Natural gas is often informally referred to simply as "gas," especially when compared to other energy sources such as oil or coal. However, it is not to be confused with synthetic gas produced from coal, oil, biomass, and waste by manmade thermal, catalytic and biological gasification processes which can be renewable or nonrenewable depending on the nature of the feedstock (see Chapters 3 through 5).

Natural gas is mainly composed of methane. After release to the atmosphere it is removed by gradual oxidation to carbon dioxide and water by hydroxyl radicals (OH) formed in the troposphere or stratosphere, giving the overall chemical reaction $CH_4 + 2O_2 \rightarrow CO_2 + 2H_2O$. While the lifetime of atmospheric methane is relatively short when compared to carbon dioxide, with a half-life of about 7 years, it is more efficient at trapping heat in the atmosphere, so that a given quantity of methane has 84 times the global-warming potential of carbon dioxide over a 20-year period and 28 times over a 100-year period. Natural gas is thus a more potent greenhouse gas than carbon dioxide due to the greater global-warming potential of methane. Current estimates by the Environment Protection Agency (EPA) place global emissions of methane at 85 billion cubic meters (bcm) (3.0×10^{12} ft^3) annually or 3.2% of global production. Direct emissions of methane represented 14.3% by volume of all global anthropogenic greenhouse gas emissions in 2004 [11–20].

During extraction, storage, transportation, and distribution, natural gas is known to leak into the atmosphere, particularly during the extraction process. A Cornell University study in 2011 demonstrated that the leak rate of methane may be high enough to jeopardize its global warming advantage over coal. This study was criticized later for its overestimation of methane leakage values. Preliminary results of some air sampling from airplanes done by the National Oceanic and Atmospheric Administration indicated higher-than-estimated methane releases by gas wells in some areas but the overall results showed methane emissions in line with previous EPA estimates [19,20].

Natural gas is an energy source often used for heating, cooking, and electricity generation. It is also used as fuel for vehicles and as a chemical feedstock in the manufacture of plastics and other commercially important organic chemicals, liquid fuels, and fuel additives. Many studies from MIT (Reference 2 of Chapter 1), and the DOE [5–7] predict that natural gas will account for a larger portion of electricity generation and heat in the future [1–8]. The usages of natural gas are discussed in detail in Chapters 9 and 10.

2.1.1 World Gas Reserve

World natural gas reserve estimate for the past 50+ years is illustrated in Figure 2.1. The Energy Information Agency (EIA), in conjunction with the *Oil and Gas Journal* and *World Oil* publications [4–7], estimates world's proven natural gas reserves to be around 6609 trillion cubic feet (tcf). Most of these reserves are located in the Middle East with 2658 tcf, or 40% of the world total, and Europe and the former USSR with 2331 tcf, or 35% of total world reserves. These numbers are constantly changing as new discoveries and new technologies are brought to attention [1,3–6].

In 2013, five top gas extractor countries were Iran 33,600 bcm (1 m³ is equal to 35.3 ft³), Russia 32,900 bcm, Qatar 25,100 bcm, Turkmenistan 17,500 bcm, and the United States 8,500 bcm. During the last 2 years, with increased shale gas production, the United States has overtaken all other countries in the total amount of natural gas extraction. The world's largest gas field is the offshore South Pars/North Dome Gas-Condensate field, shared between Iran and Qatar. It is estimated to have 51 tcm of natural gas and 50 billion barrels of natural gas condensates [1,3–8].

It is estimated that there are about 900 tcm of "unconventional" gas such as shale gas, of which 180 tcm may be recoverable [14]. The number for recoverable gas will of course change as new and improved technologies for gas extraction are developed for all types of unconventional gas.

Due to constant announcements of shale gas *recoverable* reserves, as well as drilling in Central Asia, South America, and Africa, and deep water drilling, estimates are undergoing frequent updates, mostly increasing. Since 2000, some countries, notably the United States and Canada, have seen large increases in proven gas reserves due to development of shale gas, but shale gas deposits in most countries are yet to be added to reserve calculations [1,3–8].

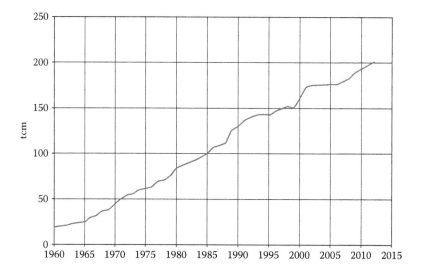

FIGURE 2.1 World's proven natural gas reserves 1960–2012. (From List of countries by natural gas proven reserves, Wikipedia, the free encyclopedia, 2015, https://en.wikipedia.org/.../List_of_countries_by_natural_gas_proven_ reserves, accessed on October 4, 2016.)

2.1.2 U.S. Natural Gas Resource Estimates

The EIA [1,3–7] estimates that there are 2543 tcf of technically recoverable natural gas in the United States. This includes undiscovered, unproved, and unconventional natural gas. Proven reserves make up a very small proportion of the total recoverable natural gas resources in the United States. In 2007, the National Petroleum Council estimated there were 1451 tcf remaining in the United States. According to this report, advances in technology could bring natural gas resources to 1887 tcf by 2017. Since then, natural gas resources have dramatically changed because of the ability to extract natural gas from shale rock found in more than 20 states in the United States [1,3–8].

Most of the natural gas that is found in North America is concentrated in relatively distinct geographical areas, or basins. Given this distribution of natural gas deposits, those states which are located on top of a major basin have the highest level of natural gas reserves. Within the United States, natural gas reserves historically have been concentrated around Texas and the Gulf of Mexico. With the recent onset of shale production, the number of states across the United States with significant resources has increased. Major shale gas producing states are New York, Pennsylvania, Arkansas, and Oklahoma along with Texas and Louisiana [1,3–8].

As shown in Figure 2.2, natural gas production in the United States has been rising over the past many decades. This production rate has increased significantly over the last several years due to the success in shale gas production. The United States relies on clean-burning natural gas for almost one-quarter of all energy used. Natural gas has proven to be a reliable and efficient energy source that burns much cleaner than other fossil fuels. In 2013, the United States produced between 85% and 90% of the natural gas it consumed [1]. Most of its other needs were imported by pipeline from Canada. This picture has changed considerably in recent years due to shale gas production. Annual U.S. natural gas consumption is projected to rise from 22.1 tcf in 2004 to 30.7 tcf in 2025 [1,3–8] because of the following reasons, among others:

1. Utilities realize advantages by using natural gas–fired generators to create electricity (lower capital costs, higher fuel efficiency, shorter construction lead times, and lower emissions).
2. The residential sector benefits from the higher fuel efficiency and lower emissions of gas appliances.

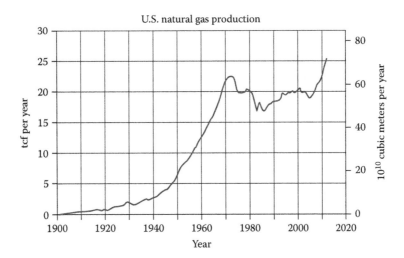

FIGURE 2.2 U.S. natural gas marketed production 1900–2012. (From Natural gas, Wikipedia, the free encyclopedia, September 2013, https://en.wikipedia.org/wiki/natural-gas, accessed on October 14, 2016.)

3. The industrial sector relies on natural gas as a feedstock or fuel for manufacturing many of the products we rely on today, including pulp and paper, metals (for computers, automobiles, and telecommunications), chemicals, fertilizers, fabrics, pharmaceuticals, and plastics.
4. The transportation sector is beginning to see natural gas as a clean and readily available alternative to other fossil fuels.

Historically, conventional natural gas deposits have been the most practical and easiest deposits to mine. However, as technology and geological knowledge advance, unconventional natural gas deposits are beginning to make up an increasingly large percentage of the supply picture. In the broadest sense, unconventional natural gas is a gas that is more difficult or less economical to extract, usually because the technology to reach it has not been developed fully, or is too expensive.

For example, prior to 1978, natural gas that had been discovered buried deep underground in the Anadarko basin was virtually untouched. It simply wasn't economical or possible to extract this natural gas. It was unconventional natural gas. However, the passage of market-based rate regulation and the passage of the Natural Gas Policy Act provided incentives toward searching for and extracting unconventional natural gas, and also spurred investment into deep exploration and development drilling, making much of the deep gas in the basin conventionally extractable [1,3–8].

Therefore, what is really considered unconventional natural gas changes over time, and from deposit to deposit. The economics of extraction play a role in determining whether or not a particular deposit may be unconventional, or simply too costly to extract. However, as mentioned earlier, at present there are six main categories of unconventional natural gas. These are: deep gas, tight gas, shale gas, coalbed methane, gas from geopressurized zones, and Arctic and subsea hydrates. Unconventional natural gas constitutes a large proportion of the natural gas that is left to be extracted in North America, and is playing an ever-increasing role in supplementing the nation's natural gas needs. As technology advances and new methods of extracting and using this natural gas are developed, the resource potential of unconventional natural gas becomes enormous [1,3–8].

2.2   METHODS FOR NATURAL GAS EXPLORATION

2.2.1   TYPES OF DATA COLLECTION

The success of natural gas industry depends heavily on the reliable exploration of various underground reservoirs and the drilling techniques used to extract gas. While conventional gas has been successfully extracted using conventional drilling techniques, which are quite extensively described in literature [21–23], new techniques used to extract gas from unconventional sources are briefly described in Section 2.3. The following description of gas exploration methods largely summarizes excellent descriptions given in References 21–25.

The practice of locating natural gas reservoirs has been transformed dramatically over the last 20–30 years with the advent of extremely advanced, ingenious technology. In the early days of the industry, the only way of locating underground natural gas deposits was to search for surface evidence of these underground formations. In more recent years, exploration for natural gas typically begins with geologists examining the surface structure of the earth, and determining areas where it is geologically likely that petroleum or gas deposits might exist. In the mid-1800s it was discovered that "anticlinal slopes" had a particularly increased chance of containing petroleum or gas deposits. Geologists have now many tools and data at their disposal, such as the outcroppings of rocks on the surface or in valleys and gorges, the geologic information attained from the rock cuttings, and samples obtained from the digging of irrigation ditches, water wells, and other oil and gas wells. This information is all combined to allow them to make inferences as to the fluid content, porosity, permeability, age, and formation sequence of the rocks underneath the surface of a particular area [21–25].

Application of basic seismology in 1921 for petroleum and natural gas exploration was a game changer. In this process, various types of underground formations were detected by observing how seismic waves interact with different geological formations. Two-, three-, and even four-dimension image processing of seismic wave reflections from underground significantly enhanced our capability to identify potential gas reservoirs. In practice, using seismology for exploring onshore areas involves artificially creating seismic waves, the reflection of which are then picked up by sensitive pieces of equipment called "geophones" that are embedded in the ground. The data picked up by these geophones are then transmitted to a seismic recording truck, which records the data for further interpretation by geophysicists and petroleum reservoir engineers. In the early days of seismic exploration, seismic waves were created using dynamite. Recently, due to environmental concerns, nonexplosive technology consists of a large heavy-wheeled or tracked-vehicle carrying special equipment designed to create a large impact or a series of vibrations. These impacts or vibrations create seismic waves similar to those created by dynamite [21–25].

The same sort of process is used in offshore seismic exploration. When exploring for natural gas that may exist thousands of feet below the seabed floor, which may itself be thousands of feet below sea level, a slightly different method of seismic exploration is used. Instead of trucks and geophones, a ship is used to pick up the seismic data and hydrophones are used to pick up seismic waves underwater. These hydrophones are towed behind the ship in various configurations depending on the needs of the geophysicist. Instead of using dynamite or impacts on the seabed floor, the seismic ship uses a large air gun, which releases bursts of compressed air under the water, creating seismic waves that can travel through the Earth's crust and generate the seismic reflections that are necessary.

In addition to using seismology to gather data concerning the composition of the Earth's crust, the magnetic properties of underground formations can be measured to generate geological and geophysical data. This is accomplished through the use of magnetometers, which are devices that can measure small differences in the Earth's magnetic field. In the early days of magnetometers, the devices were large and bulky and only able to survey a small area at a time [21–25].

Besides examining the Earth's magnetic field, geophysicists can also measure and record the difference in the Earth's gravitational field to gain a better understanding of what is underground. Different underground formations and rock types all have a slightly different effect on the gravitational field that surrounds the earth. By measuring these minute differences with very sensitive equipment, geophysicists are able to analyze underground formations and develop clearer insight into the types of formations that may lie below ground, and whether or not the formations have the potential for containing hydrocarbons [21–25].

The best way to gain a full understanding of subsurface geology and the potential for natural gas deposits to exist in a given area is to drill an exploratory well. This consists of digging into the Earth's crust to allow geologists to study the composition of the underground rock layers in detail. In addition to looking for natural gas and petroleum deposits by drilling an exploratory well, geologists also examine the drill cuttings and fluids to gain a better understanding of the geologic features of the area. This is a time-consuming and expensive process and is used only when there is other evidence for oil and gas reservoirs in a particular location [21–25].

Logging refers to performing tests during or after the drilling process to allow geologists and drill operators to monitor the progress of the well drilling and to gain a clearer picture of subsurface formations. While they can be as large as 100 different logging tests, their main purpose is to test and record the true composition and characteristics of the different layers of rock that the well passes through. Logging provides the information whether or not correct drilling equipment is used and when drilling should be discontinued if unfavorable conditions develop. Most prolific and often used logging methods are standard logging and electronic logging [21–25].

Standard logging examines and records nature of subsurface rock cut by the drilling process. This includes lifting a sample of underground rock intact to the surface and examining its nature and thickness by a highly powerful microscope. Examination under microscope allows the geologist to examine the porosity and fluid content of the subsurface rock being drilled [21–25].

Electric logging measures electric resistance of the rock layers in the "down hole" portion of the well by running an electric current through the rock formation and measuring the resistance that it encounters along its way. This gives geologists an idea of the fluid content and characteristics. A new technique called *induction electric logging* provides the same type of information more easily and the ones that are simple to interpret [21–25].

2.2.2 Computer-Assisted Exploration and Data Interpretation

As shown earlier, there are many sources of data and information for geologists and geophysicists to use in the exploration for hydrocarbons. However, these raw data alone would be useless without careful and methodical interpretation. Some techniques, including seismic exploration, lend themselves well to the construction of a hand- or computer-generated visual interpretation of an underground formation. Other sources of data, such as that obtained from core samples or logging, give good information for the subsurface geological structures.

Seismic imaging can view underground rock formation in two, three, and four dimensions (which includes time as one of the variables). While addition of each dimension adds more certainty to identifying locations for hydrocarbon reservoir, it also adds significantly to the cost of data gathering and analysis. Generally, progression from two- to three- to four-dimensional data gathering and analysis is carried out in more and more focused areas of hydrocarbon reservoir possibility. Often three-dimensional imaging is used first to narrow possible hydrocarbon reservoir region and then cheaper two-dimensional imaging is used to further pinpoint the reservoir location [21–25].

One of the greatest innovations in the history of petroleum exploration is the use of computers to compile and assemble geologic data into a coherent "map" of the underground. Use of this computer technology is referred to as "CAEX," which is short for "computer-assisted exploration." With the development of the microprocessor, it has become relatively easy to use computers to assemble seismic data that are collected from the field. This allows for the processing of very large amounts of data, thus increasing the reliability and informational content of the seismic model.

There are three main types of computer-assisted exploration models: two-dimensional (2-D), three-dimensional (3-D), and, most recently, four-dimensional (4-D). These imaging techniques, while relying mainly on seismic data acquired in the field, are becoming more and more sophisticated. Computer technology has advanced so far that it is now possible to incorporate the data obtained from different types of tests, such as logging, production information, and gravimetric testing, which can all be combined to create a "visualization" of the underground formation. Thus geologists and geophysicists are able to combine all of their sources of data to compile one clear, complete image of subsurface geology [21–25].

Two-dimensional seismic imaging gives a cross-sectional picture of the underground rock formations. The geophysicist interprets the seismic data obtained from the field, taking the vibration recordings of the seismograph and using them to develop a conceptual model of the composition and thickness of the various layers of rock underground. This process is normally used to make estimates of the possible location for deposits based on the geologic structures. In the mid-1970s a technique called "direct detection" was used where white bands, called "bright spots," often appeared on seismic recording strips. These white bands could indicate deposits of hydrocarbons. The nature of porous rock that contains natural gas could often result in reflecting stronger seismic reflections than normal, water-filled rock. This technique was, however, found not to be completely reliable [21–25].

Two-dimensional (2-D) computer-assisted exploration includes generating an image of subsurface geology much in the same manner as in normal 2-D data interpretation described earlier. However, with the aid of computer technology, it is possible to generate more detailed maps more quickly than by the traditional method. In addition, with 2-D CAEX it is possible to use color graphic displays generated by a computer to highlight geologic features that may not be apparent using traditional 2-D seismic imaging methods.

It is interesting to note that 3-D imaging techniques were developed prior to 2-D techniques. Thus, although it does not appear to be the logical progression of techniques, the simpler 2-D imaging techniques were actually an extension of 3-D techniques. Because it is simpler, 2-D imaging is much cheaper, and more easily and quickly performed, than 3-D imaging. Because of this, as mentioned before, 2-D CAEX imaging may be used in areas that are somewhat likely to contain natural gas deposits, but not likely enough to justify the full cost and time commitment required by 3-D imaging [21–25].

Three-dimensional imaging utilizes seismic field data to generate a three-dimensional "picture" of underground formations and geologic features. This, in essence, allows the geophysicist and geologist to see a clear picture of the composition of the Earth's crust in a particular area. This is tremendously useful in allowing for the exploration of petroleum and natural gas, as an actual image could be used to estimate the probability of formations existing in a particular area, and the characteristics of that potential formation. This technology has been extremely successful in raising the success rate of exploration efforts. In fact, using 3-D seismic has been estimated to increase the likelihood of successful reservoir location by 50%. Although this technology is very useful, it is also very costly and time-consuming. The technique is generally used along with 2-D computer modeling.

In addition to broadly locating petroleum reservoirs, 3-D seismic imaging allows for the more accurate placement of wells to be drilled. In fact, 3-D seismic can increase the recovery rates of productive wells to 40%–50%, as opposed to 25%–30% with traditional 2-D exploration techniques. Three-dimensional seismic imaging has become an extremely important tool in the search natural gas. In 1996, in the Gulf of Mexico, one of the largest natural gas-producing areas in the United States, nearly 80% of wells drilled in the Gulf were based on 3-D seismic data [21–25].

One of the latest breakthroughs in seismic exploration and the modeling of underground rock formations has been the introduction of four-dimensional (4-D) seismic imaging. This type of imaging is an extension of 3-D imaging technology. However, instead of achieving a simple, static image of the underground, in 4-D imaging the changes in structures and properties of underground formations are observed over time. Since the fourth dimension in 4-D imaging is time, it is also referred to as 4-D "time lapse" imaging.

Various seismic readings of a particular area are taken at different times, and this sequence of data is fed into a powerful computer. The different images are amalgamated to create a "movie" of what is going on under the ground. By studying how seismic images change over time, geologists can gain a better understanding of many properties of the rock, including underground fluid flow, viscosity, temperature, and saturation. Although very important in the exploration process, 4-D seismic images can also be used by petroleum geologists to evaluate the properties of a reservoir, including how it is expected to deplete once petroleum extraction has begun. Using 4-D imaging on a reservoir can increase recovery rates above what can be achieved using 2-D or 3-D imaging. Where the recovery rates using these two types of images are 25%–30% and 40%–50%, respectively, the use of 4-D imaging can result in recovery rates of 65%–70% [21–25].

The historical development of various methods of natural gas exploration has made exploration more of a science than art. Since drilling a well (even exploratory well) is very expensive, gas industry has used these tools to minimize cost and time for the determination of different types of gas reservoirs underground.

2.3   METHODS FOR DRILLING AND PRODUCTION OF NATURAL GAS

Once natural gas reservoirs are identified, gas well is drilled to extract the gas out of ground. There are several methods of drilling.

2.3.1 VERTICAL DRILLING

In this method, a hole is drilled straight down using drilling mud, which cools the drill, carries the rock cuttings back to the surface, and stabilizes the wall of the well bore [26–29]. Once the hole extends below the deepest freshwater aquifer, the drill pipe is removed and replaced with a steel pipe called "surface casing."

Next, cement is pumped down the casing and then back up between the casing and the borehole wall, where it sets. This cement bond prevents the migration of any fluids vertically between the casing and the hole. In doing so, it creates an impermeable protective barrier between the well bore and any freshwater sources. Tests are undertaken to confirm the success of this process before drilling further.

Typically, depending on the geology of the area and the depth of the well, additional casing sections will be run and, like surface casing, are then cemented in place to ensure that there can be no movement of fluids or gas between those layers and the groundwater sources [26–29].

2.3.2 HORIZONTAL DRILLING

Technological advances in drilling process allow drillers to deviate from vertical drilling and steer the drilling equipment to a location that is not directly underneath the point of entry. This is in contrast to "slant drilling," where the well is drilled at an angle instead of directly vertical. New technology allows for the drilling of tightly curved well holes where 90° turns can be accomplished within several feet underground. Vertical drilling continues to a depth called the "kick off point" [29]. This is where the well bore begins curving to become horizontal. Traditional directional drilling takes several thousand feet to turn 90° [29–34]. These new technologies are aided by borehole telemetry to gain real-time information from steerable drilling motors [29–34]..

Conventional vertical wellbore suffers from a lack of exposure to the shale formation in comparison to horizontal wellbores. Horizontal drilling is particularly useful in shale formations that do not have sufficient permeability to produce economically; therefore, it is becoming more and more pervasive, especially in North America [25]. In the United States, tight reservoirs such as the Bakken (North Dakota and Montana), Montney, Barnett, and Haynesville (Texas/Oklahoma) Shale and most recently Marcellus Shale (New York, Pennsylvania, and Ohio) are drilled, completed, and fractured using this method [29–34].

One of the advantages of horizontal drilling is that it's possible to drill several laterals from only one surface-drilling pad, minimizing the impact on the surface environment. In many cases, drilling can also be done as "directional drilling," which can follow the nature of the shallow reservoir. When the targeted distance is reached, the drill pipe is removed and additional steel casing is inserted through the full length of the well bore and once again, the casing is cemented in place. Once the drilling is finished and the final casing has been installed, the drilling rig is removed and preparations are made for the next step: well completion [29–34].

2.3.3 WELL COMPLETION

The first step in completing a well is the creation of a connection between the final casing and the rock holding the oil and gas. A specialized tool, called a perforating gun, is lowered to the rock layer. This perforating gun is then fired, creating holes through the casing and the cement and into the target rock. These perforating holes connect the rock holding the oil and gas and the well bore [29–37].

Since these perforations are only a few inches long and are performed more than a mile underground, the entire process is imperceptible on the surface. The perforation gun is then removed in preparation for the next step, hydraulic fracturing [29–37].

2.3.4 HYDRAULIC FRACTURING

The first fracking operation in the United States was performed in 1947 in the Hugoton Kansas gas fields by Halliburton. However, hydraulic fracking did not become economical for commercial use for several decades. Significant R&D was necessary before hydraulic fracturing could be commercially applied to shale gas deposits, due to shale's high porosity and low permeability. In the 1970s, the federal government initiated both the Eastern Gas Shale Project and the Gas Research Institute project which further explored this process. During that time, Sandia contributed its geologic micro-mapping software, which proved to be crucial for the commercial recovery of natural gas from shale [1,38–50].

In the late 1970s, the Department of Energy (DOE) pioneered a massive effort on hydraulic fracturing, which was later improved for the economic recovery of shale gas. In 1986, a joint DOE–private venture completed the first successful multifracture horizontal well in shale. The DOE later subsidized Mitchell Energy's first successful horizontal drilling in the North Texas Barnett Shale in 1991. Mitchell Energy engineers developed the hydraulic fracturing technique known as "slick water fracturing [47]." The addition of chemicals to water was to increase fluid flow. This innovation was implemented in 1996, and it started the modern shale gas boom [1,35,38,39,42,43,45,46,48–50].

It is estimated that hydraulic fracturing will eventually account for nearly 70% of natural gas development in North America. Worldwide hydraulic fracturing has been used to stimulate approximately 1 million oil and gas wells. Hydraulic fracturing is not a "drilling process" but a process used after the drilled hole is completed. Hydraulic fracturing, or "fracking," is either the propagating of fractures in a rock layer caused by the perforating air guns (during well completion process) or fracturing of rock by the pressurized fluid in deep, underground geological formations to liberate oil or natural gas. This process is used to release petroleum, natural gas (including shale gas, tight gas, deep gas, and coal seam gas), or other substances for extraction, via a technique called induced hydraulic fracturing. While the fracking process used to recover natural gas generally uses water as the fracking fluid, in general, as shown in Table 2.2 [38,42,45,46,48,49], different types of fracking fluids are required for creations of short, medium, or long fractures under different geology. In all cases, however, fracking process includes steps to protect groundwater contained in intervening aquifers. These steps involve an insertion of steel surface or intermediate casing to prevent the fluid from entering the water supply. Normally the depths of the insertions are between 1000 and 4000 ft. Also, cement needs to be filled into the annulus, the space between the casing strings and the drilled hole. Once the cement has set, then the drilling continues from the bottom of the surface or intermediate cemented steel casing to the next depth. This process is repeated, using smaller steel casing each time, until the oil and gas-bearing reservoir is reached (generally 6,000–10,000 ft) [1,35,38,39,42,43,45,46,48–50].

TABLE 2.2

Fracking Fluids Required for Short, Medium, and Long Fractures

Fluid Plus Ingredients	Function
Water with HPG, HEC, CMHPG, Guar	Low temperature and short fractures
Oil with gelling agent	Short fractures in water sensitive formations
Acid with HPG or Guar	Short fractures in carbonate formations
Water with surfactant and electrolyte	Moderate temperature and moderate fractures
Water with oil and emulsifier	Moderate length fractures with good fluid loss control
Acid with oil and emulsion	Moderate length fractures in carbonate formations
Water with crosslinker and CMHPG, CMHEC, Guar, or HPG	High temperature and long fractures
Oil with gelling agent and crosslinker	Long fractures in water sensitive formations
Acid with crosslinker and Guar or HPG	Longer and wider fractures in carbonate formations

Source: Modified and adapted from Fracturing fluid and additives, Petrowiki, A report from SPE, 2015.

In order to fracture the formation to recover unconventional gas, fracturing fluids—water and sand, and proprietary chemical mixes—are injected down the well bore into the formation. The fluid, injected under pressure, causes the rock to fracture along weak areas and propagates the fractures created by perforated air guns during the well completion process. The fluids that create the initial fractures are then mixed with thicker fluids that include sand and gelatin. These thicker fluids lengthen the openings in the rock. When the fractures are complete, and pressure is relieved, a portion of the fluids flows back up the well, where it is captured and stored for later treatment or disposal. As the fluids flow back up, sand remains in the fractures and props the rock open, maintaining an open pathway to the well. This allows the oil and gas to seep from the rock into the pathway, up the well and to the surface for collection [1,38,42,45–50].

A distinction can be made between low-volume hydraulic fracturing used to stimulate high-permeability reservoirs, which may consume typically 20,000–80,000 gal of fluid per well, with high-volume hydraulic fracturing, used in the completion of tight gas and shale gas wells which may consume enormous amounts of water; up to 5 million gal of water for a single well [1,38,45–50]. After the fracturing procedure is complete, 15%–80% of the fluid returns to the surface as wastewater, often contaminated by fracturing chemicals and subsurface contaminants, including toxic organic compounds, heavy metals, and naturally occurring radioactive materials [44]. Additionally, a significant amount of brine, containing salts and other minerals, may be produced with the gas. If the chemicals in the recovered wastewater are not treated, they can have harmful environmental and health effects. In any circumstances, this wastewater should not be allowed to be mixed with potable groundwater resources [1,38,42,45–50]. Figure 2.3 illustrates an overview of the hydraulic fracking water cycle [51].

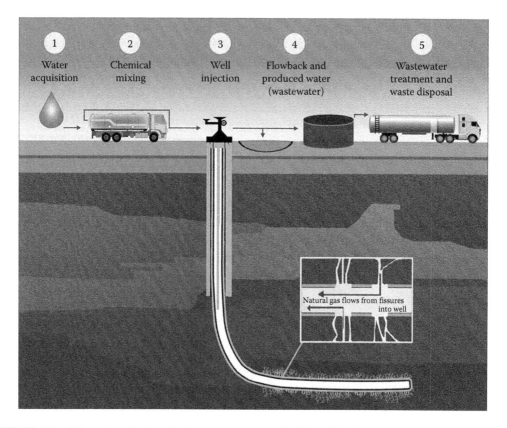

FIGURE 2.3 Schematic of hydraulic fracturing water cycle. (From The hydraulic fracturing water cycle, EPA's study of hydraulic fracturing and its potential impact on drinking water resources, EPA, Washington, DC, June 1, 2015, https://www.epa.gov/.../h...)

The purpose of the chemicals during fracking is lubrication, to keep bacteria from forming and to carry the sand. The level and strength of chemicals and additives used depends on the geological nature of the reservoir. In the United Kingdom, Cuadrilla Resources were granted approval from the Environment Agency (EA) to use the following nonhazardous amounts of chemicals at their Preese Hall exploration [1,38,42,45–50].

- Polyacrylamide (friction reducer)
- Sodium salt (for tracing fracturing fluid)
- Hydrochloric acid (highly diluted with water—for improved efficiency)
- Glutaraldehyde biocide (used to cleanse water and remove bacteria)

These chemicals were nonhazardous and ranged in concentrations from 0.1% to 0.5% by volume. They helped to improve the performance of the hydraulic fracturing. In actual tests, only a small amount of polyacrylamide friction reducer (0.04%) and a miniscule amount of salt were used. There was no need to use biocide as the water supplied had already been treated to remove bacteria, and diluted hydrochloric acid was not necessary. Polyacrylamide is a nonhazardous, nontoxic substance, which is also used extensively in other industries to remove suspended solids in drinking and waste-water plants, and pollutants or contaminants from soils. The simulated fluid was pumped by high-pressure tanks mounted on the movable skid platforms at the well bore. Once the fluid opened up the rock structure, it was withdrawn, leaving sand behind. Thus, the initial stimulation segment was then isolated with a specially designed plug and the perforating guns were used to perforate the next horizontal section of the well. This stage was then hydraulically fractured in the same manner. This process was repeated along the entire horizontal section of the well, which extended 1–2 miles [1,38,42,45–50].

While tests performed in the United Kingdom used mild water solutions, other more toxic compositions have also been tested for perforating more difficult rock structure; particularly for shale plays in the United States. Contamination of surface waterways and groundwater with fracking fluids in these cases can be problematic. There have also been instances of methane migration, improper treatment of recovered wastewater, and pollution via reinjection wells reported in the literature [1,38,42,45–50]. Shale gas deposits are generally several thousand feet below ground.

In shale gas plays, water typically makes up more than 98% of the fluids used for hydraulic fracturing. In addition to water, the fracturing process uses a proprietary mix of chemicals and other fluids, with each serving a specific, engineered purpose. Additionally, more than 1 million pounds of "proppants" may be used in hydraulic fracturing a well to prop the newly created fractures open and allow formation fluids to flow into the borehole. Proppants are compression-resistant particles, originally mainly fine-grained sand but now also including aluminum or ceramic beads, sintered bauxite, and other materials [52]. In a representative example from a Fayetteville well, water and sand made up more than 99% of the volume, with various chemicals making up the rest as shown in Figure 2.4 [52].

As mentioned earlier, over its lifetime, an average shale gas well can require 5 million gallons of water for the initial hydraulic fracturing operation and possible re-stimulation. The large volumes of water required have raised concerns about fracking in water shortage areas such as Texas, which has been in a multiple-year drought. Chemical additives used in fracturing fluids typically make up less than 2% by weight of the total fluid. Nonetheless, over the life of a typical well, this may amount to 100,000 gal of chemical additives [1,38,42,45–50]. A summary of functions of various chemical additives to fracking fluids is given in Table 2.3 [52,53]. These additives may include some that are known carcinogens, some that are toxic, and some that are neurotoxins. These include benzene, lead, ethylene glycol, methanol, boric acid, and 2-butoxyethanol. Besides these chemical additives, high levels of iodine-131 (a radioactive tracer used in hydraulic fracturing) is the major contributor to the generally elevated radiation levels found near hydraulic fracturing sites.

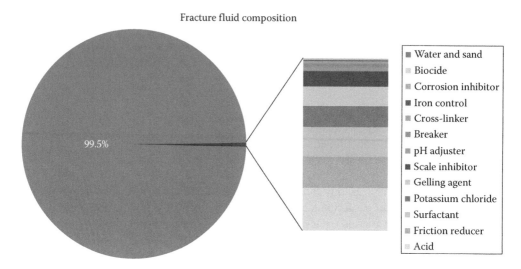

FIGURE 2.4 Volumetric composition of a hydraulic fracturing fluid. (From Zammerilli, A. et al., Environmental impacts of unconventional natural gas development and production, NETL report, DOE NETL contract number DE-FE0004001, Office of Fossil Energy, Department of Energy, Washington, DC, May 29, 2014; Department of Energy (DOE), Modern shale gas development in the United States: A primer, U.S. Department of Energy, National Energy Technology Laboratory (NETL), Strategic Center for Natural Gas and Oil, Morgantown, WV, 96pp., April 2009.)

2.3.4.1 Fracking Hot Spots in the United States

While the locations of gas-bearing shale have been known for some time, the confluence of advanced technology and market demand for clean-burning fuel has made development of these resources more urgent. About 20 states can expect to feel the effects of fracking exploration. The densest deposits are the Marcellus and Utica shale belts stretching from New York through Pennsylvania, Ohio, and Indiana into Illinois. Some states, like Texas, Oklahoma, Colorado, and Wyoming, are also experiencing resource extraction near populations. Others, like North Dakota and Montana, are largely rural [1,38,42,45–50].

2.3.4.2 Negative Effects of Hydraulic Fracturing

Various environmental impacts of hydraulic fracturing have been reported in the literature [41], including enhanced radioactivity near fracking sites [44]. Hydraulic fracturing routinely produces microseismic events much too small to be detected, except by sensitive instruments. These microseismic events are often used to map the horizontal and vertical extent of the fracturing. However, as of late 2012, there have been three known instances worldwide of hydraulic fracturing, through induced seismicity, triggering quakes large enough to be felt by people. Beginning in 2001, the average number of earthquakes occurring per year of magnitude 3 or greater increased significantly, culminating in a sixfold increase in 2011 over twentieth-century levels [1,38,40,50]. Over 109 small earthquakes (M_w 0.4–3.9) were detected during January 2011 to February 2012 in the Youngstown, Ohio, area, where there were no known earthquakes in the past. These shocks were close to a deep fluid injection well. Significant numbers of small earthquakes are also reported in the state of Oklahoma where hydraulic fracturing is practiced regularly [1,38,40,50].

2.3.5 PRODUCTION

Once the stimulation is complete, the isolation plugs are drilled out and production begins. Initially fracture fluid, and then natural gas or oil, flow into the horizontal casing and up the well bore.

TABLE 2.3

Fracturing Fluid Additives, Main Compounds, and Purposes

Additive	Compound(s)	Purpose	Percentage of Fluid (% of Volume)	
			DOE [54]	FracFocus [53]
Dilute acid	Hydrochloric or muriatic acid	Helps dissolve minerals and initiates cracks in the rock	0.123	0.07
Friction Reducer	Polyacrylamide or mineral oil	Minimizes friction between fluid and pipe	0.088	0.05
Surfactant	Isopropanol	Used to increase the viscosity of the fracture fluid	0.085	No data
KCl	Potassium chloride	Creates a brine carrier fluid	0.060	No data
Gelling agent	Guar gum or hydroxyethyl cellulose	Thickens water to suspend sand	0.056	0.5
Scale inhibitor	Ethylene glycol	Prevents scale deposits in the pipe	0.043	0.023
pH adjusting agent	Sodium or potassium bicarbonate	Maintains effectiveness of other components, such as crosslinkers	0.011	No data
Breaker	Ammonium persulfate	Allows a delayed break down of the gel polymer chains	0.01	0.02
Cross-linker	Borate salts	Maintains fluid viscosity as temperature increases	0.007	0.032
Iron control	Citric acid	Prevents precipitation of metal oxides	0.004	0.004
Corrosion inhibitor	N,N-Dimethyl formamide	Prevents the corrosion of the pipe	0.002	0.05
Biocide	Glutaraldehyde	Eliminates bacteria in the water that produce corrosive by-products	0.001	0.001
Oxygen scavenger	Ammonium bisulfate	Removes oxygen from the water to protect pipe from corrosion	No data	No data
Clay control	Choline chloride, sodium chloride	Minimizes permeability impairment	No data	0.034
Water and proppant	Proppant: silica or quartz sand	Allows fractures to remain open so gas can escape	99.51	99.2

Sources: Zammerilli, A. et al., Environmental impacts of unconventional natural gas development and production, NETL report, DOE NETL contract number DE-FE0004001, Office of Fossil Energy, DOE, Washington, DC, May 29, 2014; FracFocus, www.fracfocus.org, Available from http://fracfocus.org/, Cited: June 25, 2013.

In the course of initial production of the well, about 25%–75% of the fracturing fluid is recovered. This fluid is either recycled for use on other fracturing operations, or safely disposed of according to government and environmental regulations. This whole process of developing a well typically takes from 3 to 5 months, which include the following:

1. A few weeks to prepare the site
2. Eight to twelve weeks to drill the well
3. One to three months of completion activities, which include between 1 and 7 days of stimulation

This initial 3- to 5-month investment could, however, result in a well that will produce oil or natural gas for 20–40 years, or more [1,3–8,23–25,35].

2.3.6 Well Abandonment

When all of the oil or natural gas that can be recovered economically from a reservoir has been produced, the land is returned to the way it was before the drilling operations started. Wells, are filled with cement and pipes cut off 3–6 ft below ground level. All surface equipment are removed and all pads are filled in with earth or replanted. The land can then be used again by the landowner for other activities, and there will be virtually no signs that a well was once there.

Today, hydraulic fracturing has become an increasingly important technique for producing natural gas in places where hydrocarbons were previously inaccessible. The technologies for the recovery of oil and gas will continue to be improved for the safe, environmentally acceptable and economic production [1,3–8,23–25,35].

2.4 LNG, CNG, AND LPG

One of the drawbacks of natural gas is its low density. In order to circumvent this problem, natural gas is often compressed (CNG) [55–66] or liquefied as LNG [67–89]. Similarly propane and butane are also often liquefied as LPG [58,90–96]. Here we briefly evaluate these two forms of natural gas and petroleum gas [55–96].

2.4.1 Liquid Natural Gas

Cooling natural gas to about −260°F at normal pressure results in the condensation of the gas into liquid form, known as liquefied natural gas (LNG). LNG can be very useful, particularly for the transportation of natural gas, since it takes up about one six-hundredth the volume of gaseous natural gas. While LNG is reasonably costly to produce, advances in technology are reducing the costs associated with the liquefaction and regasification of LNG. For stranded gas where construction of pipeline is uneconomical, LNG provides an easy alternative for gas transport. LNG also allows easy transportation of natural gas across sea and oceans [67–89].

LNG, when vaporized to gaseous form, will only burn in concentrations of between 5% and 15% mixed with air. In addition, LNG, or any vapor associated with LNG, will not explode in an unconfined environment. Thus, in the unlikely event of an LNG spill, the natural gas has little chance of igniting an explosion. Liquefaction also has the advantage of removing oxygen, carbon dioxide, sulfur, and water from the natural gas.

Typically, LNG is 85–95-plus percent methane, along with a few percent ethane, even less propane and butane, and trace amounts of nitrogen. The exact composition of natural gas (and the LNG formed from it) varies according to its source and processing history. Like methane, LNG is odorless, colorless, noncorrosive, and nontoxic. Just one shipload of LNG can provide nearly 5% (roughly 3 billion cubic feet (bcf)) of the U.S. average daily demand for natural gas, or enough energy to heat more than 43,000 homes for an entire year [67–70].

Efforts to liquefy natural gas for storage began in the early 1900s, but it wasn't until 1959 that the world's first LNG ship carried cargoes from Louisiana to the United Kingdom, proving the feasibility of transoceanic LNG transport. Five years later, the United Kingdom began importing Algerian LNG, making the Algerian state-owned oil and gas company Sonatrach the world's first major LNG exporter. The United Kingdom continued to import LNG until 1990, when British North Sea gas became a less expensive alternative.

Japan first imported LNG from Alaska in 1969 and moved to the forefront of the international LNG trade in the 1970s and 1980s with a heavy expansion of LNG imports. These imports into Japan helped to fuel natural gas–fired power generation to reduce pollution and relieved pressure from the oil embargo of 1973. Japan currently imports more than 95% of its natural gas requirements, and serves as the destination for about half the LNG export worldwide. To this date, Japan remains the world's largest LNG consumer, although its share of global LNG trade has fallen slightly over the past decade as the global market has grown. Japan's largest LNG suppliers are Indonesia and Malaysia, with substantial volumes also imported from Qatar, the United Arab Emirates, Australia, Oman, and Brunei Darussalam. Early in 2004, India received its first shipment of LNG from Qatar at the newly completed facility at Dahej in Gujarat [67–70,72–78].

The United States first imported LNG from Algeria during the 1970s, before the regulatory reform and rising prices led to a rapid growth of the domestic natural gas supply. The resulting supply–demand imbalance (known as the "gas bubble" of the early 1980s) led to reduced LNG imports during the late 1980s and eventually to the mothballing of two LNG import facilities. In the 1990s, natural gas demand grew rapidly, and the prospect of supply shortfalls led to a dramatic increase in U.S. LNG deliveries. In 1999, a liquefaction plant became operational in Trinidad and Tobago, supplying LNG primarily to the United States. In the United States, imported LNG accounts for slightly more than 1% of the natural gas used in the United States. According to the EIA, the United States imported 0.41 tcf of natural gas in the form of LNG in 2010 [74,76,77]. However, due to increased domestic production, LNG imports are expected to decrease by an average annual rate of 4.1%, to levels of 0.14 tcf of natural gas by 2035.

Russia, Iran, and Qatar hold 58.4% of the world's natural gas reserves, yet consume only about 19.4% of worldwide natural gas. Such countries tend to "monetize" their gas resource—converting it into a salable product. LNG makes this possible. The world's major LNG-exporting countries hold about 25% of the total natural gas reserves. Two countries with significant reserves (Russia and Norway) are currently building their first liquefaction facilities. At least seven more are considering the investment to become LNG exporters in the near future [67–70,72–78].

In some cases, conversion to LNG makes use of natural gas that would once have been lost. For example, Nigeria depends on its petroleum exports as a primary source of revenue. In the process of oil production, natural gas was flared [12–14]—a wasteful practice that adds carbon dioxide to the atmosphere. Converting this natural gas to LNG provides both economic and environmental benefits.

International trade in LNG centers on two geographic regions [67–70,75]:

- The Atlantic Basin, involving trade in Europe, northern and western Africa, and the U.S. Eastern and Gulf coasts.
- The Asia/Pacific Basin, involving trade in South Asia, India, Russia, and Alaska.

LNG is the preferred form for long-distance, high-volume transportation of natural gas, whereas pipeline [87] is preferred for transport for distances up to 4000 km (2485 miles) over land and approximately half that distance offshore. More details on LNG storage and transportation are given in Chapter 8. The use of LNG for heating and transportation industry is described in Chapter 10. As shown in these chapters, LNG demand in vehicle industries is rapidly expanding [38,67–70,78–83].

Floating liquefied natural gas (FLNG) is an innovative technology [71,85,86] designed to enable the development of offshore gas resources that would otherwise remain untapped. Due to environmental or economic reasons these resources cannot be developed via a land-based LNG operation. FLNG technology [71,85,86,89] also provides a number of environmental and economic advantages:

1. Since all processing is done at the gas field, there is no requirement for long pipelines to shore, compression units to pump the gas to shore, dredging and jetty construction, and onshore construction of an LNG processing plant, all of which significantly reduce the environmental footprint; avoiding construction also helps preserve marine and coastal environments. In addition, environmental disturbance will be minimized during decommissioning because the facility can easily be disconnected and removed before being refurbished and redeployed elsewhere.

2. When pumping gas to shore becomes prohibitively expensive, FLNG makes development economically viable. As a result, it will open up new business opportunities for countries to develop offshore gas fields that would otherwise remain stranded, such as those offshore East Africa.

Many gas and oil companies are considering the economic and environmental benefits of FLNG. However, currently the only FLNG facility being built is by Shell, which is due for completion around 2017 [67–72,85,86,89].

2.4.2 Compressed Natural Gas

Methane stored at high pressure can be used in place of gasoline (petrol), diesel fuel, and propane/LPG. CNG combustion produces fewer undesirable gases than the fuels mentioned earlier. It is safer than other fuels in the event of a spill, because natural gas is lighter than air and disperses quickly when released. CNG may be found above oil deposits, or may be collected from landfills or wastewater treatment plants where it is known as biogas [55–66]. As shown in Chapters 8 and 10, CNG is becoming very popular in vehicle industries due its safety and reliability. CNG can be used in existing vehicles after only few modifications in car design [56,57,59–66]. CNG is preferred in larger vehicles due to its large space requirement for storage.

CNG is made by compressing natural gas (which is mainly composed of methane, CH_4), to less than 1% of the volume it occupies at standard atmospheric pressure. It is stored and distributed in hard containers at a pressure of 20–25 MPa (2900–3600 psi), usually in cylindrical or spherical shapes. CNG is odorless, colorless, and tasteless. It is drawn from domestically drilled natural gas wells or in conjunction with crude oil production. Compressors and decompression equipment are less capital-intensive and may be economical in smaller unit sizes than liquefaction/regasification plants. CNG trucks and carriers may transport natural gas directly to end users, or to distribution points such as pipelines. More details on storage and transport of CNG are given in Chapter 8. The use of CNG for mobile applications is described in detail in Chapter 10.

Compressed natural gas is often confused with LNG (liquefied natural gas). While both are stored forms of natural gas, the key difference is that CNG is a gas that is stored (as a gas) at high pressure, while LNG is stored at very low temperature, becoming liquid in the process. CNG has a lower cost of production and storage compared to LNG as it does not require an expensive cooling process and cryogenic tanks. CNG requires a much larger volume to store the same mass of gasoline or petrol and the use of very high pressures (3000–4000 psi or 205–275 bar). As a consequence of this, LNG is often used for transporting natural gas over large distances, in ships, trains or pipelines, and the gas is then converted into CNG before distribution to the end user.

CNG is being experimentally stored at lower pressure in a form known as adsorbed natural gas (ANG) tank, at 35 bar (500 psi, the pressure of gas in natural gas pipelines) in various sponge-like materials, such as activated carbon and metal-organic frameworks (MOFs) (for more details see Chapters 8 and 10). The ANG fuel is stored at similar or greater energy density than CNG. This means that vehicles can be refueled from the natural gas network without extra gas compression, and the fuel tanks can be slimmed down and made of lighter, weaker materials. CNG is sometimes mixed with hydrogen (HCNG), which increases the H/C ratio (heat capacity ratio) of the fuel and gives it a flame speed up to eight times higher than CNG. More details on the storage and transport of CNG are given in Chapter 8.

2.4.3 Liquefied Petroleum Gas

In 1910, Dr. Walter Snelling, a chemist with the U.S. Bureau of Mines, discovered that propane was a component of liquefied gas. Soon afterward, he discovered a means to store and transport propane and butane. Snelling received a patent for LPG in 1913, which he then sold to Frank Phillips, founder of Phillips Petroleum Company [1,96]. Initially, LPG was used to fuel metal-cutting torches, but by 1927, manufacturers were making gas cooking ranges fueled by LPG. Soon after World War II, propane was used as a transportation fuel in buses and cars [58,90–96]. Its use in vehicle industry is rapidly expanding.

LPG becomes a liquid at normal pressure and a temperature of −42°C, or at normal temperatures under a pressure of about eight atmospheres (standard units equivalent to ordinary atmospheric pressure at sea level and 0°C). *Liquefied petroleum gas* (LPG) is a term describing a group of hydrocarbon-based gases derived from crude oil and or natural gas. Natural gas purification produces about 55%–60% of all LPG, while crude oil refining produces about 40%–45%. Varieties of LPG bought and sold include mixes that are primarily propane (C_3H_8), primarily butane (C_4H_{10}) and, most commonly, mixes including both propane and butane. The chemical composition of LPG can vary, but is usually made up of butane and propane with a 30%–99% propane mix [58,90–96].

In winter, the mixes contain more propane, while in summer, they contain more butane [1,58,90–96]. In the United States, primarily two grades of LPG are sold: commercial propane and HD-5. These specifications are published by the Gas Processors Association (GPA) [90–94] and the American Society of Testing and Materials (ASTM) [93]. Propane–butane blends are also listed in these specifications. They also include ethane, ethylene, propylene, butylene, isobutene, and isobutylene; these are used primarily as chemical feedstocks rather than fuel. HD-5 limits the amount of propylene that can be placed in LPG to 5%, and is utilized as an autogas specification. A powerful odorant, ethanethiol, is added so that leaks can be detected easily. The internationally recognized European Standard for LPG is EN 589. In the United States, tetrahydrothiophene (thiophane) and amyl mercaptan are also approved odorants, although neither is currently being utilized [58,90–96].

As its boiling point is below room temperature, LPG will evaporate quickly at normal temperatures and pressures, and is usually supplied in pressurized steel vessels. They are typically filled to 80%–85% of their capacity to allow for thermal expansion of the contained liquid. The ratio between the volumes of the vaporized gas and the liquefied gas varies depending on composition, pressure, and temperature, but is typically around 270:1. The pressure at which LPG becomes liquid likewise varies depending on composition and temperature; for example, it is approximately 220 kPa (32 psi) for pure butane at 20°C (68°F), and approximately 2200 kPa (320 psi) for pure propane at 55°C (131°F). LPG is heavier than air (unlike natural gas) and thus will flow along floors and tend to settle in low spots, such as basements. There are two main dangers from this. The first is a possible explosion if the mixture of LPG and air is within the explosive limits and if there is an ignition source. The second is suffocation due to LPG displacing air, causing a decrease in oxygen concentration [58,90–96].

LPG currently provides about 3% of all energy consumed, and burns relatively cleanly with no soot and very few sulfur emissions. As it is a gas, it does not pose ground or water pollution hazards, but it can cause air pollution. LPG has a typical specific calorific value of 46.1 MJ/kg compared with 42.5 MJ/kg for fuel oil and 43.5 MJ/kg for premium grade gasoline [6]. However, its energy density per volume (26 MJ/L) is lower than either that of gasoline or fuel oil. Its density (about 0.5–0.58 kg/L) is lower than that of gasoline (about 0.71–0.77 kg/L) [90–94].

According to 2010–2012 estimates, proven world reserves of natural gas, from which most LPG is derived, stand at 300 tcm (10,600 tcf). Added to the LPG derived from cracking crude oil, this amounts to a major energy source that is virtually untapped and has massive potential. Production continues to grow at an average annual rate of 2.2%, virtually assuring that there is no risk of demand outstripping supply in the foreseeable future. Bio-LPG or bio-propane is a clean-burning, renewable fuel which can replace autogas and liquefied petroleum gas in all consumer and industrial applications. It is molecularly identical to propane produced from petroleum refining. Several chemical process pathways are being explored to produce bio-propane on a commercial scale [58,67–70,95,96].

In 2002, Texas produced 3.5 billion gal of propane, or 36% of the national total [96]. Because LPG is a by-product of oil and gas, the amount available is directly tied to the amount of oil and gas available. Texas' crude oil reserves in 2006 represented almost one-fourth, or 23.3%, of the total proven U.S. reserves [96]. Texas' proven gas reserves in 2006 accounted for 29.2% of all proven natural gas reserves in the United States [96]. LPG is a nonrenewable fuel source, as are the natural gas and crude oil from which it is produced. LPG is a cleaner alternative to many fuels, but its combustion does produce pollutants. These include particulate matter, sulfur dioxide, nitrogen oxides, nitrous oxide, carbon monoxide, carbon dioxide, methane, and nonmethane total organic carbon [96]. Again, since LPG is a by-product of oil and natural gas production, its water consumption and quality implications are similar to those of oil and gas [58,67–70,95,96].

There are federal and state regulations on the production, transportation, and storage of LPGs and other pressurized gases to minimize risks. Though rare, LPG, particularly propane and butane, poses a risk of sudden depressurization and explosion during storage and transport (see Chapters 8 and 10). The Texas Railroad Commission (RRC) administers and enforces state laws and rules related to LPG. The RRC also licenses LPG activities in the state, including its sale, transportation, and storage; the manufacture, repair, sale, and installation of LPG containers; and the installation, servicing, and repair of LPG-fueled appliances. Drivers and dealers of LPG vehicles also must obtain a fuels tax permit from the Texas Comptroller of Public Accounts. There are also regulations by EPA and DOT at the federal level on the use of LPG. For more details on storage and transportation of LPG, see Chapter 8.

Texas is the nation's largest consumer of LPG for all sectors combined. Other states, however, exceed Texas in some sectors. Texas was the second-largest consumer of residential LPG in 2005, accounting for 6.3% of the nation's total; Michigan led the states with 9.7%. Similarly, Texas was the second-largest consumer of commercial LPG in 2005, again accounting for 6.3% of the nation's total commercial use. Michigan was again first, with 9.7% of the national total. Texas was also the largest consumer of industrial LPG in the nation in 2005, accounting for nearly three-quarters (71.2%) of all industrial LPG used in the nation [58,67–70,95,96].

In Texas, chemical feedstock usages account for 90% of the state's LPG use, with nearly all of the remaining 10% used to produce energy [1–8,58,67–70,90–96]. While only 0.1% of LPG in 2005 was used for transportation in Texas, propane was nevertheless the most common alternative transportation fuel in the United States, used by public transportation fleets as well as many state and federal agency vehicles ([58,90–96], see also Chapter 10). Propane has a lower energy output than gasoline, producing 84,000 British thermal units (Btu) per gallon, or about 74% of gasoline's energy potential [90–94]. LPG also ranks third in the United States, behind gasoline and diesel. When specifically used as a vehicle fuel, it is often referred to as autogas (see Chapter 10). LPG vehicles emit around a third less reactive organic gas, which reacts with other pollutants in sunlight to create ozone, and about 50% less of the vapors that create smog, than do gasoline vehicles. LPG vehicles also release 20% less nitrogen oxide and 60% less carbon monoxide than gasoline vehicles. Finally, LPG contributes very little to acid rain because of its low sulfur content [90–94].

LPG has a very high energy content per ton compared to most other oil products and burns easily in the presence of air. This has made LPG a popular fuel for domestic heating and cooking, commercial use, agricultural and industrial processes, and increasingly as an alternative transport fuel (see Chapters 8 and 10). LPG, primarily propane, is widely used as a fuel for heating and cooking in rural America and other areas where natural gas lines are unavailable. Its transportability and easy storage have boosted its popularity. Although relatively few urban residences depend upon large propane tanks for heating and cooking, smaller tanks for outdoor grills are extremely common throughout the nation. It is increasingly used as an aerosol propellant and a refrigerant replacing chlorofluorocarbons in an effort to reduce damage to the ozone layer [15–17,20].

Propane is also used to generate electricity through micro-turbines and combined heat and power (CHP) technology. Micro-turbines are very small turbines intended to generate electricity for homes or commercial establishments, as well as for vehicles such as hybrid buses. More use of micro-turbines still continues to be developed. For more details on the use of LPG for vehicles and micro-turbines, see Chapter 10.

2.4.4 COMPARISON OF LPG WITH NATURAL GAS

LPG is composed primarily of propane and butane, while natural gas is composed of the lighter methane and ethane. At atmospheric pressure, vaporized LPG has a higher calorific value (94 MJ/m^3 equivalent to 26.1 kWh/m^3) than natural gas (methane) (38 MJ/m^3 equivalent to 10.6 kWh/m^3), which means that LPG cannot simply be substituted for natural gas. In order to allow the use of the same burner controls and to provide for similar combustion characteristics, LPG can be mixed with air to produce a synthetic natural gas (SNG) that can be easily substituted. LPG/air mixing ratios average 60/40, though this is widely variable based on the gases making up LPG. The method for determining the mixing ratios is by calculating the Wobbe index of the mix. Gases having the same Wobbe index are considered to be interchangeable [67–70,72,90–93,96].

LPG-based SNG is used in emergency backup systems for many public, industrial, and military installations, and many utilities use LPG peak shaving plants in times of high demand to make up shortages in natural gas supplied to their distribution systems. LPG–SNG installations are also used during initial gas system introductions, when the distribution infrastructure is in place before gas supplies can be connected. Developing markets in India and China (among others) use LPG–SNG systems to build up customer bases prior to expanding existing natural gas systems [67–70,72,90–93,96].

2.5 COALBED METHANE

Many coal seams contain natural gas, either within the seam itself or the surrounding rock. This coalbed methane is trapped underground and is generally not released into the atmosphere until coal mining activities unleash it. Historically, coalbed methane has been considered a nuisance and hazardous in the coal mining industry. Once a mine is built, and coal is extracted, the methane contained in the seam usually leaks out into the coal mine itself. This poses a safety threat, as too high a concentration of methane in the well creates dangerous conditions for coal miners. In the past, the methane that accumulated in a coal mine was intentionally vented into the atmosphere. Today, however, coalbed methane has become a popular unconventional form of natural gas. The present coalbed methane fields are illustrated in Figure 2.5.

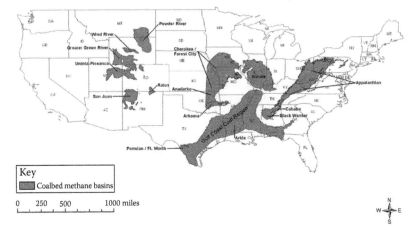

2010 Coalbed methane detailed study report

FIGURE 2.5 Current coalbed methane fields in lower United States. (From Technical development document for the coalbed methane [CBM] extraction industry, EPA report no. EPA-820-R-13-009, U.S. Environmental Protection Agency Office of Water (4303T), Washington, DC, April 2013; Energy Information Administration [EIA].)

TABLE 2.4
Current CBM-Producing Basins and Locations

Basin	States
Appalachian	Virginia, West Virginia, Pennsylvania
Black Warrior	Alabama
Cahaba	Alabama
Greater Green River	Wyoming
Powder River Basin (PRB)	Montana, Wyoming
Raton	Colorado, New Mexico
San Juan	New Mexico
Uinta-Piceance	Utah, Colorado
Anadarko	Oklahoma
Arkoma	Oklahoma, Arkansas
Cherokee/Forest City	Kansas
Arkla	Louisiana
Permian/Ft. Worth	Texas
Illinois	Illinois, Indiana
Wind River	Wyoming

Source: Technical development document for the coalbed methane (CBM) extraction industry, EPA report no. EPA-820-R-13-009, U.S. Environmental Protection Agency Office of Water (4303T), Washington, DC, April 2013.

Current CBM-producing basins and locations are listed in Table 2.4. At present time, coalbed methane is an important source of natural gas in Wyoming, Colorado, Montana, New Mexico, and Alabama [37,97–114].

Coalification, the geologic process that progressively converts plant material to coal, generates large quantities of natural gas, which are subsequently stored in coal seams. The increased pressures from water in the coal seams force the natural gas to adsorb to the coal. This natural gas contained in and removed from the coal seams consists of approximately 96% methane, 3.5% nitrogen, and trace amounts of carbon dioxide [37,109,110], and it is called coalbed methane (CBM) [106].

Since coal has a large internal surface area, it can store surprisingly large volumes of methane-rich gas, six or seven times as much gas as a conventional natural gas reservoir of equal rock volume can hold. In addition, much of the coal, and thus much of the methane, lie at shallow depths, making their exploration costs low, wells easy to drill and inexpensive to complete. With greater depth, increased pressure closes fractures (cleats) in the coal, which reduces permeability and the ability of the gas to move through and out of the coal [97,100,101,115].

The amount of available methane in coal varies with coal's hardness (the resistance to scratching). Level of hardness is known as "rank." The softest coals (peats and lignite) are associated with high porosity, high water content, and biogenic methane. In higher-rank coals (bituminous), porosity, water, and biogenic methane production decreases, but the heat associated with the higher-rank coals breaks down the more complex organics to produce methane. The highest-rank anthracite coals are associated with low porosity, low water content, and little methane generation [115]. The most sought-after coal formations for CBM development, therefore, tend to be mid-rank bituminous coals. Coal formations in the eastern United States tend to be higher-rank, with lower water content than western coal formations. They also tend to have more methane per ton of coal than western coal formations in the key basins, but can require fracturing to release the methane because of their low porosity [115]. Conservative estimates [116] suggest that in the lower United States, more than 700 tcf of coalbed methane exists, with perhaps 100 tcf economically recoverable with the existing technology. In June 2009, the Potential Gas Committee estimated that 163 tcf of technically recoverable coalbed methane existed in the United States, which made up 7.8% of the total natural gas resource base [37,97–116].

The recovery of coalbed methane also has environmental benefits. As mentioned before, the emission of methane in atmosphere is even more harmful to environment than the emission of carbon dioxide. Only 40% of methane emission is caused by natural activities of forest decomposition and wetlands. The remaining 60% of methane emission is caused by human activities. These include rice cultivation (19%), biomass burning (11.5%), livestock (11.5%), landfills (8%), coal mining (6%), and oil and gas venting (4%) [97,100,101,115,116]. This list does not include the methane leakage from gas hydrates from the Arctic environment or from the bottom of the sea. Of all the sources listed here, methane release from coal mining has the best chance to be recovered. The recovery of coalbed methane has also a very positive environment impact [97–107].

The recovery of methane from coalbed often uses enhanced coalbed methane recovery (ECBM) method. Permeability is a key factor for the recovery of coalbed methane, with almost all the permeability created due to fractures. Increasing fractures via hydraulic fracturing is common in the oil and gas industry and can be used effectively with coal seams. ECBM involves injecting gases, including nitrogen and carbon dioxide. These gases will displace the methane that permeates coal fields and drive it toward the production well. In the case of unmineable coal, the use of CO_2 has the dual benefit of extracting usable methane while sequestering the CO_2. Considerable R&D effort is under way to combine ECBM and carbon sequestration projects [97–107].

Extraction of CBM requires drilling and pumping the water from the coal seam, which reduces the pressure and allows CBM to release from the coal [107,117]. This extraction process is schematically described in Figure 2.6 [98]. CBM extraction often produces large amounts of water

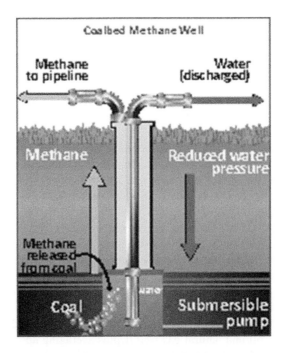

FIGURE 2.6 A schematic of coalbed methane extraction process. (From Nuccio, V., Coalbed methane: Potential and concerns, A report from USGS, USGS Fact Sheet FS-123-00, Denver, CO, October 2000, pubs. usgs.gov/fs/fs123-00/fs123-00.pdf.)

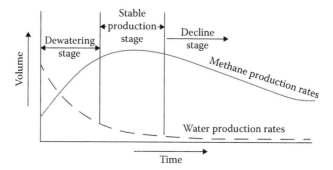

FIGURE 2.7 Typical production curves for a coalbed methane well showing relative volumes of methane and water through time. (Modified from Kuuskraa, V.A. and Brandenberg, C.F., *Oil Gas J.*, 87(41), 49, 1989; also from Nuccio, V., Coalbed methane: Potential and concerns, A report from USGS, USGS Fact Sheet FS-123-00, Denver, CO, October 2000, pubs.usgs.gov/fs/fs123-00/fs123-00.pdf.)

(see Figure 2.7) [98], which is generally recorded. The methane then flows to a compressor station, where the gas is compressed and then shipped via pipeline [97–107]. The produced water is a by-product of the gas extraction process, requiring some form of management (i.e., use or disposal). As shown in Figure 2.7, CBM wells go through the following three production stages:

1. An early stage, in which large volumes of groundwater are pumped from the seam to reduce the underground pressure which facilitates the natural gas to be released from the coal seam.
2. A stable stage, in which the amount of natural gas produced from the well increases as the amount of groundwater pumped from the coal seam decreases.
3. A late stage, in which the amount of gas produced declines and the amount of groundwater pumped from the coal seam remains low [98].

A CBM well's typical lifespan is between 5 and 15 years, with maximum methane production often achieved after 1–6 months of water removal [118]. Both scientific understanding and production experience for coalbed methane are in the early stages of learning. Additional information for the geologic, geochemical, engineering, technological, and economic factors affecting the occurrence and recoverability of coalbed methane are needed. Also, the environmental implications of developing the resource need to be ascertained [111].

Well construction for any well drilling operation—including a CBM well—usually follows one of two basic types: open hole or cased [111]. In open-hole completions, the well is drilled but no lining material is installed, so any gas can seep out all along the well into the wellbore for removal to the surface. In cased completions, a lining is installed through all or most of the wellbore. These casings need to be perforated or slotted to allow gas to enter the wellbore for removal to the surface. Open-hole completions, which are less expensive than perforated or slotted completions, are used more often in CBM production than in conventional oil and gas production, which use open-hole completion only under certain limited circumstances [36,37,97–110]. For example, open-hole completion is widely used in Wyoming's Powder River Basin (PRB) [115].

Operators drill wells into coal-bearing formations that are often not as deep as those containing conventional hydrocarbon reserves, particularly in western regions. In the PRB, for example, some of the methane-bearing formations are shallow, at hundreds to one thousand feet below land surface, compared to conventional oil and natural gas well depths averaging approximately 6000 ft [119]. CBM wells can often be drilled using water well drilling equipment, rather than rigs designed for conventional hydrocarbon extraction, which are used to drill several thousands of feet into typical conventional reservoirs [120].

2.5.1 CBM PRODUCTION HISTORY

Interest in producing methane gas from coal seams began in the 1970s, but little development occurred until the early 1980s. In 1983, the Gas Research Institute began a field study investigating the potential for producing methane gas from coalbed strata [121]. By the end of that year, 165 wells were drilled, producing about 6 bcf of gas, less than 1% of the amount produced in 2008.

The first area to be developed was the Black Warrior Basin in Alabama, followed by the San Juan Basin in New Mexico and Colorado, which began development in the latter part of the 1980s. For many years, CBM was almost exclusively produced from these three states [121]. However, the older basins, such as San Juan and Black Warrior, have not seen growth in CBM production during the 2000s. San Juan production appears to have peaked in 2002, with some decline since then. Black Warrior production was leveled in the 2000s [105].

Production in the PRB began in earnest in the early 1990s, and the PRB quickly became a major source of CBM by the end of the 1990s (Wyoming Oil and Gas Conservation Commission [122]). Since that time, production has risen fairly steadily. By 2008, Wyoming was producing approximately a third of all CBM [37,105]. CBM production (2 tcf) in 2008 totaled nearly 8% of all natural gas produced [37,105]. In 2008, approximately 56,000 CBM wells operated in the United States in 15 basins located in 16 states [37].

More than two-thirds of all CBM produced in 2008 was produced in the San Juan and Powder River Basins (69%). About 88% of all CBM was produced by the five largest producing basins (San Juan, Powder River, Appalachian, Raton, and Black Warrior). Wyoming and New Mexico produce the largest amount of coalbed methane. In all five largest basins, produced water was discharged to surface waters or publicly owned treatment works (POTWs). Some basins followed the practice of zero discharge.

The basins that have been developed to date are those with mid-rank coals; although some evidence suggests that CBM exists in North Dakota lignite. Because of the existing pipeline infrastructure, coal rank, and coal volume, the most likely basin to produce commercial quantities of CBM over the next 10 years is the Black Mesa Basin [102].

2.5.2 PRODUCED WATER CHARACTERISTICS AND MANAGEMENT

Water within the coal seam usually must be removed before and during CBM production. The quantity and quality of this produced water vary from basin to basin, and even within the basin itself. The quality of produced water depends, in part, on the hardness of the coal found within the formation. The quantity of produced water depends on the type of coal and the overall production history of the basin. Basins with a longer production history, such as the San Juan basin, produce less total water and less water per well than the more recently developed basins, such as the PRB. EPA estimated that, in 2008, more than 47 billion gallons of produced water were pumped out of coal seams and approximately 22 billion gallons of that produced water (or about 45%) were discharged to surface waters [115–119,121–155].

CBM produced water is generally characterized by elevated levels of salinity, sodicity (i.e., level of sodium), and trace elements (e.g., barium and iron) [115]. Other trace pollutants that may be present in produced water include potassium, sulfate, bicarbonate, fluoride, ammonia, arsenic, and radionuclides. The characteristics of the produced water depend on the geography and location (e.g., naturally occurring elements). All of these parameters can cause adverse environmental impacts and also affect the potential for beneficial use of produced water.

Salinity represents the total concentration of dissolved salts in the produced water, including magnesium, calcium, sodium, and chloride. Salinity can be measured as electrical conductivity (EC), expressed in deci Siemens per meter (dS/m), as well as total dissolved solids (TDS). TDS includes any dissolved minerals, salts, metals, cations, or anions in the water. The salinity of CBM produced water also relates to the measured sodicity value. Sodicity is excess sodium present in produced water that can deteriorate soil structure (i.e., swell and disperse clays reducing pore size), which reduces the infiltration of produced water through the soil. The sodicity of produced water is expressed as the SAR, which is the ratio of sodium (Na) present in the water to the concentration of calcium (Ca) and magnesium (Mg) [115–119,121–155].

CBM well operators use a variety of methods to manage, store, treat, and dispose of CBM produced water. The produced water from the project might be managed using various storage, treatment, and disposal methods, and each CBM project can use several different management methods. The water can be directly discharged to surface water or indirectly discharged to POTW. The water may also not be discharged but handled by evaporation/infiltration [115,123,155], underground injection, and land application with no crop production. This method is called zero discharge with no beneficial use. Finally water may be used for beneficial purposes such as land application, wildlife watering, or other beneficial usages. This method is called zero discharge with beneficial use.

The produced water management methods used in a particular basin depend on a variety of factors such as water quantity, water quality, availability of receiving waters, availability of formations for injection, landowner interests, and state regulations. The management practices for each project are divided into two major groups: discharging practice (direct discharge to surface waters or indirect discharge) or zero discharge practice (land application, evaporation/infiltration pond, underground injection, beneficial use, transport to a commercial disposal facility, or no water generated). The basins in which direct or indirect discharge is practiced are called "discharging basins," and the basins where produced water is managed without discharging any portion of it directly or indirectly to surface waters are called "zero discharge basins."

In zero discharge basins, the primary produced water management practices are underground injection and hauling for commercial disposal. In discharging basins, in addition to direct and indirect surface water discharge, operators also use zero discharge methods. In these basins, evaporation/infiltration ponds and beneficial use (livestock and wildlife watering) are common zero discharge practices; underground injection and hauling are less common. Land application, another practice that can be considered zero discharge, is relatively rare. More details on these methods of water discharge are given in References 115–119,121–155.

Operators may treat the CBM produced water prior to discharge or use other methods of management. The level of CBM produced water treatment depends on the pollutants present in the water and the final destination. EPA identified and investigated technologies for treating produced water, including aeration, chemical precipitation, reverse osmosis, ion exchange, electrodialysis, thermal distillation, and combination technologies. These technologies reduce or eliminate pollutants in the produced water, allowing beneficial use or surface water discharge.

Using survey responses and other data, EPA evaluated the following: the quality and quantity of produced water generated from CBM extraction; the available methods of management, storage, treatment, and disposal options; and the potential environmental impacts of surface discharges. The conclusions were as follows [115–119,121–155]:

1. Approximately 45% of all produced water is discharged to waters of the United States.
2. Various pollutants such as sodium, calcium, and magnesium (used to calculate the sodium adsorption ratio [SAR]), total suspended solids (TSS), and metals (e.g., selenium, chromium) are present in discharges. Pollutants from CBM discharges may negatively affect fish populations over time.
3. Surface water discharges of produced water can increase stream volume, streambed erosion, suspended sediment, and salinity.
4. Surface impoundment and land application of produced waters may impact groundwater from infiltration and the concentration and/or bioaccumulation of CBM-associated pollutants. Advanced water treatment options are being used in the field in some operations to remove pollutants in produced water.
5. Widely practiced zero discharge options may be available depending on well location.

2.5.3 Current Economics of CBM Production

CBM is generally produced from relatively shallow coalbeds. These coalbeds underlie the surface in broad areas, often covering many hundreds of square miles. Large amounts of produced water are typically generated initially; over time, the amount of water produced generally diminishes. In contrast, conventional gas is often contained within sharply defined geological formations, which can be accessed only from a relatively small area using deeper wells, typically, than those required for CBM production. Extracting conventional gas often generates relatively little water at first, but the production of water can increase over time. These differences in production between conventional gas and CBM lead to a very different economic profile in terms of production economics and, in some cases, firm economics. Because produced water management costs are a significant portion of operating costs in either type of gas production [37,100,101,103–105,109,110,115,123], CBM projects often begin with high operating costs that tend to diminish over time, while operating costs for conventional oil and gas often rise over time.

CBM wells are rarely operated as single units responsible for their own production costs, because operators realize economies of scale in operating several wells together as an economic unit. Given that CBM production requires numerous wells distributed over the coalbed, operators tend to include a large number of wells (1000) in each economic production unit, or project. According to EPA's screener survey [37,109,110], a total of about 56,000 CBM wells, organized into approximately 750 projects, produced gas and/or water in 2008 [37,102,104,105,109,110]. Of these projects, a minority (approximately 180 projects) discharged some produced water.

The zero discharge basins have a relatively small number of wells (18,600 or 33%) but account for about 50% of production, because average production per well is greater in the zero discharge basins than in the discharging basins. Projects that discharged at least some produced water to surface waters averaged gas production of 27 million cubic feet (mcf) per well and 4.4 bcf per project in 2008, while those that discharged no produced water averaged greater gas production per well (45 MMcf) but lower production per project (2.1 bcf) than projects discharging to surface waters. The higher per-project production in discharging basins results from the higher average number of wells per project in the discharging basins [36,37,97–110].

2.5.4 The Future of CBM

As mentioned earlier, it is estimated that the United States has coalbed methane resources of at least 700 tcf; about 100 tcf appears to be economically recoverable with the existing technology. The U.S. DOE EIA [104,105] predicts that CBM production will remain roughly steady through 2035. In the longer term, natural gas production of all types is expected to rise, contributing to a slight decline in the percentage of natural gas attributable to CBM production. The largest growth categories of natural gas types are shale gas and conventional natural gas from Alaska (the result of predicted pipeline construction completion). Shale gas is by far the largest growth category and by 2035, might be close to total conventional onshore volumes [37,104,105].

In short term, costs of production tend to rise and fall as gas prices rise and fall. Key long-term factors affecting costs of production include availability of pipelines to transport CBM to central distribution hubs, the number of years over which development has occurred in a region, technology changes, and project-specific trends, such as potential decreases in produced water production over time and regulatory requirements.

Years of development in a basin also can affect long-term production costs. As easy-to-reach coalbeds are tapped, future development relies on producing from deeper coalbeds, thinner coalbeds (e.g., those that are only a few feet thick), "tighter" coalbeds (those with fewer spaces that allow gas to escape easily), or coalbeds with lower-rank coals (with less gas), all of which can be more expensive to produce and/or generate lower revenues. Deeper coalbeds require deeper wells, taking longer to drill and requiring more piping and often more energy to bring the gas to the surface

[118,119,121,130–154]. Tighter coalbeds might require special treatment (hydro-fracturing—a method of opening up additional cracks in the coal seam to allow gas to escape more readily) incurring an additional expense [37,100,105–110]. Costs will also be affected by the changes in technologies in CBM production techniques (e.g., multiseam completions and horizontal drilling).

2.6 TIGHT GAS

Tight gas refers to natural gas reservoirs locked in extraordinarily tight, nonporous, impermeable, hard rock, sandstone, or limestone formations. While a conventional gas formation can be relatively easily drilled and extracted from the ground, tight gas requires more effort to pull it from the ground because of the extremely tight formation in which it is located. In other words, permeability for the gas to escape from the pores in the rock formation is very low due to overly narrow, badly connected, and irregular capillaries within the rock. Tight gas makes up a significant portion of the nation's natural gas resource base, with the EIA estimating that, as of January 1, 2000, 253.83 tcf of technically recoverable tight natural gas existed in the United States [156,157]. This represents over 21% of the total recoverable natural gas in the United States, and represents an extremely important portion of natural gas resources. Major tight gas plays in the United States are described in Figure 2.8.

Without secondary production methods, gas from a tight formation would flow at very slow rates, making production uneconomical. While conventional gas formations tend to be found in the younger tertiary basins, tight gas formations are much older. Deposited some 248 million years ago, tight gas formations are typically found in Paleozoic formations. Over time, the rock formations have been compacted and have undergone cementation and recrystallization, which all reduce the level of permeability in the rock. Typical conventional natural gas deposits boast a permeability level of 0.01–0.5 darcy, but the formations trapping tight gas reserves portray permeability levels measuring in the millidarcy or microdarcy range [156–168].

"Tight gas" lacks a formal definition, and usage of the term varies considerably. Law and Curtis [163] defined low-permeability (tight) reservoirs as having permeability less than 0.1 millidarcy (mD).

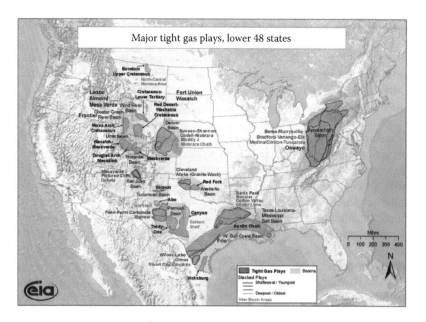

FIGURE 2.8 Major tight gas plays in the United States. (From Major tight gas plays, lower 48 states—EIA, U.S. Energy Information, Washington, DC, 2010, https://www.eia.gov/oil_gas/rpd/tight_gas.pdf.)

Recently the German Society for Petroleum and Coal Science and Technology (DGMK) announced a new definition for tight gas reservoirs, which includes reservoirs with an average effective gas permeability less than 0.6 mD. While there are differences in the definition of tight gas, industry generally defines tight gas reservoir as a gas-bearing sandstone or carbonate matrix (which may or may not contain natural fractures), which exhibits an in situ permeability of gas less than 0.10 mD. Many "ultratight" gas reservoirs may have in situ permeability down to 0.001 mD. In tight gas reservoirs, the pores are poorly connected by very narrow capillaries, resulting in very low permeability. Gas flows through these rocks generally at low rates and special methods are necessary to produce this gas [158–160,163,165].

Four criteria that define basin-centered (generally involves more than one well) gas accumulations are (1) low permeability, (2) abnormal pressure, (3) gas saturated reservoirs, and (4) no down dip water leg. While "tight gas sands" are an important type of basin-centered gas reservoir, not all of them are basin-centered gas accumulations (BCGAs). A concerted technology effort to both better understand tight gas resource characteristics and develop solid engineering approaches is necessary for significant production increases from this low-permeability, widely dispersed resource. Gas production from a tight gas well will be low on a per-well basis compared with gas production from conventional reservoirs. A lot of wells should be drilled to get most of the tight gas out of the underground reservoirs.

Based on the criteria defined above, basin-centered and deep basin (>15,000 ft) accumulations are defined by law [161,162] as "an abnormally-pressured, gas-saturated accumulation in low permeability reservoirs lacking a down-dip water contact." They are characterized by regionally pervasive gas-saturated reservoirs, containing abnormally pressured gas accumulations. The up-dip boundary of the deep basin is somewhat nebulous, as each reservoir unit may have its own up-dip edge.

Many scientists think of tight or low-permeability reservoirs as occurring only within basin-centered, or deep basin settings. However, tight gas reservoirs of various ages and types are produced where structural deformation creates extensive natural fracture systems, whether it is basin margin or foothills or plains. Fractured, tight, and unconventional reservoirs can occur in tectonic settings dominated by extensional, compressional, or wrench faulting and folding. Late burial diagenesis of the sandstone may also result tight reservoirs. Although "tight gas sands" are an important type of basin-centered gas reservoir, not all of them are BCGAs [161,162,164,166].

There could be a number of reasons for making a reservoir tight. Basically the permeability that determines the ease at which a fluid can flow is a multivariable function governed by the Darcy's law of fluid flow in porous media. Effective porosity, viscosity, fluid saturation, and the capillary pressure are some of the import parameters controlling the effective permeability of a reservoir. Besides the factors relating to the fluid nature, the rock parameters are equally important. These are controlled by depositional and post-depositional environments of the reservoir.

Understanding gas production from low permeability rocks requires an understanding of the petrophysical properties–lithofacies associations, facies distribution, in situ porosities, saturations, effective gas permeability at reservoir conditions, and the architecture of the distribution of these properties. Petrophysics is a critical technology required for understanding low-permeability reservoirs. Improvements in completion and drilling technology will allow well-identified geologic traps to be fully exploited, and improvements in product price will allow smaller accumulations or lower-rate wells to exceed economic thresholds, but this is true in virtually every petroleum province. Well clusters and onsite waste management are the key components of new technology concepts for tight gas development.

Major tight gas reservoir sites in the United States are [156–168] as follows:

1. Wind River Basin—Wyoming
2. Greater Green River Basin—Wyoming
3. San Juan Basin—New Mexico
4. Piceance Basin—Colorado
5. Uinta Basin—Utah
6. Denver Basin—Colorado, Nebraska

2.6.1 METHODS FOR TIGHT GAS RECOVERY

In order to overcome the challenges that the tight formation presents, there are a number of procedures that can be applied to help produce tight gas. Changes in drilling practices and acquiring more specific seismic data and their computer analysis (as discussed earlier in Section 2.2) can help in tapping tight gas. While vertical wells may be easier and less expensive to drill, they are not the most conducive to developing tight gas. In a tight gas formation, it is important to expose as much of the reservoir as possible, making horizontal and directional drilling a must. Here, the well can run along the formation, opening up more opportunities for the natural gas to enter the wellbore.

A common technique for developing tight gas reserves includes drilling more wells. The more the formation is tapped, the more the gas will be able to escape the formation. This can be achieved through drilling myriad directional wells from one location, lessening the operator's footprint and lowering costs. After seismic data has illuminated the best well locations, and the wells have been drilled, production stimulation by both fracturing and acidizing the wells can be employed on tight gas reservoirs to promote a greater rate of flow.

Once the well is drilled, hydraulic fracturing described earlier of tight reservoir should be achieved by pumping the fracture fluids under high pressure in the well to break the rocks in the reservoir apart and improve their permeability. Additionally, the technique of acidizing the well is employed to further improve permeability and production rates of tight gas formations. Acidization [169] involves pumping the well with acids that dissolve the limestone, dolomite, and calcite cement between the sediment grains of the reservoir rocks [30–34,38–50]. This form of production stimulation helps to reinvigorate permeability by reestablishing the natural fissures that were present in the formation before compaction and cementation.

The removal of water from the tight gas wells can help to overcome some production challenges. In many tight gas formations, the reservoirs also contain small amounts of water. This water can collect and undermine production processes. Its removal can be achieved through artificial lift techniques, such as using a beam pumping system to remove the water from the reservoir. This method is, however, not always found to be most effective. Engineers continue to develop new techniques and technologies to better produce tight gas. The gas from "ultratight gas" reservoirs can also be similarly recovered using multidimensional drilling, hydraulic fracturing, and acidizing techniques [30–34,38–50,156–169].

2.7 SHALE GAS

Natural gas can also exist in shale deposits, which formed 350 million of years ago. Shale is a very fine-grained sedimentary rock, which is easily breakable into thin, parallel layers. It is a very soft rock, but it does not disintegrate when it becomes wet. These shales can contain natural gas, usually when two thick, black shale deposits "sandwich" a thinner area of shale. The extraction of natural gas from low permeability shale deposits is more difficult and expensive than recovery of conventional natural gas from sandstone rocks with high permeability where the tiny pores within the rock are well connected. Shale basins are scattered across the United States. The Federal Energy Regulatory Commission (FERC) estimated that as of November 2008 [170], there were 742 tcf of technically recoverable shale gas resources in the United States. Shale represents a large and growing share of the U.S. recoverable resource base. As shown in Figure 2.9 [171,172], shale gas is found in many parts of the United States.

Although the basic principles of shale gas formation are fairly well understood, generation of the gas within individual shales may differ significantly. Better knowledge is needed, for example, on basin modeling, petrophysical characterization, or gas flow in shales, for an improved understanding

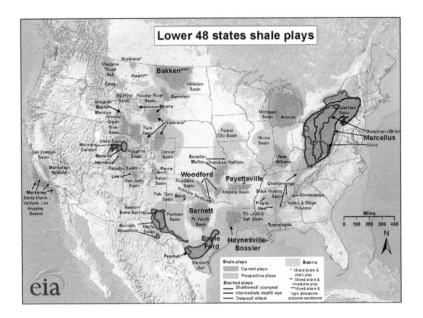

FIGURE 2.9 Regions of shale gas reservoirs in the United States. (From Shale Gas, Wikipedia, the free encyclopedia, 2015; Shale Gas in the United States, Wikipedia, the free encyclopedia, 2015; Modern Shale gas development in the United States: An update, A report from NETL, Strategic Center for Natural Gas and Oil, DOE, Washington, DC, September 2013; Energy Information Administration based on data from various published studies, Updated May 9, 2011.)

of these unconventional reservoirs. In Europe, this knowledge is acquired by research conducted within GASH, the first European interdisciplinary shale gas research initiative [173].

Shale rock characteristically has small pores which are relatively impermeable to gas flow unless they are naturally or artificially fractured to create channels connecting the pores. Shale gas areas are often known as *resource plays* (as opposed to *exploration plays*). The geological risk of not finding gas is low in resource plays, but the potential profits per successful well are usually also lower [174–197].

Shales that host economic quantities of gas have a number of common properties. They are rich in organic material (0.5%–25%), and they are usually mature petroleum source rocks, where high heat and pressure have converted petroleum to natural gas. They are sufficiently brittle and rigid enough to maintain open fractures. Natural gas within shale rock appears in three forms. Some of the gas produced is held in natural fractures, some in pore spaces, and some is adsorbed onto the organic material. The gas in the natural fractures is produced immediately; the gas adsorbed onto organic material is released as the formation pressure is drawn down by the well. The gas in pores can be released as shale structure is fractured resulting in connections of these pore spaces. In order to open up the pores, horizontal drilling (accompanied by hydraulic fracturing) [30–34,38–50,190,195] is often used with shale gas wells, with lateral lengths up to 10,000 ft (3,000 m) within the shale, in order to create maximum borehole surface area in contact with the shale.

2.7.1 Brief History of Shale Gas Development

Shale gas was first extracted as a resource in Fredonia, New York, in 1821, in shallow, low-pressure fractures. Horizontal drilling began in the 1930s, and in 1947 a well was first fracked in the United

States. Along with horizontal drilling and hydraulic fracturing, a micro-seismic imaging tool developed by Sandia National Laboratory helped the recovery of shale gas from shale rock. Although the Eastern Gas Shales Project had increased gas production in the Appalachian and Michigan basins, shale gas provided only 1.6% of U.S. gas production in 2000. George P. Mitchell is regarded as the father of the shale gas industry, by making it commercially viable in the Barnett Shale by getting costs down to $4 per million Btu. Mitchell Energy achieved the first economical shale fracture in 1998 using slick-water fracturing. Since then, natural gas from shale has been the fastest growing contributor to total primary energy in the United States, and has led many other countries to pursue shale deposits. According to the IEA [182], by 2010 shale gas contributed over 20% of the total gas production and by 2035, 46% of the U.S. natural gas supply will come from shale gas. Some analysts expect that shale gas will greatly expand worldwide energy supply. Following the success in the United States, shale gas exploration is beginning in countries such as Poland, China, and South Africa. The increase in shale gas production has made the United States the number one natural gas producer in the world. China is estimated to have the world's largest shale gas reserves [171,172,176].

2.7.2 SHALE GAS RESOURCE

Shale gas is present across much of North America in basins of both extreme and moderate size. As of 2013, there were at least 22 major shale plays in the United States, spread diversely over more than 20 states. The production of shale gas has been diversified across the country, bringing supply closer to areas where it is consumed. New reserves are discovered so frequently that industry and government authorities have trouble keeping map of shale plays up to date [170–172,181, 184–186,192]. Some of the major shale reserves include the following:

1. Marcellus (Pennsylvania, New York, Ohio, West Virginia, Kentucky)
2. Bakken (North Dakota, Wyoming)
3. Haynesville (Texas, Louisiana)
4. Barnett (Texas)
5. Eagle Ford (Texas)
6. Fayetteville (Arkansas)
7. Antrim (Michigan, Illinois, Ohio)
8. Woodford (Oklahoma)

The geographic diversity of U.S. shale gas resources and advances in technology has helped ensure a stable and deliverable natural gas supply.

The best estimate of global supply of shale gas based on data collected by the EIA [32] is outlined in Table 2.5, where numbers for the estimated amount of "technically recoverable" shale gas resources are provided alongside numbers for proven natural gas reserves. The data in this table are constantly evolving. The U.S. EIA had made an earlier estimate of total recoverable shale gas in various countries in 2011, which for some countries differed significantly from the 2013 estimates. The total recoverable shale gas in the United States, which was estimated at 862 tcf in 2011, was revised downward to 665 tcf in 2013. Recoverable shale gas in Canada, which was estimated to be 388 tcf in 2011, was revised upward to 573 tcf in 2013. The numbers in Table 2.5 should just be considered as only crude estimates.

For the United States, the EIA estimated [187,198,199] a total "wet natural gas" (methane plus natural gas liquids) resource of 2431 tcf, including both shale and conventional gas, with shale gas as 27% of the total resource. For the rest of the world (excluding the United States), the EIA estimated [187,198,199] a total wet natural gas resource of 20,451 tcf (579.1×10^{12} m^3), with shale gas as 32% of the total resource. While Europe has estimated shale gas reserves of 639 tcf (18.1×10^{12} m^3), it is more expensive to extract (three and a half times more than one in the United States due to its geology) [171,172].

TABLE 2.5
Best Estimates of Recoverable and Proven Shale Gas across World

Country		Estimated Technically Recoverable Shale Gas (tcf)	Proven Natural Gas Reserves of All Types (tcf)	Date of Report
1	China	1115	124	2013
2	Argentina	802	12	2013
3	Algeria	707	159	2013
4	United States	665	318	2013
5	Canada	573	68	2013
6	Mexico	545	17	2013
7	South Africa	485	—	2013
8	Australia	437	43	2013
9	Russia	285	1688	2013
10	Brazil	245	14	2013
11	Indonesia	580	150	2013

Source: Modified from Shale Gas, Wikipedia, the free encyclopedia, 2015, https://en.wikipedia.org/wiki/shale-gas, accessed August 18, 2016.

2.7.3 ENVIRONMENTAL IMPACTS OF SHALE GAS RECOVERY

The extraction and use of shale gas can affect the environment through the leaking of extraction chemicals and waste into water supplies, the leaking of greenhouse gases during extraction, and the pollution caused by the improper processing of natural gas. As mentioned before, there are also evidence that hydraulic fracturing method used in shale gas recovery can also cause minor earthquakes in the local region. The prevention of pollution by shale gas recovery is challenging because of wide variations in shale gas and its extraction process among different wells even in the same project. This means that the processes that reduce pollution sufficiently in one extraction may not be enough in another [171,172,177,178,189].

In late 2010, the EPA concluded that shale gas emits larger amounts of methane, a potent greenhouse gas, than does conventional gas, but still far less than coal [200]. On the other hand, a 2013 review by the United Kingdom's DOE and Climate Change as well as the University of Manchester noted that most studies of the subject have estimated that life-cycle greenhouse gas (GHG) emissions from shale gas are similar to those of conventional natural gas, and are much less than those from coal, usually about half the GHG emissions of coal. The most recent study of 2012 by Lawrence Cathles [18,201] concludes that "shale gas has a GHG footprint that is half and perhaps a third that of coal." Methane is a powerful greenhouse gas, although it stays in the atmosphere for only one-tenth as long a period as carbon dioxide. Recent evidence suggests that methane has a global warming potential (GWP) that is 28-fold greater than carbon dioxide (on equal mass basis) when viewed over a 100-year period [11–14,18,171,172,177,178,189,192].

Several studies that have estimated lifecycle methane leakage from shale gas development and production have found a wide range of leakage rates, from less than 1% of the total production to nearly 8%. Using data from the EPA's most recent Greenhouse Gas Inventory yields a methane leakage rate of about 1.4%. The most comprehensive study by EPA in 2013 indicates a methane leakage rate of 0.42% for shale gas production [11–14,18,171,172,177,178,189,192].

As shown earlier, during hydraulic fracturing chemicals are added to the water to facilitate the underground fracturing process that releases natural gas. Fracturing fluid is primarily water and approximately 0.5% (and can be as high as 2%) chemical additives (friction reducer, agents countering rust, agents killing microorganism, acids, etc.). Since (depending on the size of the area) millions of liters of water are used, hundreds of thousands of liters of chemicals are often injected into the subsurface [184,190,191,193,195]. About 50%–70% of the injected volume of contaminated water is recovered and stored in aboveground ponds to await removal by tanker. The remaining volume remains in the subsurface. Hydraulic fracturing opponents fear that it can lead to contamination of groundwater aquifers, though the industry indicates this to be "highly unlikely." Foul-smelling odors and heavy metals contaminating the local water supply aboveground have been reported. Hydraulic fracturing was exempted from the Safe Drinking Water Act in the Energy Policy Act of 2005. Besides using water and industrial chemicals, it is also possible to frack shale gas with only liquefied propane gas. This reduces the environmental degradation considerably. The method was invented by the company GasFrac, Alberta, Canada.

A 2011 study by the Massachusetts Institute of Technology [192,200] concluded, "The environmental impacts of shale development are challenging but manageable." The study addressed groundwater contamination, noting, "There has been concern that these fractures can also penetrate shallow freshwater zones and contaminate them with fracturing fluid, but there is no evidence that this is occurring." This study blames known instances of methane contamination on a small number of substandard operations, and encourages the use of industry best practices to prevent such events from recurring. There have been mixed reports on methane leakage and its influence on drinking water in numerous shale plays reported by EPA and studied by numerous universities.

2.7.4 WATER CONSUMPTION BY COAL INDUSTRY VERSUS SHALE GAS INDUSTRY

Coal-fired power plants consume two to five times more water as compared to natural gas plants. While 520–1040 gal of water are required per MWh of coal, gas-fired combined cycle power requires 130–500 gal per MWh. The environmental impact of water consumption at the point of power generation depends on the type of power plant: plants either use evaporative cooling towers to release excess heat or discharge water to nearby rivers [62]. Natural gas combined cycle power plant (NGCC), which captures the exhaust heat generated by combusting natural gas to power a steam generator, are considered the most efficient large-scale thermal power plants. One study found that the life-cycle demand for water from coal power in Texas could be more than halved by switching the fleet to NGCC.

Water usage for shale gas development in the United States represents less than half a percent of total domestic freshwater consumption, although this portion can reach as high as 25% in particularly arid regions [184,190,191,193,195]. As mentioned before, over its lifetime, an average shale gas well can require 5 million gal of water for the initial hydraulic fracturing operation and possible re-stimulation. The large volumes of water required have raised concerns about the impact of fracking in water shortage areas such as Texas, which has been in a multiple-year drought.

2.7.5 ECONOMICS OF SHALE GAS

North America has been the leader in developing and producing shale gas. The economic success of the Barnett Shale play in Texas in particular has spurred the search for other sources of shale gas across the United States and Canada. A Visiongain research report calculated the 2011 worth of the global shale gas market as $26.66 billion. In 2012, natural gas prices went down to $3/MMBtu due to shale gas. Advances in hydraulic fracturing and horizontal completions have made shale gas wells more profitable [196,197]. Improvements in moving drilling rigs between nearby locations, and the use of single well pads for multiple wells, have increased the productivity of drilling shale gas wells [66]. Shale gas tends to cost more to produce than gas from conventional wells, because of the expense of the massive hydraulic fracturing treatments, and expenses that incur due to horizontal drilling.

One of the by-products of shale gas exploration is the opening up of deep underground shale deposits to "tight oil" or shale oil production. By 2035, according to PricewaterhouseCoopers (PwC), shale oil production could boost the world economy by up to $2.7 trillion. It has the potential to reach up to 12% of the world's total oil production, reaching 14 million barrels a day [196]. As America demands more and more energy, Marcellus and Utica plays are shaping up to be key suppliers for domestic natural gas [174,185,186,194].

2.8 DEEP GAS

Deep natural gas exists in deposits very far underground, beyond "conventional" drilling depths. This gas is typically 15,000 ft or deeper underground, quite a bit deeper than conventional gas deposits, which are traditionally only a few thousand feet deep at most. In recent years, deep gas has become more conventional due to improvement in exploration and extraction techniques. However, deep gas is still more expensive to produce than conventional natural gas, and therefore economic conditions have to be such that it is profitable for the industry to extract from these sources.

Dyman and others [202–214,267] showed that more than 5000 productive deep gas wells had been drilled in the United States as of 1997. More than half of these are in the Gulf Coast, with the Anadarko and Permian basins accounting for much of the rest. Based on 1988 data, cumulative gas production from deep wells exceeded 21 tcf, primarily from the Permian (12.4 tcf), the Gulf Coast (6.2 tcf), and the Anadarko (2.4 tcf) basins. In 1995, the U.S. Geological Survey (USGS) estimated that 114 tcf of technically recoverable conventional and nonconventional deep gas remains to be discovered in the Rocky Mountains (57 tcf), Gulf Coast (27 tcf), Alaska (18 tcf), West Texas/New Mexico (4 tcf), and Midcontinent (3 tcf), among others.

Worldwide estimates of deep gas are also high. The world estimates are divided into eight regions. A total of 274 deep assessment units and plays in 123 petroleum provinces have been identified based on the U.S. Geological Survey World Petroleum Assessment [208]. These deep assessment units and plays contain a mean undiscovered conventional gas resource of 844 tcf (about 17% of total world gas resource) below 4.5 km. Of this resource, about 23 tcf of gas occurs below 7.5 km (about 25,000 ft). Of the eight regions, the former Soviet Union (Region 1) contains the largest estimated volume of undiscovered deep gas with a mean resource of 343 tcf. The second-largest estimated deep gas volume (142 tcf) occurs in Europe (Region 4), and the third-largest estimated deep gas volume (131 tcf) occurs in the Middle East and North Africa (Region 2).

Exploration for deep natural gas resources deserves special attention because these resources are widespread and occur in diverse geologic environments. Efficiently locating and developing deep undiscovered natural gas depends on improving our knowledge of the geology and reservoir characteristics of deep sedimentary basins, continued advances in exploration, drilling, and completion technologies, and improved economics. During the 1990s, deep natural gas exploration and development were influenced strongly, both by advances in technology and by lower unit costs. This progress in technology and costs helped spur development of frontier plays such as the deep Norphlet play in the eastern Gulf Coast basin, the low-permeability deep cretaceous plays of the Green River Basin, and the deep Madison Play on Madden anticline in the Wind River Basin [214–231].

Based on the IHS WHCS data [222], as of December 1998, deep gas wells made up nearly 76% of the total deep wells (more than 15,000 ft deep) producing gas or oil. At depths higher than 25,000 ft, all wells produced gas only. For the entire United States, 10 ultra-deep wells (more than 25,000 ft deep) were from Permian reservoirs in the Permian Basin, 6 wells were from the Ordovician Ellenberger Formation in the Permian Basin, and 5 wells were reported as producing from Hunton Group reservoirs in the Anadarko Basin. A historic success ratio of more than 50% ultradeep wells producing gas was achieved [207,210,213,215–230,232,233].

The evaluation of deep gas basin process is important because natural gas may form in the deep central portions of basins and migrate into shallower regions where it is trapped. An understanding of deep basin processes will aid in understanding the occurrence of natural gas in shallow basin environments as well as deep ones [231–266]. The opportunities for undiscovered accumulations in deep basins remain strong. Finally, economic conditions strongly affect the development of deep reservoirs. Economic and (or) technologic improvements could produce conditions appropriate for increased exploration. It is therefore important to maintain a database of information on deep natural gas resources and reservoirs. Current information on various possible locations for deep gas are described in detail in various USGS studies and other related works [231–266].

2.8.1 HISTORICAL PERSPECTIVE AND ORIGIN OF DEEP GAS

The literature [205,222] indicates that by 1998, more than 20,000 deep wells were distributed widely and were drilled into rocks of various ages and lithologies, but they represent less than 1% of more than 3 million oil and gas wells drilled in the United States. The Anadarko Basin of Oklahoma, where record-breaking deep wells were drilled in the 1970s and early 1980s, has been viewed traditionally as the center of deep drilling in the United States. The first deep gas discovery was in 1956 in the Carter-Knox field in the southern part of the Anadarko Basin. This well produced an open-flow potential of 31 mcf of gas per day (MMcfd) at a depth of 15,300 ft. Other deep discoveries followed in southern Oklahoma and in the Texas Panhandle.

Table 2.6 contains identification and location information for the seven deepest wells drilled in the United States, regardless of completion classification [222]. All of these wells were drilled in Oklahoma and Texas in the Anadarko and Permian basins. The deepest well drilled in the United States, the Bertha Rogers No. 1, was completed in 1974 in the Anadarko Basin in Oklahoma to a depth of 31,441 ft. Wells 4, 6, and 7 in Table 2.6 were drilled and abandoned, and well 5 was completed as gas well.

TABLE 2.6
Seven Deepest Exploration and Production Wells in the United States in Decreasing Order of Total Depth

Well Name	Completed Depth (ft)	State[a]
1. #1 Bertha Rogers 1974	31,441	A-OK
2. #1 Earnest R. Baden 1972	30,050	A-OK
3. 1-9 Cerf Ranch 1983	29,670	P-TX
4. Cerf Ranch 1994	29,670	P-TX
5. 2 Emma Lou Unit #1 1980	29,622	P-TX
6. #1-3 Duncan 1983	29,312	A-OK
7. #1-1 Robinson 1984	29,241	A-OK

Sources: Modified and adapted from IHS Energy Group, PI-Dwights WHCS (through October and December 1998), Available from IHS Energy Group, Denver, CO, 1998; Dyman, T.S. and Cook, T.A., Summary of deep oil and gas wells in the U.S. through 1998, in: Dyman, T.S. and Kuuskraa, V.A., eds., Studies of deep natural gas, U.S. Geological Survey Digital Data Series 67, one CD-ROM, USGS, Washington, DC, 2001; Wells listed regardless of completion class.

[a] A, Anadarko Basin; P, Permian Basin; OK, Oklahoma; TX, Texas.

Ultradeep gas is defined as gas occurring below 25,000 ft. Few penetrations, and even fewer successful completions, have exceeded 25,000 ft. Except for a few ultra-deep wells in the Anadarko, Gulf Coast, Rocky Mountain, and Permian basins, little is known about ultra-deep gas resources. Assuming ultra-deep reservoirs exist, significant natural gas production could be achieved in the future from such great depths, assuming technological difficulties and high costs can be countered with highly productive wells.

Deep gas occurs in either conventionally trapped or unconventional (continuous-type) basin-centered accumulations. The term "continuous type" accumulation is used by the U.S. Geological Survey to describe large accumulations of gas having spatial dimensions that usually exceed those of conventional fields. The number of deep conventional plays decreases with increasing depth. Of the 101 deep conventional plays, 73 plays have maximum depths ranging from 15,000 to 20,000 ft and only three plays exceed 30,000 ft in depth [202,203]. While at lower depth, conventional plays are larger in number than basin-centered plays, more gas is recovered in basin-centered plays.

Continuous accumulations are geologically diverse and fall into several categories, including coalbed gas, shallow biogenic gas, fractured shale gas, and basin-centered gas. Only basin-centered gas comprises a significant portion of the deep continuous gas resource. Basin-centered gas was described as "pervasive gas" by Davis [237] and documented for accumulations in Western Canada [223] and the United States. Such basins are defined by pressure profiles that are either subnormal or supernormal and have no free water within or down dip from the gas package. Other common geologic and production characteristics of basin-centered accumulations include large in-place hydrocarbon volumes, relatively low matrix permeability, gas down dip from water, source rocks within the gas package, and a lack of obvious traps or seals.

The key to producing gas from basin-centered accumulations is to locate "sweet spots" (enhanced permeability) regardless of structure. Once a sweet spot is identified, be it from fractures, clean conglomerates, stimulation of the low permeability matrix or other factors, it should produce gas without concern for water production. For continuous accumulations, source and reservoir rocks are related closely to each other, and migration distances are short, whereas for conventional accumulations, long-distance migration is possible [231–266].

The smaller number of deep gas wells compared to shallow depth wells has restricted our understanding of the origins of the deep gas. We do know that much of the shallow depth gas mass has its origins in the deeper parts of basins and has migrated to shallower depths during basin evolution [209]. The factors that may have contributed to the origin and accumulation of deep gas may be initial concentration of organic matter and the role of minerals, water, and nonhydrocarbon gases on the kinetics of natural gas generation, thermal cracking of oil to gas and thermal stability of methane, and finally porosity loss with increasing depth and thermal maturity and source rock potential based on thermal maturity and kerogen type [204,206,219,234,235,240,241].

Recent experimental simulations using laboratory pyrolysis methods have provided much information on the origins of deep gas. These studies have indicated that basins with slow heating rates, where source rocks subside slowly through low thermal gradients, are more likely to yield deep gas from kerogen than basins with fast heating rates and rapid subsidence rates. Type-III kerogen will yield the largest deep gas of the three kerogen types irrespective of heating rate, implying that basins with deeply buried coals are most likely to contain deep gas [219,234,236,255,262].

Type-I kerogen has the lowest potential for deep gas generation, implying that basins with deeply buried hydrogen-rich source rocks are not likely to contain deep gas. Thermal cracking of oil will generate the most deep gas irrespective of heating rate. Therefore, the main requirement for deep gas accumulation from the cracking of oil is that the original oil trap remains competent throughout the burial history. The Gulf Coast offshore and the Anadarko Basin may serve as examples of this geologic setting. Finally, the kinetic model derived from hydrous pyrolysis indicates that oil cracking to gas requires higher thermal maturity than those predicted by the anhydrous-pyrolysis model [219,234,236,255,262].

More studies that include an inventory of heating rates and gas accumulations of different basins, cracking of oil in the presence of water, and the catalytic effects of commonly occurring reservoir minerals and their surfaces are needed [267,268].

2.8.2 GENERAL GEOLOGIC AND TECHNOLOGIC FRAMEWORK

In a series of papers and reports Dyman and coworkers [202–214,217,267,268] have presented an excellent analysis of deep gas reservoirs, both from the geologic and technologic framework. In the following paragraphs, we briefly summarize their assessments. From a geological perspective, deep natural gas is generally defined as occurring in reservoirs below 15,000 ft. From an operational point of view, "deep" usually is thought of in a relative sense based on the geologic and engineering knowledge of gas resources in a particular area. For example, in the Anadarko and Gulf Coast basins, many wells have been drilled to depths exceeding 15,000 ft, and deep production is well established [208,209]. We know a lot about resources in the Gulf Coast and Anadarko basins below 15,000 ft. Conversely, in the San Juan Basin, relatively few wells have exceeded 10,000 ft, and the basin itself barely exceeds 15,000 ft. We know less about potential petroleum resources in the San Juan Basin in the 10,000–15,000 ft interval.

The problems associated with drilling and completing wells differ significantly for different depth intervals in different basins. Wells may encounter overpressured intervals, sour gas, or special drilling problems at any depth. In the longer term, it may be useful to view deep resources in a relative sense by understanding the geologic changes that occur with increasing depth regardless of the absolute depth of those resources [202–214,231–268]. Exploration and drilling strategies for deep gas need to differ depending on whether a deep gas play is conventional or unconventional in nature. On average, deep conventional fields need to be larger than shallow conventional fields because deep fields must justify the increased cost of drilling [203]. For moderately sized deep accumulations, well stimulation and completion practices utilizing new technologies may increase gas recoveries.

Deep gas plays are diverse as implied by the geologic age and location of the existing deep gas basins and wells. Some deep gas plays are merely down dip extensions of shallower plays where drilling has progressed beyond 15,000 ft. A good example of this situation is Wilcox sandstone and other Cenozoic plays in the western Gulf of Mexico. Plays of this type generally require "incremental" technology advancements for successful development. "True" deep plays, where the bulk of the gas play occurs at depths greater than 15,000 ft, are usually geologically older, either Mesozoic or Paleozoic in age, represent a different reservoir setting than the geologically younger plays, and require a unique assemblage of technologies for successful development [202–214,267].

Exploration and drilling strategies also need to be modified from conventional thinking as deep basin-centered prospects are considered. Abnormal pressures (either high or low) may be associated with pervasively gas-charged (continuous-type) reservoirs. Exploration strategies for these tight gas plays require the identification of relatively high porosity and permeability zones. Even low-matrix reservoir porosity can be enhanced by natural fractures associated with anticlinal flexure or faulting [202–214,267]. For cretaceous reservoirs at fields such as Madden in the Wind River Basin, which is considered as a "sweet spot" in a larger basin-centered accumulation, fracture permeability is best developed along the crest but decreases significantly off structure. For other fields, such as Elmworth field in the Western Canada Basin in Alberta, the distribution of high-permeability conglomerate zones plays an important role in gas deliverability and recovery [231–268]. Horizontal drilling is effective especially in accessing fractured reservoirs, particularly when vertical fractures are abundant [231–268].

Because deeper formations generally are tighter, special care should be taken to minimize formation damage while drilling. Careful selection of drilling fluids should be a priority; in particular, aqueous fluids can be damaging because of inhibition of water into the formation [202–214,231–268]. A drilling strategy using underbalanced drilling procedures could lead to significant improvement in the success rate. Not only would this procedure reduce reservoir impairment, it would allow for a continuous evaluation of the formations being penetrated [231–268]. Drilling rates could be increased, and a greater penetration between round trips would help to reduce costs and ensure that all productive zones would be recognized [224].

A lack of geological and geophysical information continues to be a major barrier to deep gas exploration [232]. Much information about deeper faults and possible traps may be interpreted from shallower wells. Improving well completion and stimulation methods for deep gas wells remains a major technologic and economic challenge due to high temperatures, pressures, and corrosive gases encountered at great depths [207,210,213,215–230,232,233]. Well completion improvements should include better control of acid-reaction times during matrix stimulation, and maintaining hydraulic fracture fluid stability while avoiding formation damage. For deep wells, hydraulically fractured wells are not necessarily more productive than less expensive acidized wells [214,263]. Research should be carried out to understand the effects of hydraulic fracturing and acidizing on rock structure. Appropriate treatment selection and design for deep reservoirs could reduce stimulation costs [202–214,267].

Management of nonhydrocarbon gases such as H_2S and CO_2 affects cost, safety, and success of deep gas drilling. The production of hot, sour gas requires a low-cost, acid-resistant coatings for drilling tubes and advanced membranes for gas separation [242]. Special attention should be given to develop packers and sealing elements that can withstand high temperatures and corrosive fluids. Advanced logging tools and sophisticated analytical techniques are also important for deep formation evaluation [231–268]. Critical petrophysical analyses can incorporate information from logs, well cuttings, drilling shows, penetration rates, and other drilling results to better characterize deep reservoirs. Technologies such as petrographic image analysis and auto fluorescent illumination can be helpful in evaluating micro fractures that may be the key to profitable production in deep tight formations [231–268].

2.8.3 ECONOMICS OF DEEP GAS DRILLING

Despite the large underlying deep gas resource, less than 5% of all gas wells drilled in 1996 had deep gas as a target. Deep gas is viewed by industry as a high-risk venture. Deep wells are expensive, and exploring and even delineating deep gas prospects entail high dry-hole rates. At over $5 million/deep gas well on average (including allocated dry-hole costs) and with a 27% dry-hole rate (almost double that of shallower wells), one can easily understand the industry's caution. Still, in 1996, the industry spent $1.3 billion (22%) of its $5.7 billion onshore drilling budget on deep wells.

During late 1900 and early 2000, in the Texas Gulf Coast, the heightened activity reduced well costs by 19% to about $3.9 million per well. Oklahoma deep well costs were already low due to experience in basin specific problems but still declined 9% during this period. The most expensive deep gas wells in active deep gas areas were in South Louisiana, where high-pressure zones, high bottom hole temperatures, and corrosive reservoir fluids required special practices [202–214,267].

A recent study performed for the Gas Research Institute (GRI) showed that even though (on average) onshore deep gas (15,000 ft) provides 6.58 bcf/well of reserves, 5 times more than for a shallower onshore well at 5,000–10,000 ft, these deep wells cost nearly 10 times more to drill [219]. The study also showed that gas recoveries and costs per well vary widely among deep plays and different basins. The wells are more expensive to drill in the Rocky Mountains than in Oklahoma. For ultra-deep plays in the Rocky Mountain region such as the deep Madden gas play in the Wind River Basin, well cost can exceed $10 million, and high recoveries are required to justify the costs [219]. A hypothetical well drilled to 40,000 ft could cost from $25 to $50 million and will require very sizeable gas accumulation.

In general, well costs, dry hole rates, well completion efficiency, and sour gas production and processing costs control deep gas economics. A study funded by the GRI indicated that rig time (day rate times days of drilling), which includes drilling and tripping, was the largest single cost item for a deep gas well [219]. Opportunities to reduce these costs include hiking penetration rates through improved bit designs and the development of more powerful and temperature-tolerant down hole motors.

High dry-hole rates for exploratory wells remain a significant barrier to deep gas development and may call for modifications to traditional gas exploration methods. One approach for finding naturally fractured settings involves combining satellite-platform imagery for near-surface analysis with aero-magnetics and gravity surveys for basement-feature analysis. This macro-exploration process can be augmented on a prospect level with high-resolution 3D seismic, a technique that has been proven successful in other tight gas sand settings [219,231–268].

Present evidence shows that deep well completion methods are not particularly effective. Some needed completion advances for deep, hot, and high-stress formations include improving acid-reaction times during matrix stimulation, maintaining hydraulic fracture fluid stability, and preventing proppant crushing/embedment in hydraulic fractures [231–268]. Stimulation costs can also be extremely high, requiring a heightened level of understanding with respect to treatment selection/design. Managing H_2S and CO_2 in deep gas is expensive and requires increased safety precautions, sophisticated gas processing, and use of special alloy materials in drilling tubes. Low-cost, acid-resisting coatings for tubes and advanced membranes for gas separation can help lower cost.

2.8.4 THE FUTURE OF DEEP GAS

The outlook for U.S. deep gas is promising and represents a substantial future resource for domestic energy supply. Despite the recent upsurge in activity, however, significant barriers still exist, constraining its development. Decreasing well costs through improved drilling practices, reducing dry-hole rates with more focused deep gas exploration technologies, optimizing completion practices, and finding more cost-effective methods for producing and processing sour gas are critical technical issues that need to be addressed [202–214,267].

Technologic problems are among the greatest challenges to deep drilling. Problems associated with overcoming hostile drilling environments (e.g., high temperatures and pressures, and acid gases such as CO_2 and H_2S) for successful well completion present the greatest obstacles to drilling, evaluating, and developing deep gas fields. Even though the overall success ratio for deep wells (producing below 15,000 ft) is about 25%, a lack of geological and geophysical information and high drilling costs continue to be the major barriers to deep gas exploration.

In spite of all challenges, deep gas holds significant promise for future exploration and development. Both basin-centered and conventional gas plays could contain and deliver significant deep undiscovered technically recoverable gas resources. Because of the highly mature state of drilling and production in many U.S. basins, exploration companies are looking for new opportunities and prospects. Exploration for deep gas continues in the United States, in the Gulf Coast, Rocky Mountains, and Permian and Anadarko basins [202–230,267].

2.9 GAS FROM GEOPRESSURIZED ZONES

Geopressure occurs when the pore pressure in a subsurface rock unit exceeds the normal hydrostatic pressure expected for the depth of burial. These areas are formed by layers of clay that are deposited and compacted very quickly on top of more porous, absorbent material such as sand or silt. Water and natural gas that are present in this clay are squeezed out by the rapid compression of the clay, and enter the more porous sand or silt deposits. The natural gas, due to the compression of the clay, is deposited in this sand or silt under very high pressure (hence the term "geopressure"). In addition to having these properties, geopressurized zones are typically located at great depths, usually 10,000–25,000 ft below the surface of the earth. The combination of all these factors makes the extraction of natural gas in geopressurized zones quite complicated [269–282,298].

Hottman's 1966 patent [283,284] apparently represents the first formal recognition of the deep gulf coast sediments, known as geopressured zones, as an energy resource. The patent claims that under certain conditions, overpressured formations (i.e., geopressured formations) are capable of producing hot water and dissolved methane gas. In the years following, the resource received the attention of the USGS and their work was documented in a series of publications by Jones and his associates during 1968–1974 [277,285–290]. After the Organization of Petroleum Exporting Countries (OPEC) oil embargo of 1973, geopressurized zones received significant attention by the DOE that supported a series of well flow tests which removed much of the speculation surrounding the resource.

Of all the unconventional sources of natural gas, geopressurized zones are estimated to hold the greatest amount of gas [289,291–297]. Most of the geopressurized natural gas in the United States is located in the Gulf Coast region (around Texas and Louisiana coasts) at depths below 10,000 ft. While there are some disputes, its technically recoverable estimates can be as high as 26,000 tcf on shore and 49,000 tcf when offshore sandstone reservoirs are included. The known reserve is estimated to be around 1110 tcf. Total global resources of geopressurized gas are believed to exceed all other conventional and unconventional gas resources put together, with the exception of methane hydrates.

Geopressurized zones may be found either under dry land or beneath seabeds. Geopressured reservoirs or aquifers are deep underground reservoirs containing brine. The brine is usually saturated with methane. A substantial fraction of the natural gas is projected to be dissolved in water having a temperature in excess of 300°F. The methane content could be in the range of 30–80 ft^3 per barrel of reservoir fluid. The formation temperatures in geopressured zones are higher than normal and constitute some geothermal potential. The magnitude of the technically and economically producible fraction of the resource depends on the life of the potential reservoirs, the temperature of the produced water, and the quantity of the producible methane contained in the solution [283,284,286–316]. The geopressured zone of the upper Gulf Coast occurs in a broad band 300–500 km wide that stretches from below the Rio Grande along the coast to the mouth of the Pearl River, a distance of more than 2000 km. The geopressured zone extends offshore at least to the shelf edge and contains an accumulation of clastic sediments that exceeds 15,000 m in thickness in some areas [269–282,285,298,317–329].

2.9.1   HISTORICAL PERSPECTIVES OF GULF COAST GEOPRESSURED SEDIMENTS

The geopressured sediments range in age from the Upper Cretaceous, approximately 70 million years old, to Pleistocene, only about 1 million years old. Three major sedimentary facies—a massive sandstone facies and an alternating sandstone, shale facies of the great deltas which shaped the coast in the geologic past, and a massive shale facies—formed offshore and now generally occupying the deeper portion of the Gulf Coast geosyncline predominate geopressurized sediments [285,317–329].

The sands may be of the transgressive type, or they may be regressive. The transgressive sands are by far the most favorable for fluid production, possessing greater porosity, permeability, continuity, and areal extent. Unfortunately, regressive-type sand bodies predominate on the Gulf Coast. When a sand body is contained in an interval of geopressured shale, it becomes charged with the geothermal fluids and thus becomes a potential reservoir. The pattern and distribution of the sand bodies are determined largely by the numerous faults that lace the coast in a subparallel trend to the present shoreline. These faults may have throws of 1000 m or more and act as effective barriers to retard the escape of geopressured waters (i.e., they may form reservoir boundaries). Sand distribution is further affected by complex diapirism and flowage of shale and salt underlying tertiary sediments [285,317–329].

The Tigre Lagoon Gas Field [271,274,281,327] in east-central Vermilion Parish, Louisiana, occupies a complexly faulted northwest-southeast trending structure. As mapped on the top of the Planulina zone, it is about four miles long and two miles wide, and is located about four miles due west of the Avery Island salt dome. Nonassociated gas is produced from several high-pressure conventional reservoirs formed by extensive sand bed systems in the geopressured zone. In the field area, geopressure occurs below a depth of about 12,000 ft; the geopressure "seal" is a shale bed only 300–500 ft thick [271,274,281,327].

The Tigre Lagoon rollover anticline is cut diagonally by arcuate branch faults which fill out westward at distances of 2–3 miles. At least five widespread sand bed aquifers, probably formed by the winnowing action of waves in ancient coastal lakes, occur between depths of 12,500 and 14,000 ft. These sand beds range in thickness up to about 250 ft in the eastern part of the Planulina basin, but generally are no more than 100 ft thick; their areal extent is commonly greater than 50 miles [271,274,281,327].

Natural gas in the producing reservoirs at Tigre Lagoon is not associated with oil. The solubility of natural gas (methane) in water of moderate to low salinity [290,292,311–313,315] is very great at the elevated pressures and temperatures of the geopressured zone; each barrel of water rising from depths of 15,000–20,000 ft in the Planulina basin may, at low salinity, contain up to 100 ft^3 of methane. As the pressure and temperature of the rising water are reduced along the path of flow and the salinity is increased by hyperfiltration, and as water escapes through the shale-bed "seal," methane comes out of solution and released vapor-phase gas accumulates in structural traps and forms commercial reservoirs. A pressure drop from 16,000 to 6,000 psi, at 400°F, for example, reduces methane solubility in freshwater by 52 cf/bbl; a temperature drop from 400°F to 200°F, at 10,000 psi, reduces it by 49 cf/bbl [283,284,286–316].

In the Tigre Lagoon structure, hyperfiltration of saline formation waters that seep through the clay bed "seal" as a consequence of the great pressure differential has concentrated the dissolved solids in the uppermost aquifers of the geopressured zone, where water salinities locally exceed 100,000 mg/L. As water salinity is increased, methane solubility is substantially reduced—as much as 30% if salinity is raised from 10,000 to 100,000 mg/L. The combined effects of pressure, temperature, and water salinity changes have resulted in methane removal from solution and accumulation in the Tigre Lagoon structure, and commercial gas reservoirs are found in six of the uppermost eight sand bed aquifers in the geopressured zones [283,284,286–316].

Geopressured-geothermal reservoir quality along the Texas Gulf coast is controlled by sandstone depositional environment, mineralogical composition, and consolidation history (compaction, cementation, and leaching). Geothermal reservoirs are not composed of simple primary porosity between grains, but consist of secondary leached porosity. The Austin Bayou Prospect in Brazoria County, Texas, is a prospective geothermal reservoir that is the product of secondary leached porosity. However, it must be noted that it is difficult to differentiate between primary and secondary porosity.

2.9.2 Assessment of Recoverable Energy in Gulf Coast and Its Usages

Natural gas is contained in geopressured aquifers in Gulf Coast in three forms: a continuous gas-phase pore volume saturation in the presence of connate water (this is the gas phase above methane saturated water), gas dissolved in the aquifer water, and a dispersed gas-phase pore volume saturation, small bubbles surrounded by water. These three types of gas saturation are called free gas saturation, solution gas, and residual gas saturation, respectively.

Free gas saturation is of most interest to the producers of geopressured gas reserves. It has yielded production in gas wells with minimal technical problems for years. The primary consideration is that a gas well produces gas essentially water-free. When water lies beneath the gas accumulation, well perforations are placed high in the free gas portion of such a reservoir and gas production proceeds, essentially water-free, until with expansion, the aquifer water invades the gas reservoir and eventually floods the entire perforated well space. After the invasion of the gas reservoir by water, some gas is trapped in rock pores after the mobile gas has been produced at the well(s). In many instances, some free gas saturation remains in the reservoir above the new water level after water has flooded the well(s). The gas dissolved in water can be recovered by depressurization and lowering of temperature. Hot brine solution, however, carries a significant amount of geothermal energy.

A preliminary estimate of the geopressured geothermal energy of the northern Gulf of Mexico basin was presented by Papadopulos et al. [273] in Circular 726. Using data from 250 wells, they estimated that 46,000 × 10^{18} J of thermal energy was contained in geopressured waters of the onshore tertiary sedimentary rocks of the Gulf Coast to depths of 6 km in Texas and 7 km in Louisiana. They also estimated that an additional 25,000 × 10^{18} J of energy was represented by methane dissolved in these geopressured waters. The total identified and undiscovered thermal and methane energy in geopressured fluids of the northern Gulf of Mexico basin was estimated to be approximately 106,000–178,000 × 10^{18} J [273].

Wallace et al. [276,321] presents an estimate of the thermal and dissolved methane energy contained in the entire northern Gulf of Mexico basin, both onshore and offshore, to depths of 22,500 ft (6.86 km). Their estimate, based on data from over 3500 wells, in general, substantiates the preliminary estimate of Papadopulos et al. [273]. The total identified thermal energy in fluids of both sandstone and shale is estimated to be 170,000 × 10^{18} J [273]. The major uncertainty in geopressured-geothermal resource assessments lies in determining the amount of fluid that can be recovered at the surface.

Samuels [298] indicated that the geopressured aquifers that extend along the northern Gulf of Mexico are a large, perhaps the largest, potential source of geothermal energy and natural gas in the United States. Because of the high cost of completing wells into these formations and their relatively low temperatures (200°F–400°F), the utilization of the geothermal energy will heavily depend on the value of methane and, in general, of secondary importance.

The economics of extracting either geothermal energy or natural gas from these aquifers strongly depends on the reservoir size, permeability, and compressibility, or specific storage coefficient. It must be noted that the high production rates noted earlier may not be necessary if natural gas were the only resource one might be attempting to recover. The recoverability of energy from geopressured reservoirs depends on the amount of water that can be produced by wells tapping these reservoirs. Garg et al. [270] indicated that energy recoverable from a geopressured-geothermal reservoir would be increased 5–10 times with reinjection into the producing reservoir.

Geopressured-geothermal water along the Texas and Louisiana Gulf coast [271–274,280–282, 297,300,301,308–310,321–323] contains three forms of energy capable of utilization through appropriate use of technology: thermal due to high temperature of brine water, kinetic due to high pressure of brine water, and dissolved methane. While in principle these forms of energy can be harnessed for heat, power, and other application purposes, special problems and unique solutions will have to be addressed for their appropriate use.

Geothermal water, as with all hot waters, can be used directly as a heat source in the warming of buildings and for some other direct heating uses; however, the distance to which such heat can be transmitted economically is limited to an estimated 50 km. The salt in water will either have to be removed or special materials will have to be used for heat exchangers. Highly corrosive or scaling brine may also require the use of a secondary fluid and heat exchange system for circulation in heating systems and equipment. Fossil fuel–fired peaking units may also be required with many of these applications. The generation of electric power produces a form of energy capable of widespread economical distribution and utilization for many purposes. Because of its many favorable characteristics, geothermal energy will be used for the generation of electric power. This can presently be done by two methods: flashing steam from the geothermal water by reducing the pressure to a predetermined point and passing the steam through a low-pressure expansion turbine connected to an electric generator and transferring heat from the geothermal water to a suitable secondary fluid, which is, as a result, vaporized and passed through an expansion turbine connected to an electric generator [283,284,286–316].

Electric power [299,300] may also be generated from the kinetic energy of the geopressured waters. It is believed that well head pressures as high as 140 kg/cm^2 (2000 psi) will be realized. This pressure may be converted to electric power by a hydraulic turbine in much the same manner as hydroelectric power is produced. This pressure, however, would decline with time. If all the potential geopressured-geothermal resources of the Texas and Louisiana Gulf coast offshore as well as onshore-could be economically exploited without adverse ecological impact in the form of small 10–100 MW(e) power plants.

A number of other possible uses of geopressured geothermal energy other than electrical power generation have been suggested, depending on the heat and kinetic energy content of this resource. It is unlikely that many of these alternatives are economically viable without base load use of the geopressured-geothermal brine for power generation and without methane extraction for additional saleable energy value [283,284,286–316].

Ecological considerations, such as possible subsidence and brine disposal, indicate that location of early sites will be remote from highly urbanized or industrialized areas, further limiting a number of these nonelectrical power generation uses. It should be noted, however, that the efficiency of use of geothermal resources for nonelectrical purposes is greater than for electrical power generation. The conversion efficiency for electrical power production approximates 8%–15%, while conversions of up to 85% energy efficiency may be reached in some nonelectrical applications such as direct contact heating [283,284,286–316].

Nonelectrical applications of geothermal resources are already of primary importance in Iceland, New Zealand, Hungary, France, Rumania, Italy, the former USSR, Japan, and several cities in the United States. Some applications are summarized in Table 2.7 [286,287,298–300].

TABLE 2.7
Nonelectrical Applications of Geopressurized Source of Energy

Industrial usages

1. Various types of heating operations; heat source for sugar cane and pulp and paper operations; various drying and evaporation operations (cement, clays, fish, or other marine products); low-level process and space heat for chemical, petroleum, petrochemical, and other industries; heat for lumber, brick, and concrete block curing kilns and water desalination by either flash steam condensation or process heat supply to distillation-type desalting units to provide industrial boiler and pure process water
2. Sulfur franchising if fluids can be obtained in reasonable proximity to salt domes containing sulfur resources
3. Steam turbine-driven natural gas and petroleum pipeline pumping and compressing
4. Injection of brine effluent for secondary recovery of petroleum
5. Mineral recovery from hydrothermal fluids (salt concentration, chemical extraction, etc.)
6. Absorption refrigeration and freeze-drying of foodstuffs
7. Gasohol plant energy source

Agricultural usages

1. Various types of heating: greenhouse heating for limited specialty crops and ornamental plants; rice and grain drying; animal husbandry, including space and water heating, cleaning, sanitizing, and drying of animal shelters; creating optimal thermal environmental conditions for maximum growth and production of agricultural products
2. Hydroponics temperature and humidity control
3. Refrigeration and frozen food preparation
4. Aquatic farming
5. Processing of agricultural products (waste disposal or conversion, drying, fermentation, canning, etc.)

Municipal and residential usages

1. Various types of heating: homes, multiunit dwellings, and buildings: closed hot water or steam space heating systems or district heating by thermal distribution systems; water (potable, hot/cold utility, etc.) heating; deicing bridges, overpasses, and driveways; and heating of swimming pools, fish hatcheries, etc
2. Waste treatment (disposal, bioconversion, etc.)
3. Absorption refrigeration and space cooling

Sources: Modified and adapted from Lindal, B., Industrial and other applications of geothermal energy, in: Armstead, H.C.H., ed., *Geothermal Energy: Review of Research and Development*, UNESCO, Paris, France, pp. 135–148, 1973; Samuels, G., Geopressure energy resource evaluation, DOE, Off. of Energy Technology, Oakridge National Laboratory, Oakridge, TN, 72pp., Available from: NTIS, Springfield, VA (1979), Schmidt, G.W., *Am. Assoc. Petrol. Geol. Bull.*, 57(3), 321, 1973; Swink, D.G. and Shultz, R.J., Conceptual study for total utilization of an intermediate temperature geothermal resource, Prepared by Aerojet Nuclear Company for Energy Research and Development Administration, Idaho Operations Office, Contract No. E(10-1)-1375, Available from: NTIS, Springfield, VA, ANCR-1260, 1976; Wilson, J.S. et al., A study of Phase O Plan for the production of electrical power from U.S. Gulf Coast geopressured geothermal waters, in: *Proceedings of the Second Geopressured Geothermal Energy Conference*, Center for Energy Studies, University of Texas, Austin, TX, Appendix B, 69, 1976, p. 30.

2.10 GAS HYDRATES

Huge quantities of natural gas (primarily methane) exist in the form of hydrates under sediment on offshore continental shelves and on land in Arctic regions that experience permafrost, such as those in Siberia. Hydrates require a combination of high pressure and low temperature to form. In 2010, the cost of extracting natural gas from crystallized natural gas was estimated to be 100%–200% the cost of extracting natural gas from conventional sources, and even higher from offshore deposits [31]. In 2013, Japan Oil, Gas and Metals National Corporation announced that they had recovered commercially relevant quantities of natural gas from methane hydrate [330,331].

Clathrate hydrates are solid crystalline "inclusion" compounds that form when water is contacted with small hydrophobic molecules such as methane, ethane, H_2S, and CO_2 [332–337,362] under certain pressure and temperature conditions. When the inclusion compound is a constituent of natural gas, clathrate hydrates are also referred to as gas hydrates [330,332–343,362,366]. The gas (or methane) hydrate composition is in general 5.75 mol of water for every molecule of methane, although these numbers do depend on the cage structure of the water ice. Various molecular structures of gas hydrate and clathrate are illustrated in Figure 2.10 [333]. The average density of methane hydrate is about 0.9 g/cm^3. Under standard conditions, the volume of methane hydrate will be 164 times less than the volume of methane gas [330–343,362,366].

Gas hydrates are formed when natural gas and water are brought together under suitable conditions of low temperatures and elevated pressures. The formation depends on (1) presence of sufficient amount of water, (2) presence of hydrate former, and (3) appropriate pressure and temperature conditions. In a gas hydrate reservoir, free gas, ice, water, and other components like ethane, propane, hydrogen sulfide, and carbon dioxide can be found at different temperatures, pressures, and depth values. Two- and three-phase equilibria curves [330,331,335–337,342,343] are used for correlation between phases where the number of components present plays a significant role; very small and large amount of water are not conducive to the formation of hydrates.

The gas hydrates are unstable compounds in which the water molecules form a sort of cage or lattice around the methane molecules and the two establish weak chemical bonds with one another. Methane from methane hydrates must be released in situ due to inherent instability of

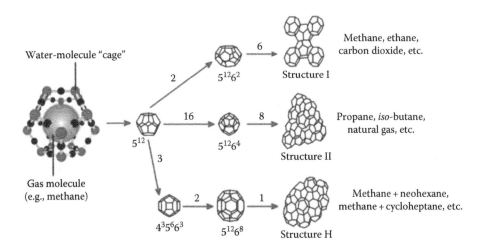

FIGURE 2.10 Structure of hydrates. (Adapted from Shah, Y.T., *Water for Energy and Fuel Production*, CRC Press, New York, 2014.)

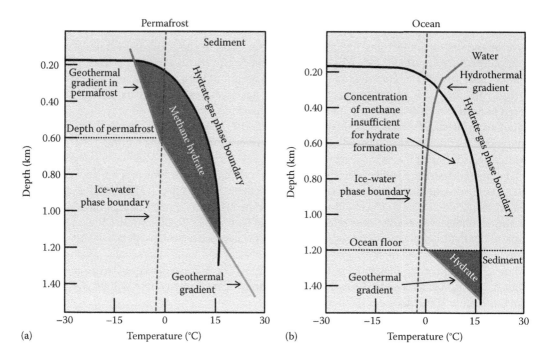

FIGURE 2.11   Hydrate phase diagrams. (Adapted from Shah, Y.T., *Water for Energy and Fuel Production*, CRC Press, New York, 2014.)

hydrate molecules. The temperature at which methane hydrate is stable depends on the prevailing pressure. For example, at 0°C, it is stable under a pressure of about 30 atm; whereas at 25°C, nearly 500 atm pressure is needed to maintain its integrity. The occlusion of other gases within the ice structure tends to add stability, whereas the presence of salts requires higher stabilizing pressures. Appropriate conditions of temperature/pressure exist on earth in the upper 2000 m of sediments in two regions (1) permafrost at high latitudes in polar regions where the surface temperatures are very low and (2) submarine continental slopes and rises where not only is the water cold but the pressures are high (greater than 30 atm). Phase boundary of methane hydrates in permafrost and deep sea regions are graphically illustrated in Figure 2.11a and b [344]. These two figures allow the estimations of regions where the stable hydrate formations are most likely to occur.

2.10.1   SOURCES, SIZES, AND IMPORTANCE OF GAS HYDRATE DEPOSITS

Gas hydrates were only discovered in the late twentieth century, and along with geopressurized zone gas, they are the best means of prolonging the carbohydrate age of energy [336,345–395,549]. As mentioned earlier, vast quantities of methane gas hydrates can be discovered in sediments and sedimentary rocks within about 2000 m of earth surface in polar and deep water regions. Furthermore, the required conditions are found either in polar continental sedimentary rocks where surface temperature is less than 0°C or in oceanic sediment at water depths greater than 300 m, where the water temperature is around 2°C. Methane hydrates can also be formed in freshwater but not in salt water.

In 1995, the USGS conducted a study to assess the quantity of natural gas hydrate resources in the United States and found that the estimated quantity exceeded known conventional domestic gas resources [332]. The USGS estimates that methane hydrates may contain more

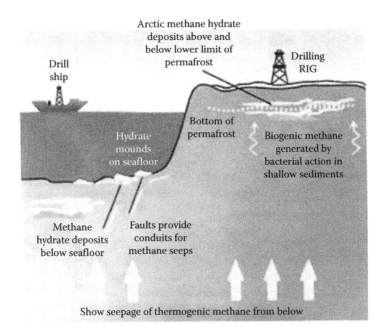

FIGURE 2.12 Types of methane hydrate deposits. (From U.S. DOE, NETL, Washington, DC; Adapted from Shah, Y.T., *Water for Energy and Fuel Production*, CRC Press, New York, 2014.)

carbon than world's coal, oil, and conventional natural gas combined. Methane hydrate estimates range anywhere from 7,000 tcf to over 73,000 tcf [344].

Types of methane hydrate deposits found on earth are graphically illustrated in Figure 2.12. Methane hydrates are believed to be formed by the migration of gas from depth along geological faults, followed by precipitation, or crystallization on contact of the rising gas stream with cold sea water. Methane hydrates are also present in deep Arctic sea cores and record a history of atmospheric methane concentrations dating to 800,000 years ago [336,345–353,369].

In polar regions, methane hydrates are found where temperatures are cold enough for onshore and offshore permafrost to be present. In offshore sediments methane hydrates are found at water depths of 300–500 m, according to prevailing bottom water temperatures. Continental deposits have been located in Siberia and Alaska in sandstone and siltstone beds at depth less than 800 m. Oceanic deposits seem to be widespread in the continental shelf and can occur within the sediments at depth or close to sediment–water interface. They may cap even larger deposits of gaseous methane. In 2008, Canadian and Japanese researchers extracted a constant stream of natural gas from Mallik gas hydrate field in the Mackenzie River delta [336,344–362,369,365–378]. This hydrate field was first discovered by Imperial Oil Co. in 1971–1972 [344,351].

The occurrence of gas hydrates on the Alaska North Slope was confirmed in 1972 in the northwest part of the PBU field [344,345,374,376,378] and the North Slope now is known to contain several well-characterized gas hydrate deposits. The methane hydrate stability zone extends beneath most of the coastal plain province and has thicknesses >1000 m in the Prudhoe Bay, Kuparuk River, and Milne Point oil fields on the North Slope of Alaska. The estimated amount of gas within these gas hydrate accumulations is approximately 37–44 tcf, which is equivalent to twice the volume of conventional gas in the Prudhoe Bay field [338]. More details on the locations of gas hydrate reservoirs in Alaska are given in various USGS reports by Ruppel and Collett [334,336,338,339,345,346, 357,360,364].

The size of the oceanic methane clathrate reservoir is poorly known. The recent estimates constrained by direct sampling suggest the global inventory occupies between 1 and 5 million km^3.

This estimate corresponds to 500–2500 gigatons carbon which is substantially larger than 230 gigatons estimated for other natural gas resources. The reservoir in Arctic permafrost has been estimated at 400 gigatons but no estimates are made for possible Antarctic reservoirs. Low concentration at most sites imply that only a small percentage of clathrate deposits may be economically recoverable [336,344–378].

There are two distinct types of oceanic deposits. The most common type is the one where methane is contained in type I clathrate and generally found in the depth of the sediment. This type is derived from microbial reduction of CO_2. These deposits are located within a mid-depth zone around 300–500 m thick in the sediments. The second, less common type is found near the sediment surface. This type is formed by the thermal decomposition of organic matter. Examples of this type are found in Gulf of Mexico and Caspian Sea. Some deposits have characteristics intermediate between the microbial and thermal source types, and they are considered to be formed from a mixture of the two.

While the sedimentary methane hydrate reservoir probably contains 2–10 times the currently known reserves of conventional natural gas, the majority of site deposits are too dispersed to recover economically. The detection of viable sources is also problematic. The technology for extraction of methane gas from hydrate is also an issue. To date, Messoyakha Gas field in the Russian city of Norilsk is the only commercially operationalized field. Japan is planning to start commercial operations by 2016 [347,358,371,372], and China has invested $100 million over 10 years to study hydrates. A possible economic reserve in the Gulf of Mexico may contain 10^{10} m^3 of gas [362,366,367,375].

Gas hydrates are of great importance for a number of reasons graphically illustrated in Figure 2.13 [344]. Naturally occurring methane gas clathrates contain an enormous amount of strategic energy reserve. In offshore hydrocarbon drilling and production operations, gas hydrates cause major and potentially hazardous flow assurance problems. The recovery of gas hydrates by carbon dioxide provides an opportunity to dispose carbon dioxide by sequestration [384–405,549]. Gas hydrates also provide an increasing awareness of the relationship between hydrates and subsea slope stability.

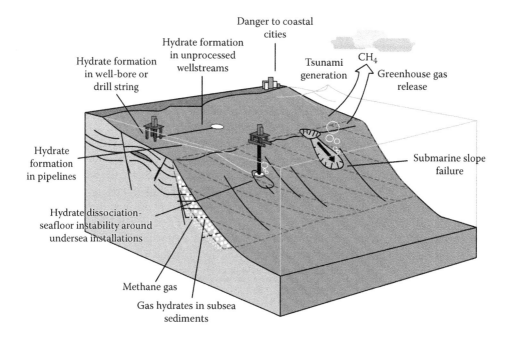

FIGURE 2.13 Locations of hydrate formations. (Adapted from Shah, Y.T., *Water for Energy and Fuel Production*, CRC Press, New York, 2014.)

Gas hydrates also pose potential danger to deep water drilling installations, pipelines, and subsea cables [371,379–422,549]. Finally, they pose long-term concerns with respect to hydrate stability, methane release, and global climate change [423–437].

2.10.2 Importance of Gas Hydrates on Offshore Oil and Gas Operations

The existence of gas hydrates affects both drilling and production of offshore oil and gas operations. These effects are briefly described here.

2.10.2.1 Drilling

Methane clathrates (hydrates) are commonly formed during natural gas production operations, when liquid water is condensed in the presence of methane at high pressure. It is known that larger hydrocarbon molecules such as ethane and propane can also form hydrates, although these are not as stable as methane hydrates. Once formed, hydrates can block pipeline and processing equipment. They are generally removed by (1) reduction of the pressure, (2) addition of heat, or (3) dissolving them using chemicals such as methanol and ethylene glycol. Care must be taken to ensure that the removal of the hydrates is carefully controlled, because as hydrates undergo phase transition, the release of water and methane can occur at very high rates. The rapid release of methane gas in a closed system can result in a rapid increase in pressure [409,410,412]. In recent years, hydrate formation during drilling operation is controlled by use of kinetic hydrate inhibitors [399–402,417–420], which dramatically slow the rate of hydrate formation and anti-agglomerates, which do not prevent hydrate formation but prevent them sticking together to block pipes and other parts of equipment.

When drilling in oil- and gas-bearing formations submerged in deep water [383–389], the reservoir gas may flow into the well bore and form gas hydrates owing to the low-temperature and high-pressure conditions found during deep water drilling. The gas hydrates may then flow upward with drilling mud or other discharged fluids. As they rise, the pressure in the annulus decreases and the hydrates dissociate into gas and water. The rapid gas expansion ejects fluid from the well, reducing the pressure further, which leads to more hydrate dissociation and further fluid ejection. The resulting violent expulsion of fluid from the annulus is one potential cause or contributor to what is referred to as a "kick" [409,410,412], which can cause blowouts. This can cause serious well safety and control problems and create hazardous conditions such as flow blockage, hindrance to drill string movement, loss of circulation, and even abandonment of the well. Since gas hydrates contain 85% water, their formation can withdraw water from drilling fluids changing the properties of the fluids, causing salt precipitation, an increase in fluid weight, or the formation of solid plug.

The condition of the hydrate formation during kick depends on the composition of the kick gas, temperature, and pressure. A combination of salts and chemical inhibitors can provide required inhibition to avoid hydrate formation, particularly at water depths greater than 1000 m [399–402,417–420].

2.10.2.2 Production by Enhanced Oil and Gas Recovery Methods

Enhanced oil and gas recovery methods increase risk of gas hydrate presence. Process equipment and multiphase transfer lines from well-head to the production platform where low temperature and high pressure conditions exist are prone to hydrate formation. The following methods are generally adopted to reduce hydrate problems in hydrocarbon transfer lines and process facilities [390–396,549].

1. Use high flow rates, which limit the time for hydrate formation in a volume of fluid, thereby reducing the kick potential [409,410,412]. Make careful measurement of line flow to detect incipient hydrate plugging [409,410,412] particularly at low gas production rate. Also, monitor the pressure rise in well casing after it is "shut in" (isolated). The hydrate formation will decrease the rate of pressure rise [409,410,412].

2. Additions of energy (e.g., the energy released by setting cement used in well completion) can raise the temperature and convert hydrates to gas, producing a "kick."
3. For a given pressure, operate at temperatures higher than the hydrate formation temperature. This can be done by insulation or heating of the equipment. At fixed temperature, operate at pressure below hydrate formation pressure.
4. Reduce water concentration to avoid hydrate formation. Change feed composition.
5. Add compounds such as methanol, salts, or other kinetic inhibitors to prevent hydrate formation. Also prevent hydrate clustering by using hydrate growth modifiers or covering working surfaces with hydrophobic substances [390–396,409,410,412,549].

With conventional oil and gas exploration methods extending into progressively deeper waters, potential hazards gas hydrates can pose to operation is becoming increasingly more important. Two possible events—(1) the release of overpressured gas (or fluids) trapped below the zone of hydrate stability and (2) destabilization of in situ hydrates—can be hazardous. Care must be taken to avoid these incidences.

2.10.2.3   Natural Gas Hydrates versus Natural Gas in Transportation

Since methane clathrates are stable at a higher temperature than liquefied natural gas (LNG) (−20°C vs. −162°C) (see Section 2.4), there is some interest in converting natural gas into clathrates rather than liquefying it when transporting it by seagoing vessels. A significant advantage would be that the production of natural gas hydrate (NGH) from natural gas at the terminal would require a smaller refrigeration plant and less energy than one required for LNG. This topic is further discussed in Chapter 8.

2.10.3   ENVIRONMENTAL IMPACTS OF GAS HYDRATES

Gas hydrates alter the physical properties of the sediment. In the absence of hydrates, fluids and gas migrate freely at seafloor. Solid hydrates reduce permeability and restrict sediment consolidation, fluid expulsion, and cementation. Hydrate dissociation leads to increased pore fluid pressure and under-consolidated sediments, with a reduced cohesive strength compared to overlying hydrate-bearing sediments, forming a zone of weakness. This zone of weakness could act as a site of failure in the event of increased gravitational loading or seismic activity. The link between seafloor failure and gas hydrate destabilization is a well-established phenomenon [330,332–343,362,366]. The exploration of hydrates from ocean floor by drilling through hydrate zones can create the problem of destabilizing support foundations for platforms and production wells. The disruption of ocean floor can also result in surface slumping or faulting, which can endanger work crews and the environment [330,332–343,362,366].

Since hydrates prevent sediment compaction, their in situ dissociation can also cause climate change and falling of sea level. If the hydrate breaks down it will weaken the sediment and may cause submarine landslides and simultaneously release methane into the atmosphere. The methane released from the reservoir to the atmosphere can contribute to the climate change. Submarine landslides can cause tsunamis and catastrophic coastal flooding. The thickness of the gas hydrate stability zone in continental margins depends on water depth (hydrostatic pressure), water temperature, the geothermal gradient, and gas composition [332,344].

Methane is a powerful greenhouse gas. Despite its short atmospheric half-life of 7 years, methane has a significant global warming potential [330,332–343,362,366]. Recent research carried out in 2008 in the Siberian Arctic has shown millions of tons of methane being released [438–443], with concentrations in some regions reaching up to 100 times above normal [330,332–343,362,366]. Past and future climate changes can be linked to methane released from gas hydrates.

Currently the link between stability of gas hydrates and global warming is being examined. Since methane warms the environment 15–20 times more than carbon dioxide, the release of methane can create a chain reaction for global warming and this leads to more hydrate instability with the additional release of methane. Methane release in air eventually (within 10 years) is converted to carbon dioxide, another greenhouse gas [438–501].

The analysis of the link between gas hydrate and climate warming can be divided into five regions [344].

Region 1: *Thick (>300 m) onshore permafrost*: Gas hydrates that occur within or beneath thick terrestrial permafrost will remain largely stable even if climate warming lasts hundreds of years. The warming could, however, cause hydrates at the top of the stability zone, about 625 ft below the earth's surface to dissociate over thousands of years [438–501]. It contributes less than 1% of total hydrates and its effect on climate change will be minimal.

Region 2: *Subsea permafrost on the circum-Arctic shelves*: The shallow water continental shelves that circle parts of the Arctic ocean were formed when sea level rise during the past 10,000 years inundated permafrost that was at the coastline. The methane hydrates in subsea permafrost, which is thawing beneath these continental shelves, is being released now. While this methane can rise to ocean surface and then to atmosphere, the amount is only considerably less than about 1% of the world gas hydrates [438–501].

Region 3: *Upper edge of stability (or deep water marine hydrates at the feather edge of GHSZ)*: Gas hydrates on upper continental slopes beneath 1000–1600 ft of water lie at the shallowest water depth for which methane hydrates are stable. The upper continental slopes that ring all the continents could host gas hydrates in zones that are roughly 30 ft thick. Within the next 100 years, warm water can completely dissociate these hydrates but they are more likely to be oxidized in water than released in the atmosphere. These hydrates contribute about 3.5% of earth's total hydrates [438–501].

Region 4: *Deepwater gas hydrates*: Ninety-five percent of earth's gas hydrates are at depths greater than 3000 ft. Even with an increase in ocean temperature, they are likely to stay stable over thousands of years. They also occur deep within the sediments and the released methane will remain in the sediments; and if they move upward, they will form new hydrates or consumed by oxidation within water [438–501].

Region 5: *Seafloor gas hydrate mounds*: At some marine seeps such as Gulf of Mexico, massive relatively pure gas hydrate occurs in seafloor mounds. While seafloor gas hydrate mounds and shallow subseafloor gas hydrate constitute only a trace amount of the global gas hydrate inventory, they can dissociate rapidly due to the expulsion of warm fluids from the seafloor and release a significant amount of methane to the atmosphere.

Based on the analysis of these five regions, a general consensus [438–501] is that catastrophic widespread dissociation of methane gas hydrates will not be triggered by continued climate warming at the contemporary rate (0.2°C per decade) over a timescale of few hundred years. In spite of this conclusion, there has been an enormous interest in studying methane release from the hydrates to the atmosphere and its effect on the environment. The vast literature [438–501] is cited here to demonstrate the significant interest on the subject.

2.10.4 PRODUCTION OF METHANE FROM GAS HYDRATE RESERVOIRS

Hydrates are known to occur at temperatures less than 295 K and pressure greater than 3000 kPa. The dissociation of these hydrates occurs as:

$$CH_4 \cdot 6H_2O(s) \rightarrow CH_4(g) + 6H_2O(l) \tag{2.1}$$

With enthalpy = 10–20 kcal/mol of gas dissociated [330,332–343,362,366]. This reaction requires an external energy source to propagate along the right-hand side [330,332–343,362,366].

In conventional gas reservoirs, natural gas migrates to the recovery point via pressure gradients. For these reservoirs, the recovery rate is a function of the formation permeability and pressure gradients established between the reservoir and extraction well(s). Production of methane from hydrate-bearing deposits requires additional energy to dissociate the crystalline water lattice that forms the gas hydrate structure. A variety of methods have been proposed for producing natural gas from hydrate deposits: (1) thermal stimulation, where the temperature is increased above the hydrate stability region; (2) depressurization, where the pressure is decreased below the hydrate stability region; (3) chemical injection of inhibitors, where the temperature and pressure conditions for hydrate stability are shifted; and (4) CO_2 or mixed CO_2 and N_2 exchange, where CO_2 and N_2 replace CH_4 in the hydrate structure; and (5) enhanced gas hydrate recovery (ERGH) methods where two-phase emulsion (of CO_2 and water) and other solution injection techniques are used to replace methane from hydrate structure. Each of these methods is briefly reviewed here.

2.10.4.1 Thermal Stimulation

The recovery of methane gas from gas hydrates via thermal stimulation has been examined both experimentally [490,491,494] and theoretically [488,489,492,493]. Technologies for implementing thermal stimulation include steam injection, cyclic steam injection, fire flooding, and hot brine injection and electromagnetic heating. The techniques of steam injection and cyclic steam injection are very similar to the ones used in the recovery of conventional and unconventional oils. Various possibilities for heating hydrates using steam or cyclic steam injections have been examined in the literature [491,494]. All of these techniques, however, suffer from high heat losses and by-products of fire flooding can dilute the produced natural gas. The energy efficiency of electromagnetic heating is also low.

A more promising approach is to inject a saline aqueous solution at an elevated temperature into gas hydrate–bearing geological reservoir. This method works on the heating process that is dominated by the advection of sensible heat carried by the brine solution. The dissolved salt depresses the dissociation temperature of the gas hydrate. The experimental evidence indicates that with the injection of brine, hydrates become colloidal and migrate advectively with the brine [495,496]. Tang et al. [490,501] showed that the energy efficiency of the hot brine injection process is dependent on the brine temperature, injection rate, and initial hydrate saturation.

The energy efficiency is defined as the ratio of combustion heat of produced gas over the heat input of the brine. The study showed that better energy efficiency was obtained at higher initial hydrate saturation and lower temperature and injection rates [488–496]. This higher energy efficiency is, however, accompanied by lower production rates. For moderate to high temperature and injection rate, about 50% of the recovered energy from methane is used to heat the brine solution. A modification of this approach was suggested by Chatterji and Griffith [502], who proposed an injection of two aqueous fluids that react and produce the heat required to release methane from the hydrates. This type of acidic and basic solution reactions will yield a hot salt solution, and this will not require the external heating of brine solution, thereby improving the energy efficiency.

2.10.4.2 Depressurization

Gas hydrate production via depressurization is considered to be the most economically promising technology [497–501,503–508]. This method has been adopted in Messoyakha field in northern Russia which contains both free natural gas and hydrates. This reservoir has been constantly producing natural gas because of dissociation of gas hydrates into gas due to depressurization. The production rate in this field is, however, controlled by the heat transfer toward the hydrate dissociation region. Moridis et al. [500,509,510] and Moridis [511,512] numerically simulated the effect of depressurization at Mallik site assuming 0.03°C/m temperature gradient in the hydrate-bearing formation. The simulation showed a vertical drop in temperature in response to depressurization and hydrate dissociation. This temperature drop can be reversed by injecting warmer water in the well, which provided the needed energy to sustain hydrate dissociation in the depressurized system.

When steam or hot methane gas was injected from a second well, natural gas production was superior in terms of the ratios of produced gas to water and fraction of produced methane from hydrates.

Several other simulation studies showed that hydrate dissociation rates and associated gas productions are controlled by the far-field reservoir pressure and temperature, via energy supplied by natural gas advected from the far field to the dissociation front [500,527,509–517,520]. Few studies have reported experimental data of gas recovery by depressurization. While depressurization is a viable option because of thermal self-regulation of gas hydrates, the method results in slow production rates. Sustained production requires a heat source, which at the Messoyakha field is supplied by thermal conduction and advection in the dissociation zone. This heat transfer ultimately controls the production rate.

There are three important mechanisms involved in the depressurization of the gas hydrates: kinetics of dissociation, heat transfer including conduction, and convective flow of fluids like gas and water. While a lot of theoretical work including a three-dimensional model in a porous media which would simulate the exact conditions of a reservoir with regard to all the mechanisms involved has been reported [500,527,509–520], so far researchers have been able to arrive at answers only with certain assumptions. Often a two-well system involving a combination of depressurization at the production well and a thermal input at the injection well where hot fluids are injected appears to be better than single vertical system [496–501,503,504].

2.10.4.3 Inhibitor Injection

Sung et al. [520], Kawamura et al. [521], and Li et al. [522,523] showed that the thermodynamic inhibitors lower the hydrate formation temperature, which can result in hydrate dissociation when injected into a gas hydrate–bearing formation. The most important thermodynamic organic inhibitors are methanol, monoethylene glycol (MEG), and di-ethylene glycol (DEG) commonly referred to as glycol [522,523]. Dissolved salts such as NaCl, $CaCl_2$, KCl, and NaBr can also be inhibitors [524–527]. While gas hydrate inhibitors are an effective methodology for preventing hydrate formation in engineering applications, their use in the production of natural gas hydrates is restrictive due to environmental impact, economic costs, and thermal self-regulation of gas hydrates. Of the inhibitors examined, methanol and glycols are the most successful ones [528]. The principles by which alcohol, glycols, and salts inhibit hydrates are the same. However, salts have corrosion problems and due to low vapor pressures, they cannot vaporize.

In adding inhibitors, besides temperature and pressure conditions, composition and amount of inhibitors are important. The inhibitor must be at or below its water dew point (i.e., must be water saturated). In addition, dehydration can be used as an alternative. An addition of an inhibitor can shift pressure–temperature diagram such that the temperature decreases at specific pressures, and this facilitates hydrate dissociation. After temperature depression due to an addition of an inhibitor, free gas will form and hydrate zone will shift to the left to lower the temperature side. Methanol has a high vapor pressure and infinite water solubility, and it can easily shift to gas phase.

In most offshore applications, hydrate formation is controlled by injection of a thermodynamic hydrate inhibitor. Inhibitor injection at a given pressure will reduce the temperature at which hydrate is formed. Overall ethylene glycol seems to be most useful inhibitor for the gas hydrates.

2.10.4.4 Gas Exchange

The concept of exchanging CO_2 with CH_4 was first advanced by Ohgaki et al. [529]. Their experimental study showed that CO_2 to be preferentially clathrated over CH_4 in the hydrated phase and they demonstrated the possibility of producing CH_4 by injecting CO_2 gas. Ohgaki et al. [529], however, noted that during the exchange process the mole fraction of CO_2 in the hydrate phase was greater than that in the gas phase.

This effect was further studied quantitatively by Seo et al. [530] and Seo and Lee [531], and they showed that CO_2 concentration in the hydrate phase was greater than 90% when the gas-phase concentration of CO_2 in the hydrate formers (i.e., CO_2 and CH_4) was above 40%. Pure CH_4 and CO_2 form structure I (sI)-type hydrates and their mixtures also form sI-type hydrates [529–542,568]. In forming mixed CH_4 and CO_2 hydrates, the CH_4 molecules occupy both the large and small cages of type sI hydrates, whereas CO_2 molecules only occupy the large cages. Without hydrate dissociation, there is an upper limit to the substitution of CO_2 for CH_4 in hydrates.

Lee et al. [536] showed that approximately 64% of CH_4 can be released by exchange with CO_2. In additions to equilibrium considerations, the heat of CO_2 hydrate formation is higher (–57.9 kJ/mol) than the heat of dissociation of CH_4 hydrate (–54.5 kJ/mol), making the overall process exothermic which favors the normal exchange of CO_2 with CH_4 hydrate.

While the exchange of CO_2 for CH_4 is thermodynamically favorable process, the kinetics of exchange mechanism is slow [529–543,568], with the induction time requiring several days. The original studies [500] also did not address the rate of CO_2 gas penetration further into gas hydrate, beyond the first few hundred manometers at the interface. The exchange of CO_2 with CH_4 at high pressure (with liquid CO_2) was also examined in the literature, but once again a slow rate of exchange was observed. The use of nitrogen instead of CO_2 gave much higher rate. For liquid CO_2 injection, thermodynamic conditions can favor either CO_2 or CH_4 cage occupation. This transition occurs where pure CO_2 and CH_4 temperature-versus-pressure equilibrium functions cross the increasing pressure above the gas–liquid CO_2 phase boundary.

Thermodynamic properties of hydrates depend on the pore size distribution in the geologic media; hydrate formation will occur in large pores first and then in small pores until equilibrium is achieved [509–512,519,544,545,554]. Porous media also affect other thermodynamic properties of hydrates. In geologic media that have distribution of pore sizes, hydrates would form and dissociate over a range of temperatures and pressures according to the distribution of pore radii and the impact of salts in the residual pore water [490,501,527]. Goel [537] and Goel et al. [515] indicated that in order to understand gas exchange technology in porous media, there is a need for quantitative estimates of formation and dissociation processes in geologic media core samples.

2.10.4.5   Enhanced Gas Hydrate Recovery Method

As shown earlier, a strict gas exchange of CO_2 for CH_4 in bulk methane hydrate is slow by several orders of magnitude to be considered an effective method of gas hydrate production. An enhanced gas hydrate recovery (EGHR) process that involves injecting a two-phase emulsion of liquid CO_2 and water at proper volumetric ratio can considerably enhance (three times or higher) production rate over injecting cool water (15°C) alone [502,543,546–552]. It is important to know the range of reservoir conditions where EGHR technique can be applied. Collett and co-workers [553–555] calculated these conditions for Alaska Northern slope (ANS) and concluded that EGHR method can be applied over a large fraction of ANS. He also found that CO_2 hydrate would be stable under almost any conditions on the ANS short of very near the ground surface. He also suggested that typical ANS reservoir conditions would inject liquid CO_2 with a density approximately 82%–94% of the water phase. ANS well log temperature data as well as carbon dioxide hydrate and vapor–liquid equilibrium data are described by Collet et al. [358,554,555].

The laboratory studies indicated that there are no signs of coagulation into macrodroplets as the emulsion moves away from the injector—the conclusion that needs to be tested at reservoir scale. Another important restriction is that temperature of the water-CO_2 emulsion remains above the equilibrium point where CO_2 hydrate could form in the wellbore or near-wellbore. Interruption of the supply of emulsion fluid during production for an extended period could result in the premature formation of CO_2 hydrate and plugging [540,546]. Provisions for temporary injection of heat may be needed to allow for flow interruptions, such as for well maintenance.

EGHR method has been tested in laboratory for continuous production of a suitable liquid carbon dioxide and water emulsion [502,540,543,546–552]. This test is largely one-dimensional. A suitable downhole tool that can work in actual field needs to be developed. The injector tool design should be compatible with downhole conditions typical of gas hydrate formations. Wellbore completion requirements such as open hole, uncased, or perforated casing influence the design parameters of the injection tool. Injection of the liquid carbon dioxide and water emulsion in the target formation is the most important requirement. A new design [540] to fit these requirements is depicted and described in my previous book [344].

EGHR technique is still being developed [540,546,550]. A number of questions such as placement of recovery wells, including the distance from the injection site and spacing to maximize the recovery of CH_4 gas, need to be determined. Identification and delivery logistics of an economic supply of carbon dioxide for a given site also need to be ascertained. Both theoretical and experimental works to address these questions still need to be carried out [543,547,552].

In summary, EGHR process has several advantages: (1) Replacing methane with carbon dioxide in gas hydrate sediments is thermodynamically favorable and the heat generated from the formation of CO_2 hydrate is approximately 20% greater than the heat consumed from the dissociation of methane. This net exothermic process allows a low-grade heat source to facilitate further dissociation of gas hydrate. (2) Once CO_2-rich fluid fills pore space and methane is extracted, the subsequent formation of carbon dioxide hydrate would mechanically stabilize the formation eliminating subsidence concerns in some production situations. (3) The overall process is carbon-neutral in terms of replacing methane with carbon dioxide which is permanently sequestered in situ as gas hydrate. Produce water can also be used to form the emulsion, eliminating a problematic disposal issue in Arctic settings [543,547,552].

2.10.4.5.1 EGHR Simulation

There are relatively few published studies of commercial production methods for gas hydrates, and all of these studies have examined essentially conventional production concepts, principally depressurization, coupled with some form of thermal stimulation [509–520,544,551,552,554,556–558]. An EGHR process that utilizes a microemulsion of liquid CO_2 and water to decompose methane hydrate in situ and produce free gas described earlier has been successfully demonstrated in laboratory-scale experiments with gas hydrate–bearing sediments. While these laboratory-based studies are extremely encouraging, a reservoir modeling assessment that compares and contrasts the EGHR process with conventional methods of gas hydrate production was carried under a DOE project [509–520,544,551,552,554,556–558] by Battelle simulation and modeling group.

Within DOE project [517,540], STOMP-HYD simulator was applied to a series of one- and two-dimensional simulations that investigated the production of CH_4 hydrates in geologic media using CO_2 injection. Brief details of this simulation work is described in my previous book [2]. One critical finding of this work was that the formation of secondary CO_2 hydrate has the potential to halt the production process by inhibiting fluid migration. A complete exchange of CO_2 and CH_4 is possible without forming excessive secondary hydrate and while maintaining elevated hydrate saturations. The pore-water salinity may play a strong role in the inhibition of secondary hydrate formation beyond certain saturation levels, an observation in agreement with the published experimental results [540].

2.11 COMMERCIAL APPLICATION

In the recent years, the mentioned production methods and computer simulations have been applied to numerous practical sites [358,544,545,553–569]. These include the north slope of Alaska and numerous sites in that region such as Malik field and Milne points. Nankat trough and Ulleung basin of Korea have also been examined [358,544,545,553–569]. Several general production strategies have also been investigated. More work on both experimental and theoretical applications of various production methods to on- and off-shore commercial sites is needed. The successful commercial operation to recover methane from gas hydrates will significantly increase our energy supply.

REFERENCES

1. Natural gas, Wikipedia, the free encyclopedia (September 2013), http://en.wikipedia.org./wiki/natural_gas, Last accessed October 14, 2016.
2. Composition of natural gas, a report by Naturalgas.org, Retrieved July 14, 2012.
3. Natural gas overview, a report by Naturalgas.org, Retrieved February 6, 2011.
4. Natural gas—Proved reserves, a report by Naturalgas.org, Accessed December 1, 2013.
5. International Energy statistics, US Energy Information Administration (EIA), Washington, DC (2016).
6. Natural Gas Gross Withdrawals and Production, Annual, 2008 a report by U.S. DOE EIA DOE, Washington, DC (2010), Available online at: http://www.eia.gov/dnav/ng/ng_prod_sum_dcu_NUS_a.htm.
7. Annual energy outlook 2012, EIA, Department of Energy, Washington, DC (2012), Available online at: http://www.eia.gov/forecasts/aeo/pdf/0383(2012).pdf, Accessed August 28, 2012.
8. List of countries by natural gas proven reserves, Wikipedia, the free encyclopedia (2015), https://en.wikipedia.org/.../List_of_countries_by_natural_gas_proven_ reserves, Last accessed October 4, 2016.
9. History of natural gas (January 1, 2011), Naturalgas.org.
10. What's the difference between wet and dry natural gas? a report by Stateimpact.npr.org, Retrieved January 13, 2014.
11. Myhre, G. et al., Anthropogenic and natural radiative forcing, in Stocker, T.F., Qin, D., Plattner, G.-K., Tignor, M., Allen, S.K., Boschung, J., Nauels, A., Xia, Y., Bex, V., and Midgley, P.M. (eds.), *Climate Change 2013: The Physical Science Basis. Contribution of Working Group I to the Fifth Assessment Report of the Intergovernmental Panel on Climate Change.* Cambridge University Press, Cambridge, U.K., p. 714 (2013). Retrieved August 22, 2014.
12. Satellite observation of flares in the world, a report by Naturalgas.org, Retrieved February 6, 2011.
13. Satellite observation of methane in earth's atmosphere a report by Naturalgas.org, 2015.
14. Revkin, A. and Krauss, C., Curbing emissions by sealing gas leaks, Energy and Environment, *NY Times* (October 14, 2009), Retrieved June 11, 2013.
15. Allen, D.T. et al., Measurements of methane emissions at natural gas production sites in the United States (PDF), *Proceedings of the National Academy of Sciences of the United States of America*, National Academy Press, 110 (44), 17768–17773 (September 16, 2013), doi:10.1073/pnas.1304880110.
16. Skone, T.J., Life cycle greenhouse gas analysis of natural gas extraction & delivery in the United States, National Energy Technology Laboratory report, DUE, Washington (May 12, 2011).
17. Jaramillo, P., Griffin, W.M., and Matthews, H.S., Comparative life-cycle air emissions of coal, domestic natural gas, LNG, and SNG for electricity generation (PDF), *Environmental Science & Technology* (July 25, 2007), doi:10.1021/es063031o.
18. Howarth, R., Santoro, R., and In graffea, A., Methane and the greenhouse-gas footprint of natural gas from shale formations (PDF), *Climate Change*, 106 (4), 679–690 (2011).
19. NOAA, Methane leaks from three large U.S. natural gas fields in line with federal estimates, Office of Oceanic and Atmospheric Research Headquarters, Silver Spring, MD (February 18, 2015).
20. Natural Gas and the Environment, a report by NaturalGas.org, Retrieved June 11, 2013.
21. Devold, H., Oil and gas production handbook—An introduction to oil and gas production, transport, refining and petrochemical industry, A report by ABB Industry, Oslo, Norway (August 2013). www04.abb.com/global/.../Oil+and+gas+production+handbook.pdf.
22. Natural gas exploration (September 2013), a report by Naturalgas.org.
23. Hyne, N.J., *Dictionary of Petroleum Exploration, Drilling & Production.* PennWell Books, Tulsa, OK, pp. 190, 625 (1991).
24. Natural gas data, US Energy and Information Administration, DOE, Washington, DC (2016), www.eia.gov/naturalgas/data.cfm.
25. Exploration and production, oil and gas overview, a report by American Petroleum Institute, Washington, DC (2015), www.api.org/Oil-and-Natural-Gas.../Exploration-and-Production.
26. Drilling, Wikipedia, the free encyclopedia (2015), https://en.wikipedia.org/wiki/Drilling, Last accessed October 13, 2016.
27. Natural gas extraction (2015), a report by Naturalgas.org, naturalgas.org/naturalgas/extraction/.
28. Pad drilling and rig mobility lead to more efficient drilling, US Energy Information Administration, DOE, Washington (September 11, 2012).

29. Lerner, N., Schaab, B., Garcia, J., Bianco, D., Thomas, S., Thompson, J., and Hollan, J., Evolution of drilling and completions in the slave point to optimize economics, *SPE Drilling and Completion*, 20 (1), 1–14 (March 2014), Document ID SPE-163816-PA.

30. Seale, R., Athans, J., and Themig, D., An effective horizontal well completion and stimulation system, *Journal of Canadian Petroleum Technology*, 46 (12), 12 (2007).

31. E&P, New tools enable CBM horizontal drilling, EPA-HQ-OW-2008-0517, DCN 07229 (2007), Available online at: http://www.epmag.com/archives/features/536.htm, May.

32. Augustine, J.R., Open hole versus cemented completions for horizontal wells with transverse fractures: An analytical comparison, in Paper SPE 142279 presented at the *2011 SPE Production and Operations Symposium*, Oklahoma City, OK (March 27–29, 2011).

33. Seale, R., An efficient horizontal open hole multi-stage fracturing and completion system, in Paper SPE 108712 presented at the *2007 SPE International Oil Conference and Exhibition*, Veracruz, Mexico (June 27–30, 2007).

34. Seale, R., Donaldson, J., and Athans, J., Multistage fracturing system: Improving operational efficiency and production, in Paper SPE 104557 presented at the *SPE Eastern Regional Meeting*, Canton, OH (October 11–13, 2006).

35. Well Completion (2004), a report by NaturalGas.org, Available online at: http://www.naturalgas.org/naturalgas/well_completion.asp, Accessed August 18, 2006.

36. U.S. DOE, Multi-seam well completion technology: Implications for Powder River Basin coalbed methane production, EPA-HQ-OW-2008-0517, DCN 07356 (September 2003).

37. Coalbed methane extraction: Detailed study report, EPA-820-R-10-022, a report by EPA, Washington, DC (December 2010).

38. Hydraulic fracturing, Wikipedia, the free encyclopedia (2015), http://en.wikipedia.org./wiki/hydraulic_fracturing, Last accessed October 14, 2016.

39. Bennett, L. et al., *The Source for Hydraulic Fracture Characterization*, Oilfield Review. Schlumberger, Houston, TX, pp. 42–57 (Winter 2005/2006). http://www.slb.com/~/media/Files/resources/oilfield_review/ors05/win05/04_the_source_for_hydraulic.pdf, Last accessed December 15, 2016.

40. How is hydraulic fracturing related to earthquakes and tremors?, a report by US Geological Survey, Department of Interior, Washington, DC (2013), www.usgs.gov/faq/categories/10132/3830, Accessed April 20, 2013.

41. Environmental impacts of hydraulic fracturing, Wikipedia, the free encyclopedia (2015), https://en.wikipedia.org/.../Environmental_impact_of_hydraulic_fracturing, last accessed October 14, 2016.

42. Gallegos, T.J. and Varela, B.A., Trends in hydraulic fracturing distributions and treatment fluids, additives, proppants, and water volumes applied to wells drilled in the United States from 1947–2010— Data analysis and comparison of literature, USGS scientific investigation report 2014-5137, U.S. geological survey, Department of Interior, Washington, D.C. (2015).

43. Morton, M.Q., Unlocking the earth: A short history of hydraulic fracturing, *Geo ExPro* (online magazine) 10 (6), (2013), www.geoexpro.com, Retrieved February 27, 2014.

44. Carus, F., Dangerous levels of radioactivity found at fracking waste site in Pennsylvania (October 2, 2013), theguardian.com, Retrieved October 9, 2013.

45. Fracturing fluid and additives, Petrowiki, A report from SPE (2015).

46. Al-Muntasheri, G.A., Fluids for fracturing petroleum reservoirs, Web Events, Society of Petroleum Engineers (2015), https://webevents.spe.org/products/fluids-for-fracturing-petroleum-reservoirs.

47. Schein, G., The application and technology of slickwater fracturing, Distinguished Lecture presented 2004–2005, SPE-108807 (2012).

48. Howard, C.C. and Fast, C.R., Optimum fluid characteristics for fracture extension, in *API Drilling and Production Practice, 24*, New York, 261 (1957).

49. Gidley, J. et al. (eds.), Fracturing fluids and additives, Chapter 7, in *Recent Advances in Hydraulic Fracturing*, SPE Monograph Series, Vol. 12, Society of Petroleum Engineers, Richardson, TX, p. 131 (1989), ISBN: 978-1-55563-020-1.

50. Natural gas extraction-hydraulic fracturing, EPA report, EPA, Washington, DC, (February 2016), https://www.epa.gov/hydraulicfracturing.

51. The hydraulic fracturing water cycle, EPA's study of hydraulic fracturing and its potential impact on drinking water resources, EPA, Washington, DC (June 1, 2015), https://www.epa.gov/.../h...

52. Zammerilli, A., Murray, R.C., Davis, T., and Littlefield, J., Environmental impacts of unconventional natural gas development and production, NETL report, DOE NETL contract number DE-FE0004001, Office of Fossil Energy, Department of Energy, Washington, DC (May 29, 2014).

53. Summary of FracFocus 1.0 hydraulic fracturing data user guide for state summaries, a report by FracFocus, www.fracfocus.org, Available from http://fracfocus.org/, Cited: June 25, 2013.

54. Department of Energy (DOE), Modern shale gas development in the United States: A primer, U.S. Department of Energy, National Energy Technology Laboratory, Strategic Center for Natural Gas and Oil, Morgantown, WV, 96pp. (April 2009).

55. Compressed natural gas, Wikipedia, the free encyclopedia, (2015), https://en.wikipedia.org/wiki/Compressed_natural_gas, Last accessed October, 09, 2016.

56. Alvi, M., Pakistan has highest number of CNG vehicles: Survey, Natural Gas Vehicle Statistics: NGV Count—Ranked Numerically as at December 2009, International Association for Natural Gas Vehicles (June 3, 2011), Retrieved April 27, 2010, Wikipedia, The free encyclopedia. https://en.www.wikipedia.org/wiki/natural_gas_vehicles, Last accessed October 5, 2016.

57. Market Studies Series, GNC (PDF), Consulate of the Argentinian Republic, Mumbai, India (in Spanish) (2009), Retrieved January 3, 2011.

58. What's the difference between CNG, LNG, LPG and Hydrogen? A report by Alternative Fuel Systems Inc., Calgary, Alberta, Canada, Retrieved May 15, 2015.

59. ISO standards–ISO/TC22/SC41, Road vehicles—Compressed natural gas (CNG) refuelling connector—Part 2: 20 MPa (200 bar) connector, size 2, Retrieved May 15, 2015, www.ISO.org/.../catalogue.tc.browse.htm.

60. ISO standards–ISO/TC22/SC41, Road vehicles—Compressed natural gas (CNG) fuel system components—Part 9: Pressure regulator, Retrieved May 15, 2015, www.ISO.org/.../catalogue.tc.browse.htm.

61. NGV Market Growth 2013 Analysis, Wuxi Banner Vessel, Retrieved May 15, 2015.

62. Fernandes, R., (August 20, 2008), Latin America NGVs: An update report, International Association of Natural Gas Vehicles (August 20, 2008), Archived from the original on November 20, 2008, Retrieved October 11, 2008.

63. Allen, R., New fuel cleans up: CNG: Compressed natural gas is rapidly gaining popularity with drivers, Surveys edition, *Financial Times*, 17 (May 11, 1999).

64. Current natural gas vehicle statics, a report by NGV global (November, 2013). www.iangv.org/current-ngv-stats/.

65. "Compressed natural gas fueling stations" a report by Alternative Fuels Data Center, Energy Efficiency and Renewable Energy, Department of Energy, Washington, DC (May 9, 2016).

66. Fuel gases—Heating values, a report by Naturalgas.org, Retrieved April 17, 2015.

67. Liquefied natural gas, Wikipedia, the free encyclopedia (2015), https://en.wikipedia.org/wiki/Liquefied_natural_gas.

68. World LNG Report—2014 edition, International Gas Union, News, views and knowledge on gas world-wide', A report, Total, Statoil, Fornebu, Norway (2014).

69. "Global LNG Report – Will new demand and new supply mean new pricing", Global natural gas and energy demand, a report by EYGM Ltd., London, U.K. (2013).

70. Liquefied natural gas: Understanding the basic facts, A report prepared by the U.S. Department of Energy (DOE) in collaboration with the National Association of Regulatory Utility, USDOE and National Energy Technology Laboratory, Washington, DC, DOE/FE-0489 (August 2005).

71. The Floating Liquefied Natural Gas (FLNG) Market 2011–2021, a Report by, Energy, Visiongain, ENE8974, https://www.visiongain.com/.../The-Floating..., Retrieved June 11, 2015.

72. Liquefied Natural Gas (LNG), a report by Naturalgas.org, Retrieved April 17, 2015.

73. Hrastar, J., *Liquid Natural Gas in the United States: A History*, 1st edn. McFarland & Company, Inc., Publishers, Jefferson, NC (2014).

74. Liquefied Natural Gas (LNG), a report by US Energy Information Administration, DOE, Washington (2015), https://www.eia.gov/.../index.cfm?...lng%20(liquefied%20natural%20gas).

75. Trends in the top five LNG—Importing nations as of 2009, a report by Naturalgas.org, Retrieved April 17, 2015.

76. The global liquefied natural gas market: Status and outlook, Appendix F, a report by Energy Information Administration (PDF), DOE, Washington, Retrieved April 17, 2015.

77. The global liquefied natural gas market: Status and outlook, a report by US Energy Information Administration, DOE, Washington (December 2003).

78. Global LNG Industry Review in 2014, a report by Naturalgas.org, Retrieved April 17, 2015.

79. High horse power off-road LNG vehicles in USA, Naturalgas.org, Retrieved April 17, 2015.

80. Frailey, M., Development of the high-pressure direct-injection ISX G Natural Gas Engine (PDF), a report by Energy Efficiency and renewable gas—Engine and vehicle research and development. Department of Energy, Washington, DC (August, 2004), DOE-102004-1940, Retrieved April 17, 2015.

81. Tosif, F., An innovative vision for LNG Fuel System for MD Diesel Dual Fuel Engine (DDF+LNG) (PDF), a report by Westport, Vancouver, Canada (September 7, 2014), Retrieved April 17, 2015.

82. Meyer Werft to build cruise ships powered by LNG, a report by Meyer Werft, Papenburg, Germany, Retrieved June 17, 2015.

83. Blomerus, P. and Oulene, P., LNG as a fuel for demanding high horsepower engine applications: Technology and approaches (PDF). Desport innovation inc., Vancover, Canada, Retrieved April 17, 2015.

84. Jensen, J., The outlook for global trade in liquefied natural gas projections to the year 2020, Prepared for: California Energy Commission, CEC-200-2007-017, (August 2007), Energy.ca.gov.

85. Shell's floating LNG plant, Natural gas.org, Retrieved April 17, 2015.

86. Shell's floating technology given green light, Naturalgas.org, Retrieved April 17, 2015.

87. Rankin, R. and Mick, M., Buried, subsea line advanced as LNG alternative a report in Oil and gas Journal (November 4, 2005), Retrieved June 22, 2012.

88. LNG: Benefits and risks of liquefied natural gas, a report by Lt. Governor.com (2013), Retrieved February 25, 2013.

89. Parfomak, P. and Flynn, A., Liquefied natural gas (LNG) import terminals: Siting, safety, and regulation (PDF), CRS report for Congress, 26 pages. Library of Congress, Washington, DC (January 28, 2004), Retrieved February 25, 2013.

90. Totten, G.E. (ed.), *Fuels and Lubricants Handbook: Technology, Properties, Performance, and Testing*, 2nd printing edn. ASTM International, West Conshohocken, PA (2003).

91. Analysis of seasonal mixtures—Propane-butane fuel mixture (Summer, Winter), a report by Unipetrol Prague, Czech Republic (2012), Retrieved April 29, 2013.

92. Liquefied petroleum gas specifications and test methods, a report by Gas Processors Association, Tulsa, OK (2012), Retrieved May 18, 2012.

93. Standard specification for liquefied petroleum (LP) gases, ASTM D1835-16, developed by subcommittee D02HO, American Society for Testing & Materials, ASTM International, West Conshohocken, PA (2016), doi:10.1520/01835-16.

94. Bauer, H. (ed.), *Automotive Handbook*, 4th edn. Robert Bosch GmbH, Stuttgart, Germany, pp. 238–239 (1996).

95. Qi, D., Bian, Y., Ma, Z., Zhang, C., and Liu, S., Combustion and exhaust emission characteristics of a compression ignition engine using liquefied petroleum gas–fuel-oil blended fuel, *Energy Conversion and Management*, 48 (2), 500 (2007).

96. Liquefied Petroleum gas, Wikipedia, the free encyclopedia (2015), https://en.wikipedia.org/wiki/Liquefied_petroleum_gas, Last accessed October, 27, 2016.

97. Coalbed methane, Wikipedia, the free encyclopedia (2015), https://en.wikipedia.org/wiki/Coalbed_methane.

98. Nuccio, V., Coalbed methane: Potential and concerns, A report from USGS, USGS Fact Sheet FS-123-00, Denver, CO, USGS, Washington, DC, (October 2000), pubs.usgs.gov/fs/fs123-00/fs123-00.pdf.

99. AKCEP (Alaska Center for Energy and Power), Coalbed methane, Alaska Energy Wiki (2012), Available online at: http://energy-alaska.wikidot.com/coalbed-methane, Accessed August 28, 2012.

100. Coal Bed Methane Primer, a report by ALL Consulting Inc., Tulsa, OK (2004), Available online at: http://www.all-llc.com/publicdownloads/CBMPRIMERFINAL.pdf.

101. ALL, Siting, design, construction, and reclamation guidebook for CBNG impoundments, EPA-HQ-OW-2008-0517, DCN 05399, Prepared for: U.S. Department of Energy, National Petroleum Technology Office (2006).

102. ARI, Technical/economic profiles of major producing coalbed methane basins in the U.S. Prepared for ARI for James Covington, EPA-HQ-OW-2008-0517, DCN 07347, EPA (2010).

103. EIA (U.S. Department of Energy, Energy Information Administration), US coalbed methane: Past, present, and future, Department of Energy, Washington, DC (2007), Available online at: http://www.eia.gov/oil_gas/rpd/cbmusa2.pdf.

104. EIA, Natural gas gross withdrawals from coalbed wells, 2002–2011 (2013), Available online at: http://www.eia.gov/dnav/ng/ng_prod_sum_a_epg0_fgc_mmcf_a.htm, Accessed March 3, 2013.

105. U.S. DOE EIA, Coalbed methane proved reserves and production (2010), Available online at: http://tonto.eia.doe.gov/dnav/ng/NG_ENR_CBM_A_EPG0_R52_BCF_A.htm.

106. U.S. DOE, Future supply and emerging resources–Coalbed natural gas, EPA-HQ-OW-2004-0032, DCN 03480 (2006), Available online at: http://www.netl.doe.gov/technologies/oil-gas/FutureSupply/CoalBedNG/CoalBed_NG.html, Accessed July 12, 2006.

107. U.S. DOE (Department of Energy), Future supply and emerging resources—Coalbed natural gas, EPAHQ-OW-2004-0032, DCN 03480 (2006).

108. Development of coal bed methane utilizing GIS technologies, Final report, DOE contract no. DE-FG26-01BC15334, prepared by ALL Consulting, Tulsa, OK (April, 2003).

109. Technical development document for the coalbed methane (CBM) extraction industry, EPA report no. EPA-820-R-13-009, U.S. Environmental Protection Agency Office of Water (4303T), Washington, DC (April 2013).

110. Economic analysis for existing and new projects in the coalbed methane industry, EPA 820-R-13-006, EPA, Washington, DC (July 29, 2013).

111. De Bruin, R.H., Lyman, R.M., Jones, R.W., and Cook, L.W., Coalbed methane in Wyoming information pamphlet 7 (revised), EPA-HQ-OW-2004-0032-1904, DCN 03070, Wyoming State Geological Survey, Laramie, WY (2001).

112. Stanford, J.A. and Hauer, F.R., Coalbed methane (CBM) in Montana: Problems and solutions, A white paper, EPA-HQ-OW-2008-0517, DCN 07229, by Flathead Lake Biological Station Division of Biological Sciences, University of Montana, Polson, MT (February 4, 2003).

113. 1999 Resource assessment of selected tertiary coal beds and zones in the Northern Rocky Mountains and Great Plains Region, Professional Paper 1625-A, USGS, Washington, DC, (1999), Available online at: http://pubs.usgs.gov/pp/p1625a/.

114. Preliminary gulf coast coalbed methane exploration maps: Depth to Wilcox, apparent Wilcox thickness and vitrinite reflectance, Report 2000-113 by USGS, Washington, DC (2000), Available online at: http://pubs.usgs.gov/of/2000/ofr-00-0113/downloads/OF00-113.pdf.

115. Handbook on coal bed methane produced water: Management and beneficial use alternatives, Prepared by ALL Consulting, Tulsa, OK EPA-HQ-OW-2004-0032-2483, DCN 03451, Prepared for Ground Water Protection Research (July 2003), Available online at: http://www.all-llc.com/publicdownloads/CBM_BU_Screen.pdf.

116. Rice, D.D., Coalbed methane—An untapped energy resource and an environmental concern, U.S. Geological Survey Fact Sheet FS-019-97, Department of Interior, Washington, DC (1997), Available online at: http://energy.usgs.gov/ factsheets/Coalbed/coalmeth.html.

117. Wheaton, J., Donato, T., Reddish, S., and Hammer, L., 2005 Annual coalbed methane regional ground-water monitoring report: Northern portion of the Powder River Basin, Open-File Report, MBMG 538, DCN 03474, Montana Bureau of Mines and Geology, a report by Montana Tech of the University of Montana, Butte, MT (2006). http://mbmg.mtech.edu/pdf-open-files/mbmg538-CBM-2005-Annual_Report.pdf, Last accessed December 15, 2016.

118. Horsley & Witten, Inc., Draft evaluation of impacts to underground sources of drinking water by hydraulic fracturing of coalbed methane reservoirs, EPA-HQ-OW-2004-0032-2543 (DCN 03489), Prepared for the U.S. Environmental Protection Agency (2001).

119. U.S. DOI, Assessment of contaminants associated with coal bed methane-produced water and its suitability for wetland creation or enhancement projects, Contaminant Report Number: R6/721C/05, EPA-HQ-OW-2004-0032-2422, U.S. Fish & Wildlife Service Region 6 (2005).

120. Apache Canada: New Brunswick exploration program, a report by Apache Corporation, New Brunswick, Canada (2010), www.apachecorp.com/.../ApacheCanada_ProjectOverview_2010-April.pdf.

121. Fisher, J.B., Environmental issues and challenges in coal bed methane production, EPA-HQ-OW-2008-0517, DCN 07229, prepared by Exponent, Inc., Tulsa, OK (2001), Available online at: http://ipec.utulsa.edu/Conf2001/fisher_92.pdf.

122. Wyoming CBM production, a report by WOGCC, Laramie, WY, also by EP, Washington, DC (2010), Available online at: http://wogcc.state.wy.us/StateCbmGraph.cfm, EPA-HQ-OW-2008-0517, DCN 07364; Yahoo Finance, Industry center—Major integrated oil and gas (2010), Available online at: http://biz.yahoo.com/ic/120_cl_all.html, EPA-HQ-OW-2008-0517, DCN 07365.

123. A guide to practical management of produced water from onshore oil and gas operations in the United States, a report prepared by Interstate Oil and gas compact commission and ALL Consulting, Tulsa, OK (Report No. DE-PS26-04NT15460-2), also EPA-HQ-OW-2008-0517, DCN 10000 (October 2006).

124. Arthur, J.D., Langhus, B., Epperly, D., Bohm, B., Richmond, T., and Halvorson, J., CBM in the Powder River Basin of Montana, Presented at the *Ground Water Protection Council's Annual Forum*, Reno, NV, EPA-HQ-OW-2008-0517, DCN 07229 (2001).

125. Buchanan, M.M., Soil water flow and irrigated soil water balance in response to Powder River Basin coalbed methane product water, MS thesis, Montana State University, Bozeman, MO, EPA-HQ-OW-2008-0517, DCN 07229 (2005).

126. BLM, Task 1B report for the Powder River Basin coal review current water resources conditions, Casper Field Office and Wyoming State Office, EPA-HQ-OW-2008-0517, DCN 07229, EPA, Washington, DC (2006).

127. Clearwater, S.J., Morris, B.A., and Meyer, J.S., A comparison of coalbed methane product quality versus surface water quality in the Powder River Basin of Wyoming, and an assessment of the use of standard aquatic toxicity testing organisms for evaluating the potential effects of coalbed methane product waters, EPA-HQ-OW-2008-0517, DCN 07229, Report submitted to Wyoming Department of Environmental Quality, University of Wyoming, Laramie, WY (cited in MacDonald, 2007) (2002).

128. Confluence Consulting, Inc., Powder River Biological survey and implications for coalbed methane development, EPA-HQ-OW-2008-0517, DCN 07229, Report prepared for Powder River Basin Resource Council, Sheridan, WY. 179pp. (cited in Confluence Consulting, 2004) (2004).

129. Davis, W.N., Bramblett, R.G., and Zale, A.V., Literature review and development of a study plan to assess the effects of coalbed natural gas activities on fish assemblages, EPA-HQ-OW-2008-0517, DCN 07229, Prepared for United States Department of the Interior Bureau of Land Management, Miles City Field Office (2006).

130. Davis, W.N., Effects of coalbed natural gas development on fish assemblages in tributary streams in the Powder River Basin, Montana and Wyoming, MS thesis, Montana State University, Bozeman, MO, EPA-HQ-OW-2008-0517, DCN 07229 (2008).

131. Doherty, M.K., Mosquito populations in the Powder River Basin, Wyoming: A comparison of natural, agricultural and effluent coal bed natural gas aquatic coalbed methane extraction: Detailed study report December 2010 5-3 (2007).

132. Forbes, M.B., Clearwater, S.J., and Meyer, J.S., Acute toxicity of coalbed methane product water and receiving waters to Fathead Minnows (*Pimephales promelas*) and Daphnia magna, EPA-HQ-OW-2008-0517, DCN 07229, Report submitted to Wyoming Department of Environmental Quality, University of Wyoming, Laramie, WY (cited in MacDonald et al., 2007) (2001).

133. Forbes, M.B., Morris, B.A., and Meyer, J.S., Acute and chronic toxicity of coalbed methane product waters and receiving waters to *Ceriodaphnia dubia*, EPA-HQ-OW-2008-0517, DCN 07229, Report submitted to Wyoming Department of Environmental Quality, University of Wyoming, Laramie, WY (cited in MacDonald et al., 2007) (2002).

134. Ganjegunte, G.K., Vance, G.F., and King, L.A., Soil chemical changes resulting from irrigation with water co-produced with coalbed natural gas, *Journal of Environmental Quality*, 34, 2217–2227 (2005), EPA-HQ-OW-2008-0517, DCN 07229.

135. Gore, J.A., Preliminary analysis of habitat loss for target biota in rivers impacted by long-term flow increases from CBM production, Coalbed methane extraction: Detailed study report December 2010 5-4, EPA-HQ-OW-2006-0771-1336, EPA-HQ-OW-2008-0517, DCN 07229 (2002).

136. Horpestad, A., Skaar, D., and Dawson, H., Water quality technical report: Water quality impacts from coal bed methane development in the Powder River Basin, Wyoming and Montana (cited in Todd, 2006), EPA-HQ-OW-2008-0517, DCN 07229 (2001).

137. Jackson, R. and Reddy, K.J., Trace element chemistry of coal bed natural gas produced water in the Powder River Basin, Wyoming, EPA-HQ-OW-2006-0771-1005 (2007).

138. Klarich, D.A. and Regele, S.M., Structure, general characteristics, and salinity relationships of benthic macroinvertebrate associations in streams draining the Southern Fort Union Coalfield Region of Southeastern Montana, USGS Grant No. 14-08-0001-G-503 final report, EPA-HQ-OW-2008-0517, DCN 07229, Water Quality Bureau, Montana Department of Health and Environmental Sciences, Billings, MO, 148pp. (cited in Regele and Stark, 2000) (1980).

139. McBeth, I.H. et al., Coalbed methane product water chemistry in three Wyoming watersheds, *Journal of the American Water Resources Association*, 39 (3), 575–585 (2003), doi:10.1111/j.1752-1688.2003.tb03676.x, also published in EPA Report, EPA-HQ-OW-2003-0074, DCN 00491, EPA, Washington (2003).

140. O'Neil, P.E., Harris, S.C., Mettee, M.F., Shepard, T.E., and McGregor, S.W., Surface discharge of wastewaters from the production of methane from coal seams in Alabama, EPA-HQ-OW-2008-0517, DCN 07229, The Cedar Cove Model, Geological Survey of Alabama, Bulletin 155, Tuscaloosa, AL (1993).

141. Osborne, T. and Adams, J., Coalbed natural gas and water management in the Powder River Basin, EPA-HQ-OW-2008-0517, DCN 07229, Southwest Hydrology (November/December 2005).

142. Ramirez, P., Assessment of contaminants associated with coal bed methane-produced water and its suitability for wetland creation or enhancement projects, Contaminant Report Number: R6/721C/05, EPA-HQ-OW-2006-0771-0928, U.S. Fish and Wildlife Service, Region 6 (2005).

143. Rice, C.A., Bartos, T.T., and Ellis, M.S., Chemical and isotopic composition of water in the Fort Union and Wasatch Formation of the Powder River Basin, Wyoming and Montana: Implications for coalbed methane development, in Schwochow, S.D. and Nuccio, V.F. (eds.), Coalbed Methane of North America, I, Rocky Mountain Association of Geologists, EPA-HQ-OW-2008-0517, DCN 07229, pp. 53–70, Washington, DC (cited in Kirkpatrick, 2005) (2002).

144. Shepard, T.E., O'Neil, P.E., Harris, S.C., and McGregor, S.W., Effects of Coalbed natural gas development on the water-quality and fish and benthic invertebrate communities of the big sandy creek drainage system, Alabama, EPA-HQ-OW-2008-0517, DCN 07229, Geological Survey of Alabama Circular 171, Tuscaloosa, AL (cited in Davis et al., 2006) (1993).

145. Skaar, D., Morris, B., and Farag, A., National pollution discharge elimination system: Toxicity of the major salt (sodium bicarbonate) from coalbed natural gas production on fish in the tongue and powder river drainages in Montana, EPA-HQ-OW-2008-0517, DCN 07229, Progress report prepared for U.S. Environmental Protection Agency, Helena (cited in Davis et al., 2006) (2004).

146. Skaar, D., Farag, A., and Harper, D., National pollution discharge elimination system, Toxicity of the major salt (sodium bicarbonate) from coalbed methane production to fish in the tongue and powder river drainages in Montana, EPA-HQ-OW-2008-0517, DCN 07229, Prepared for the U.S. Environmental Protection Agency, Washington, DC (2005).

147. Todd, A.L., Forage quality characteristics of barley irrigated with coalbed methane discharge water, MS thesis, Montana State University, Bozeman, MO, EPA-HQ-OW-2008-0517, DCN 07229, Coalbed methane extraction: Detailed study report December 2010 5-7 (2006).

148. USGS, Water produced with coal-bed methane, USGS, Fact Sheet FS-156-00 (November), DCN 03487, Coalbed methane extraction: Detailed study report December 2010 5-8 (2000).

149. USGS, Coalbed methane extraction and soil suitability concerns in the Powder River Basin, Montana and Wyoming, Fact Sheet 2006-3137 (November), EPA-HQ-OW-2008-0517, DCN 07229 (2006).

150. USGS, Toxicity of sodium bicarbonate to fish from coal-bed natural gas production in the tongue and powder river drainages, Montana and Wyoming, EPA-HQ-OW-2006-0771-0943 (2006).

151. U.S. EPA, Evaluation of impacts to underground sources of drinking water by hydraulic fracturing of coalbed methane reservoirs, Attachment 11, Office of Groundwater and Drinking Water a report by EPA, Washington, EPA 816-R-04-003 (2004), Available online at: http://water.epa.gov/type/groundwater/uic/class2/hydraulicfracturing/wells_coalbedmethanestudy.cfm; www.epa.gov/safewater.

152. Veil, J.A., Puder, M.G., Elcock, D., and Redweik, R.J. Jr., A white paper describing produced water from production of crude oil, natural gas, and coal bed methane, EPA-HQ-OW-2008-0517, DCN 07229, Prepared for the U.S. Department of Energy (January 2004).

153. Vickers, D.T., Disposal practices for waste waters from coalbed methane extraction in the Black Warrior Basin, Alabama, in *Proceedings of the First International Symposium on Oil and Gas Exploration and Production Waste Management Practices*, New Orleans, LA (September 10–13, 1990), Natural Service Center for Environmental Publications, EPA-HQ-OW-2008-0517, DCN 07229, EPA, Washington, DC (1990).

154. Wang, X., Assefa, M.M., McClain, M.E., and Yang, W., Water quality changes as a result of coalbed methane development in a Rocky Mountain watershed, *Journal of the American Water Resources Association*, 43 (6), 1383–1399 (17) (2007), EPA-HQ-OW-2008-0517, DCN 07229.

155. *Handbook on Coal Bed Methane Produced Water: Management and Beneficial Use Alternatives.* Prepared by ALL Consulting, Tulsa, OK for Ground Water Protection Research Foundation U.S. Department of Energy National Petroleum Technology Office Bureau of Land Management, Washington, DC (July 2003).

156. Staub, J., The growth of U.S. natural gas: An uncertain outlook for U.S. and world supply, in *2015 EIA Energy Conference*, June 15, 2015, DOE, Washington, DC (2015), www.eia.gov/conference/2015/pdf/presentations/staub.pdf.

157. Sieminski, A., Outlook for North American natural gas, in *LDC Natural Gas Forum*, November 11, 2014, Toronto, Ontario, Canada, EIA, DOE, Washington, DC (2014), www.eia.gov/pressroom/presentations/sieminski_01042014.pdf.

158. Hart, B.S., Engler, T., Pearson, R., and Robinson, R.L., 3-D seismic horizon-based approaches to fracture-swarm sweet spot, definition in Tight Gas Reservoirs, *The leading Edge*, 21, 28–35 (2002).

159. Dutton, S. et al., Major low-permeability sandstone gas reservoirs in the Continental United States, Bureau of Economic Geology, University of Texas, Austin, TX (1993).

160. Kaush, R., Hume, D., and Barson, D., Hydrodynamic model for biogenic gas trapped in low permeability sands—Western Plains of North America, AAPG: Tight Gas Reservoirs (2002).

161. Law, B.E., Basin-centered gas systems, *AAPG Bulletin*, 86, 1891–1919 (2002).

162. Law, B.E., A review of basin-centered gas systems with a focus on the Greater Green River basin: Rocky Mountain, in *Association of Geologists and Rocky Mountain Region of Petroleum Technology Transfer Council Petroleum Systems and Reservoirs of Southwest Wyoming Symposium*, Denver, CO (September 19, 2003).

163. Law, B.E. and Curtis, J.B., Introduction to unconventional petroleum systems, *AAPG Bulletin*, 86, 1851–1852 (2002).

164. Masters, J.A., Deep basin gas trap, Western Canada, *AAPG Bulletin*, 63 (2), 152–181 (1979).

165. Shanley, K.W., Cluff, R.M., and Robinson, J.W., Factors controlling prolific gas production from low-permeability sandstone reservoirs: Implications for resource assessment, prospect development, and risk analysis, *AAPG Bulletin*, 88 (8), 1083–1122 (2004).

166. Surdam, R.C., Wyoming and the 21st century: The age of natural gas, in *Proceedings of a Workshop on the Future of Natural Gas in Wyoming*, Institute for Energy Research, Laramie, Wyoming (1995).

167. Tight gas, Wikipedia, the free encyclopedia (2015), https://en.wikipedia.org/wiki/Tight_gas.

168. Tight gas reservoirs, Petowiki, SPE, Washington, DC (2013), petrowiki.org/Tight_gas_reservoirs.

169. Gunawan, H., Susanto, H., Widyantoro, B., and Noguera, J.A., Fracture assisted sandstone acidizing, alternative approach to increase production in tight sandstone reservoir, in *SPE Oil and Gas India Conference and Exhibition*, March 28–30, Mumbai, India, SPE-154947-MS (2012).

170. Caldwell, C. et al., 2008 State of the market report, Federal Energy Regulatory Commission, Washington, DC (August 2009).

171. Shale Gas, Wikipedia, the free encyclopedia (2015).

172. Shale Gas in United States, Wikipedia, the free encyclopedia (2015).

173. GASH-Gas Shales in Europe, European Gas Shale Database, TNO, Utrecht, the Netherlands (2010), https://www.energydelta.org/.../c99d7dab-69fa-425f-9088-34f12d40384f.

174. Stevens, P., The 'Shale Gas Revolution': Developments and changes, a publication Chatham House, The Royal Institute of International Affairs, London, U.K. (August 2012), Retrieved August 15, 2012, https://www.chathamhouse.org/publications.

175. Staff, World shale gas resources: An initial assessment of 14 regions outside the United States, US Energy Information Administration, Analysis and Projections, Department of Energy, Washington, SC (April 5, 2011), Retrieved August 26, 2012.

176. Statement on US China shale gas resource initiative, arrangement by White House, Washington, DC (November 17, 2009), America.gov, Retrieved August 6, 2013, LLP digital. usembassy.gov/st/english/2009/...

177. Carey, J.M., Surprise side effect of shale gas boom: A plunge in U.S. greenhouse gas emissions, a guest article in *Forbes Magazine*, New York (December 7, 2012), Retrieved February 21, 2013.

178. Howarth, R., Sontaro, R., and Ingraffea, A., Methane and the greenhouse-gas footprint of natural gas from shale formations (PDF) (November 12, 2010), Springerlink.com, Retrieved March 13, 2011.

179. Burnham, A. et al., Life-cycle greenhouse gas emissions of shale gas, natural gas, coal, and petroleum, *Environmental Science and Technology*, 46 (2), 619–627 (January 17, 2012).

180. Logan, J., Heath, G., Macknick, J., Paranhos, E., Boyd, W., and Carlson, K., Natural gas and the transformation of the U.S. energy sector: Electricity, Technical report NREL/TP-6A50-55538, Golden, CO (November 2012).

181. Wang, Z. and Krupnick, A., A retrospective review of shale gas development in the United States, Resources for the Futures, discussion paper, Washington, DC (April 2013).

182. International Energy Agency (IEA), World energy outlook special report on unconventional gas: Golden rules for a golden age of gas? (PDF), Retrieved August 6, 2013.

183. Jarvie, D., Worldwide shale resource plays, PDF file, *NAPE Forum* (August 26, 2008), also presented at AAPK European Regional Conference, Kiev, Ukraine (2010).

184. Modern shale gas development in the United States, A Primer Prepared by Ground Water Protection Council Oklahoma City, ALL Consulting, Work Performed Under DE-FG26-04NT15455, National Energy Technology Laboratory, Pittsburgh, PA, US Department of Energy, Washington, DC (April 2009).

185. North America leads the world in production of shale gas, US Energy Information Administration, Department of Energy, Washington, DC (October 23, 2013), https://www.eia.gov/todayinenergy/detail.cfm?id=13491.

186. Technically recoverable shale oil and shale gas resources: An assessment of 137 shale formations in 41 countries outside the United States, Analysis and projections, United States Energy Information Administration, Department of Energy, Washington, DC (June 13, 2013).

187. World Shale Gas Resources, US Energy Information Administration, Department of Energy, Washington, DC (April 2011).

188. Jiang, M. et al., Life cycle greenhouse gas emissions of Marcellus shale gas, *Environmental Research Letters* 6 (4), 034014 (9pp), IOP Publishers, Bristol, U.K. (2011), doi:10.1088/1748-9326/6/3/034014.

189. Hultman, N. et al., The greenhouse impact of unconventional gas for electricity generation, *Environmental Research Letters* 6 (4), 049504 (1pp), IOP Publishers, Bristol, U.K. (2011), doi:10.1088/1748-9326/6/4/044008.

190. Brino, A., Shale gas fracking without water and chemicals, Dailyyonder.com (2013), Retrieved August 6, 2013.

191. Kerr, R.A., Study: High-tech gas drilling is fouling drinking water, *Science Now*, 332, 775 (May 13, 2011).

192. The future of natural gas: An interdisciplinary MIT study (PDF), a report by MIT Energy Initiative, MIT, Cambridge, MA, 7, 8 (2011), Retrieved July 29, 2011.

193. Warner, N.R., Christie, C.A., Jackson, R.B., and Avner, V., Impacts of shale gas wastewater disposal on water quality in Western Pennsylvania, *Environmental Science and Technology* (American Chemical Society) 47 (20), 11849–11857 (October 2, 2013).

194. US Environmental Protection Agency, Natural gas drilling in the Marcellus shale, NPDES program frequently asked questions, Attachment to memorandum from James Hanlon, Director of EPA's Office of Wastewater Management to the EPA Regions (March 16, 2011).

195. Jenkins, J., Friday energy facts: How much water does fracking for shale consume?, a report in *The Energy Collective* (April 6, 2013), www.theenergycollective.com/jessejenkins/.../fiday-energy-facts.

196. Mason, C., Muehlenbachs, L., and Olmstend, S., "The economics of shale gas production", discussion paper by Resources for the Future, RFF DP 14-42-REV, Washington, DC (2015), http://www.rff.org.

197. The shale gas market Report 2014–2024, a report by Visiongain (April 9, 2014), Product Code: ENE0083, https://www.visiongain.com/Report/.../Shale-Gas-Market-2014-2024.

198. Kuuskraa, V.A., EIA/ARI world shale gas and shale oil resource assessment, in *EIA 2013 Energy Conference*, U.S. Department of Energy, U.S. Energy Information Administration, Washington, DC (June 17, 2013).

199. World shale resource assessment—Analysis and projections, US Energy Information Administration, Department of Energy, Washington, DC (September 24, 2015), https://www.eia.gov/analysis/studies/worldshalegas/.

200. O'Sullivan, F. and Paltsev, S., Shale gas production: Potential versus actual GHG emissions, MIT joint program on the science and policy of global change report no. 234, MIT, Cambridge, U.K. (November 2012).

201. Cathles, L. II, Brown, L., Taam, M., and Hunter, A., A commentary on "The greenhouse—Gas footprint of natural gas in shale formations" by Howarth, R., Santoro, R. and Ingraffea, A., *Climate Change*, 113, 525–535 (2012), doi: 10.1007/s10584-011-0333-0.

202. Dyman, T.S. and Schmoker, J.W., Comparison of national oil and gas assessments, Open-File Report 97-445, U.S. Geological Survey, Department of Interior, Washington, DC, 30pp. (1996).

203. Dyman, T.S., Schmoker, J.W., and Root, D.H., Assessment of deep conventional and continuous-type (unconventional) natural gas plays in the U.S., Open-File Report 96-529, U.S. Geological Survey, Department of Interior, Washington, DC, 22pp. (1996).

204. Dyman, T.S., Rice, D.D., and Westcott, P.A. (eds.), Geologic controls of deep natural gas resources in the United States, U.S. Geological Survey Bulletin 2146, Department of Interior, Washington, DC, 239pp. (1997).

205. Dyman, T.S. and Cook, T.A., Summary of deep oil and gas wells in the U.S. through 1998, in Dyman, T.S. and Kuuskraa, V.A. (eds.), Studies of deep natural gas, U.S. Geological Survey Digital Data Series 67, one CD-ROM, USGS, Washington, DC (2001).

206. Dyman, T.S. and Kuuskraa, V.A. (eds.), Studies of deep natural gas, U.S. Geological Survey Digital Data Series 67, one CD-ROM, Department of Interior, Washington, DC (2001).

207. Dyman, T.S., Wyman, W.E., Kuuskraa, V.A., Lewan, M.D., and Cook, T.A., Deep and ultra-deep natural gas resources hold big-time reserves, *American Oil and Gas Reporter*, 45 (6), 35–47 (2002).

208. Dyman, T.S., Crovelli, R.A., Bartberger, C.E., and Takahashi, K.I., Worldwide estimates of deep natural gas resources based on the U.S. Geological Survey World Petroleum Assessment 2000, *Natural Resources Research*, 11 (3), 207–218 (2002).

209. Price, L.C., Origins, characteristics, evidence for and economic viabilities of conventional and unconventional gas resource bases, in Dyman, T.S., Rice, D.D., and Westcott, P.A. (eds.), Geologic controls of deep natural gas resources in the United States, U.S. Geological Survey Bulletin 2146, Department of Interior, Washington, DC, pp. 181–207 (1997).

210. Dyman, T.S. and Cook, T.A., Summary of deep oil and gas wells in the United States through 1998, Chapter B, in Dyman, T.S. and Kuuskraa, B.N. (eds.), Geologic studies of deep natural gas resources, U.S. Geological Survey Digital Data Series DDS-67, Version 1.00, Department of Interior, Washington, DC, pp. B1–B9 (2001).

211. Dyman, T.S. et al., Worldwide estimates of deep natural gas resources based on the US Geological Survey World Petroleum Assessment 2000, *Natural Resource Research*, 11 (6), 207–218 (2002).

212. Dyman, T.S., Litinsky, V.A., and Ulmishek, G.F., Geology and natural gas potential of deep sedimentary basins in the Former Soviet Union, Chapter C, in Dyman, T.S. and Kuuskraa, V.A. (eds.), U.S. Geological Survey studies of deep natural gas resources, U.S. Geological Survey Digital Data Series DDS-67, Version 1.00, Department of Interior, Washington, DC, pp. C1–C29 (2001).

213. Dyman, T.S. et al., Geologic and production characteristics of deep natural gas resources based on data from significant fields and reservoirs, Chapter C, in Dyman, T.S., Rice, D.D., and Westcott, W.A. (eds.), Geologic controls of deep natural gas resources in the United States, U.S. Geological Survey Bulletin 2146-C, Department of Interior, Washington, DC, pp. 19–38 (1997).

214. Spencer, C.W. and Wandrey, C.J., Initial potential test data from deep wells in the United States, in Dyman, T.S., Rice, D.D., and Westcott, W.A. (eds.), Geologic controls of deep natural gas resources in the United States, U.S. Geological Survey Bulletin 2146, Department of Interior, Washington, DC, pp. 63–69 (1997).

215. Cochener, J.C. and Brandenburg, C., Expanding the role of onshore deep gas, *Gas TIPS* (Gas Research Institute), 4 (3), 4–8 (1998).

216. Cochener, J.C. and Hill, D.G., Deep for gas, *American Oil and Gas Reporter*, 44 (3), 69–77 (2001).

217. Crovelli, R.A., A probabilistic method for subdividing resources into depth slices, in Dyman, T.S. and Kuuskraa, V.A. (eds.), Geologic studies of deep natural gas resources, U.S. Geological Survey Digital Data Series 67, Department of Interior, Washington, DC, pp. F1–F3 (2001).

218. Erskine, R.D., String of drilling successes puts El Paso out front in deep gas, *American Oil and Gas Reporter*, 44 (12), 44–54 (2001).

219. Gas Research Institute, GRI baseline projection of U.S. energy supply and demand, 2000 edn., Gas Research Institute, Topical Reporter 00/0005, Chicago, IL, 104pp. (2000).

220. Gautier, D.L., Dolton, G.L., Takahashi, K.I., and Varnes, K.L. (eds.), 1995 National Assessment of United States Oil and Gas Resources—Results, methodology, and supporting data, U.S. Geological Survey Digital Data Series DDS-30, Release 2, one CD-ROM, Department of Interior, Washington, DC (1996).

221. Hill, D.G., Contribution of unconventional gas to U.S. supply continues to grow, *Gas TIPS* (Gas Research Institute), 7 (3), 4–8 (2002).

222. IHS Energy Group, PI-Dwights WHCS (through October 1998), Available from IHS Energy Group, Denver, CO (1998).

223. Masters, J.A. (ed.), Elmworth—Case study of a deep basin gas field, *American Association of Petroleum Memoirs*, 38, 189–203 (1984).

224. McLennan, J., Carden, R.S., Curry, D., Stone, R.C., and Wyman, R.E., Underbalanced drilling manual: Gas research, Institute report 97/0236 (unpaginated), Chicago, IL, Available from Society of Petroleum Engineers (1997).

225. National Petroleum Council, The potential for natural gas in the United States, Vols. I and II, National Petroleum Council, Washington, DC, 520pp. (combined) (1992).

226. National Petroleum Council, Meeting the challenges of the Nations growing natural gas demand, Vol. 1, Summary report, National Petroleum Council, Washington, DC, 53pp. with Appendices (1999).

227. Oil and Gas Journal, Advances, needs highlighted for deep U.S. gas drilling, *Oil and Gas Journal*, 96 (44), 82–84 (1998).

228. Perry, W.J. Jr., Structural settings of deep natural gas accumulations in the conterminous United States, in Dyman, T.S., Rice, D.D., and Westcott, P.A. (eds.), Geologic controls of deep natural gas resources in the United States, U.S. Geological Survey Bulletin 2146, Department of Interior, Washington, DC, pp. 41–46 (1997).

229. Potential Gas Committee, Potential supply of natural gas in the United States—Report of the Potential Gas Committee (December 31, 2000), Potential Gas Agency, Colorado School of Mines, Golden, CO, 346pp. (2001).

230. Rothwell, N.R. et al., Gyda: Recovery of difficult reserves by flexible development and conventional reservoir management, in *SPE26778, Proceedings of the Offshore Europe Conference*, pp. 271–280 (1993).

231. Cluff, R.M. and Cluff, S.G., The origin of Jonah field, Northern Green River Basin, Wyoming, in Robinson, J.W. and Shanley, K.W. (eds.), Jonah field: Case Study of a Tight-Gas Fluvial Reservoir, Vol. 52, AAPG Studies in Geology, Ch. 8, pp. 127–145 (2004), http://www.discovery-group.com/pdfs/Cluff,%20Cluff,%202004-AAPG%20Jonah%20origins.pdf.

232. Wyman, R.E., Challenges of ultradeep drilling, in Howell, D.G. (ed.), The future of energy gases, U.S. Geological Survey Professional Paper 1570, Department of Interior, Washington, DC, pp. 205–216 (1993).

233. Barker, C. and Takach, N.E., Prediction of natural gas composition in ultradeep sandstone reservoirs, *AAPG Bulletin*, 76(12), 1859–1873 (1992).

234. Williams, L.B. et al., The influence of organic matter on the boron isotope geochemistry of the gulf coast sedimentary basin, USA, *Chemical Geology*, 174, 445–461 (2001).

235. Andersen, E.E., Maurer, W.C., Hood, M., Cooper, G., and Cook, N., Deep drilling basic research—Deep drilling activity, Report 90/0265.2, Gas Research Institute, Chicago, IL, 78pp. (1990).

236. Behar, F., Vandenbroucke, M., Tang, Y., Marquis, F., and Espitalie, J., Thermal cracking of kerogen in open and closed systems: Determination of kinetic parameters and stoichiometric coefficients for oil and gas generation, *Organic Geochemistry*, 26 (5–6), 321–339 (1997).

237. Davis, T.B., Subsurface pressure profiles in gas-saturated basins, in Masters, J.A. (ed.), Elmworth—Case study of a deep basin gas field, *American Association Petroleum Geologists Memoir*, 38, 189–203 (1984).

238. Horsfield, B., Schenk, H.J., Mills, N., and Welte, D.H., An investigation of the in-reservoir conversion of oil to gas—Compositional and kinetic findings from closed-system programmed-temperature pyrolysis, *Advances in Organic Geochemistry*, 19 (1–3), 191–204 (1992).

239. Knauss, K.G., Copenhaver, S.A., Braun, R.L., and Burnham, A.K., Hydrous pyrolysis of New Albany and Phosphoria shales—Production kinetics of carboxylic acids and light hydrocarbons and interactions between the inorganic and organic chemical systems, *Organic Geochemistry*, 27 (7–8), 477–496 (1997).

240. Pepper, A.S. and Corvi, P.J., Simple kinetic models of petroleum formation—Part I, Oil and gas generation from kerogen, *Marine and Petroleum Geology*, 12 (3), 291–319 (1995).

241. Pepper, A.S. and Dodd, T.A., Simple kinetic models of petroleum formation—Part II, Oil-gas cracking, *Marine and Petroleum Geology*, 12 (3), 321–340 (1995).

242. Reeves, S.R., Kuuskraa, J.A., and Kuuskraa, V.A., Deep gas poses opportunities, challenges to U.S. operators, *Oil and Gas Journal*, 96 (18), 133–140 (1998).

243. Shirley, K., Independents lead charge into deep gas frontiers, *American Oil and Gas Reporter*, 43 (5), 61–71 (2000).

244. Shirley, K., Making hay with deep gas, *American Oil and Gas Reporter*, 44 (6), 57–67 (2001).

245. Pang, X.-Q., Jia, C.-Z., and Wang, W.-Y., Petroleum geology features and research developments of hydrocarbon accumulation in deep petroliferous basins, *Petroleum Science*, 12, 1–53 (2015).

246. Barker, C., Calculated volume and pressure changes during the thermal cracking of oil to gas in reservoirs, *AAPG Bulletin*, 74, 1254–1261 (1990).

247. Burruss, R.C., Stability and flux of methane in the deep crust a review, in The future of energy gases, US Geological Survey Professional Paper, USGS, Washington, DC, pp. 21–29 (1993).

248. Davies, G.R. and Smith, L.B., Structurally controlled hydrothermal dolomite reservoir facies: An overview, *AAPG Bulletin*, 90 (11), 1641–1690 (2006).

249. Domine, F., Dessort, D., and Brevart, O., Towards a new method of geochemical kinetic modelling: Implications for the stability of crude oils, *Organic Geochemistry*, 28, 576–612 (1998).

250. Durand, B., Understanding of HC migration in sedimentary basins (present state of knowledge), *Organic Geochemistry*, 13(1–3), 445–459 (1988).

251. Hirner, A., Graf, W., and Hahn-Weinheimer, P., A contribution to geochemical correlation between crude oils and potential source rocks in the eastern Molasse Basin (Southern Germany), *Journal of Geochemical Exploration*, 15(1–3), 663–670 (1981).

252. Jemison, R.M., Geology and development of Mills Ranch complex—World's deepest field, *AAPG Bulletin*, 63 (5), 804–809 (1979).

253. Lin, C.S., Li, H., and Liu, J.Y., Major unconformities, tectonostratigraphic framework, and evolution of the superimposed Tarim Basin, northwest China, *Journal of Earth Science*, 23(4), 395–407 (2012).

254. Magara, K., *Compaction and Fluid Migration: Practical Petroleum Geology*. Elsevier, London, U.K., p. 319 (1978).

255. Mango, F.D., The stability of hydrocarbon under the time/temperature conditions of petroleum genesis, *Nature*, 352, 146–148 (1991).

256. Masters, J.A., Deep basin gas trap, Western Canada, *AAPG Bulletin*, 63 (2), 152–181 (1979).

257. Moretti, I. et al., Compartmentalization of fluid migration pathways in the sub-Andean zone, Bolivia, *Tectonophysics*, 348 (1–3), 5–24 (2002).

258. Mukhopadhyay, P.K., Wade, J.A., and Kruge, M.A., Organic facies and maturation of Jurassic/Cretaceous rocks, and possible oil-source rock correlation based on pyrolysis of asphaltenes, Scotian Basin, Canada, *Organic Geochemistry*, 22 (1), 85–104 (1995).

259. Nelson, R.A., *Geologic Analysis of Naturally Fractured Reservoirs*. Gulf Publishing Company, Houston, TX, pp. 304–351 (1985).

260. Odden, W., Patience, R.L., and Van Graas, G.W., Application of light hydrocarbons (C4-C13) to oil/source rock correlations: A study of the light hydrocarbon compositions of source rocks and test fluids from offshore Mid-Norway, *Organic Geochemistry*, 28 (12), 823–847 (1998).

261. Pape, H. et al., Anhydrite cementation and compaction in geothermal reservoirs: Interaction of pore-space structure with flow, transport, P-T conditions, and chemical reactions, *International Journal of Rock Mechanics and Mining Sciences*, 42 (7–8), 1056–1069 (2005).

262. Seifert, W.K., Steranes and terpanes in kerogen pyrolysis for correlation of oils and source rocks, *Geochimica et Cosmochimica Acta*, 42 (5), 473–484 (1978).

263. Spencer, C.W., Review of characteristics of low-permeability gas reservoirs in western United States, *AAPG Bulletin*, 73 (5), 613–629 (1989).

264. Surdam, R.C., Boese, S.W., and Crossey, L.J., The chemistry of secondary porosity, *AAPG Memoirs*, 37, 127–149 (1984).

265. Walker, R.G. and James, N.P., Facies models: Response to sea level change, *Geological Association of Canada*, 4 (5), 153–170 (1992).

266. Waples, D.W., Time and temperature in petroleum formation: Application and detachment free deformation, *Journal of Structural Geology*, 12 (3), 355–381 (1990).

267. Henry, A.A. and Lewan, M.D., Comparison of kinetic-model predictions of deep gas generation, Chapter D, in Dyman, T.S. and Kuuskraa, V.A. (eds.), Geologic studies of deep natural gas resources, U.S. Geological Survey Digital Data Series, USGS, Department of Interior, Washington, DC, 67 (2001).

268. Lewan, M.D. and Henry, A.A., Gas: Oil ratios for source rocks containing type-I, -II, -IIS, and -III kerogens as determined by hydrous pyrolysis, Chapter E, Geologic studies of deep natural gas resources, in Dyman, T.S. and Kuuskraa, V.A. (eds.), U.S. Geological Survey Digital Data Series, USGS, Washington, DC 67 (2001).

269. Bebout, D.G., Geopressured geothermal fairway evaluation and test-well site location, Frio Formation, Texas Gulf Coast, in Meriwether, J. (ed.), *Third Geopressured-Geothermal Energy Conference*, November 16–18, 1977, Lafayette, LA, Vol. 1, Center for Energy Studies, University of Southwestern Louisiana, Lafayette, LA, pp. GI-251–GI-313 (1977).

270. Garg, S.K., Pritchett, J.W., Rice, M.H., and Riney, T.D., U.S. Gulf Coast geopressured-geothermal reservoir simulation: SSS-R-77-3147, Prepared by Systems, Science and Software for the University of Texas at Austin, University of Texas, Austin, TX, 112pp. (1977) (U.S. Energy Research and Development Administration contract E 40-1-5400).

271. Hankies, B.E. and Karkalits, O.C., Geopressured-geothermal test of the Edna Delcambre No. 1 Well, Tigre Lagoon Field, Vermilion Parish, Louisiana: Analysis of water and dissolved natural gas. Final report submitted to Department of Energy, Division of Geothermal Energy, Contract No. EY-76-S-05937, McNeese State University, Lake Charles, LA, 144pp. Available from: NTIS, Springfield, VA, ORO-4937-T1 (1978).

272. Loucks, R.G. and Moseley, M.C., Factors controlling geopressured geothermal reservoir quality-Frio sandstone facies, Texas Gulf Coast, in Meriwether, J. (ed.), *Third Geopressured-Geothermal Energy Conference*, November 16–18, 1977, Lafayette, LA, Vol. 1, Center for Energy Studies, University of Southwestern Louisiana, Lafayette, LA, pp. GI-315–GI-349.29 (1977).

273. Papadopulos, S.S., Wallace, R.H. Jr., Wesselman, J.B., and Taylor, R.E., Assessment of onshore geopressured geothermal resources in the Northern Gulf of Mexico Basin, in White, D.E. and Williams, D.L. (eds.), Assessment of geothermal resources of the United States—1975, U.S. Geological Survey Circular 726, U.S. Geological Survey, Reston, VA, USGS, Washington, DC, pp. 125–146 (1975).

274. Bassiouni, Z., Evaluation of potential geopressure geothermal test sites in Southern Louisiana, Progress Report, DOE Contract No. DE-AS05-76ET28465 (April 1980).

275. Bebout, D.G. and Gutierrez, D.R., Geopressured geothermal resource in Texas and Louisiana— Geological constraints, in Bebout, D.G. and Bachman, A.L. (eds.), *Proceedings of the Fifth Geopressured-Geothermal Energy Conference*, Baton Rouge, LA (October 1981).

276. Wallace, R.H. et al., Assessment of geopressured-geothermal resources in the Northern Gulf of Mexico Basin, in White, D.E. and Williams, D.L. (eds.), Assessment of geothermal resources of the United States—1978, U.S. Geological Survey Circular, USGS, Washington, DC, 790, pp. 132–155 (1979).

277. Jones, P.H., Natural gas resources of the geopressured zones in the Northern Gulf of Mexico Basin, in *Natural Gas from Unconventional Geologic Sources*, National Research Council, Board of Mineral Resources Commission on Natural Resources, National Academy of Sciences, Washington, DC, pp. 17–33 (1976).

278. Brown, W.M., 100,000 quads of natural gas?, Research memorandum #31, Report HI-2415/2-P, Hudson Institute, Inc., Croton-on-Hudson, New York (July 1976).

279. Hise, B.R., Natural gas from geopressured aquifers, in *Natural Gas from Unconventional Geologic Sources*. National Research Council, Board of Mineral Resources, Commission on Natural Resources, National Academy Press, Washington, DC, pp. 41–63 (1976).

280. Dorfman, M.H., Potential reserves of natural gas in the United States Gulf coast geopressured zones, in *Natural Gas from Unconventional Sources*, National Research Council, Board of Mineral Resources, Commission on Natural Resources, National Academy Press, Washington, DC, pp. 34–40 (1976).

281. Swanson, R.K. and Osaba, J.S., Production behavior and economic assessment of geopressured reservoirs in Texas and Louisiana Gulf Coast, in *Proceedings of the Third Geothermal Conference and Workshop*, EPR, Report, Palo Alto, CA, EPRI-WS-79-166 (1980).

282. Gustavson, T.C., McGraw, M.M., Tandy, M., Parker, F., and Wohlschlag, D.E., Potential environmental impacts arising from geopressured-geothermal energy development, Texas-Louisiana Gulf Coast Region, in Meriwether, J. (ed.), *Third Geopressured-Geothermal Energy Conference*, November 16–18, 1977, Vol. 1, Lafayette, LA, Center for Energy Studies, University of Texas, Lafayette, LA, pp. E-1–E-40 (1977).

283. Hottman, C.E., Method for producing a source of energy from an overpressured formation, U.S. Patent No. 3,258,067, filed February 1963, granted June 1968, assigned to Shell Oil Company, New York.

284. Hottman, C.E., Apparatus for using a source of energy from an overpressured formation, U.S. Patent No. 3,330,356, filed February 1966, granted July 1967, assigned to Shell Oil Company, New York.

285. Burst, J.F., Diagenesis of Gulf Coast clayey sediments and its possible relation to petroleum migration, *American Association of Petroleum Geologists Bulletin*, 53(1), 73–93 (1969).

286. Lindal, B., Industrial and other applications of geothermal energy, in Armstead, H.C.H. (ed.), *Geothermal Energy: Review of Research and Development*. UNESCO, Paris, France, pp. 135–148 (1973).

287. Schmidt, G.W., Interstitial water composition and geochemistry of deep Gulf Coast shales and sandstones, *American Association of Petroleum Geologists Bulletin*, 57(3), 321–337 (1973).

288. Parmigiano, J.M., Geohydraulic energy from geopressured aquifers, MS thesis, Petroleum Engineering Department, Louisiana State University, Baton Rouge, LA (1973).

289. Sultanov, R.C., Skripka, V.E., and Namiot, A., Solubility of methane in water at high temperatures and pressures' *Gazova Promyshlennost*, p. 17 (May 1972) (in Russian) from Blount, USGS open file report, pp. 81–129, http://pubs.USGS gov/of/1981/1214/report.pdf. (1981).

290. Collins, A.G. and Crocker, M.E., Exploitation of minerals in disposal brines, in Paper SPE 3453 presented at the *1971 Annual Fall Technical Conference and Exhibition*, New Orleans, LA (October 3–6, 1971).

291. Doscher, T.M. et al., The technology and economics of methane production from geopressured aquifers, *Journal of Petroleum Technology*, 31, 1502–1514 (December 1979).

292. Marsden, S.S., Natural gas dissolved in Brine—A major energy resource of Japan, in Paper SPE 8355 presented at the *1979 Annual Fall Technical Conference and Exhibition*, Las Vegas, NV (September 23–25, 1979).

293. Suzanne, K., Hamon, G., Billiotte, J., and Trocme', V., Experimental relationships between residual gas saturation and initial gas saturation in heterogeneous sandstone reservoirs, in Paper SPE 84038 presented at the *2003 Annual Technical Conference and Exhibition*, Denver, CO (October 5–8, 2003).

294. Culberson, O.L. and McKetta, J.J., Phase equilibrium in hydrocarbon—Water systems, III, The solubility of methane in water at pressures to 10,000 psia, *AIME Petroleum Transactions*, 192, 223–226 (1951).

295. Price, L.C., Aqueous solubility of methane at elevated pressures and temperatures, *AAPG Bulletin*, 63, 1527–1533 (1979).

296. Quong, R., Otsuki, H.H., Locke, F.E., and Netherton, R., High pressure solvent extraction of methane from geopressured fluids, in Bebout, D.G. and Bachman, A.L. (eds.), *Proceedings of the Fifth Geopressured-Geothermal Energy Conference*, Baton Rouge, LA (October 1981).

297. Buckley, S.E., Hocott, C.R., and Taggart, M.S., Distribution of dissolved hydrocarbons in subsurface waters (Gulf Coastal Plain), in *Habitat of Oil—A Symposium*, American Association of Petroleum Geologists, Tulsa, OK, pp. 850–882 (1958).

298. Samuels, G., Geopressure energy resource evaluation, Department of Energy, Off. of Energy Technology, Oakridge National Laboratory, Oakridge, TN, 72pp., Available from: NTIS, Springfield, VA (1979).

299. Swink, D.G. and Shultz, R.J., Conceptual study for total utilization of an intermediate temperature geothermal resource, Prepared by Aerojet Nuclear Company for Energy Research and Development Administration, Idaho Operations Office, Contract No. E(10-1)-1375, Available from: NTIS, Springfield, VA, ANCR-1260 (1976).

300. Wilson, J.S., Michael, H.K., Shepherd, B.P., Ditzler, C.C., Thomas, L.E., Bradford, B.B., and Steanson, R., A study of Phase O Plan for the production of electrical power from U.S. Gulf Coast geopressured geothermal waters, in *Proceedings of the Second Geopressured Geothermal Energy Conference*, Center for Energy Studies, University of Texas, Austin, TX, Appendix B, 69, p. 30 (1976).

301. Samuels, G., An evaluation of the geopressured energy resource of Louisiana and Texas, SPE/DOE 8848, Presented at the *1980 SPE/DOE Symposium on Unconventional Gas Recovery*, Pittsburgh, PA (May 18–21, 1980).

302. Wrighton, F., An economic overview of geopressured solution gas, in Bebout, D.G. and Bachman, A.L. (eds.), *Proceedings of the Fifth Geopressured Geothermal Energy Conference*, Baton Rouge, LA (1981).

303. McMullan, J.H. and Bassiouni, Z., Prediction of maximum flow rates from geopressured aquifers, in Paper SPE 10282 presented at the *1981 Annual Fall Technical Conference and Exhibition*, San Antonio, TX (October 5–7, 1981).

304. Isokari, O.F., Natural gas production from geothermal geopressured aquifers, in Paper SPE 6037 presented at the *1976 Annual Fall Technical Conference and Exhibition*, New Orleans, LA (October 3–6, 1976).

305. Knapp, R.M. et al., An analysis of production from geopressured geothermal aquifers, in Paper SPE 6825 presented at the *1977 Annual Fall Technical Conference and Exhibition*, Denver, CO (October 9–12, 1977).

306. Doscher, T.M. et al., The numerical simulation of the effect of critical gas saturation and other parameters on the productivity of methane from geopressured aquifers, in Paper SPE 8891 presented at the *1980 Annual California Regional Meeting of the Society of Petroleum Engineers of AIME*, Los Angeles, CA (April 9–11, 1980).

307. Randolph, P.L., Natural gas from geopressured aquifers?, Presented at the *1977 Annual Fall Technical Conference and Exhibition*, Denver, CO (October 9–12, 1977).

308. Zinn, C.D., The economics of producing methane and electrical energy from the Texas Gulf Coast geopressured resource, in Paper SPE 7542 presented at the *1978 Annual Fall Technical Conference and Exhibition*, Houston, TX (October 1–3, 1978).

309. Kharaka, Y.K. et al., Predicted formation and scale-formation properties of geopressured geothermal waters from the Gulf of Mexico Basin, *Journal of Petroleum Technology*, 319–324 (February 1980).

310. Janssen, J.C. and Carver, D.R., A computer program for predicting surface subsidence resulting from pressure depletion in geopressured wells: Subsidence prediction for the DOW Test Well No.1, Pacrperdue, Louisiana, in Bebout, D.G. and Bachman, A.L. (eds.), *Proceedings of the Fifth Geopressured-Geothermal Energy Conference*, Baton Rouge, LA (October 1981).

311. Tomsor, M.B., Rogers, L.A., Varughese, K., Prestwich, S.M., Waggett, G.G., and Salimi, M.H., Use of inhibitors for scale control in brine-producing gas and oil wells, in Paper SPE 15457 presented at *1986 Annual Technical Conference and Exhibition*, New Orleans, LA (October 5–8, 1986).

312. Silva, P. and Bassiouni, Z., Accurate determination of geopressured aquifer salinity from the SP Log, in Bebout, D.G. and Bachman, A.L. (eds.), *Proceedings of the Fifth Geopressured-Geothermal Energy Conference*, Baton Rouge, LA (October 1981).

313. McGee, K.A., Susak, N.J., Sutton, A.J., and Haas, J.L. Jr., The solubility of methane in sodium chloride brines, United States Department of the Interior Geological Survey, Open-File Report 81-1294 (1981).

314. Blount, C.W., Gowan, D.M., Wenger, L., and Price, L.C., Methane solubility in brines with application to the geopressured resource', in Bebout, D.G. and Bachman, A.L. (eds.), *Proceedings of the Fifth Geopressured-Geothermal Energy Conference*, Baton Rouge, LA (October 1981).

315. Morton, R.A., Posey, J.S., and Garrett, C.M. Jr., Salinity of deep formation waters, Texas Gulf Coast—Preliminary results', in Bebout, D.G. and Bachman, A.L. (eds.), *Proceedings of the Fifth Geopressured-Geothermal Energy Conference*, Baton Rouge, LA (October 1981).

316. Swanson, R.K., Geopressured energy availability, Final report, EPRI AP-1457, Electric Power Research Institute, Palo Alto, CA (1980).

317. Brown and Root, Inc., Gulf Coast geopressured geothermal energy study, in *Proceedings of the Second Geopressured Geothermal Energy Conference*, February 23–25, 1976, Vol. 4, Austin, TX, Center for Energy Studies, University of Texas, Austin, TX, Appendix A, 93pp. (1976).

318. Dorfman, M.H. and Deller, R.W., Summary and future projections, in *Proceedings of the Second Geopressured Geothermal Energy Conference*, February 23–25, 1976, Vol. 4, Austin, TX, Center for Energy Studies, University of Texas, Austin, TX, pp. 27–29 (1976).

319. Underhill, G.K. et al., Surface technology and resource utilization, in *Proceedings of the Second Geopressured Geothermal Energy Conference*, February 23–25, 1976, Vol. 4, Austin, TX, Center for Energy Studies, University of Texas, Austin, TX, 203pp. (1976).

320. Loucks, R.G., Richman, D.L., and Milliken, K.L., Factors controlling porosity and permeability in geopressured frio sandstone reservoirs, General crude oil/ department of energy pleasant bayou test wells, Brazoria County, Texas, Presented at the *Fourth Gulf Coast Geopressured Geothermal Energy Conference*, Austin, TX (October 29–31, 1979), Published by Texas University Bureau Economics Geology (1980) OSTI No. 6060550.

321. Wallace, R.H. Jr., Kraemer, T.F., Taylor, R.E., and Wesselman, J.B., Assessment of geopressured-geothermal resources in the northern Gulf of Mexico basin, in Muffler, L.J.P. (ed.), Assessment of geothermal resources of the United States—1978, U.S. Geological Survey Circular 790, U.S. Geological Survey, Reston, VA, pp. 132–155 (1979).

322. Swanson, R.K., Bernard, W.J., and Osoba, J.S., A summary of the geothermal and methane production potential of U.S. Gulf Coast geopressured zones from Test Well Data, *Journal of Petroleum Technology*, 1365–1370 (December 1986).

323. Eaton, B.A., Discussion of 'A summary of the geothermal and methane production potential of U.S. Gulf Coast geopressured zones from test well data, *Journal of Petroleum Technology*, 483 (April 1987).

324. Lombard, D.B. and Wallace, R.H. Jr., Discussion of A summary of the geothermal and methane potential of U.S. Gulf Coast geopressured zones from test well data, *Journal of Petroleum Technology*, 484–486 (April 1987).

325. Chacko, J.J., Maciasz, G., and Harder, B.J., Gulf Coast 439 geopressured-geothermal program summary report compilation, Vols. I, IIA, IIB, III and IV, Basic Research Institute, Louisiana State University, Baton Rouge, LA, Work performed under U.S. Department of Energy, Contract No. DE-FG07-95ID13366, DOE, Washington, DC (June 1998).

326. Durham, C.O. Jr., Background and Status of the sweet lake geopressured-geothermal test, Cameron Parish, Louisiana, in Bebout, D.G. and Bachman, A.L. (eds.), *Proceedings of the Fifth Geopressured-Geothermal Energy Conference*, Baton Rouge, LA (October 1981).

327. Lamb, J.P. and Rhode, D.L., Wellbore flow simulations for Texas-Louisiana geopressured reservoirs, in Paper SPE 7543 presented at the *1978 Annual Fall Technical Conference and Exhibition*, Houston, TX (October 1–3, 1978).

328. White, W.A., McGraw, M., and Gustavson, T.C., Environmental analysis of geopressured geothermal prospect areas, Brazoria and Kenedy Counties, Texas, Report No. ORO-5401-T2, Bureau of Economic Geology, University of Texas, Austin, TX (1978).

329. Phillippi, G.T., On the depth-time mechanism of petroleum generation, *Geochimica et Cosmochimica Acta*, 29, 1021–1049 (1965).

330. Methane clathrate, Wikipedia, the free encyclopedia (2012).

331. Makogon, Y.F., Holditch, S.A., and Makogon, T.Y., Natural gas-hydrates—A potential energy source for the 21st Century, *Journal of Petroleum Science and Engineering*, 56 (1–3), 14–31 (2007).

332. Englezos, P., Clathrate hydrates, *Industrial and Engineering Chemistry Research*, 32(7), 1251–1274 (1993).

333. Methane hydrates, A communication by Center for gas hydrate research, Hariot Watt University, Edinburgh, Scotland, The Hydrate forum Org. (2012).

334. Collett, T.S., Johnson, A.H., Knapp, C.C., and Boswell, R., Natural gas hydrates: A review, in Collett, T., Johnson, A., Knapp, C., and Boswell, R. (eds.), *Natural Gas Hydrates—Energy Resource Potential and Associated Geologic Hazards*, AAPG Memoir 89, pp. 146–219 (2009).

335. McIver, R., Gas hydrates, in Meyer, R. and Olson, J. (eds.), *Long-term Energy Resources*. Pitman, Boston, MA, pp. 713–726 (1981).

336. Collett, T.S., Gas hydrates as a future energy resource, *Geotimes*, 49 (11), 24–27 (2004).

337. Sloan, E.D. and Koh, C., *Clathrate Hydrates of Natural Gases*, 3rd edn. Taylor & Francis, Inc., Boca Raton, FL (2008).

338. Ruppel, C., Methane hydrates and the future of natural gas, MITEI natural gas report, Supplement paper 4 (2011).

339. Ruppel: MITEI natural gas report, Supplementary paper on methane hydrates, 19 (2011).

340. Energy Information Administration (EIA), Natural gas, International Energy Outlook 2010, U.S. Department of Energy, Washington, DC (2010).

341. Holder, G.D., Kamath, V.A., and Godbole, S.P., The potential of natural gas hydrates as an energy resource, *Annual Review of Energy*, 9, 427–445 (1984).

342. Methane hydrate phase diagram, Wikipedia, the free encyclopedia (May 10, 2010).

343. Clathrate hydrate, Wikipedia, the free encyclopedia, pp. 1–7 (2012).

344. Shah, Y.T., *Water for Energy and Fuel Production*. CRC Press, New York, 359, 366, 369 (2014).

345. Collett, T.S., Energy resource potential of natural gas hydrate, *American Association of Petroleum Geologists Bulletin*, 86, 1971–1992 (2002).

346. Ruppel: MITEI natural gas report, Supplementary paper on methane hydrates, 18, 19, 22, 23, 24 (2011).

347. Fujii, T. et al., Resource assessment of methane hydrate in the Nankai Trough, Japan, in *Offshore Technology Conference*, Houston, TX, Paper 19310 (2008).

348. Gornitz, V. and Fung, I., Potential distribution of methane hydrates in the world's oceans, *Global Biogeochemical Cycles*, 8, 225–347 (1994).

349. Milkov, A., Global estimates of hydrate-bound gas in marine sediments: How much is really out there? *Earth-Science Reviews*, 66, 183–197 (2004).

350. Trofimuk, A., Cherskiy, N., and Tsarev, V., Accumulation of natural gases in zones of hydrate—Formation in the hydrosphere, *Doklady Akademii Nauk SSR*, 212, 931–934 (1973) (in Russian).

351. Kvenvolden, K.A., Ginsburg, G.D., and Soloviev, V.A., Worldwide distribution of subaquatic gas hydrates, *Geo-Marine Letters*, 13(1), 32–40 (1993).

352. Kvenvolden, K.A., Natural-gas hydrate occurrence and issues, *Sea Tech*, 36(9), 69–74 (1995).
353. Sloan, E.D. Jr., *Clathrate Hydrates of Natural Gases*. Marcel Dekker Inc., CRC Press, New York (1998).
354. Klauda, J.B. and Sandler, S.I., Global distribution of methane hydrate in ocean sediment, *Energy & Fuels*, 19 (2), 459–470 (2005).
355. Borowski, W.S., A review of methane and gas hydrates in the dynamic, stratified system of the Blake Ridge region, offshore southeastern North America, *Chemical Geology*, 205, 311 (2004).
356. Cherskiy, N.V., Tsaarev, V.P., and Nikitin, S.P., *Petroleum Geology*, 21, 65 (1982); Collett, T., Natural gas hydrates of the Prudhoe Bay and Kuparuk River area, North Slope, Alaska, *American Association of Petroleum Geologists Bulletin*, 77 (5), 793–812 (1993).
357. Collett, T. and Ginsburg, G., Gas hydrates in the Messoyakha gas field of the West Siberian Basin—A re-examination of the geologic evidence, *International Journal of Offshore and Polar Engineering*, 8 (1), 22–29 (1998).
358. Dallimore, S.R. and Collett, T.S. (eds.), Scientific results from the Mallik 2002 gas hydrate production research well program, MacKenzie Delta, Northwest Territories, Canada, Geological Survey of Canada Bulletin 585 published by Natural Resources of Canada, p. 140 (2005), Geoscam ID: 220702.
359. Uchida, T., Lu, H., Tomaru, H., and the MITI Nankai Trough Shipboard Scientists, Subsurface occurrence of natural gas hydrate in the Nankai Trough area: Implication for gas hydrate concentration, *Resource Geology*, 54, 35–44 (2004).
360. Collett, T., Riedel, M., Boswell, R., Cochran, J., Kumar, P., Sethi, A., and Sathe, A., International Team Completes Landmark Gas Hydrate Expedition in the Offshore of India, *Fire in the Ice, NETL Methane Hydrates R&D Program Newsletter*, Fall, Pittsburgh, PA; also DoE, Washington, DC (2006).
361. Park, K.P., Gas hydrate exploration in Korea, in *Proceedings of the Second International Symposium on Gas Hydrate Technology*, Daejeon, Korea (November 1–2, 2006).
362. Boswell, R. and Collett, T., The gas hydrates resource pyramid, *Fire in the Ice*, US Department of Energy, Office of Fossil Energy, National Energy Technology Laboratory, 6 (3), 5–7 (2006).
363. Boswell, R. et al., Joint Industry Project Leg II discovers rich gas hydrate accumulations in sand reservoirs in the Gulf of Mexico, *Fire in the Ice*, US Department of Energy, Office of Fossil Energy, National Energy Technology Laboratory, 9 (3), 1–5 (2009).
364. Collett, T., Agena, W., Lee, M., Zyrianova, M., Bird, K., Charpentier, T., Houseknecht, D., Klett, T., Pollastro, R., and Schenk, C., Assessment of gas hydrate resources on the North Slope, Alaska, 2008, U.S. Geological Survey Fact Sheet 2008-3073, 4pp. (2008).
365. Collett, T., Riedel, M., Cochran, J.R., Boswell, R., Kumar, P., and Sathe, A.V., Indian continental margin gas hydrate prospects: Results of the Indian National Gas Hydrate Program (NGHP) expedition 01, in *Proceedings of the Sixth International Conference on Gas Hydrates*, Vancouver, British Columbia, Canada (2008).
366. Boswell, R. and Collett, T.S., Current perspectives on gas hydrate resources, *Energy and Environmental Science*, 4, 1206–1215 (2011).
367. Dai, J., Snyder, F., Gillespie, D., Koesoemadinata, A., and Dutta, N., Exploration for gas hydrates in the deepwater northern Gulf of Mexico: Part I. A seismic approach based on geologic model, inversion, and rock physics principles, *Marine and Petroleum Geology*, 25, 830–844 (2008).
368. Dai, J., Banik, N., Gillespie, D., and Dutta, N., Exploration for gas hydrates in the deepwater northern Gulf of Mexico: Part II. Model validation by drilling, *Marine and Petroleum Geology*, 25, 845–859 (2008).
369. Frye, M., Preliminary evaluation of in-place gas hydrate resources, Gulf of Mexico Outer Continental Shelf: Minerals Management Service Report, OCS Report mMS 2008-004, U.S. Department of Interior, Minerals Management Service, Resource Evaluation Division, Washington, DC (February 1, 2008).
370. Ryu, B.-J., Riedel, M., Kim, J.-H., Hyndman, R.D., Lee, Y.-J., Chung, B.-H., and Kim, I.S., Gas hydrates in the western deep-water Ulleung Basin, East Sea of Korea, *Marine and Petroleum Geology*, 26, 1483–1498 (2009).
371. Tsuji, Y., Ishida, H., Nakamizu, M., Matsumoto, R., and Shimizu, S., Overview of the MITI Nankai Trough wells: A milestone in the evaluation of methane hydrate resources, *Resource Geology*, 54, 3–10 (2004).
372. Tsuji, Y. et al., Methane-hydrate occurrence and distribution in the Eastern Nankai Trough, Japan: Findings of the Tokai-oki to Kumano-nada methane-hydrate drilling program, in Collett, T., Johnson, A., Knapp, C., and Boswell, R. (eds.), *Natural Gas Hydrates—Energy Resource Potential and Associated Geologic Hazards*, AAPG Memoir American Association of Petroleum Geologists, 89, pp. 228–249 (2009), doi: 10.1306/13201142M891602, archives.datapages.com.

373. Wu, N. et al., Preliminary discussion on gas hydrate reservoir system of Shenhu area, north slope of South China Sea, in *Proceedings of the Sixth International Conference on Gas Hydrates (ICGH 2008)*, July 6–10, 2008, Vancouver, British Columbia, Canada, 8pp. (2008).
374. Zhang, H., Yang, S., Wu, N., Su, X, Holland, M., Schultheiss, P., Rose, K., Butler, H., Humphrey, G., and GMGS-1 Science Team, Successful and surprising results for China's first gas hydrate drilling expedition, a Report by NETL, US Department of Energy, Office of Fossil Energy, National Energy Technology Laboratory, 7 (3), 6–9 (2007).
375. Paull, C., Reeburgh, W.S., Dallimore, S.R., Enciso, G., Green, S., Koh, C.A., Kvenvolden, K.A., Mankin, C., and Riedel, M., Realizing the energy potential of methane hydrate for the United States, National Research Council Report, The National Academy Press, Washington, DC, pp. 204, doi:10.17226112831, ISBN 978-0-309-14889-4 (2010).
376. Cook, A., Goldberg, D., and Kleinberg, R., Fracture-controlled gas hydrate systems in the Gulf of Mexico, *Marine and Petroleum Geology*, 25 (9), 932–941 (2008).
377. Collett, T.S. et al., Permafrost associated natural gas hydrate occurrences on the Alaskan North Slope, *Marine and Petroleum Geology*, 28, 279–294 (2011).
378. Dallimore, S.R., Uchida, T., and Collett, T.S. (eds.), Scientific results from the JAPEX/JNOC/GSC Mallik 2L-38 gas hydrate research well, Mackenzie Delta, Northwest Territories, Canada, Geological Survey of Canada Bulletin 544, Natural Resources Canada (1999), NTS 107C/06NW, GEOSCAN ID 210744.
379. Ruppel, C., MITEI natural gas report, Supplementary paper on methane hydrates, 21 (2011).
380. Walsh, T., Stokes, P., Panda, M., Morahan, T., Greet, D., MacRae, S., Singh, P., and Patil, S. Characterization and quantification of the methane hydrate resource potential associated with the Barrow Gas Field, in *Proceedings of the Sixth International Conference on Gas Hydrates (ICGH 2008)*, Vancouver, British Columbia, Canada (2008).
381. Park, K.P. et al., Korean National Program expedition confirms rich gas hydrate deposits in the Ulleung Basin, East Sea, *Fire in the Ice*, US Department of Energy, Office of Fossil Energy, National Energy Technology Laboratory, 8 (2), 6–9 (2008).
382. Why are gas hydrates important, A publication by Institute of Petroleum Engineering, Hariot Watt University, Edinburgh, Scotland (2011).
383. Collett, T.S. and Ladd, J., Detection of gas hydrate with downhole logs and assessment of gas hydrate concentrations (saturations) and gas volumes on the Blake Ridge with electrical resistivity log data, in Paull, C.K., Matsumoto, R., Wallace, P.J., and Dillion, W.P. (eds.), *Proceedings of ODP, Scientific Results*, 164, 179–191 (2000).
384. Hato, M., Matsuoka, T., Inamori, T., and Saeki, T., Detection of methane-hydrate-bearing zones using seismic attributes analysis, *The Leading Edge*, 25, 607–609 (2006).
385. Holbrook, W.S., Gorman, A.R., Hornbach, M., Hackwith, K.L., and Nealon, J., Direct seismic detection of methane hydrate, *The Leading Edge*, 21, 686–689 (2002).
386. Hovland, M. and Gudmestad, O.V., Potential influence of gas hydrates on seabed installations, in Paull, C. and Dillon, W. (eds.), *Natural Gas Hydrates—Occurrence, Distribution and Detection*, American Geophysical Union, Washington, DC, pp. 307–315 (2001).
387. Lee, J.Y., Santamarina, J.C., and Ruppel, C., Parametric study of the physical properties of hydrate-bearing sand, silt, and clay sediments: 1. Electromagnetic properties, *Journal of Geophysical Research*, 115, B11104 (2010).
388. Lee, J.Y., Francisca, F.M., Santamarina, J.C., and Ruppel, C., Parametric study of the physical properties of hydrate-bearing sand, silt, and clay sediments: 2. Small-strain mechanical properties, *Journal of Geophysical Research*, 115, B11105 (2010).
389. Moridis, G.J., Reagan, M.T., and Zheng, K., On the performance of Class 2 and Class 3 hydrate deposits during co-production with conventional gas, in *Offshore Technology Conference*, OTC 19435-MS (2008).
390. Paull, C.K. et al., in *Proceedings of the Ocean Drilling Program, Initial Reports*, Vol. 164, Ocean Drilling Program, College Station, TX (1996).
391. Rutqvist, J. and Moridis, G., Evaluation of geohazards of in situ gas hydrates related to oil and gas operations, *Fire in the Ice*, US Department of Energy, Office of Fossil Energy, National Energy Technology Laboratory, 10 (2), 1–4 (2010).
392. Ruppel, C., Collett, T., Boswell, R., Lorenson, T., Buckzowski, B., and Waite, W., A new global gas hydrate drilling map based on reservoir type, *Fire in the Ice*, US Department of Energy, Office of Fossil Energy, National Energy Technology Laboratory, 11 (1), 15–19 (2011).
393. Satyavani, N., Sain, K., Lall, M., and Kumar, B.J.P., Seismic attribute study of gas hydrates in the Andaman, Offshore India, *Marine Geophysical Researches*, 29, 167–175 (2008).

394. Ruppel, C., Boswell, R., and Jones, E., Scientific results from Gulf of Mexico gas hydrates joint industry project Leg 1 drilling: Introduction and overview, *Marine and Petroleum Geology*, 25 (2008), doi:10.1016/j.marpetgeo.2008.02.007.

395. Ruppel, C., MITEI natural gas report, Supplementary paper No. on methane hydrates, USGS, Woods Hole, MA, 20, 25 (2011).

396. Birchwood, R.A. et al., Modeling the mechanical and phase change stability of wellbores drilled in gas hydrates by the Joint Industry Participation Program (JIP) Gas Hydrates Project Phase II, in *SPE Annual Technical Conference*, November 11–14, 2007, SPE 110796 (2007).

397. Ameripour, S., Prediction of gas-hydrate formation conditions in production and surface facilities, MS thesis, Chair of Advisory Committee: Dr. Maria A. Barrufet, Texas A&M University, p. 79 (August 2005) .

398. Dalmazzone, D., Kharrat, M., Lachet, V., Fouconnier, B., and Clausse, D., DSC and PVT measurements—Methane and trichlorofluoromethane hydrate dissociation equilibria, *Journal of Thermal Analysis and Calorimetry*, 70, 493–505 (2002).

399. Edmonds, B., Moorwood, R.A.S., and Szczepanski, R., A practical model for the effect of salinity on gas hydrate formation, in *European Production Operations Conference & Exhibition*, Stavanger, Norway, SPE 35569 (1996).

400. Grigg, R.B. and Lynes, G.L., Oil-based drilling mud as a gas-hydrates inhibitor, *SPE Drilling Engineering*, 32–38 (March 1992).

401. Kotkoskie, T.S., Al-Ubaidi, B., Wildeman, T.R., and Sloan, E.D. Jr., Inhibition of gas hydrates in water-based drilling muds, *SPE Drilling Engineering*, 130–136 (June 1992).

402. Lai, D.T. and Dzialowski, A.K., Investigation of natural gas hydrates in various drilling fluids, in *SPE/IADC Drilling Conference*, New Orleans, LA, SPE 18637 (February 1989).

403. Tohidi, B., Østergaard, K.K., Danesh, A., Todd, A.C., and Burgass, R.W., Structure-H gas hydrates in petroleum reservoir fluids, *The Canadian Journal of Chemical Engineering*, 79, 384–391 (June 2001).

404. Yousif, M.H., Dunayevsky, V.A., and Hale, A.H., Hydrate plug remediation: Options and applications for deep water drilling operations, in *SPE/IADC Drilling Conference*, Amsterdam, the Netherlands, SPE 37624 (March 1997).

405. Kim, N., Bonet, E., and Ribeiro, P., Study of hydrate in drilling operations: A review, in *Fourth PDPETRO*, Campinas, Brazil, (October 21–24, 2007).

406. LaBelle, R., Hydrates-Hazard/safety issues, TAR program, Natural Gas Hydrate Research, a personal communication (2012).

407. Tohidi, B., Gas hydrates challenges in oil and gas industry, Report by Center for gas hydrate research, Institute of Petroleum Engineering, Heriot Watt University, Edinburgh, U.K. (2012).

408. Bagirov, E. and Lerche, I., Hydrate represent gas source, drilling hazard, A report by *Oil and Gas Journal* (1997), www.ogj.com/articles/print/.../hydrates-represent-gas-source-drilling-hazard.html.

409. Helgeland, L., Kinn, A., Kvalheim, O., and Wenaas, A., Gas kick due to hydrates in the drilling for offshore natural gas and oil, A report by Department of Petroleum Engineering and Applied Geophysics, NTNU, Trondheim, Norway (November 2012).

410. Skalle, P., Pressure control during oil well drilling, Chapter 7, in *Special Offshore Safety Issues*, 2nd edn. Pål Skalle & Ventus Publishing ApS (2011), http://bookboon.com/en/textbooks/geoscience/pressure-control-during-oil-well-drilling, Accessed October 3, 2012.

411. Dillon, W.P. and Max, M.D., Oceanic gas hydrates (2003), doi:10.10071/978-94-011-4387-5-6.

412. Skalle, P., Chapters 1 and 4, *Pressure Control During Oil Well Drilling*; Oilfield Glossary, Blowout, Schlumberger, Pal Skalle & Ventus Publishing ApS [online]. http://www.glossary.oilfield.slb.com/Display.cfm?Term=blowout, Accessed October 19, 2012.

413. Khabibullin, T., Drilling through gas hydrate formation: Managing wellbore formation risks, M.S. Thesis, Texas A&M University, College Station, TX (August 2010).

414. Qadir, M.I., Gas hydrates: A fuel for future but wrapped in drilling challenges, SPE-156516, in Paper presented at *SPE/PAPG Annual Technical Conference*, Islamabad, Pakistan (November 22–23, 2011).

415. Williamson, S.C., McConnell, D.R., and Bruce, R.J., Drilling observations of possible hydrate-related annular flow in the deepwater Gulf of Mexico and Implications on Well Planning, OTC 17279, in Paper presented at the *2005 Offshore Technology Conference*, Houston, TX (May 2–5, 2005).

416. Hannegan, D., Todd, R.T., Pritchard, D.M., and Jonasson, B., MPD—Uniquely applicable to methane hydrate drilling, SPE-91560, in Paper presented at the *SPE/IADC Underbalanced Technology Conference and Exhibition*, Houston, TX (October 11–12, 2004).

417. Ebeltoft, H., Yousif, M., and Sægråd, E., Hydrate control during deepwater drilling: Overview and new drilling-fluids formulations, SPE Drilling & Completion 16(1), 19–26 (March 2001), http://dx.doi.org/10.2118/68207-PA.

418. Halliday, W., Clapper, D.K., and Smalling, M., New gas hydrate inhibitors for deepwater drilling fluids, IADC/SPE-39316, in Paper presented at the *1998 SPE/IADC Drilling Conference*, Dallas, TX (March 3–6, 1998).

419. Ravi, K. and Moore, S., Cement slurry design to prevent destabilization of hydrates in deepwater environment, SPE-113631, in Paper presented at the *2008 Indian Oil and Gas Technical Conference and Exhibition*, Mumbai, India (March 4–6, 2008).

420. Halliday, W., Clapper, D., and Smalling, M., New gas hydrate inhibitors for deepwater drilling fluids, SPE 39316, in Paper by Baker Hughes Inteq, presented at *IADC/SPE Drilling Conference*, Dallas, TX (March 3–6, 1998).

421. Amodu, A., Drilling through gas hydrate formations: Possible problems and suggested solutions, MS thesis, Department of Petroleum Engineering, Texas A&M University, College Station, TX (August 2008).

422. Catak, E., Hydrate dissociation during drilling through in-situ hydrate formations, MS thesis, Department of Petroleum Engineering, Louisiana State University, Baton Rouge, LA (May 2006).

423. Ruppel, C., Methane hydrates and contemporary climate change, *Natural Education Knowledge*, 3 (10), 29 (2011).

424. Buffett, B. and Archer, D., Global inventory of methane clathrate: Sensitivity to changes in environmental conditions, *Earth and Planetary Science Letters*, 227, 185–199 (2004).

425. Dutta, N.C., Utech, R.W., and Shelander, D., Role of 3D seismic for quantitative shallow hazard assessment in deepwater sediments, *The Leading Edge*, 930–942 (2010).

426. Ellis, M., Evans, R.L., Hutchinson, D., Hart, P., Gardner, J., and Hagen, R., Electromagnetic surveying of seafloor mounds in the northern Gulf of Mexico, *Marine and Petroleum Geology*, 25, 969–968 (2008).

427. Hadley, C., Peters, D., and Vaughan, A., Gumusut-Kakap project: Geohazard characterisation and impact on field development plans, in *International Petroleum Technology Conference*, Kuala Lumpur, Malaysia, 12554, 15pp. (December 3–5, 2008).

428. Weitemeyer, K., Constable, S., and Key, K., Marine EM techniques for gas-hydrate detection and hazard mitigation, *The Leading Edge*, 25 (5), 629–632 (2006).

429. Bunz, S. and Meinert, J., Overpressure distribution beneath hydrate-bearing sediments at the Storegga Slide on the Mid-Norwegian margin, Paper 3007, in *Proceedings of ICGH 2005*, Trondheim, Norway (June 13–16, 2005) Vol. 3, pp. 755–758 (2005), ISBN 9781615670666; also published by Curran Associates Inc., Red Hook, NY (August 2009).

430. Archer, D. et al., Ocean methane hydrates as a slow tipping point in the global carbon cycle, *Proceedings of the National Academy of Sciences of the United States of America*, 106, 20956–20601 (2009).

431. Biastoch, A. et al., Rising Arctic Ocean temperatures cause gas hydrate destabilization and ocean acidification, *Geophysical Research Letters*, 38, L08602 (2011).

432. Bock, M. et al., Hydrogen isotopes preclude marine hydrate CH_4 emissions at the onset of Dansgaard-Oeschger events, *Science*, 328, 1686–1689 (2010).

433. Bohannon, J., Weighing the climate risks of an untapped fossil fuel, *Science*, 319, 1753 (2008).

434. Bowen, R.G. et al., Geomorphology and gas release from pockmark features in the Mackenzie Delta, Northwest Territories, Canada, in Kane, D.L. and Hinkel, K.M. (eds.), *Proceedings of the Ninth International Conference on Permafrost*, Institute of Northern Engineering, Fairbanks, AK, pp. 171–176 (2008).

435. Dickens, G.R. et al., Dissociation of oceanic methane hydrate as a cause of the carbon isotope excursion at the end of the Paleocene, *Paleoceanography*, 10, 965–971 (1995).

436. Dickens, G.R., Down the rabbit hole: Toward appropriate discussion of methane release from gas hydrate systems during the Paleocene-Eocene thermal maximum and other past hyperthermal events, *Climate of the Past*, 7, 831–846 (2011).

437. Harvey, L.D.D. and Huang, Z., Evaluation of potential impact of methane clathrate destabilization on future global warming, *Journal of Geophysical Research*, 100, 2905–2926 (1995).

438. Maslin, M. et al., Linking continental-slope failures and climate change: Testing the clathrate gun hypothesis, *Geology*, 32, 53–56 (2004).

439. Renssen, H. et al., Modeling the climate response to a massive methane release from gas hydrates, *Paleoceanography*, 19, PA2010 (2004).

440. Schmidt, G.A. and Shindell, D.T., Atmospheric composition, radiative forcing, and climate change as consequence of a massive methane release from gas hydrates, *Paleoceanography*, 18, 1004 (2003).
441. Semiletov, I. et al., Methane climate forcing and methane observations in the Siberian Arctic Land-Shelf system, *World Resource Review*, 16, 503–543 (2004).
442. Solomon, E.A. et al., Considerable methane fluxes to the atmosphere from hydrocarbon seeps in the Gulf of Mexico, *Nature Geoscience*, 2, 561–565 (2009).
443. Westbrook, G.K. et al., Escape of methane gas from the seabed along the West Spitsbergen continental margin, *Geophysical Research Letters*, 36, L15608 (2009).
444. Hesselbo, S.P. et al., Massive dissociation of gas hydrate during a Jurassic anoxic event, *Nature*, 406, 392–395 (2000).
445. Hinrichs, K.-U. and Boetius, A., The anaerobic oxidation of methane: New insights in microbial ecology and biochemistry, in Wefer, G. et al. (eds.), *Ocean Margin Systems*, Springer-Verlag, Berlin, Germany, pp. 457–477 (2002).
446. Hu, L., Yuon-Lewis, A., Kessler, J., and McDonald, J., Methane fluxes to the atmosphere from deep-water hydrocarbon seeps in the northern Gulf of Mexico, *Journal of Geophysical Research*, 117, C1, 2156–2202 (January 2012), doi:10.1029/2011JC007208.
447. IPCC (Intergovernmental Panel on Climate Change), *Climate Change 2001: The Scientific Basis.* Cambridge University Press, New York (2001).
448. IPCC (Intergovernmental Panel on Climate Change), *Climate Change 2007: The Physical Basis.* Cambridge University Press, New York (2007).
449. Jiang, G. et al., Stable isotope evidence for methane seeps in Neoproterozoic postglacial cap carbonates, *Nature*, 426, 822–826 (2003).
450. Judge, A.S. and Majorowicz, J.A., Geothermal conditions for gas hydrate stability in the Beaufort-Mackenzie area: The global change aspect, *Palaeogeography, Palaeoclimatology, Palaeoecology*, 98, 251–263 (1992).
451. Kennett, J.P. et al., *Methane Hydrates in Quaternary Climate Change—The Clathrate Gun Hypothesis.* American Geophysical Union, Washington, DC (2003).
452. Kessler, J.D. et al., A persistent oxygen anomaly reveals the fate of spilled methane in the deep Gulf of Mexico, *Science*, 331, 312–315 (2011).
453. Krey, V. et al., Gas hydrates: Entrance to a methane age or climate threat? *Environmental Research Letters*, 4, 034007 (2009).
454. Lachenbruch, A.H., Permafrost, the active layer, and changing climate, United States Geological Survey, Open File Report 94-694, USGS, Washington, DC (1994).
455. Lachenbruch, A.H. and Marshall, B.V., Changing climate: Geothermal evidence from permafrost in the Alaskan Arctic, *Science*, 234, 689–696 (1986).
456. Lammers, S. et al., A large methane plume east of Bear Island (Barents Sea): Implications for the marine methane cycle, *Geologische Rundschau*, 84, 59–66 (1995).
457. Lelieveld, J. et al., Changing concentration, lifetime and climate forcing of atmospheric methane, *Tellus*, 50B, 128–150 (1998).
458. Liro, C.R. et al., Modeling the release of CO_2 in the deep ocean, *Energy Conversion and Management*, 33, 667–674 (1992).
459. Macdonald, G., Role of methane clathrates in past and future climate, *Climatic Change*, 16, 247–281 (1990).
460. Macdonald, I.R. et al., Thermal and visual time-series at a seafloor gas hydrate deposit on the Gulf of Mexico slope, *Earth and Planetary Science Letters*, 233, 45–59 (2005).
461. Macdonald, I.R. et al., Gas hydrate that breaches the seafloor on the continental slope of the Gulf of Mexico, *Geology*, 22, 699–702 (1994).
462. Majorowicz, J.A. et al., Onset and stability of gas hydrates under permafrost in an environment of surface climatic change-past and future, in *Proceedings of the Sixth International Conference on Gas Hydrates*, ICGH, Vancouver, British Columbia, Canada (2008).
463. Mascarelli, A.L., A sleeping giant? *Nature Reports Climate Change*, 3, 46–49 (2009).
464. Maslin, M. et al., Gas hydrates: Past and future geohazard, *Philosophical Transactions of the Royal Society A: Mathematical, Physical & Engineering Sciences*, 368, 2369–2393 (2010).
465. Mau, S. et al., Dissolved methane distributions and air-sea flux in the plume of a massive seep field, Coal Oil Point, California, *Geophysical Research Letters*, 34, L22603 (2007).
466. McGinnis, D.F. et al., Fate of rising methane bubbles in stratified waters: How much methane reaches the atmosphere? *Journal of Geophysical Research*, 111, C09007 (2006).

467. Niemann, H. et al., Novel microbial communities of the Haakon Mosby mud volcano and their role as a methane sink, *Nature*, 443, 854–858 (2006).

468. Paull, C. et al., Tracking the decomposition of submarine permafrost and gas hydrate under the shelf and slope of the Beaufort Sea, in *Proceedings of the Seventh International Conference on Gas Hydrates*, ICGH, Edinburgh, Scotland (2011).

469. Petrenko, V. et al., $^{14}CH_4$ measurements in Greenland ice: Investigating last glacial termination CH_4 sources, *Science*, 324, 506–508 (2009).

470. Rachold, V. et al., Near-shore arctic subsea permafrost in transition, *Eos, Transactions of the American Geophysical Union*, 88, 149–156 (2007).

471. Reagan, M.T. and Moridis, G.J., Dynamic response of oceanic hydrate deposits to ocean temperature change, *Journal of Geophysical Research*, 113, C12023 (2008).

472. Reeburgh, W.S., Oceanic methane biogeochemistry, *Chemical Reviews*, 107, 486–513 (2007).

473. Röhl, U. et al., On the duration of the Paleocene-Eocene thermal maximum, *Geochemistry, Geophysics, Geosystems*, 8, Q12002 (2007).

474. Ruppel, C. et al., Degradation of subsea permafrost and associated gas hydrates offshore of Alaska in response to climate change, *Sound Waves*, 128, 1–3 (2010).

475. Shakhova, N. et al., Geochemical and geophysical evidence of methane release over the East Siberian Arctic Shelf, *Journal of Geophysical Research*, 115, C08007 (2010).

476. Shakhova, N. et al., Extensive methane venting to the atmosphere from sediments of the East Siberian Arctic Shelf, *Science*, 327, 1246–1250 (2010).

477. Sowers, T., Late Quaternary atmospheric CH4 isotope record suggests marine clathrates are stable, *Science*, 311, 838 (2006).

478. Suess, E. et al., Sea floor methane hydrates at Hydrate Ridge, Cascadia Margin, in Dillon, W.P. and Paull, C.K. (eds.), *Natural Gas Hydrates-Occurrence, Distribution and Detection*. American Geophysical Union, Washington, DC, pp. 87–98 (2001).

479. Treude, T. et al., Anaerobic oxidation of methane at Hydrate Ridge (OR), *Geochimica et Cosmochimica Acta*, 67, A491 (2003).

480. Tryon, M.D. et al., Fluid and chemical flux in and out of sediments posting methane hydrate deposits on Hydrate Ridge, OR, II: Hydrological processes, *Earth and Planetary Science Letters*, 201, 541–557 (2002).

481. Walter, K.M. et al., Methane bubbling from northern lakes: Present and future contributions to the global methane budget, *Philosophical Transactions of the Royal Society A: Mathematical, Physical & Engineering Sciences*, 365, 1657–1676 (2007).

482. Wang, J.S. et al., A 3-D model analysis of the slowdown and interannual variability in the methane growth rate from 1988 to 1997, *Global Biogeochemical Cycles*, 18, GB3011 (2004).

483. Yvon-Lewis, S.A. et al., Methane flux to the atmosphere from the Deepwater Horizon oil disaster, *Geophysical Research Letters*, 38, L01602 (2011).

484. Zachos, J. et al., Trends, rhythms, and aberrations in global climate 65 Ma to present, *Science*, 292, 686–693 (2001).

485. Zachos, J. et al., Rapid acidification of the ocean during the Paleocene-Eocene thermal maximum, *Science*, 308, 1611–1615 (2005).

486. Gas hydrates and climate warming, USGS report, USGS, Washington, DC, pp. 1–9 (January 24, 2012), https://www.usgs.gov/.../usgs../gas hydrates_and_climate_change.

487. Kennett, J., Role of methane hydrates in climate change: Compelling evidence and debate, Department of Geological sciences and Marine Science Institute, University of California, Santa Barbara, CA, personal communication (2012).

488. Hancock, S., Collett, T.S., Dallimore, S.R., Satoh, T., Huenges, E., and Henninges, J., Overview of thermal stimulation production test results for the Japex/JNOC/GSC Mallik Gas Hydrate Research Well, in Dallimore, S.R. and Collett, T.S. (eds.), Scientific results from Mallik 2002 gas hydrate production research well program, Mackenzie Delta, Northwest Territories, Canada, Geological Survey of Canada Bulletin 585, pp. 140 (2005), published by Natural Resources of Canada, GEOSCAN ID: 220702.

489. Circone, S., Kirby, S.H., and Stern, L.A., Thermal regulation of methane hydrate dissociation: Implications for gas production models, *Energy & Fuels*, 19 (6), 2357–2363 (2005).

490. Tang, L.G., Xiao, R., Huang, C., Feng, Z.P., and Fan, S.S., Experimental investigation of production behavior of gas hydrate under thermal stimulation in unconsolidated sediment, *Energy & Fuels*, 19 (6), 2402–2407 (2005).

491. Kawamura, T., Ohtake, M., Sakamoto, Y., Yamamoto, Y., Haneda, H., Komai, T., and Higuchi, S., Experimental study on steam injection method using methane hydrate core samples, in *Proceedings of the Seventh (2007) ISOPE Ocean Mining Symposium*, Lisbon, Portugal, pp. 83–86 (July 1–6, 2007).

492. Li, G., Li, X., Tang, L.-G., and Li, Q.-P., Control mechanisms for methane hydrate production by thermal stimulation, in *Proceedings of the Sixth International Conference on Gas Hydrates* (*ICGH 2008*), Vancouver, British Columbia, Canada (July 6–10, 2008).

493. Computer Modeling Group, Steam, thermal, and advanced processes reservoir simulator (STARS), Computer Modelling Group Ltd., Houston, TX (2015), online: www.cmgroup.com/software/stars.htm.

494. Kamath, V.A., Study of heat transfer characteristics during dissociation of gas hydrates in porous media, PhD dissertation, University of Pittsburgh, Pittsburgh, PA (1984).

495. Ruppel, C., Dickens, G., Castellini, D., Gilhooly, W., and Lizzarralde, D., Heat and salt inhibition of gas hydrate formation in the northern Gulf of Mexico, *Geophysical Research Letters*, 32 (4), L04605 (2005).

496. Bai, Y., Li, Q., Li, X., and Du, Y., The simulation of nature gas production from ocean gashydrate reservoir via depressurization, *Science in China Series E: Technological Sciences*, 51, 1272–1282 (2008).

497. Hong, H. and Pooladi-Darvish, M., Simulation of depressurization for gas production from gas hydrate reservoirs, *Journal of Canadian Petroleum Technology*, 44 (11), 39–46 (2005).

498. Ji, C., Ahmadi, G., and Smith, D.H., Natural gas production from hydrate decomposition by depressurization, *Chemical Engineering Science*, 56 (20), 5801–5814 (2001).

499. Kono, H.O., Narasimhan, S., Song, F., and Smith, D.H., Synthesis of methane gas hydrate in porous sediments and its dissociation by depressurizing, *Powder Technology*, 122 (2–3), 239–246 (2002).

500. Moridis, G.J., Kowalsky, M.B., and Pruess, K., Depressurization-induced gas production from class 1 hydrate deposits, *SPE Reservoir Evaluation & Engineering*, 10 (5), 458–481 (2007).

501. Tang, L., Li, X., Feng, Z., Li, G., and Fan, S., Control mechanisms for gas hydrate production by depressurization in different scale hydrate reservoirs, *Energy & Fuels*, 21 (1), 227–233 (2007).

502. Chatterji, J. and Griffith, J.E., Methods of decomposing gas hydrates, Patent No. 5,713,416 (1998).

503. Yousif, M.H., Li, P.M., Selim, M.S., and Sloan, E.D., Depressurization of natural gas hydrates in Berea sandstone cores, *Journal of Inclusion Phenomena and Molecular Recognition in Chemistry*, 8, 71–88 (1990).

504. Hong, H. and Pooladi-Darvish, M., Simulation of depressurization for gas production from gas hydrate reservoirs, *Journal of Canadian Petroleum Technology*, 44 (11), 39–46 (2005).

505. Fan, S.S., Zhang, Y.Z., Tian, G.L., Liang, D.Q., and Li, D.L., Natural gas hydrate dissociation by presence of ethylene glycol, *Energy & Fuels*, 20 (1), 324–326 (2006).

506. Kamath, V.A., Mutalik, P.N., Sira, J.H., and Patil, S.L., Experimental study of Brine injection and depressurization methods for dissociation of gas hydrate, *SPE Formation Evaluation*, 6 (4), 477–484 (1991).

507. Kawamura, T., Yamamoto, Y., Ohtake, M., Sakamoto, Y., Komai, T., and Haneda, H., Dissociation experiment of hydrate core sample using thermodynamic inhibitors, in *15th International Offshore and Polar Engineering Conference* (*ISOPE 2005*), Seoul, South Korea, pp. 346–350 (June 19–24, 2005).

508. Li, X., Zhang, Y., Li, G., Chen, Z., and Wu, H., Experimental investigation into the production behavior of methane hydrate in porous sediment by depressurization with a novel three-dimensional cubic hydrate simulator, *Energy & Fuels*, 25 (10), 4497–4505 (2011).

509. Moridis, G., Collett, T.S., Boswell, R., Kurihara, M., Reagan, M., Koh, C., and Sloan, E.D., Toward production from gas hydrates: Current status, assessment of resources, and simulation-based evaluation of technology and potential, in *SPE Unconventional Reservoirs Conference*, SPE 114163 (2008), also as Eseolas publication by Lawrence Berkeley National Laboratory, University of California (February 12, 2008), http://escholarship.org/uc/item/7hm710jd.pdf.

510. Moridis, G.J., Collett, T.S., Dallimore, S.R., Satoh, T., Hancock, S., and Weatherill, B., Numerical studies of gas production from several CH_4 hydrate zones at the Mallik Site, Mackenzie Delta, Canada, *Journal of Petroleum Science and Engineering*, 43 (3–4), 219–238 (2004).

511. Moridis, G.J., Numerical studies of gas production from methane hydrates, *SPE Journal*, 8 (4), 359–370 (2003).

512. Moridis, G.J., Numerical studies of gas production from class 2 and class 3 hydrate accumulations at the Mallik Site, Mackenzie Delta, Canada, *SPE Reservoir Evaluation and Engineering*, 7 (3), 175–183 (2004).

513. Pooladi-Darvish, M., Gas production from hydrate reservoirs and its modeling, *Journal of Petroleum Technology*, 56 (6), 65–71 (2004).

514. Sun, X., Nanchary, N., and Mohanty, K.K., 1-D modeling of hydrate depressurization in porous media, *Transport in Porous Media*, 58 (3), 315–338 (2005).

515. Goel, N., Wiggins, M., and Shah, S., Analytical modeling of gas recovery from in situ hydrates dissociation, *Journal of Petroleum Science and Engineering*, 29 (2), 115–127 (2001).
516. Moridis, G.J., Collett, T.S., Boswell, R., Kurihara, M., Reagan, M.T., Koh, C., and Sloan, E.D., Toward production from gas hydrates: Current status, assessment of resources, and simulation-based evaluation of technology and potential, *SPE Reservoir Evaluation & Engineering*, 12 (5), 745–771 (2009).
517. Tsypkin, G.G., Mathematical models of gas hydrates dissociation in porous media, in *Gas Hydrates: Challenges for the Future, Annals of the New York Academy of Sciences*, Vol. 912, pp. 428–436 (2000).
518. White, M.D. and Oostrom, M., STOMP: Subsurface transport over multiple phases, Version 4.0, User's Guide, PNNL-15782, Pacific Northwest National Laboratory, Richland, Washington, DC (2006).
519. Moridis, G.J., Numerical studies of gas production from methane hydrates, *SPE Journal*, 32 (8), 359 (2003).
520. Sung, W.M., Lee, H., and Lee, C., Numerical study for production performances of a methane hydrate reservoir stimulated by inhibitor injection, *Energy Sources*, 24 (6), 499–512 (2002).
521. Kawamura, T., Sakamoto, Y., Ohtake, M., Yamamoto, Y., Haneda, H., Yoon, J.H., and Komai, T., Dissociation behavior of hydrate core sample using thermodynamic inhibitor, *International Journal of Offshore and Polar Engineering*, 16 (1), 5–9 (2006).
522. Li, G., Li, X., Tang, L., and Zhang, Y., Experimental investigation of production behavior of methane hydrate under ethylene glycol stimulation in unconsolidated sediment, *Energy & Fuels*, 21 (6), 3388–3393 (2007).
523. Li, G., Li, X., Tang, L., Zhang, Y., Feng, Z., and Fan, S., Experimental investigation of production behavior of methane hydrate under ethylene glycol injection, *Huagong Xuebao/Journal of Chemical Industry and Engineering* (*China*), 58 (8), 2067–2074 (2007).
524. Kawamura, T., Yamamoto, Y., Ohtake, M., Sakamoto, Y., Komai, T., and Haneda, H., Experimental study on dissociation of hydrate core sample accelerated by thermodynamic inhibitors for gas recovery from natural gas hydrate, in *The Fifth International Conference on Gas Hydrate*, Trondheim, Norway, pp. 3023–3028 (June 12–16, 2005).
525. Li, G., Tang, L., Huang, C., Feng, Z., and Fan, S., Thermodynamic evaluation of hot brine stimulation for natural gas hydrate dissociation, *Huagong Xuebao/Journal of Chemical Industry and Engineering* (*China*), 57 (9), 2033–2038 (2006).
526. Li, X.S., Wan, L.H., Li, G., Li, Q.P., Chen, Z.Y., and Yan, K.F., Experimental investigation into the production behavior of methane hydrate in porous sediment with hot brine stimulation, *Industrial & Engineering Chemistry Research*, 47 (23), 9696–9702 (2008).
527. Tang, L.G., Li, G., Hao, Y.M., Fan, S.S., and Feng, Z.P., Effects of salt on the formation of gas hydrate in porous media, in *The Fifth International Conference on Gas Hydrate*, Trondheim, Norway, pp. 155–160 (2005).
528. Sira, J.H., Patil, S.L., and Kamath, V.A., Study of hydrate dissociation by methanol and glycol injection, in *Proceedings—SPE Annual Technical Conference and Exhibition*, Society of Petroleum Engineers of AIME, Richardson, TX, pp. 977–984 (1990).
529. Ohgaki, K., Takano, K., Sangawa, H., Matsubara, T., and Nakano, S., Methane exploitation by carbon dioxide from gas hydrates—Phase equilibria for CO_2-CH_4 mixed hydrate system, *Journal of Chemical Engineering of Japan*, 29 (3), 478–483 (1996).
530. Seo, Y.T., Lee, H., and Yoon, J.H., Hydrate phase equilibria of the carbon dioxide, methane, and water system, *Journal of Chemical and Engineering Data*, 46 (2), 381–384 (2001).
531. Seo, Y.T. and Lee, H., Multiple-phase hydrate equilibria of the ternary carbon dioxide, methane, and water mixtures, *Journal of Physical Chemistry B*, 105 (41), 10084–10090 (2001).
532. Ersland, G., Husebo, J., Graue, A., Baldwin, B., Howard, J.J., and Stevens, J.C., Measuring gas hydrate formation and exchange with CO_2 in Bentheim sandstone using MRI tomography, *Chemical Engineering Journal*, 158, 25–31 (2010).
533. Farrell, H., Boswell, R., Howard, J., and Baker, R., CO_2-CH_4 exchange in natural gas hydrate reservoirs: Potential and challenges, *Fire in the Ice*, US Department of Energy, Office of Fossil Energy, National Energy Technology Laboratory, 10 (1), 19–21 (2010).
534. Graue, A., Kvamme, B., Baldwin, B.A., Steven, J., Howard, J., Aspenes, E., Ersland, G., Husebo, J., and Zornes, D., MRI visualization of spontaneous methane production from hydrates in sandstone core plugs when exposed to CO_2, *SPE Journal*, 13, 146–152 (2008).
535. Jung, J.W., Espinoza, D.N., and Santamarina, J.C., Properties and phenomena relevant to CH_4-CO_2 replacement in hydrate-bearing sediments, *Journal of Geophysical Research*, 115, B10102 (2010).
536. Lee, H., Seo, Y., Seo, Y., Moudrakovski, I., and Ripmeester, J., Recovering methane from solid methane hydrate with carbon dioxide, *Angewandte Chemie—International Edition*, 42 (41), 5048–5051 (2003).

537. Goel, N., In situ methane hydrate dissociation with carbon dioxide sequestration: Current knowledge and issues, *Journal of Petroleum Science and Engineering*, 51 (3–4), 169–184 (2006).

538. Smith, D.H., Seshadri, K., and Wilder, J.W., Assessing the thermodynamic feasibility of the conversion of methane hydrate into carbon dioxide hydrate in porous media, in *First National Conference on Carbon Sequestration*, Washington, DC (May 14–17, 2001), organized by National Energy Technology Laboratory, Pittsburgh, PA (2001).

539. Uchida, T., Takeya, S., Ebinuma, T., and Narita, H., Replacing methane with CO_2 in clathrate hydrate: Observations using Raman spectroscopy, in Williams, D.J., Durie, R.A., McMullan, P., Paulson, C.A.J., and Smith, A.Y. (eds.), *Proceedings of the Fifth International Conference on Greenhouse Gas Control Technologies*, CSIRO Publishing, Collingwood, Melbourne, Australia, pp. 523–527 (2001).

540. McGrail, B., Schaef, H., White, M., Zhu, T., Kulkarni, A., Hunter, R., Patil, S., Owen, A., and Martin, P., Using carbon dioxide to enhance recovery of methane from gas hydrate reservoirs: Final summary report, Prepared for US Department of Energy under contract no. DE-AC06-76RLO 1830, Pacific Northwest National Laboratory (PNNL 17035), PNNL, Richland, WA (September 2007).

541. White, M. and McGrail, P., Numerical simulation of methane hydrate production from geologic formations via carbon dioxide injection, in *Society of Petroleum Engineers, Offshore Technology Conference*, Houston, TX (May 5–8, 2008), Paper # OTC-19458 (2008), ISBN: 978-1-55563-224-3.

542. Yezdimer, E., Cummings, P., and Chalvo, A., Extraction of methane from its gas clathrate by carbon dioxide sequestration—Determination of the Gibbs Free Energy of gas replacement and molecular simulation, *Journal of Physical Chemistry A*, 106, 7982–7987 (2002).

543. Comparative assessment of advanced gas hydrate production methods, DOE/NETL methane hydrates projects, DE-FC26-06NT42666, DOE, Washington, DC (September 2009).

544. Moridis, G.J., Kowalsky, M.B., and Pruess, K., TOUGH+HYDRATE v1.0 User's Manual: A code for the simulation of system behaviour in hydrate-bearing porous media, Report LBNL-149E, Lawrence Berkeley National Laboratory, Berkeley, CA (2008).

545. Moridis, G.J., Reagan, M.T., Kim, S.J., Seol, Y., and Zhang, K., Evaluation of the gas production potential of marine hydrate deposits in the Ulleung Basin of the Korean East Sea, *SPE Journal*, 14 (4), 759–781 (2009).

546. Ota, M., Morohashi, K., Abe, Y., Watanabe, M., Smith, R., and Inomata, H., Replacement of CH_4 in the hydrate by use of liquid CO_2, *Energy Conversion and Management*, 46 (11–12), 1680–1691 (2005).

547. Hirohama, S., Shimoyama, Y., Wakabayashi, A., Tatsuta, S., and Nishida, N., Conversion of CH_4-hydrate to CO_2-hydrate in liquid CO_2, *Journal of Chemical Engineering of Japan*, 29 (6), 1014–1020 (1996).

548. Cortis, A. and Ghezzehei, T.A., On the transport of emulsions in porous media, *Journal of Colloid Interface Science*, 313 (1), 1–4 (2007).

549. Ruppel, C., Tapping methane hydrates for unconventional natural gas, *Elements*, 3 (3), 193–199 (2007).

550. Tegze, G., Gránásy, L., and Kvamme, B., Phase field modeling of CH_4 hydrate conversion into CO_2 hydrate in the presence of liquid CO_2, *Physical Chemistry Chemical Physics*, 9 (24), 3104–3111 (2007).

551. Yan, L., Thompson, K.E., and Valsaraj, K.T., A numerical study on the coalescence of emulsion droplets in a constricted capillary tube, *Journal of Colloid Interface Science*, 298 (2), 832–844 (2006).

552. Kurihara, M. et al., Analysis of the JOGMEC/NRCAN/Aurora Mallik gas hydrate production test through numerical simulation, in *Proceedings of the Sixth International Conference on Gas Hydrates*, Vancouver, British Columbia, Canada (2008).

553. Inks, T.L., Lee, M.W., Agena, W.F., Taylor, D.J., Collett, T.S., Zyrianova, M.V., and Hunter, R.B., Seismic prospecting for gas hydrate and associated free gas prospects in the Milne Point area of northern Alaska, in Collett, T., Johnson, A., Knapp, C., and Boswell, R. (eds.), *Natural Gas Hydrates—Energy Resource Potential and Associated Geologic Hazards*, AAPG Memoir 89, USGS Publication, US Geological Survey, Washington, DC, pp. 555–583 (2009).

554. Moridis, G.J. and Collett, T.S., Strategies for gas production from hydrate accumulations under various geologic conditions, Report LBNL-52568, Lawrence Berkeley National Laboratory, Berkeley, CA (2004).

555. Walsh, M.R., Hancock, S.H., Wilson, S.J., Patil, S.L., Moridis, G.J., Boswell, R., Collett, T.S., Koh, C.A., and Sloan, E.D., Preliminary report on the commercial viability of gas production from natural gas hydrates, *Energy Economics*, 31, 815–823 (2009).

556. Moridis, G.J., Collett, T., Dallimore, S., Satoh, T., Hancock, S., and Weatherhill, B., Numerical studies of gas production from several methane hydrate zones at the Mallik Site, Mackenzie Delta, Canada, *JPSE* 43, 219 (2004).

557. Pawar, R.J., Zyvoloski, G.A., Tenma, N., Sakamoto, Y., and Komai, T., Numerical simulation of gas production from methane hydrate reservoirs, in *Proceedings of Fifth International Conference on Gas Hydrates*, Vol. 1, June 13–16, Trondheim, Norway, Paper 1040, pp. 258–267 (2005).

558. Wilder, J.W. et al., An international effort to compare gas hydrate reservoir simulators, in *Proceedings of the Sixth International Conference on Gas Hydrates (ICGH 2008)*, Vancouver, British Columbia, Canada (2008).

559. Dallimore, S.R., Collett, T.S., Uchida, T., Weber, M., and Takahashi, H., Mallik Gas Hydrate Research Team; Overview of the 2002 Mallik gas hydrate production research well program, in *Proceedings of the Fourth International Conference on Gas Hydrates*, Yokohama Synopsis, Yokohama, Japan (May 19–23, 2002), Domestic Organizing Committee, ICGH-4, Vol. 1, pp. 36–39 (2002).

560. Hancock, S., Collett, T., Pooladi-Darvish, M., Gerami, S., Moridis, G., Okazawa, T., Osadetz, K., Dallimore, S., and Weatherill, B., A preliminary investigation on the economics of onshore gas hydrate production based on the Mallik Field discovery, in *American Association of Petroleum Geologists Hedberg Conference Proceedings*, Vancouver, British Columbia, Canada (2004).

561. Howe, S.J., Production modeling and economic evaluation of a potential gas hydrate pilot production program on the North Slope of Alaska, MS thesis, University of Alaska Fairbanks, Fairbanks, AK, 138pp. (2004).

562. Nakano, S., Yamamoto, K., and Ohgaki, K., Natural gas exploitation by carbon dioxide from gas hydrate fields—High-pressure phase equilibrium for an ethane hydrate system, *Proceedings of the Institution of Mechanical Engineers*, 212, 159–163 (1998).

563. Ohgaki, K., Takano, K., and Moritoki, M., Exploitation of CH_4 hydrates under the Nankai trough in combination with CO_2 storage, *Kagaku Kogaku Ronbunshu*, 20, 121–123 (1994).

564. Takahashi, H., Fercho, E., and Dallimore, S.R., Drilling and operations overview of the Mallik 2002 production research well program, in Dallimore, S.R. and Collett, T.S. (eds.), Scientific results from Mallik 2002 gas hydrate production research well program, Mackenzie Delta, Northwest Territories, Canada, Vol. Bulletin 585, Geological Survey of Canada, Vancouver, British Columbia, Canada (2005).

565. Sung, W.M., Lee, H., Kim, S., and Kang, H., Experimental investigation of production behaviors of methane hydrate saturated in porous rock, *Energy Sources*, 25 (8), 845–856 (2003).

566. Yousif, M.H., Abass, H.H., Selim, M.S., and Sloan, E.D., Experimental and theoretical investigation of methane-gas-hydrate dissociation in porous media, *SPE Reservoir Engineering*, 6 (4), 69–76 (1991).

567. Moridis, G.J. and Reagan, M., Strategies for production from oceanic Class 3 hydrate accumulations, OTC 18865, p. 29 (2007), www.osti.gov/scitech/servlets/purl/918823.

568. Park, Y., Kim, D., Lee, J., Huh, D., Park, K., Lee, J., and Lee, H., Sequestering carbon dioxide into complex structures of naturally occurring gas hydrates, *Proceedings of the National Academy of Sciences of the United States of America*, 103–134, 12690–12694 (2006).

569. Tohidi, B., Anderson, R., Clennell, M.B., Burgass, R.W., and Biderkab, A.B., Visual observation of gas-hydrate formation and dissociation in synthetic porous media by means of glass micromodels, *Geology*, 29 (9), 867–870 (2001).

570. Kuuskraa, V.A. and Brandenberg, C.F., Coalbed methane sparks a new energy industry, *Oil and Gas Journal*, 87 (41), 49–56 (1989).

571. Modern Shale gas development in the United States: An update, A report from NETL, Strategic center for natural gas and oil, Department of Energy, Washington, DC (September 2013).

572. Major tight gas plays, lower 48 states—EIA, U.S. Energy information, Washington, DC (2010). https://www.eia.gov/oil_gas/rpd/tight_gas.pdf.

3 Synthesis Gas by Thermal Gasification

3.1 INTRODUCTION

As discussed in Chapter 2, natural gas, both conventional and unconventional, largely contains methane. The percentage of methane in different sources of natural gas can, however, vary. For example, while conventional gas contains 90%–95% methane, shale gas contains around 50%–55% of methane, the remaining being carbon dioxide and some volatile hydrocarbons. With the use of appropriate purification treatments (described in Chapter 7), all types of natural gas can, however, use same storage and transportation infrastructure and can be used for heat and power productions as well as for the productions of liquid fuels and fuel additives with the help of reforming processes described in Chapter 6.

All forms of natural gas are nonrenewable because they come from underground and it took millions of years of geological transformation of buried plants and animals to produce them. Due to its limited availability, natural gas can only be considered as a "transition fuel." Natural gas is a more pure form of fossil chemical energy and in general less harmful to the environment compared to coal or oil. The leakage of methane in the atmosphere can, however, produce significant negative impacts on the environment because methane has a stronger negative effect on global warming than carbon dioxide over a shorter period of time.

Synthetic gas is man-made, and with available technology, it can be produced for unlimited time because the feedstock required for their production can be renewable like water, biomass, and waste, as well as nonrenewable like coal and oil. Unlike natural gas, man-made synthetic gas can significantly vary in its composition depending on the nature of feedstock and its method of generation.

Fundamentally, there are three different methods for synthetic gas generation: thermal gasification, anaerobic digestion, and hydrothermal gas production. Thermal gasification is a process of producing gas from a wide variety of feedstocks like coal, oil, biomass, and waste by thermal decomposition in the presence of air (or oxygen) or carbon dioxide. The process generates synthetic or synthesis gas, which is a mixture of methane, hydrogen, carbon monoxide, carbon dioxide, water, and maybe some impurities. Coal and biomass/waste or their mixtures are the most commonly used feedstock. Thermal gasification is a well-established commercial technology and it is being practiced all over the world. Many modifications of the conventional thermal gasification technology such as underground gasification, catalytic gasification, plasma gasification, solar gasification, indirect gasification, and molten media gasification are also being explored. This chapter examines all of these technologies in full detail.

The second method for synthetic gas production is anaerobic digestion, a process that produces gas from landfills all over the world. The gas produced by this fermentation process is called "biogas." In recent years, this process has been managed and optimized to obtain biogas from different types of aqueous biological waste streams. While most man-made biogas generation facilities produce methane (about 50%–55%, the remaining being largely carbon dioxide), with the use of different types of microbes and operating conditions, this process can also generate hydrogen. Chapter 4 evaluates this method of synthetic gas production in full detail.

Hydrothermal process involves water (either as steam, water, or supercritical water) to produce synthetic gas from all types of carbonaceous substances. Steam gasification and hydrothermal gasification in subcritical and supercritical water are very closely aligned with steam reforming, aqueous phase reforming, and supercritical water reforming. Sometimes hydrothermal gasification also uses hydrogen to improve hydrogen concentration in the product. All hydrothermal processes generate gas that predominantly contains hydrogen or syngas. Hydrothermal process also involves direct dissociation of water by numerous technologies to produce hydrogen. Chapter 5 examines both types of hydrothermal processes in full details.

Thermal gasification process was discovered in 1699 by Dean Clayton as mentioned in Reference 1. It was implemented during the nineteenth century in factories for producing town gas. The first gas plant was established in 1812 in London. With the discovery of the Fischer–Tropsch (FT) process in 1923 by Franz Fischer and Hans Tropsch, it became possible to convert coal to liquid fuel like gasoline, diesel, and jet fuel. During World War II, the German Army needed to improve the use of the gasification process for fuel and chemical production. They used FT process along with gasification to produce liquid fuels needed for war.

Gasification is an incomplete oxidation of organic compounds after a pyrolysis decomposition step. The oxygen contained in the oxidizing agent used for the gasification (air, oxygen, CO_2, or steam) reacts with carbon to achieve a combustible gas, called "producer gas." This producer gas is mainly composed of carbon monoxide (CO) and hydrogen (H_2) with low quantities of carbon dioxide (CO_2), water (H_2O), methane (CH_4), hydrogen sulfide (H_2S), ammonia (NH_3), and, under certain conditions, solid carbon (C), nitrogen (N_2), argon (Ar), and some tar traces. Nitrogen and argon form from the use of air as the reactant or are due to their use as plasma gas. The composition of gas depends on the temperature and pressure of the gasification process. At high temperatures and in the presence of pure oxygen (instead of air), gasification produces "syngas" that mainly consists of carbon monoxide and hydrogen with some carbon dioxide, water, and other impurities depending on the feedstock. While producer gas can contain significant amount of methane, syngas contains very little methane [1–19].

The conversion of carbonaceous materials into syngas involves complex chemical reactions. Heterogeneous reactions take place in gas–solid phase, while homogeneous reactions occur in gas–gas phase. The main chemical reactions of gasification occurring after the pyrolysis of the carbonaceous materials like coal, oil shale, biomass, and waste are given in Table 3.1 [1–19].

The homogeneous reactions (reactions 1–4 in Table 3.1) are almost instantaneous in high-temperature conditions in contrast to the heterogeneous reactions (reactions 5–12 in Table 3.1). A very large number of gasification reactions take place in the reactor, but we can differentiate three of them that are independent gasification reactions: water gas reaction (10), Boudouard reaction (11), and hydrogasification (12). In the gas phase, these reactions can be reduced to only two reactions: water gas shift reaction (henceforth

---

TABLE 3.1

Main Chemical Reactions of Gasification

(1) Carbon monoxide oxidation	$CO + (1/2)O_2 \rightarrow CO_2$	Exothermic
(2) Hydrogen oxidation	$H_2 + (1/2)O_2 \rightarrow H_2O$	Exothermic
(3) WGS reaction	$CO + H_2O \rightarrow CO_2 + H_2$	Exothermic
(4) Methanation	$CO + 3H_2 \rightarrow CH_4 + H_2O$	Exothermic
(5) $C_nH_mO_k$ partial oxidation	$C_nH_m + (n/2)O_2 \rightarrow (m/2)H_2 + nCO$	Exothermic
(6) Steam reforming	$C_nH_m + nH_2O \rightarrow (n + m/2)H_2 + nCO$	Endothermic
(7) Dry reforming	$C_nH_m + nCO_2 \rightarrow (m/2)H_2 + 2nCO$	Endothermic
(8) Carbon oxidation	$C + O_2 \rightarrow CO_2$	Exothermic
(9) Carbon partial oxidation	$C + (1/2)O_2 \rightarrow CO$	Exothermic
(10) Water gas reaction	$C + H_2O \rightarrow CO + H_2$	Endothermic
(11) Boudouard reaction	$C + CO_2 \rightarrow 2CO$	Endothermic
(12) Hydrogasification	$C + 2H_2 \rightarrow CH_4$	Exothermic

Sources: Prepared from information in References 1–19.

denoted as WGS reaction), (3) which is the combination of the reactions (10) and (11), and methanation (4), which is the combination of the reactions (10) and (12).

It is important to notice that all these gasification reactions, except the oxidation ones, are equilibrium reactions. The final composition of the crude producer gas or syngas will be determined by reaction rates that in turn depend on temperature, pressure, and catalysts, if present. Tar decomposition is aided by temperature, pressure, and catalyst.

There are presently several tens of different gasification processes that differ by the feedstock, configuration of the reactors, and oxidizing medium. These different configurations are fully described in numerous books [2–6,11,17,19], reports [7,8,13], and reviews [9,12,14–16,18]. While the main reactor configurations are fixed-bed (or moving), FB (bubbling, turbulent, or circulating), and entrained-bed reactors, numerous modifications in each category are also implemented. Along with reactor configuration, feed pretreatment, process configuration, and operating parameters affect the production rate and quality of syngas. This chapter briefly assesses all of these interactions.

Gasification systems are a promising approach for clean and efficient power generation as well as for polygeneration of a variety of products, such as steam, sulfur, hydrogen, methanol, and ammonia [8]. As of 1996, there were 354 gasifiers located at 113 facilities worldwide. The gasifiers use solid fuels (petroleum residuals, petroleum coke, refinery wastes, coal, biomass, municipal solid waste [MSW], and other fuels) as inputs and produce a synthesis gas containing carbon monoxide (CO), hydrogen (H_2), and other components. The syngas can be processed to produce liquid and gaseous fuels, chemicals, and electric power by using it in gas turbine or fuel cell. In recent years, gasification has received increasing attention as an option for repowering at oil refineries, where there is currently a lack of markets for low-value liquid residues and coke [1–27]. With increasing use of gasification for biomass and waste, efforts are also being made to process mixtures of coal and biomass/waste. Novel approaches such as solar, plasma, and indirect gasification are also being investigated. This chapter addresses all of these topics.

3.2 PARAMETERS AFFECTING SYNTHESIS GAS PRODUCTION AND ITS QUALITY

Thermal gasification can produce a wide variety of gaseous fuels that can vary in composition of methane, hydrogen, carbon monoxide, carbon dioxide, and other impurities. Generally, at low temperatures (<1000°C), gasification generates producer gas of different Btu contents. The Btu content is largely dictated by the concentration of methane and the concentration of diluent like nitrogen in the producer gas; the higher the methane content and the lower the diluent concentration, the higher the Btu content of the gas. At high temperatures (>1200°C), thermal gasification produces syngas that largely contains hydrogen and carbon monoxide with some carbon dioxide but very little or no methane. While producer gas is useful for heat and power applications, syngas is more versatile and can be used for heat, power, and production of numerous liquid fuels and fuel additives. It should be noted, though, that producer gas can be converted to syngas by the process of gas reforming, a subject covered in great detail in Chapter 6.

3.2.1 FEEDSTOCK

Currently, thermal gasification process is largely applied to coal, biomass/waste, and their mixtures. In this chapter, we evaluate our state of knowledge for all three feedstocks. Of the three feedstocks, coal gasification is the most mature commercial technology. The gasification of biomass and waste is continually developed at a larger scale. The industry is also very much interested in the processing mixtures of coal, biomass, and waste in large-scale coal gasification plants. In some parts of the world (e.g., Estonia, Turkey, Russia), efforts are also being made to produce syngas from oil shale. This subject is briefly covered in Section 3.3.

Gasification of petroleum-derived or bio-derived oil can also result in the production of synthetic gases. This is, however, not normally done because oil can be directly used for various downstream operations. Pyrolysis of naphtha to generate ethylene and propylene is a commercialized process. Ethylene and propylene are good feedstocks for the productions of polyethylene (PE) and polypropylene. Pyrolysis of biomass can also result in biooil, which can be gasified to produce synthetic gas. This chapter will mostly emphasize the production of synthetic gas from coal and biomass/waste.

Coal and biomass (or organic waste) are very different feedstocks, with different degrees of reactivity, different compositions, and different volatile matter (VM) contents [2–6,11,17,19,21–36]. For example, coal contains a significant amount of organic ash, low VM (depending on rank), and in some cases, a high amount of sulfur. Biomass, on the other hand, contains high oxygen and VMs, inorganic ash, other inorganic matters, and, in some cases, more chlorine and less sulfur and more alkaline matters and heavy metals [2–6,11,17,19,21–36]. Biomass is more reactive than coal and decomposes at lower temperature than coal. The basic elemental composition of coal, which includes carbon, oxygen, nitrogen, hydrogen, sulfur, etc., is very different from that of biomass and waste. Coal is more organic and aromatic, while except for lignin, biomass is more cellulosic in its composition. Within coal, the composition and reactivity of coal depends on its rank (i.e., lignite, subbituminous, bituminous, anthracite). Biomass is very heterogeneous and its composition can vary widely depending on its source. The thermal gasification of coal and biomass thus results in different compositions of synthetic gas and requires somewhat different operating conditions within the gasification reactors.

Thermal decomposition of coal is an established commercial technology and it is being practiced on a large scale for several decades to produce high-, medium-, and low-Btu synthetic (producer) gas [2,5,7,19]. Coal gasification can also be used to generate syngas, which in turn can be used to generate heat, power, and other liquid fuels. In recent years, the carbon dioxide generated by coal gasification/combustion plants has led to a slowdown in the coal industry. Efforts are being made to install carbon capture technology in coal combustion and gasification plants. Most coal gasification plants use bituminous or subbituminous coals. While lignite coals are more reactive, they are less energy efficient due to high moisture content [2,5,7,19,21]. Anthracite has high heating value but low hydrogen content. It is, therefore, largely used to generate heat and not syngas. Oil shale contains kerogen that can generate shale oil that is a good raw material for jet fuel. High temperature and pressure and the presence of hydrogen and catalyst can help in the production of syngas from oil shale [37–46].

Biomass and waste gasification plants are also aggressively developed in many parts of the world [3,9–16]. Biomass combines solar energy and carbon dioxide into chemical energy in the form of carbohydrates via photosynthesis. The use of biomass as a fuel is a carbon-neutral process since the carbon dioxide captured during photosynthesis is released during its combustion/gasification. Biomass includes agricultural and forestry residues, wood, by-products from processing of biological materials, and the organic parts of municipal and sludge wastes. Photosynthesis by plants captures around 4000 EJ/year in the form of energy in biomass and food [3,16].

The estimates of potential global biomass energy vary widely in literature [3,9–16]. The variability arises from the different sources of biomass and the different methods of determining estimates for those biomasses. Fischer and Schrattenholzer [16] estimated the global biomass potential to be 91–675 EJ/year for the years 1990–2060. Their biomass included crop and forestry residues, energy crops, and animal and municipal wastes. Hoogwijk et al. [15] estimated these to be 33–1135 EJ/year for energy crops on marginal and degraded lands, agricultural and forestry residues, animal manure, and organic wastes. Parikka [9,10] estimated the total worldwide energy potential from biomass on a sustainable basis to be 104 EJ/year, of which woody biomass, energy crops, and straw constituted 40.1%, 36%, and 16.6%, respectively. Only about 40% of potential biomass energy is currently utilized. Only in Asia does the current biomass usage slightly exceed the sustainable biomass potential. Currently, the total global energy demand is about 470 EJ/year. Perlack et al. [13] estimated that, in the United States, without many changes in land use and without interfering with the production of food grains, 1.3 billion tons of biomass can be harvested each year on a sustainable basis for biofuel production. This amount of biomass is equivalent to 3.8 billion barrels of oil in energy content. U.S. equivalent energy consumption is about 7 billion barrels/year [20]. However, harvesting, collecting, and storing of biomass adds another dimension of technical challenges to the use of biomass for the production of fuels, chemicals, and biopower [3,16].

High oxygen content in biomass reduces the energy density of the biomass. The production of hydrocarbons, similar to petroleum transportation fuels, requires the removal of oxygen from the carbohydrate structure. Oxygen may be removed in the form of CO_2 and H_2O. Thermochemical conversion of biomass to syngas is an attractive route to extract oxygen from carbohydrate structures to produce intermediate compounds having C_1 (CO and CH_4), which can be further synthesized into hydrocarbons by catalysis or fermentation. Other thermochemical schemes of decarboxylation (CO_2 removal) and dehydration (H_2O removal) from carbohydrates result in higher hydrocarbons (higher than C_2) having undesired properties that require further conversion to be compatible with transportation fuels [3,16]. In general, at low temperatures, biomass and waste generate producer gas with large amount of tars (polyaromatic compounds), and at high temperatures, they generate syngas with very low tar content [3,16,21–23].

Thermochemical conversion technologies have certain advantages and disadvantages over biochemical conversion technologies. The main advantages are that (1) the feedstock for thermochemical conversion can be any type of biomass, including agricultural residues, forestry residues, nonfermentable by-products from biorefineries, by-products of food industry, by-products of any bioprocessing facility, and even organic municipal wastes and (2) the product gases can be converted to a variety of fuels (H_2, FT diesels, synthetic gasoline) and chemicals (methanol, urea) as substitutes for petroleum-based chemicals; the products are more compatible with existing petroleum-refining operations.

The major disadvantages are the high cost associated with the cleaning of the product gas from tar and undesirable contaminants like alkali and S, N, and Cl compounds, inefficiency due to the high temperatures required, and the unproven use of products (biosyngas and biooil) as transportation fuels. On a larger scale, however, research on the optimization of gasifier operating conditions and heat recovery, syngas cleaning, biooil stabilization, and efficient product utilization can make the process important for sustainable production of biofuels. Expensive feed pretreatment for gasifiers can also be an issue. With life cycle assessment, Wu et al. [18] concluded that the use of cellulosic biofuels (ethanol via gasification and fermentation, FT diesel and dimethyl ether [DME] from biomass, etc.) in light duty locomotives results in significant savings of fossil fuel resources and reduction in greenhouse gases (GHGs). Coproduction of cellulosic biofuels and power generation by GTCC consumes the least fossil fuel resources and results in the greatest reduction in GHG emissions on a per-mile basis of the thermochemical conversion techniques [21].

Biomass gasification differs from coal gasification. Biomass is a carbon-neutral and sustainable energy source unlike coal. Because biomass is more reactive and has higher volatile content than coal, biomass gasification occurs at a lower temperature. Lower temperature reduces the extent of heat loss, emissions, and material problems associated with high temperatures. Biomass also has low sulfur content, which results in lower SO_x emission. But the high alkali contents in biomass, like sodium and potassium, cause slagging and fouling problems in gasification equipment [3,9–16,21–23].

Biomass gasification can be considered as one of the competitive ways of converting distributed and low-value lignocellulosic biomass to fuel gas for combined heat and power (CHP) generation and fuel cell and synthetic diesel production. However, from the collection of biomass to the utilization of fuel gas for downstream application, the process suffers numerous problems that slow down the commercial exploitation of biomass-based energy technology [3,9–16,21–23].

One method to take advantage of large-scale operation of coal and carbon-neutral conversion of biomass/waste is to process coal/biomass/waste mixture for gasification. This method has gained significant momentum in recent years. This chapter also evaluates the opportunities and challenges of this method of gasification. Processing of coal/biomass mixtures requires different process configurations and can result in different compositions of synthesis gas. The chapter briefly addresses these issues as well [21].

3.2.2 NATURE OF GASIFICATION PROCESS

Thermal destruction of coal and biomass/waste can be carried out in excess of oxygen (combustion), below stoichiometric oxygen requirement environment (gasification) or in the absence of oxygen (pyrolysis). The destruction can also be carried out in the presence of carbon dioxide, methane, hydrogen, and steam as well as a mixture of them. These differences result in the different types of gasification processes and different product distributions. The combustion of coal and biomass/waste results in flue gases with little or no heating value. This process is outside the scope of this book. Thermal gasification in the presence of steam or hydrogen is considered here as a part of the hydrothermal process, which is covered in detail in Chapter 5. While gasification can be carried out in the presence of oxygen, carbon dioxide, methane, etc., this chapter mainly focuses on oxidative (in the presence of oxygen or air) gasification since that produces the best quality of syngas and it is most practical to implement. The gasification in the presence of methane can be very expensive. Also coal is not very reactive in the presence of carbon dioxide [19].

It should be noted that gaseous fuel (mixtures of hydrocarbons, hydrogen, and carbon monoxide) can also be produced by pyrolysis (i.e., decomposition in the absence of oxygen) under certain operating conditions of residence time, temperature, and heating rate. The pyrolysis process is generally carried out by subjecting the coal or biomass to a high temperature under an inert or oxygen-deficient atmosphere. The pyrolysis of coal results in the gas that are very high in VOC and not suitable for syngas production. In the past, biomass has been used for the production of hydrogen and medium-Btu gas [3,9–16,23]. Extensive studies have been done on the pyrolysis of cellulose, wood, and biomass materials [3,9–16,23–25].

The fast pyrolysis process of biomass generally gives three products, namely, gas, biomass-derived oil (BDO), and char. The biooil thus produced contains unsaturated hydrocarbons and is thus highly unstable. This BDO has found a variety of applications in various areas. Unlike fossil fuels, BDO is renewable, cleanly burns, and is GHG neutral. It does not produce any SO_x (sulfur dioxide) emission during combustion but produces approximately half the NO_x (nitrogen oxide) emission in comparison with fossil fuels. Therefore, it is a potential raw material for renewable fuel and can be used as a fuel oil substitute [3,9–16,23–25]. However, several challenges are identified in biooil applications resulting from their properties. Biooil contains large amount of oxygen and tends to be unstable over long period.

Some research has been done on catalytic upgrading of BDO [3,9–16,23–25]. The product gas consisted of H_2 CO, CO_2, CH_4, C_2–C_4, and higher hydrocarbons. The BDO has been converted to hydrogen via catalytic steam reforming followed by a shift conversion step [3,9–16,23–25]. The hydrogen yield was as high as 85%. An attempt has been made to produce clean fuels including hydrogen, high-Btu gaseous fuel, and synthesis gas by pyrolysis of BDO in the absence of a catalyst and relatively high temperatures (650°C–800°C). Thus, pyrolysis of BDO can generate gaseous fuel. In this process, the product gases essentially consisted of H_2, CO, CO_2, CH_4, C_2H_4, C_2H_6, C_3H_8, and C_{4+} components. This process can be utilized for producing hydrocarbons and synthesis gas for various applications. By adjusting parameters such as inert gas flow rate and reactor temperature, the composition of the product gas from BDO can be tuned in the desired direction [3,22,23].

Gasification is a process related to pyrolysis, and the most significant difference between the two processes is that gasification occurs in the presence of oxygen. Gases and liquids produced in the pyrolysis process can be used for heat and electricity or chemical production. Gasification mostly produces gases and solids. While the gas produced by the pyrolysis process consists mainly of H_2, CO, CO_2, CH_4, and other low-molecular-weight hydrocarbons and is used as fuel gas, gaseous

products obtained by gasification consist of hydrogen (H_2), carbon monoxide (CO), carbon dioxide (CO_2), and methane. If the objective is to produce syngas, gasification is a better process than pyrolysis [3,22,23].

It should be, however, noted that valuable gases, such as H_2 and CO, can also be generated by pyrolysis. These gases can be useful, among other applications, in chemical synthesis and high-efficiency combustion systems such as fuel cells. The quantity and composition of gases produced in the pyrolysis process significantly depend on numerous operating parameters such as temperature, pressure, presence of a catalyst, heating rate, and residence time. Pyrolysis is also an endothermic process and requires a significant supply of external heat, particularly if operated at high temperatures [3,22,23]. Gasification, on the other hand, is an exothermic process. The composition of synthesis gas produced by gasification depends not only on the temperature, pressure, presence of a catalyst, residence time in the reactor, and nature of gaseous environment but also on the nature of the gasification reactor and the gasification process. These effects on gasification are described in detail in the rest of this chapter.

3.3   COAL AND OIL SHALE GASIFICATION

While coal is not as heterogeneous as biomass, based on its hydrogen-to-carbon ratio and age, coal is generally divided into four ranks. Anthracite is the most aged coal with the lowest hydrogen-to-carbon ratio. While it has a very high heating value, low hydrogen does not make it suitable for the production of syngas by gasification. On the other extreme, lignite has high hydrogen-to-carbon ratio but it also has high amount of oxygen content. It is a very reactive coal, but once again due to high moisture content, its gasification is not very energy efficient and produces low yield of syngas per unit mass. Bituminous and subbituminous coals are in the average rate and they make good candidates for coal gasification to produce syngas. Bituminous coal is perhaps the most widely used coal for gasification. The sulfur and organic ash contents in the coal are also important parameters for a particular coal effectiveness for gasification. More details on coal characterization are given by Lee [19], Shah [21,26], and others [1,4–8,27,28].

In coal gasification reactors, the feedstock is converted into a synthesis gas (syngas), a mixture of H_2, CO, and CO_2, which enables the production of a variety of downstream energy carriers. A large experience exists on coal gasification worldwide as the so-called town gas was produced from coal as early as 1792, a high-temperature fluidized-bed (FB) gasifier was patented in 1921 by Winkler, and synfuel production from coal was common practice in Germany during World War II [1,2,4–8,19,34]. According to the Gasification Technologies Council, in 2007, some 144 gasification plants and 427 gasifiers were in operation worldwide, adding up to an equivalent thermal capacity of 56 GWth, of which coal gasification accounted for approximately 31 GWth [1,2,4–8,19,34].

Gasification takes place under oxygen shortage. Coal is first heated in a closed reaction chamber where it undergoes a pyrolysis process at temperatures above 400°C. During pyrolysis, hydrogen-rich VM is released, along with tar, phenols, and gaseous hydrocarbons. Then, char is gasified, with the release of gases, tar vapors, and solid residues. The dominant reactions consist of partial oxidation of char, which produces a syngas with high fractions of H_2 and CO. The process takes place at temperatures between 800°C and 1800°C. Specific operating conditions depend on coal type, on properties of the resulting ash, and on the gasification technology. The most important variable in a gasification process is the oxidant. It can be either air (with its nitrogen component) or pure oxygen if the process includes an air separation unit (ASU) for oxygen production [1,2,4–8,19,34].

The use of oxygen instead of air facilitates the partial combustion of coal but involves higher investment costs due to costly additional equipment. As gasification takes place under stoichiometric shortage of oxygen, the reaction mechanism in the gasification chamber has to be adjusted with appropriate energy balance. The direct partial oxidation of carbon to CO, for instance, is strongly exothermic, leading to a high release of energy in the form of sensible heat. However, steam gasification of coal (forming both CO and H_2) is strongly endothermic. As a consequence, a steam/oxygen mixture is commonly used. In the gasification practice, the basic equipment can be grouped in three main categories: moving-bed gasifiers, FB gasifiers, and entrained-flow gasifiers. When gasification is performed at an elevated pressure, the resulting syngas is typically at a higher pressure and not diluted by nitrogen, allowing for much easier, efficient, and less costly removal of CO_2 [1,2,4–8,19,34].

If the coal is heated by external heat sources, the process is called "allothermal," while "autothermal" process assumes heating of the coal via exothermal chemical reactions occurring inside the gasifier itself. Oxygen and water molecules oxidize the coal and produce a gaseous mixture of carbon dioxide (CO_2), carbon monoxide (CO), water vapor (H_2O), and molecular hydrogen (H_2) and methane. Some by-products like tar and phenols are also possible end products, depending on the specific gasification technology utilized. The desired end product is usually syngas (i.e., a combination of H_2 + CO), but the produced coal gas may also be further refined to produce additional quantities of H_2 [19,34]:

$$3C \text{ (i.e., coal)} + O_2 + H_2O \rightarrow H_2 + 3CO \tag{3.1}$$

If hydrogen is the desired end product, the coal gas (primarily the CO product) undergoes the WGS reaction where more hydrogen is produced by additional reaction with water vapor [19,34]:

$$CO + H_2O \rightarrow CO_2 + H_2 \tag{3.2}$$

3.3.1 SYNTHETIC GAS (PRODUCER GAS) CLASSIFICATION BASED ON ITS HEATING VALUE

Depending on the heating values of the resultant producer gases generated by gasification processes, product gases are typically classified as three types of gas mixtures [1,2,4–9,19,34]:

1. *Low-Btu gas* consisting of a mixture of carbon monoxide, hydrogen, and some other gases with a heating value typically less than 300 Btu/scf.
2. *Medium-Btu gas* consisting of a mixture of methane, carbon monoxide, hydrogen, and various other gases with a heating value in the range of 300–700 Btu/scf.
3. *High-Btu gas* consisting predominantly of methane with a heating value of approximately 1000 Btu/scf. It is also referred to as "synthetic or substitute natural gas" (*SNG*).

Coal gasification involves the reaction of coal carbon (precisely speaking, macromolecular coal hydrocarbons) and other pyrolysis products with oxygen, hydrogen, and water to provide fuel gases.

3.3.1.1 Low-Btu Gas

For the production of low-Btu gases, air is typically used as a combusting (or gasifying) agent. As air, instead of pure oxygen, is used, the product gas inevitably contains a large concentration of undesirable constituents such as nitrogen or nitrogen-containing compounds. Therefore, it results in a low heating value of 150–300 Btu/scf [1,2,4–9,19,34].

Sometimes, this type of gasification of coal may be carried out in situ, that is, underground, where mining of coal by other techniques is not economically favorable. For such in situ gasification, low-Btu gas may be a desired product. Low-Btu gas contains five principal components with around 50% v/v nitrogen, some quantities of hydrogen and carbon monoxide (combustible), carbon dioxide, and some traces of methane. The presence of such high contents of nitrogen classifies the product gas as low Btu. The other two noncombustible components (CO_2 and H_2O) further lower the heating value of the product gas. The presence of these components limits the applicability of low-Btu gas to chemical synthesis. The two major combustible components are hydrogen and carbon monoxide; their ratio varies depending on the gasification conditions employed. One of the most undesirable components is hydrogen sulfide (H_2S), which occurs in a ratio proportional to the sulfur content of the original coal. It must be removed by gas-cleaning procedures before product gas can be used for other useful purposes such as further processing and upgrading. Methane concentration can also affect the heating value [1,2,4–8,19,34]. Underground gasification is further discussed in Section 3.10.

3.3.1.2 Medium-Btu Gas

In the production of medium-Btu gas, pure oxygen rather than air is used as the combusting agent, which results in an appreciable increase in the heating value, by about 300–400 Btu/scf [19,34]. The product gas predominantly contains carbon monoxide and hydrogen with some methane and carbon dioxide. It is primarily used in the *synthesis of methanol*, higher hydrocarbons via *FT synthesis*, and a variety of other chemicals. It can also be used directly as a fuel to generate steam or to drive a gas turbine. The *H_2-to-CO ratio* in medium-Btu gas varies from 2:3 (CO rich) to more than 3:1 (H_2 rich). The increased heating value is attributed to higher contents of methane and hydrogen as well as to lower concentration of carbon dioxide, in addition to the absence of nitrogen in the gasifying agent [1,2,4–8,19,34].

3.3.1.3 High-Btu Gas

High-Btu gas consists mainly of pure methane (>95%) and, as such, its heating value is around 900–1000 Btu/scf [19,34]. It is compatible with natural gas and can be used as an SNG. This type of syngas is usually produced by the catalytic reaction of carbon monoxide and hydrogen, which is called the "methanation reaction." Feed syngas usually contains carbon dioxide and methane in small amounts. Further, steam is usually present in the gas or added to the feed to alleviate carbon fouling, which alters the catalytic effectiveness. Therefore, the pertinent chemical reactions in the methanation system include the following [1,2,4–8,19,27,28,34]:

$$3H_2 + CO = CH_4 + H_2O \tag{3.3}$$

$$2H_2 + 2CO = CH_4 + CO_2 \tag{3.4}$$

$$4H_2 + CO_2 = CH_4 + 2H_2O \tag{3.5}$$

$$2CO = C + CO_2 \tag{3.6}$$

$$CO + H_2O = CO_2 + H_2 \tag{3.7}$$

Among these, the most dominant chemical reaction leading to methane is the first one. Therefore, if methanation is carried out over a catalyst with a syngas mixture of H_2 and CO, the desired H_2-to-CO ratio of the feed syngas is around 3:1. The large amount of H_2O produced is removed by condensation and recirculated as process water or steam. During this process, most of the exothermic heat due to the methanation reaction is also recovered through a variety of energy integration processes. Whereas all the reactions listed earlier are quite strongly exothermic except the forward WGS reaction, which is mildly exothermic, the heat release depends largely on the amount of CO present in the feed syngas. For each 1% of CO in the feed syngas, an adiabatic reaction will experience a 60°C temperature rise, which may be termed as "adiabatic temperature rise" [1,4–8,19,27,28].

A variety of metals exhibit catalytic effects on the methanation reaction. While the order of catalytic activity is Ru > Ni > Co > Fe > Mo, nickel is by far the most commonly used catalyst in commercial processes because of its relatively low cost and also of reasonably high catalytic activity. Nearly all the commercially available catalysts used for this process are, however, very susceptible to sulfur poisoning and efforts must be taken to remove all hydrogen sulfide (H_2S) before the catalytic reaction starts. It is necessary to reduce the sulfur concentration in the feed gas to lower than 0.5 ppm in order to maintain adequate catalyst activity for a long period of time. Therefore, the objective of the catalyst development has been aimed at enhancing the *sulfur tolerance* of the catalyst.

Some of the noteworthy commercial methanation processes include Comflux, HICOM, and direct methanation. Comflux is a Ni-based, pressurized FB (PFB) process converting CO-rich gases into SNG in a single stage, where both methanation and WGS reaction take place simultaneously. The HICOM process developed by the British Gas Corporation is a fixed-bed process that involves a series of methanation stages using relatively low H_2-to-CO ratio syngas. Direct methanation is a process developed by the Gas Research Institute, which methanates equimolar mixtures of H_2 and CO, producing CO_2 rather than H_2O (steam) in addition to methane [1,4–8,19,27,28,34]:

$$2H_2 + 2CO = CH_4 + CO_2 \qquad (3.8)$$

The catalyst developed is claimed to be unaffected by sulfur poisoning and, as such, the process can be used to treat raw, quenched gas from a coal gasifier with no or little pretreatment [1,4–8,19,27,28,34].

3.3.2 GENERAL ASPECTS OF COAL GASIFICATION

The kinetic rates and extents of conversion for various gasification reactions are typically functions of temperature, pressure, gas composition, and the nature of the coal being gasified. The rate of reaction is intrinsically higher at higher temperatures, whereas the equilibrium of the reaction may be favored at either higher or lower temperatures depending on the specific type of gasification reaction. The effect of pressure on the rate also depends on the specific reaction. Thermodynamically, some gasification reactions such as carbon–hydrogen reaction producing methane are favored at high pressures (>70 atm) and relatively lower temperatures (760°C–930°C), whereas low pressures and high temperatures favor the production of syngas (i.e., carbon monoxide and hydrogen) via steam or carbon dioxide gasification reaction.

Supply and recovery of heat is a key element in the gasification process from the standpoints of economics, design, and operability. Partial oxidation of char with steam and oxygen leads to the generation of heat and synthesis gas. Another way to produce a hot gas stream is via the cyclic reduction and oxidation of iron ore. The type of coal being gasified is also important to the gasification and downstream operations. Only suspension-type gasifiers such as entrained-flow reactor can handle any type of coal, but if caking coals are to be used in fixed bed or FB, special measures must be taken so that the coal does not agglomerate (or cake) during gasification [1,4–8,19,27,28,34].

If such agglomeration does happen, it would adversely affect the operability of the gasification process. In addition to this, the chemical composition, the VM content, and the moisture content of coal also play important roles in coal processing during gasification. The S and N contents of coal seriously affect the quality of the product gas, as well as the gas-cleaning requirements. The sulfur content of coal typically comes from three different sources of coal sulfur, namely, pyritic sulfur, organic sulfur, and sulfatic sulfur. The first two are more dominant sulfur forms, whereas weathered or oxidized coals have more sulfatic forms than fresh coals. Sulfurous gas species can be sulfur dioxide, hydrogen sulfide, or mercaptans, depending on the nature of the reactive environment. If the reactive environment is oxidative, sulfur dioxide is the most dominant sulfur-containing species in the product gas [1,4–8,19,27–34].

In coal gasification, four principal reactions are crucial:

1. Steam gasification
2. Carbon dioxide gasification
3. Hydrogasification
4. Partial oxidation reaction

The subject of steam gasification and hydrogasification with and without steam is discussed in detail in Chapter 5 under hydrothermal gasification. In most gasifiers, several of these reactions, along with the WGS reaction, occur simultaneously. Equilibrium constants (K_p's) for these reactions as functions of temperature (T) are described by Lee [19]. He draws the following conclusions:

1. The plots of log10 K_p versus $1/T$ are nearly linear for all reactions.
2. The exothermicity of reaction is on the same order as the slope of the plot of log10 K_p versus $1/T$ for each reaction.
3. By the criterion of $K_p > 1$ (i.e., log10 $K_p > 0$), it is found that hydrogasification is thermodynamically favored at lower temperatures, whereas CO_2 and steam gasification reactions are thermodynamically favored at higher temperatures.
4. The equilibrium constant for the WGS reaction is the weakest function of the temperature among all the compared reactions. This also means that the equilibrium of this reaction can be reversed relatively easily by changing the imposed operating conditions [19]. The subject of WGS reaction is discussed in detail in Chapter 6.

3.3.3 CARBON DIOXIDE GASIFICATION

The reaction of coal with CO_2 may be approximated or simplified as the reaction of carbon with carbon dioxide, for modeling purposes. Carbon dioxide reacts with carbon to produce carbon monoxide and this reaction is called "Boudouard reaction." This reaction is also endothermic in nature, similar to the steam gasification reaction [19].

$$C(s) + CO_2(g) = 2CO(g) \quad \Delta H^{\circ}_{298} = 172.5 \text{ kJ/mol} \tag{3.9}$$

The reverse reaction is a carbon deposition reaction that is a major culprit of carbon fouling on many surfaces, such as process catalyst deactivation. This gasification reaction is thermodynamically favored at high temperatures ($T > 680°C$), which is also quite similar to steam gasification. The reaction, if carried out alone, requires high temperature (for fast reaction) and high pressure (for higher reactant concentrations) for significant conversion. However, this reaction in practical gasification applications is almost never attempted as a solo chemical reaction, because of a variety of factors, including low conversion, slow kinetic rate, low thermal efficiency, and unimpressive process economics.

There is a general agreement that experimental data on the rate of carbon gasification by CO_2 fit an empirical equation of the form (3.10)

$$r = \frac{k_1 p_{CO_2}}{(1 + k_2 p_{CO} + k_3 p_{CO_2})} \tag{3.10}$$

where p_{CO} and p_{CO_2} are the partial pressures of CO and CO_2 in the reactor. This rate equation is shown to be consistent with at least two mechanisms whereby carbon monoxide retards the gasification reaction [19,35]:

Mechanism A

$$C_f + CO_2 \rightarrow C(O)_A + CO \qquad (3.11)$$

$$C(O)_A \rightarrow CO \qquad (3.12)$$

$$CO + C_f \leftrightarrow C(CO)_B \qquad (3.13)$$

Mechanism B

$$C_f + CO_2 \leftrightarrow C(O)_A + CO \qquad (3.14)$$

$$C(O)_A \rightarrow CO \qquad (3.15)$$

In both the mechanisms, carbon monoxide retards the overall reaction rate. The retardation is via carbon monoxide adsorption to the free sites in the case of Mechanism A, whereas it is via the reaction of chemisorbed oxygen with gaseous carbon monoxide to produce gaseous carbon dioxide in Mechanism B.

As mentioned earlier, when discussing steam gasification, the CO_2 gasification rate of coal is different from that of the carbon–CO_2 rate for the very same reason. Generally, the carbon–CO_2 reaction follows a global reaction order on the CO_2 partial pressure that is around one or lower, that is, $0.5 < n < 1$, whereas the coal–CO_2 reaction follows a global reaction order on the CO_2 partial pressure that is one or higher, that is, $1 < n < 2$. The observed higher reaction order for the coal reaction is also based on the high reactivity of coal for the multiple reasons described earlier [19,35].

3.3.4 PARTIAL OXIDATION

Combustion of coal involves reaction with oxygen, which may be supplied as pure oxygen or as air, and forms carbon monoxide and carbon dioxide. Principal chemical reactions between carbon and oxygen involve [19]

$$C(s) + O_2(g) = CO_2(g) \quad \Delta H^\circ_{298} = -393.5 \text{ kJ/mol} \qquad (3.16)$$

$$C(s) + \frac{1}{2}O_2(g) = CO(g) \quad \Delta H^\circ_{298} = -111.4 \text{ kJ/mol} \qquad (3.17)$$

If sufficient air or oxygen is supplied, combustion proceeds sequentially through vapor-phase oxidation and ignition of VM to eventual ignition of the residual char. Certainly, it is not desirable to allow the combustion reaction to continue too long, because it is a wasteful use of carbonaceous resources.

Even though the combustion or oxidation reactions of carbon may be expressed in terms of simple stoichiometric reaction equations, partial oxidation involves a complex reaction mechanism that depends on how fast and efficiently combustion progresses. The reaction pathway is further complicated because of the presence of both gas-phase homogeneous reactions and heterogeneous reactions between gaseous and solid reactants. The early controversy involving carbon oxidation reaction centered on whether carbon dioxide is a primary product of the heterogeneous reaction of carbon with oxygen or a secondary product resulting from the gas-phase oxidation of carbon monoxide [19,35]. Oxidation of carbon involves at least the following four carbon–oxygen interactions, of which only two are stoichiometrically independent [19,35]:

$$C + \frac{1}{2}O_2 = CO \tag{3.18}$$

$$CO + \frac{1}{2}O_2 = CO_2 \tag{3.19}$$

$$C + CO_2 = 2CO \tag{3.20}$$

$$C + O_2 = CO_2 \tag{3.21}$$

Based on a great deal of research work, including isotope labeling studies, it is generally agreed concerning the carbon–oxygen reaction that the following apply [19,35]:

1. CO_2, as well as CO, is a primary product of carbon oxidation.
2. The ratio of the primary products, CO–CO_2, is generally found to increase sharply with increasing temperature.
3. There is disagreement in that the magnitude of the ratio of the primary products is the sole function of temperature and independent of the type of carbon reacted.

Further details on the carbon oxidation can be found from the classical work done by Walker et al. [35]. Combustion or oxidation of coal is much more complex in its nature than oxidation of carbon. Coal is not a pure chemical species; rather, it is a multifunctional, multispecies, heterogeneous macromolecule that occurs in a highly porous form (typical porosity of 0.3–0.5) with a very large available internal surface area (typically in the range of 250–700 m^2/g). As pointed out by Lee [19], the internal surface area of coal is usually expressed in terms of specific surface area, which is an intensive property that is a measure of the internal surface area available per unit mass. Therefore, coal combustion involves a very complex system of chemical reactions that occur both simultaneously and sequentially. Further, the reaction phenomenon is further complicated by the transport processes of simultaneous heat and mass transfer. The overall rate of coal oxidation, both complete and partial, is affected by a number of factors and operating parameters, including the reaction temperature, O_2 partial pressure, coal porosity and its distribution, coal particle size, types of coal, types and contents of specific mineral matter, and heat and mass transfer conditions in the reactor [19]. Kyotani et al. [36] determined the reaction rate of combustion for five different coals in a very wide temperature range between 500°C and 1500°C to examine the effects of coal rank (i.e., carbon content) and catalysis by coal mineral matter. Based on their experimental results, the combustion rates were correlated with various char characteristics.

It was found that in a region where chemical reaction rate is controlling the overall rate, that is, typically in a low-temperature region where the kinetic rate is much slower than the diffusional rate of the reactant, the catalytic effect of mineral matter is a determining factor for coal reactivity. It was also found that for high-temperature regions where the external mass transfer rate controls the overall rate, the reactivity of coal decreased with increasing coal rank. When the external mass transfer rate limited (or controlled) the overall rate of reaction, the mechanistic rate of external mass transfer is the slowest of all mechanistic rates, including the surface reaction rate and the pore diffusional rate of the reactant and product. Such a controlling regime is experienced typically at a high-temperature operation, whereas the intrinsic kinetic rate is more strongly dependent on temperature than external mass transfer rate [19,36].

3.3.5 OIL SHALE GASIFICATION

Unlike coal, the gasification of oil shale is not widely practiced. Retorting or pyrolysis of oil shale is often the preferred mode. Some literature for thermal, catalytic, and hydrogasification of oil shale is available [37–46]. Oil shale has also been co-gasified with coal [21].

Oil shale is a carbonate mineral that rather tenaciously holds oil-yielding hydrocarbons called kerogen. The production of pipeline gas from this kerogen requires the addition of sufficient hydrogen (hydrogasification) or subtraction of enough carbon (pyrolysis) to convert it into methane or a mixture of methane, ethane, and hydrogen having burning properties similar to methane.

The most attractive present technique for producing pipeline gas from oil shale is by hydrogasification. A major processing problem with oil shale arises from the fact that the pipeline gas forming reactions are accompanied by side reactions that occur at significant rates. The major side reactions are mineral carbonate (calcite and dolomite) decomposition, liquid formation, and carbon deposition. Maximum heat economy and kerogen utilization can be obtained if these side reactions are minimized to the greatest extent possible.

The hydrogasification of oil shale depends on important process variables such as the hydrogen-to-shale feed ratio, feed gas composition, shale space velocity, pressure, temperature, gas–solid contacting scheme, and oil shale feedstock. In principle, the production of pipeline gas from oil shale is possible over a wide range of operating conditions. Literature data [37–42] indicate that the conversion of the organic carbon in the shale is a function mainly of the hydrogen-to-shale feed ratio and independent of pressure and diluents in the feed gas. A slight effect of temperature on the conversion was noted; however, the conversion of kerogen to gaseous hydrocarbons was reduced at the higher temperature level due to increased residual carbon formation and at the lower temperature due to increased liquid formation. Particle residence had a minor effect on the conversion. A set of reaction paths for oil shale thermal decomposition can be illustrated by the following equation (3.22) [37]:

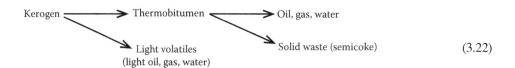

$$(3.22)$$

The production of oil shale gas from kukersite oil shale is a successfully implemented industrial process. This process was urgently introduced in the former Soviet Union (FSU) due to the energy crisis after World War II. During the second half of the twentieth century, oil shale processing technologies and material sciences developed very fast, although the success depended on economics of oil shale and environmental issues posed by oil shale pyrolysis or gasification. Kukersite oil shale in Estonian state was one of the profitable developments.

The literature shows [37] that the degradation of kukersite oil shale begins at 170°C–180°C while water evaporates at temperatures over 100°C. At 170°C–180°C structurally related gaseous components are released, at 270°C–290°C the formation of retort water from kerogen oxygen and hydrogen begins and part of CO_2 and H_2S is released, and at 325°C–350°C the generation of oil and hydrocarbon gases starts. At 450°C–500°C a solid residue or semicoke is formed and at temperatures over 500°C an intensive decomposition of primary oil begins. The chemistry of kerogen degradation is complicated. Already at 250°C an intensive dissolution of kerogen to oil begins. At a temperature of about 300°C, 70% of kerogen has already dissolved and the product formed is thermobitumen. Being liquid and viscous, thermobitumen may cause clogging of the retort equipment, which may result in serious breakdowns. At temperatures over 350°C, the decomposition of thermobitumen continues, hence giving oil, gas, and semicoke. Thus, the gasification of oil shale requires much lower temperature than that for coal.

From the point of view of thermal decomposition of oil shale, two technological processes are notable [37]:

1. In the temperature range of 500°C–550°C, retorting or semicoking occurs.
2. Over 550°C, coking dominates.

Zhunko [45] reported the composition of gas from the oil shale semicoke of Gdov deposit. According to his data, at 1000°C the gas comprises mainly CO (56.7%), H_2 (18.8%), CO_2 (16.5%), and saturated hydrocarbons (8.0%). The gas yield was 122 mN/ton.

The research work by Hisin et al. reported by Zhunko [45] showed that with increasing temperature, the gas yield increased, but that of oil decreased. In the gas composition, the content of CO_2, H_2S, and unsaturated and saturated hydrocarbons decreased and simultaneously the content of CO and H_2 increased. In the temperature range of 1000°C–1100°C, the gas was obtained that consisted mainly of CO and H_2 only. The investigation also demonstrated that while high temperature is important, other variables like the reaction time, semicoke activity, and the presence of water vapor also played a certain role [45].

Qiuan and Yin [46] briefly dealt with the gasification of oil shale semicoke formed as a product of pyrolysis of Chinese Fushun oil shale. In the selected temperature range of 650°C–750°C, the semicoke was gasified with the air/water vapor mixture, which resulted in two different reaction steps:

1. In the first step, additional amounts of CO, CO_2, and H_2 were formed.
2. In the second step, the produced CO reacted with water vapor forming CO_2 and H_2 (WGS reaction).

The authors concluded that the mineral matter in semicoke could act as a catalyst. They also found that the reacting surface played an important role in the process. The gasification was more intensive with finer grain of semicoke [46]. Qiuan and Yin [46] also described the work carried out by Thompson and Nuttall et al. in early 1980 for the gasification of Colorado oil shale coke in the environment of oxygen, as well as CO_2 and water vapor. The reaction of CO_2 with coke is an endothermic reaction and produces CO that reacts with O_2, thus giving additional heat. These results were similar to the ones reported by Mallon and Brau [46]. The positive results obtained for Kukersite oil shale can be expanded to oil shale found in Jordan and China.

In recent years, efforts have been made by Yang et al. [43] and Qian et al. [44] to develop an integrated oil shale refinery process with coal gasification for hydrogen production. As shown in Section 3.5, cogasification of coal and oil shale has also been examined in some details. This approach has some very positive features for generating syngas.

3.4 GASIFICATION OF BIOMASS AND WASTE

Biomass is a nonfossilized and biodegradable organic material originating from plants, animals, and microorganisms. It also include products, by-products, residues, and waste from agriculture, forestry, and related industries as well as the nonfossilized and biodegradable organic fractions of industrial and municipal wastes. Biomass has high but variable moisture content and is made up of carbon, hydrogen, oxygen, nitrogen, sulfur, chlorine, and inorganic elements.

A biomass fuel contains higher hydrogen and oxygen compared to coal. Lignin within biomass would generally have lower oxygen and higher carbon compared to cellulose or hemicellulose. Biomass, in general, is more reactive than coal at lower temperature and produces more hydrogen and oxygenated compounds than coal during gasification. Biomass gasification is therefore a good source for hydrogen production. An extensive research has been carried out on biomass gasification and this is well summarized in numerous publications [47–60]. These publications have evaluated all aspects of biomass gasification in details.

During biomass gasification, carbonization or slow pyrolysis moves the product toward carbon through the formation of solid char; fast pyrolysis moves the product toward hydrogen and away from oxygen, which implies higher liquid product. Oxygen gasification produces more oxygen in the product and steam gasification generates product less abundant in carbon and finally hydrogasification process increases the hydrogen in the product.

The physical and chemical characteristics of biomass, the capacity of the gasifier, and the intended application of the product gas dictate the choice of gasification system and the oxidizing environment. The downdraft gasifier is suitable for both thermal and engine applications [47–60]. The gasification of low-density biomass such as rice husk presents fewer problems in a throatless downdraft gasifier (see Section 3.7 for more details). The tar content can be minimized by separating pyrolysis and gasification zones as in two-stage gasifier. The commercial installations for fixed-bed gasification systems have been carried out in many countries [47–60].

Biomass gasification can be considered as one of the competitive ways of converting distributed and low-value lignocellulosic biomass to fuel gas for CHP generation and fuel cell and synthetic diesel production. Biomass gasification can meet the small-scale (1–10 MW) power requirement in the remote areas in an economical way, where biomass is abundant and long-range transportation is not required. The mixed gasifying agent, for instance, air and steam, could provide suitable gas composition for gas engines with higher thermal efficiency. The utilization of a catalyst, especially a cheap and active catalyst, for gas cleaning can provide the required gas quality for gas engines.

The biomass gasification technology, however, faces economic and other nontechnical barriers when trying to compete in large-scale energy markets. Large-scale biomass gasification plants face challenges with biomass feed preparation and storage and transportation issues to maintain a sustainable operation. In recent years, efforts have been made to achieve benefits of the economy of scale by using coal/biomass/waste mixture gasification (see Section 3.5).

3.4.1 Types of Biomass

Biomass can be divided into two broad groups: (1) virgin biomass and (2) waste. Primary or virgin biomass comes directly from plants or animals. It can be further subdivided into two categories: terrestrial biomass, which includes forest biomass, grasses, energy crops, and cultivated crops, and aquatic biomass, which includes algae and water plants. Waste or derived biomass comes from different biomass-derived waste products that include (1) municipal wastes like MSW, biosolids, sewage, and landfill gas; (2) agricultural solid wastes that include livestock and manures and agricultural crop residues, (3) forestry residues that include bark, leaves, and floor residues; and (4) industrial wastes that include black liquor, demolition wood, waste oil, or fat [3,10–12,47–60].

Energy crops, which belong in the virgin biomass group, are grown especially for the purpose of producing energy encompassing short-rotation or energy plantations: they comprise herbaceous energy crops, woody energy crops, industrial crops, agricultural crops, and aquatic crops. Typical examples of these are eucalyptus, willows, poplars, assorghum, sugarcane, soya beans, sunflowers, cotton, etc. These crops are suitable to be used in combustion, pyrolysis, and gasification for the production of biofuels, synthesis gas, and hydrogen. Large quantities of agricultural plant residues are produced annually worldwide and are vastly underutilized. The most common agricultural residue is the rice husk, which makes up 25% of rice by mass. Biomass gasification is somewhat complex due to the heterogeneous nature of different types of biomass.

3.4.2 Components of Biomass

Cellulose, hemicellulose, and lignin and extractives are found to be the major components of biomass. Raveendran et al. [53], among others [47–51], have reported the composition of biomass in terms of these components. Table 3.2 presents lignin, cellulose, and hemicellulose components of several different biomasses. Woody plant species are typically characterized by slow growth and

TABLE 3.2
Typical Components of Different Types of Biomass

Plant	Lignin (wt%)	Cellulose (wt%)	Hemicellulose (wt%)
Subabul wood	24.7	39.8	24.0
Wheat straw	16.4	30.5	28.9
Bagasse	18.3	41.3	22.6
Corn cob	16.6	40.3	28.7
Groundnut shell	30.2	35.7	18.7
Coconut shell	28.7	36.3	25.1
Millet husk	14.0	33.3	26.9
Rice husk	14.3	31.3	24.3
Scots pine			
Stem wood	27.0(0)	40.7(0.7)	26.9(0.6)
Bark	13.1(5.4)	22.2(3.2)	8.1(0.4)
Stump	19.5	36.4	28.2
Roots	29.8	28.6	18.9
Branches	21.5(5.9)	32.0	32.0
Needles	6.9(0.8)	29.1	24.9
Norway spruce			
Stem wood	27.4(0.7)	42.9(1.2)	27.3(1.6)
Bark	11.8(0.9)	26.6(1.3)	9.2(1.1)
Stump	29.4(1.8)	42.9	27.9
Roots	25.5	29.5	19.2
Branches	22.8(1.7)	29.0	30.0
Needles	8.4(2.1)	28.2	25.4
Silver/Downy birch			
Stem wood	20.2(0.8)	43.9(2.7)	28.9(3.7)
Bark	14.7(3.9)	10.7(0.3)	11.2(0.3)
Stump	13.4	29.5	19.4
Roots	27.1	26.0	17.1
Branches	20.8(3.9)	33.3	23.4
Leaves	11.1(0.0)	N/A	N/A

Sources: Modified and adapted from References 3,9–18,47–72.

are composed of tightly bound fibers, giving a hard external surface, whereas herbaceous plants are usually perennial, with more loosely bound fibers, indicating a lower proportion of lignin, which binds together the cellulosic fibers.

The relative proportions of cellulose and lignin are two of the determining factors in identifying the suitability of plant species for subsequent processing as energy crops. Table 3.2 indicates a wide variation in the cellulose and lignin concentrations of the different types of biomass. Similar compositions of many more biomass are described in the literature [3,9–18,47–72].

3.4.3 COMPOSITION OF BIOMASS

Every biomass type has carbon, hydrogen, and oxygen as major chemical constitutive elements. These element fractions can be quantified with the ultimate analysis. Ultimate analyses are reported using the $C_xH_yO_z$ formula where x, y, and z represent the elemental fractions of C, H, and O, respectively. To fully describe biomass characteristics, it is customary to provide the proximate analysis.

Proximate analysis gives the composition of the biomass in terms of gross components such as moisture (M), volatile matter (VM), ash (ASH), and fixed carbon (FC). It is a relatively simple and inexpensive process. The ultimate analyses and proximate analysis of few biomass feedstocks are reported in Tables 3.3 and 3.4, respectively. Similar compositions for many more biomass are available in the literature [3,9–18,47–72]. Ultimate analysis is relatively difficult and expensive compared to proximate analysis.

TABLE 3.3
Typical Proximate Analysis of Various Types of Biomass

Biomass	VM (wt%)	Ash (wt%)	Fixed Carbon (wt%)
Bagasse	84.2	2.9	15.8
Coconut shell	80.2	0.7	19.8
Corn stalks	80.1	6.8	19.9
Groundnut shell	83.0	5.9	17.0
Rice husk	81.6	23.5	18.4
Subabul	85.6	0.9	14.4
Wheat straw	83.9	11.2	16.1
Bean straw	75.3	5.93	18.77
Barley straw	68.8	10.3	20.90
English walnut	80.82	1.08	18.10
Black walnut	80.69	0.78	18.53
Tokay	76.53	2.45	21.02
Grandis	82.55	0.52	16.93
Poplar	82.32	1.33	16.35
Sudan grass	72.75	8.65	18.60
Zinfandel	76.99	3.04	19.49
Globulus	81.60	1.10	17.30
Cardinal	78.17	2.22	19.61
Ribier	76.97	3.03	20.0
Hazelnut shell	69.80	1.40	28.50
Akhrot shell	79.98	1.20	18.78
Red wood	79.72	0.56	19.92
Peach pit	79.10	1.10	19.80

Sources: Modified and adapted from References 3,9–18,47–72.

TABLE 3.4
Typical Ultimate Analyses of a Diverse Variety of Biomass Compositions (in wt%)

	Rice Straw	Beech	Maple	Tokay	Globulus	Grandis	Poplar
C (%)	39.2	51.6	50.64	47.77	48.18	48.33	48.45
H (%)	5.1	6.26	6.02	5.82	5.92	5.89	5.85
N (%)	0.6	0.0	0.25	0.75	0.39	0.15	0.47
O (%)	35.8	41.45	41.74	42.63	44.18	45.13	43.69
S (%)	0.1	0.0	0	0.03	0.01	0.01	0.01
CR (%)	0.34	0	0.07	0.20	0.08	0.10	—
Ash (%)	19.2	0.65	1.35	2.93	1.12	0.41	1.43

Sources: Modified and adapted from References 3,9–18,47–72.

Biomass energy conversion technologies, especially pyrolysis and gasification, have been substantially studied to promote renewable energy utilization and solve partially the environmental issues. Various types of gasification systems have been developed and some of them are commercialized. Yang et al. [71] concluded that fixed-bed gasification is the most common technology for the energy use of biomass and solid municipal wastes. During the biomass gasification process, this renewable material undergoes different subprocesses. In the first step, biomass is dried up. Then, as the temperature increases, biomass is pyrolyzed and the lignin and cellulose are decomposed into volatile molecules such as hydrocarbons, hydrogen, carbon monoxide, and water. Finally, the remaining solid fraction, which is called vegetal char, is oxidized when an excess of oxygen is available (combustion).

Vegetal char can be gasified by the pyrolysis and partial oxidation. This process is governed by the chemical reduction of hydrogen, carbon dioxide, and water by char. The inorganic components in the biomass are not volatilized and remain in solid state as ash. Overall, generating energy from biomass is expensive due to its low conversion efficiency and issues related to its storage and transportation and feed preparation [3,21,23,73,82].

3.4.4 REACTION ZONES

During biomass gasification, the carbonaceous material undergoes several different processes like drying, pyrolysis, combustion, and gasification. The dehydration or drying process occurs at around 100°C. Typically, the resulting steam is mixed into the gas flow and may be involved with subsequent chemical reactions, notably the water gas reaction if the temperature is sufficiently high. Pyrolysis (or devolatilization) process occurs at around 200°C–300°C. Volatiles are released and char is produced, resulting in up to 70% weight loss for biomass. The process is dependent on the properties of the carbonaceous material and determines the structure and composition of the char, which will then undergo gasification reactions. Combustion process occurs as the volatile products and some of the char reacts with oxygen to primarily form carbon dioxide and small amounts of carbon monoxide, which provides heat for the subsequent gasification reactions. Thus, different reaction zones during gasification can be identified as follows.

3.4.4.1 Drying Zone

Wood entering the gasifier has moisture content of 10%–30%. Various experiments on different gasifiers in different conditions have shown that on an average, the condensate formed is 6%–10% of the weight of the gasified wood [3,9–16]. Some organic acids also come out during the drying process. These acids give rise to the corrosion of gasifiers.

3.4.4.2 Pyrolysis Zone

Wood pyrolysis is an intricate process that is still not completely understood [3,9–16]. The products depend upon temperature, pressure, residence time, and heat losses. In this zone, up to the temperature of 200°C, only water is driven off. Between 200°C and 280°C, carbon dioxide, acetic acid, and water are given off. Real pyrolysis, which takes place between 280°C and 500°C, produces large quantities of tar and gases containing carbon dioxide. Besides light tars, some methyl alcohol is also formed. Between 500°C and 700°C, the gas production is small and contains hydrogen.

Updraft gasifier will produce much more tar than downdraft one because in downdraft gasifier the tars have to go through combustion and reduction zone and are partially broken down. Since majority of fuels like wood and biomass residue produce large quantities of tar, downdraft gasifier is preferred over others.

3.4.4.3　Reduction Zone

The products of partial combustion (water, carbon dioxide, and uncombusted partially cracked pyrolysis products) now pass through a red-hot charcoal bed where the following reduction reactions take place [3,9–16]:

$$C + CO_2 = 2CO \quad (-164.9 \text{ MJ/kg mol}) \tag{3.23}$$

$$C + H_2O = CO + H_2 \quad (-122.6 \text{ MJ/kg mol}) \tag{3.24}$$

$$CO + H_2O = CO_2 + H_2 \quad (+42 \text{ MJ/kg mol}) \tag{3.25}$$

$$C + 2H_2 = CH_4 \quad (+75 \text{ MJ/kg mol}) \tag{3.26}$$

$$CO_2 + H_2 = CO + H_2O \quad (-42.3 \text{ MJ/kg mol}) \tag{3.27}$$

Reactions 3.25 and 3.26 are the main reduction reactions and, being endothermic, have the capability of reducing gas temperature. Consequently, the temperatures in the reduction zone are normally 800°C–1000°C. The lower the reduction zone temperature (~700°C–800°C), the lower is the calorific value of gas.

3.4.4.4　Combustion Zone

The combustible substance of a solid fuel is composed of the elements carbon, hydrogen, and oxygen. In complete combustion, carbon dioxide is obtained from carbon in fuel and water is obtained from hydrogen, usually as steam. The combustion reaction is exothermic and yields a theoretical oxidation temperature of 1450°C [3,9–16]. The main reactions, therefore, are

$$C + O_2 = CO_2 \quad (+393 \text{ MJ/kg mol}) \tag{3.28}$$

$$2H_2 + O_2 = 2H_2O \quad (-242 \text{ MJ/kg mol}) \tag{3.29}$$

3.4.5　PROPERTIES OF BIOPRODUCER GAS

The producer gas is affected by various processes as outlined earlier; hence, one can expect variations in the gas produced from various biomass sources. Table 3.5 lists the composition of gas produced from few typical sources. The gas composition is also a function of a gasifier design, and

TABLE 3.5

Product Composition for Various Biomass Gasifications in a Downdraft Gasifier

Composition of Producer Gas from Various Fuels	Volume Percentage				
	Fuel	CO	H$_2$	CH$_4$	CO$_2$
Wheat straw pellets	14–17	17–19	—	11–14	—
Coconut husks	16–20	17–19.5	—	10–15	—
Corn cobs	18.6	16.5	6.4	—	—
Rice hulls pelleted	16.1	9.6	0.95	—	—
Cotton stalks cubed	15.7	11.7	3.4	—	—

Sources: Modified and adapted from References 3,9–16.

thus, the same fuel may give different calorific value as when used in two different gasifiers. The composition significantly depends on the temperature of gasification. As shown in the next section, biosyngas is generally produced at high temperatures (temperature greater than approximately 1200°C) [3,9–16].

The gasification process is applicable to biomass having moisture content lesser than 35%. Direct gasification (or autothermal gasification) occurs when the oxidant gasification agent partially oxidizes the feedstock and provides the heat for the process. Low-temperature direct gasification (below 900°C) can be carried out in a fixed bed, a FB, or a circulating FB (CFB) and high-temperature direct gasification (over 1200°C) in an entrained-flow gasifier [73].

3.4.6 Biosyngas

The gasification of biomass and waste produces two types of synthetic gas: bioproducer gas and biosyngas. The *bioproducer gas generated* by low-temperature biomass gasification (below 1000°C) contains CO, H_2, CH_4, C_xH_y aliphatic hydrocarbons, benzene, toluene, and tars (besides CO_2 and H_2O). The syngas components H_2 and CO typically contain only ~50% of the energy in the gas, while the remainder is contained in CH_4 and higher (aromatic) hydrocarbons (often known as tars). *Biosyngas* is produced by high temperature (above 1200°C) or catalytic biomass gasification. Under these conditions the biomass is completely converted into H_2 and CO (besides CO_2 and H_2O). Biosyngas is chemically similar to syngas derived from fossil sources and can replace its fossil equivalent in all applications. Biosyngas can also be made from producer gas by thermal cracking or catalytic reforming. Both types of gases need additional gas cleaning and conditioning to prepare a gas with the correct composition and specifications for the final application, for example, synthesis or power production.

Slagging entrained-flow gasification is the most suitable technology for high-temperature gasification. Bioproducer gas is preferred for utilization for power generation and SNG synthesis. Biosyngas is the feedstock for the more advanced applications like FT synthesis, ammonia, and hydrogen production and in processes like olefin hydroformylation and mixed alcohol synthesis. The conversion of bioproducer gas into biosyngas can be carried out by numerous technologies to destroy tar components [61–66]. These technologies can be thermal cracking, catalytic cracking, reforming, or destruction by plasma, among others.

Besides temperature, gasification processes can also be differentiated as direct (autothermal) and indirect (or allothermal) processes. The direct processes are typically operated with air as gasification medium, although for biosyngas production often oxygen is used as an oxidizing medium. The main direct and indirect biomass gasifiers are

- Fixed-bed updraft and downdraft
- FB (bubbling and recirculating, that is, BFB and CFB)
- Indirect FB (steam blown)

For the production of biosyngas where high temperature is required, often entrained bed is preferred. These gasifiers are described in details in Section 3.7. The gasification process is also improved by the use of indirect gasification. This process is described in details in Section 3.14.

3.4.7 Gasification of Waste

While MSW largely contains cellulosic waste, the technology used for its gasification is somewhat different than the ones used for conventional biomass. This is partly due to the heterogeneous properties of feedstock, the scale of operations, environmental regulations for MSW treatment, and various economic factors that affect its operation. In recent years, waste to energy has become a very viable alternative to the disposal of MSW in landfill. The process of MSW gasification is well described in some excellent literature publications [67–73].

3.5 COGASIFICATION

Thermochemical transformation of coal, oil, and biomass to synthetic gases has been in commercial practice for a long time. The synthetic gases produced in this way have been the important part of the energy portfolio of the world, and they are very important sources of heat, power, transportation fuels, and production of chemicals and materials. While this type of transformation has so far been restricted to single raw materials like coal, oil, or biomass, in recent years, more interests in using multiple feedstock like coal and biomass, coal and waste, and biomass and waste have evolved. This chapter also assesses our state of knowledge of the cogasification of multiple fuels [21,58–60,73–83].

Gasification is fundamentally a partial oxidation process where solid or liquid fuels are oxidized in below stoichiometric requirements of oxygen. Like combustion, the process is exothermic and produces a mixture of synthetic fuel gas, which has significant heating value and can be used to generate heat and power in a downstream process if desired. The synthetic gas is also capable of producing liquid fuels and chemicals by processes such as FT synthesis, oxysynthesis, and isosynthesis (see Chapter 9). While the process of combustion (complete oxidation) produces heat and flue gas, which largely contains CO_2, water, and other gaseous impurities, the process of gasification produces gaseous fuel, which largely contains methane, hydrogen, and carbon monoxide along with CO_2, water, and other impurities.

Coal gasification is a well-established industry. Large-scale coal gasification plants exist all over the world. In recent years, these plants have come under close scrutiny due to large emissions of GHGs. The modifications of plants to reduce GHG are expensive and difficult to implement. Biomass gasification, on the other hand, is a relatively young industry. Large-scale biomass and waste gasification plants face challenges of unsustainable raw material supply, expensive storage and transportation, and energy-intensive feed preparation. For these reasons, biomass and waste gasification plants are generally of smaller size and distributed in nature than centralized large-scale commercial coal gasification plants. This does not allow them to take full advantage of economy of scale [21,73].

Biomass in general has a high content of hydrogen (H), making it suitable as a blend to low-H-content coal. Biomass as gasification feedstock, although giving a high hydrogen yield, has the disadvantage of low energy density because of its high oxygen and moisture contents. This shortcoming is compensated for when blended with a higher-energy-content coal. Other challenges such as the seasonal limitation of biomass are somewhat mitigated through coconversion with coal. The higher tar release (due to excessive volatile release and low gasification temperature from biomass gasification) is also reduced as blending with coal increases the temperature and enhances tar cracking. Blending biomass and coal as feedstock can reduce the shortcomings of each fuel and boost the efficacy of the overall system.

The most attractive benefit of coal and biomass cogasification is the reduction of GHG emissions and environmental pollution. Biomass, compared to coal, has lower sulfur, N, and heavy organic metals. It is also carbon neutral if produced sustainably; thus, cogasification of biomass and coal can make significant contributions in mitigating GHG and other emissions. Compared to carbon capture and storage (CCS), cogasification of biomass with coal is relatively cost effective, given the fact that CCS can incur high energy penalty, which can range from 15% to 40% of the energy output [21,73,83]. Even if CCS were to be implemented as a means for GHG mitigation, it would not, alone, be able to meet the 50% emission reduction target by 2050 suggested by the International Panel on Climate Change [21,73,83].

Just as in combustion, the use of multifuels for synthetic gas production has gained significant momentum in recent years. Some of the arguments for this increased activity are the same as that for combustion, that is, reduced CO_2 emissions, more use of renewable feedstock, and diversion of waste from landfills to a more constructive use. In recent years, the use of coal, biomass, and waste to generate synthetic gas has gained more popularity than simple combustion of these materials due to the versatility of the applications of synthetic gas in energy industry. Synthetic gas can be used for heat and power but it can also generate liquid fuels and chemicals that are needed for transportation and in industrial sectors. We also have an infrastructure for natural gas, which can be used for synthetic gas. Thus, gaseous fuel is one of the most important and versatile sources of heat, power, liquid fuels, and chemicals. It has also more sustainable future because it can be obtained from natural (conventional and unconventional) sources and it can be produced from coal, oil, biomass, and waste [21,73–83].

The use of coal and biomass or coal and waste can be assessed in a number of different ways. In cogasification, biomass can be considered as a partial feed (cofeed) to large-scale coal-fed gasification process to avoid the key problems of stand-alone biomass gasification plants such as high cost, low efficiency, and shutdown risks if there is a biomass shortage. While large-scale refineries produce 500,000 barrels/day (bbl/day) oil, biofuel refineries can only be 1/10th of this size. This size difference applies to both power and fuel productions. One way to avoid all the issues encountered with stand-alone biomass gasification plant is to cofeed biomass in large-scale coal gasification plants. Oxy-cogasification allows increased efficiency and reduced environmental impact [21,58–60,73–83]. Also, while steam gasification of coal is not thermodynamically favorable, steam gasification of biomass is a very attractive way to produce hydrogen. Thus, cogasification of coal and biomass in the presence of oxygen and steam appears to be an easier way to increase the role of renewable source of energy in the fossil-dominated industry.

Another way to look at the importance of cogasification is that it is the method by which the use of coal in energy industry is reduced. This helps the environment, particularly for CO_2 emission, and also preserves our valuable coal resources for a long-term purpose. The cogasification will be more acceptable by the supporters of the environment and politicians for the future use of coal. It will also gain more social acceptance for the use of coal.

Yet another positive argument for cogasification of coal and waste or cogasification of biomass and waste is that this allows less waste to be diverted to already overcrowded landfills and makes more constructive use of waste. While converting waste into fuel carries enormous acceptance by the proenvironment advocates and politicians, stand-alone waste gasification plants suffer from the same disadvantages that biomass plants do. Coal–waste gasification plants offer many positive benefits. In recent years, coutilization of coal and waste at large scales has received considerably more attention than coutilization of coal and biomass. One example is the large-scale Lurgi plant in the United Kingdom that uses coal and waste. The waste investigated has been MSW that has had minimal presorting or refuse-derived fuel (RDF) that has had significant pretreatment such as mechanical screening, shredding, and torrefaction ([21,74], see Section 3.6).

Just as for combustion of multiple fuels, the gasification of multiple fuels offers several advantages and some drawbacks. In this chapter, we focus on the multifuel systems that involve coal, biomass, wastes, and water. The advantages and disadvantages of cogasification can be summarized as follows [21,73,83].

3.5.1 ADVANTAGES

1. The use of biomass and/or waste with coal reduces the use of coal in the synthetic fuel industry. This reduction also produces less CO_2, which fits well for the green environment objectives.
2. The use of waste reduces the need for landfill space.
3. The use of coal/biomass mixture allows the use of biomass for the production of synthetic fuels without starting new stand-alone plants for biomass. In general, stand-alone plants will be more expensive than retrofitting existing coal gasification plants to accommodate biomass and waste.
4. The use of mixed fuel plant will allow the thermochemical conversion plants to be on the larger scale, thus using the economy of scale for synfuel production. The stand-alone biomass and/or waste plants for synfuel production cannot be operated on large scale due to

difficulty in providing feedstock in a sustainable manner. Unlike coal, biomass and waste are difficult to transport and store. Also, they require much more complex and expensive feed preparation systems for gasification. The long-distance transportation of biomass and waste can also be very expensive because of its low density, fibrous nature, and high moisture content.

5. In many cases, if properly chosen, mixture can provide synergy among its constituents, which may improve the conversion rate and quality of the product.

6. As shown in my previous book [21], the mixture of coal and biomass/waste can be processed in a number of different ways, which gives more flexibility to the overall operations. Due to its flexibility the process has higher long-term sustainability.

7. If mixture is carefully chosen, the quality of gaseous, liquid, or solid products can be significantly improved over single component feed. Hydrogen production can be optimized. The overall thermal efficiency can also be increased.

8. Cogasification can help improve the public attitude displayed toward the use of renewable feedstock and the development of multifuel supply network.

9. The project risk reduction by cogasification process as opposed to stand-alone biomass process may provide enough security for project financiers [80].

10. Mixture provides more stable and reliable feed supply to the gasification process. Thus, mixture provides more security, less risk, and potential for large-scale operation.

3.5.2 DISADVANTAGES

1. The processing of mixtures will require different feed pretreatment processes (it may be one for each fuel), different designs of gasifiers, or other processing equipment and downstream product cleaning processes. Separate equipment for all three cases may become expensive.

2. The feeding of biomass and waste to the gasifier may become difficult depending on the nature of biomass and waste. Depending on the type of gasification technology employed, pretreatment of biomass may be complex and energy intensive. For example, the particle size required by entrained-flow gasifier is less than 1 mm, which will require energy-intensive milling. The circulating FB, however, may not require such small-diameter particles [21,73,74].

3. An effective integration of biomass feeding with coal feeding may be difficult. The capacity of the biomass handling and feed system may be quite large and expensive compared to the energy content of biomass fed to the gasifier.

4. There may be impacts on the gasifier injection system relative to feeding a single mixture of coal–biomass versus utilizing separate injectors/burners. The dry feeding of biomass against the pressure (in case of pressurized thermochemical processes) may become challenging. Two separate feed injectors versus a single feed injector may affect the gasifier performance.

5. Not much is known about gasification and pyrolysis of mixed fuel with different burning characteristics, particularly when physical and chemical properties of the biomaterials and waste are significantly different from that of coal.

6. Depending on the N, S, Cl, alkali, and heavy metal contents of the two or more feedstock, more treatments of gaseous emissions may be required. Generally, coal contains less chlorine and alkali and heavy metals compared to some biomass like straw and switchgrass. The high chlorine content in some biomass and waste may cause more corrosion and erosion problems in the process equipment. These would require more assessment of the use of right materials of construction for process equipment during processing of mixtures [21,73,83].

7. Cogasification of coal and different types of biomass may negatively impact the slagging behavior of the combined ash in the gasifier, depending on its design and type. The slagging

behavior will also be affected by the nature and rank of coal. The ash composition of the mixed gasification process may be significantly different from the one for pure coal alone. The significant amount of inorganic metal content in the ash (which may be present in biomass) may limit its downstream usefulness.

8. Generally, biomass gasifies at a lower temperature than coal. When both are gasified together, selection of a temperature for the optimum reactor performance may be challenging.

9. For combined thermochemical transformation, it is good to have uniform particle size for all constituents of the feedstock. This may not be possible when using biomass like straw or miscanthus or shredded waste or the feed with different shape particles. Predictions and control of the reactor performance for nonuniform particle size and shape may become challenging.

10. Cogasification of biomass may contribute to tar and oil formation in the raw syngas depending on the type of the gasification technology employed. This requires their separations from the syngas or their further conversion. Tar and oils can be avoided by operating gasification reactor at high temperatures (>1200°C). The tar and oils can also be converted to commodity products like phenols, an approach being taken by Dakota Gasification Company for lignite-based gasification plant [21,73,83].

3.5.3 Typical Examples of Cogasification

3.5.3.1 Cogasification of Coal and Biomass in Intermittent Fluidized Bed

Wang and Chen [84] examined cogasification of coal and biomass in an intermittent FB reactor to investigate the effects of temperature (T), steam-to-biomass ratio (SBMR), and biomass-to-coal ratio (BMCR) on hydrogen-rich gas production. The results showed that H_2-rich gas, free of N_2 dilution, is produced and the H_2 yield was in the range of 18.25–68.13 g/kg. The increase of T, SBMR, and BCMR is all favorable to the production of hydrogen. Both hydrogen and carbon monoxide contents and hydrogen yield are increased with temperature. While gas composition was not strongly affected by BCMR, the yield and content of hydrogen increased with BCMR, reaching a maximum at BCMR = 4. While hydrogen content and yield increased with SBMR, carbon monoxide showed a maximum with respect to SBMR. The study showed that the order of the influence of the operation parameters on hydrogen production efficiency is T > SBMR > BCMR.

3.5.3.2 Cogasification of Coal and Biomass in High-Pressure Fluidized Bed

McLendon et al. [85] examined cogasification of coal and biomass in a jetting, ash-agglomerating, FB pilot-scale-sized gasifier. The biomass used was sanding waste from a furniture manufacturer. Powder River Basin subbituminous and Pittsburgh No. 8 bituminous coals were mixed with sawdust. Feed mixture ranged up to 35 wt% biomass. The results with subbituminous coal/sawdust mixture showed few differences in operations compared to only subbituminous coal tests. The coal/sawdust mixture showed a marked difference from bituminous coal-alone data. The transport properties of coal/biomass mixtures were greatly improved compared to coal only.

3.5.3.3 Cogasification of Coal and Polyethylene Mixture

Yasuda et al. [86,87] examined hydrogasification of coal/PE mixtures at 1073 K under 7.1 MPa of hydrogen. The reaction time varied from 1 to 80 s. Coal/PE mixtures in the ratios 90:10 and 75:25 were used in the study. Both product distribution and temperature profiles were analyzed. For PE alone tests, the yield (carbon basis) of methane reached 90%. A significant synergistic effect (even with 10% PE in the mixture) was found when coal and PE were mixed and used in hydrogasification.

The study indicated that the compensation of endothermic coal pyrolysis process by heat evolved from hydrogenation of PE may be the reason for synergy. The early drop in coal hydrogasification was prevented during cogasification, and the gasification of the mixture resulted in increased carbon conversion to methane. The study recommended this mixture for future hydrogasification processes.

3.5.3.4 High-Pressure Cogasification of Coal with Biomass and Petroleum Coke

Fermoso et al. [88] examined the effects of temperature, pressure, and gas composition on gas production, carbon conversion, cold gas efficiency, and high heating value during the steam–oxygen gasification of a bituminous coal. The temperature and oxygen concentration were the most important variables during gasification process. Cogasification was studied with biomass concentration up to 10% and petroleum coke up to 60%. The ternary mixture of coal–petcoke–biomass with 45:45:10 was also studied to evaluate the effect on gas production and carbon conversion. Cogasification produced very positive results.

3.5.3.5 Cogasification of Woody Biomass and Coal with Air and Steam

Kumambe et al. [89] examined the cogasification of woody biomass and coal with air and steam to produce syngas for light fuels. The experiments were carried out in a downdraft fixed-bed reactor at 1173 K. The study varied BMCR from 0 to 1 on carbon basis. The gas production increased with an increased concentration of biomass, whereas the production of char and tar decreased. With the increase in biomass concentration, hydrogen concentration decreased and carbon dioxide concentration increased. However, CO concentration in the gas phase was independent of biomass concentration in the feed. A low biomass ratio produced gas-phase composition more suitable for methanol and hydrocarbon fuel synthesis, and a high biomass concentration produced gas favorable for DME synthesis. The synergy due to cogasification may be observed in the extent of the WGS reaction. The gasification conditions in the study provided a cold gas efficiency ranging from 65% to 85%.

3.5.3.6 Cogasification of Coal, Biomass, and Plastic Wastes with Air/Steam Mixtures in Fluidized Bed

Pinto et al. [90] studied optimum conditions for cogasification of coal and waste with respect to gas using only air, only steam, and mixtures of them. An increase in temperature increased hydrogen and decreased tars. Increasing temperature from 750°C to 890°C for a mixture of 60:20:20 coal/pine/PE waste (w/w) led to a decrease in methane and other hydrocarbon concentration of about 30% and 63%, respectively, while hydrogen concentration increased around 70%. An increase in air flow rate decreased hydrocarbon and tar production. The presence of air also decreased higher heating value of the product gas. The increase in steam increased reforming reaction and thereby increased hydrogen production.

3.5.3.7 Coutilization of Biomass and Natural Gas in Combined Cycles

Power production from biomass can occur through external combustion (e.g., steam cycle, organic Rankine cycles, Stirling engines) or internal combustion after gasification or pyrolysis (e.g., gas engines, integrated gasification combined cycle [IGCC]). External combustion has the disadvantage of low efficiency (30%–35%). Internal combustion, on the other hand, has the potential of high efficiency, but it always needs a more severe and mostly problematic gas cleaning.

De Ruyck et al. [91] examined an alternate route where advantages of external firing are combined with the potential high efficiency of the combine cycles through coutilization of natural gas and biomass. Biomass is burned to provide heat for partial reforming of the natural gas feed. In this way, biomass energy is converted into chemical energy contained in the produced syngas. Waste heats from the reformer and from biomass combustor are recovered through a waste heat recovery system.

The study showed that in this way, biomass can replace up to 5% of the energy in the natural gas feed. The study also showed that in the case of combined cycles, this alternative route allowed for external firing of biomass without important drop in cycle efficiency.

3.5.3.8 Steam Gasification of Coal–Biomass Briquettes

Yamada et al. [92] examined a biobriquette made by mixing low-grade Chinese coal and larch bark with $Ca(OH)_2$ as the desulfurizing agent to measure sulfur evolution in the mixture of nitrogen and steam at 1173 K. The briquettes were more effective in early pyrolysis condition than subsequent steam gasification condition for H_2S removal.

3.5.3.9 Syngas Production by Coconversion of Methane and Coal in a Fluidized-Bed Reactor

Wu et al. [93] examined the concept of coconverting methane reforming and steam gasification of coal at 1000°C in a FB reactor without a catalyst. This concept is applicable where coalbed methane can be used to reform methane and gasify coal by steam simultaneously in situ. In addition, the integrated coal gasification and gas reforming offers an advantage that the H_2/CO ratio in the produced syngas can be adjusted between 1 and 3 by varying the ratio of coal/gas in the feedstock. The study showed some initial success of the concept in which over 90% conversion of natural gas in a laboratory FB reactor was obtained with favorable quality of produced syngas (i.e., high H_2 and CO and low CH_4 and CO_2).

3.5.3.10 Cogasification of Coal and Biomass in a Dual Circulating Fluidized-Bed Reactor

Seo et al. [94] and others [95,96] examined the effects of temperature (750°C–900°C), steam-to-fuel ratio (0.5–0.8), and biomass ratio (0, 0.25, 0.5, 0.75, 1.0) on the cogasification of coal and biomass in dual circulating fluidized reactor (combustor/gasifier). Indonesian Tinto subbituminous coal and *Quercus acutissima* sawdust were used as coal and biomass, respectively. With increasing temperature and steam/fuel ratio, the product gas yield, carbon conversion, and cold gas efficiency of the mixtures were higher than the ones for coal alone. After pyrolysis, surface area, pore volume, and micropores of coal/biomass blend char increased. The maximum increase in gas yield can be obtained with a biomass ratio of 0.5 at the given reaction temperature. Calorific values of the product gas were 9.89–11.15 MJ/m³ with the coal, 12.10–13.19 MJ/m³ with biomass, and 13.77–14.39 MJ/m³ with coal/biomass blends at 800°C. The synergistic effects on the basis of calorific value and cold gas efficiency were pronounced with the coal/biomass blends. This work also presents a summary of previous cogasification studies.

3.5.3.11 Cogasification of Coal and Chicken Litter

Priyadarsan et al. [97] examined cogasification of coal and chicken litter in a 10 kW capacity fixed-bed countercurrent atmospheric pressure gasifier at air flow rate of 1.3 and 1.7 m³/h under batch mode of operation. An increased in air flow rate decreased heating value of product gas from 5.2 to 4.9 MJ/m³, which resulted in a minor decrease in gasification efficiency. For both air flow rates, the product gas composition was CO at 30% dry basis and H_2 at 10% dry basis. The presence of coal completely inhibited ash agglomeration in the bed, which was observed during gasification of pure chicken litter. This can be attributed to reduced amount of Na and K in the blended fuel as compared to pure chicken litter biomass ash.

3.5.3.12 Cogasification of Biomass and Waste Filter Carbon

Sun et al. [98] examined cogasification of waste filter carbon and char of wood chips with steam at atmospheric pressure. The effects of temperature 600°C–850°C and partial pressure of steam from 0.3 to 0.9 atm on the gasification rate were examined. The modified volumetric reaction model was used to evaluate kinetic data. The gasification rate of waste filter carbon was compared with that of cogasification rate. The activation energies of filter carbon and wood chips were determined to be 89.1 and 171.4 kJ/mol, respectively.

3.5.3.13 Cogasification of Low-Rank Fuel–Biomass, Coal, and Sludge Mixture in a Fluidized Bed in the Presence of Steam

Ji et al. [99] examined the cogasification of low-rank fuels in an FB reactor. Within the range of experimental conditions examined, the highest amount of hydrogen and carbon monoxide was observed at 900°C and steam partial pressure of 0.95 atm. Temperature and steam were the most important variables in the system. High temperature favored hydrogen production and gas yield but did not always favor heating value. Sludge, oil, and coal mixture showed great potential for gas production. As reaction temperature and steam partial pressure increased, the heating value of product gas increased. Kurkela et al. [100] examined the gasification of peat and biosolid.

3.5.3.14 Cogasification of Residual Biomass/Poor Coal in a Fluidized Bed

Pan et al. [101] examined cogasification of residual biomass/poor coal blends and gasification of individual feedstock used in the blends in a bench-scale, continuous FB reactor working at atmospheric pressure. Two types of blends were prepared: mixing pine chips (from Valcabadillo, Spain) with black coal, a low-grade coal from Escatron, Spain, and Sabero coal, a refuse coal from Sabero, Spain, in the ratio range of 0/100–100/0. Experiments were carried out using mixtures of air and steam at gasification temperatures of 840°C–910°C and superficial fluidized gas velocities of 0.7–1.4 m/s. Feasibility studies were very positive, showing that blending effectively improved the performance of FB cogasification of the low-grade coal and the possibility of converting the refuse coal to a low-Btu fuel gas. This study indicated that a blend ratio with no less than 20% pine chips for the low-grade coal and 40% pine chips for the refuse coal is the most appropriate. The dry product gas low heating value augmented with increasing blend ratio from 3700 to 4560 kJ/Nm^3 for pine chips/low-grade coal and from 4000 to 4750 kJ/Nm^3 for pine chips/refuse coal. Dry product gas yield rose with the increase of the blend ratio from 1.80 to 3.20 Nm^3/kg (pine chips/low-grade coal) and from 0.75 to 1.75 Nm^3/kg (pine chips/refuse coal), respectively. The study indicated that about 50% cogasification overall process thermal efficiency can be achieved for the two types of blend.

3.5.3.15 Cogasification of Biomass and Coal for Methanol Synthesis

Chmielniak and Sciazko [102] examined the economy of methanol production through coal–biomass gasification by linking it with modern gas–steam power systems. The essence of linking is the full utilization of the capacity of coal–biomass gasification installations. The paper describes the up-to-date experience of coal–biomass gasification including processing toward syngas production and methanol production. A conceptual flow diagram of pressurized and oxygen-fed cogasification of coal and biomass integrated with combined-cycle and parallel methanol production is evaluated. The effect of methanol production rate on the economy of power production is assessed.

3.5.3.16 Underground Cogasification of Coal and Oil Shale

Zhao et al. [103] tested the feasibility of in situ cogasification of coal and oil shale. Based on the specification analysis of coal and oil shale through simulating the occurrence state and characteristics of coal and oil shale, the underground coal gasification (UCG) model test was carried out. The experiments were carried out for different oxygen-to-steam ratios of 0.3, 0.35, 0.4, 0.45, and 0.5. The effect of temperature on the quality of gas was studied. The results showed that for oxygen/steam ratio of 0.4–0.45, the temperature rising rate was 7°C/min and the extended rate of gasification was 0.036 m/h, the extend of temperature field was continuous and stable, and both oil shale and coal temperature changes were uniform. The high temperature of 1000°C achieved here satisfied the requirement for oil gas production. The heating value of syngas improved by 26.37% due to cogasification.

3.5.3.17 Cogasification of Petcoke and Coal/Biomass Blend

Khosravi and Khadse [96] presented an excellent review of petcoke and coal/biomass blend gasification. The following discussion is a brief summary of their review. A number of studies have investigated synergetic effects of cogasification of petcoke and coal blend. These include cogasifications of anthracite and petcoke in thermogravimetric analyzer (TGA), which exhibited positive synergy due to catalytic effects of alkali and alkaline earth metals (AAEMs) [83,96]. Similarly, due to catalytic effects, CO_2 gasification of petcoke was found to be less than that of several coals [83,96]. The addition of a catalyst lowered the gasification temperature in the petcoke. Numerous studies [83,94–96,104] examined the effects of coal residue that had AAEMs and transition metals [83,96], the effects of transition metal and iron species [83,96], and the effects of calcium-promoted potassium carbonate catalyst on gasification. In all cases, gasification rate was significantly enhanced due to the presence of a catalyst.

Biomass was also found to have a positive effect of gasification if blended with petcoke [105]. Nemanova et al. [105] studied cogasification of biomass and petcoke in a TGA and FB reactor. They found that higher biomass content led to shorter gasification times. Fermoso et al. [88] performed cogasification of binary and ternary mixtures of biomass, coal, and petcoke in a highly pressurized fixed-bed reactor and found that addition of biomass to coal up to 10% raised the cold gas efficiency and carbon conversion, whereas addition of more than 10% biomass into the blend did not have any effect on gas composition. The gas composition was highly altered in all blends by temperature and concentrations of oxygen and steam. Higher pressure slightly decreased CO and H_2 productions. Fermoso et al. [88] also found interactions among different components of the blends during cogasification. Nemanova et al. [105] studied cogasification of petroleum coke and biomass in an atmospheric bubbling FB reactor and a TGA at KTH Royal University of Technology. Biomass ash in the blends was found to have a catalytic effect on the reactivity of petroleum coke during cogasification. Furthermore, this synergetic effect between biomass and petcoke was observed in the kinetic data. The activation energy E_a determined from the Arrhenius law for pure petcoke steam gasification in the TGA was 121.5 kJ/mol, whereas for the 50/50 mixture, it was 96.3, and for the 20/80 blend, 83.5 kJ/mol. Several other studies on cogasification are also reviewed by Brar et al. [95] and others [84–112].

3.5.4 Barriers and Potentials for Future Growth in Cogasification

Cogasification of coal and biomass allows for compensating the shortcomings of one fuel by another, since both fuels seem to be complementary in their drawbacks and advantages. In many specific cases, synergy among components enhances performance. GHG mitigation is one of the most attractive benefits of the cogasification of coal and biomass. Lifecycle assessment on cogasification of coal and biomass has shown that CO_2 emission declines proportionally to the amount of coal offset by biomass, considering biomass as a carbon-neutral source produced in a sustained manner [21–23,73,75]. The literature [21–23,73,75] has shown that 70/30 mixture of coal and biomass will be CO_2 neutral to the environment. With further reduction in CH_4 and N_2O emission, the overall outcome of cogasification is that a percentage reduction of global warming potential is higher than the percentage of biomass in the blend.

Entrained-flow gasification produces a relatively clean gas compared to fixed-bed and FB gasification. However, size reduction of biomass to the order of hundreds of microns, required in entrained-flow systems, may be expensive and difficult to achieve for some biomass feedstock. This may also require preprocessing steps like torrefaction or fast pyrolysis.

In general, biomass contains high concentrations of AAEMs and low concentrations of aluminosilicates. Coal contains high concentrations of aluminosilicates. Since AAEMs can be catalysts for coal gasification [113–116], some coal/biomass mixtures can exhibit improved performance during cogasification.

While there is no consensus on synergistic effects between coal and biomass during cogasification, some believe that free radical formation and hydrogen transfer from biomass to coal can create synergy during cogasification. This synergy is more pronounced at lower temperatures and not noticeable at higher temperatures [83,117,118]. Because of the low melting temperature of AAEMs and high melting temperature of aluminosilicate metals, biomass ash tends to melt at lower temperatures than coal ash. It is therefore important to know the behavior of ash mixture from coal and biomass in the design of cogasification reactors. In the absence of this knowledge, the best strategy for the detailed design of cogasification process is to handle the design on a case-by-case basis because of a wide variation of mineral matter content and quality in both biomass and coals. The knowledge of formation rate and characteristics of coal–biomass ash mixtures under different gasification conditions is needed.

Based on thermodynamic equilibrium of cogasification, the amount of gas produced, its gross calorific value, and cold gas efficiency increase with the concentration of biomass in the coal/biomass mixture. Higher temperature favors the extent of endothermic reaction and formation of hydrogen and carbon monoxide. However, the cold gas efficiency and the gross calorific value first increase and then decrease with an increase in temperature. Increasing the pressure has an opposite effect to that of temperature. At higher temperatures, the pressure effect is more dominant.

Coal gasification is a proven technology that has been operated successfully at commercial scale for heat, power generation, and production of synthetic fuels. On the other hand, based on the literature information, stand-alone biomass or cellulosic waste gasification and subsequent production of biofuels appear to be more complex and carried out at small scale. The share of biofuel in the transportation fuel market is likely to grow rapidly in the next decade due to numerous benefits, including sustainability, reduction of GHG emissions, regional development, reduction of rural poverty, and energy security [21,83]. The economics of scale is important for biomass gasification because only large biomass-to-liquid (BtL) plants with synfuel production capacities of at least 1,000,000 tons/annum would produce profit [21,83,112]. The desired scale is best achieved by processing coal and biomass mixtures.

Biomass, like most existing renewable energy resources, is both dispersed and variable over time. Regular and constant supply of a huge amount of biomass that is required for large-scale operation, coupled with processing and handling difficulties, makes a dedicated biomass power and heat plant or BtL plant highly improbable; but this limitation can be alleviated if coal and biomass are both used as feedstock. The backup storage capacity needed for sustainable power or fuel generation can also be enhanced using mixture as feedstock [21,73,83].

There are several barriers to the future growth of coal/biomass or coal/waste cogasification. Some of these relate to public image and perception. While gasification has a better public perception than combustion, waste is an area of public concern and has low image. The large-scale coal/biomass and coal/waste plants will require transportation of biomass and waste at a significant level. Such transportation will need public support. Transport of fuel to a power station is always a contentious issue. The transportation of waste on a larger scale is also problematic [21–23,73,83].

There is also a cultural issue of the acceptance of biomass and waste into an industry that has been single focused on coal and the promotion of its usage. We have developed a culture and know-how of single-dimensional industry like coal, oil, and gas. Cogasification thinking at the industrial scale requires a paradigm shift to the energy and fuel system integration. Overall, due to its necessity and potential for success, the future of cogasification is very certain.

3.6 FEED PREPARATION

One of the most important topic in coal, biomass/waste, or coal–biomass/waste mixture gasification is the feed pretreatment that can affect the selection and performance of gasifiers and the production rate and quality of a producer or syngas. The nature of feed pretreatment depends on the nature of feedstock and the selection of gasifier. Pretreatment is also required for feedstock storage and transportation. In almost all cases of gasification, feed pretreatment is a necessity. Here, we briefly discuss the feed pretreatment issues and options related to coal and biomass/waste.

3.6.1 COAL PREPARATION

Coal preparation is both a science and an art. Coal preparation, which is also called washing, cleaning, processing, and beneficiation, is the method by which mined coal is upgraded in order to satisfy size and purity specifications dictated by a given market and subsequent application. The upgrading, which occurs after mining and before transport of the cleaned product to market, is achieved using low-cost, solid–solid, and solid–liquid separation processes that remove waste rock and water from the mined coal. This can also be followed by appropriate physical methods to reduce coal particle size and a number of chemical cleaning processes to remove sulfur, ash, and other impurities within coal. The processing is driven by a desire to reduce freight costs, improve utilization properties, and minimize environmental impacts without destroying the physical and chemical identity of coal.

Coal preparation is required because freshly mined coals contain a heterogeneous mixture of organic (carbonaceous) and inorganic (mineral) matter. The inorganic matter includes noncombustible materials such as shale, slate, and clay. These impurities reduce coal heating value, leave behind an undesirable ash residue, and increase the cost of transporting coal to the market. The presence of unwanted surface moisture also reduces heating value and can lead to handling and freezing issues for consumers. Therefore, essentially all coal supply agreements with electric power stations impose strict limitations on the specific energy (heat), ash, and moisture contents of purchased coal.

Coal preparation operations make it possible to meet coal quality specifications by removing impurities from run-of-mine coals prior to shipment to power stations or other users of syngas. Moreover, as the first step in the power cycle, coal preparation plants improve the environmental acceptability of coal by removing impurities that may be transformed into harmful gaseous or particulate pollutants when burned or gasified. These pollutants typically include particulates (fly ash) and sulfur dioxide (SO_2), as well as air toxins such as mercury. The presence of mineral impurities can also influence the suitability of coal for high-end uses such as the manufacture of metallurgical coke or generation of petrochemicals and synthetic fuels. Coal preparation is typically needed to achieve the levels of coal purity demanded by these secondary markets.

The capability of coal preparation to improve coal quality varies widely from site to site. The most significant part of this variation occurs because of inherent differences in the liberation characteristics of run-of-mine coals. The degree of liberation is determined by the relative proportion of composite particles (i.e., particles of coal and rock that are locked together) that are present in a particular coal. The presence of composite particles makes it impossible to physically separate all of the organic matter from all of the inorganic matter. Consequently, plant operators purposely sacrifice coal recovery by discarding some composite particles as waste to improve coal quality to a level that can meet customer specifications. This loss often accounts for 10%–15% of the heating value contained in the source coal.

Coal washability has a tremendous impact on how effectively a preparation plant can upgrade a particular run-of-mine coal. Separating densities in a plant are often set in response to changes in coal washability to ensure that product coal continues to meet quality specifications. However, the types of processes employed and practices used for operation and maintenance can also greatly influence the performance of the preparation facility. This effectiveness is typically reported as organic efficiency, which is defined as the yield of coal product produced by the separation divided by the theoretical maximum yield of coal that could be achieved at the same ash content according to a washability analysis. Organic efficiencies may be in the high ninetieth percentile for well-designed and well-run operations, although lower values are not uncommon for problematic plants. These inefficient processes or practices misplace significant amounts of potentially recoverable coal into waste and rock into the washed product. Although this misplacement is typically small in comparison to losses created by washability constraints, these inefficiencies have a large impact on plant profitability [119].

The large financial impact of poor efficiency has pushed the industry to abandon old technology and to develop and adopt new processes and practices for cleaning run-of-mine coals.

Labor-intensive methods such as manual sorting were soon replaced by simple mechanical separation processes that reduced misplacement and provided higher levels of productivity. Many decades of technology development ultimately led to the design and operation of relatively efficient plants that are capable of complete or partial upgrading of the entire size range of mined coals. Many modern coal preparation plants now in operation in the United States are as complex as industrial facilities once employed only by the chemical processing industry.

Coal preparation plant flowsheets can be generically represented by a series of sequential unit operations for particle sizing, cleaning, and dewatering. This sequence of operations, commonly called a circuit, may be repeated several times. The repetition is needed to maintain efficiency, because the processes employed in preparation plants each have a limited range of applicability in terms of particle size. In the United States, modern plants may include as many as four separate processing circuits for treating the coarse (greater than 10 mm), small (between 1 and 10 mm), fine (between 0.15 and 1 mm), and ultrafine (less than 0.15 mm) material. Although many commonalities exist, the final selection of what number of circuits to use, which types of unit operations to employ, and how they should be configured is highly subjective and dependent on the characteristic properties of the feed coal in terms of size, composition, and washability.

In a typical plant, feed coal is sorted into narrow particle size classes using vibrating screens for coarser particles and classifying cyclones for fine particles. The coarse fraction is usually cleaned using a chain-and-flight dense medium vessel, while the smaller fraction is upgraded using dense medium cyclones. These processes use a dense medium suspension to separate coal from rock based on differences in particle densities. The fine fraction is usually cleaned by water-only cyclones, spirals, or a combination of these separators. These water-based processes exploit differences in particle size, shape, and density to separate coal from rock.

Unfortunately, conventional density separators cannot be used to upgrade the ultrafine fraction because of the low mass of the tiny particles. This fraction is usually upgraded using a process known as froth flotation, which separates coal from rock based on differences in the surface wettability of organic and inorganic matter. In many cases, the ultrafine fraction is resized ahead of flotation to remove particles under 40 μm that are detrimental to flotation and downstream dewatering [120]. Finally, the water used in processing is removed from the surfaces of coarse particles using combinations of screens and centrifugal basket-type dryers. Screen-bowl centrifuges or vacuum filters are usually employed to dewater fine coal that tends to retain larger amounts of moisture. Abundant literature is available on various aspects of coal preparation technologies [121–123].

For the downstream applications of coal gasification, efforts are also made to remove organic sulfur, inorganic sulfur, and sulfates from coal. This process is often called "chemical cleaning." While inorganic sulfur (pyrites) and sulfates can be relatively easily removed by various flotation, sedimentation, and density and size separation techniques [121–123], it is very difficult to remove organic sulfur because organic sulfur is chemically bonded to coal matrix and cannot be removed without losing coal itself. Significant efforts have been made [121–123] to remove organic sulfur by various chemical oxidation or fermentation methods. The removal of sulfur during feed preparation reduces the need for the removal of sulfur compounds (SO_2, H_2S, and COS) from the product gas to meet the specification of downstream operations.

Once the coal is prepared for its required size, water content, and tolerable impurity level, coal is easy to store, transport, and process in the gasification reactor either in dry form or in wet form as a slurry. The technologies for the feed systems for different types of gasification reactors are commercially well proven.

3.6.2 Technological and Environmental Barriers

There are several barriers associated with coal preparation that may limit future coal production in the United States. These barriers differ in the eastern and western states because of regional variations in the characteristics of the coal resources and industry activities in these regions.

The steady decline in the quality of U.S. coal reserves will require processing of feed coals with increasingly difficult washing characteristics. Therefore, continued development of improved solid–solid and solid–liquid separation technologies for coal preparation is needed to help offset the adverse effects of these changes to coal quality and recovery. Examples of improvements may include the development of advanced processes for fine coal cleaning, dewatering, and reconstitution or the stepwise integration of some coal preparation functions within mine extraction operations. Potential examples of evolutionary technology may include the construction of small-scale gasifiers that obviate the need for dewatering by utilizing fine coal slurry at existing preparation plant sites as well as nontraditional processing strategies at end-user sites. A new generation of online systems for real-time characterization of coal size, density (washability), and quality will also be advantageous to deal with future declines in feedstock consistency [123].

Western coal operations face even greater challenges from a decline in reserve quality, because coals in this region have traditionally not required preparation other than size reduction. Dry cleaning processes, such as pneumatic separators and electronic sorters, which can efficiently upgrade coals over a wide range of particle sizes, need to be developed for use in western states with scarce water resources. The remoteness of western resources may also dictate the need for next-generation upgrading facilities, such as mild conversion plants, which can reduce moisture and increase the heating value of low-rank coal so the existing energy transportation system can be better utilized.

Another option would be the construction of mine-mouth gasification or power plants close to coal production facilities, which would eliminate transportation barriers and improve the cost effectiveness of utilizing the large reserve base of low-sulfur western coals.

Several environmental issues represent significant challenges to expanded utilization of U.S. coal preparation facilities. Although these impediments vary from state to state, the most significant challenge facing the industry is the management of coal wastes. The declining quality of reserves has contributed to the expansion of waste storage repositories such as slurry impoundments. Well-publicized events, such as impoundment failures, have raised serious questions as to whether new regulations, better practices, and improved technologies are needed to eliminate the possibility of future disasters. New methodologies need to be developed for dewatering, handling, and permanently disposing waste slurry. New techniques are also needed for locating and assessing the stability of impounded slurry over abandoned workings. In addition, the development of new processing technology that is specifically designed to re-treat and recover coal resources from existing or abandoned impoundment areas is an attractive approach for reducing waste.

Issues are also being raised regarding the environmental effects of chemical additives used in coal preparation. To address the concerns, new processes or chemical additives need to be developed that minimize, and preferably eliminate, the use of processing reagents that have potential risk to the ecosystem [121–123].

3.6.3 Pretreatment and Feeding Options for Biomass/Waste

As shown earlier, the major elements of feed preparation of coal involve washing, solid–solid and solid–liquid separation, particle sizing, sulfur and ash removals, and VM (including water) removals irrespective of the nature of the coals. Biomass or waste on the other hand is a very nonhomogeneous material and harvested, stored, and transported differently depending on its nature and physical and chemical properties. Most biomass have higher water contents, higher VMs, and lower mass and energy densities and are softer than coals. Biomass contains higher amount of VMs (particularly oxygen), chlorine, alkali matters, and inorganic ash compared to coal. Size reduction of biomass is much more difficult. Biomass can also go through biochemical decay during storage and transportation [21,22,73,74]. Constraints and role of biomass pretreatment in biomass or coal–biomass cogasification or co-firing are well reviewed by Maciejewska et al. [117].

In all types of gasifiers, a consistent supply of the feedstock in the appropriate grade and quality is the single most critical factor for good operational performance, which leads to the desired production and outgoing gas composition targets. Coal preparation for coal gasification is a basically well-established technology. Biomass, on the other hand, is very heterogeneous and require different types of technologies for particle sizing and feed preparation depending upon the nature of biomass and the conditions required for the gasification [124–126]. Waste (MSW) can also be very heterogeneous and may require significant pretreatment (like preparation of RDF and palletization) before gasification [73,74].

While coal gasification can be carried out in three basic types of reactors, namely, moving bed (of different types), FB (including circulating), and entrained bed, about 85% of existing and new coal gasification plants are based on the entrained-bed technology. Biomass/waste gasification also currently uses either fixed-bed, moving-bed, or FB and entrained-bed reactors depending on the requirement of product gas quality (i.e., either producer gas that contains significant amount of methane or syngas that largely contains hydrogen and carbon monoxide). While FB reactors are suitable for smaller-scale biomass gasification, for large-scale coal–biomass mixture gasification, industry prefers entrained-bed reactors [21,22,73]. Unfortunately, the existing coal feeding systems for this type of reactor are usually not suitable for biomass. A number of issues related to this matter are being assessed and handled.

The method adopted for feeding biomass to entrained-bed reactor depends on the nature of the feedstock. Woody biomass can be milled to 1 mm and then feed to the reactor by screw feeder with piston compressor. In general, all types of biomass can be subjected to the torrefaction followed by milling process and then feed to the entrained-bed reactor either by pneumatic feeder and piston compressor or by screw feeder and piston compressor. All types of biomass can first be gasified in a PFB gasifier, and the product gas and char can be fed to entrained-bed gasifier for further gasification. Straw-like biomass, which is difficult to mill, can be pyrolyzed to form bioslurry that can then be fed to entrained-bed reactor by feeder and injector. Waste can be pelletized and form RDF before feeding into the reactor but this is a very expensive process. These pretreatment and feeding options are not competitive but complementary or an alternative to each other. Table 3.6 describes various characteristics of the options for biomass feed preparation [21–23,73,74].

Two techniques that are widely used to prepare biomass for size reduction and required feed processing are torrefaction and fast pyrolysis. These techniques are briefly described as follows.

TABLE 3.6
Options for Biomass Feed Preparation

Drying/Grinding	Torrefaction/Grinding	Pyrolysis
Physical	Physical/chemical	Chemical
Lowest energy	Medium energy	High energy
Dry feed	Dry feed	Slurry feed
Pelletization possible	Pelletization possible	Not applicable
Storage and transportation are problematic because of moisture absorption and fungal attack	Storage and transportation easy	Slurry storage and transportation require an inert atmosphere
Grindability, fluidization properties, and combustibility for entrained-flow fluidization are superior	Same as for drying	Pyrolysis can be considered as a first stage of a two-stage gasification process

Source: Shah, Y. and Gardner, T., Biomass torrefaction: Application in renewable energy and fuels, in: Lee, S. ed., *Encyclopedia of Chemical Engineering*, Taylor & Francis Group, New York, pp. 1–18, April 24, 2012, published online.

3.6.4 TORREFACTION

Torrefaction is a process of mild pyrolysis at lower temperature (around 200°C–250°C) and longer contact time (between 30 and 60 min). This process dehydrates and depolymerizes the long polysaccharide hydrocarbon chains present in biomass. This results in a product that is hydrophobic and has a higher energy density and improved grinding and combustion capabilities [21,22,73,74,114–116,127–171]. These improved properties of the torrefied biomass allow for easy and more energy-efficient use in existing coal-fired power plants.

The process of torrefaction is best illustrated through the Van Krevelen plot shown in Figure 3.1 [21,22,73,74,114–116,127–171]. This figure illustrates that torrefaction results in the reduction of oxygen content and correspondingly increased heating value of the treated biomass. Generally, during torrefaction, an increase in both mass and energy density occurs because approximately 30 wt% of the biomass is transformed into volatile gases. These gases carry 10% of the original biomass energy content [21,22,73,74,114–116,127–171]. This indicates that during torrefaction, a substantial amount of chemical energy is retained from the raw starting material to the product state. This differs from conventional pyrolysis where the energy yield varies from 20% to a maximum of 65%, even in very advanced pyrolysis technologies [21,74]. The torrefaction of biomass results in gases, such as H_2, CO, CO_2, CH_4, and C_xH_y; liquids such as toluene, benzene, H_2O, sugars, polysugars, acids, alcohol, furans, ketones, terpenes, phenols, fatty acids, waxes, and tannins, and solids comprised of char and ash.

Generally, the torrefaction process is operated under conditions that minimize the production of liquid products. As shown in Table 3.7 [21,74], torrefied biomass possesses very valuable properties. It has lower moisture content [124] and therefore a higher heating value when compared to untreated biomass. The storage and the transportation capabilities of the torrefied biomass are superior over those of untreated and dried biomass. Torrefied biomass is hydrophobic and does not gain humidity in storage and transportation. It shows little water uptake on immersion, between 7 and 20 wt%, and it is more stable and resistant to fungal attack compared to charcoal and untreated biomass. Pelletization, by itself, produces biomass with higher mass density; however, the pellets are not hydrophobic and remain susceptible to fungal attack. Process of palletization can also be improved when it is preceded by the torrefaction of biomass.

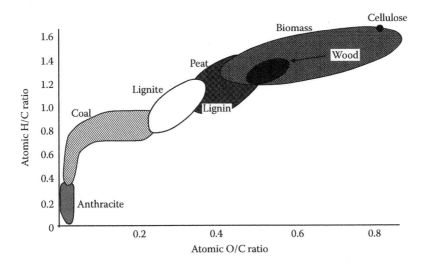

FIGURE 3.1   Van Krevelen plot. (From Shah, Y., *Energy and Fuel Systems Integration*, CRC Press, Taylor & Francis Group, New York, May 2014; Shah, Y. and Gardner, T., Biomass torrefaction: Application in renewable energy and fuels, in: Lee, S. ed., *Encyclopedia of Chemical Engineering*, Taylor & Francis Group, New York, April 24, 2012, pp. 1–18, published online.)

TABLE 3.7

Aspects of Torrefied Biomass for Gasification and Other Applications of Torrefied Products

1. Has lower moisture content and higher heating value
2. Is easy to store and transport
3. Is hydrophobic and does not gain humidity in storage and transportation
4. Is less susceptible to fungal attack
5. Is easy to burn, forms less smoke, and ignites faster
6. Significantly conserves the chemical energy in biomass
7. Has heating value (11,000 Btu/lb) that compares well with coal (12,000 Btu/lb)
8. Generates electricity with a similar efficiency to that of coal (35% fuel to electricity) and considerably higher than that of untreated biomass (23% fuel to electricity)
9. Has grindability similar to that of coal
10. Requires grinding energy 7.5–15 times less than that for untreated biomass for the same particle size
11. Has mill capacity 2–6.5 times higher compared to untreated biomass
12. Possess better fluidization properties in the gasifiers
13. Is suitable for various applications in heating, fuel, steel, and new materials in manufacturing industries

Sources: Adapted from Bergman, P.C.A. and Kiel, J.H.A., Torrefaction for biomass upgrading, in: *Proceedings of the 14th European Biomass Conference and Exhibition*, Paris, France, October 2005; ETA—Renewable Energies, Florence, Italy, 2005, pp. 17–21; Bergman, P.C.A. et al., Torrefaction for entrained flow gasification of biomass, Report no. ECN-RX-04-046, in *Proceedings of the Second World Biomass Conference on Biomass for Energy, Industry and Climate Protection*, Rome, Italy, May 10–14, 2004; Van Swaaij, W.P.M. et al., (eds.), *Proceedings of Second World Conference on Biomass for Energy Industry and Climate Protection*, Rome, Italy, May 10–14, 2004, pp. 679–682, see also report ECN-C-05-26, Petten, the Netherlands. Bergman, P.C.A., Combined torrefaction and pelletization: The TOP process, Report no. ECN-C-05-073, Energy Research Centre of the Netherlands (ECN), Petten, the Netherlands, 2005; Bergman, P.C.A. et al., Torrefaction for entrained flow gasification of biomass, Report no. ECN-C-05-067, Energy Research Centre of the Netherlands (ECN), Petten, the Netherlands, 2004; Bergman, P. et al., *Torrefaction for Biomass Co-Firing in Existing Coal-Fired Power Stations (BIOCOAL)*, ECN-C-05-013, ECN, Petten, the Netherlands, 2005, 72pp.

Torrefied biomass significantly conserves the chemical energy present in the biomass. The heating value of torrefied wood is approximately 11,000 Btu/lb and is nearly equal to that of a high-volatility bituminous coal that is 12,000 Btu/lb. It generates electricity with an efficiency comparable to that of coal of approximately 35%, on a fuel to electricity basis [21,22,73,74,114–116,127–171], and much higher than that of untreated biomass that has an efficiency of 23%, on a fuel to electricity basis [21,22,73,74,114–116, 127–171]. Bergman et al. [21,73,74,130,132,135,171] showed that torrefied biomass has better fluidization properties than that of untreated biomass, but similar to that of coal. Bergman et al. [130,132,134,135,171] also showed that torrefied willow ignites quicker and heats faster than refined coal particles.

Untreated biomass requires many times the grinding energy, by a factor of 7.5–15, to achieve a similar particle size compared to torrefied biomass (see Figure 3.2). This energy difference is significantly larger than the energy loss of biomass and energy supplied during torrefaction. The torrefied biomass is suitable for various applications such as working fuel, residential heating, and new materials for the manufacture of fuel pellets, as reducer in the steel smelting industry [21,73,74], in the manufacture of charcoal and active carbon, and in the gasification and co-firing with other fuels in gasifiers and boilers. Such a wide usefulness makes torrefied biomass a valuable and marketable product.

3.6.5 FAST PYROLYSIS

Fast pyrolysis is a rapid heating process at about 500°C in an inert atmosphere. Rapid condensation of the vapors in one to a few seconds occurs, and more than half of the biomass mass can be obtained as liquid pyrolysis oil at the expense of low char and gas yield. In these conditions, the

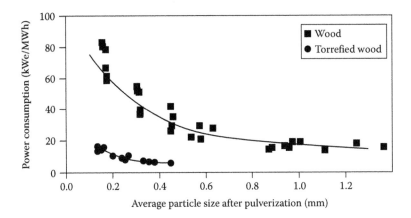

FIGURE 3.2 Power consumption for size reduction: untreated versus torrefied wood. (From Shah, Y. and Gardner, T., Biomass torrefaction: Application in renewable energy and fuels, in: Lee, S. ed., *Encyclopedia of Chemical Engineering*, Taylor & Francis Group, New York, April 24, 2012, pp. 1–18, published online.)

pyrolysis gases contain less than 10% of the biomass energy, which is consumed on-site as energy input to the pyrolysis process. The brittle char obtained is ground to a fine powder and suspended in the liquid pyrolysis oil to form the bioslurry, with an overall energy content of up to 80%–90% of the original biomass energy [3,21–23,73,74] and considerable volume reduction.

The dispersed nature of the biomass feedstock requires the pretreated fuel to be transported to a central plant where it is gasified or co-gasified with coal. Important parameters in deciding which of the two pretreatment options to consider are the energy density and postprocessing of the pretreated fuel.

Bioslurry is denser (and less bulky) than torrefied biomass. Thus, besides feeding bioslurry under pressure being easier, pneumatically feeding powdered torrefied wood requires pressurizing with an inert gas that reduces the calorific value and complicates the gas-cleaning process. Another advantage of the fast pyrolysis process is that to avoid soil nutrient depletion, pyrolysis can be designed to optimize and obtain only an energy-rich biooil containing 60%–70% of the original biomass energy [3,21–23,73,74], while the remaining char may be immediately returned to the soil. This approach may well be more efficient since it avoids transporting the ash from the gasification plant to the forest. Moreover, it ensures that almost all the ash extracted by biomass is returned to the soil and prevents possible toxic elements that might be present in coal ash to be applied to the soil. Additional savings could result from the simplicity associated with handling biooil at the plant site. Handling biooil at the plant site will require fewer and comfortable metering and handling equipment compared to handling solid feedstock. The most important factor in final decision between torrefaction and fast pyrolysis is whether or not one needs to have dry feed or slurry feed into the gasifier. Most entrained-bed gasifiers use dry feed [3,21–23,73,74].

3.6.6 SIZE REDUCTION

The reduction of coal particles and their handling are well-established technologies, and they are relatively easy and reliable. In general, the reduction of coal particles is a relatively easy task. On the other hand, size reduction of biomass is much more complex and may require a separate system. The degree of size reduction required depends on the nature of biomass/waste and the nature of the gasification technology. Many current size reduction and handling systems lack reliability [21,73,74,114–116,127–171]. Fuel variability, especially for softer biomass like straw, poses particular difficulties for the size reduction and handling systems. In cogasification systems, two extreme options for feed handling are (1) separate reception, comminution, conveying, and gasification systems for biomass and coal and (2) separate reception followed by combined comminution, conveying, and gasification systems for biomass and coal. Numerous other alternatives in between these two extremes are also possible. In general, grinding equipment appropriate for a given feedstock must be selected carefully. Woody biomass can be easily grinded, while straw biomass needs chopping.

The required degree of size reduction depends on the choice of the gasifier. FBC provides the most flexibility for particle size. Even for easily grindable fuel such as wood, significant fine grinding is required to achieve satisfactory burnout (<1 mm). The use of entrained-bed gasifier for multifuel requires expensive milling process. This also requires biomass to be dried and pelletized so that it can be milled to sufficient fine particles. As shown in Figure 3.2, the power required for the grinding biomass increases exponentially below 1 mm particle size. Often, the process of torrefaction [21,73,74,114–116,127–171] is used to facilitate the grinding process. Figure 3.2 also shows that power requirement for size reduction is significantly decreased for torrefied wood compared to untreated wood. The torrefied biomass also resists water absorption.

Hey et al. [172] outlined laboratory evaluations of fuel preparation methods for multifuel pyrolysis, gasification, and combustion of coal and biomass. They examined granular feeding in FBC and grinding sewage sludge for co-firing with coal in pulverized coal (PC) combustors. They also found that grinding of fibrous materials like straw and miscanthus in vibration or hammer mills was accompanied by stickiness or dust evolution and is therefore not a suitable approach. As long as conditions were carefully optimized, the grinding of coal and chopped barley and wood pieces in hammer mills was possible. However, such grinding in ball mill was problematic. For all of these cases, Hey et al. [172] concluded that grinding energies were of the order of 1%–2% of the feed calorific values.

For size reduction and handling purposes, biomass can be broken into four groups [21,173]:

1. Wood and wood waste, which are easy to grind and handle, but grinding and handling of fines may be an issue
2. MSW and industrial waste, which are very nonhomogeneous and difficult to process uniformly
3. Sewage sludge, which can be broken into fine particles, but its fibrous nature can cause handling problems
4. Straw and miscanthus whose particle size can only be reduced by chopping action and are difficult to feed in a mixture form particularly if they are wet

In the following paragraphs, we briefly examine size reduction and associated handling issues of these four distinct groups of biomass/waste for gasification or cogasification.

3.6.6.1 Wood/Waste Wood

In Europe, several studies [173–178] using wood/waste wood and coal showed that with the use of secondary air, hogged wood of less than 12 mm can be blended with crushed coal on the feed belts. Some studies [173–178] have, however, suggested that 1 mm is the maximum acceptable size for wood gasification system.

3.6.6.2 Municipal/Industrial Waste

Many gasifiers use RDF prepared from MSW and industrial wastes. A major drawback with the use of MSW as a fuel feedstock is its heterogeneous composition, which can present problems in a multifuel system. This can be partially handled by separating impurities like glass and metals and converting MSW into RDF pellets. This, however, is an expensive process. In order to maintain pellet structure integrity, water contamination to the pellets should be avoided. Often, the process of torrefaction precedes the pelletization process. The torrefied wood is more resistant to further water absorption.

In the United Kingdom, fiber fuel or packaging-derived fuel (PDF) is mixed with coal in FB boilers of Slough Heat and Power Ltd. [173,179–183]. Unlike RDF, PDF is made from segregated paper, packaging materials, plastic, and board and has a higher heating value. The multifuel feed can also include industrial waste such as tires, postconsumer carpets, auto shredder residues, and clinical wastes. Generally, tires are chipped to produce tire-derived fuel, carpet is shredded and premixed with coal, and auto shredder residues are usually conveyed pneumatically. The latter feedstocks are more often used for combustion than gasification.

3.6.6.3 Sewage Sludge

Dried sewage sludge can be ground down more to smaller particle sizes than most biomass materials. The presence of fine particles, fibrous nature, and wide particle size distribution can, however, cause handling problems. The heat liberated during grinding can make plastic pellets, which in turn block the screens [177,178]. The dried sewage sludge can be well pulverized using a table mill. Probst and Wehland [184] successfully processed a mixture of PC (40%) and dried sewage sludge (60%) along with some limestone in a 10 MW CFB plant. Similar processing is possible in CFB gasification plant.

3.6.6.4 Straw and Miscanthus

The fibrous and soft nature of this type of waste requires the chopping action for size reduction. Fortunately, the high reactivity of these materials allows the use of large sizes. Some data are available for co-firing. In Vestkraft coal-fired power station at Esbjerg, Denmark, some 15,000 tons of straw were combusted with wood and hard coal in 125 MW unit [21,173]. The feed was prepared by feeding baled straw to a shredder followed by a cutter. Straw feeding was difficult and below optimum. The straw moisture content above 25% caused failure in the process. The excessive moisture content caused problems in the straw feeding rates in Grenaa coal/straw dual-fuel CFB plant in Denmark [21,173]. In the 250 MW Amager power plant in Denmark, coal was mixed with 20% straw using cylindrical pellets of straw with 9% moisture content. The pellets were mixed with coal before feeding to the mills for cogrinding. These experiences can be applied to cofeeding in gasification plants.

The chopping, handling, and feeding of straw have been found difficult [185]. Meschgbiz and Krumbeck [186] used a mobile straw mill designed for agricultural use, but the system was hazardous. A screw feeder was used to supply straw to coal feed; however, segregation between biomass and coal in the bunker resulted in uncontrolled feed rates. In general, however, feeding of straw in CFB combustion or gasification system by pneumatic feeding can work well due to longer residence time and technology being more tolerant of feed moisture or size. For PFB combustion system, Andries et al. [187] fed up to 20% of crushed and pelletized straw and miscanthus and up to 20% of granular mixture of milled straw with PC. The results indicated that granular mixture was difficult to process through a lock hopper system. The same conclusion can be applied to feeding a pressurized gasifier.

3.6.7 EFFECTS OF FEED TREATMENT FOR STORAGE AND TRANSPORTATION OF BIOMASS

Unlike coal, the transportation of biomass can be very difficult, time consuming, and expensive. Low mass and energy densities and the fibrous nature of some biomass make the long-distance transportation very expensive. Grassy materials need to be rolled into bales before transportation. For these reasons, multifuel plants are often situated at the locations of easy access to biomass and waste. Otherwise, the transportation costs can limit the size of the sustainable gasification operation.

In most biomass, waste, or multifuel plant consisting of coal and biomass and waste, the concentration of biomass or waste ranges around 10%, thereby requiring less materials for transportation. Most biomass requires a combination of air-drying before storage (in an open environment) followed by further drying immediately prior to use. Biomass should be stored in an inert environment to avoid its biochemical degradation.

Bauer [188] described the storage issues during summer and winter for wood chips used in a commercial-scale 100 MW plant that used both wood chips and lignite at the Lubbenau power station in Brandenburg, Germany. While feedstock with 30% of moisture content was dried in this plant both during summer and winter within 5 weeks with no spontaneous combustion, this may not be always possible in humid summer weather. For storage of softer materials like straw, moisture control in the storage tank is very important for consistent quality of fuel supply to the plant. Very wet materials will be almost impossible to feed in the combustion or gasification plant. The ignition temperature for various biomass is generally higher than that for coal [21,173,189]. Biomass will have a higher tendency for spontaneous explosion than that for coal.

Unlike in coal, the nature of storage vessels also depends on the nature of the biomass. For example, dried sewage sludge requires storage in mass flow silos with stationary storage time being limited to 5 days in order to prevent sticking [21,173,189]. The discharge of this biomass from silos requires wide silo outlet openings, steep sides, and use of discharge aids such as air guns [21,173,189]. The storage of straw on the other hand requires an inert environment to avoid biological decay. The storage of soft biomass such as miscanthus can be problematic since this material tends to be unusable if stored at high moisture contents. The storage of coal on the other hand is a relatively easy task.

3.6.8 General Issues Regarding Particle Size, Shape, and Density of Biomass and Feeding Gasifier during Cogasification

The high reaction rates of entrained-flow gasifiers demand very small feedstock particle size (~100 μm), which is easily achievable for friable materials like coal but is more challenging and energy consuming for biomass due to its fibrous structure and hygroscopic nature. Milling biomass to the same size has a five times higher electricity consumption. Figure 3.2 shows the particle size versus the power (or energy) required for biomass. It is clear from this figure that power required for biomass increases exponentially as the particle size is reduced below 1 mm. Buggenum plant has shown [21,73,74] that for wood, a particle of 1 mm is sufficient to get complete conversion. Thus, for woody biomass, downsizing to 1 mm is sufficient, and for this case, the electricity consumption is similar to the coal milling. For straw and miscanthus, on the other hand, reducing particle size to 1 mm may not be sufficient.

In case of straw, its long narrow needle-type shape may require special type of chopping machines, and fibrous nature can create additional issues with feeding. The viability of entrained-flow gasification of this biomass relies on its pretreatment. The fibrous nature of the biomass also prevents it from fluidizing and fluffs are formed that may plug piping. Among the pretreatment options available to alleviate these biomass shortcomings are pelletization, torrefaction, and fast pyrolysis [21,73,74]. Pelletization can be expensive and fast pyrolysis leads to liquids that can be mixed with coal for slurry feeding in the gasifier. For dry feeding, torrefaction has been found an attractive alternative. Because of their wide acceptance, we briefly discuss here torrefaction and fast pyrolysis alternatives for the feed pretreatment [21,22,73,74,114–116,127–171].

In General Electric (GE) gasifier, coal is fed into the gasifier as a slurry of coal and water. Milled biomass of 1 mm can be mixed with the coal–water slurry. As discovered by Polk plant [21,23,73,74], the way to make this work effective is to use automated method of milling, pneumatic transport, and feeding into the coal slurry system. It is absolutely critical that the larger particles are effectively screened and recycled for further milling. Often, the milling of particles to a smaller diameter will require the use of processes such as torrefaction [21,22,73,74,114–116,127–171] and briquetting or pelletizing to achieve small particles without excessive power consumption. Figure 3.2 also indicates the reduction in power requirement of terrified wood as a function of particle size.

The density difference between coal and biomass particles also creates some issues with effective feeding. Mixtures of coal and switchgrass have shown very poor flow characteristics from sloped storage bunkers due to difference in bulk density. This applies to mixtures of only 5% by volume of switchgrass and progressively gets worse as the switchgrass percentage increases [21,22,73,74,114–116,127–171]. The problem can be partially alleviated by preprocessing switchgrass into pellets before blending with coal. This, however, can be expensive. Other alternatives are to modify storage bunker design or separate coal and switchgrass feed systems. The recommended approach depends on the nature of switchgrass delivered to the plant and the desire to maintain single fuel processing train. The fibrous nature of the biomass like switchgrass, straw, and miscanthus and their low density also cause problems with the pneumatic feeding system with inert carrier gas. The fibrous and compressible nature of these biomass materials can aggregate and plug the feeding line. Larger 1 mm switchgrass or straw needles are not suitable for pneumatic feeding due to their shape [21,22,73,74,114–116,127–171].

As mentioned earlier, the most desirable reactor technology of the future is an entrained-bed, oxygen-blown, high-pressure (10–50 bar) technology [21–23,73,74]. Unlike coal, the feeding of biomass in a pressurized reactor through an alternating pressurized and depressurized lock system poses several problems. While this periodic procedure is suited for high-bulk-density fuels and moderate pressures, at low bulk densities (e.g., biomass) and higher pressures, the amount of lock gas can exceed the fuel weight. Therefore, a well-designed feeding system to withstand back pressure from the gasifier is required, and this may be as expensive as the gasifier itself [21–23,73,74].

Pressurization of biomass in conventional lock hopper system requires much more inert gas than that for coal. Even when this system works for biomass, the consumption of inert gas on energy basis is approximately twice as high for biomass compared to coal due to its lower energy density. This will significantly lower the efficiency of the gasifier, as all the inert gas has to be heated to the gasification temperature. It also results in the dilution of the biosyngas. The extra inert gas has to be compressed, which results in additional electricity consumption.

For compression of biomass, a piston compressor has been developed in Europe in which approximately 50 times less inert gas is consumed [21–23,73,74]. A solid feed system developed jointly by the Department of Energy (DOE) and Stamet appears to offer a solid feed alternative system to standard lock hopper system for feeding dry coal into pressurized gasifiers [97,98]. The Stamet system often called Stamet Posimetric pump was originally developed to feed oil shale into the gasifier systems. The system provides positive flow control. The device consists of a single rotating element that is made up of multiple disks and hub that are installed inside a stationary housing. The pump has been successfully used for lignite, bituminous, and PRB coal up to 560 psi pressure. The applicability of this and other devices for coal and biomass mixture dry feeding in pressurized gasifiers is still being improved [21,73].

3.6.9 System Design Issues for Cogasification

As mentioned in Section 3.4, compared to coal, biomass and waste feedstock are more difficult to prepare and handle because of their (1) low heating value, (2) high moisture content, (3) low bulk density, (4) high fibrosity, and (5) great size variability. Several of these factors affect the transportation, storage and design, and operation of the feeding system adopted. For example, low bulk density can affect the transportation costs, size of conveyors, storage and feed bins, screw feeder size, etc.

In cogasification, often, a separate feeding system for biomass component of multifuel system is used. Both in Europe and the United States, most grain-producing countries or states produce substantial amount of surplus straw as a by-product. The straw available is a low-grade nonhomogeneous fuel source, characterized by high volatile content plus high chlorine and alkali levels. Because of its abundant availability, straw is still widely used in multifuel systems for several CFB gasification plants. Feeding systems adopted tend to be relatively complex and involve a number of separate stages. Sorting, blending, and preparing of fuels derived from wood and RDF can also be equally complex. While RDF is used in many waste gasification plants, palletization process can be very expensive.

Unlike in coal-fired power plants, the gasification and combustor using multifuel system of coal and straw results in excessive corrosion in downstream components of the plant due to high chlorine and alkali levels in straw. This issue requires a careful selection of construction materials for the plant components. The work carried out in Italy [21,73] with multifuel systems involving eucalyptus and poplar with coal required careful handling to achieve stable plant operating conditions. While both of these biomass are fast growing and in ample supply, the bark of their trees proved very difficult to handle, transport, and feed [21,73]. Furthermore, moisture content of the wood significantly affected its grinding characteristics and flow ability.

The gasifiers are designed for certain particle size distribution of feed, dust, etc. For multifuel processing, R&D work is needed to economically prepare biomass/waste feed for the gasifier. For example, for entrained-bed gasifiers biomass needs to be finely and evenly pulverized. This may be a very difficult task for softer and fibrous materials such as straw and miscanthus. Feeding of biomass/waste into a pressurized gasifier at predictable feeding rates is problematic and requires further investigation. Straw has proven to be very difficult to process and handle, and its unpredictable behavior needs further assessment. For gasification of pure biomass, often FB is preferred over entrained bed due to its flexibility in handling larger particles and particles with a larger size distribution.

3.7 GASIFICATION REACTORS

There are basically three types of reactors: fixed- or moving-bed (updraft, downdraft, cross draft, grated, etc.), FB (bubbling, turbulent, or recirculating), and entrained-bed (or transport bed) reactors used for coal and biomass gasification (see Figure 3.3). The basic difference in the design of these reactors is the status of solids and the degree of mixing in gas and solid phases. In fixed and moving bed, solids are either stagnant or move together and gas is generally in plug flow. In FB, solids are suspended and are backmixed. The gas phase also has significant amount of backmixing. In entrained bed, solids are carried by the gas phase and both gas and solids are close to plug flow. The residence time in entrained bed is the shortest of all three types of reactors. Coal gasification

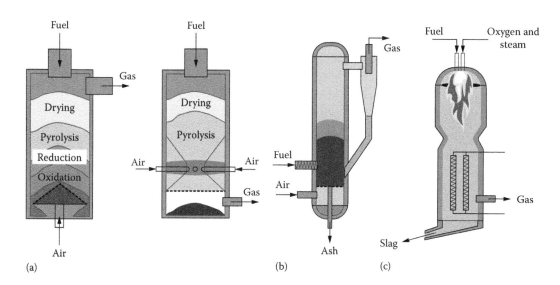

FIGURE 3.3   Three major types of gasifiers. (a) Fixed bed up-flow and down-flow, (b) fluidized bed, and (c) entrained bed. (From Gasification, Wikipedia, the free encyclopedia, 2015.)

reactors are predominantly (85%) entrained-bed reactors. Besides these, rotary kiln reactors are often used for the gasification of waste. Special types of reactors such as solar, plasma and others are discussed in Sections 3.9 through 3.14.

Both coal and biomass gasification can generate either producer gas (low temperature; heat and power applications) or syngas (high temperature; feedstock for liquid fuel applications). In general, the choice of the gasification reactors depends on the following factors:

1. How the coal or biomass is fed into the gasifier and is moved around within it: Biomass is either fed into the top of the gasifier or into the side and is then moved around either by gravity or air flows. Coal is sometimes kept on the grate or moved downward by gravity.
2. Whether oxygen, air, or steam is used as an oxidant: Using air dilutes the syngas with nitrogen, which adds to the cost of downstream processing. Using oxygen avoids this, but this method is expensive, and so oxygen-enriched air can also be used. The choice of oxidant also affects the reactor technology. Normally, in an entrained-bed reactors, pure oxygen is used as an oxidant.
3. The temperature range in which the gasifier is operated: Syngas is generally produced at higher temperatures. Producer gas is generated at lower temperatures.
4. Whether the heat for the gasifier is provided by partially combusting some of the coal or biomass in the gasifier (directly heated) or from an external source (indirectly heated), such as circulation of an inert material or steam.
5. Whether or not the gasifier is operated at above atmospheric pressure: Pressurized gasification provides higher throughputs, with larger maximum capacities, promotes hydrogen production, and leads to smaller, cheaper downstream cleanup equipment. Furthermore, since no additional compression is required, the syngas temperature can be kept high for downstream operations and liquid fuel catalysis. However, at pressures above 25–30 bar, costs quickly increase, since gasifiers need to be more robustly engineered, and the required feeding mechanisms involve complex pressurizing steps. Entrained-bed reactors are generally operated at pressures higher than atmospheric pressure.
6. How feed is injected in the gasifier (dry or slurry): A GE coal gasifier uses slurry feeding. Feeding in a pressurized reactor is more complex than in an atmospheric reactor.
7. The particle size and shape and its variation dictates the nature of the gasifier: Fixed or moving bed can take a larger particle size, FB can take more distribution of particle size, and entrained bed requires uniform and fine particles.

3.7.1 Fixed- or Moving-Bed Reactors

In the fixed-bed coal gasification reactor, coal is supported by a grate and the gasifying media (steam, air, or oxygen) pass upward through the supported bed, whereby the product gases exit from the top of the reactor. Lurgi developed an atmospheric fixed-bed reactor in 1927 and a pressurized version in 1931. An important feature of the Lurgi dry bottom gasifier is the low consumption of oxygen and the high steam demand. Only noncaking coals can be used in the fixed-bed reactor. The Lurgi dry-ash gasifier has been extensively used for town gas production and in South Africa for the production of chemicals from coal [19,21,34,73,190,191].

In the moving-bed reactor, coal and gaseous streams move countercurrently (updraft), that is, coal moves downward by gravity while gas passes upward through the coalbed. The temperature at the bottom of the reactor is higher than that at the top. Because of the lower temperature at the top for coal devolatilization, relatively large amounts of liquid hydrocarbons are also produced in this type of gasifier. Moving-bed gasifiers need graded coal in the range 6–50 mm. Highly caking coals cannot be processed in moving-bed gasifiers. Mildly caking coals require the assistance of a stirrer in order to avoid pasting up of the bed. Tars and other oxygenated compounds are produced as by-products.

In both fixed- and moving-bed reactors, the residence time of the coal is much longer than that in a FB or entrained-bed reactors, thus providing ample contact time between reactants. Ash is removed from the bottom of the reactor as dry ash or slag. Besides Lurgi gasifiers, Wellman-Galusha gasifiers belong in this category. It should be clearly understood that a moving-bed reactor is classified as a kind of fixed-bed reactor because solids in the bed stay together regardless of the movement of the hardware that supports the bed [19,21,34,73,190,191].

An advanced variant of the original Lurgi pressure gasifier was developed jointly by British Gas and Lurgi during the 1950s and 1960s. The British Gas/Lurgi (BGL) *slagging gasifier* incorporates a molten slag bath. Much lower steam and somewhat lower oxygen consumption of the slagging gasifier results in much higher syngas production per unit of coal intake and much lower yield of pyrolysis products compared with the dry bottom unit. Furthermore, the CO_2 content of the gas is lower and the methane content is halved [19,21,34,73,191].

The moving-bed reactors are also used for the gasification of MSW where drying and feed particle size reduction is not carried out. These reactors can operate at temperatures as high as 1000°C, but the syngas produced from these reactors carries significant amount of tars that need to be removed as a part of syngas-cleaning process. Many waste industries are making transformation of waste combustion to waste gasification using this type of reactors. The most notable commercial application of this type is the British Lurgi gasifier for coal–waste mixture [21,73,191].

The moving-bed biomass gasification reactors can be operated as updraft, downdraft, or cross draft reactors. In these reactors, large biomass particles move slowly down the bed while gases move up or down (or cross) through the bed. Biomass moisture content controls the product gas temperature. These reactors are generally operated at low (near atmospheric) pressures. All moving-bed reactors have low oxidant requirement, need to handle biomass with high mineral matters and VMs, and have limited ability to handle fines. They produce tar and oils and have high *cold gas* thermal efficiency when heating value of the hydrocarbon liquids is included. There are three commercial moving-bed gasifier technologies [21,73,192–212].

The selection of updraft or downdraft gasifier would mainly depend upon the available fuel and the requirements of the application. Downdraft wood gasifier requires wood chips, biomass briquettes, or wood-like agro-residues (such as stems and stalks). If fuel flexibility is required and the need for fuel switching (say, from biomass to coal or from one biomass to another) is anticipated, the updraft gasifier would be suitable. For gasification of coal, updraft gasifier is suitable. Downdraft gasifier cannot accept coal. For direct heating applications, such as firing in kiln, furnaces, boilers, etc., updraft gasifier would make a better choice. However, if the contaminants in fuel gas are likely to affect the end-product quality, the choice of downdraft gasifier makes better sense. There are situations where the effect on end-product quality is critical, but fuel flexibility is also an important consideration. In such situations, updraft gasifier with gas scrubbing arrangement may be selected. For running IC engines, normally downdraft gasifiers are selected. However, for large installations where flexibility may be an additional requirement, updraft gasifiers with extensive gas cleanup system could be an option. In addition to the general criteria mentioned earlier, there would be site-specific requirements that need to be analyzed in detail while making the final selection [3,21,73,192–212].

In *updraft (or countercurrent) moving-bed gasifiers* (see Figure 3.3), the wastes are fed at the top of the reactor. The oxidizing agent, which may be air, oxygen, CO_2, or steam, is fed into the bottom of the reactor. The gasification reaction takes place in the bottom of the reactor between the downcoming material and the ascending gas. The reaction temperature is normally between 1030°C and 1430°C. The rise of the hot gas starts waste pyrolysis at lower temperatures and dries it. The tar levels in the crude gas with this reactor configuration are between 10% and 20%, which makes them difficult to clean for electricity applications. In this case, the oxidizer passes through coke and (more likely) ashes to the reaction zone where it interacts with coal. The hot gas produced then passes fresh fuel and heats it while absorbing some products of thermal destruction of the fuel, such as tars and phenols. Thus, the gas requires significant refining before being used in the FT reaction.

Products of the refinement are highly toxic and require special facilities for their utilization. As a result, the plant utilizing the described technologies has to be very large to be economically efficient, such as Sasol in South Africa [19,34,73,194].

A typical syngas exit temperature for a moving-bed updraft gasifier is approximately 600°C–700°C. At this temperature, some of the heavier volatilized hydrocarbon compounds, such as tars and oils, will not be cracked and can easily condense in downstream syngas cooling equipment. Because fuel is introduced at the top of the gasifier where the syngas is exiting, this type of gasifier cannot handle fine fuel particles. Such particles would be entrained with the exiting syngas and would not be converted to syngas in the reactor bed. Cyclones are typically used to capture fine particles in the syngas, which are often sent to a briquetting facility to form larger particles and then recycled to the gasifier for another attempt at conversion. An overall measure of gasifier performance is the cold gas efficiency. The cold gas efficiency is the ratio of the heating value of "cold" syngas, at standard temperature, to the heating value of the amount of fuel consumed/required to produce the syngas. The cold gas efficiency does not take into account recovery of energy in the gasifier such as through steam generation or associated with sensible heat of the syngas at high temperatures. Moving-bed gasifiers tend to have very high cold gas efficiencies, with values in the range of 80%–90% [19,34,73].

In *downdraft (cocurrent) gasifiers* (reverse blowing), the wastes are fed at the top part of the reactor. As shown in Figure 3.3, the oxidizing agent, which may be air, oxygen, CO_2, or steam, comes in the middle part of the gasifier. In this case, there is no chemical interaction between coal (and biomass) and oxidizer before the reaction zone. The major part of the tars is burned for the pyrolysis of the wastes. This process is called "flaming pyrolysis." Thus, the tar levels in this reactor configuration are very low, around 0.1%, as the major part of tars is burned to supply the energy for the pyrolysis/gasification reactions of the wastes. The gas produced in the reaction zone passes solid products of gasification (coke and ashes), and CO_2 and H_2O contained in the gas are additionally chemically restored to CO and H_2. As compared to the "direct blowing" technology, no toxic by-products are present in the gas: those are disabled in the reaction zone. The rate of gas production in it is significantly lower than that in *direct blowing* [3,34,73,192,193,195–212].

This reactor configuration is particularly suitable for the production of clean gas requiring low posttreatment for their use in electricity production with gas turbines. However, the operation generally requires a long residence time (1–3 h) [3,9–16,34,73]. This configuration is considered most attractive to small units of 80–500 kWe and has the disadvantage to have low energy efficiency but with low tar concentrations [3,9–16,34,61–66,73]. There are two types downdraft gasifiers: (1) single throat and (2) double throat. Single-throat gasifiers are mainly used for stationary applications, whereas double-throat gasifiers are for varying loads as well as automotive purposes. A notable process using this technology was developed by the Soviet Union as "TERMOKOKS-S" process [213]. Industrial plants utilizing this technology are now known to function in Ulaanbaatar (Mongolia) and Krasnoyarsk (Russia).

The downdraft technology has been used for gasification of coal and biomass mixtures. Zainal et al. [192] performed an experimental study on a downdraft biomass gasifier using wood chips and charcoal and varied the equivalence ratio from 0.259 to 0.46. It was found that the calorific value increased with equivalence ratio and reached a peak value of 0.388, for which the calorific value is reported to be 5.34 MJ/Nm^3. It was also observed that complete conversion of carbon to gaseous fuel was not taken place even for the optimum equivalence ratio.

Pérez et al. [211] developed a steady, one-dimensional model of the biomass gasification process, which was validated with biomass of different sizes and varying air superficial velocities. This model allowed evaluating the effect of the physical, chemical, and energy properties of biomass (size, density, proximate and ultimate analysis, and heating value) on the gasification process. Moreover, it enabled the study of the gasifier geometry, the heat exchange, and the different injection points of the gasifying agent. Dogru et al. [196] carried out gasification studies using hazelnut shell as a biomass. Jayah et al. [197] investigated the gasification of chips of rubber wood of varying moisture contents (12.5%–18.5%) and chip sizes (3.3–5.5 cm) in an 80 kW downdraft throated gasifier, which was double walled with an air gap in between.

Sheth and Babu [198,210] varied the range of air-to-fuel ratio 1.37–1.64 Nm³/kg and that of equivalence ratio 0.262–0.314 and reported that optimum operation of the gasifier was found to be between 1.44 and 1.47 Nm³/kg of air-to-fuel ratios at the values of 4.06 and 4.48 kg/h of wet feed rate, which generated the producer gas with a calorific value of about 5 MJ/m³. Olgun et al. [199] designed and constructed a small-scale fixed-bed downdraft gasifier system that used agricultural and forestry residues as feed. The air-to-fuel ratio was adjusted to produce a gas with acceptably high heating value and low pollutants. Heating values above 4 MJ/Nm³ were achieved in less than 10 min with hazelnut shells and similar quality gas was obtained for about 45 min. With wood chips, 15 min were required to reach 4 MJ/Nm³ and the delivery of this quality gas was limited to 30 min. The obtained heating values were acceptable. The equivalence ratio resulting in the highest heating value product gas was found as 0.35 for both biomass species investigated in this study.

Biosyngas is produced by high-temperature (>1200°C) gasification. In principal, the (oxygen-blown) downdraft and the entrained-flow gasification (slagging and nonslagging) processes are suitable for this purpose. However, downdraft fixed-bed gasifiers are limited in scale and require a well-defined fuel, making them not fuel flexible. Therefore, the preferred process to produce biosyngas is entrained-flow gasification [21,34,73].

Cross draft moving-bed reactors are not as commonly used as updraft and downdraft moving-bed reactors. They have short design height, fast response time, and flexible gas production. They also have high pressure drop and high sensitivity to slag formation. They are not used as often as downdraft or updraft reactors. The updraft and downdraft gasifiers are schematically described in Figure 3.3 [1].

3.7.2 Fluidized-Bed Reactors

For coal gasification, this type of reactor uses PC particles. The feedstock fuel, oxidant, and steam are introduced at the bottom of the reactor. The gas (or gasifying medium) flows upward through the bed and fluidizes the coal particles. Better contact between gas and suspended coal particles allows more exposure of coal surface area to gas, which in turn promotes the gas–solid chemical reaction, resulting in enhanced carbon conversion. Bubbling FB reactors feature rapid mixing and excellent heat and mass transfer of fuel particles in a 0.1–10 mm size range with both the oxidant and steam in a FB [3,9–16,21,73,214–224]. In these reactors, backmixing of incoming feedstock fuel, oxidant, steam, and fuel gas takes place resulting in a uniform distribution of solids and gases in the reactors. The gasification takes place in the central zone of the reactor (see Figure 3.3).

This type of reactor allows intimate contact between gas and solid coal, at the same time providing relatively longer residence times than entrained-flow reactor. On the other hand, individual particles have widely varying residence time in the bed volume. Therefore, unreacted carbon particles are inevitably removed from the bed along with fully reacted particles (ash). The best existing FB devices offer a carbon conversion of 97%. In comparison, both moving-bed and entrained-flow processes offer carbon conversions of 99%. Either dry ash is removed continuously from the bed or the gasifier is operated at such a high temperature that it can be removed as agglomerates. Such beds, however, have limited ability to handle caking coals, owing to operational complications in fluidization characteristics. At low gas flow rate, FB is called bubbling FB. As the fuel gas flow rate increases and bed becomes turbulent, it is called turbulent FB. Winkler and Synthane processes use this type of reactor [3,9–16,19,21,34,73].

The reactors generally operate in a narrow temperature range of 950°C–1050°C. The FB is maintained at a nearly constant temperature, which is well below the initial ash fusion temperature to avoid clinker formation and possible defluidization of the bed. Unconverted coal in the form of char is entrained from the bed and leaves the gasifier with the hot raw gas. This char is separated from the raw gas in the cyclones and is recycled to the hot ash-agglomerating zone at the bottom of the gasifier. The temperature in that zone is high enough to gasify the char and reach the softening temperature for some of the eutectics in the ash. The ash particles stick together, grow in size, and become dense until they are separated from the char particles and then fall to the base of the gasifier where they are removed.

The processes in these reactors are restricted to reactive, noncaking coals to facilitate easy gasification of the unconverted char entering the hot ash zone and for uniform backmixing of coal and fuel gas. The cold gas efficiency is approximately 80%. These reactors have been used for Winkler gasification process and high-temperature Winkler (HTW) gasification process [224]. A key example of fluidized gasification design is the KRW gasifier.

For the biomass gasification, FB reactors are widely used because they provide the most flexibility in the range of operating conditions. FB gasifiers can handle a wide range of feedstock including solid waste, wood, pulp sludge, MSW, RDF, corn stover, and high-ash coals with pressure range from 1 to 33 bar and average reactor temperature range from 725°C to 1400°C [19,34,73,214–225]. For complete tar and methane conversions, the reactor needs to be operated at temperatures greater than 1200°C–1300°C. FB offers more flexibility in particle size variations. For most BFB gasifiers, biomass will have to be dried to increase operating temperatures. BFBs generally have uniform moderate temperature with good mixing and moderate oxygen and steam requirement and extensive char recycling [1–6,19,34,73,214–225]. FB gasifiers may differ in ash conditions (dry or agglomerated) and in design configurations for improving char use. There are relatively few large FB gasifiers in operation.

The tar rate in a FB reactor is at an intermediate level between the updraft and downdraft reactors: between 1% and 5%. Gasification reactions are homogenized by suspended grounded wastes in the reactor. This method optimizes the temperature along the reactor and has a high reaction rate for short residence time (less than 30 min). The disadvantage of this configuration is the high proportion of particulates (tars) in the exhaust gas that requires high gas treatment and has low mass and energy yields [1–6,19,21,34,61–66,73,214–225].

In CFB, a bed of fine inert material has air, oxygen, or steam blown upward through its fast enough (5–10 m/s) velocity to suspend material throughout the gasifier. Biomass is fed in from the side, is suspended, and combusts providing heat or reacts to form syngas. A circulating FB recycles unreacted solid back into the reactor. Directly heated circulating FB gasification of biomass is not as widely used as BFB. Very few have been operated at high pressures and most have operated at temperatures below 1000°C to prevent sticking of ash. CFBs have not demonstrated the use of pure oxygen and steam as reactants. The mixture of syngas and particles are separated using a cyclone, with the material returned into the base of the gasifier. The reported CO_2 content and H_2/CO ratios of the product coming from CFB are low [1–6,19,21,34,73,214–225].

Dual-FB system has two chambers—a gasifier and a combustor. Biomass is fed into the CFB/BFB gasification chamber and converted to nitrogen-free syngas and char using steam. The char is burnt in air in the CFB/BFB combustion chamber, thus heating the accompanying bed particles. This hot bed material is then fed back into the gasification chamber, provided that the indirect reaction heat cyclones remove any CFB chamber syngas or flue gas operates at temperatures below 900°C to avoid ash melting and sticking. This system can be pressurized. More details on this type of system often known as "indirect gasification" is given in Section 3.14.

3.7.3 ENTRAINED-BED REACTORS

While entrained-bed reactors are often not used for biomass/waste gasification due to its fine particle size requirement, more than 85% of commercial current coal gasification reactors are entrained-bed reactors in which fine coal particles gasify with large amount of oxygen and steam [1–6,19,21,34,73,225–227].

The entrained-bed reactors (see Figure 3.3) are plug flow reactors where the fine feedstock fuel particles (less than 0.1 mm) flow cocurrently with oxidant and/or steam. The feedstock, oxidant, and steam are introduced at the top of the reactor. Because of the entrainment requirement, high space velocity of gas stream and fine powdery coal particles are very essential to the operation of this type of reactor. Due to high gas flow rate and fine particle size, both gas and solids move in plug flow (with no backmixing). The residence time of fuel particles in the reactor is short (of the order of seconds). Solid fuel must be fine and homogeneous and the reactors operate well above the ash-slagging conditions to assure high carbon conversion. This type of reactor is able to gasify all coals regardless of coal rank, caking characteristics, or amount of coal fines. The coals with lower ash are preferred.

Entrained-bed reactors can use dry feed or wet slurry feed. The gasification takes place rapidly in pure oxygen environment at temperatures in excess of 1200°C. The feedstock is converted primarily to H_2, CO, and CO_2 with no methane and liquid hydrocarbons in the syngas. The raw gas leaves from the bottom of the reactor at high temperatures of 1200°C and greater. The gasifier design can also vary in their internal designs to handle the very hot reaction mixtures and heat recovery configurations [1–6,19,21,34,73,225–227].

The advantage of entrained-flow gasifiers is their ability to handle any coal feedstock and produce a clean, tar-free gas. Additionally, ash is produced in the form of inert slag or frit. This is achieved with the penalty of additional effort in coal preparation and high oxygen consumption, especially in the case of coal–water slurries or coals with a high moisture or ash content. The majority of the coal gasification processes that have been developed after 1950 are based on entrained-flow, slagging gasifiers operating at pressures of 20–70 bar and at high temperature (≥1400°C). Entrained-flow gasifiers have become the technology of choice for hard coals and have been selected for the majority of commercial-sized IGCC plants. Due to high temperatures, entrained-flow gasifiers use more oxidants than the other designs. The cold gas efficiency is approximately 80% [1–6,19,21,34,73,225–227].

Typical examples of such reactors are GE gasifiers and E-Gas gasifiers. GE gasification is a specialized form of entrained-flow gasification in which coal is fed to the gasifier in a water slurry. Because of the water in the slurry, which acts as heat moderator, the gasifier can be operated at higher pressures than other types of entrained-flow gasifiers. Higher operating pressure leads to increased gas production capability per gasifier of a given size [7,228]. Entrained-bed gasifiers have been the choice for all coal- or oil-based IGCC plants under construction [1–7,19,21,34,73,225–227]. Besides GE gasifier, commercial gasifiers include two variants of Shell gasifier and the Prenflo gasifier. Both GE and Shell oil gasifiers have more than 100 units in operation worldwide.

In a slagging gasifier, the ash-forming components melt in the gasifier. The molten particles condense on the relatively cold walls and ultimately form a layer being solid close to the wall and liquid on the inner side. This slag layer serves as a protective layer for the wall. The liquid slag is removed from the bottom of the gasifier. In order to generate a liquid slag with the right viscosity at the given temperature, generally so-called fluxing material must be added. For coal-fired plants, this often is limestone or another Ca-rich material. Slagging entrained-flow gasifier manufacturers are Shell, Texaco, Krupp-Uhde, Future Energy (formerly Noell and Babcock Borsig Power), E-Gas (formerly Destec and Dow), MHI (Mitsubishi Heavy Industries), Hitachi, and CHOREN (formerly UET). Examples of commercial gasifiers that use this type of reactor include the Koppers–Totzek gasifier and Texaco gasifier [1–7,19,21,34,73,225–227].

In most cases, EF gasifiers are operated under pressure (typically 20–50 bar) and with pure oxygen and with capacities in the order of several hundreds of MWth. Several commercial EF gasifiers exist and have proven their availability at large scales (700 MWth). EF gasifiers are operated at sufficient high temperatures (1200°C–1500°C) to ensure complete conversion of the biomass (>99.5% carbon conversion) with a high coal-to-syngas or biomass-to-biosyngas efficiency.

The slagging entrained-flow gasifier can convert all types of biomass materials, coal, and waste. EF is suitable for wood, alkaline-rich biomass like straw and grasses, high-ash streams like sludges and manure, and wastes like RDF and plastics besides coal and oil shale. Due to the high temperature in the gasifier, the biosyngas is absolutely free of organic impurities (i.e., tars) and can easily be cleaned from small traces of inorganic impurities with conventional proven technologies.

The minerals from the biomass are recovered in the slag and the fly ash. The slag can be used as construction material; this in contrast to other (low-temperature gasification) processes that yield a carbon-containing ash that has to be disposed of as chemical waste. The carbon-free fly ash can be used for mineral recycling and fertilization of biomass production areas [1–7,19,21,34,73,225–227].

3.7.4 COMPARISON OF DIFFERENT TYPES OF REACTORS

Table 3.8 compares capacity of different types of gasifier [7,190–228]. A comparison among three major types of reactors (gasifiers) discussed earlier is shown in Table 3.9 [21,73]. These tables thus show the capacity and pluses and minuses of each type of reactor.

3.7.4.1 Additional Industrial Perspectives

1. Feedstock requirements vary considerably between gasifiers: entrained-flow reactors having the most stringent requirements. However, the cost estimates show that the costs of additional on-site pretreatment needed for EF do not result in higher total plant costs than the other technologies. Similarly, the costs of achieving the sizing and moisture requirements for CFB

TABLE 3.8
Gasifier Capacity versus Nature of Gasifier

Type of Gasifier	Range of Capacity (odt/day Biomass Input)
Downflow fixed bed	0.2–10
Updraft fixed bed	7–80
Atmospheric BFB	10–100
Plasma	15–250
Atmospheric CFB and dual	90–550
Pressurized BFB, CFB, and dual	150–1,500
Entrained bed	750–10,000

Sources: Prepared from the data obtained in References 7,190–228.

TABLE 3.9
Comparison of Three Major Gasifiers

Characteristics of Different Gasifier Types[a]	Fixed Bed	Fluidized Bed	Entrained Flow
Outlet temperature	Low (425°C–600°C)	Moderate (900°C–1050°C)	High (1250°C–1600°C)
Oxidant demand	Low	Moderate	High
Ash conditions	Dry ash or slagging	Dry ash or agglomerating	Slagging
Size of coal feed	6–50 mm	6–10 mm	<100 μm
Acceptability of fines	Limited	Good	Unlimited
Other characteristics	Methane, tars, and oils present in syngas	Low-carbon conversion	Pure syngas, high-carbon conversion

Sources: Shah, Y., *Energy and Fuel Systems Integration*, CRC Press, Taylor & Francis Group, New York, May 2014; Ratafia-Brown, J. et al., Assessment of technologies for co-converting coal and biomass to a clean syngas—Task 2 report (RDS), DOE/NETL-403.01.08 Activity 2 Report, Department of Energy, Washington, DC, May 10, 2007.

[a] Adapted from Reference 10.

and BFB do not have a large impact on the syngas production costs. Feedstock tolerance is unlikely to be a determining factor in the choice of gasifier technology, as all types can ultimately accept a range of feedstocks with little implication on overall production cost.

2. All of the gasifiers can achieve the required syngas quality for fuel production, albeit with varying levels of syngas cleanup and conditioning.

3. All of the gasifiers can be scaled up to achieve the minimum economic scale for FT synthesis, either as a single gasifier or as a combination of a small number of gasifier modules. Modular systems may not have the same economies of scale as single systems but could have benefits in terms of use of different feedstocks and of availability.

4. Based on the data available on gasification plant costs and the uncertainty in these data, it is not possible to differentiate clearly between the gasifier types on the basis of syngas production costs. For all gasifier types, a more detailed analysis of a particular system concept would be needed to give an accurate comparison of the economics, paying particular attention to pretreatment costs, plant efficiency (as this has an impact on feedstock costs), and syngas cleanup steps.

5. Entrained-flow gasification is the most advanced toward commercialization, with developers having pilot plants in operation for fuel production and larger-scale demonstration plants operating currently or planned to operate in the very near term (CHOREN). The developers involved in entrained-flow gasification and their partners have significant commercial and technical experience in gasification and liquid fuel production. Despite having high pretreatment costs in some cases, entrained flow has the greatest potential for scale up to very large plants and therefore potentially low costs due to economies of scale.

6. BFB gasification benefits from a longer history of biomass gasification than that of entrained flow. There are several commercially focused players in BFB gasification, with pressurized and oxygen-blown systems in development (Carbona, EPI, Enerkem). These are aimed at fuel production, and the developers have planned biofuel demonstration projects, either alone or with biofuel companies. It is anticipated that these should provide the first performance data for large-scale BFB processes.

7. CFB gasification also has a relatively long history of biomass gasification, but much of the experience is not with the pressurized and oxygen-blown systems needed for fuel production. Nevertheless, there are several players involved in CFB gasification for fuels, including the strong VTT and Foster Wheeler collaboration, used in the NSE Biofuels (Stora Enso/Neste Oil joint venture) project.

8. Dual-FB gasification benefits from the experience gained with BFB and CFB, although is at an earlier stage of development than EF, BFB, and CFB. Dual-FB systems are only currently operating in small-scale heat and power applications, and they still need to be demonstrated at high pressure—however, if developed, these pressurized systems have the potential to produce low-cost, nitrogen-free syngas. The players involved have a shorter track record of experience but have successfully operated plants at high availabilities, and some have plans for liquid fuel production in the future.

3.7.5 Best Option of Reactor Technology for Multifuel Coal–Biomass Gasification

As mentioned earlier, BGL moving-bed reactor is being used for coal–waste gasification in the United Kingdom. In the Royal Institute of Technology in Stockholm, Sweden, an investigation of coal–biomass blend gasification in an oxygen-containing environment in a PFB reactor was carried out. It was noted that char from woody biomass was very sensitive to thermal annealing effect, which occurred at relatively low temperature (around 650°C) and short 8 min of soak time. The study showed high gasification rate and lower yields of tar and ammonia in oxygen-rich environment for fuel mixture of birch and coal [1–7,19,21,34,73,225–227]. Also, higher oxygen content of biomass reduced gasifier oxygen consumption proportional to the quantity of coal displaced.

While all three types of reactors can handle coal–biomass mixtures, the industry prefers entrained-flow reactor for coal–biomass mixture gasification. The main reasons are that (1) it provides high fuel flexibility; (2) high temperature allows it to operate under slagging conditions; (3) it mainly produces syngas, which can be easily used for a variety of downstream operations; and (4) it can be easily designed for large-scale, high-throughput operations [1–7,19,21,34,73,225–227].

A slagging entrained-bed reactor is preferred for coal–biomass gasification because mineral matters from biomass can end up in the slag. Improved ash handling also occurs with entrained-flow gasifiers under slagging conditions. In order to obtain the proper slag properties, fluxing materials are often used. Sometimes, slag recycling may be necessary to obtain enough slag inside the reactor to ensure sufficient wall coverage [1–7,19,21,34,73,225–227].

At entrained-bed reactor temperatures of 1300°C–1700°C, a gas with very low concentration of tar, methane, and carbon dioxide is obtained, which significantly decreases the gas-cleaning cost. Modern entrained-flow gasifiers operate at a pressure of between 15 and 60 bar. These elevated pressures are advantageous for gas to liquid synthesis. High flow rate and small particle size requirement for the entrained bed will necessitate biomass feed pretreatment such as torrefaction or pyrolysis.

Since entrained-bed reactors are widely used in coal industry and much is known about its commercial operation, its use for coal–biomass mixture (with small concentration of biomass) makes most sense. This choice will also allow easy retrofitting of existing large-scale entrained-bed coal gasification reactors for coal–biomass mixtures. All indications are that future coal–biomass commercial gasification reactors would be entrained-flow reactors.

A cogasification model for a large-scale entrained-flow gasifier was developed by the Center of Research for Energy Resources and Consumptions, University of Zaragoza, Spain. The model was validated with a large number of operating data from Elcogas IGCC power plant in Puertollano, Spain. The model has been successfully applied to a mixture of coal and coke and up to 10% biomass. The model predicts how oxygen and steam requirements should be changed and optimized with the change in the nature of the feedstock [21,73].

3.7.6 ROTARY KILN GASIFIER FOR WASTE

One of the gasifiers not covered in the earlier discussion is the rotary kiln gasifier often used for gasification of waste materials. Rotary kiln gasification systems use a rotary kiln similar to those commonly found in use in the cement, lime, and hazardous waste industries. Fuel is fed into the upper end of a slowly rotating kiln along with a controlled amount of air. As the fuel travels through the kiln, the tumbling action causes mixing with air and the gasification process occurs. Syngas is captured within the kiln and is removed from the high side of the kiln [229–233].

Rotary kilns have significant advantages over fixed-bed gasifiers in that fuel type and particle size are not design dependent; thus, the kiln has the ability to use a variety of fuels over time. The chief challenge in the use of rotary kiln gasifiers is the ability to control air. Syngas from rotary kiln gasifiers is typically fed to a secondary combustion chamber (SCC), or oxidizer, where the syngas is combusted to create hot exhaust gases that are used to create steam in a heat recovery boiler.

3.7.6.1 Rotary Kiln Gasification

A rotary kiln gasifier is a slight modification of a traditional rotary kiln whereby the amount of air to the kiln is restricted to allow gasification to occur. Instead of combustion flue gases, syngas is produced and is subsequently burned in a SCC or thermal oxidizer.

The principal challenge with creating an effective rotary kiln gasifier is the control of air. It is important to create effective end seals to prevent air entering. Effective seals include overlapping steel collars, as gaskets tend to burn or deteriorate over time. The process must have a method for fuel feed to prevent unwanted air entering with the fuel. Effective fuel feeds are a continuous auger feed, whereby the fuel acts as the plug to prevent air infiltration. The air traditionally enters a kiln through the low end, which tends to cause full combustion and hot spots near the air entrance. Some method of uniform and controllable air introduction throughout the length of the kiln has been difficult and so far not well achieved [229–233].

A control of fuel/air mixing is also an issue. The rotational pattern causes some fuel on the bottom of the kiln to be minimally exposed. Past efforts have included adding fuel lifters to cause more suspension and mixing of the fuel. This method tends to cause more combustion and creation of particulates that needs to be handled. Also, a method to prevent air infiltration at the ash discharge needs to be devised. Good methods include ash falling into a water bath that both cools the ash and prevents air from entering the kiln. Prevention of ash slagging from high temperatures is also needed. Proper control of the air flow in the kiln keeps slagging to a minimum.

All rotary kiln gasifier designs strive to find the right balance to minimize combustion, optimize production of syngas, minimize operator inputs (automation), provide low maintenance, and perform with predictability and high reliability.

3.7.6.2 Callidus Technologies Rotary Kiln Gasifiers

Callidus Technologies (currently a division of Honeywell) is a global manufacturer of thermal oxidizers and historically produced a rotary kiln gasification system that is in use at several North American wood product facilities today.

The Callidus gasification systems were built in 1999–2002 and are used for gasification of mill wastes to provide thermal heat for steam generation, raw product drying, and subsequent destruction of dryer and mill exhaust gases through a thermal oxidizer system. These systems are currently operational at *Norbord* oriented strand board (OSB) mills in South Carolina and Alabama, a Del-Tin OSB mill in Arkansas, and other mills in Canada.

The Callidus system at Norbord mill in Kinards, South Carolina, has three vertical thermal oxidizers at each rotary kiln and the wood fuel feed conveyors. The CLGS is a 300 MMBtu/h unit that uses natural gas as a secondary fuel source for start-up. The CLGS consists primarily of a fuel feed system, three rotary gasifiers, three SCCs, hot air heat exchanger, recuperative heat exchanger, hot oil heat exchanger, heat recovery boilers, ESP, induced draft fan, and exhaust vent stack. The system is considered closed loop in that the hot air supplied to the dryers returns to the system as combustion air and is not released to the atmosphere until after the VOCs are destroyed and the particulate captured. The press vent and dryer exhaust gases, laden with VOCs, pass through a recuperative heat exchanger to be heated to 875°F before being routed into the rotary gasifier and SCC.

The rotary gasifier is operated in the starved-air mode, where the oxygen level is too low for complete combustion to occur. As a result, combustible gases are formed, primarily carbon monoxide (CO) and hydrogen (H_2). The gasifier has a wood fuel residence time of approximately 60 min. The flue gas leaving the gasifier is oxidized in a vertical SCC to assure complete destruction of any remaining VOC, fine wood particles, and CO. The remainder of the dryer exhaust gas is used as an oxygen source to complete the combustion process in the SCC. The hot flue gases leaving the SCC pass through a series of heat exchangers to extract heat required for the different processes of manufacturing. The cooled flue gases pass through the ESP to remove any suspended particulate. The flue gases are then exhausted out the vent stack with the aid of an induced draft fan [233].

3.8 COMMERCIAL GASIFICATION PROCESSES

3.8.1 HISTORICAL TOWN GAS PROCESS

Town gas is a flammable gaseous fuel made by the destructive distillation of coal and contains a variety of calorific gases including hydrogen, carbon monoxide, methane, and other volatile hydrocarbons, together with small quantities of noncalorific gases such as carbon dioxide and nitrogen, and is used in a similar way to natural gas. This technology is not usually economically competitive with others used today.

Most old town "gashouses" were simple by-product of coke ovens that heated bituminous coal in air-tight chambers. The gas driven off from the coal was collected and distributed through networks of pipes to residences and other buildings where it was used for cooking and lighting. The coal tar (or asphalt) that are collected in the bottoms of the gashouse ovens was often used for roofing and other waterproofing purposes. It was also mixed with sand and gravel and used for paving streets [1–6,19,34].

3.8.2 TYPICAL COMMERCIAL COAL GASIFICATION PROCESSES

There are a large number of widely varying commercial coal gasification processes. The gasification processes can be classified basically in two general ways: (1) by the Btu content of the product gas [1–6,19,34] and (2) by the type of the reactor configuration and the level of the pressure. The Btu content of the product gas is directly dependent on the methane content and the level of diluent in the gas. The most famous medium- and high-Btu gasification processes deal with Lurgi gasifiers (in their different forms). Widely used medium- to low-Btu gasification processes are Koppers–Totzek, Texaco, and Shell Gasification processes. Here, we briefly summarize the descriptions of these processes given by Lee [19]. Numerous other processes for coal, oil shale, and biomass/waste gasification are also described in References 40,45,46,80,234–244. Large-scale cogasification reactors and processes are described in References 21 and 73.

3.8.2.1 Lurgi Gasification

The Lurgi gasification process is one of the several processes for which commercial technology has been fully developed [9]. The process is now widely used throughout the world. This process produces low- to medium-Btu gas as product gas and it uses a fixed-bed reactor concept. The older version of Lurgi process is the *dry-ash gasification* process that differs significantly from the more recently developed *slagging gasification* process.

The dry-ash Lurgi gasifier is a pressurized vertical reactor that only uses noncaking coals [19,34]. In this gasifier, coal sized between 1.5 in. and 4 mesh reacts with steam and oxygen in a slowly moving bed. The process is operated semicontinuously. The coal feed is supported at the base of the reactor by a revolving grate through which the steam and oxygen mixture is introduced and the ash removed. This process takes place at around 24–31 atm and in the temperature range of 620°C–760°C.

The residence time in the reactor is about 1 h. The product gas from a high-pressure reactor has a relatively high methane content compared to a nonpressurized gasifier. The high methane content of the product gas is a result of the relatively low gasification temperature. If oxygen is used as an injecting (and gasifying) medium, the exiting gas has a heating value of approximately 450 Btu/scf. More than 80% of the coal fed is gasified, the remainder being burned in the combustion zone.

The crude gas leaving the gasifier contains a substantial amount of condensable products including tar, oil, and phenol, which are separated in a devolatilizer. The gas is then subjected to methanation ($CO + 3H_2 = CH_4 + H_2O$) to produce a high-Btu gas (pipeline quality).

Recent modification of the Lurgi process called "slagging Lurgi gasifier" has been developed to process caking coals [1,2,19,34,191,234]. Therefore, the operating temperature of this gasifier is kept higher and the injection ratio of steam is reduced to 1–1.5 mol/mol of oxygen. These two factors cause the ash to melt easily and, therefore, the molten ash is removed as a slag. Coal is fed to the gasifier through a lock hopper system and distributor. The molten slag formed during the process is quenched with water and removed through a slag lock hopper. The high operating temperature and fast removal of product gases lead to higher output rates in a slagging Lurgi gasifier than a conventional dry-ash Lurgi unit. A typical product composition for oxygen-blown operation has H_2-to-CO ratio higher than 2:1 and it contains large amount of CO_2.

3.8.2.2 Koppers–Totzek Gasification

This gasification process uses entrained-flow technology, in which finely pulverized coal is fed into the reactor with steam and oxygen. The process operates at atmospheric pressure and the space time in the reactor is very short. The reactor typically operates at a temperature of about 1400°C–1500°C. At this high temperature, the reaction rate of gasification is extremely high and about 90% of the carbonaceous matter is gasified in a single pass, depending on the type of coal. Lignite is the most reactive coal, for which gasification approaches nearly 100%.

The reactor can be used for caking coals and coals with a wide variety of mineral matters and it is not restricted just to coals or dry solid materials. Because of very high operating temperatures, the ash agglomerates and drops out of the combustion zone as molten slag and subsequently gets removed from the bottom of the reactor. The hot effluent gases are quenched and cleaned. This gas product contains no tar, ammonia, or condensable hydrocarbons and is predominantly synthesis gas. It has a heating value of about 280 Btu/scf and can be further upgraded by reacting with steam to form additional hydrogen and carbon dioxide via WGS reaction. The Koppers–Totzek process has high capacity, versatility, flexibility, simplicity of construction, ease of operation, low maintenance, and high safety and efficiency measures. This process is capable of instantaneous shutdown with full production resumable in only 30 min. The only moving parts in the gasifiers are the moving screw feeders for solids or pumps for liquid feedstocks [1–6,19,30,34].

Control of the gasifiers is achieved primarily by maintaining carbon dioxide concentration in the clean gas at a reasonably constant value. Slag fluidity at high process temperatures may be visually monitored. The process has a track record of over 50 years of safe operation. The overall thermal efficiency of the gasifier is 85%–90%. The time on stream or availability is better than 95%.

3.8.2.3 Shell Gasification

The Shell coal gasification process uses a pressurized, slagging entrained-flow reactor for gasifying dry PC. Similar to the Koppers–Totzek process, it has the potential to gasify widely different ranks of coals. Unlike other gasifying processes, it uses *pure oxygen* as the gasifying medium. Shell Global Solutions licenses two versions of gasification technologies, that is, one for liquid feedstock applications and the other for coal and petroleum coke. The process achieves almost 100% conversion of a wide variety of coals. By the production of high-pressure superheated steam, the process achieves about 75%–80% of thermal efficiency. The process produces clean gas with little by-products.

Coal is crushed and ground before feeding to the gasifier vessel. The process is operated at 1800°C–2000°C (flame) temperature and 30 atm pressure and in the presence of oxygen and steam, and it produces product gas that consists of 62%–63% carbon monoxide, 28% hydrogen, and some carbon dioxide. While some ash is entrained in synthesis gas, majority is collected in water-filled collection box at the bottom. Particulate matters in the product gas are removed by *cyclones and scrubbers* [1–6,19,30,34].

3.8.2.4 Texaco Gasification

The Texaco process also uses vertical entrained-flow reactor for gasification of coal/oil or coal/water slurry under pressure. The slurry reacts with either air or oxygen at high temperature. The product gas contains primarily carbon monoxide, carbon dioxide, and hydrogen with some quantity of methane. This process is basically used to manufacture *CO-rich synthesis gas* [1–6,19,30,34].

The reactor is provided with a slag quench zone at the bottom, where the resultant gases and molten slag are cooled down. In the latter operation, large amounts of high-pressure steam can be obtained, which boosts the thermal efficiency of the process. The thermal efficiency can also be enhanced by reducing the water content of the feed coal/water slurry. This gasifier favors high-energy dense coals so that the water-to-energy ratio in the feed is small. The gasifier operates at around 1100°C–1370°C and a pressure of 20–85 atm.

The product gases and slag are cooled and are then processed for further treatment. The gas, after being separated from slag and cooled, is treated to remove carbon fines and ash. These fines are then recycled to the slurry preparation system, while the cooled gas is treated for acid gas removal and elemental sulfur is recovered from the hydrogen sulfide (H_2S)-rich stream.

3.8.3 Typical Commercial Biomass/Waste Gasification Processes

3.8.3.1 Twin-Rec
The Twin-Rec is a circulating FB system coupled with a slagging ash melter. MSW is shredded and screened to sizes 12 in. or less and fed into the air-blown, FB gasifier, which contains a fluidizing medium (sand) and operates at 930°F–1200°F [80]. Heavy inert material, such as glass, ceramics, and metals, as well as some of the ash, is extracted from the bed along with some of the sand. The sand is mechanically separated from the other materials and returned to the FB. Fine char, fly ash, and fuel gases, generated in the FB, are carried over into the cyclonic melting chamber where the gases and char are combusted in air at 2400°F–2600°F. The hot gases leaving the cyclonic melting chamber are fed into a boiler to produce steam for electricity generation and process heat. Molten slag leaving the bottom of the cyclonic melter chamber is quenched to form granules. Both the gasifier and melter operate at atmospheric pressure, without auxiliary fuel (except during start-up) or oxygen. The system requires a steam generator. The waste materials are converted to electricity and/or heat. EBARA [245] claims 14 commercial Twin-Rec facilities treating waste plastics, shredder residues, sludges, industrial wastes, and MSWs [243–246].

3.8.3.2 Schwarze Pumpe
Schwarze Pumpe is a location in Germany that houses one of the largest German plants handling solid and liquid wastes including plastics, MSWs sewage sludges, shredder residues, oils, tars, and solvents. Sekundärrohstoff Verwertungs Zentrum is a large-scale operator of the three types of gasifiers at the location—entrained flow (GSP), slagging (BGL), and rotating grate (FDV, Lurgi). The site opened in 1955 as a briquetting factory and power station. Schwarze Pumpe began generating town gas in the late 1960s. Their experience with MSWs started in the mid-1990s with an emphasis on methanol production [234].

3.8.3.3 Thermal Converter
Global Energy Solution's "Thermal Converter" resembles a slagging, downdraft gasifier. MSW is shredded and screened to a lumped size no greater than a 3 in. cube. It has three distinct sections. Preheated air is introduced into the top of the fixed bed to heat and dry the MSW in the upper portion of the vessel. The middle section of the vessel is a rotated section where MSW is pyrolyzed at temperatures ranging from 2600°F to 2730°F. In the lower section of the vessel, the inorganic fraction of the MSW is converted into a molten bed of slag that drips from the bottom of the vessel through an afterburner section and into a quench tank where the slag is cooled and solidified. Gases passing through the molten slag enter the afterburner where additional air is introduced to convert any remaining combustible gases. This zone reaches temperatures of 2900°F. Heat is recovered from the hot exhaust gases in a heat exchanger where inlet air is preheated and a recuperator where, presumably, steam can be produced for electric power generation and process heat. Preheated air for the gasifier can also be produced in the recuperator [234–248]. This gasifier, as currently configured, is closely coupled with a combustion zone (afterburner) that precludes the production of a syngas, even if purified oxygen were to be used instead of air. The largest unit appears to have a capacity of about 75 tons/day.

3.8.3.4 Thermoselect Gasifier

The Thermoselect Gasifier handles raw MSW. The feedstock is preshredded to 20 in. or less, compressed into plug that is ram fed into a long degassing chamber that moves the MSW horizontally to the gasification chamber that is vertically oriented. Heat radiated from the gasification chamber slowly pyrolyzes the compacted MSW as it slowly moves through the degassing chamber (approximately 1–1.5 h solids residence time). Gases and vapors produced in the degassing chamber pass upward through the gasification chamber where they are exposed to temperatures of approximately 2190°F for at least 2 s.

Char and inorganic solids enter the gasification chamber and drop to its bottom, where the char and supplemental natural gas are combusted with purified oxygen producing temperatures as high as 3600°F causing the inorganic materials to form a molten slag. The molten slag, consisting of separate metal and mineral layers, is quenched in a water quench basin forming granules. The granules are sorted according to their properties to recover the metals. The product gas is quenched and cleaned in a multistage system in preparation for power generation or chemical synthesis (City of Los Angeles Department of Public Works 2005). Thermoselect's pilot plant was operated in Italy. There are seven facilities in Japan, three of which handle MSW. The remaining four facilities gasify industrial waste. Two facilities export fuel gas. All facilities handling MSW use the fuel gas for gas engines. The MSW facilities use two to three lines to treat 120–300 tons/day [240,249].

3.8.3.5 TPS Termiska Process

TPS Termiska is a Swedish R&D company that focuses on the combustion and gasification of solid waste and biomass. TPS Termiska licensed its waste gasification technology to Ansaldo Aerimpianti SpA in 1989. In 1992, Ansaldo installed a commercial, two-bed unit in Greve in Chianti, Italy, which is operated by TPS. The gasification plant has a total capacity of 200 metric tons of RDS per day. The two units have a combined capacity of 30 MWth. The TPS technology uses a starved-air gasification process in a combined bubbling and circulating FB reactors operated at 850°C and near atmospheric pressure. RDF is fed to the FB. Air is used as the gasification/fluidizing agent. Part of the air is injected into the gasifier vessel through the bottom section and the remainder higher up in the vessel. This pattern of air distribution causes a density gradient in the vessel. The lower part maintains bubbling fluidization that allows coarse fuel particles adequate residence time for good gasification reactions. The secondary air introduced higher up in the vessel increases the superficial velocity of air through the reactor so that smaller, lighter particles are carried away in the gas flow [241,250].

The process gas from each gasifier passes through two stages of solids separation before being fed to a furnace/boiler. The flue gas exiting the boiler is then cleaned in a three-stage dry scrubber before being exhausted through the stack. Alternatively, some of the raw gas stream can be sent to a nearby cement factory, without cleaning, to be used as fuel in the cement kilns.

TPS Termiska has developed a patented catalytic tar-cracking system. Immediately downstream of the gasification vessel, a dolomite (mixed magnesium–calcium carbonate)- containing vessel catalyzes most of the tars formed in the gasification process and breaks them down into simpler compounds with lower molecular weights and melting points. The dolomite also will absorb acids in the flue gas, including HCl and sulfur oxides. The product gas can then be cooled and passed through conventional scrubbing systems without operational problems. After cooling, the syngas can be compressed and is clean enough to be used with a combined-cycle turbine. This gas-cleaning technology has been demonstrated successfully at a biomass gasification plant as part of a joint venture with ARBRE Energy Ltd. in England [251]. TPS Termiska has also been selected by the World Bank to build the first commercial power station in the world using wood-fuelled, combined gasification and gas turbine technology. This plant is currently being installed in the state of Bahia, in northeastern Brazil.

3.8.4 TYPICAL COMMERCIAL COGASIFICATION PROCESSES

3.8.4.1 250 MWe IGCC Plant, Nuon Power, Buggenum, B.V. Willem-Alexander Centrale

Due to Dutch government policy decision called "Dutch Coal Covenant," 253 MWe Nuon Power Plant at Buggenum, Netherlands, was converted from coal plant to coal–biomass plant to make CO_2 emission reduction of 200,000 metric tons/year. This amounted to about 30 wt% of biomass use with relative biomass and coal feeds as 185,000 and 392,000 metric tons/year, respectively. This plant was started in 1993 and uses Shell dry feed gasification technology. The gasifier is an oxygen-blown, continuous slagging, entrained-flow reactor. It is designed to accept a wide range of imported coals and contains several design features that differ from the U.S. IGCC plants. The ASU and the gas turbine are tightly coupled where the gas turbine compressor supplies all the air to ASU. This increases the plant efficiency but also makes plant more complex and difficult to start. The plant efficiency based on lower heating value is about 43%.

The process involves pulverized and dried coal being pressurized in lock hoppers and fed into the gasifier with a transport gas via dense phase conveying. The carrier gas is nitrogen or product gas when nitrogen in the product gas is undesirable. The coal is oxidized by preheated, 95% pure oxygen mixed with steam. The coal is oxidized in the temperature range of 1500°C–1600°C and pressure range of 350–650 psi to produce a syngas principally composed of hydrogen and carbon monoxide with little carbon dioxide. Very-high-temperature gasification eliminates any hydrocarbon gases and liquids in the product gas [21,73,235]. The solid ash in the form of slag runs down the refractory-walled gasifier and is collected into a water bath as slurry. The gas leaving at about 1400°C–1650°C contains a small amount of char and about half of the molten ash. The hot gas is cooled by a couple of cooling stages where waste heat is recovered to generate steam that can be used in other parts of the processes. The solid particles from cooled gas are removed by cyclones, and the cooled gas then goes through a series of gas-cleaning processes to remove sulfur (H_2S, COS) and chlorine before using the clean syngas for downstream purposes of heat, power, or liquid fuels and chemicals.

The plant processed mixed feedstock of coal and biomass first from 2001 to 2004 with about 18% by weight of pure and mixed biomass. More recently, biomass concentration has increased up to 30 wt%. In addition of demolition wood, tests were also performed with chicken litter and sewage sludge. The test program evaluated the effects of biomass on product gas and ash quality. The Nuon/IGCC plant takes about 30 wt% biomass, most of which is waste wood to provide about 17% of energy input to the gasifier [73,235].

3.8.4.2 250 MWe IGCC Plant, Tampa Electric's Polk Power Station

This is another cogasification commercial plant that uses slurry-fed GE (formerly Texaco) gasification technology that also uses oxygen-blown, continuous entrained-bed gasification reactor [21,73]. This process was brought online in 1996. In 2001/2002, the DOE sponsored a project to demonstrate if Polk Unit # 1 can coprocess biomass as a fraction of its primary coal/coke feedstock (98% particle less than 12 mesh in size) without significant impact on its performance. The biomass used was a 5-year-old locally grown eucalyptus grove with feed concentration up to 1.2% by weight for about 8.5 h. The original system was not designed to handle softer fibrous biomass. The results showed that biomass did not impede the performance of the plant and it yielded 860 kW (7700 kW h total) of electricity during the test period based on the relative heating value and flow rates of biomass and base fuel.

The Polk Power Station used old Chevron Texaco IGCC technology that is now owned by General Electric. In this process, 60%–70% of coal–water slurry is fed to the gasifier at the rate of 2200 tons (on dry basis) of coal per day. The normal feed is a blend of coal and petcoke, the solid residue from crude oil refining. The fresh feed is mixed with unconverted recycled solids and finely ground in rod mills until 98% of the particles are less than 12 mesh in size. The slurry passes through a series of screens before being pumped into the gasifier. The slurry and oxygen are mixed in the gasifier process injector. The gasifier is designed to convert 95% of carbon per pass, and it produces syngas of 250 Btu/scf heat content.

A schematic of the Polk Power Station plant is shown in my previous book [21]. In this process, the syngas coming out of gasifier is cooled in a series of steps, each recovering heat in the form of saturated high-pressure steam. The first syngas cooler, called the "radiant syngas cooler" (RSC), produces 1650 psig saturated steam. The gas from RSC is split into two streams, and they are sent to parallel convective syngas coolers where the process of cooling and generating additional high-pressure steam (at lower temperature) is repeated. The gases then further go through a simultaneous cooling and impurity removal (particulates, hydrogen chloride) process. A final trim cooler reduces the syngas temperature to around 100°F for the cold gas cleanup (CGCU). The CGCU system is a traditional amine scrubber system, and it removes sulfur, which is then converted to sulfuric acid and sold to the local phosphate industry.

The eucalyptus feedstock used in this power plant contained about 1/3 of heating value per pound at about half the density of coal. The characteristics of the mixed feedstock for the Polk Power Station are described by Shah [21] and Ratafia-Brown et al. [73], and they indicate that even a modest concentration of this biomass will require a massive and expensive feed system. Although the combined characteristics of the mixed feedstock are not considerably different from the baseline, it increases hydrogen, oxygen, and ash content by 4.6%, 11%, and 3.4%, respectively. The CO_2 discharge is reduced by 0.87%. Biomass used in the Polk plant did not lend itself to size separation and screening, and it caused minor plugging of the suction to one of the pumps [21,73,235]. The results indicate that for a slurry system, feed preparation must be tailored to the nature of the biomass in order to prevent any malfunction by the slurry pump as well as downstream gas cleaning and turbine operation. Typical experimental results for the Polk Power Station are described by Ratafia-Brown et al. [21,73,235]. The experience of the Polk Power Station can be extended to coal and other materials. Based on the experiences obtained with Polk Power Station and Nuon Power Station, a number of observations regarding commercial cogasification processes are identified by Ratafia-Brown et al. [21,73,235].

3.8.4.3   High-Temperature Winkler Gasifier

The HTW plant in Berrenrath, Germany, is commercial (for lignite) and has been in operation since 1986. Dried lignite, oxygen, and steam are fed to a FB reactor that operates at 145 psi and 1750°F. Product gas from the gasifier is cooled and cleaned in a ceramic candle filter and a water scrubber. Solid residues are combusted in an adjacent power plant. Dust collected from the filter is used in an adjacent wastewater treatment plant. The synthesis gas undergoes carbon monoxide conversion to obtain a hydrogen-to-carbon monoxide ratio suitable for methanol synthesis.

Hydrogen sulfide and carbon dioxide are removed. The resulting syngas is sent to DEA Mineralol AG where it is synthesized into methanol. A series of tests were performed, using the HTW system, to successfully co-gasify RDF pellets (at rates up to 50% of the feed input) in 1998. The demonstration was completed that year [1–6,19,34,73,80,252].

3.8.4.4   Potential Issues and Options for the Commercial Process
Configurations of Cogasification

Coal–biomass gasification complete plant can have a number of configurations, which are influenced by [21,73]

1. Overall scale of the plant and composition of feedstock
2. Required biomass capacity and its geographical dispersion relative to central gasification facility
3. Biomass pretreatment requirement
4. Levels of integration of various equipment
    a. At pretreatment stage
    b. Within the gasification stage
    c. Within the product cleaning and separation stage

This is to provide maximum flexibility, operability, and reliability of the entire plant operations.

In order to take the maximum advantage of the economy of scale, the size of the cogasification plant should be as large as possible. This will allow the most efficient gasification and syngas cleanup operations. Once the total plant size is fixed, the most important parameters that affect plant configuration are (1) the biomass concentration in the feed, (2) nature of biomass (e.g., wood vs. straw), (3) size requirement of biomass based on the chosen gasification reactor technology, (4) the availability of biomass locally or the transport distance for the biomass feedstock, and (5) required flexibility for processing of coal and biomass for stable and sustainable operation.

An analysis of the last item sets up the degree of common versus segregated pretreatment, gasification production, and syngas cleanup systems. Another important issue is the location of the pretreatment configurations: on-site or off-site. This issue is very important when biomass is not available locally and it has to travel long distance to the gasification plant. Low-density biomass is expensive to move to long distance. Often, it makes sense to carry out pretreatment at the site of the biomass and then transport the prepared biomass to the plant. This saves money if the density, particle size, shape, and structure are significantly altered during the pretreatment process like torrefaction, pelletization, or flash pyrolysis to generate feedstock in a slurry form. The original nature of biomass plays an important role in the pretreatment process. The pretreatment enhances both mass and energy densities of biomass, which significantly reduces biomass transportation costs.

If the plant is designed for a multiple types of biomass, each coming at significant distance from the central gasification facility, geographically dispersed biomass pretreatment facilities make significant sense. The transportation of coal at any distance is not a significant issue.

Process configuration is more important and more complex for gasification than combustion because the products are used for downstream upgrading as well as power generation. Process configuration has at least the following five options:

1. Gasify both coal and biomass together in the same gasifier and have a unified downstream operation. In this case, it is also possible to feed biomass and coal separately in the same gasifier but at different locations. This is to take advantage of the different reactivities of coal and biomass.
2. Gasify coal and biomass in separate gasifiers with separate pretreatment systems and then combine the product gas for a unified downstream operation. This allows the optimum gasifier design for each type of feedstock.
3. Same as (2) in pretreatment and gasifier systems but have separate cleaning steps and then combine the cleaned syngas coming from coal and biomass.
4. Any options from (2) or (3) but have different forms of feed for biomass (dry or slurry).
5. Any options from (2), (3), or (4) but share a common or separate ASU to produce oxygen for the biomass or coal gasifiers.

These different options, and there can be others, lead to different process configurations. Ratafia-Brown et al. [21,73] considered the following six possible configurations and their advantages and disadvantages. This analysis will allow others to consider other possible configurations:

Configuration 1: Cofeeding coal and biomass to the gasifier as a mixture in dry or slurry form.
Configuration 2: Cofeeding biomass and coal to the gasifier using separate gasifier feed systems, either in dry form or in slurry form.
Configuration 3: Pyrolyzing as received biomass followed by cofeeding pyrolysis char and coal to the gasifier and separately feeding pyrolysis gas to the syngas cleanup system.
Configuration 4: Biomass and coal are coprocessed in separate gasifiers followed by a combined syngas cleanup.

Configuration 5: Biomass and coal are coprocessed in separate gasifiers followed by separate syngas cleanup trains, and the syngas feeds are combined prior to sulfur and CO_2 removal unit operations.

Configuration 6: Same as (4) and (5) but share common ASU for oxygen feed to the separate gasifiers.

Ratafia-Brown et al. [73] gave an extensive assessment of these six process configurations. A summary assessment and advantages and disadvantages of each option is described in my previous books [21,22]. The analysis can also apply to other mixtures such as coal and waste and biomass and waste of widely differing properties. The important thing here is that the process options for cogasification are many and are significantly more complex than the ones discussed earlier for cocombustion. While cogasification is more complex than cocombustion, it also provides more flexible product distribution.

3.8.5 COMMERCIAL GASIFICATION OF OIL SHALE

The first plant tests on oil shale were carried out in coal gasification retorts [40,45,46,236–238]. In those days, there was no knowledge about the specific nature of oil shale decomposition and therefore only 38%–40% of kerogen was gasified. The chamber furnace for oil shale gasification designed by Zhunko and Zaglodin [236] opened a new way to designing and testing chamber furnace. The chamber furnaces were constructed from both firebricks and regular bricks. A 9.3 m high chamber battery was divided into 23 chambers by vertical partitions. The smooth downward flow of oil shale was carried out in a vertical downward widening flue-type structure. In order to reduce the content of organic matter in the solid residue, air was injected into the bottom section of chambers (70–90 mN/ton oil shale). Tar, gasoline, and retort water were separated from the gas produced in the chamber furnaces. The gas contained sulfur compounds, naphthalene, and ammonia along with some other compounds. The calorific value of gas from industrial processing was 3752 kcal/mN.

These and many other similar results indicated the following:

1. The production of oil shale gas is an industrially successful process while the production capacity of chamber furnaces (the actual oil shale throughput) was 16–17 tons/day. The calorific value of coke gas can be 4700–4800 kcal/mN.
2. Depending on the parameters of the process, the yield of products may vary significantly: coke oven gas 257–361 mN/ton, tar 2.8%–7.0%, and gasoline 1.57%–2.1%.
3. Introduction of 80 mN/ton of air and 90 kg/ton of water vapor into the central section of the chamber increased the productivity of chamber furnaces significantly [253]. The gas yield increased 27% and that of liquid products 2.2%.
4. Industrial processing of the heavier fractions of generator oil in the chamber furnace into high-calorific gas enabled the production of gasoline in greater volumes, containing 90% aromatic compounds.
5. While the commercial success of kukersite oil shale gasification will depend on the competitive cost of the process with other fuel availability, oil shale gasification can be technically carried out.

3.9 UNDERGROUND COAL GASIFICATION

In situ gasification is a technology for recovering the energy content of coal deposits that cannot be exploited either economically or technically by conventional mining (or ex situ) processes. Coal reserves that are suitable for in situ gasification have low heating values, thin seam thickness, great depth, high ash or excessive moisture content, large seam dip angle, or undesirable overburden properties. UCG is carried out in nonmined coal seams using injection of a gaseous oxidizing agent, usually oxygen, air, or steam (or their mixture) and bringing the resulting product gas to the surface through production wells drilled from the surface. The technique can be applied to resources that are otherwise not economical to extract and also offers an alternative to conventional coal mining methods for some resources [254].

A considerable amount of investigation has been performed on UCG in the former USSR and in Australia, but it is only in recent years that the concept has been revived in Europe and North America as a means of fuel gas production. In addition to its potential for recovering deep, low-rank coal reserves, the UCG process may offer some advantages with respect to its resource recovery, minimal environmental impact, operational safety, process efficiency, and economic potential [19,21,22,254–297].

The in situ coal combustion process by oxygen, air, or steam can be handled in either forward or reverse mode. Forward combustion involves movement of the combustion front and injected air in the same direction, whereas in reverse combustion, the combustion front moves in the opposite direction to the injected air. The process involves drilling and subsequent linking of the two boreholes to enable gas flow between the two [273–276]. Combustion is initiated at the bottom of one borehole called injection well and is maintained by the continuous injection of air.

In the initial reaction zone, carbon dioxide is generated by the reaction of oxygen (air) with coal, which further reacts with coal to produce carbon monoxide by the Boudouard reaction ($CO_2 + C = 2CO$) in the reduction zone. Further, at such high temperatures, the moisture present in the seam may also react with carbon to form carbon monoxide and hydrogen via the steam gasification reaction ($C + H_2O = CO + H_2$). In addition to all these basic gasification reactions, coal decomposes in the pyrolysis zone owing to high temperatures to produce hydrocarbons and tars, which also contribute to the product gas mix [19,21,22,254].

The heating value from the air-blown in situ gasifier is roughly about 100 Btu/scf. The low heat content of the gas makes it uneconomical for transportation, making it necessary to use the product gas on-site. An extensive discussion on in situ gasification can be found in references by Thompson [282], Gregg and Edgar [262], and others [256,257,283,285,286,288,297]. A noteworthy R&D effort in UCG has also been conducted by the Commonwealth Scientific and Industrial Research Organization (CSIRO) in Australia. CSIRO researchers have developed a model to assist with the implementation of this technology [283,285,286,296]. A number of other trials and trial schemes were evaluated in Europe, China, India, South Africa, and the United States [19,261].

Coal gasification generally requires construction of special plants, including large coal storage facilities and gasifiers. Meanwhile, in UCG process injection and production wells are drilled from the surface and linked together in a coal seam. Once the wells are linked, air or oxygen is injected, and the coal is ignited in a controlled manner. Water present in the coal seam or in the surrounding rocks flows into the cavity formed by the combustion and is utilized in the gasification process. The produced gases (primarily H_2, CO, CH_4, and CO_2) flow to the earth's surface through one or more production wells. After being cleaned, these gases can be used to generate electric power or synthesize chemicals (e.g., ammonia, methanol, and liquid hydrocarbon fuels).

UCG has numerous advantages over conventional underground or strip mining and surface gasification including the following [19,254–297]:

1. Conventional coal mining is eliminated with UCG, hence reducing operating costs and surface damage and eliminating mine safety issues such as mine collapse and asphyxiation. No coal is transported at the surface, thus reducing cost, emissions, and local footprint associated with coal shipping and stockpiling. UCG eliminates much of the energy waste associated with moving waste as well as usable product from the ground to the surface.
2. Coals that are unmineable (too deep, low-grade, thin seams) are exploitable by UCG, thereby greatly increasing domestic resource availability.

3. No surface gasification systems are needed; hence, capital costs are substantially reduced.
4. Most of the ash in the coal stays underground, thereby avoiding the need for excessive gas cleanup and the environmental issues associated with fly ash waste stored at the surface.
5. There is no production of some criteria pollutants (e.g., SO_x, NO_x) and many other pollutants (mercury, particulates, sulfur species) are greatly reduced in volume and easier to handle. UCG, compared to conventional mining combined with surface combustion, produces less GHG and has advantages for geologic carbon storage. The well infrastructure for UCG can be used subsequently for geologic CO_2 sequestration operations. It may be possible to store CO_2 in the reactor zone underground as well in adjacent strata.

The UCG process, however, needs some further R&D and improvement in linking of injection and production wells within a coal seam [264,266,273–280,288,294,295], minimization of variation in the composition of the produced gas, and prevention of any degradation of potable groundwater supplies [259,260]. In many cases, the coal seam has low permeability, and a linkage technology is necessary. After testing different methods for linking the injection and production wells, relatively inexpensive technologies were developed in the FSU, such as hydraulic fracturing of the coal seam by pressurized air (or water) (this technology is common in the oil and gas industry) and so-called reverse combustion linking (ignition near the production well and countercurrent flame propagation toward the injection well). It should be noted that directional in-seam drilling has been successfully competing with these technologies for many decades. Nevertheless, hydraulic fracturing and reverse combustion linking remain attractive because of their relatively low costs, and they can be used either alone or in combination with drilling [19,264,266,273–280,288,294,295].

The coal seam thickness plays an important role in UCG process. It has been shown that a decrease in the seam thickness can reduce the heating value of the produced gas, which is associated with heat loss to the surrounding formation. For example, for one particular UCG plant, the gas heating value decreased significantly as the seam thickness fell below 2 m [269].

The UCG process usually consumes water contained in the coal seam and adjacent strata. Also, water can be pumped as steam, along with air or oxygen, into the injection well. In any case, some amount of water will remain unreacted, which potentially can lead to contamination of groundwater by harmful by-products of the UCG process. To avoid this, environmental monitoring during and after the UCG process needs to be conducted. The results of environmental monitoring in the FSU indicated that water contamination during UCG was of a local nature and at admissible concentrations of harmful compounds.

R&D of underground gasification technology has been conducted in the FSU using mathematical modeling to simulate gasification processes and products. A steady-state model was developed for coal gasification in a long channel with a constant cross section, where air and water flow into the channel and react with the coal [19,21,22,264,266,273–280,288,294,295]. This model involves the following heterogeneous chemical reactions:

$$C + O_2 \rightarrow CO_2 \tag{3.30}$$

$$2C + O_2 \rightarrow 2CO \tag{3.31}$$

$$C + CO_2 \rightarrow 2CO \tag{3.32}$$

$$C + H_2O \rightarrow CO + H_2 \tag{3.33}$$

$$C + 2H_2O \rightarrow CO_2 + 2H_2 \tag{3.34}$$

$$C + 2H_2 \rightarrow CH_4 \tag{3.35}$$

Along with the following reactions in the gas phase:

$$2CO + O_2 \rightarrow 2CO_2 \tag{3.36}$$

$$2H_2 + O_2 \rightarrow 2H_2O \tag{3.37}$$

$$CH_4 + 2O_2 \rightarrow CO_2 + 2H_2O \tag{3.38}$$

$$CO + H_2O \rightarrow CO_2 + H_2 \tag{3.39}$$

$$CO + 3H_2O \rightarrow CH_4 + H_2O \tag{3.40}$$

One can assume that the flow is turbulent and that the gas is radially well mixed (no gradients over the channel cross section). A possible mathematical model should include balance equations for gas species (O_2, CO_2, CO, H_2O, H_2, CH_4, and N_2), equations for momentum, and energy, as well as a thermal conduction equation for the coal. Kinetic parameters for the involved reactions can be taken from the literature [19,264,266,273–280,288,294,295]. Gas compositions and temperatures along the channel axis can be calculated for various parameters, such as the entrance pressure, air flow rate, and water-to-coal ratio. Such a modeling effort was carried out in the FSU to design and optimize UCG process.

Yang et al. [255,295] studied the product distribution from underground coal gasification processes in China. They found that with pure oxygen gasification the hydrogen volume percentage in product gas varied from 23.63% to 30.24% and the carbon monoxide volume percentage varied from 35.22% to 46.32%. When oxygen/steam mixture was used for the gasification, the gas compositions virtually remained stable and $CO + H_2$ volume percentage was basically between 61.66% and 71.29%. Moving point gasification improved the changes in the cavity in the coal seams or the effect of roof in-break on gas quality. For steep seams, during oxygen/steam mixture gasification, the composition of $CO + H_2$ remained within the volume percentage of 58% and 72%. The average oxidation zone temperature reached 1200°C and it was higher for forward gasification than backward gasification. In general, for both types of seams, hydrogen concentration increased and carbon monoxide concentration decreased with an increase in steam-to-oxygen ratio. The hydrogen concentration reached about 60% at steam/oxygen ratio of about 3.

3.9.1 Underground Gasification Reactors

A typical underground gasification reactor is illustrated in Figure 3.4 [19]. In this type of reactor, the combustion process can be handled in either forward or reverse mode. The forward combustion involves the movement of the combustion front and injected air in the same direction. In the reverse combustion, the combustion front moves in the opposite direction to the injected air. The process involves drilling and subsequent linking of the two boreholes to enable gas flow between the two. Combustion is initiated at the bottom of the one borehole (called injection well) and is maintained by the continuous injection of air and steam. A typical underground reaction system involves linking of a series of such a unit reactor system.

3.9.2 Methods for Underground Gasification

There are two principal methods for underground steam gasification that have been tried successfully: shaft methods and shaftless methods (and a combination of two) [19,255–297]. Selection of a specific method depends on the parameters such as natural permeability of coal seam; the geochemistry of coal deposit; the seam thickness, depth, width, and inclination; closeness to the metropolitan

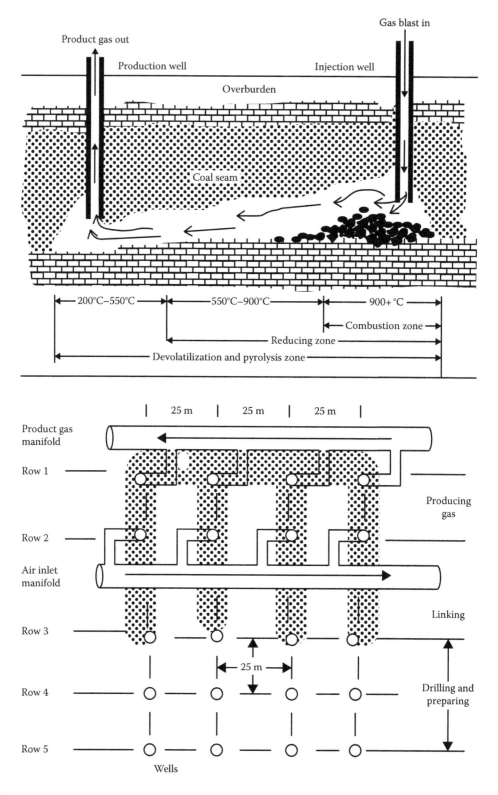

FIGURE 3.4 Details of underground coal gasification process. (From Lee, S., Gasification of coal, Chapter 2, in: Lee, S. et al., eds., *Handbook of Alternative Fuel Technology*, Taylor & Francis Group, New York, 2007, pp. 26–78.)

areas; and the amount of mining desired. Shaft methods involve driving of shafts and drilling of other large diameter openings that require underground labor, and shaftless methods use boreholes for gaining access to the coal seam and it does not require any underground labor [19].

3.9.2.1 Shaft Methods

The shaft method can be further divided into three subdivisions: (1) chamber or warehouse method in which underground galleries are prepared and the coal panels are isolated with brick wall, (2) borehole producer method in which parallel underground galleries are created about 500 ft apart within the coalbed, and (3) stream method in which inclined galleries following the dip of the coal seam of steeply pitched coalbeds are constructed parallel to each other [19,255–298].

3.9.2.1.1 Chamber or Warehouse Method

This method requires the preparation of underground galleries and the isolation of coal panels with brick wall. The blast of air for gasification is applied from the gallery at the previously ignited face of one side of the panel, and the gas produced is removed through the gallery at the opposite side of the panel. This method relies on the natural permeability of the coal seam for air flow through the system.

Gasification and combustion rates are usually low, and the product gas may have variable composition from time to time. To enhance the effectiveness, coal seams are precharged with dynamites to rubblize them in advance of the reaction zone by a series of controlled explosions.

3.9.2.1.2 Borehole Producer Method

This method typically requires the development of parallel underground galleries that are located about 500 ft apart within the coalbed. From these galleries, about 4 in. diameter boreholes are drilled about 15 ft apart from one gallery to the opposite one. Electric ignition of the coal in each borehole can be achieved by remote control.

This method was originally designed to gasify substantially flat-lying seams. Variations of this technique utilize hydraulic and electric linking as alternatives to the use of boreholes.

3.9.2.1.3 Stream Method

This method can be applied to steeply pitched coalbeds. Inclined galleries following the dip of the coal seam are constructed parallel to each other and are connected at the bottom by a horizontal gallery or "fire drift." A fire in the horizontal gallery initiates the gasification, which proceeds upward with air coming down one inclined gallery and gas leaving through the other. One obvious advantage of the stream method is that ash and roof material drop down, tend to fill void space, and do not tend to choke off the combustion zone at the burning coal front. However, this method is structurally less suitable for horizontal coal seams because of roof collapse problems [19,255–298].

3.9.2.2 Shaftless Methods

The shaftless method carries out gasification through a series of boreholes drilled from the surface to the coal seam. The coalbeds are made more permeable between inlet and outlet boreholes by a chosen linking method, ignite the coal seam, and gasify it by passing air and steam from inlet to outlet borehole. In percolation or filtration method, multiple boreholes, at a distance that depends on the seam permeability, are used to gasify the underground coal [19,255–298].

3.9.2.2.1 Percolation or Filtration Methods

This is the most direct approach to accomplish shaftless gasification of a coal seam using multiple boreholes. The distance required between boreholes depends on the seam permeability. Lower-rank coals such as lignites have a considerable natural permeability and, as such, can be gasified without open linking. However, higher-rank coals such as anthracites are far less permeable, and it becomes necessary to connect boreholes by some efficient linking techniques that will increase the permeability and fracture of the coal seam so that an increased rate of gas flow can be attained. Air or air/steam is blown through one borehole, and product gas is removed from another borehole. Either forward or reverse combustion can be permitted by this method. As the burn-off (a combination of combustion and gasification) progresses, the permeability of the seam also increases and the compressed air blown through the seam helps enlarge cracks or openings in the seam. When the combustion of a zone nears completion, the process is transferred to the next pair of boreholes and continues. In this operation, coal ash and residues should be structurally strong enough to prevent roof collapse [19,264,266,273–280,284,285,288,294,295].

An ideal underground steam gasification system must be [19,21,22]

1. Operable on large scales
2. Such that no large deposit of coal remains ungasified
3. Such that process is controllable and product gases quantity and quality are constant and uniform
4. Mechanically stable and removed from any leakages to the groundwater
5. Such that the process requires a minimal or no underground work

3.9.3 POTENTIAL PROBLEM AREAS WITH IN SITU GASIFICATION

The potential problems in all of these methods include [256,257,262,282,283,285,286,297,298]:

1. High and constant quality of product gas
2. High percentage recovery of coal energy
3. Control of groundwater contamination
4. Combustion control
5. Roof structure control
6. Product gas leakage control
7. Proper control of permeability, linking, and fracturing
8. Proper monitoring of underground processes

Combustion control is essential for controlling the product gas quality as well as the extent of coal conversion. The contact between the coal and the gasifying agent should be such that the coal is completely in situ gasified, all oxygen in the inlet gas is consumed, and the production of fully combusted carbon dioxide and water is minimized. In a typical in situ coal gasification process, the heating value of product gas decreases with an increase in the processing time due to either poor contact of gas with coal face or large void volume that can cause roof collapse.

Uncontrolled roof collapse can cause numerous problems such as leakage of reactant gases, seepage of groundwater into the coal seam, loss of product gas, and surface subsidence above the coal deposit. These issues hinder combustion control and successful overall gasification process [256,257,262,282,283,285,286,297,298].

The loss of substantial amount of product gas can adversely affect the recovered amount of the product gas as well as the gasification economics. Further, the inlet reactant gases should not be wasted. Influx of water can also affect the control of the process. Leakage varies from site to site and also depends on a number of factors including geologic conditions, depth of coal seam, types of boreholes and their seals, and permeability of coalbed.

Low permeability of coal seams does not permit the passage of oxidizing gases through it without a serious pressure drop. Methods such as pneumatic, hydraulic, and electric, as well as fracturing with explosives, do not result in a uniform increase in permeability throughout the coalbed. They can also disrupt the surrounding strata and worsen the leakage problems. Therefore, the use of boreholes is proven to provide a more predictable method of linking and is a preferred technique.

Proper monitoring of the underground processes is a necessary component of successful operation and design of an underground gasification system. A priori knowledge of all the parameters affecting the gasification is required so that adequate process control philosophy can be adopted and implemented for controlling the operation. These factors include the location, shape, and temperature distribution of the combustion front, the extent and nature of collapsed roof debris, the permeability of coal seam and debris, the leakage of reactant and product gases, the seepage of groundwater, and the composition and yield of the product gases.

UCG should never be attempted in a severely fractured area, in shallow seams, or in coal seams adjoining porous sedimentary layers. It is also essential to prevent roof collapse and to properly seal inlet and outlet boreholes after operation [256,257,262,282,283,285,286,297,298].

3.9.4 CONTROLLED RETRACTING INJECTION POINT TECHNIQUE

An important result of prior UCG work in the United States is the development of the Controlled Retracting Injection Point (CRIP) process by researchers of the Lawrence Livermore National Laboratory (LLNL). In the CRIP process [258,265], a production well is drilled vertically, and an injection well is drilled using directional drilling techniques to connect it to the production well. Once the connection, or channel, is established, a gasification cavity is initiated at the end of the injection well in the horizontal section of the coal seam.

The CRIP technique involves the use of a burner attached to coiled tubing. The device is used to burn through the borehole linear or casing and ignite the coal. The ignition system can be moved to any desired location in the injection well. The CRIP technique enables a new reactor to be started at any chosen upstream location after a deteriorating reactor has been abandoned. Once the coal near the cavity is used up, the injection point is retracted (preferably by burning a section of the linear), and a new gasification cavity is initiated. In this manner, precise control over the progress of gasification is obtained.

The CRIP technique and clean-cavern concept were used in the Rocky Mountain 1 trial (Carbon County, Wyoming), which is considered to be the most successful UCG test in the United States. Oxygen and steam were injected into a subbituminous coal seam (thickness, 10 m; depth, 130 m). Along with CRIP, another linking technology, the so-called extended linked well (ELW) [256–262], was tested. The ELW test lasted 57 days producing gas with an average heating value of 9.7 MJ/m^3. The CRIP trial lasted a total of 93 days producing average gas heating values of 10.7 MJ/m^3. Environmental impact of these tests were minimized by maintaining pressure in the UCG cavity below hydrostatic to minimize the loss of organic-laden gases and by ensuring a small but continuous influx of groundwater into the gasification cavity [256–262,265,274,275,298].

3.9.5 CRITERIA FOR UNDERGROUND COAL GASIFICATION SITE SELECTION

The determination of selection criteria for UCG locations is an important problem. The criteria for underground mining, including technological and land-use restrictions, are well known, but in some cases, the criteria for UCG are expected to be different. For example, the UCG process has specific requirements for the depth and thickness of coal seams that differ from those applicable to mining.

3.9.5.1 Thickness, Depth, and Dip of Coal Seam

Based on coal seams of the Powder River Basin, Wyoming, which are mainly from 10 to over 30 m in thickness, GasTech [299,300] indicated that the optimal coal seam thickness should be more than 10 m. On the other hand, Ergo Exergy states that UCG can be used in coal seams as thin as 0.5 m [271,275]. A more reliable lower limit based on UCG work at FSU (mentioned earlier) is 2 m, below which heating value of produced gas decreases significantly.

UCG literature shows that the depth of coal seams is not a critical parameter. The depth varied from 30 to 350 m in the FSU developments and U.S. experiments, whereas Western European trials were conducted in much deeper coals (600–1200 m). While the LLNL experts indicated that the minimum depth should be 12 m [283,285,296,299,300], 60 m is the typically applied depth limit for surface mining [19,254–297]. Furthermore, the proximity of potable and potentially potable groundwater supplies at shallow depths discourages UCG activity for those coals that are located near the ground surface. UCG is therefore definitely not advisable for coals less than 60 m deep.

Burton et al. [257] recommend operational depths of >200 m to reduce risk of subsidence. Depths of more than 300 m require more complicated and expensive drilling technologies, but they also have advantages such as minimized risk of subsidence and the possibility of conducting the UCG process at higher pressure, which increases the heating value of the produced gas. Also, deeper seams are less likely to be hydrologically linked with potable aquifers, thus avoiding drinkable water contamination problems. Further, if the product gas is to be used in gas turbines, additional compression might not be necessary. Finally, UCG cavities at depths of more than 800 m could be used for CO_2 sequestration. The choice of an optimum depth for UCG depends on the trade-off between additional cost for deeper seams versus higher heating value and higher pressure and more suitable product gas composition that can be useful for downstream applications.

Sury et al. [259,260] have indicated that shallow dipping coal seams are preferable. Such seams facilitate drainage and the maintenance of hydrostatic balance within the gasifier; they also minimize potential damage to the down dip production well from material that is moved in association with the UCG process. A report by GasTech [299,300] recommends dip angles of 0°–20°. However, UCG has been successfully carried out in steeply dipping seams [255]; thus, dip is not a critical constraining factor for selecting and operating UCG sites.

3.9.5.2 Amount of Coal, Coal Rank, and Other Coal Properties

Since the gas produced from UCG operation can be used for various heat, power, and mobile applications, the amount of coal is an important parameter for determining the size and sustainability of product gas supply for these downstream operations. The productive lifetime of the site must also be determined as a function of the required gas yield. For illustration, for the 20-year continuous operation of a 300 MW UCG-based combined-cycle power plant (efficiency, 50%), it is necessary to produce 75.6×10^9 Nm^3 of syngas with a heating value of 5 MJ/m^3. This may require 33×10^6 metric tons of coal gasification. This number can be decreased by a factor of 2 by using oxygen and steam as injection gases, which, however, increases the cost.

With the present state of knowledge, low-rank, high-volatility, noncaking bituminous coals are preferable. UCG might work better on lower-rank coals because such coals tend to shrink upon heating, hence enhancing permeability and connectivity between injection and production wells [271–278,298]. Also, the impurities in lower-rank coals might improve the kinetics of gasification by acting as catalysts for the burn process. For coals of the same rank, the heating value of the UCG gas is directly related to the heating value of the coal.

While it is difficult to determine the porosity and permeability of coal seams, in general, better permeability gives better pore structure for injection and production process and faster transport of reactants and higher rate of gasification. However, higher porosity and permeability also increases the influx of water and increases product gas losses. Sury et al. [259,260] stated that, in general, reverse combustion works well in shallow nonswelling coal but is not recommended for use at great depths and in swelling coals. This contradicts, however, the opinion of Burton et al. [257] who noted that the FSU methods demonstrated minimum sensitivity to coal swelling: the large-dimension channels formed in the linkage process employed in those operations did not appear to be plugged by coal swelling. Areas of seams that are free of major faulting in the vicinity (<45 m) of the proposed UCG location and the ones that could potentially provide a pathway for water inflow or gas migration should be preferentially targeted [19,283,285,286,296].

3.9.5.3 Groundwater and Land Use Requirements and Restrictions

Water is an essential component of the UCG process, and thus its availability either from within a coal seam or from a source adjoining the seam is an important characteristic. The adjoining rocks must contain saline water (>10,000 ppm total dissolved solids, as per U.S. Environmental Protection Agency regulations) and have a significant deliverable volume. In many cases, the coal itself serves as the principle aquifer within the stratigraphic section and is bounded by impermeable shales and low-density rock. In some cases, permeable sandstones form the roof rock and therefore are in hydrological connectivity with strata outside the coal seam. Sury et al. [259,260] recommended using coal seams with no overlying potable aquifers within a distance of 25 times the seam height. Trials have been successfully carried out in seams in closer proximity to potable underground aquifers, but the potential risk of contamination increases in such a setting.

There is no indication in the literature that UCG should be farther from towns, roads, and other objects than underground mines, assuming that the process design and environmental monitoring eliminate water contamination and air pollution. Thus, the land-use restrictions for underground mining can be applied to potential UCG sites [19,254–300].

3.10 MOLTEN MEDIA GASIFICATION PROCESSES

In this process, coal, biomass, or waste is fed along with steam or oxygen in the molten bath of salt or metal operated at 750°C–1400°C. Ash and sulfur are removed as slag. Generally speaking, molten media may mean one of the following: molten salt, molten metal, or molten slag. The molten medium not only catalyzes the gasification reaction but also supplies the necessary heat and serves as a heat exchange medium [19,301–316]. There have been several distinct small-scale, pilot-scale, or commercial processes developed over the years. Here, we briefly examine the following five processes [19,301–312]:

1. Kellogg–Pullman molten salt process
2. Atgas molten iron gasification process
3. Rummel–Otto molten salt gasification
4. Rockwell molten salt gasification process
5. Alden process for biomass gasification using molten salts

3.10.1 KELLOGG–PULLMAN MOLTEN SALT PROCESS

In this process, gasification of coal is carried out in a bath of molten sodium carbonate (Na_2CO_3) through which steam is passed [19,314]. The molten salt produced by this process offers several advantages. The steam–coal reaction is strongly catalyzed by sodium carbonate, resulting in complete gasification at a relatively low temperature. Molten salt disperses coal and steam throughout the reactor, thereby permitting direct gasification of caking coals without carbonization. Finally, a salt bath can be used to supply heat to the coal undergoing gasification. Owing to the uniform temperature throughout the medium, the product gas obtained is free of tars and tar acids.

Crushed coal is picked up from lock hoppers by a stream of preheated oxygen and steam and carried into the gasifier. In addition, sodium carbonate recycled from the ash rejection system is also metered into the transport gas stream, and the combined coal, salt, and carrier are admitted to the gasifier. Along with the usual gasification reactions, sulfur entering with the coal accumulates as sodium sulfide (Na_2S) to an equilibrium level. At this level, it leaves the reactor according to the following reaction:

$$Na_2CO_3 + H_2S \rightarrow Na_2S + CO_2 + H_2O$$

Ash accumulates in the melt and leaves along with the bleed stream of salt, where it is rejected and sodium carbonate is recycled. The bleed stream of salt is quenched in water to dissolve sodium carbonate (Na_2CO_3) and this permits separation of coal ash by filtration. The dilute solution of sodium carbonate is further carbonated for precipitation and recovery of sodium bicarbonate ($NaHCO_3$). The filtrate is recycled to quench the molten salt stream leaving the reactor. Sodium carbonate is regenerated from sodium bicarbonate filtrate cake and recycled back to the gasifier. The gas stream leaving the gasifier is processed to recover the entrained salt and the heat and is further processed for conversion to the desired product gas such as synthesis gas, pipeline gas, or SNG.

3.10.2 ATGAS MOLTEN IRON COAL GASIFICATION

This process is based on the molten iron gasification concept in which coal is injected with steam or air into a molten iron bath. Steam dissociation and thermal cracking of coal VM generate hydrogen and carbon monoxide, that is, the principal ingredients of synthesis gas. Coal sulfur is captured by iron and transferred to a lime slag from which elemental sulfur can be recovered as a by-product. The coal dissolved in the iron is removed by oxidation to carbon monoxide with oxygen or air injected near the molten iron surface.

The Atgas process uses coal, steam, or oxygen to yield product gases with heating values of about 900 Btu/scf. The Atgas molten iron process has several inherent advantages over the fixed-bed or FB reactors [19,313]. In this process, gasification is carried out at low pressures; hence, the mechanical difficulty of coal feeding in a pressurized zone is eliminated. Unlike conventional gas–solid reactors, caking properties, ash fusion temperatures, and generation of coal fines are not problems in Atgas process.

The sulfur content of coal does not cause any environmental problem in Atgas process as it is retained in the system and recovered as elemental sulfur from the slag. Elemental sulfur by-product helps the overall process economics. The product gas is essentially free of sulfur compounds. The system is very flexible with regard to the physical and chemical properties of the feed coal. Relatively coarse size particles can be handled without any special pretreatment. Shutdown and start-up procedures are greatly simplified compared to fixed-bed or FB reactors. Finally, the formation of tar is suppressed owing to very-high-temperature operation [19].

In the process, coal and limestone are injected into the molten iron through tubes using steam as a carrier gas. The coal goes through devolatilization with some thermal decomposition of the volatile constituents, leaving the fixed carbon and sulfur to dissolve in iron whereupon carbon is oxidized to carbon monoxide. The sulfur, in both organic and pyritic forms (FeS_2), migrates from the molten iron to the slag layer where it reacts with lime to produce calcium sulfide (CaS). The product gas, which leaves the gasifier at approximately $1425°C$, is cooled, compressed, and fed to a shift converter (WGS reactor) in which a portion of carbon monoxide is reacted with steam via WGS reaction to attain a CO-to-H_2 ratio of 1:3.

The carbon dioxide produced is removed from the product gas, and the gas is cooled again. It then enters a methanator in which carbon monoxide and hydrogen react to form methane via $CO + 3H_2 = CH_4 + H_2O$. Excess water is removed from the methane-rich product. The final gaseous product has a heating value of around 900 Btu/scf.

3.10.3 RUMMEL–OTTO MOLTEN SALT GASIFICATION

In this concept, the combustion of coal and air to heat the molten salt is separated from gasification of coal that is heated by the molten salt. The molten salt is thus recycled back and forth from combustion chamber to gasification chamber. This method separates the product gas from combustion gas, the former comprising mainly CO, H_2, CO_2, and CH_4. The process is thus a form of *indirect gasification* discussed in Section 3.14.

An early version of this type of process was the Rummel–Otto double-shaft slag bath gasifier [19,315,317,318] in which the heart of the reactor was a circular bath of slag, the contents of which were rotated by the tangential entry of the reactants through tuyeres to the chamber above the melt. The chamber was divided into two sections by baffles so that melt could pass from one section to another without gases above it being able to mix. PC and steam were injected beneath the melt in one chamber where they reacted to give the product gas. The melt was heated by the reaction of coal and air injected in the other section.

The ash from the coal fused and mixed with the slag, the excess melt being removed from the reactor. The slag, to which lime was added to decrease its viscosity, acted not only as a heat transfer medium and sealing fluid but also as a reactant as it transferred carbon to the chamber. As no refractories were available that would withstand the erosion and corrosion at 1550°C under both oxidizing and reducing conditions, the slag was contained in a frozen-wall reactor. Since throughputs were not very high, this resulted in large heat losses, thus rendering the process uneconomical.

With the use of Rummel–Otto slag bath gasifier (also known as Saaberg–Otto gasifier), a 264 tons/day plant was designed to operate at 25 atm. Coal was fluidized with oxygen/air and steam and the flame injected tangentially onto the slag at 1500°C–1700°C. The slag acted as a heat shield and in the event of interruption of coal supply, the oxygen reacted with the carbon suspended in the melt, thus avoiding the possibility of explosions downstream. The coal can contain up to 40% ash and the gross overall efficiency was reported to be 88%–94%; this included high-pressure steam raised by the sensible heat of the product gases. The wall of the reactor was water cooled and initially lined with refractory that was gradually replaced by frozen slag as it eroded and corroded.

3.10.4 ROCKWELL MOLTEN SALT GASIFICATION PROCESS

Rockwell International has extensively developed molten salt processes for the treatment of combustible wastes and for the gasification of coal [19,311,312,314,317,318]. Waste treatment was demonstrated in coal-, gas-, or waste-fueled systems operating at 750°C–1000°C in single-stage processes using metal or ceramic-lined vessels. The processes were demonstrated in air-breathing systems with off-gas filtering and continuous gas discharge. Scales were in the range from 1 ton/day to 1 ton/h. The melt mixture was generally based on sodium carbonate, which accumulated sulfates and halides depending on the wastes. Destruction and removal efficiencies in excess of 99.9999% were demonstrated in a series of independent experiments between 1978 and 1983. NO_x generation was insignificant at these operating temperatures. Dioxins and furans were not detected by CG/MS in off-gases from chlorinated hydrocarbon treatment [19,310–312,314,317,318].

A generalized three-step salt recovery process was developed that separated the salt intermediate product into a water-washed ash, a soda-rich fraction that was dried and returned to the molten salt vessel, and a chloride-rich product representing the stoichiometric equivalent of chloride in the original waste stream. The products of this process represented a nearly theoretical mass reduction by any oxidative process that fixes halogens as neutral salts.

Carbonate/halide salt mixtures were chosen because of their alkalinity, fluidity, and aggressive wetting and dissolving capabilities. These same properties limit materials of construction. Dense alpha-alumina bricks were found to withstand the corrosive effects of the oxygenated chloride/carbonate melt, with erosion rates measured to be about 2 mm/year under agitation and normal thermal cycles. Inconel 600 and certain other high-Cr alloys were found sufficiently stable for vessel design [19,310–312,314,317,318]. Low-cost synthetic sapphire was available for application to specialized components. Refractory life depended on duty cycle, morphology, and trace impurities.

One potentially important problem was encountered when coal containing silica and alumina was oxidized in soda melt above 900°C [19,310–312,314,317,318]. At concentrations above 15% and with excess oxygen, the alumina and silica combined to form solid deposits on the vessel walls. The phenomena was not observed during oxygen-deficient conditions (i.e., coal pyrolysis) or in the presence of silica or alumina alone. Consequently, waste treatment must avoid critical conditions for this instability by (1) operation below critical ash concentrations under oxidizing conditions with carbonate recycle, (2) queuing of wastes to avoid critical ash buildup, (3) sequential oxygen-deficient treatment (pyrolysis) followed by oxidation of the off-gas, and (4) operation at lower temperatures.

3.10.5   ALDEN PROCESS FOR BIOMASS GASIFICATION USING MOLTEN SALTS

The use of molten salts as a catalyst for the breakdown of biomass is a venue with potential that has not been explored extensively. Alden [308] analyzed the effects of molten salts with varying compositions of Li_2CO_3, Na_2CO_3, K_2CO_3, KOH, and NaOH on the biomass gasification. It is believed that the highly ionic molten salt media acts to break down the lignin binding the cellulose and hemicellulose and to further break down the long-chain cellulose strands into smaller sugars in a manner that does not alter the molten salt. In addition, the molten salt increases the surface interactions between the deconstructed carbonaceous materials with gaseous carbon dioxide and water vapor that catalyzes the formation of carbon monoxide and hydrogen (syngas). Experimental results indicated the production of syngas from the proposed process. In addition, experimental results suggested that the molten salts catalyzed the reaction between carbon dioxide, water, and the char left over from pyrolysis to produce additional carbon monoxide [308,309].

The salt composition used by Alden [308] was both catalytic and regenerative. It occasionally had a grayish tint, but this was due to biomass particles trapped in the salt and the possibility of it absorbing some water. This was important because if the salt was meant to be used for extended periods of time, it was necessary for it to be catalytic.

The Alden [308,309] process produced syngas. Hydrogen was produced in large amounts and was the main product in many of the experiments run. It was postulated that by running the reactor with continuous gas flow and minimizing the side reactions and reducing the production of water, the concentrations of hydrogen and carbon monoxide could be greatly increased. The results also indicated possible production of biooil.

3.11   PLASMA GASIFICATION

Plasma gasification is a process that converts organic matter into synthetic gas and solid slag using plasma. A plasma torch powered by an electric arc is used to ionize gas and catalytically convert organic matter into synthetic gas and solid waste (slag). It is used commercially as a form of waste treatment and has been tested for the gasification of biomass/waste and solid hydrocarbons, such as coal, oil sands, and oil shale [319,320].

Plasmas processes have been used and developed during the nineteenth century by the metalworking industry to provide extremely high temperatures in furnaces. During the early twentieth century, plasma processes were used in the chemical industry to manufacture acetylene from natural gas. Since the early 1980s, plasma technology is considered as a highly attractive route for the processing of MSW. Plasma creates tar-free, pollutant-free syngas that can be very valuable for downstream applications. It is this thinking that has taken significant momentum in recent years. Plasma has been extensively used to either remove tar from conventional gasification process [62,321–328] or apply destruction of waste to produce synthetic gas [67,248,329–349].

Plasma is often mentioned as the fourth state. A plasma torch itself typically uses an inert gas such as argon. The electrodes vary from copper or tungsten to hafnium or zirconium, along with various other alloys. A strong electric current under high voltage passes between the two electrodes as an electric arc. Pressurized inert gas is ionized passing through the plasma created by the arc. The torch's temperature ranges from 4,000°F to 25,000°F (2,200°C–13,900°C) [319,320]. The temperature at a location several feet away from the torch can be as high as 5000°F–8000°F, a temperature high enough to completely destroy all carbonaceous materials into their elemental forms. The temperature of the plasma reaction determines the structure of the plasma and composition of the product gas. With the application of plasma, carbonaceous material is heated, melted, and finally vaporized. At these conditions molecular dissociation can occur by breaking down molecular bonds. Complex molecules are separated into individual atoms. The resulting elemental components are in a gaseous phase. Molecular dissociation using plasma is referred to as "plasma pyrolysis" [319,320].

While the feedstock for plasma waste treatment is most often MSW, organic waste, coal, or oil shale, feedstocks may also include biomedical waste and hazardous materials. The content and consistency of the waste directly impacts the performance of a plasma facility. Presorting and recycling useful material before gasification provides consistency. Too much inorganic material such as metal and construction waste increases slag production, which in turn decreases syngas production. However, a benefit is that the slag itself is chemically inert and safe to handle (certain materials, however, may affect the content of the gas produced) [67,248,319,320,329–349]. Shredding waste before entering the main chamber helps to increase syngas production. This creates an efficient transfer of energy that ensures more materials are broken down. For better processing, air and/or steam is added into plasma gasifier. As shown in Chapter 6, plasma is also used for steam or dry reforming in the presence or absence of an externally added heterogeneous catalyst.

High-Btu synthetic gas produced from plasma gasification consists predominantly of carbon monoxide (CO), H_2, and CH_4, among other components. The conversion rate of plasma gasification exceeds 99% [319,320]. Nonflammable inorganic components in the waste stream are not broken down. This includes various metals. A phase change from solid to liquid adds to the volume of slag. Plasma processing of waste is ecologically clean. The lack of oxygen prevents the formation of many toxic materials. The high temperatures in a reactor also prevent the main components of the gas from forming toxic compounds such as furans, dioxins, nitrogen oxides, or sulfur dioxide. Water filtration removes ash and gaseous pollutants. While the gas product contains no phenols or complex hydrocarbons, circulating water from filtering systems is toxic. The water removes toxins (poisons) and the hazardous substances that must be cleaned [319,320]. Product gas from plasma gasification can be cleaned by conventional technologies, including cyclone, scrubbers, and ESPs. Cyclone/scrubber effluents can normally be recycled for further processing. Metals resulting from plasma pyrolysis can be recovered from the slag and eventually sold as a commodity. Inert slag is granulated and this can be used in construction industries. A portion of the syngas produced feeds on-site turbines, which power the plasma torches and thus support the feed system. Thus, power requirement for plasma process is internally generated.

3.11.1 Energy from MSW by Plasma Gasification

While plasma can gasify any feedstock, in recent years, it is mainly used for MSW gasification. Numerous commercial activities for MSW plasma gasification are now a reality in the United States, Canada, and Norway, among other countries, and they are described in the attached references [350–362]. In the United States, Westinghouse has taken a lead role in the commercial development of this technology. Plasma technology is considered as a highly attractive route to generate energy from MSWs (among other types of wastes such as heavy oil, used car tires, and medical wastes) via syngas production. The high enthalpy, the residence time, and the high temperature in a plasma can advantageously improve the conditions for gasification, which are inaccessible in other thermal processes and can allow better net electrical efficiency than autothermal processes.

When applied to waste materials such as MSW, plasma gasification possesses unique advantages for the protection of air, soil, and water resources through extremely low limits of air emissions and leachate toxicity. Because the process is not based on the combustion of carbonaceous matters, generation of greenhouse chemicals, in particular carbon dioxide, is far less than from any other conventional gasification technology. Furthermore, air emissions are typically orders of magnitude below the current regulations. The slag is monolithic and the leachate levels are orders of magnitude lower than the current EP-toxicity standard [320,350–362]. Slag weight and volume reduction ratios are typically very large. Even though the data for a variety of coals and coal wastes are not readily available in the literature, both the mass reduction ratio and the volume reduction ratio for coals and coal wastes are believed to be significantly higher than those for nonplasma gasification technology, thus substantially reducing the burden of waste and spent ash disposal problem.

Since the end of World War II, all developed countries generate more and more domiciliary and industrial wastes per capita at a level that is becoming unmanageable, causing permanent damages to the environment. For example, in Japan the total quantity of MSW is about 5.2×10^7 tons/year. This number is 250 million tons per year in United States and it will reach 1 billion tons per year globally. Public and political awareness to environmental issues has led to plan more conversion of this waste to energy and less for landfill disposal [22]. In recent year, plasma gasification process is considered as one of the viable options compared to other thermochemical processes [22,320,350–362].

MSW is a heterogeneous fuel containing a very wide variety of solid wastes. Due to the presence of some postrecycling materials, such as paper fiber and plastics, its heating value can be high. The chemical composition of MSW can be compared to any solid organic fuel like coal or biomass. The element composition of MSW is in the following range (wt%): C (17–30), H_2 (1.5–3.4), O_2 (8–23), H_2O (24–34), and ashes (18–43). Also, the average specific combustion heat of MSW is in the range from 5 to 10 MJ/kg [22,67,248,319,320,329–349]. Just like all gasification processes described in this chapter, a plasma gasifier associated with a gas turbine combined-cycle power plant can target up to 46.2% efficiency [22,67,248,319,320,329–349]. Moreover, this synthesis gas produced by gasification, mainly composed of CO and H_2, can also be used as feedstock for the production of synthetic liquid fuels in processes such as FT process.

Conventional methods based on autothermal gasification present some limitations that might be overcome through plasma [319,320], particularly in terms of material yield, syngas purity, energy efficiency, dynamic response, compactness, and flexibility. Injected plasma power can be adjusted independently of the heating value of the treated material. The enthalpy provided by the plasma can easily be adjusted by the tuning of the electric power supplied to the system, making the process independent of the ratio O/C and the H/C. Contrary to the autothermal gasification processes, the nature of the plasma medium (neutral, oxidizing, or reducing atmosphere) can also be independently adjusted. It is possible to achieve gas temperature up to 15,000 K with a thermal plasma [319,320]. Such a temperature in plasmas can allow synthesizing or degrading chemical species in some conditions unreachable by conventional combustion and can greatly accelerate the chemical reactions.

Thermochemistry of combustion does not allow precise control of the enthalpy injected into the reactor. Plasma process allows an easiest enthalpy control by adjusting the electric power. The reactive species produced by the plasma, such as atomic oxygen and hydrogen or hydroxyl radicals, is an additional advantage for the use of plasma. In the literature, it is reported that these species enhance strongly the degradation of the tars with greater efficiency than conventional processes [319,320].

The current market for high-power plasma torches is mainly shared by four companies: Westinghouse, Europlasma, Tetronics, and Phoenix Solutions Company (PSC). The technologies developed by Westinghouse, Europlasma, and PSC [320,350–362] are based on transferred and nontransferred DC torches with water-cooled metal electrodes, while Tetronics torch [248,337–339,343,360] is based on a transferred DC torch with two graphite electrodes not water cooled. Advanced Plasma Power and Tetronics have a collaboration agreement for the development and commercialization of plasma gasification WTE plants based on the technology of transferred DC torch [248,337–339,343,360].

The strategies used by Westinghouse and Europlasma are different since they have each developed a plasma gasification process based on their own DC torch technology and market turnkey plants through subsidiaries (Alter NRG for Westinghouse and CHO-Power for Europlasma, respectively) [353,355–358,361]. In parallel of these developments of industrial plasma gasification WTE plants, some companies also develop their own facility based on Westinghouse, Europlasma, or PSC DC torches [350,352,356,359,361].

Often, there is very few information on homemade torch technologies developed but it seems to be mainly based on DC torches. Although not yet validate for the waste gasification at an industrial scale, other plasma torch technologies (RF and AC) are being developed at a pilot scale in several research laboratories such as Applied Plasma Technologies (United States); PERSÉE, MINES Paris Tech (France); and Institute for Electrophysics and Electric Power, Russian Academy of Sciences (Russia).

There are two commercial waste processing facilities and one commercial ash vitrification plant in Japan, all of which were established between 1995 and 2002. One waste processing facility is a 242 short ton per day mixed auto shredder and MSW operation for power generation. The other is a 30 short ton per day mixed MSW and sewage sludge operation for heat production, which is then used in the wastewater treatment facility [352–354,361]. Westinghouse Plasma Corporation claims that capacities between 500 and 750 tons/day can be handled in a single reactor vessel [352–354,361]. The torch can be oxygen blown and the company claims that it can make synthesis gas as a product [352–354,361].

For the industrial-scale waste to energy plasma processes, the position of torches in the reactor is mainly based on the waste to be treated. For the gasification of waste with low organic matter content, it is necessary to treat the waste at high temperature in order to melt the inorganic part. The products obtained are syngas from the organic part of the waste and slag from the nonorganic part of the waste. In this case, the plasma torches are placed in the reactor body closest to the molten bath and the torches are nontransferred arc or transferred arc (the bath playing the role of anode for the plasma torch).

In the case of waste with a high proportion of organic matter, it is not necessary to raise the temperature of the reactor above 1800 K and in this case, the waste gasification in the reactor can be made by either autothermal or allothermal ways (plasma, dual FBs; see Section 3.14). In this case, the plasma torch is placed at the outlet of the gasification reactor before the cooling of the crude syngas in the aim to treat the tar content in the syngas at an optimized energy cost (primary method).

In summary, plasma treatment for waste gasification provides purified gas, high in hydrogen by limiting the production of tars and supporting WGS reaction. Plasma methods have also the advantages to be able to operate at very high temperatures and to be retrofitted to existing installation. These high temperatures in plasma processes accelerate chemical reactions and degrade chemicals to the extent that are not possible in conventional gasification processes. Plasma process allows an easy enthalpy control within the reactor by adjusting the electric power; such control is not possible in conventional combustion process. The atomic oxygen and hydroxyl radicals formed in the plasma process strongly enhance the degradation of the tars with greater efficiency than conventional processes. Plasma gasification technologies for the disposal of MSW at an industrial scale is a growing market, and the efficiency of the waste gasification by plasma is validated but the economic viability of this technology must be proven further before it is accepted.

3.12   SOLAR GASIFICATION

Solar thermochemistry refers to a number of process technologies like thermal or thermochemical splitting of water, solar electrolysis, solar gasification, and reforming or cracking of water and other carbonaceous materials [363–385]. Many of these endothermic reactions are carried out by energy harnessed by concentrated solar beams. Solar gasification generally deals with upgrading and decarbonization of fossil fuels. Such gasification is often carried out in the presence of steam. Successful solar gasification of carbonaceous materials was first reported in the 1980s in which coal, activated carbon, coke, and coal/biomass mixtures were employed in a fixed-bed windowed reactor. Charcoal, wood, and paper were gasified with steam in a fixed-bed reactor. More recently, steam gasification of oil shale and coal, waste tires and plastics, and coal in a FB reactor and also steam gasification of petroleum coke and vacuum residue in fixed-bed, FB, and entrained-bed reactors were examined [363,364,366–370,373–381]. In the last type of reactor dry coke particles, coal–water slurries and vacuum residues were tested for the steam gasification.

In a conceptual solar gasification process using steam, biomass is heated rapidly in a solar furnace to achieve flash pyrolysis at temperatures of about 900°C [363]. Some steam is added to the pyrolyzer to increase the gas yield relative to char. The char constituting about 10%–20% of the biomass by weight is steam gasified with external heating at temperatures of 900°C–1000°C; all of the volatile hydrocarbons are then steam reformed in a solar reformer. The steam for the process is generated from heat recovered from the product gas. The composition of the syngas is adjusted to the user's needs utilizing conventional operation involving the WGS reaction and CO_2 stripping.

A number of gasification experiments were carried out using small quantities of biomass, coal, oil shale, and residual oil with external heat supplied by the sun [363–381]. These experiments included cellulose gasification and oil shale gasification with carbon recovery approaching nearly 100% at temperature of 950°C and short residence times [367,382,383]. While these experiments confirm the applicability of the flash pyrolysis approach, they did not provide data for the design and scale-up of a solar gasification process [363–381].

3.12.1 SOLAR GASIFICATION REACTORS AND PROCESSES

The use of solar energy as the heat carrier for pressurized coal gasification process has the following advantages: saving available coal reserve and reducing coal-specific emissions, particularly CO_2. In this process, finely powdered coal is fed by a specially designed injection system. The oxidizing and fluidizing agent is superheated steam. The heat required for the endothermic gasification reaction is introduced by means of a tubular heat exchanger assembly immersed in the FB. The technical feasibility of a solar power tower and pressurized gasifier integration can be demonstrated in the form of a small pilot plant [363,370]. As shown in these references, coal was gasified in this process at a temperature of 859°C with an 87% conversion rate.

A number of different types of solar steam gasification reactors have also been examined in the literature [363–377,380]. The reactor configuration examined by Z'Graggen et al. [373] at ETH in Zurich consists of a cylindrical cavity receiver that is 21 cm in length and 12 cm in inside diameter length and contained 5 cm diameter aperture for solar beams. The cavity-type geometry was designed to effectively capture the incident solar radiation and its apparent absorption is estimated to exceed 0.95. The cavity was made of Inconel 601, lined with Al_2O_3, and insulated with an Al_2O_3/ZrO_2 ceramic foam. The aperture is closed by 0.3 cm thick clear fused quartz window mounted in a water-cooled aluminum ring that also serves as a shield for spilled radiation. The window is actively cooled and kept away from particles and condensable gases.

Steam and particles are injected separately into the reactor cavity, permitting the separate control of mass flow rates and stoichiometry. Steam is introduced through several ports. The carbonaceous material feed unit is positioned on the top of the reactor vessel with its inlet port located at the same plane as the primary steam injection system, allowing for the immediate entrainment of particles by the steam flow. The reactor temperature was measured at 12 separate locations by thermocouples inserted in the Inconel walls. Both inlet and exit temperatures were also measured by the thermocouples. The dry, slurry, and liquid feeding of raw materials was carried out by different devices.

Piatkowski et al. [368] used a packed bed solar steam gasification reactor. This reactor is specially designed for beam-down incident solar radiation as obtained through a Cassegrain optical configuration that makes use of a hyperbolic reflector at the top of the solar tower to redirect the sunlight collected by a heliostat field to a receiver located at the ground level. The reactor has two cavities in series. The upper one absorbs the solar radiation and contains a small aperture to gather concentrated solar radiation. The lower cavity contains carbonaceous materials on top of a steam injector. An emitter plate separates the two cavities.

A 3D compound parabolic concentrator is incorporated in the aperture of the reactor further augmenting the incident solar flux before passing it through a quartz window in the upper cavity.

The emitter plate acts as a transmitter of the radiation to the lower cavity, thus avoiding direct contact between the quartz window and the reactants and products. This setup also provides uniform temperature in the lower cavity and a constant supply of radiant heat through the upper cavity that can act as energy storage needed due to intermittent supply of radiant heat. This type of batch, two-cavity solar reactor has been successfully used for the carbothermal reduction of ZnO and the detoxification of solid waste. The reactor can be operated with a wide variety of particle sizes and as the reaction proceeds, both the particle size and the packed bed reactor volume decrease. The detailed dimensions and operation of this type of reactor are given by Piatkowski and Steinfeld [366]. Piatkowski et al. [368] and Perkins et al. [367] also showed an effective use of such a reactor to produce syngas from coal, biomass, and other carbonaceous feedstock. Z'Graggen [369] and Z'Graggen et al. [373] produced hydrogen from petcoke using a solar gasification process.

Solar gasification of petcoke project (SYNPET) (2003–2009) was tested at the Plataforma Solar de Almería, Spain. The reactor has reached the hydrogen production efficiency of 60% working at 1500 K [363,370,373,382–384]. The solar energy is also used as the heat carrier for pressurized coal gasification/reforming process. In this process, finely powdered coal is fed by a specially designed injection system. The oxidizing and fluidizing agent is superheated steam. The heat required for the endothermic gasification/reforming reaction is introduced by means of a tubular heat exchanger assembly immersed in the FB. The technical feasibility of a solar power tower and pressurized gasifier integration has been demonstrated in small pilot plant [365,370,373]. Solar energy has also been used to gasify biomass in different types of reactors [367,374,375,377,379,380].

3.13   CATALYTIC GASIFICATION

Catalysts can be used to enhance the reactions involved in gasification. Many gasifiers must operate at high temperatures so that the gasification reactions will proceed at reasonable rates. Unfortunately, high temperatures can sometimes necessitate special gasifier materials and extra energy input and cause efficiency losses if heat cannot be reclaimed.

In general, studies on the catalysis of coal or biomass/waste gasification have threefold objectives [63,386–443]: (1) to understand the kinetics of coal gasification that involves active mineral matter, (2) to design possible processes using these catalysts, and (3) to design new catalysts that can further reduce gasification temperature, improve carbon conversion efficiency, and improve quality of product. The quality of product involves concentration of syngas and degree of presence of tar components. The catalyst is often used to improve concentration of syngas and reduce tar concentration. The last objective has to also weigh in the cost, life, and recoverability of the catalyst. Since the use of catalysts lowers the gasification temperature, product composition under equilibrium conditions and thermal efficiency are generally improved. However, under normal conditions, a catalytic process cannot compete with a noncatalytic one unless the catalyst is quite inexpensive or highly active and stable at low temperatures. Recovery and reuse of catalyst in the process is undesirable and unattractive in coal gasification because of the expensive separation efforts and the low cost of coal and coal gas. Gasification is a bulk commodity production process and requires a cheaper catalyst.

Research on catalysis covers mainly three subjects: basic chemistry, application-related problems, and process engineering. Juntgen [411] published an extensive review article on catalytic gasification. Nishiyama [412] also published a review article, which features some possibilities for a well-defined catalytic research effort. The article contains the following observations:

1. Salts of AAEMs as well as transition metals are active catalysts for gasification.
2. The activity of a particular catalyst depends on the gasifying agent as well as the gasifying conditions.

3. The main mechanism of catalysis using AAEM salts in steam and carbon dioxide gasification involves the transfer of oxygen from the catalyst to carbon through the formation and
   decomposition of the C–O complex, that is, C(O).

The improvement of carbon conversion in the presence of carbon dioxide is particularly important
because it utilizes an environment pollutant as a reactant in the gasification process. Both alkali
matters and transition metal catalysts have been examined for this purpose. The mechanism of
hydrogasification reactions catalyzed by iron or nickel is still not very clear. But a possible explanation is that the active catalyst appears to be in the metallic state and there are two main steps for
the mechanism. These are hydrogen dissociation and carbon activation [413–417]. For the latter
case, carbon dissolution into and diffusion through a catalyst particle seems logical. Gasification
proceeds in two stages, each of which has a different temperature range and thermal behavior, so
that a single mechanism cannot explain the entire reaction. Thus, the catalyst is still assumed to
activate the hydrogen.

Alkali metal salts of weak acids (like potassium carbonate [K_2CO_3], sodium carbonate [Na_2CO_3],
potassium sulfide [K_2S], and sodium sulfide [Na_2S]) can catalyze steam gasification of coal. In the
early 1970s, research confirmed that 10%–20% by weight K_2CO_3 could lower acceptable bituminous coal gasifier temperatures from 925°C to 700°C and that the catalyst could be introduced to the
gasifier impregnated on coal or char. Catalysts can also be used to favor or suppress the formation
of certain components in the syngas product. The primary constituents of syngas are hydrogen (H_2)
and CO, but other products like methane are formed in small amounts. Catalytic gasification can be
used to either promote methane formation or suppress it.

Besides sodium and potassium, calcium as a catalyst has also been studied by several investigators [418–426]. This catalyst has a very high activity in the initial period when it is well
dispersed in the other promoter catalyst, but with increasing conversion, the activity drops.
The chemisorption of carbon dioxide, x-ray diffraction, and some other analytical techniques
confirmed the existence of two or more states of calcium compounds, as well as the formation
of a surface oxygen complex.

Disadvantages of catalytic gasification include increased materials costs for the catalyst itself,
as well as diminishing catalyst performance over time. Catalysts can be recycled, but their performance tends to diminish with age. The relative difficulty in reclaiming and recycling the catalyst
can also be a disadvantage. For example, the K_2CO_3 catalyst described earlier can be recovered
from spent char with a simple water wash, but some catalysts may not be so accommodating. In
addition to age, catalytic activity can also be diminished by poisoning. Many catalysts are sensitive
to particular chemical species that bond with the catalyst or alter it in such a way that it no longer
functions. Sulfur, for example, can poison several types of catalysts including nickel, iron, palladium, and platinum.

As pointed out by Lee [19], compared to other heterogeneous catalytic systems, the catalysis in
gasification is complex because of the following reasons:

1. Catalyst is only effective while in contact with substrate that itself changes with time. The
   main properties of the substrate related to the activity are (1) the reactivity of the carbonaceous constituents, (2) the catalytic effect of minerals, and (3) the effect of minerals on
   the activity of an added catalyst. All of these are difficult to model simultaneously and as
   a function of process time.
2. The definition of catalyst activity is not straightforward. For an alkali catalyst, the reaction
   rate can increase due to change in catalyst dispersion and also due to increase in catalyst/
   carbon ratio in the later stages of gasification. The activity can also be increased due to the
   change in the chemical state of the catalyst or the change in the effective catalyst surface
   area due to pore opening. The activity of the catalyst also depends on the nature of the time
   variable substrate and gasifying conditions.

3. The catalyst is very short lived. The catalyst deactivation occurs due to a number of processes such as agglomeration of catalyst particles, coking, and chemical reaction with sulfur and other trace elements. Coking causes fouling on the catalyst surface as well as sintering the catalyst, whereas reaction with sulfur poisons the catalytic activity. The simultaneous occurrence of all these processes makes modeling of deactivation process also complex.

The following general trends have been observed in reference to the factors affecting the activity of the catalysts and its management for coal gasification:

1. Nickel catalysts are more effective toward lower-rank coals because they can be more easily dispersed into the coal matrix owing to higher permeability of the coal, whereas the efficiency of potassium catalyst is independent of the rank. In any case, the coal rank alone, as given by the carbon content, cannot predict catalyst activity.
2. The internal surface area of coal char relates to the overall activity of the catalyst. It can be related to the number of active sites in cases when the amount of catalyst is large enough to cover the available surface area. For an immobile catalyst, the conversion is almost proportional to the initial surface area.
3. Pretreatment of coal before the catalytic reaction often helps in achieving higher reaction rates. Although the pretreatment of coal may not be directly applicable as a practical process, a suitable selection of coal types or processing methods could enhance the activity of catalysts.
4. The effect of coal mineral matter on the catalyst effectiveness is twofold. Some minerals such as AAEMs catalyze the reaction, whereas others such as silica and alumina interact with the catalyst and deactivate it. In general, demineralization results in enhancement of activity for potassium catalysts, but only slightly so for calcium and nickel catalysts.
5. The method of catalyst loading is also important for activity management. The catalyst should be loaded in such a way that a definite contact between both solid and gaseous reactants is ensured. It was observed that when the catalyst was loaded from an aqueous solution, a hydrophobic carbon surface resulted in finer dispersion of the catalyst when compared to a hydrophilic surface.

The most common and effective catalysts for steam gasification are oxides and chlorides of AAEMs, separately or in combination [427]. Xiang et al. [427] studied the catalytic effects of the Na–Ca composite on the reaction rate, methane conversion, steam decomposition, and product gas composition, at reaction temperatures of 700°C–900°C and pressures from 0.1 to 5.1 MPa. A kinetic expression was derived with the reaction rate constants and the activation energy determined at elevated pressures. Alkali metal chlorides such as NaCl and KCl are very inexpensive but their activities are quite low compared to the corresponding carbonates because of the strong affinity between alkali metal ion and chloride ion. Takarada et al. [428,429] have attempted to make Cl-free catalysts from NaCl and KCl by an ion-exchange technique using ammonia as a pH-adjusting agent and removed Cl ions completely by water washing. This Cl-free catalyst markedly promoted the steam gasification of brown coal. This catalyst was found to be catalytically as active as alkali carbonate in steam gasification. During gasification, the chemical form of active species was found to be in the carbonate form and was easily recovered.

The direct contact between K-exchanged coal and higher-rank coal also resulted in enhancement of gasification rate [429]. Potassium was found to be a highly suitable catalyst for catalytic gasification by the physical mixing method. Weeda et al. [430] studied the high-temperature gasification of coal under product-inhibited conditions whereby they used potassium carbonate as a catalyst to enhance the reactivity. They found that the physical mixing method is likely to be neither practical nor economical for large-scale applications.

Some researchers [431] have recovered the catalysts used, in the form of a fertilizer of economic significance. They used a combination of catalysts consisting of potassium carbonate and magnesium nitrate in the steam gasification of brown coal. The catalysts along with coal ash were recovered as potassium silicate complex fertilizer. In addition to the commonly used catalysts such as AAEMs for catalytic gasification, some less-known compounds made of rare earth metals as well as molybdenum oxide (MoO_2) have been successfully tried for steam and carbon dioxide gasification of coal [432–434]. Some of the rare earth compounds used were $La(NO_3)_3$, $Ce(NO_3)_3$, and $Sm(NO_3)_3$. The catalytic activity of these compounds decreased with increasing burn-off (i.e., conversion) of the coal. To alleviate this problem, coloading with a small amount of Na or Ca was attempted and the loading of rare earth complexes was done by the ion-exchange method.

Coal gasification is facilitated by the catalytic enhancement of steam decomposition and carbon/steam reaction. A process developed by Battelle Science & Technology International [435] indicated that calcium oxide catalyzes the hydrogasification reaction and calcium content of the catalyst and the reactivity of coal chars with carbon dioxide are well correlated. Other alkali metal compounds, notably chlorides and carbonates of sodium and potassium, can also enhance the gasification rate by as much as 35%–60%. In addition to the oxides of calcium, iron, and magnesium, zinc oxides are also found to substantially accelerate gasification rates by 20%–30%.

Some speculative mechanisms have been proposed by Murlidhara and Seras [435] as to the role of calcium oxide in enhancing the reaction rate. They proposed three different schemes. In Scheme 1, coal organic matter may function as a donor of hydrogen, which then may be abstracted by calcium oxide to form CaH_2 and water. Calcium oxide is regenerated by reaction between carbon dioxide and CaH_2. Scheme 2 explains the mechanism of generating oxygen-adsorbed CaO sites and subsequent desorption of nascent oxygen, which in turn reacts with organic carbon of coal to form carbon monoxide. In Scheme 3 direct interaction between CaO and coal organics results in liberation of carbon monoxide along with an intermediate. The scheme also explains an oxygen exchange mechanism that brings the reactive intermediates back to CaO.

ExxonMobil has reported that impregnation of 10%–20% of potassium carbonate lowers the optimum temperature and pressure for steam gasification of bituminous coals, from 980°C to 760°C and from 68 to 34 atm, respectively [436]. In their commercial-scale plant design, the preferred form of make-up catalyst was identified as potassium hydroxide. This catalyst aids the overall process in several ways. First, it increases the rate of gasification, thereby allowing a lower gasification temperature. Second, it prevents swelling and agglomeration when handling caking coals, which is another benefit of a lower gasification temperature. Most importantly, it promotes the methanation reaction because it is thermodynamically more favored at a lower temperature. Therefore, in this process, the production of methane is thermodynamically and kinetically favored in comparison to synthesis gas. A catalyst recovery unit is provided after the gasification stage to recover the used catalyst.

Just like coal gasification, biomass gasification can be improved with the use of catalysts. The catalyst can reduce temperature requirement, remove tar, and improve the quality of syngas. Catalysts for biomass gasification have been reviewed in some excellent publications by Sutton et al. [403] and Li et al. [390]. Elimination of the condensable organic compounds (tars) and methane by a suitably cheap catalyst can enhance the economic viability of biomass gasification. Another objective is to use catalyst to improve hydrogen production. The literature indicates that three groups of catalysts; olivine and dolomite, alkali metals, and nickel have been successful. Disposable biomass char in the form of catalyst filters have also been found to improve reactor performance. In recent years, the use of nanocatalysts and zirconia support has also been examined. While some interesting results [63,386–410,437–443] have been reported, the issue of cost and catalyst separation and regeneration still remains to be solved. The use of catalyst is particularly important for the production of high-quality biosyngas.

The production of syngas at lower temperature with the use of a cheap catalyst can significantly add value to the biomass gasification process. Lv et al. [387,399] showed that a promising application for biomass syngas is the production of methanol or DME. The conversion of syngas to liquid fuel significantly depends on the H_2/CO ratio in the syngas. Catalysts can be used to get improved quality of syngas with proper H_2/CO ratio at lower temperatures.

In the study by Lv et al. [387,399], a FB gasifier and a downstream fixed bed were employed as the reactors. Two kinds of catalysts, namely, dolomite-based catalyst and nickel-based catalyst, were used in the FB and fixed bed, respectively. The gasifying agent used was an air–steam mixture. The main variables studied were temperature and weight hourly space velocity in the fixed-bed reactor. Over the ranges of operating conditions examined, the maximum H_2 content reached 52.47 vol%, while the ratio of H_2/CO varied between 1.87 and 4.45, which is favorable for methanol or DME synthesis. However, besides H_2 and CO, there is also a lot of CO_2 and CH_4 in the gases, which needs to be decreased through modifying the catalysts and controlling operating conditions. The results indicated that an appropriate temperature (750°C for the current study) and more catalyst are favorable for getting a higher H_2/CO ratio. Using a simple first-order kinetic model for the overall tar removal reaction, the apparent activation energies and preexponential factors were obtained for nickel-based catalysts. A single one lump model was found perfect for tar destruction analysis. Applying this model, activation energy (E) and arrhenius constant (A) were determined as 51 kJ/mol and 14,476 (m^3 (Tb,wet)/kg h), respectively. The results indicated that biomass gasification can produce syngas for liquid fuel synthesis after further processing.

Asadullah et al. [386,437–443] showed that the GHGs, especially CO_2 and particulate matters, can be considerably reduced with the use of biomass for the production of syngas and hydrogen. The traditional process of biomass gasification is, however, problematic because of tar and char formation during the gasification process even at very high temperature. In order to get the higher energy efficiency, when the process is carried out at lower temperature (<1123 K), more tar and char are produced. The catalytic gasification of biomass seems to be promising to reduce tar amount in the product gas even at low temperature; however, the traditional Ni-based or dolomite catalysts hardly reduce the tar content in the product gas. These catalysts are suddenly deactivated in the in-bed reaction system due to deposition of carbon on the surface [386,437–443]. Highly efficient catalyst and a suitable reactor are necessary to overcome the problems. They [386,437–443] examined a process for the cellulose gasification using the $Rh/CeO_2/SiO_2$ catalyst in a continuous-feeding FB reactor at as low as 500°C. The combination of the $Rh/CeO_2/SiO_2$ [416] catalyst with the FB reactor provided the novel system for the hydrogen and syngas production from biomass at low temperature with a high energy efficiency. In the FB reactor, the catalyst circulated in the oxidizing and reducing atmosphere. Under this situation, the surface of the active site can be kept clean by this in situ treatment.

Catalytic gasification of coal–biomass mixture has also been investigated. AAEMs in biomass can provide catalytic activity during coal–biomass mixture gasification. Some of this work is reviewed by Ratafia-Brown et al. [73].

3.14 INDIRECT GASIFICATION

The conversion of biomass to a low- or medium-heating-value gaseous fuel (biomass gasification) generally involves two processes. The first process, pyrolysis, releases the volatile components of the fuel at temperatures below 600°C (1112°F) via a set of complex reactions. Included in these volatile vapors are hydrocarbon gases, hydrogen, carbon monoxide, carbon dioxide, tars, and water vapor. Because biomass fuels tend to have more volatile components (70%–86% on a dry basis) than coal (30%), pyrolysis plays a proportionally larger role in biomass gasification than in coal gasification.

The by-products of pyrolysis that are not vaporized are referred to as char and consist mainly of fixed carbon and ash. In the second gasification process, that is, char conversion, the carbon remaining after pyrolysis undergoes the classic gasification reaction (i.e., steam + carbon) and/or combustion (carbon + oxygen). It is this latter combustion reaction that provides the heat energy required to drive the pyrolysis and char gasification reactions. Due to its high reactivity (as compared to coal and other solid fuels), all of the biomass feed, including char, is normally converted to gasification products in a single pass through a gasifier system.

Depending on the type of gasifier used, the reactions given earlier can take place in a single reactor vessel or be separated into different vessels. In the case of direct gasification, pyrolysis, gasification, and combustion take place in one vessel, while in indirect gasification, pyrolysis and gasification occur in one vessel and combustion in a separate vessel. In direct gasification, air and sometimes steam are introduced directly to the single gasifier vessel. In indirect gasification, an inert heat transfer medium such as sand carries heat generated in the combustor to the gasifier to drive the pyrolysis and char gasification reactions.

Knoef [345] promoted the concept of indirect gasification and indicated that one solution to improve syngas composition (less dilution) and thermal efficiency of autothermal gasification process is to use a dual-FB reactor. This type of reactor is designed to separate the gasification from the combustion. The high temperatures are conveyed between the two reactors with sand (indirect gasification). The interest of this technology is to prevent syngas dilution by the nitrogen of the air and the combustion of the wastes by separating the gasification from the combustion. The high temperature obtained in the reactor without using the combustion process allows producing a synthesis gas with high purity and high calorific value.

The concept of indirect gasification was applied to the gasification of Victorian brown coal by the Energy Research Centre of the Netherlands (ECN) [444–448]. For this coal, in a single-stage gasifier, part of the fuel is combusted to provide heat for the gasification of the rest of the fuel. The produced gas is used for power, heat, or other high-end applications. Compared to combustion, this method is more efficient because small gas engine or gas turbine performs better than a steam cycle. The lower temperature in gasification also has a positive effect on the ash behavior related to the fouling propensity. However, lower temperature can have a downside relating to the fuel conversion. For BFB or CFB gasifiers this conversion stops at 97% [444–455] or less.

While fuel conversion in direct gasification can be improved by increasing temperature, this can cause increased agglomeration and corrosion hazard. The high temperature also lowers the efficiency. ECN noticed that indirect gasification offers the unique possibility to reach 100% fuel conversion, without the need to go to high temperatures. The twist was that the combustion and pyrolysis/gasification processes need to be separated and a bed material was used to exchange heat between the two processes.

Besides complete fuel conversion, ECN noticed that indirect gasification also offers the following other advantages [444–448]:

1. No ASU needed: The combustion takes place in a separate vessel; hence, air can be used and the product gas is essentially N_2-free.
2. No carbon loss: The char/coke that remains after gasification is completely burned. The ashes are carbon-free.
3. High efficiency: Because the gasification is decoupled from the combustion, the gasification temperature can be much lower than one for direct gasification.
4. Medium calorific gas: The product gas does not contain nitrogen, which increases the heating value substantially. This also offers the possibility to use the gas for more high-end applications.
5. It also provides the possibility to capture CO_2 in a highly concentrated gas, thus reducing the emissions to the environment.

ECN [444–448,450,453,454] developed this indirect gasification technology using an indirect gasifier, MILENA, which can operate in two modes. In one mode (MILENA), the mixture of wood and brown coal was gasified in the riser reactor and char/coke was combusted to produce heat in the BFB reactor. In the second mode, the process was reversed (i-MILENA) [444,448]. The first mode did not work well because the low residence time in the riser left too much coke to provide heat for the process.

In the second mode (i-MILENA), a vertical tube (riser) was placed at the center of the BFB with an opening at the bottom, through which the bed material containing coke was transported. The riser was operated with air to transport the solids to the top and to combust the char from the gasification process. The combustion heated up the bed material (solids) and this was returned to the BFB where it provides the heat for the gasification process [446]. The residence time of about 15 min in BFB gave sufficient time for good conversion of char/coke to gas. Since these two processes take place in separate reactors but integrated in one reactor, the heat loss was minimized. It also allowed for complete fuel conversion at much lower temperatures compared to direct gasification. The indirect gasification resulted in a low N_2-containing product gas, which in turn allowed more advanced applications of the product gas, such as CO_2 removal [447] that was identified as a key development issue.

Another application of indirect gasification is the biomass heat pipe reformer design that focuses on small-scale CHP systems with hot gas cleaning and microturbines. Hydrocarbons and tars condense at temperatures below 200°C–250°C forming tar layers in the piping or in the engine. Conventional internal combustion engines require fuel gas inlet temperatures below 100°C, whereby the condensation of tars cannot be avoided. Appropriate gas-cleaning technologies are too expensive for small-scale systems and cause additional environmental problems.

Possible solutions are systems with hot gas cleaning and microturbines. Hot gas cleaning avoids quenching of the product gas, associated efficiency losses, and the condensation of tars. However, microturbines require heating values above 10,000 kJ/kg. The required heating values are only achievable by means of allothermal gasification in FB gasifiers. A new concept—indirectly heating of a gasifier by means of high-temperature heat pipes—promised to improve the performance of indirect heated gasifiers significantly.

The expected heating value of the product gas allowed its combustion in standardized microturbines without significant modification of the combustion chamber. The hot gas cleaning not only avoided the condensation of tars, it also allowed the reduction of the gasifier dimensions. The tar content depended not only on the reaction conditions like excess steam ratio and temperature and pressure but also on the retention time of the product gas in the reactor. Accepting higher tar concentrations therefore allowed the reduction of the height of the reactor and reduced the necessity for costly catalysts [444–448].

Simple schematics of direct and indirect gasifications are depicted in Figure 3.5. Currently, indirect gasification systems operate near atmospheric pressure. Direct gasification systems have been demonstrated at both elevated and atmospheric pressures. Any one of the gasifier systems can be utilized in the larger system diagrammed earlier [444–453].

There are several practical implications of each gasifier type. Due to the diluent effect of the nitrogen in air, fuel gas from a direct gasifier is of low heating value (5.6–7.5 MJ/Nm³). This low heat content in turn requires an increased fuel flow to the gas turbine. Consequently, in order to maintain the total (fuel + air) mass flow through the turbine within design limits, an air bleed is usually taken from the gas turbine compressor and used in the gasifier. This bleed air is either boosted slightly in pressure or expanded to near atmospheric pressure depending on the operating pressure of the direct gasifier.

Since the fuel-producing reactions in an indirect gasifier take place in a separate vessel, the resulting fuel gas is free of nitrogen diluent and is of medium heating value (13–18.7 MJ/Nm³). This heat content is sufficiently close to that of natural gas (approx. 38 MJ/Nm³) that fuel gas from an indirect gasifier can be used in an unmodified gas turbine without air bleed.

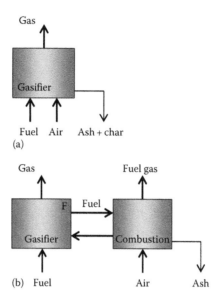

FIGURE 3.5 Direct versus indirect gasification. (a) Direct gasification and (b) indirect gasification. (From Vreugdenhil, B.J. et al., Indirect gasification: A new technology for a better use of Victorian Brown Coal, Report by ECN-M-14-012, ECN, Petten, the Netherlands, March 2014; Van der Drift, A. et al., Indirect versus direct gasification, ECN report ECN-L-13-063, ECN, Petten, the Netherlands, September 2013; Bhattacharya, S.P., *Process Saf. Environ. Prot.*, 84(6), 453, 2006; Campisi, A. and Woskoboenko, F., *Brown Coal R&D Scoping Study*, HRL Pty Ltd., Mulgrave, Australia, 2009; Vreugdenhil, B.J. et al., Indirect gasification: A new technology for a better use of victorian brown coal, ECN-M–14-02 report, ECN, Petten, the Netherlands, March, 2014.)

Other examples of indirect gasification processes, where heat is transported by inert material, are the fast internal circulating FB process first developed by the Vienna University of Technology [454,455] and the SilvaGas process based on the Battelle development [449–453] described below.

3.14.1 BATTELLE INDIRECTLY HEATED GASIFIER

The Battelle gasification process is an indirectly heated, atmospheric pressure, circulating FB system designed to gasify wood, biomass, and potentially MSW by reacting it with steam in a fluidized sand bed. Hot medium-Btu gas leaves the gasifier with the sand and a small amount of char. The sand is captured and recycled to the gasification vessel, while the char is combusted in a FB combustor that provides heat to reheat the sand, generate steam, and dry any wet feedstock. Because heat used during the endothermic gasification process is supplied externally from the sand circulating between the gasification vessel and the char combustor, the resulting fuel gas is free of nitrogen and has a heating value of 13–18 mJ/m³, using wood as the feedstock.

Battelle has experimented with a RDF feedstock in their Process Research Unit (PRU) that has throughputs between 0.22 and 9.1 Mg/day of dry RDF. The PRU has logged over 10,000 operating hours with a variety of feedstocks. The longest continuous operation with RDF was approximately 100 h at 9.1 metric tons per day. It was concluded from these studies that higher throughputs exceeding 19.5 Mg/h m² could be accomplished.

The Battelle gasifier [449–453] generates a product gas with a heating value greater than the TPS system because its indirect heating system prevents nitrogen from entering the gasification vessel. As a result, the volume of gas generated by the Battelle process is also significantly lower than TPS system. The addition of steam results in higher concentrations of CO and CH₄ formation and lower concentrations of heavy organic gases and carbon dioxide in the Battelle process. Thus, the addition of steam does encourage better gas quality; however, this benefit must be judged against the energy penalty associated with water's heat of vaporization (44 kJ/mol).

3.15 GLOBAL GASIFICATION EXPERIENCE

There is a large experience with coal gasification covering several decades: town gas was manufactured from coal as early as 1792; the first process to produce methanol from syngas was installed in 1913 (BASF); an improved high-temperature FB gasifier was patented in 1921 by Winkler; during World War II, Germany produced large amounts of synthetic fuels from coal. As mentioned before, according to the Gasification Technologies Council, in 2007 there were 144 gasification plants and 427 gasifiers in operation worldwide, adding up to an equivalent thermal capacity of some 56 GWth. Coal gasification accounted for approximately 31 GWth, with the remaining gasification plants running on petroleum, gas, petcoke, biomass, and waste feedstock [456,457].

A large part of the world's coal-based syngas is produced in 97 gasifiers of Sasol's plants in South Africa: in 2008, an estimated conversion capacity of 14 GWth enabled the conversion of some 43 million tons/year of coal into 74 million tons/year of transport fuels and chemicals (South Africa Synthetic Oil Process (Sasol), 2008). Most of the remaining coal-based syngas produced in other regions of the world is used for ammonia or methanol production and in China for the production of town gas. China has become the global test case for large-scale coal conversion activities. In 2008, China held licenses from Shell for the installation of 18 coal gasification plants; among these, 11 commercial size coal gasification plants were already in operation, most of them for the industrial production of methanol or ammonia.

While syngas is the primary product of the gasification plants, marketable products obtained from syngas include chemicals (45%), FT liquid fuels (28%), gaseous fuels (8%), and electric power (19%). Among other products, gaseous fuels include SNG. In the IGCC power plant of the Great Plains in the United States, the syngas is used to produce SNG [456,457].

While coal gasification is a commercial technology, research aims to further increase product yields, reduce consumption of catalysts and energy, and lower capital and operation costs. In IGCC plants with CCS, reducing the energy input to produce oxygen represents an essential research area.

The world's largest gasification manufacturing facility is the Gasifier and Equipment Manufacturing Corporation in the Philippines. They produce about 3000 units/year ranging in size from 10 to 250 kW. Besides, they have recently started producing gasifiers for direct heat applications. Their primary applications have been for irrigation pumps and power generating sets. To date, about 1000 units have been installed within the Philippines running on charcoal, wood chips, and briquettes. Brazil [458] is another country where large-scale gasification manufacturing programs have been undertaken, wherein about 650 units of various sizes and applications have been installed. In both the Brazilian and Philippines program, the gasifiers are mostly charcoal powered. In this case, a strict quality control of the fuel has to be maintained. Thus, the companies involved in gasifier manufacturing also supply the quality fuel. Inadequate fuel quality is the biggest problem in running these gasifiers [459].

In Europe there are many manufacturers especially in Sweden, France, West Germany, and the Netherlands who are engaged in manufacturing gasification systems for stationery applications. Most of the markets for these European manufacturers have been in developing countries. The United States and North American manufacturing activities have been summarized by Goss [457,460]. In other countries in Asia and Africa, the work is being carried out in research institutions and few prototypes have been made and tested. Currently, Japan is not very active in coal gasification largely because Japan does not have significant coal of their own. It appears to be engaged in building or managing coal gasification facilities using Australian coal.

Most of the gasifiers (up to 100 kW range) being sold by different manufacturers show a leveling off price of \$380/kWe for plant prices and about \$150 kWe for basic gasifier prices. This leveling off comes at about 100 kW system. However, for small systems the prices are extremely high. Thus, a 10 kWe gasifier plant costs about \$840/kWe, while the basic gasifier is \$350/kWe. To this must be added the transportation costs (especially for shipment to developing countries). These prices therefore can make the gasifiers uneconomic. This explains the big gasifier manufacturing push being given in countries like the Philippines and Brazil [461].

One example of a large-scale entrained-flow gasifier is the 600 MWth coal-fired Shell gasifier in Buggenum, the Netherlands [21,73,235]. It is owned by the utility company Nuon and produces electricity with a net efficiency of 43%. As mentioned earlier, tests have been performed with different kinds of biomass like wood, sewage sludge, and chicken manure up to approximately 10% on energy basis (corresponding to 18% on weight basis). It had planned to co-fire 25% biomass on energy basis in 2005 [21,73,235]. Shell has signed several contracts to make similar coal gasifiers for fertilizer industries in China. Another example is a 130 MWth gasifier (25 bar) made by Future Energy, which is operating on waste oil and sludges on the premises of the Schwarze Pumpe in Germany [234].

REFERENCES

1. Gasification, Wikipedia, the free encyclopedia (2015), https://en.wikipedia.org/wiki/gastification, Accessed July 2, 2016.
2. Bell, D., Towler, B., and Fan, M., *Coal Gasification and Its Applications*. Elsevier, Amsterdam, Netherlands (December 8, 2010).
3. Basu, P., *Biomass Gasification, Pyrolysis and Torrefaction: Practical Design and Theory*, 2nd edn. Academic Press, New York (July 18, 2013).
4. Rezalyan, J. and Chermisinoff, N., *Gasification Technologies: A Primer for Engineers and Scientists*. CRC Press, New York (April 8, 2005).
5. Higman, C. and van der Burgt, M., *Gasification*, 2nd edn. Gulf Publishing Co., Houston, TX (April 30, 2011).
6. Nikrityuk, P. and Meyer, B., *Gasification Processes: Modelling and Simulation*. John Wiley & Sons, New York (June 18, 2014).
7. Simbeck, D.R. and Dickenson, R.L., The EPRI coal gasification guidebook—A review of the recent update, in *Proceedings of 1996 Gasification Technologies Conference*, Electric Power Research Institute, Inc., Palo Alto, CA (October 1996).
8. Philcox, J.E. and Fenner, G.W., Gasification—An attraction for chemical reactions, in *Proceedings of 1996 Gasification Technologies Conference*, Electric Power Research Institute, San Francisco, CA (October 2–4, 1996).
9. Parikka, M., Global biomass fuel resources, *Biomass and Bioenergy*, 27, 613–620 (2004).
10. Parikka, M., *Biosims—A Model for Calculation of Woody Biomass in Sweden*, Vol. 27. Swedish University of Agricultural Sciences, Silvestria, Uppsala, Sweden (1997).
11. Hakkila, P., *Utilization of Residual Forest Biomass*. Springer, Berlin, Germany (1989).
12. Thrän, D. and Kaltschmitt, M., Biomass for a sustainable energy provision systems—State of technology, potentials and environmental aspects, in Sayigh, A. (ed.), *Workshop Proceedings, World Renewable Energy Congress*, June 29–July 5, 2002, Cologne, Germany (2002).
13. Perlack, R.D., Wright, L.L., Turhollow, A.F., Graham, R.L., Stokes, B.J., and Erbach, D.C., Biomass as feedstock for a bioenergy and bioproducts industry: The technical feasibility of a billion-ton annual supply, Prepared by Oak Ridge National Laboratory, managed by UT-Battelle, LLC for the U.S. Department of Energy under contract DE-AC05-00OR22725, DOE/GO-102005-2135, ORNL/TM-2005/66, USDA, Washington, DC (April 2005).
14. Kaltschmitt, M., Utilization of biomass in the German energy sector, in Hake, J.-F., Bansal, N., and Kleemann, M. (eds.), *Strategies and Technologies for Greenhouse Gas Mitigation*. Ashgate, Aldershot, U.K. (1999).
15. Hoogwijk, M., Faaija, A., Eickhoutb, B., de Vriesb, B., and Turkenburga, W., Potential of biomass energy out to 2100, for four IPCC SRES land-use scenarios, *Biomass and Bioenergy*, 29, 225–257 (2005).
16. Fischer, G. and Schrattenholzer, L., Global bioenergy potentials through 2050, *Biomass and Bioenergy*, 20, 151–159 (2001).

17. Kaltschmitt, M. and Dinkelbach, L., Biomass for energy in Europe—Status and prospects, in Kaltschmitt, M. and Bridgwater, A.V. (eds.), *Biomass Gasification and Pyrolysis—State of the Art and Future Prospects*. CPL Scientific, Newbury, U.K. (1997).

18. Wu, C.Z., Huang, H., Zheng, S.P., and Yin, X.L., An economic analysis of biomass gasification and power generation in China, Review paper, *Bioresource Technology*, 83, 65–70 (2002).

19. Lee, S., Gasification of coal, Chapter 2, in Lee, S., Speight, J., and Loyalka, S. (eds.), *Handbook of Alternative Fuel Technology*. Taylor & Francis Group, New York, pp. 26–78 (2007).

20. World Energy Outlook by International Energy Agency (IEA), Brussels, Belgium, 1998 edn. (1998), www.iea.org.

21. Shah, Y., *Energy and Fuel Systems Integration*. CRC Press, Taylor & Francis Group, New York (May 2014).

22. Shah, Y.T., *Water for Energy and Fuel Production*. CRC Press, New York (2014).

23. Lee, S. and Shah, Y.T., *Biofuels and Bioenergy: Technologies and Processes*. CRC Press, Taylor & Francis Group, New York (September 2012).

24. Bridgwater, A.V., Renewable fuels and chemicals by thermal processing of biomass, *Chemical Engineering Journal*, 91, 87–102 (2003).

25. Kayhanian, M., Tchobanoglous, G., and Brown, R.C., Biomass conversion processes for energy recovery, in Kreith, F. and Goswami, D.Y. (eds.), *Handbook of Energy Conservation and Renewable Energy*. Taylor & Francis Group, pp. 25-1–25-68 (2007).

26. Shah, Y.T., *Reaction Engineering in Direct Coal Liquefaction*. Addison Wesley Publishing Co., Reading, MA (1981).

27. Chang, R. and Smith, R., Coal gasification technology in China, PEP review 2012-03, SRI Consulting IHS Inc., Menlo Park, CA (February 2012).

28. Samuel, N., Anyigor, C.M., and Ogah, S.P.I., Coal gasification A road-map for power generation: A review, *American Journal of Scientific and Industrial Research*, 4 (5), 420–428 (2013).

29. Environmental footprints and costs of coal-based integrated gasification combined cycle and pulverized coal technologies, Final report prepared for the U.S. EPA (July 2006), http://www.epa.gov/airmarkets/articles/control.html, Retrieved March 24, 2010.

30. Higman, C., *Reliability of IGGC Power Plants Gasification Technologies Conference*, San Francisco, CA (2005).

31. Khan, M.R. and Kurata, T., The feasibilities of mild gasification of coal: Research needs, DOE/METC-85/4019, NTRS/DE85013625, Washington, DC, p. 73 (1985).

32. Penner, S.S. (ed.), Coal gasification: Direct applications and synthesis of chemicals and fuels, Elsevier, Amsterdam, the Netherlands, p. 296 (October 1987).

33. Ratafia-Brown, J. et al., Major Environmental aspects of gasification based power generation technology, Prepared for DOE/NETL, Washington, DC (2002).

34. Coal gasification, Wikipedia, the free encyclopedia (2015), https://en.wikipedia.org/wiki/coal-gasification, Last accessed October 17, 2016.

35. Walker, P.L., Rusinko, F., and Austin, L.G., Gas reactions in carbon, in Eley, D.D., Selwood, P.W., and Weisz, P.B. (eds.), *Advances in Catalysis*, Vol. XI. Academic Press, New York, pp. 133–221 (1959).

36. Kyotani, T., Kubota, K., and Tomita, A., *Fuel Processing Technology*, 36, 209–217 (1993).

37. Kann, J., Raukus, A., and Sirde, A., About the gasification of kukersite oil shale, *Oil Shale*, 30 (25), 283–293 (2013).

38. Raukas, A., Estonian state development plan of oil shale utilization, *Oil Shale*, 25 (1), 1–3 (2008).

39. Raukas, A. and Siirde, A., New trends in Estonian oil shale industry, *Oil Shale*, 29 (3), 203–205 (2012).

40. Raukas, A. and Kann, J., Estonian know-how by gasifying of kukersite oil shale (Eesti oskusteave kukersiitpõlevkivi gaasistamisel), in *Estonian Combustible Natural Resources and Wastes*, pp. 9–11 (2011) (in Estonian, summary in English).

41. Cottingham, P. and Carpenter, H., Catalytic gasification of shale oil, Fuel gasification, *Advances in Chemistry Series*, 69, 180–189 (1967).

42. Feldman, H. and Huebler, J., Hydrogasification of oil shale in a continuous flow reactor, *ACS Preprint*, 90–106 (1966).

43. Yang, Q., Qian, Y., Zhou, H., and Yang, S., Development of a coupling oil shale retorting process of gas and solid heat carrier technologies, *Energy Fuels*, 29 (9), 6155–6163 (2015).

44. Qian, Y., Yang, Q., Zhang, J., Zhou, H., and Yang, S., Development of an integrated oil shale refinery process with coal gasification for hydrogen production, *Industrial & Engineering Chemistry Research*, 53 (51), 19970–19978 (2014).

45. Zhunko, V.I., *Combustion Gases from Baltic Oil Shale (Goryuchie gazy iz Pribaltijskikh slantsev)*. Publishing House, Gostoptekhizdat, Leningrad, Russia, 63pp. (1948) (in Russian).

46. Qiuan, J. and Yin, L. (Editors in Chief), *Oil Shale—Petroleum Alternative.* China Petrochemical Press, Beijing, China, 619pp. (2010).

47. Kumar, A., Jones, D.D., and Hanna, M.A., Thermochemical biomass gasification: A review of the current status of the technology, *Energies*, 2, 556–581 (2009).

48. Rajvanshi, A., Biomass gasification, Chapter 4, in Yogi Goswami, D. (ed.), *Alternative Energy in Agriculture*, Vol. II. CRC Press, New York, pp. 83–102 (1986).

49. Worley, M. and Yale, J., Biomass gasification technology assessment consolidated report, NREL technical monitor: Abhijit Dutta, subcontract report NREL/SR-5100-57085, contract no. DE-AC36-08G028308, NREL, Golden, CO (November 2012).

50. Taylor, R., Howes, J., and Bauen, A., Review of technologies for gasification of biomass and wastes, Final report, NNFCC, E4 Tech, London, U.K. (June, 2009).

51. A report on Renewable gases supply infrastructure, Final Draft. Talent with Energy, TWE Pty Ltd., Sydney, New South Wales, Australia (March 2013), Available at: http://www.sydney2030.com.au/wpcontent/uploads/RENEWABLE-ENERGY-MASTER-PLAN-TECHNICAL-APPENDIX-2-RENEWABLE-GASES-SUPPLY-INFRASTRUCTURE.pdf.

52. Lasa, H., Salaices, E., Mazumder, J., and Lucky, R., Catalytic steam gasification of biomass: Catalysts, thermodynamics and kinetics, *Chemical Reviews*, 111, 5404–5433 (2011).

53. Raveendran, K., Ganesh, A., and Khilar, K.C., Influence of mineral matter on biomass pyrolysis characteristics, *Fuel*, 74, 1812–1822 (1995).

54. Clarke, S.J., Thermal biomass gasification, *Agricultural Engineering*, 62, 14–15 (1981).

55. Dasappa, S., Paul, P.J., Mukunda, H.S., Rajan, N.K.S., Sridhar, G., and Sridhar, H.V., Biomass gasification technology—A route to meet energy needs, *Current Science*, 87, 908–916 (2004).

56. Subramanian, P., Kirubakaran, V., Sivaramakrishnan, V., Nalini, R., Sekar, T., and Premalatha, M., A review on gasification of biomass, *Renewable and Sustainable Energy Reviews*, 13, 179–186 (2009).

57. Knoef, H., Practical aspects of biomass gasification, Chapter 3, in Knoef, H. (ed)., *Handbook Biomass Gasification.* BTG-Biomass Technology Group (BTG), Enschede, the Netherlands (2005).

58. Reed, T. and Gaur, S., *A Survey of Biomass Gasification*, 2nd edn. U.S. Department of Energy, National Renewable Energy Laboratory and the Biomass Energy Foundation, Golden, CO, p. 180 (2001).

59. Bain, R., *Biomass Gasification Overview.* NREL, Golden, CO (January 28, 2004).

60. IEA State of art of biomass gasification, prepared by European Concerted action, analysis and coordination of the activities concerning gasification of biomass AIR3-CT94-2284 and IEA bioenergy, biomass utilization, task XIII, thermal gasification of biomass activity, Sweden and Canada country reports , IEA, Paris, France (1997).

61. Fagbemi, L., Khezami, L., and Capart, R., Pyrolysis products from different biomasses: Application to the thermal cracking of tar, *Applied Energy*, 69, 293–306 (2001).

62. Han, J. and Kim, H., The reduction and control technology of tar during biomass gasification/pyrolysis: An overview, *Renewable and Sustainable Energy Reviews*, 12, 397–416 (2008).

63. Zhang, R., Brown, R.C., Suby, A., and Cummer, K., Catalytic destruction of tars in biomass derived producer gas, *Energy Conversion and Management*, 45, 995–1014 (2004).

64. Kinoshita, C.M., Wang, Y., and Zhou, J., Tar formation under different biomass gasification conditions, *Journal of Analytical and Applied Pyrolysis*, 29, 169–181 (1994).

65. Olivares, A., Aznar, M.P., Caballero, M.A., Gil, J., Frances, E., and Corella, J., Biomass gasification: Produced gas upgrading by in-bed use of dolomite, *Industrial & Engineering Chemistry Research*, 36, 5220–5226 (1997).

66. Devi, L., Ptasinski, K.J., and Janssen, F.J.J.G., A review of the primary measures for tar elimination in biomass gasification process, *Biomass and Bioenergy*, 24, 125–140 (2003).

67. Arena, U., Process and technological aspects of municipal solid waste gasification. A review, *Waste Management*, 32, 625–639 (2012).

68. Malkow, T., Novel and innovative pyrolysis and gasification technologies for energy efficient and environmentally sound MSW disposal, *Waste Management*, 24, 53–79 (2004).

69. Psomopoulos, C.S., Bourka, A., and Themelis, N.J., Waste-to-energy: A review of the status and benefits in USA, *Waste Management*, 29, 1718–1724 (2009).

70. Review of technologies for gasification of biomass and wastes, Final report, NNFCC project 09/008, a project funded by DECC, project managed by NNFCC and conducted by E4Tech, London, U.K. (June 2009).

71. Yang, W., Ponzio, A., Lucas, C., and Blasiak, W., Performance analysis of a fixed-bed biomass gasifier using high-temperature air, *Fuel Processing Technology*, 87, 235–245 (2006).

72. Song, T., Wu, J., Shen, L., and Xiao, J., Experimental investigation on hydrogen production from biomass gasification in interconnected fluidized beds, *Biomass and Bioenergy*, 36, 258–267 (2012).

73. Ratafia-Brown, J., Haslbeck, J., Skone, T., and Rutkowski, M., Assessment of technologies for co-converting coal and biomass to a clean syngas—Task 2 report (RDS), DOE/NETL-403.01.08 Activity 2 Report, DOE, Washington, DC (May 10, 2007).

74. Shah, Y. and Gardner, T., Biomass torrefaction: Applications in renewable energy and fuels, in Lee, S. (ed.), *Encyclopedia of Chemical Processing*. Taylor & Francis Group, New York, pp. 1–18, published online (April 24, 2012).

75. Williams, R.H., Larson, E.D., and Haiming, J., Synthetic fuels in a world with high oil and carbon prices, in *Proceedings of the Eighth International Conference on Greenhouse Gas Control Technologies (GHGT-8)*, Trondheim, Norway (June 19–22, 2006).

76. Antal, M.J., Biomass pyrolysis: A review of the literature. Part II: Lignocellulose pyrolysis, in Boer, K.W. and Duffie, J.A. (eds.), *Advances in Solar Energy*, Vol. 2. American Solar Energy Society, Boulder, CO, pp. 175–255 (1985).

77. Callis, H.P.A., Haan, H., Boerrigter, H., Van der Drift, A., Peppink, G., Van den Broek, R., Faaij, A., and Venderbosch, R.H., Preliminary techno-economic analysis of large-scale synthesis gas manufacturing from imported biomass, in *Proceedings of an Expert Meeting on Pyrolysis and Gasification of Biomass and Waste*, Strasbourg, France, pp. 403–417 (2003).

78. Kavalov, B. and Peteves, S.D., Status and perspectives of biomass-to liquid fuels in the European Union, Report no. EUR 21745 EN, European Commission, Joint Research Centre, ECN, Petten, the Netherlands (January 2005).

79. Rickets, B., Hotchkiss, R., Livingston, B., and Hall, M., Technology status review of waste/biomass co-gasification with coal, in IChemE (ed.), *Fifth European Gasification Conference*, Noordwijk, the Netherlands, p. 13 (April 2002).

80. Hotchkiss, R., Livingston, W., and Hall, M., Waste/biomass co-gasification with coal? Report no. COAL R216, DTI/Pub URN 02/867 (2002); Prickett, B., Hotchkiss, R., Livingston, B., and Hall, M. in Technology status review of waste/biomass co-gasification with coal, in IchemE (ed.) *Fifth European Gasification Conference,* Noordwijk, the Netherlands (April 8–10, 2012).

81. Valero, S. and Uson, S., *Oxy-Co-Gasification of Coal and Biomass in an Integrated Gasification Combined Cycle (IGCC) Power Plant*. Center for Research of Energy Resources and Consumptions (CIRCE), University of Zaragoza, Zaragoza, Spain (April 2005).

82. Shah, Y.T., Biomass to liquid fuel via Fischer–Tropsch and related syntheses, Chapter 12, in Lee, J.W. (ed.), *Advanced Biofuels and Bioproducts*. Springer Book Project, Springer Publ. Co., New York, pp. 185–207 (September 2012).

83. Tchapda, A. and Pisupati, S., A review of thermal co-conversion of coal and biomass/waste, *Energies*, 7, 1098–1148 (2014).

84. Wang, L. and Chen, Z., Hydrogen rich gas production by co-gasification of coal and biomass in an intermittent fluidized bed, *Scientific World Journal*, 276823 (2013), doi: 101155/2013/276823, published online September 15, 2013; PMC3791810 (Courtesy of Hindawi Publishing Corporation).

85. McLendon, T., Lui, A., Pineault, R., Beer, S., and Richardson, S., High pressure cogasification of coal and biomass in a fluidized bed, *Biomass and Bioenergy*, 26 (4), 377–388 (April 2004).

86. Yasuda, H., Yamada, O., Kaiho, M., and Nakagome, H., Effect of polyethylene addition to coal on hydrogasification enhancement, *Journal of Material Cycles and Waste Management*, 16 (1), 151–155 (February 2014).

87. Yasuda, H., Yamada, O., Zhang, A., Nakano, K., and Kaho, M., Hydrogasification of coal and polyethylene mixture, *Fuel*, 83 (17–18), 2251–2254 (December 2004).

88. Fermoso, J., Arias, B., Plaza, M., Pevida, C., Rubiera, F., Pis, J., Peria, G., and Casero, P., High pressure co-gasification of coal with biomass and petroleum coke, *Fuel Processing Technology*, 90 (7–8), 926–932 (July–August 2009).

89. Kumambe, K., Hanaoka, T., Fujimoto, S., Minowa, T., and Sukanishi, K., Co-gasification of woody biomass and coal with air and steam, *Fuel*, 86, 684–689 (2007).

90. Pinto, F., Franco, C., Andre, R., Tavares, C., Dias, M., Gulyurtlu, I., and Cabrita, I., Effect of experimental conditions on co-gasification of coal, biomass and plastic wastes with air/steam mixtures in a fluidized bed system, *Fuel*, 82 (15–17), 1967–1976 (October–December 2003).

91. De Ruyck, J., Delattin, F., and Bram, S., Co-utilization of biomass and natural gas in combined cycles through primary steam reforming of the natural gas, *Energy*, 32, 371–377 (2007).

92. Yamada, T., Akano, M., Hashimoto, H., Suzuki, T., Maruyama, T., Wang, Q., and Kamide, M., Steam gasification of coal–biomass briquettes, *Nippon Enerugi Gakkai Sekitan Kagaku Kaigi Happyo Ronbunshu*, 39, 185–186 (2002).
93. Wu, J., Fang, Y., and Wang, Y., Production of syngas by methane and coal co-conversion in fluidized bed reactor. Institute of Coal Chemistry, Chinese Academy of Science, Taiyuan, China, personal communication (2013).
94. Seo, M., Goo, J., Kim, S., Lee, S., and Choi, Y., Gasification characteristics of coal/biomass blend in a dual circulating fluidized bed reactor, *Energy Fuels*, 24 (5), 3108–3118 (2010).
95. Brar, J., Singh, K., Wang, J., and Kumar, S., Co-gasification of coal and biomass: A review, *International Journal of Forestry Research*, 2012, 1–10 (2012).
96. Khosravi, M. and Khadse, A., Gasification of petcoke and coal/biomass blend: A review, *International Journal of Emerging Technology and Advanced Engineering*, 3 (12), 167–173 (December 2013).
97. Priyadarsan, S., Holtzapple, M., Annamalai, K., and Mukhtar, S., Co-gasification of coal and chicken litters, in *17th National Heat and Mass Transfer Conference and 6th ASME Conference*, College Station, TX (2011).
98. Sun, H., Song, B., Jang, Y., and Kim, S., The characteristics of steam gasification of biomass and waste filter carbon, Report from Department of Chemical Engineering, Kunsan National University, Gunsan, South Korea (2012).
99. Ji, K., Song, B., Kim, Y., Kim, B., Yang, W., Choi, Y., and Kim, S., Steam gasification of low rank fuel-biomass, coal, and sludge mixture in a small scale fluidized bed, in *Proceedings of the European Combustion Meeting*, Vienna, Austria, pp. 1–5 (April 14–17, 2009).
100. Kurkela, E., Stahlberg, P., Simell, P., and Leppalahti, J., Updraft gasification of peat and biosolid, *Biosolid*, 19, 37–46 (1989).
101. Pan, Y., Velo, E., Roca, X., Manya, J., and Puigianer, L., Fluidized bed co-gasification of residual biomass/poor coal blends for fuel gas production, *Fuel*, 79, 1317–1326 (2000).
102. Chmielniak, T. and Sciazko, M., Co-gasification of biomass and coal for methanol synthesis, *Applied Energy*, 74, 393–403 (2003).
103. Zhao, L., Liang, J., and Qian, L., Coal and oil shale model test, underground cogasification, *Applied Mechanics and Materials*, 295–298, 3129 (2013).
104. Williams, R., Larson, E., and Jin, H., Comparing climate-change mitigating potentials of alternative synthetic liquid fuel technologies using biomass and coal, in *Fifth Annual Conference on Carbon Capture and Sequestration*, DOE/NETL, Pittsburgh, PA (May 8–11, 2006).
105. Nemanova, V., Abedini, A., Liliedahl, T., and Engvall, K., Co-gasification of petroleum coke and biomass, *Fuel*, 117, 870–875 (2014).
106. Nakano, K. and Kaho, M., Hydrogasification of coal and polyethylene mixture, *Fuel*, 83 (17–18), 2251–2254 (December 2004).
107. Lazar, M., Jasminska, N., and Lengyelova, M., Experiment of gasification of the synthetically mixed sample of waste in nitrogen atmosphere, Report by Mechanical Engineering Department, Technical University of Kostice, Kostice, Slovak Republic (2012).
108. Li, M., Jie, L., and Lu, X., Model test study of underground co-gasification of coal and oil shale, *Applied Mechanics and Materials*, 295–298, 3129–3136 (2013).
109. Indrawati, V., Manaf, A., and Purwadi, G., Partial Replacement of non renewable fossil fuels energy by the use of waste materials as alternative fuels, CP1169, in Handoko, L. and Siregar, M. (eds.), *International Workshop on Advanced Material for New and Renewable Energy*. American Institute of Physics, 978-0-7354-0706-0/07, pp. 179–184 (2009).
110. McDaniel, J., Tampa electric polk power station integrated gasification combined cycle project. Final Technical report under DOE contract No. DE-FC-21-91MC27363, Department of Energy, Washington, DC (August 2002).
111. Krupp, U., *A report on Gasification Technology: Shell Gasification Process*. Gasification Brochure, ThyssenKrupp Technologies, Essen, Germany (1999).
112. Henrich, E., Dahmen, N., and Dinjus, E., Cost estimate for biosynfuel production via biosyncrude gasification, *Biofuels, Bioproducts and Biorefining*, 3, 28–41 (2009).
113. Hakkou, M., Petrissans, M., Geradin, P., and Zoulalian, A, Investigations of the reasons for fungal durability of heat-treated beech wood, *Polymer Degradation and Stability*, 91 (2), 393–397 (2006).
114. Kamdem, D.P., Pizzi, A., and Jerrannaud, A., Durability of heat-treated wood, *Holz Roh Werkstoff.*, 60 (1), 1–6 (2002).
115. Weiland, J.J. and Guyonnet, R., Study of chemical modifications and fungi degradation of thermally modified wood using DRIFT spectroscopy, *Holz ala Roh und Werkstoff* (*European Journal of Wood and Wood Products*), 61 (3), 216–220 (2003).

116. White, R.H. and Dietenberger, M.A., Wood products: Thermal degradation and fire, in Buschow, K.H.J., Cahn, R.W., Flemings, M.C., Ilschner, B., Kramer, E.J., and Mahajan, S. (eds.), *The Encyclopedia of Materials: Science and Technology.* Elsevier, Amsterdam, the Netherlands, pp. 9712–9716 (2001).

117. Maciejewska, A., Veringa, H., Sanders, J., and Peteves, S., *Co-firing of Biomass with Coal: Constraints and Role of Biomass Pre-Treatment.* DGJRC Institute for Energy, European Commission (EUR 22461 EN) ECN report Petten, the Netherlands (2006), http://ie.jrc.cec.eu.int/publications/scientificpublications/2006/EUR22461EN.pdf.

118. Björkman, E. and Strömberg, B., Release of chlorine from biomass at pyrolysis and gasification conditions, *Energy Fuels*, 11, 1026–1032 (1997).

119. Akers, D. and Dospoy, R., Use of coal cleaning to reduce air toxics, in *Reprints Society for Mining, Metallurgy, and Exploration Annual Meeting*, Phoenix, AZ (no. 92-113), 19pp. (1992).

120. Bethell, P. and Luttrell, G., Effects of ultrafine deslimimg on coal flotation circuits, in *Proceedings Century of Flotation Symposium*, Brisbane, Queensland, Australia, p. 43 (2005).

121. Hardings, B., *Coal Preparation.* Society of Mining, Metallurgy and Exploration, Technology and Engineering, Amazon (1991).

122. Leonard, J. and Leonard, J., *Coal Preparation*, FastShip-CustomerFocus, Fulfilled by Amazon. Society of Mining, Metallurgy and Exploration Inc., Littleton, CO (1991).

123. Arnold, B.J., Klima, M.S., and Bethell, P.J., *Designing the Coal Preparation Plant of the Future.* SME (Society of Mining, Metallurgy and Exploration), Littleton, CO (2007).

124. Acharjee, T.C., Coronella, C.J., and Vasquez, V.R., Effect of thermal pretreatment on equilibrium moisture content of lignocellulosic biomass, *Bioresource Technology*, 102, 4849–4854 (2011).

125. Chiang, K.-Y., Chien, K.-L., and Lu, C.-H., Characterization and comparison of biomass produced from various sources: Suggestions for selection of pretreatment technologies in biomass-to-energy, *Applied Energy*, 100, 164–171 (2012).

126. Agbor, V.B., Cicek, N., Sparling, R., Berlin, A., and Levin, D.B., Biomass pretreatment: Fundamentals toward application. *Biotechnology Advances*, 29, 675–685 (2011).

127. Prins, M.J., Ptasinski, K.J., and Janssen, F.J.J.G., Torrefaction of wood: Part 1. Weight loss kinetics, *Journal of Analytical and Applied Pyrolysis*, 77 (1), 28–34 (2006).

128. Shafizadeh, F., Pyrolytic reactions and products of biomass, in Overend, R.P., Mime, T.A., and Mudge, L.K. (eds.), *Fundamentals of Biomass Thermochemical Conversion.* Elsevier, London, U.K., pp. 183–217 (1985).

129. Shafizadeh, F., Thermal conversion of cellulosic materials to fuels and chemicals, in *Wood and Agricultural Residues.* Academic Press, New York, pp. 183–217 (1983).

130. Bergman, P.C.A. and Kiel, J.H.A., Torrefaction for biomass upgrading, in *Proceedings of the 14th European Biomass Conference and Exhibition*, Paris, France (October 2005).

131. Proceedings of 14th European Conference on Biomass for Energy, Industry and Climate Protection, Paris, France (October 17–21, 2005) organized by *ETA—Renewable Energies*, Florence, Italy (2005).

132. Bergman, P.C.A., Boersma, A.R., Kiel, J.H.A, Prins, M.J., Ptasinski, K.J., and Janssen, F.J.J.G., Torrefaction for entrained flow gasification of biomass, Report no. ECN-RX-04-046; in *Proceedings of the Second World Biomass Conference on Biomass for Energy, Industry and Climate Protection*, Rome, Italy (May 10–14, 2004).

133. Van Swaaij, W.P.M., Fjällstrom, T., Helm, P.T., and Grassi, P. (eds.), Energy Research Centre of the Netherlands (ECN), Petten, the Netherlands, pp. 679–682 (2004).

134. Bergman, P.C.A., Combined torrefaction and pelletization: The TOP process, Report no. ECN-C-05-073, Energy Research Centre of the Netherlands (ECN), Petten, the Netherlands (2005).

135. Bergman, P.C.A., Boersma, A.R., Kiel, J.H.A., Prins, M.J., Ptasinski, K.J., and Janssen, F.J.J.G., Torrefaction for entrained flow gasification of biomass, Report no. ECN-C-05-067, Energy Research Centre of the Netherlands (ECN), Petten, the Netherlands (2004).

136. Prins, M.J., Ptasinski, I.G., and Janssen, F.J.J.G., More efficient biomass gasification via torrefaction, in Rivero, R., Monroy, L., Pulido, R., and Tsatsaronis, G. (eds.), *Proceedings of the 17th Conference on Efficiency, Costs, Optimization, Simulation and Environmental Impact of Energy Systems (ECOS'04)*, Guanajuato, Mexico, July 7–9, 2004. Elsevier, Amsterdam, the Netherlands, pp. 3458–3470 (2004).

137. Prins, M.J., Thermodynamic analysis of biomass gasification and torrefaction, PhD thesis, Technische Universiteit Emdhoven, Eindhoven, the Netherlands (2005).

138. Williams, P.T. and Besler, S., The influence of temperature and heating rate on the slow pyrolysis of biomass, *Renewable Energy*, 7 (3), 233–250 (1996).

139. Arcate, J.R., Torrefied wood, an enhanced wood fuel, in *Bioenergy*, Boise, ID, Paper # 207 (September 22–26, 2002).
140. Arias, B., Pevida, C., Fermoso, J., Plaza, M.G., Reubiern, F., and Pis, J.J., Influence of torrefaction on the grindability and reactivity of wood biomass, *Fuel Processing Technology*, 89 (2), 169–175 (2008).
141. Bourgois, J. and Guyonnet, R., Characterization and analysis of torrefied wood, *Wood Science and Technology*, 22 (2), 143–155 (1988).
142. Duijn, C., Torrefied wood uit resthout en andere biomassastromen, in *Proceedings of Praktijkdag Grootschalige Bioenergie Projecten*, SenterNovem (June 2004).
143. Li, J. and Gifford, J., *Evaluation of Woody Biomass Torrefaction*. Forest Research, Rotorua, New Zealand (September 2001).
144. Pach, M., Zanzi, R., and Bjømbom, E., Torrefied biomass a substitute for wood and charcoal, in *Proceedings of the Sixth Asia-Pacific International Symposium on Combustion and Energy Utilization*, Kuala Lumpur, Malaysia (May 20–22, 2002).
145. Zwart, R.W.R., Boerrigter, H., and Van der Drift, A., The impact of biomass pre-treatment on the feasibility of overseas biomass conversion to Fischer–Tropsch products, *Energy Fuels*, 20 (5), 2192–2197 (August 29, 2006).
146. Brooking, E., *Improving Energy Density in Biomass through Torrefaction*. A report by National Renewable Energy Laboratory, Golden, CO (2002), http://www.nrel.gov/education/pdfs/e_ brooking.pdf.
147. Reed, T.B. and Bryant, B., *Densified Biomass: A New Form of Solid Fuel*. U.S. Department of Energy, National Renewable Energy Laboratory, Golden, CO, p. 35 (1978).
148. Pentanunt, R., Mizanur Rahman, A.N.M., and Bhattacharya, S.C., Updating of biomass by means of torrefaction, *Energy*, 15 (12), 1175–1179 (1990).
149. Kirk, J.T.O., *Light and Photosynthesis in Aquatic Ecosystems*, 2nd edn. Cambridge University Press, Cambridge, U.K. (1994).
150. Raven, P.H., Evert, R.F., and Eichhorn, S.E., *Biology of Plants*, 6th edn. W.H. Freeman/Worth Publishers, New York (1999).
151. Weststeyn, A. and Essent Energie, B.V, First torrefied wood successfully co-fired, PyNe (biomass pyrolysis network), *Newsletter*, Issue 17 (April 2004).
152. Shafizadeh, F. and McGinnis, G.D., Chemical composition and thermal analysis of cottonwood, *Carbohydrate Research*, 16 (2), 273–277 (1971).
153. Alén, R., Kotilainen, R., and Zaman, A., Thermochemical behavior of Norway spruce (*Picea abies*) at 180–225°C, *Wood Science and Technology*, 36 (2), 163–171 (2002).
154. Alves, S.S. and Figueiredo, J.L., A model for pyrolysis of wet wood, *Chemical Engineering Science*, 44 (12), 2861–2869 (1989).
155. Di Blasi, C. and Lanzetta, M., Intrinsic kinetics of isothermal xylan degradation in inert atmosphere, *Journal of Analytical and Applied Pyrolysis Species*, 40 (41), 287–303 (1997). `
156. Varhegyi, G., Antal, M.J., Jakab, E., and Szabó, P., Kinetic modeling of biomass pyrolysis, *Journal of Analytical and Applied Pyrolysis*, 42 (1), 73–87 (1997).
157. Yang, H., Yan, R., Chen, H., Lee, D.H., and Zheng, C., Characteristics of hemicelluloses, cellulose and lignin pyrolysis, *Fuel*, 86 (12–13), 1781–1788 (2007).
158. Bradbury, A.G.W., Sakai, Y., and Shafizadeh, F.J., A kinetic model for pyrolysis of cellulose, *Journal of Applied Polymer Science*, 23 (11), 3271–3280 (1979).
159. Branca, C. and Di Blasi, C., Kinetics of the isothermal degradation of wood in the temperature range 528–707 K, *Journal of Analytical and Applied Pyrolysis*, 67 (2), 207–219 (2003).
160. Bridgeman, T.G., Jones, J.A., Shield, I., and Williams, P.T., Torrefaction of reed canary grass, wheat straw and willow to enhance solid fuel qualities and combustion properties, *Fuel*, 87 (6), 844–856 (2008).
161. Broido, A., Kinetics of solid phase cellulose pyrolysis, in Shafizadeh, F., Sarkanen, K., and Tillman, D.A. (eds.), *Thermal Uses and Properties of Carbohydrates and Lignins*. Academic Press, New York, pp. 19–36 (1976).
162. Hakkou, M., Pétrissans, M., El Bakali, I., Gérardin, P., and Zoualian, A., Wettability changes and mass loss during heat treatment of wood, *Holzforschung*, 59 (1), 35–37 (2005).
163. Hakkou, M., Pétrissans, M., Gérardin, P., and Zoualian, A., Investigation of wood wettability changes during heat treatment on the basis of chemical analysis, *Polymer Degradation and Stability*, 89 (1), 1–5 (2005).
164. Hakkou, M., Petrissans, M., Geradin, P., and Zoualian, A., Investigations of the reasons for fungal durability of heat-treated beech wood, *Polymer Degradation and Stability*, 91 (2), 393–397 (2006).
165. Voufo Panos, C.A., Maschio, G., and Lucehesi, A., Kinetic modeling of the pyrolysis of biomass and biomass components, *Canadian Journal of Chemical Engineering*, 67 (1), 75–84 (1989).

166. Bioenergy, a new process for torrefied wood manufacturing, *General Bioenergy*, 2 (4), 1–3 (2000).
167. Pétrissans, M., Gérardin, P., El Bakali, I., and Serraj, M., Wettability of heat-treated wood, *Holzforschung*, 57 (3), 301–307 (2003).
168. Bourgois, J.P. and Doat, J., Torrefied wood from temperate an tropical species, in Egnéus, H. and Ellegård, A. (eds.), *Advantages and Prospects*, Vol. III. Elsevier, London, U.K.; *Bioenergy*, 84, 153–159 (1984).
169. Jannasch, R., Quan, Y., and Samson, R., A process and energy analysis of pelletizing switchgrass—Final report, Resource Efficient Agricultural Production (REAP-Canada) for Natural Resources, Canada, pp. 1–16 (2001).
170. Hustad, J. and Barrio, M., What is biomass, in *Combustion Handbook*, International Flame Research Foundation (IFRF), File No. 23, Version No. 2 (October 17, 2000). http://www.handbook.ifrf.net/handbook/cf.html?id=2.
171. Bergman, P., Boersma, A., Zwart, R., and Kiel, J., *Torrefaction for Biomass Co-Firing in Existing Coal-Fired Power Stations* (*BIOCOAL*). ECN-C-05-013, ECN, Petten, the Netherlands, 72pp. (2005).
172. Hey, W., Pelz, W., Romey, I., and Sowa, F., Combined combustion of biomass/sewage sludge and coals of high and low rank in different systems of semitechnical scale, Final report: European Commission APAS Clean Coal Technology Program on Co-Utilization of Coal, Biomass and Waste, APAS contract COAL-CT92-0002, Vol. II: Final reports, Lisbon, Portugal (1995).
173. Opportunities and markets for co-utilization of biomass and waste with fossil fuels for power generation, Report from European Union, London, U.K. (2000).
174. Turnbull, J.H., Co-combustion of solid biomass, in Paper presented at *NOVEM/IA Symposium Co-Combustion of Biomass*, Nijmege, the Netherlands (November 8, 1995).
175. Boylan, D.M., Southern company tests of wood/coal co-firing in pulverized coal units, *Biomass and Bioenergy*, 10 (2–3), 139 (1996).
176. Gast, C.H. and Visser, J.G., The co-combustion of waste wood with coal. Effects on emissions and fly-ash composition. Final report: European Commission APAS Clean Coal Technology Program on Co-Utilization of Coal, Biomass and Waste, APAS contract COAL-CT92-0002, Vol. II: Final reports, Lisbon, Portugal (1995).
177. Abbas, T., Costen, P., Glaser, K., Hassan, S., Lockwood, F., and Ou, J.J., Combine combustion of biomass, municipal sewage sludge and coal in a pulverized fuel plant, Final report: European Commission APAS Clean Coal Technology Program on Co-Utilization of Coal, Biomass and Waste, APAS contract COAL-CT92-0002, Vol. II: Final reports, Lisbon, Portugal (1995).
178. Abbas, T., Costen, P., Kandamby, N.H., Lockwood, F.C., and Ou, J.J., The influence of burner injection mode on pulverized coal and bio-solid co-fired flames, *Combustion and Flame*, 99, 617–625 (1994).
179. McGowin, C.R. and Hughes, E.E., Efficient and economical energy recovery from waste by co-firing with coal, in Kahn, M.R. (ed.), ACS Symposium Series No. 515: *Clean Energy from Waste and Coal*, 1992. Power/Internet (1996).
180. McGowin, C.R. and Wiltsee, G.A., Strategic analysis of biomass and waste fuels for electric power generation, *Biomass and Bioenergy*, 10 (2–3), 167 (1996).
181. Lutge, C. et al., New applications for High Temperature Winkler Gasification (HTW) for thermal treatment of municipal solid waste and sewage sludge in combination with Catalytic Extraction Process (CEP), in *Gasification Technologies Conference* (October 1996).
182. Sundermann, B., Rubach, T., and Rensch, H.P., Feasibility study on the co-combustion of coal/biomass/sewage sludge and municipal solid waste for plants with 5 and 20 MW thermal power using fluidized bed technology, Final report: European Commission APAS Clean Coal Technology Program on Co-Utilization of Coal, Biomass and Waste, APAS contract COAL-CT92-0002, Vol. II: Final reports, Lisbon, Portugal (1995).
183. Tillman, D.A., Rossi, A.J., and Kitto, W.D., *Wood Combustion: Principles, Processes and Economics*. Academic Press, New York (1981).
184. Probst, H.H. and Wehland, P., Combined combustion of biomass/sewage sludge and coals, Final report: European Commission APAS Clean Coal Technology Program on Co-Utilization of Coal, Biomass and Waste, EC-research project, APAS contract COAL-CT92-0002, Vol. II: Final reports, Lisbon, Portugal (1995).
185. Morgan, D.J. and van de Kamp, W.L., The co-firing of pulverized bituminous coals with biomass and municipal sewage sludge for application to the power generation industry, Final report: European Commission APAS Clean Coal Technology Program on Co-Utilization of Coal, Biomass and Waste, APAS contract COAL-CT92-0002, Vol. II: Final reports, Lisbon, Portugal (1995).

186. Meschgbiz, A. and Krumbeck, M., Combined combustion of biomass and brown coal in a pulverized fuel and fluidised bed combustion plant, Final report: European Commission APAS Clean Coal Technology Program on Co-Utilization of Coal, Biomass and Waste, APAS contract COAL-CT92-0002, Vol. II: Final reports, Lisbon, Portugal (1995).

187. Andries, J., Vegelin, R.J., and Verloop C.M., Co-combustion of biomass and coal in a pressurized fluidized bed combustor, Final report: European Commission APAS Clean Coal Technology Program on Co-Utilization of Coal, Biomass and Waste, APAS contract COAL-CT92-0002, Final reports, Lisbon, Portugal (1995).

188. Bauer, F., Combined combustion of biomass/sewage sludge and coals of high and low rank in different systems of semi-industrial and industrial scale, Final report: European Commission APAS Clean Coal Technology Program on Co-Utilization of Coal, Biomass and Waste, APAS contract COAL-CT92-0002, Vol. II: Final reports, Lisbon, Portugal (1995).

189. Sampson, G.R., Richmond, A.P., Brewster, G.A., and Gasbarro, A.F., Co-firing of wood chips with coal in interior Alaska, *Forest Products Journal*, 41 (5), 53–56 (1991).

190. Fixed (moving) bed gasifiers-Lurgi dry-ash gasifiers, Report from NETL, Department of Energy, Washington, DC (2015). www.netl.doe.gov/research/coal/energy-systems/gasification/.../lurgi.

191. Fixed (moving) bed gasifiers-British gas/lurgi gasifier, Report from NETL, Department of Energy, Washington, DC (2015), www.netl.doe.gov/research/coal/energy-systems/gasification/.../bgl.

192. Zainal, Z.A., Rifau, A., Quadir, G.A., and Seetharamu, K.N., Experimental investigation of a downdraft biomass gasifier, *Biomass and Bioenergy*, 23, 283–289 (2002).

193. Bhavanam, A. and Sastry, R., Biomass gasification processes in downdraft fixed bed reactors: A review, *International Journal of Chemical Engineering and Applications*, 2 (6), 425–433 (December 2011).

194. Yang, Y.B., Sharifi, V.N., and Swithenbank, J., Effect of air flow rate and fuel moisture on the burning behaviours of biomass and simulated municipal solid wastes in packed beds, *Fuel*, 83, 1553–1562 (2004).

195. Martinez, J.D., Lora, E.E.S., Andrade, R.V., and Jaen, R.L., Experimental study on biomass gasification in a double air stage downdraft reactor, *Biomass and Bioenergy*, 35, 3465–3482 (2011).

196. Dogru, M., Howarth, C.R., Akay, G., Keskinler, B., and Malik, A.A., Gasification of hazelnut shells in a downdraft gasifier, *Energy*, 27, 415–427 (2002).

197. Jayah, T.H., Aye, L., Fuller, R.J., and Stewart, D.F., Computer simulation of a downdraft wood gasifier for tea drying, *Biomass and Bioenergy*, 25, 459–469 (2003).

198. Sheth, P.N. and Babu, B.V., Experimental studies on producer gas generation from wood waste in a downdraft biomass gasifier, *Bioresource Technology*, 100, 3127–3133 (2009).

199. Olgun, H., Ozdogan, S., and Yinesor, G., Results with a bench scale downdraft biomass gasifier for agricultural and forestry residues, *Biomass and Bioenergy*, 35, 572–580 (2011).

200. Pathak, B.S., Patel, S.R., Bhave, A.G., Bhoi, P.R., Sharma, A.M., and Shah, N.P., Performance evaluation of an agricultural residue based modular throat type down draft gasifier for thermal application, *Biomass and Energy*, 32, 72–77 (2008).

201. Reed, T.B., Walt, R., Ellis, S., Das, A., and Deutche, S., Superficial velocity—The key to downdraft gasification, Presented at *Fourth Biomass Conference of the Americas*, Oakland, CA (September 1999).

202. Liinanki, L., Svenningsson, P.J., and Thessen, G., Gasification of agricultural residues in a downdraft gasifier, Presented at *the Second International Producer Gas Conference*, Bangdung, Indonesia (1985).

203. Sasidharan, P., Murali, K.P., and Sasidharan, K., Design and development of ceramic-based biomass gasifier—An R and D study from India, *Energy for Sustainable Development*, 2, 49–52 (1995).

204. Warren, T.J.B., Poulter, R., and Parfitt, R.I., Converting biomass to electricity on a farm-sized scale using downdraft gasification and a spark-ignition engine, *Bioresource Technology*, 52, 95–98 (1995).

205. Panwar, N.L., Rathore, N.S., and Kurchania, A.K., Experimental investigation of open core downdraft biomass gasifier for food processing industry, *Mitigation and Adaptation Strategies for Global Change*, 14, 547–556 (2009).

206. Jain, A.K., Sharma, S.K., and Singh, D., Designing and performance characteristics of a throat less paddy husk gasifier, *Journal of Agricultural Issues*, 5, 57–67 (2000).

207. Tiwari, G., Sarkar, B., and Ghosh, L., Design parameters for a rice husk throatless gasifier reactor, *Agricultural Engineering International: The CIGR Journal of Scientific Research and Development*, 8 (2006).

208. Son, Y., Yoon, S.J., Kim, Y.K., and Lee, J.-G., Gasification and power generation characteristics of woody biomass utilizing a downdraft gasifier, *Biomass and Bioenergy*, 35, 4215–4220 (2011).

209. Tinaut, F.V., Melgar, A., Pérez, J.F., and Horrillo, A., Effect of biomass particle size and air superficial velocity on the gasification process in a downdraft fixed bed gasifier. An experimental and modeling study, *Fuel Processing Technology*, 89, 1076–1089 (2008).

210. Sheth, P.N., Babu, B.V., and Ummadisingu, A., Experimental studies on gasification of pine wood shavings in a downdraft biomass gasifier, Presented at the *Annual Meeting of AIChE*, Salt Lake City, UT (November 7–12, 2010).

211. Pérez, J.F., Tinaut, F.V., Melgar, A., and Horrillo, A., Effect of biomass particle size and air superficial velocity on the gasification process in a downdraft fixed bed gasifier. An experimental and modelling study, *Fuel Processing Technology*, 89, 1076–1089 (2008).

212. Mae, K., Chaiwat, W., and Hasegawa, I., Examination of the low-temperature region in a downdraft gasifier for the pyrolysis product analysis of biomass air gasification, *Industrial & Engineering Chemistry Research*, 48, 8934–8943 (2009).

213. TERMOKOKS—Clean coal conversion into coke and energy, TERMOKOKS, Riga, Latvia (2016), www.termokoks.com/en/technology/.

214. Kaewluan, S. and Pipatmanomai, S., Gasification of high moisture rubber wood chip with rubber waste in a bubbling fluidized bed, *Fuel Processing Technology*, 92, 671–677 (2011).

215. Kurkela, E., Stahlberg, P., and Laatikainen, J., Pressurized fluidized bed gasification experiments with wood, peat and coal at VTT in 1991–1992, VTT publications no. 161, VTT Finland (1993), info@vtt.fi.

216. Narvaez, I., Orío, A., Aznar, M.P., and Corella, J., Biomass gasification with air in an atmospheric bubbling fluidized bed. Effect of six operational variables on the quality of the produced raw gas, *Industrial & Engineering Chemistry Research*, 35, 2110–2120 (July 1996).

217. Garcia-Ibanez, P., Cabanillas, A., and Sanchez, J.M., Gasification of leached orujillo (olive oil waste) in a pilot circulating fluidized bed reactor. Preliminary results, *Biomass and Bioenergy*, 27, 183–194 (August 2004).

218. Hanping, C., Bin, L., Haiping, Y., Guolai, Y., and Shihong, Z., Experimental investigation of biomass gasification in a fluidized bed reactor, *Energy Fuels*, 22, 3493–3498 (2008).

219. Gil, J., Corella, J., Aznar, M.P., and Caballero, M.P., Biomass gasification in atmospheric and bubbling fluidized bed: Effect of the type of gasifying agent on the product distribution, *Biomass and Bioenergy*, 17, 389–403 (1999).

220. Mathieu, P. and Dubuisson, R., Performance analysis of a biomass gasifier, *Energy Conversion and Management*, 43, 1291–1299 (2002).

221. Li, X.T., Grace, J.R., Lim, C.J., Watkinson, A.P., Chen, H.P., and Kim, J.R., Biomass gasification in a circulating fluidized bed, *Biomass and Bioenergy*, 26, 171–193 (2004).

222. Arena, V., Zaccariello, L., and Mastellone, M.L., Fluidized bed gasification of waste-derived fuels, *Waste Management*, 30 (7), 1212–1219 (July 2010).

223. Arena, U., Zaccariello, L., and Mastellone, M.L., Fluidized bed gasification of waste derived fuels, *Waste Management*, 30, 1212–1219 (2010).

224. Adlhoch, W., Sato, H., Wolff, J., and Radtke, K., High-temperature winkler gasification of municipal solid waste, in *2000 Gasification Technologies Conference*, San Francisco, CA (October 8–11, 2000).

225. Cieplik, M., Coda, B., Boerrigter, A., Drift, V., and Kiel, J., Characterization of slagging behavior of wood as upon entrained flow gasification conditions, Report C-04-016 also report RX-04-082, ECN, Petten, the Netherlands (July 2004).

226. Drift, A., Boerrigter, H., Coda, B., Cieplik, M., and Hemmes, K., Entrained flow gasification of biomass; ash behavior, feeding issues, system analyses, Report C-04-039, ECN, Petten, the Netherlands, 58pp. (April 2004).

227. Zhao, Y., Sun, S., Tian, H., Qian, J., Su, F., and Ling, F., Characteristics of rice husk gasification in an entrained flow reactor, *Bioresource Technology*, 100, 6040–6044 (2009).

228. Simbeck, D.R., Dickenson, R.L., and Oliver, E.D., Coal gasification systems: A guide to status, applications, and economics, AP-3109, Prepared by Synthetic Fuel Associates, Inc. for Electric Power Research Institute, Palo Alto, CA (1983).

229. Richard, L. and Fosgitt, P.E., Gasification using rotary kiln technology, Report for Cirque Energy, LLC, Ithaca, MI (June 4, 2010).

230. Hatzilyberis, K.S., Design of an indirect heat rotary kiln gasifier, *Fuel Processing Technology*, 92 (12), 2429–2454 (December 2011).

231. Rotary Kiln, Wikipedia, the free encyclopedia (2015), https://en.wikipedia.org/wiki/Rotary_kiln, Last accessed October 5, 2016.

232. Li, S., Ma, L., Wan, W., and Yao, Q., A mathematical model of heat transfer in a rotary kiln thermoreactor, *Chemical Engineering and Technology*, 28 (12), 1480–1489 (December 2005).

233. UOP Callidus Oxidizers for waste destruction, Report by UOP, subsidiary of Honeywell Inc., Morristown, NJ (2016), www.uop.com/wp.../uop-callidus-thermal-oxidizer-brochure.pdf.

234. Kamka, F., Jochmann, A., and Picard, L., Development status of BGL gasification, in *International Freiberg Conference on IGCC and XtL Technologies*, June 16–18, 2005, Freiberg, Germany (2005).

235. Hanneman, F., Schiffers, U., and Karg, J., Buggenum experience and improved concepts for syngas applications, in Presentation to *Gasification Technology*, Vol. 92 (2002).

236. Zhunko, V.I. and Zaglodin, L.S., Author's Certificate of the USSR, No. 42027 (Avtorskoje svidetelstvo no. 42027) (1935) (in Russian).

237. Rikk, E., Processing technology of oil shale in chamber furnaces from fire-resisting bricks (Põlevkivi ümbertöötamise tehnoloogia dinastellistest kamberahjudes), Candidate's (PhD) thesis, Tallinn University of Technology, Tallinn, Estonia, 256pp. (1966) (manuscript in Estonian, printed abstract in Russian).

238. Help, K., Investigation of oil shale processing in chamber furnace (Kamberahjus põlevkivi ümbertöötamise protsessi uurimine), Candidate's (PhD) thesis, Tallinn University of Technology, Tallinn, Estonia, 230pp. (1960) (manuscript in Estonian, printed abstract in Russian).

239. Valkenburg, C., Gerber, M.A., Walton, C.W., Jones, S.B., Thompson, B.L., and Stevens, D.J., Municipal solid waste (MSW) to liquid fuels synthesis, Vol. 1: Availability of Feedstock and Technology, Prepared for the U.S. Department of Energy under contract DE-AC05-76RL01830, PNNL-18144, Pacific Northwest, Richland, WA (December 2008).

240. Campbell, F., President, An overview of the history and capabilities of the thermoselect technology, Report by Interstate Waste Technologies, Malvern, PA (November 12, 2008), www.swananys.org/pdf/Thermoselect.pdf.

241. Fjällström, T., Combustion and gasification of biomass with TPS technology, in *International Conference on Bioenergy Utilisation and Environment Protection Technology*, Dalian, China (September 24–26, 2003), also published as Proceedings, pp. 49–51.

242. Morris, M. and Waldheim, L., Energy recovery from solid waste fuels using advanced gasification technology, *Waste Management*, 18 (6–8), 557–564 (October 1998).

243. Selinger, A. and Steiner, C., TwinRec fluidized bed gasification and ash melting—Review of four years of commercial plant operation, in *IT3'04 Conference*, Phoenix, AZ, pp. 1–12 (2004).

244. Selinger, A., Steiner, Ch., Shin, K., TwinRec—Bridging the gap of car recycling in Europe, Presented at *International Automobile Recycling Congress*, March 12–14, Geneva (CH), Switzerland (2003).

245. EBARA, TwinRec fluidized bed gasification and ash melting (2007), http://www. ebara.ch/, Accessed May 2011.

246. Bosmans, A., Vanderreydt, I., Geysen, D., and Helsena, L., The crucial role of waste-to-energy technologies in enhanced landfill mining: A technology review, *Journal of Cleaner Production*, 20, 1–14 (2012).

247. Wilson, B., Williams, N., Liss, B., and Brandon Wilson, P., A comparative assessment of commercial technologies for conversion of solid waste to energy, Prepared for Enviro Power Renewable, Inc., Boca Raton, FL (October 2013).

248. Fabry, F., Rehmet, C., Rohani, V.-J., and Fulcheri, L., Waste gasification by thermal plasma: A review, *Waste and Biomass Valorization*, 4 (3), 421–439 (2013).

249. Themelis, N.J., Developments in thermal treatment technologies, NAWTEC16-1927, in *Proceedings of NAWTEC16 16th Annual North American Waste-to-Energy Conference*, Philadelphia, PA (May 19–21, 2008).

250. Klein, A., Gasification: An alternative process for energy recovery and disposal of municipal solid wastes, MS thesis, Department of Earth and Environmental Engineering, Fu Foundation School of Engineering and Applied Science Columbia University, New York (May 2002).

251. Morris, M. and Waldheim, L., Energy recovery from solid waste fuels using advanced gasification technology. *Waste Management*, 18 (6–8), 557–564 (December 1998).

252. Gasification technologies, Report from ThyssenKrupp Industrial Solutions, Dortmund, Germany (2015), www.thyssenkrupp-industrial-solutions.com/.../gasification_technologies.pdf.

253. Help, K., Investigation of oil shale processing in chamber furnace (Kamberahjus põlevkivi ümbertöötamise protsessi uurimine), Candidate's (PhD) thesis, University of Technology, Tallinn, Estonia, 230pp. (1960) (manuscript in Estonian, printed abstract in Russian).

254. Underground coal gasification, Wikipedia, the free encyclopedia (2015), https://en.wikipedia.org/wiki/Underground_coal_gasification, Last accessed October 1, 2016.

255. Yang, L., Liu, S., Yu, L., and Zhang, W., Underground coal gasification field experiment in the high dipping coal seams, *Energy Sources, Part A*, 31, 854–862 (2009).

256. Shafirovich, E. and Varma, A., Underground coal gasification: A brief review of current status, *Industrial & Engineering Chemistry Research*, 48, 7865–7875 (2009).

257. Burton, E., Friedmann, J., and Upadhye, R., Best practices in underground coal gasification, Contract W-7405-Eng-48, Lawrence Livermore National Laboratory, Livermore, CA (2006).

258. Creedy, D.P., Garner, K., Holloway, S., Jones, N., and Ren, T.X., Review of underground coal gasification technological advancements; Report COAL R211; DTI/Pub URN 01/1041, Department of Trade and Industry Technology (DTI), London, U.K. (2001).

259. Sury, M., White, M., Kirton, J., Carr, P., Woodbridge, R., Mostade, M., Chappell, R., Hartwell, D., Hunt, D., and Rendell, N., Review of environmental issues of underground coal gasification; Report COAL R272; DTI/Pub URN 04/1880, Department of Trade and Industry Technology (DTI), London, U.K. (2004).

260. Sury, M., White, M., Kirton, J., Carr, P., Woodbridge, R., Mostade, M., Chappell, R., Hartwell, D., Hunt, D., and Rendell, N., Review of environmental issues of underground coal gasifications best practice guide; Report COAL R273; DTI/Pub URN 04/1881, Department of Trade and Industry Technology (DTI), London, U.K. (2004).

261. Lazarenko, S.N. and Kreinin, E.V., *Underground Coal Gasification in Kuzbass: Present and Future*. Institute of Coal, Siberian Branch of the Russian Academy of Sciences, Nauka, Novosibirsk, Russia (1994) (in Russian).

262. Gregg, D.W. and Edgar, T.F., Underground coal gasification, *AIChE Journal*, 24, 753–781 (1978).

263. Kreinin, E.V., Fedorov, N.A., Zvyagintsev, K.N., and Pyankova, T.M., *Underground Gasification of Coal Seams*. Nedra, Moscow, Russia (1982) (in Russian).

264. Kreinin, E.V. and Shifrin, E.I., Mathematical model of coal combustion and gasification in a passage of an underground gas generator, *Combustion, Explosion, and Shock Waves*, 29, 148–154 (1993).

265. Hill, R.W., Review of the CRIP process, in *Proceedings of the 12th Annual Underground Coal Gasification Symposium*, DOE, Washington, DC (report DOE/FE/60922-H1) (1986).

266. Britten, J.A. and Thorsness, C.B., A model for cavity growth and resource recovery during underground coal gasification, *In Situ*, 13, 1–53 (1989).

267. Morris, J., Vorobiev, O., Antoun, T., and Friedmann, S.J., Geomechanical simulations related to UCG activities, Presented at the *Twenty-Fifth Annual International Pittsburgh Coal Conference*, Pittsburgh, PA, Paper 32-3 (September 29–October 2, 2008).

268. Blinderman, M.S. and Friedmann, S.J., Underground coal gasification and carbon capture and storage: Technologies and synergies for low-cost, low-carbon syngas and secure storage, Report UCRL-ABS-218560, Lawrence Livermore National Laboratory, Livermore, CA (2006).

269. Gadelle, C., Lessi, J., and Sarda, J.P., Underground coal gasification at great depth. The French Field Test of Bruay-En-Artois, *Oil & Gas Science and Technology—Revue d'IFP Energies nouvelles*, 37, 157–181 (1982).

270. Van Batenburg, D.W., Biezen, E.N.J., and Bruining, J., A new channel model for underground gasification of thin, deep coal seams, *In Situ*, 18, 419–451 (1994).

271. Kuyper, R.A., van der Meer, T.H., and Bruining, J., Simulation of underground gasification of thin coal seams, *In Situ*, 20, 311–346 (1996).

272. Coeme, A., Pirard, J.P., and Mostade, M., Modeling of the chemical processes in a longwall face underground gasifier at great depth, *In Situ*, 17, 83–104 (1993).

273. Kreinin, E.V., Two-stage in-situ coal gasification, *Khim. Tverd. Topl.*, 6, 76–79 (1990) (in Russian).

274. Blinderman, M.S. and Klimenko, A.Y., Theory of reverse combustion linking, *Combustion and Flame*, 150, 232–245 (2007).

275. Blinderman, M.S., Saulov, D.N., and Klimenko, A.Y., Exergy optimisation of reverse combustion linking in underground coal gasification, *Journal of the Energy Institute*, 81, 7–13 (2008).

276. Blinderman, M.S., Saulov, D.N., and Klimenko, A.Y., Forward and reverse combustion linking in underground coal gasification, *Energy*, 33, 446–454 (2008).

277. Perkins, G. and Sahajwalla, V., A mathematical model for the chemical reaction of a semi-infinite block of coal in underground coal gasification, *Energy Fuels*, 19, 1679–1692 (2005).

278. Perkins, G. and Sahajwalla, V., A numerical study of the effects of operating conditions and coal properties on cavity growth in underground coal gasification, *Energy Fuels*, 20, 596–608 (2006).

279. Perkins, G. and Sahajwalla, V., Modelling of heat and mass transport phenomena and chemical reaction in underground coal gasification, *Chemical Engineering Research and Design*, 85 (A3), 329–343 (2007).

280. Aghalayam, P., Sateesh, D., Naidu, R., Mahajani, S., Ganesh, A., Sapru, R.K., and Sharma, R.K., Compartment modeling for underground coal gasification cavity, Presented at the *Twenty-Fifth Annual International Pittsburgh Coal Conference*, Pittsburgh, PA, Paper 32-4 (September 29–October 2, 2008).

281. Montegomery, S. and Morzenti, S., Underground coal gasification nears commercialization, *OGJ* (*Oil and Gas Journal*) *Newsletter/MarketWatch* (April 3, 2006).

282. Thompson, P., Gasifying coal underground, *Endeavour*, 2 (2), 93–97 (1978).

283. Sajjad, M. and Rasul, M.G., Review on the existing and developing underground coal gasification techniques in abandoned coal seam gas blocks: Australia and global context, in *First International e-Conference on Energies*, Brisbane, Queensland, Australia (March 14–31, 2014), http://sciforum.net/conference/ece-1.

284. Klimenko, A.Y., Early ideas in underground coal gasification and their evolution, *Energies*, 2, 456–476 (2009).

285. Walker, L., Underground coal gasification: A clean coal technology ready for development, *The Australian Coal Review*, 19–21 (1999).

286. Moran, P.C., Costa, P.J.D., and Cuff, E.P.C., Independent scientific panel report on underground coal gasification pilot trials, Queensland Independent Scientific Panel for Underground Coal Gasification (ISP), Brisbane, Queensland, Australia (2013).

287. Daggupati, S. et al., Laboratory studies on cavity growth and product gas composition in the context of underground coal gasification, *Energy*, 36, 1776–1784 (2011).

288. Bhutto, A.W., Bazmi, A., and Zahedi, G., Underground coal gasification: From fundamentals to application, *Progress in Energy and Combustion Science*, 39, 189–214 (2013).

289. Upadhye, R., Burton, E., and Friedmann, J., *Science and Technology Gaps in Underground Coal Gasification*. A report by U.S. Department of Energy, University of California, Lawrence Livermore National Laboratory, Livermore, CA (2006).

290. Yang, L., Numerical study on the underground coal gasification for inclined seams, *AIChE Journal*, 51 (11), 3059–3071 (November 2005), doi:10.1002/aic.10554.

291. Janoszek, T. et al., Modelling of gas flow in the underground coal gasification process and its interactions with the rock environment, *Journal of Sustainable Mining*, 12 (2), 8–20 (2013).

292. Biezen, E.N.J. and Molenaar, J., An integrated 3D model for underground coal gasification, in *Society of Petroleum Engineers Annual Technical Conference & Exhibition*, Dallas, TX (1995).

293. Luo, Y., Coertzen, M., and Dumble, S., Comparison of UCG cavity growth with CFD model predictions, in *Seventh International Conference on CFD in the Minerals and Process Industries*. CSIRO, Melbourne, Victoria, Australia (2009).

294. Self, S.J., Reddy, B.V., and Rosen, M.A., Review of underground coal gasification technologies and carbon capture, *International Journal of Energy and Environmental Engineering*, 3–16 (December, 2012), doi:10.1186/2251-6832-3-16.

295. Yang, L.H., A review of the factors influencing the physicochemical characteristics of underground coal gasification, *Energy Sources, Part A: Recovery, Utilization, and Environmental Effects* 30 (11), 1038–1049 (April 2008), doi:10.1080/15567030601082803.

296. Walker, L.K., The future role for underground coal gasification in Australia, in *Energy at the Crossroads*, Australian Institute of Energy National Conference, Melbourne, Victoria, Australia (2006).

297. Khan, Md.M., Mmbaga, J.P., Shirazi, A.S., Trivedi, J., Liu, Q., and Gupta, R., Modelling underground coal gasification—A review, *Energies*, 8, 12603–12668 (2015).

298. Su, F., Nakanowataru, T., Itakura, K., Ohga, K., and Deguchi, G., Evaluation of structural changes in the coal specimen heating process and UCG model experiments for developing efficient UCG systems, *Energies*, 6, 2386–2406 (2013).

299. GasTech, Inc., Viability of underground coal gasification in the "deep coals" of the Powder River Basin, Wyoming. Prepared for the Wyoming Business Council (2007), http://www.wyomingbusiness.org/program/ucg-viability-analysis-powder-river-/1169, Accessed September 19, 2011.

300. A report by GasTech Inc., Tulsa, OK, UCG in the "deep coals" of the Powder River Basin, Wyoming Progress 2007–2008. Prepared for Zeus Development Corp., Houston, TX (2008).

301. Yosim, S.J. and Nealy, C.L., Bench-scale optimization tests for destruction of PCBs using the molten salt process, Report for the Canadian Electrical Association, Research and Development, Montreal, Quebec, Canada; Contract 144D270 (November 1, 1982).

302. Johanson, J.G., Yosim, S.J., Kellogg, L.G., and Sudar, S., Elimination of hazardous wastes by the molten salt destruction process, in Paper presented to the American Society of Testing Materials Committee D-27, Philadelphia, PA (March 22–24, 1982).

303. Barclay, K.M., Gay, R.L., Newcomb, J.C., Yosim, S.J., Lorenzo, D.K., and Van Cleve, Jr., J.E., Disposal of simulated intermediate level radioactive waste by molten sail combustion, Report of Consolidated Fuel Processing Program, ORNIVCFRP-79/9 (April 1979).

304. Gay, R.L., Barclay, K.M., Grantham, L.F., and Yosim, S.J., Destruction of toxic wastes using molten salts, in Paper presented to the American Institute of Chemical Engineers, Anaheim, CA (April 21, 1981).

305. Kohl, A.L., Harty, R.B., Johanson, J., and Naphthali, L., The molten salt gasification process. *Chemical Engineering Progress*, 7 (1), 73 (1978).

306. Gay, R.L., Newcomb, J.C., Pard, A.G., Barclay, K.M., and Yosim, S.J., Development of a metal molten salt combustor for volume reduction of intermediate-level liquid waste, Report ORNL.sub-81/40461/1, by Rockwell International, Inc., Energy Systems Group, Canoga Park, CA (July 1981).

307. McKenzie, D.E., Grantham, L.F., and Paulson, R.B., Volume reduction of radioactive beta-gamma and TRU waste by molten salt combustion, in Paper presented at *ERDA/Georgia Tech Radwaste Management Symposium*, Atlanta, GA (May 1977).

308. Alden, N., Molten salt gasification of biomass, major qualifying project, WPI, Worcester, MA (April 29, 2009).

309. Jin, G., Iwaki, H., and Arai, N., Study on the gasification of wastepaper/carbon dioxide catalyzed by molten carbonate salts. *Energy* 30 (7), 1192–1203 (2005).

310. Eatwell-Hall, R.E.A., Sharifi, V.N., and Swithenbank, J., Hydrogen production from molten metal gasification, *International Journal of Hydrogen Energy*, 35 (24), 13168–13178 (2010).

311. Kohl, A., Rockwell International Corp., Black liquor gasification process, U.S. Patent US4773918 A, also published as CA1266355A, CA1266355A (September 27, 1988).

312. Cover, A., Schreiner, W., and Skaperdas, G., The Kellog coal gasification process, *ACS Preprint*, 23, 1–7 (1995).

313. LaRosa, P. and McGarvey, R.J., in *Clean Fuels from Coal Symposium Papers*, Institute of Gas Technology, Chicago, IL, pp. 285–300 (September 10–14, 1973).

314. Cover, A.E., Schreiner, W.C., and Skaperdas, G.T., Kellogg's coal gasification process, *Chemical Engineering Progress*, 69 (3), 31–36 (1973).

315. Probstein, R.F. and Hicks, R.E., *Synthetic Fuels*. McGraw-Hill, New York (1982).

316. Lloyd, W.G., Synfuels technology update, in Thumann, A. (ed.), *The Emerging Synthetic Fuel Industry*. Fairmont Press, Atlanta, GA, pp. 19–58 (1981).

317. Speight, J., *The Chemistry and Technology of Coal*. Marcel Dekker, New York, pp. 461–516 (1983).

318. Kohl, A.L., Slater, M.H., and Xiller, K.J., Status of the molten salt coal gasification process, in *Proceedings of the Tenth Synthetic Pipeline Gas Symposium*, Chicago, IL (October 30, 1978).

319. Plasma gasification, Wikipedia, the free encyclopedia (2015), https://en.wikipedia.org/wiki/plasma_gasification, Last accessed September 15, 2016.

320. Plasma gasification commercialization, Wikipedia, the free encyclopedia (2015), https://en.wikipedia.org/wiki/Plasma_gasification_commercialization, Last accessed September 15, 2016.

321. Woolcock, P.J. and Brown, R.C., A review of cleaning technologies for biomass-derived syngas, *Biomass and Bioenergy*, 52, 54–84 (2013).

322. Kiel, J.H.A. et al., Primary measures to reduce tar formation in fluidized-bed biomass gasifiers: Final report SDE project P1999-012, ECN BM, ECN-C-04-014, Petten, the Netherlands (2004).

323. Hasler, P., Bühler, R., and Nussbaumer, T., Evaluation of gas cleaning technologies for biomass gasification, in *Biomass for Energy and Industry: 10th European Conference and Technology Exhibition*, Würzburg, Germany (1998).

324. Pemen, A.J.M., Nair, S.A., Yan, K., van Heesch, E.J.M., Ptasinski, K.J., and Drinkenburg, A.A.H., Pulsed corona discharges for tar removal from biomass derived fuel gas, *Plasmas and Polymers*, 8 (3), 209–224 (2003).

325. Chang, J.-S., Recent development of plasma pollution control technology: A critical review, *Science and Technology of Advanced Materials*, 2 (3–4), 571–576 (2001).

326. Vandenbroucke, A.M., Morent, R., De Geyter, N., and Leys, C., Non-thermal plasmas for non-catalytic and catalytic VOC abatement, *Journal of Hazardous Materials*, 195, 30–54 (2011).

327. Nair, S.A., Corona plasma for tar removal, Technische Universiteit Eindhoven, Eindhoven, the Netherlands, MEK-ET-2005-05 (2005).

328. Bosmans, A., Wasan, S., and Helsen, L., Waste to clean syngas: Avoiding tar problems, in *Second International Enhanced Landfill Mining Symposium*, Houthalen-Helchteren, Belgium, pp. 1–21 (October 14–16, 2013).

329. Gomez, E., Amutha Rani, D., Cheeseman, C., Deegan, D., Wisec, M., and Boccaccini, A., Thermal plasma technology for the treatment of wastes: A critical review, *Journal of Hazardous Materials*, 161, 614–626 (2009).

330. Dodge, E., Plasma gasification: Clean renewable fuel through vaporization of waste, *Waste Management World*, 10 (4) (2009).

331. Katou, K., Asou, T., Kurauchi, Y., and Sameshima, R., Melting municipal solid waste incineration residue by plasma melting furnace with a graphite electrode, *Thin Solid Films*, 386, 183–188 (2001).

332. Tzeng, C.-C., Kuo, Y.-Y., Huang, T.-F., Lin, D.-L., and Yu, Y.-J., Treatment of radioactive wastes by plasma incineration and vitrification for final disposal, *Journal of Hazardous Materials*, 58, 207–220 (1998).

333. Krasovskaya, L.I. and Mossé, A.L., Use of electric-arc plasma for radioactive waste immobilization, *Journal of Engineering Thermophysics*, 70, 631–638 (1997).
334. Byun, Y., Cho, M., Chung, J.W., Namkung, W., Lee, H.D., Jang, S.D., Kim, Y.-S., Lee, J.H., Lee, C.R., and Hwang, S.M., Hydrogen recovery from the thermal plasma gasification of solid waste, *Journal of Hazardous Materials*, 190, 317–323 (2011).
335. Byun, Y., Namkung, W., Cho, M., Chung, J.W., Kim, Y.S., Lee, J.H., Lee, C.R., and Hwang, S.M., Demonstration of thermal plasma gasification/vitrification for municipal solid waste treatment, *Environmental Science & Technology*, 44, 6680–6684 (2010).
336. Minutillo, M., Perna, A., and Bona, D.D., Modelling and performance analysis of an integrated plasma gasification combined cycle (IPGCC) power plant, *Energy Conversion and Management*, 50, 2837–2842 (2009).
337. Heberlein, J. and Murphy, A.B., Thermal plasma waste treatment, *Journal of Physics D: Applied Physics*, 41, 053001 (2008).
338. Camacho, S.L., Industrial worthy plasma torches: State-of-the-art, *Pure and Applied Chemistry*, 60, 619–632 (1988).
339. Park, J.M., Kim, K.S., Hwang, T.H., and Hong, S.H., Three-dimensional modeling of arc root rotation by external magnetic field in non-transferred thermal plasma torches, *IEEE Transactions on Plasma Science*, 32, 479–487 (2004).
340. Hur, M. and Hong, S.H., Comparative analysis of turbulent effects on thermal plasma characteristics inside the plasma torches with rod- and well-type cathodes, *Journal of Physics D: Applied Physics*, 35, 1946–1954 (2002).
341. Zhang, Q., Dor, L., Fenigshtein, K., Yang, W., and Blasiak, W., Gasification of municipal solid waste in the plasma gasification melting process, *Applied Energy*, 90, 106–112 (2012).
342. Leal-Quirós, E., Plasma processing of municipal solid waste, *Brazilian Journal of Physics*, 34, 1587–1593 (2004).
343. Cheng, H. and Hu, Y., Municipal solid waste (MSW) as a renewable source of energy: Current and future practices in China, *Bioresource Technology*, 101, 3816–3824 (2010).
344. Nishikawa, H., Ibe, M., Tanaka, M., Ushio, M., Takemoto, T., Tanaka, K., Tanahashi, N., and Ito, T., A treatment of carbonaceous wastes using thermal plasma with steam, *Vacuum*, 73, 589–593 (2004).
345. Dodge, E., Plasma-gasification of waste. Clean production of renewable fuels through the vaporization of garbage. Cornell University-Johnson Graduate School of Management (2008); Knoef, Ir.H.A.M., BIG biomass gasification, Report from BTG Corporation, the Netherlands (April 2008).
346. Morrin, S., Lettieri, P., Mazzei, L., and Chapman, C., Assessment of fluid bed + plasma gasification for energy conversion from solid waste, in *Third International Symposium on Energy from Biomass and Waste*, Venice, Italy, November 8–11, 2010. CISA Publisher, Italy (2010).
347. Lemmens, B., Elslander, H., Vanderreydt, I., Peys, K., Diels, L., Osterlinck, M., and Joos, M., Assessment of plasma gasification of high caloric waste streams, *Waste Management*, 27, 1562–1569 (2007).
348. An'shakov, A.S., Faleev, V.A., Danilenko, A.A., Urbakh, E.K., and Urbakh, A.E., Investigation of plasma gasification of carbonaceous technogeneous wastes, *Thermophysics and Aeromechanics*, 14, 607–616 (2007).
349. Galeno, G., Minutillo, M., and Perna, A., From waste to electricity through integrated plasma gasification/fuel cell (IPGFC) system, *International Journal of Hydrogen Energy*, 36, 1692–1701 (2011).
350. Taylor, R., Chapman, C., and Faraz, A., Transformations of syngas derived from landfilled wastes using the Gasplasma® process, in Jones, P.T. and Geysen, D. (eds.), *Enhanced Landfill Mining Symposium*, Houthalen-Helchteren, Belgium (2013).
351. ADEME, Technical, environmental and economic assessment of plasco energy group gasification process using plasma torches, *Etude réalisée pour le compte de l'ADEME (French Environment and Energy Management Agency) par ENVALYS* (2009).
352. APP, Gasplasma Technology, *Advanced Plasma Power*, http://www.advancedplasmapower.com/, Accessed on June 2013.
353. Juniper Consultancy Services Ltd., Bisley, U.K., Nippon steel gasification-full process review by Independent waste technology report (2007).
354. Willis, K.P., Osada, S., and Willerton, K.L., Plasma gasification: Lessons learned at Eco-Valley WTE facility, in *Proceedings of the 18th Annual North American Waste-to-Energy Gas Emissions Model, Hyder Consulting Conference* (*NAWTEC'18*), Orlando, FL, pp. 1–8 (2010).
355. Feasby, D.M., Tsangaris, A.V., Carter, G.W., Campbell, K., Mills, W.J., Shen, J.Z., Guo, H., and Liu, F., Resorption Canada limited proprietary plasma gasification process chemical simulator, in *IT3'03 Conference*, Orlando, FL (May 12–16, 2003).

356. Juniper Rating Gasification Report, by Juniper Waste reports.com, Bisley, England (January 19, 2009), Juniper Consulting Services, Bisley, U.K., Available on www.juniper.co.uk.
357. Juniper.com, The alter NRG/Westinghouse plasma gasification process: Independent waste technology report in Juniper.com, Bisley, England, Westinghouse-Plasma Company (2008), Available: http://www.westinghouse-plasma.com, Accessed May 20, 2011.
358. Europlasma Company, Paris, France, Available: http://www.europlasma.com, Accessed May 17, 2011.
359. Phoenix Solutions Company, Plymouth, MI, Available: http://www.phoenixsolutionsco.com/psctorches.html, Accessed May 16, 2011.
360. Tetronics Company, Tetronics International Ltd., Waste Management Services in England, London, U.K., Available: http://www.tetronics.com, Accessed May 19, 2011.
361. Juniper.com, Plasma technologies for waste processing applications: Juniper ratings report, Bisley, England (2007).
362. Willis, K.P., Osada, S., and Willerton, K.L., Plasma gasification: Lessons learned at Ecovalley WTE Facility, in *Proceedings of the 18th Annual North American Waste-to-Energy Conference (NAWTEC'18)*, May 11–13, 2010, Orlando, FL (2010).
363. Spiewak, I., Tyner, C., and Langnickel, U., Applications of solar reforming technology, SANDIA report SAND93-1959, UC-237, Sandia National Laboratory, Albuquerque, NM (November 1993).
364. Muller, W.D., Solar reforming of methane utilizing solar heat, in Baker, M. (ed.), *Solar Thermal Utilization, German Studies on Technology and Applications*, Solar Thermal Energy for Chemical Processes, Vol. 3, Springer-Verlag, Berlin, Germany, pp. 1–179 (1987).
365. Becker, M. and Bohmer, M. (eds.), *Proceedings of the Final Presentation GAST, the Gas Cooled Solar Tower Technology Program*, Lahnstein, FRG, May 30–31, 1988, Springer-Verlag, Berlin, Germany (1989).
366. Piatkowski, N. and Steinfeld, A., Solar driven coal gasification in a thermally irradiated packed-bed reactor, *Energy and Fuels*, 22, 2043–2052 (2008).
367. Perkins, C.M., Woodruff, B., Andrews, L., Lichty, P., Lancaster, B., Bringham, C., and Weimer, A., Synthesis gas production by rapid solar thermal gasification of corn stover, Report of Midwest Research Institute, Kansas, MO, under contract no. DE-AC36-99GO10337, Department of Energy, Washington, DC (2009).
368. Piatkowski, N., Wieckert, C., and Steinfeld, A., Experimental investigation of a packed bed solar reactor for the steam gasification of carbonaceous feedstocks, *Fuel Processing Technology*, 90, 360–366 (2009).
369. Z'Graggen, A., Solar gasification of carbonaceous materials—Reactor design, modeling and experimentation, Doctorate thesis, Dissertation, ETH no. 17741, ETH, Zurich, Switzerland (2008).
370. Meier, A. and Sattler, C., Solar fuels from concentrated sunlight, Solar Paces, IEA Solar Paces implementing agreement, Platforma solar de Almería, Almería, Spain (August 2009).
371. Petrasch, J. and Steinfeld, A., Dynamics of a solar thermochemical reactor for steam reforming of methane, *Chemical Engineering Science*, 62, 4214–4228 (2007).
372. Moller, S., Solar reforming of natural gas, Report by Deutsches Zentrum, DLR fur Luft-und Raumfahrt e.V., Mgheb-Europ Project, Lyon, France (June 14, 2006).
373. Z'Graggen, A., Haueter, P., Trommer, D., Romero, M., De Jesus, J., and Steinfeld, A., Hydrogen production by steam-gasification of petroleum coke using concentrated solar power—II: Reactor design, testing, and modeling, *International Journal of Hydrogen Energy*, 31 (6), 797–811 (2006).
374. Gordillo, E. and Belghit, A., A bubbling fluidized bed solar reactor model of biomass char high temperature steam-only gasification, *Fuel Processing Technology*, 92 (3), 314–321 (2010).
375. Gordillo, E. and Belghit, A., A downdraft high temperature steam-only solar gasifier of biomass char: A modelling study, *Biomass and Bioenergy*, 35 (5), 2034–2043 (2011).
376. Hathaway, B.J., Davidson, J.H., and Kittelson, D.B., Solar gasification of biomass: Kinetics of pyrolysis and steam gasification in molten salt, *Journal of Solar Energy Engineering*, 133 (2), 021011 (2011).
377. Lichty, P., Perkins, C., Woodruff, B., Bingham, C., and Weimer, A., Rapid high temperature solar thermal biomass gasification in a prototype cavity reactor, *Journal of Solar Energy Engineering*, 132 (1), 011012 (2010).
378. Klein, H.H., Karni, J., and Rubin, R., Dry methane reforming without a metal catalyst in a directly irradiated solar particle reactor, *Journal of Solar Energy Engineering*, 131 (2), 021001 (2009).
379. Flechsenhar, M. and Sasse, C., Solar gasification of biomass using oil shale and coal as candidate materials, *Energy*, 20 (8), 803–810 (1995).
380. Weimer, A., Perkins, C., Mejic, D., Lichty, P., and inventors. WO Patent WO/2008/027,980, as signee. Rapid solar-thermal conversion of biomass to syngas, U.S. Patent WO2008027980, University of Colorado, Boulder, CO (June 3, 2008).

381. Zedtwitz, P. and Steinfeld, A., The solar thermal gasification of coal—Energy conversion efficiency and CO_2 mitigation potential, *Energy*, 28 (5), 441–456 (2003).

382. Sattler, C. and Raeder, C., SOLREF-solar steam reforming of methane rich gas for synthesis gas production, Final activity report, DLR, Cologne, Germany (June 2010).

383. Suarez-Gonzalez, M., Blanco-Marigorta, A., and Peria-Quintana, A., Review on hydrogen production technologies from solar energy, in *International Conference on Renewable Energies and Power Quality (ICREPQ'11)*, Los palmas de Gran Canaria, Spain (April 13–15, 2011).

384. Ogden, J.M., Review of small stationary reformers for hydrogen production, Report for IEA, Agreement on the production and utilization of hydrogen, Task 16, hydrogen from carbon containing materials, Report No. IEA/H2/TR-02/002, IEA, Brussels, Belgium (2002).

385. Padban, N. and Becher, V., Clean hydrogen rich synthesis gas. Literature and state of art review (Re: Methane Steam Reforming), Report no. CHRISGAS, WP11 D89, Under contract no. SES6_CT_2004_502387 (October 2005).

386. Asadullah, M., Barriers of commercial power generation using biomass gasification gas: A review, *Renewable and Sustainable Energy Reviews*, 29, 201–215 (2014).

387. Lv, P., Yuan, Z., Wu, C., Ma, L., Chen, Y., and Tsubaki, N., Bio-syngas production from biomass catalytic gasification, *Energy Conversion and Management*, 48, 1132–1139 (2007).

388. Delgado, J., Aznar, M.P., and Corella, J., Calcined dolomite, magnesite and calcite for cleaning hot gas from a fluidized bed biomass gasifier with steam: Life and usefulness, *Industrial & Engineering Chemistry Research*, 35 (10), 3637–3643 (1996).

389. Courson, C., Makaga, E., Petit, C., and Kiennemann, A., Development of Ni catalysts for gas production from biomass gasification reactivity in steam- and dry-reforming, *Catalysis Today*, 63, 427–437 (2000).

390. Li, J., Xiao, B., Yan, R., and Liu, J., Catalyst for biomass gasification, *BioResources*, 4 (4), 1520–1535 (2007).

391. Baker, E.G., Mudge, L.K., and Brown, M.D., Steam gasification of biomass with nickel secondary catalysts, *Industrial & Engineering Chemistry Research*, 26 (7), 1335–1339 (1987).

392. Caballero, M.A., Aznar, M.P., Gil, J., Martin, J.A., Frances, E., and Corella, J., Commercial steam reforming catalysts to improve biomass gasification with steam-oxygen mixtures. 1. Hot gas upgrading by the catalytic reactor, *Industrial & Engineering Chemistry Research*, 36 (12), 5227–5239 (1997).

393. Chang, F.W., Hsiao, T.J., and Shih, J.D., Hydrogenation of CO_2 over a rice husk ash supported nickel catalyst prepared by deposition-precipitation, *Industrial & Engineering Chemistry Research*, 37 (10), 3838–3845 (1998).

394. Devi, L., Ptasinski, K.J., and Janssen, F.J.J.G., Pretreated olivine as tar removal catalyst for biomass gasifiers: Investigation using naphthalene as model biomass tar, *Fuel Processing Technology*, 86 (6), 707–730 (2005).

395. Furusawa, T. and Tsutsumi, A., Comparison of Co/MgO and Ni/MgO catalysts for the steam reforming of naphthalene as a model compound of tar derived from biomass gasification, *Applied Catalysis A: General* 278 (2), 207–212 (2005).

396. Li, J., Yan, R., Xiao, B., Liang, D.T., and Lee, D.H., Preparation of nano-NiO particles and evaluation of their catalytic activity in pyrolyzing biomass components, *Energy Fuels*, 22 (1), 16–23 (2008).

397. Łamacz, A., Krztoń, A., and Djéga-Mariadassou, G., Steam reforming of model gasification tars compounds on nickel based ceria-zirconia catalysts, *Catalysis Today*, 176, 347–351 (2011).

398. Li, J., Yan, R., Xiao, B., Liang, T.D., and Du, L., Development of nano-NiO/Al_2O_3 catalyst to be used for tar removal in biomass gasification, *Environmental Science & Technology*, 42 (16), 6224–6229 (2008).

399. Lv, P., Chang, J., Wang, T., Fu, Y., and Chen, Y., Hydrogen-rich gas production from biomass catalytic gasification, *Energy & Fuels*, 18, 228–233 (2004).

400. Ma, L. and Baron, G.V., Mixed zirconia–alumina supports for Ni/MgO based catalytic filters for biomass fuel gas cleaning, *Powder Technology*, 180, 21–29 (2008).

401. Martinez, R., Romero, E., Garcia, L., and Bilbao, R., The effect of lanthanum on Ni–Al catalyst for catalytic steam gasification of pine sawdust, *Fuel Processing Technology*, 85 (2–3), 201–214 (2004).

402. Rapagna, S., Provendier, H., Petit, C., Kiennemann, A., and Foscolo, P.U., Development of catalysts suitable for hydrogen or syn-gas production from biomass gasification, *Biomass and Bioenergy*, 22 (5), 377–388 (2002).

403. Sutton, D., Kelleher, B., and Ross, J.R.H., Review of literature on catalysts for biomass gasification, *Fuel Processing Technology*, 73 (3), 155–173 (2001).

404. Swierczynski, D., Libs, S., Courson, C., and Kiennemann, A., Steam reforming of tar from a biomass gasification process over Ni/olivine catalyst using toluene as a model compound, *Applied Catalysis B: Environmental*, 74 (3–4), 211–222 (2007).
405. Wang, T., Chang, J., and Lv, P., Novel catalyst for cracking of biomass tar, *Energy & Fuels*, 19 (1), 22–27 (2005).
406. Huang, Y., Yin, X., Wu, C., Wang, C., Xie, J., Zhou, Z., Ma, L., and Li, H., Effects of metal catalysts on CO_2 gasification reactivity of biomass char, *Biotechnology Advances*, 27, 568–572 (2009).
407. Domazetics, G., Liesegang, J., and James, B.D., Studies of inorganics added to low-rank coal for catalytic gasification. *Fuel Processing Technology*, 86, 463–486 (2005).
408. Fu, P., Hu, S., Xiang, J., Sun, L.Y., and Zhang, J.Y., Biomass catalytic gasification kinetics using nonisothermal TGA, *Power Systems Engineering*, 23, 14–16 (2007).
409. Moulijn, J.A., Cenfontain, M.B., and Kapteijn, F., Mechanism of the potassium catalyzed gasification of carbon in CO_2, *Fuel*, 63, 1043–1047 (1984).
410. Zhou, J.H., Kuang, J.P., Zhou, Z.J., Lin, M., and Liu, J.Z., Research on alkali-catalysed CO_2-gasification of coal black liquor slurry char and coal water slurry char, *Proceedings of CSEE*, 26 (12), 149–155 (2006).
411. Juntgen, H., Application of catalysts to coal gasification processes, incentives and perspectives, *Fuel*, 62, 234–238 (1983).
412. Nishiyama, Y., Catalytic gasification of coals—Features and possibilities, *Fuel Processing Technology*, 29, 31–42 (1991).
413. Asami, K. and Ohtuska, Y., Highly active iron catalysts from ferric chloride for the steam gasification of brown coal, *Industrial & Engineering Chemistry Research*, 32, 1631–1636 (1993).
414. Yamashita, H., Yoshida, S., and Tomita, A., Local structures of metals dispersed on coal. 2. Ultrafine FeOOH as active iron species for steam gasification of brown coal, *Energy Fuels*, 5, 52–57 (1991).
415. Matsumoto, S., Catalyzed hydrogasification of yallourn char in the presence of supported hydrogenation nickel catalyst, *Energy Fuels*, 5, 60–63 (1991).
416. Srivastava, R.C., Srivastava, S.K., and Rao, S.K., *Fuel*, 67, 1205–1207 (1988).
417. Haga, T. and Nishiyama, Y., *Fuel*, 67, 748–752 (1988).
418. Ohtuska, Y. and Asami, K., Steam gasification of high sulfur coals with calcium hydroxide, in *Proceedings of the 1989 International Conference on Coal Science*, Vol. 1, pp. 353–356 (1989).
419. Salinas Martinez, C. and Lineras-Solano, A., *Fuel*, 69, 21–27 (1990).
420. Joly, J.P., Martinez-Alonso, A., and Marcilio, N.R., *Fuel*, 69, 878–894 (1990).
421. Muhlen, H.J., *Fuel Processing Technology*, 24, 291–297 (1990).
422. Levendis, Y.A., Nam, S.W., and Gravalas, G.R., Catalysis of the combustion of synthetic char particles by various forms of calcium additives, *Energy Fuels*, 3, 28–37 (1989).
423. Zheng, Z.G., Kyotani, T., and Tomita, A., *Energy Fuels*, 3, 566–571 (1989).
424. Muhlen, H.J., *Fuel Processing Technology*, 24, 291–297 (1986).
425. Haga, T., Sato, M., and Nishiyama, Y., Influence of structural parameters on coal char gasification: 2. Ni-catalyzed steam gasification, *Energy Fuels*, 5, 317–322 (1991).
426. Pareira, P., Somorajai, G.A., and Heinemann, H., Catalytic steam gasification of coals, *Energy Fuels*, 6, 407–410 (1992).
427. Xiang, R., You, W., and Shu-fen, L., *Fuel*, 66, 568–571 (1987).
428. Takarada, T., Nabatame, T., Ohtsuka, Y., and Tomita, A., Steam gasification of brown coal using sodium chloride and potassium chloride catalysts, *Industrial & Engineering Chemistry Research*, 28, 505–510 (1989).
429. Takarada, T., Ogirawa, M., and Kato, K., *Journal of Chemical Engineering of Japan*, 25, 44–48 (1992).
430. Weeda, M., Tromp, J.J., and Moulijn, J.A., *Fuel*, 69, 846–850 (1990).
431. Chin, G., Liu, G., and Dong, Q., *Fuel*, 66, 859–863 (1987).
432. Carrasco-Marin, F., Rivera, J., and Moreno, C., *Fuel*, 70, 13–16 (1991).
433. Lopez, A., Carrasco-Marin, F., and Moreno, C., *Fuel*, 71, 105–108 (1992).
434. Suzuki, T., Nakajima, S., and Watanabe, Y., Catalytic activity of rare earth compounds for the steam and carbon dioxide gasification of coal, *Energy Fuels*, 2, 848–853 (1988).
435. Murlidhara, H.S. and Seras, J.T., Effect of calcium on gasification, in *Coal Processing Technology*, Vol. IV, A CEP Technical Manual. AIChE, New York, pp. 22–25 (1978).
436. Gallagher, J.E. and Marshall, H.A., SNG from coal by catalytic gasification, in *Coal Processing Technology*, Vol. V, A CEP Technical Manual. AIChE, New York, pp. 199–204 (1979).
437. Asadullah, M., Technical challenges of utilizing biomass gasification gas for power generation: An overview, *Journal of Energy Technologies and Policy*, 3 (11), 137–143 (2013).

438. Asadullah, M., Tomishige, K., and Fujimoto, K., A novel catalytic process for cellulose gasification to synthesis gas, *Catalysis Communications*, 2, 63–68 (2001).
439. Asadullah, M., Fujimoto, K., and Tomishige, K., Catalytic performance of Rh/CeO$_2$ in the gasification of cellulose to synthesis gas at low temperature, *Industrial & Engineering Chemistry Research*, 40, 5894–5900 (2011).
440. Asadullah, M., Miyazawa, T., Ito, S., Kunimori, K., and Tomishige, K., Demonstration of real biomass gasification drastically promoted by effective catalyst, *Applied Catalysis A: General*, 246, 103–116 (2003).
441. Asadullah, M., Ito, S., Kunimori, K., Yamada, M., and Tomishige, K., Biomass gasification to hydrogen and syngas at low temperature: Novel catalytic system using fluidized-bed reactor, *Journal of Catalysis*, 208, 255–259 (2002).
442. Asadullah, M., Ito, S., Kunimori, K., Yamada, M., and Tomishige, K., Energy efficient production of hydrogen and syngas from biomass: Development of low-temperature catalytic process for cellulose gasification, *Environmental Science & Technology*, 36, 4476–4481 (2002).
443. Asadullah, M., Miyazawa, T., Ito, S., Kunimori, K., Koyama, S., and Tomishige, K., A comparison of Rh/CeO$_2$/SiO$_2$ catalysts with steam reforming catalysts, dolomite and inert materials as bed materials in low throughput fluidized bed gasification systems, *Biomass and Bioenergy*, 26, 269–279 (2004).
444. Vreugdenhil, B.J., van der Drift, A., van der Meijden, C.M., and Grootjes, A.J., Indirect gasification: A new technology for a better use of Victorian Brown Coal, Report by ECN-M-14-012, ECN, Petten, the Netherlands (March 2014).
445. Van der Drift, A., Aranda Almansa, G., Vreugdenhil, B., Visser, H., Mourao Vilela, C., and van der Meijden, C., Indirect versus direct gasification, ECN report ECN-L-13-063, ECN, Petten, the Netherlands (September 2013).
446. Bhattacharya, S.P., Gasification performance of Australian lignites in a pressurized fluidized bed gasifier process development unit under air and oxygen enriched air blown conditions, *Process Safety and Environmental Protection*, 84 (6), 453–460 (2006).
447. Campisi, A. and Woskoboenko, F., *Brown Coal R&D Scoping Study*. A report by HRL Pty Ltd., Victoria, Australia (2009).
448. Vreugdenhil, B.J., van der Drift, A., and van der Meijden, C.M., *Co-Gasification of Biomass and Lignite in the Indirect Gasifier Milena*. Curran Associates, Inc., Rad Hook, New York, (January 2009).
449. Spath, P., Aden, A., Eggeman, T., Ringer, M., Wallace, B., and Jechura, J., Biomass to hydrogen production detailed design and economics utilizing the Battelle Columbus laboratory indirectly-heated gasifier, Technical report, NREL/TP-510-57408, Contract no. DE-AC36-99-GO10337, NREL, Golden, CO (May 2005).
450. Kinchin, C.M. and Bain, R.L., Hydrogen production from biomass via indirect gasification: The impact of NREL process development unit gasifier correlations, Technical report NREL/TP-510-44868, Contract no. DE-AC36-08-GO28308, NREL, Golden, CO (May 2009).
451. Paisley, M.A. and Overend, R.P., A biomass gasification for power generation, in *EPRI 13th Conference on Gasification Power Plants*, Palo Alto, CA (1994).
452. Paisley, M.A., Farris, G., Slack, W., and Irving, J., A commercial development of the Battelle/FERCO biomass gasification process: Initial operation of the McNeil gasifier, in Overend, R.P. and Chornet, E. (eds.), *Making a Business from Biomass in Energy, Environment, Chemicals, Fibers and Materials. Proceedings of the Third Biomass Conference of the Americas*, Montreal, Quebec, Canada, August 24–29. Pergamon, Elsevier Science, Oxford, U.K., pp. 579–588 (1997).
453. Phillips, S., Aden, A., Jechura, J., Dayton, D., and Eggeman, T., Thermochemical ethanol via indirect gasification and mixed alcohol synthesis of lignocellulosic biomass, Technical report NREL/TP-510-41168, NREL, Golden, CO (April 2007).
454. Hofbauer, H. and Rauch, R., Stoichiometric water consumption of steam gasification by the FICFB-gasification process. Also published in book "Presented at the conference progress in thermochemical biomass conversion, Innsbruck, Austria (September 17–22, 2000)", in *Progress in Thermochemical Biomass Conversion*, Blackwell Science Ltd., U.K., pp. 199–208 (2001).
455. Larsson, A., Pallarès, D., Neves, D., Seemann, M., and Thunman, H., Zero-dimensional modeling of indirect fluidized bed gasification, in Kim, S., Kang, Y., Keun, J., and Seo, V. (eds.), *Refereed Proceedings of the 13th International Conference on Fluidization—New Paradigm in Fluidization Engineering*, Gyeongju, South Korea, Art. 112 (May 16–21, 2010), dc.engconfintl.org/cgi/viewcontent.cgi?article=1106...fluidization_xiii.
456. A report on Industrial size gasification for syngas, substitute natural gas and power production, DOE/NETL-401/040607, FR-53769211, DOE, Washington, DC (April 2007).

457. Goss, J.R., An Investigation of the down draft gasification characteristics of agricultural and forestry residues, Interim report, California Energy Commission P 500-79-0017, Sacramento, CA (1979).

458. Santos, D.A., Patenting trends in green technology of gasification in Brazil: A current analysis by patent statistics, in a presentation in the conference on *"Patent Statics for Decision Makers"*, OECD, Rio de Janeiro, Brazil (November 12–13, 2013), www.oecd.org/site/stipatents/PSDM2013_5_3_Santos.pdf.

459. Gasification plant databases, a report by NETL, Department of Energy, Washington, DC (2015), www.netl.doe.gov/research/.../gasification/gasification-plant-databases.

460. Kaupp, A. and Goss, J., State of the art report for small scale (to 50 kW) gas producer-engine systems, Report prepared under U.S. Department of Agriculture, Forest Service contract no. 53-319R-0-141, Funded by U.S. Agency for International Development Bioenergy Systems & Technology project no. 936-5709 through PASA no. AG/STR-4709-6-79, Washington, DC (March 1981).

461. World Gasification Database, A report by GSTC-gasification and syngas technology coaxial, Arlington, VA (2011), www.gasification-syngas.org/resources/world-gasification-database.

4 Biogas and Biohydrogen Production by Anaerobic Digestion

4.1 INTRODUCTION

Methanogenic archaea are responsible for all biological sources of methane such as landfill gas (biogas produced from natural fermentation process) and man-made synthetic biogas. Both landfill gas and synthetic biogas are produced by the anaerobic decay of cellulosic matter (biomass). Natural sources of biogas include swamps, marshes, and landfills, in addition to enteric fermentation, particularly in cattle. Synthetic biogas can be produced and optimized from agricultural waste materials such as sewage sludge and manure by way of anaerobic digesters [1–4].

In landfill, biogas (or landfill gas) is generated by natural fermentation of waste materials. Excluding water vapor, about half of landfill gas is methane and most of the rest is carbon dioxide, with small amounts of nitrogen, oxygen, and hydrogen, and variable trace amounts of hydrogen sulfide and siloxanes. If the gas is not removed, the pressure may get so high that it works its way to the surface, causing damage to the landfill structure, unpleasant odor, vegetation die-off, and explosion hazard. The gas can be vented to the atmosphere or flared or burned to produce electricity or heat [1–7].

Biogas can also be synthetically produced by separating organic materials from waste that otherwise goes to landfills or other primary, secondary, or tertiary treatment systems. This method is more efficient than just capturing the landfill gas it produces. The method is more controlled and flexible compare to the natural biogas generation process in landfill materials. The method can also be used to produce hydrogen instead of methane by controlling various steps of the anaerobic digestion (AD) process and utilizing process conditions and microbes that favor the production of hydrogen. Anaerobic lagoons produce biogas from manure, while biogas reactors can be used for manure or plant parts. In normal conditions, like landfill gas, synthetic biogas is mostly methane and carbon dioxide, with small amounts of nitrogen, oxygen, and hydrogen. However, with the exception of pesticides, there are usually lower levels of contaminants [1–7].

As shown in this chapter, unlike methane produced in landfill, methane produced by synthetic processes can be optimized by manipulations of substrates, digester design, operating conditions, and the nature of microbes. In recent years, the use of multiple substrates (codigestion) has optimized methane production rate and efficiency of space use within the treatment plants. As shown later, codigestion has become a standard commercial practice in synthetic biogas production from numerous different types of substrates [1–7]. Unlike landfill gas, synthetic AD process can also produce hydrogen with the use of suitable microbes and a proper manipulation and control of digestion process.

Landfill gas or synthetic biogas cannot be distributed through utility natural gas pipelines unless it is cleaned up to less than 3% CO_2 and a few parts per million H_2S, because CO_2 and H_2S corrode the pipelines. The presence of CO_2 will lower the energy level of the gas below requirements for the pipeline. Siloxanes in the gas will also form deposits in gas burners and need to be removed prior to entry into any gas distribution or transmission system. Consequently, it may be more economical to burn the gas on-site or within a short distance of the landfill using a dedicated pipeline. Water vapor is often removed, even if the gas is burned on-site. If low temperatures condense water out of the gas, siloxanes can be lowered as well because they tend to condense out with the water vapor [1–7]. Other nonmethane components may also be removed to meet emission standards, to prevent fouling of the equipment, or to ensure environmental considerations. Co-firing landfill gas with natural gas improves combustion, which lowers emissions. Biogas can also be improved in quality by the methods described in Chapter 7 and can produce biomethane that is of pipeline-quality natural gas and can be transported in conventional natural gas pipeline. Biomethane has same degree of usefulness as conventional natural gas and can be stored and transported by various methods described in Chapter 8.

Synthetic biogas, landfill gas, and biomethane are already sources of energy for heating and power production. In some areas, their use, however, could be greatly expanded in future. Gas generated in sewage treatment plants is commonly used to generate electricity. For example, the Hyperion sewage plant in Los Angeles burns 8 million cubic feet (mcf) (230,000 m^3) of gas per day to generate power. New York City utilizes gas to run equipment in the sewage plants and to generate electricity and heat from boilers [1–7]. Using sewage gas to make electricity is not limited to large cities. The city of Bakersfield, California, uses cogeneration at its sewer plants. California has 242 sewage wastewater treatment plants, 74 of which have installed anaerobic digesters. The total biopower generation from the 74 plants is about 66 MW [1–7].

The global energy demand is growing rapidly because of increasing demands from countries like China, India, Russia, Brazil, and Mexico to rapidly develop their economy and quality of life. Currently, about 85% of energy is supplied by fossil energy such as oil, natural gas, and coal. However, reports from the IEA [8] predict that during this century, energy demand will increase by twofold. The supply for fossil energy will decrease somewhat, particularly for oil. The major suppliers of conventional oil and gas are also in the politically unstable regions of the Middle East. Furthermore, fossil energy is also causing more environmental problems due to emissions of greenhouse gases (GHGs) such as carbon dioxide and lower volatile hydrocarbons. According to IPCC [9], GHG emissions must be reduced to less than half of global emissions level of 1990.

Renewable energy from biomass has a significant potential to be an alternate for fossil energy. Biogas from wastes, residues, energy crops, and many other organic materials is a versatile renewable energy source. Methane-rich biogas can be used for heat and power applications as well as fuel for vehicles and for the production of a variety of chemicals and materials. Fehrenbach et al. [10] showed that biogas generated by ADs of numerous different types of biomass and effluent wastes is the most energy efficient and environmentally friendly source of bioenergy. Like natural gas, methane in biogas can be used in a variety of ways. Biogas will drastically reduce the emission of GHG compare to fossil fuels by utilizing locally available sources of wastes and other forms of biomass. The digestate is an improved fertilizer in terms of its availability to crops than conventional mineral fertilizers. Microbiology of anaerobic digester is reviewed in details by Gerardi [11].

In Europe, biogas is the fastest growing bioenergy and it reached 6 million tons of oil equivalent (mtoe) in 2007 [12] with a yearly increase of more than 20%. Germany is the biggest producer of biogas with about 4000 agricultural biogas production plants by the end of 2008. Within the agriculture sector of European Union, 1500 million tons of biomass could be digested anaerobically each year, half of which will come from energy crops [13,14]. Besides the agriculture sector, biogas coming from landfills is also becoming a more and more important source of power generation. Biogas is not only generated from various wastewater effluent streams (with solids concentration less than 10 wt%), but also from various high solids concentration streams (25–35 wt%) like municipal solid waste (MSW) and animal wastes. Since the amount of waste we create increases with the population, biogas is an effective way to convert waste into a useful and environmentally acceptable form of energy for the growing waste industry. Since every country in the world has waste problem, synthetic biogas industry is universally applied [13]. The effect of biomethane for future mobility was examined by Ahrens and Weiland [15].

The literature on biogas deals with both biomethanation and biohydrogenation. As shown in the following sections, hydrolysis of organic waste followed by AD can produce hydrogen or methane depending upon the nature of operating conditions, the nature of microorganisms present, and the nature of feedstock. It should be noted that methane can be converted to hydrogen by reforming reactions (see Chapter 6).

4.2 MECHANISM OF ANAEROBIC DIGESTION FOR METHANE

The anaerobic process is the degrading of organic substrates in the absence of oxygen to carbon dioxide and methane with only a small amount of bacterial growth [16]. The digestion process consists of several interdependent, complex, sequential, and parallel biological reactions. During these reactions, the products from one group of microorganisms serve as the substrates for the next [17]. The overall conversion process is often described as a three- or four-stage process, which occurs simultaneously within the anaerobic digester [18]. The first is the hydrolysis of insoluble biodegradable organic matter; the second is the production of acid, acetate, and hydrogen from smaller soluble organic molecules; and the third is methane generation. The three-stage scheme involving various microbial species can be described as (1) hydrolysis and liquefaction, (2) acidogenesis, and (3) methane fermentation. Some investigators [13] have broken overall mechanism for methanation into four steps: hydrolysis, acidogenesis, acetogenesis/dehydrogenation, and methanation. The difference is in the second stage where acidogenesis and acetogenesis/dehydrogenation steps are separated. This alternate four-step method was described in my previous book [19] and it can also be articulated as shown in Figure 4.1 [1]. The argument for three steps is that clear distinction between acidogenesis and acetogenesis reactions is not always possible [16–18,20–22]. Here we give more details on the three-step process in order to present an alternate picture compared to the one presented in my previous book [19].

4.2.1 Hydrolysis and Liquefaction

Hydrolysis and liquefaction are the breakdown of large, complex, and insoluble organics into small molecules that can be transported into the microbial cells and metabolized [20]. Hydrolysis of complex molecules such as proteins, carbohydrates, and lipids is catalyzed by extracellular enzymes. Some of the enzymes present include cellulase, amylase, protease, and lipase [18]. Essentially, organic waste stabilization does not occur during hydrolysis, and the organic matter is simply converted into a soluble form that can be utilized by the bacteria [21,22].

As shown by Weiland [13], the degradation of complex polymers like polysaccharides, proteins, and lipids results in the formation of monomers and oligomers such as sugars, amino acids, and long-chain fatty acids. The individual degradation steps are carried out by different consortia of microorganisms, which place different requirements on the environment [13,19]. Hydrolyzing and fermenting microorganisms are responsible for the initial attack on polymers and monomers and

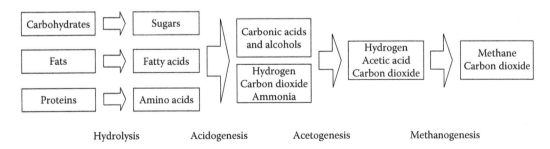

FIGURE 4.1 Alternate four-step method for anaerobic digestion. (From Biogas, Wikipedia, the free encyclopedia, 2015.)

produce mainly acetate, hydrogen, and varying amounts of fatty acids such as propionate and butyrate [13]. Hydrolytic microorganisms excrete hydrolytic enzymes such as cellulase, cellobiase, xylanase, amylase, lipase, and protease [13,19]. Thus, a complex consortium of microorganisms, most of which are strict anaerobes such as *Bacteroides*, *Clostridia*, and *Bifidobacterium*, participate in the hydrolysis and fermentation of an organic material [13]. The higher volatile fatty acids (VFAs) are converted into acetate and hydrogen by obligate hydrogen-producing acetogenic bacteria. The maintenance of an extremely low partial pressure of hydrogen is very important for the acetogenic and hydrogen-producing bacteria.

4.2.2 ACIDOGENESIS

The acidogenesis stage is a complex phase involving acid-forming fermentation, hydrogen production, and an acetogenic step. As shown in Figure 4.1 and in my previous book [19], often the acetogenic and acidogenesis steps are separated making an overall process a four-step process. In acidogenesis step, *sugars, long-chain fatty acids, and amino acids from hydrolysis* are used as substrates. Microorganisms produce organic acids (acetic, propionic, butyric, and others), alcohols, hydrogen, and carbon dioxide [18,20–22]. The products formed vary with the types of bacteria as well as environmental conditions. Bacteria responsible for acid production include facultative anaerobic bacteria, strict anaerobic bacteria, or both (i.e., *Bacteroides*, *Bifidobacterium*, *Clostridium*, *Lactobacillus*, and *Streptococcus*) [23]. Hydrogen is produced by the acidogenic bacteria and hydrogen-producing acetogenic bacteria.

Organisms that produce fermentation products, such as propionate, butyrate, lactate, and ethanol, generally exhibit obligate proton-reducing metabolism (i.e., they produce hydrogen as a fermentation product). This mechanism is commonly referred to as interspecies hydrogen transfer. The organisms are referred to as syntrophs and may be obligate as in the case of S organisms, *Syntrophomonas wolfei*, and *Syntrophobacter wolinii* or facultative as with many other syntrophs [24]. Acetogenic microorganisms can also tolerate a wide range of environmental conditions [21,25].

The main pathway of acidogenesis is through acetate, carbon dioxide, and hydrogen [18]. The accumulation of lactate, ethanol, propionate, butyrate, and higher VFAs is the response of the bacteria to increased hydrogen concentration in the medium [26]. In the absence of methanogens to utilize these substrates, hydrogen backs up the overall degradative process and organic acids accumulate causing a decrease in pH that ultimately inhibits and stops the fermentation unless controlled. The overall performance of the AD system is affected by the concentration and proportion of individual VFAs formed in the acidogenic stage because acetic and butyric acids are the preferred precursors for methane production [27]. These reactions are shown as follows with glucose as the substrate [28]. A theoretical maximum of 4 mol of hydrogen is obtained from acetic acid and 2 mol of hydrogen from butyric acid:

$$C_6O_6H_{12} + 2H_2O \rightarrow 2CH_3COOH + 4H_2 \tag{4.1}$$

$$C_6O_6H_{12} \rightarrow CH_3CH_2CH_2COOH + 2CO_2 + 2H_2 \tag{4.2}$$

$$C_6H_{12}O_6 + 2H_2O \rightarrow 2CH_3COOH + 4H_2 + 2CO_2 \tag{4.3}$$

In the acetogenic stage of acidogenesis, bacteria will degrade organic acids such as propionic, butyric, and valeric acids to acetate, carbon dioxide, and hydrogen. This intermediate conversion is important for proper AD and methane production because methanogens do not utilize these VFAs directly [18]. During acidogenesis, a clear distinction between acetogenic and acidogenic reactions is not always present [29].

The present state of knowledge indicates that hydrogen may be a limiting substrate for methanogens [13,19]. This is because an addition of hydrogen-producing bacteria to the natural biogas-producing consortium increases the daily biogas production. The studies have shown that only two groups of methanogenic bacteria produce methane from acetate, hydrogen, and carbon dioxide. These bacteria are strictly anaerobes and require a lower redox potential for growth than most other anaerobic bacteria.

4.2.3 METHANOGENESIS

The third and final stage is methane fermentation, which is the ultimate product of anaerobic treatment. Formic acid, acetic acid, methanol, and hydrogen can be used as energy sources by the various methanogens. The methane bacteria are such a unique group of organisms that they have been placed into a new evolutionary domain (separate from eukaryotic plants and animals and prokaryotic bacteria) referred to as Archaea [30]. The majority of methanogenic bacteria belong to the genera *Methanobacterium*, *Methanosarcina*, *Methanospirillum*, and *Methanococcus* [16]. Methanogens are unique because of the very different cell morphologies found between the species. Most have simple nutritional requirements, carbon dioxide, ammonia, and sulfide. The primary route of methanogenesis is the fermentation of acetic acid to methane and carbon dioxide. The bacteria that utilize acetic acid are classified as acetoclastic bacteria or acetate-splitting bacteria [23]. About two-thirds of methane gas is derived from acetate conversion by acetoclastic methanogens. Some methanogens are also able to use hydrogen to reduce carbon dioxide to methane (hydrogenophilic methanogens) with an overall reaction as shown here [25,31,32]:

$$\text{Hydrogen: } 4H_2 + CO_2 \rightarrow CH_4 + 2H_2O \tag{4.4}$$

$$\text{Acetate: } CH_3COOH \rightarrow CH_4 + CO_2 \tag{4.5}$$

$$\text{Formate: } 4HCOOH \rightarrow CH_4 + 3CO_2 + 2H_2O \tag{4.6}$$

$$\text{Methanol: } 4CH_3OH \rightarrow 3CH_4 + CO_2 + 2H_2O \tag{4.7}$$

$$\text{Carbon monoxide: } 4CO + 2H_2O \rightarrow CH_4 + 3H_2CO_3 \tag{4.8}$$

$$\text{Trimethylamine: } 4(CH_3)_3N + 6H_2O \rightarrow 9CH_4 + 3CO_2 + 4NH_3 \tag{4.9}$$

$$\text{Dimethylamine: } 2(CH_3)_2NH + 2H_2O \rightarrow 3CH_4 + CO_2 + 2NH_3 \tag{4.10}$$

$$\text{Monomethylamine: } 4(CH_3)NH_2 + 2H_2O \rightarrow 3CH_4 + CO_2 + 4NH_3 \tag{4.11}$$

$$\text{Methyl mercaptans: } 2(CH_3)_2S + 3H_2O \rightarrow 3CH_4 + CO_2 + H_2S \tag{4.12}$$

$$\text{Metals: } 4Me^0 + 8H^+ + CO_2 \rightarrow 4Me^{2+} + CH_4 + 2H_2O \tag{4.13}$$

The microbial ecology of biomethanogenesis is difficult to study. The organisms are fastidious, slow-growing anaerobes and many species will not even grow in pure culture [33]. When grown in pure culture, isolates may produce fermentation products different from those produced in the presence of hydrogen and acetate-metabolizing bacteria that are present in their natural environment [34]. Each anaerobic environment may differ in the types of bacteria involved in methanogenesis, depending on differing factors such as substrate, retention time, temperature, pH, and fluctuations in other environmental parameters. Although certain general properties are common from one environment to another, each environment may have its own unique population of bacteria and associated microbial activities [13,19].

Methane-producing microorganisms are very sensitive to environmental changes [35]. The hydrogenophilic methanogens are more resistant to environmental changes than acetoclastic methanogens [18]. Research has shown the metabolic rates of acetoclastic methanogens, which are responsible for the majority of methane production, are lower than those of acid-forming bacteria [36]. Therefore, methane production is generally the rate-limiting step in AD [37].

Only few species are able to degrade acetate into CH_4 and CO_2, for example, *Methanosarcina barkeri*, *Metanonococcus mazei*, and *Methanotrix soehngenii*, whereas all methanogenic bacteria are able to use hydrogen to form methane [13]. The first and second groups of microbes and third and fourth groups of microbes are linked closely with each other [13,19] allowing the overall process to be divided into two stages. A balanced AD process demands that in both stages, the rates of degradation must be equal in size. If the first degradation step runs too fast, the acid concentration rises and pH drops below 7.0 that inhibits methanogenic bacteria. If the second phase runs too fast, methane production is limited by the hydrolytic stage. Thus, the rate-limiting step depends on the compounds of the substrate that is used for biogas production. Undissolved compounds such as cellulose, proteins, and fats take several days to crack, whereas soluble carbohydrates crack in few hours. Therefore, the process design must be well adapted to the substrate properties for achieving complete degradation without process failure.

Numerous studies have been reported for AD to produce either methane or hydrogen from a variety of waste streams. Some of these studies are summarized in Table 4.1. Maximum gas yields and theoretical methane contents that can be generated from carbohydrates, raw protein, raw fat, and lignin along with associated reactions are summarized in Table 4.2 [13]. Finally, some typical average methane yields from various energy crops, plants and plant materials are illustrated in Table 4.3 [13,19]. The biogas generated from landfills generally contains about 50%–55% methane, and the remaining composition consists of largely CO_2 and traces of water, hydrogen, and other impurities. It is clear from these studies that AD is a very widely used process to generate methane or hydrogen from a variety of organic wastes, both of which are important gaseous biofuels.

4.2.4 Microbes and the Effects of Operating Conditions

While the success of anaerobic treatment depends upon the effectiveness of various microbes, very little is known how they work and the interactions between these microorganisms. Several investigators [82,83] have attempted to measure microorganism in an anaerobic bioreactor. Karakashev et al. [84] examined the effects of environmental conditions on methanogenic compositions in an anaerobic bioreactor. In general, the lack of knowledge sometimes results in malfunction and failure of biogas digestive process. Furthermore, only a few percent of bacteria and archaea have been isolated. Weiland [13] points out that with new molecular techniques, more information can be obtained about the community structure in the anaerobic processes [16–29,31–37,47,85–96].

The quantification of methanogens and the assessment of a wide range of samples from agriculture biogas can be carried out by fluorescence in situ hybridization technique. Klocke et al. [86,87] detected 68 taxonomic groups by 16SrDNA analysis of samples from agricultural biogas plants. They showed that while high share of acetogenic methanogens can be found in biogas plants operated at low ammonium concentration, hydrogenotrophic methanogens dominate most agricultural biogas plants.

The digestive process can be operated at a lower temperature, that is, mesophilic conditions (temp. range of 35°C–42°C) or at a high temperature, that is, thermophilic conditions (temp. range

TABLE 4.1
Some Typical Literature Studies on Anaerobic Digestion of Single and Multiple Substrates

Types of Waste	Authors
Swine waste	Chen et al. [38]
Coir path	Kunchikannan et al. [39]
Wastewater and organic kitchen waste	Weichgrebe et al. [40]
Distillery spent wash	Pathe et al. [41]
Biodiesel by-products	Kolesarova et al. [42]
Whey (a component of dairy product or an additive for food product)	Beszedes et al. [43]
Palm oil mill effluent	Yusoff et al. [44]
Tofu wastewater	Zheng et al. [45]
Starch of food waste	Ding et al. [46]
Solid organic waste and energy crops	Angelidaki et al. [47–50]
Food residuals	Shin et al. [51] and Haug et al. [52]
Dairy effluent	Desai et al. [53]
Household organic waste	Narra et al. [54]
Distillery spent waste	Nandy et al. [55]
Long fatty acids	Alves et al. [56]
Horse and cow dung	Yusuf et al. [57]
Agricultural and industrial wastes	Kujawski and Steinmetz [58]
MSW/FOG (fats, oils, and greases) wastes	Martin-Gonzalez et al. [59]
Nonedible oil cake and cow dung	Singh and Mandal [60]
Food wastes	Zhu et al. [61]
Maize grains and maize silage	Hutnan et al. [62]
Codigestion of olive mill wastewater and SM	Azaizeh and Jadoun [63]
Cow dung and water hyacinth	Yusuf and Ify [64]
Cattle manure, slaughterhouse waste	Bagge et al. [65]
Co-digesting SM with three crop residues	Wu et al. [66]
Biomass (IFBB) and whole crop digestion (WCD)	Buhle et al. [67]
Organic solid poultry slaughterhouse waste	Salminen and Rintala [68]
MSW/agricultural waste/dairy CM	Macias-Corral et al. [69]
Animal manure	Holm-Nielsen et al. [70]
Maize hybrids	Oslaj et al. [71]
By-products of sugar production/CM	Fang et al. [72]
Biomass	Gunaseelan [73]
Fruit waste	Kaparaju and Rintala [74] and Lopez et al. [75]
Food waste and sewage sludge	Iacovidou et al. [76]
Algal sludge and wastepaper	Yen and Brune [77]
FOG and wastewater	Long et al. [78]
Food waste and wastewater	Cheerawit et al. [79]
Oily waste and pet food wastewater	Acharya and Kurian [80]
Crude glycerol and chicken processing wastewater	Foucault [81]

Sources: Shah, Y.T., *Water for Energy and Fuel Production*, CRC Press, New York, 2014; Shah, Y.T., *Energy and Fuel Systems Integration*, CRC Press, New York, 2015.

IFBB, integrated generation of solid fuel and biogas from biomass.

of 45°C–60°C). Generally, temperature fluctuation decreases biogas productivity and methanogenic diversity is lower in thermophilic conditions [13,19]. In general, thermophilic operations growth rate of methanogenic bacteria is higher, thus making the process more efficient and faster; however, these bacteria are more temperature sensitive and have difficult time adjusting to temperature variations. The faster rate allows the operations to run at a lower hydraulic retention time (HRT) than in

TABLE 4.2
Maximal Gas Yields and Theoretical Methane Contents

Substrate	Biogas (Nm^3/tons TS)	CH$_4$/CO$_2$
Carbohydrates (not including inulins and single hexoses)	790–800	1/1
Raw proteins	700	Approx. 70/30
Raw fat	1200–1250	Approx. 67/33
Lignin	0	Both 0

Sources: Weiland, P., *Appl. Microbiol. Technol.*, 85, 849, 2010; Shah, Y.T., *Water for Energy and Fuel Production*, CRC Press, New York, 2014, also Baserga, U., Landwirtschaftliche Co-vergarungs-Biogasanlagen, FAT-Berichte No. 512, Tanikon/Switzerland, 1998; referred by Weiland [13].

TABLE 4.3
Average Methane Yields from Various Energy Crops, Plants, and Plant Materials

Materials	Average Methane Yield (m^3/kg VS)
Barley	0.56
Triticale	0.49
Leaves	0.48
Alfalfa	0.46
Wheat (grain)	0.45
Peas	0.43
Grass	0.42
Hemp	0.42
Clover	0.38
Potatoes	0.37
Sorghum	0.37
Rapeseed cake	0.36
Maize (whole crop)	0.36
Sugar beet	0.34
Kale	0.31
Straw	0.31
Sunflower	0.30
Oats (grain)	0.30
Sudan grass	0.28
Flax	0.23
Miscanthus	0.22

Sources: Braun, R. et al., Biogas from energy crop digestion, IEA Bioenergy, Task 37—Energy from Biogas and Landfill Gas, IEA, Paris, France, 2011. With permission; Braun, R., Potential of codigestion, 2002, http://www.novaenergie.ch/iea-bioenergy-task37/Dokumente/final.PDF, Accessed on November 7, 2007; Braun, R. and Wellinger, A., Potential for codigestion, IEA bioenergy report—Task 37—Energy from Biogas and Landfill Gas, IEA, Brussels, Belgium 2002. With permission; Shah, Y.T., *Water for Energy and Fuel Production*, CRC Press, New York, 2014; Shah, Y.T., *Energy and Fuel Systems Integration*, CRC Press, New York, 2015.

Note: These data are calculated from the arithmetic averages of the ranges.

mesophilic operations. Mesophilic bacteria, on the other hand, can tolerate temperature fluctuation of $\pm 3°$ variations without significant variation in methane production.

Since ammonia toxicity increases with temperature, thermophilic operations are more susceptible to ammonia inhibition (particularly for the ammonia concentration above 80 mg/L) and also a larger degree of imbalance due to possible washout of microbial population. An increase in ammonia concentration is, however, accompanied by an increase in VFA concentration [13]. This can lower the pH and thus counterbalance the effect of ammonia. Many strategies to reduce the ammonia inhibition effects have been examined [88,89]; the most stable recovery process was observed when biomass was diluted with reactor effluents.

pH of the reacting solution also has significant effects on the effectiveness of bacteria and methane production. The AD process best operates between a pH of 6.5 and 8.5, with an optimum value between 7 and 8 [13,19]. The process is severely affected when the pH drops below 6 or when it increases above 8.5 [13,19]. While ammonia accumulation increases the pH and VFA accumulation decreases the pH, the latter is not always valid because of the buffer capacity of some substrate. For example, animal manure has surplus alkalinity that counteracts the increase in VFA concentration. While acetic acid is always present in a larger amount than VFAs, only propionic and butyric acids are more inhibitory to methanogens [13,19]. Since inhibition effect is more associated with the undissociated form, the inhibition effect of VFA is higher in the reacting systems with lower pH values.

Besides temperature and pH, availability of several macro- and micronutrients is also very important for the growth and survival of specific groups of microorganisms. Very low amount of macronutrients such as carbon, phosphorous, and sulfur are needed (C:N:P:S = 600:15:5:1) because only a small amount of biomass is developed. Micronutrients such as iron, nickel, cobalt, selenium, molybdenum, and tungsten are important for the growth rate of microorganisms and they must be added, particularly if the energy crops are the only substrate for biogas production. Nickel is important for all methanogenic bacteria because it is necessary for the cell component cofactor F430, which is involved in the methane formation [13,19].

For optimum growth, the cell requires cobalt to build up the Co-containing corrinoid factor III. The growth of only few methanogens depends on the trace elements selenium, molybdenum, and tungsten. The required concentration is only 0.05–0.06 mg/L [13,19]. The iron is, however, necessary in higher concentrations of 1–10 mg/L [19,90]. These micronutrients are very important for the stable process and high loading for energy crops [19,91]. While the addition of manure reduces the need for micronutrients addition, even with 50% manure in the reaction medium, the addition of micronutrients can increase the biogas production rate [13,19].

As indicated by Weiland [13], each step of the AD process is complex and requires an independent assessment. For example, hydrolysis of complex insoluble substrate depends on parameters such as particle size, productions of enzymes, pH, and temperature. Angelidaki et al. [47] and Gavala et al. [92] have given a systematic assessment of complex kinetic models for organic waste digestion. The degradation can be described by a simple first-order reaction that can be applied knowing the yield of substrate and the specific reaction rate [13]. The kinetic models also depend on the nature of feedstock and the temperature range of the digestion process. The kinetic of biogas production from energy crops and manure was reported extensively by Mahnert [93], while several kinetic models were developed for low-temperature (35°C–42°C) mesophilic conditions as well as high-temperature (45°C–60°C) thermophilic conditions by Andara and Esteban [94], Linke [95], and Biswas et al. [96].

4.3 HYDROGEN PRODUCTION BY ANAEROBIC DIGESTION

Hydrogen is considered to be an alternative fuel of great potential use. In 1976, the first World Hydrogen Conference identified hydrogen as a clean energy carrier for the future [97]. Instead of GHGs, water with trace amounts of nitric oxide is produced when hydrogen is combusted with air. It also has a high energy yield of 122 kJ/g, which is 2.75 times greater than gasoline [98]. Hydrogen has the potential to lessen the world's dependency on fossil fuels, but further research and technology is needed before a sustainable hydrogen economy can be established. As mentioned in Chapter 1, hydrogen makes gaseous fuel as an "ultimate fuel" instead of just "transition fuel."

Biological production of hydrogen by anaerobic fermentation is one such area of research that shows great potential but requires further study. The biological method of hydrogen production provides a pollution free and energy-saving process, and it produces hydrogen that is less energy intensive than chemical or electrochemical methods because biological methods are normally carried out at ambient temperature and pressure [99]. Hydrogen production by fermentative bacteria is technically a simpler process over other biological processes because it proceeds at higher rates and does not require light sources [100] and it is a very energy-efficient process.

4.3.1 Mechanism of Fermentative Hydrogen Production

Equations 4.1 and 4.2 show that hydrogen along with acetic and butyric acids can be produced by dark fermentation processes using anaerobic and facultative anaerobic chemohetrotrophs [13]. Gujer and Zehnder [101] examined various conversion processes in anaerobic digestion. Production of hydrogen and methane from wastewater sludge using anaerobic fermentation was also examined by Ting et al. [102]. Different types of waste materials can also be used for hydrogen fermentation. As mentioned earlier, hydrogen production is highly dependent on a number of factors including the pH, retention time, and gas partial pressure along with the nature of microbes and substrates [13,19]. Hydrogen production is important for its use in fuel cell or microbial electrolytic cell. Wang et al. [103] described the use of low-cost cathode catalysts for high-yield biohydrogen production in microbial electrolytic cell [103–123]. Cheng and Logan [121–123] examined the roles of catalysts and membranes for high-yield biohydrogen production via electrohydrogenesis in microbial electrolytic cells. Possible pathways of fermentative hydrogen evolution and other by-products during biohydrogen fermentation are schematically described by several investigators [103–123].

In dark fermentation, different groups of bacteria are known to be responsible for hydrogen production such as *Enterobacter*, *Clostridium*, and *Bacillus*. Fang and Liu [124] reported that in a mixed culture study where hydrogen was produced, about 70% of the population was of the genus *Clostridium* and 14% belonged to genus *Bacillus*. In general, hydrogen production was correlated to the presence of *Clostridium* species in the bioreactor [13,124,125].

Clostridium is an anaerobic spore former. In response to hostile conditions, such as the presence of oxygen, heat, low or high pH, alcohol, and toxic compounds, *Clostridium* species are changed from vegetative cells to endospores, which is a stress-resistant state with greatly reduced metabolic activity. Due to endospore production, the members of genus *Clostridium* can be isolated from mixed cultures by heating the cultures to 70°C for 10 min to kill vegetative *Clostridium* cells and non-spore-forming organisms [126]. In addition to the selection process of *Clostridium* by heat treatment, some species require heat activation for endospore prior to germination [127].

According to its catabolism, *Clostridium* can be classified as two groups—saccharolytic and proteolytic—of fermenting bacteria. Saccharolytic clostridia ferment carbohydrates, consisting of simple sugars, disaccharides, oligosaccharides, and cellulose, whereas proteolytic clostridium hydrolyzes protein and ferments amino acids [128]. However, most proteolytic clostridia can also ferment carbohydrates, and hence, carbohydrates are very common substrates for the genus of *Clostridium*.

Miyake [129] examined the improvement of bacterial light–dependent hydrogen production by altering photosynthesis pigment ratio. Miyake et al. [130] also evaluated research on efficient light conversion for biohydrogen production. With regard to dark fermentation, saccharolytic *Clostridium* species metabolize simple sugars using the Embden–Meyerhof–Parnas (EMP) glycolytic pathway, where glucose is converted to pyruvate, an intermediate of hexose metabolism [131]. Pyruvate is then oxidized by the enzyme pyruvate–ferredoxin oxidoreductase to yield acetyl CoA, carbon dioxide, and reduced ferredoxin [132]. The reoxidation of reduced ferredoxin is catalyzed by the enzyme hydrogenase and generates hydrogen gas. The kinetic study of dark H_2 fermentation from sucrose and xylose using H_2-producing indigenous bacteria is given by Lo et al. [133].

Using glucose as a model substrate, hydrogen production is accompanied with either acetate formation or butyrate formation [132]. In acetate fermentation, four ATP molecules are produced, whereas three ATP molecules are produced in butyrate fermentation. Thus, for the microorganisms, it seems that the acetate fermentation is energetically more favorable than the butyrate fermentation. Zheng and Hu [134] investigated inhibition effect of butyrate on biological hydrogen production with mixed anaerobic cultures. However, based on Gibb's free energy change, butyrate fermentation is the more dominant reaction, where it only yields 3.3 ATP molecules and the maximum hydrogen production of 2.5 mol H_2/mol glucose stoichiometrically [135]. Possible explanations as to why clostridia use the butyrate fermentation pathway during hydrogen production are (1) the formation of one equivalent butyrate leads to less acidification of microorganisms' environment than the two equivalents of acetate and (2) the generation of a higher amount of butyrate may deplete excess reducing equivalents [128]:

$$C_6H_{12}O_6 + 2H_2O \rightarrow 2CH_3COOH + 2CO_2 + 4H_2 \quad \Delta G° -184 \text{ kJ} \tag{4.14}$$

$$C_6H_{12}O_6 \rightarrow CH_3CH_2CH_2COOH + 2CO_2 + 2H_2 \quad \Delta G° -257 \text{ kJ} \tag{4.15}$$

$$C_6H_{12}O_6 + 0.5H_2O \rightarrow 0.75CH_3CH_2CH_2COOH + 0.5CH_3COOH + 2CO_2 + 2.5H_2 \tag{4.16}$$

As evident from equations given earlier, saccharolytic *Clostridium* species can convert glucose not just into hydrogen and organic acids but solvents as well. In a batch culture, *Clostridium* sp. produced hydrogen and organic acids at an exponential growth phase, whereas the metabolism shifted to solvent production during the stationary growth phase [136–138]. *Clostridium acetobutylicum*, a species known to produce hydrogen and organic acids, is favored to produce organic solvents at pHs below 5 [138]. In addition, the metabolic pathway of *Clostridium pasteurianum* showed an abrupt shift from hydrogen and acid production to solvent production under iron and phosphate limitations, high substrate concentrations (125 g glucose/L), and the appearance of carbon monoxide (an inhibitor of hydrogenase) [139].

4.3.2 Environmental Factors Affecting Hydrogen Production

There are several factors that can heavily impact the performance of biological hydrogen production. These include the inocula and pretreatment utilized during hydrogen production; the nature of substrate(s); pH; HRT; temperature; the effects of products, metal ions, nitrogen, and phosphorous; and the design of the digester used. Here we briefly examine their effects.

4.3.2.1 Inocula and Pretreatment

A lot of pure cultures of bacteria have been used to produce hydrogen from various substrates. Liu and Shen [140] examined the effects of culture and medium conditions on hydrogen production from starch. *Clostridium* and *Enterobacter* have been the most widely used as inoculum for fermentative hydrogen production [13]. Most of the studies using pure cultures of bacteria for fermentative hydrogen production have been conducted in batch mode and used glucose as substrate; however, it is more desirable to produce hydrogen from organic wastes using pure cultures in continuous mode, because continuous fermentative hydrogen production from organic wastes is more feasible for industrialization to realize the goal of waste reduction and energy production [13,19]. A comparison of different pretreatment methods for enriching hydrogen producing cultures from digested sludge is given by Wang and Wan [141].

In order to carry out biological hydrogen production, the abundance of *Clostridium* species could be extracted from soil, anaerobic digested sludge, compost, etc. [142–145], and used as inoculum for fermentative hydrogen production [13,19,145]. Fermentative hydrogen production processes using mixed cultures are more practical than those using pure cultures, because the former are simpler to operate and easier to control, and may have a broader source of feedstock [13,19,145,146]. However, in a fermentative hydrogen production process using mixed cultures, the hydrogen produced by hydrogen-producing bacteria may be consumed by hydrogen-consuming bacteria. In order to harness hydrogen from a fermentative hydrogen production process, the mixed cultures [147–154] can be pretreated by certain methods to suppress as much hydrogen-consuming bacterial activity as possible while still preserving the activity of the hydrogen-producing bacteria [51,143–145,155–160]. Guo et al. [161] examined the impacts of sterilization, microwave, and ultrasonic pretreatment on hydrogen production using waste sludge. For cattle manure sludge, Cheong and Hansen [162] showed that bacterial stress enrichment enhances anaerobic hydrogen production. Zhu and Beland [163] examined alternative methods preparing hydrogen-producing seeds from digested waste water sludge.

The pretreatment methods reported for enriching hydrogen-producing bacteria from mixed cultures mainly include heat shock, acid, base, aeration, freezing and thawing, chloroform, sodium 2 bromoethane sulfonate or 2-bromoethanesulfonic acid, and iodopropane [51,143,144,155–160]. Different pretreatment methods have different properties and comparison of different pretreatment methods to obtain a better pretreatment method for a given fermentative hydrogen production process was conducted by many studies [51,143,144,155–160]. For example, baking compost or soil is a heat shock method used when the inoculum is obtained from the solid phase [144,160,164]. In addition to heat shock treatment, acid and base treatments have been developed as alternative processes for the selection of *Clostridium*. Research has reported that adjusting pH of inocula to 3 or 10 efficiently selected *Clostridium* and inhibited methanogens [165]. Meanwhile, a successful hydrogen production was observed from different types of pre-acidified inocula including activated sludge, anaerobic digested sludge, refuse compost, watermelon soil, kiwi soil, and lake sediment [166]. Another study adopted acid-pretreated anaerobic digested sludge for investigating the effect of sludge immobilization by ethylene-vinyl acetate copolymer to achieve the goal of preventing biomass washout at a low HRT [167,168].

There exists certain disagreement on the optimal pretreatment method for enriching hydrogen-producing bacteria from mixed cultures [51,143,144,155–160]. The possible reason for this disagreement may be the difference among these studies in terms of inoculum, pretreatment method studied, specific condition of each pretreatment method, and the kind of substrates used. Even though heat shock has been the most widely used pretreatment method, it is not always effective for enriching hydrogen-producing bacteria from mixed culture inoculum compared with other pretreatment methods, for it may inhibit the activity of some hydrogen-producing bacteria [51,143,144,155–160]. Since most studies have been carried out in batch mode and for glucose as substrate, more comparison in continuous mode operation and for organic substrate is needed.

4.3.2.2 Substrate

Anaerobic fermentation is also considered a simpler option because it allows the production of hydrogen by relatively straightforward procedures and can utilize substrates from many different sources [13,28,145]. Many different types of substrates have been examined for the hydrogen production. For dairy wastewater as a substrate, Mohan et al. [169] examined the effect of various pretreatment methods on anaerobic mixed microflora to enhance biohydrogen production. The major criteria for substrate selection are the availability, cost, carbohydrate content, and biodegradability [170]. Commercially produced food products, such as corn and sugar, are not yet economical for hydrogen production. Alternatively, wastewaters with organic waste such as food processing and animal waste have great potential as substrate sources [171]. Utilizing wastewaters from agriculture land food processing industries, which are generally high in carbohydrates, can provide the essential nutrients required for hydrogen production and reduce treatment and disposal costs currently needed for these particular waste streams. Treating these waste streams to protect public health and the environment while producing a clean energy source makes biological hydrogen production an attractive alternative to fossil fuels [170]. Kargi and Kapdan [172] examined biohydrogen production from waste materials.

Most studies on substrate evaluation have been restricted to batch mode. It has been demonstrated that in an appropriate range, increasing substrate concentration could increase the ability of hydrogen-producing bacteria to produce hydrogen during fermentative hydrogen production, but substrate concentrations at much higher levels could decrease it with increasing levels [12,95]. There is, however, some discrepancy on the reported optimum conditions [12,95], perhaps due to differences in inoculum and substrate concentration range. Some complex substrates are not ideal for fermentative hydrogen production due to their complex structures; however, after being pretreated by some methods, they can be easily used by hydrogen-producing bacteria [109,166].

Waste activated sludge from wastewater treatment plants contains high levels of organic matter and thus is a potential substrate for hydrogen production. After appropriate pretreatments such as ultrasonication, acidification, freezing and thawing, sterilization, methanogenic inhibitor, and microwave, the ability of hydrogen-producing bacteria to produce hydrogen from it can be improved [116,117]. Different substrate pretreatment methods have different properties and comparison of various substrate pretreatment methods have been conducted by several studies [116,117]. It is worth noting that when using *Clostridium bifermentans* as inoculum, freezing and thawing were the optimal pretreatment methods for waste activated sludge [116,117], while when using *Pseudomonas* sp. GZ1 as inoculum, sterilization was the optimal pretreatment methods for waste activated sludge [166]. This demonstrates that the optimal pretreatment methods for waste activated sludge may be dependent on the inoculum used for fermentative hydrogen production. More work on continuous mode of operation and for mixed cultures as inoculum and other substrates is needed.

Hwang et al. [173] examined biohydrogen production from ethanol fermentation. Biohydrogen production from rice slurry was examined by Fang et al. [174]. Okamoto et al. [175] studied biohydrogen potential of material characteristics of the organic function of municipal solid wastes. In addition to testing the availability of hydrogen production from different feedstock, Lay et al. [142] compared hydrogen production from carbohydrate-rich, protein-rich, and fat-rich organic solid wastes. Their batch study results indicated that hydrogen-producing microbes could evolve much more hydrogen from carbohydrate-rich organic waste. The same conclusion was also obtained from the study of converting bean curd manufacturing waste (protein-rich waste), rice, and wheat bran (carbohydrate-rich waste) into hydrogen, where rice and wheat bran were more favorable for hydrogen fermentation [176,177]. On the other hand, many studies have already concentrated on studying the utilization of carbohydrate-rich organic wastes, for example, rice winery wastewater [153] and starch-manufacturing wastewater [151,178,179], to carry out hydrogen fermentation. The effects of catalysts and membranes for high-yield biohydrogen production via electrohydrogenesis in microbial electrolysis cells were evaluated by Cheng and Logan [180]. The effect of cow dung compost on biohydrogen from wheat straw waste was examined by Fan et al. [181]. They found compost made the conversion more efficient.

4.3.2.3 pH

As mentioned earlier, research reported that pH could affect the metabolic pathway of dark fermentation in a pure culture. Hence, there is a need to investigate the effect of pH on hydrogen production in mixed cultures. Many studies determined the optimal pH from various types of substrates. Batch studies indicated that the optimum initial pH for hydrogen production using sucrose as a limiting substrate ranged from 5.5 to 5.7 [103,143–145,182]. Zhang et al. [148,149,183] reported that the optimal initial pH for converting starch to hydrogen was found at 6.0 under thermal conditions. In addition, a study showed that an initial pH 6.0 was favorable for hydrogen production from cheese whey [145,184] and found that a better performance of hydrogen production from rice slurry was obtained at an initial pH of 4.5. Based on these studies, it can be concluded that an initial pH at slightly acidophilic conditions helps to enhance hydrogen production. There are also some conflicting reports on optimum pH requirement. The optimal initial pH for fermentative hydrogen production reported by Khanal et al. was 4.5 [144], while that reported by Lee et al. was 9.0 [110,145,183,185–187]. The possible reason for this disagreement may be the difference in inoculum, substrate, and initial pH range used. The pH was also found to have a profound effect on the generation of by-products [144,145].

Even though there have been many investigations of the initial pH on hydrogen production in batch studies, the optimal pH determined from continuous operation is still limited. In a continuous operation, a pH of 5.5 was found to be the optimum for hydrogen production from glucose [188]. Lay [189,190] optimized hydrogen production by controlling the pH at 5.2 in a starch-synthetic wastewater. In addition, Lay and his coworkers determined the optimum pH of 5.8, based on the statistical contour plot analysis in a complete mixed bioreactor converting beer processing wastes into hydrogen [142,160,191]. The optimal pH for fermentative hydrogen production reported by Mu et al. was 4.2 [154,192,193], while that reported by Zhao and Yu was 7.0 [194]. More work on the determination of optimum pH for organic wastes is needed. Also, the chemicals used in laboratory experiments to control the pH are expensive and cause safety issues. A more economical way to control the pH must be found before large-scale production can successfully take place.

4.3.2.4 Hydraulic Retention Time

A kinetic study of hydrogen production using sucrose as a limiting substrate showed the maximum specific growth rate of 0.172 h^{-1} for hydrogen producers, which allowed them to retain a continuous stirred tank reactor (CSTR), operating at a short HRT [88,165,195–197]. An investigation of the effects of HRT on hydrogen production indicated that the maximum hydrogen yield of 1.76 mol H_2/mol glucose was obtained from a CSTR operated at HRT of 6 h [198,199]. In addition to the hydraulic effect on a CSTR, hydrogen fermentation could be carried out at a further shorter HRT in a high-rate bioreactor, which can maintain the biomass with an unlimited sludge retention time. In a three-phase fluidized-bed bioreactor, HRT could be reached as short as 2 h to accomplish the best hydrogen yield of 2.67 mol H_2/mol sucrose [159,167,168]. On the other hand, a maximal hydrogen yield of 3.03 mol H_2/mol sucrose was found at an HRT of 0.5 h in a carried-induced granular sludge bed bioreactor [186,187,200,201].

Since varying HRT altered the organic loading rate (OLR) simultaneously, there is a concern with the ambiguity between the effect of HRT and the OLR on hydrogen production. Therefore, a study, examining the influence of HRT and substrate concentration on continuous hydrogen production by the granular acidogenic sludge at a constant OLR, reported that the maximum yield occurred at HRT of 13.7 h with a sucrose concentration of 14.3 g/L [124,188,202].

At short HRT, hydrogen consumers, primarily methanogens, could essentially be washed out or depleted. Control of HRT could be another strategy to limit the hydrogen consumers without a pretreatment of seed sludge. Lin and Jo [203] operated a hydrogen-producing anaerobic sequencing batch reactor with a gradually reducing HRT until 8 h to completely inhibit methane production. More work on HRT effect in a continuous mode of operation (with one- or two-stage digesters) with different kinds of organic waste substrates is needed to determine optimum HRT condition.

4.3.2.5 Temperature

Temperature is one of the most important factors that influence the activities of hydrogen-producing bacteria and fermentative hydrogen production. It has been demonstrated that in an appropriate range, increasing temperature could increase the ability of hydrogen-producing bacteria to produce hydrogen during fermentative hydrogen production, but temperature at much higher levels could decrease it with increasing levels [13,19,145,193,204]. This could be due to change in reaction mechanism at different temperatures for different feed mixtures. Even though the optimal temperature reported for fermentative hydrogen production has not always been the same, it has fallen into the mesophilic range (around 37°C) and thermophilic range (around 55°C), respectively [13,19,205]. Glucose and sucrose were the most widely used substrate during the investigation of the effect of temperature on fermentative hydrogen production. Wang and Wan [206] examined the effect of temperature on fermentative hydrogen production by mixed culture. Leven et al. [207] examined the effect of process temperature on bacterial and archael communities in methanogenic bioreactors treating organic household waste. More studies using continuous mode and organic wastes are needed.

4.3.2.6 Effects of Products, Metal Ions, Nitrogen, and Phosphorous

During hydrogen fermentation, acetate and butyrate productions are always accompanied with hydrogen production. These acetate and butyrate products can result in the feedback of product inhibition to the microbes' activities. In an early study, Heyndrickx and his coworkers [208] reported no significant difference of hydrogen production was found when adding acetic acid up to 18.0 g/L.

However, the addition of butyric acid higher than 17.6 g/L began to inhibit the activity of *Clostridium butyricum*. In addition, van den Heuvel et al. [209] agreed with these results that only butyric acid up to 17.6 g/L inhibited the mixed culture acidogenic bacteria growth, but acetic acid did not inhibit bacteria growth.

Hydrogen partial pressure in the liquid phase is another factor that might interfere with biological hydrogen production. Lamed et al. [210] studied the effect of stirring on hydrogen production from cellulose and cellobiose. They found a threefold hydrogen content with less hydrogen production in the unstirred culture broth when compared to that in the stirred culture. Accordingly, Lay [189,190] demonstrated that increasing the agitation speed from 100 to 700 rpm in a lab-scale completed mixed reactor could double the daily hydrogen production rate from starch. The other approach reducing the hydrogen content showed that the process of nitrogen gas sparging could help enhance hydrogen yield from 0.85 mol H_2/mol glucose to 1.43 mol H_2/mol glucose [211].

Even though at a higher concentration, metal ion may inhibit the activity hydrogen-producing bacteria, a trace level of metal ion is required for fermentative hydrogen production [20,88,142,145,166,212]. Fe^{2+} was the most widely investigated metal ion for fermentative hydrogen production, probably because its presence is essential for hydrogenase [20,88,142,145,166,212]. Thus, more investigations of the effect of other metal ion concentration besides Fe^{2+} concentration [150] on fermentative hydrogen production are recommended. In addition, there exists certain disagreement on the optimal Fe^{2+} concentration for fermentative hydrogen production. For example, the optimal Fe^{2+} concentration for fermentative hydrogen production reported by Liu and Shen was 10 mg/L [140,145], while that reported by Zhang et al. was 589.5 mg/L [148,149,183,213]. The possible reason for this disagreement may be the difference in experimental conditions used in these studies. Since glucose and sucrose have been the most widely used substrate to evaluate the effect of metal ion on fermentative hydrogen production, more studies with organic wastes as substrate in a continuous mode of operation are needed.

Several studies also investigated the toxicity of heavy metals to fermentative hydrogen production. For example, Li and Fang [214] and Lin and Shei [215] reported that the relative toxicity of six electroplating metals to fermentative hydrogen production was in the following order: Cu > Ni–Zn > Cr > Cd > Pb [145,166,212,214–216], while Lin and Shei reported that the relative toxicity of three heavy metals to fermentative hydrogen production was in the following order: Zn > Cu > Cr [145,166,212,214–216].

Since nitrogen is a very important component for proteins, nucleic acids and enzymes that are of great significance to the growth of hydrogen-producing bacteria, it is one of the most essential nutrients needed for the growth of hydrogen-producing bacteria. Thus, an appropriate level of nitrogen addition is beneficial to the growth of hydrogen-producing bacteria and to fermentative hydrogen production accordingly [145,211,217]. Ammonia nitrogen has been the most widely investigated nitrogen source for fermentative hydrogen production. Thus, more investigations of the effect of other nitrogen source concentration besides ammonia concentration on fermentative hydrogen production are needed.

There is also some disagreement on the optimal ammonia nitrogen concentration for fermentative hydrogen production. For example, the optimal ammonia nitrogen concentration for fermentative hydrogen production reported by Bisaillon et al. was 0.01 g N/L [145,217], while that reported by Salerno et al. was 7.0 g N/L [145,166,218] possibly due to different experimental conditions used in these studies. More work with continuous systems is needed.

Phosphate is needed for hydrogen production due to its nutritious value as well as buffering capacity. It has been demonstrated that in an appropriate range, increasing phosphate concentration could increase the ability of hydrogen-producing bacteria to produce fermentative hydrogen, but phosphate concentrations at much higher levels could decrease it with increasing levels [145,217,218]. It had been shown that an appropriate carbon-to-nitrogen (C/N) and carbon-to-phosphorus (C/P) are important for fermentative hydrogen production. There is, however, some disagreement on the optimal C/N and C/P for fermentative hydrogen production. For example, the optimal C/N and C/P for fermentative hydrogen production reported by Argun et al. [145,219] were 200 and 1000, respectively, while those reported by O-Thong et al. were 74 and 559, respectively [145,220], possibly due to different experimental conditions. More work in this regard in a continuous mode is needed.

4.3.2.7 Design of Digester

As pointed out several times earlier, more studies in continuous mode for fermentative hydrogen production are needed. In a conventional CSTR, biomass is well suspended in the mixed liquor, which has the same composition as the effluent. Since biomass has the same retention time as the HRT, washout of biomass may occur at shorter HRT. The effect of HRT on hydrogen production has been described in Section 4.3.2.4.

In a CSTR, biomass concentration in the mixed liquor and the hydrogen production are limited. Immobilized cell reactors provide an alternative to a conventional CSTR, because they are capable of maintaining higher biomass concentrations and could operate at shorter HRT without biomass washout [145,148,149,157,183,213]. Biomass immobilization can be achieved through forming granules, biofilm, or gel-entrapped bioparticles [145,149,167,214,221,222]. For example, Zhang et al. found that the formation of granular sludge facilitated biomass concentration up to 32.2 g VSS/L and enhanced hydrogen production [145,148,149,183,213]. Mizuno et al. [211] showed enhancement of hydrogen production from glucose by nitrogen gas sparging. Zhang et al. concluded that the granule-based reactor was better than the biofilm-based reactor for continuous fermentative hydrogen production, because the granule-based reactor has a better ability of biomass retention [145,148,149,157,183,213]. Since most literature studies are for glucose and sucrose, more study using organic waste is needed.

Different reactors have different properties and comparison of various reactors was conducted in several studies [145,149,153,198,200,201,210,222,223]. Several obstacles must be overcome before hydrogen from biological processes can be produced economically. In the anaerobic process, there are several stages that occur simultaneously. The last stage, methanogenesis, utilizes the intermediate products from the preceding stages and converts them into methane, carbon dioxide, and water [13,18]. Under normal anaerobic conditions, the majority of hydrogen produced is consumed by methanogens. Therefore, to extract hydrogen from this process, the methanogenic bacteria must be inhibited to prevent the hydrogen from being used to form methane. A procedure must be established that inhibits the methanogenic bacteria in a continuous process over time while remaining economical and efficient. Once accomplished, the hydrogen formed by the process can be collected and utilized as an energy source.

A solution to utilizing hydrogen in the short-to-medium term until the infrastructure can be established is through hydrogen–methane mixtures. Methane produces less atmospheric pollutants and carbon dioxide per unit energy than other hydrocarbon fuels and already has a distribution network in place [224]. When combined with hydrogen, it has been shown to improve engine performance, extend operability ranges, and reduce pollutant emissions [225]. Hydrogen–methane mixtures are a potential immediate solution to a cleaner fuel supply.

Another approach is to develop a two-stage AD system to produce hydrogen and methane quantities necessary for these mixtures. The two-stage process is ideally set up to produce both hydrogen and methane while further degrading waste streams. Although promising in theory, the two-stage AD has not been widely accepted because of increased complexity and higher investment and operational costs [226–230].

4.3.3 BRIEF LITERATURE ASSESSMENT ON BIOHYDROGEN PRODUCTION

Ding et al. [46] evaluated the effect of protein on biohydrogen production from carbohydrates, in particular starch. They used two model compounds, rice as starch rich and soybean as protein-rich food waste (FW). They found that the maximum hydrogen production potential to be 0.99 mol H_2/mol of initial starch as glucose and the maximum hydrogen production rate occurred at a starch/protein ratio of 1.7. The protein content in the FW increased hydrogen production in two ways. First, it provided buffering capacity to neutralize the VFAs as concurrent products. Second, it provided readily available organic nitrogen such as soluble proteins and amino acids to microorganisms.

Thus, the existence of protein in the substrate of biohydrogen production is important. In order to get maximum hydrogen production from carbohydrates, protein content in feedstock should be optimized.

Organic nitrogen in proteins is transformed into inorganic ammonia nitrogen in anaerobic degradation. Ammonia and amino groups released from proteins neutralizes the potential pH decrease imposed by VFAs. Thus, proteins can maintain suitable pH by the production of bicarbonate. The pH stabilization by these two counteracting effects require protein to starch ratio to be at least 2 to have pH decrease within 0.5 limit. Lay [190] showed that pH-window for optimal hydrogen production from carbohydrates may be so narrow that a half unit decrease in pH can cause 50% decrease in hydrogen production from optimum.

Sung et al. [231] examined hydrogen production by anaerobic microbial communities that are exposed to repeated heat treatments. Biological hydrogen production by anaerobic mixed communities was studied in batch systems and in continuous flow bioreactors using sucrose as the substrate. The systems were seeded with anaerobically digested municipal biosolids that had been heat treated at 100°C for 15 min. During operation, repeated heat treatments of the biomass in the reactors at 90°C for 20 min were performed. Results indicated that both initial heat treatment of the inoculum and repeated heat treatments of the biomass during operation promoted hydrogen production by eliminating non-spore-forming hydrogen-consuming micro-organisms and by selecting for hydrogen-producing spore-forming bacteria. An operational pH of 5.5 was shown to be optimal for hydrogen production. The conversion efficiency and hydrogen yield were 0.0892 L H_2/g chemical oxygen demand (COD) and 1.5291 mole of H_2/mole of sucrose, respectively. Terminal restriction fragment length polymorphism analysis showed that *Clostridium* and *Bacillus* species were dominant populations in the bioreactors. A positive correlation was observed between the total abundance of *Clostridium* species and hydrogen production during part of an operational run.

Premier et al. [232,233] examined integration of biohydrogen, biomethane, and bioelectrochemical systems (BESs). Anaerobic bioprocesses such as AD, fermentative biohydrogen (BioH$_2$), and BES, converting municipal, agro-industrial wastes and crops to energy have attracted increasing interest. AD, however, still requires optimization of conversion efficiency from biomass to methane. Augmenting methane energy production with simultaneous BioH$_2$ and bioelectrochemical stage(s) would increase process efficiencies while meeting posttreatment effluent quality. Pretreatment of feedstock can increase bacterial accessibility to biomass, thus increasing the conversion yield to target product, but an alternative is separating the acidogenic/hydrolytic processes of AD from methanogenesis. Acidogenesis can be combined with BioH$_2$ production, prior to methanogenesis. Depending on operating conditions and without further treatment after digestion, the methanogenic stage may discharge a digestate with significant organic strength including VFAs. In order to meet wastewater discharge requirements, to apply adequate use of digestate on land, to minimize environmental impact, and to enhance recovery of energy, VFAs should be low. Simultaneously connected BESs-producing hydrogen and/or electricity can facilitate effluent polishing and improved energy efficiency. Various configurations of the BioH$_2$, methanogenesis, and BES are plausible and should improve the conversion of wet biomass to energy.

Kvesitadze et al. [228] examined the two-stage anaerobic process for the combined production of biohydrogen and biomethane from biodegradable solid wastes. The overall goal of the work was to increase of energy output from biodegradable fraction of MSW. For these purposes, development of cost-effective two-stage anaerobic fermentation process with biohydrogen and biomethane combined production was introduced. Anaerobic hydrogen-producing bacteria convert a large variety of carbohydrate sources (frequently obtained as refuse or waste products) mainly to hydrogen, CO_2, and VFAs. Since in natural environments, hydrogen-producing anaerobes coexist with methane forming ones that consume intermediate products of hydrogen fermentation, the most common obstacle to obtain hydrogen as the final product of hydrogen fermentation is associated with the partition of hydrogen and methane fermentation processes in a cost-effective way. The strategies that can be used to handle this issue include seeking out low-cost chemicals for partition of hydrogen and methane fermentations; isolating, studying, and revealing highly productive hydrogen-producing bacteria; and developing optimum conditions for biohydrogen and biomethane sustainable production from biodegradable fraction of MSW.

Kvesitadze et al. [228] found that the use of 10% solution of sodium hydroxide provided practically full inhibition of methane-forming bacteria. In thermophilic conditions and pH 9.0, cumulative biohydrogen productions made 82.5 L/kg dry organic matter. Percentage of hydrogen and carbon dioxide in the produced biohydrogen were 50.1% and 49.8%. Content of methane in biohydrogen didn't exceed 0. 1%. Adding of inoculums of hydrogen-producing bacteria promoted the increase of biohydrogen cumulative production up to 104 L/kg dry organic matter. Cumulative biomethane production from the metabolite products of hydrogen fermentation made 520 L/kg dry organic matter. Methane and carbon dioxide percentage determined in obtaining biomethane was 78.6% and 21.4%, respectively.

Comparing energy data for two-stage AD with those for single-stage process for biomethane production showed that energy output from biodegradable fraction of MSW increased by 23%–26% under two-stage anaerobic process. Biohydrogen and biomethane combined production can be implemented at large-scale biogas plants improving process economy. Introduction of such technology at the existing biogas plants needs low investments.

Gottardo et al. [234] examined dark fermentation optimization by anaerobic digested sludge recirculation: effects on hydrogen production. The addition of small amounts (5%–10%) of H_2 to rich CH_4 biogas improves the quality of gas combustion while reducing CO_2 emissions. For this reason, many researchers have recently been focused to optimize the two-phase AD process (dark fermentation, DF and anaerobic digestion, AD) to concurrently produce H_2 and CH_4 from biowaste and organic residues. In this study, the results of a two-phase thermophilic AD process treating biowaste for H_2 and CH_4 production are presented: in this process, neither physical nor chemical pretreatment of inoculum was used, but recirculation of anaerobic digested sludge to the DF reactor was exploited to control the pH in the optimal H_2 production range of 5–6. The experiment was carried out in bench scale using a stirred reactor (CSTR), applying an OLR of 16 kg TVS/m^3 day, and an HRT of 3 days in DF phase. Four different recirculation conditions were tested keeping constant the ammonia concentration (about 500 mg/L) through a separation process by evaporation. The aim was to investigate the influence of ammonia in the biological process of hydrogen production via dark fermentation and to research the recirculation condition that allow to reach the best yields and a stable process. The optimal hydrogen production (SHP of 0.03 m^3/kg TVS fed, 30% H_2 in the off-gas) was found using a recirculated sludge ratio of 0.66. This allowed for a stable pH value around 5.5.

Mudhoo and Kumar [216] examined the effects of heavy metals as stress factors on AD processes and biogas production from biomass. Heavy metals affect the biochemical reactions that take place during AD processes of organic matter. In this review, the different effects observed in AD processes and during the production of biomethane and biohydrogen from several substrates contaminated with and/or inheriting heavy metals from the substrates themselves were discussed. It has been found that heavy metals exert important roles in biochemical reactions. Heavy metals like copper, nickel, zinc, cadmium, chromium, and lead have been overwhelmingly reported to be inhibitory and under certain conditions toxic in biochemical reactions depending on their concentrations. Heavy metals like iron may also exhibit stimulatory effects, but these effects have been scantily observed. This review also concludes that the severity of heavy metal inhibition depends upon factors like metal concentration in a soluble, ionic form in the solution, type of metal species, and amount and distribution of biomass in the digester or chain of biochemical reactions that constitute the AD process.

A majority of studies have demonstrated that the toxic effect of heavy metals like chromium, cadmium, and nickel is attributable to a disruption of enzyme function and structure by the binding of the metal ions with thiol and other groups on protein molecules or by replacing naturally occurring metals in enzyme prosthetic groups. Published data for the effects of heavy metals on the hydrolysis stage of AD process chemistry, are lacking and hence, further studies are required any changes in this area.

Nasr et al. [226,235] examined biohydrogen production from thin stillage using conventional and acclimatized anaerobic digester sludge (AADS). To assess the viability of biohydrogen production from thin stillage, a comparative evaluation of anaerobic digester sludge (ADS) and AADS for bio-hydrogen production over a wide range of S_0/X_0 ratio (0.5/8 g COD/g VSS) was performed. A maximum hydrogen yield of 19.5 L H_2/L thin stillage was achieved for the AADS, while tests with ADS achieved a maximum yield of only 7.5 L H_2/L thin stillage. The optimum range of S_0/X_0 ratio for hydrogen production was found to be 1–2 g COD/g VSS using conventional ADS and 3–6 g COD/g VSS using AADS. The biomass specific hydrogen production rate for the AADS was 3.5 times higher than the rate for the ADS throughout the range of S_0/X_0 ratio examined in this study. The DGGE profiles of the 16S rDNA gene fragments confirmed the superior performance of the AADS over the ADS, showing that the widely known hydrogen producers *C. acetobutylicum, Klebsiella pneumoniae, C. butyricum,* and *C. pasteurianum* were the predominant species.

Noebauer and Schnitzhofer [230] examined digestion processes for the conversion of sucrose-containing substrates. A new designed carrier-based bioreactor was used for thermophilic hydrogen fermentation for a two-stage digestion process. Thick juice, a preproduct of sugar production and molasses, a by-product of it were used as a substrate to compare hydrogen productivity, hydrogen yield, and acid spectrum. Both substrates were efficiently converted to hydrogen and organic acids up to organic loads of 1.75 g/L/h and HRTs of 10 h. At similar levels of organic load, molasses yielded better overall hydrogen production yield (30%), which is contributed to the more complex composition and higher trace element content of molasses. Further products formed were acetate and lactate in a ratio of approximately 70:30, which are preferred substrates for fast and efficient conversion to methane.

Zheng et al. [223] examined production of biohydrogen using a membrane anaerobic reactor to evaluate limitations due to diffusion. Mu et al. [154] examined the maximum specific hydrogen-producing activity (SHAm) of anaerobic mixed cultures, in particular their definition and determination. Fermentative hydrogen production from wastes has many advantages compared to various chemical methods. Methodology for characterizing the hydrogen-producing activity of anaerobic mixed cultures is essential for monitoring reactor operation in fermentative hydrogen production; however, there is lack of such kind of standardized methodologies. In the present study, a new index, that is, SHAm of anaerobic mixed cultures, was proposed, and consequently, a reliable and simple method, named SHAm test, was developed to determine it. Furthermore, the influences of various parameters on the SHAm value determination of anaerobic mixed cultures were evaluated. Additionally, this SHAm assay was tested for different types of substrates and bacterial inocula. The results demonstrate that this novel SHAm assay was a rapid, accurate, and simple methodology for determining the hydrogen-producing activity of anaerobic mixed cultures. Thus, application of this approach is beneficial to establishing a stable anaerobic hydrogen-producing system.

Fuentes et al. [222] examined modeling of hydrogen production in biofilm reactors using anaerobic digestion model 1 (ADM1). Hydrogen can be produced substantially at a high rate by anaerobic fermentation from organic waste or wastewater. The aim of the work is to model the hydrogen production in anaerobic biofilm reactors using a modeling methodology previously published. The global model combines the dynamics of the three phases present in the reactor including biochemical, physicochemical, and hydrodynamic processes. The ADM1 was selected to calculate the global yield of hydrogen produced from a sugar-based substrate source. As model application, an example based on the start-up and operational performance of an anaerobic fluidized-bed reactor is presented. Several details of reactor design, biofilm and substrate characteristics, and operational conditions are required for model adjustment and validation. Only the parameter related to biomass transport phenomena was estimated by simulation, and no other parameters had to be modified. Values around 1×10^{-22} dm d^2/g were calculated. A good agreement between experimental and predicted values of soluble metabolic products and hydrogen production rate is obtained.

Toscano et al. [227] examined AD of residual lignocellulosic materials to biogas and biohydrogen. Arundo donax (giant reed) hydrolysate was exploited as a substrate for AD aimed to the production of biogas and biohydrogen. A mixed culture adapted from a primary sludge digester was used as inoculum. Besides the biogas, the anaerobic fermentation products were ethanol, acetic acid, and butyric acid. Together, the soluble products accounted for about 51% of degraded carbon. The produced biogas was 5% H_2, 41% CO_2, and 54% CH_4 by volume. Fermentation kinetics was slower and biogas yields were lower than those found with glucose fermentation with same mixed culture. Further optimization of the process is envisaged in order to improve biohydrogen yield.

Babson et al. [236] examined AD for methane generation and ammonia reforming for hydrogen production. During AD, organic matter is converted to carbon dioxide and methane, and organic nitrogen is converted to ammonia. Generally, ammonia is recycled as a fertilizer or removed via nitrification–denitrification in treatment systems; alternatively, it could be recovered and catalytically converted to hydrogen, thus supplying additional fuel. To provide a basis for further investigation, a theoretical energy balance for a model system that incorporates AD, ammonia separation and recovery, and conversion of the ammonia to hydrogen is reported. The model anaerobic digestion-bioammonia to hydrogen (ADBH) system energy demands, including heating, pumping, mixing, and ammonia reforming, were subtracted from the total energy output from methane and hydrogen to create an overall energy balance. The energy balance was examined for the ADBH system operating with a fixed feedstock loading rate with C:N ratios (gC/gN) ranging from 136 to 3 that imposed corresponding total ammonia nitrogen concentrations of 20–10,000 mg/L. Normalizing total energy potential to the methane potential alone indicated that at a C:N ratio of 17, the energy output was greater for the ADBH system than from AD generating only methane. Decreasing the C:N ratio increased the methane content of the biogas comprising primarily methane to >80% and increased the ammonia stripping energy demand. The system required 23%–34% of the total energy generated as parasitic losses with no energy integration, but when internally produced heat and pressure differentials were recovered, parasitic losses were reduced to between 8% and 17%.

Finally, Kapdan and Kargi [170] presented a review of biohydrogen production from waste materials. The review indicated that biological production of hydrogen gas has significant advantages over chemical methods. The major biological processes utilized for hydrogen gas production are biophotolysis of water by algae and dark and photofermentation of organic materials, usually carbohydrates by bacteria. Sequential dark and photofermentation process is a rather new approach for biohydrogen production. One of the major problems in dark and photofermentative hydrogen production is the raw material cost. Carbohydrate rich, nitrogen deficient solid wastes such as cellulose and starch containing agricultural and food industry wastes and some food industry wastewaters such as cheese whey, olive mill, and baker's yeast industry wastewaters can be used for hydrogen production by using suitable bioprocess technologies. Utilization of aforementioned wastes for hydrogen production provides inexpensive energy generation with simultaneous waste treatment. This review article summarizes biohydrogen production from some waste materials and addresses the roles of types of potential waste materials, bioprocessing strategies, microbial cultures to be used, bioprocessing conditions, and the recent developments on biohydrogenation with specific references to their relative advantages.

The management of sulfides during biohydrogen production has also been studied by Hilton and Archer [237] and Haghighat Afshar [238]. The literature on biohydrogen production is also well reviewed in several publications by Wang and Wan [145], Guo et al. [239], Thompson [229], Liu et al. [240], and Massanet-Nicolau et al. [152], among others.

4.4 CODIGESTION

Codigestion is the simultaneous digestion of a homogeneous mixture of multiple substrates. The most common situation is when a major amount of basic substances is mixed and digested together with minor amounts of a single or a variety of additional substrates. Recent research by Wu et al.

[66,241] demonstrates that using co-substrates in anaerobic digestive systems improves biogas yields through positive synergisms established in the digestion medium and the supply of missing nutrients by the co-substances.

Historically, AD was carried out for animal manure and sewage sludge from aerobic wastewater treatment. In recent years, agricultural biogas plants use pig, cows, and chicken manure with co-substrate, which increases organic content of total substrate. Cowastes can be organic wastes from agriculture-related industries, FWs, collected municipal biowastes from households, energy crops and top leaves of sugar beets, etc. Fats provide the largest biogas yield but require high retention time. Carbohydrates and proteins have faster conversion rates but lower yields. If pathogens or other organisms are present, pasteurization at 70°C and sterilization at 130°C of feed materials is needed prior to fermentation. The C/N ratio should be between 15 and 30 to avoid process failure by ammonia accumulation. The fermentation residue should be used as fertilizer.

The codigestion process adds another substrate to the existing digesters to utilize the availability of free capacities. The most common situation is when the major amount of main basic substrate (e.g., sewage sludge) is mixed with and digested together with minor amounts of a single or a variety of additional substrates. The word codigestion is used for digestion of all types of mixed substrates independent of its composition [242–248].

Until recently, just like single component combustion, gasification, liquefaction, and pyrolysis, methane production AD was carried out for a single substrate. For example, manure was digested to produce energy; sewage sludge was anaerobically stabilized, and industrial wastewater was pretreated before final treatment in a wastewater treatment plant. While these were convenient approaches, each AD unit did not fully utilize its capacity, nor it generated methane production to the fullest level. The concept of codigestion allows sharing of the equipment for different types of waste, thus reducing any required transportation costs for individual waste and making an optimum and most economical use of every waste treatment plant. Codigestion also requires a careful choice of multiple substrates that can give the most methane production through synergistic reaction [242–248].

Codigestion can result in an important increase of the biomethane potential (BMP) when the substrates mixture is prepared with proper percentages of the different organic substrates to be digested [242–248]. The beneficial effect of codigestion is mainly due to the optimization of the nutrient balance in the substrates mixture when codigesting nitrogen-rich substrates with carbon-rich substrates.

The higher specific methane yield of 686 Nm^3/tons volatile solid (VS) is achieved by co-digesting organic fraction of municipal solid waste (OFMSW) and vegetable oil (83/17 on dry weight). Also, OFMSW with animal fat (83/17 on dry weight) and cow manure (CM) with fruit and vegetable waste (FVW) (60/40 on weight) can give very high methane yield of 490 and 450 Nm^3/tons VS, respectively [249–265]. Codigestion can be synergistic, antagonistic, or neutral depending on the selection of co-substrates.

4.4.1 Advantages and Disadvantages

The codigestion process offers several advantages and disadvantages [242–265].

4.4.1.1 Advantages

1. It improves nutrient balance and digestion. The digestion of a variety of substrates can improve the nutrient ratio of total organic carbon/nitrogen/phosphorous, which should be ideally 300:5:1. It also maintains a reasonable mix of minerals such as Na, K, Mg, Mn, and Fe as well as balanced composition of trace metals. These types of nutrient balances help maintain stable and reliable digestion performance. It also provides a larger quantity and good quality fertilizer as the digestate.

2. In general, codigestion enhances methane (or biogas) production rate. Increased biogas productivity (m^3 biogas/m^3 digester volume/day) compensates for high investment and running costs of small- and medium-sized digesters.

3. Wastes with poor fluid dynamics, aggregating wastes, particulate or bulking materials, and floating wastes can be much easily digested after improving its rheological properties through homogenization with dilute substrate such as sewage sludge or liquid manure. Good rheological properties also help the performance of a digester.

4. Codigestion allows the mechanism to avoid overloading and underloading of a single waste that may be seasonal dependent. Through the use of multiple substrates, digestion process can be maintained at a constant rate throughout a year.

5. In agricultural digesters, the process of codigestion can considerably improve the overall economics (payback time) of a plant. Gate fee creates a win–win situation. A provider pays significantly lower price at a farm-scale AD plant than at an incineration or composting facility (usually by a factor of 3 or 4). A farmer takes the credit of the increased biogas production along with the income from gate fee.

6. If there is a sufficient farmland available, the digestate from codigestion can be directly recycled as a fertilizer at reasonable cost.

7. In a limiting situation of industrial wastes, energy crops may become an interesting alternative especially when plants are grown on fallow or set-aside land that attracts subsidies. Traditionally, all C4 plants have very good growth yields. Corn has become a very good co-substrate in Germany. In order to make sure that it is grown in a set-aside land and is taken out of the nutrition chain, it is treated after harvest with manure before it is ensiled. In Austria, Sudan grass, another C4 plant, is grown as energy crop for codigestion. It grows well in the dry soils of the southeast Austria [242–265].

8. Codigestion reduces fossil fuel consumption and minimize carbon footprint.

9. Codigestion increases a plant's value to the community by recycling difficult liquid "wastes." It can be a true water resource recovery facility.

10. Codigestion reduces operating and energy costs.

4.4.1.2   Disadvantages

1. Codigestion increases a digester effluent COD that requires more expensive treatment before it can be discarded.

2. Some substrates used in codigestion require additional pretreatment. This adds to the overall cost.

3. More substrates in the digester require additional mixing. This increases the power cost for the mixing.

4. Codigestion also increases wastewater treatment requirement. It increases waste water treatment plant (WWTP) direct loads.

5. Some co-substrates create additional hygienization requirement. Potentially infectious materials must be hygienized by law. The hygienization can be (a) sterilization (20 min at 2 bars, 121°C), (b) pasteurization (between 70°C and 90°C for 15–60 min), and (c) sanitation (heat treatment at lower temperatures over an extended time period) [265,266].

6. The land use for digestate can also be a restriction in some areas.

7. Codigestion economics very strongly depend on crop or the type of co-substrate. Flexibility in choosing the right co-substrate may be limited.

8. Substrates must be trucked in, which adds to the cost.

9. C/N ratio changes in influent when high-C, low-N substrate is directed to digesters.

10. Codigestion of manure with organic wastes can increase nitrogen by 57% compared to baseline operation, with smaller increases in ammonia and phosphorus. The additional nutrients to the farm can exacerbate existing nutrient management systems, particularly for large operations. There are also concerns that contaminants (such as materials containing prions) or pathogens arriving with the organic waste can inhibit AD process, limit the use of AD products, or create a possibility for negative public health impacts.

4.4.2 Major Applications, Users, and Related Issues with Codigestion

Codigestion is particularly applicable for agricultural biogas production from manures (which has a relatively low gas yield), which is economically not viable. The addition of co-substrates with high methane potential not only increases biogas yield (which generates more income) but also increases income through tipping fees.

Another major application for codigestion is the digester in waste treatment plant [267–271], which is usually oversized. The addition of co-substrate uses its unfilled capacity and produces more gas and consequently more electricity at only marginal additional cost. The extra electricity produced can be used for the energy needs of wastewater treatment plant at a competitive cost [242–265].

In most cases, codigestion is applied in wet single-step processes such as CSTR. The substrate is used in a slurry of 8%–15% concentration. This does not significantly add to the cost of mixing. Wet systems are particularly useful when digestate slurry (without removal of water) can be directly used on field and green lands for soil conditioner or fertilizers.

In recent years, the percentage of codigestion plants in waste treatment process is steadily increasing. In fact, a number of existing municipal sewage sludge digesters are already using co-substrates. New sewage plants or extension of existing plants use different types of organic and FWs such as bio-waste, food leftovers, fat wastes, and floatation sludges as co-substrate to improve biogas production. Codigestion plants also provide controlled organic waste disposal mechanism for fat trap contents and food leftovers for the communities.

Besides improving biogas yield, the codigestion plants also improve the energy balance over single substrate plant. On average, for sewage sludge plant, the self-sufficiency of electricity is below 50%. With the use of a co-substrate, this number can be improved as high as 80%, and in some cases, the codigestion plant becomes a net energy producer [242–265].

In agriculture industries, codigestion has become a standard norm. Many small and medium-sized farm-scale digesters use mixed co-substrates with manure. According to Braun and Wellinger [248], in 2002, most of about 2000 agricultural plants in Germany used co-substrates. In other parts of Europe, such as Austria (110), Switzerland (71), Italy (>100), Denmark (>30), Portugal (>25), Sweden, France, Spain, and England, all use codigestion process.

4.5 FEEDSTOCK EFFECTS ON BIOGAS PRODUCTION

4.5.1 Single Substrate

All substrates containing carbohydrates, proteins, fats, cellulose, and hemicellulose as major components can produce biogas by AD. The composition of biogas and methane yield depends on the content of carbohydrates, proteins, and fat, the digestion system (and its pH), and the retention time and temperature. Wood undergoes very slow anaerobic decomposition and is therefore not suitable for anaerobic decomposition. The solubilization of CO_2 in the digestate further increases methane yield.

Recently, energy crops are extensively used as pure or co-substrate for AD. The most important parameter in the choice of the energy crop is energy yield per hectare. Maize and forage beets have the highest gross energy potential and different cereal crops, perennial grasses, and forage crops also have potentials for biogas production. In general, easily degradable biomass results in biogas production. A comparison of biogas and methane potentials for various types of biomass is illustrated in Table 4.4. More discussion on biogas production for some of the selected feedstock is briefly outlined as follows.

4.5.1.1 Coir Pith

Coir Pith is a lignocellulosic agro residue that is produced as a by-product in coir industry in large quantity. Kunchikannan et al. [39] examined production of methane from this waste material by AD. The study indicated that methane yield of 38.1% per kg of dry pith weight in 44 days; the yield can be improved by about 1.5 times by the reduction of particle size.

TABLE 4.4
Mean Biogas and Methane Potentials of Different Crops, Organic Wastes, and Biomass

Substrate	Mean Biogas Yield	Range of Methane Yield
	(Nm³/ton VSs[a] Added)	
Sugar beet	750	236–381
Fodder beet	775	420–500
Maize	605	200–450
Corn cob mix	670	
Chaff		270–316
Straw		242–324
Leaves		417–454
Kale		240–334
Oats		250–295
Wheat	660	384–426
Triticale	605	337–555
Miscanthus		179–218
Sorghum	550	235–372
Grass	565	298–467
Clover grass		290–390
Red clover	575	300–350
Clover		345–350
Sunflower	480	154–400
Wheat grain	725	384–425
Rye grain	670	283–492
Reed canary grass		340–430
Nettle		120–420
Ryegrass		390–410
Hemp		355–409
Flax		212
Oilseed rape		240–340
Jerusalem artichoke		300–370
Peas		390
Potatoes		276–400
Barley		353–658
Alfalfa		340–500
Sudan grass		213–303
Sorghum		295–372
Triticale		337–555
CM (9% TS)	255	
Pig manure (7% TS)	385	
Chicken manure (15% TS)	290	
Corn (whole crop)	546	
Harvest residue (straw, stems)	375	
Sewage sludge	300	
Expired animal feed	575	
Food industry waste	500	
Stillage from breweries	600	
Green wastes (markets)	550	

(Continued)

TABLE 4.4 (*Continued*)
Mean Biogas and Methane Potentials of Different Crops, Organic Wastes, and Biomass

Substrate	Mean Biogas Yield	Range of Methane Yield
	(Nm^3/ton VSs[a] Added)	
Biowastes (source separated)	450	
Floatation sludge/animal fat	775	
Waste fat	1000	
Municipal solid waste	326	
Sunflower		154–400
Rhubarb		320–490
Turnip		314
FVWs	380	
Municipal solid waste	480	
FVW and abattoir wastewater	770	
SM	306	
Municipal solid waste	181	
Food waste leachate	267	
RS	317	
Maize silage and straw	283	
Jatropha oil seed cake	383	
Palm oil mill waste	553	
Household waste	317	
Lignin-rich organic waste	181	
SM and winery wastewater	316	
Food waste	359	

Sources: Braun, R. et al., Biogas from energy crop digestion, IEA Bioenergy, Task 37—Energy from Biogas and Landfill Gas, Denmark, 2009; Braun, R., Anaerobic digestion—A multifaceted process for energy, environmental management and rural development, in: Ranalli, P., ed., *Improvement of Crop Plants for Industrial End Users*, Springer, 2007; Murphy, J. et al., Biogas from crop digestion, IEA Bioenergy, Task 37—Energy from Biogas, Denmark, September 2011; Braun, R., Potential of codigestion, IEA Task 37 Report, IEA, Brussels, Belgium, 2002, http://www.novaenergie.ch/iea-bioenergytask37/Dokumente/final.PDF, Accessed on November 7, 2007; Braun, R. and Wellinger, A., Potential of codigestion, IEA Bioenergy, Task 37—Energy from Biogas and Landfill Gas, IEA, Brussels, Belgium, 2004; Braun, R. et al., *Biotechnol. Lett.*, 3, 159, 1981; Braun, R. et al., *Appl. Biochem. Biotechnol.*, 109, 139, 2003; Shah, Y.T., *Water for Energy and Fuel Production*, CRC Press, New York, 2014; Shah, Y.T., *Energy and Fuel Systems Integration*, CRC Press, New York, 2015.

[a] VSs, volatile solids.

The increase in acidity during digestion process decrease the methane yield, while an increase in alkalinity did not significantly changed the methane yield.

4.5.1.2 Whey

The waste from various food industries is capable of generating methane because of their high organic content. Whey is normally used as a component of dairy products or as an additive for food product. Beszedes et al. [43] examined biogas generation from membrane separated fractions, that is, permeate and concentrate of whey. The study examined the effects of pH, thermal, microwave pretreatment, and their combinations on the biogas yield. The pretreatment had a significant effect on the biogas yield. The hydrolysis of large molecules enhanced the biodegradability of whey thereby increasing the productions of biogas and methane. The long time classical heat treatment and the microwave radiation in acidic medium significantly increased the methane production. The concentrate of whey was more adaptable to AD than permeate or the whole whey.

4.5.1.3 Distillery Spent Wash

Distillery spent wash is a major pollutant in water. In recent years, methane is generated from AD of distillery spent wash by fixed-film systems and two-phase anaerobic systems. Pathe et al. [41] showed that a treatment option that involved a two-stage aerobic oxidation processes (activated sludge and extended aeration) followed by a physical chemical treatment using lime, polyaluminum chloride, polyelectrolyte, and carbon adsorption as tertiary treatment can be the most efficient method for methane generation from distillery spent wash. The treated effluent can be used for green belt development and in the agriculture industry.

Nandy et al. [55] treated a high-strength distillery spent wash in fixed-film, fixed-bed two-stage anaerobic reactors using cheaper and abundantly available pebbles as media. The experiments were carried out in laboratory as well as in pilot-scale operation. The results showed that the overall COD removal was about 80% with a specific biogas yield of 0.3 me CH_4/kg COD. For the two-stage system, while the packed-bed reactor can be easily fed with the spent wash, the detention period for each reactor has to be increased to obtain COD removal close to 80%. In both reactors, the biogas yield decreased for HRT beyond 2.43 days. Feeding to the reactors can only be stopped for maximum 3 days and reactors become sour and need to be reenergized for pH level less than 6.0. Most efficient operation was obtained at temperature of 35°C–40°C and pH of 7.0. Greater depth of reactor gave poorer performance and no clogging of the reactor was observed for 18 months.

4.5.1.4 Swine Waste

Chen et al. [38] examined various engineering options of conversion of swine waste to biomethanol. They applied target costing method in the development of marketable and environmentally friendly products like biomethanol from swine waste. Biomethanol is produced from methane that is once again generated by AD of swine waste.

4.5.1.5 By-Products of Biodiesel Productions

The process of biodiesel production is predominantly carried out by catalyzed transesterification. Besides desired methyl esters, the process produces several by-products such as crude glycerol, oil-pressed cakes, and washing water. Crude glycerol or g-phase is heavier separate liquid phase, composed mainly of glycerol. Numerous types of oil cakes such as canola, rapeseed, coconut, cottonseed, groundnut, mustard, olive, palm kernel, sesame, soybean, and sunflower are also created in this process. Although their composition widely varies depending on the parameters and substrates used for biodiesel production, all these by-products provide valuable feedstock for biogas production. The study by Kolesarova et al. [42] leads to the following conclusions:

1. Crude glycerol from biodiesel production is a valuable substrate for anaerobic degradation and for the production of biogas using g-phase as a single substrate.
2. g-Phase also has a great potential to become a co-substrate by anaerobic treatment of different types of organic wastes such as OFMSW, mixture of olive mill wastewater, slaughterhouse wastewater, corn maize waste, maize silage, and swine manure (SM).
3. Olive cakes and olive meals along with rapeseed and sunflower oil cakes can be used for AD to produce biogas and methane. High stability of the AD of sunflower oil cake under mesophilic conditions was obtained. With the increased amount of oil gained from a rapeseed meal by the extraction process, the possible biogas production from a rapeseed cake

decreases. No significant effects of pretreatment (thermal and chemical) of sunflower and rapeseed residues enhanced methane yield.

4. Washing water from biodiesel production is a good candidate for anaerobic degradation because of high content of biodegradable organic substances.
5. Specific inhibition effects resulting from the substrate composition should be considered during anaerobic treatment of biodiesel by-products. In the case of AD of crude glycerol, high salinity of the substrates may negatively affect the methanogenic microorganisms. The concentration of ammonium should also be monitored. Since nitrogen is an essential nutrient for microorganisms, the low concentration in the crude glycerol and washing water has to be compensated by ammonium supplement. On the other hand, nitrogen-rich substances have high concentration in rapeseed cake that may cause ammonium accumulation in the reactor, thereby inhibiting the digestion process.

The use of by-products of biodiesel process as potential source of energy producer than wastes makes the process of biodiesel more economically attractive.

4.5.1.6 Palm Oil Mill Effluent

Yusoff et al. [44] examined the effects of HRT and VFAs produced during fermentation on biohydrogen production from Palm oil mill effluent. Both HRT and VFA concentration played a vital role in the biohydrogen concentration, biohydrogen rate, and biohydrogen yield. The results were obtained for HRT 2, 3, and 5 days and 2 days gave the optimum operation with maximum biohydrogen yield and rate with maximum biohydrogen concentration of 30%. The VFA as soluble metabolites reduced the amount of biohydrogen production by 8%–10%. The study concluded that HRT and VFA affect the biohydrogen production and should be considered in biohydrogen fermentation.

Cowan [272] points out that one way to reduce the emission of CO_2 and increase energy production from bioprocess technologies for wastewater treatment is to use an integrated algae pond systems to address the range of wastewater treatment problems. The growth of algae requires the use of CO_2 that minimizes emission to the environment. Furthermore, IAPS produces a quality effluent suitable for irrigation, negates the food versus fuel debate, and reduces the demand for fossil fuel–derived energy and fertilizers.

Ryan et al. [273] addressed the issue of wastewater treatment from ethanol-producing biorefineries. They suggested that in order to treat the effluent from these refineries efficiently and economically to meet local requirements and minimizing the net water consumption, a process integration wherein (1) improving existing secondary (i.e., biological) treatment to maximize COD reduction, (2) incorporating a tertiary "polishing" stage to remove color, and (3) using reverse osmosis membrane technology to recover process water would be desirable. Ryan et al. [273] showed that the energy required for secondary and tertiary treatment stages can be obtained from biogas derived power from the anaerobic digester. Thus, this type of an integrated approach of post–biological treatment of ethanol stillage can address the issues of efficient refinery operation with minimum net water consumption.

4.5.1.7 Long-Chain Fatty Acids

As shown in Table 4.3 [19], potentials for biogas and methane productions from lipids are much higher than that from proteins and carbohydrates. Long-chain fatty acid (LCFA) commonly found in wastewaters include lauric acid, myristic acid, palmitic acid, palmitoleic acid, stearic acid, oleic acid, and linoleic acid, among others [19]. An extensive number of studies for the treatment of wastewater containing lipids and LCFA in different types of anaerobic reactors have been reported. These studies are evaluated by Alves et al. [56].

The high-rate anaerobic technology (HR-AnWT) for wastewater treatment requires the expansion of suitable substrates, in particular better treatment of the wastewater with high lipid content.

Waste lipids are good candidates for substrates needed to improve biogas and methane productions, as compared to proteins and carbohydrates.

Alves et al. [56] presented a review of how LCFA degradation is accomplished by syntrophic communities of anaerobic bacteria and methanogenic archaea. For optimal performance, these syntrophic communities need to be clustered in compact aggregate that is often difficult to achieve with wastewater that contains fats and lipids. Alves et al. [56] proposed a new reactor concept that provides primary biomass retention through floatation and secondary biomass retention through settling.

The types of bacteria involved in methanogenic conversion of LCFA are known and the biochemical mechanism of LCFA degradation by beta-oxidation is well understood. The initial steps in the anaerobic conversion of unsaturated LCFA are, however, unclear. Besides obligate hydrogen-producing acetogens (OHPA) that degrade unsaturated LCFA, bacteria exist that have the ability to hydrogenate unsaturated LCFA to saturated LCFA. This conversion can be coupled to growth and these bacteria may compete with hydrogenotrophic methanogens for hydrogen.

LCFA require the syntrophic cooperation of OHPA and methanogens. These synthropic communities perform optimally when they are organized in microcolonies; the interspecies hydrogen transfer is enhanced with a short intermicrobial distance. It is yet not clear how microcolonies are developed in a fatty matrix and what is the effect of hydrogen transfer. Since hydrogen is poorly soluble in water, hydrogen transfer is increased when the matrix is LCFA. More work in this area is needed.

4.5.1.8 Organic Wastes and Energy Crops

As the interest in using AD technique to generate biogas and biomethane increases due to economical and environmental reasons, it is important to determine the ultimate methane potential for a given solid substance. In fact, this parameter determines to some extent both design and economic analysis of a biogas plant. The ultimate methane potential thus identifies "thermodynamic limit" for a given substance. Furthermore, in order to compare potentials of various substrates, the definition of common units to be used in anaerobic assays is becoming increasingly important. Angelidaki et al. [47–50] presented some guidelines for biomethane assays of the AD prepared by the specialists group of International Water Association. The guidelines include the considerations of biodegradability, bioactivity, inhibition, and matrices for biostability.

Significant efforts have been made [47–50] to generate biogas (biomethane) from different types of organic wastes. AD is a preferred method for energy resource recovery from organic residuals because this method (1) generates biomethane, (2) reduces the volume of the waste, and (3) stabilizes the waste. Shin et al. [274] showed how this method has been successfully applied to FW from restaurants, markets, institutions, and households. They described a multi-step sequential batch two phase anaerobic composting (MUSTAC) process as stable, reliable, and effective in treating food residuals. The process can remove 82.4% of VSs and convert 84.4% of biomethane potential into methane in 10 days. The output from the posttreatment can be used as a soil amendment. MUSTAC process was simple to operate and had high performance. Haug et al. [52] described the use of Los Angeles Hyperion Treatment plant to anaerobically digest the food residual from LA airport and surroundings serving airline industry and passengers. The plant was cost effective and handled waste in an environmentally acceptable way.

4.5.1.9 Dairy Effluent

Energy generation potential from dairy effluent was recently evaluated by Desai et al. [53]. India is the largest milk producer in the world (100 MMT). In an organized sector, which produces only 30% of total milk generated in the country, the 140 dairy processing plants generate a very significant amount of effluent that is rich in organic waste. Desai et al. [53] described an AD system for one dairy processing 100,000 L/day milk to generate biogas (biomethane) that can provide energy for the aerators of the existing aerobic treatment system (mostly activated sludge system). The paper presents the details of the anaerobic filter system. The 40 million L of milk handled by the organized sector of milk industry in India has the potential of generating 11 MW power from methane produced by AD filter system.

4.5.1.10 Tofu Wastewater

Zheng et al. [45] examined hydrogen production from organic wastewater from tofu production by photobacteria. While this is a very useful process, NH_4^+, which is normally the integrant in organic wastewater, is the inhibitor for hydrogen production with photobacteria. Zheng et al. [45] showed that the concentration of NH_4^+ at 2 mmol/L or above significantly affected hydrogen production of wild type sphaeroides because NH_4^+ concentration inhibited nitrogenase activity. Zheng et al. [45] generated mutant named AR-3 that can produce hydrogen in the medium containing even 4 mmol/L NH_4^+ due to releasing the inhibition of NH_4^+ to nitrogenase activity. Under suitable conditions, they showed that hydrogen generation rate of AR-3 from tofu wastewater could reach 14.2 mL/L/h. It was increased by more than 100% compared with that of wild-type R sphaeroides.

4.5.1.11 Conversion of Lignocellulose to High-Value Products Employing Microbes

Methane can be produced from a range of substrates anaerobically by methanogenic bacteria. The process can be applied to the generation of methane from waste orange peel. The thermophilic AD of industrial orange waste pulp and peel with subsequent aerobic posttreatment of the digestate has been successfully demonstrated by Kaparaju and Rintala [74]. In this study, in anaerobic batch cultures, a methane production rate of 0.49 m^3/kg VSs and, in a semicontinuous methane, a production rate as high as 0.6 m^3/kg VS were generated. This does require the pH adjustment from 3.2 to 8.0 by $CaCO_3$ addition. An aerobic follow-up treatment with activated sludge produced CO_2 and water and converted ammonia into nitrate.

The removal of nitrogen required additional denitrification step. The process can be adapted to other FVWs like mango, pineapple, tomato, jackfruit, banana, and whole orange. The methane production rate from these fruits can be improved by using selected strains of *Sporotrichum*, *Aspergillus*, *Fusarium*, and *Penicillium* [13,74,75]. These fungal pretreatment enhanced the availability of nutrient in the medium, decreased the concentrations of antimicrobial components, and enabled the higher loading rate utilization.

Besides other fruits and vegetable wastes, wastewater from pressing of this by-product is also good substrate for methane production [75]. This wastewater is generated by pressing the rind of orange peel and it contains large amount of organic matter and alkalinity because in the pressing process $Ca(OH)_2$ is used as binder. Before the anaerobic treatment, the waste is pretreated by aluminum phosphate flocculent to remove solids that can hinder anaerobic treatment and to reduce pH from 11.21 to 5.5. In the batch process, this treatment removed 84% of soluble COD and that generated 295 mL of methane/g of COD removal. The presence of antimicrobial components reduced methane production when COD loadings were high [13,75].

Weichgrebe et al. [40] examined energy and CO_2 reduction potentials of anaerobic treatment of wastewater and organic kitchen wastes. They considered three different scenarios:

1. The classical waste treatment and the composting of the organic waste fraction
2. The anaerobic treatment of wastewater combined with deammonification and the digestion of the organic waste fraction
3. The mutual anaerobic treatment of wastewater and waste as codigestion with deammonification

Scenario 2 was found to be best. With today's state of the art concerning wastewater and waste treatment, energy surplus of 56.91 kWh/(person.annual) could be realized and at the same time, the CO_2 emission can be reduced by 7.97 kg of CO_2/(person.annual) for scenario 2 at 20°C without the use of the dissolved methane into the reactor's effluent. If in the future an economical process for the usage of dissolved methane is developed, emission of GHGs can be further lowered. A further positive effect of scenario 2 is that the dissolved nutrients can be reused. Since a small part of these is needed for anaerobic metabolism (<20%), a majority of mineral fertilizer can be substituted by using the effluent for irrigation. Furthermore, energy and GHGs would be additionally saved and the wastewater treatment costs will be reduced.

4.5.2 MULTIPLE SUBSTRATES (CODIGESTION)

Worldwide, anaerobic stabilization of sewage sludge is probably the most important AD process. While agricultural digesters are more in quantity, the volume treated is far smaller than that for sewage sludge. AD of manure is largely done or energy production. A digester in the farm has to be financed by the energy produced, whereas sewage sludge is a waste product of wastewater treatment that has to be stabilized. In this case, the digester is paid by the polluter or local government agency. The limiting availability of the industrial waste makes it clear that codigestion of industrial waste has only limited potential. The best gain for codigestion is the digestion of manure with energy crops.

Today, the inclusion of waste product for codigestion is economically essential. There are several organic substances that are anaerobically degraded without major pretreatment such as leachates, slops, sludges, oils, fats, or whey. Some waste can form inhibiting metabolics like NH_3 during AD, which requires higher dilutions with substrates like manure and sewage sludge. The feedstock that requires pretreatments are source-separated municipal biowastes, food leftovers, expired food, market wastes, and harvest residue. There are also limited stocks of organics such as straw, lignin, rich yard waste, and category 1 slaughterhouse animal by-products (ABPs) that are not very suitable for AD due to the high cost of pretreatment, inhibiting components, poor biodegradability, hygienic risks, or expensive transport. Approximate biogas or methane yields for various crops and organic wastes and from codigestion of organic wastes are summarized in Table 4.5. The literature on codigestion of various mixtures of substrates is very large and cannot be covered here in its entirety. References 54,57–67,69–73,76–80,205,275–383 give illustrations of some typical published literature on codigestion. More details are given by Shah [205].

One substrate ideal for codigestion is animal manure. One of the advantages of using manure for digestion is that the bacteria required for the digestion process are already present in the manure. A disadvantage of using manure alone is that it has low-energy content; therefore, a low amount of gas production is achieved per unit volume of manure. Cattle manure only contains around 3.6 GJ/tons of dry matter, and pig manure contains around 5.3 GJ/tons dry matter. Since dry matter is only 10% of the total volume, the energy content per unit volume of manure is very low (in the order 0.3–0.5 GJ/me manure). This is equivalent to about 18–30 m^3 of biogas/m^3 manure. Assuming the residence time in the digester of 30 days, this results in the production of 0.6–1 m^3 biogas/m^3 digester/day. Such low amount makes this process highly uneconomical. Codigestion with energy crops or substrates with high organic content can make them useful biogas generators.

In the following discussion, examples of feedstock effects on the codigestion process are separated into three parts: sludge/wastewater treatments, treatments of different types of manures, and other codigestion systems.

4.5.2.1 Codigestion with Different Types of Sludge and Wastewater

4.5.2.1.1 Organic Waste with Municipal Sludge

Zupancic et al. [271,289,339] studied codigestion of organic waste of domestic refuse (swill) with municipal sludge. The results were very successful and showed that the organic waste was virtually completely degraded, there is no increase in volatile sold sludge (VSS) during the process, and there are no adverse impacts on the environment. The degradation efficiency increased from 71% to 81%, and an 80% increase in biogas quantity was observed. Biogas production rate increased from 0.32 to 0.67 m^3/m^3/day. Solid by-products increased from 0.39 to a peak of 0.89 m^3/kg VSS inserted.

TABLE 4.5
Relative Methane Yield from the CoDigestion of Organic Waste

Substrate/Co-Substrate	Biogas Yield	Methane Yield
	(L/kg VSs[a])	
Cattle excreta/olive mill waste		179
Cattle manure/agricultural waste and energy crops		620
FVW/abattoir wastewater		611
Municipal solid wastes/fly ash		222
Fat, oil, and grease/waste from sewage treatment plants		350
Pig manure/fish and biodiesel waste		620
PW/sugar beet waste		680
Primary sludge/FVW		600
Sewage sludge/municipal solid waste		532
Slaughterhouse waste/municipal solid waste		500
Dairy manure	378	235
30/70 corn stillage/dairy manure	465	305
30/70 corn stillage/dairy manure (repeat)	527	301
40/60 corn stillage/dairy manure	630	402
30/70 waste grease/dairy manure	511	358
40/60 waste grease/dairy manure	569	398
30/70 whey/dairy manure	433	303
30/70 switchgrass/dairy manure	479	285

Sources: Braun, R. et al., Biogas from energy crop digestion, IEA Bioenergy, Task 37—Energy from Biogas and Landfill Gas, Denmark, IEA, Brussels, Belgium, 2009; Braun, R., Anaerobic digestion—A multi faceted process for energy, environmental management and rural development, in: Ranalli, P., ed., *Improvement of Crop Plants for Industrial End Users*, Springer, 2007; Murphy, J. et al., Biogas from crop digestion, IEA Bioenergy, Task 37—Energy from Biogas, Denmark, September 2011; Braun, R., Potential of codigestion, IEA Task 37 Report, 2002, http://www.novaenergie.ch/ieabioenergytask37/Dokumente/final.PDF, Accessed on November 7, 2007; Braun, R. and Wellinger, A., Potential of codigestion, IEA Bioenergy, Task 37—Energy from Biogas and Landfill Gas, 2004; Braun, R. et al., *Biotechnol. Lett.*, 3, 159, 1981; Braun, R. et al., *Appl. Biochem. Biotechnol.*, 109, 139, 2003; Crolla, A. et al., Anaerobic codigestion of corn thin stillage and dairy manure, in: *Canadian Biogas Conference and Exhibition*, London Convention Center, London, Ontario, Canada, March 4–6, 2013; Shah, Y.T., *Energy and Fuel Systems Integration*, CRC Press, New York, 2015.

[a] VSs, volatile solids.

Davidsson et al. [352] successfully performed the codigestion of sludge from grease traps and sewage sludge. While codigestion could not reach stable methane production in the continuous digestion tests, the addition of grease trap sludge when digesting sewage sludge increased the methane potential and methane yield (amount of methane per added amount of VS). In the pilot-scale tests, the increase in methane yield was 9%–27% for GS amounts corresponding to 10%–30% of the total VS added.

Agdag and Sponza [278] examined the feasibility of anaerobic codigestion of industrial sludge with MSW in three simulated landfilling bioreactors during a 150-day period. They noted that codigestion stabilized the waste and the treatment of leachate release. The addition of industrial sludge to MSW gave biogas with 72% methane content while improving the leachate quality.

Gomez et al. [284] examined codigestion of primary sludge and fruit and vegetable fraction of MSW under mesophilic conditions. The addition of fruit and vegetable fraction of MSW in primary sludge increased the biogas production. The specific gas production and biogas yield, however, did not change by codigestion. The application of a sudden increase in the organic load of the codigestion system led to higher gas production accompanied by downgrading of the performance of the digester.

Neves et al. [355] examined codigestion of sewage sludge and coffee wastes from the production of instant coffee substitutes. Methane yields in the range 0.24–0.28 m^3/kg VS were obtained except for barley-rich waste where methane yield was 0.02 m^3/kg VS. Four out of five wastes gave high reduction in TS (50%–73%) and VS (75%–80%).

Fernandez et al. [281] evaluated potential of mesophilic AD for the treatment of fats from different origins through codigestion with the OFMSW. No change in performance was observed when animal fat was changed to vegetable fat with a completely different LCFA profile. This indicated that no important metabolic changes are implied in the degradation of different LCFAs with an acclimatized sludge.

Fezzani and Ben Cheikh [282,321,322] investigated codigestion of olive mill wastewater (OMW) with olive mill solid waste under thermophilic condition (at 55°C). They concluded that OMW can be successfully degraded in codigestion with olive mill waste solids without previous dilution and addition of chemical nitrogen substance. The best methane productivity was 46 L CH_4/(L OMW fed) day. The process also produced the best net energy. However, the COD removal efficiency in thermophilic condition was lower than that obtained using mesophilic conditions.

4.5.2.1.2 Kitchen Waste and Sewage Sludge

Sharom et al. [356] examined different compositions of kitchen waste (KW) and sewage sludge for codigestion at 35°C and pH of 7. Five different *compositions were* examined. The cumulative biogas production increased with codigestion; the highest production occurred for 75% KW/25% activated sludge that produced 59.7 mL. The biogas production rate was also obtained at this composition. Pure activated sludge gave the least production rate of biogas.

4.5.2.1.3 Activated Sludge and OFMSW

Bolzonella et al. [341,376] studied the performances of full-scale anaerobic digesters co-digesting waste activated sludge from biological nutrients removal in wastewater treatment plants, together with different types of organic wastes (solid and liquid). Results showed that biogas production can be increased from 4,000 to some 18,000 m^3/month when treating some 3–5 tons/day of organic MSW together with waste-activated sludge. On the other hand, the specific biogas production was improved from 0.3 to 0.5 m^3/kg VS fed the reactor when treating liquid effluents from cheese factories. The addition of the co-substrates gave minimal increases in the OLR, while the HRT remained constant. Further, the potential of the struvite crystallization process for treating anaerobic supernatant rich in nitrogen and phosphorus was studied; an 80% removal of phosphorus was observed in all the tested conditions.

In a separate study, Bolzonella et al. [341,376], also examined two full-scale applications of the anaerobic codigestion of activated sludge with OFMSW. The studies were carried out at Viareggio and Treviso wastewater plants in Italy. In the first plant, 3 tons/day of source-sorted OFMSW was co-digested with activated sludge. This process increased the OLR from 1.0 to 1.2 kg TVS/(m^3 day) and a 50% increase in biogas production. At Treviso WWTP, which has been working for 2 years, some 10 tons/day of separately collected OFMSW was treated using a low-energy consumption sorting line, which allowed the removal of 99% and 90% of metals and plastics, respectively. In these conditions, the biogas yield increased from 3,500 up to 17,500 m^3/month. The payback time for the new codigestion process was estimated to be 2 years. Delia and Agdag [381] examined anaerobic codigestion of industrial sludge with MSWs in anaerobic simulated landfilling reactors.

4.5.2.1.4 Sewage Sludge and Orange Peel Waste

Serrano et al. [335] examined codigestion of sewage sludge and orange peel waste in a proportion of 70:30 (wet weight), respectively. Mesophilic AD of sewage sludge gave low methane yield, poor biodegradability, and nutrient imbalance. The codigestion improved the viability of the process. The stability was maintained within correct parameters throughout the process, while methane yield coefficient and biodegradability were 165 L/kg VS (0°C, 1 atm) and 76% (VS), respectively. The OLR increased from 0.4 to 1.6 kg VS/(m³ day). Nevertheless, the OLR and methane production rate decreased at the highest loads, suggesting the occurrence of an inhibition phenomenon.

4.5.2.1.5 Codigestion of Municipal Organic Waste and Waste Treatment Plant Sludge

Pretreatment to enhance biogas production from wastewater treatment plant sludge.

Li [372] examined codigestion of organic waste with wastewater treatment plant sludge in four phases. The final aim of the study was to evaluate optimum codigestion conditions by adding selected co-substrates and by incorporating optimum pretreatment strategies for the enhancement of biogas production from anaerobic codigestion using wastewater treatment plant sludge as the primary substrate.

In the first phase, the feasibility of using municipal organic wastes (synthetic KW and fat, oil, and grease [FOG]) as co-substrates in anaerobic codigestion was investigated. KW and FOG positively affected biogas production from anaerobic codigestion, with ideal estimated substrate/inoculum (S/I) ratio ranges of 0.80–1.26 and 0.25–0.75, respectively. Combined linear and nonlinear regression models were employed to represent the entire digestion process and demonstrated that FOG could be suggested as the preferred co-substrate.

The effects of ultrasonic and thermochemical pretreatments on the biogas production of anaerobic codigestion with KW or FOG were investigated in the second phase. Nonlinear regressions fitted to the data indicated that thermochemical pretreatment could increase methane production yields from both FOG and KW codigestion. Thermochemical pretreatments of pH = 10, 55°C, provided the best conditions to increase methane production from FOG codigestions. In the third phase, using the results obtained previously, anaerobic codigestions with FOG were tested in bench-scale semicontinuous-flow digesters at Ravens view Water Pollution Control Plant, Kingston, ON. The effects of HRT, OLR, and digestion temperature (37°C and 55°C) on biogas production were evaluated. The best biogas production rate of 17.4 ± 0.86 L/day and methane content of about 67.9% were obtained with thermophilic (55°C) codigestion at HRT of 24 days and OLR of about 2.43 g TVS/(L day).

In the fourth phase, with the suitable co-substrate, optimum pretreatment method and operational parameters identified from the previous phases and anaerobic codigestions with FOG were investigated in a two-stage thermophilic semicontinuous-flow codigestion system modified to incorporate thermochemical pretreatment of pH = 10 at 55°C. Overall, the modified two-stage codigestion system yielded of 25.14 ± 2.14 L/day (with 70.2% ± 1.4% CH_4) biogas production, which was higher than that obtained in the two-stage system without pretreatment.

4.5.2.1.6 Food Waste with Sewage Sludge

Iacovidou et al. [76] examined FW codigestion with sewage sludge in the United Kingdom. They identified the following constraints of FW codigestion with sewage sludge. Codigestion of FW with sewage sludge can be limited due to the high variability of FW. A stable digestion performance depends on the composition of FW added to sewage sludge, which if changed can cause instability in the anaerobic population and consequently in the digestion process. This is because microorganisms are acclimatized in a specific mixture, and changes in this mixture may result in changes in the process reactions. In addition, CH_4 yields may also vary due to seasonal variations in FW [384,385].

The concentration of light metal ions and biodegradation intermediates plays an important role in the smooth process performance, as they can be the most potent causes of toxicity in AD.

A compound can be described as toxic or inhibitory when it causes an adverse change in the microbial population or halts bacterial growth [88]. By increasing the fraction of FW, the risk of increasing the concentration of light metal ions and/or biodegradation intermediates at levels that can be toxic to the anaerobic population becomes greater.

Light metal ions, also known as cations, are present in many types of FW and, although essential for the growth of anaerobic microorganisms, at high concentrations can be a cause of toxicity in the digestion process due to the effect of osmosis [88,386]. Codigestion of sewage sludge with FW rich in vegetables is likely to show an increase in the potassium (K) content that may inhibit the digestion process [387]. Sodium (Na) inhibition is also likely to occur since FW is a source of Na [386,387]. VFAs, LCFAs, and ammonia (NH_3) are the main biodegradation intermediates of AD. The accumulation of these intermediates beyond certain levels in the anaerobic digester can be a cause of toxicity in the codigestion process [50,388].

The addition of FW to sewage sludge results in an initial increase in VFAs concentration due to the rapid acidification of soluble organic compounds found in FW [230,389,390]. VFAs are subsequently decreased as a result of their uptake by the anaerobic microorganisms [386,389–392]. However, when the production rate surpasses the uptake rate, the excessive levels of VFAs produced are accumulated in the digester. This accumulation can cause acidity in the digester that if not restored can lead the pH to drop to such a level that stops the digestion process from occurring [367,389,393–395]. Levels at which VFAs become toxic have not been documented; however, it was reported that VFAs can be present in the digester at concentrations up to 6000 mg/L without being toxic, provided that the pH is maintained in the optimal range [21].

A high lipidic content in FW may also affect the codigestion process with sewage sludge, due to the excessive production of LCFAs. LCFAs have been shown to be toxic to the anaerobic population, and the higher their concentration in the digester, the more toxic their effect is [88,386,395]. The levels at which LCFAs can be toxic vary widely depending on the predominant form of acid present in the digester [88,386].

The degradation of protein-rich mixtures of FW and sewage sludge is the primary cause of NH_3 production and accumulation [396,397]. NH_3 and ammonium ion (NH_4^+) concentrations both exist in anaerobic digesters. NH_4^+ can inhibit the activity of methanogens and, hence, CH_4 production. However, NH_3 was reported to be more inhibitory than NH_4^+ because of its capability to penetrate through cell membranes [88,367,398,399]. However, there is an uncertainty as to the range at which NH_3 concentrations can be inhibitive [88]. This uncertainty is mainly because of the differences in operational conditions, such as alkalinity, temperature, substrate composition, and acclimation period [50,88,367]. Mixtures with a higher proportion of FW than sewage sludge have a limited possibility for NH_3 inhibition, mainly because of the higher carbon availability [367].

Operational constraints associated with FW handling are extremely important. Impurities found in FW such as plastics, metals, glass, and other packaging parts are likely to cause tremendous technical problems in the wastewater treatment line and codigestion performance [394,400–402]. Plastics in the form of plastic bags can be wrapped around the stirring equipment in storage and reactor tanks, wear out the pumps, and form a top layer in the reactors. Furthermore, plastic contamination in the form of phthalates can change the quality of the digestate and make it unacceptable for application to agricultural land [394,403]. Metals are toxic for the bacterial biomass, whereas lignocellulosic materials such as wood and paper are digested slowly and are troublesome [402]. Mechanical problems such as clogging in the conveyor line may also arise from metal contamination [404]. These can lead to increases in operational costs and also to the loss of a relatively important fraction of FW during the pretreatment process. This loss can be associated with a reduction in the benefits for the water industry, as less biogas will be produced per ton of waste delivered [394]. Therefore, it is of critical importance that FW is free of impurities and exogenous material prior to codigestion with sewage sludge.

Overloading the digesters must also be avoided in order for the process to operate successfully. Overloading episodes can cause a decrease in CH_4 yields, and even in a failure of the whole process, while blocking of the pipes and foaming incidents in the digester are also possible [361,367,405,406]. Increased foaming, blocking of pipes, and insufficient mixing of the substrates resulting from digester overloading can ultimately lead to a loss in biogas production.

In the United Kingdom, the current operational and regulatory framework makes anaerobic codigestion of FW with sewage sludge a rather complicated matter. Sewage sludge and FW are both covered by different regulatory regimes. When FW is co-digested with sewage sludge, regulation becomes more complex and unclear, with the process standing between two sets of regulations [407]. Waste management license requirements, the quality of the co-digestate, and renewable energy generation credits are some of the regulatory constraints and uncertainties that currently prevent the adaptation of codigestion of sewage sludge with FW by the water industry.

The processing of FW that contains or has been contaminated by meat and any other animal materials falls into the ABP regulations (ABPR). The ABPR defines three categories of ABPs, with category three being the least harmful. This category includes FW originating from households, restaurants, and catering facilities [408,409]. As such, the water industry must have an ABP permit to be able to process FW. To meet the requirements set by the ABPR, FW pretreatment is required first to ensure the removal of packaging material and other impurities that may be present in FW and second to pasteurize the FW before being added to the digester [407]. The aim of the ABPR is to sanitize FW and prevent pathogen transfer. The use of co-digestate may also be covered by the ABPR although this is not yet clearly defined [410]. This also applies to other European countries that make the application of the ABPR a rather complicated matter.

The co-digestate produced from the digestion of mechanical-segregated FW with sewage sludge is considered a waste digestate in the revised Waste Framework Directive, and thus, it cannot be applied directly to land [408]. This would leave the water industry to deal with a huge amount of co-digestate if not otherwise managed.

Economic complications related to the planning and operation of the codigestion process can be a significant barrier to its adaptation. The building of new digesters, the upgrade of existing ones, and the installation of facilities for the delivery, pretreatment, and storage of FW [387,394,411], required for codigestion, cannot currently be easily provided due to economic barriers and policy restrictions. These regulations impose a level of complexity that exceeds the potential of codigestion, and initiatives to implement this process have been severely hindered. This is not surprising as individual approaches in waste management are challenging and bound to become unsustainable.

4.5.2.1.7 Algal Sludge and Wastepaper

Yen and Brune [77] examined codigestion of algal sludge and wastepaper. The unbalanced nutrients of algal sludge (low C/N ratio) were regarded as an important limitation factor to AD process. Adding high carbon content of wastepaper in algal sludge feedstock to have a balanced C/N ratio was undertaken in this study. The results showed adding 50% (based on VS) of wastepaper in algal sludge feedstock increased the methane production rate of about 1170 mL/L day, as compared to about 573 mL/L day of algal sludge digestion alone, both operated at 4 g VS/L day, 35°C and HRT of 10 days. The maximum methane production rate of about 1607 mL/L day was observed at a combined 5 g VS/L day loading rate with 60% (VS based) of paper adding in algal sludge feedstock. Results suggested an optimum C/N ratio for codigestion of algal sludge, and wastepaper was in the range of 20–25/L.

4.5.2.1.8 Fat, Oil, Grease, and Wastewater Treatment Facility

Long et al. [78] examined codigestion of high-organic-strength fat, oil, grease (FOG) from restaurant grease abatement devices with wastewater treatment facility. Addition of FOG to waste water treatment facility substantially increased biogas production. Codigestion of FOG with municipal biosolids at a rate of 10%–30% FOG by volume of total digester feed caused a 30%–80% increase in digester gas production in two full-scale wastewater biosolid anaerobic digesters. However, AD of high lipid wastes has been reported to cause inhibition of acetoclastic and methanogenic bacteria, substrate, and product transport limitation, sludge floatation, digester foaming, blockage of pipes and pumps and clogging of gas collection, and handling systems. Long et al. [78] also reviewed the scientific literature on biogas production, inhibition, and optimal reactor configurations and highlighted future research needed to improve gas production and overall efficiency of anaerobic codigestion of FOG with biosolids from municipal wastewater treatment.

4.5.2.1.9   Wastewater Using Decanter Cake

Kaosol and Sohgrathok [412] examined the codigestion of wastewater using decanter cake. The wastewater from agro-industry cannot produce biogas by biological treatment because of its low COD level and low organic content. The study examined the effect of three parameters, type of wastewater, mixing, and mesophilic temperature on the codigestion process. The study measured the biogas production of wastewater alone along with various mixtures of decanter cake and wastewater. The codigestion of decanter cake with rubber block wastewater of the R4 (wastewater 200 mL with decanter cake 8 g) produced the highest biogas yield of 3.809 L CH_4/g COD removal with maximum methane gas of 66%. The study also showed that the mixing and mesophilic temperature did not have significant effect on the biogas potential production. The codigestion of decanter cake with rubber block wastewater provided the highest biogas yield potential production at an ambient temperature. The decanter cake can be a potential source for biogas production.

4.5.2.1.10   Grease Interceptor Waste in Wastewater Treatment Plant

Aziz et al. [336] showed that grease interceptor waste (GIW) could make an ideal codigestion feedstock. GIW is comprised of the FOG, food solids, and water collected from food service establishment grease interceptors. Two ongoing research projects were attempted to address different issues related to GIW codigestion at municipal WWTFs. The first project was an experimental evaluation of the limits of biogas production from GIW codigestion with municipal sludge. The second project is developing a life-cycle decision support tool to explore the environmental and economic implications of GIW codigestion. Two lab-scale anaerobic digesters were operated under mesophilic conditions. One reactor served as a control digesting only thickened waste activated sludge (TWAS), while the other reactor received a combined feed of TWAS and GIW at varying volume fractions. At 20% GIW by volume, the second reactor showed an increase of biogas production of 317% from the control. However, at 40% GIW by volume, the digester showed a drop of methane production and signs of process failure.

These preliminary results suggested great promise for enhanced biogas production with GIW as a codigestion feed. The anaerobic codigestion decision support tool provided an economic and life-cycle assessment framework to explore the implications of enhanced biogas production during the anaerobic codigestion of GIW with municipal sludge. Results indicated that while the codigestion of GIW is environmentally favorable due to the enhanced offset of natural gas and fossil fuel-derived energy, the economics are sensitive to regional effects and available infrastructure. Despite this fact, codigestion substantially reduces the start-up cost for AD at facilities that presently use alternative solids handling.

4.5.2.1.11   Food Waste and Wastewater

Cheerawit et al. [79] investigated the potential of biogas production from the codigestion of domestic wastewater and FW. Batch experiments were carried out under various substrate ratios of domestic wastewater and FW at 10:90, 25:75, 50:50, and 70:30 at room temperature. The results showed that codigestion of domestic wastewater with FW was very promising for the production of methane gas. The biochemical methane production and COD removal efficiency were 61.72 mL CH_4/g COD and 75.77%, respectively. Moreover, the addition of FW to the AD of domestic wastewater showed an increasing trend of the biogas production. The laboratory batch study revealed that the use of FWs as co-substrate in the AD of domestic wastewater also has other advantages, that is, the improvement of the balance of the C/N ratio and efficient process stability.

4.5.2.1.12 Olive Mill Wastewater and Swine Manure

Azaizeh and Jadoun [63] examined codigestion of OMW and SM using upflow anaerobic sludge blanket (UASB) reactor. Swine wastewater (SW) and OMW are two problematic wastes that have become major causes of health and environmental concerns. The main objective of this work was to evaluate the efficiency of the codigestion strategy for treatment of SW and OMW mixtures. Mesophilic batch reactors fed with mixtures of SW and OMW showed that the two adapted sludges, namely, Gadot and Prigat, exhibited the best COD removal capacity and biogas production; therefore, both were selected to seed UASB continuous reactors. During the 170 days of operation, both sludges, Gadot and Prigat, showed high biodegradation potential. The highest COD removal of 85%–95% and biogas production of 0.55 L/g COD were obtained at a mixture consisting of 33% OMW and 67% SW. Under these conditions, an organic load of 28,000 mg/L COD was reduced to 1,500–3,500 mg/L. These results strongly suggest that codigestion technology using UASB reactors is a highly reliable and promising technology for wastewater treatment and biogas production.

4.5.2.1.13 Codigestion of Swine Wastewater with Switchgrass and Wheat Straw

Liu [413] examined anaerobic codigestion of swine wastewater (SW) with switchgrass (SG) and wheat straw (WS) for methane production. The effects of different TS concentrations (2%, 3%, 4%) on the methane yield from the codigestion with SG and WS were investigated in batch mode at mesophilic temperature (35°C). The culture from a completely mixed and semicontinuously fed anaerobic reactor treating SW and corn stover was used as the inoculum for the batch tests. The reactor had a working volume of 14 L and was operated at 35°C with an HRT of 25 days and an OLR of 0.924 kg VS/(m^3 day). Batch reactors were operated in triplicates, each with a working volume of 500 mL. Reactors were kept in thermostatic water bath maintained at mesophilic temperature (35°C) with an agitation speed of 270 rpm. The volume of methane produced in the experiment was measured by gas meters. COD, pH, total Kjeldahl nitrogen (TKN), total organic carbon, and TS and VS analyses of reactor contents were performed at the beginning and the end of the experiment.

The results indicated that with the addition of SG, methane production substantially increased. The methane yields at 2%, 3%, and 4% TS were 0.137, 0.117, and 0.104 m^3/kg VS added, respectively. The addition of WS in the batch reactors resulted in higher accumulated methane production, and the methane yield was 0.133 m^3/kg VS added at 2% and 3% TS concentrations. However, when the TS increased to 4%, methane production decreased because VFA accumulation increased rapidly and pH dropped to below 5.5. The first-order kinetic model was evaluated for the methane production. It was found that the model fitted the experimental data well. The study concluded that batch anaerobic codigestion of swine waste with SG and WS at low TS concentration is a commercially viable process.

4.5.2.1.14 Single-Source Oily Waste and High-Strength Pet Food Wastewater

Acharya and Kurian [80] focused on the revival of a formerly failed digester of 1800 m^3 volume. The most obvious cause of failure was identified to be due to capping caused by foam and scum, as a result of attempting to treat oil-rich, high-strength wastewater. The revival was affected by implementing codigestion in a three-step remedial procedure. Though, in the typical sense, codigestion involved separate waste streams, here, a single waste stream was manipulated to apply the concept of codigestion. Through the virtual codigestion, the digester succeeded to treat the daily plant effluent flow of 50 m^3 with COD > 45 g/L and around 10 tons of sludge/day that had around 20% oil and fat, exhibiting a COD removal >90%. The digester operated at 35°C and HRT of 30 days with loading rates of 3.4 kg COD/(m^3 day) and 1.3 kg O&G/(m^3 day). The biogas generated from this digester was sufficient to operate a 40 hp boiler at 100 psi.

4.5.2.1.15 Codigestion of Chicken Processing Wastewater and Crude Glycerol from Biodiesel

Foucault [81] examined codigestion of chicken processing wastewater and crude glycerol from biodiesel. The main objective of the study was to examine the AD of wastewater from a chicken processing facility and of crude glycerol from local biodiesel operations. The AD of these substrates was conducted in bench-scale reactors operated in the batch mode at 35°C. The secondary objective was to evaluate two sources of glycerol as co-substrates for AD to determine if different processing methods for the glycerol had an effect on CH_4 production. The biogas yields were higher for codigestion than for digestion of wastewater alone, with average yields at 1 atm and 0°C of 0.555 and 0.540 L/(g VS added), respectively. Another set of results showed that the glycerol from an on-farm biodiesel operation had a CH_4 yield of 0.702 L/(g VS added) and the glycerol from an industrial/commercial biodiesel operation had a CH_4 yield of 0.375 L/(g VS added). Therefore, the farm glycerol likely had more carbon content than industrial glycerol. It was believed that the farm glycerol had more impurities, such as free fatty acids, biodiesel, and methanol. The codigestion, of chicken processing water and crude glycerol, thus increased the production of methane-rich biogas.

4.5.2.1.16 Excess Brewery Yeast Codigestion in a Full-Scale Expanded Granular Sludge Bed Reactor

Zupancic et al. [271,289,339] examined the anaerobic codigestion of brewery yeast and wastewater. The study showed that such additional loading of the anaerobic process did not destroy nor damage the operation of the full-scale system. Full-scale codigestion at concentration of vol. 0.7% ± 0.05% showed no negative impacts. With additional brewery yeast (0.7%), a 38.5% increase of biogas production was detected, which resulted in an increase of the biomethane/natural gas substitute ratio in the brewery from 10% to 16%.

4.5.2.2 Codigestion with Different Types of Manure

4.5.2.2.1 Cow Manure and Highly Concentrated Food Processing Waste

Yamashiro et al. [324] examined anaerobic codigestion of dairy CM and highly concentrated food processing waste, under thermophilic (55°C) and mesophilic (35°C) conditions. Two types of feedstock were studied: 100% DM and 7:3 mixture (wet weight basis) of DM and FPW. The contents of the FPW as feedstock were 3:3:3:1 mixture of cheese whey, animal blood, used cooking oil, and residue of fried potato. Four continuous digestion experiments were carried out in a 10 L digester. The results showed that codigestion under thermophilic conditions increased methane production per unit digester volume. However, under mesophilic condition, codigestion was inhibited. TKN recovered after digestion ranged from 73.1% to 91.9%, while recoveries of ammonium nitrogen (NH_4–N) exceeded 100%. The high recovery of NH_4–N was attributed to mineralization of influent organic N. The mixtures of DM and FPW showed greater recoveries of NH_4–N than DM only, hence reflecting its greater organic N degradability. The ratios of extractable to total calcium, phosphorus, and magnesium were slightly reduced after digestion. These results indicated that codigestion of DM and FPW under thermophilic temperature enhances methane production and offers additional benefit of organic fertilizer creation.

Jepsen [250] examined codigestion of animal manure with organic household waste. He found that codigestion significantly increased biogas production. Gas yield from manure was only 15–20 m³/tons. On average, 60% gas production increased from the addition of waste. Along with household waste, industrial waste and sludge can also be used to improve biogas production. He also noted that among organic waste to improve gas yield, an order follows:

1. Concentrated fat, fish silage, etc.: 200–1000 m³/tons
2. Fish waste, fat, flotation sludge, slaughterhouse waste, dairy waste, and organic household waste: 50–200 m³/tons
3. FVW, industrial wastewater, and sewage sludge: 10–50 m³/tons

Jepsen [250] concluded that codigestion with animal manure results in a stable process.

Zhang et al. [334,362,414] evaluated anaerobic digestibility and biogas and methane yields of the FW in order to examine its suitability as co-substrate in codigestion. The tests were performed at 50°C in a batch fermenter. The daily average moisture content (MC) and the ratio of VSs to TSs were 70% and 83%, respectively. The FW contained well-balanced nutrients. The methane yield was 348 mL/g VS after 10 days and 435 mL/g VS after 28 days. The average methane content of biogas was 73%. The average VS destruction at the end of 28 days was 81%. All these data indicate that FW is a good co-substrate for AD.

4.5.2.2.2 Wastepaper and Cow Dung and Water Hyacinth

Yusuf and Ify [64,276] carried out the codigestion of cow dung and water hyacinth in a batch reactor for 60 days with the addition of various portions of wastepaper. The biogas production was measured keeping the amount of cow dung and water hyacinth fixed and the variable amounts of wastepaper. Maximum biogas volume of 1.1 L was observed at a wastepaper amount of 17.5 g, which corresponds to 10% of TSs of biomass in a 250 mL solution.

4.5.2.2.3 MSW and Cow Manure

Samani et al. [382] examined codigestion of OFMSW and dairy CM and found that while OFMSW produced 62 m³ CH_4/tons while digesting alone and dairy CM produced 37 m³/tons while digesting alone, the codigestion produced 172 m³ CH_4/tons of dry waste. Thus, codigestion gave higher methane yields. Hartmann et al. [270,394,403] studied codigestion of OFMSW and manure under thermophilic conditions at 55°C. Various concentrations of OFMSW and manure were examined at HRT of 14–18 days and OLR of 3.3–4.0 g VS/(L day) over a period of 6 weeks. The experiments were started with OFMSW–manure ratio of 1:1, and this ratio was gradually increased with time over 8 weeks. Use of recirculated process liquid to adjust organic loading had a stabilizing effect. When the pH raised to 8, the reactor showed stable performance with high biogas yield and low VFA. Biogas yield was 180–220 m³ biogas/tons of OFMSW both in codigestion configuration and in the treatment of 100% OFMSW with process liquid recirculation. A VS reduction of 69%–74% was achieved when treating 100% OFMSW. None of the processes showed signs of inhibition at the free ammonia concentration of 0.45–0.62 g N/L.

4.5.2.2.4 Dairy Manure and Food Waste

A typical comparison of biogas/methane yields from various manure and biomass mixture is illustrated in Table 4.5. El-Mashad et al. [379] and El-Mashad and Zhang [280] examined biogas production of cattle manure using sunlight and biogas production of different mixtures of unscreened dairy manure and FW, respectively. In the latter study, the effect of manure screening on the biogas yield of dairy manure was also evaluated. This study showed that two mixtures, (1) unscreened manure (68%) and FW (32%) and (2) unscreened manure (52%) and FW (48%), produced methane yields of 282 and 311 m³/kg VS, respectively, after 30 days of digestion. After 20 days, approximately 90%–95% of final biogas was obtained. The average methane content was 62% and 59% for the first and second mixtures, respectively. The predicted results from the model showed that adding the FW into manure digester at levels up to 60% of the initial VSs significantly increased the methane yield for 20 days of digestion.

Crolla et al. [415] studied the benefits of the addition of co-substrates such as energy crops, industrial wastes, or food industry wastes to manure. They noted that the addition of co-substrate to the manure can improve C/N balance that results in a stable and sustainable digestion process. The optimum C/N ratio appeared to be somewhere between 20:1 and 30:1. Codigestion also improved the flow qualities of the co-digested substrates. The optimum HRT for dairy manure was 12–25 days and for cattle manure 15–35 days. The optimum OLR was around 3.5–5.5 kg VS/(m³ day). The optimum pH for manure was also between 6.8 and 7.2. The optimum temperature for mesophilic operation was 35°C–40°C and for thermophilic operation was 55°C–63°C. Codigestion with manure also improved biogas production rate and generated additional tipping fees for the use of additional wastes. The study also presented the data on optimum biogas and methane yields associated with codigestion of manure with corn silage, SG, canola seed cake, whey, waste grease, FW, and corn silage. Some of the mean biogas and CH_4 yield data for these systems are shown in Table 4.5. Crolla and Kinsley [378] concluded that codigestion of manure and co-substrates resulted in

1. Reduced CH_4 gas emission from storage reservoirs holding digestate
2. Reduced N_2O gas emissions from the land application of digestate
3. Reduced odors in both storage reservoirs and during land application
4. Reduced pathogens and weed seeds in the digestate
5. Improved fertilizer value of the digestate by transforming nutrients into more readily available inorganic forms

Although nutrients in the digestate were readily available for plant uptake, it can be lost in the absence of plants. Cover crops were used to hold the nutrients.

A company called ANTARES in Wyoming County, New York, also examined codigestion of manure with substrates like biomass crops, agricultural residues, and FOG in local situation. These results can be obtained by directly contacting ANTARES.

4.5.2.2.5 Pig Slurry and Organic Waste from Food Industry

The main inhibitor in the AD of pig slurry is the release of free ammonia. Campos et al. [416] examined the codigestion of pig slurry with organic waste from food industry such as wastes from fruit and olive oil refineries (pear waste and oil bleaching earth). Batch experiments at both mesophilic (35°C) and thermophilic (55°C) conditions were performed. Due to large inhibition by ammonia in thermophilic conditions, the data for mesophilic conditions were better than those for thermophilic conditions. In both temperature conditions, however, methane production was improved by the addition of a co-substrate. Higher methane production was obtained from the codigestion of slurry and oil bleaching earth (95% and 5%, respectively). The methane yield was 344 mL CH_4/(g $VS_{initial}$), which was 2.4 times the methane yield for slurry (144 mL CH_4/(g $VS_{initial}$)).

4.5.2.2.6 Goat Manure with Three Crop Residues

Zhang et al. [414] examined codigestion of goat manure (GM) with WS, corn stalks (CSs), and rice straw (RS). GM is an excellent material for AD because of its high nitrogen content and fermentation stability. The experiments were performed under mesophilic conditions. With a TS concentration of 8% and different mixing ratios, results showed that the combination of GM with CS and RS significantly improved biogas production at all C/N ratios. GM/CS (30:70), GM/CS (70:30), GM/RS (30:70), and GM/RS (50:50) produced the highest biogas yields from different co-substrates after 55 days of fermentation. Biogas yields of GM/WS 30:70 (C/N 35.61), GM/CS 70:30 (C/N 21.19), and GM/RS 50:50 (C/N 26.23) were 1.62, 2.11, and 1.83 times higher than that of CRs, respectively. These values were determined to be optimum C/N ratios for codigestion. However, compared to GM/CS and GM/RS treatments, biogas generated from GM/WS was only slightly higher than the singledigestion GM or WS. This result was caused by the high total carbon content (35.83%) and lignin content (24.34%) in WS, which inhibited biodegradation.

4.5.2.2.7 Water Hyacinth with Poultry Litter versus Water Hyacinth with Cow Dung

Patil et al. [371] compared the performance of two codigestion systems: water hyacinth with poultry litter and water hyacinth with cow dung in mesophilic conditions with temperature range from 30°C to 37°C in a batch digester with a retention period of 60 days. The TS concentration was 8% in each sample. The results showed that codigestion of water hyacinth with poultry litter produced more biogas than codigestion of water hyacinth with cow dung. The overall results showed that blending water hyacinth with poultry waste had significant improvement on the biogas yield.

4.5.2.2.8 Slaughterhouse Waste with Various Crops, MSW, and Manure

Siripong and Dulyakasem [325] examined the codigestion of different agro-industrial wastes. The potential of methane production and the effects of a second feed were determined in batch AD experiments. It was shown that codigestion of SB/VC (slaughterhouse waste and various crops), SB/VC/MSW, and SB/M (manure) provided high methane potentials. The highest methane yields obtained were 592, 522, and 521 mL/g VS, respectively, in these samples. Moreover, the second feeding could increase the methane yield of some of the substrate mixtures, due to building up an active microbial consortium. In contrast, decreasing yields or inhibition was detected in some other substrate mixtures.

The study also examined long-term effects during codigestion of slaughterhouse waste in four continuously stirred tank reactors. In a continuous process, the start-up stage is really important, the OLR should be low, and then it should be slightly increased gradually to avoid overload in the system and for the adaptation of microorganisms to the substrate. VFAs, alkalinity, and ammonium–N concentrations were used as control parameters for the operation of the continuous systems. The methane content of the produced biogas during the digestion and codigestion of slaughterhouse waste was obtained between 60% and 85% (lower in the beginning and higher toward the end), and the highest methane content of 76% was found from codigestion of SB/M toward the end of the operation. Toward the end of the investigation period, average methane yields of 300, 510, 587, and 426 mL/g VS were obtained in the digestion of SB and codigestion of SB/M, SB/VC, and SB/VC/MSW, respectively. The highest average methane potential of 587 mL/g VS was found in codigestion of SB/VC, and it is comparable to the result of 592 mL/g VS obtained from the batch digestion of the same mixture.

4.5.2.2.9 Codigestion of Dairy Manure with Chicken Litter and Other Wastes

Canas [343] studied codigestion of dairy manure with chicken litters and other wastes. The following conclusions were drawn by the author:

1. Chicken litter can be added into a digester treating dairy manure to increase the OLR leading to a higher methane production rate. Chicken litter can be safely added up to a 33% as VS in the feedstock increasing methane production by 49.3%. Other researches [185,417] found a similar maximum chicken manure percentage in feedstocks for continuous and batch reactors.
2. No synergistic effects were detected when co-digesting chicken litter with dairy manure. However, chicken litter required water to be digested.
3. The selection of the initial OLR is related to inoculum acclimation and waste composition and depends on a wide number of factors.
4. For dairy manure, two retention times seemed to be enough to reach stable conditions. However, previously, three or four retention times were suggested to reach steady-state conditions.
5. Perhaps, a combined effect of high ammonia concentration and overloading resulted in the reactor's collapse. More research needed to be done evaluating the influence of high ammonia concentration under different OLRs.
6. Because of the large total alkalinity in the system, pH and VFAs were not good indicators of instability. Instead, gas production should be followed closely in order to detect any symptom of imbalance.
7. Total and free ammonia tolerance could be improved just by simply combining dairy manure with chicken litter. However, microbial adaptation for free ammonia occurred when increasing free ammonia concentrations in reactors.
8. By establishing the retention time at 20 days, it is possible to recover up to 90% of methane from substrates. In addition, this large retention time allowed the microbial population to better develop free ammonia adaptation.

9. Codigestion seemed to have no influence in pathogen indicator (*Escherichia coli*) removal. Removal values reached typical values ranging from 68.4% to 97.2%.
10. The microbial population can adapt to lower temperatures down to 19°C, but at longer retention times, this became economically unattractive for continuous reactors. At 20 days of retention time, methane production decreases by 10% when temperature decreases from 35°C to 25°C.
11. Filtered solids from dairy can be co-digested up to a maximum percentage in the feedstock of 70% VSs to increase methane production by 114.2% as a consequence of an increase in organic loading, but the efficiency (methane yield) decreased by 59.14%. Antagonistic effects were also found. GTW can be co-digested, thus improving methane yield (efficiency) and VS removal of dairy manure alone by 111.5% and 76.4%, respectively.
12. Codigestion of sawdust with dairy manure was unsuccessful.
13. By storing the substrates at a 4°C, both samples and feedstock were preserved properly.

4.5.2.2.10   Swine Manure with Energy Crop Residues

Cuetos et al. [328–330] examined codigestion of swine manure (SM) with energy crop residues (ECRs) that contain Mz, rapeseed (Rs), and sunflower (Sf) residues. The behavior of reactors and methane productions in both batch and continuous-flow reactors were examined. Three different proportions of ECRs were tested in batch experiments for codigestion with SM: 25%, 50%, and 75% VS. On the basis of results obtained from the batch study, 50% ECR content was selected as the mixture for the second stage of the study.

This stage experiments were performed under mesophilic conditions in semicontinuous reactors with HRT of 30 days, and the reactors were kept under these operational conditions over four HRTs. The results showed that the addition of ECR to the codigestion system resulted in a major increase in the biogas production. The highest biogas yield was obtained when co-digesting Rs (3.5 L/day), although no improvement was observed in specific gas production from the addition of the co-substrate.

4.5.2.2.11   Food Waste and Human Excreta

Dahunsi and Oranusi [374] examined codigestion of food waste (FW) and human excreta in Nigeria where there exists no centralized sewage system and both of these wastes end up in the septic tank of each home. An investigation was launched into the design and construction of an anaerobic digester system from locally available raw materials using local technology and the production of biogas from FWs and human excreta generated within a university campus. The experiment lasted for 60 days using a 40 L laboratory-scale anaerobic digester. The volume of gas generated from the mixture was 84,750 cm^3 and comprised of 58% CH_4, 24% CO_2, and 19% H_2S and other impurities.

The physicochemistry of the feedstock in the digester revealed an initial drop in pH to a more acidic range and a steady increase of 4.52–6.10. The temperature remained relatively constant at a mesophilic range of 22.0°C–30.5°C throughout the study. The C/N ratio of the feedstock before digestion was within 139:1. Population distributions of the microflora showed aerobic and anaerobic bacteria to include *Klebsiella* spp., *Bacillus* spp., *E. coli*, *Clostridium* spp., and a methanogen of the genus *Methanococcus*. The study concluded that in most developing nations of sub-Saharan Africa where biomass is abundant and where biogas technology is in its infant stage, anaerobic codigestion can be a solution.

4.5.2.2.12   Poultry Manure and Straw

Babaee et al. [375] examined the effects of organic loading and temperature on the anaerobic slurry codigestion of poultry manure and straw. In order to obtain basic design criteria for the AD of a mixture of poultry manure and WS, the effects of different temperatures and OLRs on the biogas yield and methane contents were evaluated. Since poultry manure is a poor substrate, in terms of the availability of the nutrients, external supplementation of carbon had to be regularly introduced in order to achieve a stable and efficient process.

The complete-mix, pilot-scale digester with a working volume of 70 L was used. The digestion operated at 25°C, 30°C, and 35°C with OLRs of 1.0, 2.0, 2.5, 3.0, 3.5, and 4.0 kg VS/(m^3 day) and an HRT of 15 days. At a temperature of 35°C, the methane yield was increased by 43% compared to the one at 25°C. Anaerobic codigestion appeared feasible with a loading rate of 3.0 kg VS/(m^3 day) at 35°C. At this state, the specific methane yield was calculated about 0.12 m^3/kg VS with a methane content of 53%–70.2% in the biogas. The VS removal was 72%. As a result of VFA accumulation and decrease in pH, when the loading rate was less than 1 or greater than 4 kg VS/(m^3 day), the process was inhibited or overloaded, respectively. Both the lower and higher loading rates resulted in a decline in the methane yield.

4.5.2.2.13 Food Waste with Dairy Manure

Lisboa et al. [418] examined codigestion of FW with dairy manure. The study showed the importance of performance anaerobic toxicity assays (ATAs) before possible codigestion food products were introduced into AD environments. This study did not give any results on biochemical methane potential assay for this codigestion system.

4.5.2.2.14 Cattle Manure and Sewage Sludge

Garcia and Perez [419] examined the influence of composition and temperature on codigestion of cattle manure and sewage sludge. Both organic wastes were from wastewater treatment stations. Codigestion of sewage sludge and cattle manure has the advantage of sharing processing facilities, unifying management methodologies, reducing operating costs, and dampening investment and temporal variations in composition and production of each waste separately.

The aim of the study was to select suitable operating conditions (both composition and temperature) of anaerobic codigestion process of cattle manure and sewage sludge to optimize the process for the biogas generation. The batch tests were developed at mesophilic and thermophilic conditions to determine the anaerobic biodegradability of three different mixtures of cattle manure and sewage sludge, in both static and stirring conditions.

The results of the study indicated that the anaerobic biodegradability of raw sludge and cattle manure mixtures was more efficient at thermophilic conditions since a greater elimination of organic matter with a greater methane yield was obtained. The most efficient process corresponded to the mixture with 25% v/v of cattle manure and 75% v/v of raw sludge with values of 62% and 75.7% of COD and DOC removals, respectively, and methane yields of 2200 mL CH$_4$/g CODr and 306 mL CH$_4$/g VSr, during the total processing time of 12 days. Also, it was verified that a higher amount of cattle manure in the mixture meant a higher alkalinity and a greater percentage of methane in biogas. The optimal composition of the mixture selected for thermophilic conditions allowed to reach values three times higher than those obtained in mesophilic conditions for all parameters examined for the generation of biogas.

4.5.2.2.15 MSW and Agricultural Waste with Dairy Cow Manure

Macias-Corral et al. [69] examined codigestion of MSW and agricultural waste with dairy cow manure (CM). Anaerobic codigestion of dairy CM, OFMSW, and cotton gin waste (CGW) was investigated with a two-phase pilot-scale AD system. OFMSW and CM were digested as single wastes and as combined wastes. The single waste digestion of CM resulted in 62 m^3 CH$_4$/tons of CM on dry weight basis. The single waste digestion of OFMSW produced 37 m^3 CH$_4$/tons of dry waste. Codigestion of OFMSW and CM resulted in 172 m^3 CH$_4$/tons of dry waste. Codigestion of CGW and CM produced 87 m^3 CH$_4$/tons of dry waste. Comparing the single waste digestions with the codigestion of combined wastes, it was shown that codigestion resulted in higher methane gas yields. In addition, codigestion of OFMSW and CM promoted synergistic effects resulting in higher mass conversion and lower weight and volume of digested residual.

4.5.2.3 Other Co-Substrates

4.5.2.3.1 *Effect of Inoculum Source on Dry Thermophilic Anaerobic Digestion of OFMSW*

Forster-Carneiro et al. [380] evaluated the effect of inoculum source on anaerobic thermophilic digestion. They used six different inoculum sources: corn silage, restaurant waste digested mixed with rice hulls, cattle excrement, swine excrement, digested sludge, and swine excrement mixed with digested sludge (1:1). The experiments were carried out at 55°C, and other conditions were 25% inoculum and 30% TSs. Results indicated that digested sludge was the best inoculum source for anaerobic thermophilic digestion of the treatment of OFMSW at dry conditions (30% TS). After 60 days of operation, the COD removal and VS removals by the digester were 44% and 43%, respectively. In stabilization stage, digested sludge showed higher volumetric biogas generated at 78.9 mL/day with a methane yield of 0.53 L CH_4/g VS. For this stage, cattle excrement and swine excrement with digested sludge were good inoculums.

4.5.2.3.2 *Codigestion of Municipal, Farm, and Industrial Organic Wastes*

Alatriste-Mondragon et al. [420] reviewed 4 years of literature on codigestion of municipal wastewater treatment plants with co-substrates like wood wastes, industrial organic wastes, and farm wastes. The review was focused on low-solids-concentration (<10%) systems for batch assays and bench-scale systems. The literature on digestibility of co-digestates, the data for performance and monitoring of codigestion, the inhibition of digestion by co-digestates, the design of the process (single or two stages), and the operation temperature (mesophilic or thermophilic) were reviewed by these authors.

4.5.2.3.3 *Fats of Animal and Vegetable Origin and Simulated OFMSW*

Fernandez et al. [281] studied the codigestion of fats of different animals and vegetable origins with OFMSW. Codigestion process was conducted at the pilot scale in semicontinuous regime under mesophilic 370°C temperature condition and for HRT of 17 days. Dry pet food was used as OFMSW and the fat used consisted of waste from food industry (animal fat) with prescribed LCFA profile. The fat concentration was raised up to 28% of the OFMSW, and then it was switched to vegetable fat. The total fat removal throughout the experiment was 88%, whereas biogas and methane yields were very similar to those simulated for OFMSW. Codigestion with fat increased the amount of biogas produced according to the applied organic loading. Authors recommended the use of codigestion for this system.

4.5.2.3.4 *Codigestion of High-Strength/Toxic Organic Liquid*

Ramsamy et al. [337] examined codigestion of high-strength/toxic organic liquid (i.e., leachate from a hazardous landfill site [Shongweni]) with textile size effluent (Frametex size effluent). The results of ATA proved the amenability of the size effluent and landfill leachate. These results suggested that codigestion of these wastes was possible. The results showed that codigestion was possible at all the sample dilutions tested, that is, 4%–40% by volume. Authors conclude that the study needed some additional work.

4.5.2.3.5 *Codigestion of Fat, Oil, and Grease Waste from Sewage-Treating Plant with Source-Collected OFMSW*

Martin-Gonzalez et al. [59] examined mesophilic codigestion of FOG waste from sewage-treating plant (STP-FOGW) with source-collected OFMSW (SC-OFMSW) at a feed ratio of 15% (VS) carried out in a 5 L lab-scale reactor that resulted in an improvement both in terms of biogas production (72% higher) and methane yield (46% higher) in comparison with anaerobic treatment of SC-OFMSW. During the codigestion process, a stable reactor performance was observed, and there was no inhibition either in LCFA accumulation or in VFA excess. VS and TS reduction percentages were stable and around 65% and 57%, respectively, and methane content in biogas was 63%.

These results suggested that anaerobic codigestion is a feasible and efficient way of managing STP-FOGW. Moreover, it is an environmentally friendly treatment in comparison with the landfill option and allows a methane potential that is presently being wasted to be recovered.

4.5.2.3.6   Biologically Pretreated Nile Perch Fish Solid Waste with Vegetable Fraction of Market Solid Waste

Kassuwi et al. [373] examined anaerobic codigestion of various organic wastes with fish wastes. Anaerobic codigestion of various organic wastes has been shown to improve biogas yield of fish wastes. The study presented the effect of pretreating Nile perch fish solid waste (FSW) using CBR-11 bacterial culture (CBR-11-FSW) and commercial lipase enzyme (Lipo-FSW), followed by batch anaerobic codigestion with vegetable fractions of market solid waste in various proportions, using potato waste (PW) and cabbage waste (CW) as co-substrates either singly or combined.

Results indicated that CBR-11-pretreated FSW co-digested with PW or CW in 1:1 ratio (substrate/inoculum) had positive effect on methane yield, while Lipo-pretreated FSW had negative effect on methane yield. Using CBR-11-FSW–PW, the highest yield was 1.58 times more than the untreated FSW, whereas using Lipo-FSW–CW, the highest yield was 1.65 times lower than the one for untreated FSW. Furthermore, the optimal mixture of CBR-11-pretreated FSW and PW and CW co-substrates resulted into higher methane yield of 1322 CH_4 mL/g VS using CBR-11-FSW (10)–PW (45)–CW (45) ratio. The ratio enhanced methane yield to 135% compared to control. The results demonstrated that optimal mixture of CBR-11-pretreated FSW with both PW and CW as co-substrates enhanced methane yield [205].

4.5.2.3.7   Energy Crops as Co-Substrates

The design of the fermenters can differ slightly, depending on the technical solutions applied. While the addition of energy crops improves the performance of AD process, it comes with the additional cost. The cultivation of energy crops requires heavy machinery, diesel fuel, and synthetic fertilizer; besides labor, all of them add a substantial cost. Commonly, energy crops are fed together with manure or other liquid substrates (codigestion), in order to keep homogenous fermentation conditions. Similar to "wet digestion," the TS content of these systems has to remain below 10% in order to enable proper reactor stirring. Recirculation of digestate is required in such digesting systems in order to maintain homogenous and well-buffered digester conditions. However, some designs of "dry fermentation" systems allow TS contents much higher than 10% TS. Without the addition of liquid, the TS content can increase above 30%. Typically, two-step, stirred tank, serial reactor designs are applied in most digestion plants. The second digester is often combined with a membrane-type gas holder. One-step digesters are rarely used [13,19,205].

The AD of energy crops requires in most cases prolonged hydraulic residence times from several weeks to months. Both mesophilic and thermophilic fermentation temperatures are commonly applied in the AD of energy crops. Complete biomass degradation with high gas yields and minimized residual gas potential of the digestate is a must in terms of proper economy, as well as the ecological soundness of the digestion process. VS degradation efficiencies of 80%–90% should be realized in order to achieve sufficient substrate use, thereby leading to negligible emissions (CH_4, NH_3) from the digestate [205].

Energy crops like Mz, Sf, grass, and beets are increasingly added to agricultural digesters either as co-substrates or as the main substrate. The cultivation of energy crops on fallow or set-aside land can reduce agricultural surpluses and provide new income for agriculture. The most popular crop today is Mz. From 1 ton of Mz (dry matter), approximately 400–600 m^3 of biogas can be produced. Approximately 8,000–12,000 m^3 biogas (50% CH_4 content) produces about 13,200–19,800 kW h electricity. Energy crop digestion is critically dependent on the obtainable price of electricity per kW h. The capital payback time of evaluated farm-scale biogas plants lies between 9 and 13 years, which is high, but reasonable. Provided that low crop production costs at high yields per land area and a high biogas yield during fermentation can be achieved, energy crop digestion can become economically viable without subsidies [13,19,205].

4.6 EFFECTS OF HARVESTING, STORAGE, AND PRETREATMENT ON BIOGAS PRODUCTION

The AD process fundamentally produces two products: biogas and usable digestate. The digestate is an important and valuable organic fertilizer, but it is only marketable if it is environmentally and hygienically safe and free of visible contaminants such as plastics, stones, and metals.

The best way to obtain usable digestate is to have clean and unpolluted co-substrates with high gas potential. If the substrate contains undesirable materials, they have to be taken out preferentially before it is used in the digestion or codigestion process. These materials can cause pipe blockages, scum formation or bottom layers in the digesters, or damage pumps and mixing devices. The removal of these impurities may require sophisticated and expensive equipment and additional operating costs. The types of collection vessel (bins, plastic bags, etc.), the region of collection, people's habits, the season, etc., affect the contents of undesirables in substrates or co-substrates [13,19,205,245–271,294–302,421,422].

The selection of a proper pretreatment process should be waste and digestion process specific and adjusted to the product quality required. For the use of organic household waste, two major digestion techniques, dry and wet digestion processes, are applied. Dry digestion technique uses solids concentration >20% and requires less pretreatment for contaminant removal [13,19,205,245–271,294–302,421,422].

Dry digestion processes retain higher concentration of particulate matters. For most of the more problematic waste, dry digestion offers the easier solution since the requirement for the separation of the materials is small as long as the dry matter content remains above 20% solids [13,19,205,245–271, 294–302,421,422]. The wet separation process, which uses less than 10% solids, requires more sophisticated approaches for the contaminant removal. It uses wet pretreatment processes to separate light solids from heavy solids. Generally, wet separation processes achieve a higher level of impurity removal compared to dry separation. However, the processes are more laborious and expensive. Only few organic wastes like biowastes from separate collection, garden and yard wastes, expired food, and leather industry wastes require extended preconditioning for wet digestion or codigestion.

In general, the pretreatment steps include size reduction of the substrate, removal of indigestible components, and hygienization. The extended pretreatment requires chopping; sieving; removal of metals; removals of glass, sand, stones, etc.; and homogenization. The hygienization includes sterilization, pasteurization, and sanitation. For most biowastes, thermophilic operation is sufficient for hygienization. Other suggestions for hygienization of biowaste are (1) using thermophilic digester operation at 55°C for at least 24 h with minimum residence time of at least 20 days; (2) in the case of mesophilic digestion, pretreating substrates at 70°C for 60 min or posttreatment of the digestate at 70°C for 60 min; and (3) composting the digestate [13,19,245–271,294–302,421,422].

Esposito et al. [251] studied codigestion of organic waste and concluded that several pretreatment methods can be applied to further increase the biogas production of a codigestion process, such as mechanical comminution, solid–liquid separation, bacterial hydrolysis, and alkaline addition at high-temperature, ensilage, alkaline, ultrasonic, and thermal pretreatments. However, other pretreatment methods, such as wet oxidation and wet explosion, can result in a decrease of the methane production efficiency [13,19,205,245–271,294–302,421,422].

You et al. [422] showed that anaerobic codigestion of corn stover with SM can be considerably improved with NaOH pretreatment. This pretreatment shortened digestion time and improved biogas yield. Different NaOH concentrations (2%, 4%, and 6%) at various temperatures (20°C, 35°C, and 55°C) and 3 h of pretreatment time were tested for corn stover pretreatment. A C/N ratio of 25:1 in the substrate of corn stover and SM was employed. The results showed that lignin removal rates of 54.57%–79.49% were achieved through the NaOH pretreatment. The highest biogas production rate was observed for corn stover pretreated at 6% NaOH at 35°C for 3 h. This was 34.59% higher than that from the untreated corn stover. The increase of methane yield was from 276 to 350 mL/g VS. The digestion time of pretreated corn stover was shortened from 18 to 12–13 days.

NaOH pretreatment thus increased biogas production, reduced digestion time, and separated lignin. The study recommended pretreatment of corn stover with 6% NaOH at 35°C for 3 h for all its future use in codigestion [13,19,205,245–271,294–302,421,422].

In general, digestate contains nondigestible and residual particulates, liquid organic and inorganic waste constituents, and bacterial biomass. In large agricultural digesters, the digested slurry is used as fertilizer for the land. The amount of nitrogen and heavy metals that can be introduced in the land and the nature of agricultural land for digestate use are often restricted [245–271,294–302,421,422].

Easy storage and conventional methods of milling are important factors in the selection of energy crops. Forage crops fit this requirement but the specific methane yield obtained from this material depends on its age [13,19,205,242]. Harvesting time and frequency of harvesting time are important for biogas yield. Crops can be grown as preceding crop, main crop, or succeeding crop, each leading to different biogas yields [13,19,292–294]. Maize crops harvested after 97 days of milk ripeness produced 37% more methane yield than those at full ripeness. Mixed cultivation such as maize and sunflower can also be used to adjust biogas and methane yields [13,292–294].

The storage of energy crops by ensiling converts soluble carbohydrates into lactic acid, acetate, propionate, and butyrate that inhibit the growth of detrimental microorganisms by a strong drop in pH between 3 and 4 [295]. An optimum process ensiling generates 5%–10% of lactic acid and 2%–4% of acetic acid within few days. Butyric acid is prevented by a rapid drop in pH. The starter cultures, enzymes, and easily degradable carbohydrates can control and accelerate acid formation. The optimum ensiling conditions are obtained by cutting particle length to 10–20 mm and for total solid contents between 25% and 35%. The storage by ensiling can be considered as a pretreatment process [13,296].

The structural polysaccharides of plant material are partly degraded during storage. They lose about 8%–20% of energy due to aerobic degradation. During storage, a plastic wrap should cover the plant material to minimize degradation. Furthermore, an addition of a heterofermentative starter culture generates acetic and lactic acids that in turn lower the pH and inhibit aerobic degradation by limiting the growth of yeasts that are responsible for heat upon exposure to oxygen [13,19]. Thus, a low pH value and continuous supply of energy crops is necessary to reduce process instabilities and fluctuating gas qualities.

Weiland [13] and Muller et al. [297] showed that thermal, chemical, mechanical, or enzymatic process as pretreatment can enhance degradation rate. Decrease in particle size accelerates biogas production rate but not methane yield [13,298]. The feed system generally includes feedstock crushing device or an ultrasonic treatment of the side stream of the fermenter [13]. Pressure hydrolysis occurs at 230°C and 20–30 atm. Pressure splits polymers into short-chain compounds, which gives better biogas yields with reduced retention time in the digesters [299,300]. Biogas yield can also be improved by about 20% by the addition of hydrolytic enzymes, which accelerates the decomposition of structural polysaccharides [13]. The effect of enzyme addition is, however, mix. The enzyme reduces viscosity of the substrate mixture and increases the degradation rates by avoiding the formation of floating layers. The protease of anaerobic microorganisms can, however, degrade the enzymes if added in excess, thus limiting its effectiveness [13,301]. For wheat grass, the addition of an enzyme improved biogas production but at the end of the digestion period, no significant improvement of methane yield or degradation rate was observed [302].

4.7 DIGESTER AND ASSOCIATED PROCESS CONFIGURATIONS FOR BIOGAS PRODUCTION

In principle, biogas is produced from wet or dry fermentation processes. This definition only suggests the level of solids concentration in the digester. In wet fermentation process, solids concentration is less than 10% and it is generally carried out in a continuous stirred slurry fermenter. Low slurry concentration allows the stirring at lower power cost. The digested material is spread on the fields for the fertilization. For energy crops, the feed must be mixed with recycled process water or liquid manure to make the slurry pumpable. Dry solids fermentation is carried out with solids concentration between 15% and 35%. Wet fermentation is mostly operated continuously, while dry fermentation is operated either batchwise or in a continuous mode. In agriculture sector, wet fermentation is the preferred mode of operation [13,19,304,305,423].

For wet fermentation, a vertical continuously stirred fermenter is the most preferred mode of operation [13,19,49,303]. Often the fermenter is covered with gas-tight single- or double-membrane roof to store the gas before utilization. The fermenter is well mixed by stirring to achieve uniform temperature and bring about the best contact between microorganisms and the feedstock. Most stirring is done mechanically and in slow or fast mode. The slow mode is operated continuously, while the fast mode is operated sequentially among several stirrers. Depending upon the nature of feedstock, height of fermentation slurry and solids content, the number, size of stirrer paddles, directions (horizontal, vertical or at an angle), and the depth are arranged to get the maximum mixing in the reactor. Besides mechanical stirrers, hydraulic and pneumatic stirrers are also used. The pneumatic stirrers use the produced biogas for mixing. Hydraulic stirring by pumps is only used in few specific types of reactors. The typical fermenter volume varies from 1000 to 4000 m^3. Horizontal digesters are generally a part of a two-stage system wherein high solids concentration flows in a horizontal plug flow mode with a low rotating horizontal paddle mixer. The reactor volume of such a reactor is limited to 700 m^3 [13,19,49,205,303].

For energy crops and processing of high solids concentration slurry, two-stage digester systems are preferred that consist of highly loaded main fermenter and a lowly loaded secondary fermenter in series to treat the digestate from the first stage. The two-stage process generally gives higher biogas production and reduced methane potential of the final digestate [13,19,205,303,307]. In the two-stage process, hydrolysis and methanation take place in both reactors. For achieving better metabolization of solid organic compounds, the two-stage reactor system with separate hydrolysis stages can be advantageous because the ideal pH required for hydrolysis (5.5–6.5) is different from the one required for methanation (6.8–7.2) [13,19,424,425]. This technology is mainly applied to municipal and industrial solid wastes and solid manure and seldom to energy crops. The control of operation and process parameters for a two-stage operation is generally difficult. If the hydrolysis stage is malfunctioned, methane and hydrogen can escape in the environment [13,19,205,426]. Thus, a gas-tight covering of the hydrolysis stage is required for preventing energy losses and emission of odorous and harmful gases.

Moist wet fermenters are operated at temperatures between 38°C and 42°C (mesophilic condition). Few biogas plants operate at a temperature of 50°C–55°C (thermophilic conditions). Higher temperature gives faster degradation rate requiring lower HRT and reactor volume but ultimate methane yield is not influenced. Lower temperature reduces the toxicity of ammonia, but the growth rate of microorganisms can drop significantly pausing the washout problem of microbial population [13,19,303]. Since energy saved by not operating under thermophilic conditions can be sold to households, thermophilic processes are less efficient. The energy crops require very high retention time (from weeks to months) and are generally fed at lower solids concentration (2–4 wt%) in a wet fermenter [13,19,205,303].

For energy crops, dry fermentation in batch processes is preferred. For these processes sometimes no mixing is required and solid inoculum up to 70% is necessary. The batch process is operated with gas-tight lids. The digester is operated for several weeks and at the end of digestion period, the digested material is unloaded and a new batch is initiated [13,19,205].

4.7.1 NOVEL DIGESTER TECHNOLOGY

The University of California at Davis developed a new anaerobic digester technology called Anaerobic Phased Solids Digester (APS Digester) for biogasification of organic waste solids that are normally difficult to process using conventional anaerobic digesters. A variety of feedstocks including crop residues, animal manures, feed processing residuals, paper sludge, and MSW can be processed by APS Digester. The digester has been used to generate power for the University of California. The first commercial APS Digester was built in Boynton Beach, Florida, to process 80 tons per day of horse stable wastes. The possible benefits of this plant are renewable energy generation, odor control, pathogen and insect control, truck traffic reduction, and production of high-quality soil amendment [13,19,205].

The APS Digester combines the favorable features of both batch and continuous operations in one system. Solids to be digested are handled in batches while biogas production is continuous. This allows solids to be loaded and unloaded without disrupting an anaerobic environment for bacteria. The typical APS Digester system consists of four hydrolysis reactors and one biogasification reactor. Liquid is recirculated intermittently between each hydrolysis reactor and the biogasification reactor. The solids are housed in the hydrolysis reactor, while the bacteria (methanogens) are housed in the biogasification reactor. The solids are broken down and liquefied in soluble compounds, which are mainly organic acids, and transferred to the biogasification reactor to generate biogas. Four hydrolysis reactors are operated in different time schedules so that biogasification reactor is constantly fed with the dissolved organic acid. High bacteria concentration in the biogasification reactor is maintained to get the optimum performance. More details on APS Digester is given by Shah [19] and Zhang [427].

Shah [19] and Narra et al. [54] evaluated a model for AD of household organic waste in high solids concentration (25 wt%) in an urban city in India and showed with pilot-scale experiments that biogas production of 209 L/kg of total solids is possible in 30 days incubation period. High solids concentration reduces the water requirement and slurry handling problems. Composting takes 35 days and yielded a quality product that can be used either as manure or a part of chemical fertilizers. Batch pilot plant was developed. Shah [19] and Abderrezaq [428] evaluated the use of anaerobic digester for the MSW generated in Jordan. They showed that the digester technology can generate the energy from waste without generating GHGs.

Applying a second-stage methanogenic fermenter in combination with the leach bed process is advantageous, because this can increase the methane yield and reduce the residual methane potential in the digestate [13,19,205,424,425]. It is also possible to use the leach bed only for hydrolysis and treat the leachate subsequently in a fixed-bed methane reactor if the recycled water is aerated slightly in order to suppress the methane bacteria that have been washed out of packings [13,19].

For dry fermentation of slurry containing more than 25% solids, a horizontal mechanically mixed fermenter or vertical plug flow reactor can also be used. These are known from anaerobic treatment of municipal organic solids [13,19,205]. In the vertical fermenter, the substrate flows from top to bottom by gravity only. The substrate fed at the top is mixed with the digestate coming from the bottom. This recycling and mixing of digestate with fresh feedstock prevents the accumulation of VFA and allows the high OLR in the fermenter. It is difficult to find a suitable and simple control parameter to control the complex fermentation process. Furthermore, only few parameters can be measured on line. In agricultural biogas plants, methane production is the only continuously measured parameter. However, complex and variable process dynamics make the interpretation of data difficult [13,19,304,305]. Only VFA can serve as an efficient indicator of process imbalances.

Weiland [13,429] and Shah [19,205] proposed that an indicator for process failure is the ratio of propionic acid/acetic acid >1. Shah [19] and Ahring et al. [430] suggested that if propionic acid concentration is greater than 1000 mg/L, the concentration of both butyrate and isobutyrate could be a reliable tool for indication of process failure. Nielsen et al. [431] suggested that propionate is the key parameter for process control and optimization. VFA analysis by manual sampling and subsequent analysis by gas chromatography or high-pressure liquid chromatography is a slow process. Online measurement is a difficult process [13,19,431]. A fast control of the process stability is possible by determining the ratio of the total VFA to the total inorganic carbonate. If this ratio is less than 0.3, the process is stable.

4.8 PROCESS-RELATED AND ECONOMIC CONSIDERATIONS FOR BIOGAS PRODUCTION

4.8.1 PROCESS MONITORING AND CONTROL

Process monitoring depends on the scale of the operation; large scale will require more attention than the small scale. The most important control parameters are the overall daily substrate flow rate (tons/day or m^3/day) and the quantity of the daily production of biogas (m^3/day). The efficiency of the digester is measured by the amount of biogas produced per unit substrate flow. For the proper control, the determination of the CH_4 concentration is highly desirable. In the case of codigestion, additional type and amount of separated impurities need to be recorded and controlled. In large-scale operations, the amount of VFAs and ammonia concentration should be monitored and controlled [13,19,68,78,283,313–318,426–462]. Mosche and Jordening [463] compared different models of substrate and product inhibition in ammonia digestion. Degradation of VFA in a highly efficient anaerobic digester was measured by Wang et al. [464]. Boe et al. [465] developed an online head space chromatographic method for measuring VFA in a biogas reactor.

The addition of co-substrate causes scum layers and bottom sediments. These need to be controlled for a stable operation. For each type and amount of waste, the time and temperature conditions for the sterilization should be monitored. Most importantly, the government regulations for the quality assurance of digestate for each type of waste should be followed [13,19,68,78,205,283,313–318,426–462].

Lin et al. [313] examined anaerobic codigestion of FVW and FW at different mixture ratios for 178 days at an organic loading of 3 kg VS/(m^3 day). The dynamics of archaeal community and the correlations between environmental variables and methanogenic community structure were analyzed by polymerase chain reactions denaturing gradient gel electrophoresis (PCR–DGGE) and redundancy analysis (RDA), respectively. PCR–DGGE results demonstrated that the mixture ratio of FVW to FW altered the community composition of Archaea. As FVW/FW ratio increased, *Methanoculleus*, *Methanosaeta*, and *Methanosarcina* became the predominant methanogens in the community. RDA results indicated that the shift of methanogenic community was significantly correlated with the composition of acidogenic products and methane production yield. The different mixture ratios of substrates led to different compositions of intermediate metabolites, which may affect the methanogenic community. These results suggested that the analysis of microbial community could be used to diagnose and control anaerobic process [13,19,68,78,205,283,313–318,426–462].

4.8.2 SCALE-UP

Shah [19,205] and Sell [316] at Iowa State University examined a scale-up procedure for substrate codigestion in anaerobic digesters through the use of substrate characterization, BMPs, anaerobic toxicity assays (ATAs), and sub-pilot-scale digesters. The digestion system utilizes cattle manure with co-substrates of numerous different types of organic wastes. Based on his study, he wrote a series of three papers. The objective of the first paper was to analyze multiple substrates using various laboratory techniques so that optimum mixture ratios could be formed. Biochemical methane potentials and ATAs were used to select and in some cases rule out substrates based on their contribution to methane production. Mixtures were created using constraints arising from the full-scale system. This included the use of all available manure, keeping total solids (TSs) below 15% to facilitate pumping, maintaining pH between 6.5 and 8.2 for microbial ecology, providing high COD concentrations to maximize methane production, and limiting ammonia levels to avoid toxicity [19,117]. The BMP and ATA results from each mixture were analyzed and compared. The mixture with the best performance was selected for subsequent testing in 100 L sub-pilot-scale anaerobic digesters.

The objective of the second paper was to analyze the performance of three 100 L sub-pilot-scale anaerobic digesters. These plug flow digesters operated at a 21-day HRT and were fed the mixture selected in the first study in a semicontinuous manner twice weekly (six loadings per HRT). Methane production was measured using submerged tipping buckets. Methane production from the sub-pilot-scale reactor was compared to that predicted by the BMP tests. After two HRTs, the BMP maximum and minimum were observed to be valid boundaries for the sub-pilot-scale anaerobic digester methane production, with some of the variability ascribed to seasonal substrate changes.

The objective of the third and final paper was to use a series of BMPs and an ATA to predict the methane production in three 100 L sub-pilot-scale anaerobic digesters that were subjected to a potential toxicant, namely, glycerin. A group of ATAs was performed with glycerin inclusion rates of 0.5%, 1.0%, 2.0%, 4.0%, 8.0%, 15%, 25%, and 35% by volume. A set of BMPs was performed where a baseline mixture was combined with glycerin such that glycerin was 0.0%, 0.5%, 1.0%, 2.0%, 4.0%, 8.0%, 15%, 25%, and 35% of the combined mixture by volume. In addition, BMPs of 100% glycerin and 50% glycerin/50% deionized water by volume were also performed [48–50]. The three 100 L sub-pilot-scale anaerobic digesters were operated at a 21-day HRT and were each fed in a semicontinuous manner twice weekly (six loadings per HRT). Each digester was fed a combination of the mixture selected in the first study with a different amount of glycerin (1%, 2%, 4% by volume) [19,205].

The ATAs showed that glycerin was toxic to methane production at all inclusion levels. The BMPs indicated no significant difference between methane productions of the 0.0%, 0.5%, 1.0%, 2.0%, and 4.0% mixture combinations; however, at 8.0%, methane production tripled. In contrast, the sub-pilot-scale reactors showed signs of toxicity for 4.0% glycerin inclusion and little to no effect on methane production for 1.0% and 2.0% glycerin inclusion. Thus, neither the ATA nor the BMP proved to be an adequate predictor for the sub-pilot-scale reactors. The most likely cause was the lack of mixing within the sub-pilot-scale digester required to keep all materials in close contact with each other to ensure adequate microbial activity and methane formation.

The aforementioned study indicates that the proper scale-up of digester requires good understanding of mixing phenomena within the digester. In order to scale up laboratory-scale performance in large-scale digesters, it is essential to obtain good contact among substrates, nutrients, and microorganisms [19,205].

4.8.3 Computer Simulation and Model

Shah [19,205] and Zaher et al. [320,456–458] examined a computer software model to assist plant design and codigestion operation. They pointed out that feeding the digester with a combination of waste streams introduces complexities in waste characterization that requires a model to simulate optimal parameters for codigestion. The general integrated solid waste codigestion (GISCOD) model was developed and tested for this purpose.

Model development overcame several challenges to achieve reliable, precise simulations. Accurate characterization of macronutrients, COD, and charge for waste streams was necessary input to the International Water Association Anaerobic Digestion Model No. 1 (ADM1) [17,68,69,436]. Particulate components of carbohydrates, proteins, and lipids vary dynamically in combined solid waste streams, making it difficult to define the waste streams for accurate input. Such waste heterogeneity could be resolved by applying a general transformer model to interface the ADM1 to practical characteristics of each waste stream. In addition, the study showed that hydrolysis rates for manure only varied considerably from hydrolysis rates for FW–manure codigestion. Thus, for codigestion applications, it is important to consider separate hydrolysis rates for each particulate component from each waste stream. Also, hydrolysis rates of solid wastes differed from that of decaying biomass, which is mainly limited by a disintegration step for cell lysis. The separate characterization and phasing of the co-digested waste hydrolysis allowed the optimization of biogas production and defined the corresponding operation settings of the digester.

As currently designed, GISCOD can support the operational decisions necessary for digesting trucked-in wastes with wastewater sludge or, generally, optimize the feedstock and operation of biogas plants. The model, however, needs to be tested and improved with the additional data from commercial codigestion plants. Recent codigestion models in literature are mainly based on ADM1, with various modifications to upgrade it from the monosubstrate to the multisubstrate system. Shah [19,205] and Esposito et al. [251] examined codigestion of organic waste. They concluded that biomethane potential data when experimentally determined are only valid for the particular co-substrates and the operating conditions (temperature, OLR, HRT, MC, S/I ratio, etc.) applied during the experiments.

However, they can be used to calibrate mathematical models capable to simulate codigestion process, and then such models can be used to predict the biomethane potential achievable with different co-substrates and under different operating conditions. For such purpose, only mathematical models specifically aimed at simulating the codigestion of different co-substrates are suitable, as monosubstrate models are not capable to take into account the peculiarities of different substrates in terms of physical and biochemical characteristics and their synergistic effects (e.g., in terms of pH, alkaline, and nutrient balance).

The ultimate aim of modeling and simulation effort is to develop a user-friendly GISCOD software package that can be made available to digestion engineering firms, wastewater treatment plants, and farm digesters [19,205].

4.8.4 PROCESS OPTIMIZATION

Alvarez et al. [205,326,344,345] and Fezzani and Ben Cheikh [205,282,321,322] developed a methodology for optimizing feed composition of anaerobic codigestion of agro-industrial wastes using linear programming method. Inoculum nutrient demand and inoculum acclimation can influence substrate biodegradation and methane potential. In spite of using granular inoculum, inoculum of different sources caused variations in substrate biodegradation. An optimization protocol for maximizing methane production by anaerobic codigestion of several wastes was carried out. A linear programming method was utilized to set up different blends aimed at maximizing the total substrate biodegradation potential (L CH_4/kg substrate) or the biokinetic potential (L CH_4/kg substrate day). In order to validate the process, three agro-industrial wastes were considered, for example, pig manure, tuna fish waste, and biodiesel waste, and the results obtained were validated by experimental studies in discontinuous assays.

The highest biodegradation potential (321 L CH_4/kg COD) was reached with a mixture composed of 84% pig manure, 5% fish waste, and 11% biodiesel waste, while the highest methane production rate (16.4 L CH_4/kg COD day) was obtained by a mixture containing 88% pig manure, 4% fish waste, and 8% biodiesel waste. Linear programming was proved to be a powerful, useful, and easy-to-use tool to estimate methane production in codigestion units where different substrates can be fed. Experiments in continuous operation were recommended to acquire broader information about the biodegradation and methane production rates of the blends determined by linear programming.

The review by Shah [205] and Khalid et al. [252] for codigestion of solid organic waste indicated that AD is one of the most effective biological processes to treat a wide variety of solid organic waste products and sludges. They, however, noted that different factors such as substrate and co-substrate composition and quality, environmental factors (temperature, pH, OLR), and microbial dynamics contribute to the efficiency of the AD process and must be optimized to achieve maximum benefit from this technology in terms of both energy production and organic waste management. The use of advanced molecular techniques can further help in enhancing the efficiency of this system by identifying the microbial community structure and function and their ecological relationships in the bioreactor.

4.8.5 ADDITIONAL NEEDS FOR CODIGESTION AND PROCESS ECONOMICS

Depending on the quality and nature of the waste to be used for codigestion and the size of the operation, some additional equipment may be required for (1) the delivery of the wastes, (2) the homogenization and mixing of co-substrates, (3) the prevention of excessive foaming and scum layer formation, and (4) the removal of sediments from the digester [53,205,281,282,308–310,466–484].

Generally, co-substrates are delivered by trucks and stored in a pressurized space to avoid odor emissions. The trucks need to be cleaned and sanitized for its reuse. The co-substrate is passed through a premixing and homogenization/pasteurization tank. The co-substrate requiring the extensive pretreatment is generally restricted for large centralized plants. In any given plant, all the information in the processing of co-substrate before, during, and after digester operations is always recorded.

Codigestion significantly improves the economics of AD processes through additional biogas production and tipping fees paid by the generator of the organic wastes to the digester owner. Various commercial-scale and laboratory-scale studies have shown that depending upon the type, concentration, and flow rate of organic wastes used, biogas production can be enhanced by as much as 25%–400% [53,205,281,282,308–310,466–484]. An economic analysis of an AD facility installed on a 700-cow dairy in northwest Washington state showed that codigestion with 16% organic wastes more than doubled biogas production and almost quadrupled annual digester revenues compared to a manure-only baseline, with 72% of all receipts directly attributable to the addition of organic waste [53,205,281,282,308–310,466–484].

4.9 PURIFICATION AND UTILIZATION OF BIOGAS AND DIGESTATE

The purification and upgrading technologies for natural and synthetic gas are examined in details in Chapter 7. Purification of biogas is examined here because some of the methods used are biogas specific and they are generally not used for natural and synthetic gas produced by thermal processes. Biogas mainly contains methane and carbon dioxide with some impurities of hydrogen sulfide and ammonia and it is generally saturated with water vapor. Before it can be used for heat and electricity generation, sulfur impurities that can be as high as 100–3000 ppm for cofermentation of manure with energy crops or harvesting residues should be reduced to a level below 250 ppm [13,19,205]. This will prevent excessive corrosion and expensive deterioration of lubrication oil. H_2S removal is carried out by biological desulfurization in which H_2S is oxidized by the injection of small amount of air (2%–5%) in biogas. For this type of desulfurization, *Sulfitobacter oxydans* bacteria must be present to convert H_2S to elemental sulfur and sulfurous acid. These bacteria are generally present in the digester. Generally, air is added in the head space of the digester. An efficient desulfurization requires specific support of wood or fabric to be installed at the top of the fermenter to achieve high contact area for microorganisms' fixation.

For biological desulfurization outside fermenter trickling filters, installations filled with plastic support materials on which the microorganisms can grow are used [19,205,485]. Raw biogas and air are injected at the bottom of the column and aqueous solution of nutrients is circulated from the top to the bottom to wash out acidic products and supply nutrients to microorganisms. The process is carried out at 35°C (mesophilic condition) and support material is washed with air/water mixture at regular intervals to prevent sulfur deposits on the filters. The direct injection of air reduces methane concentration. Desulfurization can also be done by adding commercial ferrous solution to the digester. Ferrous binds sulfur to produce insoluble compounds in the liquid phase preventing the production of gaseous hydrogen sulfide. This method is expensive [13,19,205].

Purified biogas can be used to generate electricity with about 43% efficiency. Often it is used in microgas turbine with lower (25%–31%) efficiency but good loading efficiency and long maintenance intervals. The exhaust heat from microturbine can be used to generate process heat. Clean biogas can also be used for fuel cell that gives higher electrical efficiency. Various fuel cells are operated at temperatures between 80°C and 800°C and the investment costs are higher than those for microgas or conventional turbines. Upgrading biogas to inject into the grid or the utilization as vehicle fuel has become more important. In order to inject biogas into a natural gas grid, strict standards of gas purity must be maintained. All gas contaminants as well as carbon dioxide must be removed and gas quality must be up to 95% methane for use in appliances. Both bacteria and molds must also be removed to make the use of biogas environmentally acceptable.

Carbon dioxide is removed either using organic solvents like polyethylene glycol or using a pressure swing adsorption operation. Less frequently carbon dioxide is also removed using monoethanolamine and diethanolamine solutions or using a membrane technology or cryogenic separation at a low temperature. In these techniques methane loss should be kept minimum because of economical as well as environmental reasons because methane is a GHG that is 23 times stronger than carbon dioxide.

AD process generates digestate that contains nitrogen and carbon, both of which have significant utilization as fertilizers. Ammonia nitrogen content can be increased by a factor of three if energy crops are the only substrate used [13,19,205]. Improved flow properties also allow faster permeation of digestate the soil, hence preventing the loss of ammonia in air. AD can also reduce up to 80% of a feedstock odor. The use of digestate as a fertilizer also inactivates weed seeds, bacteria, viruses, fungi, and parasites. The decay rate depends on the temperature, treatment time, pH, and VFA concentration. The sanitation is best achieved at a higher temperature (above 50°C). A 90% reduction in *Salmonella* population is achieved at a thermophilic temperature of 53°C within 0.7 h. At 35°C, at least 2.4 days are required to get the same results [13,19,205]. A separate pasteurization at 70°C is also carried out for a certain type of wastes. Pasteurization after digestion is effective but digestate is particularly prone to recontamination. More details on biogas and digestate treatments and usages are given by Shah [19,205] and Weiland [13].

4.10 TYPICAL LARGE-SCALE PLANTS FOR BIOGAS PRODUCTION

Since in recent years codigestion has become a norm in AD industry, here we outline few examples of commercial codigestion processes. Commercial AD plants using single substrates would be very similar in nature.

Shah [19,205] and Braun and Wellinger [244] indicated that large-scale codigestion plants usually have more favorable economics. In other words, the economy of scale also applies to codigestion processes. Braun and Wellinger examined large-scale centralized plants in Denmark with digester volumes of about 4650–6000 m^3 and payback times between 3 and 10 years. They concluded that careful design, layout, and operation and gate fees are essential for economic success. The plant economics also depend on restrictions on usable wastes and options for reuse of digestate. Regulations on hygienization and cost of equipment to satisfy these regulations along with process variables such as mixing requirements, level of heat treatment, and the degree of contaminant removals can also affect the economics of large-scale codigestion process. Pictures of two large-scale centralized codigestion plants in Denmark are shown in Figures 4.2 and 4.3.

In North America, significant experience with codigestion has been obtained at East Bay Municipal Utility District (EBMUD), Riverside, Inland Empire Utility Agency (IEUA), Watsonville, Millbrae, and CMSA in California, West Lafayette in Indiana, Pendleton in Oregon, and Lethbridge and Edmonton in Canada. Two prominent California water agencies, EBMUD and IEUA, have implemented and tested codigestion of FW in existing wastewater treatment biodigesters at a pilot scale [205].

A company called BioConversion Solutions (BCS) of Boston, MA [205,383], through its patented Advanced Fluidized Codigestion and Co-Generation technology improves commercial anaerobic codigestion plant profitability through the following:

1. Enhanced biodegradability of feedstock organic solids that results in up to 80%–90% conversion of VSs even for difficult-to-digest feedstock.
2. Higher biogas production yield from feedstock conversion that increases plant output, thereby allowing for the acceptance of more feedstock and higher tip fees.

FIGURE 4.2 Large-scale centralized codigestion plant in Grindsted, Denmark. (From Braun, R. and Wellinger, A., Potential of codigestion, IEA Bioenergy, Task 37—Energy from Biogas and Landfill Gas, IEA, Brussels, Belgium 2004.)

FIGURE 4.3 Large-scale centralized codigestion plant in Lemvig, Denmark. (Photo courtesy of Teodorita Al Seadi, Denmark; From Braun, R. and Wellinger, A., Potential of codigestion, IEA Bioenergy, Task 37—Energy from Biogas and Landfill Gas, IEA, Brussels, Belgium 2004.)

3. Near-complete recovery of nutrients (nitrogen, phosphorus, potassium, etc.) contained in the feedstock. (With BCS biodegradability enhancement system, these nutrients can be recovered in solid [struvite], concentrated liquid, or a combined form depending on the prevailing wholesale prices and fertilizer customer preference.)
4. Up to 80% reduction of residual sludge that otherwise incurs a disposal fee.
5. The processing of mixed feedstock "recipe" including easy-to-digest and more-difficult-to-digest feedstock.
6. The creation of feedstock by cultivating and harvesting high-energy crops on-site or nearby the owner/operator.
7. The discharge of final water output from the plant into local sewer system or waterways without violating nutrient management regulations. (The output water may also be reusable for plant or other purposes, such as irrigation or industrial use, resulting in additional potential revenue or reduced water purchase expenditures.)

The company has developed a number of commercial plants in the United States, which achieve one or more of these seven objectives [38].

Shah [205] and Murphy et al. [245] described an agricultural plant for codigestion of crops with manure. The plant (see Figure 4.4) was built in 2003, and it is located on a pig breeding farm in Austria, where 20 m³/day of manure is used as co-substrate and helps to achieve homogenization of the solid crop feedstock. The crop consists of Mz silage and crushed dry crops, together with minor amounts of residue from vegetable processing. Approximately 11,000 tons/year of crops is processed together with 7,300 tons/year of manure and leachate from the silage clamps.

Two parallel digesters are fed hourly through an automatic dosing unit. The reactors are operated at 39°C with a 77-day residence time; this corresponds to a volumetric loading rate of 4.4 kg VS/(m³ day). Reactor mixing is performed by the mechanical stirrers. Dilution of the substrate mixture to a DS content below 10% is required for sufficient mixing. The plant produces 4,020,000 m³ of biogas annually. Hydrogen sulfide is removed by the addition of air into the head space of digesters. The biogas is collected in an integrated gas holder inside the second digester as well as in an external gas holder. Power and heat are produced in two combined heat and power

FIGURE 4.4 General view of a 1 MWe plant. Crop codigestion plant using two parallel digesters (left background) and a covered final digestate storage tank (center foreground). Gas storage is integrated in digester 2 (background) and in the final storage tank (foreground); further storage capacity is provided in a dry gas storage tank (background right). (From Murphy, J. et al., Biogas from crop digestion, IEA Bioenergy, Task 37—Energy from Biogas, Denmark, September, IEA, Brussels, Belgium 2011.)

(CHP) units with a total capacity of 1 MWe and 1.034 MWth. The electricity is fed to the national power grid, and the heat is used in a local district heating network.

The digestate is collected in a gas-tight final storage tank before it is used as fertilizer in a neighborhood farm. Additionally, the biogas collected from the final digestate storage tank is used in the two CHP units. As an annual mean value, it is possible to achieve 98% of the theoretical capacity of the CHP. Of the energy content in the original substrate, 37% is converted into electricity. Electricity demand on-site is 7% of the produced electricity. Only 7.8% of the energy content of the substrate is used as heat. Heat loss equates to 50.9% of the substrate energy. Methane loss in the CHP facility is 1.8%.

Shah [205] and Ek et al. [361] outlined 15 years of slaughterhouse waste codigestion experience at Tekniska Verken plant in Linkoping AB, Sweden. Experiences from research and development and plant operations lead to several processes improving technological/biological solutions. The improvements had positive effects on energy saving, better odor control, higher gas quality, increased OLRs, and higher biogas production with maintained process stability. The study also described how much of the process stability in AD of slaughterhouse waste depends on the plant operation that allows microbiological consortia to adapt to the substrate. The study also showed that the long retention time of the plant, accomplished by low dilution of the substrate, is a vital component of the process stability when treating high-protein substrate like slaughterhouse waste. Frear et al. [486] and Steyer et al. [487] also described long-term experiences on small-scale and commercial-scale digesters.

REFERENCES

1. Biogas, Wikipedia, the free encyclopedia (2015), https://en.wikipedia.gov/wiki/biogas, Last updated October 12, 2016.
2. Biomethane, Wikipedia, the free encyclopedia (2015).
3. Anaerobic digestion, Wikipedia, the free encyclopedia (2015).
4. Landfill gas, Wikipedia, the free encyclopedia (2015).
5. Bio-hydrogen, Wikipedia, the free encyclopedia (2015).
6. Co-generation, Wikipedia, the free encyclopedia (2015).
7. Renewable natural gas, Wikipedia, the free encyclopedia (2015).
8. IEA, World energy outlook, International Energy Agency, Paris, France (2006).
9. Davidson, O., Mertz, B., co-chairs, IPCC Special report on emission scenarios, Intergovernmental panel on climate change (IPCC), "Summary for policy makers", working group III, UNEP (2000).
10. Fehrenbach, H., Giegrich, J., Reinhardt, G., Sayer, U., Gretz, M., Lanze, K., and Schmitz, J., Kriterien einer nachhaltigen Bioenergienutzung im globalen Mabstab, *UBA-Forschungsbericht*, 206, 41–112 (2008).
11. Gerardi, M.H., *The Microbiology of Anaerobic Digesters*. John Wiley, Hoboken, NJ (2003).
12. Chabrillat, R., Gillett, W., and Liebard, A., THe state of renewable energies in Europe, 13th EurObserv'ER report, Observ'ER, Paris, France (2013).
13. Weiland, P., Biogas production: Current state and perspectives, *Applied Microbiology and Biotechnology*, 85, 849–860 (2010).
14. Amon, T., Hackl, E., Jeramic, D., Amon, B., and Boxberger, J., Biogas production from animal wastes, energy plants and organic wastes, in van Velsen, A. and Verstraete, W. (eds.), *Proceedings of the Ninth World Congress on Anaerobic Digestion*, Antwerp, Belgium, pp. 381–386 (September 2–6, 2001).
15. Ahrens, T. and Weiland, P., Biomethane for future mobility, *Landbauforschung Völkenrode*, 57 (1), 71–79 (2007).
16. Gray, N.F., *Biology of Wastewater Treatment*. Imperial College Press, London, U.K. (2004).
17. Noykova, N., Muller, T.G., Gyllenberg, M., and Timmer, J., Quantitative analysis of anaerobic wastewater treatment processes: Identifiability and parameter estimation, *Biotechnology and Bioengineering*, 78, 89–103 (2002).
18. Parawira, W., Anaerobic treatment of agricultural residues and wastewater, PhD thesis, Lund University, Lund, Sweden (2004).
19. Shah, Y.T., *Water for Energy and Fuel Production*. CRC Press, New York (2014).
20. Droste, R.L., *Theory and Practice of Water and Wastewater Treatment*. John Wiley & Sons, New York (1997).

21. Parkin, G.F. and Owen, W.F., Fundamentals of anaerobic digestion of wastewater sludges, *Journal of Environmental Engineering*, 112, 867–920 (1986).
22. McCarty, P.L. and Smith, P.D., Anaerobic wastewater treatment, *Environmental Science & Technology*, 20, 1200–1206 (1986).
23. Cheong, D.Y., Studies of high rate anaerobic bio-conversion technology for energy production during treatment of high strength organic wastewaters, PhD dissertation, Utah State University, Logan, UT (2005).
24. Zinder, S.H., *Physiological Ecology of Methanogens, Methanogenesis: Ecology, Physiology, Biochemistry and Genetics.* Chapman & Hall, New York (1993).
25. Novaes, R.F., Microbiology of anaerobic digestion, *Water Science & Technology*, 12, 1–14 (1986).
26. Schink, B., Energetics of syntrophic cooperation in methanogenic degradation, *Microbiology and Molecular Biology Reviews*, 61, 262–280 (1997).
27. Hwang, S., Lee, Y., and Yang, K., Maximization of acetic acid production in partial acidogenesis of swine wastewater, *Biotechnology and Bioengineering*, 75, 521–529 (2001).
28. Nath, K. and Das, D., Improvement of fermentative hydrogen production: Various approaches, *Applied Microbiology*, 65, 520–529 (2004).
29. Fox, P. and Pohland, F.G., Anaerobic treatment applications and fundamentals: Substrate specificity during phase separation, *Water Environment Research*, 66, 716–724 (1994).
30. Woese, C.R., Kandler, O., and Wheelis, M.L., Towards a natural system of organisms: Proposal for the domains Archaea, Bacteria, and Eucarya, *Proceedings of the National Academy of Sciences of the United States of America*, 87 (12), 4576–4579 (1990).
31. Morgan, J.W., Evison, L.M., and Forster, C.F., Changes to the microbial ecology in anaerobic digesters treating ice cream wastewater during start-up, *Water Research*, 25, 639–653 (1991).
32. Chynoweth, D.P., Environmental impact of biomethanogenesis, *Environmental Monitoring and Assessment*, 42, 3–18 (1995).
33. Chynoweth, D.P. and Isaacson, R., *Anaerobic Digestion of Biomass.* Elsevier Applied Science, Ltd., London, U.K. (1987).
34. Wolin, M.J. and Miller, T.L., Interspecies hydrogen transfer. 15 years later, *ASM News*, 48, 561–565 (1982).
35. Rozzi, A., Di Pinto, A., Limmoni, N., and Tomei, M., Start up and operation of anaerobic digesters with automatic bicarbonate control, *Bioresource Technology*, 48, 215–219 (1994).
36. Mosey, F.E. and Fernandes, X.A., Patterns of hydrogen in biogas from the anaerobic digestion of milk sugars, *Water Science & Technology*, 21, 187–196 (1989).
37. Speece, R.E., *Anaerobic Biotechnology for Industrial Wastewaters.* Archae Press, Nashville, TN (1996).
38. Chen, Y., Zuckerman, G. and Zering, K., Applying target costing in the development of marketable and environmentally friendly products from the swine wastes, *The Engineering Economics*, 53, 156–170 (2008).
39. Kunchikannan, L.K.N.V., Mande, S.P., Kishore, V.V.N., and Jain, K.L., Coir Pith: A potential agro residue for anaerobic digestion, *Energy Sources Part A*, 29, 293–301 (2007).
40. Weichgrebe, D., Urban, I., and Friedrich, K., Energy and CO_2 reduction potentials by anaerobic treatment of wastewater and organic kitchen wastes in consideration of different climate conditions, *Water Science & Technology*, 58 (2), 379–384 (2008).
41. Pathe, P.P., Rao, N.N., Kharwade, M.R., Lakhe, S.B., and Kaul, S.N., Performance evaluation of a full scale effluent treatment plant for distillery spent wash, *Journal of Environmental Studies*, 59 (4), 415–437 (2002).
42. Kolesarova, N., Hutnan, M., Bodik, I., and Spalkova, V., Utilization of biodiesel by-products for biogas production, *Journal of Biomedicine and Biotechnology*, article id 126798, 1–15 (2011).
43. Beszedes, S., Laszlo, Z., Szabo, G., and Hodur, C., *Journal of Agricultural Science and Technology*, 4 (1, serial no. 26), 62–68 (February 2010).
44. Yusoff, M., Rahman, N., Abd-Aziz, S., Ling, C.M., Hassan, M., and Shirai, Y., The effect of hydraulic retention time and volatile fatty acids on biohydrogen production from palm oil mill effluent under nonsterile condition, *Australian Journal of Basic and Applied Sciences*, 4 (4), 577–587 (2010).
45. Zheng, G.H., Kang, Z.H., Qian, Y.F., Wang, L., Zhou, Q., and Zhu, H.G., Biohydrogen production from tofu wastewater with glutamine auxotrophic mutant of *Rhodobacter sphaeroides*, in Tohji, K., Tsuchiya, N., and Jeyadevan, B. (eds.), *Fifth International Workshop on Water Dynamics*, vol. 987. American Institute of Physics, Sendai, Japan (September 25–27, 2007), pp. 143–148 (2008).
46. Ding, H.B., Liu, X.Y., Stabnikova, O., and Wang, J.-Y., Effect of protein on biohydrogen production from starch of food waste, *Water Science & Technology*, 57 (7), 1031–1036 (2008).

47. Angelidaki, I., Ellegard, L., and Ahring, B., A comprehensive model of anaerobic bioconversion of complex substrates to biogas, *Biotechnology and Bioengineering*, 63, 363–372 (1999).
48. Angelidaki, I., Ellegaard, L., and Ahring, B.K., A mathematical model for dynamic simulation of anaerobic digestion of complex substrates: Focusing on ammonia inhibition, *Biotechnology Bioengineering*, 42, 159–166 (1993).
49. Angelidaki, I., Ellegaard, L., and Ahring, B., Application of the anaerobic digestion process, in Ahring, B. (ed.), *Advances in Biochemical Engineering/Biotechnology*, Biomethanation II, Springer, New York, Ch. 1, pp. 2–33 (2003).
50. Angelidaki, I., Alves, M., Bolzonella, D., Borzacconi, L., Campos, L., Guwy, A., Kalyuzhnyi, S., Jenicek, P., and van Lier, J.B., Defining the biomethane potential (BMP) of solid organic wastes and energy crops: A proposed protocol for batch assays, *Water Science & Technology*, 59 (5), 927–934 (2009). ISSN: 0276-5055.
51. Shin, H.S., Youn, J.H., and Kim, S.H., Hydrogen production from food waste in anaerobic mesophilic and thermophilic acidogenesis, *International Journal of Hydrogen Energy*, 29, 1355–1363 (2004).
52. Haug, R.T., Hernandez, G., Sarullo, T., and Gerringer, F., Using wastewater digesters to recycle food residuals into energy, *Biocycle*, 41 (9), 74–77 (September 2000).
53. Desai, H., Nagori, G., and Vahora, S., Energy generation potential of dairy effluent—A case study of Vidya dairy, in *Proceedings of the International Conference on Energy and Environment*, Chandigarh, India, pp. 457–459 (March 19–21, 2009).
54. Narra, M., Nagori, G.P., and Pushalkar, S., Biomethanation of household organic waste at high solid content a package for waste disposal and energy generation, in *Proceedings of the International Conference on Energy and Environment*, Chandigarh, India, pp. 477–480 (March 19–21, 2009).
55. Nandy, T., Kaul, S.N., Pathe, P.P., Deshpande, C.V., and Daryapurkar, R.A., Pilot plant studies on fixed bed reactor system for biomethanation of distillery spent wash, *International Journal of Environmental Studies*, 41, 87–107 (1992).
56. Alves, M., Pereira, M., Sousa, D., Cavaleiro, A.J., Picavet, M., Smidt, H., and Stams, A., Waste lipids to energy: How to optimize methane production from long-chain fatty acids (LCFA), Minireview, *Microbial Biotechnology*, 2 (5), 538–550 (2009).
57. Yusuf, M., Debora, A., and Ogheneruona, D., Ambient temperature kinetic assessment of biogas production from co-digestion of horse and cow-dung, *Research in Agricultural Engineering*, 57 (3), 97–104 (2011).
58. Kujawski, O. and Steinmetz, H., Development of instrumentation systems as a base for control of digestion process stability in full-scale agricultural and industrial biogas plants, *Water Science & Technology*, 60 (8), 2055–2063 (2009).
59. Martin-Gonzalez, L., Castro, R., Pereira, M., Alves, M., Font, X., and Vicent, T., Thermophilic co-digestion of organic fraction of MSW with FOG wastes from a sewage treatment plant: Reactor performance and microbial community monitoring, *Bioresource Technology*, 102 (7), 4734–4741 (April 2011).
60. Singh, R. and Mandal, S., The utilization of non-edible cake along with cow dung for methane enriched biogas production using mixed inoculum, *Energy Resources, Part A: Recovery, Utilization and Environmental Effects*, 33 (5), 449–458 (2011).
61. Zhu, H., Parker, W., Basnar, R., Proracki, A., Falletta, P., Beland, M., and Seto, P., Buffer requirements for enhanced hydrogen production in acidogenic digestion of food wastes, *Bioresource Technology*, 100 (21), 5097–5102 (November 2009).
62. Hutnan, M., Spalkova, V., Bodik, I., Kolesarova, N., and Lazor, M., Biogas production from maize grains and maize silage, *Polish Journal of Environmental Studies*, 19 (2), 323–329 (2010).
63. Azaizeh, H. and Jadoun, J., Co-digestion of olive mill wastewater and swine manure using up-flow anaerobic sludge blanket reactor for biogas production, *Journal of Water Resource and Protection*, 2, 314–321 (2010).
64. Yusuf, M. and Ify, N., Effect of waste paper on biogas production from co-digestion of cow dung and water hyacinth in batch reactors, *Journal of Applied Sciences and Environmental Management*, 12 (4), 95–98 (2008).
65. Bagge, E., Peterson, M., and Johansson, K.E., Diversity of spore forming bacteria in cattle manure, slaughterhouse waste and samples from biogas plants, *Journal of Applied Microbiology*, 109 (5), 1549–1565 (November 2010).
66. Wu, X., Yao, W., Zhu, J., and Miller, C., Biogas and CH_4 productivity by co-digesting swine manure with three crop residues as an external carbon source, *Bioresource Technology*, 101 (11), 4042–4047 (June 2010).
67. Buhle, L., Stulpagel, R., and Wachendorf, M., Comparative life cycle assessment of the integrated generation of solid fuel and biogas from biomass (IFBB) and whole crop digestion (WCD) in Germany, *Biomass and Bioenergy*, 35 (1), 363–373 (January 2011).

68. Salminen, E.A. and Rintala, J.A., Semi-continuous anaerobic digestion of solid poultry slaughterhouse waste: Effect of hydraulic retention time and loading, *Water Research*, 36 (13), 3175–3182 (2002).
69. Macias-Corral, M., Samani, Z., Hanson, A., Smith, G., Funk, P., Yu, H., and Longworth, J., Anaerobic digestion of municipal solid waste and agricultural waste and the effect of co-digestion with dairy cow manure, *Bioresource Technology*, 99 (17), 8288–8293 (November 2008).
70. Holm-Nielsen, J., Seadi, T., and Oleskowicz-Popiel, P., The future of anaerobic digestion and biogas utilization, *Bioresource Technology*, 100 (922), 5478–5484 (November 2009).
71. Oslaj, M., Mursec, B., and Vindis, P., Biogas production from maize hybrids, *Biomass and Bioenergy*, 34 (11), 1538–1545 (November 2010).
72. Fang, C., Boe, K., and Angelidaki, I., Anaerobic co-digestion of by products from sugar production with cow manure, *Water Research*, 45 (11), 3473–3480 (May 2011).
73. Gunaseelan, V.N., Anaerobic digestion of biomass for methane production: A review, *Biomass and Bioenergy*, 13, 83–114 (1997).
74. Kaparaju, P. and Rintala, J., Anaerobic co-digestion of potato tuber and its industrial by-products with pig manure, *Resources, Conservation and Recycling*, 43, 175–188 (2005).
75. Lopez, J., Li, Q., and Thompson, I., Biorefinery of waste orange peel, Review article, *Critical Reviews in Biotechnology*, 30 (1), 63–69 (2010).
76. Iacovidou, E., Ohandja, D., and Voulvoulis, N., Food waste co-digestion with sewage sludge-realizing its potential in UK, *Journal of Environmental Management*, 112, 267–274 (2012).
77. Yen, H.-W. and Brune, D.E., Anaerobic co-digestion of algal sludge and waste paper to produce methane, *Bioresource Technology*, 98, 130–134 (2007).
78. Long, J., Aziz, T., de Los Reyes, F., III, and Ducoste, J. Anaerobic co-digestion of fat, oil and grease (FOG): A review of gas production and process limitations, *Process Safety and Environmental Protection*, 50 (3), 231–245 (May 2012).
79. Cheerawit, R., Thunwadee, T., Duangporn, K., Tanawat, R., and Wichuda, K., Biogas production from co-digestion of wastewater and food waste, *Health and the Environmental Journal*, 3 (2), 1–9 (2012).
80. Acharya, C. and Kurian, R., Anaerobic co-digestion of a single source oily waste and high strength pet food wastewater: A study of failure and revival of a full scale digester, WEFTEC, Water Environment Foundation, pp. 5066–5073 (2006).
81. Foucault, L., Anaerobic co-digestion of chicken processing wastewater and crude glycerol from biodiesel, MS thesis, Texas A&M University, College Station, TX (August 2011).
82. Elferink, S., van Lis, R., Heilig, H., Akkermans, A., and Stams, A., Detection and quantification of microorganisms in anaerobic bioreactors, *Biodegradation*, 9, 169–177 (1998).
83. Yu, Y., Lee, C., Kim, J., and Hwangs, S., Group specific primer and probe sets to detect methanogenic communities using quantitative real time polymerase chain reaction, *Biotechnology and Bioengineering*, 89, 670–679 (2005).
84. Karakashev, D., Bastone, D., and Angelidaki, I., Influence of environmental conditions on methanogenic compositions in anaerobic biogas reactors, *Applied and Environmental Microbiology*, 71, 331–338 (2005).
85. Ange Abdoun, E. and Weiland, P., *Bermimer Agrartecnisclre Berichte*, 68, 69–78 (2009).
86. Klocke, M., Netterman, E., and Bergmann, I., Monitoring der methanbildenden Mikroflora in Praxis-Biogasanlagen im landlichen Raum: Analyse des Ist-zustandes und Entwicklung eines quantitativen Nachweissystems, Bornimer Agrartechnische Berichte No. 67 (2009).
87. Klocke, M., Nettman, E., Bergmann, I., Mundt, K., Souidiu, K., Mumme, I., and Linke, B., Characterization of the methanogenic Archaea within two-phase biogas reactor systems operated with plant biomass, *Systematic and Applied Microbiology*, 31, 190–205 (2008).
88. Chen, Y., Cheng, J., and Creamer, K., Inhibition of anaerobic digestion process: A review, *Bioresource Technology*, 99 (10), 4044–4064 (2008).
89. Nielsen, H. and Angelidaki, I., Strategies for optimizing recovery of biogas process following ammonia inhibition, *Bioresource Technology*, 99, 7995–8001 (2008).
90. Bischoff, M., Erkenntnisse beim Einsatz von Zusatz-und Hilfsstoffen sowie von Spurene lementen in Biogasanalangen, VDI-Ber 2057, 111–123 (2009).
91. Friedmann, H. and Kobe, J., Optimierung der Biogasproduktionaus nachwachsenden Rohstoffen durch den Einsatz von Mikronahrstoffen-ein Erfahrungsbericht, in *Tagungsband 17, Jahrestagung des Fachverbandes Biogas*, Nurnberg, Germany, pp. 125–130 (2008).
92. Gavala, H., Angelidaki, I., and Ahring, B., Kinetics and modeling of anaerobic digestion process, in Scheper, T. and Ahring, B.K. (eds.), *Biomethanation I*. Springer, Berlin, Germany (2003).
93. Mahnert, P., Kinetic der Biogas produktion aus nachwachsenden Rohstoffen und Gulle, Dissertation Humboldt-Universitat, Berlin, Germany, p. 202 (2007).

94. Andara, A.R. and Esteban, J.M.B., Kinetic study of the anaerobic digestion of the solid fraction of piggery slurries, *Biomass and Bioenergy*, 17, 435–443 (1999).

95. Linke, B., Kinetic study of thermophilic anaerobic digestion of solid wastes from potato processing, *Biomass and Bioenergy*, 30, 892–896 (2006).

96. Biswas, L., Chowdhury, R., and Battacharya, P., Mathematical modeling for the prediction of biogas generation characteristics of an anaerobic digester based on food/vegetable residues, *Biomass and Bioenergy*, 31, 80–86 (2007).

97. Lattin, W.C. and Utgikar, V.P., Transition to hydrogen economy in the United States: A 2006 status report, *International Journal of Hydrogen Energy*, 32, 3230–3237 (2007).

98. Antonopoulou, G., Gavala, H.N., Skiadas, I.V., Angelopoulos, K., and Lyberatos, G., Biofuels generation from sweet sorghum: Fermentative hydrogen production and anaerobic digestion of the remaining biomass, *Bioresource Technology*, 99 (1), 110–119 (February 2008).

99. Jo, J.H., Jeon, C.O., Lee, D.S., and Park, J.M., Process stability and microbial community structure in anaerobic hydrogen-producing microflora from food waste containing kimchi, *Journal of Biotechnology*, 131 (3), 300–308 (September 15, 2007). doi:10.1016/j.jbiotec.2007.07.492.

100. Han, S. and Shin, H., Biohydrogen production by anaerobic fermentation of food waste, *International Journal of Hydrogen Energy*, 29, 569–577 (2003).

101. Gujer, W. and Zehnder, A.J.B., Conversion processes in anaerobic digestion, *Water Science & Technology*, 15, 127–167 (1983).

102. Ting, C.H., Lin, K.R., Lee, D.J., and Tay, J.H., Production of hydrogen and methane from wastewater sludge using anaerobic fermentation, *Water Science & Technology*, 50, 223–228 (2004).

103. Wang, A., Liu, W., Cheng, S., Xing, D., Zhou, J., and Logan, B.E., Source of methane and methods to control its formation in single chamber microbial electrolysis cells, *International Journal of Hydrogen Energy*, 34, 3653–3658 (2009).

104. Rader, G. and Logan, B., Multi-electrode continuous flow microbial electrolysis cell for biogas production from acetate, *International Journal of Hydrogen Energy*, 35, 8848–8854 (2010).

105. Liu, H., Grot, S., and Logan, B.E., Electrochemically assisted microbial production of hydrogen from acetate, *Environmental Science & Technology*, 39, 4317–4320 (2005).

106. Selembo, P.A., Merrill, M.D., and Logan, B.E., The use of stainless steel and nickel alloys as low-cost cathodes in microbial electrolysis cells, *Journal of Power Sources*, 190, 271–278 (2009).

107. Call, D.F., Merrill, M.D., and Logan, B.E., High surface area stainless steel brushes as cathodes in microbial electrolysis cells (MECs), *Environmental Science & Technology*, 43, 2179–2183 (2009).

108. Clauwaert, P. and Verstraete, W., Methanogenesis in membraneless microbial electrolysis cells, *Applied Microbiology and Biotechnology*, 82, 829–836 (2009).

109. Parameswaran, P., Zhang, H., Torres, C.I., Rittmann, B.E., and Krajmalnik-Brown, R., Microbial community structure in a biofilm anode fed with a fermentable substrate: The significance of hydrogen scavengers, *Biotechnology and Bioengineering*, 105 (1), 69–78 (2010).

110. Lee, H.-S. and Rittman, B.E., Significance of biological hydrogen oxidation in a continuous single-chamber microbial electrolysis cell, *Environmental Science & Technology*, 44, 948–954 (2010).

111. Ye, Y., Wang, L., Chen, Y., Zhu, S., and Shen, S., High yield hydrogen production in a single-chamber membrane-less microbial electrolysis cell, *Water Science & Technology*, 61 (3), 721–727 (2010).

112. Call, D.F. and Logan, B.E., Hydrogen production in a single chamber microbial electrolysis cell lacking a membrane, *Environmental Science & Technology*, 42, 3401–3406 (2008).

113. Chae, K.-J., Choi, M.-J., Kim, K.-Y., Ajayi, F.F., Chang, I.-S., and Kim, I.S., Selective inhibition of methanogens for the improvement of biohydrogen production in microbial electrolysis cells, *International Journal of Hydrogen Energy*, 35 (24), 13379–13386 (December 2010).

114. Lalaurette, E., Thammannagowda, S., Mohagheghi, A., Maness, P.-C., and Logan, B.E., Hydrogen production from cellulose in a two-stage process combining fermentation and electrohydrogenesis, *International Journal of Hydrogen Energy*, 34 (15), 6201–6210 (2009).

115. Selembo, P.A., Perez, J.M., Lloyd, W.A., and Logan, B.E., High hydrogen production from glycerol or glucose by electrohydrogenesis using microbial electrolysis cells, *International Journal of Hydrogen Energy*, 34, 5373–5381 (2009).

116. Hu, H., Fan, Y., and Liu, H., Hydrogen production using single-chamber membrane-free microbial electrolysis cells, *Water Research*, 42, 4172–4178 (2008).

117. Zhang, Y., The use and optimization of stainless steel mesh cathodes in microbial electrolysis cells, an internal report of Environmental Engineering, University Park, Pennsylvania State University, State College, PA (2010).

118. Liu, H., Cheng, S., Huang, L., and Logan, B.E., Scale-up of membrane-free single-chamber microbial fuel cells, *Journal of Power Sources*, 179, 274–279 (2008).

119. Shimoyama, T., Komukai, S., Yamazawa, A., Ueno, Y., Logan, B.E., and Watanabe, K., Electricity generation from model organic wastewater in a cassette-electrode microbial fuel cell, *Applied Microbiology and Biotechnology*, 80, 325–330 (2008).

120. Tartakovsky, B., Manuel, M.-F., Wang, H., and Guiot, S.R., High rate membrane-less microbial electrolysis cell for continuous hydrogen production, *International Journal of Hydrogen Energy*, 34 (2), 672–677 (2009).

121. Cheng, S. and Logan, B.E., Ammonia treatment of carbon cloth anodes to enhance power generation of microbial fuel cells, *Electrochemistry Communications*, 9 (3), 492–496 (2007).

122. Logan, B., Cheng, S., Watson, V., and Estadt, G., Graphite fiber brush anodes for increased power production in air-cathode microbial fuel cells, *Environmental Science & Technology*, 41, 3341–3346 (2007).

123. Logan, B.E. et al., Microbial electrolysis cells for high yield hydrogen gas production from organic matter, *Environmental Science & Technology*, 42 (23), 8630–8640 (2008).

124. Fang, H.H. and Liu, H., Effect of pH on hydrogen production from glucose by a mixed culture, *Bioresource Technology*, 82 (1), 87–93 (2002).

125. Duangmanee, T., Chyi, Y., and Sung, S., Biohydrogen production in mixed culture anaerobic fermentation, in *Proceedings of the 14th World Hydrogen Energy Conference*, June 9–13, Montreal, Quebec, Canada (2002).

126. Sneath, P.H.A., Mair, N.S., Sharpe, M.E., and Holt, J.G., *Bergey's Manual of Systematic Bacteriology*, 1st ed., Vol. 2, Williams and Wilkins, Baltimore, MD (1986).

127. Mead, G.C., Principles involved in the detection and enumeration of clostridia in foods, *International Journal of Food Microbiology*, 17, 135–143 (1992).

128. Ljungdahl, L.G., Hugenholtz, J., and Wiegel, J., Proteolytic clostridium hydrolysis protesis and ferments ammino acid—catabolic pathway of Clostridium in hydrolysis, Chapter 5, in Minton, N.P. and Clarke, D.J. (eds.), *Biotechnology Handbooks*, Vol. III: *Clostridia*. Plenum Press, New York (1989).

129. Miyake, J., Improvement of bacterial light dependent hydrogen production by altering the photosynthetic pigment ratio, Chapter 2, in Zaborsky, O.R. (ed.), *BioHydrogen*. Plenum Press, New York, pp. 81–86 (1998), ISBN:ISBN: 978-0-306-46057-9.

130. Miyake, J., Miyake, M., and Asada, Y., Biotechnological hydrogen production: Research for efficient light conversion, *Journal of Biotechnology*, 70, 89–101 (1999).

131. Schroder, C., Selig, M., and Schonheit, P., Glucose fermentation to acetate, CO_2 and H_2 in the anaerobic hyperthermophilic eubacterium *Thermotoga maritima*: Involvement of the Embden-Meyerhof pathway, *Archives of Microbiology*, 161, 460–470 (1994).

132. Hallenbeck, P.C. and Benemann, J.R., Biological hydrogen production; fundamentals and limiting processes, *International Journal of Hydrogen Energy*, 27, 1185–1193 (2002).

133. Lo, Y.C., Chen, W.M., Hung, C.H., Chen, S.D., and Chang, J.S., Dark H_2 fermentation from sucrose and xylose using H_2-producing indigenous bacteria: Feasibility and kinetic studies, *Water Research*, 42, 827–842 (2008).

134. Zheng, X.J. and Yu, H.Q., Inhibition effects of butyrate on biological hydrogen production with mixed anaerobic cultures, *Journal of Environmental Management*, 74, 65–70 (2005).

135. Wood, W.A., *Fermentation of Carbohydrates and Related Compounds. The Bacteria. II.* Academic Press, New York (1961).

136. Afschar, A.S., Schaller, K., and Schugerl, K., Continuous production of acetone and butanol with shear-activated *Clostridium acetobutylicum*, *Applied Microbiology and Biotechnology*, 23, 315–322 (1986).

137. Brosseau, J.D., Yan, J.Y., and Lo, K.V., The relationship between hydrogen gas and butanol production by *Clostridium saccharoperbutylacetonicum*, *Biotechnology and Bioengineering*, 28, 305–310 (1986).

138. Gottwald, M. and Gottschalk, G., The internal pH of *Clostridium acetobutylicum* and its effect on the shift from acid to solvent formation, *Archives of Microbiology*, 143, 42–46 (1985).

139. Debrock, B., Bahl, H., and Gottschalk, G., Parameters affecting solvent production by *Clostridium Pasteurianum*, *Applied Environmental Microbiology*, 58 (4), 1233–1239 (1992).

140. Liu, G.Z. and Shen, J.Q., Effects of culture and medium conditions on hydrogen production from starch using anaerobic bacteria, *Journal of Bioscience and Bioengineering*, 98, 251–256 (2004).

141. Wang, J.L. and Wan, W., Comparison of different pretreatment methods for enriching hydrogen-producing cultures from digested sludge, *International Journal of Hydrogen Energy*, 33, 2934–2941 (2008).

142. Lay, J.J., Fan, K.S., Chang, J.I., and Ku, C.H., Influence of chemical nature of organic wastes on their conversion to hydrogen by heat-shock digested sludge, *International Journal of Hydrogen Energy*, 28, 1361–1367 (2003).

143. Van Ginkel, S., Sung, S., and Lay, J., Biohydrogen production as a function of pH and substrate concentration, *Environmental Science & Technology*, 35, 4726–4730 (2001).
144. Khanal, S., Chen, W.H., Li, L., and Sung, S., Biological hydrogen production: Effects of pH and intermediate products, *International Journal of Hydrogen Energy*, 29, 1123–1131 (2004).
145. Wang, J. and Wan, W., Factors influencing fermentative hydrogen production: A review, *International Journal of Hydrogen Energy*, 34, 799–811 (2009).
146. Fang, H.H.P. and Liu, H., Effect of pH on hydrogen production from glucose by a mixed culture, *Bioresource Technology*, 82, 87–93 (2002).
147. Li, C.L. and Fang, H.H.P., Fermentative hydrogen production from wastewater and solid wastes by mixed cultures, *Critical Reviews in Environmental Science and Technology*, 37, 1–39 (2007).
148. Zhang, M.L., Fan, Y.T., Xing, Y., Pan, C.M., Zhang, G.S., and Lay, J.J., Enhanced biohydrogen production from cornstalk wastes with acidification pretreatment by mixed anaerobic cultures, *Biomass and Bioenergy*, 31, 250–254 (2007).
149. Zhang, Z.P., Show, K.Y., Tay, J.H., Liang, D.T., and Lee, D.J., Biohydrogen production with anaerobic fluidized bed reactors—A comparison of biofilm-based and granule-based systems, *International Journal of Hydrogen Energy*, 33, 1559–1564 (2008).
150. Wang, J.L. and Wan, W., Effect of Fe^{2+} concentrations on fermentative hydrogen production by mixed cultures, *International Journal of Hydrogen Energy*, 33, 1215–1220 (2008).
151. Yokoi, H., Tokushige, T., Hirose, J., Hayashi, S., and Takasaki, Y., H2 production from starch by mixed culture of *Clostridium butyricum* and *Enterobacter aerogenes*, *Biotechnology Letters*, 20, 143–147 (1998).
152. Massanet-Nicolau, J., Dinsdale, R., and Guwy, A., Hydrogen production from sewage sludge using mixed microflora inoculum: Effect of pH and enzymatic pretreatment, *Bioresource Technology*, 99, 6325–6331 (2008).
153. Yu, H.Q., Zhu, Z.H., Hu, W.R., and Zhang, H.S., Hydrogen production from rice winery wastewater in an upflow anaerobic reactor by using mixed anaerobic cultures, *International Journal of Hydrogen Energy*, 27, 1359–1365 (2002).
154. Mu, Y., Wang, G., and Yu, H.Q., Response surface methodological analysis on biohydrogen production by enriched anaerobic cultures, *Enzyme and Microbial Technology*, 38, 905–913 (2006).
155. Van Ginkel, S., Oh, S.E., and Logan, B.E., Biohydrogen gas production from food processing and domestic wastewaters, *International Journal of Hydrogen Energy*, 30, 1535–1542 (2005).
156. Zhang, H., Bruns, M.A., and Logan, B.E., Biological hydrogen production by *Clostridium acetobutylicum* in an unsaturated flow reactor, *Water Research*, 40, 728–734 (2006).
157. Ueno, Y., Haruta, S., Ishii, M., and Igarashi, Y., Characterization of a microorganism isolated from the effluent of hydrogen fermentation by microflora, *Journal of Bioscience and Bioengineering*, 92, 397–400 (2001).
158. Wang, X.J., Ren, N.Q., Xiang, W.S., and Guo, W.Q., Influence of gaseous end-products inhibition and nutrient limitations on the growth and hydrogen production by hydrogen-producing fermentative bacterial B49, *International Journal of Hydrogen Energy*, 32, 748–754 (2007).
159. Wu, J.H. and Lin, C.Y., Biohydrogen production by mesophilic fermentation of food wastewater, *Water Science & Technology*, 49, 223–228 (2004).
160. Lay, J.J., Fan, K.S., Hwang, J.I., Chang, J.I., and Hsu, P.C., Factors affecting hydrogen production from food wastes by *Clostridium*-rich composts, *Journal of Environmental Engineering*, 131, 595–602 (2005).
161. Guo, L. et al., Impacts of sterilization, microwave and ultrasonication pretreatment on hydrogen producing using waste sludge, *Bioresource Technology*, 99, 3651–3658 (2008).
162. Cheong, D.Y. and Hansen, C.L., Bacterial stress enrichment enhances anaerobic hydrogen production in cattle manure sludge, *Applied Microbiology and Biotechnology*, 72, 635–643 (2006).
163. Zhu, H.G. and Beland, M., Evaluation of alternative methods of preparing hydrogen producing seeds from digested wastewater sludge, *International Journal of Hydrogen Energy*, 31, 1980–1988 (2006).
164. Fan, K.S. and Chen, Y.Y., H2 production through anaerobic mixed culture: Effect of batch S0/X0 and shock loading in CSTR, *Chemosphere*, 57, 1059–1068 (2004).
165. Chen, C.C., Lin, C.Y., and Lin, M.C., Acid-base enrichment enhances anaerobic hydrogen production process, *Applied Microbiology and Biotechnology*, 58, 224–228 (2002).
166. Montgomery, L.F.R. and Bochmann, G., Pretreatment of feedstock for enhanced biogas production, IEA Bioenergy report, Petten, the Netherlands (2014).
167. Wu, S.Y. et al., HRT-dependent hydrogen production and bacterial community structure of mixed anaerobic microflora in suspended, granular and immobilized sludge systems using glucose as the carbon substrate, *International Journal of Hydrogen Energy*, 33, 1542–1549 (2008).

168. Wu, S.Y. et al., Dark fermentative hydrogen production from xylose in different bioreactors using sewage sludge microflora, *Energy Fuels*, 22, 113–119 (2008b).
169. Mohan, S.V., Babu, V.L., and Sarma, P.N., Effect of various pretreatment methods on anaerobic mixed microflora to enhance biohydrogen production utilizing dairy wastewater as substrate, *Bioresource Technology*, 99, 59–67 (2008).
170. Kapdan, I.K. and Kargi, F., Bio-hydrogen production from waste materials, *Enzyme and Microbial Technology*, 38, 569–582 (2006).
171. Benemann, J., Hydrogen biotechnology: Progress and prospects, *Nature Biotechnology*, 14, 1101–1103 (1996).
172. Kargi, F. and Kapdan, I.K., Biohydrogen production from waste materials, in *International Hydrogen Energy Congress and Exhibition*, Istanbul, Turkey (2005).
173. Hwang, M.H., Jang, N.J., Hyun, S.H., and Kim, I.S., Anaerobic bio-hydrogen production from ethanol fermentation: The role of pH, *Journal of Biotechnology*, 111, 297–309 (2004).
174. Fang, H.H.P., Li, C.L., and Zhang, T., Acidophilic biohydrogen production from rice slurry, *International Journal of Hydrogen Energy*, 31, 683–692 (2006).
175. Okamoto, M., Noike, T., Miyahara, T., and Mizuno, O., Biological hydrogen potential of materials characteristic of the organic fraction of municipal solid wastes, *Water Science & Technology*, 41 (3), 25–32 (2000).
176. Mizuno, O., Ohara, T., Shinya, M., and Noike, T., Characteristics of hydrogen production from bean curd manufacturing waste by anaerobic microflora, *Water Science & Technology*, 42, 345–350 (2000).
177. Noike, T., Mizuno, O., and Miyahara, T., Method of producing hydrogen gas by using hydrogen bacteria, Patent # 6,860,996 B2, EP 1457566A1, EP1457566AA, 4S2004005 0078, W02003052112A1, Japan Science and Technology Corporation, Tokyo, Japan (2003).
178. Hussy, I., Hawkes, F.R., Dinsdale, R., and Hawkes, D.L., Continuous fermentative hydrogen production from sucrose and sugar beet, *International Journal of Hydrogen Energy*, 30, 471–483 (2005).
179. Yokoi, H., Tokushige, T., Hirose, J., Hayashi, S., and Takasaki, Y., Hydrogen production by immobilized cells of aciduric *Enterobacter aerogenes* strain HO-39, *Journal of Fermentation and Bioengineering*, 83, 481–484 (1997).
180. Cheng, S. and Logan, B.E., Evaluation of catalysts and membranes for high yield biohydrogen production via electrohydrogenesis in microbial electrolysis cells (MECS), *Water Science & Technology*, 58 (4), 853–857 (2008).
181. Fan, Y., Zhang, Y., Zhang, S., Hou, H., and Ren, B., Efficient conversion of wheat straw wastes into biohydrogen gas by cow dung compost, *Bioresource Technology*, 97 (3), 500–505 (February 2006).
182. Wang, C.C., Chang, C.W., Chu, C.P., Lee, D.J., Chang, B.V., and Liao, C.S., Producing hydrogen from wastewater sludge by *Clostridium bifermentans*, *Journal of Biotechnology*, 102, 83–92 (2003).
183. Zhang, Z.P., Show, K.Y., Tay, J.H., Liang, D.T., Lee, D.J., and Jiang, W.J., Rapid formation of hydrogen-producing granules in an anaerobic continuous stirred tank reactor induced by acid incubation, *Biotechnology and Bioengineering*, 96, 1040–1050 (2007).
184. Ferchichi, M., Crabbe, E., Gil, G.H., Hintz, W., and Almadidy, A., Influence of initial pH on hydrogen production from cheese whey, *Journal of Biotechnology*, 120, 402–409 (2005).
185. Fachverband Biogas, biogas dozentral erzeugen, regional profitieren, international gewinnen, in *Proc. 18. Jahrestagung des Fachverbandes Biogas*, Hannover, Germany (2009).
186. Lee, Y.J., Miyahara, T., and Noike, T., Effect of pH on microbial hydrogen fermentation, *Journal of Chemical Technology and Biotechnology*, 77, 694–698 (2002).
187. Lee, K.S., Hsu, Y.F., Lo, Y.C., Lin, P.J., Lin, C.Y., and Chang, J.S., Exploring optimal environmental factors for fermentative hydrogen production from starch using mixed anaerobic microflora, *International Journal of Hydrogen Energy*, 33, 1565–1572 (2008).
188. Xu, K., Liu, H., and Chen, J., Effect of classic methanogenic inhibitors on the quantity and diversity of archaeal community and the reductive homoacetogenic activity during the process of anaerobic sludge digestion, *Bioresource Technology*, 101 (8), 2600–2607 (2010).
189. Lay, J.J., Biohydrogen generation by mesophilic anaerobic fermentation of microcrystalline cellulose, *Biotechnology and Bioengineering*, 74 (4), 280–287 (2001).
190. Lay, J.J., Modeling and optimization of anaerobic digested sludge converting starch to hydrogen, *Biotechnology and Bioengineering*, 68, 269–278 (2000).
191. Lay, J.J., Lee, Y.J., and Noike, T., Feasibility of biological hydrogen production from organic fraction of municipal solid waste, *Water Research*, 33, 2579–2586 (1999).
192. Mu, Y., Yu, H.Q., and Wang, Y., The role of pH in the fermentative H2 production from an acidogenic granule-based reactor, *Chemosphere*, 64, 350–358 (2006).

193. Mu, Y., Zheng, X.J., Yu, H.Q., and Zhu, R.F., Biological hydrogen production by anaerobic sludge at various temperatures, *International Journal of Hydrogen Energy*, 31, 780–785 (2006).

194. Zhao, Q.B. and Yu, H.Q., Fermentative H2 production in an upflow anaerobic sludge blanket reactor at various pH values, *Bioresource Technology*, 99, 1353–1358 (2008).

195. Chen, W.M., Tseng, Z.J., Lee, K.S., and Chang, J.S., Fermentative hydrogen production with *Clostridium butyricum* CGS5 isolated from anaerobic sewage sludge, *International Journal of Hydrogen Energy*, 30 (10), 1063–1070 (2005).

196. Chen, C.C., Lin, C.Y., and Chang, J.S., Kinetics of hydrogen production with continuous anaerobic cultures utilizing sucrose as limiting substrate, *Applied Microbiology and Biotechnology*, 57, 56–64 (2001).

197. Chen, W., Biological hydrogen production by anaerobic fermentation, PhD thesis, Civil Engineering, Iowa State University, Ames, IA (2006).

198. Chang, F.Y. and Lin, C.Y., Biohydrogen production using an up-flow anaerobic sludge blanket reactor, *International Journal of Hydrogen Energy*, 29, 33–39 (2004).

199. Chang, F.Y. and Lin, C.Y., Calcium effect on fermentative hydrogen production in an anaerobic up-flow sludge blanket system, *Water Science & Technology*, 54, 105–112 (2006).

200. Lee, K.S., Lo, Y.S., Lo, Y.C., Lin, P.J., and Chang, J.S., H2 production with anaerobic sludge using activated-carbon supported packed bed bioreactors, *Biotechnology Letters*, 25, 133–138 (2003).

201. Lee, K.S., Lo, Y.S., Lo, Y.C., Lin, P.J., and Chang, J.S., Operation strategies for biohydrogen production with a high-rate anaerobic granular sludge bed bioreactor, *Enzyme and Microbial Technology*, 35, 605–612 (2004).

202. Liu, H. and Fang, H.H.P., Hydrogen production from wastewater by acidogenic granular sludge, in *Proceedings of IWA Asia Environmental Technology 2001*, October 30–November 6, Singapore (2001).

203. Lin, C. and Jo, C., Hydrogen production from sucrose using anaerobic sequencing batch reactor process, *Journal of Chemical Technology and Biotechnology*, 78 (6), 678–684 (June 2003).

204. Wang, G., Mu, Y., and Yu, H.Q., Response surface analysis to evaluate the influence of pH, temperature and substrate concentration on the acidogenesis of sucrose-rich wastewater, *Biochemical Engineering Journal*, 23, 175–184 (2005).

205. Shah, Y.T., *Energy and Fuel Systems Integration*. CRC Press, New York (2015).

206. Wang, J.L. and Wan, W., Effect of temperature on fermentative hydrogen production by mixed cultures, *International Journal of Hydrogen Energy*, 33, 5392–5397 (2008).

207. Leven, L., Eriksson, A., and Schnurer, A., Effect of process temperature on bacterial and archael communities in two methanogenic bioreactors treating organic household waste, *FEMS Microbiology Ecology*, 59, 683–693 (2007).

208. Heyndrickx, M., De Vos, P., Thibau, B., Stevens, P., and De Ley, J., Effect of various external factors on the fermentative production of hydrogen gas from glucose by *Clostridium butyricum* strains in batch culture, *Systematic and Applied Microbiology*, 9, 163–168 (1987).

209. van den Heuvel, J.C., Beeftink, H.H., and Verschuren, P.G., Inhibition of the acidogenic dissimilation of glucose in anaerobic continuous cultures by free butyric acid, *Applied Microbiology and Biotechnology*, 29, 89–94 (1988).

210. Lamed, R.J., Lobos, J.H., and Su, T.M., Effect of stirring and hydrogen on fermentation products of *Clostridium thermocellum*, *Applied and Environmental Microbiology*, 54, 1216 (1988).

211. Mizuno, O., Dinsdale, R., Hawkes, F.R., Hawkes, D.L., and Noike, T., Enhancement of hydrogen production from glucose by nitrogen gas sparging, *Bioresource Technology*, 73 (1), 59–65 (2000).

212. Salih, F.M. and Maleek, M.I., Influence of metal ions on hydrogen production by photosynthetic bacteria grown in *Escherichia coli* pre-fermented cheese whey, *Journal of Environmental Protection*, 1, 426–430 (2010), published online (December 2010).

213. Zhang, Y.F., Liu, G.Z., and Shen, J.Q., Hydrogen production in batch culture of mixed bacteria with sucrose under different iron concentrations, *International Journal of Hydrogen Energy*, 30, 855–860 (2005).

214. Li, C.L. and Fang, H.H.P., Inhibition of heavy metals on fermentative hydrogen production by granular sludge, *Chemosphere*, 67, 668–673 (2007).

215. Lin, C.Y. and Shei, S.H., Heavy metal effects on fermentative hydrogen production using natural mixed microflora, *International Journal of Hydrogen Energy*, 33, 587–593 (2008).

216. Mudhoo, A. and Kumar, S., Effects of heavy metals as stress factors on anaerobic digestion processes and biogas production from biomass, *International Journal of Environmental Science and Technology*, 10, 1383–1398 (2013).

217. Bisaillon, A., Turcot, J., and Hallenbeck, P.C., The effect of nutrient limitation on hydrogen production by batch cultures of *Escherichia coli*, *International Journal of Hydrogen Energy*, 31, 1504–1508 (2006).

218. Salerno, M.B., Park, W., Zuo, Y., and Logan, B.E., Inhibition of biohydrogen production by ammonia, *Water Research*, 40, 1167–1172 (2006).
219. Argun, H., Kargi, F., Kapdan, I.K., and Oztekin, R., Biohydrogen production by dark fermentation of wheat powder solution: Effects of C/N and C/P ratio on hydrogen yield and formation rate, *International Journal of Hydrogen Energy*, 33, 1813–1819 (2008).
220. O-Thong, S., Prasertsan, P., Intrasungkha, N., Dhamwichukorn, S., and Birkeland, N.K., Optimization of simultaneous thermophilic fermentative hydrogen production and COD reduction from palm oil mill effluent by *Thermoanaerobacterium*-rich sludge, *International Journal of Hydrogen Energy*, 33, 1221–1231 (2008).
221. Hu, B. and Chen, S.L., Pretreatment of methanogenic granules for immobilized hydrogen fermentation, *International Journal of Hydrogen Energy*, 32, 3266–3273 (2007).
222. Fuentes, M., Scenna, N., and Aguirre, P., Modelling of hydrogen production in biofilm reactors: Application of the anaerobic digestion model 1, Cuerto Congreso Nacional-Tercer Congresso Iberoamericano Hidrogeno y Fuentes Sustainables de Energia-HYFUSEN (2011).
223. Zheng, H., O'Sullivan, C., Mereddy, R., Zheng, R., Duke, M., and Clarke, W., Production of bio-hydrogen using a membrane anaerobic reactor: Limitations due to diffusion, in *Proceedings of the Environmental Research Event*, NOOSA, Queensland, Australia (2009).
224. Bauer, C.G. and Forest, T.W., Effect of hydrogen addition on performance of methane fueled vehicles. Part I: Effect on S.I. engine performance, *International Journal of Hydrogen Energy*, 26, 55–70 (2001).
225. Sarli, V. and Benedetto, A., Laminar burning velocity of hydrogen–methane/air premixed flames, *International Journal of Hydrogen Energy*, 32, 637–646 (2006).
226. Nasr, N., Elbeshbishy, E., Hafez, H., Nakhla, G., and El Nagger, M., Comparative assessment of single stage and two stage anaerobic digestion for the treatment of thin stillage, *Bioresource Technology*, 111, 122–126 (2012).
227. Toscano, G., Ausiello, A., Micoli, L., Zuccaro, G., and Pirozzi, D., Anaerobic digestion of residual lignocellulosic materials to biogas and biohydrogen, *Chemical Engineering Transactions*, 32, 487–492 (2013).
228. Kvesitadze, G., Sadunishvili, T., Dudauri, T., Metreveli, B., and Jobava, M., Two-stage anaerobic process for bio-hydrogen and bio-methane combined production from biodegradable solid wastes, a personal communication from Durmishidze Institute of Biochemistry and Biotechnology, Tbilisi, Georgia (2015).
229. Thompson, R., Hydrogen production by anaerobic fermentation using agricultural and food processing wastes utilizing a two-stage digestion system, MS thesis, Biological Engineering, Utah State University, Logan, UT (2008).
230. Noebauer, S. and Schnitzhofer, W., Efficient hydrogen fermentation for 2-stage anaerobic digestion processes: Conversion of sucrose containing substrates, personal communication, with Innovative energy systems, Im Stadtgut, A2, 4407 Steyr-Gleink, A., an integrated project HYVOLUTION funded by Framework program 6 of EC (2015).
231. Sung, S., Raskin, L., Duangmanee, T., Padmasiri, S., and Simmons, J., Hydrogen production by anaerobic microbial communities exposed to repeated heat treatments, in *Proceedings of the 2002 U.S. DOE Hydrogen Program Review*, NREL/CP-610-32405, Golden, CO (2002).
232. Premier, G.C., Kim, J.R., Michie, I., Dinsdale, R.M., and Guwy, A.J., Automatic control of load increases power and efficiency in a microbial fuel cell, *Journal of Power Sources*, 196, 2013–2019 (2011).
233. Premier, G., Kim, J., Massanet-Nicolau, J., Kyazze, G., Esteves, S., Penumathsa, B., Rodriguez, J., Maddy, J., Dinsdale, R., and Guwy, A., Integration of biohydrogen biomethane and bioelectrochemical systems, *Renewable Energy*, 35, 1–5 (2012).
234. Gottardo, M., Cavinato, C., Bolzonella, D., and Pavan, P., Dark fermentation optimization by anaerobic digested sludge recirculation: Effects on hydrogen production, *Chemical Engineering Transactions*, 32, 997–1002 (2013).
235. Nasr, N., Elbeshbishy, E., Hafez, H., Nakhla, G., and El Naggar, M., Bio-hydrogen production from thin stillage using conventional and acclimatized anaerobic digester sludge, *International Journal of Hydrogen Energy*, 36, 12761–12769 (2011).
236. Babson, D., Bellman, K., Prakash, S., and Fennell, D., Anaerobic digestion of methane generation and ammonia reforming for hydrogen production: A thermodynamic energy balance of a model system to demonstrate net energy feasibility, *Biomass and Bioenergy*, 56, 493–505 (2013).
237. Hilton, M. and Archer, D., Anaerobic digestion of a sulfate rich molasses wastewater: Inhibition of hydrogen sulfide production, *Biotechnology and Bioengineering*, 31, 885–888 (1988).

238. Haghighat Afshar, S., Management of hydrogen sulfide in anaerobic digestion of enzyme pretreated marine macro-algae, MS thesis, Water and Environmental Engineering, Department of Chemical Engineering, Lund University, Lund, Sweden (June 2012).

239. Guo, X., Trably, E., Latrille, E., Carrere, H., and Steyer, J., Hydrogen production from agricultural waste by dark fermentation: A review, *International Journal of Hydrogen Energy*, 35, 1–14 (2010).

240. Liu, D., Angelidaki, I., Zeng, R.J., and Min, B., Bio-hydrogen production by dark fermentation from organic wastes and residues, DTU Environment, Kongens Lyngby, Denmark (2008).

241. Wu, W., Anaerobic co-digestion of biomass for methane production: Recent research achievements, personal communication (2013).

242. Braun, R., Weiland, P., and Wellinger, A., Biogas from energy crop digestion, IEA Bioenergy, Task 37—Energy from Biogas and Landfill Gas, IEA, Paris, France (2011).

243. Braun, R., Potential of co-digestion (2002), http://www.novaenergie.ch/iea-bioenergy-task37/Dokumente/final.PDF, Accessed on November 7, 2007.

244. Braun, R. and Wellinger, A., Potential for co-digestion, IEA Bioenergy report, Task 37—Energy from Biogas and Landfill Gas, International Energy Agency (IEA) (2002).

245. Murphy, J., Braun, R., Weiland, P., and Wellinger, A., Biogas from crop digestion, IEA Bioenergy, Task 37—Energy from Biogas, Denmark (September 2011), also International Energy Agency (IEA), Brussels, Belgium (2011).

246. Braun, R., Weiland, P., and Wellinger, A., Biogas from energy crop digestion, IEA Bioenergy, Task 37—Energy from Biogas and Landfill Gas, Denmark (2009), also International Energy Agency (IEA), Brussels, Belgium (2009).

247. Braun, R., Anaerobic digestion—A multi faceted process for energy, environmental management and rural development, in Ranalli, P. (ed.), *Improvement of Crop Plants for Industrial End Users*. Springer, New York, (2007).

248. Braun, R. and Wellinger, A., Potential of co-digestion, IEA Bioenergy, Task 37—Energy from Biogas and Landfill Gas, IEA, Brussels, Belgium (2004).

249. Nges, I.A., Escobar, F., Fu, X., and Björnsson, L., Benefits of supplementing an industrial waste anaerobic digester with energy crops for increased biogas production, *Waste Management*, 32 (1), 53 (2012).

250. Jepsen, S., Co-digestion of animal manure and organic household waste—The Danish experience, Ministry of Environment and Energy, Danish EPA, Copenhagen, Denmark (2011).

251. Esposito, G., Frunzo, L., Giordano, A., Liotta, F., Panico, A., and Pirozzi, F., Anaerobic co-digestion of organic wastes, in *Reviews in Environmental Science and Biotechnology*, 11 (4), 325–341 (December 2012), doi:10.1007/s11157-012-9277-8.

252. Khalid, A., Arshad, M., Anjun, M., Mahmood, T., and Dawson, L., The anaerobic digestion of solid organic waste, *Waste Management*, 31, 1737–1744 (2011).

253. Arsova, L., Anaerobic digestion of food waste—Current status, problems and an alternative product, MSc thesis, Department of Earth and Environmental Engineering, Columbia University, New York (2010).

254. Klemeš, J., Smith, R., and Kim, J.-K., *Handbook of Water and Energy Management in Food Processing*. Woodhead Publishing, Salt Lake City, UT (2008).

255. Luque, R., Campelo, J., and Clark, J., *Handbook of Biofuels Production—Processes and Technologies*. Woodhead Publishing, Salt Lake City, UT (2011).

256. Monnet, F., An introduction to anaerobic digestion of organic wastes, Remade Scotland, Scotland (2003).

257. Persson, M., Jonsson, O., and Wellinger, A., Biogas upgrading to vehicle fuel standards and grid injection, IEA Bioenergy, Test 37, IEA, Brussels, Belgium (2006).

258. Polprasert, C., *Organic Waste Recycling—Technology and Management*, 3rd edn. IWA Publishing, London, U.K. (2007).

259. Frey Berg, T. (ed.), Co-digestion charge: Is waste water's new best friend? A publication by Water and Waste Water International, Tulsa, OK (2013).

260. Increasing anaerobic digester performance with co-design. A report from AgSTAR, U.S. Environment Protection Agent, Office of Air and Radiation, Washington, DC (September 2012), https://www.epa.gov/agstar.

261. Kangle, K., Kore, S., Kore, V., and Kulkarni, G., Recent trends in anaerobic co-digestion: A review, *Universal Journal of Environmental Research and Technology*, 2 (4), 210–219 (2012).

262. Co-digestion of bio-waste, A report from California Sustainability Alliance, California Public Utility Commission, Sacramento, CA (2012).

263. Froom, M., Co-digestion: A sustainable solution for sewage sludge, a report by *Renewable Energy World Magazine*, Nashua, NH (November 14, 2011).

264. Co-digestion: A developing trend and market—Areport by applied technologies, Brookfield, WI (March 10, 2009). info@atiae.com.

265. Wallis, M.J., Ambrose, M.R., and Chan, C.C., Climate change: Charting a water course in an uncertain future, *Journal of the American Water Works Association*, 100, 70–79 (2008); Groenewold, H., Anaerobic co-digestion in the Netherlands—A system analysis on greenhouse gas emissions from Dutch co-digesters, MS thesis, University of Groningen, Groningen, the Netherlands (2013).

266. Burton, C.H. and Turner, C., Anaerobic treatment options for animal manures, Chapter 7, in *Manure Management—Treatment Strategies for Sustainable Agriculture*, 2nd edn. Silsoe Research Institute, as part of the EU Accompanying Measure Project, Wrest Park, Silsoe, Bedford, U.K. (2003).

267. Romano, R.T. and Zhang, R., Co-digestion of onion juice and wastewater sludge using an anaerobic mixed biofilm reactor, *Bioresource Technology*, 99 (3), 631–637 (March 2008), DOI:10.1016/j.biortech.2006.12.043.

268. Smith, K., *Opportunities and Constraints of Co-Digestion of Sewage Sludge with Other Organic Waste Streams in Existing Waste Water Treatments Plants*. Imperial College, London, U.K. (2013).

269. Gabel, D., Co-digestion case studies enhancing energy recovery from sludge, in *MWRD PWO Seminar*, CH2M HILL, Englewood, CO (May 23, 2012).

270. Hartmann, H. and Ahring, B.K., Anaerobic digestion of the organic fraction of municipal solid waste: Influence of co-digestion with manure, *Water Research*, 39, 1543–1552 (2005).

271. Zupancic, G.D., Uranjek-Zevart, N., and Ros, M., Full-scale anaerobic co-digestion of organic waste and municipal sludge, *Biomass and Bioenergy*, 32, 162–167 (2008), available online (August 20, 2007).

272. Cowan, A.K., Bio-refineries: Bioprocess technologies for waste water treatment, energy and product valorization, CP1229, in Tarasenko, O. (ed.), *Proceedings of the Fourth Bionanotax Conference*, American Institute of Physics, pp. 80–86 (2010).

273. Ryan, D., Gadd, A., Kavanagh, J., and Barton, G., Integrated biorefinery wastewater design, *Chemical Engineering Research and Design*, 87, 1261–1268 (2009).

274. Shin, H.K., Han, S., Song, Y., and Hwan, E., Bio gasification of food residuals, *Biocycle*, 41 (8), 82–86 (August 2000).

275. Mata-Alvarez, J., Mace, S., and Llabres, P., Anaerobic digestion of organic solid wastes. An overview of research achievements and perspectives, *Bioresource Technology*, 74, 3–16 (2000).

276. Yusuf, M. and Ify, N., The effect of waste paper on the kinetics of biogas yield from the co-digestion of cow dung and water hyacinth, *Biomass and Bioenergy*, 35 (3), 1345–1351 (March 2011).

277. Salminen, E. and Rintala, J., Anaerobic digestion of organic solid poultry slaughterhouse waste—A review, *Bioresource Technology*, 83 (1), 13–26 (May 2002).

278. Agdag, O.N. and Sponza, D.T., Co-digestion of mixed industrial sludge with municipal solid wastes in anaerobic simulated landfilling bioreactors, *Journal of Hazardous Materials*, 140, 75–85 (2007).

279. Alvarez, R. and Liden, G., Semi-continuous co-digestion of solid slaughterhouse waste, manure, and fruit and vegetable waste, *Renewable Energy,* 33 (7), 726–734 (2008), available online (June 15, 2007).

280. El-Mashad, H. and Zhang, R., Biogas production from co-digestion of dairy manure and food waste, *Bioresource Technology*, 101 (11), 4021–4028 (June 2010).

281. Fernandez, A., Sanchez, A., and Font, X., Anaerobic co-digestion of a simulated organic fraction of municipal solid wastes and fats of animal and vegetable origin, *Biochemical Engineering*, 26, 22–28 (2005).

282. Fezzani, B. and Ben Cheikh, R., Thermophilic anaerobic co-digestion of olive mill wastewater with olive mill solid wastes in a tubular digester, *Chemical Engineering Journal*, 132 (1–3), 195–203 (2007).

283. Gelegenis, J., Georgakakis, D., Angelidaki, I., and Mavris, V., Optimization of biogas production by co-digesting whey with diluted poultry manure, *Renewable Energy*, 32, 2147–2160 (2007).

284. Gomez, G., Cuetos, M., Cara, J., Moran, A., and Garcia, A., Anaerobic co-digestion of primary sludge and the fruit and vegetable fraction of the municipal solid wastes: Conditions for mixing and evaluation of the organic loading rate, *Renewable Energy*, 31, 2017–2024 (2006).

285. Lehtomäki, A., Huttunen, S., and Rintala, J., Laboratory investigations on co-digestion of energy crops and crop residues with cow manure for methane production: Effect of crop to manure ratio, *Resources, Conservation and Recycling*, 51, 591–609 (2007).

286. Murto, M., Björnsson, L., and Mattiasson, B., Impact of food industrial waste on anaerobic co-digestion of sewage sludge and pig manure, *Journal of Environmental Management*, 70, 101–107 (2004).

287. Neves, L., Oliveira, R., and Alves, M., Anaerobic co-digestion of coffee waste and sewage sludge, *Waste Management*, 26, 176–181 (2006).

288. Romano, R.T. and Zhang, R., Co-digestion of onion juice and wastewater sludge using an anaerobic mixed biofilm reactor, *Bioresource Technology* 99 (3), 231–237 (February 2008), available online (June 1, 2007).

289. Zupancic, G.D., Uranjek-Zevart, N., and Ros, M., Full-scale anaerobic co-digestion of organic waste and municipal sludge, *Biomass and Bioenergy*, 32 (2), 162–167 (February 2008), available online (August 20, 2007).

290. Alvarez, J., Otero, L., and Lema, J., A methodology for optimizing feed composition for anaerobic co-digestion of agro-industrial wastes, *Bioresource Technology*, 101, 1153–1158 (2010).

291. McDonald, N., Experiences with anaerobic digestion of energy crops and high solids manures, OWS report, AgSTAR, EPA, Washington, DC (May 2011).

292. Dohler, H, Eckel, H., and Frisch, J., *Energiepflanzen*. KTBL, Darmstadt, Germany (2006).

293. Baey-Ernsten, H. and Schure, A., *Faustazahlen Biogas*. A report by Kuratorium fur Technik und Bauwesen in der Landwirtschaft, (KTBL/FNR), Darmstadt, Germany (2013).

294. Karpenstein-Machan, M., *Energiepflanzenbau fur Biogasanlagenbetreiber*. DLG-Verlag, Frankfurt, Germany (2005).

295. Weinberg, Z., Muck, R., and Weimer, P., The survival of silage inoculant lactic acid bacteria in rumen fluid, *Journal of Applied Biochemistry*, 93, 1066–1071 (2003).

296. Banemann, D. and Nelles, M., Von der Emte bis in den Fermenter, VDI-Ber 2057, 29–46 (2009).

297. Muller, J., Tiehm, A., Eder, B., Gunhert, F., Hruschka, H., Kopp, J., Kunz, P., Otte-Witte, R., Schmelz, K., and Seiler, K., Thermische, Chemische, und biochemische Desintegrationsverfahren, *Korresp Abwasser*, 50, 796–804 (2003).

298. Mahandete, A., Bjornsson, L., Kivaisi, A., Rubindamayugi, M.S.T., and Matthiasson, B., Effect of particle size on biogas yield from sisal fiber waste, *Renewable Energy*, 31, 2385–2392 (2006).

299. Prechtel, S., Anzer, T., Schneider, R., and Faulstich, M., Biogas production from substrate with high amount of organic nitrogen, in *Proceedings of the 10th World Congress—Anaerobic Digestion 2004*, Montreal, Quebec, Canada, pp. 1809–1812 (2004).

300. Mladenovska, Z., Hartmann, H., kvist, T., Sales-Cruz, M., Gani, R., and Ahring, B., Thermal pretreatment of the solid fraction of manure: Impact of the biogas reactor performance and microbial community, *Water Science & Technology*, 53, 59–67 (2006).

301. Romano, R., Zhang, R., Teter, S., and McGarry, J., The effect of enzyme addition on anaerobic digestion of Jose Tall Wheat Grass, *Bioresource Technology*, 100, 4564–4571 (2009).

302. Morgavi, D., Beauchemin, K., and Nsereko, L., Resistance of feed enzymes to proteolytic inactivation by rumen microorganisms and gastrointestinal proteases, *Journal of Animal Science*, 79, 1621–1630 (2001).

303. Gemmeke, B., Rieger, C., and Weiland, P., Biogas-Messprogramm II, 61 Biogaanlangen im Vergleich, FNR, Gulzow, Germany (2009).

304. Schon, M., *Verfahren zur Vergarung organischer Ruckstande in der Abfallwirtschaft*. Erich Schmidt Verlag, Berlin, Germany (1994).

305. De Baere, L. and Mattheeuws, B., State of the art 2008—Anaerobic digestion of solid waste, *Waste Management World*, 9, 1–8 (2008).

306. Kusch, S., Oechsner, H., and Jungbluth, T., Vergarung landwirtschaft-licher substrtae in diskontinui-erlichen Feststoffermentern, *Agratechnische Forschung*, 11, 81–91 (2007); *Biogashandbuch Bayern-Materialband*. Bayerisches Landesamt fur Umwelt, Augsburg, Germany (2005).

307. Lethomaki, A., Biogas production from energy crops and crop residues, *JyväskyläStudies in Biological and Environmental Science*, 163, 1–91 (2006).

308. Kang, M.S., Srivastava, P., Tyson, T., Fulton, J.P., Owsley, W.F., and Yoo, K.H., A comprehensive GIS-based poultry litter management system for nutrient management planning and litter transportation, *Computers and Electronics in Agriculture*, 64, 212–224 (2008).

309. Komilis Dimitris, P. and Ham Robert, K., Carbon dioxide and ammonia emissions during composting of mixed paper, yard waste and food waste, *Waste Management* (New York), 26, 62–70 (2006).

310. Desai, M., Patel, V., and Madamwar, D., Effect of temperature and retention time on biomethanation of cheese whey–poultry waste–cattle dung, *Environmental Pollution*, 83, 311–315 (1994).

311. Mladenovska, Z., Dabrowski, S., and Ahring, B.K., Anaerobic digestion of manure and mixture of manure with lipids: Biogas reactor performance and microbial community analysis, *Water Science & Technology*, 48, 271–278 (2003).

312. Møller, H.B., Sommer, S.G., and Ahring, B.K., Methane productivity of manure, straw and solid fractions of manure, *Biomass and Bioenergy*, 26, 485–495 (2004).

313. Lin, J., Zuo, J., Ji, R., Chen, X., Liu, F., Wang, K., and Yang, Y., Methanogenic community dynamics in anaerobic co-digestion of fruit and vegetable waste and food waste, *Environmental Biology*, 24, 1288–1294 (2012). doi:10.1016/s1001-0742(11)60927-3.

314. Navaneethan, N. and Zitorer, D., Anaerobic co-digestion increases net biogas production by increasing microbial activity, Central State Water Environment Association (CSWEA) meeting, Technical Program, Marquette University, Milwaukee, WI (May 14–17, 2012).

315. Liu, K., Tang, Y.Q., Matsui, T., Morimura, S., Wu, X.L., and Kida, K., Thermophilic anaerobic co-digestion of garbage, screened swine and dairy cattle manure, *Journal of Bioscience and Bioengineering*, 107 (1), 54–60 (2009).

316. Sell, S., A scale-up procedure for substrate co-digestion in anaerobic digesters through the use of substrate characterization, BMPs, ATAs, and sub pilot-scale digesters, MS thesis, Agricultural Engineering, Iowa State University, Ames, IA (2011).

317. Pagés Díaz, J., Pereda Reyes, I., Lundin, M., and Sárvári Horváth, I., Co-digestion of different waste mixtures from agro-industrial activities: Kinetic evaluation and synergetic effects, *Bioresource Technology*, 102 (23), 10834 (2011).

318. Palatsi, J., Viñas, M., Guivernau, M., Fernandez, B., and Flotats, X., Anaerobic digestion of slaughterhouse waste: Main process limitations and microbial community interactions, *Bioresource Technology*, 102 (3), 2219–2227 (2011).

319. Ghaly, A., A comparative study of anaerobic digestion of acid cheese whey and dairy manure in a two-stage reactor, *Bioresource Technology*, 58, 61–72 (1996).

320. Zaher, U., Li, R., Pandey, P., Ewing, T., Frear, C., and Chen, S., Development of codigestion software models to assist plant design and co-digestion operation. Climate Friendly Farming, Co-digestion modeling, Chapter 5, CSANR Research Report, pp. 1–15 (2010).

321. Fezzani, B. and Ben Cheikh, R., Implementation of IWA anaerobic digestion model No. 1 (ADM1) for simulating the thermophilic anaerobic co-digestion of olive mill wastewater with olive mill solid waste in a semi-continuous tubular digester, *Chemical Engineering Journal* (Amsterdam, the Netherlands), 141, 75–88 (2008).

322. Fezzani, B. and Ben Cheikh, R., Optimisation of the mesophilic anaerobic co-digestion of olive mill wastewater with olive mill solid waste in a batch digester, *Desalination*, 228, 159–167 (2008).

323. Astals, S., Ariso, M., Galí, A., and Mata-Alvarez, J., Co-digestion of pig manure and glycerine: Experimental and modelling study, *Journal of Environmental Management*, 92, 1091–1096 (2011).

324. Yamashiro, T., Lateef, S., Ying, C., Beneragama, N., Lukic, M., Mashiro, I., Ihara, I., Nishida, T., and Umetsu, K., Anaerobic co-digestion of dairy cow manure and high concentrated food processing waste, *Journal of Material Cycles and Waste Management*, 15, 539–547 (2013).

325. Siripong, C. and Dulyakasem, S., Continuous co-digestion of agro-industrial residues, MS thesis, School of Engineering, University of Boras, Boras, Sweden (June 2012).

326. Alvarez, R. and Lidén, G., Semi-continuous co-digestion of solid slaughterhouse waste, manure, and fruit and vegetable waste, *Renewable Energy*, 33 (4), 726–734 (2008).

327. Callaghan, F.J., Wase, D.A.J., Thayanithy, K., and Forster, C.F., Continuous co-digestion of cattle slurry with fruit and vegetable wastes and chicken manure, *Biomass and Bioenergy*, 22 (1), 71–77 (2002).

328. Cuetos, M.J., Fernández, C., Gómez, X., and Morán, A., Anaerobic co-digestion of swine manure with energy crop residues, *Biotechnology and Bioprocess Engineering*, 16 (5), 1044–1052 (2011).

329. Cuetos, M.J., Gómez, X., Otero, M., and Morán, A., Anaerobic digestion and co-digestion of slaughterhouse waste (SHW): Influence of heat and pressure pre-treatment in biogas yield, *Waste Management*, 30 (10), 1780–1789 (2010).

330. Cuetos, M.J., Gómez, X., Otero, M., and Morán, A., Anaerobic digestion of solid slaughterhouse waste (SHW) at laboratory scale: Influence of co-digestion with the organic fraction of municipal solid waste (OFMSW), *Biochemical Engineering Journal*, 40 (1), 99–106 (2008).

331. Kacprzak, A., Krzystek, L., and Ledakowicz, S., Co-digestion of agricultural and industrial wastes, *Chemical Papers*, 64 (2), 127–131 (2009).

332. Panichnumsin, P., Nopharatana, A., Ahring, B., and Chaiprasert, P., Production of methane by co-digestion of cassava pulp with various concentrations of pig manure, *Biomass and Bioenergy*, 34 (8), 1117–1124 (2010).

333. Xie, S., Wu, G., Lawlor, P.G., Frost, J.P., and Zhan, X., Methane production from anaerobic co-digestion of the separated solid fraction of pig manure with dried grass silage, *Bioresource Technology*, 104, 289–297 (2012).

334. Zhang, P., Zeng, G., Zhang, G., Li, Y., Zhang, B., and Fan, M., Anaerobic co-digestion of biosolids and organic fraction of municipal solid waste by sequencing batch process, *Fuel Processing Technology*, 89, 485–489 (2008).

335. Serrano, A., Lopez, J., Chica, A., Martin, M., Karouach, F., Mesfiour, A., and Bari, H., Mesophilic anaerobic co-digestion of sewage sludge and orange peel waste, *Environmental Technology*, 35 (7), 898–906 (2014).

336. Aziz, T., Wang, L., Long, H., Sawyer, H., and Ducoste, J., *Sustainable Energy from Grease Interceptor Waste Co-Digestion*. An Internal Publication by North Carolina State University, Department of Environmental Engineering, Raleigh, NC (2013).
337. Ramsamy, D., Rakgotho, T., Naidoo, V., and Buckley, C., Anaerobic co-digestion of high strength/toxic organic liquid effluents in a continuously stirred reactor: Start-up, in Paper presented at the *Biennial Conference of the Water Institute of Southern Africa (WISA)*, Durban, South Africa (May 19–23, 2002), www.wisa.co.za, Accessed December 2002.
338. Chiu-Yue, L., Feng-Yi, B., and Jen, C., Anaerobic co-digestion of septage and landfill leachate, *Bioresource Technology*, 68, 275–282 (1999).
339. Zupancic, G., Ros, M., Klemencic, M., Oset, M., and Logar, R., *Excess Brewery Yeast Co-Digestion in a Full Scale EGSB Reactor*. Department of Microbiology and Microbial Biotechnology, University of Ljubljana, Ljubljana, Slovenia (2013).
340. Martin-Gonzalez. L.. Colyurato, L., Font, X., and Vicent, T., Anaerobic co-digestion of organic fraction of municipal solid waste with FOG waste from a sewage treatment plant: Recovering a wasted methane potential and enhancing the biogas yield, *Waste Management*, 39, 615 (2010).
341. Bolzonella, D., Battistoni, P., Susini, C., and Cecchi, F., Anaerobic codigestion of waste activated sludge and OFMSW: The experiences of Viareggio and Treviso plants (Italy), *Water Science & Technology*, 53 (8), 203–211 (2006).
342. Luostarinen, S., Luste, S., and Sillanpää, M., Increased biogas production at wastewater treatment plants through co-digestion of sewage sludge with grease trap sludge from a meat processing plant, *Bioresource Technology*, 100, 79–85 (2009).
343. Canas, E., Technical feasibility of anaerobic co-digestion of dairy manure with chicken litter and other wastes, MS thesis, University of Tennessee, Knoxville, TN (2010).
344. Alvarez, R., Villca, S., and Lidén, G., Biogas production from llama and cow manure at high altitude, *Biomass and Bioenergy*, 30, 66–75 (2006).
345. Alvarez, R. and Lidén, G., Anaerobic co-digestion of aquatic flora and quinoa with manures from Bolivian Altiplano, *Waste Management*, 28, 1933–1940 (2008).
346. Amon, T., Amon, B., Kryvoruchko, V., Zollitsch, W., Mayer, K., and Gruber, L., Biogas production from maize and dairy cattle manure—Influence of biomass composition on the methane yield, *Agriculture, Ecosystems and Environment*, 118, 173–182 (2007).
347. Callaghan, F.J., Luecke, K., Wase, D.A.J., and Thayanithy, K., Co-digestion of cattle slurry and waste milk under shock loading conditions, *Journal of Chemical Technology and Biotechnology*, 68, 405–410 (1997).
348. Callaghan, F.J., Wase, D.A.J., Thayanithy, K., and Forster, C.F., Co-digestion of waste organic solids: Batch studies, *Bioresource Technology*, 67, 117–122 (1999).
349. Riggle, D., Anaerobic digestion for municipal solid waste and industrial wastewater, *Biocycle*, 37 (11), 77 (1996).
350. Sacks, J., Anaerobic digestion of high-strength or toxic organic effluents, Masters thesis, School of Chemical Engineering, University of Natal, Glenwood, Durban, South Africa (1997).
351. Callaghan, F.J., Wase, D.A.J., Thayanithy, K., and Forster, C.F., Continuous co-digestion of cattle slurry with fruit and vegetable wastes and chicken manure, *Biomass and Bioenergy*, 27, 71–77 (2002).
352. Davidsson, A., Lövstedt, C., la Cour Jansen, J., Gruvberger, C., and Aspegren, H., Codigestion of grease trap sludge and sewage sludge, *Waste Management*, 28, 986–992 (2008).
353. Demirel, B., Yenigun, O., and Onay, T.T., Anaerobic treatment of dairy wastewaters: A review, *Process Biochemistry*, 40, 2583–2595 (2005).
354. Misi, S.N. and Forster, C.F., Batch co-digestion of multi-component agro-wastes, *Bioresource Technology*, 80, 19–28 (2001).
355. Neves, L., Oliveira, R., and Alves, M.M., Co-digestion of cow manure, food waste and intermittent input of fat, *Bioresource Technology*, 100, 1957–1962 (2009).
356. Sharom, Z., Malakahmad, A., and Noor, B., Anaerobic co-digestion of kitchen waste and sewage sludge for producing biogas, in *Second International Conference on Environmental Management*, Bangi, Malaysia (2004).
357. Misi, S.N. and Forster, C.F., Batch co-digestion of two-component mixtures of agrowastes, *IChemE, Part B*, 79, 365–371 (2001).
358. Mshandete, A., Kivaisi, A., Rubindamayugi, M., and Mattiasson, B., Anaerobic batch codigestion of sisal pulp and fish wastes, *Bioresource Technology*, 95, 19–24 (2004).
359. Kaparaju, P., Luostarinen, S., Kalmari, E., Kalmari, J., and Rintala, J., Co-digestion of energy crops and industrial confectionery by-products with cow manure: Batch-scale and farm-scale evaluation, *Water Science & Technology*, 45 (10), 275–280 (2002).

360. Kuang, Y., Enhancing anaerobic degradation of lipids in wastewater by addition of cosubstrates, PhD thesis, Murdoch University, Perth, Western Australia, Australia (2002).

361. Ek, A., Hallin, S., Vallin, L., Schnurer, A., and Karlsson, M., Slaughterhouse waste codigestion—Experiences from 15 years of full scale operation, in *Biogas International Conference*, Linkoping, Sweden (May 8–13, 2011).

362. Zhang, L., Woo Lee, Y., and Jahng, D., Anaerobic co-digestion of food waste and piggery wastewater: Focusing on the role of trace elements, *Bioresource Technology*, 102, 5048–5059 (2011).

363. Zhu, Z., Hsueh, M.K., and He, Q., Enhancing biomethanation of municipal waste sludge with grease trap waste as a co-substrate, *Renewable Energy*, 36, 1802–1807 (2011).

364. Ersahin, M.V, Gomec, C.V., Dereli, R.K., Arikan, O., and Ozturk, I., Biomethane production as an alternative: Bioenergy source from codigesters treating municipal sludge and organic fraction of municipal solid wastes, *Journal of Biomedicine and Biotechnology*, 8, 1–8 (2011).

365. Bouallagui, H., Lahdheb, H., Ben Romdan, E., Rachdi, B., and Hamdi, M., Improvement of fruit and vegetable waste anaerobic digestion performance and stability with co-substrates addition, *Journal of Environmental Management*, 90, 1844–1849 (2009).

366. Buendía, I.M., Fernández, F., Villasenor, J., and Rodriguez, L., Feasibility of anaerobic co-digestion as a treatment option of meat industry wastes, *Bioresource Technology*, 100, 1903–1909 (2009).

367. Heo, N., Park, S., Lee, J., Kang, H., and Park, D., Single stage anaerobic co-digestion for mixture wastes of simulated Korean food waste and waste activated sludge, *Applied Biochemistry and Biotechnology*, 107, 567–579 (2003).

368. Ponsá, S., Gea, T., and Sánchez, A., Anaerobic co-digestion of the organic fraction of municipal solid waste with several pure organic co-substrates, *Biosystems Engineering*, 108, 352–360 (2011).

369. Braun, R., Brachtl, E., and Grasmug, M., Codigestion of proteinaceous industrial waste, *Applied Biochemistry and Biotechnology*, 109, 139–153 (2003).

370. Anaerobic co-digestion dairies in Washington State, A report from Washington State University Extension Fact Sheet—FS040E, Pullman, WA (2013).

371. Patil, J., AntonyRaj, M., Gavimath, C., and Hooli, V., A comprehensive study on anaerobic co-digestion of water hyacinth with poultry litter and cow dung, *International Journal of Chemical Sciences and Applications*, 2 (2), 148–155 (June 2011).

372. Li, C., Using anaerobic co-digestion with addition of municipal organic wastes and pretreatment to enhance biogas production from wastewater treatment plant sludge, PhD thesis, Department of Civil Engineering, Queen's University, Kingston, Ontario, Canada (September 2012).

373. Kassuwi, S., Mshandete, A., and Kivaisi, A., Anaerobic co-digestion of biological pretreated Nile perch fish solid waste with vegetable fraction of market solid waste, *ARPN: Journal of Agricultural and Biological Science*, 7 (12), 1016–1031 (December 2012).

374. Dahunsi, S. and Oranusi, U., Co-digestion of food waste and human excreta for biogas production, *British Biotechnology Journal*, 3 (4), 485–499 (2013).

375. Babaee, A., Shayegan, J., and Roshani, A., Anaerobic slurry co-digestion of poultry manure and straw: Effect of organic loading and temperature, *Journal of Environmental Health Science & Engineering*, 11, 15 (2013).

376. Bolzonella, D., Pavan, P., Battistoni, P., and Cecchi, F., Anaerobic co-digestion of sludge with other organic wastes and phosphorous reclamation in wastewater treatment plants or biological nutrients removal, *Water Science & Technology*, 53 (12), 177–186 (2006).

377. Kübler, H., Hoppenheidt, K., Hirsch, P., Kottmair, A., Nimmrichter, R., and Nordsleck, H., Full-scale co-digestion of organic waste, *Water Science & Technology*, 41, 195–202 (2000).

378. Crolla, A. and Kinsley, C., Background on anaerobic digestion on the farm, Ontario Rural Wastewater Center, Info Sheet, University of Guelph, Guelph, Ontario, Canada (2013).

379. El-Mashad, H.M., Wilko, K.P., Loon, V., and Zeeman, G., A model of solar energy utilisation in the anaerobic digestion of cattle manure, *Biosystems Engineering*, 84, 231–238 (2003).

380. Forster-Carneiro, T., Pérez, M., Romero, L.I., and Sales, D., Dry-thermophilic anaerobic digestion of organic fraction of the municipal solid waste: Focusing on the inoculum sources, *Bioresource Technology*, 98, 3195–3203 (2007).

381. Delia, T. and Agdag, N., Co-digestion of industrial sludge with municipal solid wastes in anaerobic simulated landfilling reactors, *Process Biochemistry*, 40, 1871–1879 (2007).

382. Samani, Z., Macias-Corral, M., Hanson, A., Smith, G., Funk, P, Yu, H., and Longworth, J., Anaerobic digestion of municipal solid waste and agricultural waste and the effect of co-digestion with dairy cow manure, *Bioresource Technology*, 99 (17), 8288–8293 (2008).

383. Bioconversion Solutions, Boston, MA, Company Established in 2003, www.bioconversinsolutions.com, Accessed May 2015.
384. Buffiere, P., Loisel, D., Bernet, N., and Delgenès, J.P., Towards new indicators for the prediction of solid waste anaerobic digestion properties, *Water Science & Technology*, 53, 233–241 (2006).
385. Saint-Joly, C., Desbois, S., and Lotti, J.-P., Determinant impact of waste collection and composition on anaerobic digestion performance: Industrial results, *Water Science & Technology*, 41, 291–297 (2000).
386. Carucci, G., Carrasco, F., Trifoni, K., Majone, M., and Beccari, M., Anaerobic digestion of food industry wastes: Effect of codigestion on methane yield, *Journal of Environmental Engineering*, 131, 1037–1045 (2005).
387. Edelmann, W., Co-digestion of organic solid waste and sludge from sewage treatment, *Water Science & Technology*, 41, 213 (2000).
388. Angelidaki, I. and Ahring, B., Effects of free long-chain fatty acids on thermophilic anaerobic digestion, *Applied Microbiology and Biotechnology*, 37, 808–812 (1992).
389. Kim, H.-W., Han, S.-K., and Shin, H.-S., The optimisation of food waste addition as a co-substrate in anaerobic digestion of sewage sludge, *Waste Management & Research*, 21, 515–526 (2003).
390. Sosnowski, P., Klepacz-Smolka, A., Kaczorek, K., and Ledakowicz, S., Kinetic investigations of methane co-fermentation of sewage sludge and organic fraction of municipal solid wastes, *Bioresource Technology*, 99, 5731–5737 (2008).
391. Converti, A., Drago, F., Ghiazza, G., Borghi, M.D., and Macchiavello, A., Co-digestion of municipal sewage sludges and pre-hydrolysed woody agricultural wastes, *Journal of Chemical Technology and Biotechnology*, 69, 231–239 (1997).
392. Demirekler, E. and Anderson, G.K., Effect of sewage sludge addition on the startup of the anaerobic digestion of OFMSW, *Environmental Technology*, 19, 837–843 (1998).
393. Cho, J.K., Park, S.C., and Chang, H.N., Biochemical methane potential and solid state anaerobic digestion of Korean food wastes, *Bioresource Technology*, 52, 245–253 (1995).
394. Hartmann, H., Moller, H.B., and Ahring, B.K., Efficiency of the anaerobic treatment of the organic fraction of municipal solid waste: Collection and pretreatment, *Waste Management & Research*, 22, 35–41 (2004).
395. Cirne, D.G., Paloumet, X., Björnsson, L., Alves, M.M., and Mattiasson, B., Anaerobic digestion of lipid-rich waste effects of lipid concentration, *Renewable Energy*, 32, 965–975 (2007).
396. Gallert, C. and Winter, J., Mesophilic and thermophilic anaerobic digestion of source sorted organic wastes: Effect of ammonia on glucose degradation and methane production, *Applied Microbiology and Biotechnology*, 48, 405–410 (1997).
397. Kayhanian, M. and Hardy, S., The impact of four design parameters on the performance of a high-solids anaerobic digestion of municipal solid waste for fuel gas production, *Environmental Technology*, 15, 557–567 (1994).
398. El Hadj, T.B., Astals, S., Gali, A., Mace, S., and Mata Alvarez, J., Ammonia influence in anaerobic digestion of OFMSW, *Water Science & Technology*, 59, 1153–1158 (2009).
399. Rittman, B.E. and McCarty, P.L., *Environmental Biotechnology: Principles and Applications*. McGraw-Hill International, New York (2001).
400. Appels, L., Lauwers, J., Degrève, J., Helsen, L., Lievens, B., Willems, K., Van Impe, J., and Dewil, R., Anaerobic digestion in global bio-energy production: Potential and research challenges, *Renewable & Sustainable Energy Reviews*, 15, 4295–4301 (2011).
401. De Baere, L., Anaerobic digestion of solid waste: State-of-the-art, *Water Science & Technology*, 41, 283–290 (2000).
402. Lebrato, J., Pérez-Rodríguez, J.L., and Maqueda, C., Domestic solid waste and sewage improvement by anaerobic digestion: A stirred digester, *Resources, Conservation and Recycling*, 13, 83–88 (1995).
403. Hartmann, H., Angelidaki, I., and Ahring, B.K., Increase of anaerobic degradation of particulate organic matter in full-scale biogas plants by mechanical maceration, *Water Science & Technology*, 41, 145–153 (2000).
404. *Food Waste Disposal Using Anaerobic Digestion, Renewable Energy.* Centre for Analysis and Dissemination of Demonstrated Energy Technologies, CADDET, Centre for Renewable Energy, Harwell, U.K.; a pert of IEA/OECD, Technical Brochure No. 66 (1998).
405. Ganidi, N., Tyrrel, S., and Cartmell, E., Anaerobic digestion foaming causes—A review, *Bioresource Technology*, 100, 5546–5554 (2009).
406. Purcell, B. and Stentiford, E.I., Co-digestion and enhancing recovery of organic waste, *ORBIT Journal*, 1, 1–6 (2006).

407. Policy paper, Anaerobic digestion strategy and action plan, defre; Department Environment, Food and Rural Affairs, London, U.K. (2011), www.defra.gov.uk.

408. European Parliament and Council, Regulation (EC) No. 1774/202 on laying down health rules concerning animal by-products not intended for human consumption, *Official Journal of the European Union* L273, 1–163 (2002).

409. Controls on animal by-products, Guidance on Regulation (EC) 1069/2009 and accompanying implementing Regulation (EC). 142/2011, enforced in England by the Animal By-Products (Enforcement, England) Regulations 2011, version 4, Thermal Archives (November 2011), www.defra.gov.uk.

410. Defra, Accelerating the uptake of anaerobic digestion in England: An implementation plan, Department of Environment Food and Rural Affairs, London, U.K. (2010).

411. Krupp, M., Schubert, J., and Widmann, R., Feasibility study for co-digestion of sewage sludge with OFMSW on two wastewater treatment plants in Germany, *Waste Management*, 25, 393–399 (2005).

412. Kaosol, T. and Sohgrathok, N., Enhancement of biogas production potential for anaerobic co-digestion of wastewater using decanter cake, *American Journal of Agricultural and Biological Science*, 8 (1), 67–74 (2013).

413. Liu, Z., Anaerobic co-digestion of swine wastewater with switch grass and wheat straw for methane production, MS thesis, Biological and Agricultural Engineering, North Carolina State University, Raleigh, NC (2013).

414. Zhang, T., Liu, L., Song, Z., Ren, G., Feng, Y., Han, X., and Yang, G., Biogas production by co-digestion of goat manure with three crop residues, *Materials and Methods*, 1–7 (June 2013). doi:10.1371/journal pone.0066845.

415. Crolla, A., Kinsley, C., Kennedy, K., and Sauve, T., Anaerobic co-digestion of corn thin stillage and dairy manure, in *Canadian Biogas Conference and Exhibition*, London Convention Center, London, Ontario, Canada (March 4–6, 2013).

416. Campos, E., Palatsi, J., and Flotats, X., Co-digestion of pig slurry and organic wastes from food industry, in *Conference on Waste*, Barcelona, Spain, pp. 192–195 (June 1999).

417. Bagi, Z., Acs, N., Balint, B., Horvath, L., Dobo, K., Perei, K., Rakhely, G., and Kovacs, K., Biotechnological intensification of biogas production, *Applied Microbiology and Biotechnology*, 76, 473–482 (2007).

418. Lisboa, M., lansing, S., and Jackson, C., On-farm co-digestion of food waste with dairy manure, A report from Department of Environmental Science and Technology, University of Maryland, College Park, MD (2013).

419. Garcia, K. and Perez, M., Anaerobic co-digestion of cattle manure and sewage sludge: Influence of composition and temperature, *International Journal of Environmental Protection*, 3 (6), 8–15 (June 2013).

420. Alatriste-Mondragon, F., Samar, P., Cox, H., Ahring, B., and Iranpour, R., Anaerobic codigestion of municipal, farm, and industrial organic wastes: A survey of recent literature, *Water Environment Research*, 78, 607–636 (2006).

421. Baserga, U., Landwirtschaftliche Co-vergarungs-Biogasanlagen, FAT-Berichte No. 512, Tanikon/ Switzerland (1998).

422. You, Z., Wei, T., and Cheng, J., Improving anaerobic co-digestion of corn stover using sodium hydroxide pretreatment, *Energy & Fuels*, 28 (1), 549–554 (2014).

423. Vieitez, E. and Gosh, S., Biogasification of solid wastes by two phase anaerobic fermentation, *Biomass and Bioenergy*, 16, 299–309 (1999).

424. Parawira, W., Read, J., Mattiasson, B., and Bjornsson, L., Energy production from agricultural residues: High methane yields in a pilot scale two-stage anaerobic digestion, *Biomass and Bioenergy*, 32, 44–50 (2008).

425. Busch, G., Grossmann, J., Sieber, M., and Burkhardt, M., A new and sound technology for biogas from solid waste and biomass, *Water, Air, & Soil Pollution: Focus*, 9, 89–97 (2009).

426. Oechsner, H. and Lemmer, A., Was Kann die Hydrolyse bei der Biogasvergarung leisten? VDI-Ber 2057, 37–46 (2009).

427. Zhang, R., Biogasification of organic solid wastes, *Biocycle*, 43, 56–59 (2002).

428. Abderrezaq, I.A., Employment of anaerobic digestion process of municipal solid waste for energy, *Energy Sources Part A*, 29, 657–668 (2007).

429. Weiland, P., Wichtige Messdaten fur den Prozessablauf und Stand der Technik in der Praxis, *Gulzower Fachgesprache*, 27, 17–31 (2008).

430. Ahring, B., Sandberg, M., and Angelidaki, I., Volatile fatty acids as indicators of process imbalance in anaerobic digesters, *Applied Microbiology and Biotechnology*, 34, 559–565 (1995).

431. Nielsen, H., Uellendahl, H., and Ahring, B., Regulation and optimization of biogas process: Propionate as a key factor, *Biomass and Bioenergy*, 31, 820–830 (2007).

432. Soroushian, F., Operational considerations for co-digestion, in *SARBS Seminar on Math, Operations and Maintenance for Biosolids Systems*, California Water Environment Association, CH2M HILL, Englewood, CO (September 2011).

433. Lossie, U. and Pütz, P., Targeted control of biogas plants with the help of FOS/TAC, a report from Hach Lange, Dusseldorf, Germany (2012).

434. Padilla-Gasca, E., López-López, A., and Gallardo-Valdez, J., Evaluation of stability factors in the anaerobic treatment of slaughterhouse wastewater, *Journal of Bioremediation & Biodegradation*, 2 (1), 114 (2011).

435. PCD, *Guideline for Utilization of Waste from Slaughter Process*. Pollution Control Department, Ministry of Natural Resources and Environment, Bangkok, Thailand (2012).

436. Salminen, E., Einola, J., and Rintala, J., Characterisation and anaerobic batch degradation of materials accumulating in anaerobic digesters treating poultry slaughterhouse waste, *Environmental Technology*, 22 (5), 577–585 (2001).

437. Sluiter, A., Hames, B., Hyman. D., Payne, C., Ruiz, R., Scarlata, C., Sluiter, J., Templeton, D., and Wolfe, J., Biomass and total dissolved solids in liquid process samples, Laboratory Analytical Procedure (LAP), National Renewable Energy Laboratory, Technical Report NREL/TP-510-42621, Golden, CO (March 2008); Methane production by anaerobic co-digestion of biomass and waste, 265.

438. Sluiter, A., Hames, B., Ruiz, R., Scarlata, C., Sluiter, J., and Templeton, D., Determination of ash in biomass, Laboratory Analytical Procedure (LAP), National Renewable Energy Laboratory, Golden, CO (2005).

439. Straka, F., Jenicek, P., Zabranska, J., Dohanyos, M., and Kuncarova, M., Anaerobic fermentation of biomass and wastes with respect to sulfur and nitrogen contents in treated materials, in *Proceedings of the 11th International Waste Management and Landfill Symposium*, CISA, Environmental Sanitary Engineering Centre, Cagliari, Italy (2007).

440. Verma, S., Anaerobic digestion of biodegradable organics in municipal solid wastes, MSc thesis, Department of Earth & Environmental Engineering, Columbia University, New York (2002).

441. Wang, L., Zhou, Q., and Li, F., Avoiding propionic acid accumulation in the anaerobic process for bio-hydrogen production, *Biomass and Bioenergy*, 30 (2), 177–182 (2006).

442. Alvarenga, P., Palma, P., Goncalves, A.P., Fernandes, R.M., Cunha-Queda, A.C., Duarte, E., and Vallini, G., Evaluation of chemical and ecotoxicological characteristics of biodegradable organic residues for application to agricultural land, *Environment International*, 33, 505–513 (2007).

443. Batstone, D.J., Keller, J., Angelidaki, I., Kalyuzhnyi, S.V., Pavlostathis, S.G., Rozzi, A., Sanders, W.T.M., Siegrist, H., and Vavilin, V.A., Anaerobic digestion model no. 1, Water Science and Technology, 45 (10), 64–73 (February 2002).

444. Bou-Najm, M. and El-Fadel, M., Computer-based interface for an integrated solid waste management optimization model, *Environmental Modelling & Software*, 19, 1151–1164 (2004).

445. Garcia-de-Cortazar, A.L. and Monzon, I.T., Moduelo 2: A new version of an integrated simulation model for municipal solid waste landfills, *Environmental Modelling & Software*, 22, 59–72 (2007).

446. Kleerebezem, R. and Van Loosdrecht, M.C.M., Waste characterization for implementation in ADM1, *Water Science & Technology*, 54, 167–174 (2006).

447. Lubken, M., Wichern, M., Schlattmann, M., Gronauer, A., and Horn, H., Modelling the energy balance of an anaerobic digester fed with cattle manure and renewable energy crops, *Water Research*, 41, 4085–4096 (2007).

448. Manirakiza, P., Covaci, A., and Schepens, P., Comparative study on total lipid determination using Soxhlet, Roese-Gottlieb, Bligh & Dyer, and modified Bligh & Dyer extraction methods, *Journal of Food Composition and Analysis*, 14, 93–100 (2001).

449. Nelder, J. and Mead, R., A simplex method for function minimization, *Computer Journal*, 7, 308–313 (1965).

450. Rosen, C., Vrecko, D., Gernaey, K.V., Pons, M.N., and Jeppsson, U., Implementing ADM1 for plant-wide benchmark simulations in Matlab/Simulink, *Water Science & Technology* 54, 11–19 (2006).

451. Shanmugam, P. and Horan, N.J., Simple and rapid methods to evaluate methane potential and biomass yield for a range of mixed solid wastes, *Bioresource Technology*, 100, 471–474 (2009).

452. Singh, K.P., Malik, A., and Sinha, S., Water quality assessment and apportionment of pollution sources of Gomti River (India) using multivariate statistical techniques—A case study, *Analytica Chimica Acta*, 538, 355–374 (2005).

453. Kumar, V., Mari, M., Schuhmacher, M., and Domingo, J.L., Partitioning total variance in risk assessment: Application to a municipal solid waste incinerator, *Environmental Modelling & Software*, 24, 247–261 (2009).

454. Vanrolleghem, P.A., Rosen, C., Zaher, U., Copp, J., Benedetti, L., Ayesa, E., and Jappsson, U., Continuity-based interfacing of models for wastewater systems described by Petersen matrices, *Water Science & Technology*, 52, 493–500 (2005).

455. Volcke Eveline, I.P., van Loosdrecht Mark, C.M., and Vanrolleghem Peter, A., Continuity based model interfacing for plant-wide simulation: A general approach, *Water Research*, 40, 2817–2828 (2006).

456. Zaher, U., Buffiere, P., Steyer, J.P., and Chen, S., A procedure to estimate proximate analysis of mixed organic wastes, *Water Environment Research*, 81, 407–415 (2009).

457. Zaher, U. and Chen, S., Interfacing the IWA Anaerobic Digestion Model No. 1 (ADM1) with manure and solid waste characteristics, in *WEFTEC'06, Conference Proceedings, 79th Annual Technical Exhibition & Conference*, Dallas, TX, October 21–25, 2006, pp. 3162–3175 (2006).

458. Zaher, U., Grau, P., Benedetti, L., Ayesa, E., and Vanrolleghem, P.A., Transformers for interfacing anaerobic digestion models to pre- and post-treatment processes in a plant wide modelling context, *Environmental Modelling & Software*, 22, 40–58 (2007).

459. Husain, A., Mathematical models of the kinetics of anaerobic digestion—A selected review, *Biomass and Bioenergy*, 14, 561–571 (1998).

460. Abouelenien, F., Kitamura, Y., Nishio, N., and Nakashimada, Y., Dry anaerobic ammonia-methane production from chicken manure, *Applied Microbiology and Biotechnology*, 82 (4), 757–764 (2009).

461. Purcell, B.E., Atkins, W., and Stentiford, E.I., Co-treatment: Fuelling recovery of organic wastes, *Waste Management*, 32–33 (August 2000).

462. Reinhart, D.R. and Pohland, F.G., The assimilation of organic hazardous wastes by municipal solid waste landfills, *Journal of Industrial Microbiology*, 8, 193–200 (1991).

463. Mosche, M. and Jordening, H., Comparison of different models of substrate and product inhibition in ammonia digestion, *Water Research*, 33, 2545–2554 (1999).

464. Wang, Q., Kuninobu, M., Ogawa, H., and Kato, Y., Degradation of volatile fatty in highly efficient anaerobic digestion, *Biomass and Bioenergy*, 16, 407–416 (1999).

465. Boe, K., Bastone, D., and Angelidaki, I., Online headspace chromatographic method for measuring VFA in biogas reactor, *Water Science & Technology*, 52, 473–478 (2005).

466. Redman, G., *A Detailed Economic Assessment of Anaerobic Digestion Technology and Its Sustainability to UK Farming and Waste Systems*, 2nd edn. The Andersons Center, A Project Funded by DECC and Managed by the NNFCC, Melton Mowbray, Leicestershire, U.K. (March 2010).

467. Bayr, S., Rantanen, M., Kaparaju, P., and Rintala, J., Mesophilic and thermophilic anaerobic co-digestion of rendering plant and slaughterhouse wastes, *Bioresource Technology*, 104, 28–36 (2012).

468. Benabdallah El Hajd, T., Astals, S., Galí, A., Mace, S., and Mata-Álvarez, J., Ammonia influence in anaerobic digestion of OFMSW, *Water Science & Technology*, 59 (6), 1153–1158 (2009).

469. Briefing: Anaerobic digestion, Friends of the Earth Limited on subject of Methane production by anaerobic co-digestion of biomass and waste, FOE Company number 1012357, pp. 1–7, London, U.K. (September 2007).

470. Garcia-Pena, E.I., Parameswaran, P., Kang, D.W., Canul-Chan, M., and Krajmalnik-Brown, R., Anaerobic digestion and co-digestion processes of vegetable and fruit residues: Process and microbial ecology, *Bioresource Technology*, 102 (20), 9447–9455 (2011).

471. Hansen, T.L., Schmidt, J.E., Angelidaki, I., Marca, E., Jansen, J.C., Mosbaek, H., and Christensen, T.H., Method for determination of methane potentials of solid organic waste, *Waste Management*, 24 (4), 393–400 (2004).

472. Hilkiah Igoni, A., Ayotamuno, M.J., Eze, C.L., Ogaji, S.O.T., and Probert, S.D., Designs of anaerobic digesters for producing biogas from municipal solid-waste, *Applied Energy*, 85 (6), 430–438 (2008).

473. Clemens, J., Trimborn, M., Weiland, P., and Amon, B., Mitigation of greenhouse gas emissions by anaerobic digestion of cattle slurry, *Agriculture, Ecosystems and Environment*, 112, 171–177 (2006).

474. Deng, W.-Y., Yan, J.-H., Li, X.-D., Wang, F., Zhu, X.-W., Lu, S.-Y., and Cen, K.-F., Emission characteristics of volatile compounds during sludges drying process, *Journal of Hazardous Materials*, 162, 186–192 (2009).

475. He, P.J., Shao, L.M., Guo, H.D., Li, G.J., and Lee, D.J., Nitrogen removal from recycled landfill leachate by ex situ nitrification and in situ denitrification, *Waste Management* (Amsterdam, the Netherlands), 26, 838–845 (2006).

476. Paillat, J.-M., Robin, P., Hassouna, M., and Leterme, P., Predicting ammonia and carbon dioxide emissions from carbon and nitrogen biodegradability during animal waste composting, *Atmospheric Environment*, 39, 6833–6842 (2005).

477. Criddle, C.S., The kinetics of co-metabolism, *Biotechnology and Bioengineering*, 41 (11), 1048–1056 (1993).

478. Mudunge, R., Comparison of an anaerobic baffled reactor and a completely mixed reactor, Masters thesis, School of Chemical Engineering, University of Natal, Durban, South Africa (2001).

479. Peres, C.S., Sanchez, C.R., Matumoto, C., and Schmidell, W., Anaerobic biodegradability of the organic components of municipal solid wastes, *Water Science & Technology*, 25 (7), 285–293 (1992).

480. Purcell, B.E. and Stentiford, E.I., Co-digestion—Enhancing recovery of organic wastes (2000), http://www.orbit-online.net/journal/archiv/01-01/0101_07_text_html, Accessed on May 24, 2001.

481. Albhin, A. and Vinnerås, B., Biosecurity and arable use of manure and biowaste—Treatment alternatives, *Livestock Science*, 112, 232–239 (2007).

482. Braun, R., Huber, P., and Meyrath, J., Ammonia toxicity in liquid piggery manure digestion, *Biotechnology Letters*, 3, 159–164 (1981).

483. Bryant, M.P., Commentary on the Hungate technique for culture of anaerobic bacteria, *The American Journal of Clinical Nutrition*, 25, 1324–1328 (1972).

484. McCarty, P.L. and McKinney, R.E., Salt toxicity in anaerobic digestion, *Journal of the Water Pollution Control Federation*, 33, 399–415 (1961).

485. Schneider, R., Quicker, P., Anzer, T., Prechtl, S., and Faulstich, M., Grundlegende Untersuchungen zur effektiven, kostengunstingen Entfernung von Schwefelwasserstoff aus Biogas, in *Biogasanlagen Anforderungen zur Luftreinhaltung*. Bayerisches Landesamt fur Umweltschutz, Augsburg, Germany (2002).

486. Frear, C., Liao, W., Ewing, T., and Chen, S., Evaluation of co-digestion at a commercial dairy anaerobic digester, *CLEAN–Soil, Air, Water,* 39 (7), 697–704 (July 2011).

487. Steyer, J.P., Bernard, O., Batstone, D.J., and Angelidaki, I., Lessons learnt from 15 years of ICA in anaerobic digesters, *Water Science and Technology: A Journal of the International Association on Water Pollution Research*, 53, 25–33 (2006).

5 Hydrothermal Gasification

5.1 INTRODUCTION

The word "hydrothermal gasification" (HTG) is coined in this chapter to illustrate the methods of gasification, which result in hydrogen or a mixture of gases with heavy concentration of hydrogen. Hydrothermal also implies the use of water (in some form, steam, water, or supercritical water [SCW]) and/or hydrogen in the conversion process. While natural gas, synthesis gas, and biogas largely contain methane or syngas, the processes described in this chapter largely produce gases with predominant concentration of hydrogen. Hydrogen is often produced by carrying out thermal gasification of carbonaceous materials in the presence of steam (water or SCW) and/or hydrogen. Hydrogen can also be produced by dissociation of water. For the sake of brevity, all of these processes are coined here as "hydrothermal gasification." Hydrogen is often considered as an "ultimate fuel." Hence, the commercial success of the processes described in this chapter can be a game changer in the energy and fuel industry. It should be noted that currently hydrogen is predominantly produced by the steam-reforming process, which is described in detail in Chapter 6.

This chapter is divided into two parts. In the first part, the production of hydrogen by gasification of all carbonaceous materials in the presence of steam and/or hydrogen and in the presence of subcritical water and SCW is examined. These methods of hydrogen production are described in Sections 5.3 through 5.5. In the second part, the productions of hydrogen by various water dissociation technologies are evaluated. Water is the most abundant source of hydrogen. The technologies considered for water dissociation are electrolysis; photochemical, photocatalytic, and photobiological dissociation of water; and thermal and thermochemical dissociation of water. These three technologies are evaluated in Sections 5.7 through 5.9, respectively. Finally, some novel methods for water dissociation are also evaluated in Section 5.10. Most of the water dissociation technologies are at the research or development stage and their commercial success can change the landscape of the energy and fuel industry. Often, water dissociation is carried out using waste heat from solar or nuclear energy. This approach has a considerable practical attractions and gaseous fuel (mainly hydrogen) generated using solar energy is often called solar fuels.

5.2 HYDROGEN MARKET AND METHODS FOR HYDROGEN PRODUCTION

It is important to understand hydrogen market for both stationary and mobile usages. This will provide the understanding of its storage and transportation needs. Hydrogen is already widely produced and used, but it is now being considered for use as an energy carrier for stationary power and transportation markets. Although hydrogen is the most abundant element in the universe, where it appears naturally on the earth's crust, it is bound with other elements such as carbon and oxygen instead of being in its molecular "H_2" form. Molecular hydrogen is produced for various uses, and this can be done in various ways, as discussed in this chapter and in Chapters 6, 9, and 10.

As of 2008, approximately 10–11 million metric tons of hydrogen was produced in the United States each year [1]. This is enough to power 20–30 million cars (using 700–1000 gal energy equivalents per car per year) or about 5–8 million homes. Globally, the production figure was around 40.5 million tons (equal to about 44.4 million short tons or about 475 billion cubic feet (bcf)) and was expected to grow 3.5% annually through 2013 [2,3]. This rate of growth has continued to this date.

Major current uses of the commercially produced hydrogen are for oil refining, where hydrogen is used for hydrotreating of crude oil as part of the refining process to improve the hydrogen-to-carbon ratio of the fuel, food production (e.g., hydrogenation), treating metals, and producing ammonia for fertilizer, and other industrial uses.

Because much of the hydrogen produced in the United States is done in conjunction with oil production, much of this hydrogen is produced in three states: California, Louisiana, and Texas. This hydrogen is produced from steam reforming of natural gas, the most prevalent method of hydrogen production. There are other significant production facilities up the Mississippi River valley and in the Northeast. A key point here is that vast amounts of hydrogen are currently produced and used in the United States and around the world. It is a widely used industrial gas, with a well-developed set of codes and standards governing its production, storage, and use as described in Chapters 8 through 10.

A growing use of hydrogen is to support emerging applications based on fuel cell technology along with other ways to use hydrogen for electricity production or energy storage. More than 50 types and sizes of commercial fuel cells are being sold, and the value of fuel cell shipments reached $498 million in 2009. Approximately 9000 stationary fuel cell systems and 6000 other commercial fuel cell units were shipped that year. The 15,000 total represented 40% growth over the previous year. In addition, 9000 small educational fuel cells were shipped [4,5]. Currently, this number is significantly higher. For fuel cell applications, a high level of hydrogen purity is typically more important than for many other industrial applications and thus can often entail higher costs of delivery [4,5].

The current "merchant" and "captive use" hydrogen market is, as noted previously, dominated by uses for oil refining, food production, metals treatment, and fertilizer manufacture. Power production uses relatively little of the hydrogen, perhaps on the order of 10–20 million kg/year in the United States, or about 0.1% of the total. This figure is based on an estimate by the Fuel Cell Energy (FCE) [4,5] that their fuel cell systems produced over 400 million kWh of electricity through 2009, using about 30 million kg of hydrogen over the past several years [4,5]. The FCE systems represent a significant part of the fuel cell market—approximately 1/3–1/2 of the total installed base in the country [2,6,7].

For the purposes of understanding the hydrogen market, it is useful to distinguish between captive hydrogen (where the hydrogen is produced and used on-site, such as at oil refineries) and merchant hydrogen production where the hydrogen is produced for delivery to other locations as an industrial gas. Further, a distinction can be drawn between on-purpose hydrogen, where hydrogen production is the main goal, and by-product hydrogen, where hydrogen is produced as a by-product from another process (e.g., chlor-alkali production).

On-purpose captive hydrogen production at oil refineries accounts for about 25% of total U.S. production (2.7 million tons per year). Production for ammonia represents about 21% of total U.S. production (2.3 million tons per year), and a small amount of captive production is used for methanol production and other uses. On-purpose merchant hydrogen production was about 1.6 million tons in 2006, or about 15% of the U.S. total. Finally, by-product hydrogen production amounted to about 3.8 million tons in 2006, or nearly 36% of total U.S. production. Most of this by-product hydrogen was from catalytic reforming at oil refineries and from chlor-alkali production [1].

Hydrogen has been identified as a potential zero-emission energy carrier for the future, primarily not only for the transport sector but also for energy storage and combined heat and power (CHP) applications. Although microscale hydrogen production systems are being developed for domestic use by some companies, it is likely that a network of hydrogen fueling stations would be required to supply fuel to motorists since hydrogen-fueled vehicles are likely to suffer the same range limitations as electric vehicles.

Small-scale hydrogen production from electrolysis or biomass would be possible at fueling stations, but large plants would otherwise be required for fossil fuel feedstock so that the CO_2 by-product could be captured and stored using carbon capture and storage. In the short term, small-scale steam methane reforming could also be deployed at refueling stations to facilitate the transition to a hydrogen economy, but these would not be compatible with the goal of decarbonization of the environment in the long term [1–8].

Hydrogen is relatively difficult to store and transport in comparison with petroleum fuels. Gaseous hydrogen (GH2) has two principle drawbacks. First, the unusually low volumetric energy density of GH2 means that the gas must be compressed to extremely high pressure to be used as a transport fuel. Second, the tiny molecules have a higher propensity to leak than other gases and require particularly complex storage materials. One method of avoiding these two difficulties is to compress the hydrogen into a liquid (LH2), but this is energetically expensive and difficult to handle because LH2 boils at around −253°C. Other forms of hydrogen, for example, metal hydrides, are currently being researched but are at an early stage of development. Hydrogen cannot be transported in natural gas pipelines when its concentration is greater than 5–15 vol%. For pure hydrogen, dedicated pipeline for hydrogen will be required. These pipelines will be more difficult to build and maintain due to high permeability of hydrogen through pipe walls and the problem of hydrogen-induced cracking (or corrosion) or hydrogen embrittlement, causing pipe failure. Hydrogen is often transported in liquid form but this method is expensive. All of these issues are discussed in detail in Chapter 8.

As mentioned in Chapter 1, hydrogen can be produced by a number of methods. Out of these methods, once considered in this chapter, are as follows:

1. Gasification of carbonaceous materials by steam with or without hydrogen
2. HTG of biomass/waste in subcritical water
3. SCW gasification of all carbonaceous materials
4. Water dissociation by electrolysis
5. Photocatalytic and photobiological dissociation of water
6. Thermal and thermochemical dissociation of water
7. Novel methods for water dissociation

The productions of hydrogen by a variety of reforming processes are described in Chapter 6. The subject of hydrogen production by anaerobic digestion was treated in Chapter 4.

5.3 STEAM AND HYDROGASIFICATION OF CARBONACEOUS MATERIALS

5.3.1 Mechanism of Steam Gasification

The steam gasification reaction is endothermic, that is, requiring heat input for the reaction to proceed in its forward direction. Usually, an excess amount of steam is also needed to promote the reaction:

$$C(s) + H_2O(g) = CO(g) + H_2(g) \quad \Delta H^{\circ}_{298} = 131.3 \text{ kJ/mol} \tag{5.1}$$

However, excess steam used in this reaction hurts the thermal efficiency of the process. Therefore, this reaction is typically combined with other gasification reactions in practical applications. The H_2-to-CO ratio of the product syngas depends on the synthesis chemistry and process engineering. Two reaction mechanisms [9,10] have received most attention for the carbon–steam reactions over a wide range of practical gasification conditions.

Mechanism A [9]

$$C_f + H_2O = C(H_2O)A \tag{5.2}$$

$$C(H_2O)A \rightarrow CO + H_2 \tag{5.3}$$

$$C_f + H_2 = C(H_2)B \tag{5.4}$$

In these equations, C_f denotes the free carbon sites that are not occupied, $C(H_2O)A$ and $C(H_2)$ B denote the chemisorbed species in which H_2O and H_2 are adsorbed onto the carbon site, "=" means the specific mechanistic reaction is reversible, and "→" means the reaction is predominantly irreversible. In Mechanism A, the overall gasification rate is inhibited by hydrogen adsorption on the free sites, thus reducing the availability of the unoccupied active sites for steam adsorption. Therefore, this mechanism may be referred to as "inhibition by hydrogen adsorption."

Mechanism B [10]

$$C_f + H_2O = C(O)A + H_2 \tag{5.5}$$

$$C(O)A \rightarrow CO \tag{5.6}$$

In this mechanism, the gasification rate is affected by competitive reaction of chemisorbed oxygen with hydrogen, thus limiting the conversion of chemisorbed oxygen into carbon monoxide. Therefore, this mechanism may be referred to as "inhibition by oxygen exchange."

Both mechanisms are still capable of producing the rate expression for steam gasification of carbon in the form of [12]

$$r = \frac{k_1 p_{H_2O}}{\left(1 + k_2 p_{H_2} + k_3 p_{H_2O}\right)} \tag{5.7}$$

which was found to correlate with the experimental data quite well. This type of rate expression can be readily derived by taking pseudo-steady-state approximation on the adsorbed species of the mechanism.

It has to be clearly noted here that the mechanistic chemistry discussed in this section is based on the reaction between carbon and gaseous reactants, not on the reaction between coal and gaseous reactants [11]. Even though carbon is the dominant atomic species present in coal, its reactivity is quite different from that of coal or coal hydrocarbons. In general, coal is more reactive than pure carbon, for a number of reasons, including the presence of various reactive organic functional groups and the availability of catalytic activity via naturally occurring mineral ingredients. It may now be easy to understand why anthracite, which has the highest carbon content among all ranks of coal, is most difficult to gasify or liquefy. Alkali metal salts are known to catalyze the steam gasification reaction of carbonaceous materials, including coals. The order of catalytic activity of alkali metals on coal gasification reaction is Cs > Rb > K > Na > Li. In the case of catalytic steam gasification of coal, carbon deposition reaction may affect the catalysts' life by fouling the catalyst active sites. This generally occurs when reaction environment is deficient in steam [11].

5.3.2 FEEDSTOCK EFFECT ON STEAM GASIFICATION

While steam gasification is a process similar to steam reforming, the latter process is often carried out in the presence of a nickel catalyst. Steam gasification can be carried out for any carbonaceous materials. The effects of feedstock on this process are briefly articulated in the following texts.

5.3.2.1 Coal

Corella et al. [13] used the following model for steam gasification of coal at low–medium (600°C–800°C) temperature with simultaneous CO_2 capture in fluidized bed at atmospheric pressure. The study also examines the effect of inorganic species on the gasification process.

The gasification of coal with steam follows the following set of reactions [13]: First, fast pyrolysis of coal follows the reactions

$$Coal(C_xH_yO_z \ ISs) \rightarrow Tar \ 1 + Char \ 1 \rightarrow Tar \ 2(CH_{0.85}O_{0.17})$$

$$+ \ Char \ 2(CH_{0.2}O_{0.13} \ ISs) + H_2 + CO$$

$$+ \ CO_2 + CH_4 + C_2H_4 + \cdots \tag{5.8}$$

Here, ISs are the inorganic species in the coal. The conversion of tar 1 and char 1 is an in-bed thermal reaction. Tar 2 and char 2 further react with steam and carbon dioxide as

$$Tar \ 2(CH_{0.85}O_{0.17}) + H_2O \rightarrow CO + H_2 + \cdots \tag{5.9}$$

$$Char \ 2(CH_{0.2}O_{0.13} \ ISs) + H_2O \rightarrow CO + H_2 + Char \ 3(C_{xx}H_{yy}O_{zz}) + Ashes(ISs) \tag{5.10}$$

$$Char \ 2(CH_{0.2}O_{0.13} \ ISs) + CO_2 \rightarrow CO + H_2 + Char \ 3(C_{xx}H_{yy}O_{zz}) + Ashes(ISs) \tag{5.11}$$

These reactions are not in stoichiometric proportions. Steam reforming of methane and light hydrocarbons that may occur simultaneously can be expressed as

$$CH_4 + H_2O \rightarrow CO + 3H_2 \tag{5.12}$$

$$C_2H_4 + H_2O \rightarrow 2CO + 3H_2 \tag{5.13}$$

along with shift reaction

$$CO + H_2O \rightleftarrows H_2 + CO_2 \tag{5.14}$$

All inorganic species with possible catalytic effects are designated as ISs. For example, iron-based species (Fe_2O_3, Fe_3O_4, etc.) affect the rate of overall steam gasification reaction. Some of the reactions, in particular, reforming and water gas shift reactions (henceforth denoted as WGS reaction), are catalyzed by nickel. Inorganic species such as indium can also have a catalytic effect. Finally, alkaline and alkaline-earth metallic species (sometimes called AAEM species) such as K, Ca, Na, Cs, and Mg significantly influence the overall gasification process. AAEM species either can be parts of char generated or can be additives in the gasification process. AAEM species affect (1) the reactivity of coal and char; (2) product distribution of H_2, CO_2, and CH_4; and (3) tar content in the product gas [321,322].

One of the AAEM species, CaO, is a good absorbent for carbon dioxide to form calcium carbonate. During the gasification reaction, coke can be generated on CaO that can be removed by steam or carbon dioxide reactions with coke producing hydrogen and carbon monoxide. Thus, steam gasification of coal is often carried out with an addition of CaO so that the gasifier simultaneously removes carbon dioxide during the gasification process. In general, carbon conversion and char gasification in a fluidized-bed reactor increases with temperature between 600°C and 900°C. While tar yield (or tar content) and CO_2 capture decrease with an increase in the temperature in the same range, high contents of alkalis during gasification can also cause the problems of agglomeration, sintering, and melting, all of which are harmful to the smooth operation of the gasifier. Besides CaO, often calcined dolomites (CaO–MgO) and magnesium-based minerals silicates such as serpentine ($Mg_3Si_2O_5(OH)_4$), olivine ($Mg_{0.9}Fe_{0.1}SiO_4$), and calcine limestones or calcites have also been tested [321,322].

The study showed that for a clean and efficient steam gasification of coal in a fluidized bed at low/medium temperatures, at atmospheric pressure, and with simultaneous capture of CO_2, the CaO/coal ratio is (1) a key parameter to obtain optimal product distribution, (2) a free parameter to be decided by the process designer, and (3) required to have relatively high values, clearly higher than 2 and perhaps as high as 80 or more. The type of coal or the types of ISs in the coal have some influence in the reaction network existing in the gasifier, but its influence is less than the effect of the temperature. The temperature controls more the product distribution and the usefulness of CaO than AAEMs, ISs, and nature of CaO [321,322].

Recently, Sharma [14,15] outlined a stepwise scheme to improve steam gasification reactivity of coal. In this scheme coal is first refined using coal-derived solvents such as anthracene oil and paraffin oil. The refined coal has a higher amount of inorganic materials that can act as catalyst for the steam gasification to produce chemicals and char. The particle size of coal has no effect on the gasification reactivity and catalytic effects of minerals follow the order Na > K > Ca > Ni [14,15]. The char is further subjected to steam gasification to produce syngas that can be further refined using steam-reforming reaction. According to Sharma [14,15], the following simplified set of reactions occur during coal/char gasification:

$$C_{90-120-240}H_{6-9-20}O_xS_yN_z + O_2 + H_2O \rightarrow C_nH_m + \text{Other products (CO, CO}_2\text{, etc.)} \tag{5.15}$$

$$C_nH_m \rightarrow nC + mH \tag{5.16}$$

$$C + O_2 \rightarrow CO_2 \quad \Delta H = -40.59 \text{ kJ/mol} \tag{5.17}$$

$$C + O_2 \rightarrow 2CO \quad \Delta H = +159.7 \text{ kJ/mol} \tag{5.18}$$

$$C + H_2O \rightarrow CO + H_2 \quad \Delta H = +118.9 \text{ kJ/mol} \tag{5.19}$$

$$CO + H_2O \rightarrow CO_2 + H_2 \quad \Delta H = -40.9 \text{ kJ/mol} \tag{5.20}$$

$$C + 2H_2 \rightarrow CH_4 \quad \Delta H = -87.4 \text{ kJ/mol} \tag{5.21}$$

Shift reaction takes place only at high concentration of steam. The last reaction is important under pressure. Sharma [14,15] concluded that the main factors for the steam reactivity of gasification are (1) refining of coal that increases the surface area of the coal; (2) the volatile matters in residual coal and char, that is, the more the volatile matter, the more the reactivity; and (3) the concentration of mineral matter in coal and char. Sharma [14,15] also studied steam gasification reactions that can be useful for the reactor design. Exxon [16] examined steam gasification of coal liquefaction residue. Exxon technology utilized steam to sequentially gasify and hydrogenate both raw coal and carbon residue left in coal gasification. The study also used calcium hydroxide or similar alkaline-earth metal compound as possible catalysts for the process.

5.3.2.2 Biomass

In recent years, the steam gasification of biomass is gaining more importance because it produces gaseous fuel with high hydrogen content that can either produce electricity with high efficiency or provide a feedstock for various chemical and fuel productions. Steam gasification also (1) provides gases with high heating value, (2) reduces the diluting effect of nitrogen from air, and (3) eliminates the need for expensive oxygen separation plant. Catalytic gasification in a fluidized bed allows (1) lower temperature, (2) variety of particle size, and (3) variety of feedstock [321].

A serious issue in the broad implementation of steam gasification is the generations of unwanted materials like tars, particles, nitrogen compounds, and alkali metals. Tar is a mixture of one- to five-ring aromatic hydrocarbons that can plug the reactor. Its removal is essential and that can be done either in the gasifier or by hot gas cleaning after the gasification process. Within the gasifier, tar can be reduced by choosing the appropriate operating parameters, inserting additive catalyst, or changing the gasifier design so it cannot plug the reactor. The removal of tar thermally requires the operation of the gasifier at temperature above 1000°C. The prevention of ash agglomeration however requires the gasifier temperature to be below 700°C. Ash frequently contains various oxides of Ca, K, Mg, P, Si, Na, and S that can agglomerate, deposit on the surface, and contribute to erosion and corrosion of the gasifier. Alkali metals can also react with silica to form silicates and with sulfur to form alkali sulfates, both of which are sticky and can cause sintering and defluidization [17–35,321]. Reforming tar using a Ni catalyst is an effective method for removing tar. The coke deposition in a reforming reaction can be reduced using excess steam [17–35]. Catalytic steam gasification of biomass is a complete network of heterogeneous reactions [17–35,321]. The reactions can be describes as follows:

Primary reactions:

$$C_xH_yO_z + H_2O \rightarrow C(x-1)CO + \left(\frac{y}{2}+1\right)H_2 \tag{5.22}$$

$$(C_xH_yO_z + H_2) \rightarrow (Heat)H_2 + CO + CO_2 + CH_4 + C_nH_{2m} + C(s) + Tars \tag{5.23}$$

Secondary reactions:

$$C_nH_{2m} + nH_2O \rightleftarrows nCO + (n+m)H_2 \tag{5.24}$$

With additional gas-phase reactions:

$$C + H_2O \rightleftarrows H_2 + CO \tag{5.25}$$

$$C + CO_2 \rightleftarrows 2CO \tag{5.26}$$

$$C + 2H_2 \rightleftarrows CH_4 \tag{5.27}$$

$$CO + H_2O \rightleftarrows H_2 + CO_2 \tag{5.28}$$

$$CH_4 + H_2O \rightleftarrows CO + 3H_2 \tag{5.29}$$

$$CH_4 + CO_2 \rightleftarrows 2CO + 2H_2 \tag{5.30}$$

Hydrogen and syngas production from biomass through thermochemical conversion, gasification in particular, have been widely studied. According to Chang et al. [36], the temperature is the most influent factor for gasification, and increasing the temperature resulted in an increase in gas yield and more hydrogen production. Particle size is another factor that affects the gasification process; therefore, carbon conversion efficiency and hydrogen yield increase with decreasing particle size. The optimization of gasification temperature is around 830°C and steam-to-biomass ratio of 0.6–0.7 (w/w). Xie et al. [37] studied syngas production by two-stage method of biomass catalytic pyrolysis and gasification. The results illustrated that higher temperature was needed in the gasification process (850°C) than in the pyrolysis process (750°C) to maximize syngas yield, and the maximum syngas yield could achieve up to 3.29 Nm^3/kg biomass (dry wt.), much higher than the previous studies. Gao et al. [38] investigated hydrogen production from biomass gasification with porous ceramic reforming. The results indicated that the hydrogen yield increased from 33.17 to 44.26 (g H_2)/(kg biomass) with the reactor temperature increase. The H_2 concentration of production gas in oxygen gasification (oxygen as gasifying agent) was much higher than that in air gasification (air as gasifying agent). The hydrogen yields in air and oxygen gasification varied in the range of 25.05–29.58 and 25.68–51.29 (g H_2)/(kg biomass), respectively [321].

Nipattummakul et al. [39] investigated syngas production from steam gasification of oil palm trunk waste, and it was observed that the high initial syngas flow rate is mainly attributed to the pyrolysis of volatile matter from the oil palm sample. Almost 50% of the syngas is produced during the first 5 min. The results showed that there is over 60% increase in hydrogen production with steam gasification as compared to that with pyrolysis. The increase in steam flow rate reduced the time duration of gasification and promoted steam-reforming reactions to result in increased hydrogen yield. Increase in steam flow rate provided negligible effect on the apparent thermal efficiency. Lv et al. [40] studied catalytic gasification of pine sawdust and the maximum gas yield reached 2.41 Nm^3/kg biomass at 850°C.

In order to operate gasification in the temperature range of 600°C–700°C, gasification is generally operated with reforming in the same reactor or in two stages. Demirbas [32] compared the hydrogen productions from conventional pyrolysis, steam gasification, and supercritical extraction. A comparison of hydrogen yield as a function of temperature for these three processes is illustrated in Figure 5.1. While the results are described in this figure for beech wood, similar results were obtained for corncob, olive waste, and wheat straw. The results show that an increase in steam-to-biomass ratio increases the hydrogen production. At low-temperature supercritical extraction is

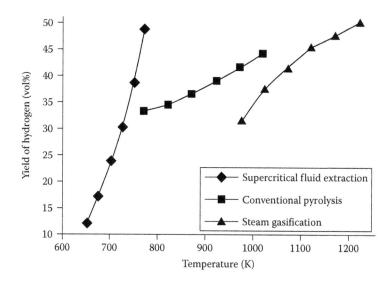

FIGURE 5.1 Comparison of various methods for hydrogen production. (Adapted from Demirbas, M., *Energy Resources*, Part A, 28, 245–252, 2006.)

the best process, while steam gasification produces the best results at higher temperatures. Inayat et al. [29] presented a model for steam gasification accompanied by CO_2 adsorption by CaO in a fluidized-bed reactor. The model indicated that high steam-to-biomass ratio gave higher hydrogen production. While an increase in temperature gave increased hydrogen production, at very high temperature, reverse WGS reaction changes the trend. The model showed that at temperature of 950 K and steam-to-biomass ratio of 3, hydrogen production was maximum [321].

Demirbas [30,33,41–45] examined different types of biomass such as hazelnut shell, tea waste, and spruce wood and once again showed that at higher temperatures, steam gasification gave higher hydrogen yield than conventional pyrolysis. Higher steam-to-biomass ratio also gave higher hydrogen production. Similar results for mosses and algae were reported by Demirbas [30]. Specific samples examined were *Polytrichum commune, Thuidium tamariscinum, Cladophora fracta, Chlorella protothecoides,* beech wood, and spruce wood. An extensive parametric modeling study for biomass steam gasification was reported by Schuster et al. [47].

Li et al. [48] examined catalytic steam gasification of municipal solid waste (MSW) in a combined (two stage) fixed-bed reactor. The catalyst used was a trimetallic catalyst (nano-Ni–La–Fe/γ-Al_2O_3) and MSW-contained kitchen garbage, wood and leaves, paper, textile, and plastics. The syngas composition was measured as functions of temperature, steam-to-MSW ratio, and catalyst-to-MSW ratio at an atmospheric pressure. The results showed more than 99% tar removal at 800°C with significant productions of hydrogen. The catalyst significantly improved hydrogen production. Higher temperature gave higher gas and hydrogen yields. While higher steam-to-MSW ratio gave better results, an excessive steam-to-MSW ratio lowered the gasification temperature and degraded the product quality. The optimum value of *S/M* was found to be 1.33 under the operating conditions. The optimum value of catalyst-to-MSW ratio was found to be about 0.5. A two-stage (pyrolysis followed by catalytic steam gasification) process for olive waste was studied by Encinar et al. [49]. The catalyst used was dolomite. The two-stage process produced gas, liquid, and solids whose yields were strongly dependent on temperature and amount of catalyst. Higher temperature and catalyst amount gave higher amount of gases and the presence of steam gave higher amount of hydrogen and carbon dioxide [321].

Hofbauer et al. [50–54] used a fast internally circulating fluidized bed (at pilot scale) to gasify biomass with steam. Using a natural catalyst as bed material, and at temperature of 750°C, tar content was significantly reduced and gas with high hydrogen content was obtained. The internal circulating bed allowed the flexibility in varying residence time needed to lower tar concentration. Herguido et al. [55] studied gasification of pine sawdust, pine wood chips, cereal straw, and thistles from energy crops in the presence of steam in a fluidized-bed reactor. The product gases were hydrogen, CO, and CO_2 and their amount and composition varied with the nature of biomass in the temperature range of 650°C–780°C.

A novel two-stage fluidized-bed approach was used by Pfeifer et al. [56] in which the first stage carried out steam gasification of solid biomass to generate heat and power and provide raw materials for downstream chemical synthesis. The residual biochar from the first stage is combusted in the second stage and the hot bed materials from the second stage provide heat needed for the first stage. This concept was also analyzed by Gopalakrishnan [57] and Matsuoka et al. [58]. The latter study showed that separating combustion zone from gasification zone resulted in high-efficiency gasification. They used gamma-alumina as particles for bed materials and tested two different types of sawdusts. Since the residence time of the bed material can be controlled in the gasifier of the circulating dual bubbling fluidized-bed system, tar captured by the porous alumina particles (coke) as well as char was effectively gasified. Since coke was preferentially gasified compared with char, higher carbon conversion and hydrogen yield can be achieved in this type of dual-bed system than in the conventional circulating fluidized bed [321].

In the studies described earlier, the process generated gases with about 40 vol% hydrogen. Furthermore, an addition of carbonate adsorbed carbon dioxide and moved carbon dioxide from gasification to combustion zone (they called it adsorption-enhanced reforming [AER]). This concept has been successfully adapted by an 8 MW CHP plant in Gussing, Austria, since 2002. A new pilot plant of 100 kW has also been built to see the effect of AER concept in improving hydrogen concentration to 75 vol% in the product gases. The possibilities for getting high hydrogen concentration along with operating the reactors at low temperature and thereby improving energy conversion efficiency make this concept very attractive. Salaices et al. [59,60] presented a very workable kinetic model for catalytic steam gasification of cellulose surrogate with Ni/alpha-alumina catalyst in a circulating fluidized bed with a riser. The model successfully predicted the productions of various gases such as hydrogen, carbon dioxide, carbon monoxide, water, and methane.

While a significant number of studies have investigated steam gasification in the presence of air (or oxygen) to improve carbon conversion and energy efficiency of the steam gasification process, Barrio et al. [61] examined the effect of hydrogen on steam gasification process. They found that hydrogen inhibits the steam gasification reaction. They also concluded that the nature of char coming from beech wood or birchwood did not significantly affect the final results.

While a major effort on steam gasification is focused at low temperature using a catalyst, Donaj et al. [62] and Gupta and Cichonski [63] examined the effectiveness of high-temperature (HT) steam gasification. Donaj et al. [62] examined the steam gasification of straw pellets at temperatures between 750°C and 950°C. The effect of steam-to-feed ratio on carbon conversion was marginal below 850°C, and in general, higher steam-to-feed ratio gave higher hydrogen production. Gupta and Cichonski [63] examined the steam gasification of paper, cardboard, and wood pellets in the temperature range of 700°C–1100°C. Once again in all cases, hydrogen production increased with temperature and steam-to-biomass ratio.

Lucas et al. [64] examined the HT air and steam gasification of densified biofuels. The experiments were carried out in a fixed-bed updraft gasifier. The results showed that an increase in feed temperature reduced productions of tar, soot, and char residues and also increased the heating value of the dry fuel gas produced. Butterman and Castaldi [65] showed that an increase in CO_2 feed rate enhanced the char conversion and the production of CO. The experiments produced a low concentration of methane and high concentration of hydrogen above 500°C for the herbaceous and nonwood samples and above 650°C for the wood biomass. The experiments also showed similarities between the gaseous products from biomass and MSW. The mass decomposition rates and gas evolution profiles showed two distinct regions with transition around 400°C. Large pyrolysis char volumes correlated well with higher lignin compositions. The biomass fuels examined included woods, grasses, and other lignocellulosic samples. These included oak, sugar maple, poplar, spruce, white pine, Douglas fir, alfalfa, cordgrass, beach grass, maple bark, pine needles, blue noble fir needles, pecan shells, almond shells, walnut shells, wheat straw, and green olive pit. The complete mass loss occurred around 900°C–1000°C [321].

Aznar et al. [66] examined biomass gasification with steam–O_2 mixtures followed by a catalytic steam reformer and a CO-shift system. The use of two CO-shift converters downstream from a fluidized-bed biomass gasifier, using steam–O_2 mixtures and a catalytic steam reformer, generated exit gas with 73% hydrogen (by volume) on dry basis and only 2.6% CO. The remaining gas contained CO_2, O_2, and CH_4. The results showed that H_2O/CO ratio in the gas phase at the inlet of the HT shift reactor is a very important parameter in the system. CO conversion up to 90% was obtained, but in order to get this conversion, the steam/CO ratio greater than 2 at the inlet of HT shift reactor was needed. Due to low tar content in the inlet gas to HT shift reactor, a significant less deactivation of the catalyst in the shift reactor occurred.

5.3.2.3 Mixed Feedstock

In recent years, significant efforts have been made to gasify the mixtures of coal and biomass, coal and waste, or biomass and waste in the presence of steam. Seo et al. [67–69] used the successful two-stage fluidized-bed model described earlier for coal/biomass blend in the temperature range of 750°C–900°C and steam/fuel ratio of 0.5–0.8. Biomass-to-coal ratio was varied from zero to one. The study showed that product gas yield, carbon conversion, and cold gas efficiency increased with increasing temperature and steam/fuel ratio. These parameters were higher for biomass gasification than those for coal gasification. Synergistic effect on gas yields was observed with a larger surface area, pore volume, and presence of micropores at biomass/total feed ratio of 0.5. The calorific values of the product gas at 800°C were 9.89–11.15 MJ/m^3 with the coal, 12.10–13.19 MJ/m^3 with the biomass, and 13.77–14.39 MJ/m^3 with the coal–biomass blend. The maximum cold efficiency was 0.45 with biomass/total feed ratio of 0.5. Sun et al. [70] examined various kinetic models for the gasification of biomass blended with waste filter carbon at temperatures around 850°C. Once again, high temperature and high flow of steam increased the gasification rate; the gasification rate of filter carbon was lower than that of wood chip. The data were taken for the steam pressure of 0.5 atm. A modified volume-reaction kinetic model best fits all the data [321,322].

Kumabe et al. [71] showed that at 900°C, the mixture of woody biomass and coal in the presence of steam and air gave favorable results. The results of this study were similar to the ones described earlier; increase in biomass gave more gases and more hydrogen was produced at higher steam-to-feedstock ratio. Higher amount of biomass also gave lower amount of char and tar. The study produced gas with composition that was favorable to methanol, hydrocarbon fuels, and dimethyl ether (DME) under high biomass feed conditions. The cogasification was carried out in a down-draft fixed-bed reactor and it provided cold gas efficiency ranging from 65% to 85%. Demirbas [44,45], on the other hand, studied the effects of co-firing MSW with pulverized coal in a bubbling fluidized-bed combustor. The results showed that the mixture produced less NO_x and SO_x in direct proportions to the MSW concentration in the mixture. Similarly, mixture produced less CO_2 than coal alone. The mixture burning can, however, bring the problems with chlorine impurities in MSW that can lead to corrosion problems and inorganic impurities such as Si, Al, Ti, Fe, Ca, Mg, Na, K, S, and P that can significantly change the composition of ash and its melting and agglomeration characteristics. This change in ash characteristics may limit the market for its downstream use [321,322].

Numerous other investigators have also examined the steam gasification (some in the presence of air or oxygen) of a variety of coal–biomass mixtures. Chmielniak and Sciazko [72] produced syngas from steam gasification of coal and biomass mixture that was subsequently transformed to methanol, DME, ethylene, and gasoline. Yamada et al. [73] also produced useful syngas from a mixture of coal and biomass briquettes. Kumabe et al. [71] examined steam gasification (with air) of a mixture of Japanese cedar and mulia coal and obtained useful syngas for the production of DME. Their gasification results were very similar to the ones described earlier and they obtained cold gas efficiency of 65%–85% during the gasification process. Pan et al. [74] examined steam gasification of residual biomass and poor coal blends. Pine chips from Spain were used as biomass and two types of coal—black coal (low grade) from Escatron, Spain, and Sabero coal from Sabero, Spain—represented poor grade coals. Once again, reasonable quality of syngas was produced with overall thermal efficiency of about 50%. Satrio et al. [75] examined steam gasification of coal and biomass mixture with the specially designed catalyst pellets with an outside shell, consisting of nickel on alumina, and a core, consisting of calcium and magnesium oxides, which can adsorb carbon dioxide. This catalyst design gave higher production of hydrogen. Finally, Ji et al. [76] studied steam gasification of a mixture of low-rank fuel mixture of biomass, coal, and sludge in a fluidized-bed reactor at 900°C temperature. Just like other studies, higher temperature gave more gas and hydrogen but not high heating value of gas. The calorific value of syngas produced from sludge mixture, sludge, wood chips, and lignite were 13, 10, 6.9, and 5.7 MJ/m^3, respectively [321,322].

An excellent review of problems associated with co-firing of coal and biomass fuel blends was given by Sami et al. [77]. This review critically assesses the effectiveness of this mixed feedstock for combustion and pyrolysis, two extreme cases of gasification. While they specifically do not discuss steam gasification and reforming, significant parts of their analysis are applicable to the process of steam gasification and reforming. Indrawati et al. [78] examined partial replacement of fossil energy by renewable sources like rice husk, palm kernel shell, sawdust, and municipal waste in the cement production. While this study also does not specifically address steam gasification and reforming of mixed feedstock, the study points to another application of the mixed feedstock [321,322].

5.3.2.4 Tar

As indicated earlier, formation of tar is a major issue with steam gasification. Tar is a complex mixture of condensable hydrocarbons and it can contain one- to five-ring aromatic compounds with other oxygen-containing hydrocarbon species [79,80]. Generally, tar is defined as C_6^+ aromatic organics produced under gasification conditions. Tar is a problem during gasification because (1) it can deposit on the outlet pipes of the gasifier and also deposit on the particulate filters, (2) it can clog fuel lines and injectors in the internal combustion engine, and (3) it reduces the gasifier's efficiency to produce additional useful fuel products such as hydrogen, carbon monoxide, carbon dioxide, and methane [321].

Baker et al. [79] illustrated the conceptual relationships between tar generation and disappearance and the temperature during thermal steam gasification of carbonaceous materials. They divided tar components in four different categories. The results showed a range based on the nature of the feedstock. At low temperatures (400°C), a significant amount of tar is produced, and for temperatures higher than around 1000°C, very little tar is produced. Elliott [80,81] showed that as temperature increases, the nature of tar undergoes transformation as shown in the following:

$$\text{Mixed oxygenates (400°C) (primary)} \rightarrow \text{Phenolic ethers (500°C) (secondary)}$$

$$\rightarrow \text{Alkyl phenolics (600°C) (tertiary-alkyl)}$$

$$\rightarrow \text{Heterocyclic ethers (700°C) (tertiary-PNA)}$$

$$\rightarrow \text{PAH* (800°C)} \rightarrow \text{Larger PAH (900°C)} \qquad (5.31)$$

Here, PAH* is high-molecular-weight polynuclear aromatic hydrocarbons. Along with temperature, tar concentration depends on the reaction time, the amount of oxygen, and the presence of a suitable catalyst during steam gasification. Higher oxygen concentration generally reduces tar concentration through the processes of cracking and oxidation, among others. The conventional steam gasification operated at 700°C–800°C produces tar with naphthalenes, acenaphthylenes, fluorenes, phenanthrenes, benzaldehydes, phenols, naphthofurans, and benzanthracenes, while HT steam gasification operating between 900°C and 1000°C produces tar that contain naphthalenes, acenaphthylenes, phenanthrenes, fluranthenes, pyrenes, acephenanthrylenes, benzanthracenes, benzopyrenes, 226 molecular weight (MW) PAHs, and 276 MW PAHs. Milne et al. [82] and Evans and Milne [83–86] further characterized tar in terms of primary, secondary, and tertiary products based on molecular beam mass spectroscopy. Some of the details of the constituents of primary, secondary, and tertiary products and their behavior with temperature are described by Evans and Milne [83–86]. Tar analysis in syngas derived from pelletized biomass in a commercial stratified downdraft gasifier was published by Gopal et al. [87]. A detailed review of catalysts needed for tar elimination in a biomass gasifier is given by El-Rub et al. [88].

5.3.2.5 Black Liquor

Huang and Ramaswamy [89] examined steam gasification of black liquor coming out of the paper and pulp industry at temperatures as high as 1500°C. Their results were in agreement with other reports. The carbon conversion was nearly complete at temperatures higher than about 750°C. Hydrogen concentration first increased with temperature but showed a maximum at high temperatures because of the dominance of reverse WGS reaction. Higher steam gave higher hydrogen concentration in the product gas. Operating with a $0.3 < SBR < 0.6$ in combination with high pressure of 30 atm, high temperature of 1000°C appears most beneficial for obtaining smelt with no C(s) and maximizing Na and S capture in the melt. Here, SBR is steam-to-dry black liquor ratio.

Black liquor gasification can be used to substitute the existing combustion process for potential higher energy efficiency, lower greenhouse gas emissions, and more safety. The steam gasification of black liquor technology can help the current paper and pulp mills technology to be extended into future biorefineries. In general, the equilibrium model examined by Huang and Ramaswamy [89] indicates that the hydrogen concentration in the product increased with a decrease in pressure and an increase in steam-to-black liquor ratio, and it showed a maximum with an increase in temperature. Demirbas et al. [41] also illustrated the conversion data for a black liquor via steam gasification with and without catalysts. Li and van Heiningen [90] studied the kinetics of steam gasification of black liquor char and Whitty [91] examined steam gasification of black liquor char under pressure.

5.3.2.6 Lignin

Bakhshi et al. [92] presented the results for steam gasification of lignin, biomass chars, and Westvaco Kraft lignin to hydrogen and high- and medium-Btu gas. Three lignins, Kraft-1, Kraft-2, and Alcell, were gasified at 600°C–800°C in a fixed-bed reactor with steam. Hydrogen contents ranged from 30 to 50 vol% in these experiments [92]. Numerous other investigators also examined hydrogen production by steam gasification of lignins [93–95]. Osada et al. [112,113] examined lignin gasification in supercritical water while Elliott et al. [114] examined organic destruction in high pressure aqueous environment. These studies indicated successful destruction of lignin in supercritical water.

5.3.3 Catalysts for Steam Gasification

Catalysts can be added to steam gasification process in two forms: (1) as active bed additives or (2) as separate heterogeneous catalysts that are used in the steam-reforming reactions. The active additives are used to (1) reduce the amount of tar formation; (2) promote several other chemical reactions to change production rate, composition, and heating value of the gas; (3) promote char gasification; (4) prevent active agglomeration of the feedstock, char, and tar that can lead to reactor choking; and (5) remove carbon dioxide through active adsorption process. The steam-reforming catalysts also reform tar and produce gas of high quality.

5.3.3.1 Dolomite, Olivine, and Alkali Metal–Based Catalysts

These are generally cheap and disposable catalysts. Dolomite, a magnesium ore with general formula $MgCO_3 \cdot CaCO_3$, is considered to be a good catalyst for biomass gasification. Dolomite is also a good adsorbent for carbon dioxide and capable of removing tar very efficiently. It is, however, very fragile substance and may quickly attrite in highly turbulent conditions within a fluidized bed. CaO additive was studied by Dalai et al. [17] and they showed that the use of this additive reduced the gasification temperature by about 150°C to get the same level of gas production; both the carbon conversion and hydrogen production increased with impregnation of CaO in cellulose, cedar, and aspen. The production rates of gas and hydrogen also depended on the nature of feedstock; cedar and aspen performed better than cellulose [321].

Hu et al. [18] tested calcined olivine and dolomite in a fixed-bed reactor and found higher activities of calcined catalysts compared to those of natural catalysts. Devi et al. [96] observed that in the presence of olivine, tar conversion increased with an increase in temperature from 800°C to 900°C, and at 900°C, all water-soluble heterocyclic compounds got converted. With 17 wt% olivine in sand at 900°C, the conversion of heavy polyaromatics increased from 48% to 71%. Calcined dolomite, however, increased the conversion up to 90%. Aznar et al. [20] showed that dolomite was very effective in removing tar coming from a blend of plastic waste with pine wood sawdust and coal in the temperature range of 750°C–880°C. The use of CaO as an absorption agent for carbon dioxide thereby improving the hydrogen concentration in the product gas at about 700°C was demonstrated by Xu et al. [23].

Monovalent alkali metals such as Li, Na, K, Rb, Cs, and Fr were also found to be catalytically active in steam gasification. Both K and Na are part of biomass and they accumulate in the ash, which in turn can act as a catalyst. This solves the problem of ash handling and ash reduces the tar content in the gas phase. The ash catalytic activity is, however, can be lost due to particle agglomeration. Sutton et al. [19] pointed out that direct addition of alkali metals can require (1) expensive recovery of catalyst, (2) increased char content after gasification, and (3) ash disposal problems. Lee et al. [24,98] found that the addition of Na_2CO_3 enhances the catalytic gasification of rice straw over a nickel catalyst and the additive increases the gas formation. They also found that the gas production rate is affected by the nature of the additive and follows the order Na > K > Cs > Li. The use of activated alumina as a secondary catalyst was found to be effective by Simell et al. [99–101]; however, this catalyst deactivated faster due to coking compare to dolomite. Sami et al. [77] showed that both zirconia and alumina promoted toluene and ammonia conversions at lower temperatures indicating enhanced oxidation activity of zirconia with alumina. Furthermore, H_2S had little effect on the activity of alumina-doped zirconia.

5.3.3.2 Nickel-Based Catalysts

As mentioned before, the gasification and reforming in the presence of steam are overlapping reactions. Tar and lower hydrocarbons produced by the gasification can be simultaneously reformed in the presence of a suitable catalyst. Rostrup-Nielsen et al. [102] presented a very good review of applicability of transition metals (group VIII) and noble metal catalysts to steam gasification/reforming process. While a number of noble metal catalysts such as Ru and Rh have superior performance for steam reforming, the cost and easy availability of these catalysts compared to that of nickel made the latter choice more practical. The literature has convincingly demonstrated the usefulness of nickel catalysts for biomass gasification [102–105].

Olivares et al. [25] showed that nickel reforming catalysts display 8–10 times more reactivity than calcined dolomite. Nickel catalysts can be, however, deactivated by the poisons such as sulfur, chlorine, and alkali metals. They can also be deactivated by the formation of coke. The coke deposition can be reduced by increasing steam/biomass ratio; however, this increases the energy cost and changes the gas-phase composition of the product. In general, Ni-gamma-alumina catalyst gave higher conversion and lower deactivation compared to Ni-alpha-alumina catalysts. The MgO/CaO addition to alumina also gives the catalyst more stability. Lanthanum-based pervoskite support also found to be very effective. The topics of coking, catalyst deactivation, and effective support for the nickel are discussed in the following sections. Suffice to say that nickel-based catalysts have gained significant support for steam gasification and reforming.

5.3.4 HYDROGASIFICATION WITH AND WITHOUT STEAM

Direct addition of hydrogen to coal under high pressure forms methane. This reaction is called "hydrogasification" and may be written as

$$Coal + H_2 = CH_4 + Carbonaceous\ matter \tag{5.32}$$

or

$$C(s) + 2H_2(g) = CH_4(g) \quad \Delta H^{\circ}_{298} = -74.8\ kJ/mol \tag{5.33}$$

This reaction is exothermic and is thermodynamically favored at low temperatures ($T < 670°C$), unlike both steam and CO_2 gasification reactions. However, at low temperatures, the reaction rate is inevitably too slow. Therefore, high temperature is always required for kinetic reasons, which in turn requires high pressure of hydrogen, which is also preferred from equilibrium considerations. This reaction can be catalyzed by K_2CO_3, nickel, iron chlorides, iron sulfates, etc. However, the use of catalyst in coal gasification suffers from serious economic constraints because of the low raw material value, as well as difficulty in recovering and reusing the catalyst.

Hydrogasification is gasification in a hydrogen-rich environment, often used for the production of synthetic natural gas (SNG) from coal or other gasifier feedstocks. Hydrogasification for SNG production from coal and biomass has been used since the 1930s. Hydrogasification tends to have low carbon conversions, low product yields, and slower reaction rates without the use of catalysts [321].

Steam hydrogasification uses both steam and hydrogen to affect the reaction. For example, steam hydrogasification of wood substantially increases the production rate of methane (the key component of SNG) over simple hydrogasification. This process requires more energy to bring the reactor to gasification temperatures because of the high latent heat of water, but if the steam produced can be used later (say, for steam methane reformation), this may not be a disadvantage. Steam hydrogasification also requires a supply of hydrogen, which can be supplied by various processes, such as reformation of natural gas or oxygen–steam gasification of char. The addition of steam into the gasifier promotes additional hydrogen formation. This hydrogen can be separated and reused, but the advantages of this may be offset by the cost of gas separation.

Hydrogasification does not require an oxygen plant, which can be a substantial cost to a gasification facility. It also minimizes exothermic reactions (better thermal balance of reactions) so it can be more thermally efficient. The addition of steam to hydrogasification significantly increases reaction rates, which lowers residence times allowing for, among other things, smaller reactors. Since the feed will be gasified with water (steam), it does not need to be dried beforehand and could potentially be fed as a slurry.

The Bourns College of Engineering–Center for Environmental Research and Technology (CE-CERT) lab in the UC–Riverside has worked to further develop the steam hydrogasification process. Their method consists of four highly coupled steps: steam hydrogasification, hot gas cleanup, steam reforming, and a fuel processing step to produce clean liquid fuels and other useful synthesized products [321].

This project carried out decomposition of various biomass feedstocks and their conversion to gaseous fuels such as hydrogen. The steam gasification process resulted in higher levels of H_2 and CO for various CO_2 input ratios. With increasing rates of CO_2 introduced into the feed stream, enhanced char conversion and increased CO levels were observed. While CH_4 evolution was present throughout the gasification process at consistently low concentrations, H_2 evolution was at significantly higher levels, though it was detected only at elevated gasification temperatures: above 500°C for the herbaceous and nonwood samples and above 650°C for the wood biomass fuels studied.

The thermal treatment of biomass fuels involves pyrolysis and gasification with combustion occurring at the higher temperatures. In the gasification environment, when combustion processes are occurring, gaseous components evolve from the fuel and react with oxygen released either from the biomass structure itself or from the injected steam and CO_2. These HT reactions are responsible for the enhanced burnout of the carbon (charcoal) structure that is produced during the low-temperature pyrolytic breakdown of the biomass. Since the lignocellulosic biomass component typically found in U.S. MSW is greater than 50%, the techniques to enhance the thermal treatment of biomass feedstocks can also aid in the processing of MSW.

The studies at CERT included oak, sugar maple, poplar, spruce, white pine, Douglas fir, alfalfa, cordgrass, beach grass, maple bark, pine needles, blue noble fir needles, pecan shells, almond shells, walnut shells, wheat straw, and green olive pit. The TGA mass decay curves showed similar behavior for the woods, grasses, and agricultural residues, where most of the mass loss occurred before 500°C. Most feedstocks exhibited two constant mass steps though several exhibited a third with completed mass loss by 900°C–1000°C. Two distinct mass decay regimes were found to correlate well with two distinct gas evolution regimes exhibited in the curves for CO, H_2, and CH_4. Most of the mass loss occurred during pyrolysis, with the remaining degradation to ash or char occurring in the HT gasification regime [321].

The above description indicates that hydrogasification is suitable for all gasification feedstocks, from coal to renewable sources like wood, agriculture residues, green wastes, MSWs, food and animal waste, and sewage sludge. Some studies have shown that coal can be mixed with dead wood, agricultural residues, and animal waste without significantly affecting the process efficiency [336].

Thus, hydrogasification of biomass can be carried out by either hydrogen or steam or both. Hydrogen, generated from renewable sources, is likely to play a major role in the future energy supply. The storage and transport of hydrogen can take place in its free form (H_2), or chemically bound, for example, as methane. However, the storage and transport of hydrogen in its free form are more complex and, probably, would require more energy than the storage and transport of hydrogen in chemical form. An additional important advantage of the indirect use of hydrogen as energy carrier is that in the future renewable energy supply, parts of the existing large-scale energy infrastructure could still be used.

The production of substitute or synthetic natural gas (SNG) by biomass hydrogasification has been assessed as a process for chemical storage of hydrogen. Thermodynamic analysis has shown the feasibility of this process. The product gas of the process has a Wobbe index, a mole percentage methane, and a calorific value quite comparable to the quality of the Dutch natural gas. With a hydrogen content below 10 mol%, the produced SNG can be transported through the existing gas net without any additional adjustment. The integrated system has an energetic efficiency of 81% (LHV).

During the pyrolysis step of biomass hydrogasification, a high percentage of biomass can be converted, with a high fraction of methane and ethane in the product gas, especially at high pressures. This can be explained by a combination of methanation and reversed shift reactions [321,322]:

$$CO + 3H_2 \rightarrow CH_4 + H_2O \qquad\qquad (5.34)$$

$$CO_2 + H_2 \rightarrow CO + H_2O \qquad\qquad (5.35)$$

In the presence of excess hydrogen, the rate of methane formation through the methanation reaction increases by increasing the operating pressure, while at the same time, carbon dioxide will react with hydrogen to produce carbon monoxide through the reversed shift reaction. The same trend is observed for the release rate of ethylene and ethane, that is, a shift from ethylene to ethane by increasing the operating pressure. This can be explained by hydrogenation of ethylene:

$$C_2H_4 + H_2 \rightarrow C_2H_6 \qquad\qquad (5.36)$$

The reaction will shift to ethane formation by increasing the operating pressure, especially in the presence of excess hydrogen.

One characteristic of biomass samples is the highly variable nature of the mineral composition. SEM/EDX analyses indicated high levels of potassium, magnesium, and phosphorus in the ash residue. The devitrification and embrittlement of the quartz furnace and balance rods were attributed to the high mineral content of many of the biomass feedstocks, with the high alkaline oxide levels of the grasses being particularly destructive. While mineral content may exert a beneficial effect through enhanced char reactivity with the possibility for a more thorough processing of the feedstock, the potential for corrosion and slagging would necessitate the judicious selection and possible pretreatment of biomass fuels. A major advantage of thermal treatment through gasification prior to combustion is the ability to remove many of the corrosive volatiles and ash elements such as potassium, sodium, and chlorine to avert damage to the process equipment [321,322].

Degradation of the lignocellulosic structure is governed by a collection of coupled mechanisms involving molecular and reactive radical intermediates. The mineral impurities inherent in the biomass feedstock confer a catalytic effect influencing pyrolysis and gasification rates and product selectivity. Subsequent to the initial drying process, initiation reactions begin at 200°C to aid in the lignin decomposition. Initial cleavage reactions produce ring structures containing 5–6 carbons; 2–6 carbon chain fragments and oxygenated species characteristic of the phenolic, aromatic, and highly cross-linked phenylpropanoid lignin structural component; and 5- and 6-carbon polysaccharide cellulose and hemicellulose structures.

Thermal treatment of hydrocarbon structures is sustained through free radical reactions, with β-carbon bond scission and H abstraction being two of the most important propagation reactions. Also significant, particularly at the onset of the biomass structural decomposition, is the cleavage of the OH radicals from the regularly repeating cellulose chain and the removal of CH_3 radicals from the methoxy groups of the lignin structure.

Though free radical reactions first become significant in the breakdown of the lignin structure above 200°C, they become increasingly more important above 300°C, when OH radicals are being cleaved off of the cellulose backbone at the onset of rapid cellulose decomposition, and above 250°C, as CH_3 and H radicals are being released due to H abstraction and β-carbon scission reactions. In addition to the OH radicals released by the cellulose structure, lignin also releases OH radicals in the 250°C–400°C temperature range through a β-aryl phenolic elimination mechanism proposed by Kawamoto et al. [106]. In this temperature interval, available CO combines with the OH radicals to produce the increased levels of CO_2 and the depressed levels of CH_4 prior to 400°C. This reaction produces H radicals that enter into subsequent propagation reactions responsible for the continual supply of low levels of CH_4 and the rise in H_2 after 500°C. With rising temperature, residual carbonyl groups are decarboxylated to CO and CO_2 as the oxygen deficient char (condensed carbon skeleton) strongly adsorbs any available oxygen that includes reactive CO. At higher gasification temperatures, this CO is liberated from the char and appears as an increase in CO gas evolution levels above 500°C [321].

Banyasz et al. [107,108] have proposed a set of slow pyrolysis depolymerization reactions that convert an active cellulose structure to tar, char, and CO_2 that can account for the rise in CO_2 in the neighborhood of the transition in regimes near 350°C–400°C. The alternative pathways proposed by Kawamoto et al. [106] for the slow pyrolysis of cellulose involve either the conversion of levoglucosan to polysaccharides via polymerization reactions that release CO and CO_2 and result in a protective carbonized layer through which further diffusion is impeded or an alternative pathway involving decomposition to low-molecular-weight volatiles. In this temperature range, radical reactions that include cleavage and decarbonylation and recombination of methylglyoxal and glyceraldehyde release CH_3 radicals and CO molecules.

While many of the HT gasification reactions governing hydrocarbon and biomass thermal treatment are similar as the final carbon skeleton is burnt out to char and residual mineral residue, the complex biomass chemical composition results in oxygenated species as well as aromatic and aliphatic fragments that are specific to the lignocellulosic structural components. The set of chemical reactions occurring during thermal treatment is highly temperature dependent. Only at very high gasification temperatures ($T > 700$°C) is the endothermic Boudouard reaction [321]

$$C + CO_2 \rightarrow 2CO \quad \Delta H = +172.5 \text{ kJ/mol} \qquad (5.37)$$

the dominant reaction that drives the rising CO levels at high temperature with increasing CO_2 injected into the stream. With rising furnace temperature, the carbon available from the biomass can react with CO_2 contained in the gasification environment. With higher rates of CO_2 injection into the system, the rate of CO production increases with rising temperature. CO formation is limited but measurable at low gasification temperatures. At lower temperatures, the highly reactive biomass char adsorbs any oxygen in the gasification medium including oxygen in the form of CO. Upon further heating, this CO is released following pyrolysis to provide a continuous supply of CO due to desorption above 450°C.

The steam-reforming reaction becomes a significant reaction above 550°C considering the high dilution rates of the biomass and the relatively high inputs of steam as compared to the concentrations of CO and H_2 that are the preferred HT products. Steam-reforming reactions are treated in great details in Chapter 6.

During pyrolysis, the steam introduced for gasification combines with any available CO in the reactor through the WGS reaction to produce H_2 that can, through direct hydrogenation methanation reactions,

$$2CO + 2H_2 \rightarrow CH_4 + CO_2 \quad \Delta H = -246.9 \text{ kJ/mol} \tag{5.38}$$

$$CO + 3H_2 \rightarrow CH_4 + H_2O \quad \Delta H = -205.7 \text{ kJ/mol} \tag{5.39}$$

$$C + 2H_2 \rightarrow CH_4 \quad \Delta H = -74.4 \text{ kJ/mol} \tag{5.40}$$

produce a continuous low-level supply of CH_4.

Several studies have examined the potential of various biomass feedstocks, particularly agricultural and forestry residues, for providing energy and chemicals. The energy potential and seasonal variability in the composition of switchgrass, corn stover, and wheat straw are discussed in Lee et al. [109]. Cool-season, rather than warm season, grasses were seen to possess lower concentrations of lignocellulosic material. New plant growth may need special pretreatment since it contains higher mineral levels than older growth. Using rice hulls

$$C + H_2O \rightarrow CO + H_2 \quad \Delta H = +131.3 \text{ kJ/mol} \tag{5.41}$$

The WGS reaction also becomes significant in this transition, between 500°C and 600°C, from low- to high-temperature steam gasification

$$CO + H_2O \rightarrow H_2 + CO_2 \quad \Delta H = -41.2 \text{ kJ/mol} \tag{5.42}$$

and is responsible for rising H_2 concentrations in this temperature interval. Both the presence of steam and high temperature are responsible for the increase in H_2 production. At high CO_2 injection ratios and at higher temperatures above 700°C, the reverse WGS reaction combines with the Boudouard reaction to create high CO concentration levels that drive the reverse steam-reforming reaction resulting in lower levels of H_2 produced.

H_2 production is depressed during steam gasification of poplar as a result of CO_2 injection. One pathway of the Demirbas mechanism involves biomass free radical stabilization through H abstraction reactions that cease once the structural component has completed mass decomposition. Though lignin begins a slow decay earlier, cellulose decomposition occurs slightly later and is rapidly completed. This H consumption for stabilization should cease earlier for high cellulosic as compared to high lignitic feedstock. Since the lignin-to-cellulose ratio is higher for woods than grasses, we would expect a more rapid decomposition in the grasses and a corresponding earlier onset of H_2 production for the grasses as compared to the woods [321].

5.4 HYDROTHERMAL GASIFICATION UNDER SUBCRITICAL CONDITIONS

The fast hydrolysis of organic molecules such as biomass at high temperature leads to a rapid degradation of the polymeric structure of biomass. A series of consecutive reactions lead to the formation of gas whose composition depends on the temperature and pressure of water, the contact time, and the catalyst if it is present. High solubility of intermediates in water, particularly at high temperature and pressure, allows further organic reactions to occur in aqueous media and prevents the formation of tar and coke. The reactive species originating from biomass (or other species) are diluted by solvation in water, thereby preventing polymerization to unwanted products. These conditions also lead to the formation of high gas yield at relatively low temperature. HTG process is thus the process of gaseous fuel generation in an aqueous medium. This, thus, differs from "steam gasification" where solids react with gaseous steam to produce a set of gaseous products. Unlike steam in "steam gasification," the water in hydrogasification is not simply a provider of oxidation environment.

The goal of HTG is to obtain high quality and yield of fuel gas. The most important components of fuel gas are hydrogen, carbon monoxide, and methane (or lower volatile hydrocarbons). Just like in conventional gasification, temperature plays an important role in the formation of methane. In principle, there are three types of HTG [110,111]:

1. *Low-temperature aqueous-phase reforming*: This occurs at low temperatures (215°C–265°C) and moderate pressures (23–65 bars) in the presence of a selective catalyst for carbohydrates with C:O ratio close to 1. This is a highly selective catalytic reaction involving sugar or sugar-derived molecules with water. The products of this reaction can be hydrogen, syngas, or lower alkanes depending on the nature of the catalyst and other operating conditions.
2. *HT catalytic gasification under subcritical water conditions*: At higher temperatures up to supercritical temperature, in the presence of a catalyst, biomass/waste or organic compounds are gasified mainly to methane and carbon dioxide. In the absence of a catalyst, this region of temperature (from 250°C to critical temperature 374°C) is also called hydrothermal liquefaction region wherein carbohydrates are liquefied to various organic products. In catalytic HTG process, the heat recovery is important for an efficient operation. The catalytic HTG process converts biomass/waste/water slurry into fuel gas. The gaseous fuel can be used for heat, power, or the generation of various chemicals. This process does not work well with coal.
3. *Gasification and reforming in SCW*: This can be carried out both in the presence and absence of a catalyst. The main products are hydrogen and carbon monoxide and carbon dioxide. This process works both for coal and biomass/waste.

The first type of reaction is a very selective catalytic reaction and mostly applied to very specific types of biomass molecules. The subject of aqueous-phase reforming is covered in my previous book [336] and in Chapter 6 in great details. This type of gaseous fuel production cannot be applied to coal. The second type of subcritical water gasification is enhanced by the use of a catalyst. It is mostly effective for biomass and waste but not for coal. Since this type of reaction also has limitations on feedstock, it is only briefly covered here. The third type of gasification applies to both coal and biomass and it can be very effective for cogasification. In the presence of a suitable catalyst, the third type of gasification is often accompanied by reforming reaction. We examine this type of reaction in some details. Reforming in SCW is discussed in Chapter 6.

5.4.1 CATALYSTS FOR HYDROTHERMAL GASIFICATION

HTG can be divided into three regions depending on the range of temperature. Osada et al. [110,111] identified region 1 as the one with temperature range of 500°C–700°C, SCW in which biomass decomposes and activated carbon can be used to avoid char formation or alkali catalyst to facilitate WGS reaction. In this region very little solids are remained and the main product of the gasification is hydrogen. In region 2 where the temperature range is 374°C–500°C, which is again in the supercritical region, biomass hydrolyzes and metal catalyst facilitates gasification. Here once again, the main product is hydrogen with some carbon dioxide, carbon monoxide, and methane.

In this section we focus on the region 3 where the temperature is below critical temperature of 374°C. In this case, biomass hydrolysis is slow and catalysts are required for gas formation. In subcritical region, the gas product distribution will be dictated by the thermodynamic equilibrium at a given temperature and pressure. In general, in subcritical region, more methane is produced compared to hydrogen. The partial pressure of water can also affect the gas composition. Higher partial pressure and lower biomass concentration can result in more steam reforming producing more hydrogen. An appropriate catalyst (such as nickel) can also reform methane to produce more hydrogen. The catalyst can also help in reducing the gasification temperature while maintaining useful kinetics. Lower temperature and pressure help in lowering the capital costs for the equipment as well as lowering possible corrosion effect on the reactor walls, thus allowing the use of less costly alloys for the reactor vessel.

An excellent review of catalysts for biomass gasification under subcritical conditions is recently given by Elliott et al. [115–118]. Their analysis is briefly described as follows.

Elliott et al. [118] examined the subcritical gasification of biomass feedstock that included cellulose, lignin, holocellulose (cellulose and hemicellulose), and a Douglas fir wood flour using nickel catalyst and added sodium carbonate cocatalyst. The results showed that at 350°C, the catalyst gave 42% of carbon fed compare to 15% of carbon fed in the absence of catalyst. Both hydrogen and methane concentrations were higher for the catalytic operations compared to the ones without catalyst. Carbon monoxide concentration was close to zero in the presence of catalyst. As regards the activity of alkali additions, the activity follows the order Cs > K > Na. The study of Elliott et al. [115–118] also indicated that conventional support for nickel, namely, alumina (other than alpha-alumina), silica, various ceramic supports, minerals such as kieselguhr, and other silica–alumina, was unstable in hot liquid water environment due to mechanisms such as dissolution, phase transition, and hydrolysis. They reported useful supports as carbon, monoclinic zirconia or titania, and alpha-alumina.

Elliott evaluated the base metal catalysis, noble metal catalysis, and activated carbon catalysis for HTG. His important conclusions are summarized as follows:

1. Of all the base metal catalysts examined by Elliott and coworkers [115–118] such as nickel, magnesium, tungsten, molybdenum, zinc, chromium, cobalt, rhenium, tin, and lead, nickel was found to be the most active and stable catalyst. Numerous other investigators [119–128] have also looked at various supports such as kieselguhr, silica–alumina, alpha-alumina, and alumina–magnesia on spinel form and carbon with varying degrees of success. The most useful promoters have been ruthenium, copper, silver, and tin impregnated at 1 wt%.

2. For noble metal catalysis, while some conflicting results are reported by various investigators [119–128], in general, platinum, palladium, and silver showed minor activities to HTG at 350°C; iridium had some activity but the best activities have been shown by ruthenium and rhodium. Rutile form of titania and carbon supports are found to be effective. Vogel et al. [127,128] found ruthenium doping on nickel catalyst on carbon to be effective for HTG.

3. While activated carbon and charcoal were found to be effective catalysts by some investigators [118–128], these results were mostly obtained under supercritical conditions.

The study of Minowa and Ogi [124] indicated that the cellulose gasification depends on the nature of support and the size of metal particles on the support. They presented the following mechanism for the cellulose gasification [321,322]:

$$\text{Cellulose} \xrightarrow{\text{Decompose}} \text{Water-soluble products}$$

$$\xrightarrow{\text{Gasification/Ni}} \text{Gases}(H_2 + CO_2) \xrightarrow{\text{Methanation/Ni}} \text{Gases}(CH_4 + CO_2) \qquad (5.43)$$

Vogel group [127,128] indicated that Raney nickel was more effective than nickel on alpha-alumina. They also studied nickel catalysts with ruthenium, copper, and molybdenum doping. Most effective results were obtained with ruthenium doping on nickel catalysts. Elliott [118] also reports that at 350°C, bimetallic Ru/Ni, Ru/C, and Cu/Ni also gave favorable gas productions by HTG of a variety of biomass. Favorable yields were obtained for lignin gasification by Ru/TiO$_2$, Ru/Al$_2$O$_3$, Ru/C, and Rh/C catalysts.

Favorable results for HTG of various biomass have been obtained both for batch and continuous systems. Ro et al. [126] showed that the subcritical HTG of hog manure feedstock can be the net energy producer for the solids concentration greater than 0.8 wt%. While the costs for gasification are higher than the ones for anaerobic digestion lagoon system, the land requirement for the gasification process and costs of transportation and tipping fees are lower. In addition, catalytic gasification process would destroy pathogens and bioactive organic compounds and will produce relatively clean water for reuse. The ammonia and phosphate by-products generated in gasification have also the potential value in the fertilizer market [321,322].

5.5 HYDROTHERMAL GASIFICATION IN SUPERCRITICAL WATER

While HTG in subcritical water produces hydrogen mainly for biomass, similar gasification in SCW can be applied to all carbonaceous materials. The main gasification reactions under SCW environment can be listed as follows:

$$C + H_2O \rightleftarrows CO + H_2 \quad \Delta H = +132 \text{ kJ/mol} \tag{5.44}$$

$$CO + H_2O \rightleftarrows CO_2 + H_2 \quad \Delta H = -41 \text{ kJ/mol} \tag{5.45}$$

$$CO + 3H_2 \rightleftarrows CH_4 + H_2O \quad \Delta H = -206 \text{ kJ/mol} \tag{5.46}$$

$$C + 2H_2O \rightleftarrows CO_2 + 2H_2 \quad \Delta H = +91 \text{ kJ/mol} \tag{5.47}$$

$$C + 2H_2 \rightleftarrows CH_4 \quad \Delta H = -87.4 \text{ kJ/mol} \tag{5.48}$$

$$C + CO_2 \rightleftarrows 2CO \quad \Delta H = +159.7 \text{ kJ/mol} \tag{5.49}$$

$$C + O_2 \rightleftarrows CO_2 \quad \Delta H = -405.9 \text{ kJ/mol} \tag{5.50}$$

Reactions (5.44) and (5.49) are important for gasification and they are endothermic. The overall process is also endothermic. Reaction (5.50) is needed to provide the heat for autothermal conditions. The reported studies for SCW gasification of complex and simple materials are briefly described here.

Li et al. [129] investigated coal gasification in the temperature range of 650°C–800°C and pressure 23–27 MPa, K$_2$CO$_3$ and Raney Ni as catalysts, and H$_2$O$_2$ as oxidant. Most experiments were performed with inlet slurry containing 16.5 wt% coal and 1.5 wt% CMC. The results showed that high temperature favors the gasification of coal in SCW, while pressure has a little effect on the gasification results. An optimum flow rate needs to be found to get the best results. Both gasification and carbon gasification efficiencies were improved by the catalysts; K$_2$CO$_3$ performed better than Raney Ni. Less char and tar were formed in the presence of catalysts. An increase in feed concentration decreased the hydrogen and gasification efficiencies. SCW desulfurizes the coal and the solid particles remained had less carbon and hydrogen than original coal.

Vostrikov et al. [130] examined coal gasification in the temperature range of 500°C–750°C, pressure of 30 MPa, and reaction time of 60–720 s with and without CO_2. Once again, the main gaseous products were CH_4, CO, CO_2, and H_2. Within the range of operating conditions examined, best carbon conversion was obtained at 750°C. The results show a significant temperature dependence on product compositions for temperatures below 650°C. Similar results were obtained by Cheng et al. [131] who studied gasification of lignite coals in the temperature range of 350°C–550°C and reaction time of 0–60 min in N_2 atmosphere.

The Battelle Pacific Northwest Laboratory demonstrated that various alkali carbonate and Ni catalysts can convert wet biomass to methane-rich gas at temperatures between 400°C and 450°C and pressure as high as 34.5 MPa. Yu et al. [132] found that glucose at low concentration (0.1 M) can be completely gasified in 20 s at 600°C and 34.5 MPa with major products being hydrogen and carbon dioxide. Higher concentration of glucose, however, reduces the product concentration of hydrogen and carbon dioxide and increases the concentration of methane. Xu et al. [133–135] showed that a wide range of carbons effectively catalyze the gasification of glucose in SCW at 600°C and 34.5 MPa pressure with nearly 100% carbon gasification efficiency. The available surface area of carbon did not affect the effectiveness of the catalyst. For concentrated organic feeds in water, in the presence of a catalyst, the temperature above 600°C is needed to achieve high gasification efficiencies. Mass transfer resistances at high concentration (if any) can affect the equilibrium of WGS reaction. In the presence of coconut shell, activated carbon, cellobiose, and various whole biomass feeds as well as depithed bagasse liquid extract and sewage sludge were completely gasified. There was some deactivation of carbon catalyst after 4–6 h of operation.

Demirbas [136,137] examined the decomposition of olive husk, cotton cocoon shell, and tea waste by water under both sub- and supercritical conditions. He also observed an increase in hydrogen production with temperature, particularly for temperatures higher than the supercritical temperature. Demirbas [136,137] observed that as temperature increased from 600°C to 800°C, hydrogen production increased from 53 to 73 vol% in reaction time of 2–6 s. She indicated that hydrogen productions can be obtained from biomass such as bionutshell, olive husk, tea waste, crop straw, black liquor, MSW, crop grain residue, pulp and paper waste, petroleum-based plastic waste, and manure slurry [321,322].

An extensive amount of work on SCW gasification of organic wastes has been reported in the literature [17–19,73,74]. The studies have shown that the gasification generally produces hydrogen and carbon dioxide mixture with simultaneous decontamination of wastes, particularly at higher temperatures. The homogeneous solution of waste and water makes it easy to pump to the high-pressure reactor without pretreatment. Guo et al. [138,139] presented an excellent review of SCW gasification of biomass and organic wastes. They as well as Lu et al. [140] showed the equilibrium effects of temperature, pressure, and feed concentration of wood sawdust on hydrogen, carbon dioxide, carbon monoxide, and methane concentrations in SCW. The data showed that the equilibrium favors the productions of hydrogen and carbon dioxide at high temperatures (see Table 5.1).

The study also showed that an increase in pressure significantly decreased the product concentration of carbon monoxide and slightly decreased the product concentration of the hydrogen. The pressure change had very little effect on the product concentrations of carbon dioxide and methane. The complex effect of pressure on the product distribution was believed to be due to complex interplay between hydrolysis and WGS reactions. Besides temperature and pressure, other parameters that affect the gas yield were feedstock concentration, oxidant, reaction time, feedstock composition, inorganic impurities in the feedstock, and biomass particle size. Guo et al. [138,139] also concluded that alkali such as NaOH, KOH, Na_2CO_3, K_2CO_3, and Ca$(OH)_2$, activated carbon, metal oxides, and metals such as noble metal catalysts (Ru/a-alumina > Ru/carbon > Rh/carbon > Pt/a-alumina, Pd/carbon, Pd-a-alumina), as well as Ni catalysts and metal oxides such as CeO_2 particles, nano-CeO_2, and nano-$(CeZr)_xO_2$, enhanced the reactivity of biomass gasification in SCW. The last two are important for the reforming under supercritical conditions.

TABLE 5.1

Equilibrium Gas Yield for 5 wt% Sawdust in SCW at 25 MPa Pressure

Temperature (°C)	Gas Yield (mol/kg)				
	Hydrogen	Carbon Dioxide	Methane	Carbon Monoxide	Methane Hydrogen
400	13	24	20	10^{-3}	1.54
500	40	31	10	2.5×10^{-3}	0.25
600	80	40	~1	3.1×10^{-3}	0.0125
700	89	43	0	1.2×10^{-3}	0.0
800	89	43	0	0.5×10^{-3}	0.0

Sources: Guo, L. et al., Supercritical water gasification of biomass and organic wastes, in: Momba, M. and Bux, F., eds., *Biomass Book*, 2010, pp. 165–182. With permission; Shah, Y.T., *Water for Energy and Fuel Production*, CRC Press, New York, 2014; [338].

Note: These data are the best estimates from the graphical data presented.

These and other studies found that the yields of H_2O and CO increased with increasing water density. Yields of H_2 were 4 times better with NaOH and 1.5 times better with ZrO_2 compared to reaction without a catalyst. Supercritical fluids gave increased pore accessibility, enhanced catalyst ability to coking, and increased desired product selectivity. While high-temperature SCW gasification produces hydrogen and carbon dioxide, Sinag et al. [141] showed that a combination of two technologies—SCW and hydropyrolysis on glucose in the presence of K_2CO_3—produces phenols, furfurals, organic acids, aldehydes, and gases. Xu and Antal [134,315] studied gasification of 7.69 wt% digested sewage sludge in SCW and obtained gas that largely contained H_2, CO_2, a smaller amount of CH_4, and a trace of CO. Other waste materials show similar behavior.

Kong et al. [122] assessed the reported work for the catalytic HTG of various types of biomass in SCW. The results show that except at low temperatures, the main product in all cases was hydrogen. Catalytic operations decrease the productions of char and tar and increase the production of hydrogen. Carbon and base catalysts play important roles in the increased gas yields and hydrogen production. Tanksale et al. [120] provided an extensive review of various catalytic and other processes to produce hydrogen from biomass. Supercritical gasification in water was one of these processes. Azadi and Farnood [119] reviewed heterogeneous catalysts for subcritical water and SCW gasification of biomass and wastes. The review provided an extensive information of carbon conversion and hydrogen and methane productions in sub- and supercritical conditions for a variety of biomass by various commercially available and laboratory-made catalysts that included supported and skeletal metal catalysts, activated carbon, metal wires, and other innovative catalysts [321,322].

The generation of hydrogen from waste has long-term and strategic implications since hydrogen is the purest form of energy and is very useful for product upgrading, fuel cell production, and many other applications. Hydrogen can be produced from waste via numerous HT technologies such as conventional or fast pyrolysis (e.g., olive husk, tea waste, crop straw), HT or steam gasification (e.g., bionutshell, black liquor, wood waste), supercritical fluid extraction (e.g., swine manure, orange peel waste, crop grain residue, petroleum-based plastic waste), and SCW gasification (e.g., all types of organic waste, agricultural and forestry waste) as well as low-temperature technologies such as anaerobic digestion and fermentation (e.g., manure slurry, agricultural residue, MSW, tofu wastewater, starch of food waste). For HT technologies, SCW gasification generates more hydrogen at a lower temperature than pyrolysis or gasification [129–166]. SCW gasification also does not require drying, sizing, and other methods of feed preparations, thereby costing less for the overall process. The temperature of the pyrolysis and gasification process can be reduced if the gases coming out of these processes are further steam reformed. This, however, adds to the overall cost. The rates for the low-temperature processes such as anaerobic digestion and fermentation can be enhanced with the use of suitable microbes and enzymes. The development of future hydrogen economy will require further research in the improvement of these technologies.

Biomass generally contains three important components: cellulose, hemicellulose, and lignin. Both cellulose and hemicellulose (collectively called homocellulose) are easy to hydrolyze, decompose, dehydrogenate, decarboxylate, and reform as shown by numerous studies mentioned earlier. Lignin component is generally toughest to convert. Yamaguchi et al. [165] studied lignin gasification in SCW. They indicated that lignin gasification involve three steps: (1) lignin decomposition to alkyl phenols and formaldehyde in SCW, (2) gasification of alkyl phenols and formaldehyde over a catalyst, and (3) formation of char from formaldehyde. They showed that SCW gasification is a promising technique to reduce the lignin gasification temperature. They also studied lignin gasification with three different catalysts at 400°C, namely, $RuCl_3/C$, $Ru(NO)(NO_3)_3/C$, and $RuCl_3/C$, and found that the order of gasification activity was $Ru/C = Ru(NO)(NO_3)_3/C > RuCl_3/C$. EXAFS analysis showed that during lignin gasification in SCW, ruthenium particle sizes in $Ru(NO)(NO_3)_3/C$ and Ru/C catalysts were smaller than that in the $RuCl_3/C$ catalyst. The study concluded that the ruthenium catalysts with smaller particle size of metal particles were more active for the lignin gasification [321,322].

Lignin is one of the major fractions of woody biomass, which is a polymer of aromatic compounds such as coniferyl alcohol, sinapyl alcohol, and coumaryl alcohol, and it constitutes about 30 wt% and 40% of energy of woody biomass. Yamaguchi et al. [151] examined the effects of various noble and transition metal catalysts and titania and activated carbon supports on lignin conversion and hydrogen production rates. The results showed that for the lignin gasification, activity order followed ruthenium > rhodium > platinum > palladium > nickel, whereas hydrogen production rate followed the order palladium > ruthenium > platinum > rhodium > nickel. Both titania and activated provided stable supports. Hydrogen production rate from lignin increased with temperature and shorter residence time.

Byrd et al. [146] examined a two-stage process to obtain clean fuels from switchgrass. In the first stage, subcritical hydrothermal liquefaction of switchgrass was carried out to obtain biocrude that did not contain some of the inorganic and other undesirable elements. In the second stage, catalytic gasification of biocrude in SCW was carried out to obtain clean syngas dominant in hydrogen. Biocrude contained many oxygenated hydrocarbons of varying molecular structure and weights, including lignin-derived products and sugars and their decomposition products. The supercritical gasification of biocrude was carried out at 600°C and 250 atm pressure. Nickel, cobalt, and ruthenium catalysts were prepared on titania, zirconia, and magnesium aluminum spinel supports. Magnesium aluminum spinel structure did not work. Over time, zirconia-supported catalyst plugged the reactor, although Ni/ZrO_2 catalyst gave the best hydrogen production. Titania-supported catalysts gave lower hydrogen conversions but did not plugged the reactor over time. All support materials suffered surface area loss due to sintering.

Glycerol $HOCH_2–CHOH–CH_2OH$ is obtained as a by-product from biodiesel manufacturing by transesterification of vegetable oils. Nine grams of biodiesel generate approximately 1 g of glycerol. With increasing production of biodiesel, glycerol production will rise and it can be used for food, oral and personal care, tobacco, polymers, pharmaceuticals, and replacements of petroleum feedstock. Kersten et al. [161] have reported gasification results for glycerol and other model compounds in a variety of catalytic and noncatalytic reactors in SCW and found that without addition of a catalyst, only very dilute concentrations of model biomass feeds could be completely gasified. The density of SCW is higher than that of steam resulting in a higher space-time yield. Higher thermal conductivity and specific heat were helpful in carrying out the endothermic reforming reactions. The formation of char and tar was also minimized because of the solubility of hydrocarbons in SCW. Importantly, hydrogen produced from SCW reforming was produced at high pressure, which can be stored directly, thus avoiding large expenses associated with compression [321,322].

The studies described here lead to some general conclusions. As the temperature increases above the critical temperature, more gases are generally produced from most carbonaceous materials. At lower temperature, higher feedstock concentration and in the absence of a catalyst, the gas production rate tends to be lower and contain more methane. At high temperature, for lower feedstock concentration and in the presence of an effective catalyst, hydrogen production rate rapidly increases. Higher temperature and the presence of a catalyst promote reforming of gas and favor reverse WGS reaction producing more hydrogen and carbon dioxide. Pressure also affects the equilibrium of WGS reaction. Higher pressure favors methane formation as opposed to hydrogen production [321,322].

5.6 WATER DISSOCIATION TECHNOLOGIES

The second general method to generate hydrogen involves various techniques to dissociate water to produce hydrogen. Fundamentally, three major techniques—electrolysis, photocatalysis/photobiological, and thermal/thermochemical—have been investigated to dissociate water. Various modifications of these techniques along with several other novel techniques are also evaluated in this chapter. These technologies are still largely at the research and development stages. The dissociation of water requires a significant amount of energy that can be provided by fossil, biomass, waste, and more importantly nuclear or solar sources. The use of solar energy for this purpose is extensively investigated. The conversion of solar energy into solar fuel (mainly hydrogen) opens up numerous possibilities [167–191]. There are basically three routes that can be used alone or in combination for producing storable and transportable fuels from solar:

- *Electrochemical*: Solar electricity made from photovoltaics (PV) or concentrating solar thermal systems followed by an electrolytic process [177–209,320]
- *Photochemical/photobiological*: Direct use of solar photon energy for photochemical and photobiological processes [210–251]
- *Thermochemical*: Solar heat at high temperatures followed by an endothermic thermochemical process [252–319]

The thermochemical route offers some thermodynamic advantages with direct economic implications. Irrespective of the type of fuel produced, higher reaction temperatures yield higher energy conversion efficiencies; but this is accompanied by greater losses by reradiation from the solar cavity receiver. The economic competitiveness of solar fuel production is closely related to two factors: the cost of fossil fuels and the necessity to control the world climate by drastically reducing CO_2 emissions. The economics of large-scale solar hydrogen production indicates that the solar thermochemical production of hydrogen can be competitive compared with the electrolysis of water using solar-generated electricity. It can even become competitive with conventional fossil fuel–based processes, especially if credits for CO_2 mitigation and pollution avoidance are applied. The applications of these three methods to generate hydrogen from water as well as other carbonaceous materials using solar energy as heating source are summarized in Figure 5.2. In this chapter, we will examine in detail all three methods of hydrogen production.

The dissociation of water to produce hydrogen reversibly requires a supply of energy as follows:

$$H_2O \rightleftarrows H_2 + O_2 \tag{5.51}$$

$$\Delta H^{\circ}_{298\,K} = 241.93 \text{ kJ/mol}, \quad \Delta G^{\circ}_{298\,K} = 228.71 \text{ kJ/mol}, \quad T\Delta S^{\circ}_{298\,K} = 13.22 \text{ kJ/mol}$$

This means that work = ΔG° and heat = $T\Delta S$ is required to split the water at 25°C and 1 atm. If the reaction does not proceed reversibly, more work is required. The energy needed for this work can be provided in a number of different ways and some of these will be evaluated in this chapter.

As mentioned before in this section, there are three major ways water can be dissociated to produce hydrogen. The first one is electrolysis wherein water is dissociated electrochemically using electrochemical cell. The cell can be operated in a number of different ways such as high temperature and high pressure, but all of them require significant amounts of energy to dissociate water. The second method is the

How it works—thermochemical routes for the production of solar fuels (H$_2$, syngas)

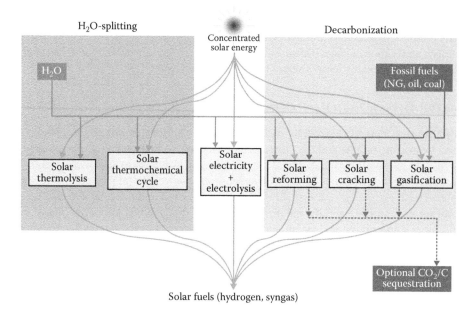

Thermochemical routes for solar hydrogen production – Indicated is the chemical source of H$_2$: H$_2$O for the solar thermolysis and the solar thermochemical cycles; fossil fuels for the solar cracking, and a combination of fossil fuels and H$_2$O for the solar reforming and solar gasification. For the solar decarbonization processes, optional CO$_2$/C sequestration is considered. All of those routes involve energy consuming (endothermic) reactions that make use of concentrated solar radiation as the energy source of high-temperature process heat.

FIGURE 5.2 Methods for generating hydrogen from water and carbonaceous materials using solar energy. (From Shah, Y. T., *Water for Energy and Fuel Production*, CRC Press, New York, 2014; also from Meier, A. and Sattler, C., Solar fuels from concentrated sunlight, Solar PACES, Solar Power and Chemical Energy System, IEA report 2009, With permission.)

use of photosynthesis and photocatalysis to dissociate water. This method also requires photonic energy with or without catalyst. The energy can however be provided using solar cell. The third method is thermal or thermochemical dissociation of water where water is dissociated either thermally or thermochemically. The latter method uses a chemical substance (or substances) to carry out dissociation using a series of chemical reactions. This method not only separates hydrogen and oxygen upon dissociation but also reduces the temperature required for the thermal dissociation. In recent years, this method has been heavily explored. Besides these three major methods, there are some miscellaneous methods such as chemical oxidation, magmalysis, and radiolysis are also explored for the water dissociation.

The use of solar energy in steam gasification, reforming, and solar cracking of fuels such as coal, biomass, and natural gas has been gaining more acceptance. Similarly, three major technologies for water dissociation such as electrochemical, photochemical/photobiological, and thermochemical can also be carried with the use of solar energy. All of these are graphically illustrated in Figure 5.2. In electrochemical process, solar electricity made from photovoltaic or concentrating solar thermal systems is used for electrolytic process. In photochemical/photobiological processes, direct use of solar photon energy carries out photochemical and photobiological processes. Finally, in thermochemical processes, solar heat at high temperature supports endothermic thermochemical water dissociation reactions. While thermochemical route offers some intriguing thermodynamic advantages over other options, in general, irrespective of the type of fuel produced, higher temperature gives higher conversion efficiency but also leads to greater losses by reradiation from the solar cavity receiver.

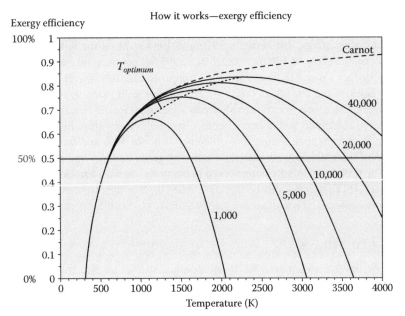

How it works—exergy efficiency

Variation of the ideal exergy efficiency as a function of the process operating temperature for a blackbody cavity-receiver converting concentrated solar energy into chemical energy. The mean solar flux concentration is the parameter (given in units of 1 sun = 1 kW/m^2): 1000; 5000; ...; 40,000 suns. Also plotted are the $\rightarrow$ Carnot efficiency and the locus of the optimum cavity temperature T$_{optimum}$.

FIGURE 5.3 Exergy efficiency as a function of temperature. (From Shah, Y. T., *Water for Energy and Fuel Production*, CRC Press, New York, 2014; Fletcher, E.A. and Moen, R.L., *Science*, 197, 1050, 1977; also from Meier, A. and Sattler, C., Solar fuels from concentrated sunlight, Solar PACES, Solar Power and Chemical Energy System, IEA report 2009, With permission.)

The recent report by Meier and Sattler [167] shows that a measure of how well solar energy is converted to chemical energy stored in solar fuels is called exergy efficiency (see Figure 5.3). The thermochemical route offers the potential of exergy efficiency to exceed 50%, a number higher than those obtained by all other methods. In solar fuel productions, half of the total investment cost is solar concentrating system. Higher exergy efficiency means lower power required to generate the same level of chemical energy in solar fuels. Thus, high exergy efficiency makes the process economically more attractive.

5.7 ELECTROLYSIS AND ITS DERIVATIVE TECHNOLOGIES

Electrolysis of water is the decomposition of water (H_2O) into oxygen (O_2) and hydrogen (H_2) by the passage of electric current through it. This process requires a large of amount of electrical energy that can be supplied by numerous sources such as hydropower, wind energy, solar energy, nuclear energy, and geothermal energy along with the electrical energy generated by fossil fuel and biomass. The electric power needed can also be supplied by the energy stored in the form of hydrogen that is generated by other sources of renewable energy [321].

Electrolysis has been known to produce hydrogen since the early nineteenth century. It gives hydrogen at 99.99% purity. Bockris and coworkers [177,194,195] showed that the cost of energy generated by electrolysis can be expressed by the following formula:

$$\text{Cost of 1 GJ in } \$ = 2.29E_c + 3 \tag{5.52}$$

which assumes 100% Faraday efficiency. E is the potential difference across the electrodes to produce current density in the order 100–500 mA/cm^2 and c is the cost of the electricity in cents per kWh of electricity. The research carried out between 1970 and 1984 reduced the value of E from 2.2 V to around 1.6 V at current density of 500 mA/cm^2. Since 1995, all commercial electrolyzers operate with 1.6 V per cell. There are several indirect methods to improve electrolytic cell performance and these are described by Bockris and coworkers [177,194,195,321]. In recent years, hydrogen is also created by coal slurry electrolysis. Recent advances can also reduce potential for the electrolysis to as low as 0.5 V.

If the electricity is obtained with a heat engine, Carnot efficiency limitation applies. Thus, if the electricity is obtained from coal, the normal efficiency is about 39% and the remaining 61% is lost as heat. This loss of thermal energy makes the electricity generated by wind or hydroelectric energy more efficient. The processes are often considered in combination with a nuclear or solar heat source. An HT electrolysis (HTE) process may be favorable when high-temperature heat is available as waste heat from other processes. The use of such waste heat makes the overall process cost efficient.

5.7.1 ALKALINE ELECTROLYSIS

Alkaline electrolyzers use an aqueous KOH solution (caustic) as an electrolyte that usually circulates through the electrolytic cells. Alkaline electrolyzers are suited for stationary applications and are available at operating pressures up to 25 bar. Alkaline electrolysis is a mature technology allowing unmanned remote operation with significant operating experience in industrial applications. The following reactions take place inside the alkaline electrolytic cell:

$$\text{Electrolyte: } 4H_2O \rightarrow 4H^+ + 4OH^- \tag{5.53}$$

$$\text{Cathode: } 4H^+ + 4e^- \rightarrow 2H_2 \tag{5.54}$$

$$\text{Anode: } 4OH^- \rightarrow O_2 + 2H_2O + 4e^- \tag{5.55}$$

$$\text{Sum: } 2H_2O \rightarrow O_2 + 2H_2 \tag{5.56}$$

Commercial electrolyzers usually consist of a number of electrolytic cells arranged in a cell stack. The major research challenges for the future are the design and manufacturing of electrolyzer equipment at lower costs with higher energy efficiency and large turndown ratios [336].

5.7.2 HIGH-TEMPERATURE ELECTROLYSIS (PROCESS)

HTE is more efficient economically than traditional room-temperature electrolysis because some of the energy is supplied by heat that is cheaper than electricity and the electrolysis reactions are more efficient at higher temperatures. While at 2500°C, thermal energy alone can dissociate water molecules, generally HTE systems operate between 100°C and 850°C [182–185]. The efficiency of HTE process can be easily estimated by assuming that the heat required comes from heat engines and heat energy required for 1 kg of hydrogen (350 MJ) at 100°C gives the efficiency of 41%. Similar calculation at 850°C gives the efficiency of 64%.

The process requires a careful use of materials for electrodes and electrolyte in a solid oxide electrolyzer cell (SOEC) is very important. Recent study [182–185] has used yttria-stabilized zirconia electrolytes, nickel-cermet steam/hydrogen electrodes, and mixed oxide of lanthanum, strontium, and cobalt oxygen electrodes. Future advances in HTE process will require materials that can withstand high temperature, high pressure, and corrosive environment. At present time, HTE process appears to be an inefficient way to store hydrogen energy. The process will become more efficient if sources such as nuclear, solar, and wind ocean energy can be the source of thermal energy [321].

When the energy is supplied in the form of heat, such as by solar or nuclear energy, the production of hydrogen by HTE is very attractive. Unlike in low-temperature electrolysis, in HTE, water converts more of initial thermal energy into chemical energy (like hydrogen) by increasing the conversion efficiency. Since energy in HTE process is supplied in the form of heat, less of the energy must be converted twice (from heat to electricity and then to chemical form) and so less energy is lost and efficiency can be doubled up to 50%.

While the heat required for the HTE process can be obtained by solar energy or nuclear energy, the latter source is more reliable and is often used. The solar form of HT heat is not consistent enough to bring down the capital cost of HTE equipment. More research into HTE and HT nuclear reactors may eventually lead to hydrogen supply that is cost competitive with natural gas steam reforming. HTE has been demonstrated in a laboratory but not at a commercial scale although the Idaho National Lab. is developing a commercial process [167,182–185,321].

5.7.3 HIGH-PRESSURE ELECTROLYSIS (PROCESS)

When electrolysis is conducted at high pressure, the produced hydrogen gas is compressed at around 120–200 bar (1740–2900 psi). By pressurizing the hydrogen in the electrolyzer, the need for an external hydrogen compressor is eliminated. The average energy consumption for internal compression is around 3% [167,177–181,194,195].

High-pressure electrolysis (HPE) is often carried out using a solid membrane (SPE) like perfluorosulfonic acid (Nafion) rather than classic liquid electrolyte (alkaline electrolyte) under high pressure. Laoun [179], LeRoy et al. [186,187], and Onda et al. [181] carried out a thermodynamic analysis of such a process and showed the importance of temperature and pressure on the entire efficiency of water electrolysis. Using the model and analysis of LeRoy et al. [186,187], Onda et al. [196] showed that a temperature change up to 250°C and pressure changes up to 70 atm can be carried out by polymer electrolytic membranes. They showed that an increase in pressure and a decrease in temperature deliver more power for water electrolysis. The increase is, however, found to be small at pressures above around 200 atm. They also found that hydrogen can be produced with about 5% less power using HPE than that required using atmospheric water electrolysis [336].

Fateev et al. [320] showed that water electrolysis using polymeric electrolyte membrane (PEM) has demonstrated its potential for high cell efficiency (energy consumption of about 4–4.2 kW/Nm^3 H$_2$) and gas purity of about 99.99%. They studied the effects of increasing operating pressure up to several hundred bars for direct storage of hydrogen in a pressurized vessel. Their study showed that while PEM water electrolyzers operating at pressures up to 70 bars can be used to produce hydrogen and oxygen of electrolytic grade with high efficiencies, an increase in cross-permeation at higher pressure, the concentration of hydrogen in oxygen and vice versa, can reach the critical levels of explosive mixtures. The cross-permeation can be reduced by surface modification of solid electrolytes using low-permeability protective layers of coating. Contaminant concentration in the produced gases can also be reduced by adding catalyst gas recombiners either directly in the electrolytic cells or along the production line. Fateev et al. [320] showed that by using gas recombiners inside the electrolysis cell, it was possible to maintain hydrogen content below 2 vol% at an operating pressure of 30 bar, with Nafion 117 as solid electrolyte.

5.7.4 PHOTOELECTROLYSIS

In this process, hydrogen and oxygen are separated in a light-driven electrolysis cell. Thus, the reactions that occur at the p-type cathode involve the evolution of hydrogen, and at the n-type anode, the evolution of oxygen. No external battery is used in the electrolysis process. While ideally current between electrodes can be used as electricity and hydrogen produced from the process can be used as fuel, the efficiency of the overall process is about 1% [210–251]. The progress in photoelectrolysis faces three major barriers: (1) there are no valid and significant theoretical analyses on the subject. The works of Scaife et al. [191,192] and Ohashi et al. [188] appear to have some deficiencies. (2) The assumption made for years that Fermi level in solution as an important aspect of the conditions under which cells would work is proven not to be true [187–195]. (3) The corrosion of semiconductor surfaces in contact with solution can be considerable. The corrosion is caused by heat as well as photoelectrochemical reactions. Photoelectrochemical reaction efficiency is currently the same as that of photosynthesis. In recent years, the increase in efficiency by photoelectrocatalysis has been achieved. Numerous metals such as $TiO_2/pGaP$, $SrTiO_2$-GaP, tin oxide, and other coatings of TiO_2 and CdS [188–191,218–220] on electrodes have been tested to improve the efficiency and life of the photoelectrolytic cell [167,188–191,220]. The work of Szklarczyk and Bockris [189,190] showed that photoelectrocatalysis is directly related to electrocatalysis. The rate determining step in photoelectrocatalysis is dependent upon the transfer of charge at the metal–solution interface and not at the semiconductor–solution interface [321].

5.7.5 PHOTOAIDED ELECTROLYSIS

One method to improve efficiency is to have light falling upon an electrode by applying to the electrode concerned a potential from an outside power source. A 30%–40% efficiency in this case is not very impressive because the overall efficiency includes light and electricity to hydrogen and not of light alone. The efficiency in this case can be improved to the level of 3%–4% [193,194]. This method is however not preferred because it does not make use of only light as source for hydrogen generation, but the overall process also includes electrical components.

5.7.6 PHOTOVOLTAIC ELECTROLYSIS

One way to avoid corrosion problem in photoelectrolysis is to use the concept of photovoltaic cell working in air and electrolyzing a distant electrolyzer [177,194]. Here, semiconductors are not in the direct contact with the solution so that corrosion problems cease to exist. This device can be effectively be used with solar energy–generated electricity. The device contains two cells.

The best setup for efficiency is, however, recorded by Murphy–Bockris cell [194] using n-on-p gallium arsenide coated with ruthenium oxide and p-on-n gallium arsenide coated with platinum. Such a cell gave about 8% conversion of light to hydrogen production at current density in the range of 10 mA/cm². The cell life was also at least as good as that of the PV in air. Two advantages of Murphy–Bockris cell are that (1) cell is in solution so that the concentration of light upon the electrode that gives high temperatures can be used to provide household heat and (2) only one device is needed compared to two that are needed in PV cell in air. There are numerous ways to use photovoltaic cell in conjunction with electrolysis. One common method is described in the following text [321].

5.7.7 SOLAR ELECTROLYSIS

The process of solar electrolysis involves generation of solar electricity via PV or concentrating solar power (CSP) followed by electrolysis of water. This process is considered to be a benchmark for other thermochemical solar processes for water splitting that offers potential for energy-efficient large-scale production of H_2. For solar electricity generated from PV cell and assuming solar thermal efficiencies at 15% or 20% and electrolyzer efficiency at 80%, the overall solar-to-hydrogen conversion efficiency will range from 12% to 16% [167,168,172,173,175,178]. If we assume solar thermal electricity cost of $0.08/kWh, the projected cost of H_2 will range from $0.15 to $0.20/kWh, that is, from $6 to $8/kg H_2. For PV electricity, costs are expected to be twice as high. HTE process can significantly reduce electricity demand if it is operated at around 800°C–1000°C via SOECs. The HT heat required for such a process can be supplied by CSP system [321].

5.8 PHOTOCHEMICAL AND ITS DERIVATIVE TECHNOLOGIES

The dissociation of water can be assisted by the photocatalysts that are directly suspended in the water [196,197]. As shown in the following, a number of photocatalysts are possible. Early work by Gray [197–200], Whitten [201,202], and Maverick and Gray [203] showed that polynuclear inorganic complexes, excited metal complexes, and surfactant ruthenium complexes can help photochemical decomposition of water to produce hydrogen. Kiwi et al. [204] presented a review of homogeneous and heterogeneous photoproduction of hydrogen and oxygen from water. The majority of the photoredox systems (heterogeneous photolysis) involve a photosensitizer, an electron acceptor, and an electron donor with the redox catalyst assisting in the gas evolution step. Excitation of the sensitizer (S) leads to an electron transfer

$$S + A \underset{}{\overset{h\nu}{\rightleftarrows}} S^+ + A^- \tag{5.57}$$

which is followed by the catalytic step

$$A^- + H_2O \xrightarrow{cat} A + OH^- + \frac{1}{2}H_2 \tag{5.58}$$

The back conversion of S^+ to S may be achieved by sacrificing a donor D added to the solution

$$S^+ + D \rightarrow S + D^+ \tag{5.59}$$

Koriakin et al. [205] used acridine dyes as sensitizers, Eu^{3+}, V^{2+} salicylates as electron acceptors, and "Adams" catalyst (PtO_2) as the redox catalysts. Numerous other sensitizers, electron acceptors, and redox catalysts are illustrated by Bockris et al. [177]. An efficiency of up to 30% at an elected wavelength for hydrogen production for a brief duration by photolytic process has been reported by Kalyasundarasam et al. [206].

5.8.1 WATER SPLITTING ON SEMICONDUCTOR CATALYSTS (PHOTOCATALYSIS)

The study by Duonghong et al. [207] was one of the first studies to investigate the splitting of water by utilizing microsystems [206–230]. In this system, the colloidal particles are made up of suitable conductor materials, for example, TiO_2. On these colloids are induced two metallic substances, for example, ruthenium oxide and platinum. When the system is irradiated, hydrogen is evolved on the platinum and oxygen on ruthenium oxide. Each colloidal particle is a micro photocell. Using small TiO_2 particles, a large area of TiO_2 can be exposed to light. The system needs to be heated to last more than several hours. There are some doubts whether or not equal production of hydrogen and oxygen is achieved and whether oxygen is engaged in side reactions. It is difficult to measure the efficiency of this system. Also, hydrogen and oxygen come off from water together and their separations add extra cost. Furthermore, the simultaneous presence of oxygen and hydrogen in water can give rise to chemical catalysis and recombination to water [206–230].

5.8.1.1 Titanium Oxide Photocatalysts

TiO_2 was the first semiconductor used in water dissociation reaction [221]. A pure and powdered TiO_2 however only absorbs UV fraction of solar light and thus not very effective for total absorption of solar light. The visible light response of TiO_2 was improved by chemical doping of TiO_2 with partially filled d-orbitals such as V_5^+, Cr_3^+, Fe_3^+, Co_2^+, and Ni_2^+ [210,226]. While these doping improved visible light response, they did not improve water dissociation reaction. Kato and Kudo [222–225] reported that TiO_2 codoped with a combination of Sb_5^+ and Cr_3^+ became active for O_2 evolution under visible light from an aqueous solution using $AgNO_3$ as sacrificial agent. The physical doping of transition metal ions into TiO_2 by the advanced ion-implantation technique also allowed modified TiO_2 to work under visible light radiation. The ion-implantation technique is however very expensive for commercial use. The visible light response can also be obtained by doping of anions such as N, S, or C [242–244] as substitutes for oxygen in the TiO_2 lattice. When TiO_2 is fused with metal oxides such as SrO, BaO, and Ln_2O_3, metal titanates and intermediate band gaps are obtained. Materials such as $SrTiO_3$, $La_2Ti_2O_7$, and $Sm_2Ti_2O_7$ have shown some promise. Promising results have also been shown when $Sm_2Ti_2S_2O_5$ is used, where sulfur anion is substituted for oxygen. Under visible light radiation, the last material works as a stable photocatalyst for the reduction of H^+ to H_2 or the oxidation of H_2O to O_2 in the presence of sacrificial electron donor Na_2S–Na_2SO_3 or methanol or acceptor Ag^+. A new class of titanium semiconductors, titanium disilicide ($TiSi_2$), which absorbs a wide range of solar light, has recently been proposed as a prototype photocatalyst for the water dissociation reaction. More description of this catalyst is given by Navarro et al. [210].

5.8.1.2 Tantalates and Niobates

Layered and tunneling structures of oxides are considered as promising materials for water dissociation reaction. Tantalates and niobates oxides with corner-sharing octahedral MO_6 (M = Ta or Nb) have been examined as photocatalysts for water dissociation. Kato and Kudo [208] observed that $MTaO_3$ (M = Li, Na, K) are effective photocatalysts for water dissociation under UV light. The oxides crystallize in pervoskite structure type. Lin et al. [209] showed that $NaTaO_3$ produced by sol–gel method gave higher activity for water dissociation than the same material prepared by the HT solid-state synthesis. The most active photocatalysts were those that achieve higher nitrogen substitution maintaining the original layered structure of $Sr_2Nb_2O_7$. More detailed discussion of these types of catalysts is given by Navarro et al. [210].

5.8.1.3 Transition Metal Oxides

Certain vanadium and tungsten compounds were found to be active in water dissociation reaction. $BiVO_4$ and scheelite structure and Ag_3VO_4 with pervoskite structure showed photocatalytic activity in visible light for oxygen evolution from and aqueous silver nitrate solution [226,230]. The WO_3 system also oxidizes water at moderately high rates in the presence of Ag^+ and Fe_3^+ ions. Under visible light, Pt-WO_3 alone with $NaIO_3$ produces oxygen at high rate but no hydrogen. Some other catalysts in this category are also examined by Navarro et al. [210].

5.8.1.4 Metal Nitrides and Oxynitrides

Navarro et al. [210] showed that nitrides and oxynitrides of transition metal cations with d10 electronic configurations (Ga_3^+ and Ga_4^+) constitute a class of photocatalysts suitable for water dissociation in visible light without sacrificial reagents. Among various cocatalysts examined, the largest improvement in activity was obtained when $(Ga_{1-x}Zn_x)(N_{1-x}O_x)$ was loaded with a mixed oxide of Rh and Cr. This semiconductor evolves hydrogen and oxygen steadily and stoichiometrically under visible light from pure water in the absence of sacrificial agent. The solid solution between ZnO and $ZnGeN_2$ ($Zn_{1+x}Ge$)– (N_2O_x) has also been found to be active oxynitride photocatalysts for pure water dissociation in visible light. Finally, $(Zn_{1+x}Ge)(N_2O_x)$ solid solution loaded with nanoparticulate RuO_2 cocatalyst is also active under visible light generating hydrogen and oxygen stoichiometrically from pure water [210].

5.8.1.5 Metal Sulfides

Small band gaps in metal sulfides make them very attractive photocatalysts for water dissociation. They are however unstable in the water oxidation reaction under visible light. A common method for the reducing photocorrosion of the sulfides under irradiation is the use of suitable sacrificial agents such as Na_2S/Na_2SO_3 salt mixture [210]. CdS with wurtzite structure is the best studied metal sulfide photocatalyst [231–233]. This catalyst property can be improved by improving preparation method that leads to CdS phases with good crystallinity and few crystal defects. Composite systems of CdS with TiO_2, ZnO, and CdO [234–236] also improved photoactivity. The incorporation of elements into the structure of CdS to make a solid solution is another strategy for improving the photocatalytic properties of CdS. A possible candidate ZnS works for this purpose. The substitution of ZnS into CdS structure improved the activity of the composite material [210].

ZnS is also another semiconductor investigated for photocatalytic activity. The chemical doping of ZnS by Cu_2^+, Ni_2^+, and Pb_2^+ [224,225,237,238] allowed ZnS to absorb visible light. These doped ZnS photocatalysts showed high photocatalytic activity under visible light for hydrogen production from aqueous solutions using SO_3^{2-}/S^{2-} as electron donor reagents. Combining ZnS with $AgInS_2$ and $CuInS_2$ to produce solid solutions $(CuAgIn)_xZn_{2(1-x)}S_2$ is another strategy for improving optical absorption in the visible light range [210,239–241]. Co catalysts such as Pt loaded on $(AgIn)_{0.22}Zn_{1.56}S_2$ showed the highest activity for hydrogen evolution. The ternary sulfides comprising In^{3+} and one type of transition metal cation (Cd_2^+, Zn_2^+, Mn_2^+, Cu^+) found to have low efficiency for water dissociation in visible light. More description of sulfide photocatalysts is given by Navarro et al. [210].

5.8.2 Photobiological Production of Hydrogen from Water

The water splitting can also be carried out photobiologically [214]. Biological hydrogen can be produced in an algae bioreactor. In the late 1990s, it was discovered that if the algae are deprived of sulfur, it will switch from the production of oxygen (a normal mode of photosynthesis) to the production of hydrogen. It seems that the production is now economically feasible by the energy efficiency surpassing 7%–10% [214].

Hydrogen can be produced from water by hydrogenase-catalyzed reduction of protons by the electrons generated from photosynthetic oxidation of water using sunlight energy. In recent years, the use of a variety of algae to produce hydrogen from water has been extensively investigated and reviewed [242,243]. These reviews mention the use of sulfur deprivation with *Chlamydomonas reinhardtii* to improve hydrogen production by algae [244,245]. Also, certain polygenetic and molecular analyses were performed in green algae [246,247]. These methods, however, did not significantly improve the rate and the yield of algal photobiological hydrogen production. Solar-to-hydrogen energy conversion using algae has efficiency less than 0.1% [248]. The rate and yield of algal photobiological hydrogen production is limited by (1) proton gradient accumulation across the algal thylakoid membrane, (2) competition from carbon dioxide fixation, (3) requirement for bicarbonate binding at photosystem II (PSII) for efficient photosynthetic activity, and (4) competitive drainage of electrons by molecular oxygen. Recently, Lee [249–251] outlined two inventions: (1) designer proton channel algae and (2) designer switchable PSII algae for more efficient and robust photobiological production of hydrogen from water. These two new inventions not only eliminate the four problems mentioned earlier but also eliminate oxygen sensitivity of algal hydrogenase and H_2–O_2 gas separation and safety issue. More work in this area is needed. The details of the two new inventions are described by Lee [214].

5.8.3 Plasma-Induced Photolysis

It has been suggested that plasma [252] can be used to produce photons of appropriate energy so that water can be dissociated in the gas phase. Thus, in a hypothetical fusion of hydrogen, it would be possible to produce a light in the region of 1800–950 A by the addition of aluminum to the plasma [177,253]. The main gain from this method is that the thermal energy absorbed would be converted to electricity in a heat engine at about 30% efficiency. A gain in efficiency is obtained because hydrogen will be produced by both photolysis and electrolysis. At present time, however, the production of high-energy protons is only possible by the injections of aluminum into plasmas. The possibility of obtaining very high efficiency (up to 90%, which is possible for electrolysis) is unlikely. Furthermore, the recombination of hydrogen and oxygen could be a major drawback of this process [177].

5.9 THERMAL AND THERMOCHEMICAL DECOMPOSITION OF WATER

The direct thermal dissociation has been examined since the 1960s [167,177,252–320]. In direct thermal decomposition the energy needed to decompose water is supplied by heat only. This requires a minimum of at least 2200°C (even for partial decomposition) and as high as about 4700°C temperature and this makes the process somewhat unrealistic. At this temperature, about 3% of all water molecules are dissociated as H, H_2, O, O_2, and OH. Other reaction products like H_2O_2 or HO_2 remain minor. At about 3200°C, about half of the water molecules are dissociated. It is well known that an initiation of thermal splitting of water even at low pressure requires 2000 K. At an atmospheric pressure, 50% dissociation requires 3500 K. This temperature can be reduced to less than 3000 K at 0.01 atm pressure. As shown later, the catalysts can accelerate the dissociation at lower temperature. The lower total pressure favors the higher partial pressure of hydrogen that makes the reactor to operate at pressures below an atmospheric pressure very difficult [167,177].

While a single-step thermolysis is conceptually simple, its realization is very challenging since it needs an HT heat source above 2200°C for achieving a reasonable degree of dissociation and an effective technique to separate hydrogen and oxygen to avoid explosive mixture. The ideas proposed to separate hydrogen from the products include effusion separation and electrolytic separation [321]. Membranes made of zirconia and other ceramics can withstand such high temperatures but they fail to absorb severe thermal shocks that often occur when working under high-flux solar radiation. Other techniques that have been evaluated are rapid quench by injecting a cold gas, expansion in a nozzle, or submersion of a solar irradiated target in liquid water. The last technique is workable and simple but a quench introduces a significant drop in energy efficiency and produces an explosive gas mixture. The efficiency can also be further decreased by reradiation and the type of temperature (e.g., 2725°C for 64% dissociation at atmospheric pressure) required creates material limitations [167,177].

One of the problems for thermal dissociation of water is the materials that can stand temperatures in the excess of at least 2200°C–2500°C. Several materials such as tantalum boride, tantalum carbide, tungsten, and graphite are possible. However, at these temperatures, only oxides are stable. Graphite is chemically unstable in the presence of hydrogen and oxygen at these high temperatures. Tungsten and tungsten carbide get oxidized at these temperatures. The effect of hydrogen on oxide catalysts at these temperatures is not known. Ceramic materials like boron nitride can also be useful if its oxidation can be controlled. Recent studies have shown that a low amount of dissociation is possible [177]. The separation of oxygen and hydrogen can be carried out in a semipermeable membrane of palladium or ZrO_2–CeO_2–Y_2O_3 that removes oxygen preferentially. Lede et al. [292,293] used a ZrO_2 nozzle through which steam is forced into a thermal stream and decomposed and unreacted water is quenched suddenly to remove water and oxygen. The resulting gas contained only a small amount (about 1.2 mol%) of hydrogen. Another possible solution is the use of heat-resistant membrane made of Pd or ZrO_2, both of which selectively permeate hydrogen. The gas can also be separated using a magnetic field. The source of heat is also an issue. Solar or nuclear sources are possibilities. They are, although at the early stages of development and at the present time, only possible on a smaller scale.

In recent years, thermal dissociation of water is achieved using nuclear and solar energy. Some prototype generation IV reactors operate at 850°C–1000°C, a temperature considerably higher than the existing commercial nuclear power plants. General Atomics predicts that hydrogen cost using HT gas-cooled reactor would cost \$1.53/kg, a cost compares well with \$1.40/kg costing by steam-reforming mechanism. One advantage of nuclear reactor producing both electricity and hydrogen is that it can shift production between the two. For example, plants can produce electricity during the day and hydrogen during night by matching the variations in electricity demand. Thus, hydrogen can act as a storage unit from which electricity can be generated when needed. The peak demand of electricity can be handled by the energy stored in hydrogen [321].

The high temperature needed to split the water can also be provided by solar energy. In Spain, a 100 kW HYDROSOL 2 pilot plant is operated at the Plataforma Solar de Almeria (PSA), which uses sunlight to get temperature at the range of 800°C–1200°C to split water [167,177]. This plant has been in operation since 2008. A megawatt plant based on this concept can be built by having several parallel reactors operated by connecting the plant to heliostat fields (field of sun-tracking mirrors) of suitable sizes [167,274,275,279]. H_2 Power Systems [167,276,279] has proposed a membrane system for solar dissociation of water at temperatures as high as 2200°C. The membrane separates hydrogen as soon as it is produced in the so-called "solar water cracker." Such a cracker with 100 m^2 concentrator can produce almost 1 kg of hydrogen per hour during full sunlight conditions.

The required scale of thermal decomposition process such that it is economical remains questionable. Large volumes may require exotic refractories. At present the choice of thermal decomposition takes second place to the thermochemical cycles described later. At laboratory scale, thermal decomposition has also been analyzed using solar energy as a source of heat. The overall efficiency of solar–thermal process for hydrogen generation is considerably higher than that of PV/electrolysis [167,274–279]. As shown in the following, solar thermal splitting of water is aided by multiple chemical steps but the following three principles govern the success of solar thermochemical reactions: (1) drive chemical reactions at the highest temperature possible, which is consistent with other pertinent constraints such as materials of construction and ability to concentrate light, (2) seek simple processes with as few steps as possible, preferably one (e.g., cracking), and (3) for multistep water-splitting thermochemical cycles, seek processes involving a highly endothermic step driven using concentrated sunlight, followed by an exothermic step that is autothermal and can run continuously [167,321].

5.9.1 Thermochemical Decomposition of Water

Thermochemical cycles have been intensely investigated over last few decades. In this method, two-, three-, or four-step chemical reactions aided by a source of heat such as nuclear or solar can dissociate water and separate hydrogen and oxygen at temperatures around 800°C–900°C. The method has some inherent issues:

1. The original concept [167,177,273] was that since the method avoided the formation of electricity by the conversion of heat to mechanical work, it would avoid the Carnot cycle, as this is the fundamental difficulty in reducing the price of hydrogen production by electrolysis method. The thermocycles was thought to produce hydrogen at a cost of about half of that for electrolytic method. This thinking was fallacious because the methods have to have reactions carried out at different temperatures in order that the entropic properties of the partial reactions in each cycle can be used to maximum advantage. Furthermore, when the individual reactions have a positive entropy change, it is desirable to carry out reactions at the highest temperature possible to minimize the overall free energy change. Conversely, if the entropy changes are negative, the reactions should be carried out at the lowest temperature. However, this requirement of changing the temperature of the reaction in various cycles gives rise to the requirements to change the pressure too and so it would be necessary to pump gases from one temperature and one pressure to another and this is similar to the Carnot cycle.
2. For three and four cycles, plant capital costs for unit hydrogen production is likely to be more than that for electrolysis due to the need for changing apparatus for each cycle. Furthermore, at temperatures like 800°C–900°C, the corrosion will cause the plant life to be short.
3. Generally, it is assumed [297,298] that the reaction would take place along the free-energy pathway but in reality they take the path down a reaction rate pathway [299,300] and not necessarily on a thermodynamic pathway. Also, because of possible side reactions, the final product may not be what was intended.
4. Because of (3), if cyclicity fails even by 1%, a considerable amount of unwanted materials buildup and calculated economics based on the cyclical nature of the process is no longer valid.

In spite of these arguments, a considerable investigation on thermochemical cycles to produce hydrogen at temperatures lower than those required for thermal dissociation has been carried out. The moderate temperatures used in these cycles, in general, also cause less material and separation problems. More than 300 different types of chemical cycles have been proposed and tested. In this section, we will evaluate some of the important ones [321].

Previously, thermochemical cycles were characterized as the ones that use process heat at temperatures below 950°C. These are expected to be available from HT nuclear reactors. These cycles required three or more chemical reaction steps and they are challenging because of material problems and inherent inefficiency involved with heat transfer and product separation in each step. One example is hybrid sulfuric acid cycle that requires two steps incorporating one electrolysis step. The leading candidates for multistep thermochemical cycles include mainly three-step sulfur–iodine cycle based on thermal decomposition of sulfuric acid at 850°C and four-step UT-3 cycle based on hydrolysis of calcium and iron bromide at 750°C and 600°C, respectively [252–297].

Recent advancement in the development of optical systems for large-scale solar concentrations capable of achieving a mean solar concentration ratio that exceeds 5000 suns allows high radiation fluxes capable of getting temperature above 1200°C. Such high temperatures allowed the development of efficient two-step thermochemical cycles using metal oxide–redox reactions. This two-step cycle along with the novel rotary reactor that was used for its study is illustrated in Figure 5.4. Some of the other important cycles are briefly described as follows.

5.9.1.1   UT-3 Cycle

UT-3 cycle is based on two pairs of chemical reactions [256,258,259]: The first pair is

$$CaO + Br_2 \rightarrow CaBr_2 + \frac{1}{2}O_2 \quad (550°C) \tag{5.60}$$

$$CaBr_2 + H_2O \rightarrow CaO + 2HBr \quad (725°C) \tag{5.61}$$

According to these reactions, a production of hydrobromic acid is accompanied by the release of oxygen. The next set of two reactions is as follows:

$$Fe_3O_4 + 8HBR \rightarrow 3FeBr_2 + 4H_2O + Br_2 \quad (250°C) \tag{5.62}$$

$$3FeBr_2 + 4H_2O \rightarrow Fe_3O_4 + 6HBr + H_2 \quad (575°C) \tag{5.63}$$

which indicates the reduction of water by a bromide, accompanied by the release of hydrogen. In the original concept, these two reactions operate separately and sequentially in two separate reactors, through the reactions among gases and solids embedded in solid matrices. The main difficulty encountered was the cycling behavior of these matrices. For example, in the first reactor during the first cycle, CaO is converted to $CaBr_2$, and in the second cycle, reverse transformation occurs and so on. The design proved difficult to extrapolate to an industrial scale. Many design issues for commercial applications are still under investigations [256–259].

5.9.1.2   Zn–ZnO Cycle

One of the most researched metal oxide redox pair is Zn/ZnO [167,256,261,276,277]. Since the product of ZnO decomposition at high temperature (viz., Zn and oxygen) readily recombines, the quenching of the product is necessary. Without heat recovery from the quench process, the estimated exergy efficiency of this cycle is around 35%. The electrothermal process to separate Zn and oxygen at high temperatures has been experimentally demonstrated in small-scale reactors. Such HT

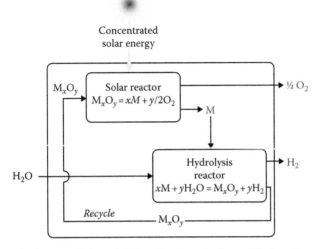

The first step of the cycle is the solar thermal release of O_2 from the metal oxide (M_xO_y). This step requires very high temperatures. The second step is the reaction of the metal (M) with H_2O to form H_2 and the corresponding M_xO_y. This step proceeds at lower temperatures and does not require additional heating in some cases. Since H_2 and O_2 are formed in different steps, the need for high-temperature gas separation is thereby eliminated. This cycle was originally proposed for an iron oxide redox system.

FIGURE 5.4 Thermochemical route based on metal oxide-redox reactions. (after Shah, Y. T., *Water for Energy and Fuel Production,* CRC press, New York, 2014; also from Meier, A. and Sattler, C., Solar fuels from concentrated sunlight, Solar PACES, Solar Power and Chemical Energy System, IEA report 2009, With permission.)

separation allows recovery of sensible and latent heats of the products to enhance the energy efficiency of the entire process. An HT solar chemical reactor (see Figure 5.5) was developed for this process and solar tests were carried out at PSI solar furnace in Switzerland [167,256,261,276,277]. These tests allowed surface temperature to reach 1700°C in 2 s with very low thermal inertia of the reactor system. The reactor was also resistant to thermal shocks. In 2010, solar chemical reactor concept for thermal dissociation of ZnO was demonstrated in a 100 kW pilot plant in a larger solar research facility [167,256,261,276,277].

In more recent studies on this cycle, a novel concept of hydrolyzer (hydrolysis step) was investigated for the hydrogen production step of $Zn + H_2O \rightarrow ZnO + H_2$. This technique showed that the water-splitting technique works at reasonable rates above 425°C. This was experimentally demonstrated in an aerosol reactor by the reaction between nano-Zn particles and water. In this reaction, Zn can be melted either by the heat generated due to the reaction or by supplying molten Zn from a nearby quencher unit of the solar plant. The heat of the reaction also generates steam for the reaction. The transportation of Zn to the reaction site eliminates the need for transportation and storage of hydrogen generated by the process. Another possible application of this process is to use energy carrier Zn directly in Zn–air batteries. This technology of redox batteries for solar energy storage is already commercially available. Some companies are pursuing a fuel cell analogue with "mechanically rechargeable" Zn–air batteries for stationary and mobile applications [167,278,286–288,321,322].

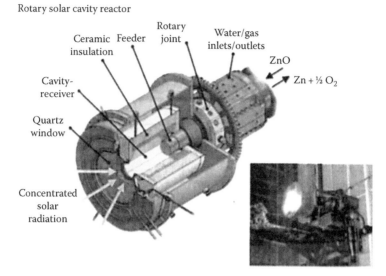

Rotary solar cavity reactor

The concept features a windowed rotating cavity-receiver lined with ZnO particles that are held by centrifugal force. With this arrangement, ZnO is directly exposed to high-flux solar irradiation and serves simultaneously the functions of radiant absorber, thermal insulator, and chemical reactant.

FIGURE 5.5 Rotary solar reactor for the thermal dissociation of Zinc oxide to Zinc and oxygen above 1700°C. (From Shah, Y. T., *Water for Energy and Fuel Production*, CRC Press, New York, 2014; also from Meier, A. and Sattler, C., Solar fuels from concentrated sunlight, Solar PACES, Solar Power and Chemical Energy System, IEA report 2009, With permission.)

5.9.1.3 SnO/SnO$_2$ Cycle

Another successful thermochemical cycle involves SnO/SnO$_2$ where exergy and energy efficiencies of 30% and 36%, respectively, can be obtained. The work carried out at Odeillo, France, in 1 kW solar reactor at atmospheric and reduced pressures has shown that SnO$_2$ reduction can be efficiently carried out at 1500°C and SnO hydrolysis can be carried out at 550°C [167,278,321,322].

5.9.1.4 Mixed Iron Oxide Cycle

Besides those mentioned earlier, manganese oxide, cobalt oxide, and iron-based mixed oxide redox pairs have also been tested [167,255,258,259,272]. The mixed iron oxide cycle was demonstrated at 10 kW level in EU's R&D project called "HYDROSOL" (2002–2005). The model for the monolithic solar thermochemical reactor (see Figure 5.6) was the catalyst converter used for automobile exhaust treatment. The reactor contained no moving parts and contained multichanneled monoliths that absorbed the solar radiation. The monolith channels were coated with mixed iron oxides and nanomaterials that can be activated by heating to 1250°C. After releasing oxygen, the water vapor was split as it is passed through the reactor trapping oxygen and releasing hydrogen as a product in the effluent gas stream at 800°C. Such a cyclic operation is established in a single closed receiver–reactor system and explosive mixture is avoided. A quasi-continuous hydrogen flow is established by alternate operation of two or more reaction chambers. "HYDROSOL 2" (2005–2009) process tested a 100 kW dual chamber pilot reactor at the PSA, Spain [167,192,258,259,272].

Inoue et al. [254] examined mixed ZnO/MnFe$_2$O$_4$ system for a two-step thermochemical cycle for the dissociation of water. This system, among many other mixed oxide systems, is workable for producing hydrogen by thermochemical cycle. At 1000°C, the mixture of ZnO and MnFe$_2$O$_4$

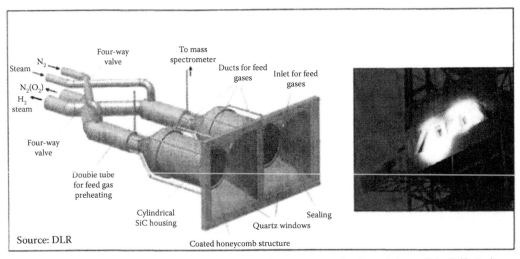

The concept features a closed receiver-reactor constructed from ceramic multi-channeled monoliths (left). Cyclic operation of the water-splitting and regeneration steps is established in two reaction chambers. Their individual temperature levels are controlled by focusing and defocusing heliostats during thermal tests at PSA, Spain (right).

FIGURE 5.6 Monolithic dual-chamber solar receiver reactor for continuous hydrogen production. (From Shah, Y. T., *Water for Energy and Fuel Production*, CRC Press, New York, 2014; also from Meier, A. and Sattler, C., Solar fuels from concentrated sunlight, Solar PACES, Solar Power and Chemical Energy System, IEA report 2009, With permission.)

reacted with water to generate hydrogen gas with 60% yield. The oxygen was produced around 1027°C completing a two-step cycle.

5.9.1.5 Carbothermal Reduction of Metal Oxides

In recent years, under EU's R&D project solar carbothermic production of zinc from zinc oxide (SOLZINC) (2001–2005), a 300 kW solar chemical reactor at the solar power research facility of WIS, in Israel, at temperatures ranging from 1000°C to 1200°C yielded up to 50 kg/h of 95% purity Zn and energy conversion efficiency of around 30% [182,270,276,302–305]. The process carried out carbothermal reduction of metal oxide (ZnO) using coke, natural gas, and other carbonaceous materials such as reducing agents. This brings down the reduction of oxides even to lower temperatures. Carbothermal reductions of metal oxides like iron oxide, manganese oxide, and zinc oxide with carbon and natural gas to produce the metals and the use of syngas were demonstrated in the solar furnaces. Such a solar chemical reactor concept as PSI's "two-cavity" solar reactor based on the indirect irradiation of ZnO and carbon (C) for producing Zn and carbon monoxide (CO) was scaled up in the SOLZINC project. The SOLZINC project provided an efficient thermochemical route for storing and transporting solar energy. If charcoal is used for ZnO reduction, the process can be CO_2 neutral [167,192,261,287–290].

5.9.1.6 Sulfur Family Thermochemical Water-Splitting Cycles

All sulfur family thermochemical water-splitting cycles (TCWSCs) depend on the concentration and decomposition of sulfuric acid for the oxygen evolution step of the cycle [256,261–271,273, 294–297]. The sulfuric acid decomposition step presents serious materials and catalyst deactivation challenges. The most active Pt catalysts deactivate very rapidly. Metal sulfate–based TCWSCs overcome this difficulty but they use thermal input, thus degrading photonic energy. T-Raissi et al. [273]

introduced FSEC's metal sulfate–ammonia (MSO_4–NH_3) hybrid photo-/thermochemical water-splitting cycle that can be represented as

$$SO_2(g) + 2NH_3(g) + H_2O(l) \rightarrow (NH_4)_2SO_3(aq) \quad \text{(chemical absorption, 25°C)} \quad (5.64)$$

$$(NH_4)_2SO_3(aq) + H_2O \rightarrow (NH_4)_2SO_4(aq) + H_2(g) \quad \text{(solar photocatalytic, 80°C)} \quad (5.65)$$

$$x(NH_4)_2SO_3 + M_2O_x \rightarrow 2xNH_3 + M_2(SO_4)_x + xH_2O \quad \text{(solar thermocatalytic, 500°C)} \quad (5.66)$$

$$M_2(SO_4)_x(s) \rightarrow xSO_2(g) + 2MO(s) + (x - 1)O_2(g) \quad \text{(solar thermocatalytic, 1100°C)} \quad (5.67)$$

where M = Zn, Mg, Ca, Ba, Fe, Co, Ni, Mn, Cu.

Chemical equilibrium calculations for the reaction between ZnO and $(NH_4)_2SO_4$ indicate that both $ZnSO_4$ and $ZnO \cdot 2ZnSO_4$ can form stable reaction products. More than 20 sulfuric acid and/or metal sulfate decomposition–based TCWSCs have been reported. Major issue remains to be electrolytic oxidation of sulfur dioxide. The use of a depolarized electrolyzer and the addition of a third process step such as S–I, S–Br, and S–Fe cycles have also been attempted. Some of these are described here [256,276,321]:

Ispra Mark 13 sulfur–bromine cycle [294]

$$Br_2(l) + SO_2 + 2H_2O(l) \rightarrow 2HBr(aq) + H_2SO_4(aq) \quad 77°C \quad (5.68)$$

$$H_2SO_4(g) \rightarrow SO_2(g) + H_2O(g) + \frac{1}{2}O_2 \quad 850°C \quad (5.69)$$

$$2HBr(aq) \rightarrow Br_2(aq) + H_2(\text{electrolytic}) \quad 77°C \quad (5.70)$$

General Atomics' sulfur–iodine cycle is described as follows [295]. Also, sulfur–iron cycle can be described as

$$Fe_2(SO_4)_3(aq) + SO_2 + 2H_2O \rightarrow 2FeSO_4(aq) + 2H_2SO_4 \quad 25°C \quad (5.71)$$

$$H_2SO_4 \rightarrow SO_2(g) + H_2O(g) + \frac{1}{2}O_2 \quad 850°C \quad (5.72)$$

$$2FeSO_4(aq) + H_2SO_4(aq) \rightarrow Fe_2SO_4(aq) + H_2 \quad 25°C \quad (5.73)$$

To make the separation of HI and H_2O easier, Sato and coworkers have proposed a nickel–iodine–sulfur version of S–I cycle [296]. Others include the following:

$$CO + H_2O \rightarrow CO_2 + H_2 \quad 550°C \quad (5.74)$$

$$CO_2 + SO_2 + H_2O \rightarrow H_2SO_4 + CO \quad 500°C \quad (5.75)$$

$$H_2SO_4(g) \rightarrow H_2O(g) + SO_2(g) + \frac{1}{2}O_2 \quad 900°C \quad (5.76)$$

$$SO_2 + H_2O + I_2 \rightarrow SO_3 + 2HI \quad 200°C \tag{5.77}$$

$$SO_3 \rightarrow SO_2 + \frac{1}{2}O_2 \quad 900°C \tag{5.78}$$

$$2HI \rightarrow H_2 + I_2 \quad 450°C \tag{5.79}$$

$$2FeSO_4 + I_2 + 2H_2O \rightarrow 2Fe(OH)SO_4 + 2HI \quad 20°C \tag{5.80}$$

$$2HI \rightarrow H_2 + I_2 \quad 450°C \tag{5.81}$$

$$2Fe(OH)SO_4 \rightarrow 2FeSO_4 + H_2O + \frac{1}{2}O_2 \quad 100°C \tag{5.82}$$

$$3FeCl_2(s) + 4H_2O \rightarrow Fe_3O_4(s) + 6HCl(g) + H_2 \quad 650°C \tag{5.83}$$

$$Fe_3O_4(s) + Fe_2O_3(s) + 6HCl + 2SO_2 \rightarrow 3FeCl_2 + 2FeSO_4 + 3H_2O \quad 100°C \tag{5.84}$$

$$2FeSO_4 \rightarrow Fe_2O_3(s) + 2SO_2(g) + \frac{1}{2}O_2 \quad 850°C \tag{5.85}$$

Although these cycles address the issue of water solubility of SO_2, they have other issues of their own. For example, efficient separation of sulfuric acid from reaction products such as HI, HBr, and $FeSO_4$ is challenging. The solution pH, particularly when other acids such as HI and HBr are formed, is a major issue. Abanades et al. [278] screened 280 TCWSCs and selected 30 as promising. There were nine metal sulfate–based TCWSCs in this selection because H_2SO_4 and M_SO_4 present an effective method for the heat-absorbing step of the TCWSCs. Some of these thermochemical cycles are also given by T-Raissi et al. [273,321].

The second approach is to introduce a metal oxide as a catalyst to convert low-concentration sulfuric acid to metal sulfate that is then decomposed to produce oxygen, sulfur dioxide, and metal oxide. Sulfur dioxide and water are sent to the acid electrolysis unit for the generation of hydrogen and sulfuric acid, thus closing the cycle. Introducing ZnO into the Westinghouse TCWSC, a new modified $ZnSO_4$ decomposition–based Westinghouse cycle can be written as [273]

$$SO_2(g) + 2H_2O(l) = H_2 + H_2SO_4(aq) \quad 77°C \text{ (electrolytic)} \tag{5.86}$$

$$H_2SO_4(aq, 50wt\%) + ZnO(s) = ZnSO_4 \cdot H_2O(s) \quad 80°C–350°C \tag{5.87}$$

$$ZnSO_4 \cdot H_2O(s) = ZnSO_4(s) + H_2O(g) \quad 450°C \tag{5.88}$$

$$ZnSO_4(s) = SO_2(g) + \frac{1}{2}O_2 + ZnO(s) \quad 850°C \tag{5.89}$$

Similarly, metal oxide catalyst can be added to sulfur–bromine, sulfur–iodine, and sulfur–iron cycles. These will give new modified metal-based TCWSCs. When the energy input for these cycles is solar energy, they can utilize only the thermal energy degrading the photonic portion of solar spectrum to lower grade heat.

5.9.1.7 Sulfur–Iodine Cycle

Sulfur–iodine cycle is one of the promising cycles for thermochemical hydrogen production [273,295,296]. It consists of three pure thermochemical steps that sum to the dissociation of water. These steps are as follows:

$$H_2O + SO_2 + I_2 = H_2SO_4 + 2HI \quad (25°C–120°C) \tag{5.90}$$

$$H_2SO_4 = H_2O + SO_2 + \frac{1}{2}O_2 \tag{5.91}$$

$$2HI = H_2 + I_2 \quad (200°C–400°C) \tag{5.92}$$

The second reaction is a two-step reaction as follows:

$$H_2SO_4 \rightarrow H_2O + SO_3 \quad (400°C–600°C) \tag{5.93}$$

$$SO_3 \rightarrow SO_2 + \frac{1}{2}O_2 \quad (800°C–900°C) \tag{5.94}$$

The first exothermic reaction is called Bunsen reaction and it is operated at 120°C. The second endothermic reaction needs a temperature of about 850°C (in two steps as shown earlier). The last endothermic reaction runs at temperatures between 300°C and 450°C. Three reactors that are a part of the cycle are called the Gibbs reactor, Bunsen reactor, and equilibrium reactor. The separation of H_2SO_4/HI mixture is the most critical part of the SI cycle [273,295,296,321].

This cycle has been investigated by several research teams because the cycle involves only liquids and gases. General Atomics has discovered that it is possible to separate two acids in the presence of excess iodine and water. However, efficient separation of HI from water and excess iodine at the outcome of Bunsen reaction still remains an issue. HT decomposition of acids is also an issue. The cycle was successfully tested in Japan to produce 45 L of hydrogen. It was also being tested in France at the capacity of 50 L/h [273,295,296,321].

5.9.1.8 Westinghouse Process

The Westinghouse process is classified under the "sulfur family" of thermochemical cycles being considered for the generation of hydrogen [256,273,274]. It is a sulfur cycle using hybrid electrochemical/thermochemical process for decomposing water into hydrogen and oxygen.

Sulfurous acid and water are reacted electrolytically to produce hydrogen and sulfuric acid. The resulting sulfuric acid is vaporized to produce steam and sulfur trioxide that is subsequently reduced at higher temperatures into sulfur dioxide and oxygen. The process may be seen as a variant of the iodine–sulfur process, wherein iodine reactions are substituted for by sulfur dioxide electrolysis as

$$SO_2 + 2H_2O \rightarrow H_2SO_4 + H_2 \quad (20°C–110°C, P = 2–10 \text{ bar}) \tag{5.95}$$

Following the separation of the water and sulfur dioxide for recycle to the electrolyzer, oxygen is available as a by-product. This has the advantage of requiring only one intermediate element. Sulfur was used because it is relatively inexpensive, its properties are well known, and it can assume a variety of valence states, thereby facilitating its use in oxidation–reduction reactions. The process requires electrical energy that restricts its efficiency. Electrolysis is carried out in a strong acid medium, leading to corrosion issues. Moreover, this would require several compartments to restrict parasitic sulfur and H_2S production at the cathode. Overall the process involves steps for oxygen generation and hydrogen generation with oxygen recovery and sulfuric acid vaporization in closing the loop for the cycle [256,273,274,321].

5.9.1.9 Copper–Chlorine Cycle

Copper–chlorine cycle is an important cycle because of its requirement for relatively low-temperature heat compared to other thermochemical water decomposition cycles [256,270,273]. It was identified by Atomic Energy of Canada Ltd. as a highly promising cycle for hydrogen production. The advantages of this cycle are (1) reduced construction materials, (2) inexpensive chemical agents, (3) minimal solids handling, and (4) reactions going to completion with few side reactions. It is well suited for energy supplied by the nuclear reactor. This cycle is unusual in that it contains five chemical steps, although efforts have been made to reduce the number of chemical steps. Just like sulfur–iodine cycle, copper–chlorine cycle has a significant potential because of lower-temperature requirements. The literature [256,270,273] has shown that the cost of hydrogen production by Cu–Cl cycle is better than electrolysis method at higher hydrogen production capacity (>30 tons/day). The five steps of this thermochemical cycle are (1) HCl production (at 400°C), (2) oxygen production (at 500°C), (3) electrochemical process (at ambient conditions), (4) flash drying (at a temperature greater than 100°C), and (5) hydrogen production (at a temperature between 430°C and 475°C) [321].

5.9.1.10 Copper–Sulfate Cycle

Copper/sulfate cycle involve two major steps: (1) hydrogen production from the reaction of water, $SO_2(g)$, and $CuO(s)$ at room temperature and (2) thermal decomposition of the products of the first step [256,263–271] to form oxygen and to regenerate reagents for the first step. The first step is performed electrolytically and the second step appears to be possible at a temperature of around 850°C. More complex versions of the copper–sulfate cycle called H-5 and H-7 involve four and six reactions. Law et al. [263–271] have given a very detailed accounting of this thermochemical cycle.

Brown et al. [255] examined efficiency of more than 100 thermochemical cycles that can use HT heat from advanced nuclear power stations. A basic requirement was the ability to deliver heat to the process interface heat exchanger at temperatures up to 900°C. They also developed a set of requirements and criteria considering design, safety, operational, economical, and development issues. Helium-cooled nuclear reactor was chosen to interface with the thermochemical cycles. Based on their and other studies, the best two-, three-, and four-step thermochemical cycles were listed by Shah [336] and they are reproduced in Table 5.2. Brown et al. [255] concluded that sulfur–iodine cycle was overall best to interlink with the helium-cooled nuclear reactor [321].

TABLE 5.2

Best Two-, Three-, and Four-Step Thermochemical Cycles

Name/Major Compound	Temperature (°C)	Details of Cycles
		Two-step cycles
Tokyo Institute of Technology/ferrite	1000	$2MnFe_2O_4 + 3Na_2CO_3 + H_2O \rightarrow 2Na_2MnFe_2O_6 + 3CO_2(g) + H_2(g)$
	600	$4Na_2MnFe_2O_6 + 6CO_2(g) \rightarrow 4MnFe_2O_4 + 6Na_2CO_3 + O_2(g)$
Westinghouse/sulfur	850	$2H_2SO_4(g) \rightarrow 2SO_2(g) + 2H_2O(g) + O_2(g)$
	77	$SO_2(g) + 2H_2O(a) \rightarrow H_2SO_4(a) + H_2(g)$
Nickel ferrite	800	$NiMnFe_4O_6 + 2H_2O \rightarrow NiMnFe_4O_8 + 2H_2(g)$
	800	$NiMnFe_4O_8 \rightarrow NiMnFe_4O_6 + O_2(g)$
Hallett Air Products/chlorine	800	$2Cl_2(g) + 2H_2O(g) \rightarrow 4HCl(g) + O_2(g)$
	25	$2HCl \rightarrow Cl_2(g) + H_2(g)$
		Three-step cycles
Ispra Mark 13/bromine/sulfur	850	$2H_2SO_4(g) \rightarrow 2SO_2(g) + 2H_2O(g) + O_2(g)$
	77	$2HBr(a) \rightarrow Br_2(a) + H_2(g)$
	77	$Br_2(l) + SO_2(g) + 2H_2O(l) \rightarrow 2HBr(g) + H_2SO_4(a)$
Ispra Mark 8/manganese/chlorine	700	$3MnCl_2 + 4H_2O \rightarrow Mn_3O_4 + 6HCl + H_2$
	900	$3MnO_2 \rightarrow Mn_3O_4 + O_2(g)$
	100	$4HCl + Mn3O_4 \rightarrow 2MnCl_2(a) + MnO_2 + 2H_2O$
Ispra/CO/Mn_3O_4	977	$6Mn_2O_3 \rightarrow 4Mn3O_4 + O_2(g)$
	700	$C(s) + H_2O(g) \rightarrow CO(g) + H_2(g)$
	700	$CO(g) + 2Mn3O_4 \rightarrow C + 3Mn_2O_3$
Ispra Mark 3/V/chlorine	850	$2Cl_2(g) + 2H_2O(g) \rightarrow 4HCl(g) + O_2(g)$
	170	$2VOCl_2 + 2HCl \rightarrow 2VOCl_3 + H_2(g)$
	200	$2VOCl_3 \rightarrow Cl_2(g) + 2VOCl_2$
Julich Center EOS/iron/sulfur	800	$2Fe_3O_4 + 6FeSO_4 \rightarrow 6Fe_2O_3 + 6SO_2 + O_2 (g)$
	700	$3FeO + H_2O \rightarrow Fe_3O_4 + H_2(g)$
	200	$Fe_2O_3 + SO_2 \rightarrow FeO + FeSO_4$
Gaz de France/KOH/K	725	$2K + 2KOH \rightarrow 2K_2O + H_2(g)$
	825	$2K_2O \rightarrow 2K + K_2O_2$
	125	$2K_2O_2 + 2H_2O \rightarrow 4KOH + O_2(g)$
Aachen University Jülich 1972/Cr/Cl	850	$2Cl_2(g) + 2H_2O(g) \rightarrow 4HCl(g) + O_2(g)$
	170	$2CrCl_2 + 2HCl \rightarrow 2CrCl_3 + H_2(g)$
	800	$2CrCl_3 \rightarrow 2CrCl_2 + Cl_2(g)$
US-Chlorine/Cu/Cl	850	$2Cl_2(g) + 2H_2O(g) \rightarrow 4HCl(g) + O_2(g)$
	200	$2CuCl + 2HCl \rightarrow 2CuCl_2 + H_2(g)$
	500	$2CuCl_2 \rightarrow 2CuCl + Cl_2(g)$
Ispra Mark 9/Fe/Cl	420	$2FeCl_3 \rightarrow Cl_2(g) + 2FeCl_2$
	150	$3Cl_2(g) + 2Fe_3O_4 + 12HCl \rightarrow 6FeCl_3 + 6H_2O + O_2(g)$
	650	$3FeCl_2 + 4H_2O \rightarrow Fe_3O_4 + 6HCl + H_2(g)$
LASL-U/uranium	25	$3CO_2 + U_3O_8 + H_2O \rightarrow 3UO_2CO_3 + H_2(g)$
	250	$3UO_2CO_3 \rightarrow 3CO_2(g) + 3UO_3(l)$
	700	$6UO_3(s) \rightarrow 2U_3O_8(s) + O_2(g)$
Ispra Mark 2 (1972)/Na/Mn	100	$Na_2O \cdot MnO_2 + H_2O \rightarrow 2NaOH(a) + MnO_2$
	487	$4MnO_2(s) \rightarrow 2Mn_2O_3(s) + O_2(g)$
	800	$Mn_2O_3 + 4NaOH \rightarrow 2Na_2O \cdot MnO_2 + H_2(g) + H_2O$
Sulfur–Iodine/S/I	850	$2H_2SO_4(g) \rightarrow 2SO_2(g) + 2H_2O(g) + O_2(g)$
	450	$2HI \rightarrow I_2(g) + H_2(g)$
	120	$I_2 + SO_2(a) + 2H_2O \rightarrow 2HI(a) + H_2SO_4(a)$

(Continued)

TABLE 5.2 (*Continued*)
Best Two-, Three-, and Four-Step Thermochemical Cycles

Name/Major Compound	Temperature (°C)	Details of Cycles
		Four-step cycles
Vanadium chloride	850	$2Cl_2(g) + 2H_2O(g) \rightarrow 4HCl(g) + O_2(g)$
	25	$2HCl + 2VCl_2 \rightarrow 2VCl_3 + H_2(g)$
	700	$2VCl_3 \rightarrow VCl_4 + VCl_2$
	25	$2VCl_4 \rightarrow Cl_2(g) + 2VCl_3$
Ispra Mark 4/Fe/Cl	850	$2Cl_2(g) + 2H_2O(g) \rightarrow 4HCl(g) + O_2(g)$
	100	$2FeCl_2 + 2HCl + S \rightarrow 2FeCl_3 + H_2S$
	420	$2FeCl_3 \rightarrow Cl_2(g) + 2FeCl_2$
	800	$H_2S \rightarrow S + H_2(g)$
Ispra Mark 6/Cr/Cl	850	$2Cl_2(g) + 2H_2O(g) \rightarrow 4HCl(g) + O_2(g)$
	170	$2CrCl_2 + 2HCl \rightarrow 2CrCl_3 + H_2(g)$
	700	$2CrCl_3 + 2FeCl_2 \rightarrow 2CrCl_2 + 2FeCl_3$
	420	$2FeCl_3 \rightarrow Cl_2(g) + 2FeCl_2$
Ispra Mark 1C/Cu/Ca/Br	100	$2CuBr_2 + Ca(OH)_2 \rightarrow 2CuO + 2CaBr_2 + H_2O$
	900	$4CuO(s) \rightarrow 2Cu_2O(s) + O_2(g)$
	730	$CaBr_2 + 2H_2O \rightarrow Ca(OH)_2 + 2HBr$
	100	$Cu_2O + 4HBr \rightarrow 2CuBr_2 + H_2(g) + H_2O$
UT-3 University of Tokyo/Fe/Ca/Br	600	$2Br_2(g) + 2CaO \rightarrow 2CaBr_2 + O_2(g)$
	600	$3FeBr_2 + 4H_2O \rightarrow Fe_3O_4 + 6HBr + H_2(g)$
	750	$CaBr_2 + H_2O \rightarrow CaO + 2HBr$
	300	$Fe_3O_4 + 8HBr \rightarrow Br_2 + 3FeBr_2 + 4H_2O$

Sources: Brown, L.C. et al., High efficiency generation of hydrogen fuels using thermochemical cycles and nuclear power, in *Spring National Meeting of AIChE*, Nuclear Engineering Session THa01 139, Hydrogen Production and Nuclear Power, New Orleans, LA, March 11–15, 2002. With permission; Schultz, K., Thermochemical production of hydrogen from solar and nuclear energy, in Presentation to the *Stanford Global Climate and Energy Project*, General Atomics, San Diego, CA, April 14, 2003. With permission; Shah, Y.T., *Water for Energy and Fuel Production*, CRC Press, New York, 2014; [336].

5.10 OTHER NOVEL TECHNOLOGIES FOR WATER DISSOCIATION

5.10.1 CHEMICAL METHODS

A number of materials react with liquid water or water containing acids to release hydrogen [177,305,307–314]. While these methods somewhat resembles steam reforming, they differ in that the reactant is liquid water instead of gaseous water and the solids involved are not naturally occurring such as coal and shale oil but the ones that require significant energy and efforts recovering metals such as zinc, aluminum, and iron.

In laboratory, zinc reacts with strong acids in Kipp's apparatus. In the presence of sodium hydroxide, aluminum and its alloys react with water to generate hydrogen [177,273,321]. This is, however, an expensive process due to the high cost of aluminum and the process also results in a large amount of waste heats that must be disposed or recovered. In relative terms, aluminum is cheaper than some other materials and it is safer and the produced hydrogen can be easily stored and transported than using other hydrogen storage materials like sodium borohydride.

The reaction between water and aluminum follows the path

$$Al + 3H_2O + NaOH \rightarrow NaAl(OH)_4 + 1.5H_2 \tag{5.96}$$

$$NaAl(OH)_4 \rightarrow NaOH + Al(OH)_3 \tag{5.97}$$

Overall reaction follows

$$Al + 3H_2O \rightarrow Al(OH)_3 + 1.5H_2 \tag{5.98}$$

The first two reactions are similar to the process that occurs inside an aluminum battery. The second reaction precipitates crystalline aluminum hydroxide. This process works well at a smaller scale and every 1 kg of aluminum can produce up to 0.111 kg of hydrogen that can be very useful in the device such as fuel cell where released hydrogen can generate electricity. Aluminum along with $NaBH_4$ can also be used as compact storage devices for hydrogen. The reaction given earlier is mildly exothermic and hence the reaction is carried out under mild temperatures and pressures providing a stable and compact source of hydrogen. The process can be a backup process for remote or marine applications. The negative effect of passivation of aluminum can be minimized by varying the temperature, alkali concentration, physical form of aluminum, and solution composition [321].

5.10.2 MAGMALYSIS

This process is another form of chemical method wherein steam is injected on a magma that is near the surface [177,309]. According to Northrup et al. [309], the following reaction would occur:

$$2FeO + H_2O \rightarrow 2FeO_{1.5} + H_2 \tag{5.99}$$

Fresh basaltic lava contains on the order of 10 wt% ferrous oxide (FeO) and 1–2 wt% ferric oxide ($FeO_{1.5}$). These components exist as dissolved constituents within the melt and in the mineral suspended in the magma. Northrup et al. [309] calculated hydrogen concentration that resulted from equilibration of water with a solid assemblage of hematite–magnetite for a total pressure of 100 MPa. This calculation agreed well with the measured data.

As water accumulates in the basaltic lava, most of the FeO is converted into $FeO_{1.5}$ resulting in the drop of hydrogen production [310,311]. Northrup et al. [309] also estimated hydrogen production at 1200°C and also showed that at lower temperature higher amount of FeO should be oxidized, hence resulting to greater hydrogen production. The estimates indicate that about 2.2×10^6 tons of hydrogen is potentially recoverable by water interacting with 1 km^3 of basalt at high temperatures at 1000 MPa. The knowledge of the rates of magma emplacement into the crust and cooling rates should permit estimations of the amount of magma area available for the reaction and whether or not this magma is a fixed or a renewable resource. Northrup et al. [309] estimated that about 10^5 km^3 of magma bodies in the areas of the United States exists where hydrogen production by this method is possible [321].

5.10.3 RADIOLYSIS

Radiolysis involves the injection of radioactive substances such as $UO_2(NO_3)_2$ into water, which emits particles that have an energy in the region of 10^6 eV [177,308,315]. This energy will decompose some 10^5 water molecules per particle and if there were no recombination, significant amounts of hydrogen and oxygen would be generated. When radioactive particles pass by water molecules, they strip a part of electron shells so that protons are produced and the oxygen becomes cationic. The mechanism of recombination in this process is unclear. The conversion efficiency is, however, low; between 1% and 5% of the radioactive energy is translated in the productions of hydrogen and oxygen [307]. The efficiency can be improved by the use of salts such as B^{10} and Li^6 compounds. The process generates hydrogen and oxygen in a mixture such as in the micromethods used in photoelectrolysis or biocatalysis. This disadvantage can be alleviated by its use in fuel cell where hydrogen and oxygen are separated by anode and cathode, respectively.

The method can be valuable if the efficiency is improved to greater than 10% and the radioactive materials used are waste. Gomberg and Gordus [308] improved the efficiency by using the nuclear fission either in a solid fuel configuration where the radiation energy/heat ratio can be about ¼ or in a fluid fuel configuration where all the energy is available as radiation [321].

5.10.4   SHOCK WAVES AND MECHANICAL PULSES

Attempts to dissociate water using shock waves and mechanical pulses have also been made [177]. The use of shock wave to dissociate diatomic molecules and organic compounds has been successful [315]. It is possible to introduce anharmonic oscillations in the molecules in such a way that the anharmonicity would introduce OH bond dissociation. A novel method could be the excitation of water molecules adsorbed on fiber optics that could be made conductive and at the same time allow part of the light wave being transmitted to interact (through the system) with the adsorbed water [336].

5.10.5   CATALYTIC DECOMPOSITION OF WATER

Another approach to the thermolysis of water is to pass water through a "getter" that will remove oxygen [177]. The "getter" then needs to be regenerated after obtaining hydrogen. Kasal and Bishop [316,317] used zeolites for this purpose. Kasal and Bishop [331,332] also described a simple two-step cycle to decompose water by cycling water over chromium- and indium-substituted aluminosilicates. For a two-step thermochemical process consisting of an endothermic step operating at lower temperature T_H, the transition between these two steps will be accompanied by a large entropy change. A large entropy change can be realized by resorting to a cycle consisting of many reaction steps or a single reaction involving many molecules. England [318] proposed a thermochemical cycle based on the results of Kasal and Bishop as [316,317,321]

$$Al_2O_3 + 4H_2O(g) + 2CrO \rightleftarrows Al_2O_3 \cdot 3H_2O + Cr_2O_3 + H_2(g) \qquad (5.100)$$

at low temperatures with an entropy change of −128.5 eu and

$$Al_2O_3 + 3H_2O + Cr_2O_3 \rightleftarrows Al_2O_3 + H_2(g) + 2CrO + \frac{1}{2}O_2 \qquad (5.101)$$

at high temperatures with an entropy change of 139.1 eu.

5.10.6   PLASMOLYSIS

The direct thermal dissociation of water by thermal means at temperatures around 3000°C suffers from the lack of durable materials for the reactor at these high temperatures [177,315]. One method by which this difficulty may be avoided is to use electrically produced plasmas [320]. Electricity generation of the plasmas involves the transformation of the energy from an electric field (microwave, radio frequency, or dc) into kinetic energy of electrons, which is further transformed into molecular excitations and to the kinetic energy of heavy particles. These discharged plasmas are divided into either hot (thermal) or cold (nonthermal) plasmas. Electron temperature can range from thousands to tens of thousands degrees in both types of discharges. The difference in energy content is a function of temperature. The low-temperature discharge has sufficient energy to dissociate water [321].

5.10.7 MAGNETOLYSIS

The idea of producing high current and low voltage was abandoned for a long time due to the fact that resistance losses are less when electricity is transmitted at high voltage over a power line than when it is transmitted at low voltage and high current [314]. However, in an electrolyzer, what is needed is low voltage and very high currents. This can be achieved by the application of a homopolar generator conceived by Farady [313]. Bockris and Gutmann [312] suggested that using this concept electrolysis can be carried out by generating the necessary potential difference by magnetic induction inside the electrolyzer [177,312–314]. Bockris et al. [177] examined the details of this method.

REFERENCES

1. Annual Energy Outlook 2008 with projections to 2030, Energy Information Administration Report, DOE/EIA 0383, Washington, DC (June 2008).
2. Adamson, K., 2008 large stationary survey, a report by *Fuel Cell Today*, Hertfordshire, U.K. (2008), Available at http://www.fuelcelltoday.com/media/pdf/surveys/2008-LS-Free.pdf.
3. World Hydrogen, Industry study with forecasts for 2013 & 2018, The Freedonia Group Report, Study Number 2605, Cleveland, OH (February 2010).
4. 2010 Fuel cell technologies market report, U.S. Department of Energy, Energy Efficiency and Renewable Energy Department, also U.S. Department of Energy Fuel Cell Technologies Program, Washington, DC (June 2011).
5. 2012 Fuel cell technology market report, U.S. Department of Energy, Energy Efficiency and Renewable Energy Department, also U.S. Department of Energy Fuel Cell Technologies Program, Washington, DC (October 2013).
6. An integrated strategies plan for research and development and demonstration of hydrogen and fuel cell technologies, The Department of Energy Hydrogen and Fuel Cells Program Plan, DOE, Washington, DC (September 2011).
7. McMurphy, K., Adamson, K., and Jerram, L., 2007 Fuel cell technologies market report, NREL report for U.S. Department of Energy, Energy Efficiency and Renewable Energy, Golden, CO (July 2009).
8. IEA, World energy outlook, International Energy Agency, Paris, France (2006).
9. Gadsby, J., Hinshelwood, C.N., and Skykes, K.W., The kinetics of the reactions of the steam-carbon system, *Proceedings of the Royal Society A*, 187, 129 (1946).
10. Johnstone, H.F., Chen, C.Y., and Scott, D.S., Kinetics of the steam-carbon reaction in porous graphite tubes, *Industrial & Engineering Chemistry*, 44, 1564–1569 (1952).
11. Lee, S., Gasification of coal, Chapter 2, in Lee, S., Speight, J., and Loyalka, S. (eds.), *Handbook of Alternative Fuel Technology*. Taylor & Francis Group, New York, pp. 26–78 (2007).
12. Walker, P.L., Rusinko, F., and Austin, L.G., Gas reactions in carbon, in Eley, D.D., Selwood, P.W., and Weisz, P.B. (eds.), *Advances in Catalysis*, Vol. XI. Academic Press, New York, pp. 133–221 (1959).
13. Corella, J., Toledo, J.M., and Molina, G., Steam gasification of coal at low–medium (600–800°C) temperature with simultaneous CO_2 capture in fluidized bed at atmospheric pressure: The effect of inorganic species. 1. Literature review and comments, *Industrial & Engineering Chemistry Research*, 45 (18), 6137–6146 (2006).
14. Sharma, D.K., Enhancing the steam gasification reactivity of coal by boosting the factors affecting the gasification reactions in the stepwise coal conversion, *Energy Sources, Part A*, 32, 1727–1736 (2010).
15. Sharma, D.K., Modelling of steam gasification reactions for reactor design, *Energy Sources, Part A*, 33, 57–71 (2011).
16. Exxon report on Coal liquefaction residue steam gasification, Linden, NJ (1997).
17. Dalai, A.K., Sasaoka, E., Hikita, H., and Ferdous, D., Catalytic gasification of sawdust derived from various biomass, *Energy & Fuels*, 17 (6), 1456–1463 (2003).
18. Hu, G., Xu, S., Li, S., Xiao, C., and Liu, S., Steam gasification of Apricot stones with olivine and dolomite as downstream catalysts, *Fuel Processing Technology*, 87 (5), 375–382 (2006).

19. Sutton, D., Kelleher, B., and Ross, J., Review of literature on catalysts for biomass gasification, *Fuel Processing Technology*, 73, 155–173 (2001).

20. Aznar, M.P., Delgado, J., Corella, J., and Aragues, J.L., Fuel and useful gas by steam gasification of biomass in fluidized bed with downstream methane and steam reforming: New results, also published in book eidted by Corassi, G., Collina, A., and Zibetta, M., Presented at *Biomass for Industry, Energy and Environment in Sixth EC Conference*, Athens, Greece (April 22–26, 1991), Elsevier Applied Science, Amsterdam, Netherlands (1992).

21. Aznar, M.P., Corella, J., Delgado, J., and Lahoz, J., Improved steam gasification of lignocellulosic residues in a fluidized bed with commercial steam reforming catalysts, *Industrial & Engineering Chemistry Research*, 32 (1), 1–10 (1993).

22. Aznar, M.P., Caballero, M.A., Gil, J., Marte, J.A., and Corella, J., Commercial steam reforming catalysts to improve biomass gasification with steam/oxygen mixtures. 2. Catalytic tar removal, *Industrial & Engineering Chemistry Research*, 37 (97), 2668–2680 (1998).

23. Xu, G., Murakami, T., Suda, T., Kusama, S., and Fujimori, T., Distinctive effects of CaO additive on atmospheric gasification of biomass at different temperatures, *Industrial & Engineering Chemistry Research*, 44 (15), 5864–5868 (2005).

24. Lee, W.J., Catalytic activity of alkali and transition metal salt mixtures for steam-char gasification, *Fuel*, 74 (9), 1387–1393 (1995).

25. Olivares, A., Aznar, M.P., Caballero, M.A., Gill, J., Franes, E., and Corella, J., Biomass gasification: Produced gas upgrading by in-bed use of dolomite, *Industrial & Engineering Chemistry Research*, 36, 5220–5226 (1997).

26. Rapagna, S., Jand, N., Kiennemann, A., and Foscolo, P., Steam gasification of biomass in a fluidized bed of olivine particles, *Biomass and Bioenergy*, 19, 187–197 (2000).

27. Ciferno, J. and Marano, J., Benchmarking biomass gasification technologies for fuels, chemicals and hydrogen production, a DOE report by National Energy Technology Laboratory, Washington, DC (June 2002).

28. Kumar, A., Jones, D., and Hanna, M., Thermochemical biomass gasification: A review of the current status of the technology, *Energies*, 2, 556–581 (2009).

29. Inayat, A., Ahmad, M., Yusup, S., and Abdul Mutalib, M., Biomass steam gasification with in-situ CO_2 capture for enriched hydrogen gas production: A reaction kinetics modeling approach, *Energies*, 3, 1472–1484 (2010).

30. Demirbas, A., Hydrogen from mosses and algae via pyrolysis and steam gasification, *Energy Sources, Part A*, 32, 172–179 (2010).

31. Li, J., Liu, J., Liao, S., Zhou, X., and Yan, R., Syn-gas production from catalytic steam gasification of municipal solid wastes in a combined fixed bed reactor, in *International Conference on Intelligent System Design and Engineering Application*, Changsha, Hunan, China (October 13–14, 2010), also published by IEE Computer Society, Los Alamitos, CA, pp. 530–534 (2010).

32. Demirbas, M., Hydrogen from various biomass species via pyrolysis and steam gasification processes, *Energy Sources, Part A*, 28, 245–252 (2006).

33. Demirbas, A., Hydrogen rich gases from biomass via pyrolysis and air-steam gasification, *Energy Sources, Part A*, 31, 1728–1736 (2009).

34. Salaices, E., Catalytic steam gasification of biomass surrogates: A thermodynamic and kinetic approach, Ph.D. thesis, Department of Chemical and Biochemical Eng., The University of Western Ontario, London, Ontario, Canada (December 2010).

35. Weiland, N., Means, N., and Morreale, B., Kinetics of coal/biomass co-gasification, NETL 2010 workshop on multiphase flow science, Pittsburgh, PA (May 4–6, 2010).

36. Chang, A.C.C., Chang, H.F., Lin, F.J., Lin, K.H., and Chen, C.H., Biomass gasification for hydrogen production, *International Journal of Hydrogen Energy*, 36, 14252–14260 (2011).

37. Xie, Q., King, S., Liu, Y., and Zeng, H., Syngas production by two-stage method of biomass catalytic pyrolysis and gasification, *Bioresource Technology*, 110, 603–609 (2012).

38. Gao, N., Li, A., Quan, C., Qu, Y., and Mao, L., Characteristics of hydrogen-rich gas production of biomass gasification with porous ceramic reforming, *International Journal of Hydrogen Energy*, 37, 9610–9618 (2012).

39. Nipattummakul, N., Ahmed, I.I., Kerdsuwan, S., and Gupta, A.K., Steam gasification of oil palm trunk waste for clean syngas production, *Applied Energy*, 92, 778–782 (2012).

40. Lv, P., Yuan, Z., Wu, C., and Ma, L., Bio-syngas production from biomass catalytic gasification, *Energy Conversion and Management*, 48, 1132–1139 (2007).

41. Demirbas, A., Karshoglu, S., and Ayas, A., Hydrogen resources conversion of black liquor to hydrogen rich gaseous products, *Fuel Science and Technology International*, 14 (3), 451–463 (1996).

42. Demirbas, A. and Caglar, A., Catalytic steam reforming of biomass and heavy oil residues to hydrogen, *Energy Education Science and Technology*, 1, 45–52 (1998).

43. Dermirbas, A., Yields of hydrogen-rich gaseous products via pyrolysis from selected biomass samples, *Fuel*, 80, 1885–1891 (2002) (Turkey).

44. Demirbas, A., Sustainable cofiring of biomass with coal, *Energy Conversion and Management*, 44 (9), 1465–1479 (June 2003).

45. Demirbas, A., Co-firing coal and municipal solid waste, *Energy Sources, Part A*, 30, 361–369 (2008).

46. Demirbas, M., Hydrogen from various biomass species via pyrolysis and steam gasification processes, *Energy Sources, Part A*, 28, 245–252 (2006).

47. Schuster, G., Loffler, G., Weigl, K., and Hofbauer, H., Biomass steam gasification ± an extensive parametric modeling study, *Bioresource Technology*, 77, 71–79 (2001).

48. Li, J., Liao, S., Dan, W., and Zhoy, X., Experimental study on catalytic steam gasification of municipal solid waste for bioenergy production in a combined fixed bed reactor, *Biomass and Bioenergy*, 46, 174–180 (October 2012).

49. Encinar, J.M., Gonzalez, J., Martinez, G., and Martin, M., Pyrolysis and catalytic steam gasification of olive oil waste in two stages, Personal communication (2010).

50. Hofbauer, H., Stoiber, H., and Veronik, G., Gasification of organic material in a novel fluidization bed system, in *Proceedings of the First SCEJ Symposium on Fluidization*, Tokyo, Japan, pp. 291–299 (1995).

51. Hofbauer, H., Veronik, G., Fleck, T., and Rauch, R., The FICFB gasification process, in Bridgwater, A.V. and Boocock, D. (eds.), *Developments in Thermochemical Biomass Conversion*, Vol. 2. Blackie Academic & Professional, Glasgow, U.K., pp. 1016–1025 (1997).

52. Hofbauer, H., Rauch, R., Löffler, G., Kaiser, S., Fercher, E., and Tremmel, H., Six years experience with the FICFB-gasification process, in Palz, W. et al. (eds.), *Proceedings of the 12th European Biomass Conference*, ETA Florence, Florence, Italy, pp. 982–985 (2002).

53. Hofbauer, H. and Rauch, R., Stoichiometric water consumption of steam gasification by the FICFB-gasification process, Presented at the conference: *Progress in Thermochemical biomass conversion* (17–22 September, 2000), Innsbruck, Austria, http://www.ficfb.at, also published in *Progress in Thermochemical biomass conversion*, vol. 1, Blackwell Science Ltd., U.K., pp. 199–208 (2001).

54. Hofbauer, H., Rauch, R., Bosch, K., Koch, R., and Aichernig, C., Biomass CHP plant guessing—A success story, in Bridgwater, A.V. (ed.), *Pyrolysis and Gasification of Biomass and Waste*, CPL Press, Newbury, U.K., pp. 527–536 (2003).

55. Herguido, J., Corella, J., and Gonzalez-Saiz, J., Steam gasification of lignocellulosic residues in a fluidized bed at a small pilot scale. Effect of the type of feedstock, *Industrial & Engineering Chemistry Research*, 31, 1274–1282 (1992).

56. Pfeifer, C., Proll, T., Puchner, B., and Hofbauer, H., H_2-rich syngas from renewable sources by dual fluidized bed steam gasification of solid biomass, in *Fluidization XII*, The Berkeley Electronic Press, Berkley, CA, pp. 889–895 (2011), also New Horizons in fluidization engineering, in Berruti, F., Bi, X., and Pugsley, T. (eds.), 12th International Conference on Fluidization, Vancouver, Canada (2007).

57. Gopalakrishnan, P., Modelling and experimental validation of biomass-steam gasification in bubbling fluidized bed reactor, PhD thesis, University of Canterbury, Christchurch, New Zealand (2007).

58. Matsuoka, K., Kuramoto, K., Murakami, T., and Suzuki, Y., Steam gasification of woody biomass in a circulating dual bubbling fluidized bed system, *Energy & Fuels*, 22 (3), 1980–1985 (2008).

59. Salaices, E., Catalytic steam gasification of biomass surrogates: A thermodynamic and kinetic approach, PhD thesis, The University of Western Ontario, London, Ontario, Canada (2010).

60. Salaices, E., Serrano, B., and de Lasa, H.I., Biomass catalytic steam gasification thermodynamics analysis and reaction experiments in a CREC riser simulator, *Industrial & Engineering Chemistry Research*, 49, 6834–6844 (2010).

61. Barrio, M., Gøbel, B., Risnes, H., Henriksen, U., Hustad, J.E., and Sørensen, L.H., Steam gasification of wood char and the effect of hydrogen inhibition on the chemical kinetics, an internal report by Norwegian University of Science and Technology, Department of Thermal Energy and Hydro Power, Trondheim, Norway (2000), orbit.dtu.dk/fedora/objects/orbit:54508/datastreams/file…/content.

62. Donaj, P., Yang, W., and Blasiak, W., High-temperature steam gasification of straw pellets, A report of KTH-ITM Division of Energy and Furnace Technology, Royal Institute of Technology, Stockholm, Sweden, pp. 1–25 (2011).

63. Gupta, A.K. and Cichonski, W., Ultra high temperature steam gasification of biomass and solid wastes, *Environmental Engineering Science*, 24 (8), 1179–1189 (2007).

64. Lucas, C., Szewczyk, D., Blasiak, W., and Mochida, S., High temperature air and steam gasification of densified biofuels, *Biomass and Bioenergy*, 27, 563–575 (December 2004).

65. Butterman, H. and Castaldi, M., CO_2 enhanced steam gasification of biomass fuels, in *Proceedings of NAWTEC 16, 16th Annual North American Waste-to-Energy Conference*, Philadelphia, PA (May 19–21, 2008).

66. Aznar, M., Caballero, M., Molina, G., and Toledo, J., Hydrogen production by biomass gasification with steam-O_2 mixtures followed by a catalytic steam reformer and a CO shift system, *Energy & Fuels*, 20, 1305–1309 (2006).

67. Seo, M., Goo, J., Kim, S., Lee, S., and Choi, Y., gasification characteristics of coal/biomass blend in a dual circulating fluidized bed reactor, *Energy & Fuels*, 24 (5), 3108–3118 (2010).

68. Seo, J., Youn, M., Nam, I., Hwang, S., Chung, J., and Song, I., Hydrogen production by steam reforming of liquefied natural gas over mesoporous $Ni-Al_2O_3$ catalysts prepared by a co-precipitation method: Effect of Ni/Al atomic ratio, *Catalysis Letters*, 130, 410–416 (2009).

69. Seo, J., Youn, M., Park, D., Jung, J., and Song, I., Hydrogen production by steam reforming of liquefied natural gas over mesoporous $Ni-Al_2O_3$ composite catalysts prepared by a single step non-ionic surfactant-templating method, *Catalysis Letters*, 130, 395–401 (2009).

70. Sun, H., Song, B., Jang, Y., and Kim, S., The characteristics of steam gasification of biomass and waste filter carbon. A report from Department of Chemical Engineering, Kunsan National University, Gunsan, South Korea (2012).

71. Kumabe, K., Hanaoka, T., Fujimoto, S., Minowa, T., and Sukanishi, K., Co-gasification of woody biomass and coal with air and steam, *Fuel*, 86, 684–689 (2007).

72. Chmielniak, T. and Sciazko, M., Co gasification of biomass and coal for methanol synthesis, *Applied Energy*, 74, 393–403 (2003).

73. Yamada, T., Akano, M., Hashimoto, H., Suzuki, T., Maruyama, T., Wang, Q., and Kamide, M., Steam gasification of coal biomass briquettes, *Nippon Enerugi Gakkai Sekitan Kagaku Kaigi Happyo Ronbunshu*, 39, 185–186 (2002).

74. Pan, Y.G., Roca, V., Manya, J., and Puigianer, L., Fluidized bed co gasification of residual biomass/poor coal blends for fuel gas production, *Fuel*, 79, 1317–1326 (2000).

75. Satrio, J., Shanks, B., and Wheelock, T., A combined catalyst and sorbent for enhancing hydrogen production from coal or biomass, *Energy & Fuels*, 21, 322–326 (2007).

76. Ji, K., Song, B., Kim, Y., Kim, B., Yang, W., Choi, Y., and Kim, S., Steam gasification of low rank fuel-biomass, coal and sludge mixture in a small scale fluidized bed, in Winter, F. (ed.), *Proceedings of Fourth European Combustion Meeting*, Vienna, Austria (14–17 April, 2009), pp. 1–5 (2009).

77. Sami, M., Annamalai, K., and Wooldridge, M., Co firing of coal and biomass fuel blends, *Progress in Energy and Combustion Science*, 27, 171–214 (2001).

78. Indrawati, V., Manaf, A., and Purwadi, G., Partial replacement of non renewable fossil fuels energy by the use of waste materials as alternative fuels, in Handoko, L. and Siregar, M. (eds.), *International Workshop on Advanced Materials for New and Renewable Energy*, American Institute of Physics, Jakarta, Indonesia (June 9–11, 2009), pp. 179–184 (2009), ISBN-13:978-0735407060, ISBN-10:0735407061.

79. Baker, E.G., Brown, M.D., Elliott, D.C., and Mudge, L.K., Characterization and treatment of tars from biomass gasifiers, Pacific Northwest Laboratory Richland, Washington, DC, pp. 1–11 (1988).

80. Elliott, D.C., Comparative analysis of gasification/pyrolysis condensates, Presented at the *Biomass Thermochemical Conversion Contractors' Meeting*, Minneapolis, MN (October 1985).

81. Elliott, D.C., Pyrolysis oils from biomass, in Soltes, E.J. and Milne, T.A. (eds.), *Relation of Reaction Time and Temperature to Chemical Composition of Pyrolysis Oils*, ACS Symposium Series, Vol. 376. Denver, CO (April 1988).

82. Milne, T.A., Evans, R.J., and Abatzoglou, N., Biomass gasification 'tars'; their nature, formation and conversion, National Renewable Energy Laboratory, Golden, CO (1998).

83. Evans, R.J. and Milne, T.A., Chemistry of tar formation and maturation in the thermochemical conversion of biomass, in Bridgwater, A.V. and Boocock, D.G.B. (eds.), *Developments in Thermochemical Biomass Conversion*, Vol. 2. Blackie Academic & Professional, London, U.K., pp. 803–816 (1997).

84. Evans, R.J. and Milne, T.A., Molecular characterization of the pyrolysis of biomass. 1. Fundamentals, *Energy & Fuels*, 1 (2), 123–138 (1987).

85. Evans, R.J. and Milne, T.A., Molecular characterization of the pyrolysis of biomass. 2. Applications, *Energy & Fuels*, 1 (4), 311–319 (1987).

86. Evans, R.J. and Milne, T.A., An atlas of pyrolysis-mass spectrograms for selected pyrolysis oils, Internal report, Solar Energy Research Institute, Golden, CO (1987).

87. Gopal, G., Adhikari, S., Thangalazhy-Gopakumar, S., Christian Brodbeck, C., Bhavnani, S., and Taylor, S., Tar analysis in syngas derived from pelletized biomass in a commercial stratified downdraft gasifier, *Bioresources*, 6 (4), 4652–4661 (2011).

88. El-Rub, Z., Bramer, E., and Brem, G., Review of catalysts for tar elimination in biomass gasification processes, *Industrial & Engineering Chemistry*, 43, 6911–6919 (2004).

89. Huang, H. and Ramaswamy, S., Thermodynamic analysis of black liquor steam gasification, *Bioresources*, 6 (3), 3210–3230 (2011).

90. Li, J. and van Heiningen, A.R.P., Kinetics of gasification of black liquor char by steam, *Industrial & Engineering Chemistry Research*, 30 (7), 1594–1601 (1991).

91. Whitty, K.J., Pyrolysis and gasification behaviour of black liquor char with H_2O under pressurized conditions, Report 93-4, Department of Chemical Engineering, thesis for degree of doctor of technology, Abo Akademi University Abo, Finland (1997), ISSN 0785-5052, ISBN 952-12-0013-8.

92. Bakhshi, N.N., Dalai, A.K., and Thring, R.W., Biomass char and lignin: Potential application, an internal report by Department of Chemical Engineering, University of Saskatchewan, Saskatoon, Saskatchewan, Canada (1999), https://web.anl.gov/PCS/acsfuel/.../44_2_ANAHEIM_03–99_0278.pdf, This work also reported in Chandhari, S., Bej, S., Bakhshi, N., and Dalal, A., Steam gasification of biomass-derived char for the production of carbon monoxide-rich synthesis gas, *Energy and Fuels*, 15 (3), 736–742 (2001).

93. Ferdous, D., Dalal, A., Bej, S., and Thring, R., Production of H_2 and medium heating value gas via steam gasification of lignins in fixed bed reactor, *The Canadian Journal of Chemical Engineering*, 79 (6), 913–922 (December 2001).

94. Kumar, V., Pyrolysis and gasification of lignin and effect of alkali addition, PhD thesis, Georgia Institute of Technology, Atlanta, GA (August 2009).

95. Kumar, V., Ilsa, K., Benerjee, S., and Frederick, W., Characterization of lignin gasification and pyrolysis, TAPPI Engineering, Pulping and Environment Conference, Portland, OR (August 24–27, 2008).

96. Devi, L., Ptasinski, K.J., and Janssen, F.J.J.G., A review of the primary measures for tar elimination in biomass gasification processes, *Biomass and Bioenergy*, 24 (2), 125–140 (2003).

97. Lopamudra, D., Catalytic removal of biomass tars, Olivine as prospective in-bed catalyst for fluidized-bed biomass gasifiers, Doctorate thesis, Technische Universiteit Eindhoven, Eindhoven, the Netherlands (2005).

98. Lee W., Nam, S., Kim, S., Lee, K., and Choi, C., The effect of Na_2CO_3 on the catalytic gasification of rice straw over nickel catalysts supported on Kieselguhr Seung, *Korean Journal of Chemical Engineering*, 17 (2), 174–178 (2000).

99. Simell, P., Stahlberg, P., Solantausta, Y., Hepola, J., and Kurkela, E., Gasification gas cleaning with nickel monolith catalysts, in Bridgewater, A. and Boocock, D.G. (eds.), *Development of Thermochemical Biomass Conversion*, Blackie Academic and Professional, London, U.K., pp. 1103–1116 (1997).

100. Simell, P., Leppalahti, J., and Bredenberg, J., Catalytic purification of Tarry fuel gas with carbonate rocks and ferrous materials, *Fuel*, 71, 211–218 (1992).

101. Simell, P., Leppalahti, J., and Kurkela, E., Tar decomposition activity of carbonate rocks under high CO_2 pressure, *Fuel*, 74 (6), 938–945 (1995).

102. Rostrup-Nielsen, J.R., Hansen, J., and Bak., H., CO_2-reforming of methane over transition metals, *Journal of Catalysis*, 144 (1), 38–49 (1993).

103. Bradford, M.C.J. and Vannice, M.A., Catalytic reforming of methane with carbon dioxide over nickel catalysts. II. Reaction kinetics, *Applied Catalysis A*, 142 (1), 97–122 (1996).

104. Bradford, M.C.J. and Vannice, M.A., CO_2 reforming of CH_4, *Catalysis Reviews—Science and Engineering*, 41 (1), 1–42 (1999).

105. Wei, J. and Iglesia, E., Isotopic and kinetic assessment of the mechanism of reactions of CH_4 with CO_2 or H_2O to form synthesis gas and carbon on nickel catalysts, *Journal of Catalysis*, 224 (2), 370–383 (2004).

106. Kawamoto, H., Ryoritani, M., and Saka, S., *Journal of Analytical and Applied Pyrolysis*, 81, 88 (2008).

107. Banyasz, J.L., Li, S., Lyons-Hart, J.L., and Shafer, K.H., Cellulose pyrolysis: The kinetics of hydroxyacetaldehyde evolution, *Journal of Analytical and Applied Pyrolysis*, 57 (2), 223–248 (February 2001).

108. Banyasz, J.L., Li, S., Lyons-Hart, J., and Shafer, K.H., Gas evolution and the mechanism of cellulose pyrolysis, *Fuel*, 80 (12), 1757–1763 (October 2001).

109. Lee, D., Owens, V.N., Boe, A., and Jeranyama, P., Composition of herbaceous biomass feedstocks, Report Sponsored by Sun Grant Initiative North Central Center South Dakota State University, Prepared by: North Central Sun Grant Center South Dakota State University, Brookings, SD, SGINC1–07 (June 2007), http://agbiopubs.sdstate.edu/articles/SGINC1-07.pdf, Accessed December 2007.

110. Osada, M., Sato, T., Watanabe, M., Shirai, M., and Arai, K., Catalytic gasification of wood biomass in subcritical and supercritical water, *Combustion Science and Technology*, 178 (1–3), 537–552 (2006).

111. Osada, M., Hiyoshi, N., Sato, O., Arai, K., and Shirai, M., Subcritical water regeneration of supported ruthenium catalyst poisoned by sulfur, *Energy & Fuels*, 22 (2), 845–849 (2008).

112. Osada, M., Hiyoshi, N., Sato, O., Arai, K., and Shirai, M. Reaction pathways for catalytic gasification of lignin in the presence of sulfur in supercritical water, *Energy & Fuels*, 21, 1854–1858 (2007).

113. Osada, M., Sato, T., Watanabe, M., Adschiri, T., and Arai, K., Low temperature catalytic gasification of lignin and cellulose with a ruthenium catalyst in supercritical water, *Energy & Fuels*, 18, 327–333 (2004).

114. Elliott, D.C., Sealock, L., and Baker, E., Chemical processing in high pressure aqueous environment. 3. Batch reactor process development experiments for organic destructions, *Industrial & Engineering Chemistry*, 33, 558–565 (1994).

115. Elliott, D.C., Peterson, K., Muzatko, D., Alderson, E., Hart, T., and Neuenschwander, G., Effects of trace contaminants on catalytic processing of biomass derived feedstocks, *Applied Biochemistry and Biotechnology*, 113–116, 807–825 (2004).

116. Elliott, D.C., Neuenschwander, G., Phelps M., Hart, T., Zacher, A., and Silva, L., Chemical processing in high pressure aqueous environment. 6. Demonstration of catalytic gasification for chemical manufacturing wastewater cleanup in industrial plants, *Industrial & Engineering Chemistry Research*, 38, 879–883 (1999).

117. Elliott, D.C., Neuenschwander, G., Hart, T., Butner, R., Zacher, A., and Englelhard, M., Chemical processing in high pressure aqueous environments & Process development for catalytic gasification of wet biomass feedstocks, *Industrial & Engineering Chemistry Research*, 43, 1999–2004 (2004).

118. Elliott, D.C., Catalytic hydrothermal gasification of biomass, *Biofuels, Bioproducts and Biorefining*, 2, 254–265 (May/June 2008).

119. Azadi, P. and Farnood, R., Review of heterogeneous catalysts for sub and supercritical water gasification of biomass and wastes, *International Journal of Hydrogen Energy*, 36, 9529–9541 (2011).

120. Tanksale, A., Beltramini, J., and Lu, G., A review of catalytic hydrogen production processes from biomass, *Renewable and Sustainable Energy Reviews*, 14, 166–182 (2010).

121. Kruse, A., Hydrothermal biomass gasification, *Journal of Supercritical Fluids*, 47, 391–399 (2009).

122. Kong, L., Li, G., Zhang, B., He, W., and Wang, H., Hydrogen production from biomass wastes by hydrothermal gasification, *Energy Sources, Part A*, 30, 1166–1178 (2008).

123. Akiya, N. and Savage, P.E., Roles of water for chemical reactions in high-temperature water, *Chemical Reviews*, 102 (8), 2725–2750 (2002).

124. Minowa, T. and Ogi, T., Hydrogen production from cellulose using a reduced nickel catalyst, *Catalysis Today*, 45, 411–416 (1998).

125. Brown, T., Duan, P., and Savage, P., Hydrothermal liquefaction and gasification of *Nannochloropsis* sp., *Energy & Fuels*, 24, 3639–3648 (2010).

126. Ro, K., Cantrell, K., Elliott, D.C., and Hunt, G., Catalytic wet gasification of municipal and animal wastes, *Industrial & Engineering Chemistry Research*, 46, 8839–8845 (2007).

127. Vogel, F., Waldner, M., Rouff, A., and Rabe, S., Synthetic natural gas from biomass by catalytic conversion in supercritical water, *Green Chemistry*, 9, 616–619 (2007).

128. Vogel, F. and Hildebrand, F., Catalytic hydrothermal gasification of woody biomass at high feed concentrations, *Chemical Engineering Transactions*, 2, 771–777 (2002).

129. Li, Y., Guo, L., Zhang, X., Jin, H., and Lu, Y., Hydrogen production from coal gasification in supercritical water with a continuous flowing system, *International Journal of Hydrogen Energy*, 35, 3036–3045 (2010).

130. Vostrikov, A., Psarov, S., Dubov, D., Fedyaeva, O., and Sokol, M., Kinetics of coal conversion in supercritical water, *Energy & Fuels*, 21, 2840–2845 (2007).

131. Cheng, L., Zhang, R., and Bi, J., Pyrolysis of a low rank coal in sub and supercritical water, *Fuel Processing Technology*, 85 (8–10), 921–932 (July 2004).

132. Yu, D., Aihara, M., and Antal, M., Hydrogen production by steam reforming glucose in supercritical water, *Energy & Fuels*, 7 (5), 574–577 (1993).

133. Xu, X.D. et al., Carbon-catalyzed gasification of organic feedstocks in supercritical water, *Industrial & Engineering Chemistry Research*, 35, 2522–2530 (1996).

134. Xu, X. and Antal, M., Kinetics and mechanism of isobutene formation from T-butanol in hot liquid water, *AIChE Journal*, 40 (9), 1524–1531 (1994).

135. Xu, X. and Antal, M., Mechanism and temperature dependent-kinetics of dehydration of tert-butyl alcohol in hot compressed liquid water, *Industrial & Engineering Chemistry Research*, 36 (1), 23–41 (1997).

136. Demirbas, A., Hydrogen production from biomass via supercritical water gasification, *Energy Sources, Part A*, 32, 1342–1354 (2010).

137. Demirbas, A., Progress and recent trends in biofuels, *Progress in Energy and Combustion Science*, 33, 1–18 (2007).
138. Guo, L., Cao, C., and Lu, Y., Supercritical water gasification of biomass and organic wastes, in Momba, M. and Bux, F. (eds.), *Biomass,* Scyo, Rijeka, Croatia, pp. 165–182 (2010), www.cafebiotecnologico. com.ar/libros/biotecnologia/Biomass.pdf.
139. Yamaguchi, A., Hiyoshi, N., Sato, O., Osada, M., and Shirai, M., EXAFS study on structural change of charcoal-supported Ruthenium catalysts during lignin gasification in supercritical water, *Catalysis Letters*, 122, 188–195 (2008).
140. Lu, Y., Guo, L., Zhang, X., and Yan, Q., Thermodynamic modeling and analysis of biomass gasification for hydrogen production in supercritical water, *Chemical Engineering Journal*, 131, 233–244 (2007).
141. Sinag, A., Kruse, A., and Rathert, J., Influence of the heating rate and the type of the catalyst on the formation of key intermediates and on the generation of gases during hydropyrolysis of glucose in supercritical water in a batch reactor, *Industrial & Engineering Chemistry Research*, 43, 502–508 (2004).
142. Takuya, Y., Yoshito, O., and Yukihiko, M., Gasification of biomass model compounds and real biomass in supercritical water, *Biomass and Bioenergy*, 26, 71–78 (2004)
143. Tang, H.Q. and Kuniyuki, K., Supercritical water gasification of biomass: Thermodynamic analysis with direct Gibbs free energy minimization, *Chemical Engineering Journal*, 106, 261–267 (2005).
144. Pinkwart, K., Bayha, T., Lutter, W., and Krausa, M., Gasification of diesel oil in supercritical water for fuel cells, *Journal of Power Sources*, 136, 211–214 (2004).
145. Kruse, A., Supercritical water gasification, *Biofuels, Bioproducts, and Biorefining*, 2, 415–437 (September/October 2008).
146. Byrd, A., Kumar, S., Kong, L., Ramsurn, H., and Gupta, R., Hydrogen production from catalytic gasification of switchgrass biocrude in supercritical water, *International Journal of Hydrogen Energy*, 36, 3426–3433 (2011).
147. Zhang, L., Champagne, P., and Xu, C., Supercritical water gasification of an aqueous by-product from biomass hydrothermal liquefaction with novel Ru modified Ni catalysts, *Bioresource Technology*, 102 (17), 8279–8287 (September 2011).
148. Michael, J.A. Jr. et al., Biomass gasification in supercritical water, *Industrial & Engineering Chemistry Research*, 39, 4040–4053 (2000).
149. Paul, T.W. and Jude, O., Composition of products from the supercritical water gasification of glucose: A model biomass compound, *Industrial & Engineering Chemistry Research*, 44, 8739–8749 (2005).
150. Peter, K., Corrosion in high-temperature and supercritical water and aqueous solutions: A review, *Journal of Supercritical Fluids*, 29, 1–29 (2004).
151. Yamaguchi, A., Hiyoshi, N., Sato, O., Bando, K., Osada, M., and Shirai, M., Hydrogen production from woody biomass over supported metal catalysts in supercritical water, *Catalysis Today*, 146, 192–195 (2009).
152. Guo, Y., Wang, S., Xu, D., Gong, Y., Ma, H., and Tang, X., Review of catalytic supercritical water gasification for hydrogen production from biomass, *Renewable and Sustainable Energy Reviews*, 14, 334–343 (2010).
153. D'Jesus, P., Boukis, N., Kraushaar-Czarnetzki, B., and Dinjus, E., Gasification of corn and clover grass in supercritical water, *Fuel*, 85, 1032–1038 (2006).
154. Kruse, A., and Gawlik, A., Biomass conversion in water at 330°C–410°C and 30–50 MPa identification of key compounds for indicating different chemical reaction pathways, *Industrial & Engineering Chemistry Research*, 42, 267–279 (2003).
155. Kruse, A. and Henningsen, T., Biomass gasification in supercritical water: Influence of the dry matter content and the formation of phenols, *Industrial & Engineering Chemistry Research*, 42, 3711–3717 (2003).
156. Lee, I.G., Kim, M.S., and Ihm, S.K., Gasification of glucose in supercritical water, *Industrial & Engineering Chemistry Research*, 41, 1182–1188 (2002).
157. Antal, M., Allen, S., Schulman, D., and Xu, X., Biomass gasification in supercritical water, *Industrial & Engineering Chemistry Research*, 39, 4040–4053 (2000).
158. Susanti, R., Nugroho, A., Lee, J., Kim, Y., and Kim, J., Noncatalytic gasification of isooctane in supercritical water: A strategy for high-yield hydrogen production, *International Journal of Hydrogen Energy*, 36 (6), 3895–3906 (March 2011).
159. Goodwin, A. and Rorrer, G., Conversion of glucose to hydrogen rich gas by supercritical water in a microchannel reactor, *Industrial & Engineering Chemistry Research*, 47 (12), 4106–4114 (May 2008).
160. Takuya, Y. and Yukihiko, M., Gasification of cellulose, xylan, and lignin mixtures in supercritical water, *Industrial & Engineering Chemistry Research*, 40, 5469–5474 (2001).

161. Kersten, S.R.A., Potic, B., Prins, W., and VanSwaaij, W., gasification of model compounds and wood in hot compressed water, *Industrial & Engineering Chemistry Research*, 45, 4169–4177 (2006).

162. Tang, H. and Kitagawa, K., Supercritical water gasification of biomass: Thermodynamic analysis with direct Gibbs free energy minimization, *Chemical Engineering Journal*, 106 (3), 261–267 (February 2005).

163. Chakinala, A., Brilman, D., van Swaaij, W., and Kersten, S., Catalytic and non catalytic supercritical water gasification of microalgae and glycerol, *Industrial & Engineering Chemistry Research*, 49 (3), 1113–1122 (February 2010).

164. Hao, X.H., Guo, L.J., Mao, X., Zhang, X.M., and Chen, X.J., Hydrogen production from glucose used as a model compound of biomass gasified in supercritical water, *International Journal of Hydrogen Energy*, 28, 55–64 (2003).

165. Sato, T., Osada, M., Watanabe, M., Shirai, M., and Arai, K., Gasification of alkyl phenols with supported nobel metal catalysts in supercritical water, *Industrial & Engineering Chemistry Research*, 42, 4277–4282 (2003).

166. Kruse, A., Meier, D., Rimbrecht, P., and Schacht, M., Gasification of pyrocatechol in super critical water in the presence of potassium hydroxide, *Industrial & Engineering Chemistry Research*, 39, 4842–4848 (2000).

167. Meier, A. and Sattler, C., Solar fuels from concentrated sunlight Solar Paces, Solar Power and Chemical Energy Systems, IEA Report, Brussels, Belgium (August 2009).

168. Smitkova, M., Janicek, F., and Riccardi, J., Analysis of the selected processes for hydrogen production, *WSEAS Transactions on Environment and Development*, 11 (4), 1026–1035 (2008).

169. Funk, J.E. and Reinstrom, R.M., Energy requirements in the production of hydrogen from water, *Industrial & Engineering Chemistry Process Design and Development*, 5 (3), 336–342 (July 1966).

170. Funk, J.E., Thermodynamics of multi-step water decomposition processes, in *Proceedings of Symposium on Non-Fossil Chemical Fuels (ACS 18) 163rd National Meeting*, Boston, MA, pp. 79–87 (April 1972).

171. Abe, I., Hydrogen productions from water, *Energy Carriers and Conversion Systems*, 1, 1–3 (2001).

172. Taylan, O. and Berberoglu, H., Fuel production using concentrated solar energy, in Rugescu, R. (ed.), Applications of solar energy, Chapter 2. INTECH, pp. 33–67 (2013), ISBN 980-953-307-937-5.

173. Klausner, J., Petrasch, J., Mei, R., Hahn, D., Mehdizadeh, A., and Auyeung, N., Solar fuel: Pathway to a sustainable energy future, Florida Energy Summit, University of Florida, Gainesville, FL (August 17, 2012).

174. Ris, T. and Hagen, E., Hydrogen production-gaps and priorities, IEA Hydrogen Implementing Agreement (HIA), a report by IEA, Brussels, Belgium, pp. 1–111 (2005).

175. Suarez, M., Blanco-Marigorta, A., and Peria-Quintana, J., Review on hydrogen production technologies from solar energy, in *International Conference on Renewable Energies and Power Quality* (ICREPQ'11), La Palmas, Spain (April 13–15, 2011).

176. Nowotny, J., Sorrell, C., Sheppard, L., and Bak, T., Solar-hydrogen: Environmentally safe fuel for the future, *International Journal of Hydrogen Energy*, 30 (5), 521–544 (2005).

177. Bockris, J., Dandapani, B., Cocke, D., and Ghoroghchian, J., On the splitting of water, *International Journal of Hydrogen Energy*, 10 (30), 179–201 (1985).

178. Naterer, G.F., Economics and Synergies of electrolytic and thermochemical methods of environmentally benign hydrogen production, in *Proceedings of WHEC*, Essen, Germany (May 16–21, 2010).

179. Laoun, B., Thermodynamics aspect of high pressure hydrogen production by water electrolysis, *Revue des Energies Renouvelables*, 10 (3), 435–444 (2007).

180. Funk, J.E., Thermochemical and electrolytic production of hydrogen, in Vezeroglu, T.N. (ed.), *Introduction to Hydrogen Energy*. International Association for Hydrogen Energy, Coral Gables, FL, pp. 19–49 (September 1975).

181. Onda, K., Kyakuno, T., Hattori, K., and Ito, K., prediction of production power for high pressure hydrogen by high pressure water electrolysis, *Journal of Power Sources*, 132, 64–70 (2004).

182. High temperature electrolysis, Wikipedia, the free encyclopedia, pp. 1–3 (2012), https://en.wikipedia.org/wiki/High-temperature_electrolysis, Last modified May 25, 2016.

183. McKellar, M., Harvego, E., and Gandrik, A., System evaluation and economic analysis of a HTGR powered high temperature electrolysis production plant, in *Proceedings of HTR*, paper 093, Prague, Czech Republic (October 18–20, 2010).

184. Herring, J., O'Brien, J., Stoots, C., Lessing, P., and Anderson, R., High temperature solid oxide electrolyzer system, Hydrogen, fuel cells, and infrastructure technologies, DOE Progress report, Washington, DC, pp. 1–5 (2003).

185. Herring, J., O'Brien, J., Stoots, C., Lessing, P., and Hartvigsen, J., High temperature electrolysis for hydrogen production, in a Paper presented by *Idaho National Laboratory for Materials Innovations in an Emerging Hydrogen Economy*, Hilton Oceanfront, Cocoa Beach, FL (February 26, 2008).

186. LeRoy, R.L., Bowen, C.T., and LeRoy, D.J., The thermodynamics of aqueous water electrolysis, *Journal of the Electrochemical Society,* 127, 1954 (1980).
187. LeRoy, R.L., Janjua, M.B.I., Renaud, R., and Leuenberger, U., Analysis of time-variation effects in water electrolysis, *Journal of the Electrochemical Society,* 126, 1674 (1979).
188. Ohashi, K., McCann, J., and Bockris, J.O'M., *Nature,* 266, 610 (1977).
189. Szklarczyk, M. and Bockris, J.O'M., *Applied Physics Letters,* 42, 1035 (1983).
190. Szklarczyk, M. and Bockris, J.O'M., *Journal of Physical Chemistry,* 88, 1808 (1984).
191. Scaife, D.E., Weller, P.F., and Fisher, W.G., Crystal preparation and properties of cesium tin(II) tri-halides, *Journal of Solid State Chemistry,* 9 (3), 308–314 (March 1974).
192. Scaife, D., Oxide semiconductors in photoelectrochemical conversion of solar energy, *Solar Energy,* 25 (1), 41–54 (1980).
193. Shyu, R., Weng, F., and Ho, C., Manufacturing of a micro probe using supersonic aided electrolysis process, DTIP of MEMS and MOEMS (April 9–11, 2008).
194. Bockris, J. and Murphy, O., One-unit photo-activated electrolyzer, U.S. Patent No. US4790916 A (December 13, 1988).
195. Bockris, J. and Veziroglu, T., A solar-hydrogen energy system for environmental compatibility, *Environmental Conservation,* 12 (2), 105–118 (1985).
196. Okada, G., Guruswamy, V., and Bockris, J., On the electrolysis of coal slurries, *Journal of Electrochemical Society,* 128, 2097 (1981).
197. Trogler, W., Geoffrey, G., Erwin, D., and Gray, H., *Journal of the American Chemical Society,* 100, 1160 (1978).
198. Erwin, D., Geoffrey, G., Gray, H., Hammond, G., Soloman, E., Trogler, W., and Zagers, A., *Journal of the American Chemical Society,* 99, 3620 (1977).
199. Tyler, D. and Gray, H., *Journal of American Chemical Society,* 103, 1683 (1981).
200. Mann, K., Lewis, N., Williams, R., Gray, H., and Gordon, J., Further studies of metal–metal bonded oligomers of rhodium (I) isocyanide complexes. Crystral structure analysis of [Rh$_2$(CNPH)$_8$] (Bph$_4$)$_2$. *Inorganic Chemistry,* 17(4), 828–834 (1978).
201. Sprintschnik, G., Sprintschnik, H., Kirsch, P., and Whitten, D., *Journal of the American Chemical Society,* 98, 2337 (1976).
202. Sprintschnik, G., Sprintschnik, H., Kirsch, P., and Whitten, D., *Journal of the American Chemical Society,* 99, 4947 (1977).
203. Maverick, A. and Gray, H., Solar energy storage reactions involving metal complexes, *Pure and Applied Chemistry,* 52, 2339 (1980).
204. Kiwi, J., Kalyanasundaram, K., and Gratzel, M., *Structure and Bonding,* Vol. 49. Springer, New York, p. 37 (1982).
205. Koriakin, B., Dshabiev, T., and Shivlov, A., *Doklady Akademii Nauk SSSR,* 298, 620 (1977).
206. Kalyasundarasam, K., Micic, O., Pramauro, E., and Gratzel, M., *Helvetica Chimica Acta,* 62, 2432 (1979).
207. Duonghong, D., Borgarello, E., and Gratzel, M., *Journal of the American Chemical Society,* 103, 4685 (1981).
208. Kato, H. and Kudo, A., Photocatalytic decomposition of pure water into H$_2$ and O$_2$ over SrTazO$_6$ prepared by a flux method, *Chemistry Letters,* 1207 (1999).
209. Lin, W., Cheng, C., Hu, C., and Teng, H., NaTaO$_3$ photocatalysts of different crystalline structures for water splitting into H$_2$ and O$_2$, *Applied Physics Letters,* 89, 211904 (2006).
210. Navarro, Yerga, R., Alvarez Galvan, M., del Valle, F., Villoria de la Mano, J., and Fierro, J., Water splitting on semiconductor catalysts under visible light irradiation, *ChemSusChem,* 2 (6), 471–485 (June 22, 2009).
211. Sato, J., Saito, N., Nishiyama, H., and Inoue, Y., Photocatalytic water decomposition by RuO$_2$ loaded antimonates, M$_2$Sb$_2$O$_7$ (M = Ca, Sr), CaSb$_2$O$_6$ and NaSbO$_3$ with d10 configuration, *Journal of Photochemistry and Photobiology A: Chemistry,* 148, 85–89 (2002).
212. Zou, Z., Ye, J., Sayama, K., and Arakawa, H., Photocatalytic hydrogen and oxygen formation under visible light irradiation with M-doped InTaO$_4$ (M = Mn, Fe, Co, Ni or Cu) photocatalysts, *Journal of Photochemistry and Photobiology A: Chemistry,* 148, 65–69 (2002).
213. Water splitting, Wikipedia, the free encyclopedia (2012).
214. Lee, J., Designer transgenic, algae for photobiological production of hydrogen from water, Chapter 20, in Lee, J. (ed.), *Advanced Biofuels and Bioproducts.* Springer Science, New York, pp. 371–404 (2012).
215. Sato, J., Kobayashi, H., Saito, N., Nishiyama, H., and Inoue, Y., Photocatalytic activities for water decomposition of RuO$_2$ loaded AlnO$_2$ (A = Li, Na) with d10 configuration, *Journal of Photochemistry and Photobiology A: Chemistry,* 158, 139–144 (2003).

216. Zou, Z. and Arakawa, H., Direct water splitting into H_2 and O_2 under visible light irradiation with a new series of mixed oxide semiconductor photocatalysts, *Journal of Photochemistry and Photobiology, A: Chemistry*, 158, 145–162 (2003).

217. Harda, H., Hosoki, C., and Kudo, A., Overall water splitting by sonophotocatalytic reaction: The role of powdered photocatalyst and an attempt to decompose water using a visible light sensitive photocatalyst, *Journal of Photochemistry and Photobiology A: Chemistry*, 141, 219–224 (2001).

218. Abe, R., Hara, K., Sayama, K., Domen, K., and Arakawa, H., Steady hydrogen evolution from water on Eosin Y-fixed TiO_2 photocatalyst using a silane coupling reagent under visible light irradiation, *Journal of Photochemistry and Photobiology A: Chemistry*, 137, 63–69 (2000).

219. Fujishima, A., Rao, T., and Tryk, D., Titanium dioxide photocatalysis, *Journal of Photochemistry and Photobiology C: Photochemistry Reviews*, 1, 1–21 (2000).

220. Navarro, R., Sanchez-Sanchez, M., Alvarez-Galvan, M., del Valle, F., and Fierro, J., Hydrogen production from renewable sources: Biomass and photocatalytic opportunities, *Energy and Environment Science*, 2, 35–54 (2009).

221. Fujishima, A. and Honda, K., Electrochemical photolysis of water at a semiconductor electrode, *Nature*, 238, 37 (1972).

222. Kato, H. and Kudo, A., *Journal of Physical Chemistry B*, 106, 5029 (2002).

223. Ishii, T., Kato, H., and Kudo, A., Hydrogen evolution from an aqueous methanol solution on $SrTiO_3$ photocatalysts codoped with chromium and tantalum ions under visible light irradiation. *Journal of Photochemistry and Photobiology A*, 163, 181 (2004).

224. Kudo, A. and Sekizawa, M., *Catalysis Letters*, 58, 241 (1999).

225. Kudo, A. and Sekizawa, M., *Chemical Communications*, 1371 (2000).

226. Konta, R., Ishii, T., Kato, H., and Kudo, A., Photocatalytic activation of noble metal ion dopped $SrTiO_3$ under light radiation, *Journal of Physical Chemistry B*, 108, 8992 (2004).

227. Anpo, M. and Takeuchi, M., *Journal of Catalysis*, 216, 505 (2003).

228. Asahi, R., Morikawa, T., Ohwaki, T., Aoki, K., and Taga, Y., *Science*, 293, 269 (2001).

229. Umebayashi, T., Yamaki, T., Itoh, H., and Asai, K., *Applied Physics Letters*, 81, 454 (2002).

230. Kudo, A., Omori, K., and Kato, H., *Journal of the American Chemical Society*, 121, 11459 (1999).

231. Asokkumar, M., *International Journal of Hydrogen Energy*, 23, 427 (1998).

232. Kalyanasundaram, M., Graetzel, M., and Pelizzetti, E., *Coordination Chemistry Reviews*, 69, 57 (1986).

233. Meissner, D., Memming, R., and Kastening, B., *Journal of Physical Chemistry*, 92, 3476 (1988).

234. Fuji, H. and Guo, L., *Journal of Physical Chemistry B*, 110, 1139 (2006).

235. Spanhel, L., Weller, H., and Hanglein, A., *Journal of the American Chemical Society*, 109, 6632 (1987).

236. Navarro, R., del Valle, F., and Fierro, J., Photocatalytic hydrogen evolution from CdS-ZnO-CdO systems under visible light irradiation: Effect of thermal treatment and presence of Pt and Ru cocatalysts, *International Journal of Hydrogen Energy*, 33, 4265 (2008).

237. Tsuji, I. and Kudo, A., *Journal of Photochemistry and Photobiology A*, 156, 249 (2003).

238. Kudo, A., Nishiro, R., Iwase, A., and Kato, H., *Chemical Physics*, 339, 104 (2007).

239. Tsuji, I., Kato, H., and Kudo, A., Visible-light-induced H_2 evolution from aqueous solution containing sulfide and sulfite over a $ZnS\text{-}CuInS_2\text{-}AgInS_2$ solid-solution photocatalyst, *Angewandte Chemie*, 117, 3631 (2005); also *Angewandte Chemie, International Edition*, 44, 3565–3568 (2005).

240. Kudo, A., Tsuji, I., and Kato, H., *Chemical Communications*, 1958 (2002).

241. Tsuji, I., Kato, H., Kobayashi, H., and Kudo, A., Photocatalytic H_2 evolution under visible-light irradiation over band-structure-controlled $(CuIn)_x Zn_{2(1-x)}S_2$ solid solutions, *Journal of Physical Chemistry B*, 109, 7323–7329 (2005).

242. Prince, R. and Kheshgi, H., The photobiological production of hydrogen: Potential efficiency and effectiveness as a renewable fuel, *Critical Reviews in Microbiology*, 31, 19–31 (2005).

243. Ghysels, B. and Franck, F., Hydrogen photo-evolution upon S deprivation stepwise: An illustration of microalgal photosynthetic and metabolic flexibility and a step stone for future biological methods of renewable H_2 production, *Photosynthesis Research*, 106, 145–154 (2010).

244. Mells, A., Zhang, L., Forestier, M., Ghirardi, M., and Seibert, M., Sustained photobiological hydrogen gas production upon reversible inactivation of oxygen evolution in the green alga *Chlamydomonas reinhardtii*, *Plant Physiology*, 122, 127–135 (2000).

245. Ghirardi, M., Zhang, L., Lee, J., Flynn, T., Seibert, M., Greenbaum, E., and Melis, A., Microalgae: A green source of renewable H_2, *Trends in Biotechnology*, 18, 506–511 (2000).

246. Nguyen, A., Thomas-Hall, S., Malnoe, A., Timmins, M., Mussgnug, J., Rupprecht, J., Krause, O., Hankamer, B., and Schenk, P., Transcriptome for photobiological hydrogen production induced by sulfur deprivation in the green alga *Chlamydomonas reinhardtii*, *Eukaryotic Cell*, 7 (11), 1965–1979 (2008).

247. Timmins, M., Thomas-Hall, S., Darling, A., Zhang, E., Hankamer, B., Marx, U., and Schenk, P., Phylogenetic and molecular analysis of hydrogen producing green algae, *Journal of Experimental Botany*, 60 (6), 1691–1702 (2009).

248. Berberoglu, H. and Pilon, L., Maximizing the solar to H_2 energy conversion efficiency of outdoor photobioreactors using mixed cultures, *International Journal of Hydrogen Energy*, 35 (2), 500–510 (2010).

249. Lee, J., Designer proton channel transgenic algae for photobiological hydrogen production, PCT international patent application publication number: WO2007/134340 A2 (2007).

250. Lee, J., Designer proton channel transgenic algae for photobiological hydrogen production, U.S. Patent 7,932,437 B2 (2011).

251. Lee, J., Switchable photosystem-II designer algae for photobiological hydrogen production, U.S. Patent 7,642,405, B2 (2010).

252. Eastlund, B. and Gough, W., Paper presented at the *163rd National Meeting of American Chemical Society*, Boston, MA (April 14, 1972).

253. McWhirter, R., in Huddlestone, R. and Leonard, S. (eds.), *Plasma Diagnostic Techniques*, Chapter 5. Academic Press, New York, pp. 201–264 (1965).

254. Inoue, M., Uehara, R., Hasegawa, N., Gokon, N., Kaneko, H., and Tamaura, Y., Solar hydrogen generation with $H_2O/ZnO/MnFe_2O_4$ system, ISES, *Solar World Congress*, Adelaide, Australia, pp. 1723–1729 (2001).

255. Brown, L.C., Besenbrauch, G.E., Schultz, K.R., Showalter, S.K., Marshall, A.C., Pickard, P.S., and Funk, J.F., High efficiency generation of hydrogen fuels using thermochemical cycles and nuclear power, in *Spring National Meeting of AIChE*, Nuclear Engineering Session THa01 139, Hydrogen Production and Nuclear Power, New Orleans, LA (March 11–15, 2002).

256. Rosen, M.A., Developments in the production of hydrogen by thermochemical water decomposition, *International Journal of Energy and Environmental Engineering*, 2 (2), 1–20 (Spring 2011).

257. Steinfeld, A. and Palumbo, R., Solar thermochemical process technology, a paper, in Meyers, R. (ed.), *Encyclopedia of Physical Science and Technology*, Vol. 15. Academic Press, pp. 237–256 (2001).

258. Funk, J.E., Conger, W.L., and Cariy, R.H., Evaluation of multi-step thermochemical processes for the production of hydrogen from water, in *THEME Conference Proceedings*, Miami Beach, FL, pp. Sll-.ll (March 1974).

259. Chao, R.E., Thermochemical water decomposition processes, *Industrial & Engineering Chemistry Product Research and Development*, 13 (2), 94–101 (June 1974).

260. Bamberger, C. and Richardson, D., Thermochemical decomposition of water based on reactions of chromium and strontium compounds, A personal communication and a report published by Oak Ridge National Laboratory, Oak Ridge, TN (2011).

261. Weimer, A., II.1.2 Fundamentals of a solar-thermal hydrogen production process using a metal oxide based thermochemical water splitting cycle, DOE contract no. DE-FC36-05G015044, Annual Progress Report (2006).

262. Roeb, M. and Sattler, C., Hycycles-materials and components of hydrogen production by sulfur based thermochemical cycles (FPS-Energy-212470), German Aerospace Center (DLR)-Solar Research Report (2010).

263. Law, V.J., Prindle, J.C., and Gonzales, R.B., Analysis of the copper sulfate cycle for the thermochemical splitting of water for hydrogen production, in Paper submitted for presentation at the *2007 AIChE National Meeting*, Salt Lake City, UT (November 2007).

264. Law, V.J., Prindle, J.C., and Gonzales, R.B., Level 1 and level 2 analysis of the copper sulfate cycle for the thermochemical splitting of water for hydrogen production, Contract no. 6F-003762, report published by Argonne National Laboratory, Oak Ridge, TN (July 2006).

265. Law, V.J., Prindle, J.C., and Gonzales, R.B., Level 3 analysis of the copper sulfate cycle for the thermochemical splitting of water for hydrogen production, Contract no. 6F-003762, report published by Argonne National Laboratory, Oak Ridge, TN (August 2006).

266. Law, V.J., Prindle, J.C., Bang, R., Hoerger, K., and Ledbetter, J., Progress report no. 1, Experimental studies of the hydrogen generation reaction for the $CuSO_4$ cycle, Contract no. 6F-01144, report published by Argonne National Laboratory, Oak Ridge, TN (October 2006).

267. Law, V.J., Prindle, J.C., Bang, R., Hoerger, K., and Ledbetter, J., Progress report no. 2, Experimental studies of the hydrogen generation reaction for the $CuSO_4$ cycle, Contract no. 6F-01144, report published by Argonne National Laboratory, Oak Ridge, TN (January 2007).

268. Law, V.J., Prindle, J.C., Bang, R., Hoerger, K., and Ledbetter, J., Progress report no. 3, Experimental studies of the hydrogen generation reaction for the $CuSO_4$ cycle, Contract no. 6F-01144, report published by Argonne National Laboratory, Oak Ridge, TN (March 2007).

269. Law, V.J., Prindle, J.C., Bang, R., Hoerger, K., and Ledbetter, J., Progress report no. 4, Experimental studies of the hydrogen generation reaction for the $CuSO_4$ cycle, Contract no. 6F-01144, report published by Argonne National Laboratory, Oak Ridge, TN (May 2007).

270. Law, V.J., Prindle, J.C., and Lupulescu, A.I., Aspen plus modeling of the three-reaction version of the copper-chloride thermochemical cycle for hydrogen production from water, Contract no. 6F-01144, Argonne National Laboratory (August 2007).

271. Law, V., Prindle, J., and Gonzales, R., Analysis of the copper sulphate cycle for the thermochemical splitting of water for hydrogen production, Tulane University, Chemical Engineering Department, New Orleans, LA (2010).

272. Roeb, M. et al., Solar hydrogen production by a two step cycle based on mixed iron oxides, *Journal of Solar Energy Engineering*, 128, 125–133 (May 2006).

273. T-Raissi, Huang, C., and Muradov, N., Hydrogen production via solar thermochemical water splitting, NASA/CR-2009-215441, Report of research for period March 2004 to February 2008 (2009).

274. Steinfeld, A. and Palumbo, R., Solar thermochemical process technology, *Encyclopedia of Physical Science and Technology*, 15 (1), 237–256 (2001).

275. Steinfeld, A. and Meier, A., *Solar Fuels and Materials*. Elsevier, Amsterdam, the Netherlands, pp. 623–637 (2004).

276. Steinfeld, A., Solar hydrogen production via a two-step water-splitting thermo-chemical cycle based on Zn/ZnO redox reactions, *International Journal of Hydrogen Energy*, 27 (6), 611–619 (2002).

277. Funke, H.H., Diaz, H., Liang, X., Carney, C.S., Weimer, A.W., and Li, P., Hydrogen generation by hydrolysis of zinc powder aerosol, *International Journal of Hydrogen Energy*, 33 (4), 1127–1134 (2008).

278. Abanades, S., Charvin, P., Lemont, F., and Flamant, G., Novel two-step SnO_2/SnO water-splitting cycle for solar thermochemical production of hydrogen, *International Journal of Hydrogen Energy*, 33 (21), 6021–6030 (2008).

279. Meier, A. and Steinfeld, A., Solar thermochemical production of fuels, *Advances in Science and Technology*, 74 (1), 303–312 (2011).

280. Chueh, W.C. and Haile, S.M., Ceria as a thermochemical reaction medium for selectively generating syngas or methane from H_2O and CO_2, *ChemSusChem*, 2 (8), 735–739 (2009).

281. Kappauf, T. and Fletcher, E.A., Hydrogen and sulfur from hydrogen sulfide—VI. Solar thermolysis, *Energy*, 14 (8), 443–449 (1989).

282. Zaman, J. and Chakma, A., Production of hydrogen and sulfur from hydrogen sulfide, *Fuel Processing Technology*, 41 (2), 159–198 (1995).

283. Steinfeld, A., Solar thermochemical production of hydrogen—A review, *Solar Energy*, 78 (5), 603–615 (2005).

284. Perret, R., Solar thermochemical hydrogen production research (STCH), thermo-chemical cycle selection and investment priority, Report No. SAND2011-3622, Sandia National Laboratories, Albuquerque, NM (2011).

285. Almodaris, M., Khorasani, S., Abraham, J.J., and Ozalp, N. (eds.), Simulation of solar thermo-chemical hydrogen production techniques, in *ASME/JSME 2011 Eighth Thermal Engineering Joint Conference*, March 13–17, 2011, ASME, Honolulu, HI (2011).

286. Aoki, A., Ohtake, H., Shimizu, T., Kitayama, Y., and Kodama, T., Reactive metal-oxide redox system for a two-step thermochemical conversion of coal and water to CO and H_2, *Energy*, 25 (3), 201–218 (2000).

287. Osinga, T., Olalde, G., and Steinfeld, A., Solar carbothermal reduction of ZnO: Shrinking packed-bed reactor modeling and experimental validation, *Industrial & Engineering Chemistry Research*, 43 (25), 7981–7988 (2004).

288. Osinga, T., Frommherz, U., Steinfeld, A., and Wieckert, C., Experimental investigation of the solar carbothermic reduction of ZnO using a two-cavity solar reactor, *Journal of Solar Energy Engineering*, 126 (1), 633–637 (2004).

289. Epstein, M., Olalde, G., Santén, S., Steinfeld, A., and Wieckert, C., Towards the industrial solar carbothermal production of zinc, *Journal of Solar Energy Engineering*, 130 (1), 014505 (2008).

290. Wieckert, C. et al., A 300 kW solar chemical pilot plant for the carbothermic production of zinc, *Journal of Solar Energy Engineering*, 129 (2), 190–196 (2007).

291. Kromer, M., Roth, K., Takata, R., and Chin, P., Support for cost analyses on solar-driven high temperature thermochemical water-splitting cycles, Report No. DE-DT0000951, TIAX, Lexington, MA, LLC2011 (February 22, 2011).

292. Lede, J., Lapicque, F., and Villermaux, J., *International Journal of Hydrogen Energy*, 8, 675 (1983).

293. Lapicque, F., Lede, J., Villermaux, J., Caler, B., Baumard, J., Anthony, A., Abdul-Aziz, G., Puechbertz, D., and Ledrix, M., *Entropie*, 19, 42 (1983).

294. Beghi, G., A decade of research on thermochemical water hydrogen at the joint research center, Ispra, *International Journal of Hydrogen Energy*, 11 (12), 761–771 (1986).
295. Besenbruch, G., General atomic sulfur-iodine thermochemical water splitting process, *American Chemical Society, Division of Petroleum Chemistry Preprint*, 271, 48 (1982).
296. Sato, S., Shimizu, S., Nakajima, N., and Ikezoe, Y., A nickel-iodine-sulfur process for hydrogen production, *International Journal of Hydrogen Energy*, 8 (1), 15–22 (1983).
297. Abanades, S., Charvin, P., Flamant, G., and Neveu, P., Screening of water-splitting thermochemical cycles potentially attractive for hydrogen production by concentrated solar energy, *Energy*, 31, 2805–2822 (2006).
298. Marchetti, C., Hydrogen and energy, *Chemical Economy and Engineering Review*, 5 (1), 7–15 (January 1973).
299. DeBeni, G. and Marchetti, C., *ACS Meeting*, Boston, MA (April 9, 1972).
300. Appleby, A. and Bockris, J., Calculation of the energy change involved in chemical reactions occurring irreversibly, *International Journal of Hydrogen Energy*, 6, 1–7 (1981).
301. Pyle, W., Hayes, M., and Spivak, A., Direct solar-thermal hydrogen production from water using nozzle/skimmer and glow discharge, IECEC96535, A report from H-ION Solar Inc., Richmond, CA (2010).
302. Baykara, S., Experimental solar water thermolysis, *International Journal of Hydrogen Energy*, 29 (14), 1459–1469 (2004).
303. Harvey, W.S., Davidson, J.H., and Fletcher, E.A., Thermolysis of hydrogen sulfide in the temperature range 1350–1600 K, *Industrial & Engineering Chemistry Research*, 37 (6), 2323–2332 (1998).
304. Perkins, C. and Weimer, A.W., Solar-thermal production of renewable hydrogen, *AIChE Journal*, 55 (2), 286–293 (2009).
305. Venugopalan, M. and Jones, R., *Chemistry of Dissociated Water Vapor and Related Systems*. Wiley, New York (1968).
306. Schultz, K., Thermochemical production of hydrogen from solar and nuclear energy, in Presentation to the *Stanford Global Climate and Energy Project*, April 14, General Atomics, San Diego, CA (2003).
307. Kerr, W. and Majumdar, D., in Veziroglu, T. (ed.), *Hydrogen Energy, Part A*. Plenum Press, New York, p. 167 (1975).
308. Gomberg, H. and Gordus, A., *Journal of Fusion Energy*, 2, 319 (1982).
309. Northrup, C. Jr., Gerlach, T., Modreski, P., and Galt, J., *International Journal of Hydrogen Energy*, 3 (1), (1978).
310. Fudali, R., *Geochimica et Cosmochimica Acta*, 29, 529 (1948).
311. Kennedy, G., *American Journal of Science*, 246, 529 (1948).
312. Bockris, J. and Gutmann, F., *Applied Physics Communications*, 1, 121 (1981–1982).
313. Farady, M., *Diary*, Vol. 1. Bell, London, U.K., p. 381 (1932).
314. Appleton, A., in Fouer, S. and Scwartz, B. (eds.), *Superconducting Machines and Devices*. Plenum Press, New York, p. 219 (1973).
315. Boyd, R. and Burns, G., in Lifshitz, A. (ed.), *Shock Waves in Chemistry*. Marcel Dekker, New York, p. 131 (1981).
316. Kasal, P. and Bishop, R. Jr., U.S. Patent No. 3,963,830 (1976).
317. Kasal, P. and Bishop, R. Jr., *Journal of Physical Chemistry*, 81, 1527 (1977).
318. England, C., in Veziroglu, T., Van Vorst, W., and Kelley, H. (eds.), *Hydrogen Energy Progress IV: Proceedings of World Hydrogen Energy Conference, IV*, Pasadena, CA (June 13–17, 1982), Vol. 2. p. 465 (1982).
319. Fletcher, E.A. and Moen, R.L., Hydrogen and oxygen from water, *Science*, 197, 1050–1056 (1977).
320. Fateev, N. et al., Water electrolysis in systems containing solid polymer electrolytes, *Electrokhimiya*, 29, 551 (1993).
321. Shah, Y.T., *Water for Energy and Fuel Productions*. CRC Press, New York (2014).
322. Shah, Y.T., *Energy and Fuel Systems Integration*. CRC Press, New York (2015).

6 Gas Reforming

6.1 INTRODUCTION

It is known that while hydrocarbons, particularly those derived from fossil fuels, are harmful to environment, hydrogen is environment-friendly and many consider it to be an "ultimate fuel" for the world. Furthermore, the mixture of hydrogen and carbon monoxide in the right proportion is a very useful feedstock for many downstream conversions to liquid fuel and useful chemicals. Thus the process of conversion of hydrocarbons to hydrogen and syngas is a "game changer" in making gaseous fuels not only "transition fuels" but also "ultimate fuels." Such a process—"gas reforming"—has become one of the most important processes in the gas industry.

The process of gas reforming is different from catalytic reforming that is used widely in the refining industry to upgrade the octane number of liquid fuels and produce other valuable chemicals. Gas reforming is focused on the production of syngas from a variety of hydrocarbons with particular emphasis on methane. This chapter strictly deals with gas reforming.

There are fundamentally three important parameters in the process of gas reforming: the nature of reactants and their environment, the method of heat provision, and the nature of the catalyst and reactor. This chapter deals with different types of gas reforming processes which address the handling of these three parameters. While several gas reforming processes are commercially established, many more are being developed to improve the overall gas reforming technology. This chapter deals with both commercial and research-based options for gas reforming.

6.2 TYPES OF GAS REFORMING

There are numerous methods for producing syngas from methane, volatile organic hydrocarbons as well as other renewable and nonrenewable feedstock such as coal, oil shale, biomass, and waste and their mixtures. The most basic form of gas reforming is the "water gas shift reaction" (henceforth denoted as WGS reaction) in which the reaction between water and carbon monoxide produces hydrogen and carbon dioxide. This reaction is always present with other forms of gas reforming, and it helps to adjust the hydrogen to carbon monoxide ratio in the final product. This is the single most important reaction in syngas production, and its kinetics and catalysis have been significantly researched [1–71]. Section 6.3 critically evaluates this basic gas reforming reaction. It is important to note that in the WGS reaction, no hydrocarbon is involved.

The next four types of gas reforming reactions depend on what reactant is used to reform hydrocarbons. The most common and commercialized form of gas reforming is steam reforming, wherein steam reacts with carbonaceous materials to produce syngas with heavy concentration of hydrogen. Steam reforming is a well-established commercial technology to produce hydrogen, which is a raw material to produce ammonia, or upgrade crude oil in standard refining operations. While steam reforming of methane is a long-standing successful commercial process, in recent years, steam reforming of other feedstock has been pursued. Section 6.4 examines the process of steam reforming in detail.

While steam reforming is a commercial process, in recent years reforming of methane and other hydrocarbons by carbon dioxide has caught increasing attention. This process, commonly known as "dry reforming," is important because it simultaneously removes two greenhouse gases—methane and carbon dioxide—thus enhancing the reduction of carbon emission in the environment.

Unlike steam reforming, dry reforming also produces syngas product with different composition of hydrogen and carbon monoxide, the one which is more suitable for downstream conversion of syngas to liquid fuels. Both steam and dry reforming are endothermic processes requiring large amount of heat, and both processes require high temperatures (higher than 500°C) for getting favorable conversion of hydrocarbons to syngas. The subject of dry reforming is treated in great detail in Section 6.5.

While steam reforming produces syngas with a hydrogen to carbon monoxide ratio of 3, and dry reforming produces syngas with a hydrogen to carbon monoxide ratio of 1, the partial oxidation (POX) of hydrocarbon produces syngas with a hydrogen to carbon monoxide ratio of 2. The product distribution of these three types of reforming processes and their downstream applications are summarized in Table 6.1. Unlike steam and dry reforming, POX is an exothermic reaction requiring no external source of heat. In recent years, this feature has encouraged the use of combine steam or dry reforming with POX in order to have reaction autothermal (self-sustained for heat requirement). When all three (steam reforming, dry reforming, and POX) are carried out simultaneously, the process is often referred as "tri-reforming" or "oxy-forming." This process allows more flexibility in the composition of syngas and requires no external heat. These obvious advantages have encouraged more investigations of this type of reforming in recent years. Section 6.6 critically examines autothermal and tri-reforming processes.

While steam reforming of hydrocarbons is an established technology, in recent years reforming in sub- and supercritical water environment is gaining momentum. Aqueous-phase reforming (APR) of carbohydrates (mainly containing equal proportion of carbon and oxygen) has been extensively researched in recent years. Section 6.7 describes this in detail. Reforming under supercritical water environment offers many distinctive advantages such as high solubility of hydrocarbons, greater reaction rate and selectivity, and less coking of catalyst. These issues, along with various other features of reforming under supercritical water environment, are also discussed in detail in Section 6.7. While APR is mainly applied to reforming with water, supercritical reforming can be applied to a combination of water, carbon dioxide, and oxygen (tri-reforming). Also, unlike APR which is largely applicable to carbohydrates and other cellulosic materials, supercritical reforming can be applied to all carbonaceous materials, including coal, oil shale, biomass, waste, and their mixtures. Both APR and supercritical water reforming produce gas with hydrogen as their main constituent.

As mentioned earlier, besides the nature of reactants and the reaction environment, another major issue in gas reforming is the availability of efficient and cheaper methods of heat supply. Fundamentally, there are three novel methods for providing efficient heat to the reforming process.

TABLE 6.1
Product Compositions and Applications for Reforming Reactions

Reforming Reaction	Product Composition (H_2/CO)	Application
Dry reforming: $CH_4 + CO_2 \rightarrow 2H_2 + 2CO$	≤ 1	Polycarbonates, formaldehyde, iron ore reduction
Partial oxidation: $CH_4 + \frac{1}{2}O_2 \rightarrow 2H_2 + CO$	2	Methanol
Steam reforming: $CH_4 + H_2O \rightarrow 3H_2 + CO$	3	H_2, NH_3, chemicals
Water gas shift: $CO + H_2O \rightleftharpoons H_2 + CO_2$	1–2	Power, fuel, and chemicals

Source: York, P.E. et al., *Catal. Rev.*, 49(4), 511, 1995.

The use of microwave to provide heat for activated carbon or other metal-catalyzed reforming process has been found to be very effective, particularly at lower temperatures. Microwave provides efficient heat without the need for heating the entire surrounding (just like in home microwave cooking) and, as shown in Section 6.8, this method has been investigated for gas reforming in the literature. This was demonstrated in my previous books [1,72] and is also briefly outlined in Section 6.8. Waste heat from nuclear reactor or from solar-based power plants can also be used to provide heat for reforming process. These processes of "co-generation" use the waste heat with very little cost. This subject was extensively covered in my previous books [1,72] and is also briefly outlined in Section 6.8. The use of microwave-, nuclear-, and solar-energy-assisted heat provision does require special types of reactors and process configurations, and these are being evaluated on a smaller scale. More work is needed to make these approaches workable on a commercial scale. Section 6.8 evaluates these novel methods for heat provision in detail. In principle, these novel methods can be applied to any types of reforming process outlined in Sections 6.3 through 6.7.

While the role of a heterogeneous catalyst has been identified in different types of reforming processes outlined in Sections 6.3 through 6.8, Section 6.9 discusses the use of plasma as a catalyst in the reforming process. Many reforming catalysts are rapidly deactivated due to the deposition of coke at high temperatures. Sections 6.3 through 6.8 describe several efforts attempted toward minimization of coking on the catalyst under reforming conditions. Plasma, the fourth state of matter, can also be used to reform hydrocarbons wherein plasma can act as a catalyst. The use of plasma as a catalyst alleviates the problem of catalyst deactivation due to coking. There are different types of plasma such as nonthermal, which includes corona discharge, dielectric barrier discharge (DBD), microwave discharge, atmospheric pressure glow discharge (APGD), and gliding arc discharge and thermal plasma. The use and effectiveness of these types of plasma for gas reforming offer some advantages and some challenges. Most of the reported work in this area is at a laboratory scale and for methane. Different types of reforming (steam, dry, etc.) have been investigated. Plasma reforming can also be carried out in the presence of a heterogeneous catalyst. Such a hybrid process has been found to give better performance at the laboratory level. More work is needed to critically analyze coking of heterogeneous catalysts in such a hybrid process.

For achieving high carbon and energy conversion efficiencies by plasma process, three key factors should be taken into account: electron density, plasma temperature, and reactor configuration. Corona discharge and DBD are nonuniform plasmas with low electron density and limited reaction volume which restrict the treatment capacity. Microwave discharge is a uniform discharge with high plasma temperature and large discharge space, but the equipment is more complicated and expensive which restricts its industrial application. Gliding arc discharge and APGD possess high electron energy and electron density, as well as proper plasma temperature, but the reactor is difficult to enlarge and the treatment capacity of both plasmas is still far from meeting industrial requirements. Thermal plasma is an efficient technology for CH_4–CO_2 reforming to synthesis gas because of its high specific enthalpy content, high temperature, high electron density, large treatment capacity as well as being easy to enlarge and due to its relatively high energy conversion efficiency. For raising the energy conversion efficiency of thermal plasma process further, the feed gases introduced into the discharge region and the synergistic effect of plasma with catalysts should be taken into consideration. These preliminary assessments are further discussed in detail in Section 6.9.

Finally, literature has shown that an effective reforming process requires careful considerations of the reactor design. Many novel reactor configurations have been considered in the literature [1,72]. The reactors must be effective and scalable. The final section, Section 6.10, evaluates various novel reactor configurations suggested for reforming process. The advantages and disadvantages of all novel reactors are briefly examined in Section 6.10.

6.3 WGS REACTION

The WGS reaction was discovered by Italian physicist Felice Fontana in 1780. It was not until much later that the industrial value of this reaction was realized. Before the early twentieth century, hydrogen was obtained by reacting steam under high pressure with iron to produce iron oxide and hydrogen. With the development of industrial processes that required hydrogen, such as the Haber–Bosch ammonia synthesis, the demand for a less expensive and more efficient method of hydrogen production arose. As a resolution to this problem, the WGS was combined with the gasification of coal to produce a pure hydrogen product [4–10].

6.3.1 Mechanism

Even though the WGS reaction is not classified as one of the principal gasification reactions, it cannot be omitted in the analysis of chemical reaction systems that involves synthesis gas. WGS reaction in its forward direction is mildly exothermic as [41–63]:

$$CO(g) + H_2O(g) = CO_2(g) + H_2(g) \quad H_{298}^{\circ} = -41.2 \, \text{kJ/mol} \tag{6.1}$$

While all the participating chemical species are in the form of a gas, scientists believe that this reaction predominantly takes place at the heterogeneous surfaces of coal and also that the reaction is catalyzed by carbon surfaces. As the WGS reaction is catalyzed by many heterogeneous surfaces and the reaction can also take place homogeneously as well as heterogeneously, a generalized understanding of the WGS reaction has been very difficult to achieve.

WGS has been extensively studied for over a hundred years. However, due to its complexity, the mechanism remains under debate. A universal rate expression and mechanistic understanding have proven elusive, reflecting the many reaction variables, the nature of the catalyst, and the proprietary nature of commercial processes [10].

The WGS reaction is a moderately exothermic reversible reaction. Therefore with increasing temperature the reaction rate increases, but the conversion of reactants to products becomes less favorable because high carbon monoxide conversion is thermodynamically favored at high temperatures. Despite the thermodynamic favorability at low temperatures, the reaction is kinetically favored at high temperatures. Over the temperature range 600–2000 K, the logarithm of the equilibrium constant for the WGS is given by the following equation (see Figure 6.1):

$$\log_{10} K_{eq} = -2.4198 + 0.0003855T + \frac{2180.6}{T} \tag{6.2}$$

The equilibrium constant first sharply decreases with an increase in temperature above around 190°C and levels off around 480°C. The relationship between the equilibrium constant and temperature has also been expressed in the literature as [70]:

$$K_p = \exp\left(\left(\frac{4477.8}{T}\right) - 4.33\right) \tag{6.3}$$

where T is in K. Thus, high forward conversion of WGS reaction is favored at low temperatures and it is essentially unaffected by the total pressure. At high temperatures, reverse WGS reaction dominates. The reaction is reversible and the forward reaction rate is strongly inhibited by the reaction products CO_2 and H_2. Low CO level can be obtained by maintaining the reactor temperature at around 200°C. At low temperatures, however, condensation of water and its contact with the

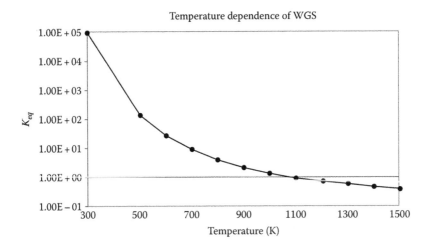

FIGURE 6.1 A plot of the temperature dependence of K_{eq}. (From WGS reaction, Wikipedia, the free encyclopedia, 2015.)

catalyst should be avoided. The equilibrium carbon monoxide concentration is also affected by the concentration of steam in the gas. Since WGS reaction is always present and it is equilibrium-controlled, in any steam reforming process, a substantial amount of carbon dioxide is present in the reaction mixture. The presence of carbon dioxide also forces "dry reforming" reaction between hydrocarbons and carbon dioxide.

Two main reaction mechanisms for WGS have been proposed [3–10,70]: an associative "Langmuir–Hinshelwood" mechanism and a regenerative "redox" mechanism. While the regenerative mechanism is generally implemented to describe the WGS at higher temperatures, at low temperatures both the redox and associative mechanisms are suitable explanations [10].

6.3.1.1 Associative Mechanism

In 1920, Armstrong and Hilditch first proposed the associative mechanism (see Figure 6.2). In this mechanism, CO and H_2O are adsorbed onto the surface of the metal catalyst followed by the formation of an intermediate and desorption of H_2 and CO_2. In the initial step, H_2O dissociates into a metal adsorbed OH and H. The hydroxide then reacts with CO to form a carboxyl or formate intermediate which subsequently decomposes into CO_2 and the metal adsorbed H, which ultimately yields H_2. While this mechanism may be valid under low-temperature shift (LTS) conditions, the redox mechanism, which does not involve any long-lived surface intermediates, is a more suitable explanation for the WGS mechanism at higher temperatures [10].

6.3.1.2 Redox Mechanism

The regenerative "redox" mechanism is the most commonly accepted mechanism for the WGS. It involves a regenerative change in the oxidation state of the catalytic metal. In this mechanism, H_2O is activated first by the abstraction of H from water followed by dissociation or disproportionation of the resulting OH to afford atomic O. The CO is then oxidized by the atomic O forming CO_2 which returns the catalytic surface back to its pre-reaction state. Alternatively, CO is directly oxidized by the OH to form a carboxyl intermediate, followed by the dissociation or disproportionation of the carboxyl. Finally H is recombined to H_2, and CO_2 and H_2 are desorbed from the metal. The principal difference in these mechanisms is the formation of CO_2. The redox mechanism generates CO_2 by reaction with adsorbed oxygen, while the associative mechanism forms CO_2 via the dissociation of an intermediate [10].

Depending on the reaction conditions, the equilibrium for the WGS can be pushed in either the forward or reverse direction. The reversibility of the WGS is important in the production of ammonia,

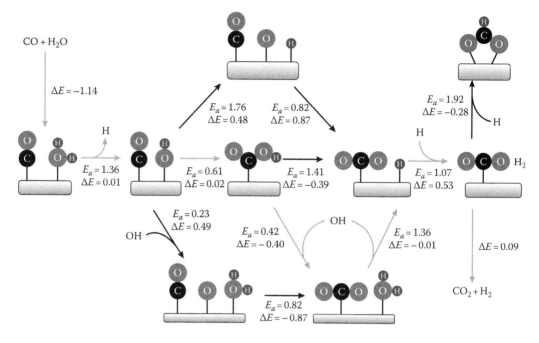

FIGURE 6.2 Associative mechanism of the WGS reaction. (From WGS reaction, Wikipedia, the free encyclopedia, 2015.)

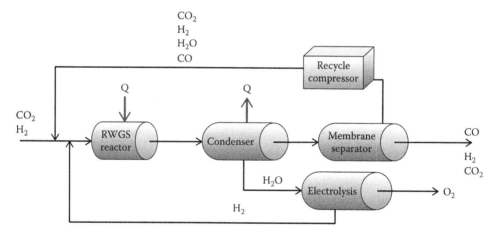

FIGURE 6.3 Reverse water gas shift flow cycle. (From WGS reaction, Wikipedia, the free encyclopedia, 2015.)

methanol, and Fischer–Tropsch (FT) synthesis where the ratio of H_2/CO is critical. Many industrial companies exploit the reverse WGS reaction (see Figure 6.3) as a source of the synthetically valuable CO from cheap CO_2. Typically, it is done using copper on aluminum catalyst [3–34,70].

6.3.2 ROLE OF WGS IN SYNGAS PRODUCTION

Syngas product from a gasifier contains a variety of gaseous species other than carbon monoxide and hydrogen. Typically, they include carbon dioxide, methane, and water (steam). Depending on the objective of the ensuing process, the composition of syngas needs to be preferentially readjusted. If the objective of the gasification were to obtain a high yield of methane, it would be preferable to have the molar ratio of hydrogen to carbon monoxide at 3:1, based on the following methanation reaction stoichiometry:

$$CO(g) + 3H_2(g) = CH_4(g) + H_2O(g) \qquad (6.4)$$

If the objective of generating syngas is the synthesis of methanol via vapor-phase, low-pressure process, the stoichiometrically consistent ratio between hydrogen and carbon monoxide would be 2:1. This ratio is generally achieved by the partial oxidation of methane. In such cases, the stoichiometrically consistent syngas mixture is often referred to as *balanced gas*, whereas a syngas composition that is substantially deviated from the principal reaction's stoichiometry is called *unbalanced gas*.

If the objective of syngas production is to obtain a high yield of hydrogen, it would be advantageous to increase the ratio of H_2 to CO by further converting CO (and H_2O) into H_2 (and CO_2) via WGS reaction. However, if the final gaseous product is to be used in fuel cell applications, carbon monoxide and carbon dioxide must be removed to acceptable levels by a process such as acid gas removal or other adsorption processes. In particular, for hydrogen proton exchange membrane (PEM) fuel cell operation, carbon monoxide and sulfurous species must be thoroughly removed from the hydrogen gas [10,11–34].

The WGS reaction is one of the major reactions in the steam gasification process, where both water and carbon monoxide are present in ample amounts. Even though all four chemical species involved in the WGS reaction are gaseous compounds at the reaction stage of most gas processing, the WGS reaction in the case of steam gasification of coal predominantly takes place heterogeneously, that is, on the solid surface of coal. If the product syngas from a gasifier needs to be reconditioned by the WGS reaction, this reaction can be catalyzed by a variety of metallic catalysts. Choice of specific kinds of catalysts has always depended on the desired outcome, the prevailing temperature conditions, the composition of gas mixture, and process economics. Many investigators have studied the WGS reaction over a variety of catalysts, including iron, copper, zinc, nickel, chromium, and molybdenum. Significant efforts have been made in developing a robust catalyst system that has superior sulfur tolerance and wider applicable temperature range. More details on catalysts for high temperature shift (HTS) and LTS reactions are given in Sections 6.3.4 and 6.3.5 respectively [4–10,70].

6.3.3 Reverse Water Gas Shift

Depending on the reaction conditions, the equilibrium for the WGS can be pushed in either the forward or reverse direction. The reversibility of the WGS reaction is important in the production of ammonia, methanol, and FT synthesis, where the ratio of H_2/CO is critical. Many industrial companies exploit the reverse water gas shift (RWGS) reaction as a source of the synthetically valuable CO from cheap CO_2. Typically, it is done using copper on aluminum catalyst. Coupling the RWGS with the water electrolysis process will yield methane and oxygen. Post electrolysis, the hydrogen produced can be recycled back into the RWGS reactor for the continued conversion of CO_2. Because this reaction is only mildly endothermic, the thermal power needed to drive this reaction can potentially be produced by a Sabatier reactor [8,10].

6.3.4 Location of WGS Reactor in Gasification Process

In applications where scrubbed syngas hydrogen/carbon monoxide (H_2/CO) ratio must be increased/ adjusted to meet downstream process requirements, the syngas is passed through a multistage, fixed-bed reactor containing shift catalysts to convert CO and water into additional H_2 and carbon dioxide (CO_2) by WGS reaction. As shown earlier, the equilibrium of the WGS reaction shows a significant temperature dependence and the equilibrium constant decreases with an increase in temperature; that is, higher carbon monoxide conversion is observed at lower temperatures. In order to take advantage of both the thermodynamics and kinetics of the reaction, the industrial-scale WGS reaction is conducted in multiple adiabatic stages consisting of a high temperature shift (HTS) followed by an LTS with intersystem cooling for the desired adjustment of H_2/CO ratio [3–10].

The shift reaction can operate with a variety of catalysts between 200°C and 480°C. The reaction does not change molar totals and therefore the effect of pressure on the reaction is minimal. However, the equilibrium for H_2 production is favored by high moisture content and low temperature for the exothermic reaction. Normally, excess moisture is present in the scrubber syngas from slurry-fed gasifiers sufficient to drive the shift reaction to achieve the required H_2-to-CO ratio. Indeed, for some slurry-fed gasification systems, a portion of the syngas feed may need to be bypassed around the sour shift reactor to avoid exceeding the required product H_2-to-CO ratio. On the other hand, additional steam injection before the shift may be needed for syngas output by dry-fed gasifiers [41–63].

In any case, the scrubber syngas feed is normally reheated to 15°C–25°C above saturation temperature to avoid catalyst damage by condensation of liquid water in the shift reactor. Shifted syngas is cooled in the low-temperature gas cooling (LTGC) system by generating low-pressure steam, preheating boiler feedwater, and heat exchanging against cooling water before going through the acid gas removal system for sulfur and chloride removals.

There is some flexibility for locating the WGS reactor in the gasification process: it can be located either before the sulfur removal step (sour shift) or after sulfur removal (sweet shift). Sour shift uses a cobalt–molybdenum catalyst and is normally located after the water scrubber, where syngas is saturated with water at about 230°C–260°C, depending on the gasification conditions and the amount of high-temperature heat recovery. An important benefit of sour shift is its ability to also convert carbonyl sulfide (COS) and other organic sulfur compounds into hydrogen sulfide (H_2S) to make downstream sulfur removal easier. Therefore, syngas treated through WGS does not need separate COS hydrolysis conditioning.

A conventional high-temperature sweet (HTS) shifting operates between 300°C and 500°C and uses chromium- or copper-promoted iron-based catalysts. Because syngas from the sulfur removal process is saturated with water at either near or below ambient temperature, steam injection or other means to add moisture to the feed is normally needed for HTS shifting. The initial HTS takes advantage of the high reaction rates, but is thermodynamically limited, which results in incomplete conversion of carbon monoxide and a 2%–4% carbon monoxide exit composition.

To shift the equilibrium toward hydrogen production, a subsequent LTS reactor is employed to produce a carbon monoxide exit composition of less than 1%. The transition from the HTS to the LTS reactors necessitates intersystem cooling. Due to different reaction conditions, different catalysts must be employed at each stage to ensure optimal activity. A conventional LTS shift, typically used to reduce residual CO content to below 1%, operates between 230°C and 260°C and uses a copper–zinc–aluminum catalyst. LTS shifting catalysts are extremely sensitive to sulfur and chloride poisoning and are normally not used in coal gasification plants [35–40].

Sweet shift is normally not used for coal gasification applications, given the problems of sulfur and chloride poisoning as mentioned earlier, in addition to the inefficiency of having to cool the syngas before sulfur removal, which condenses out all of the moisture gained in the water scrubber, and then reheating and reinjecting the steam into the treated gas after H_2S removal to provide moisture for shift. Sour shift is normally preferred for coal gasification applications since the moisture gained in the water scrubber is used to drive the shift reaction to meet the required H_2/CO ratio.

Thus in coal gasification reactors where syngas often contains more sulfur and very little chloride, WGS reactor follows a little different configuration. Here the initial HTS applied to sour gas takes advantage of the high reaction rates, but is thermodynamically limited, which results in incomplete conversion of carbon monoxide and a 2%–4% carbon monoxide exit composition. The commercial HTS catalyst used is the iron oxide–chromium oxide catalyst which is more robust and resistant toward poisoning by sulfur compounds. An important limitation for the HTS is the H_2O/CO ratio, where low ratios can lead to side reactions such as the formation of metallic iron, methanation, carbon deposition, and FT reaction. This can be avoided by insertion of steam in HTS reactor [64–69].

To shift the equilibrium toward hydrogen production, a subsequent LTS reactor is employed to produce a carbon monoxide exit composition of less than 1%. The transition from the HTS to the LTS reactors necessitates intersystem cooling. The LTS catalyst is a copper-based catalyst which is susceptible to poisoning by sulfur that may remain after the steam reformation process. This necessitates the removal of the sulfur compounds prior to the LTS reactor by a guard bed in order to protect the copper catalyst. While both HTS and LTS catalysts are commercially available, their specific composition varies based on vendor. More details on new catalyst development and preparation for HTS and LTS are given in the next two sections [3–70].

6.3.5 HIGH-TEMPERATURE SHIFT CATALYSTS

High-temperature catalysts can operate in the temperature range of 310°C–450°C–550°C and are called ferrochrome catalysts because of their composition [57]. They can be damaged if they are exposed to very high temperatures. In industrial practice, the reactor is operated adiabatically with inlet temperature around 350°C and outlet temperature not exceeding 450°C–550°C. The temperature increases along the length of the reactor due to exothermic nature of the reaction. Newsome [6] has reported the typical composition of HTS catalyst as 74.2% Fe_2O_3, 10.0% Cr_2O_3, 0.2% MgO, and the remaining being volatiles. Cr_2O_3 acts as a stabilizer and prevents the sintering of the iron oxide and, while its optimal content in the catalyst is 14%, in industry only 8% is used to prevent the compromise on surface area [24]. Several inorganic salts, boron, oils, phosphorous compounds, water, and sulfur compounds with concentrations greater than 50 ppm are poisons for iron chromia catalysts [71].

With inlet CO concentration ranging from 3% to 80%, at 8375 kPa pressure a HTS reactor can reduce CO concentration to 3% level [6]. Before the WGS reaction, the catalyst is activated by process gas mixture by partially reducing Fe_2O_3 in HTS catalyst to Fe_3O_4 [57]. Since, overreduction of the catalyst can cause damage to the catalyst pellets, reaction contact time of 3–9 s is recommended [17].

New catalysts for HTS are constantly being developed [11–34,64–69]. Grenoble and Estadt [47] found that the support for the metal or metal oxide should be acidic in nature to help the reaction. The turnover number for group VIIB, VIII, and IB decreased in the order of Cu, Re, Co, Ru, Ni, Pt, Os, Au, Fe, Pd, Rh, and Ir supported on alumina. Salmi et al. [26] and Hakkarainen et al. [19] reported that the commercial cobalt–molybdenum oxide catalyst was capable of catalyzing the reaction above 350°C and can be sulfur-tolerant. Rhodes and Hutchings [25] showed that addition of 2 wt% of B, Pb, Cu, Ba, Ag, and Hg as promoters in the temperature range of 350°C–440°C enhanced the reaction in the order Hg > Ag, Ba > Cu > Pb > unpromoted > B. While various studies have showed the promotion of the iron chromia catalyst by Cu, Au, and Ru as promising, the replacement of chromia by other less toxic components has not been commercially successful so far.

6.3.6 LOW-TEMPERATURE SHIFT CATALYSTS

A number of efforts are also being made to improve LTS catalysts. The LTS shift reactor operates at a range of 200°C–300°C. These low temperatures also reduce the occurrence of side reactions that are observed in the case of the HTS. The typical composition of the catalyst is 68%–73% ZnO, 15%–20% CuO, 9%–14% Cr_2O_3, 2%–5% Mn, Al and Mg oxides [6] and 32%–33% CuO, 34%–53% ZnO, and 15%–33% Al_2O_3 [17,57].

New catalyst development can operate at medium temperature around 300°C. The active species in the catalyst is Cu metal crystallites, while ZnO and Cr_2O_3 provide structural support and Al_2O_3 (largely inactive) helps dispersion and prevents pellet shrinkage. Cu should not be used at high temperatures due to its thermal sintering [31], while ZnO reduces sulfur poisoning of Cu, in general.

Sulfur, halogens, and unsaturated hydrocarbons in feed streams need to be avoided [32,71]. LTS catalysts give about 2–3 years of life and reduce CO concentration at the level of 0.1% in exit stream with minimum side reactions [71].

LTS catalysts are activated by reducing CuO to Cu using process stream gas with dilute H_2 [57]. The reduction should be carried out only at 230°C since high temperatures sinter the catalyst [17]. Steam should be avoided since its condensation can affect the catalyst. Tanaka et al. [34] found $CuMn_2O_4$ and $CuAl_2O_4$ mixed oxide catalyst having conversion of CO more than the commercial catalyst. Kusar et al. [22] have found the copper ceria catalyst to be nonpyrophoric and stable. Studies on Mn-promoted Cu/Al_2O_3 by Yeragi et al. [37] showed that with 8.55 wt% of Mn, 240°C, and space time of 5.33 h, CO conversion of 90% could be achieved. While a variety of catalysts have been investigated, for all catalysts copper remains the main constituent [17,27,29,32–34]. Noble metals such as Pt supported on ceria have also been extensively used for LTS [11,14,15,18,23,30].

6.3.7 Role of WGS in Fuel Cells

The WGS can aid in the efficiency of fuel cells by increasing hydrogen production. The WGS is considered a critical component in the reduction of carbon monoxide concentrations in cells that are susceptible to carbon monoxide poisoning such as the PEM fuel cell [2,94–111]. In this type of cell, the WGS not only reduces the concentration of carbon monoxide but also increases efficiency of the fuel cell by increasing hydrogen production [2]. Unfortunately, current commercial catalysts that are used in industrial water gas shift processes are not compatible with fuel cell applications [2,3]. With the high demand for clean fuel and the critical role of the WGS reaction in hydrogen fuel cells, the development of WGS catalysts for application in fuel cell technology is an area of current research interest [94–111].

Catalysts for fuel cell application would need to operate at low temperatures. Since the WGS is slow at lower temperatures where equilibrium favors hydrogen production, WGS reactors require large amounts of catalysts, which increases their cost and size beyond practical application [1]. The commercial LTS catalysts used in large-scale industrial plants are also pyrophoric in their inactive state and therefore present safety concerns for consumer applications [3]. Developing a catalyst that can overcome these limitations is crucial in the implementation of a hydrogen economy [94–111].

For compact fuel cell applications, several investigators have tried the use of noble metals as catalysts [2,3]. In the transportation sector, an integration of fossil fuel reforming and WGS with the fuel cell technology is required. The commercial iron-based catalysts are prone to coke formation in the presence of excess fuel from the reformer [62]. The use of the commercial copper-based catalysts requires excessive volume due to kinetic limitations. Also, they are pyrophoric in reduced state and get deactivated in the presence of condensed water due to leaching of active component or formation of surface carbonates [22,52,62].

The catalysts should also be capable of withstanding thermal cycling and faster response during start-up and shutdown. More efforts are being made to evaluate the effectiveness of transition metals over ceria-supported catalysts for WGS [15,20]. Wheeler et al. [62] studied the possibility of the WGS using noble metals and metals with Ceria in the temperature range of 300°C–1000°C and found the activity of the metals in the order Ni > Ru > Rh > Pt > Pd. Phatak et al. [23] carried out experiments on Pt supported on alumina and ceria at varying compositions and reported their order of reactions and activation energy. Pt exhibited lower turnover rate than Cu.

Catalyst support also plays an important role in catalyst performance. Gonzaleza et al. [18] found that Pt deposited on composite of Ceria and Titania performed better than Pt on either ceria or titania alone. Most of the recent studies have been directed in using any of the precious metals like Pt, Rh, Pd, and Au deposited on oxides of ceria, zirconia, alumina, titania, thoria, or magnesia supports or their mixtures [14]. Ratnasamy and Wagner [7] and others [4–6,8,9] have recently published comprehensive reviews on the developments on the kinetics and catalysts for the WGS reaction. An extensive literature on kinetics, catalysts, and reactor modeling for fuel cell applications has been reported [41–63,94–111].

6.4 STEAM REFORMING

Currently, an important method to produce hydrogen and syngas is the catalytic steam reforming of hydrocarbons, a process that involves the breaking of H–C and C–C bonds. The process of steam reforming is performed at lower temperatures (150°C–700°C) in the presence of catalysts [72–152], and these are divided into two groups: nonprecious metals (Ni) and precious metals of Group VIII (Platinum—Pt or Rhodium—Rh) [83,90,92,112,115,118,125]. Most studies are using nickel (Ni) supported on alumina (α-Al$_2$O$_3$) [87–89,123,124,133,140]. The nature of the metal components has a significant influence on performance and distribution of gaseous products. In steam reforming, it is important to control the formation of coke, as coke formation results in deactivation of the catalyst.

The steam reformation of organic compounds such as biomass is one of the most widely used processes for hydrogen production. In this process, the substrate reacts with steam in the presence of a catalyst to produce hydrogen, carbon dioxide, and carbon monoxide [125]. The catalysts are mainly used to increase the reaction rate and selectivity of hydrogen. This process is highly endothermic and the low pressure favors the selectivity to hydrogen. The reforming process involves mainly the breakdown of hydrocarbons in the presence of water along with the WGS reaction to produce a large amount of hydrogen and carbon dioxide (see Equation 6.1). Moreover, the methanation of carbon monoxide (Equation 6.4) also occurs [125].

Steam reforming of methane is very attractive because methane contains largest H/C ratio in any hydrocarbon. Unfortunately methane molecule is very stable with C–H bond energy of 439 kJ/mol. Such high bond energy makes methane resistant to many reactants and reactions. Furthermore, carbon-hydrogen bond in methane is very strong. Methane molecule can, however, be activated by Group VIII to X transition metals and can be oxidized to produce syngas (CO and H$_2$). Further conversion of CO by WGS reaction generates the final product with large concentrations of hydrogen and carbon dioxide. Once the carbon dioxide is removed from the mixture of carbon dioxide and hydrogen by physical (zeolite absorption or pressure swing adsorption [PSA] adsorption), chemical (absorption by amine solution), or membrane separation process (generally by use of Pd membrane), pure hydrogen is obtained. The PSA process allows the purity of hydrogen of about 999.999% at 25 bar feedstock pressure. In recent years, the use of ceramic ion transport membranes (ITM) with reformers has opened up the possibilities of the production of high-quality and low-cost hydrogen [1,72].

Methane reforming by steam is an endothermic reaction and it is favored at lower pressures. While noble metal catalysts have been tested and used in the past, most commercial operations use nickel catalyst because of its low cost and high activity, stability, and selectivity. The activity of the catalyst depends on the catalyst surface area and the temperature (around 400°C–1000°C) for steam pressure up to 30 atm. The activity of the catalyst is usually described by turnover frequency (TOF), which is generally 0.5 s^{-1} at around 450°C. This number corresponds to about 10% methane conversion. High conversion rate demands higher temperature because the reaction is limited by thermodynamics which is favored at the higher temperature. Very high conversion requires the reactor to be operated at temperatures higher than around 900°C. Often the catalysts in the reformer are poorly used because heat transfer between gas and solid is a limiting factor in the reaction. As shown later, the reactor design plays an important role in the performance of the reactor.

Numerous studies on mechanism of methane reforming have been reported, and these are well reviewed by Wei and Iglesia [82] and Rostrup-Nielsen et al. [83]. They have shown that the rate limiting step for steam reforming is C–H bond activation and proposed the following mechanism [1,72]:

$$H_2O + * \rightarrow O*(a) + H_2(g) \tag{6.5}$$

$$CH_4(g) + 2* \rightarrow CH_3*(a) + H*(a) \tag{6.6}$$

$$CH_3{*}(a) + {*} \rightarrow CH_2{*}(a) + H{*}(a) \tag{6.7}$$

$$CH_2{*}(a) + {*} \rightarrow CH{*}(a) + H{*}(a) \tag{6.8}$$

$$CH{*}(a) + O{*}(a) \rightarrow CO{*}(a) + H{*}(a) \tag{6.9}$$

$$CO{*}(a) \rightarrow CO(g) + {*} \tag{6.10}$$

$$2H{*}(a) \rightarrow H_2(g) + 2{*} \tag{6.11}$$

In the previous equations, * denotes Ni (or catalyst in general) surface atom. In this mechanism, methane adsorbs dissociatively on the Ni surface producing methyl group and water molecule reacts with surface Ni atoms to produce adsorbed oxygen and gaseous hydrogen. The methyl group goes through further stepwise dehydrogenation. The final product of this dehydrogenation, CH–, reacts with adsorbed oxygen to produce syngas (CO and H_2).

Along with the main reactions outlined earlier, the reforming reactions are accompanied by the following carbon-forming reactions:

$$2CO \rightarrow C + CO_2 \quad H^{\circ}_{298K} = -172.5\,kJ/mol \tag{6.12}$$

$$CH_4 \rightarrow C + 2H_2 \quad H^{\circ}_{298K} = +74.9\,kJ/mol \tag{6.13}$$

These two reactions deposit carbon on the catalyst in filament forms which ultimately deactivate catalyst. The carbon-forming reactions are also counterbalanced by carbon-consuming reactions:

$$C + CO_2 \rightarrow 2CO \tag{6.14}$$

$$C + H_2O \rightarrow CO + H_2 \tag{6.15}$$

Both of these reactions also depend on the operating conditions and the nature of the reactor design. Generally, at low temperature the Ni catalyst surface is covered with hydrocarbons which degrade into a polymeric layer. On the other hand, at high temperatures, cracking of olefinic and aromatic hydrocarbons produce coke which deposits on the catalyst surface. Since NiC is not stable, carbon is formed in the form of filaments which grow on the catalyst surface. The size of Ni particles has a direct bearing on the location of filaments on the nickel surface. Smaller and more dispersed Ni particles reduce the formation of carbon filaments. Thus Ni dispersion is an important variable in catalyst activity and stability (degradation). The literature has shown [76–84] that size and location of Ni particle ensemble is an important variable for controlling the coke formation on the catalyst. The coke formation can also be controlled by controlling the carbide formation. While alloys reduce carbide formation, they also hide active sites of nickel for the reforming reactions. The literature [76–84] has also shown that the addition of a small amount of dopants (e.g., Sn) reduces coking without affecting the activity for reforming reaction. Carbon formation can also be reduced by alloys of copper–nickel, sulfur–nickel, nickel–tin, and nickel–rhenium.

6.4.1 Catalysts for Steam Reforming

In general for steam reforming two types of states are needed in the catalyst: one for hydrogenation and dehydrogenation, and the second one for acidic sites. The acidic sites promote the formation of carbonium ions. For aromatization and isomerization reactions, two types of sites are necessary. While, as mentioned before, Ni catalysts on oxide supports have been most extensively used in the industry, recent studies show that bimetallic catalysts such as Ni/Ru, and Pt/Re are more effective catalysts. Again, due to economic reasons, one of the catalysts needs to be nickel. Trimetallic catalysts of noble metal alloys have also been tested. In general, bi- and trimetallic catalysts give better stability (with low sintering at high temperatures) and increased catalyst activity and stability. Coke deposition on the catalysts has been the main reason for catalyst decay; however, coke can be removed by the oxidation at high temperatures. The coke deposition can vary from 15% to 25% on the catalyst [1,72].

The coke formation can occur by Equations 6.12 and 6.13 or one of the following reactions:

$$CO(g) + 2H_2(g) \leftrightarrow H_2O(g) + C(s) \tag{6.16}$$

$$CO_2(g) + 2H_2(g) \leftrightarrow 2H_2O(g) + C(s) \tag{6.17}$$

Thus, coke can be formed from CO, CO_2 as well as from CH_4.

Coke can also be formed from ethylene through polymerization reaction as

$$C_2H_4 \rightarrow \text{polymers} \rightarrow \text{coke} \tag{6.18}$$

Coke deposition at a sustained level should be avoided because it leads to several undesirable side reactions, loss in catalyst activity, and poor heat transfer between the catalyst and the gas phase. If the coke deposition becomes very extensive, it can block the open surface area causing excessive pressure drop within the reactor and it can also cause localized "hot spots," which can induce "runaway" conditions for the reactor. Coke formation can be minimized by the use of an excess steam. The catalyst can also be regenerated periodically, by burning off the deposited coke by the oxidation reactions [1,72].

6.4.2 Feedstock Effects

6.4.2.1 Ethanol

Alcohols and in particular ethanol can be easily obtained by the process of fermentation of sugar, glucose, fructose, and many lignocellulosic biomasses [72]. In Brazil, ethanol is extensively produced using sugarcanes. Ethanol is easier and safer to store and transport because of its low toxicity and volatility and biodegradable characteristics. Ethanol can also be produced from various energy plants, waste materials from agroindustries or forestry residue materials as well as cellulosic and organic fractions of municipal solid waste. Easy availability of ethanol makes it a good candidate for steam reforming to produce hydrogen [72].

Unlike methanol and gasoline derived from fossil fuel sources, ethanol derived from biosources is carbon neutral to the environment. The carbon dioxide produced from the steam reforming of ethanol can be used to regenerate additional biomass. Bioethanol, generally containing about 12% ethanol in an aqueous solution, can be directly subjected to steam reforming, thus eliminating the distillation step required to produce pure ethanol. Since both water and ethanol can be converted to hydrogen, the process of steam reforming avoids the separation stage. The thermal efficiency of steam reforming of aqueous ethanol solution is very high (>85%), and this makes the process economically very attractive. The steam reforming of ethanol is carried out in the following reaction [72]:

$$C_2H_5OH + 3H_2O \leftrightarrow 2CO_2 + 6H_2 \qquad (6.19)$$

This reaction follows a number of steps. The steps involve the dehydrogenation of ethanol to form acetaldehyde, which decomposes to produce methane and carbon monoxide. Further reforming of methane and WGS reaction leads to the formation of hydrogen. Since ethanol has high hydrogen content, the process produces a significant amount of hydrogen. There are, however, side reactions such as dehydration and decomposition of ethanol that produce methane, diethyl ether, and acetic acid which reduce the production of hydrogen. These side reactions can be minimized by the use of selective catalysts. In addition, the formation of large amounts of carbon monoxide reduces the hydrogen yield and it also requires complex gas cleanup process. Overall, ethanol is still one of the best raw materials for steam reforming to produce hydrogen.

An extensive number of studies to develop different types of catalysts for ethanol steam reforming have been reported in the literature [73,86–93]. Mas et al. [86] used Ni(III)–Al(III) lamellar double hydroxide as catalyst precursor. They developed a Langmuir–Hinshelwood type of kinetic model for steam reforming of ethanol for this catalyst. A general model was found to be valid for a wide range of water/ethanol feed ratio and temperatures. Biswas and Kunzru [87] examined the effects of copper, cobalt, and calcium doping on Ni–CeO_2–ZrO_2 catalysts for steam reforming of ethanol. The data were obtained in the temperature range of 400°C–650°C. The nickel loading was kept fixed at 30 wt%, while Cu and Co loading varied from 2 to 10 wt% and Ca loading varied from 5 to 15 wt%. For Cu- and Ca-doped catalysts, activity increased significantly; however, Co-doped catalysts showed poor activity. The catalyst activity was in the order $N > Ncu_5 > Nca_{15} > Nco_5$. For steam reforming reaction, the highest hydrogen yield was obtained on the undoped catalyst at 600°C. With calcium doping, in the temperature range of 400°C–550°C, higher hydrogen yield was obtained compared to the ones for undoped catalysts. Akdim et al. [88] compared the steam reforming of nonnoble metal (Ni–Cu) with noble metals (Rh or Ir) supported over neutral SiO_2, amphoteric Al_2O_3, and redox CeO_2. The data showed that for each domain of temperature, quite different mechanistic routes were governing for the three tested systems. The data suggested some methods of improving catalyst formula for the steam reforming of ethanol. Finally, the effect of support on catalytic behavior of nickel catalysts in the steam reforming of ethanol for hydrogen production was investigated by Fajardo et al. [89], who studied Al_2O_3, MgO, SiO_2, and ZnO supported nickel catalysts and showed that the catalyst behavior can be influenced by the experimental conditions and chemical composition of the catalysts.

The steam reforming of ethanol by different types of Co catalysts were investigated by Sekine et al. [90], Song et al. [92], and He et al. [93]. Sekine et al. [90] examined steam reforming of ethanol over Co/$SrTiO_3$ with an addition of another metal—Pt, Pd, Rh, Cr, Cu, or Fe. Ethanol conversion and H_2 yield improved significantly by adding Fe and Rh at 823 K; however, Rh addition promoted CH_4 formation. Within Fe loading of 0.33–1.33 mol%, Fe addition increased selectivity of steam reforming of ethanol. The addition of Fe on Co/SiO_2 catalyst was not very effective. High activity of Fe/Co/$SrTiO_3$ catalyst came from interaction among Fe, Co, and $SrTiO_3$. Song et al. [92] showed that the use of novel synthesis methods such as solvothermal decomposition, colloidal crystal templating, and reverse microemulsion to prepare CeO_2-supported Co catalysts gave better performance than the catalysts prepared using conventional incipient wetness impregnation for ethanol steam reforming. The improvement can be attributed to a better cobalt dispersion and a better Co–CeO_2 interaction for the catalysts prepared using these novel methods. He et al. [93] examined a series of Co–Ni catalysts prepared from hydrotalcite (HT)-like materials by co-precipitation for steam reforming of ethanol. The results showed that the particle size and reducibility of the Co–Ni catalysts are influenced by the degree of formation of HT-like structures and increasing Co content. All catalysts were active and stable at 575°C. The activity decreased in the order 30Co–10Ni >

40Co–20Ni > 20Co > 10Co–30Ni > 40Ni. The 40Ni showed strongest resistance to deactivation, while all Co-containing catalysts showed higher activity than 40Ni catalyst. The highest hydrogen yield was found for 30Co–10Ni catalyst in which xCo and Ni are intimately mixed and dispersed in the HT-derived support [72].

Dong et al. [91] examined hydrogen production by steam reforming of ethanol using potassium-doped $12CaO–7Al_2O_3$ catalysts. The conversion of ethanol and H_2 yield over $C_{12}A_7–O^-/x\%K$ catalyst mainly depended on the temperature, K-doping amount, steam-to-carbon ratio, and contact time. Based on numerous types of catalyst analysis, the authors concluded that the active oxygen species and doped potassium play important roles in the steam reforming of ethanol over $C_{12}A_7–O^-/27.3\%K$ catalyst.

The steam reforming of ethanol undergoes several reaction pathways depending on the catalysts and the reaction conditions. Therefore the choice of the catalyst plays a vital role in the reforming process. Navarro et al. [73] pointed out that the reactions to avoid are C_4 and C_2H_4 inductive of carbon deposition on the catalyst surface. Thus, the catalysts that selectively produce hydrogen must (a) dehydrogenate ethanol, (b) break the carbon–carbon bonds of surface intermediates to produce CO and CH_4, and (c) reform these C_1 products to generate hydrogen. As shown earlier, various oxide catalysts, metal-based catalysts (Ni, Co, Ni/Cu), and noble metal-based catalysts (Pt, Pd, Rh) have proven to be active for steam reforming of ethanol. The metallic function and the acid-based properties play important roles in the steam reforming. A good review of hydrogen selectivity and coking resistance of various types of catalysts is given by Navarro et al. [73].

6.4.2.2 Methanol

Methanol is an abundant chemical often produced from fossil fuels as well as biomass. Industrially, it is produced at 250°C–300°C temperature and 80–100 atm pressure using a copper–zinc-based oxide catalyst. Methanol is an important feedstock for the production of hydrogen and hydrogen-rich syngas. While methanol can be decomposed as:

$$CH_3OH \rightarrow CO + 2H_2 \quad H^\circ_{298K} = +90.1 \, kJ/mol \qquad (6.20)$$

and this reaction is endothermic and can be catalyzed by a number of catalysts, including Ni and Pd, in this chapter we mainly focus on steam reforming of methanol. Methanol is a good feedstock because of its easy availability, high energy density, and easy storage and transportation. Currently significant work is being carried out for low-temperature steam reforming to produce high purity hydrogen for power generation in fuel cells in automobiles. The steam reforming of methanol follows the reaction [72]:

$$CH_3OH + H_2O \rightarrow CO_2 + 3H_2 \quad H^\circ_{298K} = +49.4 \, kJ/mol \qquad (6.21)$$

While a number of catalysts have been examined, commercial Cu/ZnO WGS reaction catalyst and methanol synthesis catalysts have been found to be effective for steam reforming of the methanol. Copper on ZrO_2 support prepared by numerous different methods, including precipitation, microemulsion, formation of amorphous aerogels, CuZr alloys, and so on, have been successfully attempted. For this catalyst a large surface area of the active metals needs to be maintained to avoid rapid deactivation. For this zirconia support should be in the amorphous state under the calcination and reaction conditions. $Cu/ZnO/ZrO_2$ catalyst [95] has been found to be active at temperature as low as 170°C but the catalyst deactivates rapidly at temperatures above 320°C. The deactivation can, however, be reduced by the incorporation of Al_2O_3, which increases the temperature of crystallization of ZrO_2, which remains amorphous at the reaction temperature. The incorporation of alumina also increases both the copper and no Brunauer-Emmett-Teller (BET) surface area, thereby increasing the catalyst activity [72].

Henpraserttae and Toochinda [94] examined a novel preparation technique of Cu/Zn catalyst over Al_2O_3 for methanol steam reforming. The study focused on the preparation methods of active Cu-/Zn-based catalysts with and without urea by incipient wetness impregnations to lower the metal loading and catalyst cost. The experimental data for methanol steam reforming were obtained in a fixed bed reactor in the temperature range of 453–523 K to lower energy costs. The data showed that the activity in the hydrogen production from the catalysts with urea was higher than that of catalysts without urea. The impregnated catalysts can show activity at temperatures as low as 453 K. The Cu/Zn catalysts prepared with incipient wetness impregnation over Al_2O_3 with urea can give a hydrogen yield of about 28%. Thus, the impregnated catalysts could be alternative catalysts for hydrogen production from methanol reforming with a lower cost of the catalyst compared with the co-precipitation method used in the commercial operation.

The partial oxidation of methanol is attractive because it is an exothermic reaction and it follows the reaction:

$$CH_3OH + \frac{1}{2}O_2 \rightarrow CO_2 + 2H_2 \quad H^{\circ}_{298K} = -192.2\,kJ/mol \tag{6.22}$$

The previous reaction starts at temperatures as low as 215°C. Both reaction rate and the selectivity for hydrogen increase very rapidly with temperature. Carbon monoxide formation in the entire temperature range is low. The literature has shown [95] that production of hydrogen and carbon dioxide increases with copper content, and it reaches maximum with 40/60 atomic percentage of copper and zinc. Unreduced copper-zinc oxide catalysts display very low activity and produce only carbon dioxide and water with very little hydrogen. The catalysts, however, become eventually reduced under high temperature reaction conditions. The apparent activation energy and TOF are higher at lower copper content and they slightly decrease with increase in copper content and then achieve a constant value. These and some other similar data show that the reaction depends on both ZnO and CuO phases. Methanol conversion increases with oxygen partial pressure up to 0.063 atm. A further increase in oxygen partial pressure precipitously decreases methanol conversion [72,95]. The incorporation of Al_2O_3 (up to 15% Al) to the Cu/ZnO system results in a lower activity, implying that aluminum has an inhibiting effect on the partial oxidation of methanol.

Besides Cu/ZnO catalyst, Pd/ZnO catalyst has also been effective in methanol POX reaction. For 1 wt%, Pd/ZnO catalyst, methanol conversion reaches 40%–80% within 230°C–270°C range. Methanol conversion along with the H_2 selectivity increases with an increase in temperature and the water production decreases. The nature of support also affects the kinetics. For example, Pd/ZrO_2 catalyst, while producing hydrogen and carbon dioxide, also shows a significant increase in the decomposition reaction.

A combination of steam reforming and POX results in an autothermal operation. Under this condition, the following reaction [72]

$$CH_3OH + (1 - 2n)\,H_2O + nO_2 \rightarrow CO_2 + (3 - 2n)H_2 \quad (0 < n < 0.5) \tag{6.23}$$

with copper base catalysts perform well. On Cu/ZnO catalyst, initially methanol is combusted by oxygen and water is produced. When oxygen is depleted, methanol conversion and production of hydrogen and carbon monoxide increase and water production goes down. When Al_2O_3 is added to the catalyst, better performance for steam reforming is obtained. Purnama et al. [96] also found the beneficial effect of oxygen addition to the feed during steam reforming of methanol on Cu/ZrO_2 catalysts. In autothermal operation, the relative ratio of oxygen, methanol, and steam plays an important role in hydrogen production. For Cu–ZnO (Al) catalyst the best feed ratio of oxygen/methanol/steam has been found to be 0.3/1/1. In general oxygen reforming of methanol is complex but it also strongly interacts with WGS reaction [72].

The autothermal operation for methanol for FC (fuel cell) application in vehicles has been adopted by Daimler-Chrysler, Toyota, and Nissan. Small-scale hydrogen production by reforming methanol has also been commercialized. For its application in refueling stations, hydrogen purification step is needed. This is generally carried out either by PSA or membrane separation technology. In general, the cost of hydrogen production from methanol reforming is higher than that of methane reforming. The MERCATOX project funded by European Commission is an integrated methanol steam reformer and selective oxidation system. The fuel cell contains a series of catalytic plates with combustion of anode off-gas on one side and steam reforming of methanol on the other side.

de Wild and Verhaak [139] reported the results of metal-supported catalyst systems for steam reforming of methanol for FC applications. Such catalysts overcome the slow heat transfer of packed-bed systems by integrating endothermic steam reforming with exothermic hydrogen combustion. A wash-coated aluminum heat exchanger showed the best performance using a suspension of commercial reforming catalysts. With an aluminum foam, 90% methanol conversion was achieved for a sustainable period of time (about 450 h). Lindström and Petterson [97–104] examined methanol reforming over copper-based catalysts for FC applications.

A novel technology of steam reforming of methanol accompanied by palladium membrane separation to produce pure hydrogen was investigated by Pan and Wang [108]. This technique provides a possibility to bypass the technical problems of storage and delivery of hydrogen by delivering methanol to forecourt hydrogen dispensing stations and onsite hydrogen production. Li et al. [111] examined a strategy wherein coal-derived methanol is used as a hydrogen carrier. The steam reforming of methanol can generate hydrogen at the desired place.

6.4.2.3 Liquid Hydrocarbons

Besides methane, methanol and ethanol, gasoline, diesel, and jet fuel can also be important feedstock for the steam reforming to produce hydrogen. These three types of fuels contain a variety of hydrocarbons and sulfur. While these components themselves can be important feedstock for steam reforming, they are not as readily available on a large scale as various types of fuels. The technical problems associated with these hydrocarbons include (1) catalyst deactivation by sulfur in the feedstock and (2) a significant amount of coke deposition on the catalyst which eventually results in its deactivation. Along with steam reforming, in recent years catalytic POX of high hydrocarbons using short contact time (milliseconds) and high temperatures (850°C–900°C) over noble metal catalysts on porous monolithic ceramic supports has been examined [112–119]. These reactions can be represented by a generalized reaction:

$$C_nH_m + \frac{n}{2}O_2 \rightarrow nCO + \frac{m}{2}H_2 \tag{6.24}$$

The previous reaction is about two times faster than steam reforming reaction and the heat of reaction generated by this reaction depends on the oxygen-to-fuel ratio. Unlike steam reforming and POX of methane, methanol and ethanol, steam reforming and POX of fuels involve dehydrogenation, C–C bond cleavages, total oxidation, steam reforming, CO_2 reforming, hydrocarbon cracking, methanation, and WGS reaction all occurring simultaneously [72]. Also these reactions occur for all different component hydrocarbons at different rates. Thus, the process is very complex and not clearly understood. In general aromatics are less reactive and are more prone to the reaction producing cokes than aliphatic components and olefins. Through a complex set of reactions, fuels also produce hydrogen, carbon dioxide, carbon monoxide, and water along with a significant amount of lower hydrocarbons. The final product distribution depends on the temperature and the residence time. Several catalysts, including nickel, platinum, rhodium and bimetallic, have been tested for hydrocarbons such as n-octane, n-heptane, and n-hexane [112–119]. In general, ceria and zirconia supports or a mixture of ceria–zirconia supports have been found to be reasonably effective in averting coke deposition [112–119].

A combined steam reforming, POX, and WGS reaction have been tested to obtain an auto-thermal operation. Generally, POX and steam reforming are carried out in separate zones, with the first one controlled by oxygen-to-carbon ratio and the second one controlled by steam-to-carbon ratio. The adiabatic temperature and the amount of hydrogen produced depend on the relative amounts of energy released in these two steps. Higher steam-to-carbon ratio reduces the carbon monoxide concentration in the product. For diesel fuel, thermodynamic equilibrium can be achieved at an oxygen-to-carbon ratio of 1 and steam-to-carbon ratio of 1.25 at 700°C temperature [72].

ATR process requires catalysts and supports with high resistance to coking at high tempera-tures. Excess steam and/or oxygen helps avoid coking. Also at high temperature, sulfur is less of a problem. The noble metal catalysts (Pt, Rh, Ru) supported on ceria or zirconium or their mixtures work well. In recent years, applications of perovskite oxides (ABO_3) for steam reforming of higher hydrocarbons and various fuels have been extensively examined [112–119]. Also a group of six metal carbides have also shown good success [112–119].

6.4.2.4 Glycerol

Glycerol has been a by-product of a number of conversion processes, in particular, trans-esterifi-cation of used oil, algae, and crop oils (there are about 350 of them) to produce diesel fuel. This by-product can also be effectively utilized to produce hydrogen by steam reforming process. Steam reforming of glycerol involves a complex set of reactions, numerous intermediates, and hydrogen that is accompanied by several other products. The hydrogen yield depends on the steam-to-glycerol ratio and follows the reaction [72]:

$$C_3H_8O_3 \rightarrow 3CO + 4H_2 \tag{6.25}$$

$$CO + H_2O \leftrightarrow CO_2 + H_2 \tag{6.26}$$

with overall reaction as

$$C_3H_8O_3 + 3H_2O \rightarrow 3CO_2 + 7H_2 \tag{6.27}$$

Simonetti et al. [125] showed that at about 275°C, glycerol can be catalytically converted to H_2/CO mixture. Because of this low temperature, the endothermic steam reforming process can be com-bined with exothermic FT process to make the overall process energy-efficient for fuel generation from glycerol. The primary products for steam reforming of glycerol are hydrogen, methane, carbon dioxide, carbon monoxide, and carbon and unreacted water and glycerol. The formation of meth-ane competes with the formation of hydrogen. According to steam reforming and decomposition reactions,

$$C_xH_yO_x + xH_2O \rightarrow xCO_2 + \left(\frac{x+y}{2}\right)H_2 \quad \text{steam reforming} \tag{6.28}$$

$$C_xH_yO_x \rightarrow xCO + \frac{y}{2}H_2 \quad \text{decomposition reaction} \tag{6.29}$$

The maximum hydrogen concentration in the product can be either 77% or 57%. A study done by Adhikari et al. [128,130] showed that at about 680°C, the upper limit of moles of hydrogen per mole of glycerol produced is 6 at an atmospheric pressure and a steam-to-glycerol ratio of 9.

While nickel on alumina is a workable catalyst for steam reforming of glycerol, the effects of numerous promoters such as Ce, La, Mg, and Zr were examined at 600°C [120–132,140]. These results indicate that all promoters improved the production of hydrogen with zirconium giving the best results. The increase in hydrogen production can be due to increased nickel concentration, an increased capacity to activate steam, and the stability of nickel phase.

Recent studies [72,120–132,140] investigated various noble metal catalysts on a variety of supports at 500°C–600°C, atmospheric pressure and steam-to-carbon molar ratio of 3.3. The results indicated the activity order Ru = Rh > Ni > Ir > Co > Pt > Pd > Fe. Among Y_2O_3, ZrO_2, CeO_2, La_2O_3, SiO_2, MgO, and Al_2O_3 supports, Y_2O_3 (along with ZrO_2 and CeO_2) support gave the best glycerol conversion and hydrogen production. Numerous investigators [120–132,140] demonstrated that at low conversion and low temperature (225°C–275°C) Pt/C and Pt-Re/C gave stable results. For CeO_2 support, Zhang et al. [122] showed that at 400°C Ir/CeO_2 gave the best glycerol conversion with 85% hydrogen selectivity, while Co/CeO_2 and Ni/CeO_2 gave 88% and 75% hydrogen selectivity at 425°C and 450°C. Glycerol has a higher tendency for coke formation compared to methane, and this coke formation can be significantly reduced by increasing steam-to-glycerol ratio in the feed. The catalytic steam reforming of glycerol (both conversion of glycerol and selectivity of hydrogen) is affected by operating parameters such as the reaction temperature, pressure, steam-to-glycerol ratio, and oxygen-to-glycerol ratio.

In a recent study, Maciel and Ishikura [121] have given an outstanding review of steam reforming of renewable feedstock for the production of hydrogen. They have considered methanol, ethanol, glycerol, glucose, and biomass as potential raw materials for steam reforming. Their overall analysis led to the following conclusions: (1) reforming should be carried out at lower temperatures and at atmospheric pressure to reduce operating costs, (2) the catalyst should provide high selectivity to hydrogen and inhibit CO and by-product formation such as methane, and (3) the catalyst must resist coke formation which reduces the number of active sites and hence the reaction rates and implies regeneration process which is costly. Feedstock issues such as supply, cost, logistics and the value of by-products are major factors in cost effectiveness of steam reforming process [72].

6.4.2.5 Biomass

Just like methane and other hydrocarbons, biomass can also undergo POX and steam reforming in the presence of oxygen and steam at temperatures above around 725°C yielding gaseous products and chars. The char can also be converted to gaseous products such as hydrogen, carbon dioxide, carbon monoxide, and methane under high temperature conditions. The overall reaction follows:

$$C_xH_yO_z + H_2O + O_2 \rightarrow H_2 + CO_x + CH_4 + HCs + char \qquad (6.30)$$

The hydrogen production for a variety of biomass under different operating conditions has been examined in the literature [133–137,141–151]. The literature data indicate that in a fluidized bed reactor, under suitable operating conditions, as high as 60 vol% hydrogen can be produced from biomass.

The major drawback of steam reforming of biomass is the tar formation which is not easily amenable to steam reforming process. The tar formation can be minimized by suitable operating conditions (i.e., operating at very high temperature), suitable gasifier design (i.e., entrained bed reactor), or incorporations of additives or promoters to the catalysts. At temperatures above around 1000°C, tar can be cracked, and for temperatures above around 1200°C, pure syngas (CO and H_2) can be obtained. Higher residence time can also help cracking of the tar. Additives such as dolomite and olivine to the nickel catalyst help in the reduction of the tar formation. Alkaline metal oxides are also used to reduce the tar formation.

Another important issue with biomass gasification and reforming is the formation of ash which can cause slagging, fouling, and agglomeration. The inorganic impurities in biomass can, however, be removed by biomass pretreatment by leaching and extraction processes. The literature

[133–137,141–151] has shown the leaching and subsequent gasification to produce hydrogen as a viable process for olive oil waste.

Pacific Northwest laboratories studied the gasification of biomass to produce a variety of gaseous fuels by use of appropriate catalysts. The earlier studies [141–151] used a process of catalytic steam gasification of biomass with concurrent separation of hydrogen in a membrane reactor that employed a perm-selective membrane to separate the hydrogen as it is produced. The process was particularly well suited for wet biomass and was conducted at temperatures as low as 300°C. One experiment was conducted at 4000 psi pressure and 450°C, though most others were at 15–30 psi. The process was named SepRx. Optimal gasification conditions were found to be about 500°C, atmospheric pressure, and steam-to-biomass ratio of 10/1. In the presence of a nickel catalyst, product hydrogen concentration of 65 vol% was produced under these optimal conditions. Rapagna [142] examined steam gasification of almond shell in the temperature range 500°C–800°C. Smaller particle size yielded more hydrogen. Rapagna and Foscolo [143] examined catalytic steam gasification in a fluidized bed reactor followed by a fixed bed catalytic reactor. Over a temperature range of 660°C–830°C, the catalytic converter using different steam reforming nickel catalysts and dolomite gave as high as 60% hydrogen yield [72].

Often steam gasification and steam reforming are coupled processes. McKinley et al. [134] examined various biomass gasification processes for the production of hydrogen. Turn et al. [141] showed that for a noncatalytic gasification of saw dust, the highest hydrogen yield was obtained at 825°C and for a steam-to-biomass ratio of 1.7. Zhou et al. [135], however, showed that for the production of hydrogen, adding steam to the gasification process was not as effective as adding steam to downstream nickel catalyzed steam reforming process.

6.4.2.6 Mixed Feedstock

In recent years, significant efforts have been made to gasify and steam reform mixed feedstock of coal and waste, coal and biomass, and various types of biomass. These studies are described in a recent publication by the author [1,72]. Gasification and steam reforming of mixed feedstock has a very bright future. De Ruyck et al. [152] examined co-utilization of biomass and natural gas in combined cycle through primary steam reforming of natural gas. The study proposed a method where external firing is combined with the potential high efficiency of combined cycles through co-utilization of natural gas with biomass. Biomass is burned to provide heat for partial reforming of the natural gas feed. In this way biomass energy is converted into chemical energy contained in the produced syngas. Waste heats from reformer and from biomass combustor are recovered through a waste heat recovery system. The paper shows that this way biomass can replace up to 5% of the energy in the natural gas feed. The paper also shows that in the case of combined cycles, this alternate path allows for external firing of biomass without important drop in cycle efficiency.

6.4.2.7 Bio-Oil

Catalytic steam reforming of bio-oil at 750°C–850°C over a nickel-based catalyst is a two-step process that includes the shift reaction:

$$\text{Bio-oil} + H_2O \rightarrow CO + H_2 \tag{6.31}$$

$$CO + H_2O \rightarrow CO_2 + H_2 \tag{6.32}$$

The overall stoichiometry gives a maximum yield of 11.2% based on wood. The overall reaction is

$$CH_{1.9}O_{0.7} + 1.26H_2O \rightarrow CO_2 + 2.21H_2 \tag{6.33}$$

For this process, bio-oil from regional facility is generally transported to central reforming facility. The process is compatible with other organic waste streams such as aqueous steam fractionation processes used for ethanol production and trap grease. Methanol and ethanol can also be produced from biomass by a variety of technologies and used for on-board reforming for transportation. Methane from anaerobic digestion could be reformed along with natural gas. A system analysis has shown that biomass gasification/shift conversion is economically unfavorable compared to natural gas reforming except for very low-cost biomass and potential environmental incentives [72].

6.5 DRY REFORMING

In recent years, there has been significant interest in the development of carbon dioxide (CO_2) utilization technologies that can use this environmentally harmful gas as chemical feedstock while reducing or limiting its emission [1,153]. One method for CO_2 utilization is dry reforming. Dry reforming is used to catalytically convert CO_2 and hydrocarbon feeds into syngas (mixture of hydrogen [H_2] and carbon monoxide [CO]) that can be used in the chemicals and fuels industries.

As shown in Figure 6.4a, 82% of GHGs emitted globally in 2012 comprised of CO_2 with 9% of emissions as methane (CH_4) [1,153]. While CO_2 emissions are derived from a variety of naturally occurring sources, anthropogenic emissions from expanding economies and population are responsible for the slow, but steady, increase in CO_2 emissions since the 1920s [154]. As shown in Figure 6.4b, CO_2 emissions are largely derived from a variety of sources with electricity generation (fossil fuel combustion of coal, natural gas, and oil), transportation and industry (such as steel production and natural gas systems [154–157]) are the largest contributors of CO_2 emission in the environment.

Environmental regulations may eventually require the capture of CO_2 and its sequestration or utilization as a chemical feedstock [154,158,159]. The main sources of CO_2 emissions in the United States are derived from the combustion of fossil fuels to generate electricity for homes, businesses, and industry. As shown in Figure 6.4b, electricity generation contributed 32% of total CO_2 emission in 2012 [154–158]. Mobile emission sources that utilize gasoline and diesel to transport people and goods represent the second-largest source of CO_2 emissions, accounting for 28% of all U.S. CO_2 emissions in 2012 [154,158,159]. The production of cement, steel, and chemicals represents another significant stationary source for CO_2 emissions. Industrial processes accounted for 20% of all U.S. CO_2 emissions in 2012 [153,154,159].

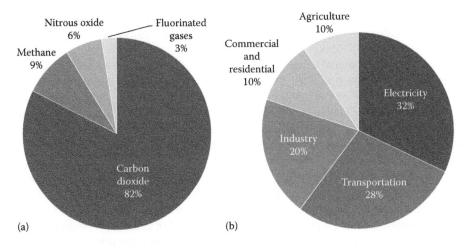

(a)   (b)

FIGURE 6.4 2012 greenhouse gas emissions by (a) gas and (b) source. (From EPA report; Shah, Y.T. and Gardner, T., *Catal. Rev. Sci. Eng.*, 54, 476, September 2014.)

TABLE 6.2

Reaction Paths and Possible Products for CO_2 Transformation

Reaction Pathway	Products
Cycloaddition	Cyclic carbonates, carbamates
Hydrogenative	Alcohols, hydrocarbons
Photochemical, electrochemical	CO, H_2CO_2, CH_4OH
Inorganic	$CaCO_3$
Biological	Alcohols, sugars, CH_3CO_2H
Dry reforming	CO, H_2, H_2O

Sources: Shah, Y.T. and Gardner, T., *Catal. Rev. Sci. Eng.*, 54, 476, September 2014; U.S. Department of State, Fourth climate action report to the UN framework convention on climate change: Projected greenhouse gas emissions, U.S. Department of State, Washington, DC, 2007; Sakakura, T. et al., *Chem. Rev.*, 107(6), 2365, 2007; Olah, G.A. et al., *J. Org. Chem.*, 74(2), 487, 2009; Darensbourg, D.J., *Chem. Rev.*, 107(6), 2388, 2007; Mio, C.X. et al., *Open Org. Chem. J.*, 2(15), 68, 2009.

The most effective way to reduce CO_2 emissions is to reduce fossil fuel consumption through increases in plant efficiency and energy conservation. Alternatively, CO_2 capture, utilization, and sequestration (CCUS) technology can potentially be developed to help reduce CO_2 emissions from stationary point sources. Dry reforming is a fuel switching technology that utilizes CO_2 and hydrocarbons, principally CH_4, to produce syngas, H_2, and CO.

Sakakura et al. [160] showed that a widely accepted idea in catalysis is that CO_2 is so thermodynamically and kinetically stable that it is rarely used as an industrial reactant [161,162]. However, thermodynamic and kinetic attributes can be overcome using thermal or irradiation energy input, high free energy co-reactants such as hydrogen and hydrocarbons, selective product removal as a condensed phase, and production of low free energy products such as organic carbonates and carbamates [163]. The various types of CO_2 conversion reactions that are possible are described in Table 6.2 [160].

A number of large-scale processes that utilize CO_2 have found industrial use in the production of soda, urea, salicylic acid, and cyclic carbonates [164]. CO_2 has been substituted into complexes containing transition metals with organic ligands whose decomposition gives industrially useful organic acids, esters, and lactones [164–166]. The bonding of CO_2 molecules as ligands in transition metal complexes is also known [167,168]. The synthesis of organic carbonates from CO_2 is a quickly developing field of CO_2 utilization.

While there are many potential routes to the conversion of CO_2, reforming reaction may have one of the greatest commercial potentials [169]. Dry reforming of methane (CH_4) is commercially utilized in two industrial processes. In the CALCOR process (caloric Gmbh), the first stage involves the production of synthesis gas with a high concentration of carbon monoxide (CO), which is produced by dry reforming liquefied petroleum or natural gas in the presence of excess CO_2 over a Ni-based catalyst [169]. In this process, carbon deposition is minimized using a graded catalyst bed with various levels of activity and catalyst geometries. The heat of reaction for dry reforming is endothermic and must be provided by burning a portion of the fuel to produce heat. The product gas comprises a mixture of CO, CO_2, H_2, water (H_2O), and a small amount of unreacted CH_4 that is further purified in the second stage to generate a product stream comprising 99.95% CO and traces of CH_4. The separated CO_2 is recycled back into the reformer and the H_2 produced is consumed as fuel or is sold as a product. CO is a toxic, but useful, raw material that is used industrially in the production of acetic acid and phosgene [169].

The dry reforming of methane has also been used in the SPARG (sulfur passivated reforming) process created by Haldor-Topsoe which was commercialized by Sterling Chemical Inc. in Texas in 1987 [170,171]. This process makes use of existing steam reforming facilities and is capable of producing a variety of syngas compositions with H_2/CO ratios as low as 1.8 [170,171] from steam reforming process which normally generates H_2-to-CO ratio of about 3.0. The SPARG process is operated between 915°C and 945°C, and the coke deposition on the Ni catalyst is minimized by catalyst pretreatment with sulfur that passivates the coke-forming sites that inhibit carbon deposition. The process uses mixtures of CO_2 and H_2O and can be considered as a combined dry and steam reforming process.

6.5.1 THERMODYNAMICS OF DRY REFORMING

Thermodynamic calculations on the reformate product distribution and carbon formation characteristics were performed using HSC Chemistry Thermodynamic Software [172] by Shah and Gardner [153]. Methane was used as the model hydrocarbon compound in these calculations. The effects of pressure, temperature, CO_2-to-CH_4 ratio, and the addition of steam were assessed. The equilibrium amount of each species formed was normalized on the basis of 1 kmol of carbon fed as fuel (i.e., CH_4).

Figure 6.5a shows the effect of temperature on the dry reforming of methane over a temperature range of 100°C–1100°C. For a constant pressure of 100 kPa and 50/50 molar ratio of CO_2

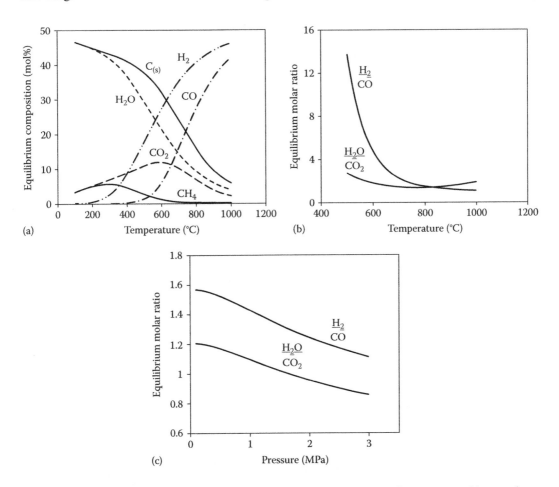

FIGURE 6.5 Effects of various variables on equilibrium molecular ratio. (a) Syngas composition as a function of temperature at 100 kPa; (b) H_2/CO and H_2O/CO_2 molar ratios as a function of temperature at 100 kPa; (c) H_2/CO and H_2O molar ratios as a function of pressure at 900°C (CO_2/CH_4 = 1/1). (From Shah, Y.T. and Gardner, T., *Catal. Rev. Sci. Eng.*, 54, 476, September 2014.)

and CH_4, H_2 and CO yields increased with the increasing temperature over this temperature range. The propensity toward carbon formation was reduced with increasing temperature. The selectivity toward combustion products, H_2O and CO_2, also diminished with increasing temperature, although both CO_2 and CH_4 exhibited maxima with respect to temperature. For constant pressure of 100 kPa, the effect of temperature on H_2-to-CO and H_2O-to-CO_2 molar ratios is described in Figure 6.5b. The results show a dramatic decrease in H_2-to-CO ratio with an increase in temperature.

For the temperature of 900°C and the 50/50 molar ratio of CO_2 and CH_4, the calculations for the effects of pressure on the carbon formation and the product distributions for the dry reforming of methane indicated that the conversion of methane was complete at pressures below 0.3 MPa and steadily declined to 50% at a pressure of 7 MPa. Carbon formation decreased with increasing pressure. The H_2 yield declined by 50%, which was more significant than the decline in CO yield over this pressure range. The selectivity toward combustion products, CO_2 and H_2O, increased marginally over this pressure range. For temperature of 900°C, the effect of pressure on H_2-to-CO and H_2O-to-CO_2 ratios is described in Figure 6.5c. As shown, both of these ratios decrease with an increase in pressure [153].

Finally, for temperature of 900°C, pressure of 20 bars and CO_2:CH_4 (1:1), the effect of steam addition (H_2O:CH_4 (1:1)) on product distribution and carbon formation during dry reforming was also analyzed. The addition of steam increased both the H_2 and CO_2 yields, while the propensity toward carbon formation was decreased. CH_4 conversion remained essentially unchanged, implying that the additional hydrogen production was due to the WGS reaction.

6.5.2 MECHANISMS AND KINETICS

The equilibrium dry reforming of CH_4 is generally expressed by:

$$CH_4 + CO_2 \rightarrow 2H_2 + 2CO \tag{6.34}$$

This reaction is generally accompanied by the RWGS reaction:

$$CO_2 + H_2 \rightarrow CO + H_2O \tag{6.35}$$

The generation of water by the WGS reaction implies that during dry reforming, some steam reforming is inevitable. Rostrup-Nielsen et al. [83] showed that the WGS reaction is extremely rapid and operates close to thermodynamic equilibrium under typical methane reforming conditions. These reactions together result in the fact that H_2-to-CO ratio during dry reforming of methane is less than 1. This is true even for dry reforming of ethane and propane. For higher hydrocarbons, however, the dominance of dehydrogenation reactions for hydrocarbons results in H_2-to-CO ratio close to or even greater than 1. When dry reforming is combined with POX and steam reforming (tri-reforming), the H_2-to-CO ratio generally exceeds 1.

The literature [82,173–180] suggests two possible mechanisms for the dry reforming of methane. In Eley–Rideal mechanism, the methane is first adsorbed on the catalyst surface and subsequently decomposes into hydrogen and adsorbed carbon on the metals of the catalyst surface. Carbon dioxide participates in the reforming reaction directly from the gas phase. In Langmuir–Hinshelwood mechanism, CH_4 is adsorbed on the catalyst surface as CH_x species and hydrogen. Also carbon dioxide is simultaneously adsorbed on the catalyst surface and is dissociated into carbon monoxide and adsorbed oxygen. The mechanism for the activation of CO_2 is unclear. One suggestion is that the CO_2 reacts with the adsorbed hydrogen and forms CO and the adsorbed hydroxide group. This OH group reacts at the metal-support interface with CH_x to form CH_xO which then dissociates into CO and H_2.

For nickel and platinum catalysts, Bradford and Vannice [175,176] proposed the following mechanism for dry reforming:

$$CH_4 + * \leftrightarrow CH*_x + \frac{1}{2}(4-x)H_2 \tag{6.36}$$

$$CO_2 + * \leftrightarrow CO*_2 \tag{6.37}$$

$$H_2 + 2* \leftrightarrow 2H* \tag{6.38}$$

$$CO*_2 + H* \leftrightarrow CO* + OH* \tag{6.39}$$

$$OH* + H* \leftrightarrow H_2O + 2* \tag{6.40}$$

$$CH* + OH* \leftrightarrow CH_xO* + H* \tag{6.41}$$

$$CH_xO* \rightarrow CO* + \frac{1}{2}xH_2 \tag{6.42}$$

$$CO* \leftrightarrow CO + * \tag{6.43}$$

As shown by Shah and Gardner [153] the previous mechanism does not quite work with the dry reforming of higher hydrocarbons where CO_2 activation is rate-controlling and dehydrogenation reactions result in numerous intermediate products as well as a significant amount of coke production.

The kinetic mechanism for dry reforming over a nickel catalyst goes through a series of steps. Carbon dioxide participates in the reaction by first producing surface OH groups through RWGS reaction. Similarly, CH_4 dissociates and adsorbs on the catalyst as CH_x intermediates. Surface OH groups then react with adsorbed CH_x species to form CH_xO intermediate which in turn decomposes to form CO and H_2. In this process, CH_4 and CH_xO decompositions are slow kinetic steps. Two main energetic roles of catalysts are to reduce activation barriers for CH_4 dissociation and CH_xO decomposition. The active site for CH_xO formation and subsequent decomposition may be at the metal–support interface. Support thus plays an important role in the reaction process. The carbon formation may be reduced by inhibition of carbon monoxide disproportionation. The catalysts that exhibit suppressed or weak carbon monoxide absorption can accomplish this task. As shown by Shah and Gardner [153], this mechanism may not be completely applicable to the dry reforming of higher hydrocarbons.

Various important kinetic models to correlate the rate data for the dry reforming of methane are summarized by Shah and Gardner [153]. The experimental data generally show that methane conversion during dry reforming is dependent on methane partial pressure but only weakly dependent on CO_2 partial pressure. Both hydrogen and CO production is dependent upon both methane and CO_2 partial pressures. The models imply that (1) methane is weakly bound on the metallic surface and its conversion depends on its gas-phase pressure, (2) CO_2 is strongly and separately adsorbed on the oxidic phase and its conversion depends both on methane conversion and the rate of RWGS reaction, and (3) both products, carbon monoxide and water, are strongly adsorbed on the active surface and they inhibit overall reforming reaction, while RWGS reaction shows a maximum with respect to its dependence on hydrogen partial pressure. Various kinetic models proposed in the literature are summarized by Shah and Gardner [153].

6.5.3 CATALYSTS FOR DRY REFORMING

Dry reforming of hydrocarbons leads to a variety of products and the transformation to syngas (CO + H_2) with different degrees of success, depending upon the operating conditions and the nature of the catalyst. The major issues with dry reforming are (1) endothermic nature of reaction requiring high energy input for the reaction process, (2) difficulty in igniting the reaction at low temperature (lower than about 500°C), and (3) requiring very high temperature (>650°C) to reduce coke deposition on the catalyst.

While Ni catalysts have been long-accepted commercial catalysts for both dry and steam reforming reactions because of their (1) high activity, (2) reasonable stability, and (3) reasonable cost and availability [181–191], in recent years significant efforts have been made to improve Ni catalysts by (1) improving the distribution of nickel on the support, (2) improving the metal–support interactions, (3) making a Ni composite (or alloy) by introducing another noble or transitional metal(s), (4) adding a promoter which affects (1) and (2) or the structure and growth rate of coke deposition, and (5) a combination of (1)–(4). Significant efforts have also been made to examine oxide and carbide catalysts. Generally, by making the catalyst bimetallic or tri-metallic, it is hoped that (1) the composite catalyst can further activate the desirable reactions, (2) reduce the nature or deposition of coke, or (3) alter the basic reaction mechanisms to improve the product selectivity and methane and carbon dioxide conversions. Different supports or promoters were examined to improve catalyst activity and life. This is particularly important when the catalyst is used for the practical systems of mixed hydrocarbons (i.e., off-gases from FT process or coal-fired power plants).

Shah and Gardner [153] have presented an extensive review of effectiveness of the different types of catalysts, supports, and promoters for dry reforming. While nickel catalysts and their composites are still the most widely used catalysts, significant research on other types of catalysts has been carried out [153,181–191]. Perspectives on some important noble metals, oxide, composite, and carbide catalysts for dry reforming are outlined in the next section. More details on all types of catalyst are given by Shah and Gardner [153].

6.5.3.1 Perspectives on Noble Metals, Composites, Oxides, and Carbide Catalysts

6.5.3.1.1 Noble Metals

Rostrup-Nielsen et al. [83] and Rostrup-Nielsen [192] gave an excellent review of the effectiveness of noble and transitional metal catalysts for dry reforming. They concluded that replacing steam by CO_2 had no significant impact on the reforming mechanism. The CO_2 reforming is slower than steam reforming. This, however, has little impact on practical reforming operation, where the RWGS reaction will soon make steam available for the faster steam reforming route.

While the nature of support significantly affects the catalysts performance, in general for noble metals the dry reforming rate follows the order Ru > Rh > Ni > Ir > Pt > Pd. At the present time, Rh catalysts on numerous supports are the most extensively investigated catalysts because of their high methane and carbon dioxide decompositions rates, high stability and low carbon deposition rates, and high selectivity for various products.

6.5.3.1.2 Composites

Besides numerous types of composites outlined by Shah and Gardner [153], following three composites appeared to be most effective. Kang et al. [193] examined trifunctional NiO–YSZ–CeO$_2$ catalyst for dry reforming reaction. The catalyst gave nearly 100% conversion for both CH$_4$ and

CO_2 above 800°C and in general higher activity than the commercial catalysts. The study used the catalyst for both dry and tri-reforming and found some NiC in dry reforming, but no NiC under tri-reforming conditions. The study concluded that tri-reforming of CH_4 is a more desirable process to produce syngas than the dry reforming. The results suggest that the tri-reforming of CH_4 over NiO–YSZ–CeO$_2$ catalyst can be applied to the production of high-value chemicals and fuel processor of SOFC and MCFC system.

Ni–Ti composite catalyst was examined by Sun et al. [194] and Guo et al. [195] for two separate applications. Sun et al. [194] showed that Ni–Ti composite xerogel catalyst with large surface area and narrow pore size distribution can be created by either (1) using a sol–gel method for the catalyst preparation and optimizing the hydrolysis and the acid-to-alkoxide ratios or (2) adding an aluminum-containing precursor. The first method avoids high-temperature calcination that results in a significant drop of surface area, collapse of pore structure, and formation of bigger crystalline particles. The second method withstands calcination temperature up to 700°C without affecting surface characteristics. The catalysts prepared by both methods showed equal or better activity and selectivity than the ones prepared by the conventional method.

Guo et al. [195] showed that $BaTi_{0.8}Ni_{0.2}O_3$ catalyst demonstrated excellent activity and stability for the partial oxidation of syngas. The study also showed that catalyst $BaTi_{1-x}Ni_xO_3$ converts methane in two steps: part of the methane is first completely oxidized to H_2O and CO_2 via combustion process, and then the unconverted methane is reformed with CO_2 and H_2O to form syngas. Choudhary et al. [196] showed methane reforming over a high-temperature stable $NiCoMgO_x$ supported on zirconia-hafnia catalyst.

6.5.3.1.3 Solid Oxide Catalysts

Besides bi- and trimetallic Ni-composite and noble metal catalysts, oxide and carbide catalysts have also been examined. Their assessments are briefly described next.

6.5.3.1.3.1 Hexaaluminate-Type Oxides

Hexaaluminates are crystalline solid oxide compounds that have several unique properties that are of interest in the design of catalysts [197–208]. Hexaaluminate compounds consist of either β-alumina or magnetoplumbite phases or mixtures of these phases and have the general formula $AO_{0.6}Al_2O_3$, where A is group I, II, or lanthanum series metals. The uniqueness of this compound lies in its molecular structure and its unit cell geometry that conducts oxygen anisotropically through the mirror plane. Oxygen conduction in the mirror plane produces an elongated crystal structure that is less susceptible to sintering at high temperatures [205–208].

Aluminum atoms that comprise the lattice are substitutable with a variety of catalytic metals. For dry reforming [198–200,202–208] and POX [197,201] reactions, Ni is the most commonly substituted metal in the hexaaluminate lattice. For catalytic combustion, Fe, Co, Mn, and Cu can be substituted into the lattice [197–208]. Collectively, this variety of metals provides a high degree of flexibility in the catalyst design. Active metals that are incorporated into the lattice initially are atomically dispersed; however, as the catalyst undergoes use, the chemical state and even the location of the catalytic metal may also undergo change [198–201]. The literature reports that Ni-substituted hexaaluminates are both active and stable for the dry reforming of methane [197–208]. The catalytic activity and the stability of hexaaluminates are a product of active metal substitution into its unique crystalline structure.

The compound, essentially, has three adjustable parameters that contribute to its activity, the first of which is the mirror cation that consists of a mono-, di-, or trivalent cation that charge balances the unit cell. The valence state and size of the mirror cation have been shown to control the amount of Ni that segregates from the bulk phase to the surface. Ni, when substituted into the lattice, has been shown to correlate with the amount of Ni on the surface and with the reforming activity of the catalyst [197–208].

The second adjustable parameter that affects catalyst activity is the extent of catalytic metal substitution into the lattice. This has also been shown to affect the activity of the catalyst. The effect is more enhanced with more Ni substitution into the lattice,

6.5.3.1.3.2 Perovskite-Type Oxides Perovskite-type oxides have been shown to possess catalytic properties that are useful for the dry reforming. Perovskite oxides with the general formula ABO_3, where A is a lanthanide series metal and B is a transition metal, have found application in the design of catalytic materials due to their high temperature thermal stability and O_{2-} ion conductivity [189,209–211]. The substitution of catalytically active metals into the perovskite structure produces defect sites that are responsible for catalytic activity and O_{2-} mobility within the crystalline lattice [189,209–211]. Reduction of weakly held –O– by B-site cations, which remain coordinated in the lattice, results in the formation of a stable and well-dispersed catalyst system.

The substitution at the A-site creates a nonstoichiometric crystal structure, which induces O_{2-} mobility. The activity of perovskite-type catalysts has been shown to be improved through partial substitution at the A-site with cations of different valence like Ce_{4+} and Sr_{2+} [189]. The unit cell structure for the perovskite phase is characterized as a face-centered cubic system [189,209–212]. In this structure, various dimensions of the unit cell are possible that affect catalytic activity by suitably choosing the A- and B-site cations. For example, two perovskite-type catalysts—$La_{1-x}Sr_xNiO_3$ ($x = 0$–0.1) and $La_{2-x}Sr_xNiO_4$ ($x = 0$–1.0)—showed markedly different performances in the dry reforming of methane. When $x = 0$, the $La_{1-x}Sr_xNiO_3$ catalyst showed high activity without coke formation, while the second catalyst was inactive. For $x > 0$, $La_{0.9}Sr_{0.1}NiO_3$ and $La_{0.8}Sr_{0.2}NiO_4$ displayed the highest catalytic activity, but exhibited coke formation. The activity of these catalysts is shown to increase with time due to the formation of lanthanum and alkaline carbonates, $La_2O_2CO_3$ and $SrCO_3$, that serve to gasify carbon deposits and result in the creation of new synergistic active sites at the Ni–La_2O_3 interface.

De Lima et al. [189] studied Ni–Fe perovskite-type oxides $LaNi_{1-x}Fe_xO_3$ ($x = 0.2$, 0.4, and 0.7). These catalysts were modified with the partial substitution of Ni by Fe. The substitution of Ni with less active Fe resulted in a less active catalyst with lower conversion; however, the composite catalyst gave greater stability and less coke deposition. The catalyst structure and properties can be optimized by choosing a suitable combination of precipitation and calcination steps. The study showed that the $LaNi_{0.8}Fe_{0.2}O_3$ catalyst was the most stable for the dry reforming reaction.

6.5.3.1.4 Metal Carbides

The relative stability of catalysts is $Mo_2C/Al_2O_3 > Mo_2C/ZrO_2 > Mo_2C/ SiO_2 > Mo_2C/TiO_2$, and calcinations of oxide precursors for short periods were found to be beneficial to catalyst stability. Although the support plays no beneficial role in the dry reforming reaction, the alumina-supported material was stable for long periods of time. Experimental evidence suggested that the differences in the stabilities may be due to interaction at the precursor stage between MoO_3 and the support, while catalyst deactivation is due to oxidation of the carbide to MoO_3, which is inactive for dry methane reforming [213–221].

Darujati and Thomson [213] observed Mo_2C/γ-Al_2O_3 to have a much higher activity than bulk phase Mo_2C due to an increase in surface area. A Ce promoter at 3 wt% was added to increase O_{2-} ion mobility that improved catalyst stability. Higher reactant conversions were obtained due to the presence of the Ce redox promoter. Naito et al. [216,217] and Tsuji et al. [219] reported that Mo_2C and WC catalyst performance and durability are strongly affected by the synthesis process, namely, nitridation and carburization routes. Supported Mo_2C catalysts have also been shown to exhibit performance differences in the following order: $Mo_2CsiO_2 < Mo_2C$–$ZrO_2 < Mo_2C$–Al_2O_3. Here, it was shown that the presence of easily reducible polymolybdate species on the surface of the precarbided catalysts coincided with higher conversions [213–221].

6.5.3.2 Catalyst Deactivation

Carbon deposition is a widely recognized mechanism of deactivation during CH_4–CO_2 reforming reactions. Its deposition onto catalytic surfaces results in the blockage of catalytically active sites which cause the reforming reaction rate to decline over time as the carbon accumulates. During methane dry reforming reactions, carbon is formed principally through two reaction pathways:

1. CH_4 decomposition at temperatures higher than 550°C

$$CH_4 \rightarrow C(s) + 2H_2 \quad H_{900°C} = +90.0 \text{ kJ/mole} \tag{6.44}$$

2. CO disproportionation at temperatures lower than 400°C

$$2CO \rightarrow C(s) + CO_2 \quad H_{900°C} = -169 \text{ kJ/mole} \tag{6.45}$$

The catalytic activity could be easily lost for many reasons such as poisoning by some ions such as sulfide, chloride, and phosphide, or it could get deactivated by carbon deposition and coke formation that occurs via pore blockage, metal encapsulation, and collapse of the catalyst support. While Ni catalyst is known for its activity for dry reforming of hydrocarbons, it is also known for its high tendency to induce coke formation. The deposited coke has different forms, all of which have their unique characteristics and different reactivity.

There are five distinct types of carbon deposits on Ni catalysts from CO and hydrocarbons [153,197]:

1. Atomic carbon (dispersed, surface carbide)
2. Polymeric films and filaments (amorphous)
3. Vermicular whiskers/fibers/filaments (polymeric, amorphous)
4. Nickel carbide (bulk)
5. Graphitic platelets and films (crystalline)

The influence of each of these carbon deposits on catalyst activity is described in detail by Shah and Gardner [153].

6.5.4 IMPORTANT CONCLUSIONS ON CATALYSIS OF DRY REFORMING OF METHANE

Based on extensive literature review, Shah and Gardner [153] have drawn the following important conclusions for catalysis of dry reforming of methane.

1. The success of the CO_2 conversion depends on three factors: catalyst activity, catalyst stability (which depends significantly on the coke formation and the nature of the coke), and efficient heat transfer operations. While there are numerous catalysts examined in the literature, it is clear that nickel catalyst is still the most practical from an economic point of view on a commercial scale. Noble metal catalysts such as Rh, Ru, and Pt are more active and perhaps more stable, but they are too expensive to be of commercial value. Future research should be focused on bimetallic catalysts such as Ni–Ru. Ru is about 40–50 times less expensive than Rh, and therefore, it will carry more practical viability for the commercial process [153]. Even Ru is an expensive catalyst. It is therefore recommended that it should be used only in a small amount.
2. The nature of the catalyst support is also very important [222–224]. Support often interacts with metals and because of that it is often considered as part of the catalyst. The best situation is the uniform distribution of very active metals in small sizes distributed along support and they do not migrate or sinter during high-temperature reforming process. Perovskite support offers special attraction because in this case metals are uniformly and tightly distributed in the support lattice. The catalyst must be basic in nature but literature has shown that too much basicity does not help the reforming process. Along with Al_2O_3, lanthanum, cerium, and zirconium oxides need to be examined. Just like mixed metals, mixed supports should also be considered [153].

6.5.5   DRY REFORMING OF HIGHER HYDROCARBONS

The literature indicates [153] that the mechanism for dry reforming of C_1–C_3 hydrocarbons is somewhat different from that of higher hydrocarbons. The same holds for steam reforming reaction. The general route in the cases of C_1–C_3 alkanes involves the dissociation of hydrocarbons and subsequent oxidation of carbon fragments; oxidative dehydrogenations of ethane and propane proceed also partially. The catalysts are in more reduced state and the activation of the hydrocarbon is the rate-controlling step. In the case of C_4 and higher hydrocarbons, the first step of the process is direct hydrogenation of alkanes. Activation of carbon dioxide, but not the activation of hydrocarbon, is the rate-controlling step. The hydrogen formed interacts with carbon dioxide and shifts the equilibrium of the dehydrogenation reaction.

Just as for methane, a required condition for dry reforming of higher hydrocarbon is that the catalyst system adsorbs and activates carbon dioxide. The acidic property of CO_2 necessitates the choice of a catalyst with basic properties. However, alkali metal and alkaline earth oxides are ineffective because of strong carbonate formation. Oxides of a moderate basicity are necessary and moreover, they must participate in redox process with CO_2 reduction. While MnO was used in the earlier studies, its modification by oxides of K, Na, Cr, and La influences both its acceptor function and the degree of surface oxidation. It controls the mechanism of hydrocarbons and alcohol transformations. Possible other good candidates are La_2O_3, cesium oxides, and praseodymium oxides. La_2O_3 showed the greatest interactions between CO_2 and hydrocarbons and alcohols. Binary oxide–based support system and dual metals can improve the performance. Promoters and the method of catalyst preparation also have an effect on the catalyst performance [153].

In more recent investigations on dry reforming, the overall objectives have been (1) to devise a process which has less coke deposition on the catalyst such that the catalyst is active and stable for a long period, (2) a process in which the catalyst ignites at as low temperature as possible, (3) a process which is heat-efficient, (4) a process in which high conversion of CO_2 and hydrocarbons is achieved, and (5) a process in which major products are carbon monoxide and hydrogen. As indicated earlier, the last objective is a particular problem without deep dehydrogenation when the hydrocarbons contain two or more carbon numbers.

Finally, three novel approaches for dry reforming—(1) an investigation of the dry reforming process under supercritical conditions, (2) microwave-assisted dry reforming, and (3) plasma reforming should be further evaluated [225,226]. Supercritical conditions help alleviate heat transfer problems. Supercritical CO_2 is a very convenient fluid because of its relatively low critical temperature and pressure. Supercritical CO_2 can enhance solubility of carbonaceous compounds, which can help catalyst activity and stability. Also, because of its relatively low pressure and temperature conditions, supercritical CO_2 operation has a commercial viability. Reforming under supercritical water is fully discussed in Section 6.7. Microwave-assisted dry reforming offers some new opportunities that should be fully explored and is described in Section 6.8. Finally, dry reforming using plasma offers a novel approach in which plasma can act as a catalyst which eliminates catalyst coking problem or plasma can be used along with a heterogeneous catalyst as a hybrid process to improve dry reforming. Plasma reforming is discussed in some detail in Section 6.9.

6.6   PARTIAL OXIDATION/AUTOTHERMAL/TRI-REFORMING

6.6.1   PARTIAL OXIDATION/AUTOTHERMAL OPERATION

Besides steam and dry reforming of methane, as shown earlier syngas can also be produced by POX of methane as:

$$CH_4 + O_2 \rightarrow CO + 2H_2 \quad \Delta H^{\circ}_{298\,K} = -38\,kJ/mol \quad \text{(Partial oxidation)} \qquad (6.46)$$

While steam and dry reforming are endothermic reactions, as shown in Section 6.4, POX is an exothermic reaction. POX of carbonaceous materials produces syngas with hydrogen-to-carbon ratio of about 2. Since POX is an exothermic reaction, when either steam reforming or dry reforming is combined with POX, one can operate reforming in an autothermal manner wherein heat generated by POX can supply the heat needed for either steam or dry reforming process without any supply of external heat.

Although thermal POX has been widely used, in catalytic POX operation, significant deposition of coke on the catalyst surface results in the catalyst deactivation [169]. As shown in Section 6.4, POX, however, offers some advantages over steam reforming. First, the reaction produces extremely high yields of synthesis gas by an exothermic reaction, and therefore, the reactor would require little or no external heating. Oxygen is often used in steam reforming to provide heat and high methane conversion. As shown in Section 6.4, POX has been extensively used with steam reforming. While POX has also been studied along with dry reforming [227–230], most studies have been restricted to one or two hydrocarbons. POX in these cases provides heat needed for dry reforming. POX also gives a better ratio of hydrogen to carbon monoxide for subsequent conversion processes to produce liquid fuels or fuel additives.

Shell uses methane POX at both Bintulu and Pearl GTL projects (see Chapter 9). There are also several POX units operating in the United States for production of hydrogen and carbon monoxide. The product gases from this reaction are low in carbon dioxide, which is commonly and mandatorily required (2%–6%) for methanol synthesis but should be further removed for some other GTL processes.

Autothermal reforming implies that the total reaction is thermoneutral [227,228]. As shown in Section 6.4, in this process, steam reforming and POX are combined [228]. The process was developed by Haldor-Topsoe A/S wherein the oxidation step is performed thermally at 1927°C, without the aid of a catalyst, and the steam reforming steps are performed with a nickel catalyst at temperatures between 927°C and 1127°C [228].

6.6.2 Tri-Reforming

It is worthwhile to recall the various types of hydrocarbon oxidation reactions that are used in the production of syngas. For methane, the relevant oxidation reactions are as follows:

$$CH_4 + H_2O \rightarrow CO + 3H_2 \quad (H^\circ_{298K} = 206\,kJ/mol) \tag{6.47}$$

$$CH_4 + CO_2 \rightarrow 2CO + 2H_2 \quad (H^\circ_{298K} = 247\,kJ/mol) \tag{6.48}$$

$$CH_4 + \frac{1}{2}O_2 \rightarrow CO + 2H_2 \quad (H^\circ_{298K} = -38\,kJ/mol) \tag{6.49}$$

Collectively, the simultaneous reactions of H_2O, CO_2, and O_2 via reactions (6.47), (6.48), and (6.49) are known as tri-reforming. They can also be characterized as autothermal reforming process, which requires an exothermic POX reaction along with steam and dry reforming (both endothermic) reactions to have overall reaction process self-sustained without any external heat input. These reactions are generally accompanied by the RWGS reaction, which is thermodynamically favorable at high temperatures:

$$CO_2 + H_2 \rightarrow CO + H_2O \quad (H^\circ_{298K} = 41\,kJ/mol) \tag{6.50}$$

While steam reforming and POX convert methane to synthesis gas, dry reforming of methane has the added advantage that it simultaneously consumes two greenhouse gases, hydrocarbons, and CO_2. While dry reforming of hydrocarbons converts CO_2 and hydrocarbons into useful syngas, tri-reforming allows the process to produce the syngas over a variety of H_2-to-CO ratios. The flexibility in H_2-to-CO ratio is very important for its downstream use. As shown in Table 6.1, different downstream applications require different H_2-to-CO ratios. For example, syngas can be converted to acetone, acetic acid, and ethylene by an exothermic reaction, while pure CO can be used for the production of acetic acid, formic acid, polyurethane, polycarbonates, methyl acrylates, and so on. H_2-to-CO ratio of about 1 is required for the production of polycarbonates, oxo alcohol, formaldehyde, iron ore reduction reaction, and so on. H_2/CO ratio of about 2 is required for methanol and FT synthesis and H_2-to-CO ratio of 3 or higher required for ammonia synthesis and hydrogen production. With a suitable manipulation of operating conditions, tri-reforming can provide this flexibility in H_2/CO ratio.

At high temperatures (greater than approximately 500°C), equilibrium favors the RWGS reaction [169]. High CO_2 concentration in the product during steam reforming also means that both steam and dry reforming reactions (Equations 6.47 and 6.48) occur simultaneously. The kinetics associated with the steam reforming reaction, in general, are faster than the dry reforming reaction, and during dry reforming, steam reforming is typically at equilibrium stage. Both dry and steam reforming are endothermic and require a significant amount of heat to operate the reaction at high-temperature (>600°C). POX is an exothermic reaction that can produce the heat required for the endothermic dry and steam reforming reactions.

As shown earlier, a major technical problem of conducting steam reforming alone is carbon deposition on the catalysts, which can lead to rapid deactivation and breakup of the catalyst. Carbon deposition can be substantially reduced by the use of an excess of water and a temperature of about 800°C. For these reasons, the stream reforming requires expensive generation of superheated steam (in excess) at high temperature. This requirement of high temperature may also be reduced by the use of tri-reforming.

An extensive study of tri-reforming was carried out by Puolakka et al. [229,230]. The study focused on the tri-reforming of five model compounds: methane, heptanes, n-dodecane, toluene, and ethanol over a number of different catalysts. It was reported that 0.25% Rh on ZrO_2 catalyst gave the best results and its performance was comparable to the results for commercial Ni catalyst. The five model compounds were chosen to represent different types of fossil/biofuels. Methane was chosen to represent natural gas, n-heptane to represent aliphatic component of gasoline, n-dodecane to represent aliphatic component of biodiesel, toluene to represent aromatic part of gasoline and diesel oil, and ethanol to represent oxygenated compounds in biofuel. The results indicated that effectiveness of tri-reforming depends on the compound, catalyst, and the operating conditions. More work with real systems is needed.

6.6.2.1 Application of Tri-Reforming to Flue Gas from Combustion Plants

A new process using tri-reforming has been proposed for effective conversion and utilization of CO_2 in the waste flue gases from fossil fuel–based power plants [231–240]. In this process, CO_2, H_2O, and O_2 in the gas need not be preseparated because they can be used as co-reactants for the tri-reforming of natural gas. In the tri-reforming process, the flue gas and natural gas are used as chemical feedstock for production of synthesis gas (CO + H) with desired H_2/CO ratio.

In the proposed process, flue gas from coal combustion unit (with large CO_2 concentration) is mixed with natural gas for tri-reforming using waste heat from the combustion-driven power plant. In principle, once the syngas with the desired H_2/CO ratio is produced from tri-reforming, it can be used to produce liquid fuels, manufacture industrial chemicals such as methanol and acetic acid, or generate electricity with either integrated gasification combined cycle (IGCC)-type generators or fuel cells.

As pointed out by Song [232,237], the proposed concept is consistent with the goals of Vision 21 EnergyPlex concept, which the U.S. Department of Energy (DOE) is developing. The proposed goals of DOE Vision 21 for power plants include more efficient power generation (>60% with coal, >75% with natural gas), higher overall thermal efficiency (85%–90%), near-zero emissions of traditional pollutants, reduction of greenhouse gases (40%–50% reduction of CO_2 emissions), and co-production of fuels [232,237].

The investigations reported in the literature [231–240] indicate that there are benefits of incorporating steam and oxygen simultaneously in CO_2 reforming of natural gas or methane. As shown earlier, adding oxygen to dry or steam reforming generates heat by POX and improves the energy efficiency of the entire process. Furthermore, adding oxygen to CO_2 reforming and/or steam reforming of methane can provide synergetic effects in processing and mitigation of coking. Inui et al. [233,235] have studied energy-efficient hydrogen production by simultaneous catalytic combustion and catalytic CO_2–water reforming of methane using a mixture of pure gases, including methane, CO_2, water, and oxygen. Choudhary et al. [236,238] have reported on their laboratory experimental study on simultaneous steam and CO_2 reforming of methane in the presence of oxygen at atmospheric pressure with Ni/CaO. Choudhary's work shows that it is possible to convert methane into syngas with high conversion and high selectivity for CO and hydrogen [236,238].

Hegarry et al. [239] and O'Connor and Ross [240] have shown that a Pt/ZrO_2 catalyst is active for steam and CO_2 reforming combined with the partial oxidation of methane. Halmann and Steinfeld [231,234] also showed that the treatment of flue gases from fossil fuel–fired power stations by tri-reforming with natural gas or by coal gasification could become an attractive approach for converting the CO_2, H_2O, O_2, and N_2 contained in these flue gases via syngas processing into useful products, such as methanol, hydrogen, ammonia, or urea.

These studies thus indicate that the proposed tri-reforming of natural gas using flue gas from power plants appears to be feasible and safe, although more detailed experimental studies, computational analyses, and engineering evaluations are still needed. Preliminary experiments [224,229–237] showed that syngas with desired H_2/CO ratios can be made by tri-reforming methane using simulated flue gas mixtures containing CO_2, water, and O_2 in a fixed-bed flow reactor. For example, tri-reforming in a fixed-bed flow reactor using gas mixtures at atmospheric pressure that simulate the cases with flue gases from coal- and natural gas–fired power plants have been reported in the literature [231–240]. Numerous catalysts have been examined, and some of them showed better resistance to coking and less deactivation.

Several technical challenges must be overcome before tri-reforming can be successfully upscaled for this concept. Flue gases contain inert nitrogen gas in high concentrations, and thus the conversion process design must consider how to dispose of nitrogen. It is possible that oxygen-enriched air or pure oxygen will be used in power plants in the future. If that becomes a reality, then the proposed tri-reforming process will be even more attractive because of much lower inert gas concentrations and higher system efficiency. Another challenge is how to deal with the small amounts of SO_x, NO_x, and other toxic substances that may be present in flue gases of power plants. These impurities should be removed before the application of tri-reforming to flue gases.

An important feature of the proposed tri-reforming process is that it is the first innovative approach to conversion and utilization of CO_2 in flue gases from power plants without separating CO_2. Many questions remain, and further research is needed to establish and demonstrate this new process concept.

6.7 SUBCRITICAL AND SUPERCRITICAL REFORMING

Reforming can also be carried out in subcritical and supercritical water. In subcritical water, APR can be carried out for carbohydrates, in particular, its sugar components. Steam, dry, and tri-reforming can also be carried out in supercritical water for all carbonaceous materials. These two types of reforming processes are fully described in two chapters of my previous book [72]. Here we present some selective brief summaries on the topics.

6.7.1 Aqueous-Phase Reforming

The pioneering work carried out by Dumesic et al. [241–264] and others [265–301] showed that carbohydrates such as sugars (e.g., glucose) and polyols such as methanol, ethylene glycol, glycerol, and sorbitol can be efficiently converted to hydrogen and carbon dioxide at 500 K by reforming under aqueous conditions. The process can be applied to all carbohydrates found in wastewater from biomass processing of cheese whey, beer brewery, sugar processing as well as carbohydrate streams from agricultural products, like corn and sugar beets and hemicellulose from any biomass [72]. Typical feedstock that can be used for APR and resulting bioforming process [72] are listed in Table 6.3.

The produced hydrogen can be used to hydrogenate many components of lignocellulosic biomass to produce glycols and other polyols (thus enlarging the feedstock possibilities for aqueous-phase processing). The hydrogen can also be used to produce ammonia and fertilizer, an additive to gasification products to produce liquid fuels via FT synthesis and fuel source for PEM fuel cells.

Besides hydrogen, APR can also produce syngas, alkanes, and monofunctional groups depending on the nature of the catalyst and the operating conditions. The production of hydrogen and syngas requires the breakage of C–C bonds within oxygenated compounds, whereas the production of alkanes and monofunctional groups requires the breakage of C–O bonds within the oxygenated compounds. With most feedstock examined so far, the alkane production is limited to six carbon atoms. More feedstock, catalysts, and reactor designs are needed to produce C_8–C_{15} alkanes from the biomass-derived reactants. The alkanes and monofunctional groups can be further upgraded catalytically by creating new C–C bondages to produce higher alkanes and liquid fuels. Light fuel additives such as pentane and hexane have limited values due to their high volatility. Various reaction paths that can be produced by APR process are schematically illustrated in Figure 6.6 [72].

The low-temperature APR to produce hydrogen has significant advantages over conventional steam reforming mentioned in Section 6.4 because of the following reasons [72]:

1. The process occurs in one liquid phase eliminating energy requirement to vaporize water and carbohydrates. Steam reforming requires high temperature and is accompanied by a phase change.
2. The raw materials for APR are nonflammable and nontoxic, allowing them to be stored and handled safely and conveniently. We have established technologies for the storage of sugar, starch, and carbohydrates.
3. The temperature and pressure used in APR favors the thermodynamics of WGS reaction allowing high conversion of CO in one reactor.

TABLE 6.3
Typical Feedstock for Aqueous-Phase Reforming

Primary Feedstock

Feedstock is generally water-soluble oxygenated hydrocarbons
(e.g., sugar, sugar alcohols, saccharides, and other polyhydric alcohols)

Secondary Feedstock

These oxygenated hydrocarbons can originate from:

   Sugar crops

   Grain crops

   Agricultural waste (corn stalks, straw, seed hulls, sugarcane leavings, bagasse, nutshells, manure from cattle, poultry and hogs)

   Wood materials (wood or bark, sawdust, timber slash, mill scrap)

   Municipal waste (waste paper, yard clippings)

   Energy crops (poplars, willows, alfalfa, switch grass, prairie bluestem, corn, soybean)

Source: Shah, Y.T., *Water for Energy and Fuel Production*, CRC Press, New York, 2014.

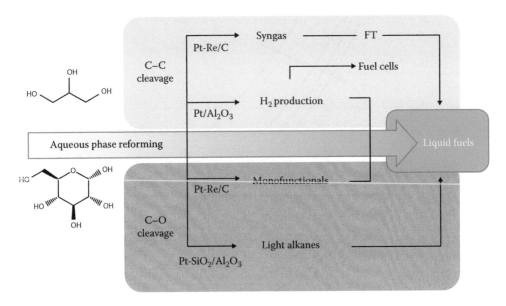

FIGURE 6.6 Various reaction paths for aqueous-phase reforming. (From Shah, Y.T., *Water for Energy and Fuel Production*, CRC Press, New York, 2014.)

4. The conventional PSA and membrane technologies are easily applicable to the product stream to separate carbon dioxide from hydrogen since pressures used in APR vary from 15 to 50 atm. Steam reforming is often carried out at low pressure, thus requiring pressurization of the product to carry out effective separation. Pure hydrogen can thus be produced more easily by APR process.

5. The low temperatures used in APR minimizes the decomposition reactions for carbohydrates and resulting coking of the catalysts. Coking of the catalyst is a significant issue in the conventional steam reforming.

6. APR can produce hydrogen in a single reactor as opposed to conventional steam reforming which will require multistage processes.

7. Since APR produces hydrogen, syngas, lower alkanes, and monofunctional groups (which can be further processed to generate different types of liquid fuels), the operating conditions and catalysts can be manipulated to obtain the desired selectivity among various products.

6.7.1.1 Thermodynamics

The prevailing thermodynamic forces for the alkanes and oxygenated compounds steam reforming along with the WGS reaction and the methanation reaction are illustrated in Figure 6.7 [72] in the form of a plots of Gibbs free energy versus temperature. The favorable thermodynamic forces require negative Gibbs free energy. The figure shows that both oxygenate reforming (of methanol, ethylene glycol, glycerol, sorbitol, and glucose) and WGS reactions are favorable at low temperatures. Also, methanation reaction is favorable at reasonably low temperatures. On the other hand, steam reforming reactions for methane and other alkanes are only favorable at higher temperatures.

The concept of APR is based on the fact that at moderate temperature and pressure oxygenated carbohydrates react with water to produce either alkanes or hydrogen and carbon monoxide by the following reforming reaction:

$$C_nH_{2n+2} + nH_2O \leftrightarrow nCO + (2n + 1)H_2 \qquad (6.51)$$

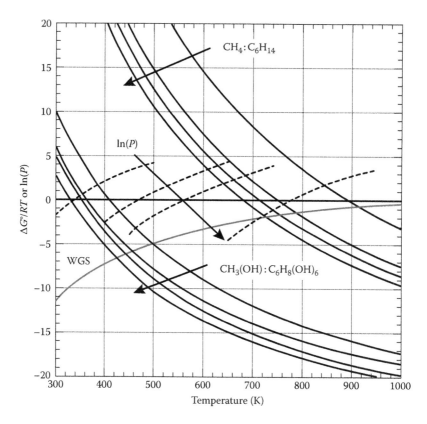

FIGURE 6.7   Thermodynamics of aqueous-phase reforming. (From Shah, Y.T., *Water for Energy and Fuel Production*, CRC Press, New York, 2014.)

Also at these temperatures and pressures water gas reaction

$$CO + H_2O \leftrightarrow CO_2 + H_2 \tag{6.52}$$

is favored. Figure 6.7 presents the Gibbs free energy ($G°/RT$) associated with the reforming of a series of alkanes such as CH_4, C_2H_6, C_3H_8, and C_6H_{14} normalized per mole of CO produced along with the one for the WGS reaction. These results show that while WGS reaction is favorable at low temperature, thermodynamics of alkanes reforming is only favorable at higher temperatures (e.g., $T > 675$ K for C_6H_{14} and $T > 900$ K for CH_4). Thus, at lower temperatures lower alkanes cannot be reformed to syngas.

The oxygenated hydrocarbons having C:O ratio of 1:1 form carbon monoxide and hydrogen according to the following reaction:

$$C_xH_{2y}O_z \leftrightarrow zCO + yH_2 \tag{6.53}$$

Figure 6.7 also shows Gibbs free energy diagram for some typical oxygenated compounds such as methanol (CH_3OH), ethylene glycol ($C_2H_4(OH)_2$), glycerol ($C_3H_5(OH)_3$), and sorbitol ($C_6H_8(OH)_6$). These results show that APR of these compounds at low temperatures is thermodynamically favorable. Sorbitol is generally obtained by the hydrogenation of glucose ($C_6H_6(OH)_6$). Thus oxygenated hydrocarbons can be reformed at much lower temperatures than the alkanes with similar carbon number. A combination of aqueous (or steam) reforming of oxygenated carbohydrates and WGS reaction will allow the production of hydrogen at low temperatures. The logarithms of vapor pressure as a function of temperature for methanol, ethylene glycol, glycerol, and sorbitol are illustrated in Figure 6.7. For first three substances, steam reforming (in the gas phase) can be carried out at temperatures 550 K and above, while for sorbitol, vapor-phase steam reforming requires temperature of at least 750 K. Thus at low temperatures (<750 K) reforming of sorbitol (and glucose) can be carried out in the aqueous phase producing hydrogen and syngas. The favorable thermodynamics for APR of oxygenated compounds illustrated in Figure 6.7 prompted significant research to evaluate favorable kinetic conditions to produce hydrogen, syngas, and alkanes via APR process [241–245,248,265,266].

Since thermodynamics of steam reforming of alkanes at low temperatures are not favorable, hydrogen and carbon dioxide formed from oxygenates at lower temperatures are not stable, and alkanes can be formed by the methanation and FT reactions between hydrogen and carbon monoxide and carbon dioxide. For example at 500 K, the equilibrium constant for methanation reaction

$$CO_2 + 4H_2 \leftrightarrow CH_4 + 2H_2O \tag{6.54}$$

is very favorable [241–245,248,265,266]. Thus, in order to selectively form hydrogen and inhibit the formation of alkanes we need a catalyst that promotes C–C scission followed by WGS reaction and inhibits C–O scission followed by the hydrogenation.

6.7.1.2 Kinetics and Catalysis of APR Process

As shown in Figure 6.6, APR process has four distinct kinetic steps depending on the desired product. APR can produce hydrogen, syngas, alkanes, or monofunctional groups depending on the catalyst and support system, promoters, and other operating conditions.

The original purpose of APR was to generate either hydrogen or alkanes by an APR of sugar, other oxygenated compounds, and polyols. For example, the aqueous-phase reforming of cellulose can form hydrogen by the following reaction:

$$C_6O_6H_{12} + 6H_2O \rightarrow 6CO_2 + 12H_2 \tag{6.55}$$

Also dehydration/hydrogenation results in the formation of alkanes as

$$C_6O_6H_{12} + 7H_2 \rightarrow C_6H_{14} + 6H_2O \tag{6.56}$$

which gives the combined reaction as:

$$1.6C_6O_6H_{12} \rightarrow C_6H_{14} + 3.5CO_2 + 2.5H_2O \tag{6.57}$$

Alkanes contained 95% of the heating value and only 30% of the mass of the biomass-derived reactant. The reforming reaction with glycerol results in

$$C_3H_8O_3 \rightarrow 3CO + 4H_2 \tag{6.58}$$

As shown in Figure 6.8, hydrogen selectivity decreases with an increase in carbon number of oxygenated compounds and increase in temperature. The literature has shown [241–245,248,265,266] that compounds like furanone and acetic acid are not amenable to the production of hydrogen by APR. The selectivity for hydrogen depends on the nature of the bond breaking in oxygenated compounds; the breaking of C–C bond favors the hydrogen formation and the breakage of C–O bond favors the formation of alkanes. Following the preferred pathway is the key to the hydrogen formation.

While the literature [72] has shown numerous ways to generate hydrogen from biomass under high-temperature conditions, the present process is unique in that it is the only process that can be carried out in liquid water. While aqueous-phase reforming can be used only for selective feedstocks,

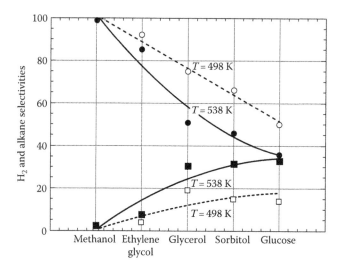

FIGURE 6.8 Temperature and feedstock effects on selectivity during aqueous-phase reforming. (From Shah, Y.T., *Water for Energy and Fuel Production*, CRC Press, New York, 2014.)

it is faster than anaerobic digestion process for generating hydrogen from cellulosic waste. Dumesic et al. [241–264], among others [265–301], have clearly shown that the nature of both metal and support has an important influence on APR reactions. The product selectivity can be tuned depending on the metal, support, and promoter. Pt-black and Pt supported on Al_2O_3, TiO_2, and ZrO_2 have been demonstrated to be active and selective for the aqueous-phase reforming of methanol and ethylene glycol to produce hydrogen. Catalysts based on Pd have shown similar activity compared to Pt analogs. Ru, Rh, and Ni, on the other hand, showed lower activity for hydrogen.

Davda et al. [242,248,263] and others [248,249,252,253,274,279,280] have examined the effectiveness of various Group VIII metal catalysts such as Ru, Rd, Pt, Ir, Pd, and Ni for APR. The studies compared the selectivity for hydrogen, alkane, and carbon dioxide by Pt, Pd, Ru, Rh, and Ni catalysts for various oxygenated compounds and at various temperatures. The results show that CO_2 selectivity was the highest for Pt and Ni catalysts and least for Rh and Pd catalysts. The alkane selectivity was the highest for Ru and Rh catalysts followed by Pt and Ni. Very little alkanes were produced by Pd catalysts. Finally, Pt and Pd (followed by Ni) showed good reforming activity and high hydrogen production rates. Good catalysts for hydrogen production by APR should show high activity for WGS reaction and for cleavage of C–C bonds. Both Pd and Pt catalysts gave poor activity for C–O scission and subsequent methanation and FT reactions. In the final analysis, since Pt catalyst gave good product distributions for all three (hydrogen, carbon dioxide, and alkanes) components, they concluded Pt to be the best catalyst. Ni catalyst, although cheap, gave preference to alkanes.

The study of Davda et al. [242,248,263] also indicated that the best support for Pt was Al_2O_3 for hydrogen production and the effect of support on reforming activity and selectivity is greater than that of metal dispersion. They also tested bimetallic catalysts and concluded that Ni–Sn catalysts show potential for APR. The selectivity for hydrogen and alkanes for different oxygenates at 225°C and 265°C using Pt/Al_2O_3 catalyst is illustrated in Figure 6.8 [72]. These results show that for all compounds hydrogen selectivity decreases with an increase in temperature. Dumesic and coworkers [241–264] also showed that an increase in pressure reduced the hydrogen selectivity. For example, for the reaction of 5 wt% sorbitol over Pt-SiAl at 498 K, hydrogen selectivity at 25.8 atm pressure was 21, while the same selectivity at a pressure between 33.1 and 52.1 atm was less than 2.

The addition of a promoter can also have some effect on the catalyst performance. One promoter, Re, was found to be an effective promoter for Pt/C catalyst. The selectivity of Pt–Re/C was found to be different from that for Pt/C. Hydrogen selectivity with promoter was lower, although hydrogen productivity was higher. An addition of base (KOH) affected glycerol APR selectivity for 3%Pt3%Re/C catalyst. Pt–Re/C catalyst following reduction was significantly more active for APR of glycerol than Pt/C catalyst. The presence of Re created surface acidity, which favored a pathway of C–O bond breaking (dehydration), resulting in lower hydrogen and $CO(CO_2)$ selectivity and higher alkane selectivity. The effects of liquid and solid acidities on carbon selectivity for sorbitol at 538 K and 57.6 bar with Pt/Al catalysts were examined by Dumesic et al. [241–264]. The results indicate that lower acidity of both liquid and solids produce higher carbon number alkanes. In general, an increase in acidity either by the use of acid catalyst support (i.e., SiO_2/Al_2O_3) or by the addition of the mineral acid like HCl to the feed increases alkane selectivity due to the increased rate of dehydration and hydrogenation pathways compared to hydrogenolysis and reforming reactions. This tendency can be exploited to produce butane, pentane, and hexane from sorbitol over $Pt-SiO_2/Al_2O_3$ catalysts. The nickel supported on SiO_2 or Ai_2O_3 was found to have low selectivity for hydrogen and it favored the formation of alkanes. On the other hand, an addition of a Sn promoter to RANEY R–Ni-based catalysts enhanced the production of hydrogen from sorbitol, glycerol, and ethylene glycol. While promoters and acidity can be used to produce alkanes, some C–C bond needs to be broken to produce hydrogen needed for the production of alkanes by the following hydrogenation and complete deoxygenation reactions for sorbitol [72].

$$C_6H_{14}O_6 + 6H_2 \rightarrow C_6H_{14} + 6H_2O \tag{6.59}$$

However, complete deoxygenation occurs as [302]

$$C_6H_{14}O_6 \rightarrow \frac{13}{19}C_6H_{14} + \frac{36}{19}CO_2 + \frac{42}{19}H_2O \tag{6.60}$$

Besides hydrogen and alkanes, reforming has also been used to produce syngas from glycerol's feedstock [276,277,282,284,285,287–290]. This once again requires the selective breakage of C–C bonds. This can be achieved with Pt catalyst in the temperature range of 498–548 K but at lower pressure. Under these conditions, Pt surface is covered by CO molecules which hinder gas-phase reaction. Pt/Ru or Pt/Re were identified as alloys that bind CO less strongly on the surface, thus mitigating reaction inhibition in the presence of products. These catalysts will produce syngas by the reaction [72].

$$C_3H_8O_3 \rightarrow 3CO + 4H_2 \tag{6.61}$$

$$C_3H_8O_3 \rightarrow \frac{7}{25}C_8H_{18} + \frac{19}{25}CO_2 + \frac{37}{25}H_2O \tag{6.62}$$

The syngas produced at these low temperatures can be easily used for the subsequent conversion of syngas to liquid fuels by FT synthesis. The increase in Re to carbon-supported Pt catalysts also promotes WGS reaction, which increases H_2-to-CO ratio and decreases CO/CO_2 ratio in syngas.

The literature results [241,242,248,273,274,277] also showed that for Pt/Re/C catalyst, an increase in pressure shifted the reaction away for reforming reaction to more in the direction of alkane production and partially deoxygenated reaction intermediates [260] such as alcohols and ketones, suggesting that it is possible to selectively deoxygenate polyols. This also suggests that it is possible to couple biomass reforming with hydrodeoxygenation to improve energy density without an external source of hydrogen. Thus Pt–Re/C catalysts operating at low temperature, high pressure, and high oxygenate feed concentration will favor C–O bond breakage and partially deoxygenate polyols to produce monofunctional intermediates which are predominantly 2-ketones, secondary

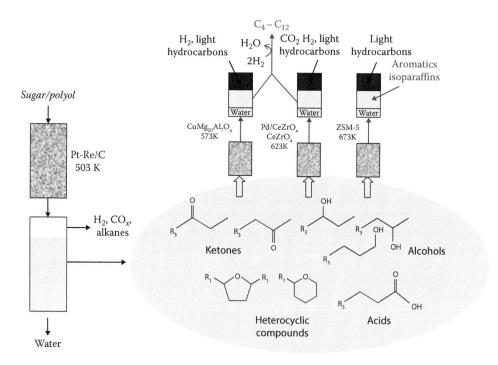

FIGURE 6.9 Ketonization and aldol condensation reactions following aqueous-phase reforming. (From Shah, Y.T., *Water for Energy and Fuel Production*, CRC Press, New York, 2014.)

alcohols, heterocycles, and carboxylic acids. As shown later, these monofunctional groups provide a platform for a variety of upgrading strategies that are not limited to the production of shorter alkanes (hexane and less) or syngas.

In fact, bioforming process developed from APR concept [294–301] show that these intermediates allow the production of fuel additives and fuels like jet fuel, diesel, and gasoline with appropriate upgrading strategies. Thus, C–C coupling is employed along with oxygen removal to obtain larger hydrocarbons starting from biomass-derived C_5 and C_6 sugar compounds. This is graphically illustrated in Figure 6.9 [72]. Ketones are coupled via aldol condensation using basic catalysts such as $MgAlO_x$, MgAl, $Pd-MgO/ZrO_2$, $MgZrO_2$, La/ZrO_2, Y/ZrO_2, and Mg/TiO_2 [72,243,248,253,248,259]. Furthermore, introduction of bifunctional metal basic catalysts allows for the coupling of secondary alcohols in the presence of hydrogen. More condensation reactions are also driven for ketones in the presence of hydrogen. C–C coupling can also be enhanced by ketonization of carboxylic acid [243]. The complete hydrogenations of monofunctional groups can also produce alcohols. The alcohols can then be converted to gasoline using MTG technology of Mobil Oil Co. which uses H-ZSM-5 catalyst [72,243,248,253,259]. Alcohols can also be dehydrated to produce olefins.

Kunkes et al. [250,260,269] designed a process of conversion of monofunctional group to pentanol and hexanol and converting these alcohols to C_6^+ gasoline by H-ZSM-5 catalyst at 673 K. In a two-step process, alcohols can also be dehydrated by acidic niobia catalyst to form olefins which can be coupled over H-ZSM-5 to form branched olefins centered around C_{12}. Less branched and more complex diesel fuel can also be created by using a mixed system of catalyst $CuMg_{10}Al_7O_x$ and $Pd/CeZrO_x$ and $CeZrO_x$ to achieve ketonization and aldol-condensation of biomass-derived monofunctional groups, as shown in Figure 6.9 [72]. All of these strategies are being further developed in providing liquid fuels by upgrading of APR products in the Virent's BioForming process described in my previous book [72] and by Held [296,298] and Cortright et al. [294,295,297,299–301].

The operating conditions and the nature of the catalyst affect not only the selectivity between hydrogen and alkanes but also the level of CO production. A low CO concentration in the product requires an ultra-shift operation wherein the reaction conditions are such that WGS reaction is favored. The lowest level of CO requires lowest partial pressure of CO_2 and H_2 in the gas phase so that forward water gas reaction is thermodynamically favored. These conditions are achieved by operating the reactor at the saturation pressure for water (at the reaction temperature) and using low feed concentration of oxygenates.

For biomass application, APR of glucose is very important because it is the basic sugar component of all starch and carbohydrates. The hydrogenation of glucose leads to the formation of sorbitol, and both glucose and sorbitol can be reformed to form carbon dioxide and hydrogen. As the glucose concentration in the feed increases, the selectivity to hydrogen decreases. Also, these reactions are favored at low temperatures. The reforming of both glucose and sorbitol can occur on Pt and Ni–Sn alloy by cleavages of C–C bonds followed by WGS reaction. The alkanes are produced on acidic sites of metals both from glucose and sorbitol. Glucose also produces acids, aldehydes, and so on through homogeneous side reactions. Since undesirable side reactions are first-order with respect to glucose and desirable reactions have a fraction order dependence on glucose, an increase in glucose concentration reduces hydrogen selectivity. The hydrogenation of glucose to sorbitol also occurs at a higher rate at low temperature (400 K) and high hydrogen pressure [72].

Davda et al. [72,248] proposed that a way to increase hydrogen selectivity from glucose is to operate in two stages: first carry out low-temperature hydrogenation step, followed by high-temperature reforming process. Other approach is to co-feed hydrogen with liquid reactant stream to the reforming reactor. This co-feeding argument lead Davda et al. to propose a process shown in Figure 6.10 [72,248] to obtain the product of desired specification using APR.

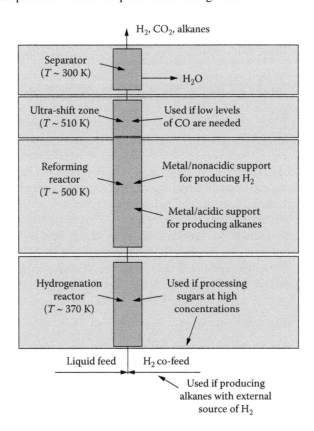

FIGURE 6.10 Process of co-feeding to optimize the product distribution during APR. (From Shah, Y.T., *Water for Energy and Fuel Production*, CRC Press, New York, 2014.)

The aforementioned discussion on thermodynamics and kinetics of the APR process gives the following conclusions about the APR process [72]:

1. The basis for the APR process is that while alkane reforming is favorable only at high temperatures, the reforming of oxygenated carbon (with C:O, 1:1) and the WGS reaction are possible at low temperatures. This allows APR to be carried out in the liquid phase.
2. The activation energy required to break up C–C bonds in oxygenated compounds is lower than the ones required in alkanes. Thus H_2 and CO_2 from oxygenated compounds can be obtained in a single reactor. This can be accomplished in liquid phase only for high boiling point compounds like glucose and sorbitol, whereas for low boiling point compounds like glycerol, ethylene glycol, and methanol, the reactions can occur both in the gas and liquid phases.
3. The choice of a catalyst can affect the products. Pt, Pd, and Ni–Sn alloys show high selectivity for hydrogen, while Ni catalysts tend to make more alkanes. Ru and Rh catalysts also make alkanes with very little hydrogen. More acidic support favors alkane production, whereas more basic/neutral support like alumina favors hydrogen production. The acidic aqueous solution similarly promotes alkane production due to acid-catalyzed dehydrogenation reactions (followed by the hydrogenation on the metal). Basic aqueous solution favors hydrogen production.
4. The type of feed and its concentration affect the product distribution. Sorbitol gives higher selectivity for hydrogen than glucose. Within polyols, hydrogen selectivity decreases with an increase in carbon number of the feed and the increase in feed concentration due to increase in side reactions.
5. Davda et al. [248] outlined a number of different pathways that can occur in the APR reactor depending on the nature of catalyst, its acidity and acidity level of aqueous solution, temperature, and pressure to obtain the desired product distributions. Generally, higher carbon number in the feed and more acidity on the catalyst or aqueous solution favor C–O scission and more alkane production. The reverse conditions promote C–C bond cleavages to form hydrogen and CO_2. The latter compounds can, however, undergo undesirable methanation and FT reactions to produce more alkanes. Some metals like Ru and Rh favor C–O scission and form more alkanes. Pt and Pd, on the other hand, favor C–C scission. More bifunctional catalysis can occur by combination of metal, support, and solutions. In general, high hydrogen selectivity requires high C–C scission, low rates of C–H scission, and low rates of methanation and FT reactions. Low CO level can be obtained by operating the reactor with low partial pressures of hydrogen and carbon dioxide.

More details and in-depth discussions regarding feedstock effect, new developments of catalysts, catalyst supports and promoters; more discussion on the production of syngas and monofunctional groups and their upgrading; and the details on the development of the demonstration-scale "bioforming process" based on the APR concept are given in my previous book [72].

6.7.2 SUPERCRITICAL WATER REFORMING

Reforming can also be carried out in supercritical water. The supercritical water (SCW) exists at pressures higher than 221 bar and temperatures above 374°C. By treatments of coal, biomass, or their mixtures in supercritical water (and in the absence of added oxidants) organics are converted into fuel gases and are easily separated from the water phase by cooling to ambient temperature. The produced high-pressure (HP) gas is generally rich in hydrogen [239].

The characteristics of supercritical water (SCW)-organics in carbonaceous materials interactions change considerably with increasing temperature. Water behaves like organic liquid and nonpolar substance at higher temperatures, and it can dissolve most organic species under supercritical conditions. With temperature increasing to 600°C, water becomes a strong oxidant and results in complete disintegration of the substrate structure by transfer of oxygen from water to the carbon atoms of the substrate. As a result of the high density, carbon is preferentially oxidized into CO_2 but also low concentrations of CO are formed. The hydrogen atoms of water and of the substrate are set free and form H_2. A typical overall reaction for glucose can be written as:

$$2C_6H_{12}O_6 + 7H_2O \Rightarrow 9CO_2 + 2CH_4 + CO + 15H_2 \tag{6.63}$$

The RSW (reforming in supercritical water) reactor operating temperature is typically between 600°C and 650°C; the operating pressure is around 300 bar. A residence time of up to 2 min is required to achieve complete carbon conversion, depending on the feedstock. Heat exchange between the inlet and outlet streams from the reactor is essential for the process to achieve high thermal efficiency. The two-phase product stream is separated in a high-pressure gas–liquid separator ($T = 25°C–300°C$), in which a significant part of the CO_2 remains dissolved in the water phase.

Possible contaminants like H_2S, NH_3, and HCl are likely captured in the water phase due to their higher solubility, and thus in situ gas cleaning is obtained. The gas from the HP separator contains mainly the H_2, CO, and CH_4 and part of the CO_2. In a low-pressure separator, a second gas stream is produced containing not only relative large amounts of CO_2 but also some combustibles. This gas can, for example, be used for process heating purposes.

The RSW process is particularly suitable for the conversion of wet organic materials (moisture content 70%–95%), which can be renewable or nonrenewable. The process is more effective for biomass (due to its higher reactivity in water) and the focus of BTG is completely on liquid biomass streams like glycerol or the aqueous phase of pyrolysis oil. In principle, slurries should also be possible, but feeding is extremely difficult and costly.

The primary gas produced by the RSW process differs significantly from the syngas that is produced in common thermal biomass gasifiers in three ways: (1) the gas is produced at high pressure, (2) the gas is rich in hydrogen, and (3) the gas does not contain diluents like nitrogen. All of these characteristics make the gas more useful for downstream applications. The produced gas is clean (no tar, or other contaminants in high-pressure gas even if they are produced in the process). The gas always contains high amounts of hydrogen; the amounts of CO and CH_4 depend on the operating conditions. Based on these process characteristics RSW can be used for (1) hydrogen production, (2) syngas production by minimizing methane production, and (3) synthetic natural gas production by minimizing CO production.

The energy efficiency for pure hydrogen is above 60%. The RSW process enables the conversion of especially wet biomass (~70–95 wt% moisture). The product gas is clean, rich in H_2 and CH_4, and available at high pressure. In a high-pressure gas–liquid separator the gases are separated from the water. Due to the high pressure, significant part of the CO_2 is dissolved in the water phase and will be released in a low-pressure gas–liquid separator. The process has now been further developed and evaluated to convert aqueous by-products from pyrolysis oil upgrading into a hydrogen-rich gas.

6.7.2.1 Feedstock Effect on Supercritical Water Reforming

Supercritical water is an ideal medium to carry out reforming reactions for both single components and complex materials. Besides all the positive features of the supercritical medium outlined earlier, supercritical water provides possibilities of lower temperature, lesser coking issues, and more active and stable catalytic reforming process.

6.7.2.1.1 Hydrocarbon Fuels

Lee et al. [303] showed that reforming of JP-8 fuel and diesel fuel can be carried out in supercritical water in the absence of a catalyst. High enthalpy level of supercritical water and high solubilities of fuel in supercritical water allowed the reforming reactions to occur in the temperature range

650°C–825°C and 220–330 atm pressure. The study examined production of hydrogen and methane as functions of reactor operating conditions and possibility of autothermal operation by simultaneously carrying out POX reaction with reforming reaction. The process handled fuel with sulfur. The results were obtained at temperatures lower than conventional reforming temperature. The autothermal operation was achieved by adding oxygen into the reacting mixture. In a noncatalytic operation, hydrogen production of 14% of theoretical maximum was obtained. In another study, supercritical water reforming of logistic diesel fuel at 550°C in the absence of a catalyst also resulted in significant hydrogen production [72].

Shah [72] and Veriansyah et al. [304–307] examined reforming of gasoline in supercritical water. They used methanol and isooctane (2,2,4-trimethylpentane) as model compounds for gasoline for experimental and simulation studies. The study presented the following conclusions:

1. Supercritical water reforming of hydrocarbons offers a possible way to convert hydrocarbons to hydrogen at a lower temperature. It does not require a steam reforming catalyst; although nickel in reactor wall can act as a catalyst. It avoids the poisoning and deactivation problems associated with the catalyst.
2. The reactor is much compact compared to conventional steam reforming reactors, and it is scalable and the reaction time is in seconds. Supercritical water provides dual functions: excellent reactant and homogeneous medium.
3. As reaction temperature, initial feed concentration, and residence time increase, hydrogen, carbon dioxide, and methane production increases while carbon monoxide and ethane yield remains stable. At high temperature, methane yield is higher than hydrogen yield because at high temperature methanation reaction is favored. In order to increase hydrogen yield, methanation reaction needs to be suppressed. High inlet feed temperature decreases the yields of hydrogen, carbon monoxide, and carbon dioxide and increases the yields of methane and ethane. High inlet temperature also forms coke in the feed line which may plug the inlet pipes.

Numerous other studies [193,303,308–336] have also examined catalytic reforming of various hydrocarbons in supercritical water. Shekhawat et al. [333] studied catalytic reforming of liquid hydrocarbon fuels for fuel cell applications. They concluded that supercritical reforming of hydrocarbons occurs at lower temperatures than those required in conventional industrial reforming process. They also showed that hydrogen yield increases when commercial catalysts are used, even if they are not optimized for these conditions. Acetone and diesel fuel produced black liquor and plugged the reactors. Pinkwart et al. [309] showed that under supercritical water, *n*-decane can be converted to hydrogen-rich gas. They also showed that reforming of diesel oil by four different commercial reforming catalysts can be carried out at a lower temperature than the conventional steam reforming process. The lower temperature also caused lower production of coke during reforming reaction. Ramaswamy and T-Raissi [326] studied hydrogen production during reforming of lube oil in supercritical water. They also examined the role of nickel, carbon, and alkali catalysts in hydrogen production. Very little catalyst deactivation was observed under supercritical conditions.

While tri-reforming of methane in supercritical water has been investigated by a number of researchers [72,229], these studies have been carried out with conventional Ni or bimetallic catalysts. The studies have shown that the supercritical conditions lower the required temperature for gasification, and at high temperatures (>600°C), hydrogen and carbon dioxide are the dominating products. The studies [72,229] have also shown that the product composition from tri-reforming under supercritical conditions depends on a number of variables, such as temperature, pressure, feedstock and oxygen concentrations, reaction time, biomass properties, presence of inorganic elements, biomass particle size, and the nature of the catalyst. In supercritical water environment, the syngas composition will heavily depend on the effectiveness of the dry reforming reaction. More catalytic studies to improve dry reforming reaction are presently being pursued. The use of nanocatalysts is also being examined in depth.

6.7.2.1.2   Biomass

A number of investigators have looked at glucose as a model for biomass reforming under supercritical water. The pertinent reaction in this case is:

$$C_6H_{12}O_6 + 6H_2O \rightarrow 6CO_2 + 12H_2 \tag{6.64}$$

Generally, hydrogen yield is smaller than predicted from the previous equation because varying amounts of methane are produced depending upon the reaction conditions. Shah [72] and Kruse [313,316,317] gave a simplified reaction mechanism for cellulose reforming. Since glucose (as well as fructose) is the main product of hydrolysis of cellulose, its reaction mechanism also applies to glucose. The reforming of glucose was accelerated by alkali catalysts such as K_2CO_3 and $KHCO_3$, both of which increased the hydrogen production and decreased coke formation. For biomass with low salt content and high protein content, these catalysts can increase the hydrogen yield.

Antal and Xu [310,311] and Antal et al. [318,334] showed the effectiveness of supercritical water reforming for the production of hydrogen for numerous different types of biomass such as wood sawdust, corn starch gel, digested sewage sludge, glycerol, glycerol/methanol mixture, poplar wood sawdust, potato starch gels, and potato waste. Once again higher temperature and catalysts gave better hydrogen production. The final product distribution did depend on the nature of the feedstock. Similar results were obtained by Boukis et al. [321] for biomass slurries and sludges. They also showed an improved heat exchange scheme in "VERENA" German pilot plant for these processes. The "VERENA" pilot facility successfully demonstrated high carbon and energy efficiency for the supercritical water reforming of ethanol and corn silage in the temperature range of 540°C–600°C for at least 10 h. On average, the hydrogen concentration in the product for these biomass was about 77 vol%.

Zhang et al. [314] examined the SCW reforming of glucose solution (50–200 g/L), a simulated aqueous organic waste (composed of glucose, acetic acid, and guaiacol) and a real aqueous organic waste stream generated from a sludge hydrothermal liquefaction process. The experiments were performed using two different types of catalysts: 0.1 RuNi/gamma-Al_2O_3 or 0.1RuNi/ activated carbon catalysts (10 wt% Ni with a Ru-to-Ni molar ratio of 0.1). While the first catalyst was very effective with glucose solutions and simulated aqueous organic waste giving hydrogen yield of 53.9 mol/kg dried feedstock at 750°C, 24 MPa and WHSV of 6 h^{-1}, it was not effective in resisting the alkali and nitrogen compounds in the real waste. The second catalyst supported on active carbon exhibited higher stability. Finally, Penninger and Rep [331] and Penninger et al. [330] examined reforming of aqueous wood pyrolysis condensate in supercritical water, and Yamaguchi et al. [328] studied hydrogen production from woody biomass over supported metal catalysts in supercritical water.

6.7.2.1.3   Glycerol

Reforming of glycerol for hydrogen production can be summarized by the following reactions.

First, the steam reforming of glycerol can be expressed as

$$C_3H_8O_3 \downarrow\downarrow \rightarrow 3CO + 4H_2 \tag{6.65}$$

followed by the WGS reaction:

$$CO + H_2O \rightarrow CO_2 + H_2 \tag{6.66}$$

The desired overall reaction is then summarized as

$$C_3H_8O_3 + 3H_2O \rightarrow 7H_2 + 3CO_2 \tag{6.67}$$

Some hydrogen is also lost via the methanation of CO and CO_2:

$$CO + 3H_2 \rightarrow CH_4 + H_2O \tag{6.68}$$

$$CO_2 + 4H_2 \rightarrow CH_4 + 2H_2O \tag{6.69}$$

As a result, the product stream is a mixture of the mentioned gases. Furthermore, the yield of hydrogen depends on several process variables, such as system pressure, temperature, and water-to-glycerol feed ratio. Guerrero-Perez et al. [324] examined recent inventions in glycerol processing and transformation.

Most recently Byrd et al. [329,335] studied the reforming of glycerol over Ru/Al_2O_3 catalyst in supercritical water conditions at temperature range of 700°C–800°C, feed concentration up to 40 wt%, and reaction time less than 5 s. Under these conditions, glycerol was completely gasified to hydrogen, carbon dioxide, and methane along with a small amount of carbon monoxide. Xu et al. [310–312,318] showed that even in the absence of a catalyst glycerol decomposes in supercritical water to a hydrogen rich gas with almost no CO after 44 s at 600°C and 34.5 MPa. Higher temperature, more active reforming catalyst, and longer residence time result in higher gas and hydrogen production. Van Bennekom et al. [336] examined reforming of crude glycerin in supercritical water to produce methanol for reuse in biodiesel plants.

6.7.2.1.4 *Ethylene Glycol*

De Vlieger et al. [327] studied catalytic reforming of ethylene glycol (5 and 15 wt%) in SCW at 450°C and 250 atm pressure. The results were obtained for Pt, Ir, and Ni containing mono- and bimetallic catalysts. The best catalyst was found to be $Pt–Ni/Al_2O_3$ having a metal loading of 1.5 wt% (molar ratio of Pt:Ni, 1:1). With this catalyst high hydrogen and carbon dioxide yields (selectivity of around 80%) were obtained by suppressing methanation reaction. The addition of Ni prevented sintering of Pt particles, thereby providing a stable performance by bimetallic catalysts. Ethylene glycol also produced more CH_4 and CO than what was produced in methanol reforming.

6.7.2.1.5 *Methanol*

Numerous studies have reported methanol reforming in supercritical water to produce hydrogen. Compared to water, which has a critical pressure of 22.1 MPa, critical temperature of 374°C, and critical density of 320 kg/m³, methanol has a lower critical temperature of 239°C, critical pressure of 8.1 MPa, and critical density of 270 kg/m³. Thus reaction of methanol in supercritical water also implies that methanol is also under supercritical conditions. Methanol reforming can be described by five chemical reactions:

$$CH_3OH \leftrightarrow CO + 2H_2 \quad H_{298}^{\circ} = +91.7\,kJ/mol \tag{6.70}$$

$$CO + H_2O \leftrightarrow CO_2 + H_2 \quad H_{298}^{\circ} = -41\,kJ/mol \tag{6.71}$$

$$CH_3OH + H_2O \leftrightarrow CO_2 + 3H_2 \quad H_{298}^{\circ} = +50.7\,kJ/mol \tag{6.72}$$

$$CO + 3H_2 \leftrightarrow CH_4 + H_2O \quad H_{298}^{\circ} = -211\,kJ/mol \tag{6.73}$$

$$CO_2 + 4H_2 \leftrightarrow CH_4 + 2H_2O \quad H_{298}^{\circ} = -223\,kJ/mol \tag{6.74}$$

While both methanation reactions and WGS reaction are exothermic, main methanol reforming reaction is endothermic and is favored at higher temperatures. Boukis et al. [321] showed that for reaction times as low as 4 s and at temperature of 600°C, and pressure of 25–45 MPa, high conversion rate of methanol can be obtained. The reaction can occur at temperatures as low as 400°C. The heavy metal of the inner surface of Inconel 625 can influence the conversion and the product composition of the reforming reaction. Boukis et al. [321] examined the feed concentration from 5 to 64 wt% methanol. Methanol conversion up to 99.9% can be obtained in the absence of a catalyst. The major product is hydrogen (up to 70%–80%) with small amounts (less than 20%–30%) of carbon dioxide, carbon monoxide, and methane. An increase in temperature increases methanol conversion, decreases CO concentration, and increases CO_2 concentration in the product. Complete methanol conversion at 600°C is achieved [321].

Taylor et al. [322] also examined reforming of methanol under supercritical water conditions in the temperature range of 550°C–700°C and at 27.6 MPa in an Inconel 625 reactor. They also reported a product rich in hydrogen and low in CH_4 and near the equilibrium ratio of CO and CO_2. A comparison of the product gas composition with equilibrium predictions indicated that the reaction occurs in two steps. First, methanol decomposes to CO and H_2 and subsequently CO is converted to CO_2 by WGS reaction. Higher steam-to-carbon ratios gave lower CO in the product gas. Both methanol decomposition and WGS reactions are kinetically limited at temperatures under 700°C. Also methanation reaction was kinetically limited. As shown by Gadhe and Gupta [323], high pressure favored formation of methane.

6.7.2.1.6 Ethanol

Wenzel [320] studied supercritical water reforming of ethanol under noncatalytic conditions for the temperature range of 618°C–710°C and pressure of 24.2 MPa. The ethanol feed rate varied from 0.17 to 2.2 g/min and water flow rate varied from 6.4 to 19.7 g/min in a 1 L 625 grade 1 alloy tubular reactor. A complete conversion of ethanol was obtained producing hydrogen, carbon dioxide, methane, ethane, and carbon monoxide in the descending order of their concentrations. Hydrogen was produced by two competing reactions: the direct reformation of ethanol into hydrogen and carbon oxides, and the pyrolytic dehydrogenation of ethanol:

$$C_2H_5OH \rightarrow C_2H_4O + H_2 \tag{6.75}$$

where acetaldehyde goes through further decarbonylation as

$$C_2H_4O \rightarrow CH_4 + CO \tag{6.76}$$

This decomposition is fast with Rh-cerium oxide catalyst at temperatures above 650°C. The net result of the two previous reactions is generation of hydrogen, methane, and carbon oxides. In this system, forward WGS reaction is active even without the presence of a water gas shift catalyst. An undesirable competing reaction of dehydration of ethanol to form ethylene occurs which is subsequently hydrogenated to form ethane. This reaction not only consumes hydrogen but also produces the coking precursor ethylene. Both pyrolytic and direct reforming reactions were first-order reactions.

Byrd et al. [329] studied supercritical reforming of ethanol over Ru/Al_2O_3 catalyst. Experiments were conducted at various temperature, pressure, residence time, and water-to-carbon ratio to evaluate their effects on the hydrogen yield. The results showed that hydrogen formation was favored at high temperature and high water-to-ethanol ratios. Under the same conditions and for an optimum residence time, methane production was suppressed. Excellent conversions were obtained for the residence time as low as 4 s. Pressure had negligible effect on hydrogen yield above the critical pressure and there was negligible coke formation for ethanol concentration in the feed less than 10 wt%. The overall reforming reaction for ethanol can be expressed as

$$C_2H_5OH + 3H_2O \leftrightarrow 6H_2 + 2CO_2 \quad \left(H_{298}^{\circ} = 174\,kJ/mol\right) \tag{6.77}$$

In the presence of Ru/Al_2O_3 catalyst, high reforming performance may be due to the fact that intermediates formed during ethanol decomposition such as dimethyl ether and acetaldehyde were also gasified in the presence of supercritical water. In the subcritical steam gasification, formation of a significant amount of carbon limits hydrogen production. Reaction products also contain acetaldehyde, diethyl ether, ethane, and ethylene. The gasification under supercritical conditions is accompanied by several complex reactions such as ethanol decomposition, steam reforming, WGS reaction, and methanation reaction. The product distribution depended on the relative rates of these reactions. It was assumed that during reforming ethanol dehydrogenates on the metal surface to give adsorbed intermediates before the cleavage of C–C and C–O bonds. The WGS reaction reduces CO concentration, and the final products predominantly contain hydrogen and carbon dioxide.

Gadhe and Gupta [323] examined the strategies for the reduction of methane formation and thereby increased production of hydrogen. Three strategies that were examined were (1) operation at a low residence time by having a smaller reactor length or a high feed flow rate, (2) addition of a small amount of K_2CO_3 or KOH in the feed, and (3) utilization of the surface catalytic activity of the reactor made of Ni–Cu alloy. All three strategies worked, resulting in lower methane production and correspondingly higher hydrogen production. The methanation reactions were favored by high pressure, high residence time and low steam-to-carbon ratio. Therdthianwong et al. [332] examined hydrogen production by reforming bioethanol using Ni/Al_2O_3 and $Ni/CeZrO$ catalysts in supercritical water.

Finally, the use of supercritical water as reaction medium for conducting the tri-reforming can be an attractive and novel method. The literature on gasification/reforming under supercritical water indicates that in general, supercritical water reduces coking lowers the required temperature for the same level of conversion and modifies the product distribution particularly in the favor of more production of hydrogen. These results imply the need for a study of tri-reforming under supercritical water (critical point 374°C and 22.1 MPa) conditions. It is expected that the supercritical conditions will bring about significant improvement in product distributions, reaction temperature severity, and catalyst activity, stability, and life. Under supercritical water gasification, syngas is produced directly at high pressure, which means a smaller reactor volume and lower energy needed to pressurize the gas in a storage tank [72].

6.8 PLASMA REFORMING

Plasma is an ionized gas produced mainly by electric field, which consists of a mixture of electrons, ions, neutral particles, and others. The term "plasma" was first introduced by Irving Langmuir in 1928 [114,371–381]. Since the mass of ions and neutral particles are much larger than that of electrons in plasma, electrons are called light particles, whereas ions and neutral particles are called heavy particles.

Depending on the energy density level, temperature, and electron density, plasma applied for reforming is classified as thermodynamic nonequilibrium plasma and thermodynamic equilibrium plasma. The thermodynamic nonequilibrium plasma is also called cold plasma, in which the thermal kinetic energy of electrons is much larger than that of heavy particles; the weighted average temperature of plasma is near room temperature [114,374,377,381]. The chemical reaction in cold plasma is mainly induced by energetic electrons. Due to the action of electric field, electrons are accelerated and energized. Through electron-impact dissociation, excitation, and ionization of gas molecules, the energetic electrons transfer their energy to the gas molecules upon inelastic collision. As a result, excited species, free radicals, ions, as well as additional electrons are produced [114,374,377,381], which induces the desired plasma chemical reactions.

The thermodynamic equilibrium plasma is simply called thermal plasma, in which the temperature of heavy particles is close to that of electrons in the range of thousands K [372,379]. Because the electron density in thermal plasma is higher than that in cold plasma, by a large amount of impact between electrons and heavy particles, the electron energy from the electric field goes into heating heavy particles to achieve the thermodynamic equilibrium in electrons and heavy particles. The chemical reactions in thermal plasma are both electron-induced reactions and thermochemical reactions.

6.8.1 Dry Reforming of Methane by Cold Plasma

Several kinds of cold plasmas have been tested in CH_4–CO_2 reforming, such as corona discharge, DBD, microwave discharge, APGD, and gliding arc discharge. Due to different discharge modes, the electron temperature of cold plasmas can vary from 1 to 10 eV, whereas the temperature of heavy particles is in the range of hundreds K; the local electron density is in the range of 10^{15}–10^{20} m^{-3}. Generally, cold plasmas are inhomogeneous in discharge space which results in limited reaction region. So the conversions and the treatment capacity are restricted. For increasing reaction conversion rate, catalytic process may be introduced in CH_4–CO_2 reforming by cold plasma. There are two ways to introduce catalysts in cold plasma: one is to place catalysts in the discharge space, and the other after the discharge space [374,381].

6.8.2 Dry Reforming of Methane by Thermal Plasma

Thermal plasma, generated in the way of electric arc, is continuous and uniform plasma. Thermal plasma with features of relatively high enthalpy content, high temperature (temperature of heavy particle and electron is 0.5–1 eV), and high electron density (10^{19}–10^{20} m^{-3}) has obvious thermal effect and chemical effect. It covers a wide range of applications and has been applied in industry [379,380].

There are several types of thermal plasma devices: direct current (DC) arc torch, alternating current (AC) arc torch, radiofrequency (RF) inductively coupled torch, and high-frequency capacitive torch. The DC arc torch is used widely. Commonly, a DC arc torch consists of a thoriated tungsten (2%–3% ThO_2) rod- or button-type cathode and one or more water-cooled copper anodes which are separated by insulating materials. The arc is established between the electrodes by a special trigger and pushed through the nozzle, resulting in a high temperature flame. The power of the torch controlled by changing the input voltage or current can reach 10^2–10^7 W [372,374,379–381]. Since thermal plasma is a source of high temperature and chemically active species, it has significant potential for CH_4–CO_2 reforming.

Compared to cold plasma, CH_4–CO_2 reforming by single-anode thermal plasma exhibits the significant advantages like large treatment capacity, little by-product, and relatively high energy conversion efficiency, which is closer to industrial application. Furthermore, in many studies, the feed gases were injected only into the plasma jet, not into the discharge region between the anode and the cathode. If the feed gases are introduced into the discharge region as plasma-forming gas directly, the treatment capacity and the energy conversion efficiency could be raised further. To verify the idea, a binode thermal plasma generator is applied to CH_4–CO_2 reforming [372,374,379–381]. The binode thermal plasma generator consists of a club-shaped cathode and two columnar anodes which are separated by insulating materials. After the discharge is first ignited between the cathode and the first anode, the ionized gas is puffed to the second anode where the DC power supply is applied. Then the discharge channel is established between the cathode and the second anode, and the binode plasma jet is formed.

It is known that the ionization degree of cold plasma is much lower than that of thermal plasma. Besides electrons, free radicals, and ions, there are still many molecules in the plasma. The ionization degree is also proportional to the input power. The higher the input power is, the higher ionization degree will be achieved. As a result, more radicals but fewer molecules can be formed and the conversions of CH_4 and CO_2 will be higher. The APGD plasma jet and thermal plasma have the highest ionization degree and appropriate electron temperature, as well as proper gas temperature compared to all types of plasmas mentioned earlier. Dry reforming of methane by them shows better energy conversion efficiency and specific energy. Between APGD plasma jet and thermal plasma, the latter possesses much larger treatment capacity and considerable energy conversion efficiency and specific energy. It is not difficult to obtain thermal plasma equipment with large power of thousands kilowatt, but its energy conversion efficiency can be enhanced by improving the thermal plasma generator, more reasonable reactor design and operation, as well as an optimal synergy of plasma with catalysts and heat recovery of the process.

As mentioned in Chapter 1, the literature review of dry reforming of methane indicates that for achieving high conversions and high energy conversion efficiency by plasma process, three key factors should be taken into account: electron density, plasma temperature, and reactor configuration. Corona discharge and DBD are nonuniform plasmas with low electron density and limited reaction volume which restrict the treatment capacity. Microwave discharge [376,380] is a uniform discharge with high plasma temperature and large discharge space, but the equipment is more complicated and expensive which restricts its industrial application. Gliding arc discharge [378] and APGD possess high electron energy and electron density, as well as proper plasma temperature, but the reactor is difficult to enlarge and the treatment capacity of both plasmas is still far from industrial requirements. Thermal plasma is an efficient technology for dry reforming of methane to synthesis gas because of its high specific enthalpy content, high temperature, high electron density, large treatment capacity, easy to enlarge, and relatively high energy conversion efficiency. For raising the energy conversion efficiency of thermal plasma process further, the nature of feed gases introduced into the discharge region and the synergistic effect of plasma with catalysts should be further evaluated.

Besides using plasma for dry reforming [379–381], several studies [114,375–377] have also examined steam reforming of methane and other hydrocarbons by plasma technology. In many studies, plasma was used jointly with heterogeneous catalysts [371,373,374,377]. Hammer et al. [373] examined hybrid catalytic steam reforming of methane. Nonthermal plasma using DBD reactor was investigated. Experiments were also conducted with dielectric packed bed catalytic reactor to examine the effect of catalyst. The investigation indicated that NTP-reforming of methane in CH_4–H_2O-mixtures with dielectric barrier discharges mainly induce formation of H_2 and C_2H_6 by decomposition of CH_4 with negligible H_2O conversion. Numerical simulations of the DBD-induced chemical kinetics showed that 38% and 7% of the energy dissipated in discharge filaments are spend for CH_4- and H_2O-dissociation respectively [114,371–373,377].

By the combination of the DPB-plasma and a Ni-catalyst high H_2O-conversion rates and selectivity toward H_2- and CO_2-formation was achieved. For temperatures above 200°C, the energy requirements dropped by an order of magnitude down to values as low as 315 kJ/mol H_2 at 600°C. Further improvements can be expected if thermal losses, for example, due to barrier heating, could be avoided, which can be well above 60% of the plasma input power.

Zhou et al. [371] examined hydrogen production by reforming methane in a corona-inducing DBD and catalyst hybrid reactor. A novel corona-inducing dielectric barrier discharge (CIDBD) and catalyst hybrid reactor was developed for reforming methane. This corona-inducing technique allows DBD to occur uniformly in a large gap at relatively low applied voltage. Hydrogen production by reforming methane with steam and air was investigated with the hybrid reactor under atmospheric pressure and temperatures below 600°C. The effects of input power, O_2/C molar ratio, and preheat temperature on methane conversion and hydrogen selectivity were investigated experimentally. It was found that higher methane conversions were obtained at higher discharge power, and methane conversion increased significantly with input power less than 50 W; the optimized molar ratio of O_2/C was 0.6 to obtain the highest hydrogen selectivity (112%); under the synergy of DBD and catalyst, methane conversion was close to the thermodynamic equilibrium conversion rate.

Aziznia et al. [374] compared dry reforming of methane in low-temperature hybrid catalytic corona with thermal catalytic reactor over Ni on gamma alumina catalyst. The hybrid effect of a corona discharge and γ-alumina-supported Ni catalysts in CO_2 reforming of methane was investigated. The study included both purely catalytic operation in the temperature range of 923–1023 K and hybrid catalytic-plasma operation of DC corona discharge reactor at room temperature and ambient pressure. The effects of feed flow rate, discharge power, and $Ni/\gamma-Al_2O_3$ catalysts were studied. When CH_4-to-CO_2 ratio in the feed was 1/2, the syngas of low H_2-to-CO ratio at about 0.56 was obtained, which is a potential feedstock for synthesis of liquid hydrocarbons. Although Ni catalyst was active only above 573 K, presence of Ni catalysts in the cold corona plasma reactor ($T \leq 523$ K) showed promising increase in the conversions of methane and carbon dioxide. When Ni catalysts were used in the plasma reaction, H_2-to-CO ratios in the products were slightly modified, selectivity to CO increased whereas fewer by-products such as hydrocarbons and oxygenates are formed.

Bromberg et al. [372] found that during steam reforming, hydrogen-rich gas (50%–75% H_2, with 25%–50% CO) can be efficiently made in compact plasma reformers. Experiments were carried out in a small device (2–3 kW) and without the use of efficient heat regeneration. For POX, it was determined that the specific energy consumption in the plasma reforming process was 40 MJ/kg H_2 (without the energy consumption reduction that can be obtained from heat regeneration from an efficient heat exchanger). A variety of hydrocarbon fuels (gasoline, diesel, oil, biomass, natural gas, jet fuel, etc.) were used. The experimental results also indicated hydrogen yields of about 100% and moderate energy consumption without the use of heat regeneration and sufficient residence time. There were no problems with soot buildup on the electrode or on the surfaces of reactor. In addition, with water injection, it was possible to combine reforming and water shift reactions in a single stage. Larger reactors, better reactor thermal insulation, efficient heat regeneration, and improved plasma catalysis could also play a major role in specific energy consumption reduction for plasma technology. With an appropriate heat exchanger to provide a high degree of heat regeneration, the projected specific energy consumption can be expected to be about 15–20 MJ/kg H_2. In addition, a system can be operated for hydrogen production with low CO content (about 2%) with power densities of about 10 kW (H_2 HHV)/L of reactor, or space velocity of about 4 m^3/h H_2 per liter of reactor. Power density should increase further with power and improved reactor design.

Plasma reformers could also be used for hard-to-use fuels, such as raw biofuels or heavy oils. Chernyak et al. [375] examined plasma reforming of liquid hydrocarbon fuels into hydrogen for the use in aerospace technology. More specifically, they examined steam reforming of ethanol by DGCLW plasma. The main conclusions of this study were the following:

1. The dynamic plasma-liquid systems with the electric discharges in the gas channels with liquid wall is quite efficient in plasma chemical reforming of ethanol into free hydrogen and synthesis gas.
2. The minimal value of power inputs in investigated discharge modes was ~2.4 kWh/m^3 at the power of output syngas of ~4.4 kWh/m^3.
3. The electric discharge in gas channel with liquid wall has high power efficiency and efficiency of the nonequilibrium plasma processes comparable to other known gas-discharge plasma sources of the atmospheric pressure such as diaphragm and arc types.

Plasma reformers are relatively inexpensive (they use relatively simple metallic or carbon electrodes). The plasma conditions (high temperatures and a high degree of ionization) can be used to accelerate thermodynamically favorable chemical reactions without a catalyst or to provide the energy required for endothermic reforming processes. Plasma reformers can provide a number of advantages: (1) compactness and low weight (due to high power density); (2) high conversion efficiencies; (3) minimal cost (simple metallic or carbon electrodes and simple power supplies); (4) fast response time (fraction of a second); (5) no need for catalysts and therefore no problems of catalyst sensitivity and deterioration; (6) operation with a broad range of fuels, including heavy hydrocarbons, and

(7) can be operated with externally added heterogeneous catalyst (hybrid operation). The technology could be used to manufacture hydrogen for a variety of stationary applications, for example, distributed, low-pollution electricity generation from fuel cells [94–111,391–395]. It could also be used for mobile applications (e.g., on-board generation of hydrogen for fuel cell–powered vehicles) and for refueling applications (stationary sources of hydrogen for vehicles).

The studies just described and others [371–381] have shown that plasma technology has the potential to significantly alleviate shortcomings of conventional means of manufacturing hydrogen. These shortcomings include cost and deterioration of catalysts, size and weight requirements, limitations on rapid response, and limitations on hydrogen production from heavy hydrocarbons. In addition, use of plasma technology could provide for a greater variety of operating modes, including the possibility of virtual elimination of CO_2 production by pyrolytic operation.

Plasma technology also has some disadvantages. It is difficult to operate under high pressure due to increase in electrode erosion and decrease in electrode lifetime. Its dependence on electrical energy results in energetics that are less favorable than the energetics of purely thermal processes, especially for endothermic reforming reactions.

More work on plasma reforming both at lab scale and demonstration level is needed. Optimum plasma configuration (to get maximum hydrogen production at the expense of minimum energy consumption) and procedures for reactor scale-up are of utmost importance for the future study.

6.9 NOVEL REFORMING PROCESSES

6.9.1 Microwave-Assisted Reforming

Microwaves are a nonionizing electromagnetic radiation that lies in the range of the electromagnetic spectrum limited by the frequencies between 300 MHz and 300 GHz (wavelength between 1 m and 1 mm). Domestic and industrial microwave applications generally operate at a frequency of 2.45 GHz in order to avoid interference with radar and telecommunication frequencies [337–341].

Dielectric heating is caused by high-frequency electromagnetic radiation, that is, radio- and microwaves. The electric field component of the electromagnetic radiation interacts with the charged particles of a material. A current is induced when these particles are free to move. However, when the particles are linked to the material, they try to align themselves with the alternating field, as a consequence of which the material heats up (dielectric polarization) [342]. Two principal dielectric polarizations are involved in microwave radiation [337,343]: (1) dipolar polarization, which occurs in dielectrics that have induced or permanent dipoles, such as water. (2) Space charge polarization, which occurs mainly in dielectric solid materials with charged particles which are free to move in a delimited region (Maxwell-Wagner polarization).

The materials which interact with microwave radiation to produce heat are called dielectrics or microwave absorbers. The ability of a material to be heated in the presence of a microwave field is described by its dielectric loss tangent $\tan \delta = \varepsilon''/\varepsilon'$. The dielectric loss tangent is composed of two parameters: the dielectric constant (or real permittivity), ε', which measures the ability to propagate microwaves into the material, and the dielectric loss factor (or imaginary permittivity), ε'', which measures the ability of the material to dissipate the energy in the form of heat [337–340]. Materials which reflect microwaves from the surface and do not heat are called conductors, and materials which are transparent to microwaves are classed as insulators. As microwave energy is transferred directly to the material that is heated (volumetric heating), the temperature inside the material is usually higher than the temperature of the surrounding atmosphere, unlike conventional heating.

Microwave heating offers a number of advantages over conventional heating, such as (1) non-contacting heating, (2) rapid heating, (3) selective heating of materials, (4) quick start-up and stoppage, (5) a higher level of safety and automation, (6) a reduction in the size of equipment and higher flexibility, and (7) reduced processing time [337,339,340,344]. In recent years, microwave heating has been applied to numerous industrial processes, including heterogeneous gas-phase catalytic systems [337,338,343–345]. In a catalytic heterogeneous system, the catalyst should be a dielectric material and acts not only as a catalyst but also as a microwave receptor [337,346]. However, some catalysts are insulator materials, and consequently, they must be used in conjunction with microwave receptors, such as carbons and certain oxides, in order to be heated easily [337,342]. Carbon materials are usually very good microwave absorbers, so they can indirectly heat materials that are transparent to microwaves.

Microwave radiation is known to have the potential to increase the rate of reaction, selectivity, and yield of catalytic heterogeneous reactions [337,338,343–345,347]. The improvement observed under microwave heating is normally attributed to various thermal effects [344], although the presence of hot spots within the catalyst bed, which are at higher temperature than the average temperature, may be the main reason for the improvement in gas–solid reactions. Hot spots are electric arcs which are caused by an uneven distribution of the electromagnetic fields and preferential heating, due to differences in dielectric properties, impurities, or geometric defects within the catalyst. The electric arcs may cause the ignition of the surrounding atmosphere. Moreover, in the case of carbons, the increase in the kinetic energy of the delocalized π-electrons, which are free to move in relatively broad regions and which try to align themselves with the alternating electric component of the microwave field, may give rise to the ionization of the surrounding atmosphere. These hot spots can be therefore considered as microplasmas both from the point of view of space and time, since they are confined to a tiny region of space and last for just a fraction of a second [337,348].

The microwave-assisted CO_2 reforming of CH_4 over carbon-based catalysts combines the catalytic and dielectric properties of carbonaceous materials with the advantages of microwave heating, which favors catalytic heterogeneous reactions due to, among other reasons, the generation of hot spots or microplasmas. Under certain operating conditions, the microwave-assisted dry reforming reaction can be considered as a combination of CH_4 decomposition and CO_2 gasification of carbon deposits, leading to the continuous regeneration of active centers. The most appropriate operating conditions to achieve high conversions for a long period of time are temperatures ranging between 700°C and 800°C and the presence of high proportions of CO_2 in the feed (at least 50%).

A suitable catalyst for the microwave-assisted dry reforming of methane can be carbonaceous material and metal catalyst. The microwave-assisted CO_2 reforming of CH_4 offers an alternative to the well-established process of steam reforming of natural gas for the production of synthesis gas and its resulting by-products. As indicated earlier, dry reforming has clear environmental benefits since it turns two greenhouse gases (CH_4 and mainly CO_2) into a valuable feedstock. Moreover, microwave-assisted dry reforming could lead to the reduction of CO_2 emissions or even to their complete elimination if the electricity consumed in the generation of microwave energy in the dry reforming process were produced from renewable sources. In addition, the dry reforming of CO_2-rich natural gas, biogas, and CO_2-rich industrial residual streams can be carried out without the need for pretreatments. Also, CO_2 reforming of CH_4 yields syngas with a H_2-to-CO ratio of 1:1 which is most suitable for downstream FT operation.

While Fidalgo and Menendez [225] successfully demonstrated dry reforming of methane on carbon and nickel catalysts in the laboratory, this needs to be demonstrated for a longer period and for a larger scale. Also, the energy consumption during dry reforming needs to be reduced in order to be able to rival the steam reforming process. A reduction in energy consumption can be achieved by the improvement of the catalysts in order to operate at large volumetric hourly space velocity but still maintain high conversions, which would yield larger syngas production per mass of catalyst. A scaling up of the process would also enhance the energetic yield. Scaling up of microwave-assisted dry reforming is a considerable challenge as it entails the designing of new microwave equipment which is able to satisfy the requirements of a large-scale process and prove the effectiveness of the heating at the industrial scale.

The effectiveness of the microwave-assisted dry reforming also needs to be proven for industrial gases. The use of microwave energy in the dry reforming of coke oven gas was carried out by Bermúdez et al. [349–352]. They obtained better yields than those obtained under conventional electric heating if the process is carried out with an activated carbon as catalyst. However, when the process is carried out with mixtures of activated carbon and Ni/Al_2O_3 as catalyst, the results obtained in the microwave oven were worse than those obtained in a conventional oven. These results differ significantly from those achieved in the dry reforming of CH_4, which are considerably higher in the case of the microwave oven, independently of the kind of catalyst, with conversions of 100% being achieved for both gases.

Durka [353] examined microwave-assisted steam reforming of alcohols using Rh/CeO_2–ZrO_2, Rh/Al_2O_3, Ni/Al_2O_3, and Ni/CeO_2–ZrO_2 catalysts. The results indicated that the microwave-assisted steam reforming of alcohols is a potential alternative to the conventional process. The benefits of microwave heating include lower operating temperature, the higher conversion and selectivity to the desired product, and the improved thermal efficiency.

These and numerous other publications in the literature have demonstrated the effectiveness of microwave-assisted heating in the reforming process using a variety of catalysts [225,346,354,355]. While the results indicate most success for reforming of methane, other feedstock for both dry and steam reforming needs to be further examined. This subject needs additional attention both at small and larger scales.

6.9.2 Solar Reforming

The high temperatures required for solar reforming effectively limit the nature of solar energy collector. The bulk energy production, whether in closed loop or open-loop configurations, probably must be carried out on a large scale to compete with fossil fuels and probably requires the tower (central receiver) solar technology. Solar reforming can be carried out using different processes such as direct and indirect, each requiring different types of reformer configuration.

The use of solar energy to create solar fuels by numerous gasification and reforming technologies is described in several excellent reviews [356–359]. Solar reforming requires novel reactors, which are described in detail in Section 6.10. More details on solar reforming are also covered in my previous books [1,72].

6.9.3 Nuclear Heat–Aided Reforming

Belghit and El Issami [360] developed a theoretical model of a moving bed chemical reactor for gasifying coal with steam. The heat was supplied by a high-temperature nuclear reactor. As shown in my previous books [1,72], high-temperature nuclear waste heat can be used for reforming of carbonaceous materials (coal, biomass, waste, natural gas, etc.). Reforming can also be carried out by dedicated small-scale nuclear heat reactors.

6.9.4 Other Novel Reforming Processes

Cypres [361] discussed metallurgical process for hydrogen production from coal and other carbonaceous materials, including coal gasification in a molten iron bath. An argument was made to place such a gasifier in the vicinity of steel manufacturing plant.

A steam-iron process is one of the oldest commercial methods for the production of hydrogen from syngas [362–369]. Various types of oxides of iron were examined. Neither chemical composition nor porosity of the ores was found to govern the efficiency. Potassium salts enhanced the activity of both natural and synthetic oxides. A number of recent studies have examined the classical steam-iron (sponge-iron) process for upgrading synthesis gas (mainly CO and H_2) to pure hydrogen for use in fuel cells and other energy devices. Friedrich et al. [365] looked at this purification of nitrogen-containing "reduction" gas from biomass gasifier using wood and wood wastes. The process involved two steps: (1) cleaning of gas from solid biomass, coal, or methane and (2) energy storage in sponge iron. This study investigated woody biomass and commercially available sponge iron. The reactions are

$$Fe_3O_4 + 4CO \rightarrow 3Fe + 4CO_2 \quad \text{(coal, biomass, or natural gas)} \tag{6.78}$$

$$3Fe + 4H_2O \rightarrow Fe_3O_4 + 4H_2 \tag{6.79}$$

This process was stated to have few risks. Jannach et al. [369] extended the sponge iron process to FeO, as well as Fe as the oxidant. The sponge iron reaction was further studied by Hacker et al. [366–368] in thermogravimetric analysis (TGA) and tube furnace devices. Other types of reactors were also examined by Fankhauser et al. [364] and Hacker et al. [366–368]. Biollaz et al. [370] explored the iron redox process to produce clean hydrogen from biomass. In the first step, iron oxide in the form of Fe_3O_4 reacts with the reducing components of wood gas to produce FeO, CO_2, and H_2. The kinetics of the second step, $3FeO + H_2O \rightarrow H_2 + Fe_3O_4$, could be improved by adding other transitional metal oxides. The reduction of iron oxide with biosyngas to sponge iron and later oxidation of sponge iron with steam offers the potential of shifting and purifying biosyngas and storing and transporting its energy. Bijetima and Tarman [362] described the steam iron process for hydrogen production operated in a large-scale pilot facility. Economic advantages of the process were also presented.

6.10 NOVEL REFORMING REACTORS

All novel types of thermal gasification reactors described in Chapter 3 are used for steam gasification (with or without oxygen). A number of novel reactors have been used for reforming process, some of which are briefly described here.

6.10.1 STEAM REFORMING REACTORS

While hydrogen production can be achieved by a number of commercially proven technologies such as gasification of coal, biomass and residue (waste), methanol decomposition and steam reforming of coal, biomass, liquid hydrocarbons and natural gas, it is the last technology that produces the largest portion of hydrogen supply. With the considerable advances in unconventional production of natural gas which include shale gas, deep gas, tight gas, coalbed methane, gas from geo-pressurized zones, and gas hydrates, the steam reforming of natural gas is likely to become even more important. An increase in natural gas production is likely to make the steam reforming of methane the choice of significant hydrogen production.

The capital cost for steam reformer is very heavily scale-dependent. For 5×10^6 Nm³/day plant, it can be as low as \$80/kW of H_2, while for 2300 Nm³/day plant the same cost would be \$4000/kW for hydrogen [72,81]. In large plants, reformer tubes can be as long as 12 m and of high cost due to expensive alloy materials for high-temperature and high-pressure operations. These are, however, unsuited for small-scale operations that require small and compact reformers at a lower cost. The small-scale operation, while expensive, is often used for niche applications such as FC technology and hydrogen refueling station.

In normal commercial reformers, the steam-to-hydrocarbon ratio is kept high enough to prevent coking and to avoid overloading the reformer duty. Generally, a ratio of 3 is used. The inlet temperature of 760°C is used and because reforming reaction is endothermic, additional heat is added as mixture flows down the catalyst-filled reformer tubes. A critical factor in the reformer heater design is keeping the tube wall temperature uniform and hot enough to promote reforming reaction. For this purpose, two types of heater design—side-firing reforming furnace and roof-fired heater design—have been employed.

In side-firing furnace, two parallel rectangular boxes are connected at the top with horizontal duct work into the vertical convection stack. Several rows (typically 4) are used to directly fire the tubes. A typical reformer furnace has 300 burners. Reformer tubes are 5 in. in diameter with wall thickness of 0.5 in. and about 34 ft of wall is exposed to the burners. The tubes are generally 25% chrome, 20% nickel, or a high-nickel steel such as HL-40.

The top-fired reformer is a rectangular box, tubes are still vertical, and inlet and outlet are pig-tails to pigtail inlet header and outlet transfer line. The burners have a pencil-shaped flame design. All burners are located above the inlet manifold. Hydrogen plants with a single reformer heater and capacity up to 100,000 ft³/day is used in the vertical down-firing approach. The outlet transfer line from the reformer is used to generate high-pressure (650 psig) steam. The reformer effluent gas exits through the transfer line at about 760°C [72].

While large commercial reformers are designed as described earlier, more compact and economical designs are used in the smaller-scale reformers. Some of these designs are briefly described next [92–110,302,395]. A good review of small-scale reformers is given by Ogden [395].

6.10.2 ANNULAR BED REFORMER

For small-scale applications such as FC technology or even independent H_2 production where the requirement is in the range 0.4–3 kW [72], the reformer normally operates at lower temperatures and pressures, and have lower cost parts. For H_2 production application, the reformer is generally operated at 3 atm and at about 700°C temperature. This is helpful in reducing the cost of the materials. For 20–200 Nm³/day capacity, estimated capital cost is about $150–$180 kW H_2 for 1000 units sold [72]. The energy conversion efficiency of such a reformer lies in the range of 70%–80%.

The annular bed reformer design for FC applications and/or H_2 production is practiced by a number of industries such as Haldor-Topsoe, Ballard Power Systems, Sanyo Electric, International Fuel Cells, Osaka Gas Co. and Praxair [72,92–110]. The technology is presently being commercialized, and it produces more compact reformers at lower costs compared to conventional reformers.

6.10.3 PLATE-TYPE REFORMERS

Another design that has been used for small-scale FC applications is plate-type reformers which are more compact than annular reformers or conventional long-tube reformers. The plates are arranged in a stack where one side of the plate is coated with a catalyst and on the other side (anode) exhaust gas from FC undergoes catalytic combustion to supply heat for the endothermic steam reforming reaction. The unit is compact, is low cost, and has good heat transfer and small heat-up period. The energy conversion efficiency of this type of the reformer lies somewhere between 70% and 77%.

Plate-type reformer design is produced by Osaka Gas Co. and GASTEC. Osaka Gas Co. design produces the reformer for the PEM FC application. In this unit, all components of an integrated design, namely, sulfur removal unit, steam reformer, water gas shift reactor and CO removal unit are made of plate deign. This makes the commercial process more compact and economical. GASTEC is applying the technology for residential type FC (20 kW) and has tested various variables such as combustion catalysts, coatings, and substrate materials to minimize the cost of the unit [72].

6.10.4 MEMBRANE REFORMERS

In this type of reactor, reforming, WGS reaction and further CO cleaning step all occur in the same unit. The reactor operates under high pressure and uses Pd membrane on one side through which H_2 permeates with high selectivity. The constant removal of hydrogen on the downstream side allows equilibrium to be shifted to achieve better conversion by reforming at a lower temperature. The membrane reactor and its analogous ion-transported membrane reactors are described later in this section [72].

6.10.5 AUTOTHERMAL REFORMERS

Steam reformers require heat because of the endothermic nature of reaction. A number of companies are involved in producing hydrogen by partial oxidation of hydrocarbons. Unlike steam reforming, POX is an exothermic reaction and does not require an external heat input. An FC operated by POX process is developed by companies like Arthur D. Little, Nuvera, Epyx, and a consortium of McDermott Technology/Catalytica and Hydrogen Burner Technology for a 50 kW FC [72,92–110]. When steam reforming is combined with the POX process, the resulting reactor is called autothermal reactor which does not require external heat for the reforming process once the oxidation reaction generates enough heat to sustain steam reforming reaction. Small-scale (10–50 kW) autothermal reactors have been developed for PEM FC by Honeywell, Daimler-Chrysler, Analytical Power, IdaTech, Hydrogen Burner Technology, Argonne National Laboratory, Idaho National Energy and Environmental Laboratory, and McDermott Technologies [72,92–110]. Autothermal reactors mostly use gasoline, diesel, and logistic fuels along with natural gas. The use of diesel and other logistic fuels makes them useful for various FC applications above ships [72].

6.10.6 ION TRANSPORT MEMBRANE REFORMERS

A large consortium headed by Air Products which include a number of companies and academic institutions is developing a ceramic membrane technology to generate hydrogen and syngas mixture. The membrane is made of nonporous multicomponent oxides which operate at temperature higher than 725°C and have high permeability and selectivity for oxygen transfer. The hydrogen production by ITM technology in the capacity range 3,000–30,000 Nm^3 of H_2/day appears to be about 27% cheaper than liquid hydrogen transported by road [72,92–110,395].

In ITM technology, one side of the membrane separates oxygen from air at around room temperature and 0.03–0.20 atm pressure. On other side, methane and steam react at high pressure (3–20 atm) to produce syngas (hydrogen and carbon monoxide). POX provides the heat for reforming reaction. The mixture of syngas can be either purified to obtain pure hydrogen or can be converted to produce fuels. The flat plate system seems to work best for ITM technology [72,92–110,395].

6.10.7 SORBENT-ENHANCED REFORMERS

In this technology, the steam reforming is accompanied by simultaneous removal of carbon dioxide and carbon monoxide by carbonation of calcium oxide [72,92–110,395]. The removal of carbon dioxide allows the reforming reaction to occur at 400°C–500°C as opposed to the normal reforming temperature of 800°C–1000°C. The reaction also produces reasonably pure hydrogen (90% H_2, 9.5% CH_4, 0.5% CO_2, and less than 50 ppm CO) and this alleviates the downstream purification processes such as WGS reaction and preferential oxidation, both of which are expensive. This technology is at the development stage and major issues to be resolved are sorbent lifetime and system design [72].

6.10.8 PLASMA REFORMERS

Thermal plasma technology is also used to generate hydrogen and hydrogen-rich gases from natural gas and other liquid hydrocarbons. Plasma provides energy to create free radicals for reforming process. The typical temperature varies from about 2,700°C to about 10,000°C. Such a high temperature accelerates the rate of reforming process. The process is very flexible and can be used for a variety of feedstock. The products generally contain ethylene and acetylene along with hydrogen, carbon monoxide, and carbon dioxide. The process can be operated in a large range of operating conditions, including reaction volume, inter-electrode gap, autothermal or steam reforming modes, variable level of sulfur impurities, and carbon deposit. The process is robust and can generate a large range of fuel power (10–40 kW) and can give up to 90% of methane conversion [114,371–381]. As described earlier in Section 6.9, numerous types of plasma reformers such as corona discharge, DBD, microwave discharge, APGD, and gliding arc discharge for cold plasma and DC arc torch, AC arc torch, RF inductively coupled torch and high-frequency capacitive torch for thermal plasma can be used. Proper design of the reactor in each case is important for best production of syngas and energy efficiency.

Plasma devices referred to as plasmatrons can generate very high temperatures (>2000°C) with a high degree of control, using electricity. The heat generation is independent of reaction chemistry, and optimum operating conditions can be maintained over a wide range of feed rates and gas composition. Compactness of the plasma reformer is ensured by high energy density associated with the plasma itself and by the reduced reaction times, resulting in a short residence time.

Plasma reforming of methane, especially plasma catalysis, has been demonstrated with sufficiently low energy consumption and with high hydrogen yields to be economically interesting for energy applications. The pyrolysis process consumes substantially higher energy consumption than could be recovered from the produced hydrogen. Substantial reduction of specific energy consumption will be required before this becomes feasible for energy applications. The results should improve with increasing reactor size and with use of uncooled graphite electrode plasmatrons. For homogeneous reforming with POX with air, improvements in plasmatron and reactor design resulted in a decrease of specific energy consumption but with values that are still relatively high [114,371–385]. There are several areas where improvements on plasmatron and reactor are possible. They include system optimization (such as heat regeneration), improved design (increased thermal insulation in the reactor chamber), and increased power operation (larger plasmas with larger volume-to-surface ratios).

Another possible application of this technology is for onboard reforming for the manufacturing of hydrogen-rich gases. The compactness of the reformer, with high power densities, makes the technology attractive for this application. Plasma reforming could also find uses for decentralized hydrogen production, such as in transportation refueling systems, either on-demand generation of hydrogen (requiring 200–500 kW plasmatrons) or small 20 kW CW plasmatrons (with stationary storage) for 30–50 vehicle fleets. The technology could also be used for industrial applications.

6.10.9 MICRO-CHANNEL REFORMERS

It is well known that reforming reactors are transport-limited. One attractive method to improve heat and mass transfer rates is to use microchannel reactor which can operate at 10 ms residence time compared to conventional 450,000 Nm^3 of H_2 per day reactor that operates at the residence time of 1 s. It is known that a significant portion of the catalyst volume in steam reformer is not effectively used. The microchannel reactor allows the reduction of plant volume to 88 Nm^3 as opposed to conventional 2700 Nm^3 required in the normal operation. Thus microchannel reactor reduces both capital and operating costs for producing hydrogen from methane. The microchannel reactors allow high reaction rates by increasing heat transfer rates. The literature study [72,92–110,395] showed that with a 10% Rh–4.5% $MgO–Al_2O_3$ catalyst at 0.5 ms, the reaction mixture was close to the equilibrium mixture. The success of microchannel reactor indicates that for highly active catalyst, methane reforming can be carried out at a contact time less than 1 ms. An increase in catalyst thickness such that it minimizes the transport resistances to the reaction can further reduce the required contact time.

6.10.10 MICROWAVE-ASSISTED REFORMER

Microwave-assisted reformer can be unimode or multimode depending on the need of the process. Generally, unimode microwave is used in the lab scale to gather fundamental data for the catalysts used in the reforming process. Large-scale microwave-assisted processes often use multimode operation. Fidalgo and Menendez [225] have given excellent descriptions of these two types of micro-wave-assisted operations, and their descriptions are largely reproduced in the following paragraphs.

Unimode microwave ovens can be employed for the lab study of microwave-assisted reforming reactions [345–347,355,386–388]. A typical experimental setup for unimode microwave device to carry out experiments of the decomposition and CO_2 reforming of methane over carbon catalysts has been described by Fidalgo and Menendez [225], among others [346,347,355]. In this case, the microwave oven can operate at a variable power from 0 to 1860 W and at a fixed frequency of 2.45 GHz. The catalyst sample is placed inside a quartz reactor, which is housed in the center of the rectangular microwave guide that directs the microwaves from the magnetron into the sample. The nonabsorbed radiation is dissipated by a water sink. The power reflected back toward the magne-tron is minimized by adjusting the manual 2-stub unit in the waveguide. The catalyst temperature is measured with an IR-pyrometer. Other unimode microwave ovens with variable output power have been described elsewhere [386–389]. There are some differences in these setups which include different maximum operating powers, the use of coaxial cable instead of a rectangular microwave guide, the measurement of temperature with optic fiber, and the use of mobile piston, short circuits, apertures, irises or posts, instead of stub units.

Although multimode microwaves are widely used in industrial applications, there are few examples of the use of lab-scale multimode microwave ovens for reforming reactions [343,354]. This lab-scale pilot plant is a multimode microwave device, which operates at a frequency of 2450 MHz over a power range of 0–1500 W. Microwaves are generated in a magnetron and directed through a rectangular waveguide to the multimode applicator, which houses the reactor charged with the catalyst. The reactor is isolated by an insulator, which is transparent to micro-waves. The catalyst temperature is measured by means of a thermocouple placed inside the cata-lyst bed. The insulator temperature is measured by means of another thermocouple. The power reflected back to the magnetron is minimized by adjusting a manual 3-stub unit, situated inside the waveguide.

A circulator prevents the reflected microwaves from reaching and damaging the magnetron. The reflected power is measured by a detector and dissipated by a water sink. The power needed to generate microwaves during the CO_2 reforming of CH_4 is recorded by an energy consump-tion meter. The instantaneous electric current and voltage are displayed on a control panel, so that the power required at any one moment can be known. In addition, the microwave pilot plant can operate in manual mode (the operating power is fixed and the carbon-microwave reactor (C/MR) is heated up to the maximum possible temperature under these conditions) and in automatic mode (the operating temperature is fixed at the desired value and controlled by a proportional-integral-derivative (PID), which adjusts the power emitted by the magnetron in order to keep the sample temperature constant).

Another multimode microwave oven has been described to investigate the microwave-assisted heterogeneous catalytic gas-phase POX of hydrocarbons [343]. In this case, a commercial micro-wave oven with two magnetrons of 900 W and rotating antennas was modified using two indepen-dent switched-mode power supplies, two metallic flanges for supporting the quartz reactor where the catalyst sample is placed, and another metallic flange to support an IR-pyrometer for measuring the temperature. A thermocouple is also used to measure the catalyst temperature after the microwave power has been turned off. The design of a microwave oven for industrial applications is very much dependent on each specific process and it must take into account critical issues such as power effi-ciency, uniform power distribution, and consistent and reliable performance over the expected range of process conditions [340,343,390].

6.10.11 Solar Steam Reforming Reactors

Solar reforming of natural gas, using either steam or CO_2, has been extensively studied in solar-concentrating facilities with small-scale solar reactor [391–406]. This solar-reforming process developed within the EU's project solar reforming process (SOLREF) (2004–2009) has been scaled up to power levels of 300–500 kWth and tested at 850°C and 8–10 bars in a solar tower [72,358]. Solar dry reforming of methane (CH_4) with CO_2 in an aerosol solar reactor with residence times around 10 ms and temperatures of approximately 1700°C, methane and CO_2 conversions of 70% and 65%, respectively, were achieved in the absence of catalysts. The high temperatures required for solar reforming effectively limit the nature of solar energy collector. The bulk energy production required for reforming must be carried out on a large scale to compete with fossil fuels and probably requires the tower (central receiver) solar technology. Solar reforming can be carried out using different processes such as direct and indirect, each requiring different types of reformer configuration [72,358,391–406].

An earlier joint Spanish–German project, Advanced Steam Reforming of Methane in Heat Exchange (ASTERIX) experiment, examined steam reforming of methane using solar-generated high-temperature process heat [72,358] using an indirectly heated reformer. The specific objectives of the ASTERIX experiments were to collect and store an amount of solar energy to obtain maximum conversion of methane and to produce consistently high-quality syngas. The experiment used a gas-cooled solar tower (GAST) system to produce hot air (up to 0.36 kg/s at 1000°C and 9 bars) to drive the separate steam reformer. This air was then fed back into the GAST cycle. The GAST technology program is described in References 72 and 358.

During normal operation, the heating medium, air, is taken from the GAST circuit (receiver) at a temperature of 1000°C over a suitable bench line and fed through the electric heater to the reforming reactor inlet. In this solar-only operating mode, air flows through the heater passively without any additional electric heating. Methane reforming is initiated at the process gas end of the reformer. A liquid natural gas storage tank directly provides the reforming unit with natural gas at the required pressure via the LNG evaporator. The process gas mixture is heated by air from 500°C to about 850°C as it passes through the catalyst bed. The endothermic reforming reaction results in the production of hydrogen and carbon monoxide with 3:1 ratio. More details of the ASTERIX experiment are given in References 72 and 358.

The Weizmann Institute of Science (WIS) operated a solar central receiver for development of high-temperature technology, including the storage and transport of solar energy via methane reforming. The WIS had designed a facility for testing reformers up to about 480 kW absorbed energy. The facility was designed for either steam or carbon dioxide reforming and can accommodate reformers that operate between 1 and 18 bars. A cavity receiver containing eight vertical reformer tubes (2 in. schedule 80) and 4.5 m long was designed. The overall dimension of the device was about 5 m high, 4.5 m wide, and 3 m deep. The reactor was designed to produce syngas at 800°C [72,358].

In the Soltox process, a parabolic dish is used to concentrate sunlight through a quartz window into an internally insulated aluminum reactor vessel where it is absorbed on a rhodium-coated reticulated ceramic foam absorber. Concentrated organic waste and steam are mixed and flow through the hot (>1000°C) catalyst bed, where they react completely in fractions of a second to produce hydrogen, carbon dioxide, carbon monoxide, and halogen acids (which are easily neutralized to simple salts). The extremely good heat and mass transfer within the reactor result in a compact, highly efficient system [72,358].

When a vaporized organic waste is mixed with steam and passed through the reactor, highly specific, irreversible, endothermic reforming reactions take place on the catalyst-coated surface of the radiantly heated absorber to quantitatively destroy the waste. For example, trichloroethylene reacts with steam to produce hydrogen, carbon monoxide, and hydrogen chloride. Because reforming is not a combustion process, neither fuel, nor air, nor oxygen needs to be supplied to the reactor. Thus, unlike incineration, solar-driven, high-temperature catalytic reforming produces neither NO_x nor products of incomplete combustion. Furthermore, variable absorber thickness and adjustable gas flow rates mean that residence times within the absorber and thus reaction times and destruction efficiency can be controlled [72,358].

The applications of open-loop solar syngas production include the following [72,358]:

1. Natural gas reforming for power plants. A number of European countries have imported natural gas via pipelines from North Africa and have reformed this gas to either syngas or hydrogen, increasing its calorific value by about 25% before combustion in gas turbine or FC power plants.
2. Syngas production from municipal, agricultural, and organic industrial waste. In Sunbelt countries, concentrated waste streams can be gasified to syngas with solar energy at potentially acceptable costs and with essentially no emissions to the atmosphere.
3. Soltox-type processing which provides an option for environmentally acceptable disposal of a number of toxic organic materials.

Open-loop syngas production can also be used for the generation of syngas that is being supplied worldwide for the production of hydrogen, methanol, ammonia, and oxyalcohols [72,358,391–406].

A number of studies have focused on the production of hydrogen by steam reforming of methane and other hydrocarbons using solar reactor [72,358,391–406]. A schematic of the solar reactor used by Seinfeld and co-workers is depicted in Figure 6.11. Shah [72] and Yeheskel et al. [407] studied chemical kinetics of high-temperature hydrocarbon reforming using a solar reactor. Watanuki et al. [403] examined methane steam reforming using a molten salt membrane reactor. In a membrane reforming reactor, the reforming reaction takes place in tubular reactors that

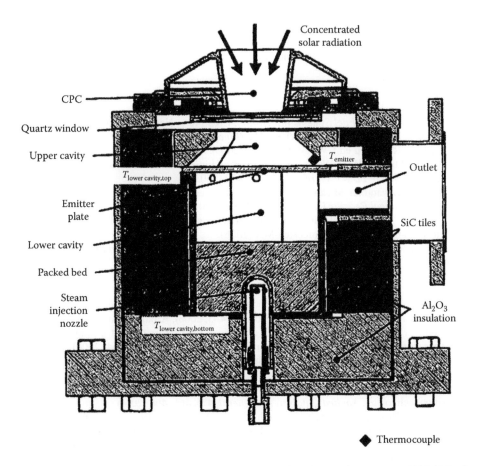

FIGURE 6.11 Solar reforming reactor used by Shah [72] and Steinfeld and co-workers [397,400]. (Section view of the packed-bed solar reactor configuration. Indicated are the locations of the thermocouples.)

consist of selective membranes, generally palladium, which separates hydrogen as it is produced. The principal advantages of a solar membrane reforming process compared to the conventional reforming process are as follows [72,358]:

1. The reforming is carried out at a lower temperature (550°C). This means a significant reduction in the energetic consumption. Low-temperature reactors also use less costing materials for the reforming reactor tubes.
2. Hydrogen is obtained with a higher purity due to highly efficient membrane separation process.
3. Methane conversions up to 90% can be reached due to high hydrogen extraction through the membrane.
4. A big part of CO–CO_2 conversion is produced inside the reactor itself.
5. Emissions are reduced by about 34%–53% due to the use of concentrated solar energy to obtain the process heat.

In this study, steam reforming of methane proceeded with the original module with palladium membrane below the decomposition temperature of molten salt (around 870 K). SOLREF process and its various options for solar reforming of natural gas by steam are also described by Moller [359]. A review of hydrogen production technologies from solar energy is also given by Suarez-Gonzalez et al. [393].

The reactor and process just described can be easily adapted to solar gasification, but for this case heavy hydrocarbons are used as feedstock. These are transformed into cleaner fuels for a combined cycle or in the process that can produce hydrogen. As mentioned earlier, a solar gasification plant using petroleum coke has been tested in the solar platform of Almeria (Spain). The reactor has reached the hydrogen production efficiency of 60% working at 1500 K [1,72,358,395].

REFERENCES

1. Shah, Y.T., *Energy and Fuel Systems Integration*. CRC Press, New York (2015).
2. Lamm, A., Vielstich, W., and Gasteiger, H., *Handbook of Fuel Cells: Fundamentals, Technology, Applications* (Reprinted ed.). Wiley, New York (2003).
3. Choi, Y. and Stenger, H.G., Water gas shift reaction kinetics and reactor modeling for fuel cell grade hydrogen, *Journal of Power Sources*, 124, 432–439 (2003).
4. Byron, S.R.J., Loganthan, M., and Shantha, M.S., A review of the water gas shift reaction, *International Journal of Chemical Reactor Engineering*, 8, 1–32 (2010).
5. Callaghan, C., Kinetics and catalysis of the water-gas-shift reaction, Ph.D. thesis, Department of Chemical Engineering, WPI, Worcester, MA (2006).
6. Newsome, D.S., The water-gas shift reaction, *Catalysis Reviews: Science and Engineering*, 21 (2), 275–318 (1980).
7. Ratnasamy, C. and Wagner, J.P., Water gas shift catalysis, *Catalysis Reviews*, 51 (3), 325–440 (September 2009).
8. Whitlow, J.E. and Parish, C.F., Operation, modeling and analysis of the reverse water gas shift process, AIP Conference Proceedings, 654, 1116 (February 4–5, 2003), Alburquerque, New Mexico.
9. Criscuoli, A., Basile, A., and Drioli, E., An analysis of the performance of membrane reactors for the water gas shift reaction using gas feed mixtures, *Catalysis Today*, 56, 53–64 (2000).
10. Water-gas shift reaction, Wikipedia, the free encyclopedia (2015). http://en.wikipedia.org/wiki/water-gas-sh.pt_reaction (last accessed June 9, 2016).
11. Jain, R., Synthesis of nano-Pt onto ceria support as catalyst for water–gas shift reaction by reactive spray deposition technology, *Applied Catalysis A: General*, 475, 461–468 (2014).
12. Giuseppe, B., New catalyst for the H_2 production by water-gas shift reaction processes, Ph.D. thesis, Departmento de chimica industriale e dei Materiali, Universita di Bologna, Bologna, Italy. (2010).
13. Barakat, T., Rooke, J.C., Genty, E., Cousin, R., Siffert, S., and Su, B.-L., Gold catalysts in environmental remediation and water-gas shift technologies, *Energy & Environmental Science*, 6 (2), 371 (January 1, 2013).
14. de Farias, A.M.D., Nguyen-Thanh, D., and Fraga, M.A., Discussing the use of modified ceria as support for Pt catalysts on water–gas shift reaction, *Applied Catalysis B: Environmental*, 93 (3–4), 250–258 (2010).

15. Apanee, L., Osuwan, S., and Gulari, E., Comparative studies of low temperature water–gas shift reaction over Pt/CeO$_2$, Au/CeO$_2$, and Au/Fe$_2$O$_3$ catalysts, *Catalysis Communications*, 4, 215–221 (2003).

16. Boon, J., van Dijk, E., Pirgon-Galin, O., Haije, W., and van den Brink, R., Water–gas shift kinetics over FeCr-based catalyst: Effect of hydrogen sulphide, *Catalysis Letters*, 131, 406–412 (2009).

17. Callaghan, C., Fishtik, I., Datta, R., Carpenter, M., Chmielewski, M., and Lugo, A., An improved microkinetic model for the water gas shift reaction on copper, *Surface Science*, 541, 21–30 (2003).

18. Gonzaleza, I.D., Navarroa, R.M., Wen, W., Marinkovic, N., Rodriguez, J.A., Rosa, F., and Fierro, J.L.G., A comparative study of the water gas shift reaction over platinum catalysts supported on CeO$_2$, TiO$_2$ and Ce-modified TiO$_2$, *Catalysis Today*, 149 (3–4), 372–379 (2010).

19. Hakkarainen, R., Salmi, T., and Keiski, R.L., Water-gas shift reaction on a cobalt molybdenum oxide catalyst, *Applied Catalysis A: General*, 99, 195–215 (1993).

20. Hilaire, S., Wanga, X., Luoa, T., Gorte, R.J., and Wagner, J., A comparative study of water-gas-shift reaction over ceria supported metallic catalysts, *Applied Catalysis A: General*, 215, 271–278 (2001).

21. Jacobs, G., Graham, U.M., Chenu, E., Patterson, P.M., Dozier, A., and Davis, B.H., Low temperature water gas shift: Impact of platinum promoter loading as the partial reduction of ceria and consequences for catalyst design, *Journal of Catalysis*, 229, 499–512 (2005).

22. Kusar, H., Hocevar, S., and Levec, J., Kinetics of the water–gas shift reaction over nanostructured copper–ceria catalysts, *Applied Catalysis B: Environmental*, 63, 194–200 (2006).

23. Phatak, A.A., Koryabkina, N., Rai, S., Ratts, J.L., Ruettinger, W., Farrauto, R.J., Blau, G.E., Delgass, W.N., and Ribeiro, F.H., Kinetics of the water–gas shift reaction on Pt catalysts supported on alumina and ceria, *Catalysis Today*, 123, 224–234 (2007).

24. Rhodes, C., Peter Williams, B., King, F., and Hutchings, G.J., Promotion of Fe$_3$O$_4$/Cr$_2$O$_3$ high temperature water gas shift catalyst, *Catalysis Communications*, 3, 381–384 (2002).

25. Rhodes, C. and Hutchings, G.J., Studies of the role of the copper promoter in the iron oxide/chromia high temperature water gas shift catalyst, *Physical Chemistry Chemical Physics*, 5, 2719–2723 (2003).

26. Salmi, T., Bostrom, S., and Lindfors, L.E., A dynamic study of the water-gas shift reaction over an industrial ferrochrome, *Catalyst Journal of Catalysis*, 112, 345–356 (1988).

27. Salmi, T. and Hakkarainen, R., Kinetic study of the low-temperature water-gas shift reaction over a Cu-ZnO catalyst, *Applied Catalysis*, 49, 285–306 (1989).

28. Sun, J., DesJardins, J., Buglass, J., and Liu, K., Noble metal water gas shift catalysis: Kinetics study and reactor design, *International Journal of Hydrogen Energy*, 30, 1259–1264 (2005).

29. Tang, Q.L., Chen, Z.-X., and He, X., A theoretical study of the water gasshift reaction mechanism on Cu (1 1 1) model system, *Surface Science*, 603, 2138–2144 (2009).

30. Olivier, T., Rachedi, K., Diehl, F., Avenier, P., and Schuurman, Y., Kinetics and mechanism of the water–gas shift reaction over platinum supported catalysts, *Topics in Catalysis*, 52, 1940–1945 (2009).

31. Twigg, M.V., *Catalyst Handbook*, 2nd edn. Wolfe Publishing Ltd. Prescott, Arizona (1989).

32. Twigg, M.V. and Spencer, M.S., Deactivation of supported copper metal catalysts for hydrogenation reactions, *Applied Catalysis A: General*, 212, 161–174 (2001).

33. Van Herwijnen, T. and De Jong, W.A., Kinetics and mechanism of the CO shift on Cu/ZnO. 1. Kinetics of the forward and reverse CO shift reactions, *Journal of Catalysis*, 63, 83–93 (1980).

34. Tanaka, Y., Utaka, T., Kikuchi, R., Sasaki, K., and Eguchi, K., Water gas shift reaction over Cu-based mixed oxides for CO removal from the reformed fuels, *Applied Catalysis A: General*, 242, 287–295 (2003).

35. Jacobs, G. and Davis, B.H., Low temperature water-gas shift catalysts, *Catalysis*, 20, 122–285 (2007).

36. Amadeo, N.E. and Laborde, M.A., Hydrogen production from the low temperature water gas shift reaction: Kinetics and simulation of the industrial reactor, *International Journal of Hydrogen Energy*, 20 (12), 949–956 (1995).

37. Yeragi, D.C., Pradhan, N.C., and Dalai, A.K., Low-temperature water-gas shift reaction over Mn-promoted Cu/Al$_2$O$_3$ catalysts, *Catalysis Letters*, 112 (3–4), 139–148 (2006).

38. Gokhale, A.A., Dumesic, J.A., and Mavrikakis, M. On the Mechanism of Low-Temperature Water Gas Shift Reaction on Copper, *Journal of American Chemical Society*, 130, 1402–1414 (2008).

39. Li, Y., Fu, Q., and Flytzani-Stephanopoulos, M., Low-temperature water-gasshift reaction over Cu- and Ni-loaded cerium oxide catalysts, *Applied Catalysis B: Environmental*, 27 (3), 179–191 (2000).

40. Singh, C.P.P. and Saraf, D.N., Simulation of low-temperature water-gas shift reactor, *Industrial & Engineering Chemistry Process Design and Development*, 19 (3), 393–396 (1980).

41. Ding, O.L. and Chan, S.H., Water-gas shift reaction—A 2-D modeling approach, *International Journal of Hydrogen Energy*, 33, 4325–4336 (2008).

42. Elnashaie, S.S.E.H. and Elshishini, S.S., *Modeling, Simulation and Optimization of Industrial Fixed Bed Catalytic Reactors*. Gordon and Breach Science Publishers, Philadelphia, PA (1993).

43. Fan, L.T., Lin, Y.-C., Shafie, S., Hohn, K.L., Bertók, B., and Friedler, F., Graph-theoretic and energetic exploration of catalytic pathways of the water-gas shift reaction, *Journal of the Chinese Institute of Chemical Engineers*, 39, 467–473 (2008).

44. Fishtik, I. and Datta, R., A UBI–QEP microkinetic model for the water–gasshift reaction on Cu(111), *Surface Science*, 512, 229–254 (2002).

45. Gideon Botes, F., Water–gas-shift kinetics in the iron-based low-temperature Fischer–Tropsch synthesis, *Applied Catalysis A: General*, 328, 237–242 (2007).

46. Grabow, L.C., Gokhale, A.A., Evans, S.T., Dumesic, J.A., and Mavrikakis, M., Mechanism of the water gas shift reaction on Pt: First principles, experiments, and microkinetic modeling, *Journal of Physical Chemistry C*112 (12), 4608–4617 (2008).

47. Grenoble, D.C. and Estadt, M.M., The chemistry and catalysis of the water gasshift reaction 1. The kinetics over supported metal catalysts, *Journal of Catalysis*, 67, 90–102 (1981).

48. Keiski, R.L., Salmi, T., and Pohjola, V.J., Development and verification of a simulation model for a non-isothermal water-gas shift reactor, *The Chemical Engineering Journal*, 48, 17–29 (1992).

49. Koryabkina, N.A., Phatak, A.A., Ruettinger, W.F., Farrauto, R.J., and Ribeiro, F.H., Determination of kinetic parameters for the water–gas shift reaction on copper catalysts under realistic conditions for fuel cell applications, *Journal of Catalysis*, 217, 233–239 (2003).

50. Levent, M., Water–gas shift reaction over porous catalyst: Temperature and reactant concentration distribution, *International Journal of Hydrogen Energy*, 26, 551–558 (2001).

51. Mao, J.H., Ming, N.Z., Xiang, P.G., and Qian, X., Mechanism of the copper catalyzed water gas shift reaction, *Acta Physico-Chimica Sinica*, 24 (11), 2059–2064 (2008).

52. Mhadeshwar, A.B. and Vlachos, D.G., Is the water–gas shift reaction on Pt simple? Computer-aided microkinetic model reduction, lumped rate expression, and rate-determining step, *Catalysis Today*, 105, 162–172 (2005).

53. Moe, J.M., Design of water-gas shift reactors, *Chemical Engineering Progress*, 58, 33–36 (1962).

54. Ovesen, C.V., Stoltze, P., Norskov, J.K., and Campbell, C.T., A kinetic model of the water gas shift reaction, *Journal of Catalysis*, 134, 445–468 (1992)

55. Ovesen, C.V., Clausen, B.S., Hammershøi, B.S., Steffensen, G., Askgaard, T., Chorkendorff, I., Nørskov, J.K., Rasmussen, P.B., Stoltze, P., and Taylor, P., A microkinetic analysis of the water–gas shift reaction under industrial conditions, *Journal of Catalysis*, 158, 170–180 (1996).

56. Podolski, W.F. and Kim, Y.G., Modeling the water-gas shift reaction, *Industrial & Engineering Chemistry Process Design and Development*, 13 (4), 415–442 (1974).

57. Rhodes, C., Hutchings, G.J., and Ward, A.M., WGS reaction: Finding the mechanistic Boundary, *Catalysis Today*, 23, 43–58 (1995).

58. Seo, Y.S., Seo, D.-J., Seo, Y.-T., and Yoon, W.-L., Investigation of the characteristics of a compact steam reformer integrated with a water-gas shift reactor, *Journal of Power Sources*, 161, 1208–1216 (2006).

59. Adams II, T.A. and Barton, P.I., A dynamic two-dimensional heterogeneous model for water gas shift reactors, *International Journal of Hydrogen Energy*, 34, 8877–8891 (2009).

60. Wang, G. et al., Investigation of the kinetic properties for the forward and reverse WGS reaction by energetic analysis, *Journal of Molecular Structure (Theochem)*, 634, 23–30 (2003).

61. Chen, W.-H., Hsieh, T.-C., and Jiang, T.L., An experimental study on carbon monoxide conversion and hydrogen generation from water gasshift reaction, *Energy Conversion and Management*, 49, 2801–2808 (2008).

62. Wheeler, C., Jhalani, A., Klein, E.J., Tummala, S., and Schmidt, L.D., The water–gas-shift reaction at short contact times, *Journal of Catalysis*, 223, 191–199 (2004).

63. Huanqi, Z., Hu, Y., and Li, J., Reduced rate method for discrimination of the kinetic models for the water-gas shift reaction, *Journal of Molecular Catalysis A: Chemical*, 149, 141–146 (1999).

64. Hakkarainen, R., Salmi, T., and Keiski, R.L., Comparison of the dynamics of the hightemperature water-gas shift reaction on oxide catalysts, *Catalysis Today*, 20, 395–408 (1994).

65. Keiski, R.L., Salmi, T., Niemisto, P., Ainassaari, J., and Pohjola, V.J., Stationary and transient kinetics of the high temperature water gas shift reaction, *Applied Catalysis A: General*, 137, 349–370 (1996).

66. Pasel, J., Samsun, R.C., Schmitt, D., Peters, R., and Stolten, D., Test of a water–gas shift reactor on a 3kWe-scale—Design points for high- and low temperature shift reaction, *Journal of Power Sources*, 152, 189–195 (2005).

67. San, S.H., Park, D., Duffy, G.J., Edwards, J.H., Roberts, D.G., Ilyushechkin, A., Morpeth, L.D., and Nguyen, T., Kinetics of high-temperature water-gas shift reaction over two iron-based commercial catalysts using simulated coal derived syngases, *Chemical Engineering Journal*, 146, 148–154 (2009).

68. Singh, C.P.P. and Saraf, D.N., Simulation of high-temperature water-gasshift reactors, *Industrial & Engineering Chemistry Process Design and Development*, 16 (3), 313–319 (1977).

69. Chen, W.-H., Lin, M.-R., Jiang, T.L., and Chen, M.-H., Modeling and simulation of hydrogen generation from high temperature and low-temperature water gas shift reactions, *International Journal of Hydrogen Energy*, 33, 6644–6656 (2008)

70. Lee, S., Speight, J., and Loyalka, S. (eds.), *Handbook of Alternative Fuel Technologies*. CRC Press, New York (2007).

71. Rase, H.F., *Chemical Reactor Design for Process Plants*, Volume Two—Case Studies and Design Data. John Wiley & Sons, New York (1977).

72. Shah, Y.T., *Water for Energy and Fuel Production*. CRC Press, New York (2014).

73. Navarro, R., Pena, M., and Fierro, J., Hydrogen production reactions from carbon feedstocks: Fossil fuels and biomass, *Chemical Reviews*, 107, 3952–3991 (2007).

74. Kechagiopoulos, P., Voutetakis, S., Lemonidou, A., and Vassalos, I., Hydrogen production via steam reforming of the aqueous phase of bio-oil in a fixed bed reactor, *Energy & Fuels*, 20, 2155–2163 (2006).

75. Ye, T., Yuan, L., Chen, Y., Kan, T., Tu, J., Zhu, X., Torimoto, Y., Yamamoto, M., and Li, Q., High efficient production of hydrogen from bio-oil using low-temperature electrochemical catalytic reforming approach over NiCuZn-Al$_2$O$_3$ catalyst, *Catalysis Letters*, 127, 323–333 (2009).

76. Barrio, M., Gobel, H., Risnes, H., Henricksen, U., Hustad, J., and Sorensen, L., Steam gasification of wood char and the effect of hydrogen inhibition on the chemical kinetics, a personal communication (2012).

77. Meijer, R., Kapteijn, F., and Moulijn, J.A., Kinetics of the alkali-carbonate catalysed gasification of carbon: H$_2$O gasification, *Fuel*, 73 (5), 723–730 (1994).

78. Sørensen, L.H. et al., Straw—H$_2$O gasification kinetics, Determination and discussion, Nordic Seminar on Thermochemical Conversion of Solid Fuels, December 3, 1997, Chalmers University of Technology, Gothenburg, Sweden (1997).

79. Li, J. and van Heiningen, A.R.P., Kinetics of gasification of black liquor char by steam, *Industrial & Engineering Chemistry Research*, 30 (7), 1594–1601 (1991).

80. Van Heek, K., General aspects and engineering principles for technical application of coal gasification, in Figuieiredo, J.L. and Moulijn, J.A. (eds.), *Carbon and Coal Gasification: Science and Technology*. Springer, Alvor, Portugal, pp. 383–402 (1986).

81. Elshout, R., Hydrogen production by steam reforming, *Chemical Engineering Processing*, 2, 34–38 (May 2010). Chemeng-processing.blogspot.com/2010/.../Hydrogen-production-by-stream-reforming.

82. Wei, J. and Iglesia, E., Isotopic and kinetic assessment of the mechanism of reactions of CH$_4$ with CO$_2$ or H$_2$O to form synthesis gas and carbon on nickel catalysts, *Journal of Catalysis*, 224 (2), 370–383 (2004).

83. Rostrup-Nielsen, J.R., Hansen, J., and Bak, H., CO$_2$-reforming of methane over transition metals, *Journal of Catalysis*, 144 (1), 38–49 (1993).

84. Jackson, S.D., Thomson, S.J., and Webb, G., Carbonaceous deposition associated with catalytic steam reforming of hydrocarbons over nickel alumina catalysts, *Journal of Catalysis*, 70, 249–263 (1981).

85. Aznar, M.P., Caballero, M.A., Gil, J., Marte, J.A., and Corella, J., Commercial steam reforming catalysts to improve biomass gasification with steam/oxygen mixtures. 2. Catalytic tar removal, *Industrial & Engineering Chemistry Research*, 37 (97), 2668–2680 (1998).

86. Mas, V., Bergamini, M., Baronetti, G., Amadeo, N., and Laborde, M., A kinetic study of ethanol reforming using a Nickel based catalyst, *Topics in Catalysis*, 51, 39–48 (2008).

87. Biswas, P. and Kunzru, D., Steam reforming of ethanol on Ni-CeO$_2$-ZrO$_2$ catalysts: Effect of doping with copper, cobalt and calcium, *Catalysis Letters*, 118 (1–2), 36–49 (October 2007).

88. Akdim, O., Cai, W., Fierro, V., Provendier, H., van Deen, A., Shen, W., and Mirodatos, C., Oxidative steam reforming of ethanol over Ni-Cu/SiO$_2$, Rh/Al$_2$O$_3$ and Ir/CeO$_2$: Effect of metal and support on reaction mechanism, *Topics in Catalysis*, 51, 22–38 (2008).

89. Fajardo, H., Longo, E., Mezalira, D., Nurenberg, G., Almerindo, G., Collasiol, A., Probst, L., Garcia, I., and Carreno, N., Influence of support on catalytic behavior of nickel catalysts in the steam reforming of ethanol for hydrogen production, *Environmental Chemistry Letters*, 8, 79–85 (2010).

90. Sekine, Y., Kazama, A., Izutsu, Y., Matsukata, M., and Kikuchi, E., Steam reforming of ethanol over cobalt catalyst modified with small amount of iron, *Catalysis Letters*, 132, 329–334 (2009).

91. Dong, T., Wang, Z., Yuam, L., Torimoto, Y., Sadakata, M., and Li, Q., Hydrogen production by steam reforming of ethanol on potassium doped 12CaO-7Al$_2$O$_3$ catalyst, *Catalysis Letters*, 119, 29–39 (2007).

92. Song, H., Tan, B., and Ozkan, U., Novel synthesis techniques for preparation of Co/CeO$_2$ as ethanol steam reforming catalysts, *Catalysis Letters*, 132, 422–429 (2009).

93. He, L., Barntsen, H., Ochoa-Fernandez, E., Walmsley, J., Blekkan, E., and Chen, D., Co-Ni catalysts derived from hydrotalcite-like materials for hydrogen production by ethanol steam reforming, *Topics in Catalysis*, 52, 206–217 (2009).

94. Henpraserttae, S. and Toochinda, P., Effects of preparation of Cu/Zn over Al_2O_3 catalysts for hydrogen production from methanol reforming, *Suranaree Journal of Science and Technology*, 16 (2), 103–112 (2009).

95. Lee, S., *Methanol Synthesis Technology*. CRC Press, New York (1989).

96. Purnama, H., Ressler, T., Jentoft, R., Soerijanto, H., Schlogl, R., and Schomacker, R., Steam reforming of methanol on Cu/ZrO_2 catalysts, *Applied Catalysis A*, 259, 83 (2004).

97. Lindström, B., Development of methanol reformer for fuel cell vehicles, Thesis, KTH-Kungliga Tekniska Hogskolan, Department of Chemical Engineering and Technology, Stockholm, Sweden (2003).

98. Lindström, B. and Pettersson, L.J., Deactivation of copper-based catalysts for fuel cell applications, *Catalysis Letters*, 74, 27–30 (2001).

99. Lindström, B. and Pettersson, L.J., Steam reforming of methanol for fuel cell applications, in *Proceedings of Ninth Nordic Symposium on Catalysis*, Lidingö, Sweden, pp. 101–102 (June 4–6, 2000).

100. Lindström, B. and Pettersson, L.J., A study of ethanol and methanol as a fuel for onboard hydrogen generation by steam reforming on copper-based catalysts, in *Proceedings of XIII International Symposium on Alcohol Fuels*, Stockholm, Sweden (July 3–6, 2000).

101. Lindström, B. and Pettersson, L.J., Steam reforming of methanol for automotive applications, in *Proceedings of 2000 Fuel Cell Seminar*, October 2000, Portland, Oregon, pp. 325–328 (2000).

102. Lindström, B., Agrell, J., and Pettersson, L.J., Combinatorial reforming of methanol for hydrogen generation over monolithic catalysts, in *Proceedings of 17th North American Catalysis Society Meeting*, June 3–8, 2001, Toronto, Ontario, Canada, p. 140 (2001).

103. Lindström, B. and Pettersson, L.J., Catalytic steam reforming of methanol for automotive fuel cell applications, in *Proceedings of Fifth European Congress on Catalysis (EUROPACAT 5)*, Limerick, Ireland (September 2–7, 2001).

104. Lindström, B. and Pettersson, L.J., Steam reforming of methanol over copper-based monoliths: The effects of zirconia doping, in *Proceedings of Seventh Grove Fuel Cell Symposium*, London, U.K. (September 11–13, 2001).

105. Kolb, G., Keller, S., Tiemann, D., Schelhaas, K., Schürer, J., and Wiborg, O., Design and operation of a compact microchannel 5 kWel,net methanol steam reformer with novel Pt/In_2O_3 catalyst for fuel cell applications, *Chemical Engineering Journal*, 207–208, 358–402 (2012). http://dx.doi.org/10.1016/j.cej.2012.06.14/

106. Lindström, B., Pettersson, L.J., and Menon, P.G., Influence of the operating conditions on the performance of a methanol reformer, in *Proceedings of 10th Nordic Symposium on Catalysis*, Helsingør, Denmark (June 2–4, 2002).

107. Zhang, L., Pan, L.-w., Ni, C.-j., Zhao, S.-s., Wang, S.-d., Hu, Y.-k., Wang, A.-j., and Jiang, K., Optimization of methanol steam reforming for hydrogen production, *Journal of Fuel Chemistry and Technology*, 41 (01), 116–122 (2013).

108. Pan, L. and Wang, S., Modeling of a compact plate-fin reformer for methanol steam reforming in fuel cell systems, *Chemical Engineering Journal*, 108, 51–58 (2005).

109. Pan, L. and Wang, S., Methanol steam reforming in a compact plate-fin reformer for fuel cell systems, *International Journal of Hydrogen Energy*, 30, 973–979 (2005).

110. Pan, L., Ni, C., Zhang, X., Yuan, Z., Zhang, C., and Wang, S., Study on a compact methanol reformer for a miniature fuel cell, *International Journal of Hydrogen Energy*, 36, 319–325 (2011).

111. Li, Z., Gao, D., Chang, L., Liu, P., and Pistikopoulos, E., Coal-derived methanol for hydrogen vehicles in China: Energy, environment and economic analysis for distributed reforming, *Chemical Engineering Research and Design*, 88, 73–80 (2010).

112. Olafadehan, O., Susu, A., and Jaiyeola, A., Mechanistic kinetic models for n-heptane reforming on Platinum/Alumina catalysts, *Petroleum Science and Technology*, 26, 1459–1480 (2008).

113. Salaun, M., Capela, S., De Costa, S., Gagnepain, L., and De Costa, P., Enhancement of 3-way CNG catalyst performance at high temperature due to the presence of water in the feed: On the role of steam reforming of methane and on the influence of ageing, *Topics in Catalysis*, 52, 1972–1976 (2009).

114. Futamura, S. and Kabashima, H., Steam reforming of aliphatic hydrocarbons with nonthermal plasma, *IEEE Transactions of Industrial Applications*, 40 (6), 1476–1481 (November/December 2004).

115. Mota, N., Alvarez, Galvan, A.C., Villoria, J.A., Rosa, F., Fierro, J., and Navarro, R., Reforming of diesel fuel for hydrogen production over catalysts derived from $LaCo_{1-x}M_x O_3$ (M = Ru, Fe), *Topics in Catalysis*, 52, 1995–2000 (2009).

116. Ambroise, E., Courson, C., Keinnemann, A., Roger, A., Pajot, O., Samson, E., and Blanchard, G., On-board hydrogen production through catalytic exhaust-gas reforming of isooctane: Efficiency of mixed oxide catalysts $Ce_2Zr_{1.5}Me_{0.5}O_8$ (Me=Co,Rh or Co-noble metal), *Topics in Catalysis*, 52, 2101–2107 (2009).

117. Almansa, G. and Kroon, P., Towards green iron and steel industry: Opportunities for MILENA biomass gasification technology in direct reduction of iron, ECM E-16-022 report ECN, Petten, The Netherlands (May 2016).
118. Kaila, R., Gutierrez, A., Korhonen, S., and Krause, A., Autothermal reforming of n-dodecane, toluene and their mixture on mono- and bimetallic noble metal zirconia catalysts, *Catalysis Letters*, 15 (1–2) 70–78 (May 2007).
119. Kiryanov, D., Smolikov, M., Pashkov, V., Proskura, A., Zatolokina, E., Udras, I., and Belyi, A., Modern state of catalytic reforming of petrol fractions. Experience of manufacture and industrial exploitation of reforming catalysts of the PR series, *Russian Journal of General Chemistry*, 77 (12), 2255–2264 (2007).
120. Encinar, J., Gonzalez, J., Martinez, G., Sanchez, N., and Sanguino, J., Hydrogen production by means pyrolysis and steam gasification of glycerol, in *International Conference on Renewable Energies and Power Quality (ICREPQ'11)*, Las Palmas de Gran Canaria, Spain (April 13–15, 2010).
121. Maciel, C. and Ishikura, S., Steam reforming of renewable feedstock's for the production of hydrogen, a report for diploma de especializacion en technologies de hidrogenoy piles de combustible, CIRCE, Universidad de Zaragoza, Zaragoza, Spain (2007).
122. Zhang, B., Tang, X., Li, Y., Xu, Y., and Shen, W., Hydrogen production from steam reforming of ethanol and glycerol over ceria supported metal catalysts, *International Journal of Hydrogen Energy*, 32 (13), 2367–2373 (September 2007).
123. Sadanandam, G., Sreelatha, N., Phanikrishna Sharma, M., Reddy, S., Srinivas, B., Venkateswarlu, K., Krishnudu, T., Subrahmanyam, M., and Kumar, V., Steam reforming of glycerol for hydrogen production over Ni/Al$_2$O$_3$ catalyst, *ISRN Chemical Engineering*, 2012, Article ID 591587, 10pp. (2012).
124. Hakim, L., Yaakob, Z., Ismail, M., Daud, W., and Sari, R., Hydrogen production by steam reforming of glycerol over Ni/Ce/Cu hydroxyapatite-supported catalysts, VERSITA, *Chemical Papers*, 67 (7), 703–712 (July 2013).
125. Simonetti, D., Kunkes, E., and Dumesic, J., Gas phase conversion of glycerol to synthesis gas over carbon supported platinum and platinum-rhenium catalysts, *Journal of Catalysis*, 247 (2) 298–306 (2007).
126. Reddy, R., Patel, S., Nair, S., and Suvikram, Y., Preparation of hydrogen from glycerol via steam reforming process, in *International Conference on Current Trends in Technology, NUICONE*, Nirma University, Institute of Technology, Ahmedabad, India (December 8–10, 2011).
127. Wang, H., Wang, X., Li, M., Li, S., Wang, S., and Ma, X., Thermodynamic analysis of hydrogen production from glycerol autothermal reforming, *International Journal of Hydrogen Energy*, 34, 5683–5690 (2009).
128. Adhikari, S., Fernando, S., Gwaltney, S., Filipito, S., Mark Bricka, R., Steele, P., and Haryanto, A., A thermodynamic analysis of hydrogen production by steam reforming of glycerol, *International Journal of Hydrogen Energy*, 32, 2875–2880 (2007).
129. Douette, A., Turn, S., Wang, W., and Keffer, V., Experimental investigation of hydrogen production from glycerin reforming, *Energy & Fuels*, 21, 3499–3504 (2007).
130. Adhikari, S., Fernando, S., and Haryanto, A., Hydrogen production from glycerol: An update, *Energy Conversion and Management*, 50, 2600–2604 (2009).
131. Slinn, M., Kendall, K., Mallon, C., and Andrews, J., Steam reforming of biodiesel by-product to make renewable hydrogen, *Bioresource Technology*, 99, 5851–5858 (2008).
132. Chen, H., Ding, Y., Cong, N., Dou, B., Dupont, V., Ghadiri, V., and Williams, P., A comparative study on hydrogen production from steam-glycerol reforming: Thermodynamics and experimental, *Renewable Energy*, 36, 779–788 (2011).
133. Milne, T.A., Elam, C., and Evans, R., Hydrogen from Biomass-State of the art and Research Challenges, A report for the International Energy Agency, Agreement on the production and utilization of hydrogen, Task 16, Hydrogen from carbon containing materials, IEA/H2/TR-02/001, pp. 1–78 IEA, Brussels, Belgium (2001).
134. McKinley, K.R., Browne, S.H., Neill, D.R., Seki, A., and Takahashi, P.K., Hydrogen fuel from renewable resources, *Energy Sources*, 12, 105–110 (1990).
135. Zhou, J., Ishimura, D.M., and Kinoshita, C.M., Effect of injecting steam on catalytic reforming of gasified biomass, in *Fourth Biomass Conference of the Americas*, Oakland, CA, pp. 991–997 (August 29–September 2, 1999).
136. Cox, J.L., Tonkovich, A.Y., Elliott, D.C., Baker, E.G., and Hoffman, E.J., Hydrogen from biomass: A fresh approach, in *Proceedings of the Second Biomass Conference of the Americas* (NREL/CP-200-8098, CONF-9508104), Portland, Oregon, pp. 657–675 (August 1995).
137. Weber, S.L., Sealock, L.J., Mudge, L.K., Rboertus, R.J., and Mitchell, D.H., Gasification of biomass in the presence of multiple catalysts for the direct production of specific products, in *AIAA Paper Symposium Pap—Energy from Biomass and Wastes 4*, Lake Buena Vista, FL, pp. 351–367 (January 21–25, 1980).

138. Holladay, J., Hu, J., King, D., and Wang, Y., An overview of hydrogen production technologies, *Catalysis Today*, 139 (4), 244–260 (January 30, 2009).

139. de Wild, P.J. and Verhaak, M.J.F.M., Catalytic production of hydrogen from methanol, ECN-RX-01-005, ECN, Petten, the Netherlands (January 2001).

140. Cheng, C. and Adesina, A., Kinetics of glycerol steam reforming catalyzed by bimetallic Co-Ni/Al$_2$O$_3$, Personal communication, University of New South Wales, School of Chemical Science and Engineering (2012).

141. Turn, S., Kinoshita, C., Zhang, Z., Ishimura, D., and Zhou, J., An experimental investigation of hydrogen production from biomass gasification, *International Journal of Hydrogen Energy*, 23 (8), 641–648 (1998).

142. Rapagna, F.P.U., Hydrogen from biomass by steam gasification, in *Hydrogen Energy Progress XI, Proceedings of the 11th World Hydrogen Energy Conference*, Stuttgart, Germany, Vol. 1, pp. 907–912 (June 23–28, 1996).

143. Rapagna, S. and Foscolo, P.U., Catalytic gasification of biomass to produce hydrogen rich gas, *International Journal Hydrogen Energy*, 23 (7), 551–557 (1998).

144. Encinar, J., González, J., RodrIguez, J., and Ramiro, M., Catalysed and uncatalysed steam gasification of eucalyptus char: Influence of variables and kinetic study, *Fuel*, 80, 2025–2036 (2001).

145. Encinar, J., González, J.F., and González, J., Steam gasification of *Cynara cardunculus* L.: Influence of variables, *Fuel Processing Technology*, 75, 27–43 (2002).

146. Kumar, A., Jones, D., and Hanna, M., Thermochemical biomass gasification: A review of the current status of the technology, *Energies*, 2, 556–581 (2009).

147. Timpe, R.C. and Hauserman, W.B., The catalytic gasification of hybrid poplar and common cattail plant chars, in *Energy from Biomass and Wastes XVI*, Institute of Gas Technology, Chicago, Illinois, March 2–6, 1992, pp. 903–919 (1993).

148. Kojima, T., Assavadakorn, P., and Furusawa, T., Measurement and evaluation of gasification kinetics of sawdust char with steam in an experimental fluidized bed, *Fuel Processing Technology*, 36, 201–207 (1993).

149. Capart, R. and Gélus, M., A volumetric mathematical model for steam gasification of wood char at atmospheric pressure, Energy from Biomass 4, in *Proceedings of the Third Contractors' Meeting*, May Paestum, Italy, pp. 580–583 (May 25–27, 1988).

150. Hemati, M. and Laguerie, C., Determination of the kinetics of the wood sawdust Steam-gasification of charcoal in a thermobalance, *Entropie*, 142, 29–40 (1988).

151. Moilanen, A. and Saviharju, K., Gasification reactivities of biomass fuels in pressurised conditions and product gas mixtures, in Bridyewater, A.V. and Boocock, D. (eds.), *Developments in Thermochemical Biomass Conversion*, Vol. 1/Vol. 2. Blackie Academic & Professional, London, U.K., pp. 828–837 (1997).

152. De Ruyck, J., Delattin, F., and Bram, S., Co-utilization of biomass and natural gas in combine cycles through primary steam reforming of natural gas, *Energy*, 32, 371–377 (2007).

153. Shah, Y.T. and Gardner, T., Dry reforming of hydrocarbon feedstocks, *Catalysis Reviews: Science and Engineering*, 54, 476–536 (September 2014).

154. EPA 430-R-12-001, Inventory of U.S. Green House Gas Emissions and Sinks: 1990–2012, U.S. Environmental Protection Agency, Washington, DC (2012).

155. Claridge, J.B., Green, M.L.H., and Tsang, S.C., Methane conversion to synthesis gas by partial oxidation and dry reforming over rhenium catalysts, *Catalysis Today*, 12 (2–3), 455–460 (1994).

156. Juan-Juan, J., Román-Martíez, M.C., and Illán-Gómez, M.J., Nickel catalyst activation in the carbon dioxide reforming of methane: Effect of pretreatments, *Applied Catalysis A*, 355 (1–2), 27–32 (2009).

157. Jager, B., Dry, M.E., Shingles, T., and Steynberg, A.P., Experience with a new type of reactor for Fischer-Tropsch synthesis, *Catalysis Letters*, 7 (1–4), 293–302 (1990).

158. NRC, *Advancing the Science of Climate Change*, National Research Council. The National Academies Press, Washington, DC (2010).

159. U.S. Department of State, Fourth climate action report to the UN framework convention on climate change: Projected greenhouse gas emissions, U.S. Department of State, Washington, DC (2007).

160. Sakakura, T., Choi, J.C., and Yasuda, H., Transformation of carbon dioxide, *Chemical Reviews*, 107 (6), 2365–2387 (2007).

161. Olah, G.A., Goeppert, S., and Prakash, G.K., Chemical recycling of carbon dioxide to methanol and dimethyl ether: From greenhouse gas to renewable, environmentally carbon neutral fuels and synthetic hydrocarbons, *Journal of Organic Chemistry*, 74 (2), 487–498 (2009).

162. Darensbourg, D.J., Making plastics from carbon dioxide: Salen metal complexes as catalysts for the production of polycarbonates from epoxides and CO$_2$, *Chemical Reviews*, 107 (6), 2388–2410 (2007).

163. Mio, C.X., Wang, J.Q., and He, L.N., Catalytic processes for chemical conversion of carbon dioxide into cyclic carbonates and polycarbonates, *Open Organic Chemistry Journal*, 2 (15), 68–82 (2009).

164. Aresta, M. and Forti, G., *Carbon Dioxide–A Source of Carbon, Biochemical and Chemical Use*, NATA ASI Series. Reidel Publishing Co., Dordrecht, Holland (1986).

165. Braunstein, P., Matt, D., and Nobel, D., Reactions of carbon dioxide with carbon–carbon bond formation catalyzed by transition metal complexes, *Chemical Reviews*, 88 (5), 747–764 (1988).

166. Keim, W., *Catalysis in C1 Chemistry*. D. Reidel Publishing Co., Dordrecht, Holland (1983).

167. Breslow, G.S., Hitchcock, P.B., and Laptest, M.F., A novel carbon dioxide complex: Synthesis and crystal structure of N_6 (N-$C_5H_4Me)_2(CH_2 SiMe_3$) (N_2-CO), *Journal of the Chemical Society, Chemical Communications*, 21, 1145–1146 (1981).

168. Fachinelti, G., Floriani, C., and Zanazzi, P.F., Bifunctional activation of carbon dioxide synthesis and structure of reversible CO_2 carrier, *Journal of the American Chemical Society*, 100 (23),7405–7407 (1978).

169. York, P.E., Xiao, T.C., Green, M.L.H., and Claridge, J.B., Methane oxyforming for synthesis gas production, *Catalysis Reviews*, 49 (4), 511–560 (1995).

170. Dibbern, H.C., Olesen, P., Rostrup-Nielsen, J.R., Tottrup, P.B., and Udengaard, N.R., Make low H_2/CO syngas using sulfur passivated reforming, *Hydrocarbon Processing*, 65 (1), 71–74 (1986).

171. Osaki, T., Horiuchi, T., Suzuki, K., and Mori, T., Suppression of carbon deposition in CO_2-reforming of methane on metal sulfide catalysts, *Catalysis Letters*, 35 (1–2), 39–43 (1995).

172. Roine, A., *HSC Chemistry 6.1*. Outotec Research Oy, Pori, Finland (2007).

173. Mhadeshwar, B. and Vlachos, D.G., A catalytic reaction mechanism for methane partial Decomposition and Oxidation on Platinum, *Industrial & Engineering Chemistry Research*, 46 (16), 5310–5324 (2007).

174. Martha, M., Quiroga, B., and Luna, A.E.C., Kinetic analysis of rate data for dry reforming of methane, *Industrial & Engineering Chemistry Research*, 46 (16), 5265–5270 (2007).

175. Bradford, M.C.J. and Vannice, M.A., Catalytic reforming of methane with carbon dioxide over nickel catalysts. II. Reaction kinetics, *Applied Catalysis A*, 142 (1), 97–122 (1996).

176. Bradford, M.C.J. and Vannice, M.A., CO_2 reforming of CH_4, *Catalysis Reviews—Science and Engineering*, 41 (1), 1–42 (1999).

177. Erdohelyi, A., Cserenyi, J., Papp, E., and Solymosi, F., Catalytic reaction of methane with carbon dioxide over supported palladium, *Applied Catalysis A*, 108 (2), 205–219 (1994).

178. Efstathiou, A.M., Kladi, A., Tsipouriari, V.A., and Verykios, X.E., Reforming of methane with carbon dioxide to synthesis gas over supported rhodium catalysts. II. A steady state tracing analysis: Mechanistic aspects of the carbon oxygen reaction pathways to form CO, *Journal of Catalysis*, 158 (1), 64–75 (1996).

179. O'Connor, A.M., Schuurman, Y., Ross, J.R.H., and Mirodatos, C., Transient studies of carbon dioxide reforming of methane over Pt/ZrO_2 and Pt/Al_2O_3, *Catalysis Today*, 115 (1–4), 191–196 (2006).

180. Múnera, J.F., Irusta, S., Cornaglia, L.M., Lombardo, E.A., Cesar, D.V., and Schmal, M., Kinetics and reaction pathway of the CO_2 reforming of methane on Rh supported on lanthanum-based solid, *Journal of Catalysis*, 245 (1), 25–34 (2007).

181. Nagavasu, Y., Asai, K., Nakavarma, A., Iwamoto, S., Yagasaki, E., and Inoue, M., Effect of carbon dioxide co feed on decomposition of methane over Ni catalysts, *Journal of the Japan Petroleum Institute*, 49 (4), 186–193 (2006).

182. Cui, Y., Zhang, H., Xu, H., and Li, W., Kinetic study of the catalytic reforming of CH_4 with CO_2 to syngas over Ni/α-Al_2O_3 Catalyst. The effect of temperature on the reforming mechanism, *Applied Catalysis A*, 318, 79–88 (2007).

183. Guo, J., Lou, H., and Zheng, X., The deposition of coke from methane on a Ni/$MgAl_2O_4$ catalyst, *Carbon*, 45 (6), 1314–1321 (2007).

184. Boukha, Z., Kacimi, M., Pereira, M.F.R., Faria, J.L., Figueiredo, J.L., and Ziyad, M., Methane dry reforming on Ni-loaded hydroxyapatite and fluorapatite, *Applied Catalysis A*, 317 (2), 299–309 (2007).

185. Olafsen, C., Daniel, Y., Schuarman, L., Raberg, B., Olsbye, U., and Mirodatos, C., Light alkanes CO_2 reforming to synthesis gas over Ni-based catalysts, *Catalysis Today*, 115 (1–4), 179–185 (2006).

186. Rezaei, M., Alavi, S.M., Sahebdelfar, S., and Yan, Z.F., Mesoporous nanocrystalline zirconia powders: A promising support for nickel catalyst in CH_4 reforming with CO_2, *Materials Letters*, 61 (13), 2628–2631 (2007).

187. Rezaei, M., Alavi, S.M., Sahebdelfar, S., Xinmei, L., Qian, L., and Yan, Z.F., CO_2-CH_4 Reforming over Ni catalysts supported on mesoporous nanocrystalline zirconia with high surface area, *Energy Fuels*, 21 (2), 581–589 (2007).

188. Murata, S., Hatanaka, N., Kidena, K., and Nomura, M., Improvement of lifetime of Ni/mordenite catalysts for CO_2 reforming of methane by support modification with alumina and Co-K loading, *Journal of the Japan Petroleum Institute*, 49 (5), 240–245 (2006).

189. De Lima, S.M., Assaf, J.M., Peña, M.A., and Fierro, J.L.G., Structural features of $La_{1-x}Ce_xNiO_3$ mixed oxides and performance for the dry reforming of methane, *Applied Catalysis A*, 311, 94–104 (2006).

190. Sahli, N., Petit, C., Roger, A.C., Kiennemann, A., Libs, S., and Bettahar, N.M., Ni Catalysts from $NiAl_2O_4$ spinel for CO_2 reforming of methane, *Catalysis Today*, 113 (3–4), 187–193 (2006).

191. Goncalves, G., Lenzi, M.K., Santos, O.A.A., and Jorge, L.M.M., Preparation and characterization of nickel-based catalysts on silica, alumina and titania obtained by sol-gel method, *Journal of Non-Crystalline Solids*, 352 (32–35), 3697–3704 (2006).

192. Rostrup-Nielsen, J.R., Aspects of CO_2-reforming of methane, *Studies in Surface Science and Catalysis*, 81, 25–41 (1994).

193. Kang, J.S., Kim, D.H., Lee, S.D., Hong, S.I., and Moon, D.J., Nickel-based tri-reforming catalyst for the production of synthesis gas, *Applied Catalysis A*, 332 (1), 153–158 (2007).

194. Sun, H., Wang, H., and Zhang, J., Preparation and characterization of nickel-titanium composite xerogel catalyst for CO_2 reforming of CH_4, *Applied Catalysis B*, 73 (1–2), 158–165 (2007).

195. Guo, C., Zhang, J., Li, W., Zhang, P., and Wang, Y., Partial oxidation of methane to syngas over BaTixNixO3 catalysts, *Catalysis Today*, 98 (4), 583–587 (2004).

196. Choudhary, V.R., Mondal, K.C., and Choudhary, T.V., Methane reforming over a high temperature stable–NiCoMgOx supported on zirconia-hafnia catalyst, *Chemical Engineering Journal*, 121 (2–3), 73–77 (2006).

197. Li, J., Wang, D., Zhou, G., and Cheng, T., *Chemistry Letters*, 39 (7), 692–694 (2010).

198. Zhang, K., Zhou, G., Li, J., Zhen, K., and Cheng, T., Effective additives of A (Ce, Pr) in modified hexaaluminate $La_xA_{1-x}NiAl_{11}O_{19}$ for carbon dioxide reforming of methane, *Catalysis Letters*, 130 (1–2), 246–253 (2009).

199. Zhang, K., Zhou, G., Li, J., and Cheng, T., The electronic effects of Pr on $La_{1-x}Pr_xNiAl_{11}O_{19}$ for CO_2 reforming of methane, *Catalysis Communications*, 10 (14), 1816–1820 (2009).

200. Wang, J., Meng, D., Wu, X., Hong, J., An, D., and Zhen, K., Catalytic properties of Mg-modified Ni-based hexaaluminate catalysts for CO_2 reforming of methane to synthesis gas, *Reaction Kinetics and Catalysis Letters*, 96 (1), 65–73 (2009).

201. Ikkour, K., Sellam, D., Kiennemann, A., Tezkratt, S., and Cherifi, O., Activity of Ni substituted Ca-La-hexaaluminate catalyst, *Catalysis Letters*, 132 (1–2), 213–217 (2009).

202. Xu, Z., Jia, S., Zhao, L., Ren, Y., Liu, Y., Bi, Y., and Zhen, K., Kinetics of carbon deposition on hexaaluminate $LaNiAl_{11}O_{19}$ catalyst during CO_2 reforming of methane, *Journal of Natural Gas Chemistry*, 12 (3), 189–194 (2003).

203. Liu, Y., Cheng, T., Li, D., Jiang, P., Wang, J., Li, W., Bi, Y., and Zhen, K., Studies on the stability of a $La_{0.8}Pr_{0.2}NiAl_{11}O_{19}$ catalyst for syngas production by CO_2 reforming of methane, *Catalysis Letters*, 85 (1–2), 101–108 (2003).

204. Liu, Y., Xu, Z., Cheng, T., Zhou, G., Wang, J., Li, W., Bi, Y., and Zhen, K., Studies on carbon deposition on hexaaluminate $LaNiAl_{11}O_{19}$ catalysts during CO2 reforming of methane, *Kinetics and Catalysis*, 43 (4), 522–527 (2002).

205. Xu, Z., Zhen, M., Bi, Y., and Zhen, K., Carbon dioxide reforming of methane to synthesis gas over hexaaluminate $ANiAl_{11}O_{19-\delta}$ (A = Ca, Sr, Ba and La) catalysts, *Catalysis Letters*, 64 (2–4), 157–161 (2000).

206. Xu, Z., Zhen, M., Bi, Y., and Zhen, K., Catalytic properties of Ni modified hexaaluminates $LaNi_yAl_{12-y}O_{19-\delta}$ for CO_2 reforming of methane to synthesis gas, *Applied Catalysis A*, 198 (1–2), 267–273 (2000).

207. Xu, Z., Liu, Y., Li, W., Bi, Y., and Chen, K., CO_2 reforming of methane to syngas over Ni-modified hexaaluminate oxides $BaNi_yAl_{12-y}O_{19-\delta}$, *Journal of Natural Gas Chemistry*, 9 (4), 273–282 (2000).

208. Xu, Z., Bi, Y., and Zhen, K., Preparation and catalytic properties of M-modified hexaaluminates $MNiAl_{11}O_{19-\delta}$, *Journal of Natural Gas Chemistry*, 9 (4), 132–138 (2000).

209. De Lima, S.M., Peña, M.A., Fierro, J.L.G., and Assaf, J.M., $La_{1-x}Ca_xNiO_3$ perovskite oxides: Characterization and catalytic reactivity in dry reforming of methane, *Catalysis Letters*, 124 (3–4), 195–203 (2008).

210. Khalesi, A., Arandiyan, H.R., and Parvari, M., Production of syngas by CO2 reforming on $M_xLa_{1-x}Ni_{0.3}Al_{0.7}O_{3-d}$ (M = Li, Na, K) catalysts, *Industrial & Engineering Chemistry Research*, 47 (16), 5892–5898 (2008).

211. Rivas, M.E., Fierro, J.L.G., Goldwasser, M.R., Pietri, E., Pérez-Zurita, M.J., Griboval-Constant, A., and Leclercq, G., Structural features and performance of $LaNi_{1-x}Rh_xO_3$ system for the dry reforming of methane, *Applied Catalysis A*, 344 (1–2), 10–19 (2008).

212. Valderrama, G., Goldwasser, M.R., Urbana de Navarro, C., Tatibou_t, J.M., Barrault, J., Batiot-Dapeyrat, C., and Martinez, F., Dry reforming of methane over Ni perovskite-type oxides, *Catalysis Today*, 107–108, 785–791 (2005).

213. Darujati, A.R.S. and Thomson, W.J., Stability of supported and promoted molybdenum carbide catalysts in dry-methane reforming, *Applied Catalysis A*, 296 (2), 139–147 (2005).

214. Bao, X. and Xu, Y., *Studies in Surface Science and Catalysis*, Vol. 147. Elsevier B.V., Amsterdam, the Netherlands (2004).

215. LaMont, D.C. and Thomson, W.J., The influence of mass transfer conditions on the stability of molybdenum carbide for dry methane reforming, *Applied Catalysis A*, 274 (1–2), 173–178 (2004).

216. Naito, S., Tsuji, M., Sakamoto, Y., and Miyao, T., in Gaigneaux, E. et al. (eds.), *Studies in Surface Science and Catalysis*, Vol. 143. Elsevier Science BV., Amsterdam, the Netherlands, pp. 415–423 (August 2002). e-book IBBN: 9780080540740.

217. Naito, S., Tsuji, M., and Miyao, T., Mechanistic difference of the CO_2 reforming of CH_4 over unsupported and zirconia supported molybdenum carbide catalysts., *Catalysis Today*, 77 (3), 161–165 (2002).

218. Sehested, J., Jacobsen, C.J.H., Rokni, S., and Rostrup-Nielsen, J.R., Activity and stability of molybdenum carbide as a catalyst for CO_2 reforming, *Journal of Catalysis*, 201 (2), 206–212 (2001).

219. Tsuji, M., Miyao, T., and Naito, S., Remarkable support effect of ZrO_2 upon the CO_2 reforming of CH4 over supported molybdenum carbide catalysts, *Catalysis Letters*, 69 (3–4), 195–198 (2000).

220. Brungs, A.J., York, A.P.E., Claridge, J.B., Márquez-Alvarez, C., and Green, M.L.H., Dry reforming of methane to synthesis gas over supported molybdenum carbide catalysts, *Catalysis Letters*, 70 (3–4), 117–122 (2000).

221. Brungs, A.J., York, A.P.E., and Green, M.L.H., Comparison of the group V and VI transition metal carbides for methane dry reforming and thermodynamic prediction of their relative stabilities, *Catalysis Letters*, 57 (1–2), 65–69 (1999).

222. Rodriguez, N.M., A review of catalytically grown carbon nanofibers, *Journal of Material Research*, 8 (12), 3233–3250 (1993).

223. Choi, S.O. and Moon, S.H., Performance of $La_{1-x}Ce_xFe_{0.7}Ni_{0.3}O_3$ perovskite catalysts for methane steam reforming, *Catalysis Today*, 146 (1–2), 148–153 (2009).

224. Swaan, H.M., Kroll, V.C.H., Martin, G.A., and Mirodatos, C., Deactivation of supported nickel catalysts during the reforming of methane by carbon dioxide, *Catalysis Today*, 21 (2–3), 571–578 (1994).

225. Fidalgo, B. and Menendez, J.A., Syngas production by CO_2 reforming of CH_4 under microwave heating-challenges and opportunities, Chapter 5, in *Syngas: Production, Applications and Environmental Impact*, Indarto, A. and Palgunadi, J. (eds.). Nova Science Publishers Hauppauge, New York (2011).

226. Fernandez, Y., Arenillas, A., Bermudez, J., and Menendez, J., Comparative study of conventional and microwave assisted pyrolysis, steam and dry reforming of glycerol for syngas production using a carbonaceous catalyst, *Journal of Analytical and Applied Pyrolysis*, 88 (2),155–159 (2010).

227. Rozovskii, Y. and Lin, G., *Theoretical Bases of Methanol Synthesis*. Khimia, Moscow, Russia (1900).

228. Gao, J., Hou, Z.Y., Shen, K., Lou, H., Fei, J.H., and Zheng, X.M., Autothermal reforming and partial oxidation of methane in fluidized reactor over highly dispersed Ni catalyst prepared from Ni complex, *Chinese Journal of Chemistry*, 24 (6), 721–723 (2006).

229. Puolakka, J., CO_2 reforming, Thesis for the Degree of Licentiate of Science and Technology, Helsinki University of Technology, Helsinki, Finland (2007).

230. Puolakka, K.J., Juutilainen, S., and Krause, A., Combined CO_2 reforming and partial oxidation of n-heptane on noble and zirconia catalysts, *Catalysis Today*, 115 (1–4), 217–221 (2006).

231. Halmann, M. and Steinfeld, A., Thermoneutral tri-reforming of flue gases from coal- and gas-fired power stations, *Catalysis Today*, 115, 170–178 (2006).

232. Song, C. and Pan, W., Tri-reforming of methane: A novel concept for catalytic production of industrially useful synthesis gas with desired H_2/CO ratios, *Catalysis Today*, 98, 463–484 (2004).

233. Inui, T., Rapid catalytic reforming of methane with CO_2 and its application to other reactions, *Applied Organometallic Chemistry*, 15, 87–94 (2001).

234. Halmann, M. and Steinfeld, A., Methanol, hydrogen, or ammonia production by tri-reforming of flue gases from coal- and gas-fired power stations, in *Proceedings of the ECOS2004 Conference*, Guanajuato, Mexico, pp. 1117–1128 (July 7–9, 2004).

235. Inui, T., Saigo, K., Fujii, Y., and Fujioka, K., Catalytic combustion of natural gas as the role of one-site heat supply in rapid catalytic CO_2-H_2O reforming of methane, *Catalysis Today*, 26 (3/4), 295–302 (2004).

236. Choudhary, V.R., Rajput, A.M., and Prabhakar, B., NiO/CaO catalyzed formation of syngas by coupled exothermic oxidation conversion and endothermic CO_2 and steam reforming of methane, *Angewandte Chemie*, 33 (20), 2104–2106 (1994).

237. Song, C., Tri-reforming: A new process for reducing CO_2 emission, *Chemical Innovation*, 31 (1), 21–26 (2001).

238. Choudhary, V.R., Rajput, A.M., and Prabhakar, B., *Catalysis Letters*, 32 (3–4), 391–396 (1995).

239. Hegarry, M.E.S., O'Connor, A.M., and Ross, J.R.H., *Catalysis Today*, 42 (3), 225–232 (1998).

240. O'Connor, A.M. and Ross, J.R.H., *Catalysis Today*, 46 (2–3), 203–210 (1998).

241. Huber, G.W., Cortright, R.D., and Dumesic, J.A., Renewable alkanes by aqueous phase reforming of biomass derived oxygenates, *Angewandte Chemie International Edition*, 43, 1549–1551 (2004).

242. Davda, R. and Dumesic, J., Catalytic reforming of oxygenated hydrocarbons for hydrogen with low levels of carbon monoxide, *Angewandte Chemie International Edition*, 42, 4068 (2003).

243. Alonso, D.M., Bond, J.Q., and Dumesic, J.A., Catalytic conversion of biomass to biofuels, *Green Chemistry—The Royal Society of Chemistry*, 12, 1493–1513 (2010).

244. Cortright, R., Davda, R., and Dumesic, J., Hydrogen from catalytic reforming of biomass derived hydrocarbons in liquid water, *Nature*, 418, 964–967 (August 2002).

245. Huber, G. and Dumesic, J., An overview of aqueous phase catalytic processes for production of hydrogen and alkanes in a biorefinery, *Catalysis Today*, 111 (1–2), 119–132 (2006).

246. Soares, R., Simonetti D., and Dumesic, J., *Angewandte Chemie International Edition*, 45, 3982–3985 (2006).

247. Huber, G., Shabaker, J., and Dumesic, J., Raney Ni-Sn catalyst for H_2 production from biomass derived hydrocarbons, *Science*, 300, 2075–2077 (2003).

248. Davda, R., Shabaker, J., Huber, G., Cortright, R., and Dumesic, J., A review of catalytic issues and process conditions for renewable hydrogen and alkanes by aqueous-phase reforming of oxygenated hydrocarbons over supported metal catalysts, *Applied Catalysis B*, 56, 171–186 (2005).

249. Shabaker, J., Huber, G., and Dumesic, J., Aqueous-phase reforming of oxygenated hydrocarbons over Sn-modified Ni catalysts, *Journal of Catalysis*, 222,180–191 (2004).

250. Kunkes, E., Simonetti, D., Dumesic, J., Pyrz, W., Murillo, L., Chen, J., and Buttrey, D., *Journal of Catalysis*, 260, 164–177 (2008).

251. Serrano-Ruiz, J. and Dumesic, J., Catalytic production of liquid hydrocarbon transportation fuels, Chapter 2, in Guczi, L. and Erdohelyi, A. (eds.), *Catalysis for Alternative Energy Generation*. Springer Sci + Business Media, New York (2012).

252. Shabaker, J., Davda, R., Huber, G., Cortright, R., and Dumesic, J., Aqueous phase reforming of methanol and ethylene glycol over alumina-supported platinum catalysts, *Journal of Catalysis*, 215, 344–352 (2003).

253. Shabaker, J., Huber, G., Cortright, R., and Dumesic, J., Aqueous-phase reforming of ethylene glycol over supported platinum catalysts, *Catalysis Letters*, 88, 1–8 (2003).

254. Chheda, J. and Dumesic, J., An overview of dehydration, aldol condensation, and hydrogenation processes for production of liquid alkanes from biomass-derived carbohydrates, *Catalysis Today*, 123 (1–4), 59–70 (2007).

255. Barrett, C., Chheda, J., Huber G., and Dumesic, J., *Applied Catalysis B*, 66, 111–118 (2006).

256. Serrano-Ruiz J. and Dumesic, J., *Green Chemistry*, 11, 1101–1104 (2009).

257. West, R., Braden D., and Dumesic, J., *Journal of Catalysis*, 262, 134–143 (2009).

258. Cortright, R., Davda, R., and Dumesic, J., *Nature*, 418, 964–967 (2002).

259. Huber, G., Chheda, J., Barrett, C., and Dumesic, J., Chemistry: Production of liquid alkanes by aqueous phase processing of biomass derived carbohydrates, *Science*, 308 (5727), 1446–1450 (2005).

260. Kunkes, E., Gurbuz, E., and Dumesic, J., *Journal of Catalysis*, 266, 236–249 (2009).

261. Gurbuz, E., Kunkes, J., and Dumesic, J., *Green Chemistry*, 12, 223–227 (2010).

262. Gūrbūz, E., Kunkes, E., and Dumesic, J., *Applied Catalysis B*, 94, 134–141 (2010).

263. Davda, R.R., Shabaker, J.W., Huber, G.W., Cortright, R.D., and Dumesic, J.A., *Applied Catalysis B: Environmental*, 43, 13–26 (2003).

264. West, R., Kunkes, E., Simonetti, D., and Dumesic, J., *Catalysis Today*, 147, 115–125 (2009).

265. Jiang, T., Chen, S., and Cao, F., Advances in H_2 production from the aqueous-phase reforming of biomass-derived polyols: A review [J], *Chemical Industry and Engineering Progress*, 31 (05), 1010–1017 (2012).

266. King, D., Hydrogen production via aqueous phase reforming, a paper by Pacific Northwest Laboratory, in Presented at *NIChE Catalysis Conference*, Washington, D.C. (September 21, 2011).

267. Cortright, R.D., Hydrogen generation from sugars via aqueous phase reforming, in *WHEC 16*, Lyon, France (June 13–16, 2006).

268. D'Angelo, M., Ordomsky, V., van der Schaaf, J., Schouten, J., and Nijhuis, T., Aqueous phase reforming in a microchannel reactor: The effect of mass transfer on hydrogen selectivity, *Catalysis Science Technology*, 3, 2834–2842 (2013).

269. Kunkes, E.L., Simonetti, D.A., West, R.M., Serrano-Ruiz, J.C., Gartner, C.A., and Dumesic, J.A., *Science*, 322, 417–421 (2008).

270. Huber, G., Iborra, S., and Corma, A., Synthesis of transportation fuels from biomass: Chemistry, catalysts, and engineering, *Chemical Reviews*, 106, 4044–4098 (2006).

271. King, D., Biomass derived liquids distributed (aqueous phase) reforming, Pacific Northwest National Laboratory, 2012 DOE Hydrogen and Fuel Cells Program Review (May 17, 2012).

272. Blommel, P.G. and Cortright, R.D., Production of conventional liquid fuels from sugars, A white paper for European Biofuel Technology Platform, Brussels, Belgium (2012).

273. Tanksale, A., Beltramini, J.N., and Lu, G., Aqueous phase reforming (APR) of biomass derived products for hydrogen generation using metal supported catalysts, Personal communication (2008).

274. Park, H. J., Kim, H.-D., Kim, T.-W., Jeong, K.-E., Chae, H.-J., Jeong, S.-Y., Chung, Y.-M., Park, Y.-K., and Kim, C.-U., Production of biohydrogen by aqueous phase reforming of polyols over platinum catalysts supported on three-dimensional bimodal mesoporous carbon, *ChemSusChem*, 5, 629–633 (2012).

275. D'Angelo, M. F. N., Ordomsky, V., Paunovic, V., van der Schaaf, J., Schouten, J.C., and Nijhuis, T.A., Hydrogen production through aqueous-phase reforming of ethylene glycol in a washcoated microchannel, *ChemSusChem*, 6, 1708–1716 (2013).

276. Wen, G., Xu, Y., Ma, H., Xu, Z., and Tian, Z., Production of hydrogen by aqueous phase reforming of glycerol, *International Journal of Hydrogen Energy*, 33, 6657–6666 (2008).

277. Luo, N., Fu, X., Cao, F., Xiao T., and Edwards, P., Glycerol aqueous phase reforming for hydrogen generation over Pt catalyst—Effect of catalyst composition and reaction conditions, *Fuel*, 87, 3483–3489 (2008).

278. Valenzuela, M., Jones, C., and Agrawal, P., Batch aqueous phase reforming of woody biomass, *Energy & Fuels*, 20, 1744–1752 (2006).

279. Chu, X., Liu, J., Sun, B., Dai, R., Pei, Y., Qiao, M., and Fan, K., Aqueous phase reforming of ethylene glycol on Co/ZnO catalysts prepared by the coprecipitation method, *Journal of Molecular Catalysis A, Chemical*, 335, 129–135 (2011).

280. Wen, G., Xu, Y., Xu, Z., and Tian, Z., Characterization and catalytic properties of the Ni/Al$_2$O$_3$ catalysts for aqueous phase reforming of glucose, *Catalysis Letters*, 129, 250–257 (2009).

281. Tanksale, A., Wong, Y., Beltramini, J.N., and Lu, G.Q., Hydrogen generation from liquid phase catalytic reforming of sugar solutions using metal-supported catalysts, *International Journal of Hydrogen Energy*, 32, 717 (2007).

282. Manfro, R., Pires, T., Ribeiro, N., and Souza, M., Aqueous phase reforming of glycerol using Ni-Cu catalysts prepared from hydrotalcite like precursors, *Catalysis Science and Technology*, 3, 1278–1287 (2013).

283. Tungal, R. and Shende, R., Subcritical aqueous phase reforming of wastepaper for biocrude and H$_2$ generation, *Energy & Fuels*, 27 (6), 3194–3203 (2013).

284. Manfro, R., de Costa, A., Ribeiro, N., and Souza, M., Hydrogen production by aqueous phase reforming of glycerol over nickel catalysts supported on CeO$_2$, *Fuel Processing Technology*, 92, 330–335 (2013).

285. Vaidya, P. and Rodrigues, A., Glycerol reforming for hydrogen production: A review, *Chemical Engineering & Technology*, 32, 1463–1469 (2009).

286. Cruz, I., Ribeiro, N., Aranda, D., and Souza, M., Hydrogen production by aqueous-phase reforming of ethanol over nickel catalysts prepared from hydrotalcite precursors, *Catalysis Communications*, 9, 2606–2611 (2008).

287. Behr, A., Eilting, J., Irawadi, K., Leschinski, J., and Lindner, F., Improved utilization of renewable resources: New important derivatives of glycerol, *Green Chemistry*, 10, 13–30 (2008).

288. Iriondo, A., Barrio, V., Cambra, J., Arias, P., Güemez, M., Navarro, R., Sanchez-Sanchez, M., and Fierro, J., Hydrogen production from glycerol over nickel catalysts supported on Al$_2$O$_{27,42,64}$ modified by Mg, Zr, Ce or La, *Topics in Catalysis*, 49, 46–58 (2008).

289. Cho, S.H. and Moon, D.J., Aqueous phase reforming of glycerol over Ni-based catalysts for hydrogen production, *Journal of Nanoscience and Nanotechnology*, 11 (8), 7311–7314 (August 2011).

290. Tuza, P., Manfro, R., Ribeiro, N., and Souza, M., Production of renewable hydrogen by aqueous-phase reforming of glycerol over Ni–Cu catalysts derived from hydrotalcite precursors, *Renewable Energy*, 50, 408–414 (February 2013).

291. Holmgren, J. and Arena, B., Solid base as catalysis in Aldol condensation, U.S. Patent, 5,254,743 (1993).

292. King, F., Kelly, G., and Stitt, E., Improved base catalysts for industrial condensation reactions, *Studies in Surface Science and Catalysis*, 145, 443–446 (2003).

293. Dooley, K., Bhat, A., Plaisance, C., and Roy, A., Ketones from acid condensation using supported CeO$_2$ catalysts: Effect of additives, *Applied Catalysis A: General*, 320, 122–133 (2007).

294. Blommel, P., Keenan, G., Rozmiarek, R., and Cortright, R., *International Sugar Journal*, 110, 672 (2008).
295. Huber, G., Cortright, R., and Dumesic, J., *Angewandte Chemie International Edition*, 43, 1549–1551 (2004).
296. Held, A., Catalytic conversion of renewable plant sugars to fungible liquid hydrocarbon fuels using Bioforming process TAPPI, IBCC Session 3, Conversion Pathways, Memphis, TN (October 15, 2009).
297. Blommel, P., Catalytic conversion of carbohydrates to hydrocarbons, DOE Biomass R&D TAC meeting, DOE, Arlington, VA (May 19, 2011).
298. Held, A., Production of renewable aromatic chemicals using Virent's catalytic bioforming process, Frontiers in Biorefining, St. Simons Island, GA (October 19–22, 2010).
299. Cortright, R.D., Bioforming process—Production of conventional liquid fuels from sugar, in *ACS/EPA Green Chemistry Conference* (June 23, 2009).
300. Cortright, R.D. and Blommel, P.G., Synthesis of liquid fuels and chemicals from oxygenated hydrocarbons, U.S. Application Number 12/044,908, Anticipated Publication date of December 4, 2008.
301. Cortright, R.D. and Blommel, P.G., Synthesis of liquid fuels and chemicals from oxygenated hydrocarbons, U.S. Application Number 12/044,876, Anticipated Publication date of December 4, 2008.
302. Yakaboylu, O., Harinck, J., Smit, K.G., and de Jong, W., Supercritical water gasification of biomass: A literature and technology overview, *Energies*, 8, 859–894 (2015).
303. Lee, S., Lanterman, H., Wenzel, J., and Picou, J., Noncatalytic reformation of JP-8 fuel in supercritical water for production of hydrogen, *Energy Sources, Part A: Recovery, Utilization and Environmental*, 31 (19), 1750–1758 (January 2009).
304. Veriansyah, B. and Kim, J.-D., Supercritical water oxidation for the destruction of toxic organic waste-waters: A review, *Journal of Environmental Sciences*, 19 (5), 513–522 (2007).
305. Veriansyah, B., Kim, J.-D., and Lee, J.-C., Destruction of chemical agent simulants in a supercritical water oxidation bench-scale reactor, *Journal of Hazardous Materials*, 147 (1–2), 8–14 (2007).
306. Veriansyah, B., Kim, J.-D., and Lee, J.-C., A double wall reactor for supercritical water oxidation: Experimental results on corrosive sulfur mustard simulant oxidation, *Journal of Industrial and Engineering Chemistry*, 15 (2), 153–156 (2009).
307. Veriansyah, B., Kim, J.-D., Lee, J.-C., and Lee, Y.-W., OPA oxidation rates in supercritical water, *Journal of Hazardous Materials*, 124 (1–3), 119–124 (2005).
308. May, A., Salvado, J., Torras, C., and Montane, D., Catalytic gasification of glycerol in supercritical water, *Chemical Engineering Journal*, 160 (2), 751–759 (June 2010).
309. Pinkwart, K., Bayha, T., Lutter, W., and Krausa, M., Gasification of diesel oil in supercritical water for fuel cells, *Journal of Power Sources*, 136, 211–214 (2004).
310. Xu, X. and Antal, M., Kinetics and mechanism of isobutene formation from t-butanol in hot liquid water, *AIChE Journal*, 40 (9), 1524–1531 (1994).
311. Xu, X., Antal, M., and Anderson, D., Mechanism and temperature dependent kinetics of the dehydration of tert-butyl alcohol in hot compressed liquid water, *Industrial & Engineering Chemistry Research*, 36 (1), 23–41 (1997).
312. Xu, X.D. et al., Carbon-catalyzed gasification of organic feedstocks in supercritical water, *Industrial & Engineering Chemistry Research*, 35, 2522–2530 (1996).
313. Kruse, A., Supercritical water gasification, *Biofuels, Bioproducts, and Biorefining*, 2, 415–437 (September/October 2008).
314. Zhang, L., Champagne, P., and Xu, C., Supercritical water gasification of an aqueous by-product from biomass hydrothermal liquefaction with novel Ru modified Ni catalysts, *Bioresource Technology*, 102 (17), 8279–8287 (September 2011).
315. Knoef, H., *Handbook Biomass Gasification*. Biomass Technology Group Press, Enschede, the Netherlands, pp. 22–23 (2005).
316. Kruse, A. and Gawlik, A., Biomass conversion in water at 330°C–410°C and 30–50 MPa identification of key compounds for indicating different chemical reaction pathways, *Industrial & Engineering Chemistry Research*, 42, 267–279 (2003).
317. Kruse, A., and Henningsen, T., Biomass gasification in supercritical water: Influence of the dry matter content and the formation of phenols, *Industrial & Engineering Chemistry Research*, 42, 3711–3717 (2003).
318. Antal, M., Allen, S., Schulman, D., and Xu, X., Biomass gasification in supercritical water, *Industrial & Engineering Chemistry Research*, 39, 4040–4053 (2000).
319. Vos, J., Reforming of crude glycerine in supercritical water to produce methanol for re-use in biodiesel plants, A report by B.T.G. Biotechnology Group BV, Enschede, the Netherlands (2007).

320. Wenzel, J., The kinetics of non-catalyzed supercritical water reforming of ethanol, PhD thesis, University of Missouri, Columbia, MO (May 2008).
321. Boukis, N., Diem, V., Habicht, W., and Dinjus, E., Methanol reforming in supercritical water, *Industrial & Engineering Chemistry Research*, 42, 728–735 (2003).
322. Taylor, J., Herdman, C., Wu, B., Walley, K., and Rice, S., Hydrogen production in a compact supercritical water reformer, *International Journal of Hydrogen Energy*, 28, 1171–1178 (2003).
323. Gadhe, J. and Gupta, R., Hydrogen production by methanol reforming in supercritical water: Suppression of methane formation, *Industrial & Engineering Chemistry Research*, 44, 4577–4585 (2005).
324. Guerrero-Perez, M., Rosas, J., Bedia, J., Rodriguez-Mirosol, J., and Cordero, T., Recent inventions in glycerol transformations and processing, Recent patents in *Chemical Engineering*, 2, 11–21 (2009).
325. Huidong, W. and Renan, W., Conversion and reforming of fossil fuel by supercritical water, *Chemical Industry and Engineering Progress*, 03, 1–7 (1999).
326. Ramaswamy, K. and T-Raissi, A., Hydrogen production from used lube oil via supercritical water reformation, A report from Florida Solar Energy Center, Cocoa, FL (2010).
327. De Vlieger, D., Chakinala, A., Lefferts, L., Kersten, S., Seshan, K., and Brilman, D., Hydrogen from ethylene glycol by supercritical water reforming using noble and base metal catalysts, *Applied Catalysis B: Environmental*, 111–112, 536–544 (2012).
328. Yamaguchi, A., Hiyoshi, N., Sato, O., Bando, K., Osada, M., and Shirai, M., Hydrogen production from woody biomass over supported metal catalysts in supercritical water, *Catalysis Today*, 146 (1–2), 192–195 (August 2009).
329. Byrd, A., Pant, K., and Gupta, R., Hydrogen production from glycerol by reforming in supercritical water over Ru/Al$_2$O$_3$ catalyst, *Fuel*, 87, 2956–2960 (2008).
330. Penninger, J., Maass, G., and Rep, M., Compressed hydrogen rich fuel gas (CHFG) from wet biomass by reforming in supercritical water, *International Journal of Hydrogen Energy*, 32 (10/11), 1472–1476 (July 2007).
331. Penninger, J. and Rep, M., Reforming of aqueous wood pyrolysis condensate in supercritical water, *International Journal of Hydrogen Energy*, 31 (11), 1597–1606 (September 2006).
332. Therdthianwong, S., Srisinwat, N., Therdthianwong, A., and Croiset, E., Reforming of bioethanol over Ni/Al$_2$O$_3$ and Ni/CeZrO catalysts in supercritical water for hydrogen production, *International Journal of Hydrogen Energy*, 36 (4), 2877–2886 (February 2011).
333. Shekhawat, D., Berry, D., Gardner, T., and Spivey, J., Catalytic reforming of liquid hydrocarbon fuels for fuel cell applications, *Catalysis*, 19, 184 (2006).
334. Antal, M.J. Jr., Manarungson, S., and Mok, W.S.-L., Hydrogen production by steam reforming glucose in supercritical water, in Bridyewater, A.V. (ed.), *Advances in Thermochemical Biomass Conversion*. Blackie Academic and Professional, London, U.K., pp. 1367–1377 (1994).
335. Byrd, A., Pant, K., and Gupta, R., Hydrogen production from ethanol by reforming in supercritical water using Ru/Ai$_2$O$_3$ catalyst, *Energy & Fuels*, 21 (6), 3541–3547 (2007).
336. Van Bennekom, J., Venderbosch, R., and Heeres, H., Supermethanol: Reforming of crude glycerine in supercritical water to produce methanol for re-use in biodiesel plants, A report by University of Groningen, Groningen, the Netherlands (2011).
337. Menéndez, J.A., Arenillas, A., Fidalgo, B., Fernández, Y., Zubizarreta, L., Calvo, E.G., and Bermúdez, J.M., *Fuel Processing Technology*, 91, 1 (2010).
338. Will, H., Scholz, P., and Ondruschka, B., *Chemical Engineering & Technology*, 27, 113 (2004).
339. Jones, D.A., Lelyveld, T.P., Mavrofidis, S.D., Kingman, S.W., and Miles, N.J., *Resources, Conservation and Recycling*, 34, 75 (2002).
340. Haque, K.E., *International Journal of Mineral Processing*, 57, 1–24 (1999).
341. Thostenson, E.T. and Chou, T.-W., *Composites Part A*, 30, 1055 (1999).
342. Jacob, J., Chia, L.H.L., and Boey, F.Y.C., *Journal of Material Science*, 30, 5321 (1995).
343. Will, H., Scholz, P., and Ondruschka, B., *Topics in Catalysis*, 29, 175 (2004).
344. Zhang, X. and Hayward, D.O., *Inorganica Chimica Acta*, 359, 3421 (2006).
345. Zhang, X., Lee, C.S.M., Mingos, D.M.P., and Hayward, D.O., *Catalysis Letters*, 88, 129 (2003).
346. Fidalgo, B., Domínguez, A., Pis, J.J., and Menéndez, J.A., *International Journal of Hydrogen Energy*, 33, 4337 (2008).
347. Domínguez, A., Fidalgo, B., Fernández, Y., Pis, J.J., and Menéndez, J.A., *International Journal of Hydrogen Energy*, 32, 4792 (2007).
348. Menéndez, J.A., Juárez-Pérez, E.J., Ruisánchez, E., Bermúdez, J.M., and Arenillas, A., *Carbon*, 49, 346 (2011).

349. Bermúdez, J., Arenillas, A., and Menéndez, J.A., Equilibrium prediction of CO_2 reforming of coke oven gas. Suitability for methanol production, *Chemical Engineering Science*, 82, 93–103 (2012).

350. Bermúdez, J.M., Fidalgo, B., Arenillas, A., and Menéndez, J.A., Dry reforming of coke oven gases over activated carbon to produce syngas for methanol synthesis, *Fuel*, 89, 2897–2902 (2010).

351. Bermúdez, J.M., Arenillas, A., and Menéndez, J.A., *International Journal of Hydrogen Energy*, 36, 13361–13368 (2011).

352. Bermúdez, J.M., Arenillasa, A., Luqueb, R., and Menéndeza, J.A., An overview of novel technologies to valorise coke oven gas surplus, *Fuel Processing Technology*, 110, 150–159 (June 2013).

353. Durka, T., Microwave effects in heterogeneous catalysis: Application to gas-solid reactions for hydrogen production, MS thesis, Department of Chemical Engineering, Warsaw Institute of Technology, Geboren Te Warsaw, Poland (2013).

354. Fidalgo, B., PhD thesis, Universidad de Oviedo, Oviedo, Spain (2010). Downloadable from: http://hdl.handle.net/10261/27755.

355. Domínguez, A., Fernández, Y., Fidalgo, B., Pis, J.J., and Menéndez, J.A., *Energy & Fuels*, 21, 2066 (2007).

356. Spiewak, I., Tyner, C., and Langnickel, U., Applications of solar reforming technology, SANDIA Report SAND93-1959, UC-237, Sandia National Laboratories, Albuquerque, NM (November 1993).

357. Muller, W.D., Solar reforming of methane utilizing solar heat, in Baker, M. (ed.), *Solar Thermal Utilization*, Solar Thermal Energy for Chemical Processes, German Studies on Technology and Applications, Vol. 3. Springer-Verlag, Berlin, Germany, pp. 1–179 (1987).

358. Meier, A. and Sattler, C., Solar fuels from concentrated sunlight, Solar Paces, IEA Solar Paces Implementing Agreement, Platforma Solar de Almeria, Almería, Spain (August 2009).

359. Moller, S., Solar reforming of natural gas, A report by Deutsches Zentrum, DLR fur Luft-und Raumfahrt e.V., Mgheb-Europ Project, Lyon, France (June 14, 2006).

360. Belghit, A. and El Issami, S., Hydrogen production by steam gasification of coal in gas–solid moving bed using nuclear heat, *Energy Conversion and Management*, 42 (1), 81 (2001).

361. Cypres, R., Modern carbochemical processes for hydrogen production from coal, *International Journal of Hydrogen Energy*, 12 (7), 451–460 (1987).

362. Bijetima, R. and Tarman, P.B., Development states of the steam-iron process for hydrogen production, *Alternative Energy Sources*, 2, 3335–3347 (1981).

363. Das, A., Chatterjae, D.S., and Mukherjee, P.N., Steam-iron process for production of hydrogen, *Indian Journal of Technology*, 15, 339–341 (1977).

364. Fankhauser, R., Hacker, V., Spreitz, B., Kleindienst, K., Fuchs, H., Rabel, A., Friedrich, K., and Faleschini, G., First operation results of a small scale sponge iron reactor (SIR) for hydrogen production, in *Hydrogen Energy Progress XII, Proceedings of the 12th World Hydrogen Energy Conference*, Buenos Aires, Argentina, 12 WHEC, Vol. 1, pp. 551–553 (June 21–26, 1998).

365. Friedrich, K., Kordesch, K., Simader, G., and Selan M., The process cycle sponge iron/hydrogen/iron oxide used for fuel conditioning in fuel cells, in *New Materials, Fuel Cell System, Proceedings International Symposium*, Montreal, Canada, pp. 239–248 (July 9–13, 1995).

366. Hacker, V., Fuchs, H., Fankhauser, R., Spreitz, B., Friedrich, K., and Faleschini, G., Hydrogen production from gasified biomass by sponge iron reactor (SIR), in *Hydrogen Energy Progress XII, Proceedings of the 12th World Hydrogen Energy Conference*, Buenos Aires, Argentina, 12 WHEC, Vol. 1, pp. 543–550 (June 21–26, 1998).

367. Hacker, V., Faleschini, G., Fuchs, H., Fankhauser, R., Simader, G., Ghaemi, M., Spreitz, B., and Friedrich, J., Usage of biomass gas for fuel cells by the SIR process, *Journal of Power Sources*, 71, 226–230 (1998).

368. Hacker, V., Fankhauser, R., Faleschini, G., Fuchs, H., Friedrich, J., Muhr, M., and Kordesch, K., Hydrogen production by steam-iron process, *Journal of Power Sources*, 86, 531–535 (2000).

369. Jannach, G., Krammer, G., Staudinger, G., and Friedrich, K., Hydrogen from biomass gasification for fuel cell systems by application of the sponge iron/iron oxide process cycle, in *Proceedings of the Second International Symposium on New Materials for Fuel Cell and Modern Battery Systems*, Montreal, Quebec, Canada, pp. 42–53 (July 6–10, 1997).

370. Biollaz, S., Sturzenegger, M., and Stucki, S., Redox process for the production of clean hydrogen from biomass, in *Proceeding of Thermochemical Biomass Conversion Abstracts*, Tyrol, Austria (September 17–22, 2000).

371. Zhou, Z.P., Zhang, J.M., Ye, T.H., Zhao, P., and Xia, W., Hydrogen production by reforming methane in a corona inducing dielectric barrier discharge and catalyst hybrid reactor, *Chinese Science Bulletin*, 56, 2162–2166 (2011).

372. Bromberg, L., Cohn, D.R., Rabinovich, A., O'Brien, C., and Hochgreb, S., Plasma reforming of methane, *Energy & Fuels*, 12, 11–18 (1998).

373. Hammer, Th., Kappes, Th., and Baldauf, M., Plasma catalytic hybrid processes: Gas discharge initiation and plasma activation of catalytic processes, *Catalysis Today*, 89, 5–14 (2004).

374. Aziznia, A., Bozorgzadehb, H.R., Seyed-Matin, N., Baghalha, M., and Mohamadalizadeh, A., Comparison of dry reforming of methane in low temperature hybrid plasma-catalytic corona with thermal catalytic reactor over Ni/γ-Al$_2$O$_3$, *Journal of Natural Gas Chemistry*, 21 (4), 466–475 (July 2012).

375. Chernyak, V., Olshewskii, S.V., Yukhymenko, V.V., Shchedrin, A.I., Levko, D.S., Naumov, V.V., Nedybalyuk, O.A., Sidoruk, S.M., Demchina, V.P., and Kudryavzev, V.S., Plasma reforming of liquid hydrocarbons into free hydrogen for the use in the aerospace technologies, in A paper presented at *Aero-Ukraine*, Kyiv, Ukraine (October 28–29, 2009).

376. Wnukowski1, M., Decomposition of tars in microwave plasma—Preliminary results, *Journal of Ecological Engineering*, 15 (3), 23–28 (July 2014).

377. Nozaki, T., Tsukijihara, H., Fukui, W., and Okazaki, K., Kinetic analysis of the catalyst and nonthermal plasma hybrid reaction for methane steam reforming, *Energy & Fuels*, 21 (5), 2525–2530 (2007).

378. Thanompongchart, P. and Tippayawong, N., Experimental investigation of biogas reforming in gliding arc plasma reactors, *International Journal of Chemical Engineering*, 2014, Article ID 609836, 9pp. (2014).

379. Tao, X., Bai, M., Li, X., Long, H., Shang, S., Yin, Y., and Dai, X., CH$_4$-CO$_2$ reforming by plasma—Challenges and opportunities, *Progress in Energy and Combustion Science*, 37, 113–124 (2011).

380. Spencer, L., The study of CO$_2$ conversion in a microwave plasma/catalyst system, PhD thesis, University of Michigan, Ann Arbor, MI (2012).

381. Gallon, H., Dry reforming of methane using non-thermal plasma catalysis, PhD thesis, University of Manchester, Manchester, U.K. (2010).

382. Plasmatrons, Wikipedia, the free encyclopedia (2015). https://en.wikipedia.org/wiki/Plasmatrons (last accessed October 17, 2016).

383. Cohn, D., Cleaner higher efficiency vehicles using Plasmatrons, Plasma Science and Fusion center, MIT, in *Presentation to Fusion Power Associates Meeting*, Washington, DC (November 21, 2003).

384. Bromberg, L., Cohn, D.R., Rabinovich, A., Surma, J.E., and Virden, J., Compact plasmatron boosted hydrogen generation technology for vehicular applications, *International Journal of Hydrogen Energy*, 24, 341–350 (1999).

385. Cohn, D.R., Rabinovich, A., Bromberg, L., Surma, J.E., and Virden, J., Onboard plasmatron reforming of biofuels, gasoline and diesel fuel, in Presented at the *Future Transportation Technology Conference & Exposition*, Costa Mesa, CA (1998), SAE-981920.

386. Perry, W.L., Katz, J.D., Rees, D., Paffet, M.T., and Datye, A.K., Kinetics of the microwave-heated CO oxidation reaction over alumina-supported Pd and Pt catalysts, *Journal of Catalysis*, 171, 431 (1997).

387. Wan, J.K.S., Chen, Y.G., Lee, Y.J., and Depew, M.C., Highly effective methane conversion to hydrocarbons by means of microwave and rf-induced catalysis, *Research on Chemical Intermediates*, 26, 599 (2000).

388. Cha, C.Y. and Kim, D.S., Microwave induced reactions of sulfur dioxide and nitrogen oxides in the char and anthracite bed, *Carbon*, 39, 1159 (2001).

389. Zhang, X., Hayward, D.O., and Mingos, D.M.P., *Chemical Communications*, 11, 975 (1999).

390. Haque, K.E., Microwave energy for mineral treatment processes—A review, *International Journal of Mineral Processing*, 57, 1–24 (1999).

391. Klein, H.H., Karni, J., and Rubin, R., Dry methane reforming without a metal catalyst in a directly irradiated solar particle reactor, *Journal of Solar Energy Engineering*, 131 (2), 021001 (2009).

392. Sattler, C. and Raeder, C., SOLREF-solar steam reforming of methane rich gas for synthesis gas production final activity report, DLR, Cologne, Germany (June 2010).

393. Suarez-Gonzalez, M., Blanco-Marigorta, A., and Peria-Quintana, A., Review on hydrogen production technologies from solar energy, in *International Conference on Renewable Energies and Power Quality ICREPQ'11*, Los Palmas de Gran Canaria, Spain (April 13–15, 2011).

394. Olsson, D., Comparison of reforming process between different types of biogas reforming reactors, Project report, 2008 MVK 160 Heat and Mass Transport, Lund, Sweden (May 2008).

395. Ogden, J.M., Review of small stationary reformers for hydrogen production, A report for IEA, Agreement on the production and utilization of hydrogen, Task 16, Hydrogen from carbon containing materials, IEA/H2/TR-02/002 (2002).

396. Becker, M. and Bohmer, M. (eds.), *Proceedings of Final Presentation GAST, the Gas Cooled Solar Tower Technology Program*, May 30–31, 1988, Lahnstein, Germany, FRG, Springer Verlag, Berlin, Germany (1989).

397. Piatkowski, N. and Steinfeld, A., Solar driven coal gasification in a thermally irradiated packed-bed reactor, *Energy & Fuels*, 22, 2043–2052 (2008).

398. Z'Graggen, Solar gasification of carbonaceous materials—Reactor design, modeling and experimentation, Doctorate thesis, Dissertation, ETH No. 17741, ETH, Zurich, Switzerland (2008).

399. Petrasch, J. and Steinfeld, A., Dynamics of a solar thermochemical reactor for steam reforming of methane, *Chemical Engineering Science*, 62, 4214–4228 (2007).

400. Piatkowski, N., Wieckert, C., and Steinfeld, A., Experimental investigation of a packed bed solar reactor for the steam gasification of carbonaceous feedstocks, *Fuel processing Technology*, 90, 360–366 (2009).

401. Z'graggen, A., Haueter, P., Trommer, D., Romero, M., De Jesus, J., and Steinfeld, A., Hydrogen production by steam-gasification of petroleum coke using concentrated solar power—II Reactor design, testing, and modeling, *International Journal of Hydrogen Energy*, 31 (6), 797–811 (2006).

402. Gordillo, E. and Belghit, A., A bubbling fluidized bed solar reactor model of biomass char high temperature steam-only gasification, *Fuel Processing Technology*, 92 (3), 314–321 (2010).

403. Watanuki, K., Nakajima, H., Hasegawa, N., Kaneko, H., and Tamaura, Y., Methane-steam reforming by molten salt membrane reactor using concentrated solar thermal energy, in *WHEC 16*, Lyon, France, pp. 13–16 (June 2006).

404. Fercher, E., Hofbauer, H., Fleck, T., Rauch, R., and Veronik, G., Two years experience with the FICFB-gasification process, in *10th European Conference and Technology Exhibition*, Würzburg, Germany (June 1998).

405. Gil, J., Aznar, M.P., Caballero, M.A., Francés, E., and Corella, J., Biomass gasification in fluidized bed at pilot scale with steam-oxygen mixtures. Product distribution for very different operating conditions, *Energy & Fuels*, 11, 1109–1118 (1997).

406. Bengtsson, S., Final activity report on CHRISGAS fuels from biomass, Clean hydrogen rich synthesis gas, 502587, 1st September 2004–28th February 2010. Reports available through the CHRISGAS website. http://www.chrisgas.com; Linnaeus University, Vaxjo, Sweden (September 2010).

407. Yeheskel, J., Rubin, R., Berman, A., and Karni, J., Chemical kinetics of high temperature hydrocarbons reforming using a solar reactor, Solar Research Facilities, Weizmann Institute of Science, Israel, Personal communication (2012).

7 Gas Processing, Purification, and Upgrading

7.1 INTRODUCTION

Both natural and synthetic gases carry impurities that must be removed before their downstream processing for heat, power, or liquid fuels. The nature of impurity depends on the source of natural and synthetic gas and its level of removal depends on the nature of the downstream applications. The removal of impurities are important for gas storage and transport as well. One big difference between impurities associated with natural gas and that from thermal gasification processes is the level of temperature of the gas. Natural gas is recovered at low temperatures, while synthesis gas produced by thermochemical processes is at higher temperatures. As shown in Section 7.10, more efforts are being made to remove impurities at higher temperatures from syngas for its better thermal integration with the downstream operations [1–5].

The natural gas coming out of wells contains a number of impurities like water, CO_2, H_2S, hydrocarbons, oils, helium, mercury, and radon that must be removed before it can be transported via natural gas pipeline and used for several downstream applications [1–44]. Similarly, thermal gasification of coal or biomass/waste yields a product gas composed of carbon monoxide, carbon dioxide, hydrogen, water, methane, and other volatile hydrocarbons. The gasification also produces particulate matter, tar, char, ash, and impurities such as HCN, NH_3, HCl, H_2S, COS, alkaline, and heavy metals. These impurities must be removed to a low level before syngas can be stored, transported, and used for numerous downstream applications. This chapter focuses on the available technologies to remove these impurities from natural and synthetic gas [1–5].

7.2 NATURAL GAS PROCESSING

More than 500 natural gas processing plants currently operate in the United States [1–44]. Most are located in proximity to the major gas-/oil-producing areas of the Southwest and the Rocky Mountain States. Not surprisingly, more than half of the current natural gas processing plant capacity in the United States is located convenient to the Federal Offshore, Texas, and Louisiana. Four of the largest capacity natural gas processing/treatment plants are found in Louisiana, while the greatest number of individual natural gas plants is located in Texas. Although Texas and Louisiana still account for the larger portion of the U.S. natural gas plant processing capability, other states have moved up in the rankings somewhat during the past 10 years as new trends in natural gas production and processing have come into play [1–6,10].

Over the past 10 years, average plant capacity increased from 76 million cubic feet (mcf) per day (mcf/d) to 114 mcf/d and decreased in only 4 of the 22 states with natural gas processing plant capacity. In Texas, although the number of plants and overall processing capacity decreased, the average capacity per plant increased from 66 to 95 mcf/d as newer plants were added and old, less-efficient plants were idled. In Alabama, Mississippi, and the eastern portion of South Louisiana, new larger plants and plant expansions built to serve new offshore production increased the average plant capacity significantly in those areas [1–6,10].

Since 1995, Wyoming's natural gas plant processing capacity increased by almost 46%, adding more than 2.2 billion cubic feet (bcf)/d. Much of the activity has been focused in the southwestern area of Wyoming's Green River Basin where one of the nation's largest gas plants, the Williams

Companies 1.1 bcf/d Opal facility, is located. In 2004, approximately 24.2 trillion cubic feet (tcf) of raw natural gas was produced at the wellhead [1–6,10,45]. A small portion of that, 0.1 tcf, was vented or flared, while a larger portion, 3.7 tcf, was reinjected into reservoirs (mostly in Alaska) to maintain pressure. The remaining 20.4 tcf of "wet" natural gas was converted into the 18.9 tcf of dry natural gas that was put into the pipeline system. All these numbers (i.e., number of plants and their capacities as well as number of states involve) are rapidly changing with increased productions of unconventional gas, in particular shale gas [1–6,10,45].

Natural gas processing begins at the wellhead. The composition of the raw natural gas extracted from producing wells depends on the type, depth, and location of the underground deposit and the geology of the area. Oil and natural gas are often found together in the same reservoir. The natural gas produced from oil wells is generally classified as "associated–dissolved," meaning that the natural gas is associated with or dissolved in crude oil. Natural gas production absent from any association with crude oil is classified as "nonassociated." In 2004, 75% of U.S. wellhead production of natural gas was nonassociated. Nonassociated gas wells typically produce only raw natural gas, while condensate wells produce raw natural gas along with other low-molecular-weight hydrocarbons. Those that are liquid at ambient conditions (i.e., pentane and heavier) are called "natural gas condensate" (sometimes also called "natural gasoline" or simply "condensate"). Thus, raw natural gas comes primarily from any one of the three types of wells: crude oil wells, gas wells, and condensate wells [1–6,10,11,17].

Natural gas is called "sweet gas" when relatively free of hydrogen sulfide; however, gas that does contain hydrogen sulfide is called "sour gas." Natural gas, or any other gas mixture, containing significant quantities of hydrogen sulfide, carbon dioxide, or similar acidic gases, is called "acid gas." Raw natural gas can also come from methane deposits in the pores of coal seams and especially in a more concentrated state of adsorption onto the surface of the coal itself. Such gas is referred to as "coalbed gas" or "coalbed methane." Coalbed gas has become an important source of energy in recent decades. The natural gas extracted from coal reservoirs and mines (coalbed methane) is essentially a mix of mostly methane and carbon dioxide (about 10%). In recent years, natural gas is also recovered from shale deposits and this is called shale gas. Shale gas only contains about 55%–60% methane, and the remaining gases are carbon dioxide (30%–35%) and some volatile hydrocarbons such as ethane, propane, butane, ethylene, and propylene [1–10,35,45].

Most natural gas production contains, to varying degrees, small (from two to eight carbons) hydrocarbon molecules in addition to methane. Although they exist in a gaseous state at underground pressures, these molecules will become liquid (condense) at normal atmospheric pressure. As mentioned before, collectively, they are called condensates or natural gas liquids (NGLs). Natural gas production from the deepwater Gulf of Mexico and conventional natural gas sources of the Rocky Mountain area is generally rich in NGLs and must be processed to meet pipeline-quality specifications. Deepwater natural gas production can contain in excess of 4 gal of NGLs per thousand cubic feet (ft^3) of natural gas compared with 1–1.5 gal of NGLs per ft^3 of natural gas produced from the continental shelf areas of the Gulf of Mexico. Natural gas produced along the Texas Gulf Coast typically contains 2–3 gal of NGLs per ft^3 [1–6,10,11,17].

The processing of wellhead natural gas into pipeline-quality dry natural gas can be quite complex and usually involves several processes to remove/recover:

1. Oil
2. Water
3. Sulfur and carbon dioxide (acid gases)
4. Nitrogen
5. Helium
6. NGLs
7. Components of NGLs

In addition to these seven processes, it is often necessary to install scrubbers and heaters at or near the wellhead. The scrubbers primarily remove sand and other large-particle impurities and the heaters ensure that the temperature of the natural gas does not drop too low and form a hydrate with the water vapor content of the gas stream. These hydrates can impede the passage of natural gas through valves and pipes. Several wells are connected to downstream gas processing facilities by small-diameter gathering pipes. When natural gas is produced of pipeline quality at wellhead or field facility, it is directly passed on to natural gas pipeline grid. In some cases, particularly for nonassociated gas, gas is dehydrated and decontaminated by custom-made "skid-mounted plants" tailored for the local gas. Acceptable pipeline-quality gas is then passed on to the pipeline gas grid. Non-pipeline-quality production is piped to natural gas processing plants [1–6,10,11,17,35].

A natural gas processing plant typically receives gas from a gathering system and sends out processed gas via an output (tailgate) lateral that is interconnected to one or more major intra- and interstate pipeline networks. Liquids removed at the processing plant usually will be taken away by pipeline to petrochemical plants, refineries, and other consumers of gas liquids. Some of the heavier liquids are often temporarily stored in tanks on-site and then trucked to customers [1–43].

7.2.1 PROCESSING DETAILS

Natural gas, as it is used by consumers, is much different from the natural gas that is brought from underground up to the wellhead. While processing of crude oil is very complex and results in numerous usable fuels and chemicals, processing of natural gas is less complex and aimed toward producing natural gas (mainly methane) that can be stored, transported by pipelines, and used for heat, power, and liquid fuel generation and mobile applications.

The natural gas used by consumers is composed almost entirely of methane. However, natural gas found at the wellhead, although still composed primarily of methane, contains a number of contaminants. Natural gas that comes from oil wells ("associated gas") can exist separate from oil in the formation (free gas) or dissolved in the crude oil (dissolved gas). Natural gas from gas wells produces raw natural gas by itself. Finally, natural gas from condensate wells produces free natural gas along with a semiliquid hydrocarbon condensate. Whatever the source of the natural gas, once separated from crude oil (if present), it commonly exists in mixtures with other hydrocarbons: principally ethane, propane, butane, and pentanes. In addition, raw natural gas contains water vapor, hydrogen sulfide (H_2S), carbon dioxide, helium, nitrogen, and other compounds.

Natural gas processing consists of separating all of the various hydrocarbons and other contaminants from the crude natural gas to produce what is known as "pipeline-quality" dry natural gas. Major transportation pipelines usually impose restrictions on the makeup of the refined natural gas that is allowed into the pipeline. That means that before the natural gas can be transported, it must be purified. While ethane, propane, butane, and pentanes must be removed from natural gas, this does not mean that they are all "waste products." In fact, *NGLs* can be very valuable by-products of natural gas processing. NGLs include ethane, propane, butane, isobutane, and natural gasoline. These NGLs are sold separately and have a variety of different uses, including enhancing oil recovery in oil wells and providing raw materials for oil refineries or petrochemical plants, as sources of energy and raw materials for various chemicals [1–6,10,11,17,35].

While some of the needed processing can be accomplished at or near the wellhead (field processing), the complete processing of natural gas takes place at a processing plant, usually located in a natural gas–producing region. The extracted natural gas is transported to these processing plants through a network of gathering pipelines, which are small-diameter, low-pressure pipes. A complex gathering system can consist of thousands of miles of pipes, interconnecting the processing plant of more than 100 wells in the area. According to the American Gas Association's Gas Facts 2000, there was an estimated 36,100 miles of gathering system pipelines in the United States in 1999 [6,44]. In addition to the processing done at the wellhead and at centralized processing plants, some

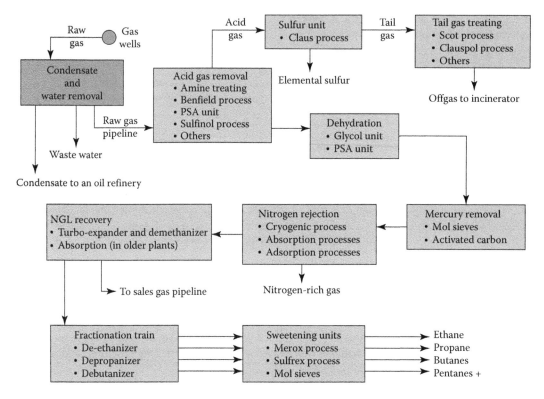

FIGURE 7.1 Schematic flow diagram of a typical natural gas processing plant. (From Natural gas, Wikipedia, the free encyclopedia, 2015; Natural gas processing, Wikipedia, the free encyclopedia, 2015.)

final processing is also sometimes accomplished at "straddle extraction plants." These plants are located on major pipeline systems to further remove small quantities of NGLs.

There are many great ways to configure the various unit processes used in the processing of raw natural gas. The block flow diagram shown in Figure 7.1 is a generalized, typical configuration for the processing of raw natural gas from nonassociated gas wells. It shows how raw natural gas is processed into pipeline-quality gas that can be used for various markets [4,5]. It also shows how processing of the raw natural gas yields several by-products:

7.2.2 PROCESS STEPS FOR THE PRODUCTION OF PIPELINE-QUALITY NATURAL GAS AND NGLS

While in general, as mentioned before, seven types of impurities from natural gas need to be removed, the number of steps and the type of techniques used in the process of creating pipeline-quality natural gas most often depend upon the source and makeup of the wellhead production stream. In some cases, several of the steps (or contaminant removal) shown in Figure 7.1 and the ones outlined earlier may be integrated into one unit or operation, performed in a different order or at alternative locations (lease/plant) or not required at all. The methods normally used for the removal of seven contaminants outlined earlier are briefly outlined in the following.

7.2.2.1 Oil and Condensate Removal

In order to process and transport associated–dissolved natural gas, it must be separated from the oil in which it is dissolved. This separation of natural gas from oil is most often done using equipment installed at or near the wellhead. The actual process used to separate oil from natural gas, as well as the equipment that is used, can vary widely. Although dry pipeline-quality natural gas is virtually identical across different geographical areas, raw natural gas from different regions may have different compositions and separation requirements. In many instances, natural gas is dissolved in oil underground primarily due to the pressure that the formation is under. When this natural gas and oil is produced, it is possible that it will separate on its own, simply due to decreased pressure. In these cases, separation of oil and gas is relatively easy, and the two hydrocarbons are sent separate ways for further processing. The most basic type of separator is known as a conventional separator. It consists of a simple closed tank, where the force of gravity serves to separate the heavier liquids like oil and the lighter gases like natural gas [1–6,10,11,17,35].

In some cases, however, a multistage gas–oil separation process is needed to separate the gas stream from the crude oil. These gas–oil separators are commonly closed cylindrical shells, horizontally mounted with inlets at one end, an outlet at the top for removal of gas, and an outlet at the bottom for removal of oil. The separation between gas and oil is accomplished by alternatively heating and cooling the flow stream in multiple steps. Some water and condensate, if present, will also be extracted as the process proceeds [1–6,10,11,17].

An example of this type of equipment is the low-temperature separator (LTX). This is most often used for wells producing high-pressure gas along with light crude oil or condensate. These separators use pressure differentials to cool the wet natural gas and separate the oil and condensate. Wet gas enters the separator, being cooled slightly by a heat exchanger. The gas then travels through a high-pressure liquid "knockout," which serves to remove any liquids into an LTX. The gas then flows into this LTX through a choke mechanism, which expands the gas as it enters the separator. This rapid expansion of the gas allows for the lowering of the temperature in the separator. After liquid removal, the dry gas then travels back through the heat exchanger and is warmed by the incoming wet gas. By varying the pressure of the gas in various sections of the separator, it is possible to vary the temperature, which causes the oil and some water to be condensed out of the wet gas stream. This basic pressure–temperature relationship can work in reverse as well, to extract gas from a liquid oil stream [1–6,10,11,17,35].

Condensates are most often removed from the gas stream at the wellhead through the use of mechanical separators. In most instances, the gas flow into the separator comes directly from the wellhead, since the gas–oil separation process is not needed. The gas stream enters the processing plant at high pressure (600 lb per square inch gauge [psig] or greater) through an inlet slug catcher where free water is removed from the gas, after which it is directed to a condensate separator. Extracted condensate is routed to on-site storage tanks [1–6,10,11,17].

7.2.2.2 Water Removal

In addition to separating oil and some condensate from the wet gas stream, it is necessary to remove most of the associated water. Most of the liquid, free water associated with extracted natural gas is removed by simple separation methods at or near the wellhead. However, the removal of the water vapor that exists in solution in natural gas requires a more complex treatment. This treatment consists of "dehydrating" the natural gas, which usually involves either absorption or adsorption.

7.2.2.2.1 Glycol Absorption

An example of absorption dehydration is known as "glycol dehydration." In this process, a liquid desiccant absorbs water vapor from the gas stream. Glycol, the principal agent in this process, has a chemical affinity for water. Essentially, glycol dehydration involves bringing a glycol solution, usually either diethylene glycol or triethylene glycol, into contact with the wet gas stream in an absorber. The glycol solution will absorb water from the wet gas. Once absorbed, the glycol particles become heavier and sink to the bottom of the absorber where they are removed. The natural gas, having been stripped of most of its water content, is then transported out of the dehydrator. The glycol solution is processed in a specialized boiler designed to vaporize only the water out of the solution. The difference in boiling point of water (212°F) and glycol (400°F) allows this to happen easily. The regenerate glycol is reused in the dehydration process [1–6,11,12,31,35].

Along with water in wet gas stream, the glycol solution occasionally carries with it small amounts of methane and other compounds found in the wet gas. In order to decrease the amount of methane and other compounds that are lost in the boiler, flash tank separator condensers are used to remove these compounds before the glycol solution reaches the boiler. Essentially, a flash tank separator consists of a device that reduces the pressure of the glycol solution stream, allowing the methane and other hydrocarbons to vaporize ("flash"). The glycol solution then travels to the boiler, which may also be fitted with air- or water-cooled condensers, which serve to capture any remaining organic compounds that may remain in the glycol solution. In practice, according to the Office of Fossil Energy of the Department of Energy (DOE), these systems have been shown to recover 90%–99% of methane that would otherwise be flared into the atmosphere [1–6,10,11,17,44].

7.2.2.2.2 Solid-Desiccant Adsorption

In this process, two or more adsorption towers containing solid desiccant are used to remove water. Typical desiccants include activated alumina or a granular silica gel material. Wet natural gas is passed through these towers, from top to bottom. As the wet gas passes around the particles of desiccant material, water is adsorbed on the surface of these desiccant particles. The dry gas exits the bottom of the adsorption tower. The multiple adsorption towers are used to make sure to capture any remaining water vapor in the gas stream from the previous tower. This happens when desiccant is saturated with water over time. Generally saturated desiccant is regenerated using hot gas, which removes water from the desiccant.

Solid desiccant is often more effective than glycol dehydrators and are usually installed as a type of straddle system along natural gas pipelines. These types of dehydration systems are best suited for large volumes of gas under very high pressure and are thus usually located on a pipeline downstream of a compressor station [1–6,10,11,17,34,44].

7.2.2.3 Sulfur and Carbon Dioxide Removal

In addition to water, oil, and NGL removal, one of the most important parts of gas processing involves the removal of sulfur and carbon dioxide (acid gases). Natural gas from some wells contains significant amounts of sulfur and carbon dioxide. This natural gas, because of the rotten smell provided by its sulfur content, is commonly called "sour gas." Sour gas is undesirable because the sulfur compounds it contains can be extremely harmful, even lethal, to breathe. Sour gas can also be extremely corrosive. In addition, the sulfur that exists in the natural gas stream can be extracted and marketed on its own. In fact, according to the USGS, U.S. sulfur production from gas processing plants accounts for about 15% of the total U.S. production of sulfur. This sulfur exists in natural gas as hydrogen sulfide (H_2S), and the gas is usually considered sour if the hydrogen sulfide content exceeds 5.7 mg of H_2S per m^3 of natural gas. The process for removing hydrogen sulfide from sour gas is commonly referred to as "sweetening" the gas [1–6,14,26,31,35].

The acid gases (CO_2 and H_2S) can be absorbed by either physical or chemical solvents. Absorption by physical solvents is mostly not recommended at low partial pressures as the compression of the gas for physical absorption is relatively uneconomical. In general, the economics of CO_2 separation is strongly influenced by the partial pressure of CO_2 in the feed natural gas [10–44]. However, if the gas is available at high pressure, physical solvents might be a better choice than chemical solvents. While physical solvents can often be stripped of impurities by reducing the pressure without the application of heat, regeneration of chemical solvents is achieved by the application of heat. Mostly, physical solvents tend to be favored over chemical solvents when the concentration of acid gases or other impurities is very high. Unlike chemical solvents, physical solvents are noncorrosive, requiring only carbon steel construction.

The concentration of heavy hydrocarbons in the feed gas also affects the choice of gas treating solvent. If the concentration of heavy hydrocarbons is high, a physical solvent may not be the best option due to higher coabsorption of hydrocarbons, particularly pentanes plus. Unlike synthesis gases, where there aren't appreciable quantities of hydrocarbons, natural gases can be a problem for physical solvents as the result of hydrocarbon coabsorption. This makes physical solvents particularly applicable to synthesis gas treating [26,31,35].

Generally preferred process for sweetening sour natural gas and removal of carbon dioxide and hydrogen sulfide uses amine solutions to remove the hydrogen sulfide and carbon dioxide. This process is known simply as the "amine process," or alternatively as the Girdler process and is used in 95% of U.S. gas sweetening operations [1–6,14,26,31,35]. The sour gas containing carbon dioxide and hydrogen sulfide is run through a tower, which contains the amine solution. This solution has an affinity for sulfur and carbon dioxide and absorbs it much like glycol absorbing water. There are two principal amine solutions used: monoethanolamine (MEA) and diethanolamine (DEA). Either of these compounds, in liquid form, will absorb sulfur compounds and carbon dioxide from natural gas as it passes through. The effluent gas is virtually free of sulfur compounds and carbon dioxide. Like the process for NGL extraction and glycol dehydration, the amine solution used can be regenerated (i.e., the absorbed sulfur and carbon dioxide are removed), allowing it to be reused to treat more sour gas (or sometimes called "acid gas"). A schematic of amine solution process is described in Figure 7.2 [46].

Besides absorption technique, adsorption technique has also been investigated. Although adsorption technique is restricted to small gas stream and moderate pressure due to complexity of the design, the use of pressure swing adsorption (PSA) technology using solid desiccants like iron sponges and molecular sieves is mostly used in shut-in natural gas wells that usually contained too much N_2. As a typical example, titanosilicate adsorbent (Engelhard Corporation) combined with a PSA process using a vacuum swing adsorption is used to remove N_2 and/or CO_2 from natural gas feed streams [14,21,34,37,38].

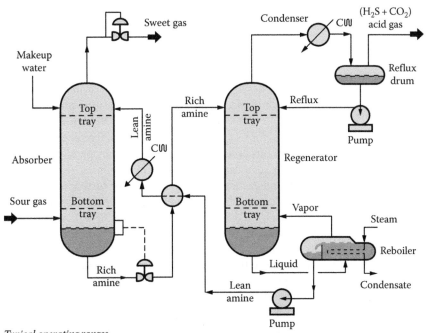

Typical operating ranges

Absorber: 35°C–50°C and 5–205 atm of absolute pressure
Regenerator: 115°C–126°C and 1.4–1.7 atm of absolute pressure at tower bottom

FIGURE 7.2 Process flow diagram of a typical amine treating process used in petroleum refineries, natural gas processing plants, and other industrial facilities. (From Amine gas treating, Wikipedia, the free encyclopedia, 2015, https://en.wikipedia.org/wiki/Amine_gas_treating.)

When the natural gas well contains high CO_2 and H_2S, the use of membrane process is most preferable under high pressure conditions. CO_2 separation is accomplished by pressure-driven selective permeation through a membrane where separation is due to the differences in permeation rate of different gas penetrants. Although the acid gas is usually recovered at low pressure, high-purity product containing approximately 95% CO_2 can be achieved with one or two stages, depending upon feed gas pressure and percent recovery. Economical considerations may dictate additional capital and incremental energy requirements to increase feed pressure and/or utilize two-stage separation with recompression of gas from the first stage.

Cryogenic process, as compared to other methods of separating CO_2, has the advantage that the CO_2 can be obtained at relatively high pressure. However, this advantage may be offset by the large refrigeration requirement. In this process, the need of special materials is also critical [11,17,31,35].

Although the choice of acid gas removal technology depends on the needs of the gas processor, the current market trend showed that membranes [13–43,47] have also proven their usefulness to compete with absorption (amines) technology. Nevertheless, the absorption technology based on amine treatment is still an efficient method even though the amine units are large and heavy.

In recent years, efforts have been made to use hybrid processes to reduce overall cost and improve efficiency. In hybrid separation processes, two separate processes are combined to create in a single unit operation [13–22,47]. A properly designed hybrid process can replace a drawback of one process with the advantage of other process with the hope that the new integrated process will be more effective with lower costs [13–22,47].

While membrane systems work well for high-pressure, high-CO_2-concentration streams, low concentration of CO_2 requires the use of amine or cryogenics system [29–34,43]. In order to achieve low CO_2 concentration, some investigators [14,15] have suggested different types of membrane-amine hybrid systems that are more economical. A study with a membrane-cryogenic distillation hybrid system has also shown favorable economics [16].

The integration of membranes with other separation processes such as PSA is also well established in the chemical and petrochemical industries [17–20,22]. Membrane permeation can be an effective aid in the pressurization and high-pressure adsorption steps of a typical PSA process; the pressure difference available from the PSA can be used for operating the membrane incorporated into the blowdown step of the PSA cycle [21]. Usually, the integration of membranes with PSA is considered in H_2 separation, while hybrid membranes + amine absorption are applied to the CO_2 separation [13–43]. Cost comparison for membrane and absorption (DEA) process showed that the membrane process is more economical for CO_2 feed concentrations in the range 5–40 mol% [14]. It does ascertain that when the feed contains other gases such as H_2S, the operating cost for removing the additional gases also increases. In such cases, use of hybrid membrane processes (membranes for bulk separation of CO_2 and H_2S and gas-absorption processes for further removal) is economically feasible [14]. A hybrid system comprising of Cynara membranes and amine absorption is operating since 1994 in Mallet (Texas, United States) [23] to perform the bulk removal of CO_2 from associated gas (90% CO_2 and heavy hydrocarbons), before downstream treating. The membrane system offered a 30% reduction in operating cost when compared with a methyldiethanolamine (MDEA) system and significantly reduced the size of the subsequent operations [13–23].

The combination of membrane processes with amine process offers an economical advantage over the use of amine process alone. However, such an integrated process can be too complex to control [42]. The removal of CO_2 for LNG production using amine-based process is outlined by Ebenezer and Gudmunsson [29]. Leal et al. [24] examined a novel approach of CO_2 adsorption by amine-bonded solid-phase adsorbent.

Sulfur can be sold and used if reduced to its elemental form. Elemental sulfur is a bright yellow powderlike material and can often be seen in large piles near gas treatment plants. In order to recover elemental sulfur from the gas processing plant, the sulfur compound contained in the discharge from the gas sweetening process is converted to the elemental sulfur by the "Claus process." This process involves thermal and catalytic reactions to extract the elemental sulfur from the hydrogen sulfide solution. The process is designed to treat high concentration of hydrogen sulfide. The chemical reactions involve partial oxidation of hydrogen sulfide to sulfur dioxide and a catalytic reaction of hydrogen sulfide and sulfur dioxide to produce elemental sulfur:

$$H_2S + 1.5O_2 \rightarrow SO_2 + H_2O \tag{7.1}$$

$$2H_2S + SO_2 \rightarrow 3S + 2H_2O \tag{7.2}$$

As shown above, the conversion of hydrogen sulfide to pure sulfur occurs in two steps. These two-step reactions are generally carried out in three catalytic stages with removal of sulfur in between stages. This method gives an overall conversion of about 97%. The product gas of Claus process is treated by "tail gas treating" before releasing to the atmosphere. Several modified Claus processes are developed to increase the economy and efficiency of the process [217,218,227–229].

After desulfurization, the gas flow is directed to the next section, which contains a series of filter tubes. As the velocity of the stream reduces in the unit, primary separation of the remaining contaminants occurs due to gravity. Separation of smaller particles occurs as gas flows through the tubes, where they combine into larger particles that flow to the lower section of the unit. Further, as the gas stream continues through the series of tubes, a centrifugal force is generated that further removes any remaining water and small solid particulate matter. Several other techniques for particulate removal are described later in this chapter.

In 2004, about 6.2 bcf of carbon dioxide was produced in seven plants in the United States. The carbon dioxide produced at these natural gas treatment plants is used primarily for reinjection in support of tertiary enhanced oil recovery efforts in the local production area. The smaller, uneconomical amounts of carbon dioxide that are normally removed during the natural gas processing and treatment in the United States are vented to the atmosphere [1–6,31,35].

7.2.2.4 Nitrogen Extraction

Once the hydrogen sulfide and carbon dioxide are processed to acceptable levels, the stream is routed to a nitrogen rejection unit (NRU), where it is further dehydrated using molecular sieve beds. In the NRU, the gas stream is subjected to a series of passes through a column and a brazed aluminum plate fin heat exchanger. Here nitrogen is cryogenically separated and vented. Another type of NRU unit separates methane and heavier hydrocarbons from nitrogen using an absorbent solvent. The absorbed methane and heavier hydrocarbons are flashed off from the solvent by reducing the pressure on the processing stream in multiple gas decompression steps. The liquid from the flash regeneration step is returned to the top of the methane absorber as lean solvent [1–6,31,35].

7.2.2.5 Helium (He) Extraction

Twenty-two natural gas treatment plants in the United States currently produce helium as a major by-product of natural gas processing. Twenty of these plants, located in the Hugoton–Panhandle Basin, produce marketable helium that is sold in the open market when profitable, while transporting the remaining unrefined helium to the Federal Helium Reserve (FHR). The FHR was created in the 1950s in the Bush salt dome, underlying the Cliffside field, located near Amarillo, Texas. Sales of unrefined helium in the United States for the most part come from the FHR [1–6,12,31,35].

Helium can be extracted from the gas stream using a PSA process. The world's supply of helium comes exclusively from natural gas production. The single largest source of helium is the United States, which produces about 80% of the annual world production of 3.0 bcf. In 2003, U.S. production of helium was 2.4 bcf, about two-thirds of which came from the Hugoton Basin in north Texas, Oklahoma, and Kansas. The rest mostly comes from the La Barge field located in the Green River Basin in western Wyoming, with small amounts also produced in Utah and Colorado. According to the National Research Council, the consumption of helium in the United States doubled between

1985 and 1996, although its use has leveled off in recent years. Helium is used in such applications as magnetic resonance imaging, semiconductor processing, and pressurizing and purging of rocket engines by the National Aeronautics and Space Administration [1–6,12,31,35].

7.2.2.6 Separation of NGLs

Natural gas coming directly from a well contains many NGLs that are commonly removed. In most instances, NGLs have a higher value as separate products, and it is thus economical to remove them from the gas stream. The removal of NGLs usually takes place in a relatively centralized processing plant and uses techniques similar to those used to dehydrate natural gas.

There are two basic steps to the treatment of NGLs in the natural gas stream. First, the liquids must be extracted from the natural gas. Second, these NGLs must be separated themselves, down to their base components [1–6,11,12,17,31,35].

One simple method to remove condensed liquids from the natural gas is cooling it down, so that the condensate (water and higher hydrocarbons) will fall out and can be fractionated. Using the Joule–Thomson effect means that the gas temperature decreases during gas expansion, for example, in an expansion valve ("self-refrigeration process"). This effect can be used at the wellhead, where the pressure has to be reduced in most cases, as a predehydration, or after the gas purification using an expansion valve and a gas–gas heat exchanger. This might require a recompression of the gas before the feed-in. The condensed liquid can be subsequently fractionated.

Refrigeration is another simple and most direct process for NGL recovery. External refrigeration is supplied by a vapor compression cycle (usually using propane as refrigerant). A cold gas/hot gas heat exchanger provides the first cooling step of the inlet gas. The chiller (mostly a shell-and-tube kettle-type unit) saves final cooling and partial condensation of liquids. The refrigerant (often propane) boils off, leaves the chiller as a saturated vapor, and is fed into the refrigeration plant. Ethylene glycol is often injected at the inlet of the gas/gas exchanger and/or the chiller to prevent hydrate formation or to prevent block exchangers. The achievable dew point depends on the cooling temperature and the pressure of the gas. This technology may not require a recompression; however, external energy for cooling is needed. The condensed liquid can also be subsequently fractionated.

Besides the simple Joule–Thomson expansion and refrigeration methods, there are two principal techniques for removing NGLs from the natural gas stream: the absorption method and the cryogenic expander process. According to the Gas Processors Association, these two processes account for around 90% of the total NGL production [1–6,11,12,31,35,44].

The absorption method of NGL extraction is very similar to using absorption for dehydration. The main difference is that in NGL absorption, oil is used as an absorbent as opposed to glycol. This oil has an "affinity" for NGLs in much the same manner as glycol has an affinity for water. Before the oil has picked up any NGLs, it is termed "lean" absorption oil. As the natural gas is passed through an absorption tower, it is brought into contact with the absorption oil that soaks up a high proportion of the NGLs. The "rich" absorption oil, now containing NGLs, exits the absorption tower through the bottom. It is now a mixture of absorption oil, propane, butanes, pentanes, and other heavier hydrocarbons. The rich oil is fed into lean oil stills, where the mixture is heated to a temperature above the boiling point of the NGLs but below that of the oil. This process allows for the recovery of around 75% of butanes and 85%–90% of pentanes and heavier molecules from the natural gas stream [1–6,11,12,31,35,44].

The basic absorption process mentioned earlier can be modified to improve its effectiveness or to target the extraction of specific NGLs. In the refrigerated oil absorption method, where the lean oil is cooled through refrigeration, propane recovery can be upward of 90%, and around 40% of ethane can be extracted from the natural gas stream. The extraction of the other heavier NGLs can be close to 100% using this process [1–6,11,12,31,35,44].

Cryogenic processes are also used to extract NGLs from natural gas. While absorption methods can extract almost all of the heavier NGLs, the lighter hydrocarbons, such as ethane, are often more difficult to recover from the natural gas stream. In certain instances, it is economical to simply leave the lighter NGLs in the natural gas stream. However, if it is economical to extract ethane and other lighter hydrocarbons, cryogenic processes are required for high recovery rates. Essentially, cryogenic processes consist of dropping the temperature of the gas stream to around $-120°F$ [1–6,11,12,31,35,44].

There are a number of different ways of chilling the gas to these temperatures, but one of the most effective is known as the turbo expander process. In this process, external refrigerants are used to cool the natural gas stream. Then, an expansion turbine is used to rapidly expand the chilled gases, which causes the temperature to drop significantly. This rapid temperature drop condenses ethane and other hydrocarbons in the gas stream, while maintaining methane in gaseous form. This process allows for the recovery of about 90%–95% of the ethane originally in the gas stream. In addition, the expansion turbine is able to convert some of the energy released when the natural gas stream is expanded into recompressing the gaseous methane effluent, thus saving energy costs associated with extraction of ethane [1–6,11,12,31,35,44].

7.2.2.7 NGL Fractionation

Once NGLs have been removed from the natural gas stream, they must be broken down into their base components to be useful. That is, the mixed stream of different NGLs must be separated out. The process used to accomplish this task is called fractionation. Fractionation works based on the different boiling points of the different hydrocarbons in the NGL stream [1–6]. Essentially, fractionation occurs in stages consisting of the boiling off of hydrocarbons one by one. The name of a particular fractionator gives an idea as to its purpose, as it is conventionally named for the hydrocarbon that is boiled off. The entire fractionation process is broken down into steps, starting with the removal of the lighter NGLs from the stream. This means the first step involves separation of ethane from NGL stream (using deethanizer), the second step involves separation of propane (using depropanizer) from NGL stream, and the third step involves separation of butane (using debutanizer) from the remaining NGL stream that contains pentanes and heavier hydrocarbons. In the fourth step, butane can be further fractionated in normal butane and isobutane (using deisobutanizer).These four steps provide different hydrocarbons in pure forms that can be sold separately.

7.3 FACTORS AFFECTING THE LEVEL OF IMPURITIES IN SYNGAS

While natural gas can be recovered by conventional and unconventional methods, as shown in the previous section, the nature of impurities contained in it basically lie in the framework described earlier. While the major component of natural gas is methane, the remaining composition can vary depending on the source of the natural gas. The technologies described earlier can be applied to natural gas coming from any sources. These technologies are mostly low-temperature technologies.

Like natural gas, biogas or "landfill gas" also mainly contains methane unless special conditions are used to generate biohydrogen. Besides methane, biogas contains a large amount of carbon dioxide and small amounts of other hydrocarbons. Methane from biogas can also be purified using the same technologies that are described earlier since biogas is also recovered at low temperatures. The recovered biomethane can then either be transported using natural gas pipeline or be used to generate heat and power locally. Biohydrogen can be injected in a natural gas pipeline in small concentration (5–15 vol%) or needs to be transported by hydrogen infrastructure.

As shown in Chapter 5, the major purpose of the hydrothermal processes is to generate either pure hydrogen or the gas containing a significant amount of hydrogen. The purification of hydrothermal gas depends on the process used to generate hydrothermal gas. If hydrogen is generated using steam or sub- and supercritical water gasification processes, the technologies required to purify gas would be the same as gases generated from conventional thermal gasification processes. Hydrogen generated from various water dissociation technologies is likely to be more in pure state.

The composition of syngas generated from thermal or hydrothermal gasification processes depends significantly on the nature of the feedstock and the operating conditions of the gasification process.

TABLE 7.1
Allowable Concentrations of Contaminants in Syngas for Catalytic Synthesis

Syngas Contaminants	Contaminant Specification
$H_2S + COS + CS_2$	<1 ppmv
$NH_3 + HCN$	<1 ppmv
HCl + HBr + HF	<10 ppbv
Alkali metals (Na + K)	<10 ppbv
Particles (soot, ash)	"Almost completely removed"
Organic components (viz. tar)	Removed to a level at which no condensation occurs upon compression to FT synthesis pressure (25–60 bar)
Hetero organic components	<1 ppmv
CO_2, N_2, CH_4, and larger hydrocarbons	"Soft maximum" of 15 vol% that has been identified (the lower, the better)

Sources: Ratafia-Brown, J. et al., Assessment of technologies for co-converting coal and biomass to a clean syngas—Task 2 report (RDS), DOE/NETL-403.01.08 Activity 2 Report, May 10, 2007; Shah, Y.T., Biomass to liquid fuel via Fischer–Tropsch and related syntheses, Chapter 12, in: Lee, J.W., ed., *Advanced Biofuels and Bioproducts*, Springer Book Project, Springer Publ. Co., New York, September 2012, pp. 185–207.

In general, at low temperature coal gasification generates producer gas with different levels of methane composition and will contain significant amounts of particulate matters, sulfur, nitrogen, and metal impurities. Biomass gasification at low temperature will also generate producer gas with methane along with syngas, but the gas can contain significant amounts of particulate matters, tar, chlorine, and alkali and heavy metals but not as much sulfur depending on the nature of the feedstock. At high temperatures (greater than around 1200°C) both coal and biomass gasification will largely generate syngas with low tar concentration. Other operating parameters such as pressure and catalyst can also affect gas composition. Thus, the level of impurities in gas produced by thermal gasification depends on the nature of feedstock and the operating conditions used for the gasification. The allowable levels of contaminant in synthetic gas depend on its downstream use. Typical allowable concentrations of contaminants in syngas for catalytic synthesis are illustrated in Table 7.1 [49–52].

Besides the nature of feedstock and nature of gasification conditions, the level of impurities in the product also depends on the level of feed pretreatment. For example, the level of sulfur impurities present in the syngas during coal gasification depends on the extent of coal pretreatment to remove sulfur. The same applies to the level of ash and its effect on the emissions. Biomass feed pretreatment is an important step in syngas quality. Feed pretreatment is thus an important factor in the product quality [49–52].

Besides all parameters discussed earlier, an important recent development of cogasification also affects the quality of syngas due to the synergy that exists in the interactions among various components in biomass and coal during cogasification. Such synergy can affect all the components of impurities discussed earlier. The effects of synergy and catalytic reactions during cogasification on the fate of various impurity actors are briefly discussed in the following sections [49–166].

7.3.1 EFFECTS OF COGASIFICATION OF COAL AND BIOMASS/WASTE ON TAR RELEASE

The literature pertaining pyrolysis under high temperature, pressure, and heating rate conditions that normally exist during gasification is sparse. The reported studies have also not examined the mechanism by which biomass components (cellulose, hemicellulose, and lignin) interact with each other under heating and the effect of mineral matter on these interactions. The knowledge of how these individual components and biomass as a whole would interact when blended with coal has also not been elucidated. Feedstock and their compositions can thus significantly influence the conversion behavior during the early stage of fuel decomposition. For an entrained-bed gasification reactor operating at temperatures higher than 1200°C, the effects of feedstock components on char gasification are also not well understood [49,53].

Tar can be defined as the organics produced under thermal or partial oxidation regimes of any organic material and generally assumed to be largely aromatic [55]. While there are different tiers of tar components [49–52], saturated polynuclear aromatics such as benzene, toluene, and naphthalene are the hardest to decompose under gasification conditions. The decomposition of all tar components generally requires high temperature (>1200°C).

The tar content of the product gases from gasification of biomass is one of the major factors affecting the subsequent process stages. The type of feedstock used and gasification temperature are the main factors affecting tar yield during gasification [56] Synergistic effects between coal and biomass particles are expected to lower the yield of tar compared to independent gasification of these feedstocks. Kumabe et al. [141] obtained a slight decrease in tar yield by varying Japanese cedar (biomass) concentration from 0 to 100 when co-gasifying with Mulia (Indonesia) coal with air and steam in a downdraft fixed-bed gasifier at 900°C. Pinto et al. [57] observed a decrease in tar yield for a mixture of 80% coal (high-ash coal from Puertollano, Spain) and 20% pine wood waste compared to coal alone in a fluidized-bed gasifier operating at atmospheric pressure and temperatures of 850°C–900°C using a mixture of oxygen and steam as the gasifying agent. However, the addition of polyethylene (PE) waste in the feed led to an increase in tar release. A probable reason for this could be that the polymeric structure of PE breaks into smaller fractions by thermal cracking, contributing to a greater amount of tars [57].

Pinto et al. [57] also noticed that a ternary blend of coal, PE, and pine resulted in less tar release than a mixture of coal and PE. They and others [34] also found that dolomite was an efficient catalyst for reducing the tar yield during cogasification. Collot et al. [144] found no synergetic effect during cogasification of Polish coal and forest residue mixture. Aznar et al. [142] and Andre et al. [143] identified an increase in tar yield with an increase in biomass content of the feedstock, suggesting that synergetic effect during coal and biomass cogasification might be highly dependent on biomass type as well as gasification conditions. Mettler et al. [60] point out that due to lack of understanding of intermediates formed during gasification, their effects on the cogasification process are unknown. Further investigations are still needed to clarify various issues of synergy during coal and biomass/waste cogasification.

7.3.2 Roles of Mineral Matters and Slagging in Cogasification

Tchapda and Pisupati [53] analyzed the role of mineral matters on slagging during cogasification. A summary of their analysis is briefly presented here.

Slagging and fouling are caused by the relatively reactive alkali and alkaline earth compounds (K_2O, Na_2O, and CaO) found in biomass ash. The alkali and alkaline earth metals (AAEMs) present and dispersed in biomass fuels induce catalytic activity during cogasification with coal. The catalytic activity is most noticeable when blended with high-rank coals. The presence of synergy during cogasification is still controversial [53].

The merits of alkali (K^+ and Na^+) and alkaline earth (Ca^{2+}) metals as catalysts for coal gasification have been extensively investigated [62–65]. Due to high content of these elements in some biomass, cogasification of coal and biomass is expected to exhibit some catalytic behavior, whose significance will depend on the type of biomass and coal being used. In fact, in lignite (low-rank coal) Na is the principal alkali metal, while the amount of K is low [66]. In bituminous coals (high rank), K is contained exclusively in illite or closely related clay structure [67], while Na is generally present as NaCl [68].

From low-rank coals to higher-rank coals, Ca is systematically changed from carboxyl bound to calcite [69], with decreasing catalytic activity [29]. The K present in the illite of bituminous coal is converted to K-aluminosilicate glass [67]. Therefore, mineral matters of high-rank coals have little catalytic activity during coal gasification. Any catalytically active ions such as Ca and Na in low-rank coals are highly dispersed [71].

As pointed out by Tchapda and Pisupati [53], the analysis presented earlier for coal is also applicable to biomass fuels. Ren et al. [58] compared cogasification of meat and bone meal (MBM) blended with high-rank (anthracite) and low-rank coals (lignite). They found that for both rank coals, acid-based MBM samples gave a lower carbon conversion compared to the raw MBM blend. Thus, biomass mineral matter (Na, K, and Ca) demonstrated catalytic influence during cogasification, as reported by other authors [54,72]. They also found that the catalytic effect of MBM minerals was not perceivable on lignite but was significant for anthracite. This is in accordance with McKee et al. [59] and Srivastava et al. [73] who demonstrated the increased catalytic activity of alkali metals on gasification with increasing coal rank.

The literature also reports [76] that the reaction rates of a blend of waste birchwood and Daw Mill coal were significantly increased, thereby reducing char in a pressurized fluidized-bed gasifier under oxygen environment. Brown et al. [72] showed that the gasification of coal char from Illinois No. 6 coal increased by eightfold at 895°C in a mixture of 10:90 of coal char and switchgrass ash. In general, biomass is more reactive than any coal [56]. Its char is therefore continuously consumed during gasification, leaving very little remains at the end of the process, whereas coal char is less reactive and continuously accumulates in the bed during the course of the gasification. Blending biomass and coal takes advantage of both the high reactivity of biomass and its catalytic effect.

Habibi et al. [74] observed a negative catalytic behavior between switchgrass and a subbituminous coal, while the mixture of switchgrass and fluid coke showed a synergy. They explain the negative behavior by a deactivation of mineral catalysts due to sequestration of the mobile alkali elements by the reaction with aluminosilicate minerals in coal to form inactive alkali aluminosilicates [53]. The catalytic effect of biomass in cogasification with coal is also expected to play an important role for the abatement of environmentally harmful species containing sulfur and nitrogen. Biomass species with a high content of K, Na, and Ca can form sulfate and capture sulfur from the gas phase when coprocessed with coal [75].

During co-firing of straw and coal, Pedersen et al. [75] observed a net decrease of NO and SO_x emission, due to the decrease of fuel-nitrogen conversion to NO and due to retention of sulfur in the ash. Sjöström et al. [76] identified a lower ammonia yield during cogasification of birchwood and Daw Mill coal. Cordero et al. [77] showed enhancement in desulfurization when blending coal with different types of biomass during co-pyrolysis. Haykiri-Acma and Yaman [78] showed that the addition of hazelnut shell to lignite contributed to the sulfur fixing potential of the resulting char in the form of CaS and $CaSO_4$ during co-pyrolysis of these feedstocks. For temperatures higher than 700°C, potassium is shown to be more trapped in Si in ash [79,80].

7.3.3 PRODUCT CLEANING AND SEPARATION MECHANISMS

Syngas produced from coal–biomass mixture gasification will likely contain nitrogen, sulfur, chlorine, alkali, and heavy metal impurities since those elements are part of the original coal and biomass. These impurities must be removed from syngas before its downstream use. Here, we briefly examine the behavior of these impurities during cogasification. The acceptable levels of these impurities in product gas are outlined in Table 7.1.

7.3.3.1 Behavior of Nitrogen, Chlorine, and Sulfur

Tchapda and Pisupati [53] gave a detailed accounting of nitrogen and their forms present in biomass and coal. They also pointed out that reaction pathways involving nitrogen during pyrolysis/gasification are complex due to the overlapping of reaction stages, the influence of reaction conditions (residence time, temperature, pressure, and gasification agent), and the differences between fuels. During pyrolysis, part of the fuel-bound nitrogen (FBN) is released with the devolatilizing gases, while the remainder is retained in the solid char to be released during subsequent gasification of the char [82,83].

The allocation of fuel N to volatiles and char is dictated by temperature [81,84–87], particle size [85], fuel type [81], residence time [85,87], heating rate [88], and pressure [88]. Temperature, fuel type, and particle size are the dominant factors in this allocation, while heating rate and pressure only have a minor effect.

Tchapda and Pisupati [53] also pointed out that any synergy during the decomposition of coal and biomass/waste blend influences the release of fuel nitrogen. Yuan et al. [90,91] found that synergy between coal and biomass enhanced the release of fuel N as volatile, thus reducing the amount of residual nitrogen left in the char during co-pyrolysis. Since synergy increases the volatile yields, it is expected that it will also augment the yield of nitrogen species in the gas phase. At higher temperatures, N_2 was the dominant nitrogen species observed by Yuan et al. [90,91].

While the literature is not clear about the effects of experimental conditions on partitioning between HCN and NH_3, there are strong evidences [90,91,145] that slow heating rate produces NH_3, while fast heating rate produces HCN as the main N compound. The literature has also shown that at low temperature more N ends up in char than at high temperatures [53]. In cogasification, the release of FBN in the volatile is desirable over its remaining in the char.

During cogasification, chlorine is released in two steps: during pyrolysis at temperatures as low as 500°C in the form of HCl and during char burnout at temperatures higher than 700°C in the form of KCl and NaCl [94–96]. K will preferentially be released in the gas phase as KCl for Cl-rich biomass fuels. Tchapda and Pisupati [53] pointed out that during cogasification, sulfur is released mainly in the form of hydrogen sulfide (H_2S) and carbonyl sulfide (COS) [94]. In coal, sulfur exists in both inorganic and organic forms. Zhao et al. [98] observed that most of the organic sulfur in the temperature range of 400°C–900°C escaped as H_2S, a conclusion supported by other studies [52,53,114–118].

In cogasification, since woody biomass and grasses contain much less sulfur than most coals, the replacement of coal by biomass will reduce H_2S and COS in the product syngas. Tchapda and Pisupati [53] pointed out that coal mineral sulfur could interact with the organic matrix during coal pyrolysis [104–108]. It has also been proven that H_2S evolved from decomposition of pyrite can be partially retained in the solid phase (char/ash) as alkali sulfides [106–108]. Si in coal can also have a beneficial effect on the sulfur retention at high temperature [79,94]. Organically associated Ca can recapture the H_2S to form sulfidic sulfur at temperatures above 600°C [108]. Given the high content of alkalis in some biomass fuels, sulfur capture by these metals could be an added advantage of cogasification of coal- and biomass-based fuels.

7.3.3.2 Bottom and Fly Ash Characteristics and Role of Alkalis

Cogasification results in the formation and deposition of ash, which can cause slagging and fouling of heat transfer surfaces. Inorganic matters and impurities in coal and biomass contribute to ash formation during cogasification. The characteristics of ash formed depend on feedstock composition and the reaction conditions.

The important factors contributing to ash characteristics and behavior are (1) the rank and type of coal and the type of biomass [109] and (2) the reaction conditions such as temperature, pressure, oxidizing or reducing atmosphere, and flow dynamics [109]. Generally, bark, straw, grasses, and grains have higher mineral contents compared to wood, and MSW and sewage sludge have higher heavy metals compared to coal and biomass [110–113].

In general, the inorganic matter of the feedstock has a major impact on the gasifier operation. This impact depends on the type of the gasifier employed and whether or not gasifier is slagging or nonslagging. The nonslagging ash interferes with moving bed [53], if the temperature is not maintained below the melting temperature. If ash fuses together and forms clinker, it can stop or inhibit the downward flow of biomass. Ash fusion temperature also changes with sodium content. The clinker formation was observed in the moving-bed Lurgi gasifiers at Dakota Gasification Great Plains plant, which forced the shutdown of the plant [49,53,161]. Even if ash does not fuse, it can lower the fuel's reaction response. In a CFB, silica in the sand bed material can agglomerate with high-sodium feedstock (like birchwood), and this agglomeration should be avoided by injecting additives.

Ash in some fresh wood does not melt even at 1300°C–1500°C due to the presence of CaO. This can cause problems with slagging entrained-flow gasifier. Despite the high-ash melting temperature in this case, slagging entrained-flow gasifier is preferred over nonslagging gasifier because (1) melt can never be avoided and (2) slagging entrained-flow gasifier is more fuel flexible. Slagging cogasification often requires fluxing material in order to obtain proper slag properties [66,81–86,111–113,122,130–136,141–144,161–163].

Biomass mineral matter is usually rich in AAEMs, which melts at low temperatures. Aluminosilicates have higher melting temperatures and their concentrations are higher in coal than in biomass. Therefore, biomass ash tends to melt at lower temperatures than coal ash. It is therefore important to know the behavior of the ash mixture from coal and biomass in the design of cogasification systems. The wide variation of mineral matter content and quality observed in biomasses as well as in coals causes the design of cogasification systems to be carried out on a case-by-case basis. More quantitative assessment of characteristics of coal–biomass ash mixtures under different gasification conditions is needed [53].

Higman and van der Burgt [163] suggest that for gasification application, the ash-melting characteristics should be determined under reducing conditions. The knowledge of possible chemical reactions among mineral matters present in coal and biomass during cogasification can be very helpful for understanding ash characteristics. Mineral matter in solid fuels is generally grouped into two factions [53]:

1. The inherent inorganic material, which is most commonly associated with the oxygen-, sulfur-, and nitrogen-containing functional groups
2. The extraneous inorganic material coming from soil or during harvesting, handling, storage, and processing of the fuel [130]

Tchapda and Pisupati [53] point out that inorganic materials in biomass can be further categorized in a number of ways. Vaporization, condensation, and coagulation/agglomeration are the main mechanisms involving biomass ash formation; alkali metals and chlorides contained in biomass vaporize to form $HCl(g)$, $KCl(g)$, $K_2SO_4(g)$, $Na_2SO_4(g)$, and $NaCl(g)$ [162]. Their condensation constitutes the main contributor to the fine ash fraction. Strand et al. [137] identified homogeneous nucleation and heterogeneous condensation as two principal routes for condensation of these vapors into particles. The inherent and extraneous inorganic materials of coal have different pathways for ash formation. Inherent inorganic materials are close to each other in the char particle; therefore, they become molten and easily coalesce and agglomerate during char burnout. Extraneous minerals mainly form ash through fragmentation. The study of Van Loo and Koppejan [131] and Wu et al. [132] partially agree with the conclusion that coarse ash particles are formed by coalescence and shedding and fine ash particles are formed through vaporization and condensation. The slagging and fouling indexes often used in the gasifier design are based on the fuel ash content and/or the ash chemical composition, and they give an indication of fuel inclination to form deposit in the gasifier. Chemical equilibrium calculation has been used in many studies to predict the behavior of ash at given pressures and temperatures [53,131,132].

There are no data available on the effects of alkalis on fly ash characteristics under cogasification conditions. Most of the studies involving co-firing coal and straw have focused on alkali metals and the reactions of alkali metal with chlorine to determine the properties of fly ash deposits [133–136]. Significant amounts of K are released in the gas phase as KCl and KOH [133] as well as K_2SO_4 [133,138–140], and they may react with other species present in the gas phase. Zheng et al. [162] concluded that the total K in the fly ash can be predicted when the ash composition of the fuel and straw is known.

Based on observations during co-combustion, Na is likely to be released as NaCl, NaOH, NaSO$_4$, and Na [133] and reacts with the gas-phase compounds. These observations corroborate with the fuel characterization based on the reactive and nonreactive ash-forming matter. This prediction method segregates the ash-forming matter in the biomass fuels in the "reactive" and "unreactive" fraction. More detailed description of this method and other matters related to roles of alkalis in ash is given in the excellent review by Tchapda and Pisupati [53].

7.3.3.3 Ash and Slag Utilizations

Gasification ash may contain high concentrations of unburned carbon and harmful soluble organic compounds restricting its usefulness. If common gasifier is used for coal–biomass mixture, additional impurities like trace metal contaminants may make that ash even less useful. In "nonslagging" biomass alone plant, dry ash produced in the process is useful for the downstream purposes. They can be recycled and used for forestry and agricultural purposes. If ash contains heavy metals such as Cd, Pb, and Zn, it cannot be used for soil amendment. Volatile metals can also gather in the fly ash, making it nonuseful.

In cogasification entrained-flow reactor, oxygen-blown slagging reactors are going to be used to avoid tar formation. Since biomass ash contains potash, phosphates, and calcium and iron oxides, these will melt at high slagging temperature and will not be useful for soil remediation. If fluxing agent with high Si and Al is added to facilitate slagging, slag will be the major solid product of cogasification. The nonleachable slag can be used for structural landfill, blasting grit, roofing tiles, asphalt aggregate, Portland cement aggregate, and other construction building products [49,53].

7.3.3.4 Assessment of Synergy during Pyrolysis

Since pyrolysis controls the initial part of gasification, it is worth examining synergy and catalytic effects during co-pyrolysis of coal and biomass. The understanding of synergy between pyrolysis of biomass and coal requires first understanding of reaction during individual pyrolysis [51]. Biomass contains three substances: cellulose, hemicellulose, and lignin. Pyrolysis at slow heating rate for these three components carried out by Williams et al. [146] showed that cellulose that is high in carbonyl groups releases CO, hemicellulose that is rich in carboxyl group releases CO_2, and lignin that is aromatic releases H_2 and CH_4 due to cracking and deformation of the aromatic C=C and C–H bonds as well as cracking of methyl groups.

For the temperatures less than 927°C, where most pyrolysis process occurs, the literature has presented conflicting reports on interactions among various components of biomass. Some [164] have claimed no interactions among biomass components, while others [165] have claimed interactions among various components of biomass during pyrolysis. In all studies, pyrolysis characteristics and product distributions were found to be significantly affected by the biomass mineral matters and biomass ash. The literature also indicates that during pyrolysis, volatiles come from cellulose and char from lignin and hemicellulose produces both char and volatiles [60].

Since inorganic matters (mineral matters) and ash from biomass have catalytic effects on the yields of volatile matters and char composition, these effects need to be separated while evaluating the synergy between pyrolysis of biomass and coal. It is known from the literature that pyrolysis conditions and the nature of biomass feedstock [146–150] dictate the yield and composition of the volatile products formed during the early stage of the pyrolysis and they influence the char composition and the environment in which char decomposes during the later stage of the pyrolysis.

Tchapda and Pisupati [53] pointed out that there are two indicators for the synergy during co-pyrolysis of coal and biomass:

1. The extent of increase in the production of total volatiles (tars and light gases) and the decrease in char yield
2. Overall decrease in pollutants (oxides of nitrogen and their precursors as well as SO_x and their precursors)

As shown earlier, synergy also plays role in the distributions of H_2S, COS, HCN, NH_3, HCl, and others in the gas phase and in the solid phase like char, ash, and slag. The discussions also indicate the roles of alkalis and mineral matters in these distributions.

During co-pyrolysis of coal and biomass, biomass breaks down early and releases volatiles due to its weak covalent bonds and high oxygen content. This breakdown results in the production of free radicals, which facilitate the decomposition of coal. The hydrogen generated from cracking of heavy and light volatile molecules of biomass can react with the free radicals generated from coal. Thus, biomass can act as a hydrogen donor for coal pyrolysis. This type of hydrogen exchange during co-pyrolysis prevents recombination reactions and reduces the production of less reactive secondary chars [53].

The oxidation of high-heating-value volatile matters coming from coal further increases the pyrolysis temperature. This facilitates the cracking of tar molecules produced from coal and biomass as well as endothermic conversion of biomass and coal chars into gases. The net effects are the increase in volatile yield, decrease in char, and increase in fuel conversion rate. This type of synergy was reported by Shah [52] for the thermal decomposition of PP, petroleum vacuum residue, and biomass blend. As pointed out by Tchapda and Pisupati [53], the synergy occurred because free radicals generated from PP cracking reacted with the thermal decomposition products from petroleum vacuum residue and biomass.

The increase in total volatiles released (tars and gases) and the corresponding decrease in char production have received the most attention for synergistic evaluation during co-pyrolysis of coal and biomass [53]. As mentioned earlier, synergy can also be manifested in the overall decrease in pollutants. This type of synergy can occur through catalytic reactions by mineral matters or through other hydrogen transfer/free radical mechanisms. Tchapda and Pisupati [53] pointed out that the existence of noncatalytic synergistic effects remains controversial. As shown by Tchapda and Pisupati [53] and Shah [52], some found the evidence of synergy and others found no synergy. Some of these synergy effects are also described in References 54–166.

The way biomass and their components are intermeshed with coal can also affect pyrolysis behavior and the product distribution [153,166]. Since blending of coal and biomass/waste leads to an increase of the volatile products (CO, CH_4, and C_nH_m), the energy content of the gas is expected to be higher than when gasifying coal alone because of the high heating value of CH_4 and C_nH_m [153,154,166]. An increase in biomass to coal ratio in the mixture further increases conversion efficiency [155,166]. As shown by Shah [52], an addition of polymeric waste like PE (in coal and pine waste mixture) also increases conversion and heating value due to its higher C and H contents. They, however, found that for mixtures of coal with equal amounts of pine and PE wastes, the gasification results were found to lie between those obtained for compositions with the same amount of only one waste.

Tchapda and Pisupati [53] pointed out that several investigators [144,161,167–183] did not find significant synergy during pyrolysis of coal and biomass mixtures. The reports include mixtures like bituminous coal (Gottelborn hard coal) and straw [180]; coal and sewage sludge [180]; coal and biomass [167]; subbituminous Collie coal with either waste wood or wheat straw [182]; Daw Mill coal (bituminous) and silver birchwood [144]; high-volatility bituminous coal (Drayton coal) and radiata pine sawdust [168]; Malaysian subbituminous coal and empty fruit bunches, kernel shell, and mesocarp fibers of palm tree [183]; lignite blends with olive kernels, forest residues, and cotton residues [181]; and high-volatility bituminous or lignite coals, blended with olive kernel and straw [170] where no synergy was found. More details on these studies are given in an excellent review by Tchapda and Pisupati [53].

Despite the lack of synergy in the co-pyrolysis of mixtures cited earlier, Tchapda and Pisupati [53] point out that many other investigators [75–78,79,184–209] found synergy during mixture pyrolysis. The mixtures include a variety of coals (Wujek, bituminous; Kaltim Prima, bituminous; Turoszow, lignite) with a variety of biomass (pinewood, cellulose, lignin, xylan, and polywax model compounds) [204]; Yallourn coal (lignite) and Taiheiyo coal (subbituminous), with fine powder of homo-PP, low-density PE, or high-density PE [205]; Seyitömer lignite and safflower seeds [206]; coals of different ranks with straw [209]; sawdust and subbituminous coal [191]; and Dayan coal (lignite) and legume straw [186]. For the last two mixtures and others [207], the synergy was pronounced at lower temperatures (around 600°C) but reduced as temperature increased and was less pronounced around 720°C. However, Ulloa et al. [208] noted interaction between subbituminous (Bitsch coal) and bituminous (Lemington coal) coals, with radiata pine sawdust at temperatures higher than 400°C and up to 1200°C. Park et al. [191], Onay et al. [206], and Gao et al. [207] noted that synergy was more pronounced at higher biomass blending ratio (around 70%). Once again, more details on these studies are outlined by Tchapda and Pisupati [53].

The aforementioned results indicate that in general, synergy seems to be more pronounced at intermediate temperatures (not too high nor too low), higher biomass concentrations, and higher concentrations of polymeric wastes. A thorough understanding of free radical formation and their evolution during pyrolysis seems to be the key to the possible interaction between coal and biomass. While extensive studies have been conducted on free radicals release during coal pyrolysis [52,53], such studies involving direct in situ observation of radicals released during biomass or biomass components (cellulose, hemicelluloses, and lignin) or polymeric wastes pyrolysis are lacking [53]. Understanding the mechanism of free radicals release during biomass/biomass components and polymeric wastes pyrolysis can pave the way to improvements in the modeling of coal/biomass/waste blend decomposition during co-pyrolysis and the synergistic effect between coal, biomass, and waste.

7.4 TECHNOLOGIES FOR PARTICULATE REMOVAL

Particulate matter in gasification product streams originates from several sources depending on the reactor types, feedstocks, and process conditions. Particulate matters consist of attrited-bed material and char that becomes entrained in the gas flow. The particle size distribution of this particulate material is a function of the initial size of the bed material. The char tends to be more friable and less dense than other bed material and typically has a smaller particle size distribution. The smallest particles exiting the gasifier tend to be alkali metal vapor condensation aerosols. The concentration of the particles is a strong function of the ash content and ash chemistry of the feedstock [52].

The requirements for a particulate removal depend on the end use of the gas. Gasification coupled with gas engines used for stationary power applications requires particulate loading below 50 mg/Nm^3 (Nm^3 is the normal cubic meter). Particulate loading less than 15 mg/Nm^3 with a maximum particle size of 5 μm is required to protect gas turbines in integrated gasification combined-cycle processes. The most stringent requirements for particulate removal are for fuel synthesis applications that require particulate loadings of less than 0.02 mg/Nm^3 to protect syngas compressors and minimize catalyst poisoning by alkali fumes and ash mineral matter.

Several technologies have been developed and are commercially available for particulate removal from high-temperature gas streams. Choosing an appropriate technology for biomass or waste gasification applications depends on the desired particle separation efficiency for the expected particle size distribution to achieve the ultimate particulate loading based on the end use of the syngas. Pressure drop through the particle removal unit operation and thermal integration are also the key design parameters. Tars produced during biomass gasification also have a significant impact on particulate removal strategies. The operating temperature of the most particulate removal devices should be above the tar dew point to avoid tar condensation and prevent particulate matter from becoming sticky and agglomerating [52].

There are primarily four types of devices that can be used to separate solid particles from a gas stream:

1. Cyclones
2. Filtration
3. Electrostatic separators or Electrostatic Precepitator (ESP)
4. Wet scrubbers

7.4.1 CYCLONES

Cyclones are used as an initial step in the gas cleanup process to remove the bulk of the char entrained in the syngas stream. This technology is standard in industry due to its low cost and high level of performance for removing particulates [52].

Commercial cyclones for particulate removal are well-known and well-proven technologies. A dirty gas stream enters a cyclone separator with high tangential gas velocity and angular momentum forces of particles close to the walls of the cyclones so that they no longer follow the gas stream lines. The cyclone efficiency is a function of gas flow, particle size, temperature, and pressure. Cyclones can be designed for optimum removal of particles of specific size distribution usually down to a lower size limit. Multiple cyclones of different designs can be used in series to achieve near submicrometer particle removal with high efficiency. Alter NRG process has followed this approach [223]. The 2010 Madison Pilot Test with Alter/Westinghouse demonstrated that the refuse-derived syngas from a plasma gasifier carries very fine particulates: 65% of the particulates are less than 1 μm.

In general, cyclones are an effective device to remove solids when particles sizes are greater than 10 μm. A properly designed cyclone with significant pressure drop (and consequently greater power consumption) may separate out small particles down to 3–5 μm. Cyclone separation efficiency is expected to be low in this particle range. In general, the efficiency of cyclones drops considerably for particle size lower than 1 μm. In addition, the hot raw syngas may deposit soot or high-melting-point metals on its surface, which will cause lower efficiency and frequent downtime for its maintenance [49,210–219].

7.4.2 FILTRATION

Operating temperature and product gas composition are the primary process parameters that need to be considered in selecting the appropriate filtering medium. Barrier filters are a technology option for high-temperature particulate removal. Filter housing design and filter media selection are key for optimizing particle capture efficiency within a manageable window of pressure drop across the filter. The initial pore size of the filter medium is the design basis for pressure drop and particulate removal; however, filtration efficiency improves as particles collect on the surface to produce a filter cake, and pressure drop across the filter increases as the thickness of the filter cake increases. Pulsing inert or clean product gas back through the filters dislodges the filter cake and in this way the pressure drop across the filter can be restored to approaching its original performance. Filter housings need to accommodate regular backpulsing to remove the filter cake and be designed so that the particulate matter can be removed before the filter element is brought back online after the backpulse. Improper design and operation can cause the material that was removed to immediately recoat the surface and such that near-original pressure drop is not restored [52].

Generally, nonfabric filters (i.e., candle type) that can sustain high temperature are made of sintered metal, metal mesh, or ceramic. Problems with nonfabric filters operated at 450°C in dirty syngas environment have been reported; breakage and blockages have caused significant downtime. The resistance of metallic filters to corrosive and erosive nature of the raw syngas at high temperatures is another significant concern with their use [49,210–219]. The basic principles of a barrier filter are graphically illustrated in Figure 7.3 [210].

7.4.2.1 Ceramic Filters

Ceramic candles are being developed for high-temperature gas filtration (above 500°C) applications. Ceramic filters can be designed for any flow requirement and can remove 90% of particulates larger than 0.3 μm. In theory, the ceramic filter elements normally made of aluminosilicate or silicon carbide powder with a sodium aluminosilicate binder have exceptional physical and thermal properties and can withstand temperatures up to 1800°F. Thermal shock from repeated backpulsing can cause filters to break, and pore blinding over extended operation can reduce long-term filter

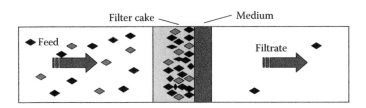

FIGURE 7.3 Principles of barrier filters. (From NREL, Equipment design and cost estimation for small modular biomass systems, synthesis gas cleanup, and oxygen separation equipment—Task 2: Gas cleanup design and cost estimates-wood feedstock, NREL/DOE subcontract report NREL/SR-510-39945, Nexant Inc., San Francisco, CA, May 2006. Also NREL, Golden, CO.)

performance. Candle filters could be used in place of cyclones for char and catalyst separation from the syngas stream. Little commercial experience exists in operating these types of filters at the cyclones temperature (1500°F+). At temperatures below 850°F, ceramic filters have demonstrated satisfactory operational reliability. A recent study performed by Nexant for the DOE NETL examined replacing a third-stage cyclone with a ceramic candle filter. The cost of this high-temperature filter, however, did not justify the change [210].

Sulfur, chlorine, and alkali metal salts can be present in the product gases generated from certain feedstocks. When these impurities contact the ceramic filters or supports, high-temperature reactions can lead to morphological changes and embrittlement, which can also reduce long-term filter performance. Optimizing the seal between the ceramic candles and the metal support plate has been a key technical challenge to overcome success of this technology [210].

7.4.2.2 Metal Candle Filters

Metal filters are used in high-temperature cleanup and it can achieve filtration level as low as 1 μm. Depending upon the material of construction, it can meet any flow requirements and it can operate over a wide range of temperatures. Metal filters made from stainless steel can be used in cleanup systems for temperatures below 650°F, while Inconel or alloy HR filters are suitable for operating temperatures up to 1100°F. Fecralloy can withstand temperatures up to 1800°F, although commercial operation at this temperature has not been demonstrated. Commercial operation of metal filters operating at maximum temperature of 915°F has been successful at a few gasification facilities in Europe. A schematic of candle filter process and candle filter element is illustrated in Figure 7.4 [220,221].

Corrosion and tar deposition on filter elements are the problem areas for these types of filters. The corrosion rate is about 10 times higher than the ones used in heat exchangers. This problem requires a regular maintenance check. The deposited tar also needs to be periodically removed [210].

7.4.2.2.1 Sintered Metal Filters

Sintered metal filter elements are an alternative to ceramic candles. The operating temperature of sintered metal filters is typically lower than that for ceramic filters to minimize sintering. At the appropriate operating temperature, sintered metal filters are more robust than ceramic candles, as the risk for cracking and rupture is much lower. Fabric filters such as flexible ceramic bags are another alternative filter medium that has been commercially proven at lower temperatures, but materials and mechanical compatibility have limited use of these materials for high-temperature syngas filtration [52,210].

7.4.2.3 Low-Temperature Baghouse Filters

Baghouse filters are not appropriate for high-temperature applications. They cannot replace cyclones as an effective solids removal option. Baghouse filters are made of a woven fabric or felted (nonwoven) material to remove particulate materials down to 2.5 μm. While for felted filter systems removal

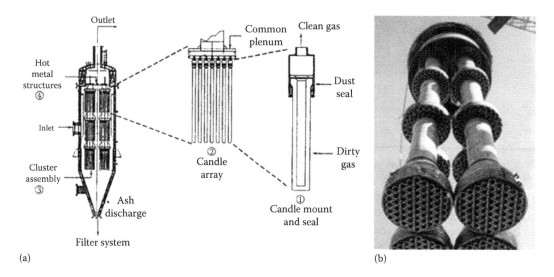

(a) (b)

FIGURE 7.4 (a, b) Candle filter system process. (From Lippert, T.E. et al., Westinghouse advanced particle filter system, Research sponsored by the U.S. Department of Energy's Morgantown Energy Technology Center, under contract DE-AC21-94MC31147 with Westinghouse Electric Corporation, Science & Technology Center, Pittsburgh, PA, 1996, www.netl.doe.gov/publications/proceedings/96/96ps/.../96ps2_1.pdf; Syngas contaminant removal and conditioning-particulate removal, NETL report, Department of Energy, Washington, DC, 2015, www.netl.doe.gov/research/coal/...systems/.../particulate-removal, accessed on April 2015.)

efficiency is constant, for other filters the efficiency can increase with the cake thickness. Baghouse filters are a modular design and thus can accommodate a wide flow range from 1,500 to 150,000 CFM. The air-to-cloth ratio or the ratio of volumetric flow to cloth area sets the size of a baghouse unit. The baghouse filters made of polyester and acrylic are suitable for temperatures below 300°F, while NOMEX, Teflon, Ryton, or fiberglass materials can take temperatures up to 500°F. Baghouse filters are suitable for low-temperature operations. They are often used downstream of cyclones so that particulate loading on the filters can be reduced [52,210].

The periodic bag replacement results in high maintenance cost. It also has a potential for bag fire or explosion. The filter fabrics can also degrade drastically with tar deposition on the fabric surface. This can be avoided by surface treatment such as Teflon coating, precoating with limestone, or other compatible filter aids [52,210].

7.4.3 ELECTROSTATIC PRECIPITATORS

Electrostatic precipitators (ESPs) have been commercially used for particulate removal in the electric power industry for many years and have found wide applications in petroleum refineries for capturing catalyst dust from fluid catalytic cracking (FCC) units. Electric charge is induced on the surface of the particles that are the removed from the gas stream as they follow the electric field lines to a grounded collector plate. ESPs can be applied to particulate removal in high-temperature, high-pressure gas streams. However, maintaining the stability of corona discharge for reliable steady state operation is a technical challenge. Another technical challenge is ensuring materials compatibility of the high-voltage discharge electrodes and other metal internal components with the syngas impurities. The overall size and capital cost of ESPs tend to make them best suited for large-scale operation [52,210].

ESPs are an effective particulate removal device for fine particles. However, ESPs are not suitable for high-temperature applications. Additionally, the soft and sticky nature of the particulates in high-temperature conditions would almost certainly cause widespread fouling and plugging of the

ESP grids. The ESP is more suitable for further use downstream in the syngas-cleaning process to remove the very fine particulates below 0.5 µm at near-ambient temperatures and after bulk particulates have already been removed in the Venturi quench scrubber. Since dry ESPs can only operate up to 750°F and wet ESP up to around 200°F, this option cannot replace cyclones for solids removal. In addition, the high cost and waste streams produced in this process make them unattractive relative to other filtration systems [210].

There are two types of ESP—dry and wet. The selection of the wet ESP as the final particulate removal unit is driven by several factors. With the specified particulate loading in the syngas, the saturation environment inherent in the wet electrostatic precipitation (WESP) reduces the possibility of igniting the syngas with the electric charge and, consequently, an energy release. The high solid loading of extremely fine particulate may cause space charge in a dry ESP. WESPs can achieve up to several times the typical corona power levels of dry precipitators, thus greatly enhancing the collection of submicron particles. In wet precipitators, reentrainment in the last field is virtually nonexistent due to adhesion between the water and collected particulate. Also, the gas stream temperature is lowered to the saturation temperature, hence promoting condensation and enhancing particulate collection. In recent years, a saturator is added before the WESP to ensure the inlet syngas is at its saturation point [52,210].

WESPs are highly efficient collectors of submicron particulate, including condensable aerosols. In the WESP, the remaining or condensed particulate in the syngas will be given an electric charge as the syngas stream is distributed to flow over and around wetted tubes with the opposite, attractive electric charge. The electrically charged particulates are attracted to the tubes by a strong electric field, where they become trapped in a sheet of water flowing over the surface of the tubes. Once captured, these particles are washed to the base of the unit where they are blown down to the process effluent treatment plant. WESPs are known to also capture mercury and organic aerosols that may condense from syngas at saturation temperatures [49,210–219].

The high-voltage system for the WESP consists of a rigid high-voltage frame suspended at four points by robust ceramic insulators. The duration and frequency of the spray depends on the severity of the application but is typically set at 2 min every 6 h. Particulates are flushed from the system and piped to the wastewater treatment system where they are fed to a clarifier and ultimately returned to the gasifier as filter cake.

7.4.4 Wet Scrubbers

As mentioned earlier, for dry feed gasification, cyclones and candle filters are often used to recover most of the particulates for recycle to the gasifiers before final cleanup with water quenching and scrubbing. In addition to fine particulates, chloride, ammonia, and some H_2S and other trace contaminants are also removed from the syngas during scrubbing process. The scrubbed gas is then either reheated for COS hydrolysis and/or sour WGS when required or cooled in the low-temperature gas cooling system by generating low-pressure steam, preheating boiler feed water, and heat exchanging it against cooling water before further processing. Spent water from the scrubber column is directed to the sour water treatment system, where it is depressurized and decanted in a gravity settler to remove fine particulates [52].

There are four major categories of wet particulate scrubbers:

1. Spray tower scrubber
2. Packed-bed or tray-type scrubbers
3. Mechanically aided scrubbers
4. Venturi scrubbers

The separation principle of wet scrubbers is that particulates are knocked out with water droplets by various means. The effectiveness of different devices depends on particle size, water droplet size, density of water droplets, etc. Residence time can be an important factor in determining efficiency, depending on design principles.

The mechanically aided scrubber is similar to a fan equipped with water sprays. Due to the high temperature and highly erosive and corrosive nature of syngas, this type of separator is not considered suitable for the application [49,210–219].

For high particulate solid loading (30,000 mg/Nm^3 dry basis) and small particle size (65% of the particulates are less than 1 μm), spray-type, tray-type, and packed-bed scrubbers are not believed to be suitable. A search of open literature publications and discussions with equipment suppliers uncovered no instances where such scrubbers have been used in similar applications. Generally, these devices are large vessels, intended to provide sufficient residence time for particle separation. However, the long residence time would allow chemical reactions in the hot syngas to take place and would likely produce undesirable chemical species by-products, such as dioxin. Additionally, tray-type and packed-bed scrubbers have intricate narrow flow passages where blockage and/or fouling can easily result, thus leading to plant shutdown for maintenance [49,210–219].

Wet scrubbing systems use liquid sprays either water or chilled condensate from the process to remove particulates that collide with liquid droplets. The droplets are then removed from the gas stream in a demister. Venturi scrubbers are the most common wet scrubbers but often require a relatively high-pressure drop to circulate quench liquid to be sprayed into the gas stream. Because wet scrubbing requires a liquid quench medium, operating temperatures need to be less than 100°C. This requires significant gas cooling for removing the sensible heat from the product gases. Heat losses from the wet scrubbing systems can adversely affect the energy efficiency of the overall process. On the other hand, for indirect gasification systems, the excess steam that is used as the gasifying agent needs to be quenched and recovered. Wet scrubbing systems are inevitable in indirect gasification systems to remove excess water vapor prior to compression and downstream syngas utilization [52].

In a typical process, the hot syngas from the gasifier first passes through a cyclonic separator for removal of particulate greater than 10 μm. The syngas is then passed through an evaporative quencher to cool the gas to saturation. After the quencher, the syngas enters a condenser/absorber (C/A) tower to subcool the syngas. The packing is wet-film, random dumped polypropylene packing. Process water is collected in the C/A sump and recirculated to the quencher and to the C/A through a fin tube cooler. A blowdown stream is taken from C/A to purge solids from the system. The subcooled syngas leaves the C/A and enters a highly efficient Venture scrubber. The Venturi scrubber removes fine particles before the gas enters a Chevron style entrainment separator for water drop removal. Finally, the gas passes through a multistage centrifugal blower to compress the gas before it goes to the engine. Recirculation liquid from the Venture is collected in the sump of the entrainment separator vessel. The liquid is recirculated to the Venturi scrubber with a recycle pump. The Venture overflow is pumped to the C/A sump. The effect of subcooling is to grow the particulate making it easier to collect them in the downstream of Venturi scrubbers [52,211].

Air Products' design [211] for syngas cleaning used Venturi scrubbing in two stages with a downstream liquid/gas separator. The Venturi quench scrubber is designed to reduce syngas temperature from 850°C to ~85°C and solid loading from 30,000 to 500 mg/Nm^3 (dry basis). In addition to cooling the syngas to near-saturation temperature, the dual Venturi scrubber will also remove a high percentage (around 99%) of solid particulate over 5 μm in size that may be carried over from the gasifier, thereby minimizing downtime. The convergent section of the Venturi is covered with a water film to eliminate particles sticking to the wall. High gas velocity in the Venturi throat causes violent agitation in the water film creating numerous fine droplets to capture the particulates. Because of the close contact between the water and hot gas, a rapid temperature drop takes place over a very short distance and time and brings all chemical reactions to an abrupt stop. The addition of caustic (NaOH) to control the resultant water stream to a pH range of 6.5–7 will remove greater than 99% of HCl and HF and two-thirds of ammonia from the syngas stream. Separation efficiencies of 99% have been commercially demonstrated in a two-stage unit. The solid-loaded water downstream of the Venturi scrubbers is introduced tangentially along the liquid/gas separator wall to enhance the separation and to ensure that no liquid (containing solids) entrained in the gas stream exits the unit. Rapid quench of the syngas will prevent recombination of organic molecules into complex organics such as dioxins or furans. The captured impurities will be removed from the initial wet scrubber system as "blowdown," which is transferred to the on-site effluent treatment system.

The cooled and partially cleaned syngas leaves the quench scrubber at a subambient pressure. Before it can be further processed to remove submicron particulates by an ESP, the pressure needs to be increased above an atmospheric level with a blower for safety reasons. The syngas exiting the Venture quench scrubber at 85°C has a moisture content of up to 59%. In order to reduce the gas volume for compression, a cooler is installed before the blower in order to condense out moisture. A direct contact cooler (DCC) is designed to cool the syngas to 46°C and reduce the syngas volume by 57% primarily due to moisture condensation.

The DCC is a packed-bed column with a circulating loop of cooled water that will condense water from the syngas stream. Hot water collected at the bottom of the DCC is pumped through the DCC heat exchanger where the circulating water is cooled before reentering the DCC. Water is purged from the system before the DCC heat exchanger and this purged water is recirculated back to the scrubber as quench/scrubbing water.

7.4.5 TECHNOLOGIES USED IN TYPICAL INDUSTRIAL PROCESSES

Removal of particulates is an important problem for the gasification of MSW. In PyroGenesis process [225], the synthesis gas coming out of the primary gasifier carries fine particles that mainly include ash fines. A refractory-lined cyclone designed to operate at high temperatures is used to capture these entrained solids and return them to the furnace for further treatment. This is then sent to secondary gasification furnace.

The gas coming out the secondary gasification process is quenched by water. The quench is designed to rapidly cool high-temperature (1000°C) gases that are essentially at thermodynamic equilibrium down to below 90°C in less than 0.5 s. This shock cooling of hot gases eliminates the possibility of reformation of dioxins and furans. The quench is designed to cool the hot gases to their saturation temperature by evaporation of an injected spray water. Typical off-gas outlet temperatures range from 70°C to 90°C.

The Plasco Energy system process [222] uses (1) dry quench, (2) main baghouse, and (3) main carbon bed filter to remove particulate matters. This is thus a low-temperature system. The purpose of the dry quench is to cool syngas down to 200°C–230°C, to reach the design temperature requirements for the rest of the gas-cleaning systems. The cooling eliminates the formation of dioxins and furans.

The main baghouse used by Plasco Energy systems contains a number of collectors that provide filtration capabilities to separate dust particles from syngas. Baghouses are one of the most efficient and cost-effective types of dust collectors available and can achieve a collection efficiency of more than 99% for very fine particulates. A baghouse captures fly ash and heavy metals from the gas stream using bags that are precoated with a mixture of feldspar and activated carbon. Particulate from the baghouse goes to bins for later disposal. Plasco Energy systems guarantee removal of 99.9% for particulate matter, 99.65% for Cd removal, 99.9% for lead removal, and 90% for mercury removal.

Alter NRG process [223] first cools gas by a quench followed by coarse particulate removal system that is followed by removal system of a fine particle and heavy metals.

7.5 TECHNOLOGIES FOR TAR REMOVAL

Tars contain polyaromatic hydrocarbons and they are the most difficult to remove during gasification process. Gasifier plants that do not use thermal tar destruction generally have lower outlet temperatures but higher energy content in the syngas. Numerous approaches, physical, chemical, and catalytic have been attempted both commercially and in pilot-scale units. Here we briefly summarize our state of the art in each of these approaches.

7.5.1 PHYSICAL

In this method, the objective of tar removal unit is twofold: (1) to get as much of the tars as possible in the condensed phase and (2) to maximize tar collection and removal, particularly for tar droplets larger than 1–3 μm. Tars into condensed phase early in the gas-cleaning system keep them for condensing in downstream process equipment that include recirculation lines, pumps, heat exchangers, and media beds. The physical methods involve wet scrubbing (Envitech) [224] and filtration. Wet scrubbing to condense tars out of product gas is an effective gas conditioning technology that is commercially available and can be optimized for the tar removal. Wet scrubbing technique does not eliminate tars but transfers it to oil or water phase that will then have to be appropriately treated. The treatments of oil or water phase in wet scrubbing technique are important. Also, when the tar is removed from the product gas stream, its fuel value is lost and the overall energy efficiency of the integrated gasification process is reduced [52].

7.5.1.1 Oil Loop Tar Removal

Envitech process [224] uses two-stage physical process for tar removal. In the first stage, it uses an oil-based scrubber to condense large particles and heavier tars. Oil loop makes sense when (1) tar concentration is high, (2) tar condenses as sticky solid rather than an oily liquid, and (3) at high degrees condensed tars don't mixed well with the water. Under these conditions, oil scrubber provides a medium where tars will be more soluble and less prone to foul the recirculation lines. Tar condensation and removal occurs in a downflow, cocurrent, direct contact condenser cooler followed by a Venturi scrubber and mist eliminator. It operates at temperatures above the water dew point for the gas. Direct contact with recirculated oil is used in conjunction with a liquid cooling circuit to cool the gas to an intermediate tar condensation temperature. The intermediate cooling temperature is determined by evaluating the fusion temperature distribution in conjunction with vapor pressure data for the specific components. A horizontal Venturi scrubber and mist eliminator are installed downstream of the condenser cooler to collect and remove condensed tars and particulates. A waste oil stream purges the system of collected tars and particulates. The waste oil can be filtered for recycling back to the oil loop. The collected sludge is fed back to the gasifier for reprocessing [210–219].

7.5.1.2 Water Loop Tar Removal

While some syngas compositions will require an organic recirculation liquid, others will be suitable for a water loop configuration. The syngas is first cooled to saturation in a vertical downflow quencher using water. This is followed by a vertical Venturi scrubber to collect the condensed tar and particulate. After the Venturi scrubber, the syngas passes through a mist eliminator to remove the water droplets in the syngas. The Venturi scrubber recirculation water passes through a water-recycle unit to separate the oil from water. The recycled water will be free of oils that either float or sink. The collected oils can be sent back to the gasifier. In Envitech process [224], oil loop and water loop are operated as two stages for tar removal.

7.5.1.3 Venturi Wet Scrubbers

Wet scrubbing is generally used to remove water-soluble contaminants from syngas by absorption into a solvent. It can be used for water-soluble tar components, particulates, alkali species, halides, soluble gases, and condensable liquids. A Venturi scrubber and mist eliminator is used to collect particulates and condensed tars in the recirculation liquid in both oil and water loop configurations. The performance will be similar in either configuration and is not dependent on the type of the recirculation liquid. Rather it depends on particle size and the pressure drop [212,213].

The literature data show that Venturi scrubber is effective at removing roughly 90%–99% of condensed tars and particulates larger than 1 μm. The flow range for a single-throat Venturi is 500–100,000 SCFM. Flow above this range requires either multiple Venturi scrubbers in series or multiple-throat Venture. Venturi scrubbers with a quench section can accommodate high-temperature gas streams up to 450°F and they can operate over a wide range of pressures. The standard materials for Venturi scrubbers are carbon steel. For corrosive high-temperature applications, stainless steel or special alloys such as fiberglass-reinforced plastics (FRPs) and Inconel are used. The disadvantages of scrubbers include high-pressure drop, the need for effluent wastewater treatment prior to disposal, and heat loss due to quenching.

7.5.2 CHEMICAL/CATALYTIC TREATMENTS

A number of chemical and catalytic treatments can also be used to remove tars from syngas.

7.5.2.1 Thermal Cracking

Thermal cracking of tars using solid acid catalysts such as silica–alumina and zeolites is well known. The catalyst with strong acid sites can crack hydrocarbons above 200°C. The cracking of aromatics results in the formations of hydrocarbons and coke. Gil et al. [216] showed that cracking of tars with spent FCC catalyst reduced the tar concentration from 20 to 8.5 g/Nm^3 at 800°C–820°C compared to the tar concentration reduction from 20 to 2 g/Nm^3 with dolomite catalysts under the same operating conditions [216]. The mechanism of cracking with these two catalysts may be different (i.e., cracking for FCC catalyst versus steam reforming for dolomite).

7.5.2.2 Hydrogenation

While hydrogenation of polyaromatic tar compounds is possible, it is thermodynamically not very favorable at higher temperatures. It is however possible to have a ring opening activity at temperatures as low as 200°C for Rh, Pt, Ir, and Ru catalysts [210].

7.5.2.3 Plasma Conversion

In PyroGenesis process [225], the soot and complex hydrocarbons coming out of cyclones and gasifier tars are converted to syngas by a secondary plasma gasifier. The secondary plasma gasifier is essentially a nozzle in which air, moisture, and synthesis gas are mixed together with a plasma jet. The plasma jet provides the activation energy for the reaction between any hydrocarbons and other organic molecules contained in the synthesis gas with the moisture and oxygen in the air. By controlling the chemistry in the secondary gasifier, all the organic molecules are converted to hydrogen and carbon monoxide. The main gasifications occurring in the secondary gasifier are as follows:

$$C + O_2 \rightarrow CO_2 \quad \text{(exothermic)} \tag{7.3}$$

$$C + H_2O \rightarrow CO + H_2 \quad \text{(endothermic)} \tag{7.4}$$

$$C + CO_2 \rightarrow 2CO \quad \text{(endothermic)} \tag{7.5}$$

$$CO + H_2O \rightarrow CO_2 + H_2 \quad \text{(exothermic)} \tag{7.6}$$

PyroGenesis has also used the plasma torch for other types of waste-to-energy applications.

7.5.2.4 Catalytic Steam Reforming

Catalytic steam reforming of tars is a developing technology with significant research interests in recent years. In this technology, tars are reformed by steam and a catalyst in a fluidized-bed reactor to produce syngas. This technique offers several advantages: (1) catalyst reactor temperatures can be thermally integrated with the gasifier exit temperature, (2) the composition of the product gas can be catalytically adjusted, and (3) steam can be added to the catalytic reactor to ensure complete reforming of tars. The subject of catalysts for steam reforming of tar and different novel approaches taken for this process are well reviewed in the literature [52,210,217,218]. Among others, the catalysts that are well tested are (1) disposable catalysts such as dolomite and olivine; (2) conventional Ni and other catalysts that include ruthenium, iron, manganese, potassium, and barium; (3) catalysts with novel supports such as perovskites and hexaaluminate; and (4) more bi- and even trifunctional catalysts. Of all the catalysts tested, disposable catalysts or commercial Ni catalysts are preferred due to their low costs, durability, and long life. A proprietary nickel monolithic catalyst has also shown considerable promise for destruction of biomass gasification tar [210,217,218].

The catalytic reformer can also remove light hydrocarbons such as benzene and ammonia in addition to tar. A few large-scale biomass gasification facilities such as Carbona in Denmark and the FERCO gasifier in Vermont have demonstrated a novel catalyst in their tar crackers since commercial catalysts are too friable for this application [210]. The FERCO tar cracker removed 90% of the tar in the syngas stream using a novel catalyst known as DN34 [210,217,218]. In both of these processes, a wet scrubber was used downstream of the tar cracker to remove residual tars and impurities.

A tar cracker known as the reverse-flow tar cracking (RFTC) reactor developed by BTG uses steam reforming process with a commercial Ni catalyst [52,210,217,218]. Since Ni catalyst is sensitive to sulfur, the process requires the removal of sulfur before the use of RFTC reactor. Due to cooling requirement of the desulfurization process, syngas is fed to the reactor at 660°F–1200°F temperature and is heated to 1650°F–1740°F in the reactor entrance section. The main reactions in RFTC reactor are as follows:

$$C_nH_m + nH_2O \leftrightarrow nCO + \left(\frac{1}{2}m + n\right)H_2 \quad \text{Hydrocarbon reforming} \tag{7.7}$$

$$2NH_3 \leftrightarrow N_2 + 3H_2 \quad \text{Reverse ammonia synthesis} \tag{7.8}$$

$$CO + H_2O \leftrightarrow CO_2 + H_2 \quad \text{Water gas shift reaction} \tag{7.9}$$

A small amount of syngas is combusted to counterbalance the endothermic tar reforming reactions:

$$H_2 + \frac{1}{2}O_2 \rightarrow H_2O \tag{7.10}$$

$$CO + \frac{1}{2}O_2 \rightarrow CO_2 \tag{7.11}$$

$$CH_4 + 2O_2 \rightarrow CO_2 + 2H_2O \tag{7.12}$$

The typical conversion for the RFTC reactor is described in Table 7.2 [52,210].

TABLE 7.2

Conversion Efficiency of Various Components by Reverse-Flow Tar Cracking

Components	Conversion (%)
Benzene	82
Naphthalene	99
Phenol	96
Total aromatic	94
Total phenols	98
Total tar	96
Ammonia	99

Sources: Shah, Y., *Energy and Fuel Systems Integration*, CRC Press, Taylor & Francis Group, New York, 2015; Equipment design and cost estimation for small modular biomass systems, synthesis gas cleanup, and oxygen separation equipment—Task 2: Gas cleanup design and cost estimates-wood feedstock, NREL/DOE subcontract report NREL/SR-510-39945, Nexant Inc., San Francisco, CA, May 2006.

TABLE 7.3

Comparison of Syngas Reforming Process Technologies

Reforming Process	H_2/CO	Comments
Tar cracking/reforming	Wide range	Developing technology; detailed information not available
Steam reforming	3 or higher	Industrially proven for hydrogen production; high efficiency
Dry reforming	Around 1	Still being developed; very important
Partial oxidation	1.7–2.0	Exothermic; low efficiency; used to upgrade heavy liquid fuels
Autothermal reforming	2.4–4.0	Hybrid (PO and SR); no external heat
Trireforming	Wide range	Hybrid (SR, DR, PO); no heat required

The partial oxidation reaction was also investigated as the possible process for tar removal. In this process, syngas enters the reactor with oxygen at 300°F and leaves at about 2500°F due to highly exothermic oxidation reactions. Tar, methane, light hydrocarbons, and benzene are converted to syngas. The disadvantage of this process is the low product heating value due to combustion of CO and H_2 as well. Also it is very difficult to selectively react methane by this process. The composition of the product is also shifted to lower H_2/CO ratio.

An autothermal reactor that includes both steam reforming and partial oxidation has also been successfully applied. Such an application would only apply to a particulate free gas since any particulate in the gas could shortly blind catalyst surface. Table 7.3 shows a comparison of various syngas reforming technologies.

7.6 TECHNOLOGIES FOR SULFUR REMOVAL

7.6.1 LOW-TEMPERATURE PROCESSES

For low H_2S concentration (<2000 ppm) and low sulfur quantities (<3 TPD), an absorption process with chemical agents that are selective toward H_2S would be most suitable. Some of the low-temperature sulfur removal processes that are widely used in industries are briefly described here:

1. THIOPAQ is a biotechnological process for removing H_2S from gaseous streams by absorption into a mild alkaline solution followed by the oxidation of the absorbed sulfide to elemental sulfur by naturally occurring microorganisms. The process was first used in 1993. Its application to syngas cleaning is not demonstrated.

   CrystalSulf is a new process to remove H_2S. It was developed for natural gas environment of high pressure and high CO_2/H_2S ratio containing heavier hydrocarbons. The process uses flammable and hazardous chemicals for absorption and sulfur oxidation. SO_2 is used as a reactant to undergo the modified Claus process for sulfur formation. The process needs to be demonstrated in a wider range of operating conditions.

2. For low temperature and high pressure, physical solvent processes such as Rectisol/Selexol are also used. The Rectisol process that uses methanol at temperatures less than 32°F can achieve a sulfur removal level as low as 0.1 ppm. The Selexol process that uses mixtures of dimethyl-ethers of PE glycol can achieve a sulfur level of 1 ppm. Selection of materials of construction depends on the solvent used. For example, Rectisol process uses mostly stainless steel contributing a significant capital cost. In Selexol process, carbon steel is the standard material of construction except for those areas with high severity where stainless steel is used.

Process 1 is a purely biological process that would be slow and will not be suitable for removal of sulfur from syngas that is unclean and will require significant cooling. Pure biological process may be affected by tars, particulate matters, and other chloride and heavy metal impurities. Process 2 is more applicable for sulfur removal in high-pressure operations [52].

In the following paragraphs, we examine four other low-temperature sulfur removal processes.

7.6.1.1 SULFURTRAP Process by Chemical Products (Oklahoma City, Oklahoma)

SULFURTRAP is a nonhazardous solid adsorbent used for the removal of hydrogen sulfide and light mercaptans from gas (or liquid) hydrocarbon streams. It can be operated as a fixed-bed process that has excellent sulfur loading capacity, minimal replacement time, and reliable performance. In typical configurations, SULFURTRAP is loaded in parallel vertical vessels where gas flows through vessels. The sulfur compounds chemically react with SULFURTRAP to form a stable, safe, and nonhazardous by-product. Depletion of the scavenger is dependent only upon the concentration of hydrogen sulfide and flow rates through the bed. SULFURTRAP process can be applied to natural gas, acid gas, syngas, landfill gas, geothermal gas, and tank vent gas. It has high sulfur removal capacity (outlet conc. <1 ppm), minimal replacement time, no caking/clumping, and good effectivity under a wide range of conditions [229].

The application of this process for sulfur removal requires good understanding of the effects of tar, particulate matter, chlorine, and heavy metals in the syngas on the adsorbent performance. The effectiveness of the process in the presence of chlorine is of particular concern. The information on how high-temperature process can operate, what maximum allowable flow rate is important, and more importantly what residence time in the column is required is all needed. This information will set the requirements of the diameter, the length of each column, and the number of columns. Since literature appears to be largely related to small flow operations, more knowledge on commercial experience on large-scale operation and flexibility in capital and operating costs for large-scale operation needs to be understood. Finally, the use of this process for COS removal needs to be determined. If COS is to be removed by hydrolysis process, should it be carried out before or after H_2S removal by this process? [229].

7.6.1.2 Low-Temperature Amine Process (Bionomic Industries, Mahwah, New Jersey)

Amine processes are proven technologies for the removal of H_2S and CO_2 from gas streams by absorption. Amine systems generally consist of an absorber, a stripper column, a flash separator, and heat exchangers. This is a low-temperature process in which gas usually enters the absorber at about 110°F. The acid gases are removed in the absorber by chemical reactions with the amine solution. Regeneration of rich amine is accomplished through the flash separator followed by a stripper column to recover H_2S and CO_2 from amine solutions. A basic simplified diagram of acid removal process is illustrated in Figure 7.2. The lean amine solution is recycled back to the absorber and stripped gases are sent to a sulfur recovery unit. Amine systems normally operate in the low- to medium-pressure range of 70–360 psi. Higher pressure makes the process more expensive. A sulfur removal level as low as 1 ppm is possible; however, this increases operating costs due to large solvent circulation rate. Commercially, available amines are as follows:

1. *MEA*: Monoethanolamine removes both H_2S and CO_2 and is generally used at low pressures and in operations that require stringent sulfur removal.
2. *DGA*: Diglycolamine is used when there is a need for COS and mercaptan removal in addition to H_2S. DGA can hydrolyze COS to H_2S; thus, COS hydrolysis unit is not needed in the cleanup system.
3. *DEA*: Diethanolamine is used in medium- to high-pressure systems (above 500 psia) and is suitable for gas stream with high ratio of H_2S to CO_2.
4. *MDEA*: Methyldiethanolamine has a higher affinity for H_2S than CO_2. MDEA is used when there is low ratio of H_2S to CO_2 in the gas stream so that H_2S can be concentrated in the acid gas effluent. If Claus plant is used for sulfur recovery, a relatively high concentration of H_2S (>15%) in the acid gas effluent is required for optimum Claus operation.

After prolonged use, MEA, DGA, and MDEA solutions accumulate impurities that reduce their H_2S removal efficiency. A reclaim unit is needed to bring back the H_2S removal efficiency. One problem with amine system is corrosion. In water, H_2S dissociates to form weak acid and CO_2 forms carbonic acid, both of which corrode metal. The equipment in amine systems may be needed to clad with stainless steel to improve equipment life.

While the capital cost for the amine process is low, the operating cost for amine can be very high. In order to remove COS, DGA would have to be used. This is even more expensive than either MEA or DEA. If MEA or DEA is used, COS must be removed by hydrolysis process. The effectiveness of amine process in the presence of high concentration of HCl is not known. It is likely that HCl would have to be removed before using the amine process.

The amine process would be able to remove H_2S to the desired level: less than 5 ppm. In principle, amine process is simple to build and operate. It has been widely used in the industry [49,210,217,218].

7.6.1.3 MV Technologies H2SPlus System

MV Technologies H2SPlus™ System is a proven superior dry chemical scrubber system that effectively and efficiently removes hydrogen sulfide from gas streams. The technology is based on their performance upgrade of commercially available iron sponge media, which itself is hydrated iron oxide on a carrier of wood fibers and has been used for more than a hundred years to treat gas streams. This enhanced iron sponge media is called BAM™ [226].

Biological agents are added to compound BAM, to remove numerous non-sulfur-bearing species such as VOCs and readily biodegradable organic compounds, to provide improved reoxidation of the iron sulfide compounds (troilite and pyrite), and to regenerate iron oxides for subsequent reactions. All of this plus other system design features add up to the extended life of the reactive media and lower cost of operation.

The basic reactions are as follows:

$$Fe_2O_3 \cdot H_2O + 3H_2S \rightarrow Fe_2S_3 + 4H_2O \tag{7.13}$$

$$Fe_2O_3 \cdot H_2O + 3H_2S \rightarrow 2FeS + S + 4H_2O \quad \text{(removal of } H_2S\text{)} \tag{7.14}$$

$$2Fe_2S_3 + 3O_2 + 2H_2O \rightarrow 2Fe_2O_3 \cdot H_2O + 6S \quad \text{(regeneration of iron oxide)} \tag{7.15}$$

The spent media is nonhazardous, passes the U.S. Environmental Protection Agency's TCLP tests, meets U.S. Occupational Safety and Health Administration definitions of a "not readily ignitable solid," and can be composted, land applied, or disposed of in landfills.

MV Technologies has applied this biologically enhanced dry-scrubber technology for odor and H_2S removal for more than a decade in a wide variety of applications for a broad base of clients in the food, beverage, and agricultural processing industries, petroleum refining and processing, and landfill and municipal waste treatment installations.

MV Technologies has incorporated H2SPlus Systems in many gas treatment/scrubbing applications for anaerobic systems and municipal wastewater treatment plants, including high-rate bioreactors, low-rate lagoon style digesters, and medium-rate solids digesters. These systems have been designed and installed on digesters generating biogas from dairy, beef and chicken manure, beer manufacturing waste products, meat-packing facility effluent streams, and municipal digesters. In all cases, longer operating life and significantly lower operating cost compared to other H_2S removal systems have been demonstrated. MV has successfully retrofitted a number of iron sponge scrubbers and demonstrated a media life almost 3× longer, hence providing a dramatic reduction in operating costs [226].

The technology claims to *provide for easy and rapid media change out*; a media change in a single vessel can typically be accomplished in less than 24 h. It *provides the lowest cost per pound of H_2S removal and operates at 100% effectiveness immediately upon start-up* and does not require any offline run-in or maturation period, as is standard in biological scrubbers. It is also not *susceptible to upset conditions* such as large swings in H_2S concentrations or changes in temperature that have impact on biological scrubbers.

The technology features a "set and forget" design; unless emission standards may demand more frequent sampling, typically operator attention is required only periodically (three times per week) to measure H_2S concentrations. The process *converts all of the H_2S in your biogas stream into pyrites and elemental sulfur*. Spent media can be land applied or disposed of as nonhazardous waste.

The process requires no additional material load to be placed on wastewater treatment system. It is *easy to install*; in most all cases, because of the low-pressure drop and simplicity of the design, the units can be added to an existing gas delivery system with little ancillary modification [226].

The process is biochemical and slow. Thus, high retention time is required. The process can deliver up to 50 ppm outlet concentration. The performance beyond that would require several modifications. Because of high retention and low-flow requirements, the required land area and the number of vessels can be very large for high-flow processes. This makes the application somewhat impractical for high-flow processes.

It isn't clear how the performance would be affected by the presence of large HCl, heavy metal concentrations, as well as some tar and particulate concentration in the syngas. The operating cost for this process is very low. The capital cost is also reasonable.

The process is more designed for low-flow (like landfill gas treatment) system. The commercial experience for syngas cleanup is lacking. So far, no waste to syngas process has used this method for sulfur removal. The process does not appear to remove COS.

7.6.1.4 Merichem LO-CAT™ Process for Sulfur Removal

Merichem's LO-CAT process for H_2S removal is based on iron chelate redox system. This is an oxidation process that uses iron catalyst held in a chelating agent to oxidize H_2S to elemental sulfur. H_2S is absorbed in the solution and sulfur is oxidized by iron reduced from +3 state to +2 state. Iron ions are held in the solution by organic chelating agents. While H_2S is the only acid gas being removed in this process, high CO_2 concentration in the feed gas requires caustic for pH adjustment. The process includes an absorber, an oxidizer for catalyst regeneration, and a sulfur-handling unit.

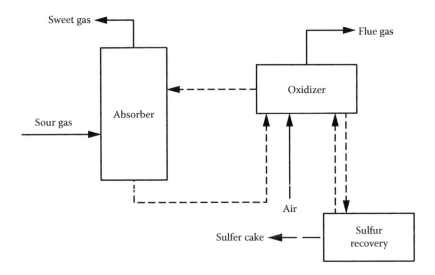

FIGURE 7.5 Typical LO-CAT system flow diagram. (From NREL Equipment design and cost estimation for small modular biomass systems, synthesis gas cleanup, and oxygen separation equipment—Task 2: Gas cleanup design and cost estimates-wood feedstock, NREL/DOE subcontract report NREL/SR-510-39945, Nexant Inc., San Francisco, CA, May 2006. Also NREL, Golden, CO.)

When the gas stream comes into contact with the LO-CAT solution in the absorber, H_2S in the gas stream is converted into sulfur. The spent catalyst is regenerated by oxygen in the oxidizer and an elemental sulfur is concentrated into a sulfur slurry. The sulfur slurry is then washed to recover any entrained catalyst in the sulfur handling unit. The sulfur recovered in the process contains some entrained residual catalyst, and this low-value sulfur is used for agricultural purposes [227]. A typical LO-CAT system is graphically illustrated in Figure 7.5 [210].

The process is most useful for small-scale applications where sulfur recovery is less than 20 TPD. The process can achieve 99.9% H_2S removal efficiency. The process can operate between atmospheric pressures to as high as 600 psi pressure. The temperature is normally maintained at about 110°F. High temperature decreases its efficiency. The process can treat a wide range of gas compositions, has a significant turndown flexibility, and has less capital cost compared to Claus process with the associated tail gas treating unit. Since the process only treats H_2S, a COS hydrolysis unit upstream of the process is needed to hydrolyze COS. Other acid gases such as HCN and mercaptans are removed by wet scrubbing. The process system uses stainless steel as material of construction. If there is a buildup of chloride ions from the feed gas, FRPs are used to provide added stability for the stainless steel components [227].

The LO-CAT process, available exclusively from Merichem [227], is a patented liquid redox system that uses a proprietary-chelated iron solution to convert H_2S to innocuous, elemental sulfur. It does not use any toxic chemicals and does not produce any hazardous waste by-products. The environmentally safe catalyst is continuously regenerated in the process.

The LO-CAT technology is applicable to all types of gas streams, including air, natural gas, CO_2, amine acid gas, biogas, landfill gas, and refinery fuel gas. Flexible design allows 100% turndown in gas flow and H_2S concentrations. With over 35 years of continuous improvement, LO-CAT units are very reliable and require minimal operator attention; many licensees report as little as 1.5 man-hours per day and over 99% on stream efficiency.

From engineering and fabrication, to installation supervision, training, and start-up, through process warranties and on-site service, Merichem provides a total sulfur recovery solution. Each system is custom designed and built to your specifications and aggressive schedules can be accommodated. Full equipment packages are provided for stick-built or modular configurations [227].

7.6.1.4.1 LO-CAT Direct Treatment Scheme

Whenever the treated gas cannot be combined with air, a direct-treat design is employed. This is achieved by use of two separate vessels, an absorber and an oxidizer. The absorber treats the sour gas, producing sweet gas in a single pass. The oxidizer serves two purposes: the regeneration of spent catalyst and the concentration of sulfur particles into a slurry. The proprietary sulfur recovery system takes the sulfur-rich slurry, washes it, and produces an elemental sulfur cake.

7.6.1.4.2 LO-CAT AutoCirc Scheme

When treating a gas that can be mixed with air, the AutoCirc design provides significant cost savings in both operating and capital expenses. By combining the absorber and oxidizer in one vessel, the solution circulation pump is eliminated resulting in reduced electricity consumption. The single vessel approach also minimizes footprint.

For treating amine acid gases and other nonexplosive, low-pressure gas streams, the patented Autocirculation system may be used. In the Autocirculation design, the absorber and oxidizer vessels are combined into a single vessel eliminating one of the vessels, the solution circulation pumps, and all associated piping and instrumentation. This design lowers both the capital costs and the operating costs when compared to conventional desulfurization systems with the two vessel design.

The sulfur produced in the LO-CAT® process is removed from the system in a variety of ways. Current users dispose of their sulfur as a 15 wt% slurry, a 30 wt% sulfur cake, and a 60 wt% cake and as molten sulfur. The sulfur filters range from low-cost bag filters to high-efficiency vacuum belt filter systems. The type of filtration is matched to the amount of sulfur produced and the users' requirements.

The LO-CAT process converts H_2S to innocuous, elemental sulfur using a patented, dual-chelated iron catalyst that is environmentally safe. The overall process reaction is

$$H_2S + \frac{1}{2}O_2 \rightarrow H_2O + S^{\circ}_{(Fe)} \tag{7.16}$$

Oxygen used in the process comes from air, which is bubbled through the catalyst solution. Because the chelated iron catalyst is not consumed during the reaction, only modest amounts of catalyst are added to the process to replace mechanical losses. A small caustic addition is required to maintain the catalyst solution in the mildly alkaline pH range. Also, additional chelates are added to replace chelates that degrade over time [210,217,218,227].

7.6.1.4.3 Process Description: LO-CAT Direct Treat Design

The following are the primary components of the unit:

1. The feedstock knockout drum
2. The absorber/gas contactor vessels that provide low-pressure drop and high-efficiency contact between the syngas and our proprietary LO-CAT liquid catalyst solution that contacts the feedstock and rapidly converts H_2S to free elemental sulfur as a precipitate
3. The oxidizer vessel that regenerates the LO-CAT catalyst solution using air from the air blowers to regenerate the catalyst solution
4. The solution circulating pumps
5. The heat exchanger that heats or cools the operating system, depending upon ambient conditions and sulfur load
6. The makeup chemical metering pumps and tanks
7. The sulfur removal filter that can be of a belt-type or bag-type design

The process can handle high flow rate and can deliver product H_2S concentration below 5 ppm. The floor space required for the process is not at large as it is for MV technology. This is because the process uses chemical system and not biological system. The chemical system does not require very high residence time. Merichem has a long-term experience in the commercialization of this process. The process operates at a low temperature. At this temperature HCl can be removed by packed-bed absorption process using caustic solution. Moreover, the process is not as sensitive to HCl concentration as MV technology and ZnO process. The process can be established as turnkey operation by Merichem Co.

7.6.2 BRIEF INDUSTRIAL ASSESSMENT OF SULFUR REMOVAL TECHNOLOGIES

Envitech syngas-cleaning process uses a media such as sponge iron. This method for H_2S removal is also used by Air Products and Alter NRG processes. PyroGenesis process uses either a regenerative or a nonregenerative technology depending on H_2S concentration. In nonregenerative technology, the contaminated syngas is passed through a bed of iron oxide–impregnated wood chips where hydrogen sulfide is converted to ferric and ferrous sulfide. These processes can achieve 90% H_2S removal. In regenerative process, a chelate iron solution is used to convert H_2S to elemental sulfur. Its environmentally safe catalyst does not use toxic chemicals and produces no hazardous by-products. The process unit can be designed for better than 99.9% H_2S removal efficiency. The decision to select either nonregenerative or regenerative technology is based on trade-off between the higher capital cost of the later and the higher operating costs of the former.

7.6.3 COS HYDROLYSIS

While COS can be removed along with H_2S in some processes, in chemical absorption processes, the degree of COS removal is dependent upon the reactivity of the solvent solution with COS. While DGA can remove all COS, MDEA has little reactivity with COS. In physical absorption processes, the solubility of COS in the physical solvent and the COS partial pressure determine the level of its removal. A COS level of 0.1 ppm is attainable in Rectisol process, while the Selexol process can achieve 10 ppm of COS. In the ZnO process, approximately 80% of COS can be removed by hydrolysis. The hydrolysis of COS follows the following reaction [52,217–219]:

$$COS + H_2O \rightarrow H_2S + CO_2 \qquad (7.17)$$

The scrubbed syngas feed is normally reheated to 30°F–50°F above saturation before entering the hydrolysis reactor to avoid catalyst damage by liquid water. COS hydrolysis uses an activated alumina-based catalyst and is normally designed to operate at 350°F–400°F. The reaction is largely independent of pressure. Due to the exothermic nature of the reaction, equilibrium is favored at low temperatures. The heat of reaction is normally dissipated throughout a large amount of nonreacting components yielding nearly isothermal reactor conditions. COS hydrolysis product is cooled before going to further acid gas removal system.

7.6.4 SULFUR RECOVERY UNIT

In a sulfur recovery unit, the acid gas stream from amine or physical solvent is recovered to elemental sulfur. For sulfur recovery of more than 20 TPD, Claus SRU is an economical approach. For a low sulfur recovery LO-CAT SRU is a more suitable process [52,217–219,227].

7.7 TECHNOLOGIES FOR NITROGEN AND CO_2 REMOVAL

Although NH_3 is not a highly stable molecule, its dissociation requires very high temperature due to its high activation energy. Krishnan et al. [215] showed that Haldor Topsoe A/S catalyst showed excellent activity even in the presence of 2000 ppm H_2S and high-temperature stability. G-65, an SRI catalyst, showed a superior activity in the temperature range of 550°C–650°C and low H_2S concentration even in the presence of HCl and HCN.

A wide variety of metal oxides/carbides/nitrides can catalyze decomposition of ammonia. Group VIII metals such Fe, Co, Ni, Ru, Rh, Pd, Os, Ir, and Pt are also active in the metal state. The activity of ammonia decomposition on a smooth metal surface follows the order Co > Ni > Cu > Zr. Groups Va (V, Nb) and VIa (Cr, Mo, W) are especially active for ammonia decomposition. Mo_2C is highly active compared to cadmium and platinum carbides. Vanadium nitrides are also very active. CaO, MgO, and dolomite are also very active at temperatures as low as 300°C.

The gas composition, particle size of the catalyst, and H_2S concentration can affect the catalyst activity. CaO is deactivated almost totally when CO, CO_2, and H_2 are present. Small quantities of H_2S (<2000 ppm) poisoned CaO. Ammonia decomposition by Fe is highly dependent on particle size (20–50 nm). Because ammonia synthesis is highly structure sensitive, the ammonia decomposition activity of Fe is highly dependent on particle size between 20 and 50 nm [49,210,217,218].

Besides catalytic processes, two methods for removing ammonia include catalytic tar reforming and wet scrubbing. Tar cracker catalysts have been demonstrated to be effective in reducing ammonia in the syngas stream by conversion to N_2 and H_2. A tar cracker can be used to remove ammonia followed by gas cooling and wet scrubbing to remove residual ammonia. This cleanup configuration should achieve complete removal of ammonia.

7.7.1 CO_2 REMOVAL IN INTEGRATED GASIFICATION COMBINED-CYCLE PLANTS

For gasification applications, or integrated gasification combined-cycle (IGCC) plants, the plant modifications required to add the ability to capture CO_2 from the syngas produced by the gasifier are minimal. The syngas produced by the gasifiers. The removal of impurities are important for gas storage and transport as well.

Syngas needs to be treated through various processes for the removal of impurities already in the gas stream, so all that is required to remove CO_2 is to add the necessary equipment, such as an absorber and a regenerator, to this process train. In combustion applications, modifications must be done to the exhaust stack and because of the lower concentrations of CO_2 present in the exhaust, much larger volumes of total gas require processing, necessitating larger and more expensive equipment [122].

Mississippi Power's Kemper Project is in the late stages of construction. It will be a lignite-fuel IGCC plant, generating a net 524 MW of power from syngas, while capturing over 65% of CO_2 generated using the Selexol process. The technology at the Kemper facility, namely, transport-integrated gasification, was developed and is licensed by KBR. The CO_2 will be sent by pipeline to depleted oil fields in Mississippi for enhanced oil recovery operations [122].

Hydrogen Energy California will be a 300 MW net, coal, and petroleum coke–fueled IGCC poly-generation plant (producing hydrogen for both power generation and fertilizer manufacture). Ninety percent of the CO_2 produced will be captured (using Rectisol) and transported to Elk Hills Oil Field for EOR, hence enabling the recovery of 5 million additional barrels of domestic oil per year.

Summit's Texas Clean Energy Project (TCEP) will be a coal-fueled, IGCC-based 400 MW power/polygeneration project (also producing urea fertilizer), which will capture 90% of its CO_2 in precombustion capture using the Rectisol process. The CO_2 not used in fertilizer manufacture will be used for enhanced oil recovery in the West Texas Permian Basin [122].

Plants such as the TCEP that employ carbon capture and storage have been touted as a partial, or interim, solution to climate change issues if they can be made economically viable by improved design and mass production. There was an opposition by utility regulators and ratepayers due to increased cost and by some environmentalists such as Bill McKibben who view any continued use of fossil fuels as counterproductive [122].

7.8 TECHNOLOGIES FOR REMOVAL OF CHLORIDES, ALKALIS, AND HEAVY METALS

For bulk removal of HCl, different sorbent materials and processes have been studied and different technologies are being developed. In the two-stage "ultraclean process," synthetic dawsonite, nahcolite, and trona ($Na_2CO_3 \cdot NaHCO_3 \cdot 2H_2O$) have been tested. The HCl concentration from stage 1 polishing at 449°C was below 3 ppm. Trona was found to be the best sorbent for HCl. Krishnan et al. [215] showed that alkali minerals reduced HCl concentration from 300 to 1 ppm in the temperature range of 550°C–650°C. Nahcolite was the best absorbent that contained 54% of chloride by weight. Stanford Research Institute (SRI), Research Triangle (RTI), and General Electric (GE) showed that nahcolite can reduce HCl concentration below 1 ppm in fixed-bed and fluidized-bed reactors in the temperature range of 400°C–650°C. They also showed high degree of absorption (70%) of nahcolite for HCl in pilot-scale circulating fluidized-bed reactor [210,215,217,218].

The literature also showed [162,210,217,218] that the equilibrium level of HCl in the presence of sodium carbonate–based sorbents depends significantly on the temperature and partial pressure of a steam. This equilibrium study indicated that low HCl concentrations are thermodynamically achievable. The diffusion through the product layer NaCl is the rate-limiting step in order to achieve such low concentration. Therefore, high surface areas are important to get high activities and sorbent absorption capacities. Commercially available chloride guards used in petroleum industries are very expensive. SRI observed that 0.3 ppm of HCl in the product is achievable using Katalco chloride guard 59-3 at 550°C. Most HCl treatment materials are once-through materials that cannot be regenerated. The naturally occurring alkali-based materials are less reactive than commercial products particularly at lower temperatures. These materials also have lower surface areas that are rapidly covered with an NaCl product layer with further reduction in their activity. While commercial materials are more active, they are also more expensive.

The halides can also be removed by wet scrubbing or purification by hydrogenation and ZnO absorption. The HCl can also be absorbed in a HCl scrubber that uses a caustic/water recirculating process liquid (maintained at pH 7.5). This guarantees the HCl outlet concentration to be 5 ppm. The packing type, depth, and nature of a spray nozzle can be changed to optimize the final results.

In PyroGenesis process [225], a packed-bed scrubber, is used to remove acid gases (mainly HCl) from the syngas stream. In order to efficiently absorb HCl, a large surface area of contact is provided. The scrubber is filled with randomly oriented saddles and rings to obtain good contact with liquid and gas phases. In a packed absorber, syngas flows from bottom up and is well mixed with a scrubber solution. The scrubber solution pH is continuously adjusted using controlled amounts of caustic soda solution. The scrubber solution is also indirectly cooled using a heat exchanger, which also cools down syngas and adds the moisture content between 5% and 10% to as high as 50% depending on the cooling water temperature. The condensed water is recirculated to the quench.

Compared to sulfur, chlorides, ammonia, and particulate removal, technologies for alkalis and heavy metals such as mercury, arsenic, selenium, and cadmium are not as well developed. Commercial technologies are heavily focused on mercury and arsenic. Adsorbents for mercury are developed by UOP LLC based on type X and Y zeolites that have been coated with elemental silver. The materials are regenerated by heating and purging with a sweep gas to recover mercury. Since the maximum regeneration temperature is below the target warm syngas temperature, this technology is not well suited for syngas cleaning. Various forms of activated carbon are also used effectively to remove mercury. A mercury removal unit based on a fixed bed of sulfur-impregnated activated carbon has been designed to reduce mercury concentration in water-saturated natural gas from 1000 to 5 ppm. Synetix also describes a process for removing mercury from natural gas using metal sulfides on inorganic supports. In this application, a fixed-bed reactor maintained at 15°C is used to reduce the concentration of mercury from 5 to 0.01 ppm. These processes operate at temperatures below the target temperature for warm syngas cleaning [49,210,217,218].

Alkali removal is normally accomplished by cooling the syngas stream below 1100°F to allow condensation of alkali species followed by barrier filtration or wet scrubbing. When using metal and ceramic candle filters, corrosion potentials due to reactions between alkali and filter material at high temperatures should be considered. Ceramic filters used in Lahti facility in Finland and Varnamo in Sweden carried out the removal of alkali with some impurities and found breakage of ceramic filters. They subsequently used sintered metal filters. Baghouse filters were used in Lahti's low-pressure gasification system and the FERCO facility in Vermont [49,210,217,218].

Alkali can easily be removed by wet scrubbing; thus, it is often the preferred method for alkali removal. Wet scrubbing or purification by hydrogenation and ZnO absorption can also be used to remove metals like nickel and iron in the air [217,218].

While some mercury is removed in the quench, scrubber, and DCC, generally a mercury removing guard is provided. Mercury removal from gas is normally achieved by adsorption. There are three types of adsorbent available commercially:

1. Activated carbon
2. Sulfur-impregnated activated carbon
3. Silver-impregnated molecular sieve

Activated carbon, whether sulfur impregnated or not, is used in nonregenerative fashion to remove mercury below 0.1 ppm. The silver-impregnated molecular sieve can remove mercury to equally low level and is regenerable. The sulfur-impregnated activated carbon has more mercury adsorption capacity than conventional activated carbon. It can also absorb other trace components such as VOCs and dioxins/furans with an estimated efficiency of 95% [49,210,217,218].

The lifetime of carbon bed is generally around 2 years. Its regeneration is often done off-site. The pressure drop across the mercury adsorption unit is generally around 34.5 kPa. Generally, the velocity of 0.3 m/s and the contact time of 20 s are used. The bed can reduce mercury concentration in the exit gas to as low as 0.1 ppm.

Besides the materials mentioned earlier, baghouse captures fly ash and heavy metals from the gas stream using bags that are precoated with the mixture of feldspar and activated carbon. In PyroGenesis process, after removal of fine particles, acid gases, and hydrogen sulfide, the syngas is passed through a deep bed scrubber after hydrogen sulfide removal system to remove fine particles, such as lead, cadmium, mercury, and total reduced sulfur.

7.9 SOME PERSPECTIVES ON LOW-TEMPERATURE AND LOW-PRESSURE SYNGAS-CLEANING PROCESS

Based on the investigations of the syngas-cleaning processes used by numerous industries at relatively low temperature and pressure, the following eight observations can be made:

1. Hot cyclones [217–231] (multistage, at least two) should be used for the particulate removal. Most industrial processes use hot cyclones for the particulate removal. It is the proven technology. Multistage operation (as either two different columns or one column with two stages) is important because it would be difficult to capture fine particles (particle size less than 1 μm) in one stage.
2. Various processes both thermal and catalytic have been used to convert tar into syngas. The popular processes include the use of plasma technology, partial oxidation, or catalytic reforming.

3. Tar conversion and hot cyclone units should be followed by wet Venturi scrubber (preferably two stages) for capturing of very fine particles (less than 1 or even 0.5 μm particles) and initial capturing of sulfur, chlorine, alkali, and heavy metals. Once again two stages can be in two separate columns or in one column. All industrial processes have preferred this unit operation in their syngas-cleaning steps.

Venturi scrubber should remove significant portions of hydrogen sulfide, hydrogen chloride, and ammonia. It will also remove some parts of alkali and heavy metals. The liquid used should have high pH (around 10) with the help of NaOH. This should remove around 70% of the hydrogen sulfide and hydrogen chloride and significant portion of COS. If concentrations of chlorine and sulfur are very high, multiple stages (in a single venture scrubber) should be considered.

4. Wet scrubbing should be followed by a WESP for final refining of particulate matters, tars, and gaseous impurities. Submicron particles in the syngas should be removed by the wet electrostatic precipitator. The use of baghouse has disadvantage that it will require high pressures. WESP will remove all fine particles, alkali and heavy metals, and some chlorine, sulfur, and nitrogen impurities.

5. WESP should be followed by packed-bed absorption by caustic solution to capture chlorides and hydrolyze the remaining COS. This is a well-proven process and will reduce HC, sulfur, and nitrogen concentrations at acceptable levels.

6. LO-CAT process for sulfur removal is very attractive. The amine process has very high operating costs even with no guarantee to remove COS. MV technology process is devised for sulfur removal from landfill gas at low flow rates and no chlorine impurities. Due to its biochemical mechanism (and slow rate operation), the process requires very large land footprints for its implementation, although its operating cost is low. LO-CAT process for the removal of H_2S should be followed by hydrolysis process to remove COS. The packed-bed scrubber can be used to remove COS and to remove 95% of HCl using caustic NaOH.

7. Alkali and heavy metals can be removed by Venturi scrubber, WESP, and final activated carbon bed. The bed should also remove some particulates, sulfur, chlorine, and nitrogen compounds. Sulfur-impregnated activated carbon would be desirable in order to get very high removal of heavy metals and alkalis. The activated carbon bed can be considered as a final polishing step. Generally, activated carbon can last 1–2 years in this final polishing stage.

8. The strategic locations of various cleaning steps should be carefully considered. For example, the presence of HCl may affect the removal of H_2S. Also, the chemicals used in each step (i.e., adsorbents) should be regenerated to the extent possible to reduce the operating costs of the cleaning process.

7.10 RTI HIGH-TEMPERATURE IMPURITY REMOVAL PROCESSES

Thermal gasification of coal and biomass generates synthetic gas at high temperatures. These gases are often used in various downstream operations also at high temperatures. For these cases, it makes sense to carry out some of the syngas-cleaning operations at high temperatures. Here, we briefly examine high-temperature cleaning operations.

Oxides of many metals (particularly transition metals) will react with H_2S as $MeO + H_2S \rightarrow MeS + H_2O$. This reaction can be effectively used to reduce H_2S concentration in the syngas. The minimum H_2S concentration in the treated syngas is determined by the equilibrium concentration based on the syngas composition and metal oxide. The reaction kinetics determine how rapidly H_2S reacts to reach the equilibrium concentration. ZnO possess one of the highest thermodynamic efficiencies for H_2S removal and the most favorable reaction kinetics of all the active oxide materials within the 149°C and 371°C temperature range [217,218,229,230].

Although a number of ZnO-based guard bed materials are commercially available, the low sulfur capacity of these materials given the large amount of sulfur to be processed makes them very costly for a once-through disposable material. However, economics become more attractive if ZnO-based materials can be regenerated and used for multiple cycles. A number of ZnO-based sorbents have been developed and are commercially available. While these materials can be used in fixed-bed and fluidized-bed reactors, transport reactor desulfurization system allows much higher throughput and smaller footprints, hence resulting in lower cost and excellent control of the highly exothermic regeneration reaction, which allows the use of neat air for regeneration.

RTI's experience with ZnO-based materials has demonstrated that these materials can effectively remove large amount of H_2S ranging from 500 ppmv to as high as 30,000 ppmv to effluent concentrations typically below 10 ppmv [170,171]. The actual effluent concentration appears to have a temperature dependence that is related to both thermodynamic equilibrium concentration and reaction kinetics. While at low temperature, thermodynamic is more favorable, the reaction rate and particularly diffusion of H_2S through the product layer drop off rapidly below 232°C [217,218,229,230].

ZnO-based materials are typically regenerated using a mixture of oxygen and nitrogen according to the reaction $ZnS + 1.5O_2 \rightarrow ZnO + SO_2$. The concentration of SO_2 in the regeneration off-gas depends on the sorbent and the reactor configuration. The possible SO_2 concentration ranges from 1 to 14 vol% depending on the O_2 concentration used in the regeneration inlet gas. The SO_2 off-gas must be further treated to produce sulfuric acid product or an elemental sulfur product either in Claus plant or RTI's direct sulfur recovery process (DSRP).

RTI has demonstrated that ZnO materials also reacts with COS and CS_2. The thermodynamics and kinetics for the reactions with COS and CS_2 are similar to those with H_2S. RTI has also observed that at temperatures below 232°C, the COS concentration in the effluent gas treated with ZnO-based materials increases. This may be due to a combination of decrease in desulfurization kinetics and increase in the rate of conversion to COS by the reaction $CO + H_2S \rightarrow COS + H_2$.

Quite often, the syngas exiting the gasifier (containing about 400 ppmv H_2S) is first treated with amine unit with a high circulation rate to reduce syngas sulfur concentration below 10 ppmv with a target of 2–3 ppmv. The ZnO beds are then used as polishing stage to reduce the sulfur concentration to be less than 0.1 ppmv. The gas exiting the amine absorber is heated to the operating temperature of ZnO beds, 750°F.

RTI has also developed novel sorbents to capture H_2S and COS at high temperature (300°C–700°C), which help maintain high thermal efficiency than conventional desulfurization options. Their DSRP follows the following reaction:

$$SO_2 + 2H_2 \text{ (or 2 CO)} \rightarrow 2H_2O \text{ (or } 2CO_2) + S \qquad (7.18)$$

The HCl concentration in syngas is of significant concern for the application of ZnO for sulfur removal. HCl must be removed to a less than 10 ppm level before ZnO process can be used for sulfur removal. This creates significant problems in the arrangement of syngas-cleaning setup. HCl can be removed by high-temperature adsorbent process. However, this means that temperature of gas has to be at least 400°C for HCl and H_2S removal and wet scrubbing process cannot be applied to syngas before removal of HCl and H_2S [217,218,230,231].

RTI has developed high-temperature syngas-cleaning processes for chlorine, nitrogen, sulfur, and CO_2 removal for the gas coming out of coal gasification. The chlorine content of this gas is generally low. While gas flow in coal gasification is extremely high, they may not contain tar and particulate matter at the levels obtained in waste gasification.

ZnO process has an advantage that it will remove both H_2S and COS. Thus, a separate hydrolysis process for the removal of COS is not needed. Since high-temperature cleaning technology for syngas was targeted for syngas from coal gasification, commercial experience for its use for biomass and waste gasification is somewhat lacking and needs to be further demonstrated.

7.10.1 RTI's Scale-Up Experience for High-Temperature Removal

As mentioned above, the most mature advanced gasification technologies in the RTI portfolio are the high-temperature desulfurization system (HTDP) and the DSRP. RTI's HTDP uses a regenerable sorbent to capture hydrogen sulfide (H_2S) and carbonyl sulfide present in the syngas coming from coal gasification process. RTI and Eastman Chemical have also completed a pilot-scale demonstration of the process at Eastman Chemical's gasification plant in Kingsport, Tennessee. RTI's HTDP is based on a transport reactor design that allows for high-throughput processing of syngas and continuous high-temperature, high-pressure operation.

According to RTI and DOE publications [217–219,230], a pilot-scale demonstration of RTI's HTDP, which treated actual syngas from Eastman's gasification plant for over 3000 h, achieved greater than 99% sulfur removal, producing a product syngas with below 10 ppm of sulfur. The demonstration also provided an abundance of operational data, which RTI will use for scaling up to a demonstration plant—to process roughly a 50 MW equivalent slipstream of syngas from a commercial gasifier.

Complementing RTI's high-temperature desulfurization technology is the DSRP. During regeneration of RTI's sorbent in the HTDP, the sulfur on the sorbent is converted to sulfur dioxide (SO_2). The DSRP uses a small slipstream of syngas to catalytically reduce this SO_2 to elemental sulfur.

The main advantage of the DSRP is that its high-temperature, high-pressure operating conditions allow direct thermal and process integration with HTDP. The DSRP also works well with low-gas mixtures containing low concentrations of SO_2 that cannot be processed by other technologies. The catalyst used in the DSRP is commercially available. RTI has demonstrated integrated operation of a HTDP and DSRP pilot plant at Eastman Chemical. During 110+ h of integrated operation, the DSRP system achieved 90%–98% removal of the inlet sulfur. The DSRP catalyst proved to be very robust, demonstrating consistent reaction rates in multiple experiments over a 3-year period.

In several independent economical studies, RTI's HTDP were compared with conventional gas cleanup technologies such as Selexol and Rectisol. This comparison indicated that RTI's technology had a net 2–3 percentage-point higher overall IGCC thermal efficiency and a significant reduction in capital cost [217–219,230].

According to RTI news publication, in order to complement RTI's high-temperature desulfurization technologies, RTI has assembled specialized fixed-bed testing systems to evaluate catalysts and sorbents for the removal of other contaminants typically found in coal gasification syngas at high temperature. RTI is focused on contaminants such as mercury, ammonia, chlorine, arsenic, cadmium, selenium, and carbon dioxide (CO_2). Some of these sorbents demonstrated contaminant removal from a slipstream of real coal-derived syngas. RTI is currently engaged in laboratory-scale demonstration of precombustion CO_2 capture technology. This CO_2 separation technology is based on regenerable mixed oxide sorbents and shows excellent CO_2 removal rates from syngas at high temperatures [217,218,229,230].

RTI is also pursuing membrane-based separation technology as an alternative approach for removing acid gases from syngas. RTI has developed a syngas-cleaning technology based on reverse-selective membranes that, unlike typical commercial membranes, separate materials based on their chemical properties rather than size. This allows impurities (e.g., CO_2, H_2S) to be removed at low pressures while maintaining the useful hydrogen-rich process gas at high pressure. A number of membrane materials possessing both high CO_2 and H_2S permeability and high selectivity for CO_2 and H_2S over H_2 have been identified, and process modeling shows significant cost incentives over conventional amine systems. RTI has completed a demonstration with a pilot-scale prototype membrane system at Eastman Chemical's gasification plant in Kingsport, Tennessee [217,218,229,230].

REFERENCES

1. Natural gas processing: The crucial link between natural gas production and its transportation to market, a report from, Energy Information Administration, Office of Oil and Gas, Washington, DC (January 2006).
2. Overview of natural gas-background, a report by Natural gas.org, Retrieved September 20, 2013.
3. Overview of natural gas-processing natural gas, a report from Naturalgas.org, Retrieved, September 25, 2013. Naturalgas.org/natural gas/processing-ng/.
4. Natural gas, Wikipedia, the free encyclopedia (2015). https://en.wikipedia.gov/wiki/natural_gas (last accessed October 22, 2016).
5. Natural gas processing, Wikipedia, the free encyclopedia (2015). https://en.wikipedia.gov/wiki/natural_gas_processing (last accessed July 19, 2016).
6. Fact Sheet: Natural Gas Processing Plants, Pipeline safety stakeholder communication, U.S. Department of Transportation, Washington, DC (July 2014), https://primis.phm.
7. Corvini, G., Stiltner, J., and Clark, K., Mercury removal from natural gas and liquids (PDF), a report from UOP LLC, Archived from the original (PDF), Des Plaines, NJ, on January 1, 2011.
8. Bourke, M.J. and Mazzoni, A.F., Desulfurization of mercury removal from natural gas, in *Laurance Reid Gas Conditioning Conference*, Norman, OK (March 1989).
9. Baker, R.W., Future directions of membrane gas separation technology, *Industrial & Engineering Chemistry Research*, 41, 1393–1411 (2002).
10. Hydrocarbon Processing; 2012 Gas Processes Handbook, UOP, LLC Honeywell Co., Gulf Publishing Co., Houston, TX (2012).
11. Turbo-expander, Wikipedia, the free encyclopedia (2015). https://en.wikipedia.org/wiki/turbo-expander (last accessed August 25, 2015).
12. Ward, D.E. and Pierce, A.P., Helium, in United States Mineral Resources, Professional Paper 820, U.S. Geological Survey, USGS, Washington, D.C., pp. 285–290 (1973).
13. Bernardo, P. et al., Membrane gas separation: A review/state of the art, *Industrial & Engineering Chemistry Research*, 48 (10), 4638–4663 (2009).
14. Bhide, B.D. et al., Hybrid processes for the removal of acid gases from natural gas, *Journal of Membrane Science*, 140 (1), 27–49 (1998).
15. Falk-Pedersen, O. and Dannstrom, H., Separation of carbon dioxide from offshore gas turbine exhaust, *Energy Conversion and Management*, 38, S81–S86 (1997).
16. Vu, D.Q., Formation and characterization of asymmetric carbon molecular sieve and mixed matrix membranes for natural gas purification, PhD thesis, The University of Texas at Austin, Austin, TX, pp. 1–362 (2001), www.intechopen.
17. Choe, J.S. et al., Process for separating components of a gas stream, Google Patents, 4701187 (1987).
18. Doshi, K.J., Enhanced gas separation process, Google Patents, 4690695 (1987).
19. Doshi, K.J. and Dolan, W.B., Process for the rejection of CO_2 from natural gas, Google Patents, 5411721 (1995).
20. Doshi, K.J. et al., Integrated membrane/PSA process and system, Google Patents, 4863492 (1989).
21. Esteves, I.A.A.C. and Mota, J.P.B., Gas separation by a novel hybrid membrane/pressure swing adsorption process, *Industrial & Engineering Chemistry Research*, 46 (17), 5723–5733 (2007).
22. Feng, X. et al., Integrated membrane/adsorption process for gas separation, *Chemical Engineering Science*, 53 (9), 1689–1698 (1998).
23. Blizzard, G. et al., Mallet gas processing facility uses membranes to efficiently separate CO_2, *Oil and Gas Journal*, 103 (14), 48–53 (2005).
24. Leal, O., Bolivar, C., Sepulveda, G., Molleja, G., Martinez, G., and Esparragoza, L., U.S. Patent no. 5,087,597 (1992).
25. Al-Juaied, M.A., Carbon dioxide removal from natural gas by membranes in the presence of heavy hydrocarbons and by Aqueous Diglycolamine®/Morpholine, PhD thesis, The University of Texas at Austin, Austin, TX, pp. 1–424 (May 2004).
26. Burr, B. and Lyddon, L., A comparison of physical solvents for acid gas removal, in Options for Acid Gas Removal (2008), available from: http://www.bre.com/portals/0/technicalarticles/A%20Comparison%20of%20Physical%20Solvents%20for%20Acid%20Gas%20Removal%20REVISED.pdf, 11.08.11.
27. Cavenati, S. et al., Removal of carbon dioxide from natural gas by vacuum pressure swing adsorption, *Energy & Fuels*, 20 (6), 2648–2659 (2006).

28. Dortmundt, D. and Doshi, K., Recent developments in CO_2 removal membrane technology, in Design Consideration (1999), available from: http://www.membrane-guide.com/download/CO2-removal-membranes.pdf, 16.08.11.

29. Ebenezer, S.A. and Gudmunsson, J.S., Removal of carbon dioxide from natural gas for LPG production, in Carbon Dioxide Removal Processes (December 2006), available from: http://www.ipt.ntnu.no/~jsg/studenter/prosjekt/Salako2005.pdf, 21.08.11.

30. Kesting, R.E. and Fritzsche, A.K., *Polymeric Gas Separation Membranes*. Wiley, New York (1993).

31. Kidnay, A.J. et al., *Fundamentals of Natural Gas Processing*. CRC, New York (2006).

32. Kohl, A.L. and Nielsen, R.B., *Gas Purification*, 5th edn. Gulf Professional Publishing, Houston, TX (1997).

33. Kovvali, A.S. and Sirkar, K.K., Carbon dioxide separation with novel solvents as liquid membranes, *Industrial & Engineering Chemistry Research*, 41 (9), 2287–2295 (2002).

34. Ritter, J.A. and Ebner, A.D., Carbon dioxide separation technology: R&D needs for the chemical and petrochemical industries, in Recommendation for Future R&D (2007), available from: http://www.chemicalvision2020.org/pdfs/CO2_Separation_Report_V2020_final.pdf, 22.06.11.

35. Rojey, A., *Natural Gas: Production, Processing, Transport*. Editions Technip, Paris, France (1997).

36. Shekhawat, D. et al., A review of carbon dioxide selective membranes: A topical report, in CO_2 Selective Membranes (2003), available from: http://www.osti.gov/bridge/product.biblio.jsp?osti_id=819990, 13.07.11.

37. Shimekit, B. et al., Ceramic membranes for the separation of carbon dioxide: A review, *Transactions of the Indian Ceramic Society*, 68 (3), 115–138 (2009).

38. Sridhar, S. et al., Separation of carbon dioxide from natural gas mixtures through polymeric membranes—A review, *Separation & Purification Reviews*, 36 (2), 113–174 (2007).

39. Tabe-Mohammadi, A., A review of the applications of membrane separation technology in natural gas treatment, *Separation Science and Technology*, 34 (10), 2095–2111 (1999).

40. Ebenezer, S.A., A project on optimization of amine base CO_2 removal process; removal of carbon dioxide from natural gas for LNG production, Institute of Petroleum Technology, Norwegian University of Science and Technology, Trondheim, Norway (December 2005).

41. Shimekit, B. et al., Prediction of the relative permeability of gases in mixed matrix membranes, *Journal of Membrane Science*, 373 (1–2), 152–159 (2011).

42. Baker, R.W., *Membrane Technology and Applications*, 2nd edn. John Wiley & Sons, West Sussex, U.K. (2004).

43. Shimekit, B. and Mukhtar, H., Natural gas purification technologies—Major advances for CO_2 separation and future directions, Chapter 9, in Al-Megren, H. (ed.), *Advances in Natural Gas Technology*. InTech, Rijeka, Croatia, pp. 235–270 (April 2012).

44. Gas facts—A report by American Gas Association, Washington, D.C. (2000). https://www.aga.org/fact…/natural-gas-facts.

45. Stell, J., Gas processing in the mighty Marcellus and Uber Utica, a report by Gas Processing, Houston, TX (2014), www.gasprocessingnews.com/…/gas-processing-in-the-mighty-marcellus (accessed December 2014).

46. Amine gas treating, Wikipedia, the free encyclopedia (2015), https://en.wikipedia.org/wiki/Amine_gas_treating, https://en.wikipedia.org/wiki/amine_gas_treating (last accessed October 24, 2016).

47. McKee, R.L. et al., CO_2 removal: Membrane plus amine, *Hydrocarbon Processing*, 70, 63 (April 1991).

48. Shah, Y.T., Biomass to liquid fuel via Fischer–Tropsch and related syntheses, Chapter 12, in Lee, J.W. (ed.), *Advanced Biofuels and Bioproducts*. Springer Book Project, Springer Publ. Co., New York, pp. 185–207 (September 2012).

49. Ratafia-Brown, J., Haslbeck, J., Skone, T., and Rutkowski, M., Assessment of technologies for co-converting coal and biomass to a clean syngas—Task 2 report (RDS), DOE/NETL-403.01.08 Activity 2 Report, Department of Energy, Washington, D.C. (May 10, 2007).

50. Lee, S. and Shah, Y., *Biofuels and Bioenergy: Technologies and Processes*. CRC Press, Taylor & Francis Group, New York (September 2012).

51. Shah, Y., *Water for Energy and Fuel Production*. CRC Press, Taylor & Francis Group, New York (May 2014).

52. Shah, Y., *Energy and Fuel Systems Integration*. CRC Press, Taylor & Francis Group, New York (2015).

53. Tchapda, A. and Pisupati, S., A review of thermal co-conversion of coal and biomass/waste, *Energies*, 7, 1098–1148 (2014).

54. Zhu, W., Song, W., and Lin, W., Catalytic gasification of char from co-pyrolysis of coal and biomass, *Fuel Processing Technology*, 89, 890–896 (2008).

55. Milne, T.A., Evans, R.J., Abatzoglou, N., *Biomass Gasifier Tars: Their Nature, Formation, and Conversion*, NREL/TP-570-25357. National Renewable Energy Laboratory, Golden, CO, p. 204 (1998).

56. Brage, C., Yu, Q., Chen, G., and Sjostrom, K., Tar evolution profiles obtained from gasification of biomass and coal, *Biomass and Bioenergy*, 18, 87–91 (2000).

57. Pinto, F., Lopes, H., Andre, R.N., Gulyurtlu, I., and Cabrita, I., Effect of catalysts in the quality of syngas and by-products obtained by co-gasification of coal and wastes. 1. Tars and nitrogen compounds abatement, *Fuel*, 86, 2052–2063 (2007).

58. Ren, H., Zhang, Y., Fang, Y., and Wang, Y., Co-gasification behavior of meat and bone meal char and coal char, *Fuel Processing Technology*, 92, 298–307 (2011).

59. McKee, D.W., Spiro, C.L., Kosky, P.G., and Lamby, E.J., Catalysis of coal char gasification by alkali metal salts, *Fuel*, 62, 217–220 (1983).

60. Mettler, M.S., Vlachos, D.G., and Dauenhauer, P.J., Top ten fundamental challenges of biomass pyrolysis for biofuels, *Energy* and *Environmental Science*, 5, 7797–7809 (2012).

61. Reed, T.B. and Bryant, B., Densified biomass: A new form of solid fuel, U.S. Department of Energy, National Renewable Energy Laboratory, Golden, CO, p. 35 (1978).

62. Radovic, L.R., Walker, P.L., Jr., and Jenkins, R.G., Catalytic coal gasification: Use of calcium versus potassium, *Fuel*, 63, 1028–1030 (1984).

63. Walker, P.L., Jr., Matsumoto, S., Hanzawa, T., Muira, T., and Ismail, I.M.K., Catalysis of gasification of coal derived cokes and chars, *Fuel*, 62, 140–149 (1983).

64. Freund, H., Kinetics of carbon gasification by CO_2, *Fuel*, 64, 657–660 (1985).

65. McKee, D.W., Spiro, C.L., Kosky, P.G., and Lamby, E.J., Eutectic salt catalysts for graphite and coal char gasification, *Fuel*, 64, 805–809 (1985).

66. Huffman, G.P., Huggins, F.E., Shah, N., and Shah, A., Behavior of basic elements during coal combustion, *Progress in Energy and Combustion Science*, 16, 243–251 (1990).

67. Huffman, G.P., Huggins, F.E., Shoenberger, R.W., Walker, J.S., Lytle, F.W., and Greegor, R.B., Investigation of the structural forms of potassium in coke by electron microscopy and x-ray absorption spectroscopy, *Fuel*, 65, 621–632 (1986).

68. Huggins, F.E., Huffman, G.P., Lytle, F.W., and Greegor, R.B., The form of occurrence of chlorine in U.S. coals: An XAFS investigation, *ACS Division of Fuel Chemistry Preprints*, 34, 551–558 (1989).

69. Huffman, G.P. and Huggins, F.E., Analysis of the inorganic constituents of low-rank coals, in Schobert, H.H. (ed.), *The Chemistry of Low Rank Coals*. American Chemical Society, Washington, DC, pp. 159–174 (1984).

70. Lang, R.J. and Neavel, R.C., Behaviour of calcium as a steam gasification catalyst, *Fuel*, 61, 620–626 (1982).

71. Kreith, F., *The CRC Handbook of Mechanical Engineering*. CRC Press, Boca Raton, FL (1998).

72. Brown, R.C., Liu, Q., and Norton, G., Catalytic effects observed during the co-gasification of coal and switchgrass, *Biomass and Bioenergy*, 18, 499–506 (2000).

73. Srivastava, S.K., Saran, T., Sinha, J., Ramachandran, L.V., and Rao, S.K., Influence of alkali on pyrolysis of coals, *Fuel*, 67, 1683–1684 (1988).

74. Habibi, R., Kopyscinski, J., Masnadi, M.S., Lam, J., Grace, J.R., Mims, C.A., and Hill, J.M., Co-gasification of biomass and non-biomass feedstocks: Synergistic and inhibition effects of switchgrass mixed with sub-bituminous coal and fluid coke during CO_2 gasification, *Energy Fuels*, 27, 494–500 (2012).

75. Pedersen, L.S., Nielsen, H.P., Kiil, S., Hansen, L.A., Dam-Johansen, K., Kildsig, F., Christensen, J., and Jespersen, P., Full-scale co-firing of straw and coal, *Fuel*, 75, 1584–1590 (1996).

76. Sjöström, K., Chen, G., Yu, Q., Brage, C., and Rosen, C., Promoted reactivity of char in co-gasification of biomass and coal: Synergies in the thermochemical process, *Fuel*, 78, 1189–1194 (1999).

77. Cordero, T., Rodriguez-Mirasol, J., Pastrana, J., and Rodriguez, J.J., Improved solid fuels from co-pyrolysis of a high-sulphur content coal and different lignocellulosic wastes, *Fuel*, 83, 1585–1590 (2004).

78. Haykiri-Acma, H. and Yaman, S., Synergy in devolatilization characteristics of lignite and hazelnut shell during co-pyrolysis, *Fuel*, 86, 373–380 (2007).

79. Khalil, R.A., Thermal conversion of biomass with emphasis on product distribution, reaction kinetics and sulfur abatement, PhD thesis, Norwegian University of Science and Technology (NTNU), Trondheim, Norway (2009).

80. Knudsen, J.N., Volatilization of inorganic matter during combustion of annual biomass, PhD thesis, Technical University of Denmark, Lyngby, Denmark (2004).

81. Kambara, S., Takarada, T., Yamamoto, Y., and Kato, K., Relation between functional forms of coal nitrogen and formation of nitrogen oxide (NOx) precursors during rapid pyrolysis, *Energy Fuels*, 7, 1013–1020 (1993).
82. Jiachun, Z., Masutani, S.M., Ishimura, D.M., Turn, S.Q., and Kinoshita, C.M., Release of fuel-bound nitrogen in biomass during high temperature pyrolysis and gasification, in *Proceedings of the 32nd Intersociety Energy Conversion Engineering Conference (IECEC)*, Vol. 3, Honolulu, HI, pp. 1785–1790 (July 27–August 1, 1997); *Energies*, 7, 1141 (2014).
83. De Jong, W., Nitrogen compounds in pressurised fluidised bed gasification of biomass and fossil fuels, PhD thesis, Technische Universiteit Delft, Delft, the Netherlands (2005).
84. Blair, D.W., Wendt, J.O.L., and Bartok, W., Evolution of nitrogen and other species during controlled pyrolysis of coal, *Symposium (International) on Combustion*, 16, 475–489 (1977).
85. Slaughter, D.M., Overmoe, B.J., and Pershing, D.W., Inert pyrolysis of stoker-coal fines, *Fuel*, 67, 482–489 (1988).
86. Solomon, P.R. and Colket, M.B., Evolution of fuel nitrogen in coal devolatilization, *Fuel*, 57, 749–755 (1978).
87. Pohl, J.H. and Sarofim, A.F., Devolatilization and oxidation of coal nitrogen, *Symposium (International) on Combustion*, 16, 491–501 (1977).
88. Cai, H.Y., Güell, A.J., Dugwell, D.R., and Kandiyoti, R., Heteroatom distribution in pyrolysis products as a function of heating rate and pressure, *Fuel*, 72, 321–327 (1993).
89. Bassilakis, R., Zhao, Y., Solomon, P.R., and Serio, M.A., Sulfur and nitrogen evolution in the argonne coals. Experiment and modeling, *Energy Fuels*, 7, 710–720 (1993).
90. Yuan, S., Chen, X., Li, W., Liu, H., and Wang, F., Nitrogen conversion under rapid pyrolysis of two types of aquatic biomass and corresponding blends with coal, *Bioresource Technology*, 102, 10124–10130 (2012).
91. Yuan, S., Zhou, Z., Li, J., Chen, X., and Wang, F., HCN and NH_3 (NOx precursors) released under rapid pyrolysis of biomass/coal blends, *Journal of Analytical and Applied Pyrolysis*, 92, 463–469 (2011).
92. Di Nola, G., de Jong, W., and Spliethoff, H., The fate of main gaseous and nitrogen species during fast heating rate devolatilization of coal and secondary fuels using a heated wire mesh reactor, *Fuel Processing Technology*, 90, 388–395 (2009).
93. Di Nola, G., de Jong, W., and Spliethoff, H., TG-FTIR characterization of coal and biomass single fuels and blends under slow heating rate conditions: Partitioning of the fuel-bound nitrogen, *Fuel Processing Technology*, 91, 103–115 (2010); *Energies*, 7, 1144 (2014).
94. Knudsen, J.N., Jensen, P.A., Lin, W., Frandsen, F.J., and Dam-Johansen, K., Sulfur transformations during thermal conversion of herbaceous biomass, *Energy Fuels*, 18, 810–819 (2004).
95. Dayton, D.C., French, R.J., and Milne, T.A., Direct observation of alkali vapor release during biomass combustion and gasification. 1. Application of molecular beam/mass spectrometry to switchgrass combustion, *Energy Fuels*, 9, 855–865 (1995).
96. Björkman, E. and Strömberg, B., Release of chlorine from biomass at pyrolysis and gasification conditions, *Energy Fuels*, 11, 1026–1032 (1997).
97. Knudsen, J.N., Jensen, P.A., and Dam-Johansen, K., Transformation and release to the gas phase of Cl, K, and S during combustion of annual biomass, *Energy Fuels*, 18, 1385–1399 (2004).
98. Zhao, J., Hu, X., and Gao, J., Study on the variations of organic sulfur in coal by pyrolysis, *Coal Conversion*, 16, 77–81 (1993).
99. Calkins, W.H., Determination of organic sulfur-containing structures in coal by flash pyrolysis experiments, *Preprints of Papers—American Chemical Society, Division of Fuel Chemistry*, 30, 450–465 (1985).
100. Cullis, C.F. and Norris, A.C., The pyrolysis of organic compounds under conditions of carbon formation, *Carbon*, 10, 525–537 (1972).
101. Winkler, J.K., Karow, W., and Rademacher, P., Gas-phase pyrolysis of heterocyclic compounds, part 1 and 2: Flow pyrolysis and annulation reactions of some sulfur heterocycles: Thiophene, benzo[b]thiophene, and dibenzothiophene. A product-oriented study, *Journal of Analytical and Applied Pyrolysis*, 62, 123–141 (2002).
102. Ur Rahman Memon, H., Williams, A., and Williams, P.T., Shock tube pyrolysis of thiophene, *International Journal of Energy Research*, 27, 225–239 (2003).
103. Huang, C., Zhang, J.Y., Chen, J., and Zheng, C.G., Quantum chemistry study on the pyrolysis of thiophene functionalities in coal, *Coal Conversion*, 28, 33–35 (2005).
104. Ibarra, J.V., Palacios, J.M., Gracia, M., and Gancedo, J.R., Influence of weathering on the sulphur removal from coal by pyrolysis, *Fuel Processing Technology*, 21, 63–73 (1989).

105. Cleyle, P.J., Caley, W.F., Stewart, L., and Whiteway, S.G., Decomposition of pyrite and trapping of sulphur in a coal matrix during pyrolysis of coal, *Fuel*, 63, 1579–1582 (1984).
106. Gryglewicz, G. and Jasieńko, S., The behaviour of sulphur forms during pyrolysis of low-rank coal, *Fuel*, 71, 1225–1229 (1992).
107. Gryglewicz, G., Sulfur transformations during pyrolysis of a high sulfur polish coking coal, *Fuel*, 74, 356–361 (1995).
108. Telfer, M.A. and Zhang, D.K., Investigation of sulfur retention and the effect of inorganic matter during pyrolysis of south Australian low-rank coals, *Energy Fuels*, 12, 135–1141 (1998).
109. Suárez-Ruiz, I. and Crelling, J.C., *Applied Coal Petrology: The Role of Petrology in Coal Utilization.* Elsevier Ltd., Burlington, MA (2008).
110. Biedermann, F. and Obernberger, I., Ash related problems during biomass combustion and possibilities for a sustainable ash utilization, in *Proceedings of the International Conference on World Renewable Energy Congress (WREC)*, Aberdeen, Scotland, U.K., p. 8 (May 22–27, 2005).
111. Baxter, L.L., Ash deposition during biomass and coal combustion: A mechanistic approach, *Biomass and Bioenergy*, 4, 85–102 (1993).
112. Rushdi, A., Sharma, A., and Gupta, R., An experimental study of the effect of coal blending on ash deposition, *Fuel*, 83, 495–506 (2004).
113. Tumuluru, J.S., Hess, J.R., Boardman, R.D., Wright, C.T., and Westover, T.L., Formulation, pretreatment, and densification options to improve biomass specifications for co-firing high percentages with coal, *Industrial Biotechnology*, 8, 113–132 (2012).
114. Paul, T.W. and Jude, O., Composition of products from the supercritical water gasification of glucose: A model biomass compound, *Industrial & Engineering Chemistry Research*, 44, 8739–8749 (2005).
115. Peter, K., Corrosion in high-temperature and supercritical water and aqueous solutions: A review, *The Journal of Supercritical Fluids*, 29, 1–29 (2004).
116. Veriansyah, B., Kim, J., and Kim, J.D., Hydrogen production by gasification of gasoline in supercritical water, a report from Supercritical Fluid Research Laboratory, Korean Institute of Science and Technology, Seoul, Korea, personal communication (2012).
117. Knoef, H., *Handbook Biomass Gasification.* Biomass Technology Group Press, Enschede, the Netherlands, pp. 22–23 (2005).
118. Lee, I.G., Kim, M.S., and Ihm, S.K., Gasification of glucose in supercritical water, *Industrial & Engineering Chemistry Research*, 41, 1182–1188 (2002).
119. Jannasch, R., Quan, Y., and Samson, R., A process and energy analysis of pelletizing switchgrass—Final report, Resource Efficient Agricultural Production (REAP-Canada) for Natural Resources, Ste. Marthe., Quebec, Canada (2001), www.reap-canada.com/online_library/feedstock_biomass/11%20A%20process.pdf.
120. Rutkowski, M.D., Schoff, R.L., and Keuhn, N.J., Analysis of Stamet pump for IGCC applications prepared for DOE/NETL, Department of Energy, Washington, D.C. (July 2005).
121. Aldred, D., Saunders, T., and Rutkowski, M., Successful continuous injection of coal into gasification system operating pressures exceeding 500 psi, in Presented at the *Gasification Technology Conference*, Orlando, FL (October 9–12, 2005).
122. Integrated gasification combined cycle, Wikipedia, the free encyclopedia (2015), https://en.wikipedia.org/wiki/Integrated_gasification_combined_cycle (last accessed August 5, 2016).
123. Zwart, R.W.R., Boerrigter, H., and Van der Drift, A., The impact of biomass pre-treatment on the feasibility of overseas biomass conversion to Fischer–Tropsch products, *Energy Fuels*, 20(5), 2192–2197 (August 29, 2006).
124. Brooking, E., A report on improving energy density in biomass through torrefaction. National Renewable Energy Laboratory, Golden, CO (2002), http://www.nrel.gov/education/pdfs/e_brooking.pdf (accessed 2002).
125. Williams, P.T. and Besler, S., The influence of temperature and heating rate on the slow pyrolysis of biomass, *Renewable Energy*, 7(3), 233–250 (1996).
126. Arcate, J.R., Torrefied wood, an enhanced wood fuel, in *Bioenergy*, Boise, ID, Paper # 207 (September 22–26, 2002).
127. Arias, B., Pevida, C., Fermoso, J., Plaza, M.G., Reubiern, F., and Pis, J.J., Influence of torrefaction on the grindability and reactivity of wood biomass, *Fuel Processing Technology*, 89(2), 169–175 (2008).
128. Bourgois, J. and Guyonnet, R., Characterization and analysis of torrefied wood, *Wood Science and Technology*, 22(2), 143–155 (1988).
129. Duijn, C., Torrefied wood uit resthout en andere biomassastromen, in *Proceedings of Praktijkdag Grootschalige Bioenergie Projecten*, SenterNovem, Berlin, Germany (June 2004).

130. Livingston, W.R., *Biomass Ash Characteristics and Behaviour in Combustion, Gasification and Pyrolysis Systems*. Doosan Babcock Energy Limited, West Sussex, U.K. (2007).
131. Van Loo, S. and Koppejan, J., *Handbook of Biomass Combustion and Co-Firing*. Earthscan, London, U.K. (2008).
132. Wu, H., Wall, T., Liu, G., and Bryant, G., Ash liberation from included minerals during combustion of pulverized coal: The relationship with char structure and burnout, *Energy Fuels*, 13, 1197–1202 (1999).
133. Wei, X., Lopez, C., von Puttkamer, T., Schnell, U., Unterberger, S., and Hein, K.R.G., Assessment of chlorine-alkali-mineral interactions during co-combustion of coal and straw, *Energy Fuels*, 16, 1095–1108 (2002).
134. Dayton, D.C., Belle-Oudry, D., and Nordin, A., Effect of coal minerals on chlorine and alkali metals released during biomass/coal cofiring, *Energy Fuels*, 13, 1203–1211 (1999).
135. Andersen, K.H., Frandsen, F.J., Hansen, P.F.B., Wieck-Hansen, K., Rasmussen, I., Overgaard, P., and Dam-Johansen, K., Deposit formation in a 150 MWe utility PF-boiler during co-combustion of coal and straw, *Energy Fuels*, 14, 765–780 (2000).
136. Blevins, L.G. and Cauley, T.H., Fine particulate formation during switchgrass/coal cofiring, *Journal of Engineering for Gas Turbines and Power*, 127, 457–463 (2005).
137. Strand, M., Bohgard, M., Swietlicki, E., Gharibi, A., and Sanati, M., Laboratory and field test of a sampling method for characterization of combustion aerosols at high temperatures, *Aerosol Science and Technology*, 38, 757–765 (2004).
138. Aho, M. and Silvennoinen, J., Preventing chlorine deposition on heat transfer surfaces with aluminium–silicon rich biomass residue and additive, *Fuel*, 83, 1299–1305 (2004).
139. Kyi, S. and Chadwick, B.L., Screening of potential mineral additives for use as fouling preventatives in Victorian brown coal combustion, *Fuel*, 78, 845–855 (1999).
140. Raask, E., *Mineral Impurities in Coal Combustion: Behavior, Problems, and Remedial Measures*. Hemisphere Publishing Corporation, Washington, DC (1985).
141. Kumabe, K., Hanaoka, T., Fujimoto, S., Minowa, T., and Sakanishi, K., Co-gasification of woody biomass and coal with air and steam, *Fuel*, 86, 684–689 (2007).
142. Aznar, M.P., Caballero, M.A., Sancho, J.A., and Frances, E., Plastic waste elimination by co-gasification with coal and biomass in fluidized bed with air in pilot plant, *Fuel Processing Technology*, 87, 409–420 (2006).
143. Andre, R.N., Pinto, F., Franco, C., Dias, M., Gulyurtlu, I., Matos, M.A.A., and Cabrita, I., Fluidised bed co-gasification of coal and olive oil industry wastes, *Fuel*, 84, 1635–1644 (2005); *Energies*, 7, 1134 (2014).
144. Collot, A.G., Zhuo, Y., Dugwell, D.R., and Kandiyoti, R., Co-pyrolysis and cogasification of coal and biomass in bench-scale fixed-bed and fluidised bed reactors, *Fuel*, 78, 667–679 (1999).
145. Nelson, P.F., Buckley, A.N., and Kelly, M.D., Functional forms of nitrogen in coals and the release of coal nitrogen as NOx precursors (HCN and NH$_3$), *Symposium (International) on Combustion*, 24, 1259–1267 (1992).
146. Williams, R.H., Larson, E.D., and Haiming, J., Synthetic fuels in a world with high oil and carbon prices, in *Proceedings of the Eighth International Conference on Greenhouse Gas Control Technologies (GHGT-8)*, Trondheim, Norway (June 19–22, 2006).
147. Kavalov, B. and Peteves, S.D., Status and perspectives of biomass-to liquid fuels in the European Union, Report no. EUR 21745 EN, European Commission, Joint Research Centre, Brussels, Belgium (2005).
148. Reed, T. and Gaur, S., *A Survey of Biomass Gasification*, 2nd edn. U.S. Department of Energy, National Renewable Energy Laboratory and the Biomass Energy Foundation, Golden, CO, p. 180 (2001).
149. Rickets, B., Hotchkiss, R., Livingston, B., and Hall, M., Technology status review of waste/biomass co-gasification with coal, in IChemE (ed.), *Fifth European Gasification Conference*, Noordwijk, the Netherlands, p. 13 (April 2002).
150. Hotchkiss, R., Livingston, W., and Hall, M., Waste/biomass co-gasification with coal? Report no. Coal R216, DTI/Pub URN 02/867, London, U.K. (2002).
151. Li, J. and Gifford, J., *Evaluation of Woody Biomass Torrefaction*. Forest Research, Rotorua, New Zealand (September 2001).
152. Pach, M., Zanzi, R., and Björnbom, E., Torrefied biomass a substitute for wood and charcoal, in *Proceedings of the Sixth Asia-Pacific International Symposium on Combustion and Energy Utilization*, Kuala Lumpur, Malaysia (May 20–22, 2002).
153. Prins, M.J., Ptasinski, K.J., and Janssen, F.J.J.G., Torrefaction of wood: Part 1. Weight loss kinetics, *Journal of Analytical and Applied Pyrolysis*, 77(1), 28–34 (2006).
154. Shafizadeh, F., Thermal conversion of cellulosic materials to fuels and chemicals, in *Wood and Agricultural Residues*. Academic Press, New York, pp. 183–217 (1983).

155. Bergman, P.C.A. and Kiel, J.H.A., Torrefaction for biomass upgrading, in *Proceedings of the 14th European Biomass Conference and Exhibition*, Paris, France, October 2005. ETA—Renewable Energies, Florence, Italy, pp. 17–21 (2005).

156. Bergman, P.C.A., Boersma, A.R., Kiel, J.H.A, Prins, M.J., Ptasinski, K.J., and Janssen, F.J.J.G., Torrefaction for entrained flow gasification of biomass, Report no. ECN-RX-04-046, in Van Swaaij, W.P.M., Fj.llstrom, T., Helm, P.T., and Grassi, P. (eds.), *Proceedings of the Second World Biomass Conference on Biomass for Energy, Industry and Climate Protection*, Rome, Italy, May 10–14, 2004. Energy Research Centre of the Netherlands (ECN), Petten, the Netherlands, pp. 679–682 (2004).

157. Bergman, P.C.A., Combined torrefaction and pelletization: The TOP process, Report no. ECN-C-05-073, Energy Research Centre of the Netherlands (ECN), Petten, the Netherlands (2005).

158. Bergman, P.C.A., Boersma, A.R., Kiel, J.H.A., Prins, M.J., Ptasinski, K.J., and Janssen, F.J.J.G., Torrefaction for entrained flow gasification of biomass, Report no. ECN-C- 05-067, Energy Research Centre of the Netherlands (ECN), Petten, the Netherlands (2004).

159. Prins, M.J., Ptasinski, I.G., and Janssen, F.J.J.G., More efficient biomass gasification via torrefaction, in Rivero, R., Monroy, L., Pulido, R., and Tsatsaronis, G. (eds.), *Proceedings of the 17th Conference on Efficiency, Costs, Optimization, Simulation and Environmental Impact of Energy Systems (ECOS'04)*, Guanajuato, Mexico, July 7–9, 2004. Elsevier, Amsterdam, the Netherlands, pp. 3458–3470 (2004).

160. Prins, M.J., Thermodynamic analysis of biomass gasification and torrefaction, PhD thesis, Technische Universiteit Emdhoven, Eindhoven, the Netherlands (2005).

161. Li, S., Chen, X., Wang, L., Liu, A., and Yu, G., Co-pyrolysis behaviors of saw dust and Shenfu coal in drop tube furnace and fixed bed reactor, *Bioresource Technology*, 148, 24–29 (2013).

162. Zheng, Y., Jensen, P.A., Jensen, A.D., Sander, B., and Junker, H., Ash transformation during co-firing coal and straw, *Fuel*, 86, 1008–1020 (2007).

163. Higman, C. and van der Burgt, M., *Gasification*. Elsevier, New York (2008).

164. Antal, M.J., Biomass pyrolysis: A review of the literature. Part II: Lignocellulose pyrolysis, in Boer, K.W. and Duffie, J.A. (eds.), *Advances in Solar Energy*, Vol. 2. American Solar Energy Society, Boulder, CO, pp. 175–255 (1985).

165. Callis, H.P.A., Haan, H., Boerrigter, H., Van der Drift, A., Peppink, G., Van den Broek, R., Faaij, A., and Venderbosch, R.H., Preliminary techno-economic analysis of large-scale synthesis gas manufacturing from imported biomass, in *Proceedings of an Expert Meeting on Pyrolysis and Gasification of Biomass and Waste*, Strasbourg, France, pp. 403–417 (2003).

166. Shafizadeh, F., Pyrolytic reactions and products of biomass, in Overend, R.P., Mime, T.A., and Mudge, L.K. (eds.), *Fundamentals of Biomass Thermochemical Conversion*. Elsevier, London, U.K., pp. 183–217 (1985).

167. Biagini, E., Lippi, F., Petarca, L., and Tognotti, L., Devolatilization rate of biomasses and coal–biomass blends: An experimental investigation, *Fuel*, 81, 1041–1050 (2002).

168. Meesri, C. and Moghtaderi, B., Lack of synergetic effects in the pyrolytic characteristics of woody biomass/coal blends under low and high heating rate regimes, *Biomass and Bioenergy*, 23, 55–66 (2002).

169. Moghtaderi, B., Meesri, C., and Wall, T.F., Pyrolytic characteristics of blended coal and woody biomass, *Fuel*, 83, 745–750 (2004).

170. Vamvuka, D., Pasadakis, N., Kastanaki, E., Grammelis, P., and Kakaras, E., Kinetic modeling of coal/ agricultural by-product blends, *Energy Fuels*, 17, 549–558 (2003).

171. Vuthaluru, H.B., Thermal behaviour of coal/biomass blends during co-pyrolysis, *Fuel Processing Technology*, 85, 141–155 (2004).

172. Lu, K.-M., Lee, W.-J., Chen, W.-H., and Lin, T.-C., Thermogravimetric analysis and kinetics of co-pyrolysis of raw/torrefied wood and coal blends, *Applied Energy*, 105, 57–65 (2013).

173. Kirtania, K. and Bhattacharya, S., Pyrolysis kinetics and reactivity of algae–coal blends, *Biomass and Bioenergy*, 55, 291–298 (2013).

174. Sadhukhan, A.K., Gupta, P., Goyal, T., and Saha, R.K., Modelling of pyrolysis of coal–biomass blends using thermogravimetric analysis, *Bioresource Technology*, 99, 8022–8026 (2008).

175. Pan, Y.G., Velo, E., and Puigjaner, L., Pyrolysis of blends of biomass with poor coals, *Fuel*, 75, 412–418 (1996).

176. Aboyade, A.O., Görgens, J.F., Carrier, M., Meyer, E.L., and Knoetze, J.H., Thermogravimetric study of the pyrolysis characteristics and kinetics of coal blends with corn and sugarcane residues, *Fuel Processing Technology*, 106, 310–320 (2013).

177. Li, S., Chen, X., Liu, A., Wang, L., and Yu, G., Study on co-pyrolysis characteristics of rice straw and Shenfu bituminous coal blends in a fixed bed reactor, *Bioresource Technology*, 155, 252–257 (2014).

178. Gil, M.V., Casal, D., Pevida, C., Pis, J.J., and Rubiera, F., Thermal behaviour and kinetics of coal/bio-mass blends during co-combustion, *Bioresource Technology*, 101, 5601–5608 (2010).

179. Franco, C., Pinto, F., Andre, R., Tavares, C., Dias, M., Gulyurtlu, I., and Cabria, I., Experience using INETI experimental facilities in co-gasification of coal and wastes, in *Proceedings of the Workshop Co-processing of Different Waste Materials with Coal for Energy*, Lisbon, Portugal, pp. 238–247 (November 23, 2001).

180. Storm, C., Rudiger, H., Spliethoff, H., and Hein, K.R.G., Co-pyrolysis of coal/biomass and coal/sewage sludge mixtures, *Journal of Engineering for Gas Turbines and Power*, 121, 55–63 (1999).

181. Vamvuka, D., Kakaras, E., Kastanaki, E., and Grammelis, P., Pyrolysis characteristics and kinetics of biomass residuals mixtures with lignite, *Fuel*, 82, 1949–1960 (2003).

182. Vuthaluru, H.B., Investigations into the pyrolytic behaviour of coal/biomass blends using thermogravi-metric analysis, *Bioresource Technology*, 92, 187–195 (2004).

183. Idris, S.S., Rahman, N.A., Ismail, K., Alias, A.B., Rashid, Z.A., and Aris, M.J., Investigation on ther-mochemical behaviour of low rank Malaysian coal, oil palm biomass and their blends during pyrolysis via thermogravimetric analysis (TGA), *Bioresource Technology*, 101, 4584–4592 (2010).

184. Lapuerta, M., Hernández, J.J., Pazo, A., and López, J., Gasification and co-gasification of biomass wastes: Effect of the biomass origin and the gasifier operating conditions, *Fuel Processing Technology*, 89, 828–837 (2008).

185. Suelves, I., Lázaro, M.J., and Moliner, R., Synergetic effects in the co-pyrolysis of samca coal and a model aliphatic compound studied by analytical pyrolysis, *Journal of Analytical and Applied Pyrolysis*, 65, 197–206 (2002).

186. Zhang, L., Xu, S., Zhao, W., and Liu, S., Co-pyrolysis of biomass and coal in a free fall reactor, *Fuel*, 86, 353–359 (2007).

187. Haykiri-Acma, H. and Yaman, S., Combinations of synergistic interactions and additive behav-ior during the co-oxidation of chars from lignite and biomass, *Fuel Processing Technology*, 89, 176–182 (2008).

188. Sonobe, T., Worasuwannarak, N., and Pipatmanomai, S., Synergies in co-pyrolysis of Thai lignite and corncob, *Fuel Processing Technology*, 89, 1371–1378 (2008).

189. Edreis, E.M.A., Luo, G., Li, A., Xu, C., and Yao, H., Synergistic effects and kinetics thermal behaviour of petroleum coke/biomass blends during H_2O co-gasification, Energy Conversion and Management, 79, 355–366 (2014).

190. Krerkkaiwan, S., Fushimi, C., Tsutsumi, A., and Kuchonthara, P., Synergetic effect during co-pyrolysis/gasification of biomass and sub-bituminous coal, *Fuel Processing Technology*, 115, 11–18 (2013).

191. Park, D.K., Kim, S.D., Lee, S.H., and Lee, J.G., Co-pyrolysis characteristics of sawdust and coal blend in TGA and a fixed bed reactor, *Bioresource Technology*, 101, 6151–6156 (2010).

192. Wei, L.G., Zhang, L., and Xu, S.P., Effects of feedstock on co-pyrolysis of biomass and coal in a free-fall reactor, *Journal of Fuel Chemistry and Technology*, 39, 728–734 (2011).

193. Blesa, M.J., Miranda, J.L., Moliner, R., Izquierdo, M.T., and Palacios, J.M., Low temperature co-pyrolysis of a low-rank coal and biomass to prepare smokeless fuel briquettes, *Journal of Analytical and Applied Pyrolysis*, 70, 665–677 (2003).

194. Haykiri-Acma, H. and Yaman, S., Interaction between biomass and different rank coals during co-pyrolysis, *Renewable Energy*, 35, 288–292 (2010).

195. Yuan, S., Dai, Z.-H., Zhou, Z.-J., Chen, X.-L., Yu, G.-S., and Wang, F.-C., Rapid copyrolysis of rice straw and a bituminous coal in a high-frequency furnace and gasification of the residual char, *Bioresource Technology*, 109, 188–197 (2012).

196. Xu, C., Hu, S., Xiang, J., Zhang, L., Sun, L., Shuai, C., Chen, Q., He, L., and Edreis, E.M.A., Interaction and kinetic analysis for coal and biomass co-gasification by TG–FTIR, *Bioresource Technology*, 154, 313–321 (2014); *Energies*, 7, 1136 (2014).

197. Coughlin, R.W. and Davoudzadeh, F., Coliquefaction of lignin and bituminous coal, *Fuel*, 65, 95–106 (1986).

198. Howaniec, N. and Smoliński, A., Steam co-gasification of coal and biomass—Synergy in reactivity of fuel blends chars, *International Journal of Hydrogen Energy*, 38, 16152–16160 (2013).

199. Feng, Z., Zhao, J., Rockwell, J., Bailey, D., and Huffman, G., Direct liquefaction of waste plastics and coliquefaction of coal-plastic mixtures, *Fuel Processing Technology*, 49, 17–30 (1996).

200. Mastral, A.M., Murillo, R., Callen, M.S., and Garcia, T., Evidence of coal and tire interactions in coal-tire coprocessing for short residence times, *Fuel Processing Technology*, 69, 127–140 (2001).

201. Mastral, A.M., Mayoral, M.C., Murillo, R., Callen, M., Garcia, T., Tejero, M.P., and Torres, N., Evaluation of synergy in tire rubber-coal coprocessing, *Industrial & Engineering Chemistry Research*, 37, 3545–3550 (1998).

202. Taghiei, M.M., Feng, Z., Huggins, F.E., and Huffman, G.P., Coliquefaction of waste plastics with coal, *Energy Fuels*, 8, 1228–1232 (1994).
203. Kanno, T., Kimura, M., Ikenaga, N., and Suzuki, T., Coliquefaction of coal with polyethylene using Fe(CO)5–S as catalyst, *Energy Fuels*, 14, 612–617 (2000).
204. Jones, J.M., Kubacki, M., Kubica, K., Ross, A.B., and Williams, A., Devolatilisation characteristics of coal and biomass blends, *Journal of Analytical and Applied Pyrolysis*, 74, 502–511 (2005).
205. Hayashi, J., Mizuta, H., Kusakabe, K., and Morooka, S., Flash copyrolysis of coal and polyolefin, *Energy Fuels*, 8, 1353–1359 (1994).
206. Onay, O., Bayram, E., and Kockar, O.M., Copyrolysis of Seyitömer lignite and safflower seed: Influence of the blending ratio and pyrolysis temperature on product yields and oil characterization, *Energy Fuels*, 21, 3049–3056 (2007).
207. Gao, C., Vejahati, F., Katalambula, H., and Gupta, R., Co-gasification of biomass with coal and oil sand coke in a drop tube furnace, *Energy Fuels*, 24, 232–240 (2009).
208. Ulloa, C.A., Gordon, A.L., and García, X.A., Thermogravimetric study of interactions in the pyrolysis of blends of coal with radiata pine sawdust, *Fuel Processing Technology*, 90, 583–590 (2009).
209. Kubacki, M.L., Co-pyrolysis and co-combustion of coal and biomass, PhD thesis, University of Leeds, Leeds, U.K. (2007).
210. NREL, Equipment design and cost estimation for small modular biomass systems, synthesis gas cleanup, and oxygen separation equipment—Task 2: Gas cleanup design and cost estimates-wood feedstock. NREL/DOE subcontract report NREL/SR-510-39945, Nexant Inc., San Francisco, CA (May 2006).
211. Air products tees valley renewable energy facility, EA/EPR/JP3331HK/A001, Environmental Permit Application, Air Products PLC, Hersham, Surrey, U.K. (May 2011).
212. Bertocci, A. and Patterson, R., Wet scrubber technology for controlling biomass gasification emissions, in *IT3'07 Conference*, Phoenix, AZ (May 14–18, 2007).
213. Bertocci, A., Wet scrubbers for gasifier gas cleaning, Paper no. 49, in *IT3'07 Conference*, Phoenix, AZ (May 14–18, 2007).
214. Energy evolved: Clean, sustainable energy recovery through plasma gasification, a report by Westinghouse Plasma Corporation, a division of Alter NRG Corp., Calgary, Canada (February 2011).
215. Krishnan, G., Wood, B., Tong, G., and Kothari, V., Removal of hydrogen chloride vapor from high temperature coal gases. Abstracts of the Papers for American Chemical Society, *Fuel*, 195, 32 (1988).
216. Gil, J. et al., Biomass gasification with air in a fluidized bed: Effect of the in-bed use of dolomite under different operating conditions, *Industrial & Engineering Chemistry Research*, 38(11), 4226–4235 (1999).
217. Turk, B.S., Merkel, T., Lopez-Ortiz, A., Gupta, R.P., Portzer, J.W., Krishnan, G.N., Freeman, B.D., and Fleming, G.K., Novel technologies for gaseous contaminants control, final report for the base program, DOE contract no. DE-AC26-99FT40675, Period of Performance: October 1, 1999 to September 30, 2001 for U.S. Department of Energy, Morgantown, WV (2001).
218. Dayton, D., Turk, B., and Gupta, R., Syngas cleanup, conditioning and utilization, Chapter 4, in Brown, R.C. (ed.), *Thermochemical; Processing of Biomass Conversion into Fuels, Chemicals and Power*, 1st edn., John Wiley & Sons, New York (April 2011).
219. Tuan, H.P., Janssen, H.-G., Cramers, C.A., Smit, A.L.C., and van Loo, E.M., Determination of sulfur components in natural gas: A review, *Journal of high Resolution Chromatography*, 17, 373–389 (June 1994).
220. Lippert, T.E., Bruck, G.J., Sanjana, Z.N., Newby, R.A., and Bachovchin, D.M., Westinghouse advanced particle filter system, Research sponsored by the U.S. Department of Energy's Morgantown Energy Technology Center, under contract DE-AC21-94MC31147 with Westinghouse Electric Corporation, Science & Technology Center, Pittsburgh, PA (1996), www.netl.doe.gov/publications/proceedings/96/96psl/.../96ps2_1.pdf (accessed December 1996).
221. Syngas contaminant removal and conditioning-particulate removal, NETL report, Department of Energy, Washington, DC (2015), www.netl.doe.gov/research/coal/...systems/.../particulate-removal.
222. The Plasco Process, a report by Plasco Energy Group, Kanata, Ontario, Canada (2015).
223. Alter-NRG process, a report by Westinghouse Plasma Corporation, Madison, PA (2015).
224. Envitech Process, a report by Envitech Co., San Diego, CA (2015).
225. PyroGenesis Canada, Montreal, Quebec, Canada (2015).
226. H_2S management, a publication by MV Technologies, Golden, CO (2016), mvseer.com/hydrogen-sulfide-removal/index.html.

227. LO-CAT process for cost effective desulfurization of all types of gas streams, a publication by Merichem Co., Schaumburg, IL (2016), www.merichem.com/images/casestudies/Desulfurization.pdf.

228. Air Product Tees valley renewable energy facilities, Air Product, Stockton, U.K. (2015).

229. SULFURTRAP® high performance hydrogen sulfide (H_2S) adsorbent, a publication by Chemical Products Industries Inc., Oklahoma City, OK (2016), www.chemicalproductsokc.com/.../SULFURTRAP%20 PDS.pdf.

230. Gupta, R. (PI), Recovery Act: Scale-Up of High-Temperature Syngas Cleanup Technology, a publication by NETL, Department of Energy, Washington, DC, RTI project, Period of Performance 7/20/2009 to 9/30/2015 (2016), www.netl.doe.gov/publications/factsheets/project/fe0000489.pdf.

231. Lesemann, M., *Syngas Cleanup and Conversion*. RTI International, Research Triangle, NC (2016), https://www.rti.org/brochures/rti_adv_gasification.pdf (accessed February 2016).

8 Gas Storage and Transport Infrastructure

8.1 INTRODUCTION

The supply and demand of natural or synthetic gas is time-dependent, and its management requires an infrastructure of storage and transportation. The downstream conversion of natural and synthetic gas to liquid fuels, chemicals, heat, power, and mobile applications also requires timely and required delivery. Thus, the peak demand and fast response to the changing needs require workable and economical storage and a transportation system that can bring natural and synthetic gas to its various markets. The nature of the storage and transportation system depends on the nature and state of gas (natural gas or synthetic gas like biogas, syngas, or hydrogen; liquids like LNG or LPG; CNG; and solids like natural gas hydrates), need of the market for its use, and the end purpose: heat, power, liquid fuels, commodity production, or mobile applications. The storage and transportation infrastructure also needs to address the issue of natural and synthetic gas to and from remote locations. This chapter examines our capability for storage and transportation of natural and synthetic gas for all of these situations.

Gas is difficult to store or transport because of its physical nature and needs. High pressures and/ or low temperatures are used to increase the bulk density and the ease for storage and transportation. Whereas oil is readily stored in large, relatively simple, and cheap tanks and then transported in huge tankers, gas, as a result of the storage difficulties, needs to be transported immediately to its destination after production from a reservoir [1–8]. Gas, on the other hand, is easier to recover compared to oil and has a longer future than oil [1–8]. The transmission of gas by pipelines requires pressure which generally moves gas at speed of about 25 mph within large transmission pipelines.

The cost of transporting natural gas per unit of energy to distant markets is much higher compared to oil (perhaps 10 times [2,3]) because of its volume–pressure behavior, and it is carried out by pipeline on land, and via liquefied natural gas (LNG) for overseas [3–5]. While LNG production at present costs around US$15/bbl oil equivalent (i.e., \$2.5/thousand scf of gas), many importing countries do not have the capital to build the huge storage and regeneration facilities. Selling small volumes of LNG is not yet economically attractive to the LNG market. Likewise, intermittent gas is also not economically attractive to the major gas buyers for LNG facilities, pipelines, or large-scale commodity manufacturers. Thus for the smaller markets, for example, islands where pipelines or LNG are not economical, and for smaller gas fields a different more flexible, cheaper, less massive storage and transport approach such as CNG (compressed natural gas) or NGH (natural gas hydrates) is needed. Otherwise, for stranded gas where the quantities can be far better regulated, gas is often used for the local needs, for example, a power station or local heating.

Natural gas is often found in the places where there is no local market for its use or no storage or transportation facilities. Sometimes it cannot be even discarded such as the case with flaring of associated gas. These location difficulties have prevented gas from reaching market and in many countries; for example, Trinidad and Qatar have prevented development of gas reserves for many years until the "dash for gas" created new markets and safe long-term contracts [1–8]. Capturing stranded gas often requires guaranteed market for its usages. The price of the gas can also have a dominating influence on the viability of the development of a hydrocarbon reservoir, and on the method chosen for the storage and transport of gas recovered from the reservoir.

Sources of gas may be nonassociated gas reservoirs (i.e., only gas within the reservoir) or associated gas from oil reservoirs, which is gas produced along with oil as pressure drops [2]. Some associated gas is always produced when crude oil is produced. Nonassociated gas is directly controllable by the producer. Associated gas is dependent on the rate of oil production and the amount of gas dissolved in the oil when it is produced; thus supply can be unreliable. Nonassociated gas is normally 95% or more methane; nonassociated gas contains some quantity of ethane, propane, and so on, which are valuable premium products in their own right, and are usually extracted for separate sale at source or other nearby market.

Worldwide, governments are mandating that producers stop flaring associated gas, as their citizens perceive that it is a waste of a valuable nonrenewable resource. There are often regulatory restrictions on when produced gas can be reinjected, or flared, with an understanding that any reinjected gas must eventually be produced. When such restrictions occur, oil production must be stopped until this associated gas can somehow be exported, reinjected in the well, or used locally [1–8].

The reinjection option can appear attractive at first sight, offering the added advantage of maintaining reservoir pressure, but the costs are high for the drilling and completion of the injection wells, the subsurface equipment, and the topsides equipment required to clean, pressurize, and inject the gas. The gas is basically making an expensive round trip from the production well to the topsides and back down the injection well. This is sunk cost, perhaps US$0.25–0.5/million scf, since no monetary value is gained from the gas until it is sold. Thus oil producers would like to have a robust cost effective way of disposing of "stranded" gas; otherwise, they could suffer shutdown [1–8].

Sources can also be unconventional such as deep gas, tight gas, coalbed methane, shale gas, gas from geopressurized zones or gas hydrates. While the natural gas from these sources may not be as pure as the one from conventional sources, generally the recovered gas is pretreated in order to be stored and transported within the infrastructure created for the conventional natural gas. As shown later, gas hydrates can be transported in the solid form.

The storage and transportation of synthetic gas productions, described in Chapters 3 through 6, require somewhat different strategies. The storage and transportation of syngas, biogas, and hydrogen are handled separately. The present chapter also addresses the details of the infrastructure needs for these gases. There are a number of possible options of storing and exporting natural gas from oil and gas fields and synthetic gases (syngas, biogas, hydrogen) from their productions sites to their available markets, and this chapter examines all of them in detail.

The discussion on storage and transportation infrastructure for natural and synthetic gas is divided into seven parts in this chapter. First, the storage and transportation of natural gas coming from underground is discussed in detail in Sections 8.2 and 8.3 respectively. Gas is mostly stored underground and transported by the pipelines. Gas can also be stored and transported as CNG or solid gas hydrates to decrease its volume and cost of storage and transportation. This subject is discussed in Section 8.4. The transport of gas as solid gas hydrate is also important while recovering gas hydrates from underwater or remote arctic locations. The natural gas is often liquefied as LNG and LPG to increase its mass and energy densities. The storage and transportation strategies for liquid natural gas are different from those required for gaseous natural gas and they are examined in Section 8.5.

The next three subjects of storage and transportation deal with three different types of synthetic gas. The storage and transport of syngas produced by thermal gasification require a different set of processes, as outlined in Section 8.6. Biogas and biomethane produced from anaerobic waste treatment are stored and distributed for their downstream use in a variety of ways, and this subject is discussed in Section 8.7. Finally, hydrogen management requires an independent storage and transportation infrastructure for pure hydrogen. While a small amount of hydrogen (5–15 vol%) can be transported by natural gas pipeline, pure hydrogen storage and transport requires its independent infrastructure. This subject is examined in Sections 8.8 and 8.9. These sections also deal with hydrogen storage and refueling needs for its vehicle application.

The final topic covered in this chapter is about natural gas and hydrogen grids necessary for better management of their supply and demand. These grids will have to be connected to heat grids and smart electrical grids. Smart gas grids will have to be a major method for managing supply and demand of gaseous fuels for heat and power applications. Grids also play an important role in the application of gaseous fuels for mobile usages. This topic is discussed in Section 8.10.

In general, the development of storage and transportation infrastructure for natural and synthetic gas requires the following considerations:

1. Gas needs to be sold to monetize it. At present, on a large scale, pipelines and LNG are the main "gas as fuel" transport routes. Pipeline costs are approximately proportional to distance. LNG transportation costs also increase proportionately with distance to market, but at a less steep rate. LNG transport requires high initial investment and causes large-volume quantum jumps when a new train is brought into the market. While the use of gas for liquid fuels, electricity, and commodity has market potential, it requires huge upfront investment and incurs large energy losses.

2. Gas energy movement, methods, and quantities will depend on market demands and distance to market. What will sell and what contracts can be negotiated for long-term gas reservoir and infrastructure development, including the necessary manufacturing and processing plant, are critical. Pipelines, LNG, and, to some extent, gas to liquid conversion have a large economy of scale, so they are proportionately expensive for developing small gas reserves. As mentioned earlier, LNG production is expensive (around US$15/bbl oil equivalent—$2.5/thousand scf) and many exporters do not have the quantities of gas to export, or importers do not have the need or the capital to build the storage and regeneration facilities for importing large quantities.

3. Small volumes of intermittent gas are not economically attractive to the major gas sellers, particularly for LNG facilities or pipelines. For the smaller markets, for example, islands where pipelines or LNG are not feasible, solid gas (i.e., gas hydrates) and CNG can be potential economical transport methods. The quantities can be far better regulated and designed for the needs of a particular power station. There could be options for handling niche markets for gas reserves which are stranded (no market) and for associated gas (on- or off-shore) which cannot be flared or reinjected, or for small reservoirs which cannot otherwise be economically exploited.

4. Transportation of natural gas as hydrate or CNG is believed to be feasible at costs less than for LNG and where pipelines are not possible. The competitive advantage of gas hydrates or CNG over the other nonpipeline transport processes is that they are intrinsically simple and so it is much easier to implement at lower capital costs, provided economically attractive market opportunities can be negotiated between the gas seller and buyer.

5. In the end, it is the cost per unit of the gas and the profit margin between the cost of gas and the sold product that will dictate the suitability of a particular form of storage and transportation infrastructure. The exception may be the need of energy-rich countries to supply gas to its energy-poor neighboring countries or to non-energy-producing countries (e.g., Trinidad and the other Caribbean nations, or the various non-energy-producing Indonesian islands). In these cases, suitable international contracts which consider humanitarian reasons will dictate the required storage and transportation infrastructure. In general, stability of long-term contracts is important for any workable storage and transportation projects. This applies to both natural and synthetic gas.

8.2 NATURAL GAS STORAGE

Natural gas, like most other commodities, can be stored for an indefinite period of time. *The underground gas storage industry is almost 100 years old.* Since the exploration, production, and transportation of natural gas take time and the natural gas that reaches its destination is not always needed right away, it is often stored in the underground storage facilities. These storage facilities are often located near market centers to supplement locally produced natural gas in times of need [1,2,9–17].

If gas is used mainly for heating, its demand can be seasonal (more needed in winter than summer). If, however, gas is used for heating as well as electricity, its demand will be even across the year. In either case, storage serves as a buffer between transportation and distribution, to ensure adequate supplies of natural gas are in place for overall demand shifts, and unexpected demand surges. Natural gas in storage also serves as insurance against any unforeseen accidents, natural disasters, or other occurrences that may affect the production or delivery of natural gas [1,2,9–17].

Natural gas storage plays a vital role in maintaining the reliability of supply needed to meet the demands of consumers. Historically, when natural gas was a regulated commodity, storage was part of the bundled product sold by the pipelines to distribution utilities. The deregulation of gas market in 1992 by Federal Energy Regulatory Commission's Order 636 increased the importance of gas storage. This regulation meant that storage was not just to meet the needs of utilities, but also to be used by industry participants for commercial reasons: storing gas when prices are low, and withdrawing and selling it when prices are high. This gave added incentives for storage. The purpose and use of storage have always been closely linked to the regulatory environment of the time. The importance of storage has also increased with the effective development and employment of smart gas grid [371] and its connection to smart electrical grid and local heat grid.

Underground gas storage (UGS) facilities largely contribute to the reliability of gas supplies to consumers. They level off daily gas consumption fluctuations and meet the peak demand in winter for heating and summer (for air-conditioning) or any other time of need. Underground storage facilities secure natural gas supplies to consumers regardless of a season, temperature, or other unforeseeable circumstances.

There are basically two uses for natural gas in storage facilities: meeting base load requirements and meeting peak load requirements. Base load facilities are capable of holding enough natural gas to satisfy long-term seasonal demand requirements. Typically, the turnover rate for natural gas in these facilities is a year; natural gas is generally injected during the summer (nonheating season), which usually runs from April through October, and withdrawn during the winter (heating season), usually from November to March. This schedule may change in warmer locations, where gas is more used for air-conditioning in summer (through electricity) than winter. These reservoirs are larger, but their delivery rates are relatively low, meaning the natural gas that can be extracted each day is limited. Instead, these facilities provide a prolonged, steady supply of natural gas. Depleted gas reservoirs are the most common type of base load storage facility.

Peak load storage facilities, on the other hand, are designed to have high-deliverability for short periods of time, meaning natural gas can be withdrawn from storage quickly should the need arise. Peak load facilities are intended to meet sudden, short-term demand increases. These facilities cannot hold as much natural gas as base load facilities; however, they can deliver smaller amounts of gas more quickly, and can also be replenished in a shorter amount of time than base load facilities. While base load facilities have long-term injection and withdrawal seasons, turning over the natural gas in the facility about once per year, peak load facilities can have turnover rates as short as a few days or weeks. While salt caverns are the most common type of peak load storage facility, the need can also be satisfied by aquifers storage facility.

8.2.1 TYPES OF UNDERGROUND STORAGE

Underground natural gas storage fields grew in popularity shortly after World War II. At the time, the natural gas industry noted that seasonal demand increases could not feasibly be met by pipeline delivery alone. In order to be able to meet seasonal demand increases, underground storage fields were the only options.

Natural gas is usually stored underground, in large storage reservoirs. There are three main types of underground storage: depleted gas reservoirs, aquifers, and salt caverns. In addition to underground storage, however, natural gas can be stored as LNG. LNG allows natural gas to be shipped and stored in liquid form, meaning it takes up much less space than gaseous natural gas. This subject is discussed in detail in Section 8.5.

Essentially, any underground storage facility is reconditioned before injection, to create a desired storage vessel underground. Natural gas is injected into the formation, building up pressure as more natural gas is added. In this sense, the underground formation becomes a sort of pressurized natural gas container. As with newly drilled wells, the higher the pressure in the storage facility, the more readily gas may be extracted. Once the pressure drops to below that of the wellhead, there is no pressure differential left to push the natural gas out of the storage facility. This means that, in any underground storage facility, there is a certain amount of gas that may never be extracted. This is known as physically unrecoverable gas; it is permanently embedded in the formation. In the United States currently, working gas capacity is mainly stored in depleted reservoir (86%), with remaining stored in aquifer (10%) and salt domes (4%) [1,2,9–17].

In addition to this physically unrecoverable gas, underground storage facilities contain what is known as "base gas" or "cushion gas." This is the volume of gas that must remain in the storage facility to provide the required pressurization to extract the remaining gas. In the normal operation of the storage facility, this cushion gas remains underground; however, a portion of it may be extracted using specialized compression equipment at the wellhead.

"Working gas" is the volume of natural gas in the storage reservoir that can be extracted during the normal operation of the storage facility. This is the natural gas that is being stored and withdrawn; the capacity of storage facilities normally refers to their working gas capacity or daily deliverability of gas. In the United States 74% of daily deliverability is imbedded in depleted reservoir, with 15% in salt caverns and 11% in aquifers. While depleted reservoirs contain both high working gas capacity and deliverability, salt caverns indicate high daily deliverability relative to working gas capacity [1,2,9–17]. At the beginning of a withdrawal cycle, the pressure inside the storage facility is at its highest, meaning working gas can be withdrawn at a high rate. As the volume of gas inside the storage facility drops, pressure (and thus deliverability) in the storage facility also decreases. Periodically, underground storage facility operators may reclassify portions of working gas as base gas after evaluating the operation of their facilities [1,2,9–17].

8.2.1.1 Depleted Gas Reservoirs

The first instance of natural gas successfully being stored underground occurred in Welland County, Ontario, Canada, in 1915. This storage facility used a depleted natural gas well that had been reconditioned into a storage field. In the United States, the first storage facility was developed just south of Buffalo, New York. By 1930, there were nine storage facilities in six different states. Prior to 1950, virtually all natural gas storage facilities were in depleted reservoirs [1,2,9–17].

Today, the most prominent and common form of underground storage consists of depleted gas reservoirs. Depleted reservoirs are those formations that have already been tapped of all their recoverable natural gas. This leaves an underground formation, geologically capable of holding natural gas. In addition, using an already developed reservoir for storage purposes allows the use of the extraction and distribution equipment left over from when the field was productive. Having this extraction network in place reduces the cost of converting a depleted reservoir into a storage facility. Depleted reservoirs are also attractive because their geological characteristics are already well known. Of the three types of underground storage, depleted reservoirs, on average, are the cheapest and easiest to develop, operate, and maintain [1,2,9–17].

The factors that determine whether or not a depleted reservoir will make a suitable storage facility are both geographic and geologic. Geographically, depleted reservoirs must be relatively close to consuming regions. They must also be close to transportation infrastructure, including trunk pipelines and distribution systems. While depleted reservoirs are numerous in the United States, they are more abundantly available in producing regions like south and northeast. In regions without depleted reservoirs, like the upper Midwest, one of the other two storage options is required.

Geologically, depleted reservoir formations must have high permeability and porosity. The porosity of the formation determines the amount of natural gas that it may hold, while its permeability determines the rate at which natural gas flows through the formation, which in turn determines the rate of injection and withdrawal of working gas. In certain instances, the formation may be stimulated to increase permeability [1,2,9–17].

In order to maintain pressure in depleted reservoirs, about 50% of the natural gas in the formation must be kept as cushion gas. However, depleted reservoirs, having already been filled with natural gas and hydrocarbons, do not require the injection of what will become physically unrecoverable gas; that gas already exists in the formation. This makes this type of reservoir a very cost-effective gas storage facility [1,2,9–17].

8.2.1.2 Aquifers

Gas fills much greater volumes than solids or fluids. It would be therefore difficult to find impermeable reservoirs for it, if nature hadn't built them already. Porous beds of sandstone in the Earth's crust, hermetically sealed with a dome of a clay layer at the top, are natural UGS facilities. Sandstone pores can contain water, but hydrocarbons also can accumulate there. In the process of making UGS facilities in an aquifer, gas accumulating under the clay cover is displacing the water downwards. *If a reservoir initially contains hydrocarbons, it is an oil or gas field. Impermeability of this structure is already proven by the fact that hydrocarbons accumulated in it.*

Aquifers are underground porous, permeable rock formations that act as natural water reservoirs. However, in certain situations, these water-containing formations may be reconditioned and used as natural gas storage facilities. As they are more expensive to develop than depleted reservoirs, these types of storage facilities are usually used only in areas where there are no nearby depleted reservoirs. Traditionally, these facilities are operated with a single winter withdrawal period, although they may be used to meet long-term peak load requirements as well [1,2,9–17].

Aquifers are the least desirable and most expensive type of natural gas storage facility for a number of reasons. First, the geological characteristics of aquifer formations are not as thoroughly known, as with depleted reservoirs. A significant amount of time and money goes into discovering the geological characteristics of an aquifer, and determining its suitability as a natural gas storage facility. Seismic testing must be performed, similar to what is done for the exploration of potential natural gas formations [1,2,9–17]. The area of the formation, the composition and porosity of the formation itself, and the existing formation pressure must all be discovered prior to development of the formation. In addition, the capacity of the reservoir is unknown, and may be determined only once the formation is further developed.

In order to develop a natural aquifer into an effective natural gas storage facility, all of the associated infrastructure must also be developed. This includes installation of wells, extraction equipment, pipelines, dehydration facilities, and possibly compression equipment. Since aquifers are naturally full of water, in some instances powerful injection equipment must be used, to allow sufficient injection pressure to push down the resident water and replace it with natural gas. While natural gas being stored in aquifers has already undergone all of its processing, upon extraction from a water-bearing aquifer formation the gas typically requires further dehydration prior to transportation, which requires specialized equipment near the wellhead. Aquifer formations do not have the same natural gas retention capabilities as depleted reservoirs. This means that some of the natural gas that is injected escapes from the formation, and must be gathered and extracted by "collector" wells specifically designed to pick up gas that may escape from the primary aquifer formation [1,2,9–17].

In addition to these considerations, aquifer formations typically require a great deal of more "cushion gas" than do depleted reservoirs. Since there is no naturally occurring gas in the formation to begin with, a certain amount of natural gas that is injected will ultimately prove physically unrecoverable. In aquifer formations, cushion gas requirements can be as high as 80% of the total gas volume. While it is possible to extract cushion gas from depleted reservoirs, doing so from aquifer formations could have negative effects, including formation damage. As such, most of the cushion gas that is injected into any one aquifer formation may remain unrecoverable, even after the storage facility is shut down. Most aquifer storage facilities were developed when the price of natural gas was low, meaning this cushion gas was not very expensive to give up. At higher gas prices, this development can become significantly expensive [1,2,9–17].

All of these factors mean that developing an aquifer formation as a storage facility can be time-consuming and expensive. In some instances, aquifer development can take 4 years, which is more than twice the time it takes to develop depleted reservoirs as storage facilities. In addition to the increased time and cost of aquifer storage, there are also environmental restrictions to using aquifers as natural gas storage. In the early 1980s the Environmental Protection Agency (EPA) set certain rules and restrictions on the use of aquifers as natural gas storage facilities. These restrictions are intended to reduce the possibility of freshwater contamination [1,2,9–17].

8.2.1.3 Salt Caverns

Underground salt formations offer another option for natural gas storage. These formations are well suited to natural gas storage in that salt caverns, once formed, allow little injected natural gas to escape from the formation unless specifically extracted. The walls of a salt cavern also have the structural strength of steel, which makes it very resilient against reservoir degradation over the life of the storage facility.

Essentially, salt caverns are formed out of existing salt deposits. These underground salt deposits may exist in two possible forms: salt domes, and salt beds. Salt domes are thick formations created from natural salt deposits that, over time, leach up through overlying sedimentary layers to form large dome-type structures. They can be as large as a mile in diameter, and 30,000 ft in height. Typically, salt domes used for natural gas storage are between 6000 and 1500 ft beneath the surface, although in certain circumstances they can come much closer to the surface. Salt beds are shallower, thinner formations. These formations are usually no more than 1000 ft in height. Because salt beds are wide, thin formations, once a salt cavern is introduced, they are more prone to deterioration, and may also be more expensive to develop than salt domes [1,2,9–17].

Once a suitable salt dome or salt bed deposit is discovered, and deemed suitable for natural gas storage, it is necessary to develop a "salt cavern" within the formation. Essentially, this consists of using water to dissolve and extract a certain amount of salt from the deposit, leaving a large empty space in the formation. This is done by drilling a well down into the formation, and cycling large amounts of water through the completed well. This water will dissolve some of the salt in the deposit, and be cycled back up the well, leaving a large empty space that the salt used to occupy. This process is known as "salt cavern" leaching [1,2,9–17].

Salt cavern leaching is used to create caverns in both types of salt deposits, and can be quite expensive. However, once created, a salt cavern offers an underground natural gas storage vessel with very high deliverability. In addition, cushion gas requirements are the lowest of all three storage types, with salt caverns only requiring only about 33% of total gas capacity to be used as cushion gas.

Salt cavern storage facilities are primarily located along the Gulf Coast, as well as in the northern states, and are best suited for peak load storage. Salt caverns are typically much smaller than depleted gas reservoirs and aquifers; in fact, underground salt caverns usually take up only one one-hundredth of the acreage taken up by a depleted gas reservoir. As such, salt caverns cannot hold the volume of gas necessary to meet base load storage requirements. However, deliverability from salt caverns is typically much higher than for either aquifers or depleted reservoirs. Therefore natural gas stored in a salt cavern may be more readily (and quickly) withdrawn, and caverns may be replenished with natural gas more quickly than in either of the other types of storage facilities. Moreover, salt caverns can readily begin flowing gas on as little as one hour's notice, which is useful

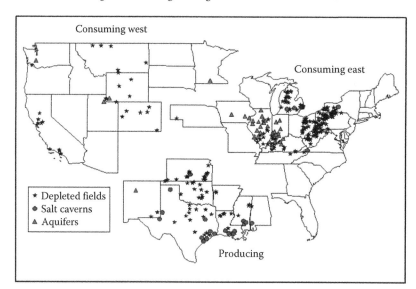

FIGURE 8.1 Locations of natural gas storage facilities: Most of them are located in Northeast and south. (From Natural gas storage, Wikipedia, the free encyclopedia, 2015; Thompson, J., U.S. Underground storage of natural gas in 1997: Existing and proposed, a report in Natural Gas Monthly, Energy Information Administration, Department of Energy, Washington, DC, September 1997.)

in emergency situations or during unexpected short-term demand surges. Salt caverns may also be replenished more quickly than other types of underground storage facilities [1,2,9–17].

Salt caves are ideal impermeable reservoirs. It is not difficult to build an underground salt cave, though it is a long process. Wells are drilled in a fitting bed of rock salt. Afterward, water is pumped into them and a cavity of the required size is washed out in the salt bed. A salt dome is not only impermeable to gas: salt is capable of "self-healing" fissures and fractures. Two storage facilities in rock salt deposits are being developed in the Kaliningrad and Volgograd Oblasts of the former Soviet Union [1,2,9–17].

In summary, storage facilities in a depleted field or aquifer feature large capacities but low flexibility. Injection and extraction of gas are much faster in the storage facilities that were built in rock salt caves. However, this type of storage is low in capacity compared to depleted fields. Locations of UGS facilities are depicted in Figure 8.1.

According to the Energy Information Administration (EIA), as of November 2015, the total design storage capacity of lower 48 states in the United States was 4658 bcf and working gas volume was 4343 bcf [11,13–16]. This number is significantly higher at the present time. The Severo-Stavropolskoye UGS facility (in Russia) is the largest in the world. Its capacity is 43 billion cubic meters (bcm) of active gas. That would be enough to meet the annual demand of France or the Netherlands [11,13–16]. The Severo-Stavropolskoye UGS facility was constructed at a depleted gas field. UGS facilities are of particular importance in Russia with its cold climate and large distances between resources and end users. Russia has the unique Unified Gas Supply System (UGSS), with the UGS system being the integral part of it. Underground storage facilities secure natural gas supplies to consumers regardless of season, temperature, or any unforeseeable circumstances.

8.3 NATURAL GAS TRANSPORT

Our dependence on natural gas is growing. Oil companies have developed many ways to transport natural gas to our homes. Moving natural gas from its source to our homes involves transporting it through elaborate pipeline systems. Many regulations on transporting natural gas are in place to ensure that it is moved in the safest and most effective manner possible. From the ground, to processing plants and ultimately to our homes, natural gas is transported by the oil and gas companies in a very precise way. This has now become a part of the development of smart gas grid where the flow of natural gas to different customers is managed and controlled in a dynamic manner [329–356]. The first step in transporting natural gas is to collect the gas from underground by a "gathering system" of pipes. The collected gas is then appropriately cleaned and processed (as shown in Chapter 7) to make it pipeline quality gas which can then be transported to the customers.

When the gas gets to the processing plant, thiols are added to it so it can be detected (by smell) during any type of leak. This is a safety measure. Once gas is processed, it is sent through intrastate and interstate pipeline networks. Intrastate pipelines carry natural gas within a state. Interstate pipelines are networks that actually cross state lines. These pipelines are constructed from rigorously tested carbon steel sections of pipelines which can withstand enormous pressure to ensure the safety of transport.

Pipelines are a very convenient method of transport but are not flexible as the gas will leave the source and arrive at its (one) destination [18–29]. Once the pipeline diameter is decided the quantities of gas that can be delivered is fixed by the pressures, although an increase in the maximum quantity can be achieved by adding compressors along the line, extra pipes in the form of loops, or by increasing the average pipeline pressure. If, for some reasons, pipeline has to be shutdown, generally both gas production and gas receiving facilities will also need to be shut down because gas cannot be readily stored, except perhaps by increasing the pipeline pressure by some percentage [18–29].

Pipeline pressures are normally 700–1100 psig (although 4000 psig lines are in operation) depending on the material of construction and the age of the pipe. Installation of pipeline costs currently, on average, US$1–5 million per mile, sometimes even higher, depending on the terrain (such as for onshore, mountains or for offshore, seabed flatness and depth) plus compressor stations, so distance becomes a major factor in the overall cost of the line, with cost being approximately proportional to distance. Overland pipelines are vulnerable to sabotage and are uneconomic for small reserves. Subsea lines over large marine distances and difficult marine environments can be hard to maintain, and generally uneconomical. Nevertheless, novel construction methods, including laser-welding techniques, are available and they allow faster pipe laying. Drag-reducing agents and treatments to the inner surface of pipes are used to increase flow and reduce pipe corrosion [18–29]. Safety is always the highest priority in pipeline installment. Export by pipeline on land is extensive throughout Europe, the United States, and soon South America [1]. Subsea lines over 2000 miles are generally considered to be uneconomical [18–29].

There are three major types of pipelines along the transportation route: the gathering system, the interstate pipeline system, and the distribution system. The gathering system consists of low-pressure, small-diameter pipelines that transport raw natural gas from the wellhead to the processing plant. If natural gas from a particular well contains high sulfur and carbon dioxide (sour gas), a specialized sour gas gathering pipe must be installed. Sour gas is corrosive, thus its transportation from the wellhead to the sweetening plant must be done carefully. While intrastate pipelines distribute gas within a particular state, interstate pipelines carry gas across state boundaries and in some case clear across state boundaries. Figure 8.2 describes interstate natural gas pipelines within the United States.

The interstate natural gas pipeline network transports processed natural gas from processing plants in producing regions to those areas with high natural gas requirements, particularly large, populated urban areas. Natural gas that is transported through interstate pipelines travels at high pressure in the pipeline, at pressures anywhere from 200 to 1500 pounds per square inch (psi). This reduces the volume of the natural gas being transported (by up to 600 times), as well as propelling natural gas through the pipeline at speed up to 25 mph. Interstate pipelines consist of a number of components that ensure the efficiency and reliability of a system that delivers gas year-round, twenty four hours a day [18–29].

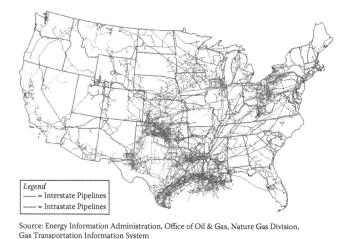

Source: Energy Information Administration, Office of Oil & Gas, Nature Gas Division,
Gas Transportation Information System

FIGURE 8.2   U.S. natural gas pipeline network. (From Storage of natural gas, Naturalgas.org, September 2013, accessed on January 2009.)

While certain component pipe sections can be as small as 0.5 in. in diameter, which are generally used only in gathering and distribution systems, mainline transmission pipes are usually between 16 and 48 in. in diameter. Lateral pipelines, which deliver natural gas to or from the mainline, are typically between 6 and 16 in. in diameter. Most major interstate pipelines are between 24 and 36 in. in diameter. While interstate main pipeline and lateral pipelines (called "line pipe") are made of carbon steel, some distribution pipes are made of highly advanced plastic, because of the need for flexibility, versatility, and the ease of replacement. Both mainline transmission pipes and "line pipe" are covered with a specialized coating to ensure that they do not corrode once placed in the ground. The purpose of the coating is to protect the pipes from moisture, which causes corrosion and rusting.

Turbine centrifugal compressors used along pipelines are operated by the natural gas they compress. Some centrifugal compressors are operated using an electric motor. Reciprocating natural gas engines are also used to power some compressor stations. In addition to compressing natural gas, compressor stations also usually contain some type of liquid separator. Usually, these separators consist of scrubbers and filters that capture any liquids or other unwanted particles from the natural gas in the pipeline. Although natural gas in pipelines is considered "dry" gas, it is not uncommon for a certain amount of water and hydrocarbons to condense out of the gas stream while in transit [18–29].

In addition to compressing natural gas to reduce its volume and push it through the pipe, metering stations are placed periodically along interstate natural gas pipelines. These stations allow pipeline companies to monitor the natural gas in their pipes. Essentially, these metering stations measure the flow of gas along the pipeline, and allow pipeline companies to "track" natural gas as it flows along the pipeline. These metering stations employ specialized meters to measure the natural gas as it flows through the pipeline, without impeding its movement. Interstate pipelines include a great number of valves along their entire length. These valves work like gateways; they are usually open and allow natural gas to flow freely, or they can be used to stop gas flow along a certain section of pipe. The valves allow replacement or maintenance of a particular segment of pipeline. Large valves can be placed every 5–20 miles along the pipeline, and are subject to regulation by safety codes [1,2,18–29].

8.3.1   Pipeline Control and Safety

Natural gas pipeline companies have customers on both ends of the pipeline—the producers and the customers. In order to manage the natural gas that enters the pipeline, and to ensure that all customers receive timely delivery of their portion of this gas, sophisticated control systems for monitoring and controlling the natural gas that is traveling through the pipeline is required. This is done by centralized gas control stations. These stations collect, assimilate, and manage data received from monitoring and compressor stations all along the pipe. Most of the data that is received by a control station are provided by Supervisory Control and Data Acquisition (SCADA) systems. These systems take measurements and collect data along the pipeline and transmit them to the centralized control station. Flow rate through the pipeline, operational status, pressure, and temperature readings may all be used to assess the status of the pipeline at any given time. These systems also work in real time [1,2,18–29].

Once the pipelines are built, pipeline companies routinely inspect their pipelines for corrosion and defects by an equipment known as "smart pigs." Smart pigs are intelligent robotic devices that are propelled down pipelines to evaluate the interior of the pipe. Some of the safety precautions associated with natural gas pipelines include aerial patrols, leak detection, pipeline markers, gas sampling, preventive maintenance, emergency response, and the one call program [1,2,18–29].

8.3.2 Natural Gas Distribution

Distribution is the final step in delivering natural gas to customers. While some large industrial, commercial, and electric generation customers receive natural gas directly from high-capacity interstate and intrastate pipelines, most other users receive natural gas from their local gas utility, also called a local distribution company (LDC).

Local distribution companies typically transport natural gas from delivery points located on interstate and intrastate pipelines to households and businesses through thousands of miles of small-diameter distribution pipe. The delivery point where the natural gas is transferred from a transmission pipeline to the local gas utility is often termed the "city gate," and is an important market center for the pricing of natural gas in large urban areas. Typically, utilities take ownership of the natural gas at the city gate, and deliver it to each individual customer's meter. This requires an extensive network of small-diameter distribution pipe. The U.S. Department of Transportation's Pipeline and Hazardous Materials Safety Administration reports that there are just over 2 million miles of distribution pipes in the United States, including city mains and service pipelines that connect each meter to the main [1,2,18–29].

While natural gas traveling through interstate pipelines may be compressed to as much as 1500 psi, natural gas traveling through the distribution network requires as little as 3 psi of pressurization and is as low as ¼ psi at the customer's meter. The natural gas to be distributed is typically depressurized at or near the city gate, as well as scrubbed and filtered (even though it has already been processed prior to distribution through interstate pipelines) to ensure low moisture and particulate content. In addition, mercaptan—the source of the familiar rotten egg smell in natural gas—is added by the utility prior to distribution. This is added because natural gas is odorless and colorless, and the familiar odor of mercaptan makes the detection of leaks much easier [1,2,18–29].

Distribution pipelines use flexible plastic and corrugated stainless steel tubing. These new types of tubing allow cost reduction, installation flexibility, and easier repairs for both local distribution companies and natural gas consumers. Another innovation in the distribution of natural gas is the use of electronic meter-reading systems. The natural gas that is consumed by any one customer is measured by on-site meters, which essentially keep track of the volume of natural gas consumed at that location.

SCADA systems, similar to those used by large pipeline companies, are also used by local distribution companies. These systems provide all the information needed to manage and control gas distribution to all customers on a dynamic basis. The management of distribution system follows the same safety codes as those for transmission pipelines. The entire gathering, transmission, and distribution pipeline system is a part of the overall smart gas grid that is managed and controlled by a sophisticated computer system [1,2,18–29].

8.4   STORAGE AND TRANSPORT OF CNG AND GAS HYDRATES

In this section, we combine hydrates and compressed natural gas due to similarities in their processes. The hydrate and CNG processes do not involve extreme temperature, either high or low, and do not require any chemical processes. They also do not feature any complex unit operations other than standard process equipment. In addition, the technology is able to cope with an intermittent and variable profile of gas production with time, as is usually the case with associated gas.

Transportation of natural gas as hydrate or CNG is believed to be feasible at costs less than for LNG and where pipelines are not possible. The competitive advantage of gas hydrates or CNG over the other nonpipeline transport processes is that they are intrinsically simple, and much easier to implement at lower capital costs, provided economically attractive market opportunities can be negotiated between gas seller and buyer.

8.4.1   Gas Hydrates

Gas can be transported as a solid, with the solid being gas hydrate. Natural gas hydrate (NGH) is the product of mixing natural gas with liquid water to form a stable water crystalline ice-like substance. NGH transport is believed to be a viable alternative to LNG or pipelines for the transportation of natural gas from source to demand [30–77].

Using gas hydrate as a means of gas transport involves three stages: production, transportation, and regasification. As shown in Chapter 2, NGH is created when certain small molecules, particularly methane, ethane and propane, stabilize the hydrogen bonds within water to form a three-dimensional cage-like structure with the gas molecule trapped within the cages. A cage is made up of several water molecules held together by hydrogen bonds. These types of structures, known as clathrates, have been well studied and are complex [30–77]. The thermodynamics and phase diagrams of water, hydrate, and various gases have all been researched and documented [30–77], and they are described in Chapter 2. Hydrates are formed from natural gas in the presence of liquid water provided the pressure is above and the temperature is below the equilibrium line of the phase diagram of the gas and liquid water. The solid has a snow-like appearance.

For gas transport, NGH can be deliberately formed by mixing natural gas and water at 80–100 bar and 2°C–10°C. It has been found that if NGH-water slurry is refrigerated to around −15°C, NGH decomposes very slowly at atmospheric pressure, so that the hydrate can be transported by ship to market in simple insulated (inexpensive compared to LNG carriers) bulk carriers, that is, a large "thermos flask" under near adiabatic conditions.

At the market, the slurry is melted back to gas and water by controlled warming for use after appropriate drying in electricity power generation stations or other applications. The hydrate yields up to 160 sm^3 of natural gas per ton of hydrate, depending on the manufacturing process [30–77]. The manufacture of the hydrate could be carried out using mobile equipment for onshore and ship for offshore using a floating production, storage, and offloading vessel (FPSO) with minimal gas processing (e.g., cleaning) prior to hydrate formation. The water can be used at the destination if there is water shortage, or returned as ballast to the hydrate generator and since it is saturated with gas it will not take more gas into solution to form new hydrates. Process operability of continuous production of hydrate in a large-scale reactor, long-term hydrate storage, and controlled regeneration of gas from storage have all been demonstrated. Reactor and process data have been obtained, and equipment design for full-scale process development has been outlined [30–77].

A pilot plant producing 1 ton of hydrate (transporting around 5000 scf gas) has been recently demonstrated [32,33]. The hydrate can be stored at normal temperatures (0°C to −10°C) and pressures (10–1 atmosphere) where 1 m^3 of hydrate should contain about 160 sm^3 gas per m^3 of water. This "concentration" of gas is attractive as it is easier to produce and safer and cheaper to store compared to the 200 sm^3 per 1 m^3 of compressed gas (high pressure of 3000 psig) or the 637 sm^3

gas per 1 m³ of LNG (low temperatures of −162°C). This efficient storing of gas in the hydrate state is due to the molar ratio of gas to water and to the exceptionally high density of gas in the hydrate state, and in fact the relative density of the gas in the hydrate lattice exceeds its liquid state density. Gas storage in hydrate form becomes especially efficient at relatively low pressures where substantially more gas per unit volume is contained in the hydrate than in the free state or in CNG when the pressure has dropped. When compared to the transportation of natural gas by pipeline or LNG, the hydrate concept has lower capital and operating costs for the movement of quantities of natural gas over adverse conditions. Overall, simplicity and flexibility of the process and cheaper cost make hydrates very plausible storage and transport alternatives for small-volume gas in remote locations [30–77].

8.4.2 COMPRESSED NATURAL GAS

CNG is typically stored in steel or composite containers at high pressure (3000–4000 psi, or 205–275 bar). These containers are not typically temperature controlled, but are allowed to stay at local ambient temperature. CNG can be stored at lower pressure in a form known as an adsorbed natural gas (ANG) tank at 35 bar (500 psi, the pressure of gas in natural gas pipelines) in various sponge-like materials, such as activated carbon and metal-organic frameworks (MOFs) [1,2,77–87]. The storage density of natural gas is graphically illustrated in Figure 8.3.

Gas can be transported in containers at high pressures, typically 1800 psig for a rich gas (significant amounts of ethane, propane, etc.) to roughly 3600 psig for a lean gas (mainly methane). Gas at these pressures is termed "compressed natural gas." CNG is used in some countries for vehicular transport as an alternative to conventional fuels (gasoline or diesel). However, the time to fill a tank with 3000 psig gas can be slow and frustrating. The filling stations can be supplied by pipeline gas but the compressors needed to get the gas to 3000 psig can be large, noisy,

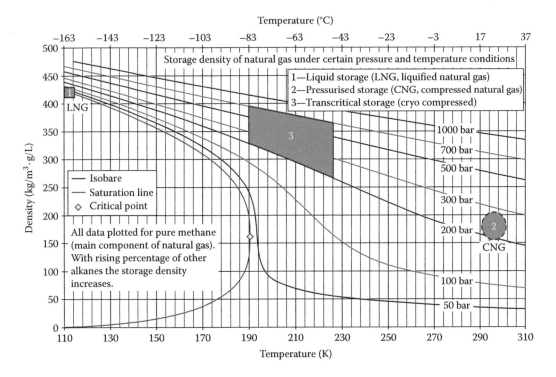

FIGURE 8.3 Storage density of natural gas. (From Natural gas storage, Wikipedia, the free encyclopedia, 2015.)

multistage, and expensive to purchase, maintain, and operate [78–87]. The thermodynamics of gas compression (heat generation), and gas expansion (significant cooling), should be considered in any gas processing operation, and the need for appropriate heat exchangers adds significant costs. A gas network is also needed.

Originally, the transport containers were heavy-walled (and hence heavy in weight) pressure vessels, but recently new lighter designs have been proposed. One design uses relatively long lengths of thin-walled tubing (6.25 in. outside diameter with a wall thickness of 0.25 in.) coiled into large diameter reels, termed by the inventors as a Coselle, "a coil in a carousel" [1,2,6–10,77–87]. The carousel structure is important since it not only protects the pipe from damage, even against total break, but it permits stacking 6–8 units high. The inventors initially proposed a Coselle of length 9.6 miles which would stand some 11 ft high with a 50 ft outside diameter and 10 ft inside diameter and contain approximately 3 million scf of gas at 3000 psig. The Coselle would have many vertical girders around the outside so that it would be a large safe pressurized gas containment system. The long-term viability of the coiled tubing under repeated loading/unloading is being tested, but no serious difficulties are anticipated. The total weight of pipes and associated structures (perhaps 500 tons) should be transported along with the gas, but the inventors claim that the lower fabrication costs for the gas containers makes this design attractive. They have also now designed Coselles for smaller markets [1,2,6–10,77–87].

An alternative approach, VOTRANS [1,2,6–10,77–87], are dedicated transport ships carrying straight long, large-diameter pipes in an insulated cold storage cargo package. The gas should be dried, compressed, and chilled for storage onboard. By careful control of temperature, more gas should be transported in any ship of a given payload capacity, subject to volume limitation and amount and weight of material of the pipe (pressure and safety considerations). Suitable compressors and chillers are needed, but they would be much less expensive than an LNG liquefier, and would be standard, so that costs could be further minimized. According to the proposers, the terminal facilities would also be simple and hence would be of low cost.

These CNG systems would make transport possible either for stranded gas (i.e., in places where there is no current market or no export pipeline) or for smaller quantities of associated gas which cannot be flared or reinjected. The number and size of Coselles or VOTRANS ships can be scaled to fit demand and would depend on daily production rates from the reservoirs (whether variable or not), and weight restrictions of transporters. Case studies by the inventors [1,2,6–10,77–87] have shown that large quantities of natural gas (~500 million scf) can be transported to markets at costs substantially below LNG costs over short distances, and probably over longer distance when the largest ships are employed. Ships capable of carrying Coselles up to 1 Bscf each and VOTRANS capable of carrying up to 2 Bscf have been proposed [2,78–87]. However, further considerations of safety check requirements and availability of dry dock for very heavy/large ships suggest that it may be a misconception to try to make the CNG ship as large as a LNG ship. An ideal CNG transport facility may be a fleet of smaller ships perhaps delivering gas daily directly into the distribution pipeline, with as many ships as the distance requires, or perhaps into a system with some backup storage in case a ship is delayed in transit. Such CNG transport systems would be more flexible and cope with variable gas supplies such as associated gas [1,2,6–10,77–87].

8.5 METHODS FOR STORAGE AND TRANSPORT OF LNG AND LPG

8.5.1 LNG STORAGE AND TRANSPORT

LNG is the liquid form of natural gas. Gas cooled to around −162°C liquefies, and has a volume ~1/600 that of gas at room temperature. Export by LNG to Japan, the United States, and Europe from distant production fields has now become less expensive due to improvements in technology and thermodynamic efficiencies of LNG facilities [1–5,88–111]. The incremental cost of transport per mile is less than that for pipeline [1–5,88–111].

LNG facilities require complex machinery with moving parts and special refrigerated ships for transporting the LNG to market [1–5,88–111]. The costs of building LNG plant have come down over the past 25 years because of greatly improved thermodynamic efficiencies, so LNG is becoming a major gas exporting method worldwide, with 15 billion scf/day (approximately tripled since 1990) and many plants being extended, or new ones built around the world, for example, Nigeria, Angola, Qatar, Egypt, and Trinidad [1–5]. The success of these projects requires long-term (20 years) committed contracts.

Huge cryogenic tanks are needed to store the LNG; typically these may be 70 m diameter, 45 m high, and hold over 100,000 m^3 of LNG. At the consumer end, an infrastructure for handling the reprocessing of vast quantities of natural gas from LNG is required, which is also very expensive and vulnerable to sabotage. Even though the cost of producing LNG has fallen by some 40% since 1985, LNG plants are large-scale, long-contract (~20 years or more), and require large 3 tscf gas reserves and ~US$1 billion investment for a train processing around 500 million scf/day [1–5,88–111].

On peak demand of winter days, LNG storage facilities prove invaluable because of their ability on short notice to regasify and deliver large amounts of natural gas into regional distribution systems. About 82% of LNG storage capacity is located in the eastern United States in major population centers such as Boston, New York, and Philadelphia [1–5].

LNG storage pressures are typically around 50–150 psi, or 3–10 bar. At atmospheric pressure, LNG is at a temperature of −260°F (−162°C); however, in a vehicle tank under pressure the temperature is slightly higher. Storage temperatures may vary due to varying composition and storage pressure. LNG is far denser than even the highly compressed state of CNG. As a consequence of the low temperatures, vacuum-insulated storage tanks typically made of stainless steel are used to hold LNG [1–5,88–111].

Prior to regasification, LNG is stored at atmospheric pressure in double-walled, insulated tanks that feature innovative, highly safe, and stable designs. The walls of the inner tank, composed of special steel alloys with high nickel content as well as aluminum and prestressed concrete, must be capable of withstanding cryogenic temperatures. LNG storage tanks are built on a base of concrete blocks with the glassy volcanic aggregate perlite added to Portland cement and special admixtures, reinforced with steel bars. These blocks insulate the cryogenic tank from the ground itself. Perlite is also used as insulation in the walls of the tank [1–5,88–111].

To safeguard against leaks, some storage tanks feature a double-containment system, in which both the inner and outer walls are capable of containing LNG. Another approach, utilized by most LNG tanks at existing U.S. import and satellite storage facilities, surrounds a single-containment tank with an earthen dam or dike that provides secondary containment, safely isolating any LNG spills [1–5,88–111].

LNG security is multifaceted. DOT's Office of Pipeline Safety provides guidelines to LNG operators for security procedures at onshore facilities. A federal security task force works to improve pipeline security practices, facilitate communications within industry and government, and lead public outreach efforts. Comprehensive safety procedures and equipment found at all LNG facilities help to maintain an outstanding record of worker safety. Precautions include avoiding asphyxiation (which can result if LNG vapors deplete breathable oxygen in a confined space), preventing lung damage (which can result if LNG vapors are inhaled), and preventing cryogenic burns (which can occur if LNG contacts human skin).

The composition of LNG received in the United States varies by country of origin, as shown in Table 8.1, and must be modified before delivery. This variation limits deliveries to certain terminals and also must be factored into the development of new facilities. LNG importing facilities deal with this problem by mixing domestic and imported gas or injecting nitrogen or air into the gas stream. At Lake Charles, Louisiana, Southern Union successfully mixes high-heat-content natural gas with relatively low-heat-content gas common to the region's substantial processing infrastructure. Therefore, LNG deliveries with high Btu content occur more often at Lake Charles. At the Everett,

TABLE 8.1

Typical Composition of LNG Imports by Country

Origin	Methane (%)	Ethane (% of Methane)
Trinidad & Tobago	96.9	2.8
Nigeria	91.6	5.0
Malaysia	89.8	5.8
Qatar	89.9	6.7
Australia	89.3	7.9
Oman	87.7	8.6
Algeria	87.6	10.3

Sources: Modified and adapted from Storage of natural gas, Naturalgas.org, September 2013; Global LNG-will new demand and new supply mean new pricing?, A report by EY, EYG no. DW0306, CSG/GSC2013/1115394, A Part of Earnest and Young Global Ltd., U.K., 2013; Liquefied natural gas: Understanding the basic facts, A report from NETL, DOE/FE-048g, Department of Energy, Washington, DC, August 2005; International Gas Union, News, views and knowledge on gas-worldwide, Sponsored by Statoil, Norway, 2014.

Massachusetts, facility, Distrigas uses in-tank blending of pipeline gas with LNG to meet standards. Btu levels can also be reduced by injecting nitrogen or air into the vaporized gas stream at sendout. This method can be expensive: approximately $18.5 million to equip a facility with air injection devices and about $28 million for nitrogen separation equipment. Installation of liquid-stripping facilities at marine terminals would also effectively allow Btu reduction, but at a cost of $30 million or more per facility [2–5,88–111].

LNG is typically transported by specialized tanker with insulated walls, and is kept in liquid form by auto-refrigeration, a process in which the LNG is kept at its boiling point, so that any heat additions are countered by the energy lost from LNG vapor that is vented out of storage and used to power the vessel. From storage, LNG is converted back into gas and fed into the natural gas pipeline system. LNG is also transported by truck to satellite storage sites for use during peak periods of natural gas demand—in the coldest weather for heating and in hot weather for fueling electric power generators, which in turn run air conditioners. LNG that is imported to the United States comes via ocean tankers. The United States gets a majority of its LNG from Trinidad and Tobago, Qatar, and Algeria, and also receives shipments from Nigeria, Oman, Australia, Indonesia, and the United Arab Emirates [1–5,88–111].

There are a number of other methods now being investigated for transporting LNG. Transportation accounts for 10%–30% of the cost of the LNG value chain. Carrier ships often are owned by LNG producers, but are also sometimes built as independent investments separate from specific LNG projects.

The evolution of LNG transport ships has been dramatic. While the first LNG carrier was a converted freighter with aluminum tanks insulated with balsa wood, modern LNG carriers are sophisticated double-hulled ships specifically designed for the safe and efficient transportation of cryogenic liquid. In May 2005, 181 LNG carriers were operating, with another 74 under construction for delivery in the 2005–2007 timeframe [2–5]. This number has been rapidly increasing every year. About half of the LNG fleet is of the *membrane* design, with the other half of the *spherical* or *Moss*® design [1–5,88–111]. As of 2004, about three-fourths of the new LNG ships under construction or planned were of the membrane design due to innovations aimed at increasing cargo capacity in a given hull size, reducing capital costs and overall construction time [2–5,88–111]. This number is significantly in more newer higher constructions.

A small number of ships in service, built by the IHI shipyard in Japan, feature a self-supporting *prismatic tank design*. Like the spherical tank, the prismatic tank is independent of the hull. Any leaking LNG evaporates or flows into a pan below the tank [88–111]. Data from the Society of International Gas Tanker & Terminal Operators (SIGGTO) show that Moss tankers represented 46% of the fleet in 2004, membrane tankers accounted for 51%, and 3% were other designs. South Korea is the world's leading builder of LNG ships, led by Hyundai Heavy Industries Co., Ltd., Samsung Heavy Industries, and Daewoo Shipbuilding & Marine Engineering Co. Japan is placed second with major firms including Mitsubishi Heavy Industries Ltd., Mitsui Engineering and Shipbuilding Co., and Kawasaki Heavy Industries Ltd. Izari in Spain and Chantiers de l'Atlantique in France are also leading builders of LNG ships [2–5].

The current largest specially built refrigerated tankers can carry 135,000 m^3 LNG, equivalent to 3.2 Bscf of gas, but are very expensive [2–5]. This makes it difficult for LNG to use smaller isolated (offshore) reserves and to serve small markets commercially because it is this large capacity, continuous running that keeps thermodynamic efficiency and costs to a minimum. Thus small volumes of intermittent gas are not economically attractive to the major gas sellers for LNG facilities. However, small well-insulated LNG container trade is being investigated, and if successful, small quantities of LNG may be able to be delivered from the LNG storage, just like the gasoline tankers of today. Even so, the LNG must be stored for periods of time (months) without significant boil-off losses. This is, in general, a difficult task [88–111].

Due to comprehensive safety and security programs for LNG tankers and receiving terminals, more than 33,000 shipments have transported in excess of 3 bcm of LNG without a serious accident at sea or in port in the past 40 years. LNG facilities and vessels feature state-of-the-art natural gas, fire, and smoke detection systems that identify hazardous situations and automatic shutdown systems that halt operations. Security measures for the waterfront portions of marine terminals and LNG ships are regulated by the U.S. Coast Guard, which prevents other ships from getting near LNG tankers while in transit or docked at a terminal. The Federal Energy Regulatory Commission (FERC) also serves as a coordinator with the Coast Guard and other agencies on issues of marine safety and security at LNG import facilities [88–111].

The Coast Guard has led the International Maritime Organization (IMO) in developing maritime security standards outside U.S. jurisdiction. These new standards, the International Ship and Port Facility Security Code (ISPS Code), contain detailed mandatory security requirements for governments, port authorities, and shipping companies as well as recommended guidelines for meeting those requirements.

In order to transfer LNG from storage tanks to warming systems, where the liquid rapidly returns to a vaporized state, *ambient temperature* systems use heat from surrounding air or from seawater to vaporize the cryogenic liquid. The *above-ambient temperature* systems add heat by burning fuel to indirectly warm the LNG via an intermediate fluid bath [2–5,88–111]. Once gas is vaporized, it is ready for delivery into the nation's network of transmission and distribution pipelines for use by residential consumers, industries, or nearby power generation plants.

The current large LNG projects have contracts typically for up to 20 years and, once they have been carefully planned, marketed and built, are expected to provide stable revenues over this time period, and would cushion shorter-term fluctuations in other parts of an oil field development (e.g., oil price variations). For green site developments and smaller niche markets, gas hydrates or CNG could be strong contenders, for example, Indonesian islands.

The gas hydrate plant is often quoted to have capital costs much lower than for LNG perhaps half [88–111] and CNG even less [78–87]. Consequently, the payback period will be shorter for the same gas throughput. A number of other factors such as higher capital borrowing for a more complicated process, differences in technical difficulty, and maintenance costs as well as local idiosyncrasies must be considered in the final decision for transport mechanism.

Calculations have been made which indicate that when the oil price is at $20/bbl, LNG is a cheaper energy source, but when at $10/bbl, oil is cheaper. Transport and liquefaction costs for LNG account for ~85% of the supply cost of delivered LNG to the customer's jetty. Additional costs occur to get the gas from the jetty via gasification to the burner tip [2–5,88–111].

8.5.2 LPG STORAGE AND TRANSPORT

Propane or liquid propane gas (LPG) exists as a liquid and a vapor. As a liquid, it is heavier than air and flows like water, collecting and pooling in low-lying areas. Transported under pressure in compact liquid form, end users value the efficiency of this clean-burning fuel. In its liquid form, propane is 270 times more compact than it is in its gaseous form. This makes it economical to store and transport, from its creation at refineries, through a vast network that can include pipelines, railway shipments, storage facilities, distribution centers, highway transport and bobtail trucks, local suppliers and retailers to the eventual consumer [1–5,112–121].

As a vapor, this highly flammable and combustible gas, while commonly associated with backyard grills, is also a portable and economical fuel source for home heating and cooling, recreational vehicles, as well as a number of agricultural and industrial uses. Propane, according to the National Propane Gas Association (NPGA), accounts for more than 4% of the total energy needs of the United States each year.

Propane is a naturally occurring by-product of domestic oil refining and natural gas processing. Since it is 270 times more compact as a liquid than as a gas, LPG is highly economical to store and transport. When propane is used as an on-road engine fuel, it is often called "propane autogas" among other names throughout the world.

Propane is stored in large tanks at various distribution points, and in smaller tanks at residential homes. Residential demand for propane tends to be seasonal, and propane and other LPGs can be stored whenever supply exceeds demand. Propane inventories often are built up during the summer months for use in the winter.

Bulk LPG vessels, on the other hand, are used in several types of facilities. They store large amounts of propane to help a supplier meet the demand of the market in the area. Bulk facilities are used to distribute propane to residential and commercial consumers. Vessels are typically designed to load bobtails. In many parts of the country, petroleum marketers are adding propane to their product mix. These facilities are installing aboveground or underground LPG vessels alongside atmospheric petroleum tanks.

In recent years, vast gas reserves trapped in the Marcellus Shale and other shale gas plays have been exploited. Naturally occurring hydrocarbons, known as NGL, are found in natural gas that is sourced from gas wells or associated with crude oil. These by-products of natural gas have increased significantly with shale gas exploration and hydraulic fracking. NGL products such as propane and butane are marketed to consumers. This has resulted in the development of new bulk facilities to store LPG [112–121]. While in most cases, bulk storage facilities are installed aboveground, more facilities are moving underground due to space limitation and for safety against fire and other natural disasters. Underground storage is also carried out in specialized HighGuard vessels (Highland tank Co.).

More than 1 million commercial establishments use propane for heating and cooling air, heating water, cooking, refrigeration, drying clothes, barbecuing, and lighting. More than 350,000 industrial sites rely on it for space heating, brazing, soldering, cutting, heat treating, annealing, vulcanizing, and many other uses. Petrochemical industries use propane to manufacture plastics. Propane is also a staple on 660,000 farms, where it is used in everything from grain drying to planting seeds, ripening fruit, and running a variety of farm equipment such as irrigation pumps and standby generators. All of these applications also use propane storage tanks [1–5,112–121].

The road and oversea transport of LPG is carried out in the same manner as LNG. LPG-based SNG can serve local community by a localized storage and piping distribution. Such a network can serve at least 5000 domestic consumers. These localized natural gas networks are successfully operating in Japan with feasibility to get connected to wider networks in both villages and cities.

FIGURE 8.4 A spherical gas container typically found in refineries. (From Liquefied Petroleum gas, Wikipedia, the free encyclopedia, https://en.wikipedia.org/wiki/Liquefied_petroleum_gas, 2015.)

In a refinery or gas plant, LPG must be stored in pressure vessels. These containers are either cylindrical and horizontal or spherical (see Figure 8.4 [93]). Large, spherical LPG containers may have up to a 15 cm steel wall thickness. Typically, these vessels are designed and manufactured according to some code. In the United States, this code is governed by the American Society of Mechanical Engineers (ASME). LPG containers have approved pressure relief valves, such that when subjected to exterior heating sources, they will vent LPGs to the atmosphere or a flare stack. If a tank is subjected to a fire of sufficient duration and intensity, it can undergo a boiling liquid expanding vapor explosion (BLEVE). This is typically a concern for large refineries and petrochemical plants that maintain very large containers. If pressure during fire cannot be released by relief valve, overpressured container may rupture violently, causing catastrophic damage to anything nearby [1,2,112–121]. Thus, safety of LPG storage vessels is of utmost importance.

8.6 STORAGE AND TRANSPORT OF SYNGAS

The technical feasibility and economic attractiveness of syngas storage can depend on the specific properties of the syngas produced from the gasification process. These properties, such as the composition, energy density, temperature, and pressure, depend on the type and rank of coal, nature of biomass, and the specific gasification process used to produce the syngas. Syngas is primarily composed of carbon monoxide and hydrogen, and is characterized by a low energy density, typically ranging from 150 to 280 Btu/scf. The low energy density of syngas, ranging from roughly one-sixth to one-third that of natural gas, means that larger amounts of syngas are required to produce an equivalent amount of heat or electricity in an IGCC facility. The implications for storage are that storage vessels must be designed to handle large volumes of gaseous syngas through large physical sizes, high working pressures, or some combination [122–202].

8.6.1 Methanation

Methanation is a process used to upgrade low energy density syngas to higher pipeline quality synthetic natural gas or SNG. In the methanation process, the calorific value and other parameters of the gas are adjusted to meet natural gas pipeline specifications [161,202].

The methanation reactions can be represented as [182]

$$CO + 3H_2 \rightarrow CH_4 + H_2O \quad Hg = -217 \text{ kJ/mol} \tag{8.1}$$

$$2CO + 2H_2 \rightarrow CH_4 + CO_2 \tag{8.2}$$

$$CO_2 + 4H_2 \rightarrow CH_4 + 2H_2O \quad Hg = -178 \text{ kJ/mol} \tag{8.3}$$

Because syngas is converted to methane during the methanation reaction, the problems associated with low energy density syngas and hydrogen-rich gases can generally be avoided. The main advantage of SNG is that its composition is nearly identical to natural gas and can therefore be used in the same manner and injected directly into natural gas pipelines [162]. The techniques and costs for natural gas handling and use are well known and can be directly applied to SNG.

Capital costs for syngas to SNG process can be obtained from data reported in the literature and from facility developers [134,160,165,190]. The reported capital costs include methanators, compressors, and water gas shift process. Data show that a facility producing 50 mscf/day of SNG from syngas would have estimated capital costs of about \$65 ± \$20 million (approximately ±30%). The literature suggests that costs for methanation process are known and are consistent among different data sources [134,160,165,190].

8.6.2 Syngas and SNG Storage

While storage options for syngas and SNG are not well reported in the literature, technical and economic aspects of hydrogen and natural gas storage are well addressed. From these related studies, costs for syngas and SNG storage can be reasonably estimated [122–202], based on the composition and properties (pressure, temperature, etc.) of the gas to be stored. Costs for SNG and syngas storage in aboveground and underground ground vessels are estimated based on existing estimates for natural gas and hydrogen storage options [122–202]. If SNG composition is close to that of natural gas, SNG storage and transport can follow the guidelines for natural gas discussed earlier.

8.6.2.1 Options for Large-Scale Storage

Options for the large-scale, bulk storage of gasses include compressed gas, cryogenic liquid, solids such as metal hydrides, and liquid carriers such as methanol and ammonia. Metal hydride storage is an emerging technology used for storing pure gases such as hydrogen. Liquid carriers such as methanol and ammonia are also useful for a pure gas. As syngas and SNG are gas mixtures of varying compositions, depending on the gasification process, solid and liquid carrier storage options are unlikely to be feasible.

8.6.2.1.1 Cryogenic Liquid Storage

Cryogenic liquid storage has been used for large-scale hydrogen storage. While the technology is largely driven by the space programs, storing liquid hydrogen presents numerous engineering challenges due to its low heat of vaporization and resultant very high loss index [179,180]. Liquid hydrogen cannot be stored in cylindrical tanks used to store LNG because the boil-off would be too high [124]. Spherical tanks are used for large-scale applications because this shape has the lowest surface area for heat transfer per unit volume. NASA uses liquid hydrogen tanks which are about 22 m in diameter [179,180]. Liquid hydrogen storage is expensive when cost includes spherical storage tanks and cost of facility required for cooling and liquefaction.

In addition to high costs, there are technical concerns related to liquid syngas storage. Syngas is a gas mixture and not pure gas. The chemical components that make up syngas liquefy and react at different temperatures and pressures. As such, it is unknown what technical difficulties may arise from liquefaction and cryogenically storing syngas. Additionally, syngas and SNG are typically used in gaseous form for an end-use process, such as combustion in a turbine or production of liquid fuels. Compressing and liquefying the gas for storage, followed by expansion and vaporization for end use, is inefficient. Because of the high capital costs, technical uncertainties, and gas-to-liquid-to-gas conversion inefficiencies, liquid storage is not particularly suited to syngas storage.

8.6.2.1.2 *Compressed Gas Storage*

Compressed gas technology is the most relevant large-scale stationary storage method for syngas production facilities. It is cheaper than cryogenic storage and it can be easily used for either hydrogen or syngas. Compressed gas storage is the simplest storage solution as the only required equipment is a compressor and a pressure vessel [124]. The main disadvantage of compressed gas storage is the low storage density, which can be increased with the storage pressure. For pure hydrogen storage, several stages of compression are required because of the low density [171]. Compressed gas can be stored both aboveground and underground.

The cost of compressed syngas storage depends on the cost of compressor and the nature of storage vessel [124,139,169].

The literature data indicate that the compressor capital costs scale linearly with the size of the compressor. A one horsepower increase in compressor size corresponds to about $492 increase in capital costs [122,124,139,169]. Capital costs also scale linearly with the vessel size and it increases approximately $62 per m^3 increase in vessel size [122,124,139,169].

Compressed gas storage requires a compressor to provide the necessary mass flow of gas into the storage vessel. No literature discusses syngas compression or compressor requirements for syngas service; however, reasonable estimates can be drawn from literature discussing compressors for natural gas and hydrogen service. The density and molecular weight of the gas to be compressed is an important consideration for compressor choice. Centrifugal compressors, which are widely used for natural gas, are not generally suitable for pure hydrogen compression as the pressure rise per stage is very small due to the low density and low molecular weight [124,172]. Large-scale hydrogen can be compressed using standard axial, radial, or reciprocating piston-type compressors with slight modifications of the seals to take into account the higher diffusivity of the hydrogen molecules [124,172].

The capital costs of compression depend on the properties of the gas to be compressed. Compressing pure hydrogen requires about three times the compressor power as natural gas and specific capital costs for large hydrogen compressors are expected to be 20%–30% higher than for natural gas [136]. Compressor costs depend on the inlet pressure, outlet pressure, and flow rate [124]. There is about a tenfold difference in capital costs of compressors reported in the literature ($650–$6600/kW). Costs for large-scale, megawatt-sized compression facilities for pipeline transport are developed by the International Energy Agency, IEA [169].

8.6.2.2 Aboveground Syngas Storage Methods

Aboveground options for syngas include storage in existing piping infrastructure, in gasometers or in cylindrical "bullets" common for LPG, LNG, and CNG storage. The choice of storage vessel depends on both technical and economic considerations, including the composition and quantity of the gas to be stored, the charge and discharge rates, as well as capital, operating, and maintenance costs.

Conventional methods of aboveground compressed gas storage range from small high-pressure gas cylinders to large, low-pressure spherical gas containers [127,171]. Compressed gas pressure vessels are commercially available at pressures of 1200–8000 psi, typically holding 6000–9000 scf per vessel. Low-pressure spherical tanks can hold roughly 13,000 Nm^3 of gas at 1.2–1.6 MPa (1700–2300 psig) [124]. High-pressure tube storage is available for larger gas volumes, typically around 500,000 scf (14,000 Nm³) [139]. Because of the relatively small storage capacity, industrial facilities typically use aboveground compressed gas storage in pressure tanks for gas storage on the order of a few million scf or less [136]. Pressure vessels are physically configured in rows or in stacks of tanks; such storage is modular, with little economy of scale [124]. Capital costs for aboveground pressure vessel storage range from approximately $22–$214/Nm³ ($0.62–$6.02/scf) [124].

Gasometers are aboveground vessels designed for storing large amounts of gas, typically at low pressure. Gasometers typically have a variable volume, through the use of a weighted movable cap, which provides gas output at a constant pressure. Gasometers operate at low pressure, with typical pressures in the range of 200–300 mm water (0.28–0.43 psig); maximum operating pressures are 1000 mm water (1.4 psig) [152]. Typical volumes for large gasometers are about 50,000–70,000 m³, with approximately 60 m diameter structures, although the largest gasholder installed by one manufacturer was 340,000 m³ [152]. Gasometers have long operating lifetimes; the structure itself can operate for over 100 years [152], while the diaphragm that seals the gasometer has a lifetime of approximately 10 years [156].

Syngas can also be stored, or packed, in piping systems. Pipelines are usually several miles long and, in some cases, may be hundreds of miles long. Because of the large volume of piping systems, a slight change in the operating pressure of a pipeline system can result in a large change in the amount of gas contained within the piping network. By making small changes in operating pressure, the pipeline can be effectively used as a storage vessel [124]. Storing gas in an existing pipeline system by increasing the operating pressure requires no additional capital expense as long as the pressure rating of the pipe and the capacity of the compressors are not exceeded [124].

Existing hydrogen pipelines are generally constructed of 0.25–0.30 m (10–12 in.) commercial steel and operate at 1–3 MPa (145–435 psig); natural gas pipelines, on the other hand, are constructed of pipes as large as 2.5 m (5 ft) in diameter and have working pressures of 7.5 MPa (1100 psig) [167]. A 30 km, 3 in. diameter hydrogen distribution pipeline can carry a flow of 5 mscf of hydrogen per day. Assuming that the pipeline operated at 1000 psi, the storage volume available in the pipeline would be 340,000 scf, or about 7% of the total daily flow rate [136].

8.6.2.3 Underground Syngas Storage Methods

In general, underground storage has a lower cost and large capacities [124] and are generally most suitable for large quantities and/or long storage times [174]. There are four underground formations in which gas can be stored under pressure: (1) depleted oil or gas field, (2) aquifers, (3) excavated rock caverns, and (4) salt caverns [139].

There is significant industrial experience in underground gas storage: natural gas has been stored underground since 1916 [139]; the city of Kiel, Germany, has been storing town gas (60%–65% hydrogen) in a gas cavern since 1971 [139]; Gaz de France has stored town gas containing 50% hydrogen in a 330 million cubic meter (mcm) aquifer structure near Beynes, France; Imperial Chemical Industries stores hydrogen at 50 atm (5 × 10⁶ Pa) pressure in three brine-compensated salt caverns at 1200 ft (366 m) near Teesside, the United Kingdom; and in Texas, helium is stored in rock strata beneath an aquifer whereby water seals the rock fissures above the helium reservoir, sealing in the helium atoms [172].

Underground storage volumes in depleted oil and gas fields can be extremely large; volumes of gas stored exceed 10^9 m³ and pressures can be up to 40 atm. Salt caverns, large underground voids that are formed by solution mining of salt as brine, tend to be smaller, typically around 10^6–10^7 m³. As shown earlier, although smaller, salt caverns offer faster discharge rates and tend to be tighter than other underground formations, reducing leakage. Hydrogen, a small molecule with high leakage rates, has been stored in salt caverns [133]. Rock caverns are usually smaller cavities, typically on the order of 1 million to 10 million m³.

Underground gas storage requires the use of a cushion gas that occupies the underground storage volume at the end of the discharge cycle. As mentioned earlier, cushion gas is a nonrecoverable base gas necessary to pressurize the storage reservoir. Cushion gas can be as much as 50% of the work-ing volume, or several hundred thousand kilograms of gas [124] and the cost of the cushion gas is a significant part of the capital costs for large storage reservoirs [139].

Capital costs for underground storage are reported in the literature. Underground storage is reported to be the most inexpensive means of storage for large quantities of gas, up to two orders of magnitude less expensive than other methods [124,127]. The only case where underground storage would not be the least cost option is with small quantities of gas in large caverns where the amount of working capital invested in the cushion gas is large compared to the amount of gas stored [124]. Capital costs vary depending on whether there is a suitable natural cavern or rock formation, or whether a cavern must be mined. An abandoned natural gas well is reported to be the least expensive; however, the likelihood of a gasification facility being near such a formation (and choosing to use it to store syngas rather than to sequester CO_2) seems small. Solution min-ing, excavating a salt formation with a brine solution, capital costs were estimated at $19–$23/m^3 ($0.54–$0.66/ft^3) [127]; hard rock mining costs were estimated at $34–$84/m^3 ($1.00–$2.50/ft^3) depending on the depth [124].

Additionally, construction times for underground storage facilities can be long and may contrib-ute to their costs. One estimate for solution mining a salt formation to create a 160 million cubic foot cavern was 2.5 years [175]. Underground compressed gas storage has been successfully used for compressed air energy storage (CAES) systems. There are currently two operating CAES systems in the world, both of which use salt caverns for air storage; 290 MW Huntorf project in Germany and 110 MW McIntosh project in the United States [175].

As with all storage technologies, the overall cost of storage depends on throughput and storage time [174]. Operating costs for underground storage are primarily for compression power and lim-ited to the energy and maintenance costs related to compressing the gas into underground storage and possibly boosting the pressure coming back out [126,174]. The cost of the electricity require-ments to compress the gas is independent of storage volume, which means the cost of underground storage is very insensitive to changes in storage time [124]. If the gasification facility is not geo-graphically located near an area with suitable underground storage, transport costs play an impor-tant role in the overall cost analysis.

From data reported in the literature, capital cost distributions were constructed for underground salt caverns, excavated rock caverns, low-pressure gasometers, and high-pressure cylindrical bullets [124,174]. These data indicated that salt cavern was the cheapest option and gasometer was the most expensive option. Excavated rock caverns and high-pressure cylindrical bullets had a similar range of cost. Because it is the lowest cost, a salt cavern is preferred if it is available.

8.6.2.4 Major Storage and Transportation Issues with Syngas

The storage and transportation of syngas faces two important technical issues: hydrogen embrittle-ment resulting in syngas leakage and biological fouling. The oil and gas industry has recognized internal and external hydrogen attack on steel pipelines, described variously as hydrogen-induced cracking (or corrosion) (HIC), hydrogen corrosion cracking (HCC), stress corrosion cracking (SCC), hydrogen embrittlement (HE), and delayed failure [172]. These issues are serious since corrosion damages cause most of the failures and emergencies of gas pipelines. Corrosion defects, such as general corrosion, pitting corrosion, and SCC, make up the majority of detected effects in pipelines [136].

If molecular hydrogen is split up in atomic hydrogen (H+) [169], it diffuses into a metal and reforms as microscopic pockets of molecular hydrogen gas, causing cracking, embrittlement, and corrosion which can ultimately lead to pipe failure. The hardness of a metal as defined by Vickers Hardness Number (VHN) correlates to the degree of embrittlement; if VHN is greater than 300, material will tend to fail due to plastic straining caused by significant absorption of atomic hydrogen [136].

Two primary mechanisms leading to HIC are wet conditions and elevated temperatures [136]. Temperature greater than 220°C and relative humidity greater than 60% can cause dissociation of molecular hydrogen into atomic hydrogen [169]. If these conditions are avoided, then HIC (or HE) may be handled without problems with standard low-alloy carbon steel irrespective of the gas pressure [169]. Options for steel pipes for 100% hydrogen service include Al–Fe (aluminum–iron) alloy, and variable-hardness pipe, with the harder material in the interior and softer material toward the exterior, so that any hydrogen which diffuses into the interior steel diffuses rapidly outward and escapes [172].

Existing natural gas pipelines can be used for less than 15%–20% hydrogen, by volume, without danger of hydrogen attack on the line pipe steel; however, further hydrogen enrichment will risk hydrogen embrittlement [172]. Existing pipelines originally designed for sour service can provide additional protection against HIC and hydrogen embrittlement due to their specific metallurgy [169]. As long as right material is used, cost is not an issue to address HIC [168]. For large-diameter pipelines and vessels, options include low-carbon steel plate, such as type X52, which is easy to make, readily available, easy to weld, and easy to fabricate. Smaller pipes can be constructed from either seamless or welded pipe. More internal stress created by heating and cooling during welding makes that area more sensitive for HE.

An additional potential problem resulting from the hydrogen content of syngas is that atomic hydrogen is a small molecule and can diffuse through most metals [169], which can cause problem over the long term. The subsurface storage of gas can cause biological fouling, such as contamination of the gas, plugging of the storage vessel, degrading its capacity, and bio-corrosion. Once again industrial experiences have suggested that that biological fouling is not an issue for short-term storage [163,166,168,178].

8.7 STORAGE AND TRANSPORTATION OF BIOGAS AND BIOMETHANE

Biogas (which generally contains less concentration of methane than natural gas) is generated by anaerobic digestion process. While biogas is generated from landfills (as "landfill gas"), most of man-made biogas is generated in farms and various types of waste treatment facilities. Depending on the size of the landfill, landfill gas can either be used locally or purified and transported through natural gas pipelines. For the latter to occur, thorough cleaning and upgrading of landfill gas is necessary. Two major impurities in landfill gas are hydrogen sulfide and carbon dioxide (see Table 8.2). These can be removed using technologies discussed in Chapter 7. In this section we assess various methods of storage and transport of biogas and biomethane generated in agriculture farms, and various types of waste treatment facilities.

Dairy manure biogas is generally used in combined heat and power applications (CHP) that combust the biogas to generate electricity and heat for on-farm use. The electricity is typically

---

TABLE 8.2
Typical Composition of Biogas

Compound	Formula	%
Methane	CH_4	50–75
Carbon dioxide	CO_2	25–50
Nitrogen	N_2	0–10
Hydrogen	H_2	0–1
Hydrogen sulfide	H_2S	0–3
Oxygen	O_2	0–0

Sources: Biogas, Wikipedia, the free encyclopedia, 2015.

TABLE 8.3
Short-Term Storage Options for Biogas or Biomethane

Pressure (psi)	Storage Device	Material	Size(ft³)
<0.1	Floating cover	Reinforced and nonreinforced plastics, rubbers	Variable volume usually less than one day's production
<2	Gas bag	Reinforced and nonreinforced plastics, rubbers	150–11,000
2–6	Water sealed gas holder	Steel	3,500
	Weighted gas bag	Reinforced and nonreinforced plastics, rubbers	880–28,000
	Floating roof	Plastic, reinforced plastic	Variable volume, usually less than one day's production

Sources: Modified and adapted from Ross, C.C. and Walsh, Jr. J.L., Impact of utility interaction on agricultural cogeneration, American Society of Agricultural Engineers Paper 86-3020, St. Joseph, MI, 1986; Ross, C.C. et al., *Handbook of Biogas Utilization*, 2nd edn., U.S. Department of Energy, Southeastern Regional Biomass Energy Program, July 1996.

psi, pounds per square inch, ambient conditions; ft³, cubic feet.

produced directly from the biogas as it is created, although the biogas may be stored for later use when applications require variable power or when production is greater than consumption [201–214].

Biogas can be upgraded to higher-value biomethane by removing the H_2S, moisture, and CO_2, and this can be used as a vehicular fuel. Since production of such fuel typically exceeds immediate on-site demand, the biomethane must be stored for future use, usually either as compressed biomethane (CBM) or liquefied biomethane (LBM). Because most farms will produce more biomethane than they can use on-site, the excess biomethane must be transported to a location where it can be used or further distributed.

Biogas or biomethane is often stored on farm temporarily for on-farm use. The least expensive and easiest to use storage systems for on-farm applications are low-pressure systems which are briefly described in Table 8.3. The long-term storage for later use or off-farm use by transportation to off-site distribution plants requires high pressure (10–2900 psig) storage tanks built of steel or ally steel of size between 350 and 2000 ft³. The energy, safety, and scrubbing requirements of medium- and high-pressure storage systems make them costly and high-maintenance options for on-farm use. Such extra costs can be best justified for biomethane, which has a higher heat content and is therefore a more valuable fuel than biogas [203–214].

8.7.1 BIOGAS STORAGE

Both biogas and biomethane can be stored for on-farm uses. In practice, however, most biogas is used as it is produced. Thus, the need for biogas storage is usually of a temporary nature, at times when production exceeds consumption or during maintenance of digester equipment. Important considerations for on-farm storage of biogas include (1) the needed volume (typically, only small amounts of biogas need to be stored at any one time); (2) possible corrosion from H_2S or water vapor that may be present, even if the gas has been partially cleaned; and (3) cost (since biogas is a relatively low-value fuel).

As shown in Table 8.3, floating gas holders on the digester form a low-pressure storage option for biogas systems. These systems typically operate at pressures up to 10 in. water column (less than 2 psi). Floating gas holders can be made of steel, fiberglass, or a flexible fabric. A separate tank may be used with a floating gas holder for the storage of the digestate and also storage of the raw biogas [123,156,203–214].

One advantage of a digester with an integral gas storage component is the reduced capital cost of the system. The least expensive and most trouble-free gas holder is the flexible inflatable fabric top, as it does not react with the H_2S in the biogas and is integral to the digester. These types of covers are often used with plug-flow and complete-mix digesters. Flexible membrane materials commonly used for these gas holders include high-density polyethylene (HDPE), low-density polyethylene (LDPE), linear low density polyethylene (LLDPE), and chlorosulfonated polyethylene covered polyester (such as Hypalon®, a registered product of DuPont Dow Elastomers LLC). Thicknesses for cover materials typically vary from 18 to 100 miles (0.5–2.5 mm) [214]. In addition, gas bags of varying sizes are available and can be added to the system. These bags are manufactured from the same materials mentioned previously and may be protected from puncture damage by installing them as liners for steel or concrete tanks.

Biogas can also be stored at medium pressure between 2 and 200 psi, although this is rarely, if ever, done in the United States. To prevent corrosion of the tank components and to ensure safe operation, the biogas must first be cleaned by removing H_2S. Next, the cleaned biogas must be slightly compressed prior to storage in tanks. Typical propane gas tanks are rated to 250 psi. Compressing biogas to this pressure range uses about 5 kWh per 1000 ft³ [214]. Assuming the biogas is 60% methane and a heating rate of 13,600 Btu/kWh, the energy needed for compression is approximately 10% of the energy content of the stored biogas. This loss of energy plus cost of H_2S removal makes this option less attractive for low-price biogas [203–214].

8.7.2 BIOMETHANE STORAGE

Biomethane is less corrosive than biogas and also is potentially more valuable as a fuel. For these reasons, it may be both possible and desirable to store biomethane for intermediate- and long-term on- or off-farm uses. Biomethane can be stored as CBM to save space. Gas scrubbing is even more important at high pressures because impurities such as H_2S and water are very likely to condense and cause corrosion. The gas is stored in steel cylinders such as those typically used for storage of other commercial gases. Storage facilities must be adequately fitted with safety devices such as rupture disks and pressure relief valves. The cost of compressing gas to high pressures between 2000 and 5000 psi is much greater than the cost of compressing gas for medium-pressure storage. Compression to 2000 psi requires nearly 14 kWh per 1000 ft³ of biomethane [214]. If the biogas is upgraded to 97% methane and the assumed heating rate is 12,000 Btu/kWh, the energy needed for compression amounts to about 17% of the energy content of the gas. Biomethane can also be stored in low- or intermediate-pressure vessels with sufficient storage capacity for about 1 or 2 days. This would suffice for a dairy with 1000 cows (30,000 cubic feet (ft³) biomethane per day). Low-pressure storage tanks are custom designed and easily available.

In most cases, biomethane cannot be all consumed locally. In this case, the compressor receives the low-pressure biomethane from the storage tank and compresses it to 3600–5000 psi. The CBM output of the compressor is fed to a number of individual high-pressure storage tanks connected in parallel and housed in a portable trailer. The trailer then transports it to the point of consumption or additional storage and/or distribution location.

Liquefied biomethane (LBM) can be transported using LNG or CNG vehicles but this requires the collection of LBM at the level of 10,000 gal, which is the normal capacity of LNG tanker trucks. LBM can also be dispensed by CNG vehicles because liquid-to-compressed natural gas (LCNG) refueling station equipment creates CNG from LNG feedstock. The low-pressure (50 psi) storage tank is a buffer for LBM after it exits the biomethane liquefaction equipment. Typical LNG storage tanks have 15,000 gal capacity which can be filled by a dairy with 1000 cows in 6 weeks. Since, most or all of the LBM must be transported to a refueling station where it can be dispensed to natural-gas-fueled vehicles, liquid biomethane is transported in the same manner as LNG, that is, via insulated tanker trucks designed for transportation of cryogenic liquids. Since there is a significant loss of LBM through the relief valve on the tank, LBM should be used fairly quickly after production. For this reason, LBM storage and transport makes most sense for larger farms, where larger number of cows can produce larger amount of LBM which allows faster transport to the end users [203–214].

8.7.3 BIOMETHANE TRANSPORT

Biogas is a low-grade, low-value fuel, and therefore, in general, it is not economically feasible to transport it for a significant distance. Also, biogas cannot be economically trucked. In contrast, biomethane can be distributed to its ultimate point of consumption by several options depending on its point of origin. These options include distribution via dedicated biomethane pipelines, distribution via the natural gas pipeline, over-the-road transport of CBM and over-the-road transport of LBM.

If the point of consumption is relatively close to the point of production (e.g., <1 mile), the biomethane would typically be distributed via dedicated biomethane pipelines (buried or aboveground). For example, biomethane intended for use as CNG vehicle fuel could be transported via dedicated pipelines to a CNG refueling station. Costs for laying dedicated biomethane pipelines may range from about $100,000 to $250,000 or more per mile [205,206,211,213,214].

If biomethane meets the requirement of natural gas pipeline and if suitable arrangements with owner (private or utility company) can be made for pipeline use, biomethane can be transported by injection in this pipeline. Once injected, it can be used as a direct substitute for natural gas by any piece of equipment connected to the natural gas grid, including domestic gas appliances, commercial/industrial gas equipment, and CNG refueling stations. Generally, natural gas pipelines resist injection of biomethane unless it is highly cleaned and monitored continuously.

If distribution of biomethane via dedicated pipelines or the natural gas grid is impractical or prohibitively expensive, over-the-road transportation of compressed biomethane (3000–3600 psi pressure) may be a distribution option. CNG or biomethane bulk transport vehicles are often referred to as "tube trailers." They are DOT-approved tanks (e.g., DOT-3AAX seamless steel cylinders) that do not exceed the rated pressure. Water vapor content in these tanks must be less than 10 ppm with a minimum methane content of 98%. The tanks must be labeled with "hazardous materials" markings. High transportation and capital equipment costs and the need for additional compression at the point of consumption make this approach attractive only to create new market for CBM [203–214].

While over-the-road transportation of liquefied biomethane is a possibility, it must use 10,000 gal LNG tankers and requires to meet all the restrictions of cryogenic liquids. The method also requires DOT-approved double-walled insulated steel tanks with two independent pressure relief systems. The tanks also must have "hazardous materials" and maximum one-way-travel-time markings. One of the most attractive features of over-the-road transportation of liquefied biomethane is that an infrastructure and market already exist. In addition to acting as a fuel for LNG vehicles, liquefied biomethane can also be used to provide fuel for CNG vehicles via LCNG refueling stations which turn LNG into CNG. As mentioned before, the major disadvantage of this method is heat loss which forces its application most meaningful only for larger farms (ones which can produce at least 3000 gal of LBM per day) [88–91,203–214].

8.8 METHODS OF HYDROGEN STORAGE

Hydrogen storage is a key enabling technology for the advancement in use of hydrogen for stationary power, portable power, and transportation. Hydrogen has the highest energy per mass of any fuel; however, it also has the lowest energy per unit volume. This requires the development of advanced storage methods that have potential for higher energy density [215–221]. While we examined some aspects of hydrogen storage and transport earlier in Section 8.6, in this section and Section 8.9, we examine some additional details pertaining to storage and transport of hydrogen for its static and mobile usages [259,260,262,264].

Clean and inexhaustible hydrogen is increasingly acknowledged as the energy carrier of choice for the twenty-first century. Its substitution for petroleum for use as an automotive fuel would largely eliminate smog in inner cities and health concerns related to airborne particulates. The acceptance of high-efficiency proton exchange membrane (PEM) fuel cells for automobiles and strategic alliances in the automotive fuel cell world (Ford/Daimler-Benz/Ballard/dbb, General Motors/Toyota) attests to the seriousness with which the automotive industry views fuel cell propulsion. Since fuel cells fundamentally depend on hydrogen, importance and prominence of hydrogen will enhance with its credible production, storage, and distribution systems. Whether for vehicular onboard storage or stationary bulk storage, the storage of hydrogen has been problematic due to its low volumetric density and resulting high cost [215–267].

Hydrogen storage technologies can be divided into two parts: (1) physical storage, where hydrogen molecules are stored via compression, liquefaction, adsorption, or absorption. (2) chemical storage, where hydrogen is stored in various chemical metal hydrides, in chemical carriers or chemical adsorbents/absorbents, among other methods. Hydrogen storage can also be divided as static storage and mobile storage. It is the mobile storage on vehicles that has gained significant momentum and research efforts in recent years. The chemical methods are of particular interest for onboard vehicle storage where weight and volume are at a premium.

Various physical and chemical methods of hydrogen storage have trade-offs with regard to the energy "penalties" involved, along with their characteristics related to safety, weight, cost, rate of energy transfer, and other factors [215–267]. Here we examine all important methods of hydrogen storage in some details. Hydrogen density dependence on pressure and temperature is illustrated in Figure 8.5.

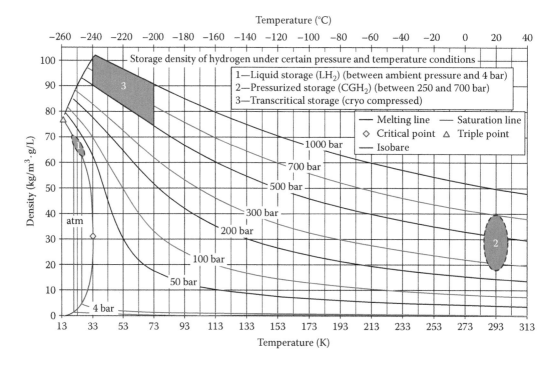

FIGURE 8.5 Net storage density of hydrogen. (From Hydrogen storage Wikipedia, the free encyclopedia, 2015.)

8.8.1 Physical Storage

Hydrogen can be stored physically as either a gas or a liquid. Storage of hydrogen as a gas typically requires high-pressure tanks (5,000–10,000 psi tank pressure). Storage of hydrogen as a liquid requires cryogenic temperatures because the boiling point of hydrogen at one atmospheric pressure is −252.8°C. Hydrogen can also be stored on the surfaces of solids (by adsorption) or within solids (by absorption).

For larger amount of hydrogen storage, underground hydrogen storage is carried out in caverns, salt domes, and depleted oil and gas fields. Large quantities of gaseous hydrogen have been stored in underground caverns by ICI for many years without any difficulties [54]. Underground static hydrogen storage (both in gaseous and liquid forms) is useful to provide grid energy storage for intermittent energy sources, like wind power, as well as providing fuel for transportation, particularly for ships and airplanes. The round-trip efficiency of this type of storage is approximately 40% and the cost is slightly higher than that for pumped hydrostorage [215,259,260,262,264].

For stationary applications, hydrogen can be stored as compressed hydrogen (CGH_2) in a hydrogen tank and liquid hydrogen (LH_2) or slush hydrogen in a cryogenic hydrogen tank. CGH_2 offers the advantages of simplicity and stable storage (no boil-off losses) but requires considerably greater volume than LH_2 depending on pressure. Even accounting for compression costs, high-pressure gaseous hydrogen is cheaper than LH_2. However, except for pipeline transmission, CGH_2 lacks the bulk transportability of LH_2. Consequently, CGH_2 will mostly be employed for storage of limited hydrogen quantities, for long-term storage, or when the cost of liquefaction is prohibitive. Remaining issues for CGH_2 include its safety perception, and the current high costs of the pressure vessels and hydrogen compressors. Compressed hydrogen costs 2.1% of the energy content [1] to power the compressor. Higher compression without energy recovery will mean more energy lost to the compression step. Compressed hydrogen storage can exhibit very low permeation leakage in right environment [215–219].

8.8.1.1 Storage Vessels and Their Transport

Hydrogen is widely used and stored as compressed hydrogen industrial gas with a well-developed set of codes and standards governing its use. Hydrogen is typically stored in steel ASME-certified vessels, or in composite vessels that are currently DOT-certified for stationary uses. Some applications of hydrogen storage are now using higher than previously used pressure levels and can be as high as 875 bar (12,500 psi). The pressure in refueling vehicles can also be as high as 700 bar (10,000 psi). These higher pressure levels are requiring additional testing and certification of high-pressure vessels, which are typically made out of composite materials for weight considerations.

Typical DOT-certified hydrogen tube trailer are 22 and 44 ft tube depending on the volume of hydrogen needed. The larger 44 ft trailers typically hold 85,000–112,000 standard cubic feet (scf) of hydrogen at 2,400–2,800 psi of pressure. More recently, 3600–5000 psi pressures are being considered for compressed hydrogen storage. These high-pressure levels for ambient temperature compressed hydrogen distribution are being explored as an alternative to the currently used methods of lower pressure (2400–2800 psi) gas and cryogenic liquid delivery [215–219,259,260,262,264].

Because hydrogen is somewhat bulky and expensive to transport, the U.S. DOE has been funding research to improve the energy efficiency and cost of transporting hydrogen by truck, pipeline, and other means. Recent research has focused on high pressure levels of 3600 psi (up from current standard industry practice of 2400–3000 psi), and testing of a hydrogen storage and transportation delivery system that would hold 600 kg of hydrogen in a total of four pressure vessels.

The 600 kg transportable in the high-pressure tank system would be enough hydrogen for approximately 150 vehicle fills or to produce about 10 MWh of electrical power with a stationary fuel cell system (or approximately enough to power 10 houses for a month) [271]. Moving to 5000 psi of pressure, which is also under investigation, would allow a total of 800 kg to be stored and transported in the same size tubes [271].

Once delivered for stationary uses, hydrogen is typically stored in what is known as "ground storage" in ASME-certified pressure vessels. These ASME pressure vessels are recertified every 5 years. When compressed gas ground storage systems are refueled by hydrogen tube trailer delivery truck, this is called "bump stop" or "field bump" delivery. For many applications this is preferable to a smaller-scale "cylinder swap" model because the frequency of deliveries can be significantly reduced with the larger storage capacity [272].

8.8.1.2 Liquid Hydrogen

Liquid hydrogen storage is currently the bulk hydrogen storage medium of choice and has a very impressive safety record. However, liquid hydrogen requires cryogenic storage and boils around 20.268 K (−252.882°C or −423.188°F). Hence, its liquefaction imposes a large energy loss (as energy is needed to cool it down to that temperature). The tanks must also be well insulated to prevent boil-off. However, adding insulation increases the cost of storage. The hydrogen is typically liquefied at the production site in large quantities (10–30 tons per day) and then trucked cross-country in 11,000 gal LH_2 tankers with no boil-off losses. Unfortunately, about 30% of the hydrogen heating value is used as the energy requirements of liquefaction leading to relatively high hydrogen cost as compared to gaseous hydrogen.

LH_2 will likely remain the main technique of bulk, stationary hydrogen storage for the foreseeable future. For applications where hydrogen demands are such that tube trailers become cumbersome, delivery of hydrogen as a cryogenic liquid becomes more attractive owing to its higher energy density. However, that requires somewhat higher amounts of energy than compression to even 3600 or 5000 psi for bulk gas delivery. Once delivered for stationary uses, hydrogen is typically stored in underground liquid hydrogen "Dewar" systems. These storage systems must be tested and certified every 5 years [215–219,259,260,262,264].

8.8.1.3 Physical Hydrogen Storage in Mobile Applications

The goal for hydrogen application research for mobile systems is to provide adequate hydrogen storage for onboard light-duty vehicles, material-handling equipment, and portable power applications to meet the U.S. DOE hydrogen storage targets. By 2020, the Fuel Cell Technologies Office (FCTO) aims to develop and verify onboard automotive hydrogen storage systems, achieving targets that will allow hydrogen-fueled vehicle platforms to meet customer performance expectations for range, passenger, and cargo space; refueling time; and overall vehicle performance. Specific system targets include the following:

1. 1.8 kWh/kg system (5.5 wt% hydrogen)
2. 1.3 kWh/L system (0.040 kg hydrogen/L)
3. $10/kWh ($333/kg stored hydrogen capacity)

The progression over the last 10 years for hydrogen storage system is described in Table 8.4 [215–219,220,221,226,245–247,253–258,261].

To overcome these challenges the FCTO is pursuing two strategic pathways, targeting both near-term and long-term solutions. The near-term pathway focuses on compressed gas storage, using advanced pressure vessels made of fiber-reinforced composites that are capable of reaching 700 bar pressure, with a major emphasis on system cost reduction. The long-term pathway focuses on both (1) cold or cryo-compressed hydrogen storage, where increased hydrogen density and insulated pressure vessels may allow for DOE targets to be met, and (2) materials-based hydrogen storage technologies, including sorbents, chemical hydrogen storage materials, and metal hydrides, with properties having potential to meet DOE hydrogen storage targets outlined in Table 8.4.

Vehicular compressed hydrogen systems consisting of 34.5 MPa (5000 psi) gaseous hydrogen in metal- or plastic-lined, carbon fiber wound pressure vessels offer simplicity of design and use. Compressed hydrogen tanks at 700 bar (10,000 psi) is based on type IV carbon-composite

TABLE 8.4

Automotive Onboard Hydrogen Storage

Storage parameter	2005	2010	2015
Gravimetric capacity (specific energy)	1.5 kWh/kg 0.045 kg H_2/kg	2.0 kWh/kg 0.060 kg H_2/kg	3.0 kWh/kg 0.090 kg H_2/kg
System weight	111 kg	83 kg	55.6 kg
Volumetric capacity (energy density)	1.2 kWh/L 0.036 kg H_2/L	1.5 kWh/L 0.045 kg H_2/L	2.7 kWh/L 0.081 kg H_2/L
System volume	139 L	111 L	62 L
Storage system cost	$6/kWh	$4/kWh	$2/kWh
System cost	$1000	$666	$333
Refueling rate	0.5 kg H_2/min	1.5 kg H_2/min	2.0 kg H_2/min
Refueling time	10 min	3.3 min	2.5 min

Source: Hydrogen storage Wikipedia, the free encyclopedia, 2015.

technology [215–219,295]. Several car manufacturers such as Honda [215–219,253–258] or Nissan have been developing this technology [215–219,245–247]. Infrastructure requirements are several fold: rapid refueling capability, excellent dormancy characteristics and minimal infrastructure impact, high safety due to the inherent strength of the pressure vessel, and little to no development risk. The disadvantages are system volume and use of high pressure. Integrating the moderate-to-large system volume will clearly challenge the automotive designer, but such a tank volume can be packaged into a "clean sheet" vehicle. Many advantageous features of compressed gas storage outweigh its larger volume. Compressed gas storage is supportable by small-scale H_2 production facilities (on-site natural gas reforming plants, partial oxidation burners, and electrolysis stations) as well as larger-scale LH_2 production facilities. Thus a plausible H_2 infrastructure transition pathway exists. For these reasons, room temperature compressed gas storage is viewed as the most appropriate fuel storage system for PEM fuel cell vehicles.

Most research into hydrogen storage is focused on storing hydrogen as a lightweight, compact energy carrier for mobile applications. For these applications, liquid hydrogen or slush hydrogen may be used, as in the Space Shuttle. Liquid hydrogen has less energy density *by volume* than hydrocarbon fuels such as gasoline by approximately a factor of four. This highlights the density problem for pure hydrogen: there is actually about 64% more hydrogen in a liter of gasoline (116 g hydrogen) than there is in a liter of pure liquid hydrogen (71 g hydrogen). The carbon in the gasoline also contributes to the energy of combustion [215–219,295].

Vehicular LH_2 systems have the highest H_2 mass fractions and one of the lowest system volumes, along with near-zero development risk, good fast fill capability, and acceptable safety characteristics. They would appear to be an excellent choice except for two adverse factors: dormancy and infrastructure impact. Dormancy concerns arise due to boil-off losses that will inevitably concern the average car owner, although daily use or proper planning for route or fleet applications can remove most, if not all, dormancy concerns [215–221,245–247,253–258,295]. Three factors affect its usefulness: first the liquefaction process is costly; second, small-scale LH_2 production is impractical; and third, low volume distribution/dispensing of LH_2 is expensive. Consequently, LH_2 systems will not easily support a transition from anemic start-up to a robust H_2 economy. Overall, LH_2

storage is a most appropriate for a mature H_2 economy where the inherent difficulties (and high cost) of large-scale remote LH_2 production and very small-scale LH_2 dispensing are least encountered.

Cryo-compressed storage of hydrogen is the only technology that meets 2015 DOE targets for volumetric and gravimetric efficiency. Furthermore, another study has shown that cryo-compressed exhibits interesting cost advantages: ownership cost (price per mile) and storage system cost (price per vehicle) are actually the lowest when compared to any other technology. For example, a cryo-compressed hydrogen system would cost $0.12 per mile (including cost of fuel and every associated other cost), while conventional gasoline vehicles cost between $0.05 and $0.07 per mile.

Like liquid storage, cryo-compressed uses cold hydrogen (20.3 K and slightly above) in order to reach a high energy density. However, the main difference is that, when the hydrogen would warm up due to heat transfer with the environment ("boil-off"), the tank is allowed to go to pressures much higher (up to 350 bars versus a couple of bars for liquid storage). As a consequence, it takes more time before the hydrogen has to vent, and in most driving situations, enough hydrogen is used by the car to keep the pressure well below the venting limit. Consequently, it has been demonstrated that a high driving range could be achieved with a cryo-compressed tank: more than 650 miles were driven with a full tank mounted on an hydrogen-fueled engine of Toyota Prius [215–219]. Research is still under way in order to study and demonstrate the full potential of the technology [215–219]. As of 2012, the BMW Group has started a thorough component and system level validation of cryo-compressed vehicle storage with ultimate goal of making it a commercial product [215–219].

8.8.2 OTHER PHYSICAL METHODS

8.8.2.1 Carbon Nanotubes and Adsorption

Hydrogen carriers based on nanostructured carbon (such as carbon buckyballs and nanotubes) have been proposed. However, at present, these balls or tables can hold maximum 3–7 wt% hydrogen at 77 K. These numbers fall short of the value set by U.S. DOE (6 wt% at nearly ambient conditions).

Gaseous hydrogen can be adsorbed onto the surface of carbon to attain storage volumetric densities greater than liquid hydrogen. Adhesion capacity is greatly increased by low temperature (particularly cryogenic temperatures) and by high pressure. Indeed significant fractions of the hydrogen contained in carbon adsorbent systems is actually held in gaseous form within the interstitial volume of the carbon adsorbent. Carbon nanofibers are a special type of carbon adsorbent systems which may exploit a fundamentally different mechanism of hydrogen storage and thereby achieve dramatically improved storage capability. This idea is at the early stage of development.

8.8.2.2 Clathrate Hydrates

In 2004, researchers from Delft University of Technology and Colorado School of Mines showed solid H_2-containing hydrates could be formed at ambient temperature and 10 s of bar by adding small amounts of promoting substances such as THF [215,248]. These clathrates have theoretical maximum hydrogen densities of around 5 wt% and 40 kg/m³. This falls a little short of DOE target.

8.8.2.3 Glass Capillary Arrays and Microspheres

A team of Russian, Israeli, and German scientists have collaboratively developed an innovative technology based on glass capillary arrays for the safe infusion, storage, and controlled release of hydrogen in mobile applications [215,249–251]. This technology has achieved the United States Department of Energy (DOE) 2010 targets for on-board hydrogen storage systems; DOE 2015 targets can also be achieved using flexible glass capillaries and cryo-compressed method of hydrogen storage [259,260,262,264].

Microsphere hydrogen storage systems consists of hollow glass spheres that are charged with hydrogen 300°C–500°C, 27–62 MPa for an hour and discharged by reducing pressure at about

200°C–250°C. The microspheres can be pumped or poured from one tank to another, making them viable for vehicular hydrogen storage. Microsphere shelf life is a concern, and the idea is in the early stages of development [215,263].

8.8.3 Chemical Storage for Static and Mobile Applications

8.8.3.1 Metal Hydrides

Metal hydrides can be subdivided into two categories: low dissociation temperature hydrides and high dissociation temperature hydrides. The low dissociation temperature hydrides suffer from low H_2 fraction (~2%). The high temperature hydrides require a heat source to generate the high temperature of dissociation (~300°C). Both systems offer fairly dense H_2 storage and good safety characteristics. They exhibit no or slow H_2 release in a crash [296–299].

Overall for vehicular hydrogen storage, metal hydrides are either very heavy or their operating requirements are poorly matched to PEM vehicle systems. Without a dramatic breakthrough achieving high weight fraction, low temperature, low dissociation energy, and fast charge time, metal hydrides will not be an effective storage medium for PEM fuel cell vehicles. For stationary storage, the high weight of metal hydride system is not an adverse factor. Consequently, their attributes of high volumetric storage density and stability make them quite attractive. Major improvements in resistance to gaseous contaminant and system cycle life are, however, needed.

Metal hydrides, such as MgH_2, $NaAlH_4$, $LiAlH_4$, LiH, $LaNi_5H_6$, $TiFeH_2$, and palladium hydride, with varying degrees of efficiency, can be used as a storage medium for hydrogen [215–219,222–226]. Some of them are liquid at room temperature, and others are solids which can be turned into pellets [265]. They possess good energy density by volume but not by weight. Most metal hydrides bind with hydrogen very strongly. As a result, high temperatures around 120°C–200°C are required to release their hydrogen content. This energy cost can be reduced by using alloys which consist of a strong hydride former and a weak one such as $LiNH_2$, $LiBH_4$, and $NaBH_4$ [215–219]. These alloys provide a happy medium between strong hydrogen affinity at moderate pressure but also its release at a reasonable temperature. The target for onboard hydrogen fuel systems is roughly <100°C for release and <700 bar for recharge (20–60 kJ/mol H_2). The reduction in hydrogen release temperature can also be achieved by doping with activators. Currently the only hydrides which are capable of achieving the 9 wt% gravimetric goal for 2015 are limited to lithium-, boron-, and aluminum-based compounds; at least one of the first-row elements or Al must be added. More research is being pursued [221–227] in this regard.

Proposed hydrides for use in a hydrogen economy include simple hydrides of magnesium [215] or transition metals and complex metal hydrides, typically containing sodium, lithium, or calcium and aluminum or boron. Hydrides chosen for storage applications provide low reactivity (high safety) and high hydrogen storage densities. Leading candidates are lithium hydride, sodium borohydride, lithium aluminum hydride, and ammonia borane.

The literature shows that for most metal hydrides, a strong relation between pressure and temperature for hydrogen adsorption exists (i.e., at high temperature, considerably high pressure is required for same level of adsorption). Using this pressure-temperature dependence, several efforts at Arizona State University, University of Pittsburgh, Georgia Tech, and Intelligent Energy (a fuel cell company) are being made to improve conditions for metal hydride hydrogen storage [215–219].

8.8.3.2 Nonmetal Hydrides

While numerous nonmetal hydrides have been pursued, as yet they have not measured up to the expected goal of DOE for hydrogen storage, particularly for automobile application. Here we briefly assess some of the reported hydrogen storage possibilities.

The Italian catalyst manufacturer Acta has proposed using liquid hydrazine (at room temperature) as an alternative to hydrogen in fuel cells. By storing it in a tank full of a double-bonded carbon-oxygen carbonyl, it reacts and forms a safe solid called hydrazone. By the flushing of the tank with warm water, liquid hydrazine hydrate is released [215–219]. Heterocyclic aromatic compounds such as N-ethylcarbazole [215] or dibenzyltoluene can be hydrogen carrier which can reach relatively high gravimetric storage densities (about 6 wt%) [215–219]. Research conducted by Dr. André Geim at the University of Manchester has shown that graphene can not only store hydrogen easily but can also release the hydrogen again, after heating to 450°C [215–219]. Cella Energy has suggested encapsulation of hydrogen gas and nano-structuring of chemical hydrides in small plastic balls, at room temperature and pressure.

A small hydrogen reformer would extract the hydrogen as needed by the fuel cell. However, these reformers are slow to react to changes in demand and add a large incremental cost to the vehicle powertrain. Without reformer, fuel cell can be operated by methanol or ethanol and SOFC can operate on light hydrocarbons such as propane and methane [228–230,232–234]. However, high temperature and slow startup time of these fuel cells are problematic for automotive applications. Ammonia can be reformed [266] to produce hydrogen with no harmful waste, or can mix with existing fuels and under the right conditions burn efficiently. In 2005, chemists from the Technical University of Denmark announced a method of storing hydrogen in the form of ammonia saturated into a salt tablet. They claim this to be an inexpensive and safe storage method [215–219].

Carbohydrate provides high hydrogen storage densities as a liquid. With mild pressurization and under cryogenic conditions, it can also be stored as a solid powder. This carbohydrate can be a high energy density hydrogen carrier with a density of 14.8 wt%. Some complex borohydrides, or aluminum hydrides, and ammonium salts [266] have an upper theoretical hydrogen yield limited to about 8.5% by weight. Amine boranes (especially ammonia borane and diborane diammoniate) have been extensively investigated as hydrogen carriers. Ignition of the amine borane(s) forms boron nitride (BN) and hydrogen gas [228–235].

Formic acid [231,236–238] can be decomposed into hydrogen and carbon dioxide in the presence of water soluble ruthenium catalysts [237]. This catalyst is stable, has good lifetime, and avoids formation of carbon monoxide [236,238]. Carbon dioxide can be hydrogenated to regenerate formic acid. Formic acid contains 53 g/L hydrogen at room temperature and atmospheric pressure. By weight, pure formic acid stores 4.3 wt% hydrogen. In 2007 DuPont and others reported hydrogen storage materials based on imidazolium ionic liquids [239]. Simple alkyl(aryl)-3-methylimidazolium N-bis(trifluoromethanesulfonyl)imidate salts that possess very low vapor pressure, high density, and thermal stability, are not inflammable, can add reversibly 6–12 hydrogen atoms in the presence of classical Pd/C or Ir0 nanoparticle catalysts, and can be used as alternative materials for on-board hydrogen storage devices. These salts can hold up to 30 g/L of hydrogen at atmospheric pressure [215].

8.8.3.3 Metal-Organic Frameworks

Metal-organic frameworks represent another class of synthetic porous materials that store hydrogen and energy at the molecular level. MOFs are highly crystalline inorganic–organic hybrid structures that contain metal clusters or ions (secondary building units) as nodes and organic ligands as linkers. When guest molecules (solvent) occupying the pores are removed during solvent exchange and heating under vacuum, porous structure of MOFs can be achieved without destabilizing the frame and hydrogen molecules will be adsorbed onto the surface of the pores by physisorption. MOFs have very high number of pores and surface area which allow high hydrogen uptake in a given volume [215–219]. Since 2003, many researches have explored what combination will provide the maximum hydrogen uptake by varying materials of metal ions and linkers within MOF. In 2006, chemists at UCLA and the University of Michigan achieved hydrogen storage concentrations of up to 7.5 wt% in MOF-74 at a low temperature of 77 K [215–219]. In 2009, researchers at University of Nottingham reached 10 wt% at 77 bar (1117 psi) and 77 K with MOF NOTT-112 [240,241,243,244]. Varying several factors such as surface area, pore size, catenation, ligand structure, spillover, and sample purity can result in different amounts of hydrogen uptake in MOFs.

8.8.4 FUTURE CHALLENGES

Multiple techniques of hydrogen storage are viable for both vehicular storage and bulk stationary storage. However, no one storage mechanism is ideal. As demand for hydrogen grows, industry must respond by supplying (and storing) hydrogen in ways suitable for the new class of consumers and must educate the public in its safe use.

The FCTO is developing onboard automotive hydrogen storage systems that allow for a driving range of more than 300 miles while meeting cost, safety, and performance requirements [215–219, 243–246,253–258]. In 2010, only two storage technologies were identified as being susceptible to meet DOE targets: MOF-177 exceeds 2010 target for volumetric capacity, while cryo-compressed H_2 exceeds more restrictive 2015 targets for both gravimetric and volumetric capacity.

High-density hydrogen storage is a challenge for stationary and portable applications, and remains a significant challenge for transportation applications. Presently available storage options typically require large-volume systems that store hydrogen in gaseous form. This is less of an issue for stationary applications, where the footprint of compressed gas tanks may be less critical.

However, fuel-cell-powered vehicles require enough hydrogen to provide the desired driving range (at least 300 miles) with the ability to quickly and easily refuel the vehicle. While some light-duty hydrogen fuel cell electric vehicles (FCEVs) that are capable of this range have emerged in the market, these vehicles will rely on compressed gas onboard storage using large-volume, high-pressure composite vessels. The required large storage volumes may have less impact for larger vehicles, but providing sufficient hydrogen storage across all light-duty platforms remains a challenge [215–219,259,260,262,264].

On a mass basis, hydrogen has nearly three times the energy content of gasoline—120 MJ/kg for hydrogen versus 44 MJ/kg for gasoline. However, on a volume basis, the situation is reversed; liquid hydrogen has a density of 8 MJ/L, whereas gasoline has a density of 32 MJ/L. Onboard hydrogen storage capacities of 5–13 kg hydrogen will be required to meet the driving range for the full range of light-duty vehicle platforms [295–323].

8.9 HYDROGEN TRANSPORT

Three methods are commonly used to deliver hydrogen to refueling stations [321]: (1) hydrogen pipeline, (2) hydrogen tube trailers, and (3) liquid hydrogen tanker. Gaseous hydrogen pipelines require large up-front capital investments but can transport large amounts of hydrogen very cheaply over short and sometimes long distances. The most cost-effective method of delivering small amounts of gaseous hydrogen over short distances is gaseous tube trailers. Finally, over longer distances, liquid hydrogen delivery by road tanker or ships becomes optimal [139,268–294].

8.9.1 PIPELINE TRANSPORT OF GASEOUS HYDROGEN (CGH$_2$)

Pipelines are the most efficient method of transporting large quantities of hydrogen, particularly over short distances. Almost 3000 km of hydrogen pipelines have been constructed since 1938 in Europe and North America [318]. As mentioned earlier, transporting hydrogen through high-pressure steel pipelines is more difficult than transporting methane because of hydrogen embrittlement, which makes strong steel pipes vulnerable to cracking, and because of hydrogen attack that allows reactions with the steel carbon atoms under certain operating conditions, again leading to cracks.

Hydrogen has a lower energy density by volume than methane but a faster flow rate; this means that the total pipe capacity is around 20% lower for hydrogen than methane but the total hydrogen stored within the pipe is only a quarter of the total methane at the same pressure in terms of its energy value. Low-pressure hydrogen pipelines are not generally used for hydrogen except in a few niches, such as hospitals. There is, however, much more flexibility over the choice of pipeline material at low pressures [295–323].

Pipeline investment costs can be split into four main categories: materials, labor, right-of-way fees, and miscellaneous. Only the material costs are likely to differ from pipelines used for methane. Pipeline costs are affected not only by the diameter but also crucially by the topography, land use, and labor costs. Material costs for hydrogen pipelines would be larger than the ones for methane transport. Pipeline costs are proportional to the length of pipeline. Since upfront investment costs for hydrogen pipeline are high, in the absence of high hydrogen throughput and its utilization factor, the decision for investment gets difficult and complex. Another complication factor is that the throughput is not fixed to the pipeline diameter. The throughput can be increased by increasing the pressure difference across the pipeline length but at a cost of greater energy consumption [310]. Increasing the throughput in a pipeline reduces the energy efficiency but also substantially increases the return on investment (ROI) of the capital cost.

Just like pipeline transport of natural gas, pipeline transport of hydrogen is distributed in four categories: high-pressure transmission lines that connect city boundaries to plant production, high-pressure distribution lines that connect transmission lines to high-demand customers and low-pressure distribution network, smaller low-pressure distribution lines that connect low-demand customers to high-pressure lines, and, finally, small service pipelines that connect customers to low-pressure distribution lines [295–323].

As of June, 2013, there were approximately 1213 miles of hydrogen pipeline in the United States, as reported by the U.S. Energy Information Administration [323]. Most of these pipelines are in Texas (847 miles), Louisiana (290 miles), Alabama (31 miles), Indiana (15 miles), and California (13 miles). Virtually all (an estimated 99%) of the transportation of hydrogen in the United States is by pipeline as a compressed gas (typically at pressures below 1000 psi), mainly for oil refinery use and ammonia production. Hydrogen transmission by pipeline dates back to the 1930s in the United States, and has had a good safety record [323].

As an example, a hydrogen pipeline in California connects Carson and nearby Wilmington, associated with oil refineries in the area. The existing pipeline was installed by Air Products and Chemicals, Inc. [269,270] to connect various refineries and to balance their capacities and needs for hydrogen for "hydro-treating" crude oil as part of the gasoline production process.

Several new or extended pipelines are being contemplated by various groups, including a proposal by Air Product and Chemicals Inc. to build a 180-mile-long segment to connect its pipeline networks in Texas and Louisiana [269,270]. Among the other hydrogen pipelines being contemplated is one between Chevron's refinery in Richmond, California, and a group of refineries in nearby Martinez [269,270].

8.9.2 Tube Trailer Transport of Gaseous Hydrogen (CGH_2)

Most existing hydrogen fueling stations dispense fuel from compressed gas canisters that are delivered to the station. This is the most economic system for locations with low fuel demand if the delivery distance is not too great [321], and it might be the most suitable in the future in some locations.

The principal advantage of tube trailer delivery is that it avoids the high liquefaction energy cost and high pipeline investment costs that affect other delivery systems. Tube trailer fueling stations can be cheaper than other hydrogen fueling stations because the hydrogen is dispensed directly from the tube trailer, so little on-site storage is required [321]. Fueling stations tend to be small because a single tube trailer can store only 250–500 kg and it is impractical to replace the trailer several times each day.

High costs are the principal disadvantages of tube trailer delivery, particularly for long-distance deliveries. The lower costs represent an average round-trip of 200 km (325–360 deliveries per year), while the higher costs represent a round-trip of 1600 km (72–150 deliveries per year) [321]. For long-distance delivery, investment costs are also much higher because a greater number of tube trailers are required to deliver the same amount of fuel. Investment cost reductions in the future will depend primarily on improvements to hydrogen storage technologies. The tube trailer capacity, which varies between 250 and 500 kg, has an important influence on the energy efficiency of delivery, particularly over long delivery distances [321]. Tube trailers are particularly inefficient for hydrogen delivery over long distances compared to road tankers because road tanker delivers 3000–4500 kg of hydrogen for the same cost as that of tube trailer.

Tube trailers are competitive at short delivery distances, but if the demand is very high then pipelines are a more economic option. This leaves two principal roles for tube trailers in a hydrogen economy. Firstly, tube trailers can be the most economic technology during the transition to a low-carbon economy when demand is low. Secondly, tube trailers can be the most economic technology for geographically isolated fueling stations in areas with low demand in a developed hydrogen economy [321]. Thus, while tube trailers play an important role in the transition to hydrogen economy, they will provide only a small niche in a developed economy.

8.9.3 Road Tanker and Ship Transports of Liquid Hydrogen

Hydrogen undergoes liquefaction at a temperature of 20 K (−253°C). Theoretically, only about 4 MJ/kg must be removed from the gas but the cooling process has a very low Carnot cycle efficiency [306], so even large plants require 30 MJ/kg to liquefy hydrogen. The principal disadvantage of liquid hydrogen (LH_2) is the substantial amount of expensive electricity that is consumed during liquefaction. Larger plants have lower investment costs and are more efficient than smaller plants.

The DOE has targeted to dramatically reduce the cost of liquefaction [312]. The energy efficiency of liquefaction varies from 68% to 84%, with larger plants being more efficient.

Road tankers are currently used worldwide for fuel deliveries. Since the energy density by volume of liquid hydrogen is three times lower than that of gasoline, a much greater number of tankers would be required to deliver the same quantity of fuel. The relative fuel demand for hydrogen would be lower, however, as a result of the greater efficiency of hydrogen fuel cells compared to internal combustion engines. It is also possible that hydrogen production could be more decentralized [312], reducing the distance between production plants and refueling stations and so enabling road tankers to perform a greater number of sorties per year [312]. Both the capital investment cost and the energy efficiency are determined by the average trip distance.

Ships could also be used to transfer hydrogen across the planet. Hydrogen ships are expected to be similar to the LNG ships that are widely used at present. All ships are forecast to have energy efficiencies of 95%, although this would depend on the transport distance [312].

8.10 NATURAL GAS AND HYDROGEN GRIDS

Gas is fundamentally a network industry. Networks are necessary to move gas from the source of production to the customers. As more gas producers and customers are attached to the network, the network itself becomes more valuable. There are also the interconnections and interdependencies between gas networks and other networks, both physical and digital [324–356].

For years, the concept of gas grid has been applied to manage supply and demand of gas, which has been controlled by elaborate gas pipeline infrastructure and storage systems (often called "gas network"). Approximately 89% of gas supply and demand is managed by gas pipeline infrastructure [324,345], which has a global network of over 1.4 million km [324,345]. About 70% of these pipelines are regional lines that support collection and distribution within their domestic markets. The balance consists of large-diameter continental scale pipelines found in places like the United States, Canada, Russia, and increasingly China [324,345]. Pressure is essential for moving gas through the pipeline system. This is achieved through compressors that are located at regular intervals along the pipeline. With compression, gas moves about 25 mph (40 km/h) through long-distance pipelines. Compared to the transmission of electricity, this is a slow process and does not allow very fast response to the long-distance demand.

The second mode of gas transport is the LNG network, which constitutes 10% of global gas trade [324,345]. This sea-based network is made possible by the invention of LNG and large special-purpose transport ships that ply long distance in the oceans linking exporters and customers. Finally, 1% of gas can also be transported using existing road and rail infrastructure. This is made possible through compression technologies like CNG and small scale LNG as well as gas hydrates. A gallon of CNG has about 25% of the energy content of a gallon of diesel fuel, and LNG has 60% of the volumetric energy density of diesel fuel. Both CNG and LNG are cost-competitive, and their market demands are rapidly expanding.

This mode of transportation is often called "virtual" pipelines. This method allows an efficient use of CNG and LNG in the transportation sector and in distributed power plants [324,345].

Natural gas grid, thus, has three subcomponents: pipelines, LNG sea network, and virtual pipelines. The cost of transportation through pipelines is much lower than that for sea-based LNG transportation, and the transportation by virtual pipeline economics plays a large role in determining which solution is selected to move gas from sources of supply to ultimate demand [324,345].

Over the years, we have developed a sophisticated transportation and storage system for gas to meet our variable needs of gas for residential, commercial, industrial, and transportation purposes. Within the United States, we have several hundred thousand miles of underground pipeline infrastructure for gas at different levels of distribution, storage, and user systems, and these are closely watched, controlled, and managed to support variable demands. While the United States always had an elaborate transportation and storage infrastructure for gas, over the past 5 years, unconventional gas activity has thrust the nation into an unexpected position. The United States is now the largest natural gas producer, at 65 billion cubic feet (bcf)/day, in the world [345]. At the same time, unconventional activity is spurring the growth of natural gas liquid (NGL) production of over 500,000 barrels of oil equivalent (boe) per day in the United States since 2008. Over the past 5 years, daily oil production of United States has also increased by 1.2 million barrels [345].

In order to accommodate the growth in U.S. natural gas, NGLs, and crude oil, the past 2 years have witnessed a rapid growth in direct capital investment toward oil and gas infrastructure assets. It is estimated that the capital spending in oil and gas midstream and downstream infrastructure has increased by 60%, from $56.3 billion in 2010 to $89.6 billion in 2013 [345]. This increase in capital spending is reshaping the U.S. oil and gas infrastructure landscape. A large portion of the projects being developed during this sustained infrastructure investment period will shift the United States toward being energy trade balance and add key infrastructure segments that enable growing energy production in the Midcontinent region to reach demand centers on the U.S. Gulf Coast and Eastern seaboard [324–356].

This newly developing infrastructure will also handle different types of unconventional gas (i.e., shale gas, coalbed methane, etc., with composition very different from that of natural gas, biogas, propane, hydrogen) [328–336,339,342–345,349] so that in future, transportation and storage infrastructure of gas is capable of handling a variety of gas to meet increasing and variable demands without significantly affecting the pricing structure. The successful penetrations of unconventional gas as well as gas coming from renewable energy sources like biomass and waste will depend on the success of our transportation and storage infrastructure. A summary of our existing natural gas infrastructure is illustrated in Table 8.5.

Besides unconventional gas and gas coming from biomass and waste (biogas), future energy landscape will significantly change with the production and use of hydrogen in a variety of ways. Hydrogen is the most abundant and noncarbon energy source that can be used either as a fuel or as a source of electrical energy. In order for hydrogen to penetrate energy market in a significant way, we will need to (1) use existing natural gas infrastructure to the extent possible or (2) develop its own transportation and storage infrastructure. At present, we can introduce 5%–15% by volume of hydrogen in natural gas pipeline. An independent development of hydrogen transportation and storage infrastructure is in the research stage [328–331,335–337,356]. Since hydrogen is also a

TABLE 8.5

Summary of Natural Gas Transportation and Storage Infrastructure in the United States [327–351,355,356]

Topic Gas Pipeline Systems	Number
Underground NG storage	210
Underground NG storage facilities	400
Hubs or interconnections market centers	24
Delivery points	11,000
Receiving points	5,000
Interconnection points	1,400
Compressor stations to maintain pipeline NG pressure	1,400
Interstate and intrastate transmission pipelines (miles)	305,000
LNG import facilities	8
LNG peaking facilities	100
Import/export locations for pipeline NG transport	49

good source for electrical energy and energy storage, hydrogen grid or natural gas grid containing hydrogen will be intertwined with the electrical grid.

Today, modern fossil fuel–based power plants (and especially natural gas combined cycles) are becoming more and more flexible. Their ramping-up speed in response to rapid changes in demand is increasing. They can provide reliable and flexible backup power for renewable energy sources like wind and solar. In the short term, however, ES is also needed to fill the gap between the ramping-down time of wind and solar and the ramping-up time of these backup plants. The challenge is to increase existing storage capacities and increase efficiencies [324,345].

Gas storage is closely linked to electricity storage. Some seasonal variations in demand for gas and electricity can be covered by natural gas storage. Gas is an important fuel for electricity production, and natural gas power plants have a very high efficiency (above 60% for the best available technology), a very high flexibility, and low CO_2 emissions (replacing an old coal-fired power plant by a natural gas–fired power plant reduces the CO_2 emissions per kWh up to 80%) [324–345]. In the future, injection of biogas and hydrogen into the natural gas grid [331–334,356], and the longer-term commercialization of carbon capture and sequestration technology, will further de-carbonize gas-powered electricity generation. Indeed, the expansion of natural gas–fired power plants, the increased efficiency, and the reduced costs of flexible combined-cycle and simple-cycle natural gas turbines combined with the strong and fast-growing interconnection of the grid on a larger scale will continue to help efficiency and security of energy supply and demand needs. These discussions once again indicate that in future, natural gas grid, hydrogen grid, and electrical grid would be strongly intertwined. The development of CHP will make localized heat grids also an integral part of this mix.

8.10.1 GRID INTEGRATIONS

Gas network or grid must be connected to other networks [342,343,346–350]. These connections are based on the needed functionality. New interconnections create more supply diversity and customer options. This also makes grid more flexible. As grid becomes more complex by adding functions of gathering and processing; storage; LNG systems at large scale; LNG, CNG, or gas hydrates systems at small scale; distribution grids; and the complex control and information and communication software systems needed to monitor, track, and trade natural gas, it also becomes more powerful and smart—a system that is far more than just interconnected pipelines. It becomes a real example of holistic integration of energy and fuel systems. In the near future, we are likely to see a natural gas grid that is complex and fully integrated both horizontally and vertically. The deepening connection between gas supply and electricity generation is an obvious example of vertical integration. Another example is the potential for gas to become more commonplace in supplying fuel to heavy vehicles. Both LNG and CNG are competitive in this market. The vertical integration of natural gas will also occur with other types of gas such as biogas, shale gas, coalbed methane, propane and hydrogen, whose compositions are considerably different from 98% to 99% pure (methane) natural gas. The use of existing pipelines for these gases will require some pretreatments. These gas supplies, with the help of new technology, will be cheaper and can be produced with lower environmental impact than ever before. Midstream and consumer technologies are also advancing [346–350]. There is a significant opportunity to further optimize and secure the sea- and land-based components of natural gas systems. Digital monitoring and control of pipelines and optimization of CNG/LNG fueling systems are two examples of innovative integrations of existing technology. The next generation of digital systems will employ technologies, including satellite, wireless, cloud storage, and software tools for remote monitoring and control and predictive analytics. This will integrate data collection, processing, reporting, and analytics in smart ways and create a grid similar to the smart grid (SG) for electricity [324,345].

Blending hydrogen into natural gas pipeline networks has also been proposed as a means of delivering pure hydrogen to markets, using separation and purification technologies downstream to extract hydrogen from the natural gas blend close to the point of end use [328–331,335–337,356]. Hydrogen can be inserted in natural gas pipelines up to the concentration of 5%–15% [345]. Anything higher than that would require its own dedicated transportation infrastructure. Hydrogen storage may also require specially designed systems. As a hydrogen delivery method, blending can defray the cost of building dedicated hydrogen pipelines or other costly delivery infrastructure during the early market development phase. This hydrogen delivery strategy also incurs additional costs, associated with blending and extraction, as well as modifications to existing pipeline integrity management systems, and these must be weighed against alternative means of bringing more sustainable and low-carbon energy to consumers. The insertion of hydrogen in natural gas grid will also enhance the interactions between gas and electricity grids.

Deeper horizontal and vertical integrations will also create the potential for greater competition across networks. Supply and demand structure will become more hybrid and interactive and provide a greater customer flexibility. As the complexity of gas systems increases, they also offer more options. Finally, deeper integration of networks, both vertically and horizontally, will enhance the overall resilience of energy systems, making them more impervious to disruption. In general, as grids become larger, they also become more efficient, more flexible, more complex, and more resilient [324,345].

Natural gas is a versatile fuel. Environmental advantages coupled with high efficiency and flexibility make natural gas a great choice for power generation and many other usages. Unlike coal or nuclear fuel, it can be cost-effective when used in both large- and small-scale applications. Furthermore, as gas networks become available, it can become an alternative to oil [324,345]. The technologies being deployed today are designed to bring flexibility to operations, help consumers capture value, and create security of supply. Effectively mobilizing these demand-side technologies is an important part of achieving the sustainability, resilience, and competitive benefits of natural gas. These technologies are also the paths for the development of more smart and improved gas grid [332–352].

Since transportation of gas is not as rapid as that of electricity, gas grid may need to be managed on a smaller scale to obtain the faster response to the changes in supply and demand. The gas grid can be distributed in satellite operations for better and faster control of supply and demand. This is not to say that various distributed (satellite) gas grids cannot be interconnected for overall better control [324]; the overriding control of supply and demand will be at regional level for gas grid. In a similar context, heat grids will also have to be managed locally, because heat cannot be transported over a long distance due to transmission losses. Thus, effective integrations among gas, heat, and electricity will have to be carried out at local and regional levels.

8.10.2 Levels of Grid Applications

The improvement in gas generation technologies at large, medium, and small scales will also improve the effectiveness of gas grid. The latest flexible combined cycle power plants are reaching thermal efficiencies such that almost two-thirds of the energy in the natural gas is converted into electricity; 750 MW gas-driven power plants can start up in fewer than 30 min and increase power output at 100 MW/min [324,345]. When necessary, they can be turned down to 14% of their base load capacity to adjust to changes in system demand or the introduction of intermittent renewables [340–346,349–351].

At a smaller scale (typically 10–120 MW), gas-fired CHP systems using gas turbines can create both electricity and steam. As shown in Chapter 10, CHP systems can be used for numerous nonelectrical applications such as heating and cooling large commercial buildings, hospitals, airports, or industrial sites, providing industrial process heat, and so on. Utilizing CHP (also called cogeneration), these power plants can achieve thermal efficiencies in excess of 80%, often with lower emissions and losses than grid-supplied options [324,340–346,349–351].

CHP operations can be designed to disconnect from the larger grid in the event of a disruption, allowing critical facilities to be operated in "island" mode during natural disasters. CHP projects, however, require customized designs and are difficult to implement in existing buildings. Additionally, coordination between electric grid operators and CHP operators is critical on a number of issues, including excess power flowing back to the grid, backup power requirements if the natural gas supply is disrupted, or safety concerns that might result from miscommunication [324,345].

For even smaller applications (0.3–10 MW), natural gas engine can have high efficiencies (up to 45% thermal efficiency) relative to other simple combustion technologies. These engines can be used in CHP configurations to achieve higher efficiency or in other distributed settings [324, 340–346,349–351]. They have low emissions relative to their diesel-fired counterparts and can run on a wide range of natural gas qualities (from rich to lean) and a variety of other gases, including biogas, landfill gas, coalbed methane, sewage gas, and combustible industrial waste gases. The ability of some gas engines to burn natural gas with higher liquid content (rich gas) is ideal for oil and gas field power generation applications such as drilling and enhanced oil recovery.

Thus, the advances in combined-cycle technology, CHP using gas turbines or gas engines, and others like fuel cells and small biogas systems will continue to expand the role of the gas grid. These improvements will also bring workings of gas grid, localized heat grid, and electricity grid more intertwined. Digital technologies and the industrial internet will also play a major role in the network improvement. Technology will thus play a critical role in supplying, securing, and growing the gas grid of the future [324,345].

8.10.3 Role of Gas Storage

The way natural gas grid is designed is somewhat different from the SG for electricity. At present, storage of compressed gas underground is the primary technology used to store natural gas [324,345,352,353]. In the natural gas industry, the most rapid consumption of natural gas occurs in winter. However, it is uneconomical to design transcontinental pipelines and natural gas treatment plants to meet peak natural gas demands. Instead, the natural gas is produced and transported at a nearly constant rate throughout the year. A variety of different types of large underground storage systems in different geologies at locations near the customer are used to store the excess natural gas produced during the summer. This practice minimizes the cost of the long-distance natural gas pipeline system and improves reliability by locating storage facilities near the customer. In the winter, these underground storage facilities provide the natural gas to meet customer demands.

While there are some differences in traditional gas grid and electricity grid, these differences become small as both grids mature and become more complex and interact with each other. In an advanced hybrid energy system, multiple sources of energy and energy storage will be combined with the use of cogeneration to generate heat, power, and other downstream applications. Thus, in an advanced hybrid energy system, gas, heat, and electricity grids are connected to simultaneously produce electricity and heat for other downstream applications [354]. The advanced hybrid system also allows the use of renewable energy for power generation at large and sustainable scale and improves thermal efficiency of the overall process. This subject is further discussed in my previous book [355].

REFERENCES

1. Natural gas storage, Wikipedia, the free encyclopedia (2015), https://en.www.wikipedia.org/wiki/Natural_gas_storage (last accessed June 29, 2016).
2. Storage of natural gas, a report by Naturalgas.org. (September 2013), Accessed January 2009.
3. Global LNG-will new demand and new supply mean new pricing?, A report by EY, EYG no. DW0306, CSG/GSC2013/1115394, A Part of Earnest and Young Global Ltd., U.K. (2013).
4. Liquefied natural gas: Understanding the basic facts, A report from NETL, DOE/FE-048g, Department of Energy, Washington, DC (August 2005).
5. News, views and knowledge on gas-worldwide, General information, International Gas Union, Statoil, Norway (2014), www.igu.org.
6. Fullerbaum, R., Fallon, J., and Flanagan, B., Oil & natural gas transportation & storage infrastructure: Status, trends, & economic benefits, Report for American Petroleum Institute, p. 68, Appendix B: Natural Gas, Prepared by IHS Global Inc., Washington, D.C. (December 2013).
7. North American midstream infrastructure through 2035: Capitalizing on our energy abundance, Interstate Natural Gas Association of America, Prepared by ICF International, Fairfax, VA (March 2013), http://www.ingau.org/Foundation/Foundation_reports/studies/14904/14889.aspx.
8. The future of U.S. nature gas supply, demand and infrastructure developments. A report for North Dakota Industrial Commission, Order no. 24665 66. Quadrennial energy review analysis, prepared by BENTEK energy, Denver, CO (July 2014). http://www.energy.gov/epsa/qer-document-library.
9. Federal Regulatory Commission, Current state of and issues concerning: Underground natural gas storage, Docket No. AD04-11-000 (September 30, 2004), http://www.ferc.gov/EventCalendar/Files/20041020081349-final-gs-report.pdf, Accessed February 2, 2014.
10. Federal Regulatory Commission, Natural gas storage—Storage fields (February 1, 2011), https://www.ferc.gov/industries/gas/indus-act/storage/fields.asp, Accessed January 28, 2015.
11. Underground natural gas storage, Energy Information Administration, Department of Energy, Washington, D.C. http://www.eia.gov/pub/oil_gas/natural_gas/analysis_publications/ngpipeline/undrgrnd_storage.html, Accessed July 2014.
12. Salt caverns and their use for disposal of oil field wastes, The National Petroleum Technology Office, Argonne National Laboratory, Lemont, IL (1999), http://web.evs.anl.gov/saltcaverns/doc/SaltCavbroch.pdf, Accessed January 28, 2015.
13. Energy Information Administration, The basics of underground natural gas storage (August 2004), http://www.eia.gov/pub/oil_gas/natural_gas/analysis_publications/storagebasics/storagebasics.html.
14. 191 Field level storage data (Annual): 2012, Natural Gas Annual Respondent Query System (EIA-191 data through 2013), Energy Information Administration, Department of Energy, Washington, D.C. (April 2014), http://www.eia.gov/cfapps/ngqs/ngqs.cfm?f_report=RP7&f_sortby=ACI&f_items=&f_year_start=2012&f_year _end=2012&f_show_compid=Name&f_fullscreen.
15. Underground natural gas storage capacity, Energy Information Administration, Department of Energy, Washington, D.C. (July 31, 2014), http://www.eia.gov/dnav/ng/ng_stor_cap_a_EPG0_SA1_Mmcf_a.htm.
16. Energy Information Administration, Monthly underground gas storage report, Form EIA-191M, DOE, Washington, D.C. (October 31, 2014) (data from 2013).
17. Natural gas infrastructure, A report by Department of Energy, Washington, DC (2014), energy.gov/sites/prod/.../Appendix%20B-%20Natural%20Gas_1.pdf.
18. The transport of natural gas, A report by Naturalgas.org. (September 20, 2013).
19. Deliverability on the interstate natural gas pipeline system, Energy Information Administration (EIA), Department of Energy, Washington, D.C. (May 1998), http://ftp.eia.doe.gov/pub/oil_gas/natural_gas/analysis_publications/deliverability/pdf/deliver.pdf.
20. Hartman, K., Making state gas pipelines safe and reliable: An assessment of state policy, National Conference of State Legislatures (NCSL), Jacquelyn Press, full report 21 pages. (March 2011), http://www.ncsl.org/research/energy/state-gas-pipelines-natural-gas-as-an-expanding.aspx.

21. Energy Information Administration, Interstate pipeline capacity on a state-to-state level (January 16, 2014), http://www.eia.gov/naturalgas/data.cfm, Accessed October 10, 2014.
22. Energy Information Administration, U.S. natural gas pipeline projects (2014), http://www.eia.gov/naturalgas/pipelines/EIA-NaturalGasPipelineProjects.xls, Accessed September 24, 2014.
23. How does the natural gas delivery system work? A report by American Gas Association, Washington, D.C. (2014). http://www.aga.org/KC/ABOUTNATURALGAS/CONSUMERINFO/Pages/NGDeliverySystem.aspx, Accessed July 30, 2014.
24. Energy Information Administration, Natural gas compressor stations on the interstate pipeline network: Developments since 1996, Department of Energy, Washington, D.C. (November 2007), http://www.eia.gov/pub/oil_gas/natural_gas/analysis_publications/ngcompressor/ngcompressor.pdf.
25. Energy Information Administration, First westbound natural gas flows begin on Rockies Express Pipeline, Department of Energy, Washington, D.C. (June 18, 2014), http://www.eia.gov/todayinenergy/detail.cfm?id=16751.
26. Energy Information Administration, Natural gas: About U.S. natural gas pipelines (2007), http://www.eia.gov/pub/oil_gas/natural_gas/analysis_publications/ngpipeline/index.html, Accessed July 28, 2014.
27. Maximum allowable operating pressure for natural gas pipelines, A report by Interstate Natural Gas Association of American Foundation, Washington, D.C. (2011), http://www.ingaa.org/File.aspx?id=12366.
28. Energy Information Administration, Distribution of natural gas: The final step in the transmission process (2008), http://www.eia.gov/pub/oil_gas/natural_gas/feature_articles/2008/ldc2008/ldc2008.pdf, Accessed July 28, 2014.
29. Natural gas pipelines: Safe, sound, and underground, A report by American Gas Association, Washington, D.C. (2014). http://www.aga.org/Kc/aboutnaturalgas/consumerinfo/Pages/NGDeliverySystemFacts.aspx, Accessed August 4, 2014.
30. Mannel, D. and Puckett, D., Transportation and storage of natural gas hydrates, Personal communication (April 2008).
31. Daimaru, T., Fujii, M., Yamasaki, A., and Yanagisawa, Y., Energy saving potential for natural gas hydrate transportation, *Preprints of Papers—American Chemical Society, Division of Fuel Chemistry*, 49 (1), 190 (2004).
32. Nazari, K., Taheri, Z., Mehrabi, M., and Khodafarin, R., Natural gas hydrate production and transportation, in *Proceedings of the Seventh International Conference on Gas Hydrates (ICGH 2011)*, Edinburgh, U.K. (July 17–21, 2011).
33. Rehder, G., Eckl, R., Elfgen, M., Falenty, A., Hamann, R., Kähler, N., Kuhs, W.F., Osterkamp, H., and Windmeier, C., Methane hydrate pellet transport using the self-preservation effect: A techno-economic analysis, *Energies*, 5, 2499–2523 (2012).
34. Davidson, D.W., Garg, S.K., Gough, S.R., Handa, Y.P., Ratcliffe, C.I., Ripmeester, J.A., Tse, J.S., and Lawson, W.F., Laboratory analysis of a naturally-occurring gas hydrate from sediment of the Gulf of Mexico, *Geochimica et Cosmochimica Acta*, 50, 619–623 (1986).
35. Yakushev, V.S. and Istomin, V.A., Gas-hydrates self preservation effect, in *Physics and Chemistry of Ice*, Hokkaido University Press, Sapporo, Japan, pp. 136–139 (1992).
36. Gudmundsson, J.S., Method for production of gas hydrates for transportation and storage, U.S. Patent 5,536,893 (July 16, 1996).
37. Gudmundsson, J.S., Natural gas hydrate: Problem solver and resource for production and transport, in *Proceedings of Gas Hydrates (Tekna) Conference*, Bergen, Norway (October 21–22, 2008).
38. Fitzgerald, A. and Taylor, M., Offshore gas-to-solids technology, in *Proceedings of Offshore Europe Conference*, Aberdeen, U.K. (2001).
39. Iwasaki, T., Katoh, Y., Nagamori, S., and Takahashi, S., Continuous natural gas hydrate pellet production (NGHP) by process development unit (PDU), in *Proceedings of the Fifth International Conference on Gas Hydrates*, Trondheim, Norway (June 13–16, 2005).
40. Kanda, H., Economic study on natural gas transportation with natural gas hydrate (NGH) pellets, in *Proceedings of the 23rd World Gas Conference*, Amsterdam, the Netherlands (2006).
41. Takahashi, M., Moriya, H., Katoh, Y., and Iwasaki, T., Development of natural gas hydrate (NGH) pellet production system by bench scale unit for transportation and storage of NGH pellet, in *Proceedings of Sixth International Conference on Gas Hydrates*, Vancouver, British Columbia, Canada (July 6–10, 2008).
42. Nakai, S., Development of natural gas hydrate supply chain, in *Proceedings of Gastech 2011 Conference & Exhibition*, Amsterdam, the Netherlands (March 21–24, 2011).
43. Stern, L.A., Circone, S., Kirby, S.H., and Durham, W.B., Anomalous preservation of pure methane hydrate at 1 atm, *Journal of Physical Chemistry B*, 105, 1756–1762 (2001).

44. Takeya, S., Shimada, W., Kamata, Y., Ebinuma, T., Uchida, T., Nagao, J., and Narita, H., In situ X-ray diffraction measurements of the self-preservation effect of CH_4 hydrate, *Journal of Physical Chemistry A*, 105, 9756–9759 (2001).

45. Takeya, S., Ebinuma, T., Uchida, T., Nagao, J., and Narita, H., Self-preservation effect and dissociation rates of CH_4 hydrate, *Journal of Crystal Growth*, 237–239, 379–382 (2002).

46. Circone, S., Stern, L.A., Kirby, S.H., Durham, W.B., Chakoumakos, B.C., Rawn, C.J., Rondinone, A.J., and Ishii, Y., CO_2 hydrate: Synthesis, composition, structure, dissociation behavior, and a comparison to structure I CH_4 hydrate, *Journal of Physical Chemistry B*, 107, 5529–5539 (2003).

47. Falenty, A. and Kuhs, W.F., Self-preservation of CO_2 gas hydrates-surface microstructure and ice perfection, *Journal of Physical Chemistry B*, 113, 15975–15988 (2009).

48. Falenty, A., Kuhs, W.F., Glockzin, M., and Rehder, G. Self-preservation of CH_4 clathrates in the context of offshore gas transport, in *Physics and Chemistry of Ice*, Hokkaido University Press, Sapporo, Japan, pp. 189–196 (2011).

49. Gudmundsson, J.S., Parlaktuna, M., and Khokhar, A.A., Storing natural gas as frozen hydrate, *SPE Production & Facilities*, 9, 69–73 (1994).

50. Gudmundsson, J.S. and Borrehaug, A., Frozen hydrate for transportation of natural gas, in *Proceedings of Second International Conference on Natural Gas Hydrate*, Toulouse, France, pp. 415–422 (June 2–6, 1996).

51. Shirota, H., Hikida, K., Nakajima, Y., Ota, S., and Iwasaki, T., Self-preservation property of hydrate pellets in bulk in ship cargo holds during sea-borne transport of natural gas, in *Proceedings of Fifth International Conference on Gas Hydrates*, Trondheim, Norway (June 13–16, 2005).

52. Kuhs, W.F., Genov, G., Staykova, D.K., and Hansen, T., Ice perfection and onset of anomalous preservation of gas hydrates, *Physical Chemistry Chemical Physics*, 6, 4917–4920 (2004).

53. Circone, S., Stern, L.A., and Kirby, S.H., The effect of elevated methane pressure on methane hydrate dissociation, *American Mineralogist*, 89, 1192–1201 (2004).

54. Takeya, S. and Ripmeester, J.A., Anomalous preservation of CH_4 hydrate and its dependence on the morphology of hexagonal Ice, *ChemPhysChem*, 11, 70–73 (2010).

55. Takeya, S., Uchida, T., Nagao, J., Ohmura, R., Shimada, W., Kamata, Y., Ebinuma, T., and Narita, H., Particle size effect of CH_4 hydrate for self-preservation, *Chemical Engineering Science*, 60, 1383–1397 (2005).

56. Stern, L.A., Circone, S., Kirby, S.H., and Durham, W.B., Temperature, pressure, and compositional effects on anomalous or self preservation of gas hydrates, *Canadian Journal of Physics*, 81, 271–283 (2003).

57. Takeya, S. and Ripmeester, J.A., Dissociation behavior of clathrate hydrates to ice and dependence on guest molecules, *Angewandte Chemie International Edition*, 54, 1276–1279 (2008).

58. Uchida, T., Shiga, T., Nagayama, M., Gohara, K., and Sakurai, T., Microscopic observations of sintering processes on gas hydrate particles, in *Proceedings of Seventh International Conference on Gas Hydrates*, Edinburgh, U.K. (July 17–21, 2011).

59. Luo, Y.-T., Zhu, J.-H., Fan, S.-S., and Chen, G.-J., Study on kinetics of hydrate formation in a bubble column, *Chemical Engineering Science*, 62, 1000–1009 (2007).

60. Tajima, H., Yamasaki, A., and Kiyono, F., Continuous gas hydrate formation process by static mixing of fluids, in *Proceedings of Fifth International Conference on Gas Hydrates*, Trondheim, Norway (June 13–16, 2005).

61. Mori, Y., Recent advances in hydrate-based technologies for natural gas storage—A review, *Journal of Chemical Industries and Engineering*, 54, 1–17 (2003).

62. Windmeier, C., Experimentelle und theoretische Untersuchungen zum Phasen-und Zersetzungsverhalten von Gashydraten, Shaker Verlag, Aachen, Germany (2009).

63. Andritz Environment and Process, Separation Krauss-Maffei vacuum and pressure filters product information, Andritz KMPT GmbH, Vierkirchen, Germany (2011); also Andritz Alg, Grag, Austria. Available online http://www.kmpt.com/uploads/media/KMFILT-DE-10_11.qxp_02.pdf, Accessed on July 1, 2011.

64. Takaoki, T., Hirai, K., Kamei, M., and Kanda, H., Study of natural gas hydrate (NGH) carriers, in *Proceedings of Fifth International Conference on Gas Hydrates*, Trondheim, Norway, pp. 1258–1265 (June 13–16, 2005).

65. IMO, International code for the construction and equipment of ships carrying liquefied gases in bulk, International Convention for the Safety of Life at Sea (SOLAS), Chapter VII/Part C, International Maritime Organization, London, U.K. (2006).

66. IMO, Sub-committee on bulk liquids and gases, Draft interim guidelines for the construction and equipment of ships carrying natural gas hydrate pellets (NGHP) in bulk, BLG Report, International Maritime Organisation, London, U.K. (2010).

67. Nakajima, Y., Takaoki, T., Ohgaki, K., and Ota, S., Use of hydrate pellets for transportation of natural gas-II-proposition of natural gas transportation in form of hydrate pellets, in *Proceedings of Fourth International Conference on Gas Hydrates*, Yokohama, Japan, pp. 987–990 (May 19–23, 2002).

68. Ota, S., Uetani, H., and Kawano, H., Use of methane hydrate pellets for transportation of natural gas-III-safety measures and conceptual design of natural gas hydrate pellet carrier, in *Proceedings of Fourth International Conference on Gas Hydrates*, Yokohama, Japan (May 19–23, 2002).

69. IMO, MSC 83/INF.2: Formal Safety Assessment—Consolidated text of the guidelines for formal safety assessment (FSA) for use in the imo rule-making process (MSC/Circ.1023–MEPC/Circ.392), International Maritime Organisation, London, U.K. (2007).

70. Braband, J., Improving the risk priority number concept, *Journal of System Safety*, 3, 21–23, 32 (2003).

71. IMO, MSC 83/INF.3: FSA—Liquefied Natural Gas (LNG) carriers details of the formal safety assessment, International Maritime Organisation, London, U.K. (2007).

72. Coselle (TM) System, A report by Sea NG Corporation, Calgary, Alberta, Canada (2011). Available online: http://www.coselle.com, Accessed on July 1, 2011.

73. Cho, J.H., Kotzot, H., de la Vega, F., and Durr, C., Large LNG carrier poses economic advantages, technical challenges, *Oil Gas Journal*, 2, 4 (2005).

74. Sethuraman, D., LNG charter rates fall 17% on increase in vessels, Blomberg report (2011). Available online: http://www.lngpedia.com/2009/03/20/lng-tanker-charter-rates-fall-17, Accessed on July 1, 2011.

75. The charter market for container ships, Available online: http://www.hanseatic-lloyd.de/ investments/schiffahrtCharter.html, Accessed on July 1 2011.

76. Gudmundsson, J.S., Mork, M., and Graff, O.F., Hydrate non-pipeline technology, in *Proceedings of Fourth International Conference on Gas Hydrates*, Yokohama, Japan (May 19–23, 2002).

77. Osokogwu, U., Ademujimi, M., and Ajienka, J.A., Economic analysis of GTP, GTL, CNG, NGH for offshore gas development in Nigeria, in *Proceedings of Nigeria Annual International Conference and Exhibition*, Abuja, Nigeria (July 30–August 3, 2011).

78. Mumtaz Alvi, Pakistan has highest number of CNG vehicles: Survey, A report in The New International (June 3, 2011).

79. CNG, A report by Naturalgas.org. (2015).

80. Bulgaria: MAN wins tender to deliver 126 CNG buses to Sofia transit agency, *NGV Journal*, www.backup.ngvjournal.com, Retrieved June 30, 2015.

81. City of Harrisburg switches city trucks, police car, to compressed natural gas, CNGnow.com, Retrieved May 15, 2015.

82. Compressed natural gas, Wikipedia, the free encyclopedia (2015).

83. Stenning, D., The Coselle CNG carrier: A new way to ship natural gas by sea, in Paper presented at the *1999 NOIA Conference*, Newfoundland, Canada (June 14–17, 1999).

84. EnerSea transport plans to commercialize marine delivery of CNG by Mid-2004, A report by EnerSea Transport, Houston, TX (October 2001). info.enersea.com. Remote Gas Strategies, 1 (October 2001).

85. Klimowski, S.R., Oceanic transport of natural gas as compressed natural gas (floating pipeline), in Paper presented at the *2001 Monetizing Stranded Gas Reserves Conference*, Denver, CO (October 10–12, 2001).

86. Wagner, J.V., Marine transport of compressed natural gas—A potential export alternative for fuel gas, in *2002 AIChE Spring Natoinal Meeting*, New Orleans, LA (March 10–14, 2002).

87. Compressed natural gas, Petrowiki, SPE International (June 2, 2015).

88. Liquefied Natural Gas (LNG) in the U.S., A report by Department of Transportation, Pipeline and Hazardous Materials Safety Administration, Washington, D.C. (2015). https://primis.phmsa.dot.gov/comm/lng.htm, Accessed January 28, 2015.

89. Distribution, transmission & gathering, LNG, and liquid annual data, A report by Department of Transportation, Pipeline and Hazardous Materials Safety Administration, Washington, D.C. (2014). http://www.phmsa.dot.gov/portal/site/PHMSA/menuitem.6f23687cf7b00b0f22e4c6962d9c8789/?vgne xtoid=a872dfa122a1d110VgnVCM1000009ed07898RCRD&vgnextchannel=3430fb649a2dc110VgnVC M1000009ed07 898RCRD&vgnextfmt=print, Accessed September 8, 2014.

90. Fullenbaum, R., Fallon, J., and Flanagan, B., Oil & natural gas transportation & storage infrastructure: Status, trends, & economic benefits, Report for: American Petroleum Institute, Submitted by: IHS Global Inc., Washington, DC (December 2013).

91. Liquefied natural gas, Wikipedia, the free encyclopedia (2015).
92. Liquefied Petroleum Gas (LPG), Liquefied Natural Gas (LNG) and Compressed Natural Gas (CNG), A report by Envocare Ltd., Nagpur, India (March 21, 2007), Retrieved September 3, 2008.
93. Liquefied Petroleum gas, Wikipedia, the free encyclopedia (2015), https://en.wikipedia.org/wiki/Liquefied_petroleum_gas.
94. Report on the investigation of the fire at the liquefaction storage, and regasification plant of the east Ohio Gas Co., Cleveland, OH, October 20, 1944, Retrieved April 17, 2015.
95. Carroll, D., *2016 World Live Report*, IGU, Statoil, Norway (2016).
96. Third Gulf Coast LNG export terminal wins conditional Nod from DOE, Department of Energy, Washington, D.C., Retrieved April 17, 2015.
97. Smith, S., China natural gas fueling station station equipment industry report 2015–2018, *PR Newswire* (September 16, 2015).
98. O'Reilly, C., Japan to introduce LNG-fueled transport, A report by LNG Industry, Palladian Publications, Farnham, OK (2015). Retrieved July 17, 2015.
99. Evaluation of LNG Technologies, A report by University of Oklahoma, Norman, OK (2008).
100. Prescott, C., Zhang, J., and Brower, D., New design uses ambient pressure insulated ING pipeline offshore, A paper presented at Offshore Technology Conference, Houston, TX (May 2–5, 2005).
101. Neandross et al., LNG opportunities for marine and rail in the great lakes, Gulf of Mexico and inland waterways, A report prepared for ANGEA by GNA (Gladstein, Neandross & Associates), (October 2014).
102. Parfomak, P. and Flynn, A., Liquefied Natural Gas (LNG) import terminals: Siting, safety, and regulation (CRS report for Congress, CRS, The Library of Congress, Washington, D.C. (January 28, 2004).), Retrieved February 25, 2013.
103. Lisowski, E. and Czyzycki, W., Transport and storage of LNG in container tanks, *Journal of KONES Powertrain and Transport*, 18 (3), 193–201 (2011).
104. Allen, M.S., Baumgartner, R.G., Fesmire, J.E., and Augustynowicz, S.D., Advances in microsphere insulation systems, in *Advances in Cryogenic Engineering, Proceedings of the Cryogenic Engineering Conference*, Vol. 49, Melville, NY, pp. 619–626 (2004).
105. Arkharow, A., Cryogenic systems: Textbook for university students majoring in engineering and low temperature physics (in Russian) in 2 vols. Vol. 1. Fundamental of theory and analysis, 3rd edn., revision and addition, Mashinostroenie, Moscow, Russia (2006).
106. Cryogel Z, *Flexible Industrial Insulation with Vapor Barrier for Sub-Ambient and Cryogenic Applications*. Aspen Aerogels, Inc., Northborough, MA (2008).
107. Górski, Z., Cwilewicz, R., Konopacki, Ł., and Kruk, K., Proposal of propulsion for liquefied natural gas tanker (LNG carrier) supplying LNG terminal in Poland, *Journal of KONES*, 15 (2), 103–108 (2008), Institute of Aviation, Warsaw.
108. Lisowski, E., Czyzycki, W., and Filo, G., Computer aided design in mechanical engineering, Transport of liquid natural gas by mobile tank container, A report from Bergen University College, Bergen, Norway (2009).
109. Lisowski, E., Czyzycki, W., and Cazarczyk, K., Using of polyamide in construction of supporting blocks of cryogenic tanks on example of LNG container, *Archives of Foundry Engineering*, 10 (3), 81–86 (2010).
110. Lisowski, E., Czyzycki, W., and Cazarczyk, K., Simulation and experimental research of internal supports in mobile cryogenic tanks, *Technical Transactions*, 2-M/2010, Wydawnictwo, PK (2010).
111. Faruque Hasan, M.M., Zheng, A.M., and Karimi, I.A., Minimizing boil-off losses in liquefied natural gas transportation, *Industrial & Engineering Chemistry Research*, 48 (21), 9571–9580 (2009).
112. Automotive LPG, the practical alternative motor fuel, A report by AEGPL, Brussels, Belgium (2003).
113. Autogas in Europe, The sustainable alternative—An LPG industry roadmap, A report by AEGPL, Brussels, Belgium (2009), Available at: http://www.aegpl.eu/media/16300/autogas%20roadmap.pdf.
114. Hanschke, C.B., Uyterlinde, M.A., Kroon, P., Jeeninga, H., and Londo, H.M., Duurzame innovatie in het wegverkeer. Een evaluatie van vier transitiepaden voor het thema Duurzame mobiliteit, ECN-E-08-076, Energy Research Centre of the Netherlands, Petten, the Netherlands (2009).
115. Roeterdink, W.G., Uyterlinde, M.A., Kroon, P., and Hanschke, C.B., Groen Tanken: Impassing van alternatieve brandstoffen in de tank- en distributie infrastructuur, ECN-E-09-082, Energy Research Centre of the Netherlands, Petten, the Netherlands (2010).
116. What is a propane vehicle? US Department of Energy, Alternative Fuels and Advanced Vehicles Data Center, DOE, Washington, D.C. (2010), Available at http://www.afdc.energy.gov/afdc/vehicles/propane_what_is.html.

117. World LP Gas Association (WLPGA), Health effects and costs of vehicle emissions, Neuilly-sur-Seine, France (2005), Available at http://www.worldlpgas.com/uploads/Modules/Publications/healtheffects_medres.pdf.
118. World LP Gas Association (WLPGA), Autogas Incentive Policies: A country-by country analysis of why and how governments promote, Autogas & what works, Neuilly-sur-Seine, France (2011), Available at http://www.worldlpgas.com/uploads/Modules/Publications/autogas-incentive-policies-2012-updated-july-2012.pdf.
119. Liquefied petroleum gas in transport, A report from Climate TechWiki (2006), www.climatetechwiki.org/technology/lpg; LPG Storage Vessels, Hoghland Tank, Lebanon, PA (2015), www.highlandtank.com.
120. Smith, M. and Gonzales, J., Costs associated with propane vehicle fueling infrastructure-factors to consider in the implementation of fueling stations and equipment, A report by U.S. Department of Energy, Energy Efficiency and Renewable Energy, Washington, D.C. (August 2014).
121. Martinsen, W.E., Johnson, D.W. and Welker, J.R., Fire safety of LPG in marine transportation, Work Performed Undw Contract Ns. AG05-78EV060M, Applied Technology Corp., Norman, Oklahoma, OK (June 1980).
122. Apt, J., Newcomer, A., Lave, L.B., Douglas, S., and Dunn, L.M., An engineering-economic analysis of syngas storage, Final report, Contract DE-AC26-04NT 41817.404.01.02, DOE/NETL-2008/1331, Department of Energy, Washington, DC (July 31, 2008).
123. Alp, T., Dames, T.J., and Dogan, B., The effect of microstructure in the hydrogen embrittlement of a gas pipeline steel, *Journal of Materials Science*, 22 (6), 2105–2112 (1987).
124. Amos, W., Costs of transporting and storing hydrogen, NREL/TP-570-25106, National Renewable Energy Laboratory, Golden, CO (1998).
125. Astaf'ev, A.A., Hydrogen embrittlement of constructional steels, *Metal Science and Heat Treatment*, 26 (2), 91–96 (1984).
126. Beghi, G. and Dejace, J., Massaro, C., and Ciborra, B., Economics of pipeline transport for hydrogen and oxygen, Springer, Berling, Germany (1974).
127. Carpetis, C., Estimation of storage costs for large hydrogen storage facilities, *International Journal of Hydrogen Energy*, 7 (2), 191–203 (1982).
128. Chernov, V.Y., Makarenko, V.D., Kryzhanivs'kyi, E.I., and Shlapak, L.S., On the causes of corrosion fracture of industrial pipelines, *Materials Science*, 38 (6), 880–883 (2002).
129. Duret, A. et al., Process design of Synthetic Natural Gas (SNG) production using wood gasification, *Journal of Cleaner Production*, 13 (15), 1434–1446 (2005).
130. Engelhard Corporation, Methanation catalyst, Carbon oxide conversion (2005), http://www.engelhard.com/documents/MET%20datasheet%20Web%20FINAL.pdf, Retrieved March 7, 2006.
131. Moeller, F.W. et al., Methanation of coal gas for SNG, *Hydrocarbon Processing*, 53 (4), 69–74 (1974).
132. Mohitpour, M. et al., *Pipeline Design and Construction: A Practical Approach*. ASME Press, New York (2000).
133. Morrow, J.M., Corrao, M., and Hylkema, S., U.S. International patent application no. PCT/US2005/009353, Submitted by PraXAIR Technology Inc., Hydrogen storage and supply method (October 27, 2005).
134. Mozaffarian, M. and Zwart, R.W.R., *Feasibility of Biomass/Waste-Related SNG Production Technologies*. Energy Research Centre of the Netherlands (ECN), Petten, the Netherlands (2003).
135. M. Mozaffarian, Zwart, R.W.R., Boerrigter, H., and Deurwaarder, E.P., "Green gas" as SNG (synthetic natural gas) a renewable fuel with conventional quality, in *Contribution to the "Science in Thermal and Chemical Biomass Conversion" Conference*, August 30–September 2, 2004, Victoria, Vancouver Island, British Columbia, Canada, ECN-RX-04-085 (August 2004); Ogden, J.M., Prospects for building a hydrogen energy infrastructure, *Annual Review of Energy and the Environment*, 24 (1), 227–279 (1999).
136. Rogante, M., Battistella, P., and Cesari, Hydrogen interaction and stress-corrosion in hydrocarbon storage vessel and pipeline weldings, *International Journal of Hydrogen Energy*, 31 (5), 597–601 (2006).
137. Seglin, L. et al., Survey of methanation chemistry and processes, in Seglin, L., (ed.), *Methanation of Synthesis Gas*, pp. 1–30, Advances in Chemistry Series 146, American Chemical Society, Washington, D.C. (1975).
138. Carbon monoxide and syngas pipeline systems, IGC Doc 120/04/E, Globally harmonised document, European Industrial Gases Association, Brussels, Belgium (2004).
139. Taylor, J.B., Alderson, J.E.A., Kalyanam, K.M., Lyle, A.B., and Phillips, L.A., Technical and economic assessment of methods for the storage of large quantities of hydrogen, *International Journal of Hydrogen Energy*, 11 (1), 5–22 (1986).

140. Barthelemy, H., Interaction of steels with hydrogen in petroleum industry pressure vessel service, in *International Conference*, Paris, France (March 28–30, 1989).

141. Loginow, A.W. and Phelps, E.H., Steel for seamless hydrogen pressure vessels, *Corrosion*, 31, 404 (1975).

142. Walter, R.J., Jewett, R.P., and Chandler, W.T., On the mechanism of hydrogen embrittlement of iron-and-nickel-base alloys, *Materials Science and Engineering*, 5 (2), 99–110 (January 1970).

143. Vennett, R.M. and Ansell, G.S., A study of gaseous hydrogen damage in certain FCC metals, *ASM Transactions Quarterly*, 62 (4), 1007–1013 (December 1969).

144. Cavett, R.H. and VanNess, H.C., Embrittlement of steel by high-pressure hydrogen gas, *Welding Journal*, Welding Research Supplement, 42, 316s–319s (July 1963).

145. Steinman, J.B., VanNess, H.C., and Ansell, G.S., The effect of high-pressure hydrogen upon the notch tensile strength and fracture model of 4140 steel, *Welding Journal*, Welding Research Supplement, 44, 221s–224s (May 1965).

146. Wachab and Nelson, H.B., *Hydrogen Effects in Metals*. AIME, Warrendate, PA (1981).

147. Walter, R.J. and Chandler, W.T., in Thompson, A.W. and Bernstein, I.M. (eds.), *Effects of Hydrogen on Behavior of Materials*. AIME, New York (1976).

148. Holbrook, J.H. and Cicione, H.J., The effect of hydrogen on low-cycle-fatigue life and subcritical crack growth in pipeline steels, Battelle Columbus Laboratories prepared for the Energy Applications and Analysis Division Department of Applied Science Under Subcontract No. 550772-S (1982).

149. Chandler, W.T. and Walter, R.J., Testing to determine the effect of high pressure hydrogen environments on the mechanical properties of metals, Hydrogen Embrittlement Testing, ASTM STP 543, 170–197 (ASTM, 1974).

150. Aiken, R., Ditzel, K.H., Morra, F., and Wilson, D.S., Coal-based integrated gasification combined cycle: Market penetration strategies and recommendations, Study prepared for DOE, NETL, GTC (Gasification Technology Council) by Booz Allen Hamilton Co., under contract no. DE-AM26-99FT40575 (September 2004).

151. Amick, P., Geosits, R., Herbanek, R., Kramer, S., and Tam, S., A large coal IGCC power plant, in *Nineteenth Annual International Pittsburgh Coal Conference*, Pittsburgh, PA (September 23–27, 2002).

152. Bennet, N., MB Engineering Services, Clayton Walker Gasholder Division (2006); Blankinship, S., Gasifier reliability and flexibility key to IGCC's future in the power sector, *Power Engineering*, 110 (7), 4–7 (2006).

153. Bloom, J.A., Generation cost curves including energy storage, in *IEEE Transactions on Power Apparatus and Systems PAS-103(7)*, pp. 1725–1731 (July 1984).

154. Booras, G. and Holt, N., Pulverized Coal and IGCC plant cost and performance estimates, in *Presentation, Gasification Technologies Conference*, Washington, DC (October 3–6, 2004), http://www.gasification.org/Presentations/2004.htm.

155. Chiang, A.C., *Elements of Dynamic Optimization*. Waveland Press, Long Grove, IL (1992); Collot, A.-G., Clean fuels from coal, IEA Clean Coal Centre, London, U.K. (2004); Colwell, R., Professor Marine Geology, Oregon State University, Corvallis, OR (2006).

156. ContiTech, Diaphragms for gas holders (2006), http://www.contitech.de/ct/contitech/themen/produkte/membranen/gasspeichermembrane n/gasspeicher_e.html, Retrieved August 21, 2006.

157. Deurwaarder, E.P. et al., Methanation of Milena product gas for the production of bio-SNG, in *14th European Biomass Conference & Exhibition*, Paris, France (2005); DOE Coal Gasification Research Needs (COGARN) Working Group, Coal gasification: Direct applications and syntheses of chemicals and fuels, U.S. Department of Energy, Office of Energy Research, Office of Program Analysis, Washington, D.C. (1987).

158. EIA, Annual energy outlook 2007, with projections to 2030, Energy Information Administration, DOE, Washington, D.C. (February 2007), http://www.eia.doe.gov/oiaf/aeo/pdf/0383(2007).pdf.

159. Energy Information Administration, Short-term energy outlook (December 12, 2006), http://www.eia.doe.gov/emeu/steo/pub/contents.html, Retrieved December 14, 2006.

160. Gray, D., Salerno, S., and Tomlinson, Potential application of coal-derived fuel gases for the glass industry: A scoping analysis, National Energy Technology Laboratory, DOE (2004).

161. Hagen, M., Polman, E., Myken, A., Jensen, J., Jönsson, O., Biomil, A.B., and Dahl, A., Adding gas from biomass to the gas grid, Contract no: XVII/4.1030/Z/99-412, Final report time period, July 1999–February 2001 (2001).

162. Collot, A.-G., Clean fuels from coal, IEA Clean Coal Centre, CCC/81, 52pp., PF 04-03, IEA, Brussels, Belgium (March 2004).

163. Colwell, R., Professor Marine Geology, Oregon State University, Personal communication (2006).

164. Farina, G. and Bressan, L., Optimizing IGCC design: Improve performance, reduce capital cost, *Foster Wheeler Review*, 1 (3), 16–21 (1999).
165. Gray, D., Salerno, S., and Tomlinson, G., Polygeneration of SNG, hydrogen, power, and carbon dioxide from Texas Lignite, National Energy Technology Laboratory, DOE, Washington, D.C. (2004).
166. Griffin, M., Executive Director Green Design Institute, Tepper School of Business, CMU, formerly BP, Personal communications (2006).
167. Hart, D., *Hydrogen Power: The Commercial Future of 'the Ultimate Fuel'*. Financial Times Energy Publishing, London, U.K. (1997).
168. Heard, R., Carnegie Mellon University, Personal communications (2006).
169. IEA GHG, Transmission of CO_2 and energy, Cheltenham, International Energy Agency Greenhouse Gas R&D Programme, Cheltenham, U.K. (2002).
170. Mathanatian for hydrogen, KATALCO 11-4 methanation catalyst, a bulletin by Johnson Matthey Process Technologies (2003), Available online at http://www.jmcatalysts.com/pct/pdfs-uploaded/Refineries%20-%20Oil%20and%20Petrochemical/11-4.pdf.
171. Korpås, M., Distributed energy systems with wind power and energy storage, Department of Electrical Power Engineering, Norwegian University of Science and Technology, Trondheim, Norway (2004).
172. Leighty, W., Hirata, M., O'Hashi, K., Asahi, H., Benoit, J., and Keith, G., Large renewables—Hydrogen energy systems: Gathering and transmission pipelines for windpower and other diffuse, dispersed sources, in *World Gas Conference 2003*, Tokyo, Japan (2003).
173. Mozaffarian, M. and Zwart, R.W.R., Production of substitute natural gas by biomass hydrogasification, in *Developments in Thermochemical Biomass Conversion*, Energy Research Centre of the Netherlands (ECN), Tyrol, Austria (2000).
174. Padró, C. and Putsche, V., Survey of the economics of hydrogen technologies, National Renewable Energy Laboratory, Golden, CO (1999).
175. Ridge Energy Storage & Grid Services L.P. and Texas State Energy Conservation Office, The economic impact of CAES on wind in TX, OK, and NM (2005).
176. Rode, D.C. and Fischbeck, P.S., The value of using coal gasification as a long-term natural gas hedge for ratepayers, Carnegie Mellon Electricity Industry Center Working Paper CEIC-06-12, Carnegie Mellon University, Pittsburgh (December 13, 2006), www.cmu.edu/electricity.
177. Herder, P.M., Stikkelman, R.M., Dijkema, G.P.J., and Correljé, A.F., Design of a syngas infrastructure, in Braunschweig, B. and Joulia, X. (eds.), *18th European Symposium on Computer Aided Process Engineering-ESCAPE 18*, Elsevier, New York (2008).
178. Stolz, J., Microbial ecologist, Department of Biological Sciences, Duquesne University, Pittsburgh, PA (2006).
179. Stoft, S., *Power System Economics: Designing Markets for Electricity*. IEEE Press/Wiley-Interscience, New York (2002).
180. Todd, D.M. and Battista, R.A., Demonstrated applicability of hydrogen fuel for gas turbines, *Proceedings of the IchemE Gasification 4 Conference*, Noordwijk, the Netherlands (2000).
181. Twigg, M.V., *Catalyst Handbook*. Wolfe, London, U.K. (1989).
182. U.S. Environmental Protection Agency, Clean air markets—Data and maps, Vintage Year NOx Allowance Price, EPA, Washington, D.C. (August 3, 2007), http://camddataandmaps.epa.gov/.
183. Valenzuela, J. and Mazumdar, M., Cournot prices considering generator availability and demand uncertainty, *IEEE Transactions on Power Systems*, 22 (1), 116–125 (2007).
184. Valenzuela, J. and Mazumdar, M., A probability model for the electricity price duration curve under an oligopoly market, *IEEE Transactions on Power Systems*, 20 (3), 1250–1256 (2005).
185. van der Drift, A. et al., MILENA gasification technology for high efficient SNG production from biomass, in *14th European Biomass Conference & Exhibition*, Paris, France (2005).
186. Wabash River Energy Ltd., Wabash river coal gasification repowering project: Final technical report, Department of Energy Technical Report, DOE, Washington, D.C. (2002).
187. Wolters, C., Operating experience at the Willem_Alexander Centrale, Nuon Power, Buggenum, Harlem, the Netherlands, NETL/DOE report (October 14, 2003), http://www.gasification.org/Docs/2003_Papers/19WOLT.pdf.
188. Wong, R. and Whittingham, E., *A Comparison of Combustion Technologies for Electricity Generation*, 2nd edn. Pembina Institute, Alberta, Canada (December 2006).
189. Walker, L.K. et al., An IGCC project at Chinchilla, Australia based on Underground Coal Gasification (UCG), in *2001 Gasification Technologies Conference*, San Francisco, CA (2001).
190. Walker, M., e3 Ventures, Personal communications (2006).
191. Robinson, S.L. and Stoltz, R.E., Toughness losses and fracture behavior of low strength carbon—Manganese steel in hydrogen, Sandia National Laboratories, Livermore, CA (July 1976).

192. Pumphrey, P.H., The effect of sulfide inclusions on the diffusion of hydrogen in steels, Internal Paper-Central Electricity Generating Board, Surrey, England (1978).

193. Mohitpour, M., Solanky, H., and Vinjamuri, G., Design basis developed for hydrogen pipelines, *Oil and Gas Journal*, 83–94 (May 28, 1990).

194. Mucek, M.W., Amoco's line pipe specs exceed industry standards, *Oil and Gas Journal*, 45–47 (June 11, 1990).

195. Holbrook, J.H., Cialone, H.J., and Scott, P.M., Hydrogen degradation of pipeline steels summary report, DOE contract no. DE-ACO2-76CH00016, Battelle Columbus Laboratories, Columbus, OH (July 1984).

196. Thompson, A.W. and Bernstein, I.M., Selection of structural materials for hydrogen pipelines and storage vessels, *International Journal of Hydrogen Energy*, 2, 163–173 (1977).

197. Brynestad, J., Iron and nickel carbonyl formation in steel pipes and its prevention—Literature survey, ORNL/TM 5499, Oak Ridge National Laboratory, Oak Ridge, TN (September 1976).

198. Leis, B.N. and Colwell, I.A., Initiation of stress-corrosion cracking in gas transmission piping, in VanDerSluys, W.A., Piascek, R.S., and Zawierucha, R. (eds.), *Effects of the Environment in the Initiation of Crack Growth*, ASTM 1298. American Society for Testing and Materials (1997).

199. Wilson, A.D., The influence of inclusions on the toughness and fatigue properties of A516-70 steel, *Journal of Engineering Materials and Technology*, 101, 265–274 (July 1979).

200. Makhnenko, V.I., The effects of residual stresses on the propagation of fatigue cracks in the component parts of welded structures, Welding Research Abroad (January 1981).

201. Zwart, R.W.R. and Boerrigter, H., High efficiency co-production of substitute natural gas (SNG) and Fischer-Tropsch (FT) transportation fuels from biomass, in Presented at the *Second World Conference and Technology Exhibition on Biomass for Energy*, May 10–14, 2004, Industry and Climate Protection, Rome, Italy, ECN-RX-04-042, ECN, Petten, the Netherlands (May 2004).

202. Mozaffarian, M., Zwart, R., Boerrigter, H., and Deurwaarder, E., Feasibility of SNG production by biomass hydrogasification, ECN-CX-01-115, ECN, Petten, the Netherlands (June 2002).

203. Biogas, Wikipedia, the free encyclopedia (2015).

204. Petersson, A. and Wellinger A., Biogas upgrading technologies—Developments and innovations, IEA Bioenergy, Task 37, IEA, Brussels, Belgium (2009).

205. Hjort, A. and Tamm, D., Transport Alternatives For Biogas in the region of Skåne, BioMil AB biogas, miljö och kretslopp, Intelligent Energy Europe, Brussels, Belgium, pp. 1–35 (November 3, 2012).

206. Persson, T. and Svensson, M., Non-grid biomethane transportation in Sweden and the development of the liquefied biogas market, Biogas in Society, A Case Story from IEA Bioenergy Task 37, Energy from Biogas, IEA, Brussels, Belgium (September 2014).

207. Gaikwad, V.R. and Katti, P.K., Design of biogas scrubbing, compression & storage system, *IOSR Journal of Electrical and Electronics Engineering*, 58–63 (2014).

208. Guide to cooperative biogas to biomethane developments, produced by Institute of Chemical Engineering, Vienna University of Technology (Austria), A project under the Intelligent Energy—Europe Programme, Brussels, Belgium, Contract Number: IEE/10/130, Deliverable Reference: Task 3.1.2, Delivery Date: (December 2012).

209. Krich, K., Augenstein, D., Batmale, J.P., Benemann, J., Rutledge, B., and Salour, D., Biomethane from dairy waste a sourcebook for the production and use of renewable natural gas in California, Prepared for Western United Dairymen Michael Marsh, Chief Executive Officer, USDA Rural Development (July 2005).

210. Small-scale Methodology AMS-III.AQ: Introduction of Bio-CNG in transportation applications, A report Version 02.0, Clean Development Mechanism CDM-EB79-A17, United Nations, Framework convention on climate change (June 1, 2014).

211. Pütz, K., Asfaw, A., Leta, B., and Müller, J., Biogas as business—Biogas transport technology and economic concept for developing countries, in *Conference on International Research on Food Security, Natural Resource Management and Rural Development Tropentag 2011*, University of Bonn, Bonn, Germany (October 5–7, 2011).

212. Bates, J., Biomethane for transport from landfill and anaerobic digestion final report for the department for transport, PPRO 04/91/63, RICARDO-AEA, Harwell, U.K. (March 6, 2015).

213. Ross, C.C. and Walsh, Jr. J.L., Impact of utility interaction on agricultural cogeneration, American Society of Agricultural Engineers Paper 86-3020, St. Joseph, MI (1986).

214. Ross, C.C., Drake, T.J., and Walsh, J.L., *Handbook of Biogas Utilization*, 2nd edn. U.S. Department of Energy, Southeastern Regional Biomass Energy Program, TVA, Muscle Shoals, AL (July 1996).

215. Hydrogen storage Wikipedia, the free encyclopedia (2015). https://en.wikipedia.gov/wiki/hydrogen_storage (last accessed September 6, 2016).

216. Hydrogen storage technologies roadmap, A report from Freedom Car & Fuel Partnership, 31 pp, EERE, DOE, Washington, D.C. (November 2005), https://www1.eere.energy.gov/vehiclesandfuels/pdfs/program/hydrogen_storage_roadmap.pdf.

217. Yang, J., Sudik, A., Wolverton, C., and Siegel, D.J., High capacity hydrogen storage materials: Attributes for automotive applications and techniques for materials discovery, *Chemical Society Reviews*, 39 (2), 656–675 (2010).

218. Hydrogen storage: Current technology, Energy.gov, Office of Energy Efficiency and Renewable Energy, DOE, Washington, D.C. (August 26, 2011), eere.energy.gov, Retrieved on January 8, 2012.

219. Ahluwalia, R.K., Hua, T.Q., Peng, J.K., and Kumar, R., System level analysis of hydrogen storage options, 2010 DOE Hydrogen Program Review, Washington, DC (June 8–11, 2010).

220. Eberle, U., Mueller, B., and von Helmolt, R., Fuel cell electric vehicles and hydrogen infrastructure: Status 2012, *Energy & Environmental Science*, 5, 8780–8798 (2012).

221. Satyapal, S. et al., The U.S. Department of Energy's National Hydrogen Storage Project: Progress towards meeting hydrogen-powered vehicle requirements, *Catalysis Today*, 120 (3), 246–256 (February 2007).

222. DOE metal hydrides, eere.energy.gov, Department of Energy, Washington, D.C. (December 19, 2008), Retrieved on January 8, 2012.

223. Christian, M. and Aguey-Zinsou, K.F., Core–shell strategy leading to high reversible hydrogen storage capacity for $NaBH_4$, *ACS Nano*, 6 (9), 7739–7751 (September 2012). Retrieved August 20, 2012.

224. Graetz, J., Reilly, J., Sandrock, G., Johnson, J., Zhou, W.M., and Wegrzyn, J., Aluminum hydride, Al_1H_3, As a hydrogen storage compound, Brook Haven National Laboratory report, BNL 77336-2006, Upton, New York (November 2006).

225. Kim, K.C., Kulkarni, A.D., Karl Johnson, J., and Sholl, D.S., Examining the robustness of first-principles calculations for metal hydride reaction thermodynamics by detection of metastable reaction pathways, *Physical Chemistry Chemical Physics*, 13, 21520 (2011).

226. Kim, K.C., Kulkarni, A.D., Karl Johnson, J., and Sholl, D.S., Large-scale screening of promising metal hydrides for hydrogen storage system from first-principles calculations based on equilibrium reaction thermodynamics, *Physical Chemistry Chemical Physics*, 13, 7218 (2011).

227. Kulkarni, A.D., Wang, L.-L., Johnson, D.D., Sholl, D.S., and Karl Johnson, J., First-principles characterization of amorphous phases of $MB_{12}H_{12}$, M = Mg, Ca, *Journal of Physical Chemistry C*, 114, 14601 (2010).

228. He, T., Pei, Q., and Chen, P., Liquid organic hydrogen carriers, *Journal of Energy Chemistry*, 24 (5), 587–594 (September 1, 2015).

229. Teichmann, D., Arlt, W., Wasserscheid, P., and Freymann, R., A future energy supply based on Liquid Organic Hydrogen Carriers (LOHC), *Energy & Environmental Science*, 4, 2767–2773 (2011).

230. Brückner, N., Obesser, K., Bösmann, A., Teichmann, D., Arlt, W., Dungs, J., and Wasserscheid, P., Evaluation of industrially applied heat-transfer fluids as liquid organic hydrogen carrier systems, *ChemSusChem* (2013).

231. Singh, A., Singh, S., and Kumar, A., Hydrogen energy future with formic acid: A renewable chemical hydrogen storage system, *Catalysis Science & Technology*, 6, 12–40 (2016).

232. Müller, B., Müller, K., Teichmann, D., and Arlt, W., Energy storage by CO_2 methanization and energy carrying compounds: A thermodynamic comparison, *Chemie Ingenieur Technik*, 2011, 11 (2002–2013) (written in German).

233. Kariya, N., Fukuoka, A., and Ichikawa, M., Efficient evolution of hydrogen from liquid cycloalkanes over Pt-containing catalysts supported on active carbons under wet–dry multiphase conditions, *Applied Catalysis A: General*, 233 (1–2), 91–102 (July 10, 2002).

234. Clot, E., Eisenstein, O., and Crabtree, R.H., Computational structure–activity relationships in H_2 storage: How placement of N atoms affects release temperatures in organic liquid storage materials, doi:10.1039/B705037B, pubs.rsc.org, Retrieved November 4, 2015.

235. Eblagon, K.M., Tam, K., and Tsang, S.C.E., Comparison of catalytic performance of supported ruthenium and rhodium for hydrogenation of 9-ethylcarbazole for hydrogen storage applications, doi:10.1039/C2EE22066K, pubs.rsc.org, Retrieved November 4, 2015.

236. Laurenczy, G., Fellay, C., and Dyson, P.J., Hydrogen production from formic acid, PCT Int. Appl., CODEN: PIXXD2 WO 2008047312 A1 20080424 AN 2008:502691 (2008).

237. Fellay, C., Dyson, P.J., and Laurenczy, G., A viable hydrogen-storage system based on selective formic acid decomposition with a ruthenium catalyst, *Angewandte Chemie International Edition in English*, 47 (21), 3966–3968 (2008).

238. Joó, F., Breakthroughs in hydrogen storage—Formic acid as a sustainable storage material for hydrogen, *ChemSusChem*, 1 (10), 805–808 (2008).

239. Stracke, M.P., Ebeling, G., Cataluña, R., and Dupont, J., Hydrogen-storage materials based on imidazolium ionic liquids, *Energy & Fuels*, 21 (3), 1695 (2007).
240. Welch, G.C., Juan, R.R.S., Masuda, J.D., and Stephan, D.W., Reversible, metal-free hydrogen activation, *Science*, 314 (5802), 1124–1126 (2006).
241. Graphene as suitable hydrogen storage substance, Physicsworld.com, Retrieved on January 8, 2012.
242. MOF-74—A potential hydrogen-storage compound, Nist.gov, Retrieved on January 8, 2012.
243. Researchers demonstrate 7.5 wt% hydrogen storage in MOFs, Green Car Congress (March 6, 2006), Retrieved on January 8, 2012.
244. New MOF material with hydrogen uptake of up to 10 wt%, A report in Career in Congress (February 22, 2009). http://www.typepad.com/services/trackback/6a00d8341c4/be53ef0111688fse22970c.
245. Other methods for the physical storage of hydrogen, Chapter 8, in *Compendium of Hydrogen Energy*, Vol. 2, Hydrogen Storage, Transportation and Infrastructure, A volume in Woodhead Publishing Series in Energy, Elsevier, Amsterdam, the Netherlands (2016).
246. Lasher, S., Analyses of hydrogen storage materials and on-board systems, in *DOE Annual Merit Review*, Department of Energy, Washington, D.C. (June 7–11, 2010).
247. Compact (L)H_2 storage with extended dormancy in cryogenic pressure vessels, Lawrence Livermore National Laboratory, Department of Energy, Livermore, CA (June 8, 2010).
248. Florusse, L.J., Peters, C.J., Schoonman, J., Hester, K.C., Koh, C.A., Dec, S.F., Marsh, K.N., and Sloan, E.D., Stable low-pressure hydrogen clusters stored in a binary clathrate hydrate, *Science*, 306 (5695), 469–471 (2004).
249. Zhevago, N.K. and Glebov, V.I., Hydrogen storage in capillary arrays, *Energy Conversion and Management* 48 (5), 1554 (2007).
250. Zhevago, N.K., Denisov, E.I., and Glebov, V.I., Experimental investigation of hydrogen storage in capillary arrays, *International Journal of Hydrogen Energy*, 35, 169 (2010).
251. Zhevago, N.K., Chabak, A.F., Denisov, E.I., Glebov, V.I., and Korobtsev, S.V., Storage of cryo-compressed hydrogen in flexible glass capillaries, *International Journal of Hydrogen Energy*, 38 (16), 6694 (2013).
252. Wicks, G.G., Heung, L.K., and Schumacher, R.F., SRNL's porous, hollow glass balls open new opportunities for hydrogen storage, drug delivery and national defense (PDF), *American Ceramic Society Bulletin*, 87 (6), 23–28 (June 2008).
253. Kurtz, J., Ainscough, C., Simpson, L., and Caton, M., Hydrogen storage needs for early motive fuel cell markets, Technical report, NREL/TP-5600-52783, Contract no. DE-AC36-08GO28308, Golden, CO (November 2012).
254. Ahluwalia, R., System level analysis of hydrogen storage options, U.S. Department of Energy, Arlington, VA (2011).
255. American Public Transportation Association, *Public Transportation Fact Book*, 62nd edn. American Public Transportation Association, Washington, DC (2011).
256. Federal Aviation Administration, Unmanned Aircraft Systems (UAS) (2010), http://www.faa.gov/news/fact_sheets/news_story.cfm?newsId=6287, Retrieved September 9, 2011.
257. Kurtz, J., Ainscough, C., and Simpson, L., Onboard energy storage performance needs for fuel cell motive markets workshop, in *Held in Conjunction with the FCHEA Conference*, National Harbor, Washington, DC (2011).
258. Kurtz, J., Ainscough, C., and Simpson, L., Onboard energy storage performance needs for material handling equipment workshop, in *Held in Conjunction with the ProMat2011 Expo*, Chicago, IL (2011).
259. Stetson, N., *2011 Annual Merit Review and Peer Evaluation Meeting*, Hydrogen Storage, U.S. Department of Energy, Arlington, VA (2011).
260. Riis, T., Hagen, E.F., Vie, P.J.S., and Ulleberg, O., Hydrogen production and storage—R&D Priority gaps, A report from IEA—hydrogen co-ordination group—Hydrogen implementing agreement, OECD/IEA, Paris, France (2006).
261. Schlapbach, L. and Züttel, A., Hydrogen-storage materials for mobile applications, *Nature*, 414, 353–357 (November 15, 2001).
262. Niedzwiecki, A., Quantum technologies): Storage, in *Proceedings for Hydrogen Vision Meeting*, US DOE, Washington, DC (November 15–16, 2001), http://www.eere.energy.gov/hydrogenandfuelcells/pdfs/hv_report_12–17.pdf.
263. Teitel, R., Hydrogen storage in glass microspheres, Report BNL 51439, Brookhaven National Laboratories, Department of Energy, Upton, NY (1981).
264. Chambers, A., Park, C., Baker, R.T.K., and Rodriguez, N.M., Hydrogen storage R&D: Priorities and gaps, *Journal of Physical Chemistry B*, 102, 4253–4256 (1998).

265. Sandrock, G., A panoramic overview of hydrogen storage alloys from a gas reaction point of view, *Journal of Alloys and Compounds*, 293–295, 877–888 (1999).

266. Autry, T., Gutowska, A., Li, L., Gutowski, M., and Linehan, J., Chemical hydrogen storage: Control of H$_2$ release from ammonia borane, DOE 2004 Hydrogram Program Review, May 24–27, 2004, Philadelphia, PA, http://www.eere.energy.gov/hydrogenandfuelcells/doe_hydrogen_program.html.

267. Lipman, T., An overview of hydrogen production and storage systems with renewable hydrogen case studies, a clean energy state alliance report, Clean Energy State Alliance, DOE contract DE-FC3608G018111, Montpelier, VT (May 2011).

268. Adamson, K.-A., 2008 Large stationary survey, *Fuel Cell Today*, available at: http://www.fuelcelltoday.com/mecha/pdf/surveys/2008. (August 2008).

269. Air products creating world's largest hydrogen pipeline supply network, news release by Air Products, Lehigh Valley, PA (October 13, 2010). www.airproducts.com/PressRoom/CompanyNews/Archived/2010/13Oct2010.htm.

270. Heydon, E., Patel, P., and Jahnke, F., Development of a renewable hydrogen energy station, A report from Air Products. Presented at National Hydrogen Association Conference and Expo, April 2, 2009, Columbia, South Carolina.

271. Baldwin, D., Design and development of high pressure hydrogen storage tank for storage and gaseous truck delivery, US Department of Energy Program Review—Hydrogen Delivery, Contract DE-FG36-08GO18062, DOE, Washington, D.C. (2009).

272. Cohen, M. and Snow, G.C., Hydrogen delivery and storage options for backup power and off-grid primary power fuel cell systems, in *IEEE Intelec 2008 Proceedings*, Telecommunications Energy Conference, 2008, INTELEC 2008, IEEE 30th Internaional Conference Dates (September 14–18, 2008).

273. Melaina, M.W., Antonia, O., and Penev, M., Blending hydrogen into natural gas pipeline networks: A review of key issues Prepared under task no. HT12.2010, Technical report NREL/TP-5600-51995, Contract no. DE-AC36-08GO28308 (March 2013).

274. Gillette, J. and Kolpa, R., Overview of interstate hydrogen pipeline systems, ANL/EVS/TM/08-2, Environmental Science Division, Argonne National Laboratory, U.S. Department of Energy, Lemont, IL (November 2007).

275. Smith, B., Frame, B., Eberle, C., Anovitz, L., Blencoe, J., Armstrong, T., and Mays, J., New materials for hydrogen pipelines, Oak Ridge National Laboratory, U.S. Department of Energy, in *Hydrogen Pipeline Working Group Meeting*, Augusta, GA (August 30–31, 2005).

276. Amos, W.A., Costs of storing and transporting hydrogen, NREL/TP-570-25106, National Renewable Energy Laboratory, Golden, CO (November 1998).

277. Au, M., Chen, C., Ye, Z., Fong, T., Wu, J., and Wang, O., The recovery, purification, storage and transport of hydrogen separated from industrial purge gas by means of mobile hydride containers, *International Journal of Energy*, 12 (1), 33–37 (1996).

278. Carpetis, C., Technology and costs of hydrogen storage, *TERI Information Digest on Energy*, 4 (1), 1–13 (1994).

279. A cyclohexane makes the grade as a carrier of hydrogen, *Chemical Engineering*, 101 (12), 19 (December 1994).

280. Cuoco, A., Sgalambro, G., Paolucci, M., and D'Alessio, L., AIs photovoltaic hydrogen in Italy competitive with traditional fossil fuels?, *Energy*, 20 (12), 1303–1309 (1995).

281. *Encyclopedia of Chemical Technology, A Hydrogen*, 4th edn., Vol. 13. Wiley, Hoboken, NJ, pp. 838–894 (1991).

282. FIBA Technologies, Inc., Price Quote (1998); Flynn, T.M., Liquification of gases, *McGraw-Hill Encyclopedia of Science & Technology*, 7th edn., Vol. 10. McGraw-Hill, New York, pp. 106–109 (1992).

283. Garret, D.E., *Chemical Engineering Economics*. Von Nostrand Reinhold, New York (1989).

284. Huston, E.L., A liquid and solid storage of hydrogen, in *Proceedings of the Fifth World Hydrogen Energy Conference*, Vol. 3, July 15–20, 1984, Toronto, Ontario, Canada (1984).

285. The storage and transmission of hydrogen: Technical note, in Johannsen, T.B., et al. (eds.), *Renewable Energy: Sources for Fuels and Electricity*, Vol. 13 (1), (1993).

286. Hydrogen, *McGraw-Hill Encyclopedia of Science & Technology*, 7th edn., Vol. 8. McGraw-Hill, New York, pp. 581–588.

287. Schwarz, J.A. and Amonkwah, K.A.G., *Hydrogen Storage Systems*. U.S. Geological Survey, Washington, DC (1993).

288. Timmerhaus, C. and Flynn, T.M., *Cryogenic Engineering*. Plenum Press, New York (1989).

289. T-Raissi, A. and Sadhu, A., A system study of metal hydride storage requirements, in *Proceedings of 1994 DOE/NREL Hydrogen Program Review*, April 18–21, 1994, Livermore, CA (1994).

290. TransCanada Pipelines, Ltd., Press release, AANG pipeline to build new gas link across Southern B.C. (November 3, 1997); Worf, J.C., A metal hydrides, *McGraw-Hill Encyclopedia of Science & Technology*, 7th edn., Vol. 11. McGraw-Hill, New York, pp. 37–41 (1992).

291. Takahashi, K., Hydrogen transportation, *Energy Carriers and Conversion Systems*, Vol. II, Encyclopedia of Life Support Systems (EOLSS) (2000), www.eolss.net/sample-chapters/c08/E3-13-08.pdf.

292. Carpetis, C., in Winter, C.-J. and Nitsch, J. (eds.), *Storage, Transport and Distribution of Hydrogen*: *Hydrogen as an Energy Carrier—Technologies, Systems, Economy*, pp. 249–289 (1988).

293. Duret, B. and Sandin, A., Microspheres for on-board hydrogen storage, *International Journal of Hydrogen Energy*, 19, 757 (1994). (Theoretical consideration of the case of hydrogen-filled microspheres, and confirmation of hydrogen adsorbing capacity by laboratory-scale experiment.)

294. Fukuda, K., WE-NET first stage summary of research and development results, in: *Proceeding of WE-NET Hydrogen Energy Symposium*, Tokyo, Japan (*Proceedings*), pp. 1–26 (February 24, 1999) (in Japanese).

295. Hanada, T., Liquid hydrogen transportation, *Energy Resource*, 13 (6), 538–545 (1992).

296. Ohsumi, Y., in Ohta, T. (ed.), *Storage and Transportation of Hydrogen Using Metal Hydride*: *Most Advanced Hydrogen Energy Technology*. NTS Publishing Co., Ltd., Austin, TX, pp. 208–254 (1995).

297. Pettier, J.D., in Yürüm, Y. (ed.), *Hydrogen Transmission For Future Energy Systems*: *Hydrogen Energy System*. Kluwer Academy Publishers, Dordrecht, the Netherlands, pp. 181–193 (1995).

298. Uehara, I., in Tamura, H. (ed.), *Production, Storage, Transportation and Utilization of Hydrogen*: *Metal hydrides*. NTS Publishing Co., Ltd., pp. 8–29 (1998).

299. Pettier, J.D. and Blondin, E., in Durum, Y. (ed.), *Mass Storage of Hydrogen: Hydrogen Energy System*. Kluwer Academic Publishers, pp. 167–179 (1995).

300. Tsar, Y., Gaseous hydrogen transport, *Energy and Resources*, 13 (6), 561–566 (1992). Takahashi, K., Hydrogen transportation, *Encyclopedia of Life Support Systems* (EOLSS).

301. Dodds, P.E. and McDowall, W., A review of hydrogen delivery technologies for energy system models, UKSHEC working paper no. 7, UCL Energy Institute, University College London, London, U.K. (2012).

302. Ahluwalia, R.K., Hua, T.Q., and Peng, J.K., On-board and Off-board performance of hydrogen storage options for light-duty vehicles, *International Journal of Hydrogen Energy*, 37, 2891–2910 (2011).

303. Balta-Ozkan, N., Kannan, R., and Strachan, N., Analysis of UKSHEC hydrogen visions in the UK MARKAL energy system model, Policy Studies Institute, London, U.K. (2007).

304. Davison, J., Performance and costs of power plants with capture and storage of CO_2, *Energy*, 32, 1163–1176 (2007).

305. Dodds, P.E. and McDowall, W., A review of hydrogen production technologies for energy system models, UKSHEC working paper no. 6, UCL, London, U.K. (2012).

306. Fichtner, M., Hydrogen storage, in Ball, M. and Wietschel, M. (eds.), *The Hydrogen Economy: Opportunities and Challenges*. Cambridge University Press, Cambridge, U.K. (2009).

307. Haeseldonckx, D. and D'haeseleer, W., The use of the natural-gas pipeline infrastructure for hydrogen transport in a changing market structure, *International Journal of Hydrogen Energy*, 32, 1381–1386 (2007).

308. IEA, Reduction of CO_2 emissions by adding hydrogen to natural gas, IEA Greenhouse Gas R&D Programme, Gastec Technology BV, Apeldoorn, the Netherlands (2003).

309. IEA, Prospects for hydrogen and fuel cells, Paris, France (2005); Kelly, B., *Liquefaction and Pipeline Costs, Hydrogen Delivery Analysis Meeting*, Columbia, MD (2007).

310. Krewitt, W. and Schmid, S.A., Fuel cell technologies and hydrogen production/distribution options, CASCADE MINTS, DLR Institute of Vehicle Concepts, Stuttgart, Germany (2004).

311. Leighty, W., Holloway, J., Merer, R., Somerday, B., San Marchi, C., Keith, G., and White, D., Compressor less hydrogen transmission pipelines deliver large-scale stranded renewable energy at competitive cost, in *16th World Hydrogen Energy Conference*, Lyon, France (2006).

312. Mytelka, L.K., Hydrogen fuel cells and alternatives in the transport sector: A framework for analysis, Chapter 1, p. 5 in Making Chores about Hydrogen Transport Tissues with Developing Countries, in: Mytelka, L. and Boyl, G. (eds.), International Development Research Center, United Nations University Press, New York (2008).

313. National Grid, Annual report and accounts 2010/11, National Grid plc, London, U.K. (2011); National Grid, Streetworks pipeline replacement, National Grid, Northampton, U.K. (2011).

314. Parker, N., Using natural gas transmission pipeline costs to estimate hydrogen pipeline costs, University of California, Davis, CA (2004).

315. Van der Zwaan, B.C.C., Schoots, K., Rivera-Tinoco, R., and Verbong, G.P.J. The cost of pipelining climate change mitigation: An overview of the economics of CH_4, CO_2 and H_2 transportation, SSRN eLibrary, "Ideas," a service hostel by Federal Bank of St. Louis, MO (2010). http://www.scienedirect.com/science/article/pii/s030626191100314x.

316. Simbeck, D. and Chang, E., Hydrogen supply: Cost estimate for hydrogen pathways—Scoping analysis, National Renewable Energy Laboratory, Golden, CO (2002).

317. Syed, M.T., Sherif, S.A., Veziroglu, T.N., and Sheffield, J.W., An economic analysis of three hydrogen liquefaction systems, *International Journal of Hydrogen Energy*, 23, 565–576 (1998).

318. van der Zwaan, B.C.C., Schoots, K., Rivera-Tinoco, R., and Verbong, G.P.J., The cost of pipelining climate change mitigation: An overview of the economics of CH_4, CO_2 and H_2 transportation, *Applied Energy*, 88, 3821–3831 (2011).

319. Weinert, J., A near-term economic analysis of hydrogen fueling stations, Institute of Transportation Studies, University of California, Davis, CA (2005).

320. West, J.E., The economics of small to medium liquid hydrogen facilities, CryoGas International (2003).

321. Yang, C. and Ogden, J.M., Determining the lowest-cost hydrogen delivery mode, *International Journal of Hydrogen Energy*, 32, 268 (2007).

322. Zaetta, R. and Madden, B., Hydrogen fuel cell bus technology state of the art review, Next HyLights. Element Energy, Cambridge, U.K. (2011).

323. EIA, Energy Information Administration, The Impact of Increased Use of Hydrogen on Petroleum Consumption and Carbon Dioxide Emissions, Energy Information Administration Office of Integrated Analysis and Forecasting Office of Coal, Nuclear, Electric and Alternate Fuels U.S. SR/OIAF-CNEAF/2008-04, Washington, DC (September 2008).

324. The 2011 oil shock—Oil and gas economy, A report in *The Economist Magazine* (March 3, 2011), Retrieved November 3, 2012.

325. EIA, Existing capacity by energy source, DOE, Washington, D.C. (January 2010).

326. Kroposki, B., Garrett, B., Macmillan, S., Rice, B., Komomua, C., O'Malley, M., and Zimmerle, D., Energy systems integration: A convergence of ideas, NREL report, NREL/TP-6A00-55649, under contract no. DE-AC36-08GO28308 (July 2012).

327. Evans, P. and Farina, M., The age of gas & the power of networks, A report from General Electric Company, New York (2013).

328. Riis, T., Sandrock, G., Ulleberg, O., and Vie, P., Hydrogen storage-gaps and priorities, HIA HCG Storage Paper, IEA HIA Task 17, IEA, Brussels, Belgium, pp. 1–13 (2005).

329. National Research Council and National Academy of Engineering, The hydrogen economy: Opportunities, costs, barriers and R&D needs, Section 4: Transportation, distribution and storage of hydrogen, National Academy Press, Washington, DC (2004).

330. Zhou, L. Progress and problems in hydrogen storage methods, *Renewable and Sustainable Energy Reviews*, 9, 395–408 (2005).

331. Melaina, M., Antonia, O., and Penev, M., Blending hydrogen into natural gas pipeline networks: A review of key issues, NREL, Golden, CO. Prepared under task no. HT12-2010, NREL/TP-5600-51995, under DOE contract no. DE-AC36-08GO28308 (March 2013).

332. Hagen, M., Polman, E., Jensen, J., Myken, A., Jonsson, O., and Dahl, A., Adding gas from biomass to the gas grid, Report SGC 118, ISSN 1102-7371, ISRN SGC-R-118-SE, Swedish gas center (July 2001).

333. Hussey, B., Biogas injection into the natural gas grid, A report by Commission for Energy Regulation, Dublin, Ireland (September 11, 2013).

334. Goellner, J., Expanding the shale gas infrastructure, CEP, 49 (August 2012).

335. Hydrogen Storage, Wikipedia, the free encyclopedia (2012).

336. Godula-Jopek, A., Jehle, W., and Wellinitz, J., *Hydrogen Storage Technologies, New Materials, Transport and Infrastructure*. Wiley Online, Hoboken, NJ (November 5, 2012).

337. Gas network simulation, Wikipedia, the free encyclopedia (2012).

338. Baldwin, J., Biomethane to grid UK project review, CNG services report (March 11, 2014).

339. Hagen, M., Polman, E., Myken, A., Jensen, J., Jonsson, O., Biomil, A.B., and Dahl, A., Adding gas from biomass to the gas grid, Final report July 1998–February 2001, under contract no. XVII/4.1030/Z/99-412 (2001).

340. Specht, D. et al., Storing bioenergy and renewable electricity in the natural gas grid, FVEE-AEE, Topics (2009). www.free.de/fileadmin/publikationen/thermenhefte/th2009-1/th2009-1_05_06.pdf.

341. Power to Gas, Wikipedia, the free encyclopedia (2012).

342. Barati, F., Seifi, H., Sepasian, M., Nateghi, A., Shafie-khah, M., and Catalao, J., Multiperiod integrated framework of generation, transmission, and natural gas grid expansion planning for large scale systems, *IEEE Transactions on Power Systems*, 99, 1–11 (October 31, 2014), IEEE Power and Energy Society.

343. EU commission task force for smart grids–expert group 4-smart grid aspects related togas, EU report EG4/SEC00601DOC, EU Commission, Brussels, Belgium (June 5, 2011).

344. Weidenaur, T., Hoekstra, S., and Wolters, M., Development options for the Dutch gas distribution grid in a changing gas market, Personal communication (2009).

345. Fullenbaum, R., Fallon, J., and Flanagan, B., Oil and gas transportation and storage infrastructure: Status, trends, & economic benefits, A report by HIS Global Inc., Submitted to American Petroleum Institute, Washington, DC (December, 2013).

346. Wolters, M., Requirements of future gas distribution networks, in *23rd World Gas Conference*, Amsterdam, the Netherlands (2006).

347. Special reliability assessment: Accommodating an increased dependence on natural gas for electric power—Phase II: A vulnerability and scenario assessment for the North American bulk power system, NERC report North American Electric Reliability Corporation, Atlanta, GA (May 2013). www.nerc.com.

348. Dehaeseleer, J., Gas industry views regarding smart gas grid, A report from Marcogaz technical association of the European Natural Gas Industry, EGATE2011, Copenhagen, Denmark (2011).

349. Weiss, J., Bishop, H., Fox-Permer, P., and Shavel, I., Partnering natural gas and renewables in ERCOT, A report by The Brattle Group, Inc., 35 pp. Prepared for The Texas Clean Energy Coalition (June 11, 2013). www.brattle.com

350. North American natural gas midstream infrastructure through 2035: A secure energy future, A report by The INGAA Foundation Inc., Washington, D.C. (June 28, 2011). http://www.ingaa.org/cm3/31/7306/7828.aspx.

351. Zwick, G., Natural gas infrastructure-papers 1–9, Prepared by the gas infrastructure subgroup of the resource and supply task group, Working document of the NPC North American Resource Development study (September 15, 2011).

352. Natural gas pipeline and storage infrastructure projections through 2030, A report submitted by ICF International to the INGAA Foundation Inc., Washington, D.C., F-2009-04 (October 2009).

353. The basics of underground natural gas storage, EIA, U.S. Energy Information Administration report, Department of Energy, Washington, D.C. (August 2004).

354. Boardman, R., Advanced energy systems-nuclear-fossil-renewable hybrid systems, A report to Nuclear Energy Agency-committee for technical and economical studies on nuclear energy development and fuel cycle, INL, Idaho Falls, ID (April 4–5, 2013).

355. Shah, Y.T., *Energy and Fuel Systems Integration*. CRC Press, New York (2015).

356. Renewable gas-vision for a sustainable gas network, A report by National Grid U.S.A. Service Co., Waltham, MA (2010).

357. Thompson, J., U.S. Underground storage of natural gas in 1997: Existing and proposed, A report in Natural Gas Monthly, Energy Information Administration, Department of Energy, Washington, DC (September 1997).

9 Natural and Synthetic Gas for Productions of Liquid Fuels and Their Additives

9.1 INTRODUCTION

The components in natural gas, methane, ethane, propane, *n*- and isobutanes, and pentanes, are all useful in their own right. Higher paraffins are particularly valuable for a wealth of chemicals and polymers such as acetic acid, formaldehyde, olefins, polyethylene, polypropylene, acrylonitrile, and ethylene glycol, as well as portable premium fuels such as, Calor Gas (propane). As discussed in Chapter 6, methane can also be converted to syngas by gas reforming. Syngas can then be converted to methanol, ammonia, syncrude, lubricant, or some precursors for chemical manufacturing, for example, dimethyl ether (DME) and urea. The technologies to manufacture some of these products directly from methane are also being developed and these will require more research into novel catalytic processes and their large-scale developments.

While methane (and other lower hydrocarbons [HCs]), syngas, and hydrogen are useful for the production of a variety of chemicals, polymers, and consumer and agricultural (like fertilizer) products, in this chapter we only examine the conversion of natural and synthetic gas to liquid (GTL) fuels and fuel additives. These synthetic naphtha, methanol, higher alcohols, gasoline, diesel, jet fuels, etc., can be substituted for the similar liquid fuels prepared from crude oils, coal, shale oil, or biomass.

We discuss here in detail the catalytic processes used to convert natural and synthetic gases to a variety of liquid fuels and fuel additives. Coal, biomass, natural gas, syngas, and hydrogen can be used to produce liquid fuels and their additives in a number of different ways [1–18]. One method is to convert coal, biomass, and natural gas into syngas (mixture of hydrogen and carbon monoxide) and then convert syngas to liquid fuels and their additives via Fischer–Tropsch (FT) and related syntheses. In Chapter 3 we examined the methods for converting coal, biomass/waste, and coal/biomass/waste mixtures to syngas. In Chapter 6 we examined the reforming process to convert natural gas into syngas. Liquid fuels like gasoline, diesel, and jet fuels are also produced from crude oils in oil refineries by various hydroprocessing operations. Hydrogen produced by various methods described in Chapters 5 and 6 plays important role in these hydroprocessing operations.

In this chapter we examine various catalytic methods for the conversion of syngas to liquid fuels and their additives as well as various ways hydrogen is used to produce liquid fuels and fuel additives in oil refineries. Specifically, we examine the following eight usages of syngas and hydrogen to produce liquid fuels and fuel additives by either direct or indirect methods:

1. Syngas to methanol
2. Methanol to DME, gasoline, and olefins
3. Mixed alcohol synthesis
4. FT synthesis
5. Isosynthesis
6. Oxosynthesis
7. Syngas fermentation
8. Role of hydrogen in the production of liquid fuels in refineries

This chapter illustrates each process in detail including available commercial processes wherever possible.

9.1.1 Paths for GTL Conversions

The general reactions for the hydrogenation of carbon monoxide are given as follows [18]:

$$(2n + 1)H_2 + nCO = C_nH_{2n+2} + nH_2O \tag{9.1}$$

$$(n + 1)H_2 + 2nCO = C_nH_{2n+2} + nCO_2 \tag{9.2}$$

$$2nH_2 + nCO = C_nH_{2n} + nH_2O \tag{9.3}$$

$$nH_2 + 2nCO = C_nH_{2n} + nCO_2 \tag{9.4}$$

$$2nH_2 + nCO = C_nH_{2n+1}OH + (n - 1)H_2O \tag{9.5}$$

$$(n + 1)H_2 + (2n - 1)CO = C_nH_{2n+1}OH + (n - 1)CO_2 \tag{9.6}$$

These reactions represent the formation of paraffins, monoolefins, and alcohols. The reactions generate water or carbon dioxide as by-products. As the concentration of carbon monoxide compared to that of hydrogen is increased, by-products change from water to carbon dioxide. However, the desired product can still be obtained during this change in the formation of the by-product [1–18].

Since the synthesis reactions lead to a smaller number of gaseous molecules, the equilibrium conversion at any given temperature increases with pressure. For example, for the reaction $20H_2 + 10CO = C_{10}H_{20} + 10H_2O$, 80% conversion can be obtained at 1 atm and 300°C. If temperature is increased to 390°C, 420°C, 445°C, or 475°C, the pressure required to maintain the same conversion would be 10, 20, 30, and 50 atm, respectively. Thus, for the formation of 1-decene (Reaction 9.3), an increase in pressure by 50 atm allows temperature to be increased by 175°C to obtain the same level of conversion [18].

Generally, the reactions forming carbon dioxide rather than water have larger equilibrium constants. The standard free energy per carbon atoms, $\Delta F°/n$, for reactions producing methane is more negative than that of the reactions forming the higher HCs. In addition, the free energy of elemental carbon formation is more negative than that for the higher HCs; therefore, the production of higher HCs for liquid fuels must depend on the selectivity of the catalyst. The nature of the catalyst affects the Schulz–Flory chain growth distribution of FT products [18]. Both the nature of the catalyst and promoters and reaction conditions (i.e., temperature and pressure) significantly affect various syngas conversions mentioned earlier. Earlier studies [18] led to following conclusions regarding the required operating conditions for various syngas conversions:

1. *Methane synthesis*: 1 atm, 250°C–500°C, with Ru and Ni catalysts and ThO$_2$ and MgO promoters
2. *Methanol synthesis*: 100–1000 atm, 200°C–400°C, with ZnO, Cu, Cr$_2$O$_3$, and MnO catalysts
3. *Higher alcohol synthesis*: 100–400 atm, 350°C–450°C, with catalysts same as in methanol and FT synthesis with alkali promoters

4. *Fischer–Tropsch synthesis (paraffinic and olefinic HCs up to wax)*: 1–30 atm, 150°C–350°C, with Fe, Co, and Ni catalysts and ThO_2, MgO, Al_2O_3, and K_2O promoters
5. *Fischer–Tropsch synthesis (high-molecular-weight paraffinic HCs)*: 100–1000 atm, 150°C–250°C, with Ru catalyst
6. *Isosynthesis*: 100–1000 atm, 400°C–500°C, with ThO_2 and ZnO + Al_2O_3 catalysts and K_2O promoter
7. *Oxosynthesis*: 100–200 atm, 100°C–200°C, with Co and Fe catalysts

In recent years, significant research on both catalysts and processes further developed these conversion processes. This chapter examines both the history and present development of all of these processes (except methanation). Oxosynthesis converts syngas into oxygenated compounds and isosynthesis converts syngas into branched HCs. These compounds can be added to liquid fuels to change their desired properties. Methanol can be dehydrated to produce DME, a fuel additive. Methanol can also be converted to gasoline by the famous Mobil methanol-to-gasoline (MTG) process and olefins by methanol-to-olefin (MTO) process. Finally, syngas can also be converted to alcohols by fermentation techniques [1–18]. All of these processes are examined in this chapter.

Besides the processes of converting syngas to liquid fuels as mentioned earlier, the role of hydrogen in upgrading crude oils or unconventional oils (like heavy oils, bitumen, oil shale, coal liquids, bio-oils) to produce gasoline, jet fuel, and diesel fuel is uncompromising. Refining processes like hydrogenation, hydrocracking, hydrodesulfurization, hydrodemetallization, and hydrodenitrogenation are essential in the productions of gasoline, diesel, and jet fuels of the right hydrogen-to-carbon ratio and the right hydrocarbon compositions and fuels with minimum impurities. We also briefly evaluate this role of hydrogen in oil-based, coal-based, and biomass-based refineries in this chapter.

9.2 CONVERSION OF SYNGAS TO METHANOL

Methanol or methyl alcohol (CH_3OH) is a colorless liquid with a boiling point of 65°C. Methanol will mix with a wide variety of organic liquids as well as with water, and accordingly, it is often used as a solvent for domestic and industrial applications. It is most familiar in homes as one of the constituents of methylated spirits. Methanol is the raw material for many chemicals such as formaldehyde, dimethyl terephthalate, methylamines and methyl halides, methyl methacrylate, acetic *acid*, and *gasoline*. In principle, methanol can be produced from coal or biomass via syngas route [19–58].

9.2.1 CHEMISTRY

Methanol synthesis has been postulated to occur via two different types of reaction steps. One mechanistic view indicates methanol to be formed by direct reaction (9.7) between carbon monoxide and hydrogen as [19–58]

$$\text{Synthesis } 2H_2 + CO \xrightarrow[\text{Cu–Zn}]{} CH_3OH \quad \Delta H_r = -92 \text{ kJ mol}^{-1} \tag{9.7}$$

Also,

$$CO_2 + H_2 = CO + H_2O \tag{9.8}$$

Overall,

$$CO_2 + CO + 5H_2 \rightarrow 2CH_3OH + H_2O + \text{Heat} \tag{9.9}$$

The second reaction (9.8) is a reverse water gas shift reaction (henceforth denoted as WGS reaction).

Experimental data containing 3%–9% carbon dioxide in the feed show a decrease in carbon dioxide concentration in the effluent stream indicating the presence of the second reaction. The first reaction (9.7) is exothermic, while the second reaction (9.8) is endothermic. In this mechanism, the presence of carbon dioxide in the feed is extremely important. Low carbon dioxide in the feed will result in lower methanol productivity. Typically, 2%–4% of carbon dioxide is present in the syngas mixture for the vapor-phase synthesis of methanol and the corresponding value for the liquid-phase methanol synthesis is 4%–8% [8,19].

The second view implies that the principal chemical reactions that lead to the synthesis of methanol are [8–11,19]

$$CO_2 + 3H_2 = CH_3OH + H_2O \tag{9.10}$$

$$CO + H_2O = CO_2 + H_2 \tag{9.11}$$

According to this view, the synthesis of methanol proceeds predominantly via direct hydrogenation of carbon dioxide and WGS reaction (9.11) in the forward direction. The WGS reaction produces carbon dioxide, which in turn increases the productivity of methanol production. Thus, methanol production from syngas involves a set of reactions among CO, H_2, and a small amount of CO_2. The process is carried out at 220°C–300°C and 50–100 bar, with the raw products fed into a distillation plant to recycle unused syngas, volatiles, water, and higher alcohols back to the reactor.

A significant literature is available to justify each of these two mechanistic views for the methanol productions. Side reactions, also strongly exothermic, can lead to formation of by-products such as methane, higher alcohols, or DME. The oldest process for the industrial production of methanol is the dry distillation of wood. Methanol is currently produced on an industrial scale exclusively by catalytic conversion of synthesis gas. Processes are classified according to the pressure used:

1. *High-pressure process*: 250–300 bar
2. *Medium-pressure process*: 100–250 bar
3. *Low-pressure process*: 50–100 bar

Both high- and low-pressure processes are available to hydrogenate carbon monoxide. The high-pressure method was commercialized by BASF [19,20,27,29,42]. This process used zinc–chromium oxide catalysts at or above 340°C and required a pressure of 300–500 bar to obtain methanol concentrations of about 5–6 vol% in the effluent gases from the reactor, together with significant amounts of methane, DME, ethanol, and higher alcohols. In 1966, Imperial Chemical Industries (ICI) introduced a low-pressure methanol synthesis that used a copper–zinc–chromium catalyst at 50 bar. The low-pressure processes differ primarily in operating pressure, catalyst, and reactor design, especially with respect to heat recovery and temperature control. The low-pressure processes are currently favored and their main advantages are lower investment and production costs, improved operational reliability, and greater flexibility in the choice of the plant size [19,36,47–50,56].

It should also be noted that although the reactor catalyst is highly specific in producing methanol, some side reactions can occur, which produce higher alcohols (ethanol, propanol, butanol) and alkanes. These may be articulated as [19,20,27,29,42]

$$nCO + 2\left(n - \frac{1}{2}\right)H_2 \rightarrow C_nH_{2n}OH + (n-1)H_2O \tag{9.12}$$

$$nCO + CH_3OH + 2nH_2 \rightarrow C_nH_{2n+3}OH + nH_2O \tag{9.13}$$

9.2.2 CATALYSTS

As pointed out by Spath and Dayton [58], the first high-temperature, high-pressure methanol synthesis catalysts were ZnO/Cr_2O_3 and were operated at 350°C and 250–350 bar. Over the years, as gas purification technologies improved (i.e., removal of impurities such as sulfur, chlorine, and metals), interest in the easily poisoned Cu catalysts for methanol synthesis was renewed. In 1966, ICI introduced a new, more active $Cu/ZnO/Al_2O_3$ catalyst that began a new generation of methanol production by a low-temperature (220°C–275°C), low-pressure process (50–100 bar). Under normal commercial operating conditions, CuO/ZnO methanol catalysts have quite long lifetimes, up to 2–5 years. The ZnO in the catalyst formulation creates a high Cu metal surface area; it is suitably refractory at methanol synthesis temperatures and hinders the agglomeration of Cu particles. ZnO also interacts with Al_2O_3 to form a spinel that provides a robust catalyst support. Acidic materials like alumina are known to catalyze methanol dehydration reactions to produce DME. By interacting with the Al_2O_3 support material, the ZnO effectively improves methanol selectivity by reducing the potential for DME formation.

Copper catalysts are extremely sensitive to site-blocking poisons such as reduced sulfur. To retain the long-term activity of Cu catalysts, it has been found empirically that the gas-phase sulfur concentration needs to be kept below 1 ppm and preferably below 0.1 ppm. If reactor temperatures are not properly controlled, the highly exothermic methanol synthesis reactions can also rapidly cause catalyst deactivation by sintering of the Cu crystallites. Reactor temperatures are maintained below 300°C to minimize sintering. The presence of Cl in syngas has also been correlated with a greatly enhanced rate of sintering of copper crystallites most likely due to the formation of volatile copper chloride compounds. The limits on HCl content to avoid catalyst poisoning are on the order of 1 ppb [19,23,25,32,38,41,43,53,54].

Catalysts are typically prepared by the co-precipitation of metal salts (such as nitrates or sulfates) with a variety of precipitating agents but avoid alkali metals because they catalyze productions of higher alcohols. Heavier alkali metals like Cs can, however, improve methanol yields. Other catalyst formulations that increase methanol yield have been presented by Klier [42]. He indicated that thoria-supported catalyst showed higher methanol yields when syngas is free of CO_2. Cu/Zr catalysts have been proven active for methanol synthesis in CO-free syngas at 5 atm and 160°C–300°C [23,25,32,58]. Supported Pd catalysts have also demonstrated methanol synthesis activity in CO_2-free syngas at 5–110 atm at 260°C–350°C [19,23,25,32,58]. Commercial catalyst formulations as reported by Spath and Dayton [58] are illustrated in Table 9.1.

TABLE 9.1
Commercial Methanol Synthesis Catalyst Formulations

Manufacturer	Cu (at%)	Zn (at%)	Al (at%)	Others	Patent Date
IFP	45–70	15–35	4–20	Zr—2–18	1987
ICI	20–35	15–50	4–20	Mg	1965
BASF	38.5	48.8	12.9		1978
Shell	71	24		Rare Earth oxide –5	1973
Sud Chemie	65	22	12		1987
DuPont	50	19	31		None found
United catalysts	62	21	17		None found
Haldor Topsoe MK-121	>55	21–25	8–10		None found

Source: Spath, P.L. and Dayton, D.C., Preliminary screening—Technical and economic assessment of synthesis gas to fuels and chemicals with emphasis on the potential for biomass-derived syngas, NREL/TP-510-34929, NREL, Golden, CO, December 2003.

9.2.3 Syngas Requirements

Methanol synthesis has very high catalyst specificity, and since syngas C–O bond remains intact, the overall process only involves a few simple chemical reactions compared to the complex reactions in an FT or a mixed alcohol process. The main requirements for syngas during methanol synthesis are as follows:

1. The relative quantities of H_2, CO, and CO_2. The stoichiometric ratio of (H_2-CO_2) to $(CO + CO_2)$ should be greater than 2 for gas reactions using alumina-supported catalysts and around 0.68 for slurry-based reactors. As an example, 11 molecules of H_2 and 4 molecules of CO to 1 molecule of CO_2 gives a stoichiometric ratio of 2.
2. Removal, to concentrations of less than 10s of ppb, of tars with dew points below the catalyst operating temperature.
3. Avoidance of alkalis and trace metals, which can promote other reactions, such as FT and mixed alcohol synthesis.

Methanol synthesis has similar syngas cleanup requirements to FT synthesis, and the overall biomass to methanol plant efficiency are generally similar to FT plants [19,23,25,32,58]. The minimum economic scale is also of the order of a few hundred tons/day output [23,25,32,58], that is, around 100,000 tons/year methanol output, equating to a biomass input of 1,520 oven dried tons (odt/day). The new process technologies in development for FT would also be applicable to methanol process.

9.2.4 Commercial Reactors and Processes

Along with pressure and temperature, a major design issue in the development of methanol synthesis reactor is the method used to remove heat and distribute gas within the reactor. Numerous approaches have been taken by different industries and they are briefly summarized in Table 9.2 [19,20,34,37,40,47–49,51,56].

Besides the ones mentioned in Table 9.2, two other novel methanol conversion processes are based on systems in which the product methanol is continuously removed from the gas phase by selective adsorption on a solid or in a liquid. The gas–solid–solid trickle flow reactor utilizes an adsorbent such as SiO_2/Al_2O_3 to trap the product methanol [24,46]. The solid adsorbent is collected in holding tanks and the methanol is desorbed. In a reactor system with interstage product removal, a liquid solvent is used to adsorb the product methanol [24].

As indicated in Table 9.2, methanol synthesis has been carried out at both high and low pressures and in both gas and liquid phases. The commercial processes that use these different types of reactors are also listed in Table 9.2. Some details of these processes are given in Table 9.3. More extensive details on these processes are given by Lee [19,20].

ICI and Lurgi offer the two most widely used low-pressure processes [57]. Low-pressure technologies are also offered by Mitsubishi Gas Chemical Company, Inc., of Japan, jointly by Haldor Topsoe of Denmark and Nihon Suido Kogyo of Japan, and by Halliburton. Table 9.4 presents the process conditions for methanol technology suppliers as reported by Spath and Dayton of NREL [58]. According to them, the world's largest producer of methanol is Methanex; they account for 17% of the total global capacity [57]. The next largest producer is SABIC, which accounts for 6.5% of the total global capacity [57].

The production of methanol from natural gas has a minimum impact on the environment. Only one major product is manufactured (methanol), which is a compound of relative low toxicity. A special methanol sewer collects any methanol wastes spilled on the plant site. This waste is burned in the reformer as fuel. A storm pond collects rainwater from the plant and is analyzed for methanol and other contaminants prior to discharge to the river. Thus, overall conversion of syngas to methanol is a very environmentally friendly process.

TABLE 9.2

Commercial Reactors for Methanol Synthesis

Company	Type of Reactor	Operating Conditions
Lurgi	Shell and tube design boiling water	Isothermal: 230°C–265°C, 50–100 bar, $Cu/ZnO/Cr_2O_3$
		With promoters, by-product steam at 40–50 bar
ICI	Low-pressure gas injected by horizontal lozenges	50–100 bar and 270°C, *Cu/ZnO/Al₂O₃ catalyst cold fresh gas and recycled syngas quenches*
ICI	ARC Converters	Distributor plates separate multiple catalyst beds
Kellogg, Brown, Root (Haliburton)	Multiple fixed-bed separated by heat exchangers	Spherical, all gas goes to first bed-less catalyst than ICI converter
Haldor Topsoe	Collect—mix—distribute (CMD)	Multiple vertical sections fixed-bed gas go into first. Section—quench at top mixes with exit gas—increased conversion per pass
Toyo Engineering Corporation	Multistage radial flow converter (MRF-ZTM)	Bayonet boiler tubes for intermediate cooling. The tubes divide the catalyst into concentric beds.
—	**Tube- cooled converter**	**Gas enters heat exchange tubes before entering catalyst bed.**
Linde	**Isothermal—Variobar converter**	**Coiled heat exchanger tubes embedded in catalyst bed**
Mitsubishi Gas Chemical	Isothermal—MGC/MHI super converter	Annular bed with cooling on both sides—gas goes up **in the center-high conversion per pass**
Chem Systems, Inc., Department of Energy Air Products	Liquid-phase slurry reactor (LPMEOH™)	225°C–265°C, 50 bar pressure **Cu/ZnO catalyst**

Source: Prepared from information in Lee, S., Methanol synthesis from syngas, Chapter 9, in: Lee, S. et al., *Handbook of Alternative Fuel Technologies*, CRC Press, New York, 2007, pp. 297–321.

The end product of all the processes described earlier is crude methanol and water. This mixture is reduced in pressure in a letdown "flash" vessel. Gas from this vessel is recycled to the furnace as fuel. The crude is then sent to the process storage tank. The methanol formed by syngas conversion may be dehydrated to give DME, which can be further dehydrated over a zeolite catalyst, ZSM-5, to give a gasoline with 80%C_{5+} hydrocarbon products. These conversions are further illustrated in Section 9.3.

9.3 CONVERSION OF METHANOL TO DIMETHYL ETHER, GASOLINE, AND OLEFINS

DME, also known as "methoxymethane," is an organic compound with the formula CH_3OCH_3, which is simplified to C_2H_6O. This is the simplest ether, which is a colorless gas that is a useful precursor to other organic compounds and an aerosol propellant and is being researched on as a future energy option [59–72]. It is an isomer of ethanol. DME is a clean, colorless gas that is easy to liquefy and transport and is made from a variety of renewable materials or fossil fuels. Physically similar to liquefied petroleum gas, DME is synthesized from dehydration of methanol as

$$2CH_3OH \rightarrow (CH_3)_2O + H_2O \qquad (9.14)$$

TABLE 9.3

Commercial Processes for Methanol Synthesis

Process/Comments

The conventional ICI's 100 atm methanol process/Cu/ZnO/Al$_2$O$_3$ catalyst system
Two parts: reforming and synthesis gas conversion

Haldor Topsoe A/S low-pressure methanol process/low capital cost
Two-stage reforming: suitable for smaller and larger plants (10,000 tons/day)

Kvaerner methanol process/similar to Haldor Topsoe process; CO$_2$ can be used as supplementary feed; 2000–3000 meter tons per day (mtpd) typical size

Krupp Uhde's methanol process/flexible feedstock, different steam
Reforming: at least 11 plants

Lurgi Ol-Gas-Chemie GMBH Process/10,000 mtpd capacity; two-stage reforming
Synetix LPM process/large capacities; improved ICI's LPM; three stages

Liquid-phase methanol process/developed by Chem Systems Inc.; slurry reactor; Cu/ZnO/Al$_2$O$_3$ catalyst at 230°C–260°C, 50–100 atm

Sources: Modified, adapted and prepared from information in Lee, S., Methanol synthesis from syngas, Chapter 9, in: Lee, S. et al., *Handbook of Alternative Fuel Technologies*, CRC Press, New York, 2007, pp. 297–321; Lee, S., *Methanol Synthesis Technology*, CRC Press, Boca Raton, FL, 1990.

TABLE 9.4

Methanol Technology Suppliers

Technology Supplier	*T* (°C)	*P* (atm)	Notes
ICI (Synetix)	210–290	50–100	Currently licenses four types of reactors: ARC, Tubular Cooled, Isothermal Linde, and Toyo (see Reactor Section 9.2.4 for details)
Lurgi	230–265	50–100	Tubular, isothermal reactor
Mitsubishi	240	77–97	Tubular, isothermal reactor
Haldor Topsoe and Nihon	260	48–300	To date, no commercial plants based on this process.
Halliburton (Kellogg, Brown and Root)	Not found	Not found	Spherical reactor geometry

Source: Spath, P.L. and Dayton, D.C., Preliminary screening—Technical and economic assessment of synthesis gas to fuels and chemicals with emphasis on the potential for biomass-derived syngas, NREL/TP-510-34929, NREL, Golden, CO, December 2003.

DME is a safe and nontoxic fuel suitable for use in diesel engines that burns cleanly without producing soot. It is, however, highly flammable. Similar to propane, DME is handled as a low-pressure liquid that can be easily shipped and stored. Global production capacity was about 11.3 million tons by 2012 [59–72].

DME is a low-temperature solvent and extraction agent that can be used for specialized laboratory procedures. While its usefulness is limited by its low boiling point (–23°C [–9°F]), the same property facilitates its removal from reaction mixtures. DME is (1) the main constituent of freezer spray used for field testing of electronic components [70], (2) a precursor to the useful alkylating agent, such as trimethyloxonium tetrafluoroborate, (3) a refrigerant with ASHRAE refrigerant designation R-E170 used in refrigerant blends with ammonia, carbon dioxide, butane, and propene [19], and (4) an agent used for the treatment of warts along with propane [70]. Unlike other alkyl ethers, DME resists autoxidation [70].

A potentially major use of DME is as a substitute for propane in LPG used as fuel in household and industry. It is also a promising fuel in diesel engines, gasoline engines (30% DME/70% LPG), and gas turbines. Its cetane number of 55 is higher than diesel fuel cetane number of 40–53 [70]. Only moderate modifications are needed to convert a diesel engine to burn DME. Its combustion leads to very low emissions of particulate matter, NO_x, and CO and sulfur-free DME passes emission regulations of Europe, United States, and Japan [70]. Mobil uses DME in their methanol-to-gasoline process. The largest use of DME is as the feedstock for the production of the methylating agent, dimethyl sulfate, which entails its reaction with sulfur trioxide as follows [59–72]:

$$CH_3OCH_3 + SO_3 \rightarrow (CH_3)_2SO_4 \tag{9.15}$$

This application consumes several thousand tons of DME annually. DME can also be converted into acetic acid by the reaction [59–72]

$$(CH_3)_2O + 2CO + H_2O \rightarrow 2CH_3COOH \tag{9.16}$$

DME vehicles are also on the rise. Auto manufacturers Shanghai Diesel Co., AB Volvo, Isuzu Trucks, and Nissan Diesel are actively developing DME-fueled heavy-duty vehicles [70]. Requirements for modifications to the fuel distribution infrastructure and vehicle engine parts to accommodate the use of DME will influence the market introduction from bus and truck fleets to passenger diesel cars. In 2009, a team of university students from Denmark won the Urban Concept/Internal Combustion class at the European Shell Eco-Marathon (the Shell Eco-Marathon is an unofficial world championship for mileage) with a vehicle running on 100% DME. The vehicle drove 589 km/L [70].

World production of DME is primarily by means of methanol dehydration. The majority of global DME production is currently in China. Japan, Korea, and Brazil have significant new production facilities, and major new capacity additions are planned or under construction in Egypt, India, Indonesia, Iran, and Uzbekistan. China's National Development and Reform Commission is calling for 20 million tons of DME production capacity by 2020. South Korea is studying all aspects of commercializing DME as a potential alternative energy source for the twenty-first century.

In Sweden, Chemrec [73] uses black liquor gasification, a waste stream from the pulping process, to produce BioDME. This synthetic second-generation biofuel offers a very high reduction of carbon dioxide emissions compared to conventional diesel fuel. By 2012, Chemrec produced up to 40 million gallons of BioDME or other renewable motor fuels a year. With the EU considering a potential biofuel mix for 2030, the market for BioDME is expected to increase dramatically.

The growing global production capacity for DME and BioDME means expanding and dynamic markets for methanol. The required methanol for the production of DME is obtained from synthesis gas (syngas). In principle, the methanol could be obtained from coal, organic waste, or biomass. Other possible improvements call for a dual catalyst system that permits both methanol synthesis and dehydration in the same process unit, with no methanol isolation and purification [70]. Both the one-step and two-step processes mentioned earlier are commercially available. Currently, there is more widespread application of the two-step process since it is relatively simple and start-up costs are relatively low. Efforts are, however, being made to improve one-step liquid-phase process [59,60,66,71].

Figure 9.1 illustrates some of processes from various raw materials to DME. As shown, DME can be produced from coal, biomass/waste, and natural gas through proper conversion processes.

9.3.1 Mobil MTG Process

Gasoline can be produced from methanol using a process developed by Mobil, which uses ZSM-5 Zeolite catalyst. The plant consists of five identical trains for conversion and has a maximum design capacity of 2200 tons per day of product gasoline. The gasoline produced by the MTG process contains durene, a substance with a high melting point (79°C). The durene produced by the MTG process is more than that permitted under product gasoline specifications. The durene content is reduced by treating the heavy gasoline produced in MTG, in heavy gasoline treatment plants prior to blending into the product gasoline [74–87].

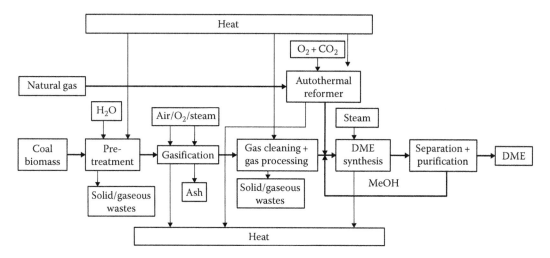

FIGURE 9.1 Various feedstocks for producing DME. (From Dimethyl ether, Wikipedia, the free encyclopedia, 2015.)

The MTG process was developed by Mobil in the early 1970s. In 1979, the New Zealand government [76,78,79,81,87] decided to employ the Mobil MTG process as an alternative in reducing the dependence on imported crude oil. A plant was built at Motunui with a production of about 14,000 barrels per day (bbl/day) of unleaded gasoline, having an octane rating of 92–94.

The MTG plant was the first commercial synthetic gasoline plant using new technology developed since World War II. The gasoline coming out from the plant was shipped to the Marsden Point refinery for blending into a gasoline pool in New Zealand. The methanol requirement for this process came from methanol plants producing two 2200 tons (water-free basis) per day [76,78,79,81,87].

9.3.1.1 Chemistry of Reactions

Starting from basic feedstock of coal, biomass, or natural gas, steam reforming process can be represented by the following general equation [74,75,80]:

$$C_nH_{2n+2} + nH_2O \rightarrow nCO + (2n + 1)H_2 \tag{9.17}$$

where C_nH_{2n+2} is any alkane and n is a positive integer. For natural gas, which is almost entirely methane, the equation given earlier becomes

$$CH_4 + H_2O \rightarrow CO + 3H_2 \tag{9.18}$$

Synthesis gas produced by these reactions can be used for methanol synthesis. The reaction of the synthesis gas can give a range of products [74,75,80,83], for example,

$$nCO + 2nH_2 \rightarrow CH_3(CH_2)_{n-1}OH + (n - 1)H_2O \tag{9.19}$$

$$nCO + (2n + 1)H_2 \rightarrow CH_3(CH_2)_{n-2}CH_3 + nH_2O \tag{9.20}$$

Hence, methanol synthesis requires proper choice of catalyst, which gives high selectivity for methanol. From this, if $n = 1$, then

$$CO + 2H_2 \rightarrow CH_3OH \tag{9.21}$$

The chemistry involved in MTG process is quite complex. A simplified reaction scheme (proposed by Chang and Silvestri [75]) is shown as follows:

$$2CH_3OH \underset{+H_2O}{\overset{-H_2O}{\rightleftharpoons}} CH_3OCH_3 \xrightarrow{-H_2O} C_2\text{--}C_5 \text{ alkenes} \rightarrow \text{Alkanes, cycloalkanes, aromatics} \qquad (9.22)$$

At a shorter contact time (in the order of 3–4 s), water and DME are the main products. When the contact time is increased, the yield of DME increases and reaches a maximum after which it further dehydrates to give C_2–C_5 alkenes. With further increase in the contact time, alkanes/C_{6+} alkenes and aromatics are obtained.

The MTG process is a selective catalytic conversion. The rate-limiting step is the conversion of DME to alkenes, a reaction step that appears to be autocatalytic. The catalytic conversion of methanol relies on the Bronsted acid sites on the zeolitic structure. The condensation of methanol molecules to DME is thought to involve the formation of surface methoxy group formed by the protonation and subsequent removal of water. The oligomerization of alkene molecules probably involves carbocation intermediates that are formed by protonation of alkene double bonds [74,75,83,84,86].

In the 1970s, Mobil's newly developed zeolite catalyst became a key element in the MTG process. Zeolites are porous, crystalline materials with three-dimensional framework composed of AlO_4 and SiO_4 tetrahedra. This catalyst, known as ZSM-5 can convert methanol to hydrocarbon products that are similar to the gasoline fraction of conventional petroleum. ZSM-5 has an intermediate pore diameter about 6 Å and has two sets of intersecting channels: elliptical, 10-membered ring channels and near-circular (sinusoidal) channels. It is this unique combination of channel shapes and sizes that make ZSM-5 so efficient and special in MTG conversion, producing gasoline range molecules (C_4–C_{10}) with practically no HCs above C_{10}. These selective properties also give ZSM-5 a reputation for high resistance to deactivation. Hence, a novel route to gasoline from either coal or natural gas can be achieved [74,75,83,84,86].

9.3.1.2 Process Description

The first step in an MTG process is to produce desulfurized natural or synthetic gas that is steam reformed to generate the mixture of hydrogen, carbon monoxide, and carbon dioxide. This mixture is cooled, compressed, and reheated to produce methanol in a methanol converter. This methanol containing 20% water becomes the feedstock for the MTG process.

There are three stages involved in an MTG process [76,78,79,81,87]:

1. Gasoline synthesis
2. Distillation
3. Heavy gasoline treating

Catalyst regeneration is also an essential part of the MTG process.

9.3.1.2.1 Gasoline Synthesis

Crude methanol is initially preheated, vaporized, and then superheated to a temperature range of 300°C–320°C in a series of heat exchangers. The vapor is then sent to a DME reactor containing a dehydration catalyst (alumina) where approximately 75% of the methanol is partially dehydrated to an equilibrium mixture of DME, water, and methanol.

$$2CH_3OH \rightarrow CH_3OCH_3 + H_2O \qquad (9.23)$$

The reaction is rapid, reversible, and exothermic. About 20% of the total heat produced is liberated in this step.

The mixture (at the temperature between 400°C and 420°C) is then mixed with recycled gas (consisting of light HCs, carbon dioxide, and hydrogen that absorbs heat of reaction) and passes to an MTG conversion reactor. In the MTG conversion reactor, DME is further dehydrated to give light alkenes that oligomerize and cyclize to give the final products by ZSM-5 catalyst with the liberation of the remainder of the heat [76,78,79,81,87].

The mixed effluent is cooled, by generating medium-pressure steam that is used to preheat the methanol feed and recycle gas. The conversion is essentially 100%. About 85%–90% of the hydrocarbon products can be used as gasoline having the remainder as fuel gas. Small amounts of CO, CO_2, and coke are formed as by-products. The recycled gas, water, and HCs then go to the product separator. The water is normally recycled to the reformer saturator, the recycled gas returns to the compressor, and the liquid HCs are sent to a distillation section.

9.3.1.2.2 Distillation
MTG hydrocarbon is refined in three distillation columns. A portion of the lighter and more volatile HCs, dissolved gases, and some water is removed by the first column. The second column removes the remaining light HCs, which are cooled to form LPG. It also recovers a high-vapor-pressure petrol blending component. Gasoline is then split into light and heavy fraction in a splicer column. Light gasoline is stored. Heavy fraction is sent to a treating facility [76,78,79,81,87].

9.3.1.2.3 Heavy Gasoline Treatment
Heavy gasoline produced in an MTG process contains a component known as durene (1,2,4,5-tetramethylbenzene) that has a high melting point (79°C). The concentration of durene is reduced in a heavy petrol treating section, by converting to low-melting-point petrol components, for example, isodurene (1,2,3,5-tetramethylbenzene) with a melting point of −23.7°C. The composition of a synthetic gasoline is quite similar to a conventional high-quality gasoline [76,78,79,81,87].

9.3.1.3 Advantages and Disadvantages of MTG Process
9.3.1.3.1 Advantages
1. Synthetic gasoline is free of sulfur and nitrogen. The product meets or exceeds existing gasoline specifications. Methanol conversion is virtually complete. Gasoline yield is high [74–87].
2. The overall energy efficiency of an MTG process including processing energy is about 92%–93% with 95% of the thermal energy of a methanol feed preserved in the hydrocarbon product. The remaining 5% is liberating as heat of reaction. The thermal efficiency of natural gas to gasoline is about 50%–60%.
3. The feed (methanol) for the MTG process can be made from the wide variety of sources, namely, natural gas, coal, and biomass. Also, the an process can convert most types of alcohol to gasoline [74–87].

9.3.1.3.2 Disadvantages
1. An MTG process is highly exothermic producing heat at 1740 kJ/kg of methanol consumed. The principal problem in reactor design is thus heat removal, which is important. The recycled gas provides a good absorbent for the heat of reaction [74–87].
2. One of the undesirable products in an MTG process is durene, which causes carburetor "icing" because of its high melting point. The synthetic gasoline contains higher concentration of durene (about 3–6 wt%) than is normally present in conventional gasoline (about 0.2–0.3 wt%). Durene can be either isomerized to give isodurene or it can potentially be used as a feedstock in the polymer industry.
3. An MTG process faces two types of catalyst deactivation. First is the aging due to coke deposition that requires regeneration every 3–4 weeks. Coke is burned off with a

heated air–nitrogen mixture. Operation of the MTG process can be kept continuous by using multiple reactors where one undergoing off-line regeneration at any given time while the others are run in parallel for MTG. The second type of irreversible coke is caused by steam (a reaction product), which leads to dealumination and loss of crystallinity of the catalyst. This can be minimized by operating at low temperatures and pressures [74–87].

4. In fixed-bed reactors, especially with fresh catalyst, the reaction only occurs over a relatively narrow band of the catalyst bed. As coke deposits, it first deactivate the front part of the bed; the active reaction zone subsequently move down the bed along the flow of the reactants (a phenomena called "band aging").

5. The major products of an MTG process are hydrocarbon and water. Therefore, any unconverted methanol will dissolve in water and be lost unless a distillation step is added to the process for recovering the methanol. Thus, essentially complete conversion of methanol is highly preferred. MTG process also does not directly produce diesel and jet fuels [74–87].

9.3.2 Methanol to Olefins Process

One of the most important classes of base chemicals are olefins and aromatics, some of which can be either directly added to liquid fuels to improve its performance or converted to chemicals that can be incorporated as additives to liquid fuels. Olefins and aromatics are also the most important base chemicals in the petrochemical industry [88–119]. Already for several decades, research is ongoing to produce these compounds directly from syngas; so far, these attempts have not been successful. The production of olefins and aromatics from methanol (MTO process) has shown more promise. Mobil has developed several processes in which methanol is converted to olefins and aromatics using zeolite catalysts. In the MTG process, the product contains approx. 15 wt% of aromatics, which can be recovered by conventional separation techniques. In a variant of the MTG process, the MTO process, olefins are produced directly from methanol. Such a process has been demonstrated and commercialized [88–119].

MTO reaction is one of the most important reactions in C_1 chemistry, which provides a chance for producing basic petrochemicals from nonoil resources such as coal and natural gas. Olefin-based petrochemicals and relevant downstream processes have been well developed for many years. Many institutions and companies have put great effort to the research on MTO reaction since it was first proposed by Mobil Corporation in 1977, and significant progress has been achieved with respect to reaction principle [91–93,113,117], catalyst synthesis [100,102,103], and process research and development (R&D) [95,111,118]. In the United States, UOP (part of Honeywell) has made significant contribution in the development of MTO process [107,109].

Methanol, which is very sensitive to a catalyst due to its high activity, could be catalyzed by acidic zeolites to form HCs. The reactant molecule is small and simple, but the reaction has been demonstrated to be very complicated with a large variety of products over different zeolite catalysts. The successful development of a commercially applicable MTO technology has addressed a number of issues [88–119]. Efficient zeolite catalyst development required the relationships among catalyst synthesis, catalyst properties, and catalyst deactivation and selectivity as well as deeper understanding of reaction mechanism. The process also required the large-scale production of catalyst from available raw materials. Once suitable catalyst was developed, a suitable reactor, optimum reaction conditions, and the method for its appropriate scale up were also pursued. Finally, both the catalyst and the reactor were integrated in a workable commercial process design. These issues along with fundamental research on MTO conversion process and process demonstration at numerous scales have been reviewed in numerous publications [88–119].

MTO is an autocatalytic reaction [88,91,92], in which the initial formation of a small amount of products leads to an enhanced methanol conversion until the efficient production period is reached.

How C–C bond forms from C_1 reactant has been debated during the past decades. Early studies proposed many direct mechanisms to explain C–C bond formation from methanol or DME, such as carbene mechanism, oxonium ylide mechanism, carbocation mechanism, and free radical mechanism [88–119]. Over fresh catalyst, the slow kinetic rate of the initial methanol conversion implies that C–C bond formation during this period possibly goes through the reaction route with relatively high energy barrier. Until now, how the first C–C bond generates is still a controversial issue.

The MTO conversion on acidic zeolites takes place through a complex network of chemical reactions. The distribution of products and thus the "selectivity" depends on the temperature, among other factors. In general, at lower temperatures methanol reacts to form DME. At higher temperatures, the desired products (olefins) are produced and the selectivity for DME decreases. While a number of zeolite catalysts have been tested, H-SAPO-34 was found to be most effective [89,93–95, 97,100,102,103,106,108]. This catalyst converted DME 100% at about 650 K. The olefin production was first observed around 523 K and the yields of lower olefins (C_2 and less) reached around 35%–45% at 650 K. Higher olefins (C_3 and C_4) were less than around 15% at this temperature.

The main focus of the published literature [89,93–95,97,100,102,103,106,108,113–118] has been to review investigations of the intermediate species involved in the MTO process. Some of the olefins, methanol, and DME react further to yield heavier species. These species, termed the "hydrocarbon pool," are occluded in the pores and catalyze the formation of olefins from methanol along with zeolite itself. "Hydrocarbon pool" are trapped because they are large relative to the size of the pores. They are formed during a kinetic induction period and are accumulated from reactions of methanol, DME, and other species produced initially. At steady state, the hydrocarbon pool continually reacts with methanol to yield olefins via elimination reaction. A simplified reaction network involving the hydrocarbon pool at steady state is illustrated in the following Equation (9.24) [89,93–95,97,100,102,103,106,108,113–118]:

$$\begin{array}{c}
C_2H_4 \\
\Updownarrow \\
CH_3OH \longrightarrow (CH_2)_n \rightleftharpoons C_3H_6 \\
\nearrow \quad \Updownarrow \quad \searrow \text{Saturated hydrocarbons} \\
\text{Hydrocarbonpool} \quad C_4H_8 \quad \searrow \text{Coke}
\end{array} \qquad (9.24)$$

Reactive species in the hydrocarbon pool include polyalkylated aromatics, large alkylated olefins, and carbenium ions. A carbenium ion has a trivalent carbocation center with three bonds and a net positive charge and is thus highly reactive. Benzenoid compounds in the hydrocarbon pool react with DME, methanol, or other species to form higher homologues such as xylenes and ethylbenzenes [102,103,106,108,113–118]. These can undergo additional reactions, including elimination to yield light olefins that can leave the catalyst pores. The original species are then realkylated to complete the catalytic cycle. An example of one possible step in this overall mechanism is the methylation of toluene by DME to produce xylene and methanol. The detailed reaction paths have been investigated by numerous researchers [102,103,106,108,113–118].

9.3.2.1 Catalyst, Reactor, and Process

The literature [89,93–95,97,100,102,103,106,108,113–118] indicates that the aromatics-based HCP reaction mechanism is dominant for the MTO reaction over SAPO-34 than on ZSM-5 and ZSM-22 [96,97,99]. Dual-cycle reactions proceed over ZSM-5 catalyst, and olefins methylation and cracking mechanism occurs in the reaction over ZSM-22. Even HCP mechanism for cage-type SAPO catalysts [97,101–103,113–119], olefin methylation, and cracking mechanism gives significant contribution in some cases. The acid density variation plays different roles in alkenes-based cycle and aromatics-based cycle. The Bronsted acid sites are very important. Due to the narrow pore opening and big super cage, methanol conversion over SAPO-34 is a characteristic of very quick coke formation and residue in the cage [97,101–103,113–119].

Haw et al. [115] reported that during methanol conversion, methylbenzenes, the most active confined organics, formed in the cage of SAPO-34 are converted with time on stream to methylnaphthalenes and polyaromatics, phenanthrene derivatives, and pyrene, which is the largest aromatic ring system that can be accommodated in the nanocages of the catalyst. The mass transport of reactants and products will be greatly reduced with the accommodation of these bulky coke species, which cause the deactivation of the catalyst. Considering the quick deactivation of SAPO-34, a fluidized-bed reactor is used in an industrial MTO process with reaction–regeneration cycle to maintain the activity of a catalyst [89,93–95,97,100,102,103,106,108,113–118].

The MTO reaction over SAPO-34 catalyst is highly exothermic, and the heat of the reaction is about −196 kcal/kg methanol feed when the reaction temperature is at 495°C. The reaction heat must be removed from the reaction bed simultaneously to keep the reaction temperature in the designed range. In addition, as mentioned earlier, the MTO reaction over SAPO-34 catalyst is featured by rapid catalyst deactivation due to coke deposition. In order to maintain the high activity of the catalyst in the reactor, the in-line combustion of coke is required. The fluidized-bed reactor is considered superior to a fixed-bed reactor because of the excellent heat transfer performance and good fluidity of the catalyst in fluidization state.

9.3.2.2 UOP/Hydro MTO Process

In UOP/hydro MTO process, the reaction for MTO can be shown in two steps [106,107]. The first step is the conversion of methanol to DME and water:

$$2CH_3OH \rightarrow CH_3OCH_3 + H_2O \tag{9.25}$$

Next, DME is converted to both ethylene and propylene. The ratio between ethylene and propylene production depends on the catalyst, reaction parameters, and the technology applied [92,106–111,119].

$$CH_3OCH_3 \rightarrow C_2H_4 + H_2O \tag{9.26}$$

$$3CH_3OCH_3 \rightarrow 2C_3H_6 + 3H_2O \tag{9.27}$$

The UOP/Hydro MTO process can achieve almost complete conversion of methanol and can provide 80% yield from methanol to ethylene and propylene [94,97]. The MTO process converts crude methanol to olefins, which results in savings for a methanol purification section.

The reaction takes place on a SAPO-34-type zeolite catalyst in a fluidized-bed reactor [97]. The catalyst is deactivated over time by the building of coke, which is why a portion of the spent catalyst is continuously removed from the reactor to a regeneration reactor. Air or oxygen is introduced to that regeneration reactor so that the catalyst can be regenerated by building carbon oxides [92,106–111,119]. The product stream leaving the reactor is fed to a separation section to remove water and to recover the nonreacted DME. An olefin-rich stream is then passed to a fractionation section that separates the mixture into the desired ethylene and propylene streams as well as the fuel gas stream and the stream that consists of medium-boiling HCs. According to the needs, the ratio between propylene and ethylene produced by the reaction can be adjusted in a range [92,106–111,119] of about 1.3–1.8.

The heavier hydrocarbon stream leaving fractionation is fed into a cracking zone to provide another source for ethylene and propylene production. The product stream of the cracking section is separated into high-boiling HCs, which are removed from the process, and an olefin-rich stream, which is rerouted to fractionation. The UOP MTO process was further expanded as MTO-OCP process for making plastics where olefin cracking is carried out in series with MTO process. This process was jointly developed by UOP and TOTAL [119].

9.3.2.3 DMTO Process

DMTO process (Dalian Institute of Chemical Physics [DICP] MTO) is a proprietary technology developed by DICP (Chinese Academy of Sciences). In 2010, the first DMTO commercial unit in the world with a production capacity of 600,000 tons of lower olefins per year was successfully initiated [93,97,100]. In this process, the large quantity of small catalyst particles provided a very large surface area for reaction and heat transfer in a fluidized bed. The temperature gradient in the fluidized bed was also readily reduced to just several degrees via optimal operation. The process was a two-stage process that included a fluidized-bed reactor and a regenerator. The fluidized catalyst particles were transported between the reactor and the regenerator allowing a continuous regeneration of catalyst [93,97,100].

Experiments in the laboratory showed that a catalyst–gas contact time of about 2–3 s is necessary in order to avoid undesired by-product in the DMTO process. Such a short catalyst–gas contact time means that an industrial DMTO reactor has to be operated at a gas velocity higher than 1 m/s, which leads to a fluidized bed with a bed height less than 3 m to achieve the desired contact time. DMTO catalyst particles have similar physical properties as FCC catalyst [120]. For these types of particles, a superficial gas velocity of 0.5–1.5 m/s would make the fluidized bed operating in the turbulent fluidized-bed regime. A turbulent fluidized-bed reactor offered the excellent catalyst–gas contact, high mass transfer efficiency, and large solids holdup. In the DMTO process, a turbulent fluidized-bed reactor operating at a superficial gas velocity of 1–1.5 m/s was used. The process has been successfully demonstrated and it is described in detail in numerous publications [93,97,100].

Tian et al. [93] suggested that while DMTO process technology should be exported to other countries with abundant coal and natural gas, MTO process can be further improved with additional research for better understanding of MTO reaction mechanism. Specifically, questions such as how the first C–C bond forms, what happens in the induction period, what exact relations among different reaction routes are used, how coke forms, and how control coking reaction in the reaction network is controlled need to be further investigated. More fundamental understanding of HCP mechanism is needed. While fundamental investigations on the SAPO molecular sieve synthesis offers strong support to the existing catalyst development, further MTO catalyst development should include the crystallization mechanism studies for the precise control of synthesis, as well as the application of new catalytic materials such as mesoporous–microporous zeolites that have been confirmed with longer lifetime for many reactions [101,113,117]. To find a molecular sieve with suitable cavity size restricting coke formation is also a potential direction for developing a better MTO catalyst. Finally, the strategy to design an MTO catalyst with similar physical properties (density, particle size distribution, etc.) to FCC catalyst, enabling the application of industrial FCC research, operation, and design experience into the future MTO process development, is critical and should be explored.

9.4 FT SYNTHESIS

In the nonselective catalytic FT synthesis, one mole of CO reacts with two moles of H_2 to form mainly paraffin straight-chain HCs (C_xH_{2x}) with minor amounts of branched and unsaturated HCs (i.e., 2-methyl paraffins and α-olefins) and primary alcohols. Undesirable side reactions include methanation, the Boudouard reaction, coke deposition, oxidation of the catalyst, or carbide formation. Typical operation conditions for FT synthesis include temperatures of 150°C–350°C and pressures between 25 and 60 bar [120–135]. High-temperature FT (HTFT) synthesis is generally operated at 25 bar. In the exothermic FT reaction, about 20% of the chemical energy is released as heat. The major reaction can be expressed as

$$CO + 2H_2 \rightarrow -(CH_2)_n + H_2O \tag{9.28}$$

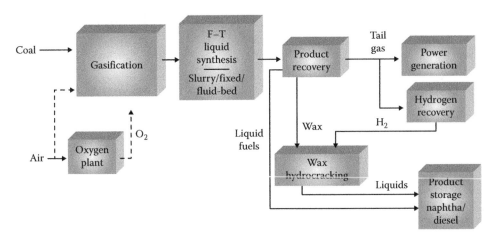

FIGURE 9.2 Simplified FT synthesis-based production scheme. (From Fischer–Tropsch synthesis-liquid fuels, NETL report, Department of Energy, Washington, DC, 2010, www.netl.doe.gov/research/coal/.../gasification/gasifipedia/ftsynthesis, accessed on March 2010.)

FT processes can be used to produce either a light synthetic crude oil (syncrude) and light olefins or heavy waxy HCs. The syncrude can be refined to high-quality sulfur and aromatic liquid product and specialty waxes or, if hydrocracked and/or isomerized, to produce excellent diesel fuel, lube oils, and naphtha, which is an ideal feedstock for cracking to olefins. For direct production of gasoline and light olefins, the FT process is operated at high temperatures (330°C–350°C) [121] and for the production of waxes and/or diesel fuel at low temperatures (220°C–250°C) [131]. CO + H_2 conversion for low-temperature FT (LTFT) ranges between 60% and 93%. The similar number for HTFT is about 85%. FT process can be applied to coal, biomass, or natural gas. The use of these three feedstocks to make liquid products is often referred to as coal to liquid (CTL), biomass to liquid (BTL), and GTL [120–150]. A simplified schematic of CTL process is described in Figure 9.2 [79].

While a variety of synthesis gas compositions can be used, the correct feed gas ratio between H_2 and CO is desirable. As shown later, FT can be carried out using iron, cobalt, or ruthenium catalysts. When using cobalt catalysts, the optimum molar ratio of H_2/CO is around 1.8–2.1. If the syngas produced by the gasifier has a lower ratio, an additional WGS reaction is used to adjust the ratio by converting a part of the CO with steam to form more H_2. Iron catalysts have intrinsic WGS activity, and so with this catalyst the H_2-to-CO ratio need not be high. The WGS reaction can be important for synthesis gas derived from coal or biomass, which tends to have relatively low H_2/CO ratios (<1). For iron catalysts, the required ratio can be between 0.6 and 1.7 depending on the presence of catalyst promoters, gas recycling, and the reactor design [120–135,151–167].

As mentioned earlier, generally, the FT process is operated in the temperature range of 150°C–300°C (302°F–572°F). Higher temperatures lead to faster reactions and higher conversion rates but also tend to favor methane production. For this reason, the temperature is usually maintained at the low-to-middle part of the range. Increasing the pressure leads to higher conversion rates and also favors formation of long-chained alkanes, both of which are desirable. Typical pressures range from one to several tens of atmospheres. Even higher pressures would be favorable, but the benefits may not justify the additional costs of high-pressure equipment, and higher pressures can lead to catalyst deactivation via coke formation. HTFT is operated at temperatures of 330°C–350°C and uses an iron-based catalyst. This process was used extensively by Sasol in their CTL plants [168–172]. LTFT is operated at lower temperatures (220°C–250°C) and uses an iron- or cobalt-based catalyst. This process is best known for being used in the first integrated GTL plant operated and built by Shell in Bintulu, Malaysia [121].

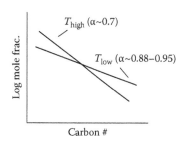

Low T	Sasol Arge	High T	Sasol Synthol
• Low C_1–C_4	13.3	• Higher C_1–C_4	43.0
• Low C_5–C_{11}	17.9	• Higher C_5–C_{11}	40.0
• Low C_{12}–C_{19}	13.9	• Less C_{12}–C_{19}	7.0
• 50%–70% wax	51.7	• Low wax	4.0
• 220°C–270°C		• 325°C–350°C	
• α: 0.87+		• α: ~0.7	
• Gasoline/diesel: 1:2		• Gasoline/diesel: 2:1	
• 80° Cetane #		• 50 – 60 Cetane #	
• 0 – 20 Octane #		• 0 – 60 Octane #	

FIGURE 9.3   High- and low-temperature product distributions by SASOL processes (dependency of Fischer–Tropsch synthesis ASF distribution on temperature. Product selectivities [in %] of the Sasol Arge [220°C] and Sasol Synthol [325°C] processes are on a C atom basis). (Reprinted from Shah, Y.T., *Energy and Fuel Systems Integration*, CRC Press, New York, 2015; Fischer–Tropsch synthesis—liquid fuels, NETL report, Department of Energy, Washington, DC, 2010, www.netl.doe.gov/research/coal/.../gasification/gasifipedia/ftsynthesis, accessed on June 2015.)

Typical product distributions of low-temperature (SASOL Arge) and high-temperature (SASOL Synthol) FT processes are briefly described in Figure 9.3 [79].

Feed syngas for FT process must have very low sulfur content (of the order of 10–100 ppb). Sulfur causes permanent loss of catalyst activity and so reduces catalyst lifetimes. There is a trade-off here between the additional costs of gas cleaning and the catalyst lifetime. In general, S, Cl, and N compounds are detrimental to catalytic conversion; hence, it is desirable to employ wet scrubbing to completely remove these contaminants before FT reactors. Cobalt catalysts have higher activities than iron catalysts but are more expensive and have lower contaminant tolerances [120–135,168–172].

The concentration of tar in syngas must be below 10s ppb and its dew point must be below the catalyst operating temperature. The heavier tars would condense onto surfaces, thus reducing the catalyst surface area and lifetimes. While this is a serious problem with fixed-bed catalysts, slurry-bed reactors can tolerate traces of aromatics without any serious problems.

The nonreactive gases such as nitrogen and methane must be in low concentration because their presence increases the size and cost of equipment needed. For all of the processes, reduction in the volume of inert components in the syngas reduces the requirements for the volume of downstream equipment, which results in the cost reduction. For these reasons, oxygen-blown or oxygen-enriched gasification is preferred by many developers currently working on liquid fuel production from syngas. Oxygen-blown entrained-bed reactor is a preferred gasifier for coal gasification [120–135].

9.4.1   Product Distribution

In general, the product distribution of HCs formed during an FT process follows an Anderson–Schulz–Flory distribution [120–135], which can be expressed as

$$\frac{W_n}{n} = (1-\alpha)^2 \alpha^{n-1} \tag{9.29}$$

where
  W_n is the weight fraction of HCs containing n carbon atoms
  α is the chain growth probability or the probability that a molecule will continue reacting to form a longer chain
  In general, α is largely determined by the catalyst and the specific process conditions.

Examination of the equation given above reveals that methane will always be the largest single product so long as alpha is less than 0.5; however, by increasing α close to one, the total amount of methane formed can be minimized compared to the sum of all of the various long-chained products. Increasing α increases the formation of long-chained HCs. The very long-chained HCs are waxes, which are solid at room temperature. Therefore, for the production of liquid transportation fuels, it may be necessary to crack some of the FT products. In order to avoid this, some researchers have proposed using zeolites or other catalyst substrates with fixed sized pores that can restrict the formation of HCs longer than some characteristic size (usually $n < 10$). This way they can drive the reaction so as to minimize methane formation without producing lots of long-chained HCs. Such optimization of product distribution is still being explored [120–135].

9.4.2 CATALYSTS

The status of the several FT processes through 1950 is given by Shah and Perrotta [18]. The yield of C_3^+ product per cubic meter of synthesis gas is very important because it directly relates to the purity of the synthesis gas. The cost of the production of purified synthesis gas can be as high as 70% of the total cost of the FT process [18]. Earlier work focused on Fe catalysts and the improvements in these catalysts that would increase their operability and selectivity and decrease the operating costs of FT process [18]. The efforts were made to prevent the reaction $2CO = CO_2 + C$ and improve the steady-state life of the catalyst. Significant developmental efforts were also made for a more active and mechanically stable catalyst to further reduce the yields of C_1 and C_2 [18]. The process improvement efforts were also focused on developing a catalyst that minimized the shift reaction.

Shah and Perrotta [18] pointed out that selectivity in FT reaction can be enhanced by either poisoning acceptable catalysts with sulfur compounds or by selecting sulfides of less frequently used catalysts. They also indicated that an addition of small amount of H_2S initially increased nickel–manganese catalyst activity. This was also confirmed by Herrington and Woodward [10] for cobalt–thoria–kisselguhr (100:18:100) catalysts. In their experiments H_2S was mixed with the synthesis gas in small batches and no H_2S was eliminated in the off-gas during the course of sulfur poisoning experiments. The first addition of H_2S increased the yield of liquid HCs at constant temperature. As sulfur addition continued, there was a decrease in the yield of gaseous HCs. The total hydrocarbon yield increased with sulfur addition to the catalyst until 8 mg of sulfur was added to each gram of the catalyst. This work suggested the advantage of stopping sulfidization at low level (1–4 mg of sulfur/g of catalyst) to obtain the benefits of increased liquid HCs yield. The results also suggested that the catalyst might show the same behavior if presulfided to the same degree before introducing the synthesis gas. Shah and Perrotta [18] also indicated that the literature showed that 69% of CO conversion can be obtained for a molybdenum disulfide catalyst alkalized with 2%–3% KOH in a feed of $2H_2 + CO$ at 530°F and 13.6 atm. Products from this synthesis were low boiling with 30% of the product C_3^+ HCs and organic oxygenated compounds. Laynes [11] indicated that by allowing the sulfur content of the iron catalyst to build up to an optimum ratio and maintain at that level will minimize CO_2 formation during hydrogenation of CO to form HCs.

The new catalyst development work indicated that group VIII transition metal oxides are generally good CO hydrogenation catalysts. Vannice reported a relative activity of these metals for FTS [165] in decreasing order of activity as Ru > Fe > Ni > Co > Rh > Pd > Pt. Ni is basically a methanation catalyst and does not have the broad selectivity of other FT catalysts. Ru has very high activity and quite high selectivity for producing high-molecular-weight products at low temperatures. Fe is also very active and has WGS activity. Fe readily forms carbides, nitrides, and carbonitrides with metallic character that also have FTS activity. Fe also has a stronger tendency than Ni or Co to produce carbon that deposits on the surface and deactivates the catalyst.

Co tends to have a longer lifetime than Fe catalysts and does not have WGS activity, which leads to improved carbon conversion to products because CO_2 is not formed. Co catalysts in FTS yield mainly straight-chain HCs (no oxygenates like Fe). Although Ru is the most active FTS catalyst, it is much more expensive than Fe. Iron is by far the least expensive FTS catalyst of all of these metals. Co catalysts are 230 times more expensive than Fe but are still an alternative to Fe catalysts for FTS because they demonstrate activity at lower synthesis pressures, so higher catalyst costs can be offset by lower operating costs [18].

The development of FT catalyst in recent years has been focused on Fe, Co, and Ru metals. The three key properties of FT catalysts are lifetime, activity, and product selectivity. Optimizing these properties for desired commercial application has been the focus of FT catalyst R&D. Each one of these properties can be affected by a variety of strategies including [10,11,18,151–167]

1. Use of promoters (chemical and structural)
2. Catalyst preparation and formulation
3. Pretreatment and reduction
4. Selective poisoning
5. Shape selectivity with zeolites

Over the last several decades, the continuous effort to improve catalyst activity, selectivity, and stability has been carried out. Both cobalt and ruthenium catalysts with different types of promoters were extensively examined by Exxon and other oil companies. Their studies indicate that while these catalysts give higher initial activity, they also tend to decay rapidly. Numerous patents on these catalysts have been reported by Exxon and other oil companies. In addition to active metals, FT catalysts typically contain a number of "promoters," such as potassium and copper [151–167].

The performance of Co catalysts is not very sensitive to the addition of promoters. Early work demonstrated that the addition of ThO_2 improved wax production at atmospheric pressure, but had little effect at higher pressures. Group 1 alkali metals, including potassium, are a poison for cobalt catalysts but are promoters for iron catalysts. Promotors also have an important influence on activity. While alkali metal oxides and copper are common promotors, their formulation and effectiveness depend on the primary metal, that is, iron versus cobalt [7]. Alkali oxides on cobalt catalysts generally cause activity to drop severely even with very low alkali loadings. C_{5+} and CO_2 selectivity increase, while methane and C_2–C_4 selectivity decrease. In addition, the olefin-to-paraffin ratio increases. With Fe catalysts, however, promoters and supports are essential catalyst components. Since the discovery of FTS, potassium has been used as a promoter for Fe catalysts to effectively increase the basicity of the catalyst surface. The objective is to increase the adsorption of CO to the metal surface, which tends to withdraw electrons from the metal, by providing an electron donor. Adding potassium oxide to Fe catalysts also tends to decrease hydrogenation of adsorbed carbon species, so chain growth is enhanced, hence resulting in a higher-molecular-weight product distribution that is more olefinic. Potassium promotion also tends to increase WGS activity leading to a faster rate of catalyst deactivation because of the increased rate of carbon deposition on the surface of the catalyst.

Copper has also been successfully used as a promoter in Fe catalysts for FT synthesis. It increases the rate of FTS more effectively than potassium but decreases the rate of the WGS reaction. Copper has been shown to facilitate iron reduction. The average molecular weight of the products increases in the presence of copper, but not as much as when potassium is used.

Catalyst preparation impacts the performance of Fe and Co catalysts. Fe catalysts can be prepared by precipitation onto catalyst supports such as SiO_2 or Al_2O_3 or as fused iron where formulations are prepared in molten iron and then cooled and crushed. Thus, FT iron catalysts need alkali promotion to attain high activity and stability (e.g., 0.5 wt% K_2O). An addition of Cu to promote reduction mechanism, an addition of SiO_2, Al_2O_3 for structural promotion, and maybe some manganese for selectivity control (e.g., high olefin concentration in the products) are recommended [58,121,151–167]. In the initial period of catalyst synthesis, catalyst is reduced with hydrogen to produce several iron carbide phases and elemental carbon along with iron oxides and some metallic iron. The control of these phase transformations can be important in maintaining catalytic activity and preventing breakdown of the catalyst particles [18,58,121,151–167].

The role of supports in Co catalysts is also important. Since Co is more expensive than Fe, precipitating the ideal concentration of metal onto a support can help reduce catalyst costs while maximizing activity and durability. The combination of light transition metal oxides such as MnO with Fe increases the selectivity of light olefins in FTS. Fe/Mn/K catalysts have shown selectivity for C_2–C_4 olefins as high as 85%–90%. Noble metal addition to Co catalysts increases FTS activity but not selectivity [18,58,121,151–167].

All catalysts for FT synthesis can be damaged with impurities such as NH_3, HCN, H_2S, and COS. These impurities poison the catalysts. HCl causes corrosion of catalysts. Alkaline metals can be deposited on the catalyst resulting in catalyst deactivation. Tars can also be deposited on the catalyst resulting in its poisoning and contamination of the products. The formation of particles (dust, soot, ash) causes fouling of the reactor. The removal limit for all impurities is based on an economic optimum determined by catalyst stand-time and investment in gas cleaning. However, in general, all these impurities should be removed to a concentration below 1 ppmv [18,58,121,151–167].

FT catalysts are particularly sensitive to poisoning by sulfur-containing compounds. Cobalt-based catalysts are more sensitive than for their iron counterparts. The FT synthesis process can be made generally selective by proper choice of operating conditions and catalysts. Most of the studies over the last several decades have been focused in this area. There is an interesting question to consider: what features have the metals nickel, iron, cobalt, and ruthenium in common to let them—and only them—be FT catalysts, converting CO/H_2 mixture to aliphatic (long chain) HCs in a "one-step reaction." The term "one-step reaction" means that reaction intermediates are not desorbed from the catalyst surface. In particular, it is also interesting that the much carbided, alkalized iron catalyst behaves similarly to the just metallic ruthenium catalyst. The kinetic principle of "selective inhibition" might be the common feature that applies, in spite of differences in catalyst composition, reaction intermediates, steps of reaction, and corresponding kinetic schemes [18,58,121,151–167].

With iron catalysts, two directions of selectivity have been pursued. One direction has aimed at a low-molecular-weight olefinic hydrocarbon mixture to be produced in an entrained-phase or fluidized-bed process (Sasol Synthol process). Due to the relatively high reaction temperature (approx. 340°C), the average molecular weight of the product is so low that no liquid product phase occurs under reaction conditions. The catalyst particles moving around in the reactor are small (particle diameter 100 mm) and carbon deposition on the catalyst does not disturb reactor operation. Thus, low catalyst porosity with small pore diameters is appropriate for this purpose. For maximizing the overall gasoline yield, the olefins C_3 and C_4 have been oligomerized at Sasol FT plant. The olefins can also be used for subsequent polymerization process.

The second direction of iron catalyst development has aimed at the highest catalyst activity to be used at low reaction temperature where most of the hydrocarbon product is in the liquid phase under reaction conditions. Typically, such catalysts are obtained through precipitation from nitrate solutions. A high content of a carrier provides mechanical strength and wide pores for easy mass transfer of the reactants in the liquid product filling the pores. The main product fraction then is a paraffin wax, which is refined to marketable wax materials at Sasol FT plant; however, it also can be very selectively hydrocracked to a high-quality diesel fuel. Thus, iron catalysts are very flexible. Iron catalysts have a higher tolerance for sulfur, are cheaper, and produce more olefin products and alcohols. The lifetime of the Fe catalysts is short and in commercial installations generally limited to 8 weeks [18,58,115,151–157]. Iron catalysts are preferred for lower-quality feedstocks such as those obtained from coal or biomass [136–149].

Cobalt catalysts have the advantage of a higher conversion rate and a longer life (over 5 years). Cobalt catalysts are used at the lower temperature range to produce waxy, long-chained products that can be cracked to diesel. Both iron and cobalt catalysts can be used in a range of different reactor types (fixed bed, slurry reactor, etc.) [129,132,136,137,139]—for example, CHOREN uses a cobalt catalyst in a fixed-bed reactor, developed by Shell, to produce FT diesel. The cobalt catalysts are also in general more reactive for hydrogenation and produce therefore less unsaturated HCs (olefins) and alcohols compared to iron catalysts.

Cobalt catalysts are more active for FT synthesis when the feedstock is natural gas. Natural gas has high hydrogen-to-carbon ratio, so the WGS is not needed for cobalt catalysts. Synthesis gases derived from hydrogen-poor feedstocks have low hydrogen content (e.g., coal) and require the WGS reaction. Iron catalysts favor WGS reaction [18,58,121,151–167].

Ruthenium is the most active at the lowest reaction temperature. It produces the highest-molecular-weight HCs ("polymethylene synthesis"). It performs the chain growth reaction in the cleanest mode. It also acts as a FT catalyst as the pure metal, without any promotors, thus providing the simplest catalytic system of FT synthesis. Just like for nickel, the selectivity for ruthenium catalyst changes to mainly methane at elevated temperature. Its high price and limited world resources prohibit its industrial application [18,58,121,136,137,139,151–167].

In recent years, metal organic framework (MOF) has been tested to improve the selectivity of FT reactions. This work is being carried out at NETL in Pittsburgh and it is still at the development stage.

9.4.3 FT Reactor Options

FT reactor can be either a fixed bed, a slurry bed, or different forms of fluidized-bed. In earlier studies and commercialization, fixed-bed reactor technology was extensively used. Since FT process is exothermic, a careful control of heat and mass transfer is very important for CO conversion and catalyst selectivity and stability. In the fixed-bed reactor, special efforts have been made to design the reactor internals to remove the heat and control the reactor temperature. Subsequently, on large-scale operations, slurry bed was preferred. In principle, four types of reactors can be used [120–135,151–167,173,174].

9.4.3.1 Multitubular Fixed-Bed Reactor

This type of reactor contains number of tubes with small diameter. These tubes contain catalyst and are surrounded by boiling water that removes the heat of reaction. Fixed-bed reactor is suitable to operate at low temperatures and has upper temperature limit of 530 K. Excess temperature leads to carbon deposition and hence blockage of the reactor. Since large amounts of products formed are in liquid state, this type of reactor can also be termed as trickle flow reactor system [174].

9.4.3.2 Entrained-Flow Reactor

An important requirement of the reactor for the FT process is to remove the heat of the reaction. This type of reactor contains two banks of heat exchangers that remove heat, the remainder of which is removed by products and recycled in the system. The formation of heavy waxes should be avoided since they condense on the catalyst and form agglomeration. This leads to incipient of fluidization regime. Hence, risers are operated over 570 K [135,173].

9.4.3.3 Fluidized-Bed and Circulating Catalyst Reactors

These are used for HTFT synthesis (nearly 340°C) to produce low-molecular-weight olefinic HCs on alkalized fused iron catalysts. The fluidized-bed technology (as adapted from catalytic cracking of heavy petroleum distillates) was introduced by Hydrocarbon Research in the years 1946–1950 and named the "Hydrocol" process. A large-scale FT Hydrocol plant (350,000 tons per annum) operated during the years 1951–1957 in Brownsville, Texas. Due to technical problems and increasing petroleum availability, this development was discontinued. Fluidized-bed FT synthesis has recently been very successfully reinvestigated by Sasol. One reactor with a capacity of 500,000 tons per annum is now in operation and even larger ones are being built (nearly 850,000 tons per annum). The process is now used for mainly olefins C_2 and C_7 production. This new development can be regarded as an important progress in FT technology [129,132,135–137,139].

A high-temperature process with a circulating iron catalyst ("circulating fluidized bed," "riser reactor," "entrained catalyst process") was introduced by the Kellogg Company and a respective plant built at Sasol in 1956. It was improved by Sasol for successful operation. At Secunda, South Africa, Sasol has operated 16 advanced reactors of this type with a capacity of approximately 330,000 tons per annum each. Now the circulating catalyst process is being replaced by the superior Sasol advanced fluidized-bed technology. Early experiments with cobalt-catalyst particles suspended in oil have been performed by Fischer. The bubble column reactor with a powdered iron slurry catalyst and a CO-rich syngas was particularly developed to pilot plant scale by Kölbel at the Rheinpreussen Company in 1953. Recently (since 1990), LTFT slurry processes are under investigation for the use of iron and cobalt catalysts [139], particularly for the production of a hydrocarbon wax, to be used as such, or to be hydrocracked and isomerized to mainly diesel fuel by Exxon and Sasol. Today, slurry-phase (bubble column) LTFT synthesis is regarded by many authors as the most efficient process for FT clean Diesel production. This FT technology is also under development by the Statoil Company (Norway) for use on a vessel to convert associated gas at offshore oil fields into a hydrocarbon liquid [120–139,151–167].

9.4.3.4 Slurry Reactors

More recently, slurry-bed reactor is preferred because it offers distinct advantages for the control of both mass and heat transfer problems that may affect the reactor performance. Slurry-bed reactor also allows more flexibility in the use of the catalyst particle size. For coal- and petroleum-derived syngas, commercial FT reactors are being operated by Shell, Exxon, Sasol, and Syntroleum, among others all over the world. Shell Pearl GTL project in Qatar produces 70,000 bbl/day. Shell also has a smaller commercial plant in Bintulu, Malaysia, which produces 14,700 bbl/day. Syntroleum operates FT process in Australia. Sasol-QP GTL ORYX-1 project in South Africa produces 34,000 bbl/day. These and many other commercial technologies can be readily used for the FT process that uses biosyngas. The size of FT process will depend on the size of the gasifier for an integrated process. For example, a BTL plant producing 2100 bbl/day will require a gasifier producing 250 MWth [57,120–137,139,151–167,173,174].

Heat removal is done by internal cooling coils. Synthesis gas is bubbled through the waxy products and finely divided catalyst, which is suspended in the liquid medium. This also provides agitation of the contents of the reactor. Catalyst particle size reduces diffusional heat and mass transfer limitations. Lower temperature in the reactor leads to more viscous product, and higher (>570 K) gives undesirable product spectrum. However, separation of the product from the catalyst is a problem.

9.4.4 PRODUCT SEPARATIONS AND UPGRADING

The products from the FT reactor are generally separated as gas, liquid, and solid (which may include catalyst for a continuous slurry-phase reactor). The off-gases can be recycled back to the gasifier or to FT reactor feed before or after a reforming stage. The catalysts from solids are separated, and the remaining solids can also be recycled to the gasifier or a reformer. Portions of hot gas and solids can be used for process heat as well as for the generation of electricity through a downstream combustion process. The main liquid product is then upgraded to produce ASTM standard biodiesel, biogasoline, or biojet fuel using standard refining and hydroprocessing operations. As mentioned earlier, the nature of the liquid produced from FT reactor will depend on the nature of the catalyst, operating conditions, and H_2/CO ratio. The upgrading strategy will, therefore, depend on the nature of the liquid and the desired end product [120–135].

9.4.5 FT COMMERCIALIZATION

Since syngas can be generated from coal, biomass/waste, or natural gas, as mentioned before, there are three types of FT conversion processes, commonly designated as CTL, BTL, or GTL. FT commercialization has largely occurred for CTL and GTL processes. These processes are large and successful for a long period. We describe few of them in brief details. We also briefly outline an integrated concept of BTL process as a case study for new emphasis on liquid production from biomass. Finally we discuss small-scale GTL processes that are being developed for stranded and remote gas reservoirs [57,120–135,173,174].

The FT process has been applied in large-scale gas–liquid and coal–liquid facilities such as Pearl GTL facility in Ras Laffan, Qatar and others. Such large facilities are susceptible to high capital costs, high operation and maintenance costs, the uncertain and volatile price of crude oil, and possible harmful impact to environment.

9.4.5.1 Sasol

The largest scale implementation of FT technology are in a series of plants operated by Sasol in South Africa, a country with large coal reserves, but little oil. The first commercial plant opened in 1952, 40 miles south of Johannesburg [121]. Sasol uses coal and now natural gas as feedstocks and produces a variety of synthetic petroleum products, including most of the country's diesel fuel.

In December 2012, Sasol announced plans to build a 96,000 barrels-a-day plant in Westlake, Louisiana, using natural gas from tight shale formations in Louisiana and Texas as feedstock. Costs are estimated to be between $11 and $12 billion with $2 billion in tax relief being contributed by the State of Louisiana. The planned complex will include a refinery and a chemical plant.

Thermal efficiency of Sasol's slurry-phase FT process is around 60%, and since it is a slurry-based process, it inherently recycles the reactants. Syngas CO conversion is 75%. Single-pass FT always produces a wide range of olefins, paraffins, and oxygenated products such as alcohols, aldehydes, acids, and ketones with water or CO_2 as a by-product. Product selectivity can also be improved using multiple step processes to upgrade the FT products. More detailed description of SASOL process is given in numerous literature [57,120–135,173,174].

9.4.5.2 PetroSA

PetroSA, a South African company that, in a joint venture, won project innovation of the year award at the Petroleum Economist Awards in 2008, has the world's largest GTL complexes at Mossel Bay in South Africa. The refinery is a 36,000 barrels-a-day plant that completed semicommercial demonstration in 2011. The technology can be used to convert natural gas, biomass, or coal into synthetic fuels [121].

9.4.5.3 Shell Middle Distillate Synthesis

One of the largest implementations of FT technology is in Bintulu, Malaysia. This Shell facility converts natural gas into low-sulfur diesel fuels and food-grade wax. The scale is 12,000 bbl/day (1,900 m³/day). This process is described in the literature [57,120–135,173,174].

9.4.5.4 Ras Laffan, Qatar

The new LTFT facility Pearl GTL, which began operation in 2011 at Ras Laffan, Qatar, uses cobalt catalysts at 230°C, converting natural gas to petroleum liquids at a rate of 140,000 bbl/day (22,000 m³/day), with additional production of 120,000 barrels (19,000 m³) of oil equivalent in natural gas liquids and ethane. The first GTL plant in Ras Laffan was commissioned in 2007 and is called Oryx GTL and has a capacity of 34,000 bbl/day. The plant utilizes the Sasol slurry-phase distillate process, which uses a cobalt catalyst. Oryx GTL is a joint venture between Qatar Petroleum and Sasol [121].

9.4.5.5 UPM (Finland)

In October 2006, Finnish paper and pulp manufacturer UPM announced its plans to produce biodiesel by the FT process alongside the manufacturing processes at its paper and pulp plants; using waste biomass resulting from paper and pulp manufacturing processes as source material, UPM constructed a biorefinery at its Kaukas mill site in Lappeenranta, Finland. It is the world's first refinery that produces second-generation, wood-based biodiesel on a large scale. The refinery produces approximately 120 million L of renewable diesel fuel a year [121].

9.4.5.6 Others

A demonstration-scale FT plant was built and operated by Rentech, Inc., in partnership with ClearFuels, a company specializing in biomass gasification. Located in Commerce City, Colorado, the facility produces about 10 bbl/day (1.6 m³/day) of fuels from natural gas. Commercial-scale facilities were planned for Rialto, California; Natchez, Mississippi; Port St. Joe, Florida; and White River, Ontario. However, Rentech closed down their pilot plant in 2013 and abandoned the work on their FT process as well as the proposed commercial facilities [121].

In the United States and India, some coal-producing states have invested in FT plants. In Pennsylvania, Waste Management and Processors, Inc., was funded by the state to implement FT technology licensed from Shell and Sasol to convert so-called waste coal (leftovers from the mining process) into low-sulfur diesel fuel. Numerous other efforts in India, Australia, and other countries are also being pursued.

Most of the large-scale commercial activities described earlier pertain to the syngas obtained from coal (CTL) or natural gas. In recent years, more emphasis has been placed on syngas conversion to liquids using biomass as original feedstock (BTL) [136–150]. CHOREN, one of the leading developers of BTLs via the FT route, estimates that the minimum economic scale for an FT plant would be around half of the scale of their Sigma plant, which corresponds to 100,000 tons/year BTL fuel output, or around 1,520 odt/day biomass input. These are smaller numbers than ones used for CTL. There are also newer process technologies in development that could reduce this minimum economic scale. For example, the Velocys technology [175,176], which was recently acquired by Oxford Catalysts, has been estimated to allow FT process to be viable at outputs of 500–2000 bbl/day, which would correspond to biomass inputs of 300–1220 odt/day. These small-scale processes are useful for conversion of stranded natural gas or small biomass footprints in remote locations. Here we briefly describe a typical BTL and small-scale GTL FT process [175,176].

9.4.5.7 An Integrated Process for Biomass to Biodiesel Oil Production via FT Synthesis (BTL)

A schematic of typical integrated BTL process and a proposed modification of this process by researchers at University of California, Riverside, are described by Shah [139]. In UC Riverside modification, conventional gasification reactor was replaced by a hydrogasification reactor to produce biosyngas with larger hydrogen concentration from biomass. In general, the integrated process has five elements: pretreatment, gasification, gas reforming, FT conversion, and product upgrading and usages. Various modifications of an integrated BTL process are possible depending on the use of preferred technologies in these basic five elements of an integrated BTL process [139].

Boerrigter et al. [15,16,137,144,147–150] point out that a design of an integrated BTL process depends on whether or not one takes a front-end approach or a back-end approach.

While existing commercial technology for GTL and CTL can be used for BTL, the scale of BTL plant is going to be important. Unlike coal and natural gas, biomass is difficult to transport and store, and the cost of feed preparation of biomass can become an important factor in the scale of BTL process. This issue has been briefly discussed later in the economy of scale of BTL operation. Since FT process is oblivious to how syngas is produced (as long as its composition is not significantly varied), gasification technology is the key to the integration of GTL, CTL, or BTL process. In order to take advantage of the economy of scale, significant efforts are being made to examine CBTL (mixture of coal and BTL) process. As shown later, this process offers some distinct advantages over CTL or BTL process [15,16,137,144,147–150].

9.4.5.7.1 Front-End Approach

In the front-end approach, the scale of the plant is dependent upon the scale of the gasifier. The principal idea in this approach is that trigeneration, FT liquid, heat, and electricity, is the desired end product. This is generally used for the gasifier of size 1–100 MWth [138]. In this approach, the off-gases from FT reactor are used to generate heat and electricity in a combined cycle. The fundamental assumptions and other elements that justify this approach are as follows [139]:

1. The gasifier is small and air blown.
2. Once pass through FT reactor operation to avoid accumulations of inert like N_2 (present at 40 vol% in biosyngas, CO_2, CH_4, and other gaseous C_1–C_4 FT products).
3. No adjustment of H_2/CO ratio via WGS reaction and CO_2 removal, thus reducing the cost of gas conditioning.
4. H_2 concentration is the limiting factor for the conversion of once-through FT synthesis. The unconverted H_2, CO, and other HCs are used to generate heat and electricity. The conversion to liquid is, therefore, low. When the purpose of the trigeneration plant is to improve the production of FT liquid, a shift step is introduced as part of the overall system. This step allows to maximize the total yield of H_2 + CO in the FT synthesis.
5. The basic gas conditioning system will include the removal of tars and BTX by OLGA unit, the removal of inorganic impurities by wet gas cleaning technique, and the removal of H_2S and remaining trace impurities by ZnO and active carbon filters.

9.4.5.7.2 Back-End Approach

The objective of this approach [139,149] is to convert feedstock into FT liquid as much as possible. The heat and particularly electricity generations are of secondary importance. This approach requires a large production capacity. Since FT process has a high fixed cost, the economic scale is important to produce FT liquid competitively with other biofuels such as bio-ethanol, biodiesel, as well as FT liquids derived from fossil fuels [146]. This means gasifiers of the size at least greater than 1000 MWth are required. Fundamental assumptions and other elements of this approach are as follows:

1. FT—off-gas is recycled to the biomass gasifier to achieve maximum syngas conversion. Electricity is produced as a "by-product" from the relatively small recycle bleed stream.
2. The yield of syngas H_2 + CO in the FT feed gas must be as high as possible.
3. Since significant energy remains in CH_4, C_2H_4, BTX, and tar, the process requires tar cracker and shift reactor or a reformer to convert all HCs into CO and H_2. The increase in feed H_2 + CO and increase in their yields result in much higher production of FT liquids and wax and higher overall efficiency of 63.5%.
4. Both gasifier and tar cracker must be oxygen blown to reduce dilution of recycle stream by N_2 and other inert materials.
5. Unlike in the front-end approach, the part of gas conditioning system, CO_2 removal step, is essential.

9.4.5.7.3 Economics of an Integrated BTL Process

Boerrigter et al. [150] have studied an economics of BTL plant to produce green biodiesel from biomass using an integrated back-end approach to the plant design. His main conclusions are as follows:

1. Fixed cost for BTL plant is generally 60% higher than one required for GTL plant of the same size. This is because of the following requirements for BTL plant: larger air separation unit, 50% more expensive gasifier because of the special solid handling, and the requirement of Rectisol unit for bulk gas cleaning.
2. BTL plants of 1000–5000 MWth (size of gasifiers) are optimal.
3. For production below 20,000 bbl/day, fixed costs increase very rapidly.

4. The heart of BTL plant is a pressurized oxygen-blown slagging entrained-flow gasifier. This technology is identified as an optimum technology for biosyngas production.
5. Torrefaction is the optimum biomass pretreatment technology for the entrained-flow gasification.
6. Commercially available technologies can be used for biosyngas cleaning and conditioning as well as for FT synthesis (both fixed bed and slurry bed).
7. Large-scale facilities are required to take advantage of economy of scale. The increase in transportation cost to provide large-scale facility is less important than decrease in fixed plant costs. For a large-scale process, the diesel generated from BTL plant is competitive with $60/bbl oil price. Large-scale BTL processes are also more environment friendly and can take advantage of the infrastructure of large-scale CTL process [79].

Other GTL processes are being developed to produce clean fuels, for example, syncrude, diesel, or many other products including lubricants and waxes, from gas but these require complex (expensive) chemical plant with novel catalyst technology [151–167]. Most of these processes are currently only in their pilot stage. At the moment such premium fuels are expensive compared to normal fuel oil, but are environmentally much cleaner due to the absence of sulfur components, which are removed during the initial processing stage of the recovery of natural gas from the reservoir. The investment and size of plants required to replace all gasoline by GTL technology, would be currently prohibitive.

The minimum syngas throughput needed to make these processes economically viable does help to determine which types of gasifier might be most suitable. At the minimum scale for conventional FT synthesis of 100,000 tons/year fuel output (1,520 odt/day biomass input), only pressurized fluidized-bed and entrained-flow systems would be appropriate. If the minimum scale is reduced to around 300 odt/day biomass input, corresponding with the minimum scale of new FT process technologies, atmospheric CFBs and plasma gasification systems might also have potential [139].

9.4.5.8   Conventional GTL versus Smaller Scale GTL

There are also newer process technologies in development that could reduce the required minimum economic scale for GTL process. For example, the Velocys technology, which was recently acquired by Oxford Catalysts, has been estimated to allow FT process to be viable at outputs of 500–2000 bbl/day [5], which would correspond to biomass inputs of 300–1220 odt/day. The use of small-scale GTL plant becomes practical only with the availability of small amount of feedstock such as "stranded gas," that is, sources of natural gas far from major cities, which are impractical to exploit with conventional gas pipelines and LNG technology or small biomass footprints in remote locations. Otherwise, the direct sale of natural gas to consumers would become much more profitable. Other small-scale FT technologies that are being developed to be commercial are CompactGTL and GreyRock energy FT processes.

In conventional GTL plants, Fischer-Tropsch process is carried out using very large fixed-bed or slurry-bed reactors. These are designed to work on a very large scale. They require a capital investment of $3 billion or more and are only economically viable for plants producing at least 30,000 bbl/day. Only about 6% of the world's known gas fields are large enough to sustain a GTL plant of this size.

Smaller scale GTL, enabled by Velocys technology, provides a cost-effective way to take advantage of smaller scale and remote gas resources. This technology makes it possible to build GTL plants that process 150–1,500 million m^3 per year (15–150 million cubic feet per day (mcf/d)) of gas, producing 1,500–15,000 bbl/day of liquid fuels. Capital costs, operating costs, and plant size are all significantly reduced relative to conventional GTL. The success of Velocys' smaller scale GTL technology lies in the combination of microchannel reactors and superactive catalysts that, together, significantly intensify the FT process.

Smaller scale GTL also provides an alternative to the flaring of associated gas, thereby unlocking oil production in fields that would otherwise be constrained by the regulation or taxation of flaring. Due to their modular construction methods, plants based on the use of microchannel reactors are very flexible and can be easily scaled to match the size of the resource. In this type of reactor, high heat flux (10 times higher than conventional reactors) allows better temperature control in the reactor. The channel depth ranges from 0.1 to 10 mm. The modules are of a standardized design and are shop fabricated in skid-mounted units, making them easier to transport to remote locations or to integrate with existing facilities on refinery or gas-processing sites. This construction approach reduces the costs and risks associated with building plants in remote locations. More details are given by the publications from Velocys [175,176].

Smaller FT processes are also useful alternative for remote area with smaller biomass or waste footprints. Since small amounts of biomass or waste are expensive to transport or store, on-site conversion of these sources to liquid fuels may be an attractive alternative. Smaller FT process can also be an alternative to heat and power for landfill or remote biogas facilities.

The use of microchannel reactors in small-scale FT process brings down the size of the reaction hardware and overcomes the heat and mass transport problems associated with conventional FT technology. Enhanced heat transfer inside the microchannels reactor allows for optimal temperature control, which maximizes catalyst activity and life.

Smaller scale GTL is suitable for use at many more locations and on many more gas fields than conventional GTL. It offers an attractive way to improve the economics and unlock production of a range of unconventional gas resources including shale gas and coalbed methane, which are too far from the existing pipeline infrastructure and markets. For refiners, it can provide diversity and security of supply and can be used to make more valuable products.

In Australia, Linc Energy commenced construction in 1999 of the world's first gas–liquid plant operating on synthesis gas produced by underground coal gasification. The GTL plant uses the FT technology and produced liquids in 2008 [121].

9.5 MIXED ALCOHOL SYNTHESIS

Mixed alcohol synthesis, also known as higher alcohol synthesis (HAS), is very similar to both FT and methanol synthesis. Just as a mixture of straight-chain HCs is produced with varying chain lengths in FT synthesis, a mixture of alcohols can also be produced via catalytic mixed alcohol synthesis. Due to large interest in bio-ethanol, the mixed alcohol synthesis has received renewed attention in the United States [177–241].

9.5.1 CHEMICAL REACTIONS AND PRODUCT DISTRIBUTIONS

The mechanism for higher alcohol synthesis (HAS) involves a complex set of numerous reactions with multiple pathways leading to a variety of products that are impacted by kinetic and thermodynamic constraints. No kinetic analysis of HAS has been published that is capable of globally predicting product compositions over ranges of operating conditions [240]. Depending on the process conditions and catalysts used, the most abundant products are typically methanol and CO_2.

The first step in HAS is the formation of a C–C bond by CO insertion into CH_3OH. Linear alcohols are produced in a stepwise fashion involving the synthesis of methanol followed by its successive homologation to ethanol, propanol, butanol, etc. [185]. Therefore, the HAS catalyst should have methanol synthesis activity because methanol can be considered a recurrent C_1 reactant. Branched higher alcohols are typically formed from modified methanol synthesis and modified FTS catalysts, and straight-chain alcohols are formed when alkalized MoS_2 catalysts are used. The mechanism for HAS over modified high-temperature methanol synthesis catalysts has been described as a unique carbon-chain growth mechanism that is referred to a carbon (adjacent to the alcohol oxygen) addition [190]. Individual reactions in HAS can be grouped into five distinct reaction types [237]:

1. Linear chain growth by C_1 addition at the end of the chain to yield primary linear alcohols
2. Beta addition between the C_1 and C_n ($n \geq 2$) to yield, for example, 1-propanol and branched primary alcohols such as 2-methyl-1-propanol (isobutanol) for $n = 2$
3. Beta addition between C_m ($m = 2$ or 3) and C_n ($n \geq 2$)
4. Methyl ester formation via carboxylic acids formed from synthesized alcohols
5. Carbonylation of methanol to yield methyl formate

Linear alcohols can proceed along the reaction path, but branched alcohols are terminal products of the aldol condensation pathways because they lack the 2α-hydrogens required for chain growth [229].

The general HAS reaction mechanism has the following overall stoichiometry [204,205]:

$$nCO + 2nH_2 \rightarrow C_nH_{2n+1}OH + (n-1)H_2O \quad (\Delta H_r = -61.2 \text{ kcal/mol}) \tag{9.30}$$

with n typically ranging from 1 to 8 [180]. The reaction stoichiometry suggests that the optimum $CO/H_2 = 2$; however, the simultaneous occurrence of the WGS reaction means that the optimum ratio is closer to 1. The major reactions in HAS are methanol synthesis, FT reactions, HAS reactions, and the WGS reaction [202]. Some of the more important main and side reactions occurring during HAS synthesis are described by Spath and Dayton [58].

Methanol formation is favored at low temperatures and high pressures [188]. At high pressures, HAS increases as the temperature is increased at the expense of methanol formation minimizing hydrocarbon formation. To maximize higher alcohols, the H_2/CO ratio should be close to the usage ratio, which is about 1. Lower H_2/CO ratios favor CO insertion and C–C chain growth. In general, the reaction conditions for HAS are more severe than those for methanol production. To increase the yield of higher alcohols, methanol can be recycled for subsequent homologation provided the catalyst shows good hydrocarbonylation activity [185,188]. Main reactions also produce H_2O and CO_2 as by-products.

WGS plays a major role, and, depending on the catalyst's shift activity, some chemical dehydration of alcohols can be undertaken in situ to produce higher alcohols, esters, and ethers [188]. Secondary reactions also produce HCs including aldehydes and ketones [187,188]. Also, frequently, substantial quantities of methane are formed [197]. Thermodynamic constraints limit the theoretical yield of HAS, and just as in other syngas-to-liquid processes, one of the most important limitations to HAS is the removal of the considerable heat of reaction to maintain control of process temperatures [187]. Compared to methanol synthesis, less alcohol product is made per mole of CO, more by-product is made per mole of alcohol product, and more heat is liberated. The main alcohol producing reactions can be summarized as

$$nCO + 2nH_2 \rightarrow C_nH_{2n+1}OH + (n-1)H_2O \tag{9.31}$$

which occurs along with WGS reaction. Some of the important reactions that occur during mixed alcohol synthesis are described in Table 9.5 [58,177–205].

Both selectivity and productivity are important parameters. For example, a catalyst system can have a high CO conversion well above 40%, but if most of that CO is converted to methane or CO_2, then the alcohol selectivity would be very low and the entire process economics would suffer.

TABLE 9.5
Important Reactions for Mixed Alcohol Synthesis

Reaction	Characterization
$CO + 2H_2 \leftrightarrow CH_3OH$	Methanol synthesis
$CO + H_2O \leftrightarrow CO_2 + H_2$	WGS
$CH_3OH + CO \leftrightarrow CH_3CHO + H_2O$	CO beta addition—aldehydes
$CH_3OH + CO + 2H_2 \leftrightarrow CH_3CH_2OH + H_2O$	Ethanol homologation
$C_nH_{2n-1}OH + CO + 2H_2 \leftrightarrow CH_3(CH_2)_nOH + H_2O$	HAS homologation
$2CH_3OH \leftrightarrow CH_3CH_2OH + H_2O$	Condensation/dehydration
$2CH_3OH \leftrightarrow (CH_3)_2CO + H_2O$	DME formation
$(CH_3)_2CO + H_2 \leftrightarrow (CH_3)_2CHOH$	Branched iso-alcohols
$2CH_3CHO \leftrightarrow CH_3COOCH_2CH_3$	Methyl ester synthesis
Competing reactions	
$nCO + 2nH_2 \rightarrow C_nH_{2n} + nH_2O$	Olefins
$nCO + (2n+1)H_2 \rightarrow C_nH_{2n+2} + nH_2O$	Paraffin

Sources: Prepared from information from References 58,177–205.

TABLE 9.6
Mixed Alcohol Product Distributions (NREL Report)

Alcohol	Dow [40] (wt%)	SRI [41] (wt%)	NREL Model (wt%)[a]
Methanol	30–70	30.77	5.01
Ethanol	34.5	46.12	70.66
Propanol	7.7	13.3	10.07
Butanol	1.4	4.14	1.25
Pentanol+	1.5	2.04	0.17
Acetates (C_1 & C_2)	2.5	3.63	
Others			10.98
Water	2.4		1.86
Total	100	100	100

Sources: Phillips, S. et al., Thermochemical ethanol via indirect gasification and mixed alcohol synthesis of lignocellulosic biomass, NREL/TP-51041168, National Renewable Energy Laboratory, Golden, CO, April 2007, http://www.nrel.gov/docs/fy07osti/41168.pdf; Aden, A. et al., The potential of thermochemical ethanol via mixed alcohols production, NREL Milestone Report, National Renewable Energy Laboratory, Golden, CO, September 30, 2005, http://devafdc.nrel.gov/bcfcdoc/9432.pdf, accessed on January, 2007.

[a] Prior to alcohol purification and methanol recycle.

Likewise, if the catalyst had a high CO conversion and selectivity, but had very low productivity, a much larger reactor would have to be built to accommodate the volume of catalyst required.

The mixed alcohol products described in literature are often high in methanol, but contain a wide distribution of several different alcohols. The product distributions described by Dow and SRI are shown in Table 9.6 along with the relative product concentrations calculated by the National Renewable Energy Laboratory (NREL) model [177,183]. The literature indicates that selectivity can be improved by recycling methanol.

9.5.2 CATALYST PERFORMANCE COMPARISONS

The mixed alcohol synthesis often uses catalysts that are modified from methanol and FT syntheses, with added alkali metals (and sometimes sulfur) to promote the mixed alcohol reaction. The process produces a mixture of alcohols such as methanol, ethanol, propanol, and butanol and some heavier alcohols. The requirements for syngas are very similar to methanol and FT processes, except that the H_2/CO ratio must be 1–1.2; hence, the need for a WGS reaction during syngas conditioning is reduced. Also, for the sulfide catalyst, some sulfur (between 50 and 100 ppmv) is actually required in syngas, rather than needing to be removed [201,203,206–240].

Since the catalysts and reactors are based on FT or methanol technology, and due to the very similar requirements in syngas cleanup to FT and methanol synthesis, the minimum economic scale for mixed alcohol synthesis is expected to be similar to that of FT synthesis, corresponding to 100,000 tons/year BTL fuel output, or 1,520 odt/day biomass input [177–205].

HAS catalysts are essentially bifunctional base hydrogenation catalysts and are typically categorized into several groups based on their composition. Common to all HAS catalysts is the addition of alkali metals to the formulation. The activating character of alkali metal promoters is a function of their basicity. Alkali metals provide a basic site to catalyze the aldol condensation reaction by activating surface adsorbed CO and enhancing the formation of the formate intermediate. Four groups of catalyst are generally evaluated in the literature [24]:

1. Modified high-pressure methanol synthesis catalysts—alkali-doped ZnO/Cr_2O_3
2. Modified low-pressure methanol synthesis catalysts—alkali-doped Cu/ZnO and Cu/ZnO/Al_2O_3
3. Modified FT catalysts—alkali-doped $CuO/CoO/Al_2O_3$
4. Alkali-doped sulfides, mainly MoS_2

One of the major hurdles to overcome before HAS becomes an economic commercial process is to improve catalysts that increase the productivity and selectivity to higher alcohols [241]. According to Spath and Dayton [58], to date modified methanol and modified FT catalysts have been more effective in the production of mixed alcohols; the sulfide-based catalysts tend to be less active than the oxide-based catalysts [190]. Rhodium (Rh)-based catalysts are another group of catalysts that are not specifically used for HAS but have been developed for selective ethanol synthesis. Other C_2 oxygenates (i.e., acetaldehyde and acetic acid) as well as increased levels of methane production are also synthesized over Rh-based catalysts [194]. The high cost and limited availability of rhodium for ethanol synthesis catalysts will impact any commercialization of these synthetic processes for converting syngas to ethanol [202].

Pacific Northwest Company [186,215–217,220] also extensively evaluated the performance of various catalysts for mixed alcohol synthesis. They reported that there were clearly differences between the different classes of catalysts in terms of the space–time yields (STYs) for C_{2+} oxygenates and total liquids as well as the selectivity to different liquid products. We briefly describe some of their reported comparison of different catalysts. More details are given in their extensive report [186,215–217,220].

The ICI methanol catalyst (copper catalyst) mainly produced methanol at low temperature. The MeOH–X catalyst (modified copper catalyst) has been operated over a temperature range of 250°C–325°C and appeared to have carbon conversions comparable to the ICI methanol catalyst under comparable operating conditions. The MeOH–X catalyst also produced C_{2+} oxygenates, predominantly ethanol. However, the selectivity to C_{2+} oxygenates was less than 17% at the conditions evaluated.

The K/Cu/Zn/Mn/Co/Cr catalyst, which was evaluated at temperatures ranging from 354°C to 398°C and a 7500 L/L$_{cat}$/h space velocity, did not perform as well as expected, producing very few oxygenates, and those were dominated by methanol [186,201,203,206–241].

The K/MoS$_2$ catalyst, which was evaluated at temperatures of 325°C–375°C at 6700 L/L$_{cat}$/h space velocity, was much less reactive than either the ICI methanol catalyst or MeOH–X catalysts with only 20.3% CO conversion achieved at the highest temperature. The highest C$_{2+}$ oxygenate at STY of about 50 g/g$_{cat}$/h was also achieved at 375°C. This is lower than the highest STY achieved for the MeOH–X catalyst. The selectivity of the K/MoS$_2$ catalyst for C$_{2+}$ oxygenates over the range of temperatures evaluated ranged from about 20% to 23%, which was much better than the MeOH–X catalyst [186,201,203,206–241].

The K/Co/MoO$_2$/C catalyst, which was evaluated at temperatures ranging from 317°C to 381°C and at 6,700 and 15,000 L/L$_{cat}$/h space velocities, appeared to be much more reactive than the KMoS$_2$ catalyst under comparable operating conditions. While the selectivity to C$_{2+}$ oxygenates appeared to be lower under comparable conditions at the higher temperatures, the K/Co/MoO$_2$/C catalyst did not produce nearly as much methanol [186,201,203,206–241].

The Rh/Mn/SiO$_2$ and Rh/Mn/Fe/SiO$_2$ catalysts appeared to be less reactive than the MeOH–X catalysts at temperatures up to ~280°C. At about 300°C, the CO conversion rates are comparable. The Rh/Mn/SiO$_2$ and Rh/Mn/Fe/SiO$_2$ catalysts, on the other hand, are very selective to C$_{2+}$ oxygenates and are not limited by equilibrium conditions at higher temperatures. Consequently, the C$_{2+}$ STYs for these two rhodium catalysts were much higher than those achieved with the MeOH–X catalyst, ranging as high as 0.11 g/L$_{cat}$/h at 280°C and 3,300 L/L$_{cat}$/h space velocity for the Rh/Mn/SiO$_2$ and 0.40 g/L$_{cat}$/h at 323°C and 11,000 L/L$_{cat}$/h space velocity for the Rh/Mn/Fe/SiO$_2$ catalyst. The Rh/Mn/SiO$_2$ and Rh/Mn/Fe/SiO$_2$ catalysts also appeared to be both more reactive and more selective to C$_{2+}$ oxygenates than the MoS$_2$ catalyst. At comparable catalyst temperatures (325°C and 323°C), the carbon conversion for the Rh/Mn/Fe/SiO$_2$ catalyst was more than 5 times greater than the K/MoS$_2$ catalyst, while the carbon selectivity to C$_{2+}$ oxygenates was about the same (24% versus 23%). However, the Rh/Mn/Fe/SiO$_2$ catalyst produced almost no C$_1$ oxygenates, whereas methanol was the major product for the K/MoS$_2$ catalyst [186,201,203,206–241].

The Rh/Mn/Fe/SiO$_2$ and FT–MeOH–Pd catalysts have similar selectivity to C$_{2+}$ oxygenates at ~350°C catalyst temperature. The selectivity to C$_{2+}$ oxygenates for Rh/Mn/Fe/SiO$_2$ decreases significantly with catalyst temperature, whereas the FT–MeOH–Pd catalyst selectivity to C$_{2+}$ oxygenates is relatively constant with temperature. Both catalysts also have relatively low selectivity to C$_1$ oxygenates, although the FT–MeOH–Pd catalyst has a distinctly higher selectivity than Rh/Mn/Fe/SiO$_2$ catalyst [186,201,203,206–241].

9.5.3 PROCESS

The alcohol synthesis reactor system is the heart of the entire HAS process. The syngas is reformed, quenched, compressed, and treated to have acid gas concentrations (H$_2$S, CO$_2$) reduced. After that, it is further compressed to the synthesis reaction conditions of 1000 psia. The compressor is a three-stage steam-driven compressor. The outlet syngas from the compressor is then mixed with recycled methanol from alcohol purification step, heated to 570°F (300°C), and sent to the reactor. The syngas is converted to the alcohol mixture across a fixed-bed catalyst. Because this is a net exothermic reaction system, water is cross-exchanged with the reactor to produce steam for the process while helping to maintain a constant reactor temperature. The product gas is subsequently cooled, allowing the alcohols to condense and separate from the unconverted syngas. The liquid alcohols are then sent to alcohol separation and purification. The residual gas stream is recycled back to the tar reformer with a small purge to fuel combustion (5%).

Although the synthesis reactor is modeled as operating isothermally, it is recognized that maintaining a constant temperature in a fixed-bed reactor system is difficult, especially since these reactions are highly exothermic. Temperature has a significant impact on the alcohol selectivity and product distribution. High pressures are typically required to ensure the production of alcohols. While MoS$_2$ catalyst at high pressure (in addition to promoting with alkali) helps to shift the syngas conversion pathways from hydrocarbon production toward alcohol production, compression requirements for achieving these pressures can be quite substantial. Thus, targeting a catalyst that achieves optimal performance at lower pressures can potentially provide significant cost savings.

The CO_2 concentration requirements for the syngas are less well known. Hermann [24] states that the presence of larger amounts of CO_2 in the synthesis gas retarded the catalyst activity. Further study showed that increasing the CO_2 concentration to 30 vol% decreased the CO conversion but did not significantly alter the alcohol to hydrocarbon ratio of the product. With CO_2 concentrations up to 6.7 vol%, the extent of CO conversion is not affected; however, higher chain alcohol yield relative to methanol does tend to decrease. The effect of CO_2 concentration on alcohol production needs to be further studied in future laboratory experiments.

The reactor effluent is cooled to 110°F (43°C) through a series of heat exchangers using cooler process streams, air and water in three stages while maintaining high pressure. A knockout drum is then used to separate the liquids (primarily alcohols) from the remaining gas, which is comprised of unconverted syngas, CO_2, and methane. From here, the liquid crude alcohols are sent to product purification while the residual syngas is superheated to 1500°F (816°C) and sent through an expander to generate additional power for the process. The pressure is dropped from 970 to 35 psia prior to being recycled to the tar reformer. A 5% purge stream is sent to fuel combustion.

If unconverted syngas is recycled to the throat of the synthesis reactor instead of to the tar reformer, it can save money on upstream equipment costs because of lower process throughput. However, it would also lower yields because the CO_2 would build up in the recycle loop. The limit to the amount of unconverted syngas that could be recycled to the reactor is less than 50% because this would cause the H_2/CO ratio to grow well above 1.2.

Mixed alcohol synthesis has been examined in detail by NREL using high surface area molybdenum disulfide (MoS_2)-based catalyst promoted with alkali metal salts (e.g., potassium carbonate) and cobalt (CoS). These promoters shift the product slate from HCs to alcohols and can either be supported on alumina or activated carbon or be used unsupported [18]. The catalyst gives relatively high ethanol selectivity, and its product slate is a mixture of linear alcohols. The reaction requires temperature of 300°C, pressure of 1000–2000 psia, H_2/CO ratio of 1.0, CO_2 concentration of 0–5 mol%, and sulfur concentration of 50 ppm.

While catalyst is expected to last 5 years, so far it has been tested for 1 year with no loss in activity. The alcohol-producing reactions assumed to occur with this catalyst are listed in Table 9.5. The product distribution obtained with NREL process [177,183] is compared with those obtained by Dow and SRI in Table 9.6. One of the benefits of this catalyst is its sulfur tolerance. It must be continuously sulfided to maintain its activity; thus, an inlet gas concentration of 50 ppmv H_2S is maintained. Concentrations above 100 ppmv inhibit the reaction rate and higher alcohol selectivity [175–205].

9.5.4 ALCOHOL SEPARATION

The mixed alcohol stream from the process is degassed, dried, and separated into three streams: methanol, ethanol, and mixed higher-molecular-weight alcohols. The methanol stream is used to back-flush the molecular sieve drying column and then recycled, along with the water removed during back flushing, to the inlet of the alcohol synthesis reactor. The ethanol and mixed alcohol streams are cooled and sent to product storage tanks.

Carbon dioxide is readily absorbed in alcohol. Although the majority of the noncondensable gases leaving the synthesis reactor are removed in the separator vessel, a significant quantity of these gases remains in the alcohol stream, especially at the high system pressure. These gases are removed by depressurizing from 970 to 60 psia. Most of the dissolved gasses separate from the alcohols in the knockout vessel. This gas stream is made up primarily of carbon dioxide with some small amounts of HCs and alcohols; it is recycled to the tar reformer. After being vaporized by cross-exchanging with steam to a 20°F (11°C) superheated temperature, the alcohol stream goes to the molecular sieve dehydrator unit operation.

The liquid mixed alcohol stream is vaporized, superheated, and then fed to one of two parallel molecular sieve adsorption columns. The adsorption column preferentially removes water and a small amount of the alcohols. While one adsorption bed is adsorbing water, the other is regenerating. In regenerating column, the adsorbed water is desorbed from the molecular sieves with a combination of depressurization and flushing with methanol. This methanol/water mixture is then recycled back to the Alcohol Synthesis section. This methanol/water mixture is cooled to 140°F (60°C) using a forced air heat exchanger and separated from any uncondensed vapor. The gaseous stream is recycled to the tar reformer, and the condensate is pumped to 1000 psia and mixed with high-pressure syngas from compressor upstream of the synthesis reactor preheater.

The dry mixed alcohol stream leaving the molecular sieve dehydrator enters into the first of two distillation columns. This column separates 99% ethanol and all methanol from all higher alcohols. The second column separates methanol from methanol/ethanol mixtures. The methanol and small quantity of ethanol exiting the overhead of second distillation column is used to flush the molecular sieve column during its regeneration step. Higher alcohols and ethanol are then separately stored.

9.5.5 PROCESS COMMERCIALIZATION

Power Energy Fuels, Inc., in Denver, Colorado, has commercialized a process to generate a product with the trade name Ecalene™ [179,200]. This started out as a collaborative project between Western Research Institute and Power Energy Fuels, Inc. [58]. The target product is a mixture of 75 wt% ethanol and the remainder being higher alcohols [24]. Other major players in commercial development are Snamprogetti (SEHT) jointly with *Enichem* and *Haldor Topsoe, DOW, and IFP* [58,177,201,203]. Generally, the H_2/CO ratio is 1–1.4 except for the Snamprogetti, which has a range of 0.5–3 [58,177,183,186,201,203]. These three companies are advanced in their commercial process development [194].

9.6 ISOSYNTHESIS

Isosynthesis (for the production of branched HCs) is interesting because (1) under proper conditions, it gives saturated branched-chain aliphatic HCs containing 4–8 carbon atoms, (2) compared to other types of synthesis, it is the only one that uses difficult reducible oxides as catalysts [18,254–257], and (3) it produces isobutane for high-octane gasoline. The oxide catalysts that are used for this synthesis are thorium oxide, aluminum oxide, tungstic oxide, uranium oxide, and zinc oxide. The thoria and alumina catalysts can be prepared by precipitation from nitrate solutions by adding a hot sodium carbonate solution. The other oxide catalysts can be prepared by addition of alkali to their respective nitrate solutions except in the case of tungsten, which can be precipitated from Na_2WO_4 solution by nitric acid addition [18,242–257].

The isosynthesis reaction was first studied by Pichler and Ziesecke [254,255] in the early 1940s using ThO_2- or zirconia-based catalysts and severe reaction conditions of 150–1000 atm and 450°C [18,251]. The goal was to optimize isobutene and isobutane production for use in high-octane gasoline. DME was the main product at higher pressures. Thorium-based catalysts are good alcohol dehydration catalysts and, in turn, are the most active isosynthesis catalysts. They have long lifetimes because they can be regenerated by oxidizing the accumulated coke that deposits on the surface and they are not poisoned by sulfur. Unfortunately, they are radioactive, making them unusable for commercial application.

Demand for methyl tert-butyl ether (MTBE) as a gasoline additive increased in the early 1990s to the point that supply shortages of petrochemical isobutene were becoming a concern [251]. This led to renewed interest in the isosynthesis reaction to convert syngas to branched HCs, particularly isobutene, which is a reactant in the synthesis of MTBE. Catalyst development was focused on

ZrO_2-based catalysts for increased activity and selectivity to i-C_4 products. Efforts were focused on finding isosynthesis catalysts that have high activity and high selectivity to isobutene at less severe process conditions, especially at lower pressure (~50 atm).

The isosynthesis reaction has often been considered a variation of FT synthesis; however, there are major differences between the two synthetic processes [244]. Isosynthesis is selective to i-C_4 HCs, whereas FTS forms a range of olefins according to the ASF distribution. Only trace amounts of oxygenates (water, methanol, isobutanol, DME, etc.) are formed under isosynthesis reaction conditions. Selective formation of branched HCs also occurs in isosynthesis.

As pointed by Spath and Dayton [58], while most laboratory studies for isosynthesis have been conducted in gas–solid fixed-bed reactors, slurry reactors are also being investigated. In fact, selectivities to C_4 products are reported to be higher in slurry reactors compared to fixed-bed reactors [242]. The WGS reaction between water and carbon monoxide produces large amount of carbon dioxide during isosynthesis reaction [252,253]. This suggests that conversion efficiency to products can be improved by recycling CO_2 back into the reactor.

Isosynthesis conditions are optimized for isobutene/isobutane production. At higher temperatures, methane and aromatics are formed. At lower temperatures, alcohols and other oxygenates are formed. The isosynthesis mechanism involves two chain growth mechanisms—a stepwise CO insertion reaction and a condensation reaction mechanism involving surface-adsorbed oxygenates. The precursor to the oxygenate products has been identified as a surface-adsorbed methoxide species [250]. The oxygenates, alcohols and ethers, are more than likely the primary reaction products that undergo dehydration and hydrogenation to form the iso-alkenes and branched alkanes. These two competing chain growth mechanisms result in a discontinuity in the ASF distribution at C_4 that explains the relatively high selectivity of the isosynthesis reaction to C_4 products.

9.6.1 CATALYSTS

The most effective catalysts for isosynthesis are the tetravalent oxides, thoria, ceria, and zirconia. Alumina and the other oxides are not as effective in producing isobutane. Other oxides such as chromia, lanthana, praseodymium oxide, magnesia, manganese oxide, titania, and berylia have been tested, but they had lower activity. Of all the catalysts tested so far, thoria or promoted thoria have been the most promising catalysts. Unlike catalysts of iron group, thoria is not poisoned by sulfur compounds. In addition, an activity decline due to carbon deposition can be easily overcome by simply passing air over the catalyst to generate CO_2 at the synthesis temperature. Some typical results for one-component catalysts at various pressures were given by Shah and Perrotta [18]. The results indicate that for thoria catalyst, water gas conversion increased from 19% to 66% when pressure increased from 30 to 300 atm. In the same pressure, range conversion increased from 9% to 36% for ZrO_2 catalyst and 10% to 44% for ZnO catalyst. For alumina catalyst, the conversion dropped from 54% to 21% in the same pressure range. For Cr_2O_3 catalyst, water gas conversion increased only from 7% to 10% with an increase in pressure from 30 to 100 atm. The hydrocarbon concentration in the product for thoria catalyst increased from 2.1% to 6.4% in the same pressure range and for ZrO_2 from 1% to 3.5% and Cr_2O_3 from 0.5% to 1% when pressure increased from 30 to 100 atm. In all three cases, avg carbon number of the product was about 2. Both alumina and ZnO gave considerably higher (about 10%) hydrocarbon concentration in the product; however, average carbon number in both cases was significantly lower than 2 (more close to 1). Thus, these early studies indicated that ThO_2, ZrO_2, and Cr_2O_3 were the best catalysts, and they worked better at higher pressure [18].

Thoria catalyze the production of HCs with relatively high carbon number from synthesis gas. Furthermore, the so-called "gasol" HCs consisting of mainly C_3–C_4 also contain large amounts of isobutane. In contrast, the trivalent oxide alumina produced mainly methane and carbon with small amount of "gasol" and traces of isobutane. The most active divalent oxide, zinc oxide, produced no liquid HCs but mainly methane and alcohols. For thoria catalyst, the best temperature has been found to be between 375°C and 475°C. Alcohol production dominated below 375°C, and methane and LPG were the main products between pressure of 300 and 600 atm. Below 300 atm, gas conversion was small; and above 600 atm, methane and DME were obtained.

Zirconium-based catalysts also have high activity for isosynthesis. Unpromoted zirconium catalysts have demonstrated 32% CO conversion at 150 atm and 450°C with much higher selectivity to isobutene [18,251] compared to the thorium-based catalysts. The overall activity of ZrO_2-based catalysts for isosynthesis is lower than ThO_2-based catalysts.

Various promoters have been investigated to improve the activity and selectivity of ThO_2 and ZrO_2 catalysts [245,246]. The most active isosynthesis catalyst is 20% Al_2O_3/ThO_2. Other promoters such as Zn, Cr, and alkali metals have also been tested. The addition of alkali metals to zirconium catalysts had a negative effect on catalyst performance [246,247]. Doped zirconium-based catalysts possess oxygen vacancies in the oxide lattice. The most active catalysts tend to have maximum ionic conductivity suggesting that vacancies in the crystal lattice play an important role in the isosynthesis reaction [248,249]. These oxygen vacancy sites are required for methoxide formation on the catalyst surface that contributes to the condensation reaction [245].

The effects of adding various promoters on the performance of thoria catalysts have also been examined. Three promoters have been studied: (1) alkali to increase molecular weight of the product, (2) phosphoric acid to convert unsaturated HCs such as propane to *n*-butene to higher branched HCs particularly to dimers, and (3) zinc oxide and alumina to enhance the formation of alcohols and subsequently the dehydration of alcohols for the formation of branched HCs. The alkali did not work well, and the activity of thoria catalyst decreased with an increase in alkali content. The most effective promoter was alumina. For thoria–alumina catalyst, the literature data show [18] that the amount of isobutene increased with an addition of alumina; best results were obtained at 20% alumina on thorium oxide. Catalysts prepared by separate precipitation and mixing of the wet precipitates produced the highest *i*-C_4 (isobutane) yields. While "gasol" increased with an increase in temperature, C_5^+ and alcohols (not shown in the table) decreased with an increase in temperature. *i*-C_4 increased with an increase in temperature. In this study, thoria were precipitated from a nitrate solution with sodium carbonate while aluminum oxide was precipitated from a sodium aluminate solution using sulfuric acid. Finally, the best thoria–alumina–zinc oxide catalysts gave lower yields of isobutane and somewhat higher yield of C_5^+ HCs than thoria–alumina catalysts. Once again, just like for FT synthesis, efforts on the catalyst development for isosynthesis have been focusing on improving product selectivity [18].

The selectivity of the isosynthesis reaction depends on the nature of the active catalyst sites, including oxygen vacancies on the surface, and the number of acidic and basic sites.

The balance between acidic and basic catalyst sites dictates overall activity and selectivity [249]. Enhancing the number of acidic sites on the ZrO_2 catalyst increased activity and selectivity to linear C_4 HCs. Increasing the number of basic sites on the catalyst increased the yield of *i*-C_4 HCs [248,258].

The acidic catalyst sites are thought to promote condensation and dehydration reactions. The basic sites are known to catalyze CO insertion reactions. The activity of promoted isosynthesis catalyst systems is related to the acid/base ratio, which can be altered and controlled by varying the preparation procedure for mixed oxide catalysts [243,244]

9.7 OXOSYNTHESIS

The oxosynthesis process is also known as hydroformylation. It involves the reaction of CO and H_2 with olefinic HCs to form an isomeric mixture of normal and iso-aldehydes. It is an industrial synthetic route for the conversion of olefins (in the C_3–C_{15} range) to produce solvents, synthetic detergents, flavorings, perfumes and other healthcare products, and other high-value commodity chemicals. Aldehydes can also be converted to alcohols, which can serve as fuels and fuel additives. Worldwide production of oxo-aldehydes and alcohols was 6.5 million tons/year in 1997.

9.7.1 CHEMISTRY

Olefins react with synthesis gas (carbon monoxide and hydrogen) in the presence of homogeneous catalysts to form aldehydes containing an additional carbon atom. This hydroformylation, also called oxosynthesis or Roelen reaction, is a commercial-scale process for the production of normal and iso-aldehydes. By catalytic addition of hydrogen and carbon monoxide to an olefin, an aldehyde is obtained under chain elongation [259–288]:

$$R-CH=CH_2 + CO + H_2 \rightarrow R-CH_2-CH_2-CHO + R-CH(CH_3)-CHO \qquad (9.32)$$

The basic oxosynthesis reaction is highly exothermic and is thermodynamically favorable at ambient pressures and low temperatures [288]. The reaction proceeds only in the presence of homogeneous metal carbonyl catalysts. The hydroformylation reaction is used on an industrial scale to produce aldehydes and alcohols. The most important oxo products are in the range $C_3–C_{19}$; with a share of roughly 75%, butanol is by far the most significant product. Propene is the olefin mostly used for this reaction. The oxo-products are converted to alcohols, carboxylic acids, aldol condensation products, and primary amines. The process typically involves the reaction of an alkene with carbon monoxide and hydrogen at temperatures between 40°C and 200°C and at high pressures between 10 and 100 atm [259–288].

Usually a 1:1 H_2/CO syngas mixture is required for oxosynthesis. The overall reaction rate has first-order dependence on the hydrogen partial pressure and inverse first-order dependence on CO partial pressure, making the reaction rate essentially independent of total pressure [287]. Higher CO partial pressures are usually required, however, to maintain the stability of the metal carbonyl catalysts. The reaction is also first order in olefin and metal concentration at the higher CO partial pressures.

The mechanism for the homogeneously catalyzed hydroformylation reaction is a function of the catalyst system used in the process. However, the general steps in the reaction paths are similar. The first step is to remove CO from the catalyst organometallic complex to give a coordinatively electron-deficient species. The olefin attaches to the vacated d-orbital at this site in the catalyst complex. The double bond in the olefin attaches to the metal atom yielding an alkyl metal carbonyl complex. The olefin inserts into the M–H bond. CO then inserts into the complex at the C–M bond followed by hydrogen insertion at the same point to yield the aldehyde product. A generalized reaction mechanism is given by Whyman [288].

A key consideration of hydroformylation is the "normal" versus "iso" selectivity. For example, the hydroformylation of propylene can afford two isomeric products, butyraldehyde or isobutyraldehyde [259–288]:

$$H_2 + CO + CH_3CH=CH_2 \rightarrow CH_3CH_2CH_2CHO \quad (\text{"normal"}) \qquad (9.33)$$

versus

$$H_2 + CO + CH_3CH=CH_2 \rightarrow (CH_3)_2CHCHO \quad (\text{"iso"}) \qquad (9.34)$$

These isomers result from the different ways of inserting the alkene into the M–H bond. Generally, normal isomer is more desirable.

When the hydrogen is transferred to the carbon bearing the most hydrogen atoms (Markovnikov addition), the resulting alkyl group has a larger steric bulk close to the ligands on the cobalt. If the ligands on the cobalt are bulky (such as tributyl phosphine), then this steric effect is greater. Hence, the mixed carbonyl/phosphine complexes offer a greater selectivity toward the straight-chain products [259–288].

(9.35)

The electronic effects that favor the Markovnikov addition to an alkene are less able to direct the hydride to the carbon atom, which already bears most hydrogen. Thus, as a result, as the metal center becomes more electron rich, the catalyst becomes more selective for the straight-chain compounds. After the alkyl formation, a second migratory insertion converts the alkyl into an acetyl ligand. The vacant site on the metal is filled by two hydrogens (from the oxidative insertion of a hydrogen molecule). One of these hydrides then takes part in a reductive elimination to form the molecule of the aldehyde and the complex [$HCo(CO)_3$]. It is important that the rate of migratory insertion of the carbonyl into the carbon–metal bond of the alkyl is fast in order to avoid the isomerization of alkenes. In systems where the migratory insertion does not occur (such as nickel hydride tristriphenyl phosphite), the reaction of the hydride with the alkene is reversible, which results in the isomerization of the alkene [259–288].

Side reactions of the olefin hydroformylation are the isomerization and hydrogenation of the olefinic double bond. While the alkanes resulting from hydrogenation of the double bond do not participate further in the reaction, the isomerization of the double bond with subsequent formation of the n-alkyl complexes is a desired reaction. The hydrogenation is usually of minor importance; however, cobalt–phosphine-modified catalysts can hydrogenate 15% of the olefins [259–288].

A usually desired consecutive reaction is the hydrogenation of the aldehydes to alcohols. Higher temperatures and hydrogen partial pressures favor the hydrogenation of the resulting aldehyde to the alcohol. In this case, aldehyde initially forms a CO–π complex with the catalyst, which is rearranged to the alkoxide complex, and by subsequent oxidative addition of hydrogen, the alcohol is produced with regeneration of the starting complex. The aldehydic carbon–oxygen double bond can also be subject to hydroformylation, which leads to formic acid and its esters. The reaction requires the carbon monoxide insertion into the oxygen–metal bond of the alkoxide complex. The resulting formyl complex can be converted into the formic acid esters and the starting complex by oxidative addition of hydrogen. Aldehyde can also react further by aldol condensation to target either product precursors or higher-molecular-weight condensation products like thick oil [259–288].

9.7.2 CATALYSTS AND ASSOCIATED COMMERCIAL PROCESSES

Oxosynthesis is a rapid reaction catalyzed by soluble cobalt or rhodium complexes. The advances made in organometallic chemistry in the 1950s and 1960s led to improved hydroformylation synthesis catalysts with higher thermal stability and greater selectivity. Three complimentary catalytic hydroformylation processes have been developed and commercialized. The choice of catalyst depends on the particular starting olefin or produce 2-ethylhexanol from propylene and syngas. Unmodified cobalt catalysts are versatile enough to be used for hydroformylation of high carbon number mixed olefins. Phosphine-modified cobalt catalysts are used for the production of higher, detergent range, alcohols rather than aldehydes. Phosphine-modified rhodium catalysts are best to produce 2-ethylhexanol, a plasticizer alcohol used to make flexible PVC, from propylene and syngas. This is the highest volume oxosynthesis process. The LP Oxo™ Low-Pressure Oxo Process is the world's leading process for the production of oxo alcohols from olefins [58,273,274,289–348].

The industrial processes vary depending on the chain length of the olefin to be hydroformylated, the catalyst metal and ligands, and the recovery of the catalyst. The original Ruhrchemie process produced propanal from ethene and syngas using cobalt tetracarbonyl hydride. Today, industrial processes based on cobalt catalysts are mainly used for the production of medium- to long-chain olefins [273,274,289–311], whereas the rhodium-based catalysts are generally used for the hydroformylation of propene [312–348]. The rhodium catalysts are significantly more expensive than cobalt catalysts. In the hydroformylation of higher-molecular-weight olefins, the separation of the catalyst from the produced aldehydes is difficult.

9.7.2.1 Cobalt Carbonyl Catalyst

The first hydroformylation catalysts were cobalt carbonyls; the specific active catalyst was found to be $HCo(CO)_4$ in equilibrium with $Co_2(CO)_8$ based on the following reaction [273,274,289–311]:

$$Co_2(CO)_8 + H_2 \leftrightarrow 2HCo(CO)_4 \qquad (9.36)$$

Cobalt metals and most cobalt salts will form cobalt carbonyl under hydroformylation conditions. The cobalt catalyzes both double-bond isomerization and oxosynthesis. Undesired competing side reactions, such as the direct hydrogenation of the starting olefin and the condensation of product aldehydes to high-boiling products, are generally avoided in the co-catalyzed process.

For cobalt carbonyl catalysts, a normal/iso ratio of 4:1 can be achieved with catalyst concentrations of 0.1%–1% metal/olefin at 200–300 atm and 110°C–200°C with a 1:1 H_2/CO ratio. Lower process temperatures and higher CO partial pressures favor the formation of the straight-chain isomer; however, the overall conversion efficiency decreases. The cobalt carbonyl catalysts are also not very stable at the higher process temperatures. Catalyst tends to deposit on reactor walls decreasing catalyst activity and reducing catalyst recovery at elevated process temperatures [58,273,274,289–311].

The BASF-oxo process starts mostly with higher olefins and relies on cobalt carbonyl-based catalyst [273,274,289–311]. The process is carried out at low temperatures, which favors the linear product. The process is carried out at a pressure of about 30 MPa and in a temperature range of 150°C–170°C. The cobalt is recovered from the liquid product by oxidation to water-soluble Co^{2+}, followed by the addition of aqueous formic or acetic acids. This process gives an aqueous phase of cobalt, which can then be recycled. Losses are compensated by the addition of cobalt salts [58,273,274,289–311].

The Exxon process, also Kuhlmann or PCUK—oxo process, is used for the hydroformylation of C_6–C_{12} olefins. The process relies on cobalt catalysts. In order to recover the catalyst, an aqueous sodium hydroxide solution or sodium carbonate is added to the organic phase. Metal carbonyl hydride is recovered by extraction with olefin and neutralization by addition of sulfuric acid solution under carbon monoxide pressure. This is stripped out with syngas and returned to the reactor. Similar to the BASF process, the Exxon process is carried out at a pressure of about 30 MPa and at a temperature of about 160°C–180°C [58,273,274,289–311].

9.7.2.2 Phosphine-Modified Cobalt

In the early 1960s, Shell Oil Company commercialized a new cobalt-based hydroformylation process. The addition of a phosphine ligand to Co resulted in a trialkylphosphine-substituted cobalt carbonyl catalyst [$HCo(CO)_3P(n\text{-}C_4H_9)_3$]. This new catalyst had higher selectivity to straight-chain aldehydes (normal/iso = 7:1) with improved thermal stability compared to the unsubstituted cobalt catalysts. Shell uses this for the hydroformylation of C_7–C_{14} olefins. The improved thermal stability allows for lower process pressures but higher process temperatures. The Shell process conditions are 50–100 atm and 160°C–200°C with H_2:CO = 1. Even though this catalyst has improved thermal stability, it has a lower hydroformylation activity than cobalt carbonyl catalysts; hence, it requires higher reaction temperature. The higher temperatures also increase the competing olefin hydrogenation reaction. The resulting aldehydes are directly hydrogenated to the fatty alcohols (C_{11}–C_{14}), which are separated by distillation, and the catalysts are recycled. The process has good selectivity to linear products, which are used as feedstock for detergents. Shell has optimized this process to produce detergent range alcohols (C_{11}–C_{14}) in a single step by capitalizing on the conversion of terminal olefins to alcohols by hydrogenation of the aldehyde hydroformylation products. A high n/i ratio results from increased isomerization rates concurrently with hydroformylation using phosphine-modified cobalt catalyst [58,273,274,289–311].

9.7.2.3 Phosphine-Modified Rhodium

Significantly lower operating pressures and temperatures and increased selectivity to linear products were demonstrated with the introduction of phosphine-modified rhodium catalysts. In 1976, Union Carbide Corporation (UCC) and Davy Process Technology commercialized the LP Oxo process based on a triphenylphosphine-modified rhodium catalyst. With a $HRh(CO)(PPh_3)_3$ catalyst, the process is maintained at 7–25 atm and 60°C–120°C and a n/i ratio of 8–12:1 is possible. Low temperatures, high carbon monoxide partial pressure, and high ligand concentration on the Rh catalysts favor the formation of the straight-chain isomers. Rh-based catalysts are mainly used for the hydroformylation of lower olefins (e.g., propylene to butyraldehyde) but have limited use for higher olefins because of thermal instability of the catalyst at the high temperatures required for distillation to separate product and catalyst. Another important factor when selecting Rh-based hydroformylation catalysts is the high cost and low availability of rhodium. This makes catalyst separation and recycle an economically important process consideration. The high cost of rhodium, however, is offset by lower equipment costs, increased activity, and higher selectivity and efficiency. The development of water-soluble Rh-based catalysts avoids some of these issues. Rhone–Poulenc commercialized an oxo process based on a water-soluble Rh catalyst in 1984 [285].

The UCC process, also known as low-pressure oxo process, relies on a rhodium catalyst dissolved in high-boiling thick oil, a higher-molecular-weight condensation product of the primary aldehydes, for the hydroformylation of propene. The reaction mixture is separated in a falling film evaporator from volatile components. The liquid phase is distilled, and butyraldehyde is removed as head product while the catalyst containing bottom product is recycled to the process. The process is carried out at about 1.8 MPa and 95°C–100°C [58,312–348].

The Ruhrchemie/Rhone–Poulenc process relies on a rhodium catalyst with water-soluble TPPTS as a ligand (Kuntz Cornils catalyst) for the hydroformylation of propene [58,259,312–348]. The trisulfonation of triphenylphosphane ligand provides hydrophilic properties to the organometallic complex. The catalyst complex carries nine sulfonate groups and is highly soluble in water (about 1 kg/L), but not in the emerging product phase [10]. The water-soluble TPPTS is used in about 50-fold excess, whereby the leaching of the catalyst is effectively suppressed. Reactants are propene and syngas consisting of hydrogen and carbon monoxide in a ratio of 1.1:1. A mixture of butyraldehyde and isobutyraldehyde in the ratio 96:4 is generated with few by-products such as alcohols, esters, and higher boiling fractions [58,259,312–348]. The Ruhrchemie/Rhone–Poulenc process is the first commercially available two-phase system in which the catalyst is present in the aqueous phase. In the progress of the reaction, an organic product phase is formed, which is separated continuously by means of phase separation, wherein the aqueous catalyst phase remains in the reactor [58,259,312–348].

The process is carried out in a stirred tank reactor where the olefin and the syngas are bubbled from the bottom of the reactor through the catalyst phase under intensive stirring. The resulting crude aldehyde phase is separated at the top from the aqueous phase. The aqueous catalyst-containing solution is reheated via a heat exchanger and pumped back into the reactor. The excess olefin and syngas are separated from the aldehyde phase and fed back to the reactor. The generated heat is used for the generation of process steam, which is used for subsequent distillation of the organic phase to separate into butyraldehyde and isobutyraldehyde [58,259,312–348]. Process does not require an elaborate removal of catalyst poisons from syngas because poisons migrate into the organic phase and are removed from the reaction with the aldehyde [58,259,312–348].

A plant was built in Oberhausen in 1984, which produced, in 1998, 500,000 tons/annum butanal. The conversion rate of propene was 98%, and the selectivity to the *n*-butanal was high. During the lifetime of a catalyst batch in the process, less than 1 ppb rhodium was lost [58,259].

9.8 SYNGAS FERMENTATION

A variety of microorganisms can use syngas as an energy and carbon source to produce ethanol, with some forming butanol, acetate, formate, and butyrate [349–359]. These include *Acetobacterium woodii, Butyribacterium methylotrophicum, Clostridium carboxidivorans* P7, *Eubacterium limosum, Moorella,* and *Peptostreptococcus productus* [349]. Current syngas fermentation efforts are predominantly focused on ethanol production. The process operates at low pressures (atmospheric to 2 bar) and low temperatures (most use near 37°C, although some species can survive and grow in temperatures ranging from 5°C to 55°C), with the exact reactor conditions and pH depending on the type of microorganism used [349–395].

The main requirement for syngas for fermentation is the avoidance of tars or HCs (to within a similar level as for FT synthesis), as they inhibit fermentation and adversely affect cell growth. The biological process is not sensitive to many of the other requirements for the chemical catalytic processes, and most of the organisms mentioned earlier grow better on CO than H_2. This means that H_2/CO ratio in the syngas can be low and the WGS reaction during or after gasification is not required. However, other requirements, such as the tolerance to sulfur, will depend on the particular type of organism used. The minimum economic scale for syngas fermentation is expected to be considerably smaller than conventional FT processes, at around 30,000 tons/year ethanol output [349], which corresponds to 290 odt/day biomass input [349–359]. The input numbers will be different if other carbonaceous sources are used to generate syngas.

Syngas fermentation can be compared with other biofuel processes. A comparison of syngas fermentation with FT process is briefly described in Table 9.7. Syngas fermentation can also be compared with lignocellulosic fermentation to produce ethanol. Lignocellulosic fermentation is a multistep process where lignocellulosic biomass is pretreated and then hydrolyzed in order to convert the carbohydrate polymers cellulose and hemicellulose to monomeric sugars, which are fermented to produce ethanol [355]. A key challenge for biochemical lignocellulosic fermentation is the recalcitrance of biomass. Lignocellulosic fermentation cannot directly utilize the lignin fraction, which typically makes up 10%–25% of biomass. This is significant, as the lignin accounts for 25%–35% of the energy content of the feedstock. To allow fermentation of the carbohydrate fraction of this resource, lignin must be separated from cellulose and hemicellulose polymers via a complex and costly pretreatment process so that these polymers may be broken down into fermentable sugars by enzymatic hydrolysis [355]. Methods used for pretreatment include physical treatments such as milling and irradiation as well as chemical treatments such as the use of oxidizing agents and strong acids [356,397]. In gas fermentation, most of these steps are unnecessary and replaced by the gasification process, which allows the entire feedstock (both the lignin and carbohydrate portions) to be converted to a fermentable syngas [412].

The unifying nature of gasification allows heterogeneous resources to be processed through a single technology for the production of a fermentable gas stream. Conversely, biochemical pretreatments are far less accepting of diverse inputs and must be calibrated for a homogenous or defined stream of feedstock [398]. Although work into consolidated bioprocessing where cellulolytic bacteria convert lignocellulose in one step has shown promise [352], standard lignocellulosic fermentation is far closer to commercial viability. The perceived technical advantages offered by gas fermentation may allow a production cost advantage over other second-generation biofuels in terms of both operating and capital costs. Furthermore, the microbial catalysts used in gas fermentation are now becoming the focus of molecular biology investigation and genetic

TABLE 9.7

Comparison of Syngas Fermentation versus FT Process

Syngas Fermentation	FT Process
Low temperature and Pressure	High temperature and pressure
Slow process (biochemical)	Fast process (thermochemical)
Flexible feedstock	Flexible feedstock
High rates of energy and carbon capture	Moderate rates of energy and carbon capture
Process robustness, catalyst flexibility, and high development potential	Limited process robustness, catalyst flexibility, and development potential
Highly selective-less downstream processing	Limited selectivity to gasoline (45%)—more downstream processing
Small scale	Small, medium, or large scale
Small H_2/CO	H_2/CO of about 1–2.15 depending on the desired final product
No gas shift required	Gas shift reaction is required.
More tolerant to contaminants in syngas—depends on microorganism	Requires sulfur and contaminants free syngas—makes process more expensive
Lower rates of deactivation of microorganisms	High catalyst deactivation—catalyst regeneration is required
Simpler and cheaper process	Complex and expensive process
Higher fuel yield and energy efficiency (57%)	Lower fuel yield and energy efficiency (45%)
Lower CO_2 emission	Higher CO_2 emission
Novel microorganisms can be developed—new genetic methods.	Use of metal catalyst is a limitation.
Commercial processes are still being developed.	Well-established commercial processes

Sources: References 349–359.

modification technologies [58,349], which offer tremendous scope for improvement in both product value and process performance.

Daniell et al. [349] also briefly compared various performance parameters of three processes for alcohol production (methanol synthesis, mixed alcohol synthesis, and syngas fermentation). Thermal efficiency of methanol synthesis is about 79%, a number higher than 62%–68% for mixed alcohol synthesis. Thermal efficiency of syngas fermentation is generally not reported. While syngas to methanol conversion per pass can be at most 25%, syngas can be converted to methanol more than 99% by recycling process. In mixed alcohol synthesis, while single pass conversion can vary between 10% and 40% (producing mainly methanol), better catalyst and recycling can give methanol, ethanol, and higher alcohol selectivity as high as 60%–90% (on CO_2 free basis). This range can be affected by the production of HCs depending on the catalyst. For syngas fermentation, syngas conversion depends significantly on gas–liquid mass transfer rates (i.e., design of fermenter), microorganism growth, and activity and use of recycling. With the use of right microorganisms 100% selectivity to ethanol can be obtained.

9.8.1 Biochemistry of Gas Fermentation

Acetogens are a group of bacteria capable of fermenting CO and/or CO_2 and H_2 into acetyl-CoA (and from there into acetic acid, ethanol, and other metabolic end products) via the reductive acetyl-CoA pathway [351,390]. The reductive acetyl-CoA pathway, also known as the Wood–Ljungdahl pathway, was first characterized by Wood and Ljungdahl in 1966 when they proposed a scheme for the synthesis of acetate from CO_2 by the organism *Clostridium thermoaceticum*, now classified as

Moorella thermoacetica [384,385]. Variations of this pathway are also found in methanogenic and sulfate-reducing organisms [351,390]; however, only acetogens are known to synthesize metabolic end products that can be used as liquid transportation fuels.

The biochemistry of this pathway has been comprehensively described in numerous reviews, including those by Wood [385] and Ljungdahl [386], Ragsdale and coworkers [350,387], and Drake et al. [351]. The reductive acetyl-CoA pathway indicates that at high CO concentrations, no or only little hydrogen uptake will occur, but it will increase once CO is utilized and its concentration drops. Ethanol and acetate can be produced according to the following reactions: with CO as the sole carbon energy source, as shown in Equations 9.37 and 9.38; with CO as a carbon source and both CO and H_2 as the energy source, according to Equations 9.39 through 9.41; and with CO_2 as carbon source and H_2 as energy source, as in Equations 9.42 and 9.43 [386]:

$$6CO + 3H_2O \rightarrow CH_3CH_2OH + 4CO_2 \quad (\Delta G^{\circ\prime} = -224 \text{ kJ/mol}) \tag{9.37}$$

$$4CO + 2H_2O \rightarrow CH_3COOH + 2CO_2 \quad (\Delta G^{\circ\prime} = -175 \text{ kJ/mol}) \tag{9.38}$$

$$3CO + 3H_2 \rightarrow CH_3CH_2OH + CO_2 \quad (\Delta G^{\circ\prime} = -164 \text{ kJ/mol}) \tag{9.39}$$

$$2CO + 2H_2 \rightarrow CH_3COOH \quad (\Delta G^{\circ\prime} = -135 \text{ kJ/mol}) \tag{9.40}$$

$$2CO + 4H_2 \rightarrow CH_3CH_2OH + H_2O \quad (\Delta G^{\circ\prime} = -144 \text{ kJ/mol}) \tag{9.41}$$

$$2CO_2 + 6H_2 \rightarrow CH_3CH_2OH + 3H_2O \quad (\Delta G^{\circ\prime} = -104 \text{ kJ/mol}) \tag{9.42}$$

$$2CO_2 + 4H_2 \rightarrow CH_3COOH + 2H_2O \quad (\Delta G^{\circ\prime} = -95 \text{ kJ/mol}) \tag{9.43}$$

Acetyl-CoA is then converted into acetate via phosphotransacetylase and acetate kinase reactions [387]. The genes for these enzymes are organized in a single operon in all sequenced acetogens to date [353,357,383,387,388]. In the acetate kinase reaction, one molecule of ATP is gained by substrate-level phosphorylation, which is used to compensate for the ATP required to activate formate. More details on the organisms used and reaction pathways for syngas conversion are described by Daniell et al. [349].

9.8.2 ACETOGENS

Acetogens are anaerobic bacteria that play key role in the global carbon cycle [351]. Acetogens have been isolated from a variety of habitats including soil, sediments, and the intestinal tracts of animals and humans and are found worldwide. All acetogens described to date produce acetate [351]. Although over 100 acetogenic bacteria from 22 genera have been isolated [396], the best characterized and researched of these fit into the genera *Acetobacterium* and *Clostridium*. Drake provides a good definition of an acetogen [351,390]: "An anaerobe that can use the acetyl-CoA pathway as a (1) mechanism for the reductive synthesis of acetyl-CoA from CO_2, (2) terminal electron-accepting, energy-conserving process, and (3) mechanism for the fixation (assimilation) of CO_2 in the synthesis of cell carbon."

Acetogenic organisms that have been investigated for use in commercial syngas fermentation are highlighted according to their product spectrum, which include acetate, ethanol, butyrate, butanol, and 2,3-butanediol. Those primarily used and characterized for the production of ethanol include *Clostridium ljungdahlii*, *Clostridium autoethanogenum*, "*Clostridium ragsdalei*," and *Alkalibaculum bacchi*. Butanol production has been observed with *C. carboxidivorans* and *B. methylotrophicum*. More details on organisms and various product distributions like ethanol, butanol, butyrate, and 2,3-butanol are given by Daniell et al. [349].

9.8.3 PARAMETERS FOR FERMENTATION PROCESS OPTIMIZATION

Daniell et al. [349] discuss in great details various parameters that need to be controlled, improved, and optimized within fermentation process to make it commercially viable. Here we briefly summarize the main conclusions of this study [397–437].

Product synthesis rates and ethanol to acetate ratios described in the literature by laboratory-scale investigations of various organisms are often poor from a commercial perspective [364,435,438]. Parameters such as nutrient media, pH, temperature, bioreactor design, and improvements in strain effectiveness through new and advanced genetic techniques need to be controlled and optimized to enhance commercial fermentation process [402,403].

Nutrient media for gas fermenting organisms must include minerals, vitamins, metal cofactors, and a reducing agent. Cotter et al. [363] found that solventogenesis was favored for cells that are in resting place in nutrient-limited media. For example, they found that nitrogen limitation led to improved ethanol production in *C. autoethanogenum* [363]. Furthermore, reducing agents increase the cellular NADH concentration, which favors alcohol production through NADH-dependent pathways [429].

Vitamins and minerals are also important to have a favorable fermentation environment. The use of lab-scale standard vitamin and mineral components is too expensive at commercial scale due to large reaction volume. Low-cost cottonseed extract or corn steep liquor serves the same purpose [401,406]. The inclusion of metal cofactors in fermentation media has also been demonstrated to improve microbial productivity. For example, nickel is an important cofactor for enzymes such as CO dehydrogenase and acetyl-CoA synthase and has been shown to improve CO uptake and ethanol production in gas fermentation [366,403].

The removal of trace metal Cu^{2+} from the medium and increasing concentrations of Ni^{2+}, Zn^{2+}, SeO^{4-}, and WO^{4-} in "*C. ragsdalei*" improved ethanol production [400]. pH has been one of the most important factors in improving the activity of gas fermenting organisms due to its effect on product composition. Studies have shown a shift from acidogenesis to solventogenesis as fermentation pH is lowered, allowing the increased production of ethanol and other highly reduced products [391,410]. Optimal pH varies depending on the organism; for example, pH 4.74 was found to be optimal for ethanol production by *C. autoethanogenum* [399], while in "*C. ragsdalei*," higher ethanol concentration was not connected to pH below 6 [406].

Temperature is also important for fermentation optimization as it influences both microbial productivity and substrate gas solubility. Optimum temperatures for mesophilic and thermophilic conditions need to be identified. Although the increased temperature required by thermophilic organisms reduces gas solubility, a decrease in liquid viscosity at higher temperature may improve GTL mass transfer coefficient.

Bioreactor design is an important variable for gas fermentation because of its role in facilitating the mass transfer of CO and H_2 from the gas phase into the microbial cell within liquid media. Good bioreactor design must give high syngas mass transfer rate in fermentation broth, easy scale-up possibility, and low operating costs. Good syngas mass transfer rate requires high solubility in broth, specific fluid flow pattern, and high interfacial area between gas and broth. This can be achieved by increased pressure and careful reactor design [380,411]. The design should also consider the power consumption along with high mass transfer rate. Thus, mass transfer performance is more usefully described by the optimum volumetric mass transfer coefficient per unit power input ($k_L a P_g^{-1}$) [411]. Munasinghe and Khanal [412] reported the highest $k_L a$ in an airlift reactor combined with a 20 μm bulb diffuser. In a study of stirred tank reactors using different impeller designs, Ungerman and Heindel found that dual impeller designs with an axial flow impeller as the top impeller had the highest $k_L a P_g^{-1}$ [411]. At commercial scale, options being considered by gas fermentation companies include bubble column, gas lift, trickle bed, immobilized cell, and microbubble reactors [367,402,404]. Coskata [439] developed a patented microporous, immobilized cell reactor in which gas directly makes contact with the fermentation organism, avoiding the need for gas to first transfer into the liquid medium.

Finally, adopting new genetic methods for improving the effectiveness of various strains in the commercial-scale bioreactors is very important. This requires using advances in genomic sequencing, developments of new genetic tools, and new biological techniques. Recently, metabolic engineering and synthetic biology techniques have been applied to gas fermentation organisms. This work strives to improve microbial productivity and robustness and to introduce pathways for the commercial production of increasingly energy-dense fuels and more valuable chemicals. Over the past two decades, a number of new genetic techniques for clostridia, such as antisense RNA strategies and others [369,370,420–424,428–432], have been developed. More recently, integration-based techniques such as ClosTron [371,372,436], as well as marker-less integration methods [437], have been applied to *Clostridium acetobutylicum* and cellulolytic *Clostridium thermocellum*. Several review articles have been published recently that give a detailed overview of the developed tools [354,392,425].

While the improvement of the robustness, productivity, and ethanol to acetate ratios of gas fermenting organisms has been achieved through random mutagenesis combined with high-throughput screening for desired characteristics, there are no published reports of improved productivity of gas fermentation organisms through targeted genetic modification; this is likely to change with recent advances in genome sequencing and developments of genetic tools for gas fermentation organisms.

The use of metabolic engineering to integrate new pathways has been reported for *C. ljungdahlii* and *C. autoethanogenum* for the production of the biofuel butanol [353,426], and its details are described in a patent from LanzaTech [353,426].

Through synthetic biology, routes to advanced biofuels such as isobutanol and farnesene have been demonstrated in *Escherichia coli* and yeast [416], and as genetic tools are further developed, there is the potential for these approaches to be implemented with acetogenic bacteria. Another intriguing strategy is to transplant the genes for gas utilization into more genetically malleable organisms. These efforts will greatly benefit from systems biology approaches and the creation of genome-scale metabolic models. While only a single acetogen had been sequenced before mid-2010, several genome sequences have since been released [353,357,388,393,427,433,440]. With advances in "omics" technologies and further reduction in the cost of sequencing, this will certainly change in the future.

9.8.4 COMMERCIALIZATION

Despite its increase in commercial interest, biomass gasification followed by syngas fermentation has yet to be achieved on a commercial scale. One of the principal technical challenges associated with commercialization is the successful scale-up of this process combination from pilot scale to a commercial level [441].

The fermentation step itself also faces a number of technical challenges including scale-up and mass transfer limitations and risks involved with the biocatalyst. For commercial viability, this fermentation must occur reliably with high conversion efficiencies and on a large scale. Compared to aerobic processes, cell concentrations in anaerobic fermentations are usually lower as ATP is limited.

While it is desirable to have most of the carbon going into products rather than biomass, a certain cell concentration is necessary to achieve high production rates. This can be a challenge, and there are several strategies to overcome this issue, starting from media optimization to allow increased biomass formation to cell recycling or cell immobilization to retain cells that have been successfully applied to other anaerobic or clostridial processes [417].

Secondly, there are risks associated with the microbial catalyst, including poisoning by oxygen or other contamination and bacteriophage infection [241]. Bacteriophage infection of biocatalyst has yet to be reported with synthesis gas fermentation. As with gas contaminants, bacteriophage infection may affect microbial productivity to different extents, with symptoms difficult to diagnose. Techniques used to avoid phage infection include improved culture practice and plant hygiene and the creation of phage-immunized strains through the serial transfer of phage-resistant mutants [394].

Almost all bacterial species investigated have been found susceptible to infection by at least one phage, and, despite most gas fermentation organisms having at least one methylation system to protect against foreign phage DNA, phage has infected other clostridia [373]. Infection could be economically damaging in a commercial process, potentially requiring an entire plant to be taken offline so that a decontamination and sterilization process can be carried out. This is therefore a noteworthy potential threat to commercialization of the gas fermentation process.

Most technologies used during downstream product separation are mature and thus are unlikely to pose technical or commercial challenges. In gas fermentation, more energy is potentially required for downstream product recovery. Mature technologies are available for the distillation of ethanol and butanol [395] and are carried out largely in the same manner as for conventional fermentation. However, the recovery of 2,3-butanediol has not yet been established at scale [374]. To improve process efficiency and reduce cost, it is important that the process is run in a continuous closed loop. Thus, the downstream product separation step must allow, for example, a water and nutrient recycle. Finally, integration of gasification, gas clean-up, fermentation, and bioreactor design and downstream processing, which includes novel technology, at a commercial scale is challenging and is a significant barrier to commercialization.

In spite of barriers, as pointed out by Daniell et al. [349], INEOS Bio, Coskata, and LanzaTech are three companies exploring the commercialization of gas fermentation for the production of liquid fuels [439,442–450]. INEOS Bio (http://www.ineosbio.com) at their pilot plant reported a production rate of 100 gallons of ethanol per dry ton of feedstock using proprietary isolates of *C. ljungdahlii* as the biocatalyst [375,402,446]. In 2011 INEOS Bio began the construction of their first commercial-scale plant, the Indian River BioEnergy Center in Florida [443], with the hope of starting operation for the third quarter of 2012. This plant is designed to produce ethanol from yard, vegetative, and household waste and is also projected to produce 6 MW (gross) of electricity from unused syngas and recovered heat [449]. The plant has a planned capacity of 300 dry tons/day, producing 8 million gallons of ethanol per year [444].

Coskata, Inc. (http://www.coskata.com) has operated its gas fermentation technology at a demonstration plant in Madison, PA, since October 2009 [447] with syngas being produced from wood biomass and municipal solid waste using a plasma gasification process developed by Westinghouse Plasma Corporation [439]. Coskata reported that they were constructing a commercial plant in Alabama for the conversion of wood chips and waste to ethanol, with a planned production capacity of 16 million gallons of ethanol per year, to be scaled up to 78 million gallons of ethanol per year [447]. This plant will not use the Westinghouse Plasma Corporation gasification technology used at the demonstration plant, but instead will use an indirect biomass gasifier [447]. Coskata reported an expected yield of 100 gallons of ethanol per bone dry ton of softwood. In July 2012 the company announced a shift in strategy toward the construction of a commercial plant funded by private strategic investors, which uses reformed natural gas as the sole feedstock [450].

LanzaTech (http://www.lanzatech.co.nz) has focused on the use of synthesis gas and CO-rich industrial off-gases to produce ethanol and 2,3-butanediol, using a proprietary strain of *C. autoethanogenum* [434]. The ability to produce chemicals such as 2,3-butanediol and butanol [383,426] as well as traditional fuels like ethanol by gas fermentation organisms is a distinct feature of the LanzaTech technology. LanzaTech has been running a pilot-scale plant using steel mill off-gases at the BlueScope Steel facility in New Zealand since 2008 [445] and has built a precommercial 100,000 gallons per year demonstration plant in partnership with BaoSteel in Shanghai, China, for the production of ethanol from steel mill off-gases [442,448]. Construction of another steel mill-based plant begun with Shougang Group, through a joint venture to develop a demonstration plant located at a Shougang steel mill [442,448].

9.9 ROLE OF HYDROGEN IN REFINERY OPERATIONS

Many believe that hydrogen is the ultimate fuel because it has least impact on environment; there is abundant supply, although in the form of compounds; and per unit mass it has the highest heating value. While, as shown in Chapter 5, synthetic production of hydrogen faces many challenges, currently hydrogen can be produced by reforming, hydrothermal gasification in sub- and supercritical water, electrolysis, and thermochemical dissociation of water, among other methods. While a significant amount of hydrogen is used in the production of ammonia, fertilizers, and urea, one major energy application is its use in refineries.

Hydrogen is essential in improving the quality of crude oil, heavy oil, bitumen, oil shale, coal liquids, bio-oils, etc., so they can be used as liquid fuel for vehicles (automobiles, ships, planes, etc.) and suitable liquid fuels for heating purposes. Liquid fuels (gasoline, diesel, jet fuel, fuel oil, etc.) require right ratio of hydrogen to carbon; right average carbon number; minimum sulfur; metal and nitrogen impurities; right balances between aliphatic, olefinic, and aromatic components; and right mix between straight and branched HCs. All of these are achieved in refineries by different set of processes, most of which (if not all) require hydrogen. The presence of hydrogen also improves catalyst life in refinery operations by reducing the coke production. Here we briefly examine five important refinery processes, which use hydrogen [414,451–454].

9.9.1 HYDROGENATION

The process of hydrogenation involves a chemical reaction between molecular hydrogen (H_2) and another compound or element, often in the presence of a catalyst such as nickel, palladium, or platinum. The process is commonly employed to reduce or saturate organic compounds. Hydrogenation typically constitutes the addition of pairs of hydrogen atoms to a molecule, generally an alkene. Catalysts are required for the reaction to be usable; noncatalytic hydrogenation takes place only at very high temperatures. Hydrogenation reduces double and triple bonds in HCs [414]. The insertion of hydrogen in crude or heavy oils increases hydrogen to carbon ratio in the oil. Hydrogenation of bio-oil improves its quality by reducing the concentration of oxygenated compounds.

Because of the importance of hydrogen, many related reactions have been developed for its use. Most hydrogenations use gaseous hydrogen (H_2), but some involve the alternative sources of hydrogen such as internal exchange of hydrogen among various reacting species or hydrogen donation from other liquid components. A reaction where bonds are broken while hydrogen is added is called hydrogenolysis. These types of reactions are important for removals of sulfur, metals, nitrogen, oxygen, etc., from liquid. Hydrogenation of unsaturated fats produces saturated fats. The addition of hydrogen to alkene produces alkane.

Hydrogenation requires an unsaturated substrate, hydrogen, and catalyst. The reduction reaction is carried out at different temperatures and pressures depending upon the substrate, nature of the reaction, and the activity of the catalyst. Hydrogenation is sensitive to steric hindrance explaining the selectivity for reaction with the exocyclic double bond but not the internal double bond.

An important characteristic of alkene and alkyne hydrogenations, both the homogeneously and heterogeneously catalyzed versions, is that hydrogen addition occurs with "syn addition," with hydrogen entering from the least hindered side [3].

In the absence of a metal catalyst, thermal hydrogenation seldom occurs at temperatures below 480°C (750 K or 900°F). In a catalytic hydrogenation, the unsaturated substrate is chemisorbed onto the catalyst, with most sites covered by the substrate. Hydrogen forms surface hydrides (M–H) from which hydrogens can be transferred to the chemisorbed substrate. Platinum, palladium, rhodium, and ruthenium form highly active catalysts, which operate at lower temperatures and lower pressures of H_2. Nonprecious metal catalysts, especially those based on nickel (such as Raney nickel and Urushibara nickel), have also been developed as economical alternatives, but they are often slower or require higher temperatures. The trade-off is activity (speed of reaction) versus cost of the catalyst and cost of the apparatus required for use of high pressures. For example, Raney nickel-catalyzed hydrogenations require high pressures [414].

Hydrogenation is a useful reaction for converting more oxidized oxygen and nitrogen compounds such as aldehydes, imines, and nitriles into the corresponding saturated compounds, that is, alcohols and amines. Primary alcohols can be synthesized from aldehydes by hydrogenation. Thus, alkyl aldehydes, which can be synthesized with the oxo process from carbon monoxide and an alkene, can be converted to alcohols, for example, 1-propanol is produced from propionaldehyde. Xylitol, a polyol, is produced by hydrogenation of the sugar xylose, an aldehyde. Primary amines can be synthesized by hydrogenation of nitriles.

Hydrogenations of coal, oil shale, and heavy oil are essential to produce liquid fuels, which can be suitable for the production of gasoline, diesel, and jet fuel. Hydrogenation allows this feedstock to improve their hydrogen to carbon ratio at the level so that they can be a reasonable feedstock for further refining and upgrading.

9.9.2 HYDROCRACKING

Hydrocracking is a catalytic cracking process assisted by the presence of added hydrogen gas. Unlike a hydrotreater, where hydrogen is used to cleave C–S and C–N bonds, hydrocracking uses hydrogen to break C–C bonds. The products of this process are saturated HCs; depending on the reaction conditions (temperature, pressure, catalyst activity), these products range from ethane, LPG to heavier HCs consisting mostly of isoparaffins. Hydrocracking is normally facilitated by a bifunctional catalyst that is capable of rearranging and breaking hydrocarbon chains as well as adding hydrogen to aromatics and olefins to produce naphthenes and alkanes [452].

The major products from hydrocracking are jet fuel and diesel, but low-sulfur naphtha fractions and LPG are also produced [8]. All these products have a very low content of sulfur and other contaminants. In the United States, fluid catalytic cracking process is more common because the demand for gasoline is higher.

The hydrocracking process depends on the nature of the feedstock and the relative rates of the two competing reactions, hydrogenation and cracking. Heavy aromatic feedstock is converted into lighter products under a wide range of very high pressures (1000–2000 psi) and fairly high temperatures (750°F–1500°F), in the presence of hydrogen and cracking catalysts [452].

The primary function of hydrogen in hydrocracking operation is, thus, as follows: (1) preventing the formation of polycyclic aromatic compounds if feedstock has a high paraffinic content, (2) reducing tar formation, (3) reducing impurities, (4) preventing buildup of coke on the catalyst, (5) converting sulfur and nitrogen compounds present in the feedstock to hydrogen sulfide and ammonia, and (6) achieving high cetane number fuel.

9.9.3 HYDRODESULFURIZATION

HDS is a catalytic chemical process widely used to remove sulfur (S) from natural gas and from refined petroleum products such as gasoline or petrol, jet fuel, kerosene, diesel fuel, and fuel oils [1,2]. The purpose of removing the sulfur is to reduce the sulfur dioxide (SO_2) emissions that result from using those fuels in automotive vehicles, aircraft, railroad locomotives, ships, gas- or oil-burning power plants, residential and industrial furnaces, and other forms of fuel combustion [451].

Another important reason for removing sulfur from the naphtha streams within a petroleum refinery is that sulfur, even in extremely low concentrations, poisons the noble metal catalysts (platinum and rhenium) in the catalytic reforming units that are subsequently used to upgrade the octane rating of the naphtha streams.

The industrial hydrodesulfurization (HDS) processes include facilities for the capture and removal of the resulting hydrogen sulfide (H_2S) gas. In petroleum refineries, the hydrogen sulfide gas is then subsequently converted into by-product elemental sulfur or sulfuric acid (H_2SO_4). In fact, the vast majority of the 64,000,000 metric tons of sulfur produced worldwide in 2005 was by-product sulfur from refineries and other hydrocarbon processing plants [3,4]. An HDS unit in the petroleum refining industry is also often referred to as a **hydrotreater**.

In the mid-1950s, the first noble metal catalytic reforming process (the Platformer process) was commercialized. At the same time, the catalytic HDS of the naphtha feed to such reformers was also commercialized. In the decades that followed, various proprietary catalytic HDS processes have been commercialized. Currently, virtually all of the petroleum refineries worldwide have one or more HDS units.

By 2006, miniature microfluidic HDS units had been implemented for treating JP-8 jet fuel to produce clean feedstock for a fuel cell hydrogen reformer [7]. By 2007, this had been integrated into an operating 5 kW fuel cell generation system [451].

Hydrogenolysis is a type of hydrogenation and results in the cleavage of the C–X chemical bond, where C is a carbon atom and X is a sulfur (S), nitrogen (N), or oxygen (O) atom. The net result of a hydrogenolysis reaction is the formation of C–H and H–X chemical bonds. Thus, HDS is a hydrogenolysis reaction. Using ethanethiol (C_2H_5SH), a sulfur compound present in some petroleum products, as an example, the HDS reaction can be simply expressed as

$$C_2H_5SH + H_2 \rightarrow C_2H_6 + H_2S \qquad (9.44)$$

The refinery HDS feedstocks (naphtha, kerosene, diesel oil, and heavier oils) contain a wide range of organic sulfur compounds, including thiols, thiophenes, organic sulfides and disulfides, and many others. These organic sulfur compounds are products of the degradation of sulfur containing biological components and present during the natural formation of the petroleum crude oil.

When the HDS process is used to desulfurize a refinery naphtha, it is necessary to remove the total sulfur down to the parts per million range or lower in order to prevent the poisoning of the noble metal catalysts in the subsequent catalytic reforming of the naphtha. When the process is used for desulfurizing diesel oils, the latest environmental regulations in the United States and Europe, requiring what is referred to as *ultralow sulfur diesel*, in turn require that very deep HDS is needed. In the very early 2000s, the governmental regulatory limits for highway vehicle diesel were within the range of 300–500 ppm by weight of total sulfur. As of 2006, the total sulfur limit for highway diesel is in the range of 15–30 ppm by weight [451].

A family of substrates that are particularly common in petroleum are the aromatic sulfur-containing heterocycles called thiophenes. Many kinds of thiophenes occur in petroleum ranging from thiophene itself to more condensed derivatives called benzothiophenes and dibenzothiophenes. Thiophene itself and its alkyl derivatives are easier to hydrogenolyze, whereas dibenzothiophene, especially its 4,6-disubstituted derivatives, is considered the most challenging substrate. Benzothiophenes are midway between the simple thiophenes and dibenzothiophenes in their susceptibility to HDS [451].

In an industrial HDS unit, such as in a refinery, the HDS reaction takes place in a fixed-bed reactor at elevated temperatures ranging from 300°C to 400°C and elevated pressures ranging from 30 to 130 atm of absolute pressure, typically in the presence of a catalyst consisting of an alumina base impregnated with cobalt and molybdenum (usually called a CoMo catalyst). Occasionally, a combination of nickel and molybdenum (called NiMo) is used, in addition to the CoMo catalyst, for specific difficult-to-treat feedstocks, such as those containing a high level of chemically bound nitrogen [451].

The main HDS catalysts are based on molybdenum disulfide (MoS_2) together with smaller amounts of other metals [10]. The nature of the sites of catalytic activity remains an active area of investigation, but it is generally assumed that basal planes of the MoS_2 structure are not relevant to catalysis, rather the edges or rims of this sheet [11]. At the edges of the MoS_2 crystallites, the molybdenum center can stabilize a coordinatively unsaturated site (CUS), also known as an anion vacancy. Substrates, such as thiophene, bind to this site and undergo a series of reactions that result in both C–S scission and C=C hydrogenation. Thus, the hydrogen serves multiple roles—generation of anion vacancy by removal of sulfide, hydrogenation, and hydrogenolysis.

Most metals catalyze HDS, but it is those at the middle of the transition metal series that are most active. Ruthenium disulfide appears to be the single most active catalyst, but binary combinations of cobalt and molybdenum are also highly active [12]. Aside from the basic cobalt-modified MoS_2 catalyst, nickel and tungsten are also used, depending on the nature of the feed. For example, Ni–W catalysts are more effective for hydrodenitrogenation. Metal sulfides are "supported" on materials with high surface areas. A typical support for HDS catalyst is γ-alumina. The support allows the more expensive catalyst to be more widely distributed, giving rise to a larger fraction of the MoS_2 that is catalytically active. The interaction between the support and the catalyst is an area of intense interest, since the support is often not fully inert but participates in the catalysis [451].

9.9.4 HYDRODEMETALLIZATION

As feedstocks become heavier and contain increasing amounts of heteroatoms, new processes for upgrading these feedstocks are important. For example, improved processes for hydrodemetallization need to be developed to remove vanadium, nickel, and other metals from crude oil. Heavier feedstocks also require hydrocracking and other hydroprocessing to yield more valuable products such as gasoline, diesel, jet fuel, and lube oil.

Hydrodemetallization catalysts can comprise various components. One component that can be used in hydrodemetallization catalysts is a magnesium aluminosilicate clay. Layered magnesium aluminosilicates can be described as a type of clay comprising alternating layers of octahedrally coordinated magnesium atoms and tetrahedrally coordinated silicon and/or aluminum atoms. Magnesium aluminosilicate clays have a negative layer charge, which can be balanced by cations. The literature contains examples of magnesium aluminosilicate clays used as catalysts or as components of catalysts.

While synthesis of clays can be difficult, particularly on a large scale, clays have received attention for use in hydroprocessing reactions. For example, U.S. pat. no. 3,844,978 discloses a layer-type, dioctahedral, clay-like mineral that is a magnesium aluminosilicate. The clay can be used as a catalyst or as a component in a catalyst composition.

9.9.5 HYDRODENITROGENATION (HDN) AND HYDRODEOXYGENATION (HDO)

Organonitrogen compounds, even though they occur at low levels, are undesirable because they are poisoning downstream catalysts. Furthermore, upon combustion, organonitrogen compounds generate NO_x, a pollutant. HDN is effected as general hydroprocessing, which traditionally focuses on HDS because sulfur compounds are even more problematic. To some extent, hydrodeoxygenation (HDO) is also effected [453].

Typical organonitrogen compounds in petroleum include quinolines and porphyrins and their derivatives. The total nitrogen content is typically less than 1%, and the targeted levels are in the ppm range. In HDN, the organonitrogen compounds are treated at high temperatures with hydrogen in the presence of a catalyst, the net transformation being [2]

$$R_3N + 3H_2 \rightarrow 3RH + NH_3 \tag{9.45}$$

The catalysts generally consist of cobalt and nickel as well as molybdenum disulfide or less often tungsten disulfide supported on alumina. The precise composition of the catalyst, that is, Co/Ni and Mo/W ratios, is tuned for particular feedstocks. A wide variety of catalyst compositions have been considered, including metal phosphides [453,454].

Using pyridine (C_5H_5N), a nitrogen compound present in some petroleum fractionation products, as an example, the hydrodenitrogenation reaction has been postulated as occurring in three steps [453,454]:

$$C_5H_5N + 5H_2 \rightarrow C_5H_{11}N + 2H_2 \rightarrow C_5H_{11}NH_2 + H_2 \rightarrow C_5H_{12} + NH_3 \tag{9.46}$$

and the overall reaction may be simply expressed as

$$C_5H_5N + 5H_2 \rightarrow C_5H_{12} + NH_3 \tag{9.47}$$

Many HDS units for desulfurizing naphtha within petroleum refineries are actually simultaneously denitrogenating to some extent as well.

Besides hydrodenitrogenation, hydrodeoxygenation is very important for upgrading bio-oils and reducing its oxygen content. Hydrogenation, hydrocracking, HDS, HDM, HDN, and HDO, in principle, all occur at the same time with a suitable catalyst and high-pressure, high-temperature operating conditions. Sometimes these processes are separated to optimize the reactor performance. Sometimes they are operated in stages. All of these processes together are generally called hydroprocessing operation, which is the heart of the refinery operation. Thus, the use of hydrogen is essential in modern-day refinery operation.

9.10 COMPARISON OF LIQUID FUEL PRODUCTION WITH OTHER METHODS

Liquid fuels like gasoline, diesel, and jet fuel have been largely produced by the process of refining crude oil, heavy oil, bitumen, etc. This petroleum-based liquid fuels are nonrenewable sources, and their composition and characteristics significantly depend on the original composition of the crude or heavy oil. Bio-oil derived from biomass and waste has significantly different composition than petroleum-based crude oil. Coal liquids are much more aromatic than crude oil, and oil shale containing kerogens is very paraffinic. Thus, the composition of liquid fuels resulting from these different original feedstock can significantly differ from each other in their composition and may not be completely miscible with each other [14,71,72].

The liquid fuels, on the other hand, produced from syngas are always of the same quality as long as syngas has the same composition of hydrogen, carbon monoxide, and other impurities and same operating conditions are used for its generation. One can reproduce syngas composition generated from coal and oil with the biosyngas coming from biomass and waste. Thus, liquid fuels coming from syngas of different original feedstocks can be very similar and miscible. This gives enormous edge to the process of liquid fuel production from syngas over the production of liquid fuels coming from coal, crude and heavy oils, oil shale, and biomass/waste.

The removal of impurities from coal, biomass, and crude oils requires feed pretreatment and extensive postrefining operations. The removal of impurities from syngas is a relatively easier task resulting in more definitive results.

REFERENCES

1. IEA Clean Coal Center, Fuels for biomass co-firing, IEA, Brussels, Belgium (2005), ISBN: 92-9029-418-3.
2. IEA: World energy outlook, 225pp., IEA, Brussels, Belgium (1999), ISBN: 92-64-17140-1.
3. Anderson, R.B., in Emmett, P. (ed.), *Catalysis*, Vol. IV. Reinhold, New York, p. 5 (1956).
4. Van Herwijnen, T., Van Doesburg, H., and Dejong, W.A., Kinetics of the methanation of CO and CO_2 on a nickel catalyst, *Journal of Catalysis*, 28, 391 (1973).
5. Lee, S., Lee, B.G., Gogate, M.R., and Parameswaran, V., Fundamentals of methanol synthesis, in *Proceedings of Annual DOE Contractors Conference for Indirect Coal Liquefaction*, US DOE/PETC, Pittsburgh, PA (1989).
6. Chinchen, G.C., Mansfield, K., and Spencer, M.S., Methanol synthesis: How does it work?, *CHEMTECH*, 29 (11), 692–699 (1990).
7. Alleman, T.L. and McCormick, R.L., Fischer–Tropsch diesel fuels—Properties and exhaust emissions: A literature review, in *2003 SAE World Congress, CI Engine Combustion Processes & Performance with Alternative Fuels (SP-1737)*, Detroit, MI (March 3–6, 2003).
8. Dalla Betta, R.A., Piken, A.G., and Shelef, M., Heterogeneous methanation: Initial rate of CO hydrogenation on supported ruthenium and nickel, *Journal of Catalysis*, 35, 54–60 (1974).
9. Storch, H.H., Golumbic, N., and Anderson, R.B., *The Fischer–Tropsch and Related Synthesis*. Wiley, New York, pp. 314, 315, 437, 439 (1951).
10. Herrington, E.F.G. and Woodward, L.A., Experiments on the Fischer-Tropsch synthesis of hydrocarbons from carbon monoxide and hydrogen, *Transactions of the Faraday Society*, 35, 958–966 (1939).
11. Laynes, E.T., To Hydrocarbon Research Inc., U.S. Patent No. 2,445,426 (August 1948).
12. Bergman, P.C.A., Combined torrefaction and pelletization—The top process, ECN report ECN-c-05-073, Petten, the Netherlands, p. 29 (2005).
13. Bergman, P.C.A., Boersma, A.R., and Kiel, J.H.A., ECN-RX-04-029, pp. 78–82, ECN, Petlen, the Netherlands (2004).
14. Shah, Y.T., *Energy and Fuel Systems Integration*. CRC Press, New York (2015).
15. Boerrigter, H., Deurwaarder, E.P., Bergman, P.C.A., Van Padsen, S.V.B., and Vann Ree Therman, R., Bio-refinery: High-efficient integrated production of renewable chemicals (transportation) fuels and products from biomass, Report ECN-RX-04-029, pp. 67–77, ECN, Petlen, the Netherlands (2004).
16. Boerrigter, H. and Van der Drift, A., Synthesis gas from biomass for fuels and chemicals, Paper for Workshop Organized by IEA Bioenergy Task 33 (Biomass Gasification) in Conjunction with the *SYNBIOS Conference* held in Stockholm, Sweden (May 2005).
17. Cavalov, B. and Peteves, S.D., Status and perspectives of biomass-to liquid fuels in the European Union, European Commission, DC JRC, Institute for Energy, Petten, the Netherlands, p. 1141 (2005).
18. Shah, Y.T. and Perrotta, A., Catalysts for Fischer–Tropsch and isosynthesis, *I&EC Product Research and Development*, 15, 123 (1976).
19. Lee, S., Methanol synthesis from syngas, Chapter 9, in Lee, S., Speight, J., and Loyalka, S. (eds.), *Handbook of Alternative Fuel Technologies*. CRC Press, New York, pp. 297–321 (2007).
20. Lee, S., *Methanol Synthesis Technology*. CRC Press, Boca Raton, FL (1990).
21. Cybulski, A., Liquid-phase methanol synthesis: Catalysts, mechanism, kinetics, chemical equilibria, vapor–liquid equilibria, and modeling—Review, *Catalysis Reviews: Science and Engineering*, 36 (4), 557–615 (1994).
22. Lee, S., Parameswaran, V., Wender, I., and Kulik, C.J., The roles of carbon dioxide in methanol synthesis, *Fuel Science and Technology International*, 7 (8), 1021–1057 (1989).
23. Sawant, A., Parameswaran, V., Lee, S., and Kulik, C.J., In-situ reduction of a methanol synthesis catalyst in a three-phase slurry reactor, *Fuel Science and Technology International*, 5 (1), 77–88 (1987).
24. Hermann, R.G., Classical and non classical route for alcohol synthesis, Chapter 7, in Guczi, L. (ed.), *New Trends in CO Activation*. Elsevier, Amsterdam, the Netherland, pp. 281–285 (1991).
25. Lee, S., Sawant, A., and Kulik, C.J., Phases in the active liquid phase methanol synthesis catalyst, *Fuel Science and Technology International*, 6 (2), 151–164 (1988).
26. Lee, S., Sawant, A., and Kulik, C.J., Process for methanol catalyst regeneration using crystallite redispersion, U.S. Patent No. 5,004,717 (April 2, 1991).
27. Lee, S., *Methane and Its Derivatives*. Marcel Dekker, New York (1997).
28. Sawant, A., Rodrigues, K., Kulik, C.J., and Lee, S., The effects of carbon dioxide and water on the methanol synthesis catalyst, *Energy & Fuels*, 3 (1), 2–7 (1989).
29. Lee, S. and Sardesai, A., Liquid phase methanol and dimethylether synthesis from syngas, *Topics in Catalysis*, 33 (1–2) (2005).

30. Gogate, M.R., Kulik, C.J., and Lee, S., A novel single-step dimethyl ether (DME) synthesis in a three-phase slurry reactor from CO-rich syngas, *Chemical Engineering Science*, 47 (13–14), 3769–3776 (1992).

31. Lee, S. and Parameswaran, V., Reaction mechanism in liquid-phase methanol synthesis, EPRI-ER/GS-6715, Electric Power Research Institute, Palo Alto, CA, pp. 1–206 (1990).

32. Liu, X.-M., Lu, G.Q., Yan, Z.-F., and Beltramini, J., Recent advances in catalysts for methanol synthesis via hydrogenation of CO and CO_2, *Industrial & Engineering Chemistry Research*, 42 (25), 6518–6530 (2003).

33. Bailey, E., The coal to methanol alternative, *Coal Technology (Houston)*, 2 (3), 149–161 (1979).

34. Bhatt, B.L., Heydorn, E.C., Tihm, P.J.A., Street, B.T., and Kornosky, R.M., Liquid phase methanol (LPMEOH) process development, *Preprints—American Chemical Society, Division of Petroleum Chemistry*, 44 (1), 25–27 (1999).

35. Chinchen, G.C., Mansfield, K., and Spencer, M.S., The methanol synthesis—How does it work, *CHEMTECH*, 20 (11), 692–699 (1990).

36. U.S. DOE, Commercial-scale demonstration of the liquid phase methanol (LPMEOHTM) process, DOE/FE-0243P, U.S. Department of Energy, Washington, D.C. (1992).

37. Dybkjaer, I., Topsoe methanol technology, *Chemical Economy and Engineering Review*, 13 (6), 17–25 (1981).

38. Dybkjaer, I. and Christensen, T.S., Syngas for large scale conversion of natural gas to liquid fuels, in Fleisch, T. et al. (eds.), *Natural Gas Conversion VI*, Studies in Surface Science and Catalysis, Vol. 136, pp. 435–440, Elsevier, Amsterdam, the Netherlands (2001).

39. Methanol, in *Ullmann's Encyclopedia of Industrial Chemistry Release 2003*, 6th edn. Wiley-VCH Verlag GmbH & Co., KGaA, Berlin, Germany (2003).

40. Haid, J. and Koss, U., Lurgi's Mega-Methanol technology opens the door for a new era in down-stream applications, in Fleisch, T. et al. (eds.), *Natural Gas Conversion VI*, Studies in Surface Science and Catalysis, Vol. 136, pp. 399–404, Elsevier, Amsterdam, the Netherlands (2001).

41. Hamelinck, C.N. and Faaij, A.P.C., Future prospects for production of methanol and hydrogen from biomass, NWS-E-2001-49, Utrecht University (2001), ISBN: 90-73958-84-9.

42. Klier, K., Methanol synthesis, *Advances in Catalysis*, 31, 243–313 (1982).

43. Kung, H.H., Deactivation of methanol synthesis catalysts—A review, *Catalysis Today*, 11 (4), 443–453 (1992).

44. Ladebeck, J., Improve methanol synthesis, *Hydrocarbon Processing, International Edition*, 72 (3), 89–91 (1993).

45. Larson, E.D. and Katofsky, R.E., Production of methanol and hydrogen from biomass, Report No. 271, Princeton University, Center for Energy and Environmental Studies, Princeton, NJ (1992).

46. Pass, G., Holzhauser, C., Akgerman, A., and Anthony, R.G., Methanol synthesis in a trickle-bed reactor, *AIChE Journal*, 36 (7), 1054–1060 (1990).

47. Pinto, A. and Rogerson, P.L., Optimizing the ICI low-pressure methanol process, *Chemical Engineering (New York)*, 84 (14), 102–108 (1977).

48. Rogerson, P.L., The ICI [Imperial Chemical Industries] low pressure methanol process, *Ingenieursblad*, 40 (21), 657–663 (1971).

49. Rogerson, P.L., The ICI low-pressure methanol process, Chapter 2-2, in Meyers, R.A. (ed.), *Handbook of Synfuels Technology*. McGraw-Hill Book Company, New York (1984).

50. Salmon, R., Economics of methanol production from coal and natural gas, ORNL-6091, Oak Ridge National Laboratory, Oak Ridge, CN (1986).

51. Supp, E. and Quinkler, R.F., The Lurgi low-pressure methanol process, in Meyers, R.A. (ed.), *Handbook of Synfuels Technology*. McGraw-Hill, New York, pp. 113–131 (1984).

52. Takase, I. and Niwa, K., Mitsubishi (MGC/MHI) methanol process, *Chemical Economy & Engineering Review*, 17 (5), 24–30 (1985).

53. Tijm, P.J.A., Waller, F.J., and Brown, D.M., Methanol technology developments for the new millennium, *Applied Catalysis A*, 221, 275–282 (2001).

54. Twigg, M.V. and Spencer, M.S., Deactivation of supported copper metal catalysts for hydrogenation reactions, *Applied Catalysis A: General*, 212 (1–2), 161–174 (2001).

55. English, A., Rovner, J., Brown, J., and Davies, S., Methanol, in *Kirk-Othmer Encyclopedia of Chemical Technology*. John Wiley & Sons, Inc., Hoboken, NJ (1995).

56. Park, C.A., Commercial-scale demonstration of the liquid phase methanol (LPMEOHTM) process, DOE/FE-0243P, Report number 543.7500, U.S. Department of Energy, Washington, D.C. (1992).

57. Davenport, B., Methanol. Chemical economics handbook marketing research report, Report number 674.5000, SRI International, Menlo Park, CA (2002).

58. Spath, P.L. and Dayton, D.C., Preliminary screening—Technical and economic assessment of synthesis gas to fuels and chemicals with emphasis on the potential for biomass-derived syngas, NREL/TP-510-34929, NREL, Golden, CO (December 2003).

59. Lee, S. and Gogate, M.R., Development of a single-stage liquid-phase synthesis process of dimethyl ether from syngas, EPRI-TR-100246, Electric Power Research Institute, Palo Alto, CA, pp. 1–179 (1992).

60. Brown, D.M., Bhatt, B.L., Hsiung, T.H., Lewnard, J.J., and Waller, F.J., Novel technology for the synthesis of dimethyl ether from syngas, *Catalysis Today*, 8 (3), 279–304 (1991).

61. Collignon, F., Loenders, R., Martens, J.A., Jacobs, P.A., and Poncelet, G., Liquid phase synthesis of MTBE from methanol and isobutene over acid zeolites and amberlyst-15, *Journal of Catalysis*, 182 (2), 302–312 (1999).

62. Ge, Q., Huang, Y., Qiu, F., and Li, S., Bifunctional catalysts for conversion of synthesis gas to dimethyl ether, *Applied Catalysis A: General*, 167 (1), 23–30 (1998).

63. Gunda, A., Tartamella, T., Gogate, M., and Lee, S., Dimethyl ether synthesis from CO_2-rich syngas in the LPDME process, in *Proceedings of the 12th Annual International Pittsburgh Coal Conference*, pp. 710–715 (1995).

64. Lidderdale, T.C.M., MTBE production economics, Energy Information Agency, Department of Energy, Washington, D.C. (2001).

65. Peng, X.D., Wang, A.W., Toseland, B.A., and Tijm, P.J.A., Single-step syngas-to-dimethyl ether processes for optimal productivity, minimal emissions, and natural gas-derived syngas, *Industrial & Engineering Chemistry Research*, 38 (11), 4381–4388 (1999).

66. Sardesai, A. and Lee, S., Liquid phase dimethyl ether (DME) synthesis: A review, *Reviews in Process Chemistry and Engineering*, 1 (2), 141–178 (1998).

67. Shikada, T., Ohno, Y., Ogawa, T., Ono, M., Mizuguchi, M., Tomura, K., and Fujimoto, K., Synthesis of dimethyl ether from natural gas via synthesis gas, *Kinetics and Catalysis (Translation of Kinetika i Kataliz)*, 40 (3), 395–400 (1999).

68. Blaszkowski, S.R. and van Santen, R.A., The mechanism of dimethyl ether formation from methanol catalyzed by zeolitic protons, *Journal of the American Chemical Society*, 118 (21), 5152–5153 (1996).

69. Jamshidi, L.C.L.A., Barbosa, C.M.B.M., Nascimento, L., and Rodbari, J.R., Catalytic dehydration of methanol to dimethyl ether (DME) using the $A_{162.2}Cu_{25.3}Fe_{12.5}$ quasi crystalline alloy, *Journal of Chemical Engineering & Process Technology*, 4, 164 (2013).

70. Dimethyl ether, Wikipedia, the free encyclopedia (2015). https://er.wikipedia.org/wiki/dimethyl_ether (last accessed Otober 13, 2016).

71. Shah, Y.T., *Water for Energy and Fuel Production*, CRC Press, New York (2014).

72. Lee, S. and Shah, Y.T., *Biofuels and Bioenergy: Technology and Processes*, CRC Press, New York (2012).

73. Chemrec, Wikipedia, the free encyclopedia (2015).

74. Lee, S., Gogate, M.R., and Fullerton, K.L., Catalytic process for production of gasoline from synthesis gas, U.S. Patent 5,459,166 (1995).

75. Chang, C.D. and Silvestri, A.J., MTG origin, evolution, operation, *CHEMTECH*, 10, 624–631 (1987).

76. Bem, J., New Zealand synthetic gasoline plant, in *Proceedings of the 20th Intersociety Energy Conversion Engineering Conference*, Miami Beach, FL (August 18–23, 1985) Vol. 1, Society of Automotive Engineers, pp. 1.517–1.522 (1985).

77. Edwards, M.S., Ulrich, W.C., and Salmon, R., Economics of producing gasoline from underground coal gasification synthesis gas, *Urja*, 6 (7), 163–178 (1979).

78. Kam, A.Y., Schreiner, M., and Yurchak, S., Mobil methanol-to-gasoline (MTG) process, Chapter 2-3, in Meyers, R.A. (ed.), *Handbook of Synfuels Technology*. McGraw-Hill Book Company, New York (1984).

79. Fischer Tropsch synthesis liquid fuels NETL report, DOE, Washington, D.C. (2010). www.netl.doe.gov/research/coal/.../gasification/gasifipediaftsysnthesis.

80. Lee, W., Maziuk, J., Weekman, V.W., and Yurchak, S., Catalytic conversion of alcohols to gasoline by the mobil process, in *Energy from Biomass and Wastes IV*, LakeBuena, Vista, FL, p. 834.

81. Maiden, C.J., New Zealand gas to gasoline plant, *APEA Journal*, 23 (1), 33–43 (1983).

82. Schreiner, M., Research guidance studies to asses gasoline from coal by methanol-to-gasoline and sasol-type Fischer–Tropsch technologies, DOE Contract No. EF-76-C-01-2447, Mobil Research and Development Corp., Princeton, NJ (1978).

83. Tabak, S.A. and Yurchak, S., Conversion of methanol over ZSM-5 to fuels and chemicals, *Catalysis Today*, 6 (3), 307–327 (1990).

84. Topp-Jorgensen, J., The Topsoe integrated gasoline synthesis, *Petrole et Techniques*, 333, 11–17 (1987).

85. Wham, R.M. and Forrester, R.C., III, Available technology for indirect conversion of coal to methanol and gasoline: A technology and economics assessment, Miami International Conference on *Alternative Energy Sources*, Miami Beach, FL (December 15, 1980); also Technical report by Oak Ridge National Lab for DOE, Oak Ridge, TN, 29 pp. (1980); report no. CONF8012101ODR; contract no. N-7405-ENG-26.|

86. Jones, S.B. and Zhu, Y., Techno-economic analysis for the conversion of lignocellulosic biomass to gasoline via the methanol-to-gasoline (MTG) process, PNNL-18481, Pacific Northwest National Laboratory, a report prepared for Department of Energy Contract DE-AC05-76RL01830 (April 2009).

87. Bem, J., New Zealand synthetic gasoline plant, in *Proceedings of the 20th Intersociety Energy Conversion Engineering Conference*, Miami Beach, FL (August 18–23, 1985), Vol. 1, pp. 1.517–1.522 (1985).

88. Chang, C.D., Hydrocarbons from methanol, *Catalysis Reviews: Science and Engineering*, 25 (1), 1–118 (1983).

89. Apanel, G. and Netzer, D., Methanol to propylene by the Lurgi MTP process, PEP Review No. 98-13, SRI Consulting, Menlo Park, CA (2002).

90. Hancock, E.G., The manufacture of gasoline and the chemistry of its components, in *Technology of Gasoline*, Critical Reports on Applied Chemistry, John Wiley, Hoboken, NJ, Vol. 10, pp. 20–56 (1985).

91. Hansen, J.B. and Joensen, F., High conversion of synthesis gas into oxygenates, in Holmen, A. et al. (eds.), *Natural Gas Conversion*, Studies in Surface Science and Catalysis, Vol. 61, Elsevier, Amsterdam, the Netherlands, pp. 457–467 (1991).

92. Bare, S., Methanol to olefins (MTO): Development of commercial catalytic process, Modern methods in heterogeneous catalysis research, FHI lecture, UOP, Des Plaines, IL (November 30, 2007).

93. Tian, P., Wei, Y., Ye, M., and Liu, Z., Methanol to olefins (MTO): From fundamentals to commercialization, *ACS Catalysis*, 5 (3), 1922–1938 (2015).

94. Funk, G., Myers, D., and Vora, B., A different game plan-new technology for the production of ethylene and propylene from methanol, UOP, Hydrocarbon Engineering, Des Plaines, IL, pp. 1–4 (December 2013).

95. Wan, V., Methanol to olefins, a private report by the process economics program, Report No. 261, SRI Consulting, Menlo Park, CA (November 2007).

96. Sun, X., Catalytic conversion of methanol to olefins over HZSM-5 catalysts, PhD thesis, Technischen Universitat Munchen, Munich, Germany (2013).

97. Ying, L., Ye, M., Cheng, Y., Li, X., and Liu, Z., A kinetic study of methanol to olefins (MTO) process in fluidized bed reactor, in Kuipers, J., Mudde, R., van Ommen, J., and Deen, N. (eds.), *The 14th International Conference on Fluidization—From Fundamentals to Products*, ECI Symposium Series (2013), http://dc.engconfintl.org/fluidization_xiv/19.

98. Chen, A.K. and Masel, R., Direct conversion of methanol to formaldehyde in the absence of oxygen on Cu(210), *Surface Science*, 343 (1–2), 17–23 (1995).

99. Park, T.-Y. and Froment, G.F., Analysis of fundamental reaction rates in the methanol-to-olefins process on ZSM-5 as a basis for reactor design and operation, *Industrial & Engineering Chemistry Research*, 43 (3), 682–689 (2004).

100. Hongxing, L., Zaiku, X., and Guoliang, Z., SINOPEC methanol-to-olefins (S-MTO) technology process, in *New Technologies and Alternative Feedstocks in Petrochemistry and Refining DGMK Conference*, October 9–11, 2013, Dresden, Germany (2013).

101. Keil, F., Methanol to hydrocarbon process technology, *Microporous and Mesoporous Materials*, 29 (1–2), 49–66 (June 1999).

102. Song, W., Fu, H., and Haw, J.F., Selective synthesis of methylnaphthalenes in HSAPO-34 cages and their function as reaction centers in methanol-to-olefin catalysis, *Journal of Physical Chemistry B*, 105 (51), 12839–12843 (2001).

103. Dubois, D.R., Obrzut, D.L., Liu, J., Thundimadathil, J., Adekkanattu, P.M., Guin, J.A., Punnoose, A., and Seehra, M.S., Conversion of methanol to olefins over cobalt-, manganese- and nickel-incorporated SAPO-34 molecular sieves, *Fuel Processing Technology*, 83 (1–3), 203–218 (September 15, 2003).

104. Park, T.-Y. and Froment, G.F., Kinetic modeling of the methanol to olefins process. 1. Model formulation, *Industrial & Engineering Chemistry Research*, 40 (20), 4172–4186 (2001).

105. Ding, J. and Hua, W., Game changers of the C3 value chain: Gas, coal, and biotechnologies, *Chemical Engineering and Technology*, 36, 83–90 (2013).

106. Chen, J.Q., Bozzano, A., Globver, B., Fuglerud, T., and Kvisle, S., Recent advancements in ethylene and propylene production using UOP/hydro MTO process, *Catalysis Today*, 106, 103–107 (2005).

107. Eng, C.N., Arnold, E.C., and Vora, B.V., Integration of the UOP/HYDRO MTO process into ethylene plants, in Presented at the *1998 AIChE Spring National Meeting, Session 16, Fundamental Topics in Ethylene Production*, New Orleans, LA, Paper 16g (March 8–12, 1998).

108. Barger, P., Methanol to olefins (MTO) and beyond, *Catalysis Science Series*, 3, 239–260 (2002).

109. Vora, B. and Pujado, P., Process for enhanced olefin production, U.S. Patent 7,317,133 B2 (November 21, 2002).

110. Kempf, R., Advanced MTO: Breakthrough technology for the profitable production of light olefins, Available online: http://www.wraconferences.com/sites/default/files/day%202%201130%20petchem-Rick%20Kempf.pdf, Accessed on September 8, 2015.

111. Jasper, S. and El-Halwagi, M.M., A techno-economic comparison between two methanol-to-propylene processes, *Processes*, 3, 684–698 (2015).

112. Chang, C.D. and Silvestri, A.J., The conversion of methanol and other O-compounds to hydrocarbons over zeolite catalysts, *Journal of Catalysis*, 47, 249–259 (1977).

113. Stocker, M., Methanol-to-hydrocarbons: Catalytic materials and their behavior, *Microporous and Mesoporous Materials*, 29, 3–48 (1999).

114. Wang, W. and Hunger, M., Reactivity of surface alkoxy species on acidic zeolite catalysts, *Accounts of Chemical Research*, 41, 895–904 (2008).

115. Haw, J.F., Song, W.G., Marcus, D.M., and Nicholas, J.B., The mechanism of methanol to hydrocarbon catalysis, *Accounts of Chemical Research*, 36, 317–326 (2003).

116. Arstad, B. and Kolboe, S., The reactivity of molecules trapped within the SAPO-34 cavities in the methanol-to-hydrocarbons reaction, *Journal of the American Chemical Society*, 123, 8137–8138 (2001).

117. Chen, D., Moljord, K., and Holmen, A., A methanol to olefins review: Diffusion, coke formation and deactivation on SAPO type catalysts, *Microporous and Mesoporous Materials*, 164, 239–250 (2012).

118. Svelle, S., Joensen, F., Nerlov, J., Olsbye, U., Lillerud, K.-P., Kolboe, S., and Bjorgen, M., Conversion of methanol into hydrocarbons over zeolite H-ZSM-5: Ethene formation is mechanistically separated from the formation of higher alkenes, *Journal of the American Chemical Society*, 128, 14770–14771 (2006).

119. Margerie, C., MTO/OCP: A strategic research project, testing innovative processes for making plastics, a report by Research and Development Department of TOTAL, Brussels, Belgium (2010).

120. Senden, M.M.G., Sie, S.T., Post, M.F.M., and Ansorge, J., Engineering aspects of the conversion of natural gas into middle distillates, in Delaga, H. et al. (eds.), *Chemical Reactor Technology for Environmentally Safe Reactors and Products*, NATO ASI Series, Series E: Applied Sciences, Vol. 225, Kluwer Academic Publishers, part of Springer-Science, Berlin, Germany, pp. 227–247 (1992).

121. Fischer–Tropsch process, Wikipedia, the free encyclopedia (2015). https://en.wikipedia.org/wiki/Fischer_Tripsch_process (last accessed September 11, 2016).

122. Davis, B.H., Fischer–Tropsch synthesis: Current mechanism and futuristic needs, *Fuel Processing Technology*, 71 (1–3), 157–166 (2001).

123. Dry, M.E., The Fischer–Tropsch process: 1950–2000, *Catalysis Today*, 71 (3–4), 227–241 (2002).

124. Dry, M.E., Fischer–Tropsch reactions and the environment, *Applied Catalysis A*, 189 (2), 185–190 (1999).

125. Dry, M.E., Practical and theoretical aspects of the catalytic Fischer–Tropsch process, *Applied Catalysis A*, 138, 319 (1996).

126. Frohning, C., Kolbel, H., Ralek, M., Rottig, W., Schnur, F., and Schulz, H., Fischer–Tropsch process, Chapter 8, in Falbe, J. (ed.), *Chemical Feedstocks from Coal*. John Wiley & Sons, New York, pp. 309–432 (1982).

127. Gradassi, M.J., Economics of gas to liquids manufacture, in Permaliana, A. et al. (eds.), *Natural Gas Conversion V*, Studies in Surface Science and Catalysis, Vol. 119, Elsevier, Amsterdam, the Netherlands, pp. 35–44 (1998).

128. Gray, D. and Tomlinson, G., A technical and economic comparison of natural gas and coal feedstocks for Fischer–Tropsch synthesis, in Permaliana, A. et al. (eds.), *Natural Gas Conversion IV*, Studies in Surface Science and Catalysis, Vol. 107, Elsevier, Amsterdam, the Netherlands, pp. 145–150 (1997).

129. Haid, M.O., Schubert, P.F., and Bayens, C.A., Synthetic fuel and lubricants production using gas-to-liquids technology, DGMK Tagungsbericht 2000-3, pp. 205–212 (2000), http://www.eia.doe.gov/emeu/cabs/safrica.html, http://www.oil barrel.com/archives/featuresarchive/2002/jan-2002/sasol310102.htm.

130. Jager, B., Developments in Fischer–Tropsch technology, in Permaliana, A. et al. (eds.), *Natural Gas Conversion V*, Studies in Surface Science and Catalysis, Vol. 119, Elsevier, Amsterdam, the Netherlands, pp. 25–34 (1998).

131. Jager, B. and Espinoza, R.L., Advances in low temperature Fischer–Tropsch synthesis, *Catalysis Today*, 23, 17 (1995).

132. Jess, A., Popp, R., and Hedden, K., Fischer–Tropsch-synthesis with nitrogen-rich syngas: Fundamentals and reactor design aspects, *Applied Catalysis A*, 186 (1–2), 321–342 (1999).

133. Marano, J.J. and Ciferno, J.P., Life-cycle greenhouse-gas emissions inventory for Fischer–Tropsch fuels, U.S. DOE National Energy Technology Laboratory, Pittsburgh, PA (2001).

134. Mills, G., Status and future opportunities for conversion of synthesis gas to liquid energy fuels: Final report, TP-421-5150, National Renewable Energy Laboratory, Golden, CO (1993).

135. Jager, B., Dry, M.E., Shingles, T., and Steynberg, A.P., Experience with a new type of reactor for Fischer–Tropsch synthesis, *Catalysis Letters*, 7, 293–302 (1990).

136. Hu, J., Yu, F., and Lu, Y., Application of Fischer–Tropsch synthesis in biomass to liquid conversion, *Catalysts*, 2, 303–326 (2012).

137. Boerrigter, H. and Rauch, R., Review of applications of gases from biomass gasification, ECN Research, Petten, the Netherlands (2006).

138. Daey, O.C., Den Uil, H., and Boerrigter, H., Trigeneration from biomass and residues, in *Proceedings of Progress in Thermodynamical Biomass Conversion (PITBC)*, Tyrol, Austria (September 17–22, 2000).

139. Shah, Y.T., Biomass to liquid fuel via Fischer–Tropsch and related syntheses, Chapter 12, in Lee, J.W. (ed.), *Advanced Biofuels and Bioproducts*. Springer Book Project, Springer Publ. Co., New York, pp. 185–207 (September 2012).

140. Babu, S., Synthesis gas from biomass gasification and its utility for biofuels, Technology report for ExCo 62, IEA Task 33, Brussels, Belgium (2008).

141. Park, C.S., Hackett, C., and Norbeck, J.M., Synthetic diesel fuel production from carbonaceous feed stocks, in Presentation to *ISAF XV, International Symposia on Alcohols and Fuels*, CERT, University of California, Riverside, CA (September 2005).

142. Opdal, O.A., Production of synthetic biodiesel via Fischer–Tropsch synthesis: Biomass-to-liquids in Namdalen, Norway, A project report, Department of Energy and Process Engineering, Faculty of Engineering and Science, Norwegian University of Science and Technology, Trondheim, Norway (2006).

143. Hamelinck, C., Faaij, A.P.C., den Uil, H., and Boerrigter, H., *Production of FT Transportation Fuels from Biomass, Technical Options, Process Analysis and Optimisation, and Development Potential*. Utrecht University, Copernicus Institute, Utrecht, the Netherlands (March 2003).

144. Boerrigter, H., den Uil, H., and Calis, H.-P., Green diesel from biomass via Fischer–Tropsch synthesis: New insights in gas cleaning and process design, in Paper presented a *Pyrolysis and Gasification of Biomass and Waste, Expert Meeting*, Strasbourg, FR (September 30, 2002).

145. Davis, S., Hay, W., and Pierce, J., Biomass in the energy industry—An introduction, A report by British Petroleum, London, U.K. (2014).

146. Callis, H.P.A., Haan, H., Boerrigter, H., Van der Drift, A., Peppink, G., van den Broek, R., Faaij, A., and Venderbosch, R.H., Preliminary techno-economic analysis of large-scale synthesis gas manufacturing from imported biomass, in *Pyrolysis and Gasification of Biomass and Waste, Expert Meeting*, September 30–October 1, 2002, Strasbourg, France, pp. 403–418; also in SDE, Amsterdam, concept report (May 5, 2002).

147. Boerrigter, H., Van der Drift, A., and Van Ree, R., Biosyngas; markets, production technologies, and production concepts for biomass-based syngas, ECN-report, ECN-CX-04-013, Petten, the Netherlands (2004); also ECN Biomass Presentation at *First International Biorefinery Workshop*, Washington, DC (July 20–21, 2005).

148. Boerrigter, H. and Van der Drift, A., Liquid fuels from solid biomass: The ECN concept(s) for integrated FT-diesel production systems, Report RX-03-060, Energy Research Centre of the Netherlands (ECN), Petten, the Netherlands (2003).

149. Boerrigter, H., Economy of biomass to liquid (BTL) plants—An engineering assessment, ECN report no. ECN-C-06-019, Petlen, the Netherlands (May 2006).

150. Boerrigter, H., Calis, H.P., Slort, D.J., Bodenstaff, H., Kaandorp, A.J., den Uil, H., and Rabou, L.P.L.M., Gas cleaning for integrated biomass gasification (BG) and Fischer–Tropsch (FT) systems, Report C-04-056, Energy Research Center of the Netherland (ECN), Petten, the Netherlands, 59pp. (November 2004).

151. Gual, A., Godard, C., Castillón, S., Curulla-Ferré, D., and Claver, C., Colloidal Ru, Co and Fe-nanoparticles. Synthesis and application as nanocatalysts in the Fischer–Tropsch process, *Catalysis Today*, 183, 154–171 (2012).

152. Jahangiri, H., Bennett, J., Mahjoubi, P., Wilson, K., and Gu, S., A review of advanced catalyst development for Fischer–Tropsch synthesis of hydrocarbons from biomass derived syn-gas, *Catalysis Science & Technology*, 4, 2210–2229 (2014).

153. Lualdi, M., Fischer–Tropsch synthesis over cobalt-based catalysts for BTL applications, PhD thesis, KTH Chemical Science and Engineering, Stockholm, Sweden (2012).

154. Zhang, Q., Kang, J., and Wang, Y., Development of novel catalysts for Fischer–Tropsch synthesis: Tuning the product selectivity, *ChemCatChem*, 2 (9), 1030–1058 (September 17, 2010).

155. Van Steen, E. and Claeys, M., Fischer–Tropsch catalysts for the biomass-to-liquid (BTL)-process, *Chemical Engineering and Technology*, 31, 655–666 (2008).

156. Bartholomew, C.H., Recent developments in Fischer–Tropsch catalysis, Chapter 5, in Guczi, L. (ed.), *New Trends in CO Activation*, Studies in Surface Science and Catalysis, Vol. 64, Elsevier, Amsterdam, the Netherlands, pp. 158–224 (1991).

157. Dry, M.E., Catalytic aspects of industrial Fischer–Tropsch synthesis, *Journal of Molecular Catalysis*, 71, 133–144 (1982).

158. Dry, M.E., The Fischer–Tropsch synthesis, in Anderson, J.R. and Boudart, M. (eds.), *Catalysis, Science and Technology*, Vol. 1. Springer-Verlag, Berlin, Germany, pp. 159–256 (1981).

159. Dry, M.E. and Hoogendoorn, J.C., Technology of the Fischer–Tropsch process, *Catalysis Reviews: Science and Engineering*, 23 (2), 265–278 (1981).

160. Oukaci, R., Singleton, A.H., and Goodwin, J.G., Comparison of patented Co F-T catalysts using fixed-bed and slurry bubble column reactors, *Applied Catalysis A*, 186, 129–144 (1999).

161. Patzlaff, J., Liu, Y., Graffmann, C., and Gaube, J., Studies on product distributions of iron and cobalt catalyzed Fischer–Tropsch synthesis, *Applied Catalysis A*, 186 (1–2), 109–119 (1999).

162. Raje, A., Inga, J.R., and Davis, B.H., Fischer–Tropsch synthesis: Process considerations based on performance of iron-based catalysts, *Fuel*, 76 (3), 273–280 (1997).

163. Rao, V.U.S., Stiegel, G.J., Cinquegrane, G.J., and Srivastave, R.D., Iron based catalysts for slurry phase Fischer–Tropsch process: Technology review, *Fuel Processing Technology*, 30, 83–107 (1992).

164. Schulz, H., Selectivity and mechanism of the Fischer–Tropsch CO-hydrogenation, *C1 Molecular Chemistry*, 1, 231 (1985).

165. Vannice, M.A., The catalytic synthesis of hydrocarbons from carbon monoxide and hydrogen, *Catalysis Reviews: Science and Engineering*, 14 (2), 153–191 (1976).

166. Wender, I., Reactions of synthesis gas, *Fuel Processing Technology*, 48 (3), 189–297 (1996).

167. Adesina, A.A., Hydrocarbon synthesis via Fischer–Tropsch reaction: Travails and triumphs, *Applied Catalysis A*, 13, 345–367 (1996).

168. Chang, T., South African company commercializes new F-T process, *Oil & Gas Journal*, 98 (2), 42–45 (2000).

169. Dry, M.E., The Fischer–Tropsch process-commercial aspects, *Catalysis Today*, 6, 183 (1990).

170. Dry, M.E., The Sasol route to chemicals and fuels, in Bibbly, D. et al. (eds.), *Methane Conversion*, Studies in Surface Science and Catalysis, Vol. 36, Elsevier, Amsterdam, the Netherlands pp. 447–456 (March 1988).

171. Espinoza, R.L., Steynberg, A.P., Jager, B., and Vosloo, A.C., Low temperature Fischer–Tropsch synthesis from a Sasol perspective, *Applied Catalysis A*, 186 (1–2), 13–26 (1999).

172. Steynberg, A.P., Espinoza, R.L., Jager, B., and Vosloo, A.C., High temperature Fischer–Tropsch synthesis in commercial practice, *Applied Catalysis A*, 186 (1–2), 41–54 (1999).

173. Davis, B.H., Fischer–Tropsch synthesis: Overview of reactor development and future potentialities, *Preprints of Papers—American Chemical Society, Division of Fuel Chemistry*, 48 (2), 787 (2003).

174. Kolbel, H. and Ralek, M., Fischer–Tropsch synthesis in the liquid phase, *Catalysis Reviews: Science and Engineering*, 21 (2): 225–274 (1980).

175. Taboada, S., Small scale GTL technologies on the brink of commercialization, a report from Nexant, Burlington, MA (February 2015).

176. Velocys announces commercial scale GTL plant gets go ahead, a report in Biomass Magazine by Velocys plc. (August 1, 2014).

177. Phillips, S., Aden, A., Jechura, J., Dayton, D., and Eggeman, T., Thermochemical ethanol via indirect gasification and mixed alcohol synthesis of lignocellulosic biomass, NREL/TP-51041168, National Renewable Energy Laboratory, Golden, CO (April 2007), http://www.nrel.gov/docs/fy07osti/41168.pdf.

178. Wyman, C.E., Bain, R.L., Hinman, N.D., and Stevens, D.J., Ethanol and methanol from cellulosic materials, Chapter 21, in Johansson, T.B. et al. (ed.), *Renewable Energy: Sources for Fuels and Electricity*. Island Press, Washington, DC (1993).

179. Ecalene, Wikipedia, the free encyclopedia (2015).

180. Forzatti, P., Tronconi, E., and Pasquon, I., Higher alcohol synthesis, *Catalysis Reviews: Science and Engineering*, 33 (1–2), 109–168 (1991).

181. Quarderer, G.J., Mixed alcohols from synthesis gas, in *Proceedings from the 78th Spring National AIChE Meeting*, New Orleans, LA (April 6–10, 1986).

182. Nirula, S., Dow/Union carbide process for mixed alcohols from syngas, PEP Review No. 851-4, SRI International, Menlo Park, CA (March 1986).

183. Aden, A., Spath, P., Atherton, B., The potential of thermochemical ethanol via mixed alcohols production, NREL Milestone Report, National Renewable Energy Laboratory, Golden, CO (September 30, 2005), http://devafdc.nrel.gov/bcfcdoc/9432.pdf.

184. Chem Systems, Biomass to ethanol process evaluation, Report prepared for NREL by Chem Systems, Tarrytown, NY (1994).

185. Quarderer, G.J. and Cochran, G.A., Process for producing alcohols from synthesis gas, U.S. Patent No. 4,749,724 (1986).

186. Gerber, M.A., Gray, M.J., Stevens, D.J., White, J.F., and Rummel, B.L., Optimization of rhodium-based catalysts for mixed alcohol synthesis—2009 Progress Report, PNNL-20115, Pacific Northwest National Laboratory, Richland, WA (2010).

187. Courty, P., Arlie, J.P., Convers, A., Mikitenko, P., and Sugier, A., C1–C6 alcohols from syngas, *Hydrocarbon Processing*, 63 (11), 105–108 (1984).

188. Courty, P., Chaumette, P., Raimbault, C., and Travers, P., Production of methanol-higher alcohol mixtures from natural gas via syngas chemistry, *Revue de l'Institut Francais du Petrole*, 45 (4), 561–78 (1990).

189. Dai, L., Chen, Z., Li, G., Li, Y., and Liu, X., Synthesis of C1–C5 alcohols from coal-based syngas, in *Proceedings of the Sixth Annual International Pittsburgh Coal Conference*, Vol. 2, Pittsburgh, PA (September 25–29, 1989), 739–746 (1989). https://www.engineering.pitt.edu/../conferences/../Pr.

190. Herman, R.G., Advances in catalytic synthesis and utilization of higher alcohols, *Catalysis Today*, 55 (3), 233–245 (2000).

191. El Sawy, A.H., Evaluation of mixed alcohol production processes and catalysts, NTIS. DE90010325. SAND89-7151, Mitre Corporation, Hampton, VA (1990).

192. Lucero, A.J., Klepper, R.E., O'Keefe, W.M., and Sethi, V.K., Development of a process for production of mixed alcohols from synthesis gas, *Preprints of Symposia—American Chemical Society, Division of Fuel Chemistry*, 46 (2), 413–419 (2001).

193. Natta, G., Colombo, U., and Pasquon, I., Direct catalytic synthesis of higher alcohols from carbon monoxide and hydrogen, in Emmett, P.H. (ed.), *Catalysis*. Reinhold, New York, pp. 131–174 (1957).

194. Nirula, S.C., Dow/union carbide process for mixed alcohols from syngas, PEP Review No. 85-1-4, SRI International, Menlo Park, CA (1994).

195. Quarderer, G.J. and Cochran, G.A., Catalytic process for producing mixed alcohols from hydrogen and carbon monoxide, PCT Int. Appl., WO, Dow Chemical Co., Midland, MI, 40pp. (1984).

196. Ricci, R., Paggini, A., Fattore, V., Ancillotti, F., and Sposini, M., Production of methanol and higher alcohols from synthesis gas, *Chemia Stosowana*, 28 (1), 155–168 (1984).

197. Roberts, G.W., Lim, P.K., McCutchen, M.S., and Mawson, S., The thermodynamics of higher alcohol synthesis, *Preprints—American Chemical Society, Division of Petroleum Chemistry*, 37 (1), 225–233 (1992).

198. Smith, K.J. and Anderson, R.B., A chain growth scheme for the higher alcohols synthesis, *Journal of Catalysis*, 85 (2), 428–436 (1984).

199. Smith, K.J. and Klier, K., An overview of the higher alcohol synthesis, *Abstracts of Papers of the American Chemical Society*, 203, 82-PETR (1992).

200. Jackson, G., Production of mixed alcohol fuels from gasified biomass (Ecalene™) Ecalene™ DE (denatural ethanol) by Power Energy Fuel Inc., altenergymag.com, online trade magazine (2006). www.altenergymag.com/content.php?issue_number06.02.01&article=ecalene.

201. Verkerk, K.A.N., Jaeger, B., Finkeldei, C.H., and Keim, W., Recent developments in isobutanol synthesis from synthesis gas, *Applied Catalysis A: General*, 186 (1, 2), 407–431 (1999).

202. Xiaoding, X., Doesburg, E.B.M., and Scholten, J.J.F., Synthesis of higher alcohols from syngas—Recently patented catalysts and tentative ideas on the mechanism, *Catalysis Today*, 2 (1), 125–170 (1987).

203. Dow develops catalytic method to produce higher mixed alcohols, *Chemical and Engineering News*, 62 (46), 29–30 (1984). doi: 10.1021/cen-v062n046.p029.

204. Wong, S.F., Patel, M.S., and Storm, D.A., Retrofitting methanol plants for higher alcohols, in *78th American Institute of Chemical Engineers, National Meeting*, New Orleans, LA (1986).

205. Hutchings, G.J., Copperthwaite, R.G., and Coville, N.J., Catalysis for hydrocarbon formation and transformations, *South African Journal of Science*, 84 (1), 12–16 (1988).

206. Park, T., Nam, I., and Kim, Y., Kinetic analysis of mixed alcohol synthesis from syngas over K/MoS$_2$ catalyst, *Industrial & Engineering Chemistry Research*, 36, 5246–5257 (1997).

207. Gunturu, A.K., Kugler, E.L., Cropley, J.B., and Dadyburjor, D.B., A kinetic model for the synthesis of high-molecular-weight alcohols over a sulfided Co–K–Mo/C catalyst, *Industrial & Engineering Chemistry Research*, 37, 2107–2115 (1998).

208. Smith, K.J., Herman, R.G., and Klier, K., Kinetic modeling of higher alcohol synthesis over alkali-promoted Cu/ZnO and MoS$_2$ catalysts, *Chemical Engineering Science*, 45 (8), 2639–2646 (1990).

209. Bashin, M.M., Bartley, M.J., Ellgen, P.C., and Wilson, T.P., Synthesis gas conversion over supported rhodium and rhodium iron catalysts, *Journal of Catalysis*, 54, 120–128 (1978).

210. Bao, J., Fu, L., Sun, Z., and Gao, C., A highly active K–Co–Mo/C catalyst for mixed alcohol synthesis from CO + H$_2$, *Chemical Communications*, 2003, 746–747 (2003).

211. Liu, Z., Li, X., Close, M.R., Kugler, E.L., Petersen, J.L., and Dadyburjor, D.B., Screening of alkali-promoted vapor-phase-synthesized molybdenum sulfide catalysts for the production of alcohols from synthesis gas, *Industrial Engineering Chemistry Research*, 36, 3085–3093 (1997).

212. Qi, H., Li, D., Yang, C., Ma, Y., Li, W., Sun, Y., and Zhong, B., Nickel and manganese co-modified K/MoS$_2$ catalyst: High performance for higher alcohols synthesis from CO hydrogenation, *Catalysis Communications*, 4, 339–342 (2003).

213. Woo, H.C., Park, K.Y., Kim, Y.G., Nam, I., Chung, J.S., and Lee, J.S., Mixed alcohol synthesis from carbon monoxide and dihydrogen over potassium-promoted molybdenum carbide catalysts, *Applied Catalysis*, 75, 267–280 (1991).

214. Zhang, Y., Sun, Y., and Zhong, B., Synthesis of higher alcohols from syngas over ultrafine Mo–Co–K catalysts, *Catalysis Letters*, 76 (3–4), 249–253 (2001).

215. Gerber, M.A., White, J.F., Gray, M.J., and Stevens, D.J., Mixed alcohol synthesis catalyst screening 2007 progress report, PNNL-17074, Pacific Northwest National Laboratory, Richland, WA (2007).

216. Gerber, M.A., White, J.F., Gray, M.J., and Stevens, D.J., Evaluation of promoters for rhodium-based catalysts for mixed alcohols synthesis, PNNL-17857, Pacific Northwest National Laboratory, Richland, WA (2008).

217. Gerber, M.A., White, J.F., and Stevens, D.J., Mixed alcohol catalyst screening, PNNL-16763, Pacific Northwest National Laboratory, Richland, WA (2007).

218. Mei, D., Rousseau, R., Kathmann, S.M., Glezakou, V.-A., Engelhard, M.H., Jiang, W., Wang, C., Gerber, M.A., White, J.F., and Stevens, D.J., Ethanol synthesis from syngas over Rh-based/SiO$_2$ catalysts: A combined experimental and theoretical modeling study, *Journal of Catalysis*, 271 (2), 325–342 (2010).

219. Hu, J., Wang, Y., Cao, C., Elliott, D.C., Stevens, D.J., and White, J.F., Conversion of biomass-derived syngas to alcohols and C2 oxygenates using supported Rh catalysts in a microchannel reactor, *Catalysis Today*, 120 (1), 90–95 (2007).

220. Albrecht, K.O. et al., Rh-based mixed alcohol synthesis catalysts: Characterization and computational report, PNNL-22697, Pacific Northwest National Laboratory, Prepared for the U.S. Department of Energy under contract DE-AC05-76RL01830 (August 2013).

221. Avila, Y., Kappenstein, C., Pronier, S., and Barrault, J., Alcohol synthesis from syngas over supported molybdenum catalysts, *Applied Catalysis A: General*, 132 (1), 97–109 (1995).

222. Bian, G.-Z., Fan, L., Fu, Y.-L., and Fujimoto, K., Mixed alcohol synthesis from syngas on sulfided K–Mo-based catalysts: Influence of support acidity, *Industrial & Engineering Chemistry Research*, 37 (5), 1736–1743 (1998).

223. Burcham, M.M., Herman, R.G., and Klier, K., Higher alcohol synthesis over double bed Cs–Cu/ZnO/Cr$_2$O$_3$ catalysts: Optimizing the yields of 2-methyl-1-propanol (isobutanol), *Industrial & Engineering Chemistry Research*, 37 (12), 4657–4668 (1998).

224. Campos-Martin, J.M., Guerreroruiz, A., and Fierro, J.L.G., Structural and surface properties of CuO–ZnO–Cr$_2$O$_3$ catalysts and their relationship with selectivity to higher alcohol synthesis, *Journal of Catalysis*, 156 (2), 208–218 (1995).

225. Courty, P., Durand, D., Freund, E., and Sugier, A., C1–C6 alcohols from synthesis gas on copper–cobalt catalysts, *Journal of Molecular Catalysis*, 17 (2–3), 241–254 (1982).

226. Dalmon, J.A., Chaumette, P., and Mirodatos, C., Higher alcohols synthesis on cobalt based model catalysts, *Catalysis Today*, 15 (1), 101–127 (1992).

227. Elliott, D.J., and Pennella, F., Mechanism of ethanol formation from synthesis gas over copper oxide/zinc oxide/alumina, *Journal of Catalysis*, 114 (1), 90–99 (1988).

228. Epling, W.S., Hoflund, G.B., and Minahan, D.M., Higher alcohol synthesis reaction study. VI: Effect of Cr replacement by Mn on the performance of Cs- and Cs, Pd-promoted Zn/Cr spinel catalysts, *Applied Catalysis A: General*, 183 (2), 335–343 (1999).

229. Hilmen, A.M., Xu, M., Gines, M.J.L., and Iglesia, E., Synthesis of higher alcohols on copper catalysts supported on alkali-promoted basic oxides, *Applied Catalysis A: General*, 169 (2), 355–372 (1998).

230. Hoflund, G.B., Epling, W.S., and Minahan, D.M., Higher alcohol synthesis reaction study using K-promoted ZnO catalysts. III, *Catalysis Letters*, 45 (1, 2), 135–138 (1997).

231. Hoflund, G.B., Epling, W.S., and Minahan, D.M., An efficient catalyst for the production of isobutanol and methanol from syngas. XI. K- and Pd-promoted Zn/Cr/Mn spinel (excess ZnO), *Catalysis Letters*, 62 (2–4), 169–173 (1999).

232. Iranmahboob, J. and Hill, D.O., Alcohol synthesis from syngas over K_2CO_3/CoS/MoS_2 on activated carbon, *Catalysis Letters*, 78 (1–4), 49–55 (2002).

233. Iranmahboob, J., Toghiani, H., Hill, D.O., and Nadim, F., The influence of clay on K_2CO_3/Co-MoS_2 catalyst in the production of higher alcohol fuel, *Fuel Processing Technology*, 79 (1), 71–75 (2002).

234. Li, X., Feng, L., Liu, Z., Zhong, B., Dadyburjor, D.B., and Kugler, E.L., Higher alcohols from synthesis gas using carbon-supported doped molybdenum-based catalysts, *Industrial & Engineering Chemistry Research*, 37 (10), 3853–3863 (1998).

235. Minahan, D.M., Epling, W.S., and Hoflund, G.B., An efficient catalyst for the production of isobutanol and methanol from syngas. VIII: Cs- and Pd-promoted Zn/Cr spinel (excess ZnO), *Catalysis Letters*, 50 (3, 4), 199–203 (1998).

236. Minahan, D.M., Epling, W.S., and Hoflund, G.B., Higher-alcohol synthesis reaction study. V. Effect of excess ZnO on catalyst performance, *Applied Catalysis A: General*, 166 (2), 375–385 (1998).

237. Nunan, J.G., Bogdan, C.E., Klier, K., Smith, K.J., Young, C.W., and Herman, R.G., Higher alcohol and oxygenate synthesis over cesium-doped copper/zinc oxide catalysts, *Journal of Catalysis*, 116 (1), 195–221 (1989).

238. Tronconi, E., Lietti, L., Forzatti, P., and Pasquon, I., Synthesis of alcohols from carbon oxides and hydrogen. 17. Higher alcohol synthesis over alkali metal-promoted high-temperature methanol catalysts, *Applied Catalysis*, 47 (2), 317–333 (1989).

239. Xiaoding, X., Doesburg, E.B.M., and Scholten, J.J.F., Synthesis of higher alcohols from syngas—Recently patented catalysts and tentative ideas on the mechanism, *Catalysis Today*, 2 (1), 125–170 (1987).

240. Beretta, A., Tronconi, E., Forzatti, P., Pasquon, I., Micheli, E., Tagliabue, L., and Antonelli, G.B., Development of a mechanistic kinetic model of the higher alcohol synthesis over a Cs-doped Zn/Cr/O catalyst. 1. Model derivation and data fitting, *Industrial & Engineering Chemistry Research*, 35 (7), 2144–2153 (1996).

241. Fierro, J.L.G., Catalysis in C1 chemistry: Future and prospect, *Catalysis Letters*, 22 (1–2), 67–91 (1993).

242. Erkey, C., Wang, J.H., Postula, W., Feng, Z.T., Philip, C.V., Akgerman, A., and Anthony, R.G., Isobutylene production from synthesis gas over zirconia in a slurry reactor, *Industrial & Engineering Chemistry Research*, 34 (4), 1021–1026 (1995).

243. Feng, Z.T., Postula, W.S., Akgerman, A., and Anthony, R.G., Characterization of zirconia-based catalysts prepared by precipitation, calcination, and modified sol–gel methods, *Industrial & Engineering Chemistry Research*, 34 (1), 78–82 (1995).

244. Feng, Z.T., Postula, W.S., Erkey, C., Philip, C.V., Akgerman, A., and Anthong, R.G., Selective formation of isobutane and isobutene from synthesis gas over zirconia catalysts prepared by a modified sol–gel method, *Journal of Catalysis*, 148 (1), 84–90 (1994).

245. Jackson, N.B. and Ekerdt, J.G., Isotope studies of the effect of acid sites on the reactions of C-3 intermediates during isosynthesis over zirconium dioxide and modified zirconium dioxide, *Journal of Catalysis*, 126 (1), 46–56 (1990).

246. Jackson, N.B. and Ekerdt, J.G., The surface characteristics required for isosynthesis over zirconium dioxide and modified zirconium dioxide, *Journal of Catalysis*, 126 (1), 31–45 (1990).

247. Li, Y.W., He, D.H., Cheng, Z.X., Su, C.L., Li, J.R., and Zhu, Q.M., Effect of calcium salts on isosynthesis over ZrO_2 catalysts, *Journal of Molecular Catalysis A—Chemical*, 175 (1–2), 267–275 (2001).

248. Li, Y.W., He, D.H., Yuan, Y.B., Cheng, Z.X., and Zhu, Q.M., Selective formation of isobutene from CO hydrogenation over zirconium dioxide based catalysts, *Energy & Fuels*, 15 (6), 1434–1440 (2001).

249. Li, Y.W., He, D.H., Yuan, Y.B., Cheng, Z.X., and Zhu, Q.M., Influence of acidic and basic properties of ZrO_2 based catalysts on isosynthesis, *Fuel*, 81 (11–12), 1611–1617 (2002).

250. Maruya, K., Takasawa, A., Haraoka, T., Domen, K., and Onishi, T., Role of methoxide species in isobutene formation from CO and H-2 over oxide catalysts—Methoxide species in isobutene formation, *Journal of Molecular Catalysis A—Chemical*, 112 (1), 143–151 (1996).

251. Sofianos, A., Production of branched-chain hydrocarbons via isosynthesis, *Catalysis Today*, 15 (1), 149–175 (1992).

252. Su, C.L., He, D.H., Li, J.R., Chen, Z.X., and Zhu, Q.M., Influences of preparation parameters on the structural and catalytic performance of zirconia in isosynthesis, *Journal of Molecular Catalysis A—Chemical*, 153 (1–2), 139–146 (2000).

253. Su, C.L., Li, J.R., He, D.H., Cheng, Z.X., and Zhu, Q.M., Synthesis of isobutene from synthesis gas over nanosize zirconia catalysts, *Applied Catalysis A—General*, 202 (1), 81–89 (2000).

254. Pichler, H. and Ziesecke, H.H., The Isosynthesis (translated by R. Brinkley and N. Golumbic), Bureau of Mines Bulletin, p. 488 (1950).

255. Pichler, H. and Ziesecke, H.H., *Breninst. Chem.*, 30, 13–22 (1949).

256. Pichler, H., Ziesecke, H.H., and Titzenthaler, E., *Breninst. Chem.*, 30, 333–347 (1949).

257. Pichler, H., Ziesecke, H.H., and Traeger, B., *Breninst. Chem.*, 31, 361–374 (1950).

258. Sastri, M., Gupta, V.C., and Viswanathan, R.B., *Journal of Catalysis (Part II)*, 32, 325 (1974).

259. Hydroformylation, Wikipedia, the free encyclopedia (2015). https://en.www.wikipedia.org/wiki/hydroformylation (last accessed October 24, 2016).

260. Pino, P. and Botteghi, C., Aldehydes from olefins: Cyclohexanecarboxaldehyde, *Organic Syntheses*, 57, 11 (1977).

261. Cornils, B., Herrmann, W.A., Rasch, M., and Roelen, O., Pioneer in industrial homogeneous catalysis, *Angewandte Chemie International Edition in English*, 33 (21): 2144–2163 (1994).

262. Hartwig, J.F., *Organo Transition Metal Chemistry—From Bonding to Catalysis*. University Science Books, Herndon, VA, pp. 753–757 (2009).

263. Cornils, B. and Herrmann, W.A. (eds.), *Aqueous-Phase Organometallic Catalysis*. VCH, Weinheim, Germany (1998).

264. Cornils, B., Herrmann, W.A., Wong, C.-H., and Zanthoff, H.-W., *Catalysis from A to Z: A Concise Encyclopedia*, 2408 Seiten. Wiley-VCH Verlag GmbH & Co. KGaA, Hoboken, NJ (2012).

265. Chan, A.S.C. and Shieh, H.-S., A mechanistic study of the homogeneous catalytic hydroformylation of formaldehyde: Synthesis and characterization of model intermediates, *Inorganica Chimica Acta*, 218 (1–2), 89–95 (1994).

266. Murata, K. and Matsuda, A., Application of homogeneous water–gas shift reaction III further study of the hydrocarbonylation—A highly selective formation of diethyl ketone from ethene, CO and H_2O, *Bulletin of the Chemical Society of Japan*, 54 (7), 2089–2092 (1981).

267. Liu, J., Heaton, B.T., Iggo, J.A., and Whyman, R., The complete delineation of the initiation, propagation, and termination steps of the carbomethoxy cycle for the carboalkoxylation of ethene by Pd–diphosphane catalysts, *Angewandte Chemie International Edition*, 43, 90–94 (2004).

268. Falbe, J. and Adams, Ch.R., *Carbon Monoxide in Organic Synthesis*. Springer Verlag, Berlin, Germany (1970).

269. Breit, B. and Seiche, W., Recent advances on chemo-, regio-, and stereo selective hydroformylation, *Synthesis*, 1, 1–36 (January 2001).

270. Frohning, C.D., Kohlpaintner, C.W., and Bohnen, K., Carbon monoxide and synthesis gas chemistry, in Cornils, B. and Herrmann, W.A. (eds.), *Applied Homogeneous Catalysis with Organometallic Compounds*, Vol. 1. VCH, Weinheim, Germany, Chapter 2.1, pp. 29–194 (1996).

271. Beller, M., Cornils, B., Frohning, C.D., and Kohlpaintner, C.W., Progress in hydroformylation and carbonylation, Review, *Journal of Molecular Catalysis A*, 104, 17–85 (1995).

272. Reinius, H., Activity and selectivity in hydroformylation: Role of ligand, substrate and process conditions, Doctoral thesis, Industrial Chemistry Publication Series, No. 9, Espoo, Finland, 52pp. (2001).

273. Falbe, J. (ed.), *New Syntheses with Carbon Monoxide*. Springer-Verlag, Heidelberg, Germany, 465pp. (1980).

274. Henrici-Olivé, G. and Olivé, S., *The Chemistry of the Catalysed Hydrogenation of Carbon Monoxide*. Springer-Verlag, Berlin, Germany, 231pp. (1984).

275. Sachtler, W.M.H. and Ichikawa, M., Catalytic sites requirements for elementary steps in syngas conversion to oxygenates over promoted Rh, *Journal of Physical Chemistry*, 90, 4752–4758 (1986).

276. Hedrick, S.A., Chuang, S.S.C., and Brundage, M.A., Deuterium pulse transient analysis for determination of heterogeneous ethylene hydroformylation mechanistic parameters, *Journal of Catalysis*, 185, 73–90 (1999).

277. Näsman, J.H., Sundell, M.J., and Ekman, K.B., Process for the preparation of a graft copolymer bound catalyst, U.S. Patent 5,326,825 (July 5, 1994).

278. Dietz, W.A., Response factors for gas chromatographic analyses, *Journal of Gas Chromatography*, 5, 68–71 (1967).

279. Van Leeuwen, P.W.N.M., Decomposition pathways of homogeneous catalysts, *Applied Catalysis A: General*, 212, 61–81 (2001).

280. Hakuli, A. and Kytökivi, A., Binding of chromium acetylacetonate on a silica support, *Physical Chemistry Chemical Physics*, 1, 1607–1613 (1999).

281. Cowie, J.M.G., *Polymers: Chemistry and Physics of Modern Materials*, 2nd edn. Blackie Academic & Professional, London, U.K., 436pp. (1991).

282. Fessenden, R.J. and Fessenden, J.S., *Organic Chemistry*, 4th edn. Brooks/Cole, Pacific Grove, CA, pp. 542–543 (1990).
283. Claridge, J.B., Douthwaite, R.E., Green, M.L.H., Lago, R.M., Tsang, S.C., and York, A.P.E., Studies of new catalyst for the hydroformylation of alkenes using C60 as a ligand, *Journal of Molecular Catalysis*, 89, 113–120 (1994).
284. Bahrmann, H. and Bach, H., Oxo synthesis, in *Ullman's Encyclopedia of Industrial Chemistry*, Wiley-VCH Verlag GmbH & Co., Hoboken, NJ (2000).
285. Billig, E. and Bryant, D.R., Oxo process, in *Kirk-Othmer Encyclopedia of Chemical Technology*. John Wiley & Sons, Hoboken, NJ (2000).
286. Bitzzari, S.N., Gubler, R., and Kishi, A., Oxo chemicals, in *Chemical Economics Handbook*. SRI International, pp. 1–121. Report number 682.7000, Menlo Park, CA (2002).
287. Pruett, R.L., Hydroformylation, *Advances in Organometallic Chemistry*, 17, 1–60 (1979).
288. Whyman, R., Industrial applications of homogeneous catalysts, in Jennings, J.R. (ed.), *Selected Developments in Catalysis*. Blackwell Scientific Publications, Oxford, U.K., p. 128 (1985).
289. Lenarda, M., Storaro, L., and Ganzerla, R., Hydroformylation of simple olefins catalyzed by metals and clusters supported on unfunctionalized inorganic carriers, Review, *Journal of Molecular Catalysis A: Chemical*, 111, 203–237 (1996).
290. Reinikainen, M., Cobalt and ruthenium–cobalt catalysts in CO hydrogenation and hydroformylation, Doctoral thesis, Technical Research Centre of Finland (VTT), Espoo, Finland, 64pp. (1998).
291. Ungváry, F., Application of transition metals in hydroformylation annual survey covering the year 2004, Review, *Coordination Chemistry Reviews*, 249, 2946–2961 (2005).
292. Hakuli, A., Preparation and characterization of supported CrO_x catalysts for butane dehydrogenation, Doctoral thesis, Helsinki University of Technology, Espoo, Finland, 48p. (1999).
293. Puurunen, R., Preparation by atomic layer deposition and characterization of catalyst supports surfaced with aluminium nitride, Doctoral thesis, Industrial Chemistry Publication Series, No. 13, Espoo, Finland, 78pp. (2002).
294. Backman, L.B., Rautiainen, A., Krause, A.O.I., and Lindblad, M., A novel Co/SiO_2 catalyst for hydrogenation, *Catalysis Today*, 43, 11–19 (1998).
295. Takeuchi, K., Hanaoka, T., Matsuzaki, T., Reinikainen, M., and Sugi, Y., Selective vapor phase hydroformylation of ethylene over cluster-derived cobalt catalysts, *Catalysis Letters*, 8, 253–261 (1991).
296. Matsuzaki, T., Hanaoka, T., Takeuchi, K., Arakawa, H., Sugi, Y., Wei, K., Dong, T., and Reinikainen, M., Oxygenates from syngas over highly dispersed cobalt catalysts, *Catalysis Today*, 36, 311–324 (1997).
297. Reuel, R. and Bartholomew, C., The stoichiometries of H_2 and CO adsorptions on cobalt: Effects of support and preparation, *Journal of Catalysis*, 85, 63–77 (1984).
298. Reuel, R. and Bartholomew, C., Effects of support and dispersion on the CO hydrogenation activity/selectivity properties of cobalt, *Journal of Catalysis*, 85, 78–88 (1984).
299. Niemelä, M.K., Backman, L., Krause, A.O.I., and Vaara, T., The activity of the Co/SiO_2 catalyst in relation to pretreatment, *Applied Catalysis A: General*, 156, 319–334 (1997).
300. Alvila, L., Pakkanen, T.A., Pakkanen, T.T., and Krause, O., Hydroformylation of olefins catalysed by rhodium and cobalt clusters supported on organic (Dowex) resins, *Journal of Molecular Catalysis*, 71, 281–290 (1992).
301. Cao, S.-K., Huang, M.-Y., and Jiang, Y.-Y., Hydroformylation of heptene-1 catalyzed by some inorganic polymer-metal complexes, *Journal of Macromolecular Science: Part A—Chemistry*, 26, 381–389 (1989).
302. Omata, K., Fujimoto, K., Shikada, T., and Tominaga, H., Vapor-phase carbonylation of organic compounds over supported transition-metal catalyst. 6. On the character of nickel/active carbon as methanol carbonylation catalyst, *Industrial & Engineering Chemistry Research*, 27, 2211–2213 (1988).
303. Tomita, A. and Tamai, Y., Hydrogenation of carbons catalysed by transition metals, *Journal of Catalysis*, 27, 293–300 (1972).
304. Matsuzaki, T., Hanaoka, T., Takeuchi, K., Sugi, Y., and Reinikainen, M., Effects of modification of highly dispersed cobalt catalysts with alkali cations on the hydrogenation of carbon monoxide, *Catalysis Letters*, 10, 193–199 (1991).
305. Halttunen, M.E., Niemelä, M.K., Krause, A.O.I., and Vuori, A.I., Some aspects on the losses of metal from the support in the hydrocarbonylation of methanol, *Journal of Molecular Catalysis A: Chemical*, 144, 307–314 (1999).
306. Kiviaho, J., Niemelä, M.K., Reinikainen, M., Vaara, T., and Pakkanen, T.A., The effect of decomposition atmosphere on the activity and selectivity of the carbonyl cluster derived Co/SiO_2 and Rh/SiO_2 catalysts, *Journal of Molecular Catalysis A: Chemical*, 121, 1–8 (1997).

307. Takahashi, N., Takeyama, T., Fujimoto, T., Fukuoka, A., and Ichikawa, J., Formation of Rh carbonyl cluster species and its conversion into metal particles during exposure of Rh/active carbon to carbon monoxide revealed by EXAFS and TPD techniques, *Chemistry Letters*, 1441–1444 (1992).

308. Guyot, A., Hodge, P., Sherrington, D.C., and Widdecke, H., Recent studies aimed at the development of polymer-supported reactants with improved accessibility and capacity, *Reactive Polymers*, 16, 233–259 (1991/1992).

309. Karinen, R.S., Krause, A.O.I., Ekman, K., Sundell, M., and Peltonen, R., Etherification over a novel acid catalyst, *Studies in Surface Science and Catalysis*, 130, 3411–3416 (2000).

310. Babich, I.V., Plyuto, Y.V., Van der Voort, P., and Vansant, E.F., Thermal transformations of chromium acetylacetonate on silica surface, *Journal of Colloid and Interface Science*, 189, 144–150 (1997).

311. Mikami, M., Nakagawa, I., and Shimanouchi, T., Far infra-red spectra and metal-ligand force constants of acetylacetonates of transition metals, *Spectrochimica Acta*, 23A, 1037–1053 (1967).

312. Ichikawa, M., Lang, A.J., Shriver, D.F., and Sachtler, W.M.H., Selective hydroformylation of ethylene on Rh-Zn/SiO$_2$. An apparent example of site isolation of Rh and Lewis acid promoted CO insertion, *Journal of the American Chemical Society*, 107 (1985), 7216–7218 (1985).

313. Evans, D., Osborn, J.A., and Wilkinson, G., Hydroformylation of alkenes by use of rhodium complex catalyst (PDF), *Journal of the Chemical Society*, 33 (21), 3133–3142 (1968).

314. Kuil, M., Soltner, T., van Leeuwen, P.W.N.M., and Reek, J.N.H., High-precision catalysts: Regioselective hydroformylation of internal alkenes by encapsulated rhodium complexes, *Journal of the American Chemical Society*, 128 (35), 11344–11345 (2006).

315. Watkins, A.L., Hashiguchi, B.G., and Landis, C.R., Highly enantioselective hydroformylation of aryl alkenes with diazaphospholane ligands, *Organic Letters*, 10 (20), 4553–4556 (2008).

316. Spencer, A., Hydroformylation of formaldehyde catalysed by rhodium complexes, *Journal of Organometallic Chemistry*, 194 (1–2), 113–123 (1980).

317. Van Leeuwen, P.W.N.M. and Clawer, C. (eds.), *Rhodium Catalyzed Hydroformylation*. Catalysis by Metal Complexes, Vol. 22. Kluwer, Dordrecht, the Netherlands, 284pp. (2000).

318. Tudor, R. and Ashley, M., Enhancement of industrial hydroformylation processes by the adoption of rhodium-based catalyst: Part II: Key improvements to rhodium process, and use in non-propylene applications, *Platinum Metals Review*, 51, 164–171 (2007).

319. Zhao, J., Zhang, Y., Han, J., and Jiao, Y., Preparation and performance of anchored heterogenised rhodium complex catalyst for hydroformylation, *Journal of Molecular Catalysis A: Chemical*, 241, 238–243 (2005).

320. Mukhopadhyay, K., Mandale, A.B., and Chaudhari, R.V., Encapsulated HRh(CO)(PPh$_3$)$_3$ in microporous and mesoporous supports: Novel heterogeneous catalysts for hydroformylation, *Chemistry of Materials*, 15, 1766–1777 (2003).

321. Feldman, J. and Orchin, M., Membrane-supported rhodium hydroformylation catalysts, *Journal of Molecular Catalysis*, 63, 213–221 (1990).

322. Ro, K.S. and Woo, S.I., Hydroformylation of propylene catalyzed over polymer-immobilized RhCl(CO)(PPh$_3$)$_2$: Effect of crosslink ratio and FTIR study, *Journal of Molecular Catalysis*, 61, 27–39 (1990).

323. Terekhova, G.V., Kolesnichenko, N.V., Alieva, E.D., Truhmanova, N.I., Teleshev, A.T., Markova, N.A., Alekseeva, E.I., Slivinsky, E.V., Loktev, S.M., and Pesin, O.Y., Rhodium carbonyl catalysts, immobilized on polymeric supports in the hydroformylation of olefins, in Maggi, R. et al. (eds.), *Preparation of Catalysts VII*, Vol. 118, Studies in Surface Science and Catalysis, Elsevier, Amsterdam, the Netherlands, pp. 255–263 (1998).

324. Pittman, C.U., Jr. and Wilemon, G.M., 1-Pentene hydroformylation catalyzed by polymer-bound ruthenium complexes, *Journal of Organic Chemistry*, 46, 1901–1905 (1981).

325. Terreros, P., Pastor, E., and Fierro, J.L.G., Hept-1-ene hydroformylation on phosphinated polystyrene-anchored rhodium complexes, *Journal of Molecular Catalysis*, 53, 359–369 (1989).

326. Junfan, W., Juntan, S., Hong, L., and Binglin, H., The stability of a polymer-supported rhodium complex in the batch hydroformylation of 1-hexene, *Reactive Polymers*, 12, 177–186 (1990).

327. Pittman, C.U., Jr. and Honnick, W.D., Rhodium-catalyzed hydroformylation of allyl alcohol. A potential route to 1,4-butanediol, *Journal of Organic Chemistry*, 45, 2132–2139 (1980).

328. Jongsma, T., Kimkes, P., Challa, G., and van Leeuwen, P.W.N.M., A new type of highly active polymer-bound rhodium hydroformylation catalyst, *Polymer*, 33, 161–165 (1992).

329. Arai, H., Hydroformylation, hydrogenation and isomerization of olefins over polymer-immobilized rhodium complexes, *Journal of Catalysis*, 51, 135–142 (1978).

330. Bonaplata, E., Ding, H., Hanson, B.E., and McGrath, J.E., Hydroformylation of 1-octene with a poly(arylene ether triaryl phosphine) rhodium complex, *Polymer*, 36, 3035–3039 (1995).
331. Hartley, F.R., Murray, S.G., and Nicholson, P.N., γ-Radiation-produced supported metal complex catalysts. IV. Rhodium(I) hydroformylation catalysts supported on phosphinated polypropylene, *Journal of Molecular Catalysis*, 16, 363–383 (1982).
332. Li, B., Li, X., Asami, K., and Fujimoto, K., Hydroformylation of 1-hexene over rhodium supported on active carbon catalyst, *Chemistry Letters*, 32, 378–379 (2003).
333. Yan, L., Ding, Y.J., Zhu, H.J., Xiong, J.M., Wang, T., Pan, Z.D., and Lin, L.W., Ligand modified real heterogeneous catalysts for fixed-bed hydroformylation of propylene, *Journal of Molecular Catalysis A: Chemical*, 234, 1–7 (2005).
334. Zhu, H., Ding, Y., Yan, L., Lu, Y., Li, C., Bao, X., and Lin, L., Recyclable heterogeneous Rh/SiO$_2$ catalyst enhanced by organic PPh$_3$ ligand, *Chemistry Letters*, 33, 630–631 (2004).
335. Zhang, Y., Zhang, H.-B., Lin, G.-D., Chen, P., Yuan, Y.-Z., and Tsai, K.R., Preparation, characterization and catalytic hydroformylation properties of carbon nanotubes-supported Rh-phosphine catalyst, *Applied Catalysis A: General*, 187, 213–224 (1999).
336. Andersson, C., Nikitidis, A., Hjortkjær, J., and Heinrich, B., Continuous liquid-phase hydroformylation of 1-hexene with a poly-TRIM bound rhodium-phosphine complex, *Applied Catalysis A: General*, 96, 345–354 (1993).
337. Davis, M.E., Butler, P.M., Rossin, J.A., and Hanson, B.E., Hydroformylation of 1-hexene by soluble and zeolite-supported rhodium species, *Journal of Molecular Catalysis*, 31, 385–395 (1985).
338. Balakos, M.W. and Chuang, S.S.C., Transient response of propionaldehyde formation during CO/H$_2$/C$_2$H$_4$ reaction on Rh/SiO$_2$, *Journal of Catalysis*, 151, 253–265 (1995).
339. Huang, L., Xu, Y., Guo, W., Liu, A., Li, D., and Guo, X., Study on catalysis by carbonyl cluster-derived SiO$_2$-supported rhodium for ethylene hydroformylation, *Catalysis Letters*, 32, 61–81 (1995).
340. Chuang, S.S.C. and Pien, S.I., Infrared study of the CO insertion reaction on reduced, oxidised, and sulfided Rh/SiO$_2$ catalysts, *Journal of Catalysis*, 135, 618–634 (1992).
341. Chuang, S.S.C., Stevens, R.W., Jr., and Khatri, R., Mechanism of C$_{2+}$ oxygenate synthesis on Rh catalysts, *Topics in Catalysis*, 32, 225–232 (2005).
342. Halttunen, M.E., Niemelä, M.K., Krause, A.O.I., Vaara, T., and Vuori, A.I., Rh/C catalysts for methanol carbonylation. I. Catalyst characterisation, *Applied Catalysis A: General*, 205, 37–49 (2001).
343. Halttunen, M.E., Niemelä, M.K., Krause, A.O.I., and Vuori, A.I., Rh/C catalysts for methanol hydrocarbonylation II. Activity in the presence of MeI, *Applied Catalysis A: General*, 182, 115–123 (1999).
344. Zhang, Y., Nagasaka, K., Qiu, X., and Tsubaki, N., Hydroformylation of 1-hexene for oxygenate fuels on supported cobalt catalysts, *Catalysis Today*, 104, 48–54 (2005).
345. Niemelä, M., Krause, O., Vaara, T., Kiviaho, J., and Reinikainen, M., The effect of precursor on the characteristics of Co/SiO$_2$ catalysts, *Applied Catalysis A: General*, 147, 325–345 (1996).
346. Etspüler, A. and Suhr, H., Deposition of thin rhodium films by plasma-enhanced chemical vapour deposition, *Applied Physics A*, 48, 373–375 (1989).
347. Flint, E.B., Messelhäuser, J., and Suhr, H., Laser-induced CVD or rhodium, *Applied Physics A*, 53, 430–436 (1991).
348. Jesse, A.C., Ernsting, J.M., Stufkens, D.J., and Vrieze, K., Vapour pressure measurements on (acac)M(substituted olefin)$_2$ and (acac)M(CO)$_2$ (M = Rh(I), Ir(I)), *Thermochimica Acta*, 25, 69–75 (1978).
349. Daniell, J., Köpke, M., and Simpson, S.D., Commercial biomass syngas fermentation, *Energies*, 5, 5372–5417 (2012).
350. Ragsdale, S.W., Life with carbon monoxide, *Critical Reviews in Biochemistry and Molecular Biology*, 39, 165–195 (2004).
351. Drake, H.L., Küsel, K., Matthies, C., Wood, H.G., and Ljungdahl, L.G., Acetogenic prokaryotes, in Dworkin, M., Falkow, S., Rosenberg, E., Schleifer, K.-H., and Stackebrandt, E. (eds.), *The Prokaryotes*. Springer, New York, pp. 354–420 (2006).
352. Olson, D.G., McBride, J.E., Shaw, A.J., and Lynd, L.R., Recent progress in consolidated bioprocessing, *Current Opinion in Biotechnology*, 23, 396–405 (2012).
353. Köpke, M., Held, C., Hujer, S., Liesegang, H., Wiezer, A., Wollherr, A., Ehrenreich, A., Liebl, W., Gottschalk, G., and Dürre, P., *Clostridium ljungdahlii* represents a microbial production platform based on syngas, *Proceedings of the National Academy of Sciences of the United States of America*, 107, 13087–13092 (2010).
354. Tracy, B.P., Jones, S.W., Fast, A.G., Indurthi, D.C., and Papoutsakis, E.T., Clostridia: The importance of their exceptional substrate and metabolite diversity for biofuel and biorefinery applications, *Current Opinion in Biotechnology*, 23, 364–381 (2012).

355. Sims, R.E.H., Mabee, W., Saddler, J.N., and Taylor, M., An overview of second generation biofuel technologies, *Bioresource Technology*, 101, 1570–1580 (2010).

356. Mabee, W.E., Gregg, D.J., Arato, C., Berlin, A., Bura, R., Gilkes, N., Mirochnik, O., Pan, X., Pye, E.K., and Saddler, J.N., Updates on softwood-to-ethanol process development, *Applied Biochemistry and Biotechnology*, 129–132, 55–70 (2006).

357. Poehlein, A. et al., An ancient pathway combining carbon dioxide fixation with the generation and utilization of a sodium ion gradient for ATP synthesis, *PLoS One*, 7, e33439 (2012).

358. Henstra, A.M. et al., Microbiology of synthesis gas fermentation for biofuel production, *Current Opinion in Biotechnology*, 18, 200–206 (2007).

359. Worden, R.M. et al., Engineering issues in syngas fermentation, Chapter 18, in *Fuels and Chemicals from Biomass*, pp. 320–335 (2006).

360. Datta, R., Maher, M.A., Jones, C., and Brinker, R.W., Ethanol—The primary renewable liquid fuel, *Journal of Chemical Technology and Biotechnology*, 86, 473–480 (2011).

361. Griffin, D.W. and Schultz, M.A., Fuel and chemical products from biomass syngas: A comparison of gas fermentation to thermochemical conversion routes, *Environmental Progress & Sustainable Energy*, 31, 219–224 (2012).

362. Abrini, J., Navean, H., and Nyris, E.-J., *Clostridium autoethanogenum*, sp. nov., an anaerobic bacterium that produces ethanol from carbon monoxide, *Archives of Microbiology*, 161, 345–351 (1994).

363. Cotter, J.L., Chinn, M.S., and Grunden, A.M., Ethanol and acetate production by *Clostridium ljungdahlii* and *Clostridium autoethanogenum* using resting cells, *Bioprocess and Biosystems Engineering*, 32, 369–380 (2009).

364. Gaddy, J.L. and Clausen, W.C., *Clostridium ljungdahlii*, an anaerobic ethanol and acetate producing microorganism, U.S. Patent 5,173,429 (December 22, 1992).

365. Girbal, L., Vasconcelos, I., Saint-Amans, S., and Soucaille, P., How neutral red modified carbon and electron flow in *Clostridium acetobutylicum* grown in chemostat culture at neutral pH, *FEMS Microbiology Reviews*, 16, 151–162 (1995).

366. Ragsdale, S.W., Nickel-based enzyme systems, *Journal of Biological Chemistry*, 284, 18571–18575 (2009).

367. Gaddy, J., Biological production of products from waste gases, U.S. Patent 6,340,581 B1 (January 22, 2002).

368. Hickey, R., Basu, R., Datta, R., and Tsai, S., Method of conversion of syngas using microorganism on hydrophilic membrane, U.S. Patent 7,923,227 (April 12, 2011).

369. Tracy, B.P., Jones, S.W., and Papoutsakis, E.T., Inactivation of σE and σG in *Clostridium acetobutylicum* illuminates their roles in clostridial-cell-form biogenesis, granulose synthesis, solventogenesis, and spore morphogenesis, *Journal of Bacteriology*, 193, 1414–1426 (2011).

370. Soucaille, P., Figge, R., and Croux, C., Process for chromosomal integration and DNA sequence replacement in clostridia, Patent WO/2008/040387 (April 10, 2008).

371. Heap, J.T., Pennington, O.J., Cartman, S.T., Carter, G.P., and Minton, N.P., The ClosTron: A universal gene knock-out system for the genus *Clostridium*, *Journal of Microbiological Methods*, 70, 452–464 (2007).

372. Heap, J.T., Kuehne, S., Ehsaan, M., Cartman, S.T., Cooksley, C.M., Scott, J.C., and Minton, N.P., The ClosTron: Mutagenesis in *Clostridium* refined and streamlined, *Journal of Microbiological Methods*, 80, 49–55 (2010).

373. Jones, D.T., Bacteriophages of *Clostridium*, in Dürre, P. (ed.), *Handbook on Clostridia*. CRC Press, Boca Raton, FL, pp. 699–719 (2005).

374. Xiu, Z.L. and Zeng, A.P., Present state and perspective of downstream processing of biologically produced 1,3-propanediol and 2,3-butanediol, *Applied Microbiology and Biotechnology*, 78, 917–926 (2008).

375. Wald, M., Yet another route to cellulosic ethanol, energy the environment and bottom line, A report in the New York Times, New York. Available online: http://nyti.ms/POnkOJ, Accessed on July 16, 2012.

376. Hacker, J., 2012 Statement—*Bioenergy—Chances and Limits*. German National Academic of Sciences Leopoldina, Halle, Germany (2012). www.leopoldina.org.

377. Mitchell, D., Kojima, M., and Ward, W., Considering trade policies for liquid biofuels, Special report, Energy Sector Management Assistance Program, Washington, DC (2007).

378. Carney, B., Can the world still feed itself? Weekend review on the Wall Street Journal (September 3, 2011), Available online: http://on.wsj.com/omahnc, Accessed on August 20, 2012. www.wsj.com/.../SB10001424053111904787404576529912

379. Thompson, A., Nestle chief calls for end to using food in biofuel production, Business News, Reuters, U.K. (August 19, 2012). Available online: http://uk.reuters.com/article/2012/08/19/uk-nestle-idUKBRE87I03W20120819, Accessed on August 20, 2012.

380. Munasinghe, P.C. and Khanal, S.K., Biomass-derived syngas fermentation into biofuels: Opportunities and challenges, *Bioresource Technology*, 101, 5013–5022 (2010).

381. Abubackar, H.N., Veiga, M.C., and Kennes, C., Biological conversion of carbon monoxide: Rich syngas or waste gases to bioethanol, *Biofuels, Bioproducts and Biorefining*, 5, 93–114 (2011).

382. Schiel-Bengelsdorf, B. and Dürre, P., Pathway engineering and synthetic biology using acetogens, *FEBS Letters*, 586, 2191–2198 (2012).

383. Köpke, M., Mihalcea, C., Liew, F., Tizard, J.H., Ali, M.S., Conolly, J.J., Al-Sinawi, B., and Simpson, S.D., 2,3-Butanediol production by acetogenic bacteria, an alternative route to chemical synthesis, using industrial waste gas, *Applied and Environmental Microbiology*, 77, 5467–5475 (2011).

384. Ljungdahl, L.G. and Wood, H., Total synthesis of acetate from CO_2 by heterotrophic bacteria, *Annual Review of Microbiology*, 23, 515–538 (1969).

385. Wood, H.G., Life with CO or CO_2 and H_2 as a source of carbon and energy, *FASEB Journal*, 5, 156–163 (1991).

386. Ljungdahl, L.G., The autotrophic pathway of acetate synthesis in acetogenic bacteria, *Annual Review of Microbiology*, 40, 415–450 (1986).

387. Ragsdale, S.W. and Pierce, E., Acetogenesis and the Wood–Ljungdahl pathway of CO_2 fixation, *Biochimica et Biophysica Acta*, 1784, 1873–1898 (2008).

388. Bruant, G., Lévesque, M.J., Peter, C., Guiot, S.R., and Masson, L., Genomic analysis of carbon monoxide utilization and butanol production by *Clostridium carboxidivorans* strain P7, *PLoS One*, 5, 1–12 (2010).

389. Ragsdale, S.W., Enzymology of the acetyl-CoA pathway of CO_2 fixation, *Critical Reviews in Biochemistry and Molecular Biology*, 26, 261–300 (1991).

390. Drake, H.L., Acetogenesis, acetogenic bacteria, and the acetyl-CoA Wood/Ljungdahl pathway: Past and current perspectives, in Drake, H.L. (ed.), *Acetogenesis*. Chapman & Hall, New York, pp. 3–60 (1994).

391. Phillips, J.R., Klasson, K.T., Claussen, E.C., and Gaddy, J.L., Biological production of ethanol from coal synthesis gas, *Applied Biochemistry and Biotechnology*, 39, 559–571 (1993).

392. Green, E.M., Fermentative production of butanol—The industrial perspective, *Current Opinion in Biotechnology*, 22, 337–343 (2011).

393. Paul, D., Austin, F.W., Arick, T., Bridges, S.M., Burgess, S.C., Dandass, Y.S., and Lawrence, M.L., Genome sequence of the solvent-producing bacterium *Clostridium carboxidivorans* strain P7T, *Journal of Bacteriology*, 192, 5554–5555 (2010).

394. Jones, D.T., Shirley, M., Wu, X., and Keis, S., Bacteriophage infections in the industrial acetone butanol (AB) fermentation process, *Journal of Molecular Microbiology and Biotechnology*, 2, 21–26 (2000).

395. Vane, L.M., Separation technologies for the recovery and dehydration of alcohols from fermentation broths, *Biofuels, Bioproducts and Biorefining*, 2, 553–588 (2008).

396. Drake, H.L., Gössner, A.S., and Daniel, S.L., Old acetogens, new light, *Annals of the New York Academy of Sciences*, 1125, 100–128 (2008).

397. Taherzadeh, M.J. and Karimi, K., Pretreatment of lignocellulosic wastes to improve ethanol and biogas production: A review, *International Journal of Molecular Sciences*, 9, 1621–1651 (2008).

398. Mosier, N., Wyman, C., Dale, B., Elander, R., Lee, Y.Y., Holtzapple, M., and Ladisch, M., Features of promising technologies for pretreatment of lignocellulosic biomass, *Bioresource Technology*, 96, 673–686 (2005).

399. Guo, Y., Xu, J., Zhang, Y., Xu, H., Yuan, Z., and Li, D., Medium optimization for ethanol production with *Clostridium autoethanogenum* with carbon monoxide as sole carbon source, *Bioresource Technology*, 101, 8784–8789 (2010).

400. Saxena, J. and Tanner, R.S., Effect of trace metals on ethanol production from synthesis gas by the ethanologenic acetogen, *Clostridium ragsdalei*, *Journal of Industrial Microbiology and Biotechnology*, 38, 513–521 (2011).

401. Saxena, J. and Tanner, R.S. Optimization of a corn steep medium for production of ethanol from synthesis gas fermentation by *Clostridium ragsdalei*, *World Journal of Microbiology and Biotechnology*, 28, 1553–1561 (2012).

402. Gaddy, J., Arora, D., Ko, C., Phillips, J., Basu, R., Wikstrom, C., and Clausen, E., Methods for increasing the production of ethanol from microbial fermentation, U.S. Patent 2012/0122173 A1 (May 17, 2012).

403. Simpson, S.D., Warner, I.L., Fung, J.M.Y., and Köpke, M., Optimised fermentation media, U.S. Patent 20110294177 (December 1, 2011).

404. Trevethick, S., Bromley, J., Simpson, S., and Khosla, V., Improved fermentation of gaseous substrates, Patent WO 2011/028137 A1 (March 10, 2011).

405. Hickey, R., Datta, R., Tsai, S.-P., and Basu, R., Membrane supported bioreactor for conversion of syngas components to liquid products, U.S. Patent 2011/0256597 A1 (October 20, 2011).

406. Kundiyana, D.K., Wilkins, M.R., Maddipati, P., and Huhnke, R.L., Effect of temperature, pH and buffer presence on ethanol production from synthesis gas by *Clostridium ragsdalei*, *Bioresource Technology*, 102, 5794–5799 (2011).

407. Fischer, C.R. et al., Selection and optimization of microbial hosts for biofuels production, *Metabolic Engineering*, 10, 295–304 (2008).

408. Heiskanen, H. et al., The effect of syngas composition on the growth and product formation of *Butyribacterium methylotrophicum*, *Enzyme and Microbial Technology*, 41, 362–367 (2007).

409. Tsai, S.P., Datta, R., Basu, R., Yoon, S.H., and Robey, R., Modular membrane supported bioreactor for conversion of syngas components to liquid products, U.S. Patent 2009/0029434 A1 (January 29, 2009).

410. Grethlein, A.J., Worden, R.M., Jain, M.K., and Datta, R., Continuous production of mixed alcohols and acids from carbon monoxide, *Applied Biochemistry and Biotechnology*, 24–25, 875–884 (1990).

411. Ungerman, A.J. and Heindel, T.J., Carbon monoxide mass transfer for syngas fermentation in a stirred tank reactor with dual impeller configurations, *Biotechnology Progress*, 23, 613–620 (2007).

412. Munasinghe, P.C. and Khanal, S.K., Syngas fermentation to biofuel: Evaluation of carbon monoxide mass transfer coefficient (kLa) in different reactor configurations, *Biotechnology Progress*, 26, 1616–1621 (2010).

413. Tsai, S.P., Datta, R., Basu, R., and Yoon, S.H., Syngas conversion system using asymmetric membrane and anaerobic microorganism, U.S. Patent 2009/0215163 A1 (August 29, 2009).

414. Hydrogenation, Wikipedia, the free encyclopedia (2015). https://en.wikipedia.org/wiki/hydrogenation (last accessed Ocotber 23, 2016).

415. Vega, J.L., Klasson, K.T., Kimmel, D.E., Clausen, E.C., and Gaddy, J.L., Sulfur gas tolerance and toxicity of CO-utilizing and methanogenic bacteria, *Applied Biochemistry and Biotechnology*, 24/25, 329–340 (1990).

416. Peralta-Yahya, P.P., Zhang, F., del Cardayre, S.B., and Keasling, J.D., Microbial engineering for the production of advanced biofuels, *Nature*, 488, 320–328 (2012).

417. Qureshi, N., Annous, B.A., Ezeji, T.C., Karcher, P., and Maddox, I.S., Biofilm reactors for industrial bioconversion processes: Employing potential of enhanced reaction rates, *Microbial Cell Factories*, 4, 24 (2005).

418. Lee, P.-H., Syngas fermentation to ethanol using innovative hollow fiber membrane, Ph.D. thesis, Civil and Environmental Engineering, Iowa State University, Ames, IA (2010).

419. Shen, Y., Attached-growth bioreactors for syngas fermentation to biofuel, Ph.D. thesis, Department of Food Science and Human Nutrition, Iowa State University, Ames, IA (2013).

420. Argyros, D.A. et al., High ethanol titers from cellulose by using metabolically engineered thermophilic, anaerobic microbes, *Applied and Environmental Microbiology*, 77, 8288–8294 (2011).

421. Tripathi, S. et al., Development of pyrF-based genetic system for targeted gene deletion in *Clostridium thermocellum* and creation of a pta mutant, *Applied and Environmental Microbiology*, 76, 6591–6599 (2010).

422. Tracy, B. and Papoutsakis, E., Methods and compositions for genetically engineering clostridia species, U.S. Patent 2010/0075424 (March 25, 2010).

423. Cartman, S. and Minton, N., Method of double crossover homologous recombination in clostridia, Patent WO/2010/084349 (July 29, 2010).

424. Desai, R.P. and Papoutsakis, E.T., Antisense RNA strategies for metabolic engineering of *Clostridium acetobutylicum*, *Applied and Environmental Microbiology*, 65, 936–945 (1999).

425. Lütke-Eversloh, T. and Bahl, H., Metabolic engineering of *Clostridium acetobutylicum*: Recent advances to improve butanol production, *Current Opinion in Biotechnology*, 22, 634–647 (2011).

426. Koepke, M. and Liew, F., Production of butanol from carbon monoxide by a recombinant microorganism, Patent WO/2012/053905 (April 26, 2012).

427. Roh, H., Ko, H.J., Kim, D., Choi, D.G., Park, S., Kim, S., Chang, I.S., and Choi, I.G., Complete genome sequence of a carbon monoxide-utilizing acetogen, *Eubacterium limosum* KIST612, *Journal of Bacteriology*, 193, 307–308 (2011).

428. Tummala, S.B., Welker, N.E., and Eleftherios, T., Development and characterization of a gene expression reporter system for *Clostridium acetobutylicum* ATCC 824, *Applied and Environmental Microbiology*, 65, 3793–3799 (1999); *Energies*, 5, 5413 (2012).
429. Girbal, L., Mortier-Barriere, I., Raynaud, F., Rouanet, C., Croux, C., and Soucaille, P., Development of a sensitive gene expression reporter system and an inducible promoter-repressor system for *Clostridium acetobutylicum, Applied and Environmental Microbiology*, 69, 4985–4988 (2003).
430. Feustel, L., Nakotte, S., and Durre, P., Characterization and development of two reporter gene systems for *Clostridium acetobutylicum, Applied and Environmental Microbiology*, 70, 798–803 (2004).
431. Cui, G., Hong, W., Zhang, J., Li, W., Feng, Y., Liu, Y., and Cui, Q., Targeted gene engineering in *Clostridium cellulolyticum* H10 without methylation, *Journal of Microbiological Methods*, 89, 201–208 (2012).
432. Dong, H., Tao, W., Zhang, Y., and Li, Y., Development of an anhydrotetracycline-inducible gene expression system for solvent-producing *Clostridium acetobutylicum*: A useful tool for strain engineering, *Metabolic Engineering*, 14, 59–67 (2012).
433. Hemme, C.L. et al., Genome announcement—Sequencing of multiple clostridia genomes related to biomass conversion and biofuels production, *Journal of Bacteriology*, 192, 6494–6496 (2010).
434. Heijstra, B., Kern, E., Koepke, M., Segovia, S., and Liew, F., Novel bacteria and methods of use thereof, Patent WO/2012/015317 (February 2, 2012).
435. Huhnke, R., Lewis, R., and Tanner, R.S., Isolation and characterization of novel clostridial species, U.S. Patent 2008/0057554 (March 6, 2008); *Energies*, 5, 5412 (2012).
436. Kuehne, S.A., Heap, J.T., Cooksley, C.M., Cartman, S.T., and Minton, N.P., ClosTron-mediated engineering of *Clostridium, Methods in Molecular Biology*, 765, 389–407 (2011).
437. Heap, J.T., Ehsaan, M., Cooksley, C.M., Ng, Y.K., Cartman, S.T., Winzer, K., and Minton, N.P., Integration of DNA into bacterial chromosomes from plasmids without a counter-selection marker, *Nucleic Acids Research*, 40, 1–10 (2012).
438. Abrini, J., Naveau, H., and Nyns, E.J., *Clostridium autoethanogenum*, sp. nov., an anaerobic bacterium that produces ethanol from carbon monoxide, *Archives of Microbiology*, 161, 345–351 (1994).
439. Coskata Inc., Coskata Inc.'s semi-commercial facility demonstrates two years of successful operation, Available online: http://www.coskata.com/company/media.asp?story=504B571C-0916-474E-BFFA-ACB326EFDB68, Accessed on July 23, 2012.
440. Pierce, E. et al., The complete genome sequence of *Moorella thermoacetica* (f. *Clostridium thermoaceticum*), *Environmental Microbiology*, 10, 2550–2573 (2008).
441. Köpke, M., Mihalcea, C., Bromley, J.C., and Simpson, S.D., Fermentative production of ethanol from carbon monoxide, *Current Opinion in Biotechnology*, 22, 320–305 (2011).
442. Burton, F. and Williams, J., *New Release—Mitsui Leads Investment in Lanza Tech's $60m, Series D Round*, Lanza Tech Publication, Roselle, IL (March 26, 2014). www.LanzaTech.com.
443. Cummings, D., McClenahar, H., and Akbazad, S., INEOS Bio, INEOS Bio JV breaks ground on 1st advanced waste-to-fuel commercial biorefinery in U.S. (February 9, 2011), Available online: http://www.ineosbio.com/76-Press_releases-15.htm, Accessed on July 16, 2012.
444. U.S. Department of Energy report, INEOS bio commercializes bioenergy technology in Florida, Available online: http://www1.eere.energy.gov/biomass/pdfs/ibr_arraprojects_ineos.pdf, Accessed on August 1, 2012.
445. LanzaTech, LanzaTech's commercialisation goes transtasman, Available online: http://www.lanzatech.co.nz/sites/default/files/imce_uploads/lanzatech_signs_with_bluescope_steel_march_2012.pdf, Accessed on July 23, 2012.
446. Williams, P., INEOS Bio takes advanced biofuel technology commercial, Available online:http://biomassmagazine.com/articles/6841/ineos-bio-takes-advanced-biofuel-technology-commercial, Accessed on July 16, 2012.
447. William, J.R., Securities and Exchange Commission, Coskata, Inc., Warrenville, IL, Available online: http://www.sec.gov/Archives/edgar/data/1536893/000119312511343587/d267854ds1.htm, Accessed on July 9, 2012.
448. LanzaTech Chinese steel miller commercializing LanzaTech's clean energy technology, Available online: http://www.lanzatech.co.nz/sites/default/files/imce_uploads/shougangprvf.pdf, Accessed on August 1, 2012.
449. INEOS Bio, INEOS Bio Names AMEC as its global license support engineering firm for its waste-to-bioenergy technology, Available online: http://www.ineosbio.com/76-Press_releases-33.htm, Accessed on July 16, 2012.

450. Lane, J., Coskata switches focus from biomass to natural gas; to raise $100M in natgas-oriented private placement, Available online: http://www.biofuelsdigest.com/bdigest/2012/07/20/coskata-switches-from-biomass-to-natural-gas-to-raise-100m-in-natgas-oriented-private-placement/, Accessed on July 23, 2012.

451. Hydrodesulfurization, Wikipedia, the free encyclopedia (2015). https://en.wikipedia.org/wiki/hydrodesulfurization (last accessed July 8, 2016).

452. Cracking (Chemistry), Wikipedia, the free encyclopedia (2015). https://en.wikipedia.org/wiki/cracking(chemistry) (last accessed September 27, 2016).

453. Hydrodeoxygenation, Wikipedia, the free encyclopedia (2015). https://en.wikipedia.org/wiki/hydrodeoxygenation (last accessed September 5, 2015).

454. Hydrodenitrogenation, Wikipedia, the free encyclopedia (2015). https://en.wikipedia.org/wiki/hydrodenitrogenation (last accessed January 12, 2015).

10 Gas for Heat, Electricity, and Mobile Applications

10.1 INTRODUCTION

In recent years, natural gas has become a more affordable heat and power source for Americans. Only about 8% of U.S. homes are on oil heat today; most of them are in the Northeastern United States and were built back in the day when oil was the cheapest way to keep warm through the long winters. Many utilities have since put gas lines into neighborhoods that didn't have them in the past, opening the door for homeowners to switch out old inefficient oil furnaces for more efficient gas units. Also, in recent years many coal-based power plants have been replaced by natural gas–based power plants due to environmental concerns regarding emissions of CO_2 and other pollutants. In general natural gas–based power plants emit more than 50%–60% lower CO_2 emission than coal-based power plants [1–8].

While gas has lower carbon emissions than oil and coal, a controversial gas extraction method (fracking, in which drillers inject water, sand, and chemicals at high pressure underground to break through rock and access the natural gas) which is increasingly employed today may take a toll on surrounding ecosystems and regional water quality. Most environmental advocates would rather see a rapid transition to truly renewable heating sources like biomass, wind, geothermal, or solar. As shown in my previous books [1,2], the capital investment in implementing renewable sources are high and overall not as cost effective as natural or synthetic gas. In the near future, renewable energies face many headwinds because of high capital cost and issues related to their scale; however, the best way to promote them is through hybrid processes that include both gas and renewable energies. Low cost and low emission of CO_2 by natural gas would provide a smooth transition to renewable energy. The use of hydrogen will also provide long-term sustainability.

10.1.1 Versatility of Natural and Synthetic Gas

Natural and synthetic gas collectively is a very versatile fuel both on supply and demand sides. On supply sides, natural gas can be obtained by both conventional and unconventional methods, and synthetic gas can be obtained from numerous feedstocks. The success of gas reforming technologies also allows interplay among methane, syngas, and hydrogen. On the demand side, besides the use of natural and synthetic gas in the productions of liquid fuels and fuel additives mentioned in the previous chapter, major usages of natural and synthetic gas are in power generation, home, and industrial heating; manufacturing of chemicals and polymers; production of other industrial and agricultural commodities; and in transportation industries. We briefly assess these applications in this chapter.

10.1.1.1 Power Generation

Most rapidly expanding use of natural gas is for the generation of power and electricity. This usage is aided by the application of concepts of combined cycle and cogeneration (to improve thermal efficiency) and the use of sophisticated gas and steam turbines both at large and small scales. Furthermore, as shown in my previous books [1,2], natural gas is also well suited for a combined use with renewable energy sources such as wind or solar and for alimenting peak-load power stations functioning jointly with hydroelectric plants. Most grid peaking power plants and some off-grid engine generators use natural gas.

The U.S. Energy Information Administration reports that in 2012, natural gas–driven power plants generated about half as much CO_2 than coal-based plants. Petroleum burning also generated 75% more CO_2 than natural gas. Coal-fired electric power generation emits around 2000 pounds of carbon dioxide for every megawatt-hour generated, which is almost double the carbon dioxide released by a natural gas–fired electric plant per megawatt-hour generated. The change from coal-based power plants to natural gas–based power plants resulted in carbon dioxide emissions in the first quarter of 2012, the lowest of any recorded for the first quarter of any year since 1992 [1–8].

Combined cycle power generation using natural gas is currently the cleanest available source of power using hydrocarbon fuels, and this technology is widely and increasingly used, as natural gas can be obtained at increasingly reasonable costs. While fuel cell technology is yet not quite cost competitive, it will eventually provide cleaner options for converting natural gas into electricity, both for static and mobile usages. Locally produced electricity and heat using natural gas–powered combined heat and power plants (CHPs) or cogeneration plant is also considered energy-efficient and a rapid way to cut carbon emissions [2,6]. Natural gas power plants are increasing in popularity and generate 22% of the world's total electricity.

Not only natural and synthetic gas are preferred over coal and oil for power production due to its less impact on environment, natural gas is easy to use at all scales: large, medium, small, and micro. The ability to design and control turbines of all sizes with natural gas combustion makes its power application very broad. The interconnections between gas grid and electrical grid also make power production (both base and peak level) at any scale easy. The use of natural gas for hybrid power production with renewable sources is very desirable at both large and small scales.

10.1.1.2 Domestic, Industrial, and Agricultural Uses

Besides generating less carbon dioxide during heating, natural gas dispensed from a simple stovetop can generate temperatures in excess of 1100°C (2000°F), making it a powerful domestic cooking and heating fuel [3–8]. In much of the developed world, it is supplied through pipes to homes, where it is used for many purposes, including ranges and ovens, gas-heated clothes dryers, heating/cooling, and central heating. Heaters in homes and other buildings may include boilers, furnaces, and water heaters. Both North America and Europe are major consumers of natural gas.

In the United States, compressed natural gas (CNG) is used in rural homes without connections to piped-in public utility services, or with portable grills. Natural gas is also supplied by independent natural gas suppliers through Natural Gas Choice programs throughout the United States. However, since CNG costs more than LPG (propane), the latter is the dominant source of rural gas. Natural gas or LPG is thus a very valuable source of heating both in urban and rural environments.

Besides applications mentioned above, gas is heavily used in the manufacturing of energy-intensive commodities such as aluminum, glass, bricks, cement, and iron bars. Here the gas is converted to thermal or electrical power, which is then used in the making of the commodity, and the commodity is then sold on the open market. Natural gas is also used in the manufacturing of fabrics, steel, plastics, paint, and other products. Natural gas is a major feedstock for the production of ammonia, via the Haber process, for its eventual use as fertilizer [3].

10.1.1.3 Transportation

For transportation, burning natural gas produces about 30% less carbon dioxide than burning petroleum. With new innovations in vehicle design, natural and synthetic gas is also becoming a more viable option to conventional gasoline, diesel, and jet fuels. CNG is a cleaner and also cheaper alternative to other automobile fuels such as gasoline (petrol) and diesel. By the end of 2012, there were 17.25 million natural gas vehicles worldwide, with Iran and Pakistan leading the pack [9]. The energy efficiency is generally equal to that of gasoline engines, but lower compared with modern diesel engines. Gasoline/petrol vehicles converted to run on natural gas suffer because of the low compression ratio of their engines, resulting in a cropping of delivered power while running on natural gas (10%–15%). CNG-specific engines, however, use a higher compression ratio due to this fuel's higher octane number of 120–130 [3–8].

Russian aircraft manufacturer Tupolev is currently running a development program to produce LNG- and hydrogen-powered aircraft. The program has been running since the mid-1970s, and seeks to develop LNG and hydrogen variants of the Tu-204 and Tu-334 passenger aircraft, and also the Tu-330 cargo aircraft. It claims that at current market prices, an LNG-powered aircraft would cost about 60% less with considerable reductions to carbon monoxide, hydrocarbon, and nitrogen oxide emissions [3–8]. Liquid methane has more specific energy than the standard kerosene-based jet fuels. Its low temperature also cools air which engine compresses for greater volumetric efficiency. It can also be used to lower the temperature of the exhaust. Finally, with the recent advancement in fuel cells, natural gas and hydrogen can play a significant role in the use of fuel cell both for stationary and mobile purposes [3–8].

Thus natural and synthetic gas is a very valuable commodity with a wide range of power, heat, and vehicle applications, both at small and large scales and for both urban and rural environments. One new development for a gas seller is the concept of gas refinery [3–8]. Designs now exist where integrated complexes use gas in an optimized manner for making aluminum, iron, cement, glass, and so on for LNG, for conversion to other products such as polymers, methanol, and ammonia, and where waste heat and by-products can be used in other associated processes [3–8]. Such refineries will enable lower costs, hence keeping the products competitive in the marketplace.

10.1.1.4 Other Factors Affecting or Enhancing Gas Utility

The use of natural or synthetic gas for heat and power can be facilitated by the natural gas grid, which can operate in harmony with smart electrical grid. With the help of these grids, electricity can be generated anywhere, particularly at or near the reservoir source and transported by cable to the required destination(s). Thus, for instance, offshore or isolated gas could be used to fuel an offshore power plant (may be sited in less hostile waters), which would generate electricity for sale onshore or to other offshore customers. Unfortunately, installing high-power lines to reach the shoreline appears to be almost as expensive as pipelines [3–8], so gas to electricity could be viewed as defeating the purpose of an alternative cheaper solution for transporting gas. There is significant energy loss from the cables along the long-distance transmission lines, more so if the power is AC rather than DC; additionally, losses also occur when the power is converted to DC from AC and when it is converted from the high voltages used in the transmission to the lower values needed by the consumers [2,10].

Some consider having the gas as energy at the consumers' end gives greater flexibility and better thermal efficiencies because the waste heat can be used for local heating, desalination, and other industrial purposes. This view is strengthened by the economics as power generation uses approximately 1 million scf/day of gas for every 10 MW of power generated, so even large-generation capacity would not consume much of the gas from larger fields, and thus not generate large revenues for the gas producers. Nevertheless, gas to electricity has been an option much considered in the United States for getting energy from the Alaskan gas and oil fields to the populated areas, particularly California. The choice between centralized and distributed use of gas for power generation will depend on the overall economics and related environmental concerns. If gas-driven power plants use cogeneration approach where waste heat is used for other purposes, it would make the most sense to generate power where waste heat is being utilized. This is because heat and electricity cannot be transported to long distances without incurring significant losses. Thus, through the use of cogeneration, gas can be effectively used in distributed manner.

The fact that gas can be used in centralized and distributed manners adds significantly to its versatility. There are many sources of distributed small-scale gas supply such as "associated gas," gas from landfills, gas from stranded islands, and small-scale synthetic gas created from local biomass fields. All of these can be used for local heating and power needs. The flexibility in the gas transport by CNG, LNG, LPG, and gas hydrates also provide additional versatility to the gas market.

The use of "associated gas" has other practical considerations. If it is only used for power generation and if there is generator shutdown, the whole oil production facility might also have to be shut down, or the gas has to be released to flare. Also, if there are operational problems within the generation plant, the generators must be able to shut down quickly (in around 60 s) to keep a small incident from escalating. Additionally, the shutdown system itself must be safe so that any plant that has complicated processes that require a purge cycle or a cool-down cycle before it can shut down is clearly unsuitable. The lack of ability for quick shutdown or rapid start-up of the generator can affect the financial retribution from power distributors. All of these factors imply that for the use of "associated gas" multiple choices and flexible applications are important.

The distributed power plants also make sense when synthetic gas is generated using biomass or waste (like landfill or MSW) because transporting biomass or waste to long distances has not proven to be economical. The long-term storage of these materials is also problematic. The best use of landfill gas is to purify and generate power and heat locally. On the other hand, synthetic gas generated from coal can be produced on a large scale on a centralized location. Thus, for synthetic gas, the source and nature of feedstock as well as the plans for use of waste heat govern the decision between centralized or distributed power generations. Natural or synthetic gas can handle both types of operations.

The transportation ease of gas also influences its use and adds to its versatility. Besides natural gas pipeline and LNG transport by sea and roads, the hydrate and CNG are very convenient tools for natural gas transport. These tools do not involve extreme temperature, either high or low, do not require an oxidant or a catalyst, and do not feature any complex unit operations other than standard process equipment. In addition, the technology is able to cope with an intermittent and variable profile of gas production with time, as is usually the case with associated gas.

It is believed (see Chapter 8) that the delivered cost of gas by hydrates or CNG can compete in the energy market at any scale with LNG transport for offshore associated gas, remote reservoirs, smaller reservoirs, or stranded onshore gas that can create markets. The actual economics of gas transportation can become one element of a much bigger scenario with the extreme example being where an oil development may not proceed if there is no means to dispose the associated gas and flaring is prohibited. Selling, and hence exporting, the gas by gas hydrates or CNG may ensure that an oil project can be commercially successful.

Gas can also satisfy small-volume needs of tourist industries of, for instance, the Caribbean, Black Sea, Indonesian, and Mediterranean islands. This industry is using increasing quantities of electricity but will have only a "small" energy demand on the world scale. A large five-star hotel which may use perhaps 11 million-kWh of electricity per year would need only some 70 million scf/year of gas (assuming that 10 MW of electricity generating capacity continuously running requires ~1 million scf/day of gas). If power is generated continuously, the hotel would need only a 1.2 MW generator (30,000 kWh/day of electricity). The actual number can be a little higher to cover variable needs during the day or seasonal differences in power needs. While this quantity may be too small to be transported by LNG train, its transportation by CNG or hydrates can provide a meaningful revenue stream for stranded or "associated gas."

While transportation of small quantity of gas by gas hydrates and CNG is relatively easy, it will require significant initial development costs. The manufacturing of hydrate could be carried out using mobile equipment for onshore and ship for offshore, or for CNG with a mobile compressor, so that the equipment can be moved on if the field ceases production. The development costs can be absorbed by initial projects. While the developments of small gas projects render small profits for local industry, they can satisfy heating and power needs for local hotels and industries. The availability of standard equipment and procedures can help these efforts. If the gas is "associated gas," the development of small projects can help in sustainable oil production [3–8].

As shown in Chapter 8, the competitive advantage of the hydrates or CNG processes over other nonpipeline gas technologies, particularly gas to liquid, is that they are intrinsically simpler processes and, as a concept, far easier to implement. These processes are feasible at lower capital costs and, hence, require smaller investments and payback times for equivalent gas sales.

Another commercial advantage that hydrates and CNG have over LNG is that to implement a project a much smaller lead time is needed, which means that hydrates and CNG are producing revenue, whereas LNG needs more years of investing (larger) capital before any return. Also, there are also smaller insurance risks concerning the former.

Finally, gas can be a better substitute for fuel oil for small applications. Fuel oil produces more CO_2 for the same energy output than gas, and the environmental cost is higher per ton of CO_2 emitted. Thus from this point of view, electricity generated by gas transported by CNG, hydrate, or even LNG in small containers could possibly be cheaper than fuel oil at, say, $30/bbl, that is, $5/thousand scf gas, particularly if there are extra costs for exhaust clean-up and increased boiler corrosion and environmental carbon tax [3–8].

10.1.2   IMPORTANT CONCEPTS AND TOOLS TO EXPAND USEFULNESS OF GAS IN ENERGY AND FUEL INDUSTRY

In the previous chapter, we examined the fact that with the use of proper catalyst or new microbes, syngas can be converted to numerous liquid fuels, fuel additives, and chemicals. These conversion processes are commercially established and are being continuously improved by new catalysts, new microbes, and new process concepts. This indicates that even if oil supply gets depleted, synthetic oil can be created from gas.

In this chapter, we examine the concepts and tools that can further advance the use of natural and synthetic gas for heating, power, and mobile applications. Five concepts that have helped advance the use of gas for power generations are (1) centralized power generation, (2) distributed power generation, (3) "grid" approach for gas and electricity management, (4) combined cycles, and (5) cogeneration or combined heat and power generation. "Grid approach" can be considered as a concept and tool to manage both centralized and distributed power systems as well as to integrate gas, heat, and electricity transport and storage systems. The chapter analyzes in detail the applications of these concepts for heat and power generations by gas.

The use of gas for power, heat, and mobile applications is also significantly advanced by constant development of four tools: (1) gas turbines; (2) micro-turbines; (3) CNG, LNG, and LPG fuels for automobiles; and (4) hydrogen-based fuel cells. The chapter examines these tools in detail and their roles in the applications of gas for power, heat, and vehicle industries. Further developments of these concepts and tools are equivalent to the additional developments of catalysts, microbes, and innovative process concepts for syngas conversions evaluated in the previous chapter.

The use of gas for power production is rapidly expanding. This expansion has created both centralized and decentralized (or distributed) power production and distribution facilities. There are pluses and minuses of both centralized and distributed power generation concepts, and each of them needs to be implemented and further developed with careful assessment of economics and environment. The development of gas and electricity grids and interactions between them are also very important for not only managing centralized and distributed power generations on dynamic basis but also integrating them. Liquefied or compressed natural gas and LPG are used more and more for vehicles replacing traditional gasoline and diesel fuels. Hydrogen-driven car or the use of hydrogen for fuel cell is rapidly expanding.

The use of natural gas or hydrogen at large- and small-scale applications has increased significantly since the advanced development of gas turbines and micro-turbines. There are numerous types of turbines, and they play an essential role in the conversion of thermal energy from combustion to electrical energy. The chapter examines the workings of these two types of turbines in detail. Micro-turbines, which use micro-combustion, have revolutionized the power generation and usages at small and micro scale both for static and mobile systems. Natural gas, syngas, and hydrogen are ideally suited for micro-combustion.

The concepts of combined cycles and cogeneration have helped improve thermal efficiencies of gas-driven power plants. The use of gas for power and heat is supported by a strong infrastructure for gas and its ease for storage and transportation as well as its connections to the power grids. This chapter examines these tools and concepts that make the use of natural and synthetic gas for heat, power, and mobile applications most attractive and viable both at large and small scales.

10.2 THE CENTRALIZED POWER SYSTEM

Since the 1990s, electricity production has been driven toward generation concentration and a higher degree of integration leading to the current centralized electricity paradigm. This move was driven by several factors [11–16].

1. *Economies of scale*: The advent of steam turbines made it possible to increase the size of the turbines while decreasing the marginal cost of electricity production. The rapid spread of this technology led to a surge in the overall plant capacities.
2. *The search for high energy efficiency*: Gains in efficiency were achieved through larger facilities capable of handling higher pressures and temperatures in steam used for electricity generation. At a certain point, the gains were, however, offset by the increase in operating and maintenance costs as materials were unable to sustain operation at high levels of operating conditions over the long run.
3. *Innovation in electricity transmission*: The use of alternative current instead of direct current permitted transmission of electricity over long distances with a significant loss reduction.
4. *The search for reliability*: In order to increase the reliability at the customer's end, large electricity production facilities were connected to the transmission networks. Pooling resources helped reduce the reliance of each customer on a particular generator as other generators were often able to compensate for the loss.
5. *Environmental constraints*: The use of transmission networks made it possible to relocate the generation facilities outside the city centers, thus removing pollution due to exhaust from coal- or gas-fired plants.
6. *Regulation favoring larger generation facilities*: Current government regulations favored large-scale power generation facilities.
7. *Advancement on smart grid and natural gas grid concepts*: Grid concepts also allowed integration of nuclear and renewable (solar, wind, geothermal biomass, etc.) power generation systems with natural gas (or coal) power plants.

In the sector's layout resulting from this move toward concentration and integration, electricity is generated and transported over long distances through the transmission network and medium distances through the distribution network to be finally used by the end customer.

In spite of its predominance, several studies were conducted to emphasize the main shortfalls of the centralized generation paradigm [17,18]. The results of these studies indicated several factors such as transmission and distribution costs and investment, energy efficiency, security and reliability, and rural electrification point to the need for distributed power generation [10–48].

On average, transmission and distribution costs can be up to 30% of the total cost of electricity delivery [28,31,45]. The lowest cost occurs when industrial customers take electricity at high to medium voltage and highest cost occurs when small customers take electricity from the distribution network at low voltage [31]. The high price for transmission and distribution results from line losses in transmission and distribution network and conversion losses required to fit the specifications of the network, for example, changing the voltage while flowing from the transmission network to the distribution network [28]. Furthermore, while larger transmission and distribution losses cause cash costs, they also contribute to larger greenhouse emissions for the electricity that is wasted due to losses.

The International Energy Agency (IEA) [32] has also projected that expansion in centralized power production with associated transmission and distribution network for OECD countries will require \$3–\$3.5 trillion of new investment up to 2030. Distributed generations which bypass the transmission and distribution network, on the other hand, will cost less than \$3 trillion (electricity generation investments remaining constant) [28,31,32,45].

Besides cost and new investment, energy efficiency of centralized power distribution indicated some concerns.

In the 1960s, the marginal gains in energy efficiency through size increase and use of higher temperature and pressure started to diminish. Higher temperatures and pressure resulted in high material wear and tear, leading to lower than expected operating life for steam turbines [30]. In order to increase energy efficiency without requiring higher pressure, cogeneration systems were developed to reuse the waste steam in a neighborhood heating system or cooling system through district heating and/or cooling. The total energy efficiency achieved when combining both electricity and heat goes up to 90% [29]. Comparatively, the sole electricity generation in a single cycle hardly goes above 40%. The main problem, however, is that steam and heat are even less easily transported than electricity, thus requiring use of a large amount of waste heat from centralized power plants to be used locally, which may not be possible in many cases.

Unlike centralized power generation, the distributed generation also contributes to the improvement of energy security and reliability in a number of ways. Distributed generation technologies can accommodate a larger range of fuels other than coal, oil, gas, and nuclear. For instance, distributed generation has been used at landfills to collect biogas and generate energy. Distributed power generation allows use of biomass/waste locally rather than shipping them expensively to a large centralized power plant. Distributed power generations are well suited for small-scale operations run by renewable fuels like solar, wind, and so on.

The main use of distributed generation is for backup capacities to prevent operational failures in case of network problems. Backup generators have been installed at critical locations such as hospitals and precincts of schools. Furthermore, in a deregulated electricity market, the reduction in required reserve margins during unplanned power outages can lead to capacity shortfalls resulting in high electricity prices to the consumers. In order to hedge against negative price impacts, large electricity consumers have acquired distributed generation capacities [10–48].

The environmental impact of the centralized energy system is significant due to the heavy reliance on coal and oil and, to a lesser extent, on natural gas. The electricity sector is responsible for one-fourth of NO_x emissions, one-third of CO_2 emissions, and two-third of the SO_2 emissions in the United States [46]. Distributed generation reduces emission attached to transmission and distribution losses by centralized plants. The use of cogeneration and renewable sources of energy in distributed power plants also reduces emissions.

Finally, distributed power generation is very important for providing electricity to rural environment and stranded islands for two reasons. (1) Large capital expenditures required to connect remote areas with small consumptions by the transmission lines are generally uneconomical. (2) The transmission and distribution losses are greater with the increased distance covered. Rural and stranded island electrification is thus costly for centralized power plants. It often proves more economical to rely on distributed generation in such cases [22]. This has often been the case for mountain areas, low population density areas, a stranded island, offshore drilling platforms, and so on which are remote from the main cities [10–48].

While, as mentioned earlier, distributed power generation offers many advantages, with the advances in smart electrical and gas grids (see Section 10.4), centralized power generation is likely to continue to have a larger share of overall power generation, particularly in urban environment.

10.3 THE DISTRIBUTED POWER SYSTEM

Under the current centralized generation paradigm, electricity is mainly produced at large generation facilities, shipped though the transmission and distribution grids to the end consumers. In recent years, however, the role of distributed power generation has increased. The recent quest for energy efficiency and reliability and reduction of greenhouse gas (GHG) emissions has led to exploration of possibilities of distributed generation where electricity is produced next to its point of use. Historically, distributed generators have been able to act as a complement to centralized generation. Price deregulation enabled distributed generators to enter particular electricity market at different price and with fewer barriers to entry. Distributed power generation also reduced emission by applying cogeneration principle to local heat and power need.

It is difficult to estimate the percentage distribution of distributed and centralized electricity because of the unclear definition of what is distributed and what is not distributed. The IEA's model [32] estimates that distributed generation will account for between 20% and 25% of additional capacities to be built up to 2030 in the reference scenario and 30%–35% in the alternative scenario. This will add up to approximately 30%–45% of power generation investments over the period 2001–2030. The strong incentive to support distributed generation will be driven, among others, by the size of investment in transmission network that needs to be avoided.

Over the long run, a significant increase in the share of distributed generation will require revamping the whole physical and regulatory architecture of the electricity network which is largely based on the centralized generation. This change will face numerous constraints. Distribution networks will have to be reinforced and partly redesigned to cope with new capacities. Besides, they will have to incorporate both control and protection software and hardware to coordinate the distributed generators and make them price competitive to centralized generators. The use of distributed generators for cogeneration and fuel cell will require more research and development to reduce the cost per kWh [10–48]. Significant work should be undertaken to alter the regulatory environment which favors the centralized power plants. Finally, in order for distributed generation to be a sustainable alternative paradigm, it will have to rely on the renewable technologies or favor cogeneration that maximizes energy efficiency and reduces emissions [25,28,33,40,45,46].

Distributed generation is often expressed as "decentralized generation," "dispersed generation," "distributed energy resources," and so on. As shown by Pepermans et al. [18], the definition varies significantly in terms of characteristics of the generators mentioned. Dondi et al. [26] define distributed generation as a generator with small capacity close to its load that is not part of a centralized generation system.

There is no consensus in the literature on the upper limit of distributed generators; this limit can range from 1 MW to over 100 MW [21] or maximum capacity of 30 kW [23]. In order to define distributed generation in a way that could encompass a wide variety of technologies, capacities, the connection type, and so forth, Ackermann et al. [21] devised a definition applicable to the vast majority of distributed generators. "Distributed generation is an electric power source connected directly to the distribution network or on the customer site of the meter" [21]. The key criteria in this definition are the connection to the distribution network and the proximity to the end consumer.

A variety of end use allows greater variety in technologies for distributed power generation [31]. Among the technologies used, a reciprocating engine is popular. This technology uses compressed air and fuel. The fuel mixture is ignited by a spark to move a piston. The mechanical energy is then converted into electrical energy. Reciprocating engines are a mature technology and largely spread due to their low capital investment requirement, fast start-up capabilities, and high energy efficiency when combined with heat recovery systems. Most reciprocating engines run either on fuel or natural gas, with an increasing number of engines running on biogas produced from biomass and waste. Besides reciprocating engines, gas turbines and micro-turbines are increasingly becoming more popular. Gas turbines are widely used for electricity generation due to regulatory incentives induced to favor natural gas, which produces low emission levels. Gas turbines are widely used in cogeneration. Micro-turbines are built with the same characteristics as gas turbines but with lower capacities and higher operating speed. As shown later, micro-turbines offer several advantages for vehicle applications. The use of micro-turbines has expanded due to successful development of micro-combustion technology.

Instead of converting mechanical energy into electrical energy, fuel cells are built to convert chemical energy of a fuel directly into electricity. The fuel used is generally natural gas or hydrogen.

Fuel cells are a major field of research, and significant effort is put in reducing capital costs and increasing efficiency, which are the two main drawbacks of this technology. Renewable technologies are also used as a way to produce distributed energy. Renewable sources include photovoltaic technologies, wind energy, and geothermal energy. These sources qualify as distributed generation only if they meet the criteria of the definition, which is not always the case. For example, offshore wind farms do not qualify as distributed generation. Power plants operated on renewable fuels with cogeneration and directly connected to distribution network and local heat grid are considered a part of distributed energy.

10.3.1 FUTURE OF DISTRIBUTED GENERATION

While in past, distributed power generation has been used to complement centralized generation, in future distributed power generation will grow because [10–48]:

1. Deregulation of electricity market will allow distributed power generation to set the price what the production and market demand.
2. Distributed market will reduce greenhouse emissions by use of renewable fuels for power production, use of cogeneration technologies, and reducing the transmission losses of electricity. Using heat for district and process heating and other applications makes it possible to reduce emissions and increase energy efficiency to high levels. Cogeneration relies heavily on distributed generation as heat transmission and storage is the source of significant energy losses.
3. Distributed power generation will provide electricity to stranded island, rural environment, offshore drilling platforms, and so on, where power line transmission costs by centralized power generation would be prohibitive. Distributed power generation can also help niche markets such as to make use of stranded gas (like associated gas) and landfill gas locally without their expensive transportation to centralized location. Distributed generation technologies make it possible to accommodate a broad range of fuel.
4. As distributed generators tend to be of smaller size and quicker to build, they will be able to benefit from price premiums. Geographical and operational flexibility make it possible to set up distributed generators in congested areas or use it only during consumption peaks. Besides, for small excess demand, it is often uneconomical to build an additional centralized generation plant, whereas with lower CAPEX and capacities, distributed generation can serve the needs [31].
5. Distributed power may be the best way to provide electricity for mobile applications in remote areas.

While distributed energy will grow over the years, it also faces numerous constraints:

10.3.1.1 Technical Constraints

Distributed energy will need technical improvements necessary to ensure high system reliability [39,40]. Adding distributed generators at the distribution level can significantly impact the amount of power to be handled by the equipment (cables, lines, and transformers). In order to avoid overload problems, reinforcement work will have to be undertaken. Distributed generators are often connected to low-voltage networks. When power is carried over long distances, voltage tends to drop due to resistance in cables. While using distributed generation, additional protection systems are required to avoid internal faults, defective distributed network, and islanding [43]. Distributed generation helps reduce transmission and distribution losses because distributed generators are not connected to the transmission grid and they distribute power locally. This often comes with condition that they have captive use for a client which limits the use of the distribution grid.

As of today, all the ancillary services positively impacting the quality of electricity delivered are provided by centralized generators. The provision of these services by distributed generators will entail additional cost. The growth of distributed generation will also require the development of "virtual power plants," where several distributed generators are coordinated to act as an integrated plant [41,44,48]. The plant is "virtual" as it is not in one place but made of the aggregation of several units. This is analogous to the use of parallel computing to perform the task of a supercomputer. The operation of such a plant requires a strong integration of information, communication, and management systems [39]. One way of integrating small-scale distributed generators is through the use of a microgrid. As of today, distributed generators are mainly integrated though medium-voltage grids. Significant research is, however, under way to facilitate the integration in low-voltage grids with local coordinating functions or microgrids [25,39,42].

10.3.1.2 Cost Competitiveness

One key hurdle to overcome in a deregulated power market is the cost competitiveness of distributed generation. This parameter varies, however, a lot from one technology to the other due to their age and current development. For example, reciprocating engines are a mature technology but potentially more effective fuel cells are still subject to significant research and development in order to become a credible source of generation. Cost competitiveness of distributed generation is heavily impacted by the capacity of the regulation to price its impact on the electricity network and on its ability to provide specific services to the end consumer such as heat generation or ancillary services.

10.3.1.3 Regulatory Barriers

The distributed generation is also negatively impacted by the regulations involving (1) network tariffs, (2) planning for costs, (3) incremental distribution costs caused by distributed generation, (4) treatment of energy losses, (5) cost associated with ancillary services for voltage control, (6) incentives for innovation for converting grids into smart grids, and (7) unbundling activities of operators toward legal independence of distributed networks. The regulations related to these issues need to be resolved in favor of a distributed network to make it more profitable.

In addition to these regulatory barriers, office of gas and electricity market (OFGEM) [38] reported that licensing requirements used for centralized power plants need to be modified and made more economical for smaller power plants. Their permissions need to be simplified and expedited. Provisions should also be made for distributed generators to sell excess power to grid and other sources to generate profit when local needs are low.

10.3.1.4 Impact on Climate Change and Global Warming

Distributed generation must use renewable sources of energy or cogeneration concept to reduce harmful emission to the environment. Diesel reciprocating engines are the worst emitters in terms of nitrogen oxides (NO_x). Combined cycle gas turbines tend to be the best performers in terms of carbon dioxide (CO_2) and sulfur dioxide (SO_2). While fuel cells are the lowest emitters when it comes to NO_x, CO, particulate matters (PM10), and hydrocarbons (HC), the lack of commercial experience and the costs of this technology limit its current usage.

The growth of distributed generation will be closely linked to the capacity of the generators to monetize their positive impact on the overall electricity sector while emitting less harmful pollutants to the environment. This will be possible only if the sector and price regulation integrates climate change and global warming effect through an adjustment in the price of electricity paid to the generators.

10.4 SMART ELECTRICAL GRID FOR CENTRALIZED AND DISTRIBUTED POWER MANAGEMENT

Both centralized and distributed power production and management can be best achieved by the grid approach. A smart electrical grid which is connected to smart gas grid and heat grid can provide the best method to manage electrical power, gas distribution, and usage of waste heat in a dynamic and two-directional manner. In Chapter 8, we examined the natural gas grid concept. Here, we briefly examine the smart electrical grid concept to manage both centralized and decentralized power distribution.

10.4.1   SMART ELECTRICAL GRID

Along with the usage of gas, the consumption of electricity is rapidly expanding. ExxonMobil [1] predicts that by 2040, 40% of our energy consumption will be electrical. The supply and demand of the electricity is managed by smart electrical grid [49]. The national electrical grid system is continuously expanding, and it is divided into three major regional parts: eastern interconnection, western interconnection, and Texas interconnection. These larger grids provide efficiency and stability in supply and demand needs. An effort is being made to interconnect these three grids into a single U.S. grid. An SG, also called smart electrical/power grid, intelligent grid, future grid, and so on [49], is an enhancement of the twentieth-century power grid. The traditional power grids are generally used to carry power from a few central generators to a large number of users or customers. In contrast, the SG uses two-way flows of electricity and information to create an automated and distributed advanced energy delivery network.

For the future electrical distribution system, grids will become more active and will accommodate bidirectional power flows and an increasing transmission of information. Some of the electricity generated by large conventional plants will be displaced by the integration of renewable energy sources. An increasing number of PV, biomass, and onshore wind generators will feed into the medium- and low-voltage grid [50,51]. Conventional electricity systems must be transformed in the framework of a market model in which generation is dispatched according to market forces and the grid control center undertakes an overall supervisory role (active power balancing and ancillary services such as voltage control).

The SG is expected to control the demand side as well as the generation side, so that the overall power system can be more efficiently and rationally operated. The SG includes many technologies such as information technology, communications and control technologies [52–106], and electrical energy storage (EES) technologies. The EES can be used in a number of different ways in the management of SG, and these are described in detail in my previous book [1].

By utilizing modern information technologies, the SG is capable of delivering power in more efficient ways and responding to wide-ranging conditions and events [60,80,107,108]. Broadly stated, the SG could respond to events that occur anywhere in the grid, such as power generation, transmission, distribution, and consumption, and adopt the corresponding strategies. For instance, once a medium-voltage transformer failure event occurs in the distribution grid, the SG may automatically change the power flow and recover the power delivery service.

Since lowering peak demand and smoothing demand profile reduce overall plant and capital cost requirements, in the peak period, the electric utility can use real-time pricing to convince some users to reduce their power demands so that the total demand profile full of peaks can be shaped to a nicely smoothed demand profile. More specifically, the SG can be regarded as an electric system that uses information, two-way, cyber-secure communication technologies [65–68], and computational intelligence in an integrated fashion across electricity generation, transmission, substations, distribution, and consumption to achieve a system that is clean, safe, secure, reliable, resilient, efficient, and sustainable.

The description mentioned earlier covers the entire spectrum of the energy system from the generation to the end points of consumption of the electricity [80,109,110]. It also presents the ultimate integration of the energy systems. While the ultimate SG is a vision, it is also a loose integration of complementary components, subsystems, functions, and services under the pervasive control of highly intelligent management and control systems. Given the vast landscape of the SG research, different researchers may express different visions [80,87,109,110] for the SG due to different focuses and perspectives [96,109–111]. The development and refinement of the SG is an ongoing task, because new functions [96,111–114] and more efficiency are constantly added and required.

10.4.2 Benefits and Requirements of Smart Electrical Grid

The initial concept of SG started with the idea of advanced metering infrastructure (AMI) with the aim of improving demand-side management [89,99–103] and energy efficiency, and constructing self-healing reliable grid protection against malicious sabotage and natural disasters [55,65,69–73]. In some way, this concept was similar to the management of natural gas grid. However, new requirements and demands expand the initially perceived scope of SG [55,70–72,102]. In spite of constantly changing final role of SG, the anticipated benefits and the basic requirements of it carry the following elements [55,66,68–106]:

1. *Improved power quality, capacity, security, and reliability*: SG improves power reliability and quality. It also enhances the capacity and efficiency of the existing electric power networks. The smart infrastructure and management system of SG will also present opportunities to improve grid security. In short, SG manages integrated, safe, secure, efficient, and sustainable power supply and demand.
2. *Provide more choices and service expansion capabilities*: It provides more choices for the consumer and enables new markets, products, and services. More recent expansion of SG will be in the management of power requirement of plug-in vehicles. The CHP will also provide new market for the SG. SG will also be required to manage new ES options.
3. *Facilitate penetration of renewable sources of power/decarbonize power industry*: By allowing more penetration of renewable energy sources in the electrical industry, SG will help to decarbonize the power industry. It will help in the reduction of GHGs. The expanded markets of hybrid and electric cars will also reduce emission of GHGs.
4. *Facilitate and automate power distribution and provide self-control maintenance*: The SG is required to accommodate distributed power sources. It will also serve the localized needs of shopping malls, hospitals, university campuses, and so on through the use of smart microgrids, which can serve as subsystems of overall SG. It will help in the reduction of oil consumption by reducing the need for inefficient generation during peak usage periods. The SG will automate maintenance and operation of the power delivery system.
5. *Optimize facility utilization*: SG will optimize the facility utilization. It will minimize the construction of backup (peak load) power plants by properly utilizing ES systems. SG will easily and effectively add or remove new or outdated sources of power generation. It will improve resilience to disruption and will allow predictive maintenance and self-healing responses to system disturbances.

These benefits and requirements will make SG development a continuous process as more functions, more demands for efficiency, and more complexities for control systems are constantly added. Smart grid has three subsystems: (1) smart infrastructure and distribution system, (2) smart customer-focused management system, and (3) smart personally secure, reliable, and failure protection system. These subsystems need to be continuously developed and upgraded [80–106].

10.4.3 Smart Grid Technology Assessment, Development, Standardization, and Optimization

Several technologies such as information, communication [55,66,73–75], and control are imbedded in the development of SG. Significant research to improve these technologies has been published in the literature. This includes the critical assessment of the basic concepts and the recommendations of technologies used in SG [52–55] and their standardizations [56,115]. For example, Vasconcelos [116] outlined the potential benefits of smart meters. Brown and Suryanarayanan [115] provided an industry perspective for the smart distribution system and identified areas of research for the improvement of technologies associated with the smart distribution system. Baumeister [117] and Chen [52] assessed SG cybersecurity and privacy issues.

The use of communication networks, wireless communication, and communication architecture for power systems has also been examined in the literature [54–91]. Gungor and Lambert [54] provided a clear analysis of the hybrid network architecture that can provide heterogeneous electric system automation application requirements. Akyol et al. [53] analyzed how, where, and what types of wireless communications are suitable for deployment in the electric power system. Wang et al. [55] provided a survey of the communication architectures in the power systems. They also discussed the network implementation considerations and challenges in the power system settings.

In order to realize the new grid paradigm, the National Institute of Standards and Technology (NIST) provided a conceptual model, which can be used as a reference for the various parts of the electric system where SG standardization work is taking place [56–67,92–99]. This conceptual model divides the SG into seven domains: customers, markets, service providers, operations, bulk generation, transmission, and distribution. Each domain encompasses one or more SG actors (such as end users, operators, service providing organizations, managers, generators of electricity and distributors), including devices, systems, or programs that make decisions and exchange information necessary for performing applications. NIST model is widely used as a standardization of various functions of SG.

The optimization of an SG depends on the ultimate objectives. Different users have different end games in mind. In general, however, the grid optimization is needed because

1. It allows maximum usage of the existing infrastructure, reduces and extends resource usage, and reduces emissions of CO_2 and other pollutants.
2. It reduces the overall cost and improves reliable delivery of power to the customers.
3. By saving costs, it allows investments in new generation, transmission, and distribution facilities.

For grid optimization, grid efficiency is very important. In 2006, a total of 1638 billion kWh of energy was lost on the U.S. power grid with 655 billion kWh lost in the distribution system alone. To put this in perspective, a 10% improvement in grid efficiency at the distribution level alone would have produced $5.7 billion in savings based on the 2006 national average price of electricity. It would also have saved over 42 million tons of CO_2 emissions [50,51,99,104].

Numerous studies [57–65,67–72] have attempted to examine different models for grid optimization. Nygard et al. [64] examined an optimization model for energy reallocation in an SG. They evaluated a self-healing problem, which takes action in near real-time to reallocate power to minimize disruption using integer linear programming models. Simmhan et al. [68] adopted an informatics approach to demand response optimization in SGs. The informatics approach allowed them to build an intelligent and adaptive grid. Ahat et al. [63] treated the SG as a complex system, locating the problems at local as well as global levels, and solving them with coordinated methods. By studying and analyzing SG, they isolated homogeneous parts with similar behaviors or objectives and applied classical optimization algorithms at different levels with coordination. The method guaranteed the flexibility in terms of system size and allowed its applicability in different scenarios and models. The optimization of each homogeneous subsystem with specific algorithms was achieved. These were then coordinated to achieve global optimization. In future, smart electrical grid will be closely intertwined with gas and heat grids, and an optimization of this holistic integration of multiple grids will also be required.

10.5 GAS TURBINE

The generation of power requires a turbine and electric generator. Generally two types of turbine—steam turbine or gas turbine—are used. In steam turbine, steam drives the turbine. The first gas turbine engine was invented by Charles Gordon Curtis in 1899 [118–120]. In gas turbine, a suitable fuel is combusted and the heat from combustion drives the turbine. In coal-based power plants, coal is burned to drive the turbine. This results in the generation of a significant amount of carbon dioxide and other pollutants which are emitted to the environment. On the other hand, when natural gas, syngas, or hydrogen is combusted to generate heat, as in gas-based power plants, significantly less carbon dioxide and other pollutants are emitted to the environment. Because of this environmental consideration along with ample supply of gas, in recent years more and more coal-based power plants have been replaced by gas-based power plants. The developments of newer and innovative gas turbines have facilitated the use of natural gas in large-scale power plants [118–143].

A *gas turbine*, also called a *combustion turbine*, is a type of internal combustion engine. It has an upstream rotating compressor coupled to a downstream turbine, and a combustion chamber in between. The basic operation of the gas turbine is similar to that of the steam power plant except that air is used instead of water to combust the fuel. Fresh atmospheric air flows through a compressor that brings it to higher pressure. An addition of fuel in air and its combustion generates high temperature gas. This high-temperature, high-pressure gas enters a turbine, where it expands down to the *exhaust pressure*, producing a *shaft work output* in the process. This shaft work drives the compressor along with an electric generator that may be coupled to the shaft. The energy that is not consumed by the shaft work comes out in the exhaust gases in the form of high velocity or high temperature. The design of gas turbine depends on its purpose such that the most desirable energy form is maximized. Besides generating static power, gas turbines are also used to power aircraft, trains, ships, electrical generators, and tanks [118,119].

10.5.1 THEORY OF OPERATION

In an ideal gas turbine, gases undergo three thermodynamic processes: an isentropic compression, an isobaric (constant pressure) combustion, and an isentropic expansion. Together, these make up the Brayton cycle for power generation. In a practical gas turbine, mechanical energy is first irreversibly transformed into heat when gases are compressed. In a subsequent combustion chamber, where heat is added, gas experiences a slight loss in pressure. During expansion amid the stator and rotor blades of the turbine, irreversible energy transformation once again occurs [118,119].

If the device has been designed to power a shaft as with an industrial generator or a turboprop, the exit pressure will be as close to the entry pressure as possible. In practice, it is necessary that some pressure remain at the outlet in order to fully expel the exhaust gases. In the case of a jet engine, only enough pressure and energy are extracted from the flow to drive the compressor and other components. The remaining high-pressure gases are accelerated to provide a jet that can, for example, be used to propel an aircraft.

Mechanically, gas turbines can be considerably less complex than internal combustion piston engines. Simple turbines might have one moving part: the shaft/compressor/turbine/alternative-rotor assembly not counting the fuel system. However, the required precision manufacturing for components and temperature-resistant alloys required for high efficiency often make the construction of a simple turbine more complex than piston engines.

A major challenge facing the turbine design is reducing the creep that is induced by the high temperatures. Turbine materials are damaged due to stress of operation. Creep becomes more important when temperature is increased to improve turbine efficiency. In practice, creep is limited by the use of thermal coatings, superalloys with solid-solution strengthening, and grain boundary strengthening for turbine blades. Creep strength can also be improved by the addition of rhenium or ruthenium to the alloy [118,119].

TABLE 10.1

Types of Gas Turbines (Prepared from Information in References 118–143)

Jet engines

Turboprop engines

Aeroderivative gas turbines

Amateur gas turbines

Auxiliary power units

Industrial gas turbines for power generation

Industrial gas turbines for mechanical drive

Turbo shaft engines

Radial gas turbines

External combustion gas turbines

Numerous types of gas turbines exist. Some of the important ones are listed in Table 10.1. These different types of gas turbines are described in detail in numerous literature publications [118–143].

10.5.2 ADVANTAGES AND DISADVANTAGES OF GAS TURBINE ENGINES

Gas turbine engines carry some advantages and some disadvantages [118–143]. They carry very high power-to-weight ratio compared to reciprocating engines. They are smaller than most reciprocating engines of the same power rating. They move in one direction only, with far less vibration than a reciprocating engine, and in general they have fewer moving parts than reciprocating engines.

Gas turbines have a greater reliability particularly in applications where sustained high power output is required. They use low operating pressures and high operation speeds. Waste heat is dissipated almost entirely in the exhaust. This results in a high-temperature exhaust stream that is very usable for boiling water in a combined cycle, or for cogeneration.

Many turbines are air cooled and in general consume less lubrication oil, resulting in lower cost. Gas turbines can run on a wide variety of fuels with very low toxic emissions of CO and HC due to excess air, complete combustion, and no "quench" of the flame on cold surfaces.

The disadvantages of gas turbines include their high cost and low efficiency compared to reciprocating engines at idle speed. Gas turbines also have a longer start-up time and are less responsive to changes in power demand compared to reciprocating engines. Finally, the characteristic whine (noise) of gas turbines can be hard to suppress.

10.5.3 GAS TURBINES IN SURFACE VEHICLES

Gas turbines are often used on ships, locomotives, helicopters, tanks, and, to a lesser extent, on cars, buses, and motorcycles. Gas turbines offer a high-powered engine in a very small and light package. However, they are not as responsive and efficient as small piston engines over the wide range of RPMs and powers needed in vehicle applications. In series hybrid vehicles, the responsiveness, poor performance at low speed, and low efficiency at low output are much less important. The turbine can be run at optimum speed for its power output, and batteries and ultracapacitors can supply additional power as needed, with the engine cycled on and off to run it only at high efficiency. The emergence of the continuously variable transmission also alleviates the problem [118–143].

Chrysler has carried out a number of experiments with gas turbine–powered automobiles. More recently, there has been some interest in the use of turbine engines for hybrid electric cars. The MTT Turbine SUPERBIKE was the first production motorcycle powered by a turbine engine producing about 283 kW (380 bhp) [126].

A turbine is theoretically more reliable and easier to maintain than a piston engine, since it has a simpler construction with fewer moving parts but in practice turbine parts experience a higher wear rate due to their higher working speeds. The turbine blades are highly sensitive to dust and fine sand, so in desert operations air filters should be fitted and changed several times daily. Gas turbines are usually multi-fuel engines, although natural gas is often preferred [118–143].

10.5.4 MARINE APPLICATIONS

Gas turbines are used in many naval vessels, where they are valued for their high power-to-weight ratio and their ships' resulting acceleration and ability to get under way quickly [118,119]. The marine gas turbine operates in a more corrosive atmosphere due to the presence of sea salt in air and fuel and use of cheaper fuels. In April 1974 Boeing launched its first passenger-carrying waterjet-propelled hydrofoil Boeing 929. It was powered by twin Allison gas turbines of the KF-501 series. Between 1971 and 1981, Seatrain Lines operated a scheduled container service with turbine-driven container ships between ports on the eastern seaboard of the United States and ports in northwest Europe [118,119].

The first passenger ferry to use a gas turbine was the GTS *Finnjet*, built in 1977 and powered by two Pratt & Whitney FT 4C-1 DLF turbines. In July 2000 the *Millennium* became the first cruise ship to be propelled by gas turbines, in a combined gas and steam turbine configuration [118–143].

10.5.5 RECENT ADVANCES IN TECHNOLOGY

Gas turbine technology has steadily advanced since its inception and continues to evolve. Development is actively producing both smaller gas turbines and more powerful and efficient engines. More advanced turbines are produced with the computer-aided design and superior high-temperature strength materials. These advances allow higher compression ratios and turbine inlet temperatures, more efficient combustion, and better cooling of engine parts. Computational fluid dynamics has contributed to substantial improvements in the performance and efficiency of gas turbine engine components. The simple-cycle efficiencies of early gas turbines are practically doubled by incorporating intercooling, regeneration (or recuperation), and reheating [118–143].

The turbine requires lower peak temperature to reduce NO_x emission with increasing inlet temperature. In May 2011, Mitsubishi Heavy Industries achieved a turbine inlet temperature of 1600°C on a 320 MW gas turbine, and 460 MW in gas turbine combined-cycle power generation applications in which gross thermal efficiency exceeded 60% [118,119]. Compliant foil bearings were commercially introduced to gas turbines in the 1990s. These can withstand over a hundred thousand start/stop cycles and have eliminated the need for an oil system. The use of a gas turbine along with a steam turbine in combined cycle power generation has significantly improved the thermal efficiency of power production. Gas turbine is thus an important tool for the expanded use of gas in the power industry, and its continuous improvement will make power generation more efficient and durable [118,119].

10.6 MICRO-TURBINES

Kurt Schreckling produced one of the world's first micro-turbines which can produce up to 22 newtons of thrust [119]. Since then, *micro-turbines* have been touted to become widespread in distributed power and combined heat and power applications. They are one of the most promising technologies for powering hybrid electric vehicles. They range from handheld units producing less than a kilowatt, to commercial-sized systems that produce tens or hundreds of kilowatts. Basic principles of micro-turbine are based on micro-combustion [118,119].

Part of micro-turbine success is due to advances in electronics, which allow unattended operation and interfacing with the commercial power grid. Electronic power switching technology eliminates the need for the generator to be synchronized with the power grid. This allows the generator to be integrated with the turbine shaft, and to double as the starter motor [144–193].

Micro-turbines are lower power machines with different applications than larger gas turbines, having typically the following characteristics: (1) variable rotation and high-frequency electric alternator, (2) reliability and simplicity, (3) compact: easy installation and maintenance, (4) air-cooled bearings, and (5) good heat recovery [184–193].

The micro-turbine variable speed is between 30,000 and 120,000 rpm depending on the manufacturer. The generator operates with a converter for AC/DC. In addition, the alternator itself is the engine starter. Some micro-turbines have operated for three years, including shutdown and maintenance. The generator is placed in the same turbine shaft, making it easy to be manufactured and maintained. It has great potential for inexpensive and large-scale manufacturing. In order to reduce noise levels during operation, micro-turbines require a specific acoustic system [144–152].

The use of air bearings for cooling avoids lubricant contamination by combustion products, and it prolongs the equipment's useful life and reduces maintenance costs. Micro-turbine manufacturers generally use heat recovery of exhaust gas to heat the air intake of the combustion chamber, thus achieving a thermal efficiency of 30% [144–193].

Micro-turbines have similar setup of small-, medium-, and large-size gas turbines, as described by Nascimento and Santos [161].

State-of-the-art micro-turbines have markedly improved in recent years. In recent years, several micro-turbines with different configurations have been developed. While the configuration depends on the application, it usually consists of a single-shaft micro-turbine, annular combustor, single-stage radial flow compressor and expander, and sometimes a recuperator or regenerator. The optimum micro-turbine rotational speeds at typical power ratings are between 60 and 90,000 rpm and pressure ratio of 3 or 4:1, in a single stage [155–162].

Gas micro-turbines have the same basic operation principle as open cycle gas turbines (Brayton open cycle). In this cycle, the incoming air is first compressed. The temperature of the compressed air is increased in the combustion chamber by fuel burning. The high-temperature working fluid is then expanded in the turbine which generates power for the compressor and the electric generator. The application of microelectronics and power switching technology have enabled the development of commercially viable electricity generation by micro-turbines for distribution and vehicle propulsion [162,170,172,173,181,187].

Micro-turbine designs usually consist of a single-stage radial compressor, a single-stage radial turbine, and a recuperator. Recuperators are difficult to design and manufacture because they operate under high pressure and temperature differentials. Exhaust heat can be used for water heating, space heating, drying processes, or absorption chillers, which create cold for air-conditioning from heat energy instead of electric energy. Typical micro-turbine efficiencies are 25%–35%. When in a combined heat and power cogeneration system, efficiencies of greater than 80% are commonly achieved [144–193].

Micro-combustion processes can result in the formation of significant amounts of nitrogen dioxide (NO_2) and carbon monoxide (CO). As shown later, some manufacturers of micro-turbines have developed advanced combustion technologies to minimize the formation of these pollutants. They have assured low emissions levels from micro-turbines fueled with gaseous and liquid fuels. Some companies in the United States, England, and Sweden have recently introduced in the world market commercial units of micro-turbines. These companies include: AlliedSignal, Elliott Energy Systems, Capstone, Ingersoll-Rand Energy Systems & Power Recuperators Works™, Turbec, Browman Power, and ABB Distributed Generation & Volvo Aero Corporation [118,119,152–154].

10.6.1 Micro-Turbines versus Reciprocating Engines

Micro-turbine systems have many claimed advantages over reciprocating engine generators, such as higher power-to-weight ratio, low emissions, and few, or just one, moving part. Micro-turbines can be designed with foil bearings, air bearings or magnetic bearings [118,119,152–154], and air-cooling, and do not require lubricating oil, coolants, or other hazardous materials. Micro-turbines also have waste heat contained in the relatively high-temperature exhaust making it simpler to capture, and use effectively in cogeneration processes. The waste heat of reciprocating engines, on the other hand, is split between its exhaust and cooling system [118,119,152–154].

The reciprocating engine generators are, however, quicker to respond to changes in output power requirement and are usually slightly more efficient, although the efficiency of micro-turbines is increasing with new developments. Micro-turbines also lose more efficiency at low power levels than reciprocating engines. With all factors considered, at present reciprocating engines are somewhat cheaper than micro-turbines. More research is being carried out to reduce the cost of micro-turbines. Micro-turbines are energy generators whose capacity ranges from 15 to 300 kW [118,119,152–154].

10.6.2 Micro-Turbines in Transportation Industry

Micro-turbines came into the automotive market between 1950 and 1970. The first micro-turbines were based on gas turbines designed to be used in generators of missile launching stations, aircraft and bus engines, among other commercial means of transport. The use of this equipment in the energy market increased between 1980 and 1990, when the demand for distributed generating technologies increased as well [159,160].

When gas turbines are used in extended-range electric vehicles, very poor throttling response does not matter because the gas turbine, which may be spinning at 100,000 rpm, is not directly, mechanically connected to the wheels. While micro-turbine can accept most commercial fuels, such as petrol, natural gas, propane, diesel, and kerosene as well as renewable fuels such as E85, biodiesel, and biogas, natural gas, or propane is preferred for micro-combustion [118,119,152–154,184].

10.6.3 State-of-the-Art Micro-Turbines

There are numerous state-of-the-art micro-turbines in the market. Here we discuss three of them.

AlliedSignal micro-turbine has shaft configuration and works with cycle Regenerative open Brayton; its bearings are pneumatic, and it has a drive direct current–alternating current (DC/AC) 50/60 Hz (the frequency is reduced from about 1200 to 50 Hz or 60 Hz), and the compressor and turbine are the radial single stage. The heat transfer efficiency of this stainless steel regenerator is 80%–90% [152–154].

Elliott Energy Systems, Stuart, Florida, has launched two commercial prototypes: a 45 kW micro-turbine (TA-45 model) and another 80 kW (TA-80), and, later, a 200 kW micro-turbine (TA-200). The TA-45 model has oil-lubricated bearings and a system starting at 24 V, which is unique to micro-turbines. The TA-80 and TA-200 micro-turbines models are similar to the TA-45 model. All three can generate electricity in 120/208/240V and can work with different fuels: natural gas, diesel, kerosene, alcohol, gasoline, propane, methanol, and ethanol [151,171,173,175–178].

The joint venture of ABB and Volvo Aero Corp. developed a micro-turbine for cogeneration. Operating on natural gas, the MT100 micro-turbine generates 100 kW of electricity and 152 kW of thermal energy (hot water). As other manufacturers of micro-turbines, the MT100 has a frequency converter that allows the generator to operate at variable speed [118,119,152–154,187].

10.6.4 Methods to Improve Gas Micro-Turbine Performance

Micro-turbines are a technology-based cycle with or without recuperation. To produce an acceptable efficiency, the heat in the turbine exhaust system must be partially recovered and used to preheat the turbine air supply before it enters the combustor, using an air-to-air heat exchanger called recuperator or regenerator. This allows the net cycle efficiency to be increased to as much as 30%, while the average net efficiency of unrecovered micro-turbines is 17% [185,186,188].

Just as in gas turbines, the maximum net power provided by a micro-turbine depends on (1) temperature the material of the turbine can support, (2) cooling technology and (3) service life required. The two main factors affecting the performance of micro-turbines are components efficiency and gases, temperature at the turbine inlet. Furthermore, micro-turbines usually employ permanent magnet variable-speed alternators generating very-high-frequency alternating current which must be first rectified and then converted to AC to match the required supply frequency.

Capstone Micro-turbines use a lean premix combustion system to achieve low emissions levels at a full-power range. Lean premix operation requires operating at high air–fuel ratio within the primary combustion zone. The large amount of air is thoroughly mixed with fuel before combustion. This premixing of air and fuel enables clean combustion to occur at a relatively low temperature. Injectors control the air–fuel ratio and the air–fuel mixture in the primary zone to ensure that the optimal temperature is achieved for the NO_x minimization. The higher air–fuel ratio results in a lower flame temperature, which leads to lower NO_x levels [118,119,152–154].

In order to achieve low levels of CO and hydrocarbons simultaneously with low NO_x levels, the air–fuel mixture is retained in the combustion chamber for a relatively long period. This process allows for a more complete combustion of CO and hydrocarbons [151,152]. Nitrogen oxides (NO_x) and carbon monoxide (CO) emission levels of Capstone micro-turbines are lower than 7 ppmv at 15%O_2 at full load when these micro-turbines are fueled with natural gas. Micro-turbines exhibit low emissions of all classes of pollutants and have environmental benefits as they release fewer emissions compared to other distributed generation technologies, like internal combustion engines [152–154].

In addition, the exhaust of micro-turbines can be used in direct heating or as an air pre-heater for downstream burners, once it has a high concentration of oxygen. Clean burning combustion is the key to both low emissions and highly durable recuperator designs.

The most effective fuel to minimize emissions is clearly natural gas. Natural gas is also the fuel choice for small businesses. Usually natural gas requires compression to the ambient pressure at the compressor inlet of the micro-turbine. The compressor outlet pressure is normally three to four atmospheres [151–154,172–181].

Capstone micro-turbine control and power electronic systems allow for different operation modes, such as grid connect, stand-alone, dual mode and multiple units for potentially enhanced reliability, operating with gas, liquid fuels, and biogas. In grid connect, the system follows the voltage and the frequency from the grid. Grid connect applications include baseload, peak shaving, and load following. One of the key aspects of a grid connect system is that the synchronization and the protective relay functions required to reliably and safely interconnect with the grid can be integrated directly into the micro-turbine control and power electronic systems. This capability eliminates the need for very expensive and cumbersome external equipment needed in conventional generation technologies [185,186]. In the stand-alone mode, the system behaves as an independent voltage source and supplies the current demanded by the load. Capstone micro-turbine when equipped with the stand-alone option includes a large battery used for unassisted start of the turbine engine and for transient electrical load management [152–154].

In both the grid connect and the stand-alone operational modes, the micro-turbine can also be designed to automatically switch between these two modes. This type of functionality is extremely useful in a wide variety of applications, and is commonly referred to as dual mode operation. Besides, the micro-turbines can be configured to operate in parallel with other distributed generation systems in order to obtain a larger power generation system. This capability can be built directly into the system and does not require the use of any external synchronizing equipment [152–154].

Some micro-turbines can operate with different fuels. The flexibility and the adaptability enabled by digital control software allow this to happen with no significant changes to the hardware. Power generation systems create large amounts of heat in the process of converting fuel into electricity.

For the average utility-size power plant, more than two-thirds of the energy content of the input fuel is converted into heat. Conventional power plants discard this waste heat; however, distributed generation technologies, due to their load-appropriate size and sitting, enable this heat to be recovered. Cogeneration systems can produce heat and electricity at or near the load side. Cogeneration plants usually have up to 85% of efficiency and operation cost lower than other applications. Small cogeneration systems usually use reciprocating engines, although micro-turbines have showed to be a good option for this application. The hot exhaust gas from micro-turbines is available for cogeneration applications. Recovered heat can be used for hot water heating or low-pressure steam applications [152–154].

MIT started its millimeter-size turbine engine project in the middle of the 1990s. Problems occurred with heat dissipation and high-speed bearings in these new micro-turbines. MIT's millimeter-size turbine delivers 500–700 Wh/kg in the near term, rising to 1200–1500 Wh/kg in the longer term [118,119]. A similar micro-turbine built in Belgium has a rotor diameter of 20 mm and produces about 1000 W [118,119,152–154].

In summary, micro-turbines are clean enough to be placed in a community with residential and commercial buildings. Micro-turbine generators have shown good perspectives for electricity-distributed generation in small scales. Once they have high reliability and simple design they show high potential for large-scale cheap manufacturing. The most profitable investment in micro-turbines is in the cogeneration case. Improved micro-turbines continue to advance the role of gas in heat, power, and vehicle industries.

10.7 COMBINED CYCLE AND IGCC TECHNOLOGY FOR POWER GENERATION

In order to improve thermal efficiency of the power generation from coal or natural gas; biomass/waste; nuclear, solar, or geothermal energy, combined cycle power plants are actively pursued. The basic principle here is to improve thermodynamic Carnot efficiency of the power production from heat removal process in gas turbine by generating additional power (in combined cycle) from waste heat coming out of turbine. When this combined cycle is combined with coal or biomass gasification, it is called integrated gasification and combined cycle system. In this section, we examine combined cycle and integrated gasification and combined cycle for power generation [194–233].

10.7.1 Combined Cycle Power Plants

In electric power generation, a *combined cycle* is an assembly of heat engines that work in tandem from the same source of heat, converting it into mechanical energy, which in turn usually drives electrical generators. The principle is that after completing its cycle (in the first engine), the working fluid of the first heat engine has enough waste heat left to generate additional power by another cycle. By combining these multiple streams of work upon a single mechanical shaft turning an electric generator, the overall net efficiency of the system may be increased by 50%–60%; that is, from an overall efficiency of, say, 34% (in a single cycle) to possibly an overall efficiency of 51% (in a mechanical combination of two [2] cycles) in net Carnot thermodynamic efficiency. This can be done because heat engines are able to use only a portion of the energy their fuel generates (usually less than 50%). In principle, cycles can be repeated multiple times if the waste heat coming out of previous cycle has high enough temperature [194–203].

Combining two or more thermodynamic cycles results in improved overall efficiency, reducing fuel costs for the same level of power generation. In stationary power plants, a widely used combination is a gas turbine (operating by the Brayton cycle) burning natural gas or synthesis gas from coal, whose hot exhaust powers a steam power plant (operating by the Rankine cycle). This is called a combined cycle gas turbine (CCGT) plant, and can achieve a real thermal efficiency of around 54% in baseload operation, in contrast to a single-cycle steam power plant which is limited to efficiencies of around 35%–42%. Many new gas power plants in North America and Europe are of this type.

Such an arrangement is also used for marine propulsion, and is called a *combined gas and steam* (*COGAS*) plant. Multiple stage turbine or steam cycles are also often used [194–203].

The efficiency of a heat engine, the fraction of input heat energy that can be converted to useful work, is limited by the temperature difference between the heat entering the engine and the exhaust heat leaving the engine. In a thermal power station, water is the working medium. High-pressure steam requires strong, bulky components. Also, high temperatures require expensive alloys made from nickel or cobalt, rather than inexpensive steel. These alloys limit practical steam temperatures to 655°C, while the lower temperature of a steam plant is fixed by the temperature of the cooling water. With these limits, a steam plant has a fixed upper efficiency of 35%–42% [194–203].

An open circuit gas turbine cycle, on the other hand, has a compressor, a combustor, and a turbine. For gas turbines the amount of metal that must withstand the high temperatures and pressures is small, and lower quantities of expensive materials can be used. In this type of cycle, the input temperature to the turbine (the firing temperature) is relatively high (900°C–1400°C). The output temperature of the flue gas is also high (450°C–650°C). This is therefore high enough to provide heat for a second cycle which uses steam as the working fluid (a Rankine cycle). In a combined cycle power plant, the heat of the gas turbine's exhaust is used to generate steam by passing it through a heat recovery steam generator (HRSG) with a live steam temperature between 420°C and 580°C. The condenser of the Rankine cycle is usually cooled by water from a lake, river, sea, or cooling towers. This temperature can be as low as 15°C [194–203].

For large-scale power generation, a typical set would be a 270 MW gas turbine coupled to a 130 MW steam turbine giving 400 MW. A typical power station might consist of between 1 and 6 such sets. Plant size is important in the cost of the plant. The larger plant sizes benefit from economies of scale (lower initial cost per kilowatt) and improved efficiency. Gas turbines of about 150 MW size are already in operation manufactured by at least four separate groups—General Electric and its licensees, Alstom, Siemens, and Westinghouse/Mitsubishi. These groups are also developing, testing, and/or marketing gas turbine sizes of about 200 MW.

The heat recovery boiler is item 5 in the COGAS (Figure 10.1) [194,195]. No combustion of fuel means that there is no need for a fuel handling plant and it is simply a heat exchanger. Exhaust enters the superheater and the evaporator and then the economizer section as it flows out from the boiler.

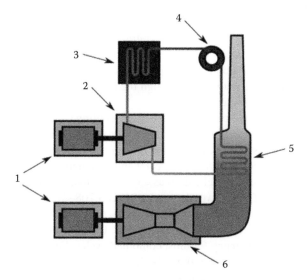

FIGURE 10.1   Working principle of a combined cycle power plant (1, electric generators; 2, steam turbine; 3, condenser; 4, pump; 5, boiler/heat exchanger; 6, gas turbine). (From Combined Cycle, Wikipedia, the free encyclopedia, 2015.)

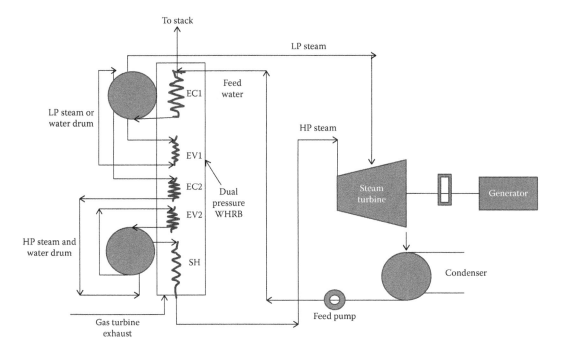

FIGURE 10.2 Steam turbine plant lay out with dual-pressure heat recovery boiler. (From Combined Cycle, Wikipedia, the free encyclopedia, 2015.)

Feed water comes in through the economizer and then exits after having attained saturation temperature in the water or steam circuit. Finally it then flows through the evaporator and superheater. If the temperature of the gases entering the heat recovery boiler is higher, then the temperature of the exiting gases is also high [195–197].

When high heat is recovered from exit gases from turbine, often a dual-pressure boiler (see Figure 10.2) is employed. It has two water/steam drums. Low-pressure drum is connected to low-pressure economizer or evaporator. The low-pressure steam is generated in low-temperature zone. The low-pressure steam is supplied to the low-temperature turbine. Superheater can be provided in the low-pressure circuit. Some part of the feed water from the low-pressure zone is transferred to the high-pressure economizer by a booster pump. This economizer heats up the water to its saturation temperature. This saturated water goes through the high-temperature zone of the boiler and is supplied to the high-pressure turbine [194–203].

The efficiency of CCGT and GT can be boosted by precooling combustion air, which is practiced in hot climates and has the effect of increasing power output. This can be achieved by evaporative cooling of water using a moist matrix placed in front of the turbine, or by using ice storage air-conditioning. The latter has the advantage of greater improvements due to the lower temperatures available. Furthermore, ice storage can be used as a means of load control or load shifting since ice can be made during periods of low power demand [194–203,208–212].

Supplementary firing may be used in combined cycles (in the HRSG) raising exhaust temperatures from 600°C (GT exhaust) to 800°C or even 1000°C. Using supplemental firing will, however, not raise the combined cycle efficiency for most combined cycles. For single boilers it may raise the efficiency if fired to 700°C–750°C; for multiple boilers, however, supplemental firing is often used to improve peak power production of the unit, or to enable higher steam production to compensate for failure of a second unit. It is possible to add more fuel to the turbine's exhaust because the turbine exhaust gas (flue gas) still contains some oxygen. Maximum supplementary firing refers to the maximum fuel that can be fired with the oxygen available in the gas turbine exhaust. The HRSG can be designed with supplementary firing of fuel after the gas turbine in order to increase the quantity or temperature of the steam generated. Supplementary burners are also called *duct burners* [194–203,208–212].

In principle, the combined cycle concept can be applied to heat generated by coal, oil, gas, biomass/waste, nuclear solar, or geothermal energy. At present, the turbines used in combined cycle plants are commonly fueled with natural gas. Due to increase in shale gas supply, it is becoming the fuel of choice. Combined cycle plants can be filled with biogas derived from agricultural and forestry waste, which is often readily available in rural areas.

10.7.1.1 Single-Shaft versus Multi-Shaft Options

A single-shaft combined cycle plant comprises a gas turbine and a steam turbine driving a common generator. In a multi-shaft combined cycle plant, each gas turbine and each steam turbine have their own generator. The single-shaft design provides slightly less initial cost and slightly better efficiency than if the gas and steam turbines had their own generators. The multi-shaft design enables two or more gas turbines to operate in conjunction with a single steam turbine, which can be more economical than a number of single-shaft units [194–203,208–212].

The primary disadvantage of single-shaft combined cycle power plants is that the number of steam turbines, condensers, and condensate systems—and perhaps the number of cooling towers and circulating water systems—increases to match the number of gas turbines. For a multi-shaft combined cycle power plant there is only one steam turbine, condenser, and the rest of the heat sink for up to three gas turbines. The use of only one large steam turbine and heat sink results in low cost. A larger steam turbine also allows the use of higher pressures and results in a more efficient steam cycle. Thus the overall plant size and the associated number of gas turbines required have a major impact on whether a single-shaft combined cycle power plant or a multishaft combined cycle power plant is more economical [194–203,208–212].

Key advantages of the single-shaft arrangement are operating simplicity, smaller footprint, and lower start-up cost. Single-shaft arrangements, however, will tend to have less flexibility and equivalent reliability than multishaft blocks. Multishaft systems have one or more gas turbine-generators and HRSGs that supply steam through a common header to a separate single steam turbine-generator. In terms of overall investment, a multishaft system is about 5% higher in costs [194–203,208–212].

10.7.2 Integrated Combined Cycles

Unlike natural gas–based combined cycle power plants, the integrated gasification combined cycle (IGCC) produces electricity from a solid (coal or biomass) or liquid fuel. First, the fuel is converted to syngas which is a mixture of hydrogen and carbon monoxide. Second, the syngas is converted to electricity in a combined cycle power block consisting of a gas turbine process and a steam turbine process which includes an HRSG. The combined cycle technology is similar to the technology used in modern natural gas–fired power plants [194–203].

Coal-based IGCC plants are still not fully commercial. A number of demonstration plants with electric output up to 300 MW have been built in Europe and the United States, all with financial support from government. The motivation for pursuing this technology is the potential for better environmental performance at a low marginal cost. This is especially true for mercury removal and CO_2 capture. In order to compete with conventional pulverized coal (PC)-fired plants under current environmental regulation, the main challenges facing the IGCC technology today are capital cost and availability [212–221].

IGCC is an advanced power generation concept with the flexibility to use coal, heavy oils, petroleum coke, biomass, and waste fuels to produce electric power as a primary product. IGCC systems typically produce sulfur as a by-product. Systems that produce many coproducts are referred to as "poly-generation" systems. IGCC systems are characterized by high thermal efficiencies and lower

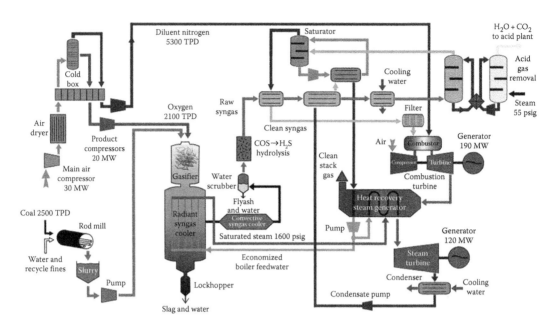

FIGURE 10.3 A schematic of IGCC plant. (From Integrated gasification combine cycle, Wikipedia, the free encyclopedia, 2015.)

environmental emissions than conventional PC-fired plants [229]. IGCC plant is the synthetic gas–based power plant and it is as efficient as combined cycle natural gas–based power plant.

A generic IGCC system is illustrated schematically in Figure 10.3. In an IGCC power plant, the feedstock to the gasifier is converted to a syngas, composed mainly of hydrogen and carbon monoxide, using a gasification process. After passing through a gas cleanup system, in which particles and soluble gases are removed via wet scrubbing and in which sulfur is removed and recovered via a selective removal process, the syngas is utilized in a combined cycle power plant. Different variations of IGCC systems exist based upon the type of coal gasifier technology, oxidant (e.g., oxygen or air), and gas cleanup system employed. A typical IGCC system includes process sections of fuel handling, gasification, high-temperature gas cooling, low-temperature gas cooling and gas scrubbing, acid gas separation, fuel gas saturation, gas turbine, heat recovery steam generator, steam turbine, and sulfur by-product recovery. The specific design of each of the process sections such as gasification and high temperature gas cooling varies in different IGCC systems. Three generic designs of gasification reactors—fixed or moving bed, fluidized bed, and entrained bed—are typically employed in IGCC systems. In all types of reactors, the feedstock fuel is converted to syngas in reactors with an oxidant and either steam or water. The oxidant is required to partially oxidize the fuel. The exothermic oxidation process provides heat for the endothermic pyrolysis reactions. Water or steam is used as a source of hydrolysis in the gasification reactions. The type of reactor used is the primary basis for classifying different types of gasifiers [194,195,206–226].

The gasification process can produce syngas from a wide variety of carbon-containing feedstocks, such as high-sulfur coal, heavy petroleum residues, and biomass waste. The plant is called *integrated* because (1) the syngas produced in the gasification section is used as fuel for the gas turbine in the combined cycle, and (2) steam produced by the syngas coolers in the gasification section is used by the steam turbine in the combined cycle. In this example, the syngas produced is used as fuel in a gas turbine which produces electrical power. In a normal combined cycle, so-called waste heat from the gas turbine exhaust is used in an HRSG to make steam for the steam turbine cycle. An IGCC plant improves the overall process efficiency by adding the higher-temperature steam produced by the gasification process to the steam turbine cycle. This steam is then used in steam turbines to produce additional electrical power. Current gasification technologies are detailed in a recent IEA Clean Coal Centre Report [200], and the principle of IGCC has been described many times [230,232]. Gas cleaning is typically undertaken by water scrubbing or the dry removal of solids, followed by hydrolysis of COS to H_2S and then scrubbing to remove H_2S. There are many possible plant configurations, because gasifier designs vary significantly and IGCC has a large number of process areas that can use different technologies. The deep cleaning needed to protect the gas turbine enables emissions of particulates and SO_2 to be very low [230]. Totally dry gas clean-up at elevated temperatures ("hot gas clean-up") may eventually be applied, with advantages in efficiency. More details on options for gasification conditions for syngas production, types of gasifiers that can be used, and technologies for syngas purifications were described earlier in Chapters 3, 5, and 7.

The concept of IGCC technology has been adopted by following four major gasification processes [195]:

1. Sasol-Lurgi Dry Ash
2. GE (originally developed by Texaco)
3. Shell
4. ConocoPhillips E-gas (originally developed by Dow)

GE, Shell, and ConocoPhillips are all perceived as the three major players with respect to future IGCC projects, which seem to concentrate on entrained flow slagging gasifiers. Fluidized bed gasifiers are less suitable for IGCC concept. In 2004, the U.S. DOE financed a $235 million grant to Southern Company's future 285 MW Orlando IGCC project in Florida which was based on the KBR Transport reactor. This type of gasifier has been developed at smaller scale, and it is potentially well suited for low-rank coals with high moisture and ash contents. IGCC is a high-efficiency power generation technology and compared with conventional PC-fired power plants IGCC has potentially many advantages, including [194–233] (1) high thermal efficiency, (2) good environmental characteristics, and (3) reduced water consumption.

Shell estimate an IGCC generation efficiency based on their gasifier of 46%–47% net, LHV basis, for bituminous coals with a gas turbine [231]. The plant's high thermal efficiency means that emissions of CO_2 are low per unit of generated power. The emission of CO_2 can be further reduced by adding carbon capture sequestration technology. In addition, emissions of SO_x and particulates are reduced by the requirement to deep-clean the syngas before firing in the gas turbine. IGCC uses less water since 60% of its power is derived from an air-based Brayton cycle reducing the heat load on the steam turbine condenser to only 40% of that of an equivalent-rated PC-fired plant. Additionally, through the direct desulphurization of the gas, IGCC does not require a large flue gas desulphurization unit which consumes large amounts of water. Further gains in reducing water use can also be achieved when CCS is incorporated into the plant [211,212,214,219,231,232].

The DOE Office of Fossil Energy considers that future gasification concepts that merit study include those that offer significant improvements in efficiency, fuel flexibility, economics, and environmental sustainability. Fuel flexibility is considered to be especially important given that future gasification plants could conceivably process a wide variety of low-cost feedstocks, in addition to coal, such as biomass, municipal, and other solid wastes, or combinations of these feedstocks [199,220,222]. DOE, is also studying the so-called transport reactor based on an advanced circulating fluidized bed reactor, in which a chemical sorbent can be added to capture sulfur impurities. New processes should also include improved technologies for carbon capture sequestration [211,212,214,219,231,232].

Another important area for research has been the development of efficient and economical oxygen separation technologies. Currently, producing oxygen involves a complex, energy-intensive cryogenic process. A number of other lower-cost alternatives based on innovations in ceramic membranes to separate oxygen from the air at elevated temperatures are being explored by the DOE.

Membrane research is also concentrating on less expensive materials that can selectively remove hydrogen from syngas so that it can be used as a fuel for turbines, future fuel cells, or refineries, or in hydrogen-powered vehicles. Other gas separation research is focused on removing carbon dioxide from syngas. Research is continuing into new types of pollutant-capturing sorbents that work at elevated temperatures and do not degrade under the harsh conditions of a gasification system. Also, new types of gas filters and novel cleaning approaches are being examined [197,219,230,231].

Gasification produces less solid wastes than other coal-based power generation options, and these wastes can have commercial value. Gasifier slag is being used for road construction and investigations are under way to use the solid material produced when coal and other feedstocks (e.g., biomass, municipal waste) are utilized in the gasification process. Some plants produce sulfur of sufficient purity for sale as a commercial product. Thus, most IGCC developments tend to be evolutionary in nature, building upon the performance of established components and materials. The requirements for future plants are therefore concentrated generally on further developments of those systems; particularly larger and more efficient gas turbines, higher duty steam cycles, more efficient oxygen separation processes including ion membrane technologies in the longer term, and improvements to ancillary components—for example, solids pumps [232,233].

A modification of conventional IGCC plant is an ISCC, INCC or IGECC plant in which a solar heat, nuclear heat or geothermal heat is integrated within combined cycle plants. In ISCC plants, solar energy is used as an auxiliary heat supply, supporting the steam cycle, which results in increased generation capacity or a reduction of fossil fuel use [8]. The first such system to come online was the Archimede solar power plant, Italy, in 2010, followed by Martin Next Generation Solar Energy Center in Florida, and in 2011 by the Kuraymat ISCC Power Plant in Egypt, Yazd power plant in Iran, Hassi R'mel in Algeria, and Ain Beni Mathar in Morocco [193,194,202]. Similarly, nuclear heat, geothermal heat can be a source of auxiliary source of heat in INCC and IGECC plants [194,195].

Combined cycles have traditionally only been used in large power plants. BMW, however, has proposed that automobiles use exhaust heat to drive steam turbines [194,195]. This can even be connected to the car or truck's cooling system to save space and weight, but also to provide a condenser in the same location as the radiator and preheating of the water using heat from the engine block. The exhaust heat can also be used to generate electricity by a thermo electric device. It may be possible to use the pistons in a reciprocating engine for both combustion and steam expansion as in the Crower engine [194,195]. A turbocharged car is also a combined cycle [194,195,223].

10.8 COMBINED HEAT AND POWER GENERATION–COGENERATION AND TRIGENERATION

Cogeneration or combined heat and power (*CHP*) is the use of a heat engineer power station to generate electricity and useful heat at the same time [234–289]. *Trigeneration* (see Figure 10.4) or *combined cooling, heat, and power* (*CCHP*) refers to the simultaneous generation of electricity and useful heating and cooling from the combustion of a fuel or a solar heat collector or heat generated by a nuclear reactor or heat recovered from enhanced geothermal system [1,234,252–265].

Cogeneration is a thermodynamically efficient use of fuel. During the production of power, waste heat is discarded in the environment through cooling towers, flue gases, or by other means. In contrast, CHP captures all the waste heat from the power plant for other useful purposes such as desalination, district heating, and industrial heating close to the power plant. For example, in Scandinavia and Eastern Europe, it is used for district heating with water temperatures ranging from approximately 80°C to 130°C. This is also called *combined heat and power district heating* (*CHPDH*). The waste heat at moderate temperatures (100°C–180°C, 212°F–356°F) can also be used in absorption refrigerators for cooling [234–289].

In 1882, Thomas Edison's Pearl Street Station, the world's first commercial power plant, was a combined heat and power plant, producing both electricity and thermal energy while using waste heat to warm neighboring buildings [234]. The use of waste energy allowed Edison's plant to

Trigeneration

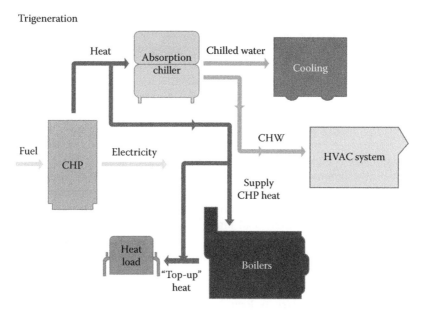

FIGURE 10.4 Trigeneration cycle. (From Cogeneration, Wikipedia, the free encyclopedia, 2015.)

FIGURE 10.5 The 250 MW Kendall Cogeneration Station plant in Cambridge, MA. (From Cogeneration, Wikipedia, the free encyclopedia, 2015.)

achieve approximately 50% overall thermal efficiency. During late 1970s cogeneration plants produced about 8% of all energy in the United States [50,51,234]. One of the first 250 MW Kendall cogeneration plant in Cambridge, Massachusetts, is shown in Figure 10.5.

The concept of cogeneration can be applied to large nuclear power plant, midsize solar power plant or large, small, or micro natural gas power plant. The concept can also be applied to high-temperature geothermal heat coming out of enhanced geothermal systems. As shown in my previous books [1,2], the application of waste heat depends on the temperature of the waste heat. At smaller scales (typically below 1 MW), a gas engine or diesel engine can be used. Trigeneration (CCHP) involves the use of waste heat for both heating and cooling, for example, by absorption refrigerators. CCHP systems can attain higher overall efficiencies than cogeneration or traditional power plants. In the United States, the application of trigeneration in buildings is called *building cooling, heating, and power (BCHP)*. Heating and cooling output may operate concurrently or alternately depending on the need and system construction [234–289].

Figure 10.4 illustrates a typical trigeneration cycle. Trigeneration has its greatest benefits when scaled to fit buildings or complexes of buildings where electricity, heating, and cooling are perpetually needed. Such installations include, but are not limited to data centers, manufacturing facilities, universities, hospitals, military complexes, and colleges. Localized trigeneration has additional benefits of lower power usage costs and the ability to sell electrical power back to the local utility. Even for small buildings such as individual family homes trigeneration systems provide benefits over cogeneration because of increased energy utilization [234–289]. This increased efficiency can also provide significant reduced GHG emissions, particularly for new communities [234–287].

Most industrial countries generate the majority of their electrical power needs in large centralized facilities with capacity for large electrical power output. These plants have excellent economies of scale, but usually transmit electricity long distances resulting in sizable losses and negatively affecting the environment. Large power plants can use cogeneration or trigeneration systems only when sufficient need exists in immediate geographic vicinity for an industrial complex, additional power plant, or a city. An example of cogeneration with trigeneration applications in a major city is the New York City steam system.

Thermal efficiency in a trigeneration system is defined as [234–289]:

$$\eta_{th} \equiv \frac{W_{out}}{Q_{in}} \equiv \frac{\text{Electrical power output} + \text{Heat output} + \text{Cooling output}}{\text{Total heat iutput}}$$

where

η_{th} is the thermal efficiency
W_{out} is the total work output by all systems
Q_{in} is the total heat input into the system

Typical trigeneration models have losses as in any system. The energy distribution below is represented as a percent of total input energy [234–289]:

Electricity = 45%
Heat + Cooling = 40%
Heat Losses = 13%
Electrical Line Losses = 2%

Thus trigeneration systems can give in excess of 80% of thermal efficiency.

Cogeneration was practiced in some of the earliest installations of electrical generation. Industries generating their own power used exhaust steam for process heating. Large office and apartment buildings, hotels, and stores commonly generated their own power and used waste steam for building heat. In the United States, Consolidated Edison distributes 66 billion kg of 350°F (180°C) steam each year through its seven cogeneration plants to 100,000 buildings in Manhattan—the biggest steam district in the United States. The peak delivery is 10 million pounds per hour (or approximately 2.5 GW) [234].

Cogeneration is still common in pulp and paper mills, refineries, and chemical plants. In this "industrial cogeneration/CHP," the heat is typically recovered at higher temperatures (above 100°C) and used for process steam or drying duties. This is more valuable and flexible than low-grade waste heat, but there is a slight loss of power generation. The increased focus on sustainability has made industrial CHP more attractive, as it substantially reduces carbon footprint compared to generating steam or burning fuel on-site and importing electric power from the grid.

An HRSG is a steam boiler that uses hot exhaust gases from the gas turbines or reciprocating engines in a CHP plant to heat up water and generate steam. The steam, in turn, drives a steam turbine or is used in industrial processes that require heat. HRSGs used in the CHP industry are different from conventional steam generators in that they are designed based on the specific features of its coupled gas turbine or reciprocating engine. Since the exhaust gas temperature and gas velocity are relatively low, its major mode of convective heat transfer requires high heat transfer surface area. Generally this is accomplished by the use of plate-fin heat exchangers.

Industrial cogeneration plants normally operate at much lower boiler pressures than utilities. Cogeneration plants face possible contamination of returned condensate. Because boiler feed water from cogeneration plants has much lower return rates than 100% condensing power plants, industries usually have to treat proportionately more boiler makeup water. Boiler feed water must be completely oxygen-free and demineralized, and the higher the pressure, the more critical the level of purity of the feed water [234–289].

Utilities, on the other hand, are typically larger scale power than industry, which helps offset the higher capital costs of high pressure. Utilities are less likely to have sharp load swings than industrial operations, which deal with shutting down or starting up units that may represent a significant percent of either steam or power demand [234–289].

CHP plants can be operated in a number of different ways depending on the local need of power and heat and the desired overall thermal efficiency. Steam turbines for cogeneration are designed for *extraction* of steam at lower pressures after it has passed through a number of turbine stages, or they may be designed for final exhaust at *back pressure* (noncondensing), or both. Some tricycle plants have used a combined cycle in which several thermodynamic cycles produced electricity; then a heating system was used as a condenser of the power plant's bottoming cycle. For example, the RU-25 MHD generator in Moscow heated a boiler for a conventional steam power plant, whose condensate was then used for space heat. A more modern system might use a gas turbine powered by natural gas, whose exhaust powers a steam plant, whose condensate provides heat. Tricycle plants can have thermal efficiencies above 80% [234–289].

The viability of CHP, especially in smaller CHP installations, depends on an on-site (or near-site) electrical demand and heat demand. A CHP plant can either meet the need for heat (*heat driven operation*) or be run as a power plant with some use of its waste heat. The viability of CHP can be greatly increased where opportunities for trigeneration exist. In such cases, the heat from the CHP plant is also used to deliver cooling. CHP is most efficient when heat can be used on-site or very close to it. A car engine becomes a CHP plant in winter when the reject heat is useful for warming the interior of the vehicle. Thermally enhanced oil recovery (TEOR) plants often produce a substantial amount of excess electricity. After generating electricity, these plants pump leftover steam into heavy oil wells so that the oil will flow more easily, increasing production. TEOR cogeneration plants in Kern County, California, produce so much electricity that it cannot all be used locally and is thus transmitted to Los Angeles [234–289].

Cogeneration can also be carried out in two ways. Topping cycle plants primarily produce electricity from a steam turbine. The exhausted steam is then condensed and the low-temperature heat released from this condensation is utilized, for example, district heating or water desalination. Bottoming cycle plants produce high-temperature heat for industrial processes; then a waste heat recovery boiler feeds an electrical plant. Bottoming cycle [281] plants are used only when the industrial process requires very high temperatures such as furnaces for glass and metal manufacturing, so they are less common [234–289].

Large cogeneration systems provide heating water and power for an industrial site or an entire town. Common CHP plant types are as follows [234–289]:

1. Gas turbine CHP plants using the waste heat in the flue gas of gas turbines. The fuel used is typically natural gas.
2. Gas engine CHP plants using a reciprocating gas engine which is generally more competitive than a gas turbine up to about 5 MW. The gaseous fuel used is normally natural gas [194–197].
3. Biofuel engine CHP plants using an adapted reciprocating gas engine or diesel engine, depending upon which biofuel is being used, and are otherwise very similar in design to a gas engine CHP plant.
4. Combined cycle power plants adapted for CHP or steam turbine CHP plants that use the heating system as the steam condenser for the steam turbine.
5. Molten-carbonate fuel cells and solid oxide fuel cells have a hot exhaust, very suitable for heating [251].
6. Nuclear power plants with waste heat, as shown in my previous book [1]. Depending on the temperature of waste heat, it can be used for numerous different applications [276].
7. Solar power plants with waste heat generating solar fuel (mainly hydrogen by numerous processes) [1,241–243] or enhanced geothermal systems that can bring high temperature to the surface can also be used for CHP systems [263].
8. Geothermal power plants with waste heat that can be used for heating and cooling buildings.

Smaller cogeneration units may use a reciprocating engine or Stirling engine [268]. The heat is removed from the exhaust and radiator. The systems are popular in small sizes because small gas and diesel engines are less expensive than small gas- or oil-fired steam-electric plants.

Some cogeneration plants are fired by biomass [30,245,253,273,282] or industrial and municipal solid waste [262]. Some CHP plants utilize waste gas as the fuel for electricity and heat generation. Waste gases can be gas from animal waste, landfill gas, gas from coal mines, sewage gas, and combustible industrial waste gas. Some cogeneration plants combine gas and solar photovoltaic generation to further improve technical and environmental performance [247,249]. Such hybrid systems can be scaled down to the building level [263] and even individual homes.

10.8.1 Micro-CHP

Micro combined heat and power or "micro cogeneration" is a so-called distributed energy resource (DER). The installation is usually less than 5 kW_e in a house or small business. Instead of burning fuel to merely heat space or water, some of the energy is converted to electricity in addition to heat. This electricity can be used within the home or business or, if permitted by the grid management, sold back into the electric power grid.

Delta energy and environment consultants stated [288,289] that in 2012, with 64% of global sales, the sale of micro-CHP fuel cell exceeded that for conventional fuel cell; 20,000 units were sold just in Japan. The development of small-scale CHP systems has provided the opportunity for in-house power backup of residential-scale photovoltaic (PV) arrays [288]. The results of a 2011 study show that a PV + CHP hybrid system not only has the potential to radically reduce energy waste in the status quo electrical and heating systems, but also enables the share of solar PV to be expanded by about a factor of five [234,247,249,288]. In some regions, in order to reduce waste from excess heat, an absorption chiller has been proposed to utilize the CHP-produced thermal energy for cooling of PV-CHP systems [288]. These trigeneration and photovoltaic systems have the potential to save even more energy and further reduce emissions compared to conventional sources of power, heating, and cooling [249].

Micro-CHP installations use five different technologies: micro-turbines, internal combustion engines, Stirling engines, closed cycle steam engines, and fuel cells. One author indicated in 2008 that micro-CHP based on Stirling engines is the most cost-effective of the so-called microgeneration technologies in abating carbon emissions; a 2013 U.K. report [288] stated that MCHP is the most cost-effective method of utilizing gas to generate energy at the domestic level; however, advances in reciprocation engine technology are adding efficiency to CHP plants, particularly in the biogas field [288] As both Mini-CHP and CHP have been shown to reduce emissions [288,289], they could play a large role in the field of CO_2 reduction from buildings, where more than 14% of emissions can be saved using CHP in buildings [288]. The ability to reduce emissions is particularly strong for new communities in emission intensive grids that utilize a combination of CHP and photovoltaic systems [288,289].

10.9   NGV (CNG AND LNG) AND LPG VEHICLES

10.9.1   NGV (CNG AND LNG) VEHICLES

One of the major application of natural and synthetic gas is the vehicle industry. Combustion of one cubic meter of natural gas yields 38 MJ (10.6 kWh). Natural gas has the highest energy/carbon ratio of any fossil fuel, and thus produces less carbon dioxide per unit of energy. Natural gas has, over the course of the 1990s, proven to be the most effective fuel for reducing emissions in an internal combustion engine. The 1993 Dodge full-size CNG vans were the first vehicles to meet the California Low Emission Vehicle (LEV) standards; the 1994 Chrysler/Dodge CNG minivans were the first to meet the Ultra-Low Emission Vehicle (ULEV) standards; the 1997 Ford CNG pickups and vans were the first to meet the Super Ultra-Low Emission Vehicle (SULEV) standards; and the CNG Honda Civic GX has been the cleanest internal combustion-engine vehicle ever tested by the EPA in every year since its 1998 introduction [9,290].

A *natural gas vehicle* (*NGV*) is an alternative fuel vehicle that uses CNG or LNG as a cleaner alternative to other fossil fuels. Worldwide, there were 14.8 million NGVs in 2011, with Iran, Pakistan, and Argentina leading the pack [9,290,291,292]. The Asia-Pacific region led the world with 6.8 million NGVs. In the Latin American region almost 90% of NGVs have bifuel engines, allowing these vehicles to run on either gasoline or CNG [9,290,291,292]. In Pakistan, almost every vehicle converted to (or manufactured for) alternative fuel use typically retains the capability to run on ordinary gasoline.

Unfortunately, according to GE, only about 250,000 NGV are being used in the United States [9,293]. The average growth rate in the United States shows a 3.7% increase per year since 2000, as contrasted with a booming global growth rate of 30.6% per year. Several manufacturers (Fiat, Opel/General Motors, Peugeot, Volkswagen, Toyota, Honda, and others) sell bifuel cars. In 2006, Fiat introduced the Siena Tetrafuel in the Brazilian market, equipped with a 1.4 L FIRE engine that runs on E100, E25 (Standard Brazilian Gasoline), ethanol, and CNG [9,294].

The number of NGV refueling stations also increased, to about 18,200 worldwide as of 2010, up 10.2% from the previous year [2]. In the United States, as of February 2011, there were 873 CNG refueling sites and 40 LNG sites, led by California with 215 CNG refueling stations in operation and 32 LNG sites. As of December 2013, the United States ranked sixth in the world in terms of the number of NGV stations [9,295–297].

NGVs are popular in regions or countries where natural gas is abundant and where the government chooses to price CNG lower than gasoline [9,290]. The use of natural gas began in the Po River Valley of Italy in the 1930s. In the United States CNG-powered buses are the favorite choice of several public transit agencies, with a fleet of more than 114,000 vehicles, mostly buses. A typical bus operated by CNG is shown in Figure 10.6. On private market, as of December 2013 Waste Management had a fleet of 2000 CNG Collection trucks and UPS had 2700 alternative fuel vehicles. India, Australia, Argentina, and Germany also have widespread use of natural gas–powered buses in their public transportation fleets [9,293,298,299].

CNG vehicles are also common in South America, with a 35% share of the worldwide NGV fleet, where these vehicles are mainly used as taxicabs in main cities of Argentina and Brazil. In these

FIGURE 10.6 Buses powered with CNG are common in the United States. (From Compressed natural gas, Wikipedia, the free encyclopedia, 2015.)

countries, gasoline vehicles are retrofitted with CNG gas cylinder in the trunk and CNG injection system and electronics. As of 2013, Argentina had about 2.49 million NGVs with about 1939 refueling stations across the nation, and Brazil had 1.78 million NGV vehicles and about 1805 refueling stations. Colombia, Bolivia, Peru, Venezuela, and Chile also had significant numbers of NGV fleets.

Existing gasoline-powered vehicles may be converted to run on CNG or LNG, and can be dedicated (running only on natural gas) or bi-fuel (running on either gasoline or natural gas). Diesel engines for heavy trucks and buses can also be converted and can be dedicated with the addition of new heads containing spark ignition systems, or can be run on a blend of diesel and natural gas, with the primary fuel being natural gas and a small amount of diesel fuel being used as an ignition source. CNG is starting to be used also in tuk-tuks and pickup trucks, transit and school buses, and trains [9,299,300,301].

NGV filling stations can be located anywhere that natural gas lines exist. Compressors (CNG) or liquefaction plants (LNG) are usually built on a large scale but with CNG small home refueling stations are possible. A company called Fuel Maker pioneered such a system called Phill Home Refueling Appliance (known as "Phill"), which they developed in partnership with Honda for the American GX model. CNG may also be mixed with biogas, produced from landfills or wastewater, which doesn't increase the concentration of carbon in the atmosphere [9,300,302–304].

Despite its advantages, the use of NGVs faces several limitations, including fuel storage and infrastructure available for delivery and distribution at fueling stations. CNG must be stored in high-pressure cylinders (3000–3600 psi operation pressure), and LNG must be stored in cryogenic cylinders (−260°F to −200°F). These cylinders take up more space than gasoline or diesel tanks that can be molded in intricate shapes to store more fuel and use less on-vehicle space. CNG tanks are usually located in the vehicle's trunk or pickup bed, reducing the space available for other cargo. This problem can be solved by installing the tanks under the body of the vehicle, or on the roof (typical for buses), leaving cargo areas free. As with other alternative fuels, other barriers for widespread use of NGVs are natural gas distribution to and at fueling stations as well as the low number of CNG and LNG stations. CNG-powered vehicles are, however, considered to be safer than gasoline-powered vehicles [9,290,296,305,306].

The cost and placement of fuel storage tanks is the major barrier to wider/quicker adoption of CNG as a fuel. In spite of these circumstances, the number of vehicles in the world using CNG has grown steadily (30% per year) [9,291,295,306]. Now, as a result of industry's steady growth, the cost of such fuel storage tanks has been brought down to a much acceptable level [9,290]. CNG's volumetric energy density is estimated to be 42% that of liquefied natural gas and 25% that of diesel fuel [9,290]. Compared to diesel fuel buses, CNG buses have greater breaking distance due to their increased fuel storage system weight.

CNG locomotives are operated by several railroads. The Napa Valley Wine Train successfully retrofit a diesel locomotive to run on CNG before 2002 [9,290,307–309]. Ferrocarril Central Andino in Peru has been running a CNG locomotive on a freight line since 2005 [9,290,309]. CNG locomotives are usually diesel locomotives that have been converted to use CNG generators instead of diesel generators to generate the electricity that drives the traction motors. CNG costs about 50% less than gasoline or diesel, and emits up to 90% fewer emissions than gasoline [9,290,300].

In many ways, CNG is like LPG. It is very easy on the engine, giving longer service life and lower maintenance costs [310–314,319]. CNG is the least expensive alternative fuel (except electricity) compared with equal amounts of fuel energy. The high octane rating of natural gas allows the CNG-powered Honda Civic GX to use a very high compression ratio and produce more power than stock gasoline versions. CNG provides high octane rating of the fuel and allows timing and mixture to be adjusted for more efficiency without causing detonation ("knocking"). As with LPG, because the fuel tanks have to withstand such enormous internal pressures, they are incredibly tough, with good results for safety. In addition, because natural gas is lighter than air and has very narrow flammability limits, if a leak develops it is very likely that the fuel will dissipate harmlessly into the air without causing a danger of ignition or explosion. CNG and LNG are largely methane. Methane is slightly soluble in water and under certain anaerobic conditions does not biodegrade. If excess amounts accumulate, the gas can bubble in water creating a possible risk of fire or explosion [9,305].

Natural gas does not auto-ignite at pressures and temperatures relevant to traditional gasoline and diesel engine design, thus providing more flexibility in the design of a natural gas engine. Methane, the main component of natural gas, has an autoignition temperature of 580°C/1076°F [9], whereas gasoline and diesel auto ignite at approximately 250°C and 210°C respectively. With a CNG engine, the mixing of the fuel and the air is more effective since gases typically mix well in a short period of time, but at typical CNG compression pressures the fuel itself is less energy dense than gas or diesel; thus, the end result is a lower energy dense air–fuel mixture. Thus for the same cylinder displacement engine, a nonturbocharged CNG-powered engine is typically less powerful than a similarly sized gas or diesel engine. For that reason turbochargers are popular on European CNG cars [9,290].

Although LNG and CNG are both considered NGVs, the technologies are vastly different. Refueling equipment, fuel cost, pumps, tanks, hazards, and capital costs are all different. One thing they share is that due to engines made for gasoline, computer-controlled valves to control fuel mixtures are required for both of them, often being proprietary and specific to the manufacturer. The on-engine technology for fuel metering is the same for LNG and CNG.

In summary, there are several other advantages and disadvantages of CNG vehicles. All of these can be briefly summarized as follows [9,290].

10.9.1.1 Advantages of CNG

CNG does not contain any lead, thereby eliminating fouling of spark plugs (unleaded fuel is lead-free, but still can cause plugs to foul). CNG-powered vehicles have lower maintenance costs than other hydrocarbon-fuel-powered vehicles. CNG tends to corrode and wear the parts of an engine less rapidly than gasoline. CNG fuel systems are sealed, thus preventing fuel losses from spills or evaporation.

CNG does not contaminate or dilute the crankcase oil, thereby increasing the life of lubricating oils. Being a gaseous fuel, CNG mixes easily and evenly in air. CNG is less likely to ignite on hot surfaces, since it has a high autoignition temperature (540°C) and a narrow range (5%–15%) of flammability [9,290].

The lifecycle GHG emissions for CNG compressed from California's pipeline natural gas has been given a value of 67.70 g of CO_2-equivalent per mega joule (gCO_2e/MJ) by CARB (the California Air Resources Board), approximately 28% lower than the average gasoline fuel in that market (95.86 gCO_2e/MJ). CNG can reduce carbon monoxide emissions by 90%–97% and nitrogen oxide emissions by 35%–60% when compared with gasoline. CNG can also potentially reduce nonmethane hydrocarbon emissions by 50%–75%, while producing fewer carcinogenic pollutants and little or no particulate matter [14]. CNG produced from landfill biogas was found by CARB to have the lowest GHG emissions of any fuel analyzed, with a value of 11.26 gCO_2e/MJ (more than 88% lower than conventional gasoline) [9,290].

The cost of CNG can be as little as half that of a gallon of gas when a home refueling device is used. At commercial stations, the cost is still significantly less than gasoline. Some research pegs on average the fuel savings at about 30%–40% less than gasoline. Although CNG car acceleration is typically slower, the car starts and drives normally. The DOE vehicles powered by natural gas are as safe as conventional gasoline or diesel vehicles, and that their pressurized tanks have been designed to withstand severe impact, temperature, and environmental exposure [9,290,305].

10.9.1.2 Disadvantages of CNG

As mentioned earlier, CNG tanks can take up trunk space unless it is installed under the body or on the roof of the vehicle. CNG tanks are quite bulky and heavy, about three times more so than LPG tanks, though modern composite technology is starting to be used to cut weight and cost (as on the Honda Civic GX). CNG vehicle range is significantly less than for an equivalent gasoline vehicle. While initial cost for a CNG vehicle is larger ($3,000–$10,000) compared to the gasoline version, low fuel and maintenance costs in a high-mileage application like a transit bus, a taxi, or a shuttle van can make up the price difference over time [9,290].

The lack of easy availability of refueling stations (except in California and New York) along with lower mileage range per tank creates real anxiety for CNG vehicle passengers. CNG refueling stations are more expensive to operate (due to high-pressure compressor requirement). The time required to refill a CNG tank is also significantly higher than filling a gasoline tank. Even "fast fill" stations take 5–10 min to fill a 10 gasoline gallon equivalent (GGE) tank. Home refueling can take up to a day. CNG is not as effective for high-powered engine as LNG. Westport recommends CNG for engines 7 L or smaller and LNG with direct injection for engines between 20 and 150 L. For engines between 7 and 20 L either option is recommended [9,290].

10.9.1.3 LNG-Fueled Vehicles

LNG, or liquefied natural gas, is natural gas that has been cooled to a point that it is a cryogenic liquid. In its liquid state, it is still more than two times as dense as CNG. LNG is dispensed from bulk storage tanks at LNG fuel stations at rates exceeding 20 GGE/min. Because of its cryogenic nature, it is stored in specially designed insulated tanks. Generally speaking, these tanks operate at fairly low pressures (about 70–150 psi) when compared to CNG. A vaporizer is mounted in the fuel system that turns the LNG into a gas (which may simply be considered low-pressure CNG). When comparing building a commercial LNG station with a CNG station, utility infrastructure, capital cost, and electricity heavily favor LNG over CNG. There are existing LCNG stations (both CNG and LNG), where fuel is stored as LNG and then vaporized to CNG on-demand. LCNG stations ironically require less capital cost than fast-fill CNG stations alone, but more than LNG stations.

LNG is in the early stages of becoming a mainstream fuel for transportation needs. It is being evaluated and tested for over-the-road trucking, off-road, marine, and railroad applications [9,293]. There are known problems with the fuel tanks and delivery of gas to the engine but despite these concerns the move to LNG as a transportation fuel has begun [9,293,296,303,308,309,315–318].

China has been a leader in the use of LNG vehicles [9,305] with over 100,000 LNG powered vehicles on the road as of September 2014. In the United States the beginnings of a public LNG fueling capability is being put in place. An alternative fueling center tracking site shows

69 public truck LNG fuel centers as of February 2015. The 2013 National Trucker's Directory lists approximately 7000 truck stops; thus, approximately 1% of U.S. truck stops have LNG available [9,296,303,305,308,309,315–318].

During 2013, several companies like Dillon Transport, Raven transportation, and Lowe made significant commitments to buy LNG trucks for their needs. UPS had over 1200 LNG fueled trucks (about 10% of its fleet) on the roads in February 2015. Numerous other refueling facilities are being established in Texas, Oklahoma, and California [9,296,303,305,308,309,315–318]. More stations are needed in Illinois and Colorado. As of December 2014, LNG fuel and NGVs were not favored within Europe and it is questionable whether LNG will ever become the fuel of choice among fleet operators [9,296,303,305,308,309,315–318].

LNG offers a unique advantage over CNG for more demanding high-horsepower applications by eliminating the need for a turbocharger. Because LNG boils at approximately −160°C, using a simple heat exchanger a small amount of LNG can be converted to its gaseous form at extremely high pressure with the use of little or no mechanical energy. A properly designed high-horsepower engine can leverage this extremely high-pressure energy dense gaseous fuel source to create a higher energy density air–fuel mixture than can be efficiently created with a CNG-powered engine. The end result when compared to CNG engines is more overall efficiency in high-horsepower engine applications when high-pressure direct injection technology is used [9,295,303,305,308,309,315–318].

High horsepower engines in the oil drilling, mining, locomotive, and marine fields are being developed. Blomerous [9,303,305,308,309,315–318] reports that as much as 40 million tons per annum of LNG (approximately 26.1 billion gal/year or 71 million gal/day) could be required just to meet the global needs of the high-horsepower engines by 2025–2030.

10.9.2 LPG Vehicles

Autogas is the common name for LPG when it is used as a fuel in internal combustion engines in vehicles as well as in stationary applications such as generators. It is a mixture of propane and butane. Autogas is widely used as a "green" fuel, as its use reduces CO_2 exhaust emissions by around 15% compared to gasoline. One liter of gasoline is equivalent to 1.33 L of autogas. It has an octane rating (MON/RON) that is between 90 and 110. The energy content (higher heating value—HHV) for pure propane is 25.5 MJ per liter and for pure butane is 28.7 MJ per liter [310–314,319–321].

Propane autogas is domestically produced, costs approximately 30%–40% less than gasoline per gallon, and does not take away the performance and power unlike other alternative fuels. It has an octane rating of 105. Additionally, autogas burns cleaner than gasoline, which means a smaller carbon footprint. The autogas market is the sleeping giant of alternative fuel, and it is often stored in Highland Tank manufactured by Highland Co. (see Chapter 8). Propane is the most commonly used alternative fuel for transportation. In 2006, Texas had 525 LPG fueling stations, or 22.9% of the national total. By contrast, Texas has about 16,500 gasoline fueling stations [319]. The Texas state government fleet had about 7400 vehicles using alternative fuels in fiscal 2006 [9,290,319], with LPG vehicles accounting for 73% of the total number of alternate fuel vehicles [319]. Autogas is on the rise in the United States. As of February 2012, the United States had more than 147,000 autogas vehicles on the road, accounting for just 2% of the world's total and as of June 2013, the United States had 2,843 autogas fueling stations, making it easier for drivers to refuel across the country [319]. While few automakers offer LPG vehicles in the United States, they are much more common in Europe and Australia. Ford and GM both offer LPG-fueled models to those markets. The popularity of these vehicles is due to tax incentives for purchasing LPG and/or tax disincentives for purchasing gasoline-driven cars [310–314,319–321].

The initial infrastructure costs required to expand the sales of LPG in the transport sector are mainly determined by the investment costs of the LPG refueling infrastructure. The costs incurred relate mainly to service-station storage and dispensing facilities. Autogas, however, generally makes use of existing service-station infrastructure for distribution of conventional fuels; therefore, additional costs for Autogas are relatively low compared to other alternative fuels; for example, the cost of installing a tank, pump, and metering equipment for autogas alongside existing gasoline or diesel facilities is about one-third the cost for equivalent CNG capacity [322].

In the United States, autogas is more commonly known by the name of its primary constituent, *propane*. In 2010, the Propane Education & Research Council adopted "propane autogas" to refer to LPG used in on-road motor vehicles [305,319]. In the United States, *LPG* and *autogas* are used interchangeably. In Australia, the common terms are *LPG* and *gas*. In Italy and France, *GPL* (an acronym for *gas di petrolio liquefatto* and *gaz de pétrole liquéfié*) is used. In Spain, the term *GLP* (*gas licuado del petróleo*) is used. In Asian countries, the terms *LPG* or *auto LPG* are more widely used by consumers, especially by taxi drivers, many of whom use converted vehicles [319]. The converted vehicles are commonly called *LPG vehicles* or *LPG cars* [319].

Toyota produced a number of LPG-only engines in their 1970s *M*, *R*, and *Y* engine families. A number of automobile manufacturers such as Citroën, Fiat, Ford, Hyundai, and General Motors have OEM *bi-fuel* models that will run equally well on both LPG and gasoline. Vialle manufactures OEM bi-fuel models, LPG-powered scooters and LPG-powered mopeds that run equally well on gasoline [310–314,319–321].

Autogas enjoys great popularity in numerous countries and territories, including Australia, the European Union, Hong Kong, India, the Philippines, the Republic of Macedonia, South Korea, Serbia, Sri Lanka, and Turkey. It is also available at larger gasoline stations in several countries. In the Republic of Armenia, for example, the Transport Ministry estimates as many as 20%–30% of vehicles use autogas, because it offers a very cheap alternative to both diesel and gasoline, being less than half the price of gasoline and some 40% cheaper than diesel [310–314,319–321].

Autogas is the third most popular automotive fuel in the world. There are more than 7 million LPG-powered vehicles in Europe, and LPG accounts for about 2% of the fuel mix of passenger cars in Europe which is likely to increase to 10% of Europe's passenger car fuel mix by 2020 [323]. Worldwide there are more than 17.4 million autogas vehicles and over 57,000 refueling sites. Table 10.2 shows a summary of the seven largest markets across the globe for LPG vehicles. LPG also has substantial reserves because of its dual origins of natural gas processing and crude oil refining. Demand between

TABLE 10.2

World and Top Seven Country Markets for LPG as Transport Fuel in 2007–2011

Country	Vehicles[a] (% of World)	LPG Consumption[a] (% of World Usage Ton per Vehicle)	Refueling[a] (% of World Sites per Vehicle)
Korea	13.2	147	21
Turkey	13.7	79	117
Russia	7.3	137	47.7
Poland	13.3	54	77.7
Italy	9.7	55	51.0
Japan	1.6	318	201
Australia	3.7	134	150

Source: Modified and Adapted from World LP Gas Association (WLPGA), Autogas Incentive Policies: A country-by country analysis of why and how governments promote Autogas & what works, 2011, Available at http://www.worldlpgas.com/uploads/Modules/Publications/autogas-incentive-policies-2012-updated-july-2012.pdf.

[a] Total world: No. of vehicles: 17,473; LPG consumption per vehicle: 1.31 (tons per vehicle); Refueling sites per vehicle: 3.27.

2000 and 2010 increased by 60%, but remains concentrated in a small number of markets with the top 5 countries accounting for 53% of world consumption in 2010.

The different autogas systems generally use the same type of filler, tanks, lines, and fittings but use different components in the engine bay. There are three basic types of autogas system. The oldest of these is the conventional *converter-and-mixer* system, which has existed since the 1940s and is still widely used today. A converter-mixer system uses a converter to change liquid fuel from the tank into vapor, and it then feeds that vapor to the mixer where it is mixed with the intake air. This is also known as a venture system or "single-point" system. The other two types are known as *injection* systems, but there are significant differences between the two [310–314,319–321].

Vapor-phase injection systems also use a converter, but unlike the mixer system, the gas exits the converter at a regulated pressure. The gas is then injected into the air intake manifold via a series of electrically controlled injectors. The injector opening times are controlled by the autogas control unit. This unit works in much the same way as a gasoline fuel injection control unit. This allows much more accurate metering of fuel to the engine than is possible with mixers, thus improving economy and/or power while reducing emissions. Because the fuel vaporizes in the intake, the air around it is cooled significantly. This increases the density of the intake air and can potentially lead to substantial increases in engine power output, to the extent that such systems are usually *detuned* to avoid damaging other parts of the engine [310–314,319–321].

Liquid injection systems use special tanks with circulation pumps and return lines similar to gasoline fuel injection systems. Liquid phase injection systems do not use a converter, but instead deliver the liquid fuel into a *fuel rail* in much the same manner as a petrol injection system. Liquid phase injection has the potential to achieve much better economy and power plus lower emission levels than are possible using mixers or vapor phase injectors [310,319–322].

Commercially available LPG is currently derived from fossil fuels. Burning LPG releases carbon dioxide, a greenhouse gas. The reaction also produces some carbon monoxide. LPG does, however, release less CO_2 per unit of energy than does coal or oil. It emits 81% of the CO_2 per kWh produced by oil, 70% of that of coal, and less than 50% of that emitted by coal-generated electricity distributed via the grid. Being a mix of propane and butane, LPG emits less carbon per joule than butane but more carbon per joule than propane. LPG can be considered to burn more cleanly than heavier molecule hydrocarbon, in that it releases very few particulates.

LPG-powered passenger cars have about 10% lower tailpipe CO_2 emission than comparable gasoline-powered cars. When compared to a diesel car, there is no significant CO_2 emission reduction per kilometer driven; however, LPG-powered vehicles do have substantially lower NO_x emissions than diesel-powered vehicles [310–314,319–321].

A study by Hanschke et al. [324] compared Well-to-Wheels (WTW) CO_2 emissions of LPG vehicles with gasoline and diesel-powered vehicles. WTW CO_2 emissions also consider the energy which is needed to extract and refine the fuels. This study concludes that the corresponding WTW CO_2 emissions per MJ of energy content are lower for LPG (73.4 g/MJ) than for gasoline (82.8 g/MJ) and diesel (85.4 g/MJ). As the energy efficiency of an LPG-powered car is comparable to the energy efficiency of a gasoline-powered car, the WTW CO_2 emissions per km are calculated to be about 11% lower for LPG (165 g/km) [324] than for a gasoline-powered car (185 g/km). The WTW CO_2 emissions per kilometer of an LPG-powered vehicle are comparable than for a diesel-powered car. This mainly due to by the higher vehicle efficiency of the diesel-powered car. The primary reason why governments in many countries actively encourage autogas use is the environment. Autogas is shown in many studies to perform better environmentally than its gasoline and diesel counterparts [322].

Other benefits of using LPG as an automotive fuel include lower maintenance costs and lower fuel costs compared to conventional gasoline and diesel. It gives a longer engine lifetime. This is due to LPG's high octane rating and low carbon and oil contamination, which puts less pressure on the engine. Using LPG may also increase energy security as it may be available domestically or, if it is imported, may diversify a country's fossil fuel sources [310–314,319–321].

An important difference between LPG and conventional vehicles is the method of fuel storage. LPG is gaseous at room temperature and a pressurized storage tank is required at the fueling station as well as in the vehicle. LPG can be used in dedicated LPG vehicles or in vehicles converted from gasoline use. The availability of dedicated LPG models is limited, and most LPG-powered passenger vehicles have a modified combustion engine. Such converted vehicles normally operate in bi-fuel mode, using either LPG or regular gasoline. Modern bi-fuel vehicles use electronically controlled gas injection systems lowering the NO_x and CO_2 emissions substantially. However, bi-fuel cars rarely achieve the full emission benefits because a compromise in engine tuning is required for the two fuels. The advantage of a bi-fuel vehicle is that the car owner is less dependent on a LPG refueling infrastructure and in areas, where LPG is not available, regular gasoline can be used. A drawback of the bi-fuel vehicle is that requirement of two fueling tanks lowers the available space in the vehicle [310–314,319–321].

LPG tank is located under the car, saving space in the trunk of the vehicle. LPG also has a lower energy density than gasoline, and therefore requires more storage volume for an equivalent drive range. The safety risk of LPG is higher than of other (alternative) fuels, although LPG liquefies at moderate pressures (just over 10 bar). Pressurized storage of liquid can give rise to a "boiling liquid expanding vapor explosion" (BLEVE). This type of explosion occurs when the storage vessel ruptures. Moreover, LPG has a higher density than air which means that in case of a leak there can be a buildup of the gas around the fueling station [325,326]. In the past, most accidents have happened during the transport of LPG. These safety problems make it generally necessary to locate an LPG fueling station well outside an urban location [319,326].

10.9.3 LPG VERSUS NGV VEHICLES

Autogas or LPG can have different chemical composition, but it is still a petroleum-based gas. It has a number of inherent and noninherent advantages and disadvantages. The inherent advantage of autogas over CNG is that it requires far less compression (20% of CNG cost), is denser as it is a liquid at room temperature, and thus requires far cheaper tanks (consumer) and fuel compressors (provider) than CNG. As compared to LNG, it requires no chilling (and thus less energy), or has no problems associated with extreme cold such as frostbite. Like NGV, it also has advantages over gasoline and diesel in cleaner emissions, along with less wear on engines over gasoline. The major drawback of LPG is its safety. The fuel is heavier than air, which causes it to collect in a low spot in the event of a leak, making it far more hazardous to use and more care is needed in handling it [9,305,319,320,323].

In places like the United States, Thailand, and India, there are 5–10 times more LPG stations than NGV stations, thus making the fuel more easily accessible. In countries like Poland, South Korea, and Turkey, both LPG stations and automobiles are widespread, while NGVs are not. In Thailand, the retail LPG fuel is considerably cheaper in cost. The cost of NGV infrastructure and vehicle tanks can be lowered by using ANG (adsorbed natural gas) technology at 500 psi in both vehicles and fueling stations [9,300,304,319].

10.10 FUEL CELL

A *fuel cell* (see Figure 10.7) is a device that converts the chemical energy from a fuel into electricity through a chemical reaction of positively charged hydrogen ions with oxygen or another oxidizing agent [327–367]. Its market, how it works, and progress on its growth are discussed in a series of reviews and books [327–340]. Fuel cells can produce electricity continuously for as long as these inputs are supplied. The first fuel cells were invented in 1838. The first commercial use of fuel cells came more than a century later in NASA space programs. Since then, fuel cells have been used in many other applications. Fuel cells are used for primary and backup power for commercial, industrial, and residential buildings [368–383] and in remote or inaccessible areas. They are also used to

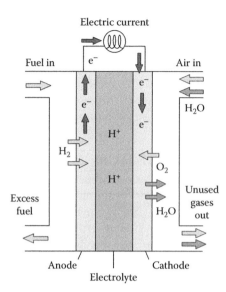

FIGURE 10.7 Scheme of a proton-conducting fuel cell. (From Fuel Cell, Wikipedia, the free encyclopedia, 2015.)

power fuel cell vehicles, including forklifts, automobiles, buses, boats, motorcycles, and submarines [384–400]. Fuel cells are also used in co- or trigeneration modes to improve its efficiency [401–413].

10.10.1 Types of Fuel Cells

Fuel cells come in many varieties; however, they all work in the same general manner [341–367]. They are made up of three adjacent segments: the anode, the electrolyte, and the cathode. Two chemical reactions occur at the interfaces of the three different segments. The net result of the two reactions is that fuel is consumed, water or carbon dioxide is created, and an electric current is created, which can be used to power electrical devices, normally referred to as the load.

At the anode a catalyst oxidizes the fuel, usually hydrogen, turning the fuel into a positively charged ion and a negatively charged electron. The electrolyte is a substance specifically designed so ions can pass through it, but the electrons cannot. The freed electrons travel through a wire creating the electric current. The ions travel through the electrolyte to the cathode. On reaching the cathode, the ions are reunited with the electrons and the two react with a third chemical, usually oxygen, to create water or carbon dioxide.

The main difference among fuel cell types is the electrolyte; fuel cells are classified by the type of electrolyte they use and by the difference in start-up time ranging from 1 s for proton exchange membrane fuel cells (PEM fuel cells, or PEMFC) to 10 min for solid oxide fuel cells (SOFC). In addition to electricity, fuel cells produce water, heat, and, depending on the fuel source, very small amounts of nitrogen dioxide and other emissions. The energy efficiency of a fuel cell is generally between 40% and 60%, or up to 85% efficient in cogeneration if waste heat is captured for use. The fuel cell market is growing, and Pike Research has estimated that the stationary fuel cell market will reach 50 GW by 2020 [331].

The most important design features in a fuel cell are the electrolyte substance, fuel (normally hydrogen), and anode and cathode catalysts. The anode catalyst breaks down the fuel into electrons and ions, and it is usually made up of very fine platinum powder. The cathode catalyst turns the ions into the waste chemicals like water or carbon dioxide, and it is often made up of nickel but it can also be a nanomaterial-based catalyst [327–340].

Voltage created by fuel cell (0.6–0.7 V) at full load decreases as current increases, due to activation loss, ohmic loss (voltage drop due to resistance of the cell components and interconnections), and mass transport loss (depletion of reactants at catalyst sites under high loads, causing rapid loss of voltage. To deliver the desired amount of energy, the fuel cells can be combined in series to yield higher voltage, and in parallel to allow a higher current to be supplied. Such a design is called a *fuel cell stack*. The cell surface area can also be increased, to allow higher current from each cell. Within the stack, reactant gases must be distributed uniformly over each of the cells to maximize the power output [327–340].

10.10.1.1   Proton Exchange Membrane Fuel Cells

In the archetypical hydrogen oxide–proton exchange membrane fuel cell design, a proton-conducting polymer membrane (the electrolyte) separates the anode and cathode sides [341–354]. This was called a "solid polymer electrolyte fuel cell" (SPEFC) in the early 1970s, before the proton exchange mechanism was well understood. On the anode side, hydrogen diffuses to the anode catalyst, where it later dissociates into protons and electrons. These protons often react with oxidants causing them to become what are commonly referred to as multi-facilitated proton membranes. The protons are conducted through the membrane to the cathode, but the electrons are forced to travel in an external circuit (supplying power) because the membrane is electrically insulating. On the cathode catalyst, oxygen molecules react with the electrons (which have traveled through the external circuit) and protons to form water. While most PEMFC operate with hydrogen, they can also operate with diesel, methanol, natural gas, and chemical hydrides [331].

The different components of a PEMFC are biploar plates, electrodes, catalyst, membrane, and the necessary hardware [331]. The bipolar plates may be made of different types of materials, such as metal, coated metal, graphite, flexible graphite, and various carbon composites. The membrane electrode assembly (MEA) is usually made of a proton exchange membrane sandwiched between two catalyst-coated carbon papers. Platinum and/or similar type of noble metals are usually used as the catalyst for PEMFC. The electrolyte could be a polymer membrane. The issues facing this type of fuel cell are (1) cost, (2) water and air management, (3) temperature management, (4) durability, and (5) limited carbon monoxide tolerance. Significant research is being carried out to address these issues [341–354].

10.10.1.2   Phosphoric Acid Fuel Cell

Phosphoric acid fuel cells (PAFC) were first designed and introduced in 1961 by Elmore and Tanner. In these cells phosphoric acid is used as a nonconductive electrolyte to pass positive hydrogen ions from the anode to the cathode. These cells commonly work in temperatures of 150°C–200°C. This high temperature will cause heat and energy loss if the heat is not removed and used properly [355]. This heat can be used to produce steam for air-conditioning systems or any other thermal energy consuming system. Using this heat in cogeneration can enhance the efficiency of phosphoric acid fuel cells from 40% to 50% to about 80% [402–413]. Phosphoric acid, the electrolyte used in PAFCs, is a nonconductive liquid acid which forces electrons to travel from the anode to the cathode through an external electrical circuit. Since the hydrogen ion production rate on the anode is small, platinum is used as catalyst to increase this ionization rate. A key disadvantage of these cells is the use of an acidic electrolyte. This increases the corrosion or oxidation of components exposed to phosphoric acid [355].

10.10.1.3   High-Temperature Fuel Cells (SOFC)

Solid oxide fuel cells (SOFCs) use a solid material, most commonly a ceramic material called yttria-stabilized zirconia (YSZ), as the electrolyte. Because SOFCs are made entirely of solid materials, they are not limited to the flat plane configuration of other types of fuel cells and are often designed as rolled tubes. They require high operating temperatures (800°C–1000°C) and can be run on a variety of fuels, including natural gas [356–359].

In SOFCs, negatively charged oxygen ions travel from the cathode to the anode. This is different from other types of fuel cells in which positively charged hydrogen ions travel from the anode to the cathode. Oxygen gas is fed through the cathode, where it absorbs electrons to create oxygen ions. The oxygen ions then travel through the electrolyte to react with hydrogen gas at the anode. The reaction at the anode produces electricity and water as by-products. Carbon dioxide may also be a by-product depending on the fuel, but the carbon emissions from an SOFC system are less than those from a fossil fuel combustion plant [331]. SOFC systems can run on any fuels that contain hydrogen atoms. However, for the fuel cell to operate, the fuel must be converted into pure hydrogen gas. SOFCs are capable of internally reforming light hydrocarbons such as methane, propane, or butane. These fuel cells are at an early stage of development [331].

Due to its high temperature operation, carbon dust buildup on the anode in SOFC slows down the internal reforming process. Research carried out at the University of Pennsylvania has shown that the use of copper-based cermet (heat-resistant materials made of ceramic and metal) can reduce coking and the loss of performance. SOFC systems have slow start-up time, making them less useful for mobile applications [331]. Despite these disadvantages, a high operating temperature provides an advantage by removing the need for a precious metal catalyst like platinum, thereby reducing cost. Additionally, waste heat from SOFC systems may be captured and reused (via cogeneration), increasing the theoretical overall efficiency to as high as 80%–85% [331,401–413].

While this type of fuel cell requires high temperature due to YSZ electrolyte whose ionic conductivity decreases with temperature, Ceres Power, a UK SOFC fuel cell manufacturer, has developed a method of reducing the operating temperature of their SOFC system to 500°C–600°C. They replaced the commonly used YSZ electrolyte with a CGO (cerium gadolinium oxide) electrolyte. The lower operating temperature allows them to use stainless steel instead of ceramic as the cell substrate, which reduces cost and start-up time of the system [331].

10.10.1.4 Hydrogen-Oxygen Fuel Cell (Bacon Cell)

The hydrogen-oxygen fuel cell was designed and first demonstrated publicly by Bacon in the year 1959. The cell consists of two porous carbon electrodes impregnated with a suitable catalyst such as Pt, Ag, and CoO. The space between the two electrodes is filled with a concentrated solution of KOH or NaOH which serves as an electrolyte. Hydrogen and oxygen are bubbled into the electrolyte through the porous carbon electrodes which results in a reaction between hydrogen and oxygen to form water. The cell runs continuously until the reactant's supply is exhausted. This type of cell operates efficiently in the temperature range 70°C–140°C and provides a potential of about 0.9 V [331,366,367].

10.10.1.5 MCFC

Molten carbonate fuel cells (MCFCs) require a high operating temperature, 650°C (1200°F), similar to SOFCs. MCFCs use lithium potassium carbonate salt as an electrolyte, and this salt liquefies at high temperatures, allowing for the movement of negative carbonate ions within the cell [331,360–363].

Like SOFCs, MCFCs are capable of converting fossil fuel to a hydrogen-rich gas in the anode, eliminating the need to produce hydrogen externally. The reforming process creates CO_2 emissions. MCFC-compatible fuels include natural gas, biogas, and gas produced from coal. The hydrogen in the gas reacts with carbonate ions from the electrolyte to produce water, carbon dioxide, electrons, and small amounts of other chemicals. The electrons travel through an external circuit creating electricity and return to the cathode. There, oxygen from the air and carbon dioxide recycled from the anode react with the electrons to form carbonate ions that replenish the electrolyte, completing the circuit. MCFCs have slow start-up times because of their high operating temperature. This makes MCFC systems not suitable for mobile applications, and this technology will most likely be used for stationary fuel cell purposes. The main challenge of MCFC technology is the cells' short life span. The high-temperature and carbonate electrolyte lead to corrosion of the anode and cathode. These factors accelerate the degradation of MCFC components, decreasing the durability and cell life [331].

MCFCs are resistant to impurities. They are not prone to carbon buildup on the anode that results in reduced performance by slowing down the internal fuel reforming process. Therefore, carbon-rich fuels like gases made from coal are compatible with the system. MCFCs also have relatively high efficiencies. They can reach a fuel-to-electricity efficiency of 50%, which is considerably higher than the 37%–42% efficiency of a phosphoric acid fuel cell plant. Efficiencies can be as high as 65% when the fuel cell is paired with a turbine, and 85% if heat is captured and used in a cogeneration system [401–413].

Fuel Cell Energy, a Connecticut-based fuel cell manufacturer, develops and sells MCFC fuel cells. The company says that their MCFC products range from 300 kW to 2.8 MW systems that achieve 47% electrical efficiency and can utilize CHP technology to obtain higher overall efficiencies. One product, the DFC-ERG, is combined with a gas turbine and, according to the company, it achieves an electrical efficiency of 65% [337,340].

A comparison of different types of fuel cell is illustrated in Table 10.3.

TABLE 10.3
Comparison of Fuel Cell Types

Fuel Cell Name	Electrolyte	Qualified Power (W)	Working Temperature (°C)	Efficiency (System)
Metal hydride fuel cell	Aqueous alkaline solution		>−20 (50% P_{peak} at 0°C)	
Electro-galvanic fuel cell	Aqueous alkaline solution		<40	
Direct formic acid fuel cell (DFAFC)	Polymer membrane (ionomer)	<50 W	<40	
Microbial fuel cell	Polymer membrane or humic acid		<40	
Upflow microbial fuel cell (UMFC)			<40	
Regenerative fuel cell	Polymer membrane (ionomer)		<50	
Direct borohydride fuel cell	Aqueous alkaline solution		70	
Alkaline fuel cell	Aqueous alkaline solution	10–100 kW	<80	62%
Direct methanol fuel cell	Polymer membrane (ionomer)	100 mW–1 kW	90–120	10%–25% [48]
Reformed methanol fuel cell	Polymer membrane (ionomer)	5 W–100 kW	250–300 (Reformer) 125–200 (PBI)	25%–40%
Direct ethanol fuel cell	Polymer membrane (ionomer)	<140 mW/cm²	>25 ? 90–120	
Proton exchange membrane fuel cell	Polymer membrane (ionomer)	1 W–500 kW	50–100 (Nafion) [50] 120–200 (PBI) [51]	30%–50% [48]
Phosphoric acid fuel cell	Molten phosphoric acid (H_3PO_4)	<10 MW	150–200	40% [48] Co-Gen: 90%
Solid acid fuel cell	H⁺-conducting oxyanion salt (solid acid)	10 W–1 kW	200–300	40%–45%
Molten carbonate fuel cell	Molten alkaline carbonate	100 MW	600–650	45%–55% [48]
Tubular solid oxide fuel cell (TSOFC)	O²⁻-conducting ceramic oxide	<100 MW	850–1100	55%–60%
Protonic ceramic fuel cell	H⁺-conducting ceramic oxide		700	
Direct carbon fuel cell	Several different		700–850	70%
Planar solid oxide fuel cell	O²⁻-conducting ceramic oxide	<100 MW	500–1100	55%–60% [48]
Magnesium-air fuel cell	Salt water		−20 to 55	

Source: Modified and adapted from Fuel Cell, Wikipedia, the free encyclopedia, 2015.

10.10.2 THEORETICAL VERSUS PRACTICAL ENERGY EFFICIENCY

According to the DOE, fuel cells are generally between 40% and 60% energy-efficient [331]. This is higher than some other systems for energy generation. For example, the typical internal combustion engine of a car is about 25% energy-efficient [109]. In CHP systems, the heat produced by the fuel cell is captured and put to use, increasing the efficiency of the system to up to 85%–90% [401–413].

The maximum theoretical energy efficiency of a fuel cell is 83%, operating at low power density and using pure hydrogen and oxygen as reactants (assuming no heat recapture) [331]. According to the World Energy Council, this compares with a maximum theoretical efficiency of 58% for internal combustion engines [110]. While these efficiencies are not approached in most real-world applications, high-temperature fuel cells (solid oxide fuel cells or MCFCs) can theoretically be combined with gas turbines to allow stationary fuel cells to come closer to the theoretical limit. A gas turbine would capture heat from the fuel cell and turn it into mechanical energy to increase the fuel cell's operational efficiency. This solution has been predicted to increase total efficiency to as much as 70% [401–413].

Solid-oxide fuel cells produce exothermic heat from the recombination of the oxygen and hydrogen. The ceramic can run as hot as 800°C. This heat can be captured and used to heat water in a micro combined heat and power (m-CHP) application. When the heat is captured, total efficiency can reach 80%–90% at the unit, but this does not consider production and distribution losses. CHP units are being developed for the European home market.

Fuel cell makes the most sense for operation disconnected from the grid, or when fuel can be provided continuously. PEM fuel cell is becoming an attractive choice [331] when (a) frequent and relatively rapid start-ups are required, (b) zero emission is required such as in close spaces like warehouses, and (c) hydrogen is considered as an acceptable reactant [331,401–413].

10.10.3 APPLICATIONS

A fuel cell has a multitude of applications. While fuel cell can be used for solar water heating, food preservation (like fish), breathalyzers for alcohols, and carbon monoxide detectors, its two major applications are in power generation (with or without cogeneration) and numerous types of vehicle (land, air, or sea) applications. We briefly evaluate these two applications in detail.

10.10.3.1 Power

Stationary fuel cells are used for commercial, industrial, and residential primary and backup power generation. Fuel cells are very useful as power sources in remote locations, such as spacecraft, remote weather stations, large parks, communications centers, rural locations including research stations, and in certain military applications. A fuel cell system running on hydrogen can be compact and lightweight, and have no major moving parts. Because fuel cells have no moving parts and do not involve combustion, in ideal conditions they can achieve up to 99.9999% reliability [331]. This equates to less than 1 min of downtime in a 6-year period [331].

Since fuel cell electrolysis systems do not store fuel in themselves, but rather rely on external storage units, they can be successfully applied in large-scale energy storage [331]. There are many different types of stationary fuel cells, so efficiencies vary, but most are between 40% and 60% energy-efficient. However, when the fuel cell's waste heat is used to heat a building in a cogeneration system this efficiency can increase to 85% [401–413]. This is significantly more efficient than traditional coal-fired power plants, which are only about one-third energy-efficient. At the same scale level, compared to other energy generation systems fuel cells could save 20%–40% on energy costs when used in cogeneration systems [401–413]. Fuel cells are also much cleaner than traditional power generation; a fuel cell power plant using natural gas as a hydrogen source would create less than one ounce of pollution (other than CO_2) for every 1000 kWh produced, compared to 25 pounds of pollutants generated by conventional combustion systems [331]. Fuel cells also produce 97% less nitrogen oxide emissions than conventional coal-fired power plants.

A pilot-scale fuel cell is operated on Stuart Island in Washington State [331] in which a complete closed loop system is created in which solar panels power an electrolyzer, which makes hydrogen. The hydrogen is stored in a 500-U.S.-gallon at 200 psig pressure. This hydrogen is supplied to a ReliOn fuel cell to provide full electric backup to the off-the-grid residence. Another closed system loop was unveiled in late 2011 in Hempstead, New York [331]. Fuel cells can be used with low-quality gas from landfills or wastewater treatment plants to generate power and lower methane emissions. A 2.8 MW fuel cell plant in California is said to be the largest of this type [331].

Portable power systems that use fuel cells can be used in the leisure sector (i.e., RVs, cabins, marine), the industrial sector (i.e., power for remote locations including gas/oil well sites, communication towers, security, weather stations), and in the military sector. SFC Energy is a German manufacturer of direct methanol fuel cells (DMFCs) for a variety of portable power systems. Ensol Systems Inc. is an integrator of portable power systems, using the SFC Energy DMFC [326–340].

Fuel cells can also be applied to numerous other applications. They can provide backup source for a peak power need and can be used as emergency power supplier for residential homes to hospitals, scientific laboratories, data centers [368,380], and so on. They can provide instant protection from a momentary power interruption. They can also be power supplier for telecommunication [331], equipment, notebook computers, smartphones, laptops, tablets, and hybrid vehicles [331,381–383].

CHP fuel cell systems, including micro combined heat and power (micro-CHP) systems, are used to generate both electricity and heat for homes, office building, and factories. The system generates constant electric power (selling excess power back to the grid when it is not consumed), and at the same time produces hot air and water from the waste heat [401–413]. A typical capacity range of home fuel cell is 1–3 kW_{el}/48 kW_{th}. CHP systems linked to absorption chillers use their waste heat for refrigeration. The waste heat from fuel cells can be diverted during the summer directly into the ground providing further cooling while the waste heat during winter can be pumped directly into the building [368–383].

Cogeneration systems can reach 85% efficiency (40%–60% electric + remainder as thermal phosphoric-acid fuel cells [PAFCs] comprise [355] the largest segment of existing CHP products worldwide and can provide combined efficiencies close to 90% [402–413]. MCFCs and SOFCs are also used for combined heat and power generation and have electrical energy efficiency around 60% [402–413]. Disadvantages of cogeneration systems include slow ramping up and down rates, high cost, and short lifetime [331]. Also their need to have a hot water storage tank to smooth out the thermal heat production is a serious disadvantage in the domestic market place where space in domestic properties is at a great premium [331]. Since 2012, more number of micro-combined heat and power systems than conventional systems were used [401–413].

10.10.3.2 Fuel Cell Vehicles

As of 2015, two fuel cell vehicles were introduced for commercial lease and sale in limited quantities: the Toyota Mirai and the Hyundai ix35 FCEV. Following the leads by Toyota and Hyundai, General Motors Honda, Mercedes-Benz, and Nissan are planning to manufacture fuel cell vehicles [331,381–400].

The DOE's Fuel Cell Technology Program claims that, as of 2011, fuel cells achieved 53%–59% efficiency at one-quarter power and 42%–53% vehicle efficiency at full power and a durability of over 75,000 mile with less than 10% degradation [331,381–400]. With a WTW simulation analysis, General Motors and its partners estimated that per mile traveled, a fuel cell electric vehicle running on compressed gaseous hydrogen produced from natural gas could use about 40% less energy and emit 45% less greenhouse gasses than an internal combustion vehicle [331,381].

As of August 2011, there were a total of approximately 100 fuel cell buses deployed around the world. This number has been rapidly increased in last few years. China alone added 333 fuel cell buses in 2015. Most buses are produced by UTC Power, Toyota, Ballard, Hydrogenics, and Proton Motor. Fuel cell buses have a 39%–141% higher fuel economy than diesel buses and natural gas buses. Fuel cell buses have been deployed around the world, including in Whistler, Canada; San Francisco, United States; and Hamburg, Germany [331,381].

A fuel cell forklift (also called a fuel cell lift truck) is a fuel cell–powered industrial forklift truck used to lift and transport materials. In 2013, there were over 4000 fuel cell forklifts used in material handling in the United States, of which only 500 received funding from DOE. The global market is 1 million fork lifts per year Fuel cell fleets are operated by various companies, including Sysco Foods and FedEx Freight. Europe demonstrated 30 fuel cell forklifts with Hylift. In 2011, Pike Research stated that fuel cell–powered forklifts will be the largest driver of hydrogen fuel demand by 2020 [331,381–400]. Fuel cell–powered forklifts can provide benefits over battery-powered forklifts as they can work for a full 8 h shift on a single tank of hydrogen and can be refueled in 3 min. Fuel cell–powered forklifts can be used in refrigerated warehouses, as their performance is not degraded by lower temperatures [331,381–400].

In 2005, a British manufacturer of hydrogen-powered fuel cells, Intelligent Energy (IE), produced the first working hydrogen-run motorcycle called the ENV (Emission Neutral Vehicle). In 2004 Honda developed a fuel cell motorcycle that utilized the Honda FC Stack [331,390]. Taiwan and Italy are also involved in IE fuel cell motor bikes [331,381–400].

In 2003, the world's first propeller-driven airplane to be powered entirely by a fuel cell was flown. The fuel cell was a unique FlatStack™ stack design, which allowed the fuel cell to be integrated with the aerodynamic surfaces of the plane [331,381–383]. In February 2008, Boeing researchers and industry partners throughout Europe conducted experimental flight tests of a manned airplane powered only by a fuel cell and lightweight batteries. This plane used a PEMFC/lithium-ion battery hybrid system to power an electric motor, which was coupled to a conventional propeller [331,381–400].

The world's first fuel cell boat HYDRA used an AFC system with 6.5 kW net output. Iceland committed to converting its vast fishing fleet to use fuel cells to provide auxiliary power has been and, eventually, to provide primary power in its boats. Amsterdam recently introduced its first fuel cell–powered boat that ferries people around the city's famous and beautiful canals [331,381–400].

The Type 212 submarines of the German and Italian navies use fuel cells to remain submerged for weeks without the need to surface. The U212A is a nonnuclear submarine developed by German naval shipyard Howaldtswerke-Deutsche Werft. The system consists of nine PEM fuel cells, providing between 30 and 50 kW each. The submarine is silent, giving it an advantage in the detection of other submarines [331,381–400].

10.10.4 MARKETS AND ECONOMICS

In 2012, fuel cell industry revenues exceeded $1 billion market value worldwide, with Asia Pacific countries shipping more than three-fourths of the fuel cell systems worldwide [327–340,382,383]. However, as of October 2013, no public company in the industry had yet become profitable [331]. There were 140,000 fuel cell stacks shipped globally in 2010 [382,383]. Tanaka Kikinzoku Kogyo K.K. expanded its production facilities for fuel cell catalysts in 2013 to meet anticipated demand [382,383].

Approximately 50% of fuel cell shipments in 2010 were stationary fuel cells, and the four dominant producers in the fuel cell industry were the United States, Germany, Japan, and South Korea [331,333,337,382,383]. While there seems to sufficient platinum available for fuel cell [331,333,337,382,383] in 2007, research at Brookhaven National Laboratory suggested that platinum could be replaced by a gold-palladium coating, which may be less susceptible to poisoning and thereby improving fuel cell lifetime [327–340]. Another method would be to use iron and sulfur instead of platinum. This would lower the cost of a fuel [382,383].

REFERENCES

1. Shah, Y.T., *Energy and Fuel Systems Integration*. CRC Press, New York (2015).
2. Shah, Y.T., *Water for Energy and Fuel Production*. CRC Press, New York (2014).
3. Natural gas, Wikipedia, the free encyclopedia (2015).
4. Natural gas overview, A report by Naturalgas.org (2011). Retrieved February 6, 2011.
5. International Statistics, EIA, DOE, Washington, D.C. (2013). Accessed December 1, 2013.
6. Understanding combined heat and power, A report by Alfagy.com (2012). www.alfagy.com. Retrieved November 2, 2012.
7. Worldwide NGV statistics, A report by Ngv Journal (2015). www.ngvjournal.com. Retrieved January 2015.
8. Cryogenic aircraft—A report in global security.org (2011). http://en.wikipedia.org/wiki/Tupolev (last accessed August 29, 2016).
9. Natural gas vehicles, Wikipedia, the free encyclopedia (2015). https://en.wikipedia.org/wiki/Natural_gas_vehicle (last accessed October 2, 2016).
10. Maki, K., Repo, S., and Jarventausta, P., Impact of distribution generator as a part of distribution network planning, A report by Tampere University of Technology, Institute of Power Engineering, Tampere, Finland (2001), https://en.wikipedia.org/wiki/Tampere_University_of_Technology.
11. Farrell, J., Challenge of reconciling a centralized versus decentralized electricity system, A report by Institute for Local Self-Reliance (October 17, 2011).
12. Martin, J., Distributed vs. centralized electricity generation: Are we withstanding a challenge of paradigm?—An introduction to distributed generation, A project Under the supervision of Antoine Hyafil, Dean of the Energy Track at HEC Paris, HEC, Paris, France (May 2009).
13. From centralized power generation to a distributed model a report by ABB, ABB SACE, Una divisione di ABB S.p.A. Interruttori B.T., Via Baioni, Italy (2014).
14. Gueissaz, N., Distributed vs. centralized energy storage for power system applications, Supervisor: Dr Damian Flynn, University College Dublin, Dublin, Ireland (2013–2014).
15. Singh, S. and Kori, A.K., Centralized and decentralized distributed power generation in today's scenario, *IOSR Journal of Electrical and Electronics Engineering (IOSR-JEEE)*, 4 (5), 40–45 (January–February 2013).
16. Jarquin-Laguna, A., Fluid power network for centralized electricity generation in offshore wind farms, The Science of Making Torque from Wind 2014 (TORQUE 2014), *IOP Publishing Journal of Physics: Conference Series*, 524, 012075 (2014).
17. El-Kattam, W. and Salama, M.M.A., Distributed generation technologies: Definitions and benefits, *Electric Power Research*, 71, 119–128 (2004).
18. Pepermans, G., Driesen, J., Haeseldonckx, D., Belmans, R., and D'haeseleer, W., Distributed generation: Definition, benefits and issues, *Energy Policy*, 33, 787–798 (2005).
19. Willis, H.L. and Scott, W.G., *Distributed Power Generation Planning and Evaluation*. Marcel Dekker Inc., New York (2000).
20. Khan, U.N., Impact of distributed generator on electrical power network, Wroclaw University of Technology, Wroclaw, Poland (2008).
21. Ackermann, T., Andersson, G., and Söder, L., Distributed generation: A definition, *Electric Power Systems Research*, 57, 195–204 (2001).
22. Carley, S., Distributed generation: An empirical analysis of primary motivators, *Energy Policy*, 37, 1648–1659 (2009).
23. Chambers, A., *Distributed Generation: A Nontechnical Guide*. PennWell, Tulsa, OK, p. 283 (2001).
24. Cossent, R., Gomez, T., and Frias, P., Towards a future with large penetration of distributed generation: Is the current regulation of electricity distribution ready? Regulatory recommendations under a European perspective, *Energy Policy*, 37, 1145–1155 (2009).
25. Costa, M.P., Matos, M.A., and Pecas Lopes, J.A., Regulation of micro generation and micro grids, *Energy Policy*, 36, 3893–3904 (2008).
26. Dondi, P., Bayoumi, D., Haederli, C., Julian, D., and Suter, M., Network integration of distributed power generation, *Journal of Power Sources*, 106, 1–9 (2002).
27. 2008 32nd power generation order survey, A report by Diesel and Gas Turbine Worldwide (2008), http://www.dieselgasturbine.com/pdf/power_2008.pdf#zoom=100.
28. Monthly energy review March 2009, A report by Energy Information Administration, DOE, Washington, D.C. (2009), http://www.eia.doe.gov/mer.

29. Metz, B., Davidson, O., Bosch, P., Dave, R., and Meyer, L. (eds.). *Climate Change 2007: Mitigation of Climate Change, Contribution of Working Groups III in the Fourth Assessment of Report of IPCC*, Cambridge University Press, Cambridge, U.K. (2007).

30. Hirsh, R.F., *Technology and Transformation in the American Electric Utility Industry.* Cambridge University Press, New York (1989).

31. IEA, Brussels, Belgium, A report by *Distributed Generation in a Liberalized Energy Market.* Jouve, France (2002).

32. Distributed generation and renewables outlook 2030, A report by IEA, Brussels, Belgium (2003), http://www.iea.org/textbase/work/2004/distgen/Birol.pdf.

33. Jarraud, A. and Steiner, A., *Climate Change 2007—Impacts, Adaptation and Vulnerability*, Fourth assessment report of IPEE, Cambridge University Press, Cambridge, U.K. (2007).

34. Jörss, W., Jorgensen, B.H., Löffler, P., Morthorst, P.E., Uyterlinde, M., Sambeek, E., and Wehnert, T., *Decentralized Power Generation in the Liberalized EU Energy Markets.* Springer, New York (2003).

35. Lehtonen, M. and Nye, S., History of electricity network control and distributed generation in the UK and Western Denmark, *Energy Policy* 37 (6), 2338–2345 (2009).

36. McDonald, J., Adaptive intelligent power systems: Active distribution networks, *Energy Policy*, 36, 4346–4351 (2008).

37. Mendez, V.H., Rivier, J., de la Fuente, J.I., Gomez, T., Arceluz, J., and Marin, J., *Impact of Distributed Generation on Distribution Network.* Universidad Pontificia Comillas, Madrid, Spain (2002).

38. Review of distributed generation, A report by Office of Gas and Electricity Markets, (OFGEM), London, U.K. (2007), www.dti.gov.uk/energy/whitepaper.

39. Pehnt, M., Micro cogeneration technology, in Pehnt, M., Cames, M., Fischer, C., Praetorius, B., Shneider, L., Schumacher, K., and Voss, J.P. (eds.), *Micro Cogeneration Towards Decentralized Energy Systems.* Springer, Berlin, Germany, pp. 197–218 (2006).

40. Schneider, L. and Pehnt, M., Embedding micro cogeneration in the energy supply system, in Pehnt, M., Cames, M., Fischer, C., Praetorius, B., Shneider, L. (eds.), *Micro Cogeneration*, Chapter 9. Springer Science, Berlin, Germany, pp. 197–218 (2006).

41. Jänig, C., Perspektive. Lokal Energie—Geschäftsbericht 2001. Stadtwerke Unna GmbH, Unna, Germany (2002).

42. Praetorius, B., Micro cogeneration—Setting of an emerging market, in Pehnt, M., Cames, M., Fischer, C., Praetorius, B., Shneider, L., Schumacher, K., and Voss, J.P., *Micro Cogeneration Towards Decentralized Energy Systems.* Springer, Berlin, Germany, pp. 145–170 (2006).

43. Jenkins, N., Embedded generation—Part 2, *IEE Power Engineering Journal*, 233–239 (1996).

44. Feldmann, W., Dezentrale Energieversorgung-zukünftige Entwicklungen, technische Anforderungen, in *Conference; Energie Innovativ 2002*, VDI Verlag, Düsserldorf, Germany (2002).

45. The potential benefits of distributed generation and rate-related issues that may impede their expansion. A Study pursuant to Section 1817 of the Energy Policy Act of 2005, US Department of Energy, Washington, D.C. (2007).

46. US EPA, Inventory of US greenhouse gas emissions and sinks: 1990–2001, EPA 430-R-03-004, US Environmental Protection Agency, Washington, DC (2003).

47. World survey of decentralized energy, 2006, A report by World Alliance for Decentralized Energy (WADE) (2006), http://www.localpower.org/nar_publications.html.

48. Arndt, U. and Wagner, U., Energiewirtschafltiche Auswirkungen eines Virtuellen Brennstoffzellen-Kraftwerks. VDI-Berichte (2003).

49. Fang, X., Misra, S., Xue, G., and Yang, D., Smart grid—The new and improved power grid—A survey, *IEEE Communications Surveys and Tutorials*, 4 (4), 944–980 (December 2011).

50. Boardman, R., Advanced energy systems-nuclear-fossil-renewable hybrid systems, A report to Nuclear Energy Agency-committee for technical and economical studies on nuclear energy development and fuel cycle, INL, Idaho Falls, ID (April 4–5, 2013).

51. Antkowiak, M., Ruth, M., Boardman, R., Bragg-sitton, S., Cherry, R., and Shunn, L., Summary report of the INL-JISEA workshop on nuclear hybrid energy systems, Prepared under Task No. 6A50.1027, NREL, Golden, CO (July 2012), Report available at: http://www.nrel.gov/docs/fy12o-sti/55650.pdf, Presentations of Workshop, Available at: https://inlportal.inl.gov/portal/server.pt?tbb=hybrid.

52. Chen, T., Survey of cyber security issues in smart grids, in *Cyber Security, Situation Management, and Impact Assessment II; and Visual Analytics for Homeland Defense and Security II (Part of SPIE DSS*; Orlando, FL (April 5–9, 2010), pp. 77090D-1–77090D-11 (2010).

53. Akyol, B., Kirkham, H., Clements, S., and Hadley, M., A survey of wireless communications for the electric power system, Prepared for the U.S. Department of Energy, DOE, Washington, D.C. (2010).

54. Gungor, V. and Lambert, F., A survey on communication networks for electric system automation, *Computer Networks*, 50 (7), 877–897 (2006).

55. Wang, W., Xu, Y., and Khanna, M., A survey on the communication architectures in smart grid, *Computer Networks*, 55, 3604–3629 (2011).

56. NIST framework and roadmap for smart grid interoperability standards, release 1.0 Office of the National Coordinator for Smart Grid Interoperability, NIST Special Publication 1108, NIST, Department of Commerce, Washington, D.C. (January 2010), http://www.nist.gov/public affairs/releases/upload/smart-grid interoperabilityfinal.pdf.

57. Neely, M., Tehrani, A., and Dimakis, A., Efficient algorithms for renewable energy allocation to delay tolerant consumers, in *IEEE SmartGridComm'10*, pp. 549–554 (2010).

58. Sortomme, E., Hindi, M., MacPherson, S., and Venkata, S., Coordinated charging of plugin hybrid electric vehicles to minimize distribution system losses, *IEEE Transactions on Smart Grid*, 2 (1), 198–205 (2011).

59. Caldon, R., Patria, A., and Turri, R., Optimal control of a distribution system with a virtual power plant, in *Bulk Power System Dynamics and Control Conference*, Cortina d'Ampezzo, Italy (August 22–27, 2004), pp. 278–284 (2004).

60. Bakker, V., Bosman, M., Molderink, A., Hurink, J., and Smit, G., Demand side load management using a three step optimization methodology, in *IEEE SmartGridComm'10*, Gaithersburg, MD (October 4–6, 2010), pp. 431–436 (2010).

61. A smart grid is an optimized grid, A white paper by ABB, Cary, NC (2013). www.abb.com.

62. Smart grid optimization, A white paper by ECHELON, Menwith Hill, U.K. (2013).

63. Ahat, M., Amor, S., Bui, M., Bui, A., Guerard, G., and Petermann, C., Smart grid and optimization, *American Journal of Operations Research*, 3, 196–206 (2013).

64. Nygard, K., Ghosen, S., Chowdhury, M., Loegering, D., and McCulloch, R., Optimization models for energy reallocation in a smart grid, in *IEEE, Information Theory Workshop*, Paraty, Brazil (October 16–20, 2011), p. 186 (2011).

65. Rahimi, F. and Ipakchi, A., Demand response as a market resource under the smart grid paradigm, *IEEE Transactions on Smart Grid*, 1 (1), 82–88 (2010).

66. Sauter, T. and Lobashov, M., End-to-end communication architecture for smart grids, *IEEE Transactions on Industrial Electronics*, 58 (4), 1218–1228 (2011).

67. Schneider, K., Gerkensmeyer, C., Kintner-Meyer, M., and Fletcher, R., Impact assessment of plug-in hybrid vehicles on pacific northwest distribution systems, in *Power & Energy Society General Meeting*, Pittsburgh, PA (July 20–24, 2008), pp. 1–6 (2008).

68. Simmhan, Y., Giakkoupis, M., Cao, B., and Prasanna, V., On using cloud platforms in a software architecture for smart energy grid, in *IEEE 2nd International Conference on Cloud computing (Cloudcom)*, Indiana, Indianapolis, IN (November 30–December 3, 2010) (December 2010).

69. Bennett, C. and Highfill, D., Networking AMI smart meters, in *Proceeding of IEEE Energy 2030 Conference'08*, Atlanta, GA (November 17–18, 2008), pp. 1–8 (2008).

70. Berthier, R., Sanders, W., and Khurana, H., Intrusion detection for advanced metering infrastructures: Requirements and architectural directions, in *IEEE SmartGridComm'10*, pp. 350–355 (2010).

71. Andersen, P., Poulsen, B., Decker, M., Træholt, C., and Østergaard, J., Evaluation of a generic virtual power plant framework using service oriented architecture, in *IEEE PECon'08*, pp. 1212–1217 (2008).

72. Baldick, R. et al., Initial review of methods for cascading failure analysis in electric power transmission systems, in *IEEE Power and Energy Society General Meeting'08*, pp. 1–8 (2008).

73. Noda, K., Japan's activity on international standardization of smart grid, a paper by Ministry of Economy, Trade and Industry, Tokyo, Japan (November 17, 2011).

74. Gungor, V., Lu, B., and Hancke, G., Opportunities and challenges of wireless sensor networks in smart grid, *IEEE Transactions of Industrial Electronics*, 57 (10), 3557–3564 (2010).

75. Parikh, P., Kanabar, M., and Sidhu, T., Opportunities and challenges of wireless communication technologies for smart grid applications, in *IEEE Power and Energy Society General Meeting'10*, Minneapolis, MN (July 26–28, 2010).

76. Bou Ghosn, S., Ranganathan, P., Salem, S., Tang, J., Loegering, D., and Nygard, K., Agent-oriented designs for a self healing smart grid, in *IEEE SmartGridComm'10*, Gaithersburg, MD (October 4–6, 2010), pp. 461–466 (2010).

77. Brown, R., Impact of smart grid on distribution system design, in *IEEE Power and Energy Society General Meeting*, Pittsburgh, PA (July 20–24, 2008)—*Conversion and Delivery of Electrical Energy in the 21st Century*, pp. 1–4 (2008).

78. Brucoli, M. and Green, T., Fault behavior in islanded micro grids, in *19th International Conference on Electricity Distribution*, Vienna, Austria (May 21–24, 2007), pp. 1–4 (2007).

79. Caron, S. and Kesidis, G., Incentive-based energy consumption scheduling algorithms for the smart grid, in *IEEE SmartGridComm'10*, pp. 391–396 (2010).

80. Coll-Mayor, D., Paget, M., and Lightner, E., Future intelligent power grids: Analysis of the vision in the European Union and the United States, *Energy Policy*, 35, 2453–2465 (2007).

81. Colson, C. and Nehrir, M., A review of challenges to real time power management of micro grids, in *IEEE Power & Energy Society General Meeting*, Calgary, Alberta, Canada (July 26–30, 2009), pp. 1–8 (2009).

82. Conejo, A., Morales, J., and Baringo, L., Real-time demand response model, *IEEE Transactions on Smart Grid*, 1 (3), 236–242 (2010).

83. Deep, U., Petersen, B., and Meng, J., A smart microcontroller based iridium satellite communication architecture for a remote renewable energy source, *IEEE Transactions on Power Delivery*, 24 (4), 1869–1875 (2009).

84. Driesen, J. and Katiraei, F., Design for distributed energy resources, *IEEE Power & Energy Magazine*, 6 (3), 30–40 (2008).

85. Driesen, J., Vermeyen, P., and Belmans, R., Protection issues in microgrids with multiple distributed generation units, in *Power Conversion Conference'07*, pp. 646–653 (2007).

86. Efthymiou, C. and Kalogridis, G., Smart grid privacy via anonymization of smart metering data, in *IEEE SmartGridComm'10*, pp. 238–243 (2010).

87. European Committee for Electrotechnical Standardization (CENELEC), Smart Meters Coordination Group: Report of the second meeting held on 2009-09-28 and approval of SM-CG work program for EC submission, Brussels, Belgium (2009).

88. Fang, X., Yang, D., and. Xue, G., Online strategizing distributed renewable energy resource access in islanded microgrids, in *IEEE Globecom'11*, Houston, TX, (December 6–8, 2011). www.ieee.globecom.org/2011.

89. Wright, D. et al., Assessment of demand response and advanced metering, Staff Report, Federal Energy Regulatory Commission, Washington, D.C. (February 2011), http://www.ferc.gov/legal/staff-reports/2010-dr-report.pdf.

90. Guan, X., Xu, Z., and Jia, Q.-S., Energy-efficient buildings facilitated by microgrid, *IEEE Transactions on Smart Grid*, 1 (3), 243–252 (2010).

91. Ibars, C., Navarro, M., and Giupponi, L., Distributed demand management in smart grid with a congestion game, in *IEEE SmartGridComm'10*, pp. 495–500 (2010).

92. Kim, Y.-J., Thottan, M., Kolesnikov, V., and Lee, W., A secure decentralized datacentric information infrastructure for smart grid, *IEEE Communications Magazine*, 48 (11), 58–65 (2010).

93. Laaksonen, H., Protection principles for future microgrids, *IEEE Transactions on Power Electronics*, 25 (12), 2910–2918 (2010).

94. Lasseter, R. Smart distribution: Coupled micro grids, *Proceedings of the IEEE*, 99 (6), 1074–1082 (2011).

95. Lasseter, R. and Paigi, P., Micro grid: A conceptual solution, in *PESC'04*, pp. 4285–4290 (2004).

96. Mohagheghi, S., Stoupis, J., Wang, Z., Li, Z., and Kazemzadeh, H., Demand response architecture: Integration into the distribution management system, in *IEEE SmartGridComm'10*, Gaithersburg, MD, pp. 501–506 (2010).

97. Nikkhajoei, H. and Lasseter, R., Microgrid protection, in *IEEE Power Engineering Society General Meeting'07*, pp. 1–6 (2007).

98. Souryal, M., Gentile, C., Griffith, D., Cypher, D., and Golmie, N., A methodology to evaluate wireless technologies for the smart grid, in *IEEE SmartGridComm'10*, pp. 356–361 (2010).

99. Taneja, J., Culler, D., and Dutta, P., Towards cooperative grids: Sensor/actuator networks for renewables integration, in *IEEE SmartGrid-Comm'10*, pp. 531–536 (2010).

100. Wang, Z., Scaglione, A., and Thomas, R., Compressing electrical power grids, in *IEEE SmartGridComm'10*, pp. 13–18 (2010).

101. You, S., Træholt, C., and Poulsen, B., Generic virtual power plants: Management of distributed energy resources under liberalized electricity market, in *The Eighth International Conference on Advances in Power System Control, Operation and Management*, Hong Kong, China (November 8–11, 2009), pp. 1–6 (2009).

102. Tumilty, R., Elders, I., Burt, G., and McDonald, J., Coordinated protection, control & automation schemes for microgrids, *International Journal of Distributed Energy Resources*, 3 (3), 225–241 (2007).

103. Vandoorn, T., Renders, B., Degroote, L., Meersman, B., and Vandevelde, L., Active load control in islanded microgrids based on the grid voltage, *IEEE Transactions on Smart Grid*, 2 (1), 139–151 (2011).

104. Lu, Z., Lu, X., Wang, W., and Wang, C., Review and evaluation of security threats on the communication networks in the smart grid, in *Military Communications Conference'2010*, pp. 1830–1835 (2010).
105. Zareipour, H., Bhattacharya, K., and Canizares, C., Distributed generation: Current status and challenges, in *NAPS'04*, pp. 1–8 (2004).
106. Balijepalli, V., Pradhan, V., Khaparde, S., and Shereef, R., Review of demand response under smart grid paradigm, in *IEEE PES Innovative Smart Grid Technologies—Kollam,* Kerala, India (December 1–3, 2011).
107. Shamshiri, M., Gan, C., and Tan, C., A review of recent development in smart grid and micro-grid laboratories, in *2012 IEEE International Power Engineering and Optimization Conference (PEOCO 2012)*, Melaka, Malaysia (June 6–7, 2012).
108. Anderson, R., Boulanger, A., Powell, W., and Scott, W., Adaptive stochastic control for the smart grid, *Proceedings of the IEEE*, 99 (6), 1098–1115 (2011).
109. Moslehi, K. and Kumar, R., A reliability perspective of the smart grid, *IEEE Transactions on Smart Grid*, 1 (1), 57–64 (2010).
110. Ipakchi, A. and Albuyeh, F., Grid of the future, *IEEE Power and Energy Magazine*, 7 (2), 52–62 (2009).
111. Mohsenian-Rad, A. and Leon-Garcia, A., Optimal residential load control with price prediction in real-time electricity pricing environments, *IEEE Transactions on Smart Grid*, 1 (2), 120–133 (2010).
112. Hassan, R. and Radman, G., Survey on smart grid, in *IEEE SoutheastCon 2010*, Charlotte-Concord, NC (March 18–21, 2010), pp. 210–213 (March 18–21, 2010).
113. Uslar, M., Rohjansand, S., Bleiker, R., GonzÅLalez, J., Specht, M., Suding, T., and Weidelt, T., Survey of smart grid standardization studies and recommendations—Part 2, in *IEEE PES'10*, pp. 1–6 (2010).
114. Rohjansand, S., Uslar, M., Bleiker, R., GonzÅLalez, J., Specht, M., Suding, T., and Weidelt, T., Survey of smart grid standardization studies and recommendations, in *IEEE SmartGridComm'10* (October 4–6, 2010), pp. 583–587 (2010).
115. Brown, H. and Suryanarayanan, S., A survey seeking a definition of a smart distribution system, in *North American Power Symposium'09*, Starkville, MS (October 5–7, 2009), pp. 1–7 (2009).
116. Vasconcelos, J., Survey of regulatory and technological developments concerning smart metering in the European Union electricity market, EUI RSCAS PP, EU, Brussels, Belgium (2008), http://cadmus.eui.eu/handle/1814/9267.
117. Baumeister, T., Literature review on smart grid cyber security, Technical report, University of Hawaii, Honolulu, HI (2010), http://csdl.ics.hawaii.edu/techreports/10-11/10-11.pdf.
118. Gas turbine, Wikipedia, the free encyclopedia (2015). https://en.www.wikipedia.org/wiki/gas_turbine (last accessed October 24, 2016).
119. Capstone turbine, Wikipedia, the free encyclopedia (2015). https://en.www.wikipedia.org/wiki/capstone_turbine (last accessed July 21, 2016).
120. Bakken, L.E. et al., Centenary of the first gas turbine to give net power output: A tribute to Ægidius elling, in *ASME*, pp. 83–88 (2004).
121. How gas turbine power plant works, A report from office of fossil energy, Department of Energy, Washington, DC (2015), energy.gov/.../how-gas-turbine-pow.
122. Gas Turbines breaking the 60% efficiency barrier, *Decentralized Energy Magazine* (January 5, 2010), Retrieved October 15, 2015. www.peimagazine.com.
123. Tamarin, Y., *Protective Coatings for Turbine Blades*. ASM International, Materials, Park, OH, pp. 3–5 (2002).
124. Latief, F.H. and Kakehi, K., Effects of Re content and crystallographic orientation on creep behavior of aluminized Ni-based single crystal superalloys, *Materials & Design*, 49, 485–492 (2013).
125. Ratliff, P., Garbett, P., and Fischer, W., The New Siemens Gas Turbine SGT5-8000H for more customer benefit (PDF), A report by VGB PowerTech, Siemens Power Generation, Erlangen, Germany (September 2007), Retrieved July 17, 2010.
126. MTT—Leading turbine innovation, A report in Marineturbine.com, Retrieved August 13, 2012. https://pinterest.com/pin/4908220593591574091/.
127. Kay, A., *German Jet Engine and Gas Turbine Development 1930–1945*. Airlife Publishing, Ramsbury, U.K. (2002).
128. Hunt, R., The history of the industrial gas turbine (part I—the first fifty years 1940–1990), A report by Thermal Power Consultant, Morpeth, U.K. (January 2011).
129. Lamb, J. and Duggan, R., Operation of a marine gas turbine under sea conditions, *Journal of the American Society for Naval Engineers*, 66, 457–466 (1954).
130. GE goes from installation to optimized reliability for cruise ship gas turbine installations, A report by GE—Aviation, Evendale, OH (March 16, 2004), Geae.com, Retrieved August 13, 2012.

131. MHI achieves 1,600°C turbine inlet temperature in test operation of world's highest thermal efficiency J-Series Gas Turbine, A report by Mitsubishi Heavy Industries, Tokyo, Japan (May 26, 2011), Archived from the original on November 13, 2013.

132. Cocco, D., Deiana, P., and Cau, G., Performance evaluation of small size externally fired gas turbine (EFGT) power plants integrated with direct biomass dryers, *Energy*, 31 (10–11), 1459–1471 (2006).

133. Evans, R.L. and Zaradic, A.M., Optimization of a wood-waste-fuelled indirectly fired gas turbine cogeneration plant, *Bioresource Technology*, 57, 117–126 (1996).

134. Ferreira, S.B. and Pilidis, P., Comparison of externally fired and internal combustion gas turbines using biomass fuel, *ASME Journal of Energy Resources Technology*, 123, 291–296 (2001).

135. Kusterer, K., Braun, R., and Bohn, D., Organic Rankine cycle working fluid selection and performance analysis for combined application with a 2 MW class industrial gas turbine, in *Proceedings of ASME Turbo Expo 2014: GT2014*, Düsseldorf, Germany (June 16–20, 2014).

136. Kautz, M. and Hansen, U., The externally-fired gas-turbine (EFGT-Cycle) for decentralized use of biomass, *Applied Energy*, 84 (7–8), 795–805 (2007).

137. Knoef, H., The indirectly fired gas turbine for rural electricity production from biomass, Project Brochure and Reports, Contract FAIR-CT95-0291 (1998); also same paper in Knoef, H., Wagenaar, B., and Reumerman, P., *Proceedings of the 10th European Biorefinery Conference*, Wurgberg, pp. 1334–1337, C.A.R.M.E.N., Rimpar, Germany (1998).

138. Arvay, P., Muller, M.R., and Ramdeen, V., Economic implementation of the organic rankine cycle in industry, in *ACEEE Summer Study on Energy Efficiency in Industry*, ACEEE, Washington, DC (2011).

139. Pantaleo, A., Camporeale, S., and Shah, N., Thermo-economic assessment of externally fired micro gas turbine fired by natural gas and biomass: Applications in Italy, *Energy Conversion and Management*, 75, 202–213 (2013).

140. Pantaleo, A., Shah, N., and Keirstead, J., Bioenergy and other renewables in urban energy systems, in Keirstead, J. and Shah, N. (eds.), *Urban Energy Systems—An Integrated Approach*. Routledge, New York (2013).

141. Riccio, G., Martelli, F., and Maltagliati, S., Study of an external fired gas turbine power plant fed by solid fuel, in *Proceeding of ASME Turbo Expo 2000*, ASME editor, Paper 0015-GT-2000.

142. Soltani, S., Mahmoudi, S.M.S., Yari, M., and Rosen, M.A., Thermodynamic analyses of an externally fired gas turbine combined cycle integrated with a biomass gasification plant, *Energy Conversion and Management*, 70, 107–115 (2013).

143. Yan, J. and Eidensten, L., Status and perspective of externally fired gas turbines, *Journal of Propulsion and Power*, 16(4), 572–596 (2000).

144. Gemmer, B., High efficiency microturbine with integral heat recovery—Improving the operating efficiency of microturbine-based distributed generation at an affordable price, U.S. Department of Energy, Energy Efficiency and Renewable Energy, Advanced Manufacturing Office, DOE/EE-0432, DOE, Washington, D.C. (June 2014).

145. Hillman, S., Elliott micro-turbines, A report by Elliott Micro-Turbines EBARA Group, Erie, PA (2015).

146. Behavior of capstone and honeywell microturbine generators during load changes, Prepared by CERT Program Office, Lawrence Berkeley National Laboratory, Contract No. 150-99-00 for Project Manager Don Kondoleon, California Energy Commission, Consultant Report, P500-04-021 (February 2004).

147. do Nascimento, M.A.R., de Oliveira Rodrigues, L., Cruz dos Santos, E., Batista Gomes, E.E., Goulart Dias, F.L., Gutiérrez Velásques, E.I., and Miranda Carrillo, R.A., Micro gas turbine engine: A review, Chapter 5, in Benini, E. (ed.), *Progress in Gas Turbine Performance*. INTECH, pp. 107–141 (2009), http://dx.doi.org/10.5772/54444, https://de_seribd.com/document/173583097/Progress.

148. ASME, Performance test code PTC-22-1997, Gas turbine power plants (1997).

149. Barker, T.M., Catalysts and electronics, power-gen international 96, *Turbomachinery*, 38 (1), 19–21 (1997).

150. Camporeale, S.M. and Pantaleo, A.M., Influence of heat demand on techno-economic performance of a biomass/natural gas micro gas turbine and bottoming ORC for cogeneration, Paper ID: 182, in *Third International Seminar on ORC Power Systems*, October 12–14, 2015, Brussels, Belgium, p. 1 (2015).

151. Bolszo, C.D. and McDonell, V.G., Emissions optimization of a biodiesel fired gas turbine, in *Proceedings of the Combustion Institute*, Vol. 32, Elsevier, Amsterdam, the Netherlands, pp. 2949–2956 (2009).

152. Capstone low emissions microturbine technology, a White paper by Capstone Turbine Corporation, Los Angeles, CA (2000).

153. Capstone microturbine model 330 system operation manual, A report by Capstone Turbine Corporation, Los Angeles, CA (2001).

154. Capstone microturbine product catalog, A report by Capstone Turbine Corporation, Los Angeles, CA (2012), http://www.capstoneturbine.com/prodsol/products/, Accessed at June 20, 2012.

155. Cohen, H., Rogers, G.F.C., and Saravanamuttoo, H.I.H., *Gas Turbine Theory*, 4th edn., in Coher et al. (eds.), Addison-Wesley, Salt Lake City, UT (1996).

156. Dunn, S. and Flavin, C., Dimensionando a Microenergia, in Estado do Mundo 2000, Brazil, UMA Ed. (2000).

157. Gomes, E.E.B., Análise Técnico-econômica e Experimental de Microturbinas a Gás Operando com Gás Natural e Óleo Diesel, Master degree thesis, Supervised by Nas-cimento, M.A.R. and Lora, E.E.S., Federal University of Itajubá, Pinheirinho, Brazil (2002).

158. Hamilton, S.L., Microturbines, Chapter 3, in Chambers, A. (ed.), *Distributed Generation: A Nontechnical Guide*. PennWell Corporation, Tulsa, OK, pp. 33–72 (2001).

159. Kincaid, D., The role of Distributed Generation in competitive energy markets, Distributed Generation Forum, Gas Research Institute (GRI), GRI-99/0054, Chicago, IL (March 1999).

160. Liss, W.E., Natural gas power systems for the distributed generation market, in *Power-Gen International '99 Conference*, CD-Rom, New Orleans, LA (1999).

161. Nascimento, M.A.R. and Santos, E.C., Biofuel and gas turbine engines, Chapter 6, in Benini, E. (ed.), *Advances in Gas Turbine Technology*. InTech (January 2011).

162. Nwafor, O., Emission characteristics of Diesel engine operating on rapeseed methyl ester, *Renewable Energy*, 29, 119–129 (2004).

163. Pierce, J.L., Microturbine distributed generation using conventional and waste fuel, *Cogeneration and On-Site Power Production*, 3 (10), 45 (January–February 2002), James & James Science Publishers.

164. Wendig, D., Bio fuel in micro gas turbines, in *Workshop: Bio-Fuelled Micro Gas Turbines in Europe—Market Opportunities and R&D Requirements*, Brussels, Belgium (September 24, 2004), www.bioturbine.org.

165. Willis, H.L. and Scott, W.G., *Distributed Power Generation: Planning and Evaluation*. CRC Press, Taylor and Francis Group, New York (January 2000).

166. Scott, W.G., *Micro Gas Turbine Cogeneration Applications*. International Power and Light Co. (2000).

167. Sierra, R.G.A., Experimental test and thermal economical analysis of biofuel used in regenerative micro gas turbine engine, Master thesis, UNIFEI, Minas Gerais, Brazil (2008).

168. Nageimento, M. et al., Micro gas turbine engine: A review, InTech, Vol. 141. *Progress in Gas Turbine Performance*, Chapter 5, 107–141 (2014).

169. Watts, J.H., Microturbines: A new class of gas turbine engine, *Global Gas Turbine News*, ASME-IGTI, 39 (1), 4–8 (1999).

170. Weston, F., Seidman, N.L., and James, C., Model regulations for the output of specified air emissions from smaller-scale electric generation resources, The Regulatory Assistance Project (2001).

171. do Nascimento, M.A.R. and dos Santos, E.C., Biofuel and gas turbine engines, in Benini, E. (ed.), *Advances in Gas Turbine Technology,* Chapter 6. InTech (2011), Available from: http://www.intechopen.com/books/advances-in-gas-turbine-technology/biofuel-and-gas-turbine-engines.

172. Biasi, V., Low cost and high efficiency make 30 to 80 kW microturbines attractive, Gas Turbine World, Southport, CT (January–February 1998).

173. Bist, S., Development of vegetable lipids derived fatty acid methyl esters as aviation turbine fuel extenders, Master thesis of Purdue University, West Lafayette, IN (2004).

174. Boyce, P.M., *Gas Turbine Engineering Handbook*, 3rd edn. Gulf Professional Publishing, Houston, TX, (April 28, 2006).

175. Schmellekamp, Y. and Dielmann, K., Rapeseed oil in a Capstone C30, in *Workshop: Bio-fuelled Micro Gas Turbines in Europe—Market Opportunities and R&D Requirements*, September 24, 2004, Brussels, Belgium (2004), www.bioturbine.org.

176. Correia, P.S., The use of biodiesel in gas fuel micro turbine: Thermal performance and emission testing. Master thesis, UNIFEI, Minas Gerais, Brazil (2006).

177. Gökalp, I. and Lebas, E., Alternative fuels for industrial gas turbines (AFTUR), *Applied Thermal Engineering*, 24 (11–12), 1655–1663 (2004).

178. Habib, Z., Parthasarathy, R., and Gollahalli, S., Performance and emission characteristics of biofuel in a small-scale gas turbine engine, *Applied Energy*, 87 (24), 1701–179 (2010).

179. Kehlhofer, R., Hannemann, F., Stirnimann, F., and Rukes, B., *Combined-Cycle Gas & Steam Turbine Power Plants*, 3rd edn. PennWell, Tulsa, OK, pp. 64–69 (2009).

180. Schmitz, W. and Hein, D., Concepts for the production of biomass derived fuel gases for gas turbine applications, in *Proceedings of ASME Turbo Expo 2000*, Munich, Germany (May 8–11, 2000).

181. Lopp, D., Tanley, D., Ropp, T., and Cholis, J., Soy-diesel blends use in aviation turbine engines, Aviation Technology Department of Purdue University, West Lafeyette, IN (1995).

182. Petrov, A.Y., Zaltash, A., Rizy, D.T., and Labinov, S.D., Study of flue gas emissions of gas microturbine-based CHP system (1999), www.uschpa.org.
183. Mimura, N., Biodiesel fuel: A next microturbine challenge, A report by Oak Ridge National Laboratory, Oak Ridge, TN (2003), www.ornl.gov.
184. Tan, E.S. and Palanisamy, K., Experimental and simulation study of biodiesel combustion in a micro turbine, in *ASME Turbo Expo 2008*, Berlin, Germany, June 9–13, 2008, ASME GT2008-51497 (2008).
185. Rodgers, G. and Saravanamutto, H., *Gas Turbine Theory*, in Saravanamutto, H. et al. (eds.), 5th edn., Prentice Hall, New York (2001).
186. Rodgers, C., Watts, J., Thoren, D., Nichols, K., and Brent, R., Microturbines, Chapter 5, in Borbely, A.-M. and Kreider, J.F., *Distributed Generation—The Power Paradigm for the New Millennium*. CRC Press LLC, New York, pp. 120–148 (2001).
187. Peirs, J., Ultra micro gas turbine generator, A report from Department of Mechanical Engineering, Katholieke Universiteit Leuven, Leuven, Belgium (2008).
188. Galanti, L. and Massardo, A.F., Thermo economic analysis of micro gas turbine design in the range 25–500 kWe, in *Proceedings of ASME Turbo Expo 2010*, Glasgow, U.K., pp. 1–11 (June 14–18, 2010).
189. do Nascimento, M.A.R., de Oliveira Rodrigues, L., Cruz dos Santos, E., Batista Gomes, E.E., Goulart Dias, F.L., Gutiérrez Velásques, E.I., and Miranda Carrillo, R.A., Micro gas turbine engine: A review, in Benini, E. (ed.), *Progress in Gas Turbine Performance*, InTech (June 19, 2013).
190. Obernberger, I., Decentralized biomass combustion: State of the art and future development, *Biomass and Bioenergy*, 14 (1), 33–57 (1998).
191. Savola, T., Tveit, T.-M., and Laukkanen, T., Biofuel indirectly fired microturbine state of the art, A report by TKK, Laboratory of Energy Engineering and Environmental Protection, Espoo, Finland (2005), http://eny.hut.fi/research/process_integration/bioifgt_Jan.pdf.
192. Hamilton, S.L., *The Handbook of Microturbine Generators*. PennWell Corporation, Tulsa, OK (2003).
193. Riccio, G. and Chiaramonti, D., Design and simulation of a small polygeneration plant cofiring biomass and natural gas in a dual combustion micro gas turbine, *Biomass and Bioenergy*, 33 (11) 1520–1531 (2009).
194. Combined Cycle, Wikipedia, the free encyclopedia (2015).
195. Integrated gasification combine cycle, Wikipedia, the free encyclopedia (2015).
196. Commercial power production based on gasification, NETL, Department of Energy, Washington, D.C. (2015), www.netl.doe.gov/research/coal/energy-systems/gasification/.../igcc.
197. IGCC process system sections, NETL report, Department of Energy, Washington, D.C. (2015), www.netl.doe.gov/research/coal/energy-systems/.../igcc-process.
198. Fortunato, B., Camporeale, S.M., and Torresi, M., A gas-steam combined cycle powered by syngas derived from biomass, *Procedia Computer Science*, 19, 736–745 (2013).
199. Franco, A. and Giannini, N., Perspectives for the use of biomass as fuel in combined cycle power plants, *International Journal of Thermal Sciences*, 44 (2), 163–177 (2005).
200. Fernando, R., Coal gasification, Report CCC/140, IEA Clean Coal Centre, London, U.K., 6pp. (October 2008).
201. Air-cooled 7HA and 9HA designs rated at over 61% CC efficiency, A report by Gas Turbine World (April 2014). www.gasturbineworld.com.
202. Iran—Yazd integrated solar combined cycle power station, A report by Helios CSP (May 21, 2011). info@heliocsp.com.
203. Singh, S. and Prasad, B., Energy and exergy analysis of steam cooled reheat gas-steam combined cycle, *Applied Thermal Engineering*, 27 (17–18), 2779–2790, (2007).
204. Maurstad, O., An overview of coal based integrated gasification combined cycle (IGCC) technology, Massachusetts Institute of Technology Laboratory for Energy and the Environment, MIT LFEE 2005-002 WP, Cambridge, MA (September 2005).
205. Farina, G. and Bressan, L., Optimizing IGCC design: Improve performance, reduce capital cost, *Foster Wheeler Review*, 1 (3), 16–21 (1999), http://www.fwc.com/publications/heat/heat_pdf/autumn99/igcc.pdf.
206. Hannemann, F. et al., Pushing forward IGCC technology at Siemens, in *Gasification Technologies Conference*, San Francisco, CA (October 13, 2003).
207. Higman, C. and van der Burgt, M., *Gasification*. Elsevier, Amsterdam, the Netherlands (2003).
208. Holt, N., EPRI, Coal-based IGCC plants—Recent operating experience and lessons learned, in *Gasification Technologies Conference*, Washington, DC (2004).
209. Holt, N., EPRI, A summary of recent IGCC studies of CO_2 capture for sequestration, in *Gasification Technologies Conference*, San Francisco, CA (2003).

210. Holt, N., EPRI, Integrated gasification combined cycle power plants, in *Encyclopedia of Physical Science and Technology*, 3rd edn. Academic Press, New York (2001).

211. IEA GHG Report PH4/19, Potential for improvements in gasification combined cycle power generation with CO_2 capture, IEA, Brussels, Belgium (2003).

212. IEA GHG report PH4/27, Canadian clean power coalition—Studies on CO_2 capture and storage (2004).

213. McDaniel, J., Tampa Electric Polk power station integrated gasification combined cycle project—Final technical report, Report for US DOE/NETL, Tampa Electric Company, DOE, Washington, D.C. (2002).

214. Phillips, J., EPRI, Integrated gasification combined cycles with CO_2 capture, in *GCEP Research Symposium on Meeting the challenge of reducing global GHG emissions through energy research*, Stanford University, Palo Alto, CA (June 13–16, 2005).

215. Ratafia-Brown, J. et al., SAIC, Major environmental aspects of gasification-based power generation technologies, NETL report, DOE, Washington, D.C. (2002).

216. Schmoe, L., Bechtel, IGCC—Expected plant availabilities and efficiencies, in *Presentation at Platts IGCC Symposium*, Pittsburgh, PA (June 2–3, 2005).

217. Shilling, N., GE power systems understanding the natural gas to IGCC conversion option, in *Presentation at the Gasification Technology Conference*, Washington, DC (2004).

218. Thompson, Clean Air Task Force, Integrated gasification combined cycle (IGCC)—Environmental performance, in *Presentation at Platts IGCC Symposium*, Pittsburgh, PA (June 2–3, 2005).

219. Rubin, E.S., Berkenpas, M.B., Christopher Frey, H., Chen, C., and McCoy, S.T., Technical documentation: Integrated gasification combined cycle systems (IGCC) with carbon capture and storage (CCS), Final report of work performed under contract no.: DE-AC21-92MC29094 Reporting Period Start, October 2003 Reporting Period End, May 2007, Carnegie Mellon University, Pittsburgh, PA (2007).

220. Craig, K.R. and Mann, M.K., Cost and performance analysis of biomass-based integrated gasification combined-cycle (BIGCC) power systems, NREL/TP-430–21657 UC Category: 1311 DE96013105, NREL, Golden, CO (October 1996).

221. Gallaspy, D.T., Johnson, T.W., and Sears, R.E., Southern Company Services' Study of a KRW-based GCC power plant, EPRI GS-6876, Electric Power Research Institute, Palo Alto, CA, Work performed by Southern Company Services Inc., Birmingham, AL (July 1990).

222. Northern States Power et al., Economic development through biomass systems integration—Sustainable biomass energy production, NREL/TP-421-20517, Golden, CO, Work performed for the National Renewable Energy Laboratory and the Electric Power Research Institute by Northern States Power, Minneapolis, MN (May 1995).

223. Craig, K.R., Mann, M.K., and Bain, R.L., Cost and performance potential of advanced integrated biomass gasification combined cycle power systems, Published in ASME Cogen Turbo Power '94, in *Eighth Congress & Exposition on Gas Turbines in Cogeneration and Utility*, Industrial and Independent Power Generation, Portland, OR (October 1994).

224. Kawabataa, M., Kurata, O., Iki, N., Furutania, H., and Tsutsumib, A., Advanced integrated gasification combined cycle (A-IGCC) by exergy recuperation—Technical challenges for future generations, *Journal of Power Technologies*, 92 (2), 90–100 (2012).

225. Iki, N., Tsutsumi, A., Matsuzawa, Y., and Furutani, H., Parametric study of advanced IGCC, in *Proceedings of ASME Turbo Expo 2009 Power for Land, Sea, and Air*, Orlando, FL, June 8–12, 2009 *(GT2009)*, GT2009-59984, ASME, New York, pp. 1–8 (2009).

226. Guan, G., Fushimi, C., Tsutsumi, A., Ishizuka, M., Matsuda, S., Hatano, H., and Suzuki, Y., High-density circulating fluidized bed gasifier for advanced IGCC/IGFC—Advantages and challenges, *Particuology*, 8 (6), 602–606 (2010).

227. Integrated Gasification Combined Cycle (IGCC), Demonstration project Polk power station—Unit No. 1 annual report, October 1993–September 1994, Work performed under contract no.: DE-FC21-91MC27363, For U.S. Department of Energy, Tampa Electric Company, Tampa, FL (May 1995).

228. Chen, H., Yogi Goswami, D., and Stefanakos, E.K., A review of thermodynamic cycles and working fluids for the conversion of low-grade heat, *Renewable and Sustainable Energy Review*, 14 (9), 3059–3067 (December 2010).

229. Bjorge, R. and Jandrisevits, M., IGCC technology for the 21st century, in *1996 Gasification Technologies Conference*, Electric Power Research Institute, Inc., Palo Alto, CA (October 1996).

230. Henderson, C., Case studies on recently constructed fossil fired plants, in *Third International Conference on Clean Coal Technologies for our Future*, May 15–17, 2007, Cagliari and Sotacarbo Coal Research Centre, Carbonia Sardinia, Italy (2007).

231. Klara, J. and Wimer, J., Shell IGCC plant with carbon capture & sequestration, IGCC Plant-bituminous coal, Cost and performance baseline for fossil energy plants, Vol. 1, DOE/NETL-2007/1281, DOE, Washington, D.C., May 2007. B_IG_Shell_CCS_051507.

232. Henderson, C., Future developments in IGCC, CCC/143, IEA Clean Coal Center, London, U.K. (December 2008).

233. Minchener, A., Coal gasification for advanced power generation, *Fuel*, 84 (17), 2222–2235 (2005).

234. Cogeneration, Wikipedia, the free encyclopedia (2015). https://en.www.wikipedia.org/wiki/cogeneration (last accessed October 15, 2016).

235. Cogeneration combined heat and power, A report by Clark Energy, Winchester, KY, www.clarke-energy.com, Retrieved November 26, 2011.

236. Combined heat and power—Effective energy solutions for a sustainable future (PDF), A report Oak Ridge National Laboratory, Oak Ridge, TN (December 1, 2008), Retrieved September 9, 2011.

237. Andrew, D., Carbon footprints of various sources of heat—Biomass combustion and CHPDH comes out lowest, "William Orchard" Claverton Energy Research, U.K. (February 7, 2009). www.claverton_energy.com.

238. Cogeneration is the most energy transformer. Since 1992 EVW has manufactured cogeneration units and contributed to energy efficient power produced, A report by E. VAN WINGEN, Evergreen, Belgium (accessed 2015).

239. Andrew, D., Finning Caterpillar Gas Engine CHP Ratings, A report by Claverton Energy Research Group, U.K. (April 23, 2014). www.claverton-energy.com/finning. Retrieved May 15, 2015.

240. High cogeneration performance by innovative steam turbine for biomass-fired CHP plant in Iislami, Finland (PDF), OPET, Retrieved March 13, 2011.

241. Oliveira, A.C., Afonso, C., Matos, J., Riffat, S., Nguyen, M., and Doherty, P., A combined heat and power system for buildings driven by solar energy and gas, *Applied Thermal Engineering*, 22 (6), 587–593 (2002).

242. Yagoub, W., Doherty, P., and Riffat, S.B., Solar energy-gas driven micro-CHP system for an office building, *Applied Thermal Engineering*, 26 (14), 1604–1610 (2006).

243. Pearce, J.M., Expanding photovoltaic penetration with residential distributed generation from hybrid solar photovoltaic + combined heat and power systems, *Energy*, 34, 1947–1954 (2009).

244. Andrews, D., What is Micro generation? And what is the most cost effective in terms of CO_2 reduction, Claverton Energy Research Group, U.K. (November 6, 2008). Retrieved May 15, 2015. www.claverton_energy.com.

245. Best Value CHP, Combined heat & power and cogeneration—Profitable greener energy via CHP, cogen and biomass boiler using wood, biogas, natural gas, biodiesel, vegetable oil, syngas and straw, A report on alfagy.com, Leeds, U.K. (2015). alfagy.com. w3cdomain.com. Retrieved May 15, 2015.

246. Pehnt, M., Environmental impacts of distributed energy systems—The case of micro cogeneration, *Environmental Science & Policy*, 11 (1), 25–37 (2008).

247. Nosrat, A.H., Swan, L.G., and Pearce, J.M., Simulations of greenhouse gas emission reductions from low-cost hybrid solar photovoltaic and cogeneration systems for new communities, *Sustainable Energy Technologies and Assessments*, 8, 34–41 (2014).

248. Lowe, R., Combined heat and power considered as a virtual steam cycle heat pump, *Energy Policy*, 39 (9), 5528–5534 (2011).

249. Nosrat, A.H., Swan, L.G., and Pearce, J.M., Improved performance of hybrid photovoltaic-trigeneration systems over photovoltaic-cogen systems including effects of battery storage, *Energy*, 49, 366–374 (2012).

250. Martin, J. et al., Trigeneration system with fuel cells, a research paper, University of Basque Country, Spain, RE & PQJ, 1 (6), 135–140 (March 2008).

251. Tri-generation success story-world's first tri-gen energy station-fountain valley, A report from Fuel Cell technology office, A report by Energy Efficiency and Renewable Energy, Department of Energy, Washington, DC (March 2013).

252. Goth, M., Sekwambane, P., Coetzer, D., Fromme, J., Sewenig, J., and Borchard, C., Good practice brochure on co/tri-generation, status QUO of the South African Market, Sonedi (South African National Energy Development Institute), Pretoria, South Africa (May 2014).

253. Henderick, P. and Williams, R.H., Trigeneration in a northern Chinese village using crop residues, *Energy for Sustainable Development*, IV (3), 26–42 (October 2000).

254. Stassen, H.E., Small-scale biomass gasifiers for heat and power: A global review, World Bank Technical Paper Number 296, Energy Series report no. WTP296, Vol. 1 (October 1995).

255. Kelly, J. and Darby, E., TRi-generation energy system technology, Prepared by: Altex Technologies Inc., Sunnyvale, CA. Prepared for: California Energy Commission, Energy Research and Development Division, Final project report, CEC-500-2015-026 (January 2015). www.energy.co.gov/.../cee5w.2015.026.pdf.

256. Lennox, Lennox commercial chillers (n.d.), www.lennox.com, Retrieved October 25, 2012.

257. Naguib, R., The changing landscape of chillers' energy efficiency, *Energy Engineering*, 108 (4), 25–45 (2011).

258. Yazaki, Yazaki steam absorption chillers (n.d.), www.yazakienergy.com, Retrieved October 25, 2012.

259. Zogg, R.A., *Guide to Developing Air-Cooled LiBr Absorption for CHP Applications*. U.S. DOE report, DOE, Washington, D.C. (2005).

260. Attri, V.K., Sachdev, A., and Kumar, M., Performance evaluation of trigeneration systems, *International Journal of Emerging Technology and Advanced Engineering*, 4 (1), 277–292 (February 2014), www.ijetae.com (Online), *International Conference on Advanced Developments in Engineering and Technology* (ICADET-14), India.

261. Teopa Calva, E., Picon Nunez, M., and Rodrıguez Toral, M.A., Thermal integration of trigeneration systems, *Applied Thermal Engineering*, 25, 973–984 (2005), Elsevier.

262. Bassols, J., Kuckelkorn, B., Langreck, J., Schneider, R., and Veelken, H., Trigeneration in the food industry, *Applied Thermal Engineering*, 22, 595–602 (2002), Pergamon Press.

263. Ziher, D. and Poredos, A., Economics of a trigeneration system in a hospital, *Applied Thermal Engineering*, 26, 680–687 (2006), Elsevier.

264. Hernandez-Santoyo, J. and Sanchez-Cifuentes, A., Trigeneration: An alternative for energy savings, *Applied Thermal Engineering*, 76, 219–227 (2003), Elsevier.

265. Minciuc, E., Le Corre, O., Athanasovici, V., Tazerout, M., and Bitir, I., Thermodynamic analysis of tri-generation with absorption chilling machine, *Applied Thermal Engineering*, 23, 1391–1405 (2003), Pergamon Press.

266. Temir, G. and Bilge, D., Thermoeconomic analysis of a trigeneration system, *Applied Thermal Engineering*, 24, 2689–2699 (2004), Elsevier.

267. Colonna, P. and Gabrielli, S., Industrial trigeneration using ammonia–Water absorption refrigeration systems (AAR), *Applied Thermal Engineering*, 23, 381–396 (2003), Pergamon Press.

268. Kong, X.Q., Wang, R.Z., and Huang, X.H., Energy efficiency and economic feasibility of CCHP driven by stirling engine, *Energy Conversion and Management*, 45, 1433–1442 (2004), Elsevier.

269. Khaliq, A., Exergy analysis of gas turbine trigeneration system for combined production of power heat and refrigeration, *International Journal of Refrigeration*, 32 (3), 534–545 (May 2009).

270. Khaliq, A. and Kaushik, S.C., Thermodynamic performance evaluation of combustion gas turbine cogeneration system with reheat, *Applied Energy*, 24, 1785–1795 (2004), Elsevier.

271. Pilatowsky, I., Romero, R.J., Isaza, C.A., Gamboa, S.A., Rivera, W., Sebastian, P.J., and Moreira, J., Simulation of an air conditioning absorption refrigeration system in a cogeneration process combining a proton exchange membrane fuel cell, *International Journal of Hydrogen Energy*, 32, 3174–3182 (2007).

272. Zamora, I., San Martín, J.I., Mazón, A.J., San Martín, J.J., Aperribay, V., and Ma Arrieta, J., Cogeneration in electrical microgrids, in *International Conference on Renewable Energy and Power Quality*, Palma de Mallorca, Spain, April 5–7, 2006 (2006).

273. Veerapen, J., Co-generation and renewables-solutions for a low carbon energy future, TEA report, OECD/IEA, Paris, France (2011).

274. IEA, *Combined Heat and Power: Evaluating the Benefits of Greater Global Investment*. OECD/IEA Publications, Paris, France (2008).

275. IEA, *Co-generation and District Energy: Sustainable Energy Technologies for Today and Tomorrow*. OECD/IEA Publications, Paris, France (2009).

276. Kelly, S. and Pollitt, M., *Making Combined Heat and Power District Heating (CHP-DH) Networks in the United Kingdom Economically Viable: A Comparative Approach*. Electricity Policy Research Group, University of Cambridge, Cambridge, U.K. (2009).

277. Trieb, F. and Muller-Steinhagen, H., Concentrating solar power for seawater desalination in the Middle East and North Africa, *Desalination*, 220 (1–3), 165–183 (2008).

278. Al-Sulaiman, F.A., Dincer, I., and Hamdullahpur, F., Thermoeconomic optimization of three trigeneration systems using organic Rankine cycles: Part I—Formulations, *Energy Conversion and Management*, 69, 199–208 (2013).

279. He, C., Liu, C., Gao, H., Xie, H., Li, Y., Wu, S., and Xu, J., The optimal evaporation temperature and working fluids for subcritical organic Rankine cycle, Elsevier, Amsterdam, the Netherlands (2012).

280. Invernizzi, C.M., Iora, P., and Sandrini, R., Biomass combined cycles based on externally fired gas turbines and organic Rankine expanders, *Proceeding of the Institution of the Mechanical Engineers Journal of Power and Energy*, 225 (8), 1066–1075 (December 2011).

281. Camporeale, S., Ciliberti, P., Torresi, M., Fortunato, B., and Pantaleo, A., Externally fired micro gas tur-bine and ORC bottoming cycle: Optimal biomass/natural gas CHP configuration for residential energy demand, in *Proceedings of ASME Turbo Expo 2015*, Montreal, Quebec, Canada (June 15–19, 2015).
282. Camporeale, S., Turi, F., Torresi, M., Fortunato, B., Pantaleo, A., and Pellerano, A., Part load perfor-mances and operating strategies of a natural gas-biomass dual fuelled microturbine for CHP operation, in *Proceedings of ASME Turbo Expo 2014*, June 16–20, 2014, Düsseldorf, Germany, GT2014-27109 (2014); *ASME Journal of Engineering for Gas Turbines and Power*.
283. Ferreira, A.C.M., Nunes, M.L., Teixeira, S.F.C.F., Leão, C.P., Silva, Â.M., Teixeira, J.C.F., and Martins, L.S.B., An economic perspective on the optimization of a small-scale cogeneration system for the Portuguese scenario, *Energy*, 45 (1), 436–444 (2012).
284. Pantaleo, A., Candelise, C., Bauen, A., and Shah, N., ESCO business models for biomass heating and CHP: Case studies in Italy, *Renewable and Sustainable Energy Reviews*, 30, 237–253 (2014).
285. Pantaleo, A., Camporeale, S., and Shah, N., Natural gas—Biomass dual fuelled microturbines: Comparison of operating strategies in the Italian residential sector, *Applied Thermal Engineering*, 71, 686–696 (2014).
286. Pantaleo, A., Ciliberti, P., Camporeale, S., and Shah, N., Thermo-economic assessment of small scale biomass CHP: Steam turbines vs ORC in different energy demand segments, in *Proceedings of Seventh International Conference on Applied Energy—ICAE*, Abu Dhabi, United Arab Emirates (March 28–31, 2015).
287. David, G., Michel, F., and Sanchez, L., Waste heat recovery projects using Organic Rankine Cycle tech-nology—Examples of biogas engines and steel mills applications, in *Geneva 2011—World Engineer's Convention*, September 4–9, 2011, Geneva, Switzerland (2011).
288. Micro CHP, Wikipedia, the free encyclopedia (2015).
289. Blair, C., Micro-CHP commercialization—European Utility Engagement, in *Advanced Energy Conference*, Delta Energy & Environment, London, U.K. (November 2008), www.aertc.org/.../AEC_Sessions%5CCopy%20of%20Session%202%5CTrack%20C-....
290. Compressed natural gas, Wikipedia, the free encyclopedia (2015). /en.www.wikipedia.org/wiki/compressed_natural_gas (last accessed October 9, 2016).
291. Mumtaz Alvi, Pakistan has highest number of CNG vehicles: Survey (June 3, 2011).
292. Natural Gas Vehicle Statistics: Summary Data 2010, International Association for Natural Gas Vehicles, (2010), Retrieved August 2, 2011. www.wikipedia.org/wiki/naturalgasvehicle (last accessed Ocotober 5, 2016).
293. Natural gas vehicle statistics: NGV Count—Ranked Numerically as at December 2009, International Association for Natural Gas Vehicles, Retrieved April 4, 2010. www.wikipedia.org/wiki/naturalgasvehicle (last accessed Ocotober 5, 2016).
294. Lepisto, C., Fiat Siena tetra power: Your choice of four fuels, Treehugger (August 27, 2006), Retrieved August 24, 2008. www.wikipedia.org/wiki/naturalgasvehicle (last accessed Ocotober 5, 2016).
295. ISO 14469-2:2007, Road vehicles—Compressed natural gas (CNG) refuelling connector—Part 2: 20 MPa (200 bar) connector, size 2, Retrieved May 15, 2015. www.wikipedia.org/wiki/naturalgasvehicle (last accessed Ocotober 5, 2016).
296. CNG fueling sites—Compressed natural gas fueling locations, Retrieved May 15, 2015. www.wikipedia.org/wiki/naturalgasvehicle (last accessed Ocotober 5, 2016).
297. AT&T orders 1,200 CNG-powered Chevrolet express vans, media.gm.com, Retrieved 15 May 2015. www.wikipedia.org/wiki/naturalgasvehicle (last accessed Ocotober 5, 2016).
298. GFS diesel-LNG for Wyoming coal trucks, HHP Insight, Retrieved January 5, 2016. www.wikipedia.org/wiki/naturalgasvehicle (last accessed Ocotober 5, 2016).
299. Pike Research, Pike Research predicts 68% jump in global CNG vehicle sales by 2016, AutoblogGreen (September 14, 2011), Retrieved September 26, 2011. www.wikipedia.org/wiki/naturalgasvehicle (last accessed Ocotober 5, 2016).
300. Iran plans to substitute compressed gas with ANG in vehicles, *AzerNews* (September 14, 2012).
301. ISO 11439:2000, Gas cylinders—High pressure cylinders for the on-board storage of natural gas as a fuel for automotive vehicles, ISO (International Organization for Standardization) (2000). www.iso.org/iso/catalogue-detail.html.
302. Gibgas, 900th CNG filling station for Germany, *NGV Global News* (December 21, 2011), Retrieved December 28, 2011. www.wikipedia.org/wiki/naturalgasvehicle (last accessed Ocotober 5, 2016).
303. Fernandes, R., Latin America NGVs: An update report, International Association of Natural Gas Vehicles (August 20, 2008), Archived from the original on November 20, 2008, Retrieved August 11, 2008. www.wikipedia.org/wiki/naturalgasvehicle (last accessed Ocotober 5, 2016).

304. Isayev, S. and Jafarov, T., *Iran Plans to Produce ANG for Vehicles*. Trend News Agency, Baku, Azerbaijan (September 14, 2012).

305. How safe are natural gas vehicles? (PDF), Clean Vehicle Education Foundation, Retrieved May 8, 2008. www.wikipedia.org/wiki/naturalgasvehicle (last accessed Ocotober 5, 2016).

306. Natural Gas Vehicles Statistics, the International Association for Natural Gas Vehicles (2010), http://www.iangv.org/tools-resources/statistics.html. www.wikipedia.org/wiki/naturalgasvehicle (last accessed Ocotober 5, 2016).

307. Liquid Propane Injection, Holden special vehicles, Retrieved April 12, 2011.

308. High horse power off-road LNG vehicles in USA, Retrieved April 17, 2015. www.wikipedia.org/wiki/naturalgasvehicle (last accessed Ocotober 5, 2016).

309. Development of the high-pressure direct-injection ISX G natural gas engine (PDF), Retrieved April 17, 2015. www.wikipedia.org/wiki/naturalgasvehicle (last accessed Ocotober 5, 2016).

310. Canadian Propane Market Review, GPMi (October 2014)

311. Zhang, L., The Development of China's Auto LPG Industry, World LP Gas Association, Brussels, Belgium (2009), Archived from the original on September 27, 2010, Retrieved October 9, 2010.

312. Over 1 million LPG vehicles now on Thai roads, *Bangkok Post* (February 23, 2013), Retrieved August 2, 2013. wikipedia/en.www.wikipedia.org/wiki/autogas (last accessed October 22, 2016).

313. Propane Vehicles, Alternative Fuels Data Center, U.S. Department of Energy, Retrieved May 30, 2015. wikipedia/en.www.wikipedia.org/wiki/autogas (last accessed October 22, 2016).

314. Alliance autogas builds hundreds of U.S. propane autogas fueling stations, Retrieved June 21, 2013. wikipedia/en.www.wikipedia.org/wiki/autogas (last accessed October 22, 2016).

315. New LNG trucking fleet launches in Houston, Retrieved April 17, 2015. www.wikipedia.org/wiki/naturalgasvehicle (last accessed Ocotober 5, 2016).

316. Blomerus, P. and Onlette, P., LNG as a fuel for demanding high horsepower engine applications: Technology and approaches (PDF), Retrieved April 17, 2015. www.wikipedia.org/wiki/naturalgas-vehicle (last accessed Ocotober 5, 2016).

317. Schuler, M., Introducing ISLA BELLA—World's first LNG-powered containership launched at NASSCO, *gCaptain* (April 19, 2015). http://gcaptain.com/isla-bella-worlds-first-lng-powered-container-ership-launched-at-nassco/ (last accessed December 9, 2016)

318. Natural gas vehicles statistics, International Association for Natural Gas Vehicles (2010), Archived from the original on January 10, 2010. www.wikipedia.org/wiki/naturalgasvehicle (last accessed Ocotober 5, 2016).

319. Autogas, Wikipedia, the free encyclopedia (2015). en/www.wikipedia.org/wiki/autogas (last accessed October 22, 2016).

320. Liquefied Petroleum Gas (LPG), Liquefied Natural Gas (LNG) and Compressed Natural Gas (CNG), Retrieved May 15, 2015. wikipedia/en.www.wikipedia.org/wiki/autogas (last accessed October 22, 2016).

321. Almost 23 million autogas vehicles will be operating on roads worldwide by 2020, Forecasts Pike Rese, Retrieved October 10, 2012. wikipedia/en.www.wikipedia.org/wiki/autogas (last accessed October 22, 2016).

322. Autogas Incentive Policies: A country-by country analysis of why and how governments promote Autogas & what works, A report by LP Gas Association (WLPGA), Brussels, Belgium (2011), Available at http://www.worldlpgas.com/uploads/Modules/Publications/autogas-incentive-policies-2012-updated-july-2012.pdf.

323. Autogas in Europe, The sustainable alternative—An LPG industry roadmap, A report by AEGPL, Brussels, Belgium (2009), Available at http://www.aegpl.eu/media/16300/autogas%20roadmap.pdf.

324. Hanschke, C.B., Uyterlinde, M.A., Kroon, P., Jeeninga, H., and Londo, H.M., Duurzame innovatie in het wegverkeer: Een evaluatie van vier transitiepaden voor het thema Duurzame Mobiliteit, ECN-E--08-076, ECN Beleidsstudies, Petten, the Netherlands (2009).

325. Roeterdink, W.G., Uyterlinde, M.A., Kroon, P., and Hanschke, C.B., Groen Tanken: Impassing van alternatieve brandstoffen in de tank- en distributie infrastructuur, ECN-E-09-082, Energy Research Centre of the Netherlands, Mei (2010).

326. Liquefied Petroleum Gas in Transport, Climate techWIKI, a clean technology platform, Energy Research Center of the Netherlands (ECN) Policy Studies, Petten, the Netherlands (2012).

327. Nice, K. and Strickland, J., How fuel cells work: Polymer exchange membrane fuel cells, How Stuff Works, autohowstuffworks.com/fuelefficiency/alternatives_fuels/fuel.cell2.html. Accessed August 4, 2011.

328. Larmine, J. and Dicks, A., *Fuel Cell Systems Explained*. John Wiley & Sons Ltd., Chichester, U.K. (2003).

329. O'Hayre, R., Cha, S.K., Colella, W., and Prinz, F.B., *Fuel Cell Fundamentals*. John Wiley & Sons, New York (2008).

330. World's Largest Carbon Neutral Fuel Cell Power Plant. A report by Fuel Cell Energy Inc., Danbury, CT (October 18, 2012). www.energydigital.com/sustainability/2043/Worlds-largest-Carbon-Neutral-Fuel-Cell-Power-Plant.

331. Fuel Cell, Wikipedia, the free encyclopedia (2015). https://en.www.wikipedia.org/wiki/fuel_cell (last accessed October 27, 2016).

332. Types of Fuel Cells, A report by Department of Energy EERE, DOE, Washington, D.C., Accessed August 4, 2011.

333. Prabhu, R.R., Stationary fuel cells market size to reach 350,000 shipments by 2022, Renew India Campaign (January 13, 2013), Retrieved January 14, 2013. www.renewindians.com/stationary_fuel_cells_market_size_to_reach350,000shipment_by_2002.

334. Larminie, J. and Dicks, A., *Fuel Cell Systems Explained*, 2nd edn., John Wiley, Inc., Hoboken, NJ (2013).

335. Vincent, B., Gangi, J., Curtin, S., and Delmont, E., 2008 fuel cell technologies market report, Breakthrough Technologies Institute, Department of Energy, Energy Efficiency and Renewable Energy, DOE, Washington, D.C. (June 2010).

336. Accomplishments and progress, Fuel Cell Technology Program, U.S. Department of Energy, DOE, Washington, D.C. (June 24, 2011).

337. Carter, D. and Wing, J., The fuel cell industry review 2013, a report in Fuel Cell today industry review, DOE, Washington, D.C., pp. 1–50 (2013).

338. Yarguddi, O. and Dharme, A.A., Fuel cell technology: A review, *International Journal of Innovative Research in Science, Engineering and Technology*, 3 (7), July 2014, 14668–14673 (An ISO 3297: 2007 Certified Organization).

339. Carrette, L., Friedrich, K., and Stimming, U., Fuel cells—Fundamentals and applications, *Fuel Cells*, 1 (1), 5–39 (2001).

340. Fuel Cell Energy Inc., Danbury, CT. Available: www.fce.com, Accessed January 5, 2014 (Online).

341. Wang, J.Y., Pressure drop and flow distribution in parallel-channel of configurations of fuel cell stacks: U-type arrangement, *International Journal of Hydrogen Energy*, 33 (21), 6339–6350 (2008).

342. Wang, J.Y. and Wang, H.L., Flow field designs of bipolar plates in PEM fuel cells: Theory and applications, *Fuel Cells*, 12 (6), 989–1003 (2012).

343. Wang, J.Y. and Wang, H.L., Discrete approach for flow-field designs of parallel channel configurations in fuel cells, *International Journal of Hydrogen Energy*, 37 (14), 10881–10897 (2012).

344. Kakati, B.K. and Deka, D., Effect of resin matrix precursor on the properties of graphite composite bipolar plate for PEM fuel cell, *Energy & Fuels*, 21 (3), 1681–1687 (2007).

345. Kakati, B.K. and Mohan, V., Development of low cost advanced composite bipolar plate for P.E.M. fuel cell, *Fuel Cells*, 08 (1), 45–51 (2008).

346. Kakati, B.K. and Deka, D., Differences in physico-mechanical behaviors of resol and novolac type phenolic resin based composite bipolar plate for proton exchange membrane (PEM) fuel cell, *Electrochimica Acta*, 52 (25), 7330–7336 (2007).

347. Notter, D.A., Kouravelou, K., Karachalios, T., Daletou, M.K., and Haberland, N.T., Life cycle assessment of PEM FC applications: Electric mobility and μ-CHP, *Energy & Environmental Science*, 8 (7), 1969–1985 (2015).

348. Harikishan Reddy, E. and Jayanti, S., Thermal management strategies for a 1 kWe stack of a high temperature proton exchange membrane fuel cell, *Applied Thermal Engineering*, 48, 465–475 (December 15, 2012).

349. Chemical could revolutionize polymer fuel cells (PDF), A report by Georgia Institute of Technology, Atlanta, GA (August 24, 2005), Retrieved November 21, 2014.

350. Kakati, B.K. and Kucernak, A.R.J., Gas phase recovery of hydrogen sulfide contaminated polymer electrolyte membrane fuel cells, *Journal of Power Sources*, 252, 317–326 (March 15, 2014).

351. Andrea, E., Mañana, M., Ortiz, A., Renedo, C., Eguíluz, L.I., Pérez, S., and Delgado, F., A simplified electrical model of small PEM fuel cell, University of Cantabria, Cantabria, Spain. RE & PQJ, 1(4), 281–284 (April 2006).

352. Badwal, S.P.S. and Foger, K., Solid oxide electrolyte fuel cell review, *Ceramics International*, 22, 257–265 (1996).

353. Friedman, D. and Moore, R., PEM fuel cell system optimization, *Proceedings of the Electrochemical Society*, paper 27, 407–423 (1998).

354. Wang, Y., Chen, K., Mishler, J., Chan Cho, S., and Cordobes Adroher, X., A review of polymer electrolyte membrane fuel cells: Technology, applications, and needs on fundamental research, US Department of Energy Publications, Paper 132, Washington, D.C. (2011). http://digitalcommons.unl.edu/usdoepub/132.

355. Phosphoric acid fuel cell technology, *Fuel Cells Today*. Available: http://americanhistory.si.edu/fuel-cells/phos/pafcmain.htm, Accessed January 4, 2014 (Online).

356. Solid oxide fuel cell (SOFC), Available: http://www.fctec.com/fctec_types_sofc.asp, Accessed January 4, 2014 (Online).

357. Stambouli, A.B. and Traversa, E., Solid oxide fuel cells (SOFCs): A review of an environmentally clean and efficient source of energy, *Renewable and Sustainable Energy Reviews*, 6, 433–455 (October 2002).

358. Kazempoor, P. et al., Modelling and evaluation of building integrated SOFC systems, *International Journal of Hydrogen Energy*, 36 (20), 13241–13249 (2011).

359. Minh, N.Q., Solid oxide fuel cell technology—Features and applications, *Solid State Ionics*, 174 (1–4), 1–318 (2004).

360. Dicks, A., Molten carbonate fuel cells, *Current Opinion in Solid State & Materials Science*, 8, 379–383 (2004).

361. Molten carbonate fuel cell technology, A report by U.S. Department of Energy, Washington, D.C. (2014). http://www1.eere.energy.gov/hydrogenandfuelcells/fuelcells/fc_types.html, Accessed January 5, 2014.

362. Molten carbonate fuel cell technology, U.S. Department of Energy, DOE, Washington, D.C. Accessed August 9, 2011.

363. Molten carbonate fuel cells (MCFC), A report by Fuel Cell Today, FCTec.com, Accessed August 9, 2011, Archived July 1, 2014 at the Wayback Machine.

364. Habermann, W. and Pommer, E.-H., Biological fuel cells with sulphide storage capacity, *Applied Microbiology and Biotechnology*, 35, 128–133 (1991).

365. Rabaey, K. and Verstraete, W., Microbial fuel cells: Novel biotechnology for energy generation, *Trends in Biotechnology*, 23 (6), 291–298 (June 2005).

366. Giddey, S., Badwal, S.P.S., Kulkarni, A., and Munnings, C., A comprehensive review of direct carbon fuel cell technology, *Progress in Energy and Combustion Science*, 38 (3), 360–399 (2012).

367. Comparison of fuel cell technologies, U.S. Department of Energy, Energy Efficiency and Fuel Cell Technologies Program, DOE, Washington, D.C. (February 2011), Accessed August 4, 2011.

368. Boyd, J., Home fuel cells to sell in Japan (2008), http://spectrum.ieee.org/energy/renewables/home-fuel-cells-to-sell-in-japan, Retrieved April 27, 2012.

369. Elmer, T. and Ruffat, S., State of art reviews: Fuel cell technologies in domestic built environment, A report from University of Nottingham, U.K. (September 9, 2014). link.springer.com/.../10.1007 %2F978-3-319-07977-6.

370. Carter, D., Fuel cell residential micro-CHP developments in Japan, ENE.FARM, F.C. Today (2012); CFCL, BlueGEN modular generator—Power and heat (2009); CFCL, E.ON UK orders additional 105 Ceramic Fuel Cells' products (2011); DECC, Climate Change Act 2008 (2008).

371. Gencoglu, M.T. and Ural, Z., Design of a PEM fuel cell system for residential application, *International Journal of Hydrogen Energy*, 34 (12), 5242–5248 (2009).

372. Gigliucci, G. et al., Demonstration of a residential CHP system based on PEM fuel cells, *Journal of Power Sources*, 131 (1–2), 62–68 (2004).

373. Hamada, Y. et al., Hybrid utilization of renewable energy and fuel cells for residential energy systems, *Energy and Buildings*, 43 (12), 3680–3684 (2011).

374. Harrison, J., E.ON—Smart homes with fuel cell micro CHP, in *Smart Hydrogen & Fuel Cell Power Conference*, Birmingham, U.K. (2012).

375. Klose, P., Baxi Innotech—Gamma 1.0 Large scale demonstration of residential PEFC systems in Germany, status and outlook, in *Fourth IPHE Workshop—Stationary Fuel Cells*, Tokyo, Japan (2011).

376. Nishizaki, K., The Japanese experience in micro CHP for residential use, Gas Industry Micro CHP Workshop, Tokyo, Japan (May 29, 2008). wenku.baidu.com/view/abe025110b4e76815acfeab1.htm.

377. Panasonic develops new fuel cell cogeneration system for home use, Panasonic (2011), http://panasonic.co.jp/corp/news/official.data/data.dir/en080414-2/en080414-2.html, Retrieved April 27, 2012.

378. Tokyo gas and panasonic to launch new improved Ene-farm home fuel cell with world-highest*1 power generation efficiency at more affordable price, Headquarters news, Tokyo Gas Co. Ltd., Panasonic Corporation, Chesapeake, VA (February 9, 2011). http://panasonic.co.jp/corp/news/official.data/data.dir/en110209-2/en110209-2.html, Retrieved April 27, 2012.

379. Peacock, A.D. and Newborough, M., Impact of micro-CHP systems on domestic sector CO_2 emissions, *Applied Thermal Engineering*, 25 (17–18), 2653–2676 (2005).

380. Sammes, N.M. and Boersma, R., Small-scale fuel cells for residential applications, *Journal of Power Sources*, 86 (1–2), 98–110 (2000).

381. Fuel cell vehicle, Wikipedia, the free encyclopedia (2015), https://en.wikipedia.org/wiki/Fuel_cell_vehicle.

382. 2013 Fuel Cell Technologies market report, Fuel cell technologies office, Energy Efficiency and Renewable Energy, Department of Energy, Washington, DC (November 2014), energy.gov/sites/prod/files/2014/11/.../fcto_2013_market_report.pdf.

383. Fuel Cell Energy 2013 annual report, Energy efficiency and renewable energy, Department of Energy, Washington, DC (2014), http://files.shareholder.com/downloads/FCEL/3376881510x0x724661/D703F85E-D67A-441D-81E2-76401A391857/Fuel_Cell__13AR_compiled.pdf.

384. Von Helmolt, R. and Eberle, U., Fuel cell vehicles: Status 2007, *Journal of Power Sources*, 165 (2), 833 (March 20, 2007).

385. Telias, G., Day, K., and Dietrich, P., RD & D cooperation for the development of fuel cell, hybrid, and electric vehicles within the International Energy Agency, a paper presented at 25th World Battery, Hybrid, Fuel Cell Electric Vehicle Symposium and Exhibition, Shenzhen, China (November 5–9, 2010), NREL/cp-5400-49105 (January 2011), NREL, Golden, CO.

386. Brinkman, N., Wang, M., Weber, T., and Darlington, T., Well-to-wheels analysis of advanced fuel/vehicle systems—A North American study of energy use, Greenhouse Gas Emissions, and Criteria Pollutant Emissions, General Motors Corporation, Argonne National Laboratory and Air Improvement Resource, Inc., Lemont, IL (May 2005), Accessed August 9, 2011.

387. GM's fuel cell system shrinks in size, weight, cost, A report by GM in Fuel Cell Works (March 16, 2010), Retrieved March 5, 2012.

388. Hyundai Motor Co., UTC fuel cells and chevron tex & co launch Hydrogen fleet and infrastructure project, A report in California hydrogen business council (February 18, 2005).

389. Kantola, K., Intelligent energy ENV H_2 motorcycle, A report in Intelligent Energy ENV, Retrieved May 27, 2007.

390. Honda develops fuel cell scooter equipped with Honda FC stack, a news release from Honda Motor Co. (August 24, 2004), Retrieved May 27, 2007.

391. Bryant, E., Honda to offer fuel-cell motorcycle (July 21, 2005), autoblog.com, Retrieved May 27, 2007.

392. Hydrogen fuel cell electric bike (December 2007), Youtube.com, Retrieved September 21, 2009.

393. Boeing successfully flies fuel cell-powered airplane, *Boeing* (April 3, 2008), Accessed August 2, 2011, Archived from the original on May 9, 2013. www.boeing.com/aboutus/environment/environmental../flash-201-2.html.

394. First Fuel Cell Microaircraft, Archived October 4, 2012 at the Wayback Machine.

395. Ford, J., Hydrogen-powered unmanned aircraft completes set of tests, A report in Engineer Centaur, Communication Ltd., London, U.K. (June 20, 2011), www.theengineer.co.uk, Accessed August 2, 2011.

396. Coxworth, B., Drone flight powered by lightweight hydrogen-producing pellets (February 8, 2016), www.gizmag.com, Retrieved February 9, 2016.

397. First hydrogen powered canal cruise boat in Amsterdam, A report in Simply Amsterdam (December, 2009), www.simplyamsterdam.n6/First_hydrogen_powered_canal_cruise_brat_in_Amsterdam. Accessed August 2, 2011.

398. Pleitgen, F., Super-stealth sub powered by fuel cell, A report in CNN Tech: Nuclear Weapons (February 22, 2011), Accessed August 2, 2011.

399. U212/U214 Attack Submarines, Germany, A report in Naval-Technology.com, Accessed August 2, 2011.

400. Goodenough, R.H. and Greig, A., Hybrid nuclear/fuel-cell submarine, *Journal of Naval Engineering*, 44 (3), 455–471 (2008).

401. Milewski, J., Miller, A., and Badyda, K., The control strategy for high temperature fuel cell hybrid systems, *The Online Journal on Electronics and Electrical Engineering*, 2 (4), 331 (2009), Accessed August 4, 2011.

402. Reduction of residential carbon dioxide emissions through the use of small cogeneration fuel cell systems—Combined heat and power systems, IEA Greenhouse Gas R&D Programme (IEAGHG), IEA, (November 11, 2008), Brussels, Belgium. Retrieved July 1, 2013.

403. Al-Sulaiman, F.A. et al., Energy analysis of a trigeneration plant based on solid oxide fuel cell and organic Rankine cycle, *International Journal of Hydrogen Energy*, 35 (10), 5104–5113 (2010).

404. Braun, R.J. et al., Evaluation of system configurations for solid oxide fuel cell-based micro-combined heat and power generators in residential applications, *Journal of Power Sources*, 158 (2), 1290–1305 (2006).

405. Brett, D., Micro-generation, A report from UCL Electrochemical Energy Conversion Research Group, University College London, Londaon, U.K. (2007), http://www.homepages.ucl.ac.uk/~ucecdbr/research.html, Retrieved May 9, 2012.

406. Calise, F., Design of a hybrid polygeneration system with solar collectors and a Solid Oxide Fuel Cell: Dynamic simulation and economic assessment, *International Journal of Hydrogen Energy*, 36 (10), 6128–6150 (2011).

407. Hawkes, A. et al., Fuel cell micro-CHP techno-economics: Part I—Model concept and formulation, *International Journal of Hydrogen Energy*, 34, 9545–9557 (2009).

408. Peht, M., Cames, M., Fischer, C., Prateorius, B., Schneider, L., Schumacher, K., and Voss, J., *Micro Cogeneration towards Dencentralized Energy Systems*. Springer, Berlin, Germany (2006).

409. Pilatowsky, I., Romero, R.J., Isaza, C.A., Gamboa, S.A., Sebastian, P.J., and Rivera, W., *Cogeneration Fuel Cell—Sorption Air Conditioning Systems*. Springer, London, U.K. (2011).

410. Riffat, S., Durable solid oxide fuel cell tri-generation system for low carbon buildings (2012); Rowe, T. and Foger, K., Market launch of BlueGen: Essential experience from real-world field trials, Fuel Cell Seminar and Exposition, Orlando, FL (2011).

411. Wu, D.W. and Wang, R.Z., Combined cooling, heating and power: A review, *Progress in Energy and Combustion Science*, 32 (5–6), 459–495 (2006).

412. Yu, Z. et al., Investigation on performance of an integrated solid oxide fuel cell and absorption chiller tri-generation system, *International Journal of Hydrogen Energy*, 36 (19), 12561–12573 (2011).

413. Santangelo, P.E. and Tartarini, P., Fuel cell systems and traditional technologies. Part I: Experimental CHP approach, *Applied Thermal Engineering*, 27 (8–9), 1278–1284 (2007).

Index

Printed and bound by CPI Group (UK) Ltd, Croydon, CR0 4YY

01/11/2024

01782601-0016